Deuterostomes

Chordata

Subphylum Vertebrata

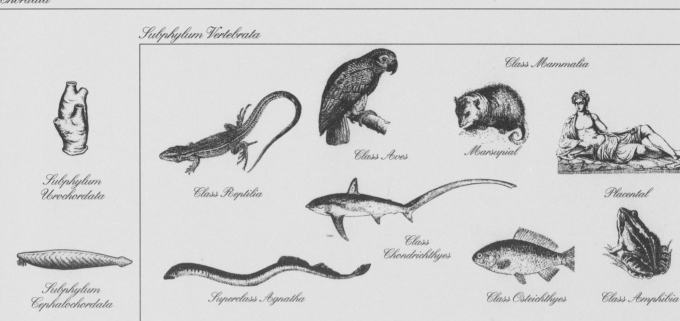

Subphylum
Urochordata

Subphylum
Cephalochordata

Class Reptilia

Class Aves

Class Mammalia

Marsupial

Placental

Class
Chondrichthyes

Superclass Agnatha

Class Osteichthyes

Class Amphibia

Echinodermata

Chaetognatha

Hemichordata

Loricifera

Entoprocta

Gastrotricha

Cnidaria

Porifera

Placozoa

Ctenophora

Mesozoa

CONCEPTS IN
ZOOLOGY

Emperor penguins (Aptenodytes forsteri) *with chick.*

JEFF McLEAN

CONCEPTS IN
ZOOLOGY

C. LEON HARRIS

State University of New York
Plattsburgh, New York

Consultant in Biology:
KAREL LIEM
Harvard University

 HarperCollins*Publishers*

SPONSORING EDITOR:	Glyn Davies
DEVELOPMENT EDITOR:	Kathleen Dolan
PROJECT EDITOR:	Karen Trost
TEXT DESIGN AND ART COORDINATION:	Michael Mendelsohn, M 'N O Production Services, Inc.
COVER DESIGN:	Teresa Delgado
COVER ILLUSTRATION/PHOTO:	David L. Pearson/Visuals Unlimited
PHOTO RESEARCHER:	Lynn Mooney
PRODUCTION MANAGER:	Kewal Sharma
COMPOSITOR:	Kingsport Press
PRINTER AND BINDER:	Arcata Graphics/Kingsport
COVER PRINTER:	The Lehigh Press, Inc.

Concepts in Zoology

Library of Congress Cataloging-in-Publication Data

Harris, C. Leon, 1943–
    Concepts in zoology/C. Leon Harris.
      p. cm.
    Includes index.
    ISBN 0-06-042659-4 (student edition)
    ISBN 0-06-500421-3 (teacher edition)
    1. Zoology. I. Title.
QL47.2.H38 1992
591—dc20   91–2850
                                               CIP

91 92 93 94 9 8 7 6 5 4 3 2 1

For Mary Jane

I think I could turn and live with animals, they are so placid and self-contain'd,
I stand and look at them long and long.

They do not sweat and whine about their condition,
They do not lie awake in the dark and weep for their sins,
They do not make me sick discussing their duty to God,
Not one is dissatisfied, not one is demented with the mania of owning things,
Not one kneels to another, nor to his kind that lived thousands of years ago,
Not one is respectable or unhappy over the whole earth.

So they show their relations to me and I accept them,
They bring me tokens of myself, they evince them plainly in their possession.

I wonder where they get those tokens,
Did I pass that way huge times ago and negligently drop them?

Walt Whitman, from "Song of Myself"

BRIEF CONTENTS

DETAILED CONTENTS

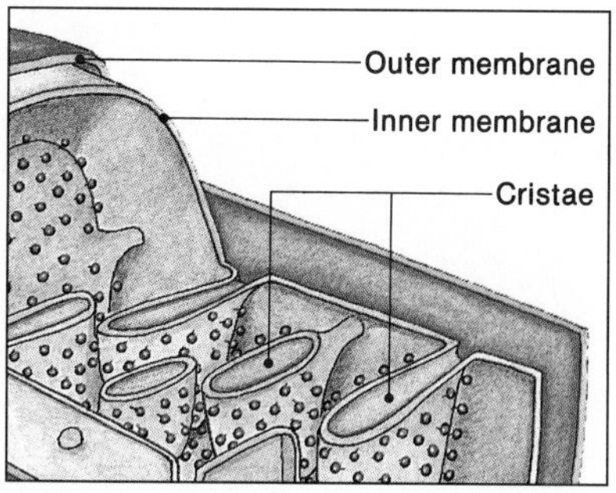

Outer membrane
Inner membrane
Cristae

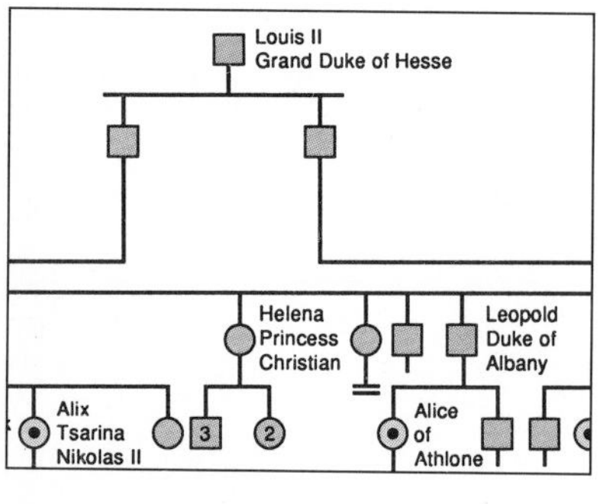

Louis II
Grand Duke of Hesse

Helena
Princess
Christian

Leopold
Duke of
Albany

Alix
Tsarina
Nikolas II

3

2

Alice
of
Athlone

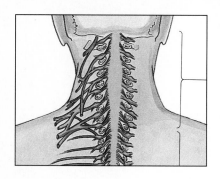

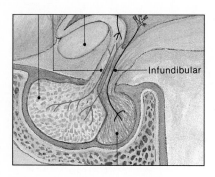

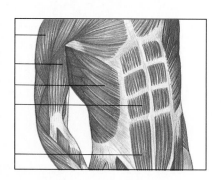

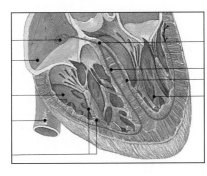

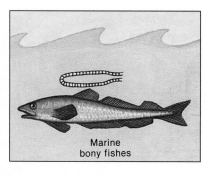

UNIT THREE: INTERACTIONS OF ANIMALS WITH THEIR ENVIRONMENTS AND WITH EACH OTHER

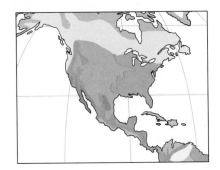

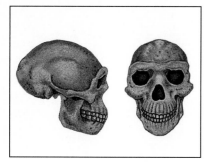

UNIT FOUR: THE DIVERSITY OF ANIMALS   436

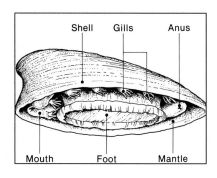

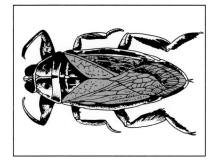

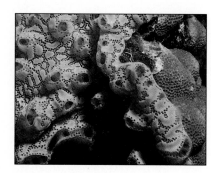

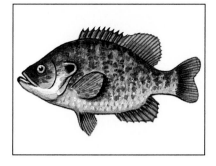

PREFACE TO INSTRUCTORS

Just as you do in the classroom, I have struggled throughout the writing of this book with the need to teach all of the information required for the understanding of concepts in zoology while, at the same time, sharing with students the enthusiasm we all feel for the fascinating ways those concepts are manifested in animals. Perhaps before we can teach students anything we must make them think, "Here is one human being who actually believes the subject of this course is important and interesting. Maybe it is." I have tried to communicate this message "between the lines" while endeavoring within the lines to express concepts as clearly as possible.

Organization and Pedagogical Aids

Unfortunately, students must often acquire quite a lot of information before concepts become clear; until they understand the material they are apt to wonder what all the excitement is about. Any device that facilitates this process will help them grasp the concept more quickly and share our enthusiasm for it. There are many such pedagogical devices in this book. I explain the functions of these pedagogical aids to students in the following Preface to Students, but unless students have changed greatly since the Jurassic period when I was in school, many will not read the preface or use the aids unless the instructor suggests they will be helpful. For that reason, I thought it would be useful to provide a synopsis of these aids so that you can recommend to your students the ones you think will be most helpful for your particular course.

I have organized the book to satisfy the greatest number of general zoology courses. However, there is such an enormous diversity of courses that the probability of this organization matching your exact preference is rather small. For that reason I have attempted to make each chapter reasonably independent of the others so that chapters can be taken in virtually any sequence, and many can be omitted entirely. However, a complete understanding of a topic may require information from a chapter not covered. For that reason I include frequent cross-references; you may wish to alert your students to the ones you consider important.

Each chapter opens with a Chapter Outline and list of Learning Objectives to help give students an overview of the chapter. The Chapter Outline is a listing of the contents of the chapter. The Learning Objectives are questions designed to encourage students to be receptive to the major concepts to be presented. These are not necessarily questions to which a student should be able to give a definitive answer after studying the material, for zoologists do not yet have the answers to all of them. Rather, they are intended to stimulate the students' curiosity. As a further aid to help students get an overall view of the material in each chapter, there is a concise Summary. Although it comes at the end of each chapter, it might not be a bad idea to point out to students that there is no rule against reading the summary before reading the rest of the chapter. Also at the end of each chapter is a Self-Test on the main concepts in the chapter. Students can test themselves or each other with these questions. In addition, you may find them helpful as models for essay questions.

One of the major barriers to learning zoology is the extensive vocabulary that is more often than not derived from Greek and Latin. I have tried to keep the use of technical terms to a minimum, but there is a limit to how many can be omitted without turning out students who are illiterate in zoology. Technical terms are written in **boldface** the first time they occur within a chapter, and the term is defined within or near the sentence in which it first occurs. Where appropriate, I have also provided pronunciation guides or derivations of the terms in context; on the back endpaper is a list of common root words and a general guide to pronunciation. The most important new words are listed as Key Terms at the end of each chapter. You might want to call students' attention to these lists as convenient places to review vocabulary. In addition, the Glossary serves as a ready reference for more than 1800 zoological terms. This is the most extensive glossary in a zoology text, and should prove valuable in subsequent biology courses that students will take. The thorough Index is another valuable resource for students who want to look up particular topics either during the introductory zoology course, or in later courses.

Finally, each chapter closes with two lists of references. The first list is Recommended Readings for students (or even instructors) who want additional information on particular subjects. The second lists some of the Additional References I used in preparing the chapter. These may be more helpful to instructors than to students. Often they include information that may not be widely known, and some of the most recent and least familiar are referred to parenthetically within the text.

Ancillaries

Concepts in Zoology is complemented by the following supplements:

Laboratory Manual. *The HarperCollins Zoology Laboratory Manual* was written by Bill Tietjen of Bellarmine College, and Leon Harris. Consisting of 26 labs, it emphasizes the living properties of animals as organisms, rather than collections of dead parts. All labs have been carefully chosen and class tested. Carefully developed art is included for each exercise, helping to clarify the experiment.

The HarperCollins Biology Encyclopedia Laser Disk. The Biology Encyclopedia Laser Disk, produced in conjunction with Nebraska Interactive Video, Inc., offers the latest in visual technology: transparencies, micrographs, slides, and film and video footage. Containing over 1500 images provided by Carolina Biological Supply, the laser disk allows instant access to any image or footage, frame by frame or moving, simply by pushing a few buttons on a hand-held remote. The disk enhances the principles of biology covered in the text much more effectively than transparencies or videos.

Instructor's Manual and Test Bank. An Instructor's Manual and Test Bank by Leon Harris and Ken Saladin (Georgia College) is available free to adopters. It provides a unique, easy-to-use guide to appropriate images on the laser disk, overhead transparencies and exercises in the laboratory manual, plus summaries of key concepts and lecture demonstrations. The test bank consists of 1500 questions in multiple-choice, true/false, matching, and short answer formats. The Instructor's manual also includes 150 transparency masters.

Testmaster. The testbank is available to adopters in a computerized form for your IBM or Macintosh.

Acetate Transparencies. A comprehensive set of 125 four-color acetates of art and photomicrographs from the text is available free to adopters.

Harper Dictionary of Biology. Written by W. G. Hale and J. P. Margham, of the Liverpool Polytechnic Institute, this dictionary contains 5600 entries that go far beyond basic definitions to provide in-depth explanations and examples. Diagrams illustrate such concepts as genetic organization, plant structure, and human physiology. The dictionary covers all major subjects (anatomy, biochemistry, ecology, etc.) and also includes biographies of important biologists.

The Student Environmental Action Handbook. HarperCollins has joined with the Student Environmental Action Coalition to bring your students a handbook of the environmental movement on campuses around the country. Using real campus examples, it contains a series of strategies for approaching the administration, the community, political leaders, and student leaders, as well as for adjusting one's own personal habits to achieve positive change. Examples include population control, transportation, water conservation, and publishing a newsletter.

Two Minutes a Day for a Greener Planet. Written by Marjorie Lamb, a veteran reporter on environmental affairs, this book provides easy, practical suggestions for individual action, on a small scale, that can make a big impact on our planet's future.

The Zoology Coloring Book. This exciting new approach is an enjoyable and effective way to learn the fundamentals of zoology. Participation by the reader through creative coloring provides significant learning reinforcement. The text accompanying each coloring plate provides supportive explanatory material and leads the reader through the plate in a step-by-step manner. When finished, the colored plates provide an excellent review that the reader has helped create.

Writing About Biology. Written by Jan A. Pechenik, professor of biology at Tufts University, this brief but straightforward guide includes sections on writing lab reports, essays, term papers, research proposals, critiques and summaries, and in-class essay examinations. It also includes special sections on effective note-taking, how to give oral presentations, and how to prepare applications for summer and permanent jobs in biology. Appendices listing commonly used abbreviations for lengths, weight, volumes, and concentrations are also featured.

Acknowledgments

Many zoologists and others have already defied the current media image of scientists by giving me their time, knowledge, and encouragement. The administration and my colleagues in the Department of Biological Sciences at SUNY Plattsburgh made possible a sabbatical during which a large part of this book was written. I am especially grateful to Lawrence Shaffer, F. Daniel Vogt, and Matthew Merrens for their support during one particular crisis. Colleagues at SUNY Plattsburgh who were especially generous with expertise were Kissu Schin, James C. McGraw, and Gerhard Gruendling. Fred A. Urquhart at the Scarborough Campus of the University of Toronto and W. D. Hummon at Ohio University also provided information. Special thanks are due Bernd Heinrich at the University of Vermont for the enlightening interview in Chapter 15.

Many at HarperCollins expertly turned all this raw information into a book. Much of the credit is due to Kathleen Dolan, who was Development Editor but could just as well be called a second author. Richard E. Morel made numerous improvements in the figures, and Karen Trost expertly guided the production. Glyn Davies, Biology Editor, oversaw the project. I would also like to thank designer Michael Mendelsohn for his beautiful and functional design, and photo researcher Lynn Mooney, whose extraordinary resourcefulness provided many of the text's images. Finally I wish to thank the following 38 reviewers, who have tried, and often succeeded, in keeping me from committing too many errors:

Mary D. Albert, *Boston College*

Joseph T. Bagnara, *University of Arizona*

David C. Brubaker, *Seattle University*

John A. Byers, *University of Idaho*

Brian T. Crother, *University of Texas–Austin*

Peter Dalby, *Clarion University*

David Gale Davis, *University of Alabama*

Dorothy Dunning, *West Virginia University*

Duwayne Englert, *Southern Illinois University at Carbondale*

Marylynn Fallon, *Connecticut College*

John N. Farmer, *University of Oklahoma*

Milton Fingerman, *Tulane University*

Jon G. Houseman, *University of Ottawa*

Patricia J. Humphrey, *Ohio University*

Gerard F. Iwantsch, *Fordham University*

Ronald Jenkins, *Samford University*

Gwilym S. Jones, *Northeastern University*

Richard N. Mariscal, *Florida State University*

Marvin Mays, *Central State University*

Grover C. Miller, *North Carolina State University*

Thomas C. Moon, *California University of Pennsylvania*

Ruthanne B. Pitkin, *Shippensburg University*

Edward B. Pivorun, *Clemson University*

Edwin Powell, *Iowa State University*

Mark Rausher, *Duke University*

John Roese, *Texas A&M University*

William Rogers, *Winthrop College*

Virginia Roth, *Duke University*

Ken Saladin, *Georgia College*

Fred Searcy, *Broward Community College*

Kimberly G. Smith, *University of Arkansas*

Robert Stiles, *Samford University*

Barbara A. Taber, *Southwest Missouri State University*

Walter Taylor, *University of Central Florida*

John Thornton, *Oklahoma State University*

John F. Tibbs, *University of Montana*

Catherine Toft, *University of California–Davis*

R. Stimson Wilcox, *SUNY Binghamton*

One final note. Throughout this text I have tried to keep students open to the tentative nature of many of our ideas, and to encourage them to think critically—though I hope not cynically. I hope you will approach this book with the same attitude. No doubt you will find many statements with which you disagree, and I hope you will feel free to point out such disagreements to your students. I don't know a better way to get them to think like scientists about concepts in zoology. I also invite you to write to me directly with any corrections, criticisms, or even compliments you may have.

C. Leon Harris

PREFACE TO STUDENTS

I might as well confess at the very beginning that I have written this book with the intention that you enjoy it. If you have been brought up to believe that good medicine must taste bad, this may seem an unpardonable sin, but I have two excuses. First, you probably learn most easily what you enjoy knowing. Second, animals *are* enjoyable, and a zoology text that obscures that fact is simply wrong from cover to cover. Most successful zoologists would agree, for they were first attracted to the field by the shapes and colors of shells, butterflies, beetles, or birds. They remain zoologists because of the pleasure of understanding animals. Zoologists are often enjoyable too, by the way, and many are as interesting as the animals they study. I want to introduce you to some of these zoologists in this text, to help you understand how they think and work; I hope to discourage the view one often gets from textbooks—that the only good zoologist is a dead zoologist.

You will find that, contrary to the popular impression, good zoologists are not gifted with an infallible vision of truth. Usually their work involves more *re*vision than vision. Most of the concepts of zoology presented here have been revised in the past, and undoubtedly many will have to be revised in the future. Often zoologists do not agree or simply do not know something. If your goal is merely to learn facts, this may discourage you. However, if your goal is to learn how to be a zoologist, you will be glad to know that there is plenty left for you to do.

Of course, not every moment of zoology will be enjoyable. One can really enjoy animals only after understanding many details about them, which requires hard work. However, I have tried to reduce the amount of work by avoiding the presentation of details for their own sake. As the title of this book indicates, the emphasis is on *concepts* in zoology. My dictionary defines a concept as "a general idea or understanding, especially one derived from specific instances or occurrences." Some "instances and occurrences" will have to be described to explain each concept, but the focus is on the concept.

You might as well abandon any hope of trying to memorize all of the "instances and occurrences" in this text; even I do not remember them all. Zoology is just too vast a subject. Even if you could remember so many facts, you still might not understand animals. The biological sciences are not about isolated facts, but about relationships. You should endeavor to understand how each organism relates to its natural environment. Try to see the world from the animal's point of view. What are its problems, and how does it solve them? That is the way to understand the organism. I have a feeling that this is also the best way to remember facts. One remembers by association with experiences, even if they are only imagined ones.

Because all of the concepts in zoology are so interrelated, finding a good place to begin learning about animals is as difficult as finding where a circle starts. After considering many alternatives, I decided to organize the book in the same hierarchical way animals are organized, from cells to organs to organisms. Unit One begins with the problems confronting individual cells as they live and attempt to reproduce within the body of the animal. Unit Two progresses to the ways in which organ systems function in animals, creating an internal environment in which cells can live and reproduce. Unit Three discusses the problems animals as a whole encounter trying to interact with each other and their external environments and ensure the survival of their own kind. Finally, Unit Four describes some of the animals that have evolved out of these diverse interactions. The descriptions of various animals in this unit provide opportunities to view the interactions of the concepts presented in the previous units within individual animals. Your professor may prefer that you go through the text in a different sequence, but try to keep this hierarchy in mind.

The topics within each chapter could also have been presented in a variety of sequences, so I begin each chapter with a Chapter Outline, which lists the contents of the chapter. The Chapter Outline is followed by a list of Learning Objectives in the form of questions. These are the big, conceptual questions that zoologists have tried and are still trying to answer—the types of questions that you would wonder about if you had the time. The Self-Test at the end of the chapter will give you a chance to find out if you can answer any of these questions and others. Also at the end of each chapter, I bring the concepts together in a Summary. You may find it helpful to read the Summary of each chapter first, as an introduction.

Often a single chapter in a general text such as this does not include all of the information you may need about a concept. By referring to cross-references and to the index you may find additional information elsewhere in the text. The references listed at the end of each chapter will provide further information. There are two kinds of references. Recommended Readings are those that I consider useful and readable background. Additional References are sources of specialized information, especially information that is not widely known. As is common practice in scientific literature, I cite many of these additional references within the text like this: (Groucho, Harpo, and Karl 1948). You might find these citations distracting at first, but I think you will soon learn to read through the ones that do not interest you.

I have tried to reduce the amount of work by introducing only the terminology needed to understand a concept. Too many textbooks assume they have covered a topic if they have used every term related to it, but defining a term is not the same as explaining a concept. This tendency is especially unfortunate in zoology, because much of the terminology was established in the days when every zoologist knew Greek and Latin. I know of one student who, after the first lecture, asked his zoology professor whether the course would ever be offered in English. Unfortunately, there is no way to avoid entirely the specialized terminology of zoology. If you want to talk and write about zoology you must use the language of zoology. Therefore, in spite of my best efforts, this book includes hundreds of terms that may be new to you. These new terms are printed in **bold** type where they first occur, and I list the most important ones at the end of each chapter. Many are defined in the Glossary at the end of the text for ready reference. With more than 1800 terms, this is by far the most complete glossary of any general zoology text, and should prove useful in other courses that you may take in the future.

A good way to learn new terms is to pronounce them aloud. This can be a source of great fun at a party. Mistakes should not be embarrassing if you remember that there is no one correct way to pronounce most zoological terms, since no one knows how ancient Greek and Latin were spoken. Zoologists usually follow the pronunciation of their teachers, which varies widely. British and American zoologists don't even agree on the pronunciation of "zoology": the British pronounce it zoo-ology, while Americans prefer zo-ology. The most difficult terms in the text are followed by a simple guide to pronunciation (pro-NUN-see-A-shun), with phonetic spelling and capitalization of accented syllables. A table inside the back cover offers general rules for pronunciation. Many of the new terms can be dissected into root words, some of which you already know because they occur in everyday English. On the inside back cover is also a list of the common root words. It would save time in the long run to learn as many of these as possible.

In addition to the specialized terminology, you will also find some funny-looking names in italics. These are the scientific names for species. (For now, we may simply define a species as a group of organisms that cannot interbreed with another group.) I include these scientific names because the common names often vary from one region to another, and because you should get used to seeing how zoologists refer to animals. In this system the name of a species consists of two italicized words, for example, *Canis familiaris*. The first word is capitalized and refers to the genus group, which usually includes several different species. The second word is not capitalized and differs for each species in a genus. These names may look imposing, but many of them make sense. *Canis familiaris*, for example, is just the familiar canine. You will see some of these names so often that you will learn them without even trying, and you will soon get into the habit of reading through the others without losing much time.

There are approximately one million of these species names for animals. This is far too many for any one person to know, so zoologists generalize about large groups of animals, such as protostomes, vertebrates, and phyla (FY-luh, plural of phylum). The 32 phyla of animals recognized in this text are represented in a figure inside the front cover. Referring to the figure frequently will be a relatively painless way to become familiar with the major animal groups.

Now plunge ahead, and don't feel guilty or surprised if you find that you actually enjoy learning about animals.

C. Leon Harris

1

Introduction

Human and camel (Camelus bactrianus)—an ancient relationship.

CHAPTER OUTLINE

LEARNING OBJECTIVES

1. What do zoologists do?

2. Is there a particular scientific method that zoologists use?

3. What are animals?

4. Where within an animal do the functions of life occur?

5. How do cells create an internal environment in which they carry out the functions essential to the life of the animal?

6. How do animals obtain the necessities for life from each other and the environment?

7. How do zoologists keep track of so many kinds of animals?

AN ANCIENT PROFESSION

For as long as there have been humans there have been those who studied animals. The evidence for this is that such slow, frail creatures as ourselves have managed to exist for two million years. That would not have been possible if our ancestors had not acquired an impressive understanding of their prey and predators. In some parts of what is now Europe, these primitive zoologists appear to have deliberately attempted to pass on their learning to future generations. With paintings as elegant as they were anatomically detailed, they revealed the strategy of the hunt and pointed out the vulnerable parts of prey. In many cases animals were portrayed not merely as potential food, but as objects of beauty and admiration (Figure 1.1).

Zoology thus began as a matter of survival and grew into a fascination with, and sometimes a reverence for, animals. So it is still. Zoology has changed, certainly, but zoologists still ask many of the same questions that are so crucial to human survival and to our understanding of life. Indeed, it seems that zoologists know even less than our ancestors did, for the more we learn, the more we realize how little we know. Now it takes more than one elder in a tribe to pass on all the knowledge about animals. The relatively little we know is already too much for a single person to master. A taxonomist who can identify beetles is an authority on beetles, but that taxonomist would probably have to seek help from a physiologist to understand the functioning of the beetle's internal organs, from an ecologist to understand its relation to other organisms, and from an embryologist, a geneticist, and an evolutionist to understand why the beetle appears as it does. These would be only a few of the many kinds of zoologists, for any biologist whose primary focus is on animals can be called a zoologist. A zoologist is not someone who understands everything about every kind of animal, but someone who out of necessity and curiosity has learned a great deal about some of them. The objective of this text is to provide a similar overview of animals.

No one would have had to explain to primitive people why they needed to understand animals. Most of us, however, no longer hunt from necessity, and even fewer have to grapple tooth and claw with predators. Now our struggle to survive is more subtle, but no less real. Legions of insects feast on crops and stored food that could feed the thousands of humans now starving. Animal parasites still weaken and kill millions through malaria, schistosomiasis, and other diseases. Our

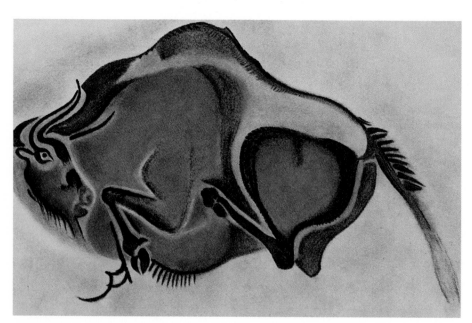

Figure 1.1
A charging bull painted tens of thousands of years ago in a cave at Altamira, Spain by a Cro-Magnon person. Although the bull may have been hunted for food, it appears that the artist was also impressed with the animal's beauty. The detail is remarkably accurate, considering that the artist undoubtedly worked from memory.

bodies and those of other animals on whom we depend for food are prey to infections from viruses, bacteria, and fungi, as well as to other disorders. Only through an understanding of the functioning of animals, including ourselves, have we been able to hold some of these threats more or less to a standoff. Because of the continuing increase in the human population and the appearance of new diseases, our survival demands that we expand our efforts to understand animals. Even if our survival did not depend on it, however, there would undoubtedly still be zoologists, simply because zoology is one of the most exciting and enjoyable endeavors many of us can imagine.

WHAT ZOOLOGISTS DO

Zoologists are employed in a variety of activities (Figures 1.2 and 1.3). Many with bachelor's degrees in zoology become technicians, attendants, research associates, conservation officers, rangers, and guides at zoos, museums, universities, game preserves, parks, and forests. Many of the same organizations also employ zoologists with graduate degrees and research experience in teaching and/or research. Zoologists are also employed in environmental conservation and wildlife management, assessing and maintaining the health and populations of animals. In these days when many people learn about animals through photographs and television, there are also opportunities in photography and cinematography for those whose knowledge allows them to get close enough to animals without disturbing them and without getting themselves killed. Other zoologists find that their interest in animals is mainly physiological, and they may ultimately go into veterinary or human medicine. This is by no means a complete list of the activities engaged in by people who think of themselves as zoologists.

Regardless of which career a zoologist chooses, it is essential to have not only a good knowledge of subjects directly related to zoology but also a broad background in the other sciences and in mathematics. Many areas of zoology also require abilities in statistics, photography, drawing, computer programming, economics, management, and other subjects. Zoologists also benefit from a broad education in many other ways, simply because they often have to interact with people of various political, economic, social, and aesthetic persuasions. Perhaps most important of all, zoologists must be able to communicate well, both orally and in writing. No matter how much knowledge a zoologist has, it is useless unless she or he can communicate it to employers, peers, or the public.

How does one choose which area of zoology to make a career in? Unfortunately, many people have little choice: they may be lucky to get *any* job, no matter how remotely related it is to their interests. Usually, however, the broader one's education, the broader the range of choices. Before you choose any career, you should remember that you will be engaged in it for a long time, and that no salary is worth being miserable for the best part of your life. Before choosing, ask yourself whether you prefer working indoors or outdoors. (Which would you enjoy more—sitting at a lab bench examining animal feces for eight hours, or tracking moose in mud while donating blood to mosquitos?) Do you prefer working with whole animals, or with their organs? Also consider whether the kinds of animals you enjoy working with includes *Homo sapiens*.

Even if you eventually choose some career unrelated to zoology, you should never consider your knowledge of animals wasted. Many of the most significant contributions to zoology have been made by people who did not earn a living as zoologists. The list ranges from Charles Darwin to Emperor Hirohito. You may also have noticed that those whose lives are the fullest and most enjoyable are those who also enjoy animals.

Figure 1.2
A zoologist, interacting closely with an animal in its environment. Here the zoologist is measuring the beak of a gull in a study of food preferences. Field research like this is often lonely and physically demanding but provides opportunities for close associations with animals.

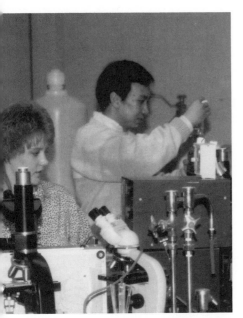

Figure 1.3
Many zoologists work in teams, taking advantage of the controlled and more comfortable environment of the laboratory.

HOW ZOOLOGISTS THINK

The Myth of Method.   We have all thrilled to the stories of zoologists who discovered exotic animals on dangerous expeditions, or zoologists who conceived of brilliant theories in dreams and then confirmed them in elegant experiments. Unfortunately, the lives of most zoologists are usually less thrilling than this. Most of the time they are feeding the animals, checking the traps, struggling to make some sense out of data, worrying about where to get the money, and dozens of other mundane matters. Almost always the ultimate goal of all their activity is to test or apply a theory that relates zoological phenomena to each other. Sometimes the theory gets lost in the details, and the zoologists must try to get it back into focus by talking to colleagues, going to conferences, or just quietly thinking. What you will seldom see a zoologist do, however, is refer to a guide book on "the scientific method." Nor are you likely to find a course on scientific method in a science curriculum. In fact, about the only place you will find discussions of "the scientific method" is in the introductions of science textbooks.

A curious change in the way such texts have presented "the scientific method" suggests that it is a myth. Once upon a time "the scientific method" was said to be induction, in which generalizations are formulated after repeated observations. A typical example was that of a supposed scientist who kept seeing black crows until at last he felt justified in concluding that all crows are black. (These texts never explained what the scientist should do about albino crows.) Then around 1960 biology texts started describing "the scientific method" as the hypothetico-deductive method. The scientist was supposed to first state or realize a problem, make relevant observations, then formulate a tentative explanation, called a theory or hypothesis. The scientist would then deduce what ought to happen in certain situations according to that hypothesis, test these deductions by experimentation, and thereby either support or disprove the hypothesis.

Did scientists around 1960 suddenly change the way they worked? Not at all. It is simply that the two "methods" are so loosely described that it is possible for almost any scientists to believe she or he is following either one. Charles Darwin, for example, wrote in his autobiography that he "distrusted greatly deductive reasoning" and developed his theory of natural selection "on true Baconian principles [of induction], and without any theory collected facts on a wholesale scale." No doubt Darwin really believed that he had followed the inductive method. It was the only reputable method in Britain at the time, and Darwin was quite sensitive to the numerous criticisms that his theory was a "mere deduction." Darwin's notebooks show, however, that he formulated his theory before he amassed the bulk of the evidence, and his theory is now presented in biology texts as if it had been developed by the hypothetico-deductive scheme.

If neither induction nor deduction is "the scientific method," how do scientists think? Just about like everyone else does. They formulate hypotheses and theories, then try them out. You do the same when you figure out why your car won't start or why the baby keeps crying. When Darwin was not trying to justify his theory he seems to have felt the same way, for he wrote in a letter, "I have often said and thought that the process of scientific discovery was identical with everyday thought, only with more care." Just as people differ in their abilities as auto mechanics and baby sitters, scientists also differ in their abilities to formulate hypotheses and theories. This brings us to the subject of creativity, about which many volumes have been written. Unfortunately, no one has ever successfully explained creativity or found a systematic method for it. The impression that emerges from the autobiographies of creative scientists is that much of the creation goes on when they are not aware of it. They find a problem so enchanting that their minds can't let it alone even when they are sleeping or working on another problem. Contrary to the popular myth of "the scientific method," the creation of theories depends on some process that is not at all methodical. It is essentially a human activity.

Although the creation of scientific theories is similar in many ways to ordinary thought, the testing of those theories almost always involves more skill and care than is required in nonscientific endeavors. Not only must scientists carefully and repeatedly test their ideas by experimentation or observations in the field, but they must convince their peers that those experiments and observations adequately support the theories. Even after peers have been sufficiently impressed by the research to recommend publication or the awarding of a grant, the theory continues to be challenged and tested by other scientists. Even Darwin's theory that evolution is mainly due to natural selection, which is probably the most widely accepted theory in biology, is continually put to the test. Few nonscientific activities lend themselves to such a rigorous approach.

Scientific Integrity. Like any complex intellectual activity, science does not always function as it should. It is a human activity, and its participants sometimes succumb to the same human weaknesses that afflict us all. Most scientists like to reap the honors and material rewards of success, or at least to feed themselves and their families. These human desires can sometimes make scientists treat other scientists like competitors rather than colleagues, and even tempt them to take credit for work others have done, or for work that has never been done. As the competition for jobs, promotions, and grants has increased, more and more "scientists" have been found to have plagiarized, fabricated, or misrepresented research. The number of dishonest scientists represents a minuscule proportion of all scientists and is vanishingly small compared with the proportion of dishonest politicians and advertisers. Nevertheless, there are enough of them to take a lot of the fun out of science for those who were attracted to it because of its integrity. Perhaps one reason why there is some dishonesty in science is that textbook authors have considered the necessity for honesty too obvious to mention.

Perhaps there is a scientific method after all. It is not a method for creating theories; no method can replace individual creativity. It is, however, a method for ensuring the honest evaluation of those theories, and it might consist of the following steps:

1. Plan experiments and observations in such a way that data are as likely to disprove as to support the theory that you hope or believe is valid.
2. Regard data as the sacred and unalterable word of nature.
3. Be as critical of your own conclusions as you are of the conclusions of others.

This self-critical approach to scientific theories is in keeping with the ideas of one of the most influential philosophers of science, Karl Popper. In the 1930s Popper formulated a **Criterion of Demarcation** to distinguish between scientific and nonscientific theories. The criterion is that a theory, in order to be regarded as scientific, must be *potentially* capable of being proved false. That is, one must be able to describe an experiment that could prove the theory to be false. This criterion of falsifiability shifts the burden of scientists from that of trying to prove theories true to that of trying to prove them false. One advantage of this outlook on science is that it removes any feeling of shame that might otherwise be attached to an incorrect theory. Theories earn respect not from being proved true, but from being interesting enough to inspire scientists to try to prove them false.

The Use of Animals in Experimentation.   Scientific theories are tested by experimentation, so by their very nature zoological theories require animal experimentation. Such experimentation often comes into conflict with another human trait: empathy—the ability to imagine oneself in the place of another. Inspired by empathy for animals, many individuals and groups have succeeded in improving the lot of research subjects. Some others have not been content with merely improving conditions but have taken legal and illegal measures toward ending all research on animals. Some have invaded and damaged laboratories and have threatened researchers with violence. Others have become increasingly successful in imposing regulations that make research more expensive and time-consuming. More and more zoologists are finding themselves compelled to justify to others and to themselves their use of animals in research.

Dishonesty in science will meet with nearly universal condemnation, but it is harder to find a consensus on animal experimentation, for it is almost impossible to take a rigid point of view on this subject without self-contradiction. There are some who believe research on animals is never justified, but who nonetheless avail themselves of foods, medicines, and surgical procedures that are based on such research. At the other extreme are physiologists who think nothing of performing

Ethical Considerations in the Use of Live Animals in Research

1. What kinds of animals feel pain? Are only humans truly conscious of pain? Can other animals feel pain too, or do they merely react with defensive or withdrawal reflexes? Can dogs feel pain? Rats? Frogs? Cockroaches? Sponges? Protozoa?

2. Is the ability to perceive pain the only criterion for judging the ethics of experimentation? If so, then is all animal experimentation ethical as long as anesthesia is used? What about experimentation on a comatose human?

3. If the ability to perceive pain is not the only criterion, then what are the other criteria? If those other criteria make it unethical to use animals for experimentation, do they also make it unethical to perform experiments on plants? Do these criteria also rule out any use of animals or plants for food and other purposes?

4. How much pain is inflicted in the name of science compared with that inflicted from carelessness and for sport, agriculture, and development? Each year an estimated 20 million animals, mostly rodents, are used in experimental research. About as many cats and dogs are abandoned by their owners to starve to death or to be "put to sleep" by humane societies. Many more pets and farm animals are neutered, often without anesthetics, than are subjected to experimental surgery. The destruction of habitat subjects countless millions of animals to starvation. Why is animal experimentation singled out? Could it be because scientists are not as politically powerful as pet owners, farmers, and developers?

5. How much pain results from research compared with that which is a natural part of the lives of wild animals, few of whom die a comfortable death in old age? If other animals inflict pain on their prey to meet their needs, why cannot humans inflict pain to fulfill their needs? Or does this commit the **naturalistic fallacy** of asserting that what occurs in nature is morally right for people?

6. Is it possible that the pain inflicted through research is outweighed by the reduction in pain made possible through advances in human and veterinary medicine that result from such research?

7. Does government regulation of animal research make sense considering that it generally protects only warm, furry animals while leaving unprotected the cold, wet ones, as well as birds, rats, mice, horses, and other farm animals? Is it rational for the government to spend millions of dollars to eradicate rats in Washington, D.C., while protecting them in laboratories?

8. Shouldn't scientists use every means to reduce the amount of unnecessary pain, by using statistics to determine the minimum number of animals required and by using isolated organs, cell cultures, mathematical models, and computer simulations when possible?

9. If scientists can inflict pain on other animals, what is to prevent scientists of future generations from adopting a similar attitude toward experimentation on humans? Does inflicting pain on an animal dull the conscience of the experimenter, making her or him less sensitive to pain in other humans? Does an unconcern for pain in animals diminish humans by blurring one distinction, empathy, that separates us from most other animals? Or does a deeper understanding of the structure and functioning of animals make us more respectful of all animals, including humans?

surgery on a live animal, but who are sickened by hunting for sport. Each person must decide individually what uses of animals are proper. The questions in the box may help.

The myth of the scientific method and the necessity of using animals in experimentation sometimes give students and nonscientists the impression that the study of animal life is itself lifeless, cold, and cruelly rational. Most zoologists do not perceive zoology this way, however. It is certainly not the way zoology is presented in this book. On the contrary, I hope to show that zoology exalts and celebrates humanity, for without it none of us would know that humanity is the product of billions of years of evolution, that we are brothers and sisters to all other living things and the children of vanished organisms that only science can memorialize, and that we are nevertheless unique in being the only animals capable of understanding all this. Zoology is itself an expression of humanity. It is conducted by human beings who share with artists, theologians, philosophers, and historians a desire to understand their lives and to find out what we are doing and ought to be doing on Earth. A considerable number of zoologists are motivated by the most humanitarian of all goals, that of improving the well-being of humans and other animals.

WHAT IS AN ANIMAL?

This may seem a silly question, but in fact biologists have often debated whether such organisms as sponges and polyps ought to be considered animals, and recent studies comparing the molecules of such organisms have reopened the debate. Until this century, organisms were considered to be animals if they moved actively and to be plants if they did not. This division of life into only two **kingdoms,** Animalia and Plantae, persisted into this century even though an enormous variety of microorganisms had been seen since 1670. Finally, biologists added a third kingdom, Monera, to accommodate the bacteria. Since the late 1950s biologists have added two more kingdoms, Fungi and Protista, for fungi and protists (protozoans and some algae). Most biologists now accept this division of all living things on this planet into these five kingdoms.

The five-kingdom classification is currently under challenge, because bacteria (kingdom Monera) differ so much from organisms in the other four kingdoms. Bacteria lack cell nuclei and are therefore said to be **prokaryotes** (Greek *pro-* before, *karyon* kernel, i.e., nucleus). All other organisms have nuclei in their cells and are called **eukaryotes** (Greek *eu-* true). Most eukaryotes are also similar to each other in having other intracellular structures that are not found in prokaryotes. Thus it seems unjustified to divide the eukaryotes into four kingdoms at an equal level with prokaryotes. Comparisons of the structures of certain molecules show, in fact, that there are actually two extremely different kinds of bacteria, the eubacteria and the archaebacteria, that differ from each other as much as they differ from protists, fungi, plants, and animals (Figure 1.4). In the future, therefore, biologists may recognize six kingdoms, or perhaps even more.

The archaebacteria live in environments that are extremely salty, hot, or lacking oxygen, so people and other animals seldom encounter them. All other kinds of organisms, however, associate intimately with animals. The eubacteria include the 5000 known species of relatively familiar bacteria. (Species are defined as groups of organisms that can breed among themselves, but not with other such groups. For bacteria and other organisms that do not reproduce sexually, however, species are distinguished on the basis of structural and other differences.) We know bacteria mainly as agents of disease and food spoilage, but probably just as many

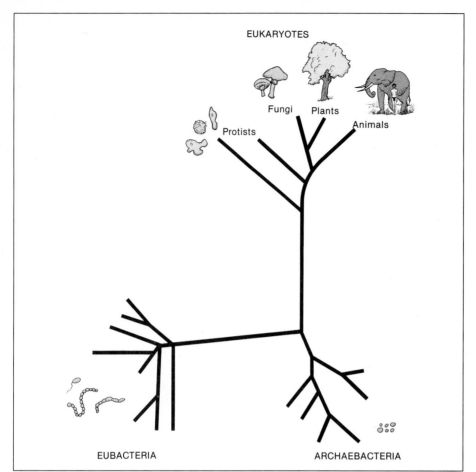

EUKARYOTES

Fungi Plants

Protists

Animals

EUBACTERIA

ARCHAEBACTERIA

Figure 1.4
The relationship of different kinds of organisms to each other, based on comparisons of molecular structure. The distance along a set of lines between any two groups of organisms represents the evolutionary distance separating the two groups. Animals are closely related to other eukaryotes, compared with the two major groups of bacteria.

species of eubacteria are beneficial to animals. Among them are the cyanobacteria (formerly called blue-green algae), which produce oxygen by photosynthesis, just as plants and algae do. In fact the cyanobacteria may have been the first photosynthesizers and might have created the oxygen-rich atmosphere that enabled animals to evolve. Some bacteria help animals by enabling food plants to grow, others help animals digest the food, and still others help eliminate the bodies of animals after they have died. No matter how clean you are, you contain about a hundred quadrillion (100,000,000,000,000,000) bacteria—about ten times as many bacterial cells as human cells (Figure 1.5).

Among the eukaryotes almost 60,000 species are neither fungi, nor plants, nor animals but are lumped together as protists. Protists are usually divided between algae and protozoa, but they are actually so diverse that one taxonomist has suggested that they be classified into 20 different kingdoms. Most are single celled and microscopic, but some consist of large colonies of more or less identical cells. The giant kelp, for example, grows to be hundreds of meters long (Figure 1.6). Many of the protists, such as kelp and other algae, are capable of photosynthesis. They are therefore important in converting solar energy into chemical nutrients that can be used by animals that feed on them. The protists also include nonphotosynthesizing organisms such as *Amoeba* and *Paramecium*, which have traditionally been discussed in zoology texts as primitive animals in the group Protozoa. Although protozoa are no longer considered to be animals, they have to be discussed in zoology texts because they are apparently the closest surviving kin of animals. Moreover, understanding how they function as single cells helps us understand how animals function with many cells.

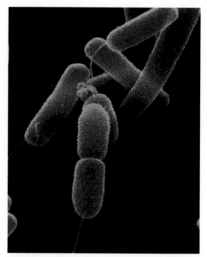

Figure 1.5
Scanning electron micrograph of Escherichia coli *bacteria on tissue from the human cheek. Each bacterium is about 1 micrometer (millionth of a meter) long. This species is found in the digestive tract of almost all mammals. In some species, such as rats and rabbits,* E. coli *contributes useful amounts of vitamins.*

Figure 1.6
Giant kelp Macrocystis pyrifera. *Even though this species grows to 60 meters long and resembles a plant, its cells identify it as a relative of single-celled protists. Fifty million dollars worth of giant kelp are harvested in California to produce emulsifiers for ice cream and other foods, as well as supplements for cattle feed. Other animals, such as sea urchins, anemones, sea otters, barnacles, snails, worms, crabs, and fishes use these protists for food and habitat.*

There are approximately 50,000 species of fungi, including yeasts, mushrooms, and organisms that cause various forms of rot in plant and animal tissue. Each fungus consists of one type of cell, each of which is enclosed in a rigid wall. Many fungi grow as fibrous networks (mycelia). Unlike protists, fungi reproduce by means of spores, which are extremely durable packets of genetic material. The fungi neither eat nor photosynthesize but obtain energy by absorbing organic matter. If the organic matter happens to belong to a human, then the fungus causes an infection such as athlete's foot and jock itch. Other fungi, such as mushrooms, are important food sources for various species of animals (Figure 1.7), and many are decomposers that allow molecules from dead tissues to be recycled back into living animals.

Figure 1.7
The leaf-cutter ant Atta *fertilizes its subterranean fungus garden with fragments of leaves. The large queen ant, a smaller soldier ant, and some workers are shown here on the spongy mass of fungi. Because fungi do not photosynthesize, they can live in the darkness of ant burrows.*

Approximately a quarter of a million species are now considered to be plants. Like fungi, their cells are enclosed in rigid walls, but unlike fungi, plants consist of various kinds of cells. Because of their ability to convert solar energy into chemical-bond energy, many plants are essentially sources of nutrients for animals. Plants also provide shelter and habitat for many terrestrial animals (Figure 1.8).

Considering the characteristics of protists, fungi, and plants enables us to formulate a definition that will answer the question that began this section: Animals are organisms with numerous and diverse eukaryotic cells that lack walls or the ability to photosynthesize. Narrow as this definition might seem, it is broad enough to embrace more than a million kinds of known and living animals, and many times that number of species that are extinct or not yet discovered.

AN OVERVIEW OF ZOOLOGY

Cells: Where Life Dwells.   Although our ancestral zoologists had an intimate knowledge of animals, they must have wondered in vain about life itself. Many probably dissected living animals (not excluding humans) in attempts to locate the place where the spirit dwells—the place where life itself lives. Probably some thought they had found it within the lungs or the heart. The Latin word *anima* meant both breath and soul, and even now we speak of being "a heartbeat away from death." What primitive zoologists did not know, and what no one could imagine until the invention of science, is that life is not a thing that can be dissected out and examined. It is a process, or many processes, that occur everywhere within an organism, in every living cell. The cells are where life dwells.

The cell doctrine—the realization that all organisms consist of cells and materials made by cells—was perhaps the single most important breakthrough in the quest to understand life. It tells us that what an animal must do to stay alive is to provide the internal conditions necessary to keep its cells alive. The beautiful part of this insight is that each cell has a reciprocal need to contribute to the maintenance of the conditions required for the survival of other cells. Animals and other multicellular organisms are therefore made of cells, by cells, and for cells. If too many of the cells fail to do their jobs, other cells will die, and the animal will die.

Cells do inevitably lose their ability to function, of course. Many of them are replaced, but there is a limit to how often that can be done. Therefore it is inevitable that each animal will die. In order for the species to survive, some cells must be capable of forming new animals of about the same kind before they die. If they did not, a zoology text would be just a collection of short stories instead of a long saga with many heroes and adventures. All aspects of zoology, whether developmental, physiological, evolutionary, ecological, or behavioral, are ultimately devoted to ensuring that the saga continues—that the cells responsible for reproducing the organism have the requirements of life.

Even though we have "cornered" life within microscopic cells, it is still not easy to identify or define life. About the best we can do is characterize it in terms of the processes associated with it. Each generation of zoologists, and perhaps each individual zoologist, could propose his own list of the characteristics of life. In Unit One of this text we shall examine the following four characteristics of cells.

1. **Living cells maintain a greater degree of order within themselves than occurs outside.** As we shall see, cells contain orderly arrangements of **organelles** ("little organs") made of molecules that are themselves orderly. Immediately following death this ordered state begins to reverse; the molecules and organelles break down, and the cells decay. In other words, death brings a return to the normal situation found among nonliving things, which decay and crumble to disordered dust.

2. **Living cells transfer energy.** The way in which cells increase order within themselves is by taking many disordered molecules from the environment and converting them into relatively few orderly molecules. They convert simple amino acids in the diet, for example, into complex proteins. This process, and many others performed by cells, requires energy. According to the First Law of Thermodynamics, energy cannot be created. The energy used by cells must therefore come from outside the cell, and ultimately from outside the animal in the form of food.

3. **The functions of living cells are regulated.** The processes occurring within cells do not happen randomly. For example, the sequence of amino acids in proteins is determined by genes. The transfer of energy is also regulated: animals stop feeding when their cells signal that their energy needs are satisfied.

4. **Living cells have means of ensuring that their genetic information survives after the cell has died.** Since every cell within an animal usually has the same genetic information, not all of them have to reproduce to form a new individual of the same species. A few reproductive cells are capable of passing to the next generation all the genetic information contained in the individual organism. By cooperating to ensure the survival of these reproductive cells, the nonreproductive cells ensure the survival of their own genetic information.

The Internal Environment. Most of an animal's activities are directed toward the maintenance of an internal environment compatible with the lives of the animal's cells. In the mid-19th century the French physiologist Claude Bernard called attention to this fact in his statement that "all the vital mechanisms, varied as they are, have only one object: that of preserving constant the conditions of life in the *milieu intérieur.*" This quotation neatly summarizes one of the most important ideas of animal physiology, but after more than a century we need to qualify it somewhat. Bernard wrote vaguely of the *milieu intérieur* as being the body fluids, but today we would more precisely say that the environment of cells is the **interstitial fluid** in which cells are bathed. Also it seems that Bernard was overlooking reproductive physiology, for reproduction of the species has little to do with maintenance of the interstitial fluid and, in fact, seems to take precedence over it.

Another exception to Claude Bernard's statement would be that not all the conditions of life are kept constant in the interstitial fluid. In fact, only the levels of nutrients, oxygen, ions, and sometimes temperature are maintained within very narrow limits. Concentrations of nutrients, such as sugars for energy and amino acids for protein synthesis, are maintained by the feeding and digestive systems. Oxygen, which is required to obtain energy from nutrients, is maintained by the respiratory system. The concentrations of ions are regulated by excretory organs such as the kidney. In birds and mammals body temperature is held within a few degrees by various systems that control the production, gain, and loss of heat. Hormone levels, heart rate, neural and muscular activities, and many other functions can change radically as they help to regulate these levels.

In 1926 the American physiologist W. B. Cannon coined the term **homeostasis** to describe the maintenance of a steady state for these conditions. These homeostatic systems will be described in Unit Two. In addition, we shall study the nervous, muscular, hormonal, and circulatory systems, which are required for homeostasis. Finally, we shall examine the reproductive system, which is certainly as important for the survival of the species as the other systems are for the survival of the individual. One could say, in fact, that reproduction is more important; that the other physiological systems have as their ultimate goal the well-being of the reproductive system. As the English writer Samuel Butler put it more than a century ago, "The hen is only the egg's way of making another egg."

Interactions of Animals with Their Environments and Each Other. The need for animals to provide their cells with nutrients, oxygen, ions, and suitable temperatures inevitably leads to interactions with their environments and with each other. These interactions are studied in the fields of ecology, evolution, and behavior, which we shall survey in Unit Three. There are three fundamental principles that characterize the scientific study of these aspects of zoology:

1. The ecological interaction of animals with their external environments is largely a struggle for energy, because energy is required to maintain the internal environment and to reproduce. That struggle manifests itself in all the varieties of ways by which animals obtain food.

2. External environments have changed over the past 3.5 billion years that life has existed on Earth. If organisms have been adapted through all that time, they too must have changed in ways that are inherited from one generation to another. This change is evolution. The record of evolution is preserved in fossils and also in comparisons of the millions of species of organisms still living. Most zoologists believe this evolution results from the large amount of genetic variability within each group of organisms, and from the competition among them to survive and successfully reproduce. Genetic variations that brought success in the competition would naturally tend to be found in higher proportions in succeeding generations. In effect, nature would tend to select the organisms that are genetically endowed with the ability to leave more offspring. The increasing proportion of individuals with those naturally selected traits would be seen as a change in the species.

3. We assume that animal behavior results from physiological mechanisms that are in part under genetic control. We therefore assume that the behavior of animals also evolves in a way that improves the ability to reproduce.

Diversity.   Unit Four will describe some of the million or so species of animals known to be alive. In an attempt to keep track of so many animals, taxonomists classify them into different taxa. (Taxa is the plural of **taxon**.) Taxonomists use at least seven levels of taxa for animals:

Kingdom

Phylum (plural, phyla)

Class

Order

Family

Genus (plural, genera)

Species (plural, species)

(You might remember these taxonomic levels by reciting the mnemonic, "Kind Professor, Carefully Observing Fauna, Guiding Students.") Each level of the series includes all the lower levels. That is, each phylum includes one or more classes, each class includes one or more orders, and so on. Phylum Chordata includes class Mammalia, which includes order Primates, which includes family Hominidae, which includes genus Homo, which includes the species *Homo sapiens.* Even these seven levels are sometimes inadequate, so they are often divided into higher and lower levels, such as superclass and subspecies. As the previous discussion of kingdoms indicates, taxa above the level of the species are not fixed and absolute, but are created solely for the convenience of biologists.

Obviously, it will not be possible to cover each species of animal or even each genus, family, or order in this text. However, some attention will be given to each phylum and most classes, and a great deal of attention to lesser taxa in which zoologists have been most interested. One chapter each will be devoted to vertebrate classes (fishes, amphibians, reptiles, birds, and our fellow mammals) even though they represent less than 5% of animal species.

Throughout this text there is repeated reference to one particular species, because zoologists, like everyone else, have always been more interested in it than in any other. This species is, of course, the one we flatter with the name *Homo sapiens*—wise man. Discussion of the way we treat our fellow animals will often make us wonder whether the name is deserved. We shall ponder whether some intelligent species coming after us will view the history of humankind as a brief zoological catastrophe, whether the little we have learned through science will prove too much for our own good, and whether it might not have been better for the other animals if we had remained in our caves painting and admiring them.

SUMMARY

Zoology, the study of animals, is an endeavor that humans have always engaged in for survival and pleasure. Today there are numerous kinds of careers in zoology that require a variety of skills and talents. Succeeding as a zoologist is more a matter of creativity than of scientific method. An indispensable trait, however, is integrity.

Zoologists study animals, which can be defined as organisms that have diverse cells with nuclei and organelles, but without cell walls, and without the ability to photosynthesize. The life of animals is carried out by cells, which cooperate to create an "internal environment" in which they can survive, and in which reproductive cells can produce new offspring. The cellular processes of life consist of creating more order within the organism than occurs outside, and in animals this requires energy derived from food. These life processes are regulated, ultimately by genetic instructions that are passed from one generation to the next. Animal cells require certain levels of nutrients, oxygen, ions, and sometimes temperature, which are maintained homeostatically by physiological systems.

Animals must obtain nutrients, oxygen, and the other requirements of life from their environments, often by consuming other animals. Interactions of animals with their environments and each other comprise the study of ecology. Since environments have changed during the history of animals, the animals themselves must also have changed, since they continue to interact successfully with their environments. These changes, called evolution, have occurred largely because animals with genes that gave them a greater ability to survive and produce offspring were more likely to pass those genes to following generations. Among the evolutionary changes are those affecting behavior.

Animals have evolved into millions of different species, approximately a million of which survive and have been described. These are classified by taxonomists into phyla, classes, orders, families, genera, and species.

KEY TERMS

prokaryote	interstitial fluid	class
eukaryote	homeostasis	order
protist	taxon	family
protozoa	kingdom	genus
organelle	phylum	species

SELF-TEST

1. Think about the following lines:

   > What am I, Life? A thing of watery salt
   > Held in cohesion by unresting cells,
   > Which work they know not why, which
   >   never halt,
   > Myself unwitting where their Master
   >   dwells?

   John Masefield, *Sonnets* 14

   a. To what extent are the lines biologically relevant?

   b. To what degree is scientific zoology capable of the questions asked by Masefield?

2. Which of the following statements from this chapter is scientific, according to the criterion of falsifiability? (*Hint:* Try to imagine experiments potentially capable of proving the statements false.) Make up a sentence that is scientific but false. Make up one that is nonscientific but true.

   a. Zoology is one of the most exciting and enjoyable endeavors many of us can imagine.

   b. All living organisms consist of cells and materials made by cells.

   c. The hen is only the egg's way of making another egg.

   d. Evolution resulted from the large amount of genetic variability within each group of organisms, and from the competition among them to survive and to leave offspring in the next generation.

   e. Zoology is itself an expression of humanity.

3. If you are taking a laboratory with this course and your instructor requires reports, discuss what you should do if you get results that differ from what lectures and the text had led you to expect. Should you report actual observations, or should you change the results to make them consistent with what you would get if the book were correct? Suppose you did not follow instructions correctly: should you "cook" the data to hide your error? Try to formulate a general rule governing when it is ethical to report data that you did not actually observe.

4. For each of four characteristics of living cells (maintenance of internal order, transformation of energy, regulation, reproduction) name a nonliving thing that also has such a characteristic. Of the nonliving things you named, are there any that have all four characteristics? A virus consists of a protein coat surrounding genetic material, and it can reproduce only with the aid of a cell it infects. Are viruses alive?

5. Explain why no animal can live in isolation from the environment and other organisms. Explain how interactions with the environment and other organisms could lead to ecological, behavioral, and evolutionary differences among animals.

6. Name one kind of bacterium, protist, fungus, and plant, and describe how it interacts with an animal.

READINGS

Alternatives to Animal Use in Research, Testing, and Education. 1986. Washington, DC: Office of Technology Assessment.

Beveridge, W. I. B. 1957. *The Art of Scientific Investigation.* London: William Heinemann Ltd.

Broad, W. J. and N. Wade. 1983. *Betrayers of the Truth.* New York: Simon & Schuster. (*On fraud in science.*)

Chubin, D. E. 1985. Research malpractice. *BioScience* 35:80–89.

Fox, M. A. 1986. *The Case for Animal Experimentation.* Berkeley: University of California Press.

Gillispie, G. C. (Ed.). 1970–1980. *Dictionary of Scientific Biography.* New York: Scribner's. (*For biographies of Claude Bernard, W. B. Cannon, and other scientists referred to throughout this text.*)

Goodfield, J. 1981. *An Imagined World: A Story of Scientific Discovery.* New York:

Harper & Row. (*Excellent account of the daily life of an immunological researcher. See especially Chapter 15 on "scientific method" and creativity.*)

Janovy, J. Jr. 1985. *On Becoming a Biologist.* New York: Harper & Row.

Kilbourne, B. K. and M. T. Kilbourne (Eds.). 1983. *The Dark Side of Science.* San Francisco: Pacific Division, American Association for the Advancement of Science. (*On fraud in science.*)

Klemm, W. R. (Ed.). 1977. *Discovery Processes in Modern Biology.* Huntington, NY: Robert E. Krieger Publishing. (*Autobiographical accounts of discovery.*)

Leroi-Gourhan, A. 1982. The archaeology of Lascaux Cave. *Sci. Am.* 246(6):104–112 (June).

March, B. E. 1984. Bioethical problems: animal welfare, animal rights. *BioScience* 34:615–620.

Medawar, P. B. 1979. *Advice to a Young Scientist.* New York: Harper & Row.

Moss, T. H. 1984. The modern politics of laboratory animal use. *BioScience* 34:621–625.

Regan, T. 1985. *The Case for Animal Rights.* Berkeley: University of California Press.

Ritvo, H. 1984. *Plus ça change:* antivivisection then and now. *BioScience* 34:626–633.

Rodd, R. 1990. *Biology, Ethics and Animals.* New York: Oxford University Press.

Sechzer, J. A. (Ed.). 1983. The Role of Animals in Biomedical Research. *Ann. NY Acad. Sci.* vol. 406.

Sperlinger, D. (Ed.). 1981. *Animals in Research: New Perspectives on Animal Experimentation.* New York: Wiley.

Starr, D. 1984. Equal rights. *Audubon* 86(6):30–35 (Nov). (*The activities of various animal rights groups.*)

Unit One:
The Cellular and Molecular Bases of Life

These blood cells in a small human artery illustrate some of the principles to be covered in this unit. One of the major principles is that animals are composed of numerous cells of a great variety, each specialized for a unique function in the life of the animal. The biconcave disc-shaped cells are red blood cells (RBCs), also called erythrocytes. Each is about 7.5 millionths of a meter wide (7.5 micrometers—μm). Erythrocytes are specialized to carry oxygen to every other cell, which depends on the oxygen for the production of energy.

The spherical, rough-coated cells are white blood cells (WBCs), also called leukocytes. Leukocytes also make an essential contribution to the animal by recognizing and helping to destroy damaged cells and invading bacteria and other parasites. Other kinds of cells make up the layered inner surface of the small artery.

The structures of cells and their abilities to contribute to the life of the animal are determined by genetic instructions encoded in large molecules within the cells. These large molecules, called nucleic acids, make up genes. Genes control structure and functions by instructing the cell in how to make various proteins. These proteins then regulate other activities within the cell and also compose most of the structure of the cell. Genes also have to make copies of themselves and direct the synthesis of new molecules that will enable each cell to divide into two new and identical cells.

Remarkably, all the many kinds of cells in an animal have the same genetic information. At some point a cell in the human embryo acquires the ability to become bone marrow but not the lining of arteries. Later the cells become even more specialized as either red blood cells or white blood cells.

Overview

(From Tissues and Organs: A Text-Atlas of Scanning Electron Microscopy. By Richard G. Kessel and Randy H. Kardon. Copyright © 1979 by W. H. Freeman and Company. Reprinted with permission.)

2

Materials
and
Mechanisms of Cells

Water and life: Australian sea lion (Neophoca cinerea).

CHAPTER OUTLINE

Water and Related Matters
Atoms and Molecules
Water as a Molecule
Water as a Solvent
Acidity
Heat
Diffusion and Osmosis
Entropy
Organic Molecules
  Carbohydrates
  Lipids
  Proteins
Mechanism, Measurement, and Microscopy
The Plasma Membrane
  Structure
  Diffusion
  Active Transport
  Exocytosis and Endocytosis
Why Do Animals Have More Than One Cell?
Cellular Shape, Movement, and Interconnections
  The Cytoskeleton
  Motility

Intercellular Connections
Cytoplasmic Membranes

LEARNING OBJECTIVES

1. What is the underlying structure of matter?

2. Why is water essential for life?

3. What kinds of molecules are unique to life?

4. What functions do these molecules serve, and what are the properties that enable them to serve those functions?

5. Why do molecules tend to diffuse until their concentrations are uniform?

6. How do plasma membranes that enclose cells defy this tendency, maintaining different concentrations inside and outside cells?

7. How do cells hold their shapes, move, and attach to other cells?

8. What internal mechanisms of cells carry out the functions of life?

WATER AND RELATED MATTERS

It may seem strange to begin the study of such a fascinating subject as life with such an ordinary subject as water. Unless you live in the desert or in one of the increasing number of areas where the water is undrinkable, you probably seldom think about water. You, yourself, are approximately three-fourths water, however, and so am I. Some animals, such as jellyfishes, are as much as 90% water. A few animals can survive almost complete drying in an inactive state, but humans and most other animals die if they lose just one-fifth of their body water. Life origi-nated in water, and the vast majority of animals still reside there. Water is the canvas on which the portrait of life is painted. Like the artist, we must begin by preparing the canvas.

Another reason why it is appropriate to begin with water is that historically that is the way the scientific study of life began. The first known scientist, Thales (pronounced THAY-leez), who lived on the Mediterranean coast around 600 B.C., thought that all matter was made of water. A few decades later Anaximander, perhaps impressed with the perpetual tide of life washed up on the Mediterranean shores, speculated that even humans originated in the sea. Aristotle (384–322 B.C.) and other philosopher-scientists of ancient Greece soon realized that there were many more phenomena of nature than could be accounted for by the properties of one substance—even water. They therefore added three other substances: earth, fire, and air.

Thales, Anaximander, and the Greek philosopher-scientists adhered to a view that still guides modern scientists, at least unconsciously. That view is called mate-rialism. Since biologists are not noted for being greedy for material wealth, per-haps we should use the term **scientific materialism.** Scientific materialism is the belief that all natural phenomena, including those of life, are due to the properties of matter. It is a view that we shall assume throughout this chapter and throughout the text. According to this view, the phenomena of life are due in large part to the properties of the matter composing living things. Scientific materialism there-fore provides a further rationale for beginning the study of life with an exami-nation of the properties of water.

ATOMS AND MOLECULES

Some of the Greek philosopher-scientists considered water, earth, fire, and air to be made of particles that could not be divided into smaller particles without destroying their properties. They called such particles **atoms,** meaning indivisible. We still use the term "atom" for such indivisible particles, but we now know that water, earth, fire, and air are not atoms at all. Each particle of water, for example, is a **molecule:** a collection of two or more atoms that are bonded together. Water is made of one atom of oxygen and two atoms of hydrogen and is therefore represented by the familiar formula H_2O. Like all other atoms, hydrogen and oxygen have nuclei made of protons and usually some neutrons. A proton is rel-atively massive and carries one positive charge. All atoms of the same type, or element, have the same number of protons in the nucleus. Hydrogen always has one proton in the nucleus, and oxygen always has eight protons in its nucleus. A neutron has about the same mass as a proton, but no charge. Hydrogen and other atoms can exist in several forms, called **isotopes,** which differ in the number of neutrons in the nucleus. The most common isotope of hydrogen has one proton and no neutron in the nucleus. Other isotopes of hydrogen have one or two neutrons in their nuclei and are called deuterium and tritium, respectively. The

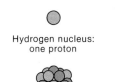

Hydrogen nucleus:
one proton

Oxygen nucleus:
8 protons + 8 neutrons

A

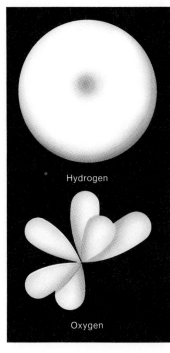

Hydrogen

Oxygen

B

Figure 2.1
(A) Models of a hydrogen (H) and an oxygen (O) nucleus. Electrons are not shown since the page would have to be about 30 feet wide to fit them into the figure. (B) Models of hydrogen and oxygen atoms showing the electrons. At the atomic scale the position of particles cannot be defined, so the cloudlike figures represent the probable locations of electrons. A hypothetical camera with its shutter left open might record such blurred images of electrons, which can be thought of as small, negatively charged particles. The nuclei are vanishingly small, somewhere near the center. One million atoms in a row would not quite reach across the period at the end of this sentence.

most common isotope of oxygen has eight protons and eight neutrons in the nucleus. Around the nucleus are electrons, which have little mass and one negative charge each. The number of electrons in a stable atom equals the number of protons, so the net charge on the atom is zero (Figure 2.1). The bonds that hold atoms together in a molecule are due to the activity of electrons.

Each electron is confined to one of several **energy levels** that correspond to different distances from the nucleus. The first energy level, which is closest to the nucleus, can hold no more than two electrons. The second and third levels can hold no more than eight electrons each. If the highest level is not completely filled, the atom can share one or more electrons with another atom. This sharing of electrons forms a type of chemical bond called **covalent bond.** Hydrogen has only one electron, so it has one vacancy in the first energy level where it can share an electron with another atom. For example, a pair of hydrogen atoms can form covalent bonds with each other, making a hydrogen molecule (H_2). Oxygen has eight electrons—two in the first energy level and six in the second—so it has two vacancies in the second energy level for electrons from other atoms to share. Like hydrogen, a pair of oxygen atoms can share electrons with each other, forming an oxygen molecule (O_2).

WATER AS A MOLECULE

Water is formed when two hydrogen atoms form covalent bonds with one oxygen atom. The chemical reaction in which hydrogen forms covalent bonds with oxygen to make water can be written as follows:

$$2H + O \longrightarrow H_2O$$

The two hydrogen atoms do not stick out of a water molecule at opposite ends, as one might expect, but actually form a 105° angle with each other (Figure 2.2). As a result the positive charges of the hydrogen nuclei are concentrated on one side of the water molecule, and the negative charges of most of the oxygen electrons are concentrated at the opposite end. This separation of charge gives water a **polar** structure, like a bar magnet with opposite poles on each end. Just as the north pole of one bar magnet tends to stick to the south pole of another, the positively charged end of a polar molecule such as water is attracted to the negatively charged region of other polar molecules. Such an attraction between two water molecules is an example of a **hydrogen bond.** A hydrogen bond is the charge attraction between a hydrogen atom that is covalently bonded to an oxygen or nitrogen atom in a polar molecule, and a negative region of the same or a different polar molecule. A single hydrogen bond is quite weak, but in large numbers hydrogen bonds account for many of the life-giving properties of water, as well as proteins and the molecules responsible for heredity.

One of the vital properties of water due to hydrogen bonds is **cohesion:** the attraction of identical molecules to each other. Cohesion accounts for the tendency of water to form drops. Were it not for cohesion, cells would lose their shapes, and body fluids would leak out of the tiny crevices in tissues. Equally important is **adhesion,** which is the attraction of different kinds of molecules to each other (Figure 2.3). Water adheres to other polar molecules because it forms hydrogen bonds with the negatively charged parts of those molecules. Without adhesion, water would not make anything wet. Life without wet cells and organs is inconceivable.

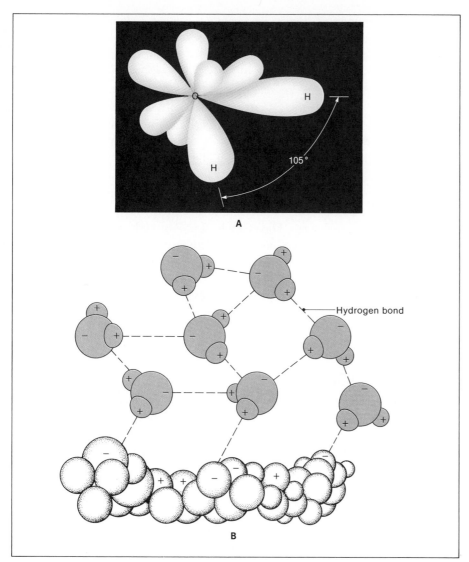

A

B

Hydrogen bond

Figure 2.2
(A) Two hydrogen atoms form covalent bonds with an oxygen atom to make water. (B) The unequal distribution of charge on water molecules causes weak attractions between them, called hydrogen bonds. Water molecules are also weakly attracted to negatively charged parts of other molecules.

WATER AS A SOLVENT

The property of adhesion explains why many substances dissolve in water. When water molecules adhere to individual molecules of a substance, the cohesion between molecules of the substance will be weakened. The molecules will then

Figure 2.3
(A) An illustration of cohesion and adhesion. The common pond snail Heliosoma *hangs from the surface of the water because molecules in its foot adhere to the water molecules. Surface tension, which is due to the cohesion of water molecules, supports the weight of the snail. (B) Cohesion and adhesion are due to hydrogen bonds, represented here as springs.*

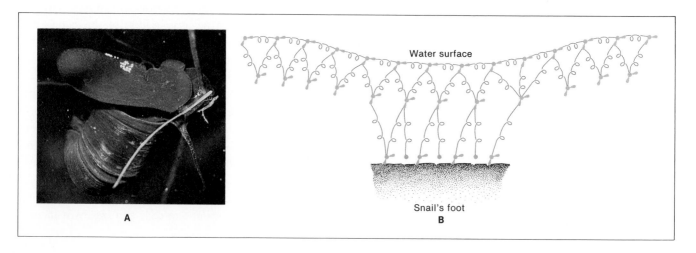

A

Water surface

Snail's foot

B

The number of molecules of a substance and the mass of the substance are related to each other by the **mole**. A mole of any substance equals 6.02×10^{23} molecules (Avogadro's number) and has a mass in grams equal to one **gram molecular mass** (= gram molecular weight). The gram molecular mass of a substance is a mass in grams approximately equal to the total number of protons and neutrons in a molecule of the substance. One mole of H_2 (one gram molecular mass) therefore has a mass of 2 grams, and one mole of O_2 has a mass of 32 grams. Concentration can therefore be expressed as the number of moles per liter of solution, or **molarity**. A 1 molar solution contains 1 mole of solute per liter of solution, or 6.02×10^{23} molecules per liter, regardless of the identity of the solute.

disperse, or dissolve, in the water. The substance that dissolves is called the **solute**, and the substance in which it dissolves is called the **solvent.** The combination of solute and solvent is called a **solution.** Water is one of the best naturally occurring solvents for polar molecules. This property is essential for life, because it allows large amounts of biologically important substances to be contained within cells and to circulate within the body fluids of organisms. The number of molecules of solute dissolved in a given volume of solution determines the **concentration** of the solution. It would be somewhat tedious to count the number of molecules in a volume of solution, however, so the number of molecules is usually determined indirectly by weighing the mass of solute.

Many molecules form chemical bonds in which electrons are not shared, as in covalent bonds, but are transferred from one atom to another. These are called **ionic bonds.** Ionic bonds are easily broken when molecules dissolve in water, allowing atoms in the molecules to dissociate. During such dissociation an electron from one atom of the molecule remains with the other part of the molecule, forming one or more **ions.** Ions are atoms or molecules that carry a net charge, because the number of electrons does not equal the number of protons. For example, when crystals of ordinary table salt, sodium chloride (NaCl), dissolve in water, they dissociate into sodium and chloride ions (Figure 2.4). During this dissociation sodium loses the single electron in its third energy level and becomes a sodium ion. Since the sodium ion has one more proton than electrons, it is written as Na^+. The lost electron takes up residence in the available place in the third energy level of chlorine, which therefore becomes the chloride ion. The chloride ion then has one extra negative charge, and it is written Cl^-. Positively charged ions such as Na^+ are called **cations.** Negatively charged ions like Cl^- are called **anions.**

ACIDITY

A small proportion of water molecules also dissociate, forming hydrogen ions (H^+ = protons) and hydroxyl groups (OH^-):

$$H_2O \longrightarrow H^+ + OH^-$$

The small concentration of hydrogen ions that results from the dissociation of water can have profound effects on biological processes. The reason is that the

Figure 2.4
(A) Atoms of sodium (Na) and chlorine (Cl) linked by ionic bonds form regular arrays in a crystal of ordinary table salt (NaCl). (B) When dissolved in water, NaCl dissociates into sodium ions (Na^+) and chloride ions (Cl^-). Because of charge attraction, water molecules orient in a shell around these ions. This phenomenon is called hydration.

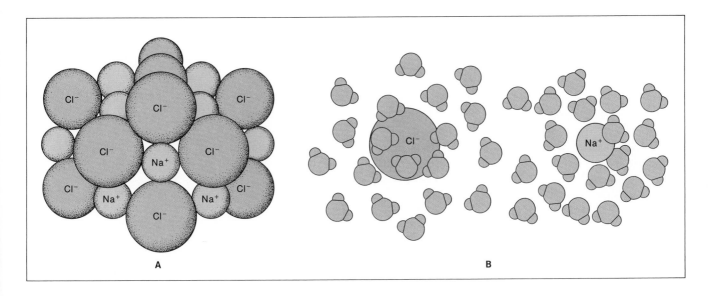

A   B

distribution of charges on proteins and other biological molecules determines their ability to function. If the negatively charged parts of such molecules are covered by hydrogen ions, they may function at a different rate, or perhaps not at all. In some cases molecules that normally adhere to each other by charge attraction split apart if the concentration of H^+ is too high.

The concentration of hydrogen ions in a solution determines its **acidity.** Acidity is usually measured by **pH,** which is the negative exponent of the H^+ concentration expressed as a power of ten. For example, the average concentration of H^+ ions in pure water is 10^{-7} molar, so the pH of pure water is 7. This is said to be neutral. By adding an acid, which contributes additional hydrogen ions, the concentration might be raised to 10^{-1} molar, for a pH of 1. This extreme acidity occurs in the human stomach when too much hydrochloric acid (HCl) is secreted (Figure 2.5). Adding a base, which is a substance that contributes hydroxyl groups, reduces the concentration of H^+ by combining them with OH^- to form water. Adding sodium hydroxide (NaOH) to water, for example, could reduce the H^+ concentration to 10^{-11} molar, for a pH of 11.

Certain molecules or combinations of molecules have the property of stabilizing, or buffering, the concentration of H^+ in water. Such molecules are called **buffers.** Buffers work by releasing H^+ when the pH begins to rise, and binding H^+ when the pH begins to fall. A buffer often consists of a weak acid and a related salt. In human blood, for example, the weak acid is carbonic acid (H_2CO_3), and the salt is sodium bicarbonate ($NaHCO_3$), which dissociates into Na^+ and bicarbonate (HCO_3^-). If the pH is too low the excess H^+ tends to bind with bicarbonate, forming carbonic acid. If the pH is too high, the carbonic acid releases H^+. Proteins also have the ability to release or bind hydrogen ions, and they therefore act as buffers in body fluids and within cells. Buffering is extremely important in preventing changes in pH from upsetting the functioning of numerous molecules essential for life. In human blood a deviation of only 0.2 pH units from the normal value of approximately 7.4 can be life-threatening.

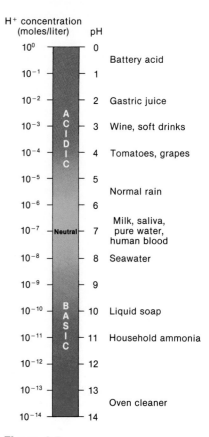

Figure 2.5
A scale showing H^+ concentration, the corresponding pH values, and some representative solutions.

HEAT

Many of the properties of water and other molecules depend on how much they are agitated by thermal energy. A measure of this thermal energy is **temperature.** The form of the energy that causes this thermal agitation is **heat.** Thus adding heat to a substance increases its temperature. One of the interesting and important properties of water is that it takes more heat to raise the temperature of a certain mass of it than for any other substance. Conversely, water has to lose a lot of heat to cool off. In other words, water has a high **heat capacity.** It takes one **calorie** of heat to raise the temperature of one gram of water by 1°C. The average adult human has from 40,000 to 50,000 grams of water in the body, making a volume of 40 to 50 liters. If you imagine trying to heat more than 10 gallons of water on a stove you will appreciate why your body temperature does not readily fluctuate. The same resistance to fluctuation in temperature also protects animals that live in lakes and oceans. Although air temperature can change drastically within a few hours, the temperatures of large bodies of water tend to remain constant for days and weeks at a time. Water has several other interesting and vital thermal properties that will be discussed in more detail in Chapter 15.

DIFFUSION AND OSMOSIS

The jostling of water and other molecules due to thermal agitation gives rise to three phenomena that lie at the heart of any understanding of life. The first such

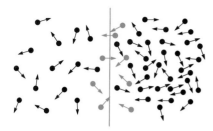

Figure 2.6
The explanation for diffusion. Each molecule is moving in a random direction (arrows) due to thermal agitation. There are more molecules in the region with high concentration, however, so more molecules will be moving toward the region with low concentration than in the reverse direction. Here a total of eight molecules are about to cross the boundary separating the two concentrations (color). Five molecules from the more concentrated solution will cross into the dilute solution, and three molecules will cross in the reverse direction. Thus, there is a net flow, or diffusion, from high to low concentrations.

phenomenon is **diffusion.** Diffusion refers to the tendency of molecules to move from a region where they are highly concentrated to a region of lower concentration until the molecules are equally distributed. If one drops a cube of sugar into a cup of coffee, for example, the sugar will dissolve and eventually become evenly distributed throughout the coffee, even without stirring. A great deal has been written about the cause of diffusion, but the concept is fundamentally simple. Thermal agitation causes each molecule to move in a random direction. It is therefore just as likely that a molecule in a region where it is highly concentrated will move to a region of low concentration as it is that a molecule in a region of low concentration will move to one with a high concentration. There are more molecules in the region with high concentration, however, so more molecules will be moving from the high to the low concentration than will be moving in the reverse direction (Figure 2.6). Eventually, so many molecules will have moved from regions of high to low concentration that the concentrations will be equal. The molecules will continue to move about in random directions, but there will be no further change in concentration. This state is called **equilibrium.**

Water obeys the same commonsense principles of probability as other molecules; it too diffuses from high to low concentrations. The more molecules of solute dissolved in a given volume of solution, the less water will fit into the volume. Therefore water diffuses from a dilute solution into one that is more concentrated. The diffusion of water from a low solute concentration into a higher one is called **osmosis.** Osmosis is usually observed only when solutions having two different concentrations are separated by a membrane or other structure that allows water, but not solute, to pass through (Figure 2.7). Because of osmosis, the more concentrated solution will accumulate water and therefore develop an increased pressure, called **osmotic pressure.** Osmotic pressure is generally proportional to the difference in concentrations in the two solutions. A pressure similar to osmotic pressure can also result from **colloids.** Colloids occur when molecules that are too large to dissolve are prevented from precipitating by hydrogen bonding with the water. The pressure due to colloids is called **colloid osmotic pressure** or **oncotic pressure.**

ENTROPY

Diffusion and osmosis are two examples of a far more general principle: Any natural process results in an increase in the total disorder of objects involved in the process. The reason is that there are many more ways for the objects to be disordered than for them to be ordered, so any change in a system is apt to increase its disorder. For example, there are many ways for sugar molecules to be distributed randomly throughout a coffee cup, but many fewer ways for them to be arranged in a cube. Diffusion or any other natural change in the distribution of sugar molecules is therefore likely to increase disorder. Sugar dissolves spontaneously; it does not form cubes spontaneously in water. The same principle applies even to mundane objects. There are countless places to lose one sock, but only a few places where the sock belongs. A sock is therefore much more likely to become misplaced than it is to spontaneously end up neatly folded in a drawer beside its mate.

This principle is so important that is has been elevated to the status of a physical law, called the **Second Law of Thermodynamics.** There are various ways of stating the Second Law. One is as follows: No process is possible in which the only result is the complete conversion of heat into work. One cannot, for example, create a steam engine that would convert all the heat in steam into useful work. If that were to happen the steam would freeze, then continue to lose heat until

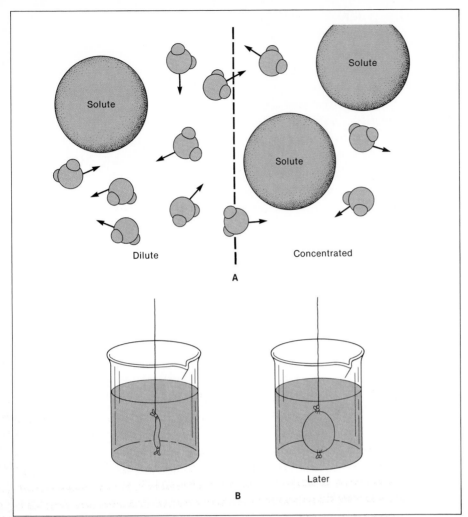

Figure 2.7
(A) In two solutions having different solute concentrations, there will be a greater concentration of water in the solution with the lower concentration of solute. Consequently, water will diffuse from the more dilute solution into the more concentrated one. This process is called osmosis. (B) If a cellophane bag partially filled with a sugar solution is suspended in water, the bag will accumulate water and begin to swell because of osmotic pressure. The same thing can happen to animal cells (see Figure 14.1).

Solute

Solute

Solute

Dilute

Concentrated

A

Later

B

all molecular vibration ceased at the temperature of absolute zero (0 kelvin = −273° Celsius). This is true even though the conversion of the heat energy into an equivalent amount of mechanical energy would not violate the First Law of Thermodynamics, which states that energy can be converted from one form to another but can never be destroyed.

The Second Law can be stated more precisely in terms of a quantity called **entropy,** which is a kind of measure of disorder. Entropy cannot be observed or measured directly, but one can calculate how much it changes. If an object at temperature T gains a small quantity of heat Q, then the object gains an amount of entropy equal to Q/T. It is a familiar fact that some heat is produced as a by-product of any natural process. Every natural process, therefore, whether it is diffusion, osmosis, operating a steam engine, or losing one's socks, increases the entropy of the universe. This is another way of stating the Second Law.

The Second Law of Thermodynamics is bad news for those who like to think very, very far ahead. It means that the universe is constantly becoming more disordered, and that there is nothing we can do about it. In fact, the only way to slow down the increase in entropy is to do absolutely nothing. Inevitably the universe will reach a state of total uniformity, with matter and temperature evenly distributed. That prospect is a long way off, however. The Second Law raises more immediate concerns having to do with our own lives and the lives of every other organism.

ORGANIC MOLECULES

At first glance life appears to violate the Second Law of Thermodynamics. Instead of increasing disorder, living organisms increase order within themselves as they grow, reproduce, and metabolize. During photosynthesis, for example, plants convert six molecules of CO_2 into a single, more-ordered molecule of the sugar glucose. Furthermore, plants and animals use the energy in glucose to create carbon-based molecules that are even more highly ordered. Such complex, carbon-based molecules are called **organic molecules,** because they now occur naturally only where there are (or were) organisms. Despite appearances, the synthesis of organic molecules and the other activities of living do not violate the Second Law. Any increase in order that results from the synthesis of organic molecules is accompanied by an even greater increase in disorder outside the organism, due to the production of heat and metabolic wastes. Consequently, living does increase the total entropy of the universe, as predicted by the Second Law.

Although it is not against the Second Law to make organic molecules, it is nevertheless noteworthy that in nature only living organisms now do it. Many respected scientists once regarded this fact as proof that living organisms have unique powers. The belief in such unique powers of life, called **vitalism,** began to weaken in 1828 after Friedrich Wöhler synthesized the simple organic substance urea from inorganic molecules. Nowadays, of course, plastics and other complex organic molecules are routinely manufactured without any organisms present. Organic molecules apparently also occurred spontaneously on Earth billions of years ago, even before life appeared (see pp. 387–390). Therefore no vitalistic power is needed to account for organic molecules.

Organic molecules are special, nonetheless. What makes them so is their enormous variety of structures and properties. This variety is due to each carbon atom's ability to form four covalent bonds with a variety of other atoms. Carbon has a total of six electrons. Therefore it has four electrons in its second energy level that it can share with other atoms, and four vacancies in that energy level that can accommodate electrons from other atoms (Figure 2.8a). In organic molecules carbons (C) commonly bond with each other, as well as with hydrogen (H), nitrogen (N), oxygen (O), phosphorus (P), and sulfur (S) (CHNOPS). These other atoms often associate with each other in common **functional groups** that affect the properties of organic substances (Figure 2.8b). Alcohols have at least one hydroxyl (—OH) group, for example. The carboxyl group (—COOH) makes an organic molecule acidic, because it can lose the H as a hydrogen ion.

The most common organic molecules in living organisms belong to one of four classes: nucleic acids, carbohydrates, lipids, and proteins. The nucleic acids, deoxyribonucleic acid (DNA) and ribonucleic acid (RNA), are involved in heredity. They consist of long sequences of sugars (either deoxyribose or ribose) linked to each other by phosphate groups, with nitrogenous bases attached to the sugars. The sequence of nitrogenous bases controls the production of proteins, thereby determining hereditary traits. The structure and functioning of nucleic acids will be described in detail in Chapter 5.

Carbohydrates. Carbohydrates include simple sugars, such as glucose, and more complex sugars such as starch and cellulose. A simple sugar has an aldehyde or ketone group and two or more hydroxyl groups and consists of C, H, and O in the ratio 1:2:1. The formulas for both glucose and fructose, for example, are $C_6H_{12}O_6$. These six-carbon sugars are called **hexoses.** Simple sugars with five carbons are **pentoses,** and those with three carbons are **trioses.** When crystallized, simple sugars take the form of carbon chains. When dissolved in water,

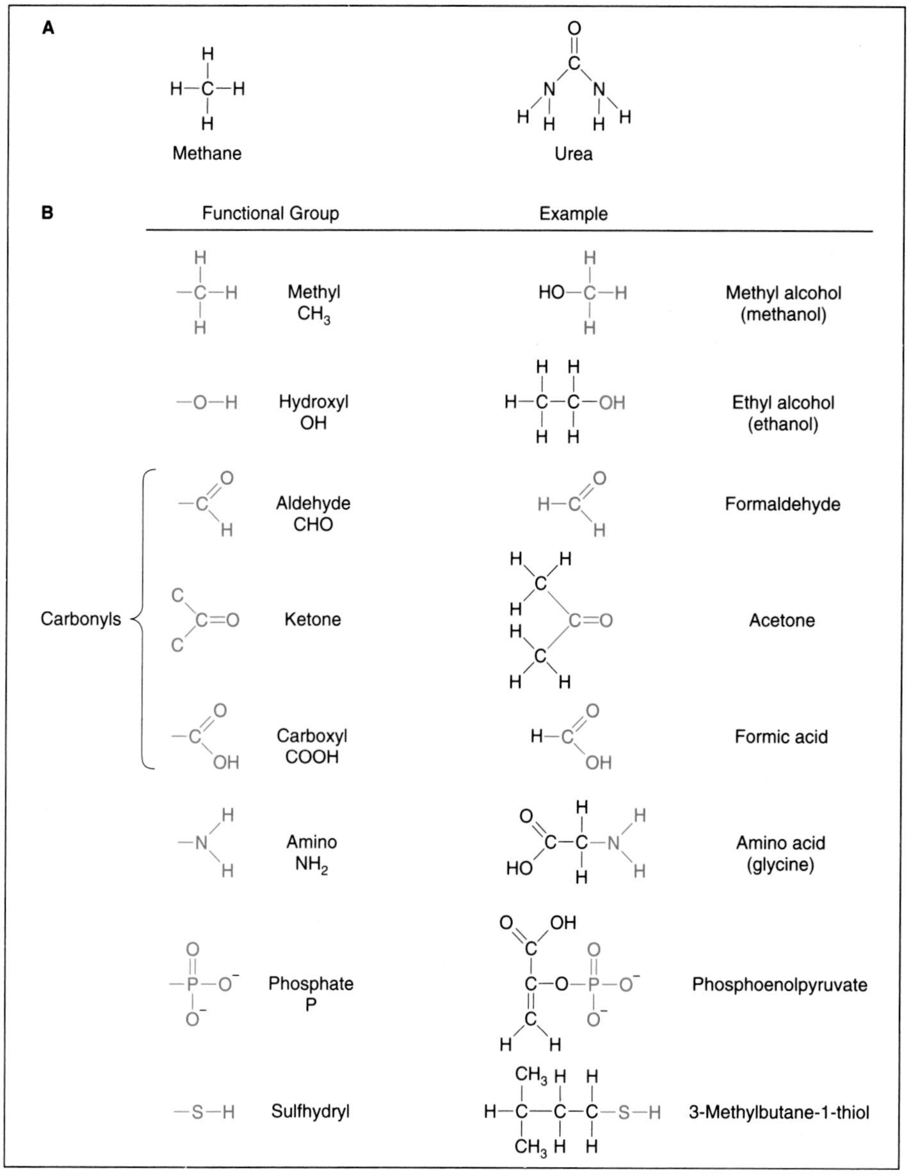

A

Methane

Urea

B

Functional Group		Example	
	Methyl CH₃		Methyl alcohol (methanol)
	Hydroxyl OH		Ethyl alcohol (ethanol)
Carbonyls {	Aldehyde CHO		Formaldehyde
	Ketone		Acetone
	Carboxyl COOH		Formic acid
	Amino NH₂		Amino acid (glycine)
	Phosphate P		Phosphoenolpyruvate
	Sulfhydryl		3-Methylbutane-1-thiol

Figure 2.8
(A) A carbon atom has six protons and (usually) six neutrons in the nucleus. Four of its six electrons occupy the second energy level, leaving four vacancies. These four electrons and vacancies allow a carbon atom to covalently bond with up to four other atoms, forming a variety of organic molecules. Two of the simplest organic molecules are methane (CH₄) and urea [CO (NH₂)₂]. In methane the one electron of each of the four hydrogens forms a single bond with the carbon. In urea the carbon shares two electrons with oxygen, forming a double bond, and forms single bonds with each of the two nitrogen (N) atoms. (B) Carbon also bonds with a variety of functional groups that affect the properties of organic molecules. Some examples are shown. Formic acid is an irritant secreted by ants in self-defense, and 3-methylbutane-1-thiol is the odorant secreted by skunks in self-defense.

however, simple sugars tend to form rings (Figure 2.9). When the covalent bonds between carbon atoms in simple sugars are broken inside cells, using the controlled processes that will be described in the next chapter, the energy from those bonds can be captured by the cell to perform other functions. Simple sugars are therefore used as immediate sources of energy for cellular functions. In humans and many other animals, glucose circulates in blood and other body fluids as a ready source of energy.

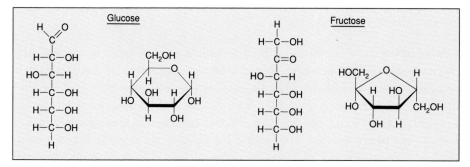

Figure 2.9
The structure of two simple sugars represented in linear form (left), and in the ring form that occurs in solution. The Cs in the ring occur at the angles and are usually not shown. Glucose circulates in the blood of humans and many other animals and is a major source of energy. Fructose occurs in semen as an energy source for sperm.

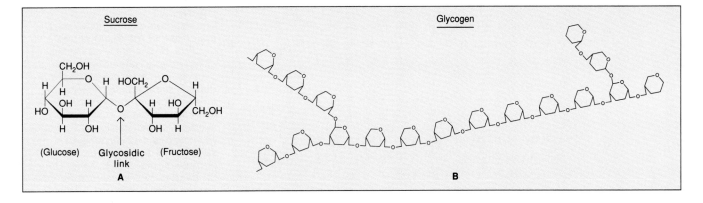

CH$_2$OH

(Glucose)   Glycosidic   (Fructose)
link
A

Glycogen

B

Figure 2.10
(A) Common table sugar, sucrose, is a disaccharide consisting of a glucose and a fructose joined by a glycosidic link. (B) The polysaccharide glycogen is a branched chain of glucose subunits. Starch is similar except that its chains have fewer branches.

Simple sugars are also called **monosaccharides.** When two simple sugars are linked together they form a **disaccharide.** Ordinary table sugar is the disaccharide sucrose, which consists of a molecule of glucose linked to a molecule of fructose (Figure 2.10A). The linkage, formed by oxygen, is called a **glycosidic link.** Digestion breaks glycosidic links, liberating simple sugars that can enter the bloodstream as sources of cellular energy. Another important disaccharide in humans and other mammals is lactose, the sugar in milk. Lactose, consisting of glucose linked to a similar hexose called galactose, provides much of the energy needed by infant mammals.

Quite long chains of simple sugars can be joined by glycosidic linkages to form **polysaccharides.** Polysaccharides are much less soluble than mono- and disaccharides and therefore generally do not circulate in blood, milk, or other body fluids. They serve mainly as long-term energy stores or for structural support. Two common polysaccharides, cellulose and starch, are made by plants. Both consist entirely of glucose, but their glycosidic linkages differ. Cellulose, which helps form the walls around plant cells, has glycosidic links that neither humans nor most other animals can digest. We and most other animals that eat plants excrete cellulose as roughage in the feces. Starch, the white material that plants synthesize to store glucose, has glycosidic links that humans and most other animals can digest. Glucose from digested starch is a major source of energy for many animals. Another important polysaccharide that is similar to starch is **glycogen** (Figure 2.10B). Certain cells in humans and other animals produce glycogen as between-meal reserves of glucose.

Lipids. Fats and several other kinds of organic molecules are called lipids. Fats—**neutral fats,** to be precise—are the most familiar lipids. A neutral fat consists of a three-carbon molecule of **glycerol** to which up to three **fatty acids** are attached (Figure 2.11). Fatty acids are long chains of carbon, with hydrogens attached to the carbons, and an acidic carboxyl group at one end. If three fatty acids are attached, the neutral fat is a **triglyceride** (= triacylglycerol). If one or two fatty acids are attached, it is a monoglyceride or diglyceride. On the same triglyceride any fatty acid can differ from the others in the number of carbons and in the number of double bonds between adjacent carbons. The fewer double bonds a fat has, the more hydrogen can attach to the carbons, and the more **saturated** the fat is said to be. The degree of saturation of a glyceride affects its fluidity at a given temperature. Saturated glycerides are fluid at higher temperatures but become solid, like lard, at room temperature. Saturated fats are more common in birds and mammals ("warm-blooded" animals), as well as in other kinds of animals and plants from warm climates. Unsaturated glycerides are fluid (oils) at room temperature and are more common in plants and animals from cold climates.

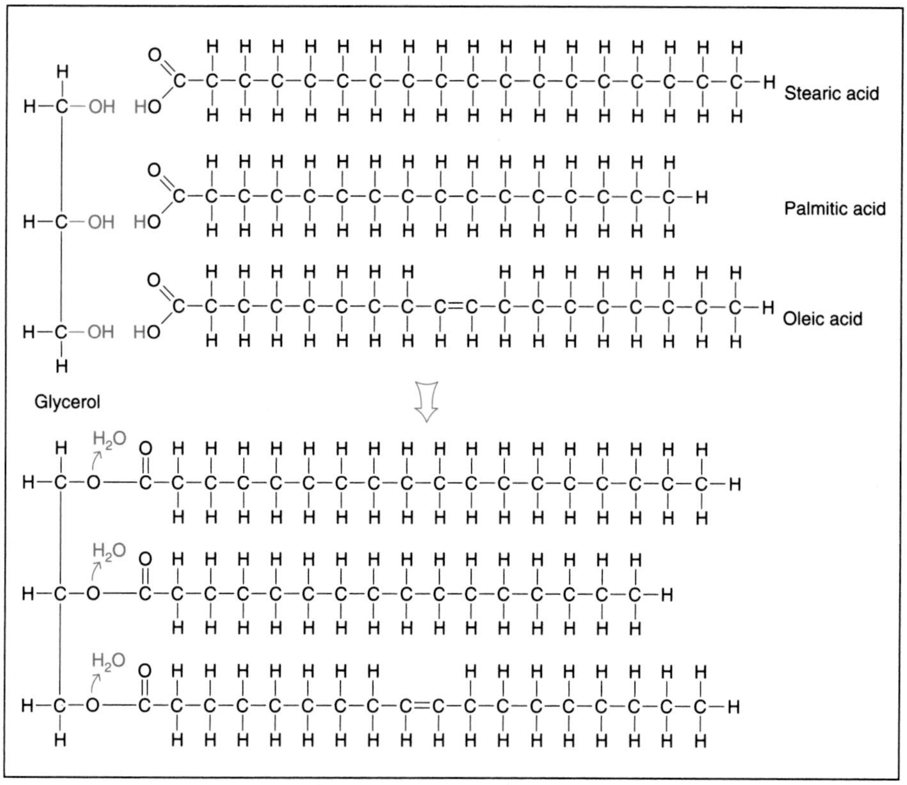

Stearic acid

Palmitic acid

Oleic acid

Glycerol

H_2O

H_2O

H_2O

Figure 2.11
The structure of a triglyceride. The upper part of the figure shows glycerol, on the left, and three different fatty acids. Stearic acid and palmitic acid are both saturated but differ in the number of carbon atoms. (The number of carbons is even because fatty acids are synthesized two carbons at a time.) Oleic acid has 18 carbons, like stearic acid, but is unsaturated because of the double bond. During formation of a triglyceride, hydroxyl groups from glycerol and an H from the fatty acid are removed as water, producing a bond through oxygen (an ester bond). Formation of a chemical bond by removal of H_2O is called a dehydration reaction. Breaking the bond by adding H_2O, as occurs when fats are digested, is called a hydrolysis reaction.

Lipids are said to be **hydrophobic** ("water-fearing") because of their insolubility in water. This insolubility explains why fats and oils form globules or a film in dishwater. Glycerides are hydrophobic because of the hydrogen nuclei that surround the fatty acid chains. These positive charges repel the positive charges of the hydrogens in water, preventing the formation of hydrogen bonds. Neutral fats do, however, dissolve in many other lipids and in liquids such as acetone and alcohol.

The insolubility of neutral fats in body fluids and their large number of carbon–carbon bonds make them ideal as long-term energy stores. In times when food is readily available animals tend to eat more than they immediately need. Fat-storing cells, called **adipose cells,** use the excess energy **(calories)** to synthesize triglycerides. When food is scarce, much of this energy can be recovered by breaking down the triglycerides. Adipose tissues are also effective in insulation against cold. In humans the amount of fat deposited depends on a variety of environmental, genetic, and behavioral factors (see pp. 276–278). The average young American man has a little over 10% of his body weight in the form of triglycerides. This is enough energy to see him through a 40-day fast. American women have an even higher proportion of body fat (approximately 18%), much of it distributed in a layer beneath the skin.

Another important lipid, **phospholipid,** is similar in structure to a triglyceride. The main difference is that in a phospholipid one fatty acid is replaced by a different kind of substance, called a **head group,** that is linked to the glycerol by a phosphate (Figure 2.12A). The head group is polar, so it is able to form hydrogen bonds with water. This makes the head group soluble, or **hydrophilic** ("water-loving").

The hydrophilic head group combined with two hydrophobic fatty acid "tails" accounts for one of the most interesting and useful properties of phospholipids. If a phospholipid such as lecithin is added to water, it spontaneously forms a double layer called a **lipid bilayer.** The hydrophobic tails avoid the water by

Figure 2.12
(A) A phospholipid is essentially a triglyceride with one fatty acid replaced by a phosphate-linked head group. The head group shown here is choline, forming the phospholipid lecithin (= phosphatidylcholine). Three other common head groups are ethanolamine, the amino acid serine, and inositol. (B) Because the head groups are hydrophilic and the fatty acid tails are hydrophobic, phospholipids in water spontaneously form a lipid bilayer.

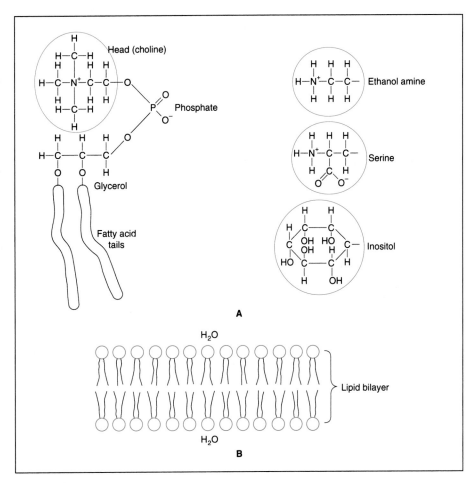

pointing toward each other into the middle of the bilayer, and the hydrophilic heads point outward toward the water on each side of the layer (Figure 2.12B). The avoidance of contact with water by the tails also causes the lipid bilayer to spontaneously form hollow spheres. We shall soon see that these properties are essential in the formation of cell membranes.

Also included among lipids are **steroids,** such as cholesterol and certain hormones (see Figure 10.4). Steroids differ from other lipids in that they are based on a four-ring structure. Differences in the functional groups attached to the rings determine how the steroids function. For example, the male sex hormone testosterone differs from the female sex hormone estradiol mainly in having a ketone group where estradiol has a hydroxyl group. Cholesterol is synthesized from saturated fatty acids and is an important constituent in membranes. Another group of lipids are **waxes,** which consist of long-chain fatty acids linked to long-chain alcohols or carbon rings. Waxes have a variety of special applications, such as the comb of beehives and as waterproof coverings in animal skins, hair, and feathers.

Proteins. The last major class of organic molecules to be described here are the proteins. Proteins consist of **polypeptides,** which are chains of **amino acids.** Polypeptides with only a few amino acids are referred to as **peptides.** As the name implies, an amino acid is a molecule with both an amino group and an acidic carboxyl group. Although an infinite variety of amino acids is possible, only 20 occur in natural proteins. These amino acids all have both the amino and carboxyl groups attached to the same carbon. Also attached to that carbon is an H, and an **R group** that is different for each of the 20 amino acids (Figure 2.13).

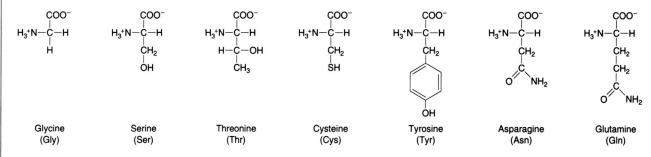

Figure 2.13
(A) The general structure of an amino acid. At the normal pH found in cells the amino groups often have an additional H^+ and the carboxyls often lose an H^+, as shown. (B) The 20 amino acids that occur in natural proteins, grouped according to their properties.

A

Nonpolar (Hydrophobic) Amino Acids

Alanine
(Ala)

Valine
(Val)

Leucine
(Leu)

Isoleucine
(Ile)

Phenylalanine
(Phe)

Tryptophan
(Trp)

Methionine
(Met)

Proline
(Pro)

Polar Uncharged Amino Acids

Glycine
(Gly)

Serine
(Ser)

Threonine
(Thr)

Cysteine
(Cys)

Tyrosine
(Tyr)

Asparagine
(Asn)

Glutamine
(Gln)

Negatively Charged Amino Acids

Positively Charged Amino Acids

Aspartic acid
(Asp)

Glutamic acid
(Glu)

Lysine
(Lys)

Arginine
(Arg)

Histidine
(His)

B

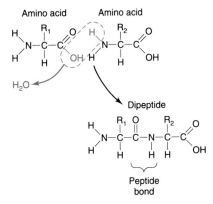

Figure 2.14
Formation of a dipeptide from two amino acids.

Differences in the R groups affect the properties of the amino acids, and the properties of the peptides and proteins in which they occur. Eight of the natural amino acids are nonpolar and therefore hydrophobic, like fats. The others are polar and therefore hydrophilic. Of the 12 polar amino acids, seven bear no net charge (at the pH normally found in cells). Three polar amino acids are positively charged, and two are negatively charged. These charged amino acids can interact electrically with each other and with other charged molecules. A large protein with many different amino acids can therefore be hydrophobic in some parts and hydrophilic in others and have different charges in different areas. These regional differences affect the overall shapes of proteins and the way they interact with other molecules.

In peptides and proteins the amino acids link to each other by **peptide bonds.** A peptide bond is formed when the OH is removed from the carboxyl of one amino acid, an H is removed from the amino group on another amino acid, and the two groups bond to each other (Figure 2.14). Two amino acids linked by a peptide bond form a dipeptide, three linked in this way form a tripeptide, and so on. Many peptides function as hormones. Some proteins function as hormones, and many others are responsible for maintaining cellular shape, for muscle contraction, for regulating chemical reactions within cells, and for such diverse structures as tendons, hair, feathers, horns, and claws. With 20 amino acids available, the number of different proteins is unimaginably large. The number of different dipeptides is $20 \times 20 = 20^2 = 400$, the number of tripeptides is $20^3 = 8000$, and so on. The number of polypeptides with 100 amino acids is 20^{100}, which equals approximately 10^{130}. Since there are only about 10^{27} molecules in a human, and only about 10^{71} atoms in the entire universe, it is obvious that no animal could ever produce all the different possible proteins. However, even the simplest animal must produce thousands of different proteins to perform all the functions of life.

The different functions of proteins are due to their different structures. One way in which they can differ is in the sequence of amino acids, called the **primary structure.** The primary structure is determined genetically, as will be described in Chapter 5. Superposed on the primary structure are characteristic patterns, called the **secondary structure,** that are due to hydrogen bonding between adjacent parts of a protein. Secondary structure takes two common forms: the α-helix and the β-sheet (alpha helix and beta sheet) (Figure 2.15). Because these secondary structures form hydrogen bonds within themselves rather than with water, they tend to be hydrophobic.

Most proteins also have a **tertiary structure** that is determined by interactions among R groups. These interactions can be due to hydrogen bonding between polar R groups, attraction or repulsion between charged R groups, association of hydrophobic R groups with each other, and covalent bonding between R groups. The latter kind of interaction is especially common between two cysteines, whose sulfhydryls (—SH) tend to form **disulfide bonds** (= S—S bridges). These disulfide bonds are especially strong reinforcements of tertiary structure. They determine the curliness of hair, for example. Finally, many proteins also have a **quaternary structure,** in which two or more polypeptides combine into one protein. The four levels of protein structure are illustrated in Figure 2.16.

Proteins tend to be either globular or fibrous in structure. **Globular proteins** are approximately spherical and usually soluble and are common in body fluids, inside cells, and incorporated into cell membranes. **Fibrous proteins** form long, insoluble strands. Examples of fibrous proteins are **collagen,** which is a reinforcement in skin, bone, and ligaments, and **keratin,** which occurs in horns, beaks, hooves, nails, feathers, and hair.

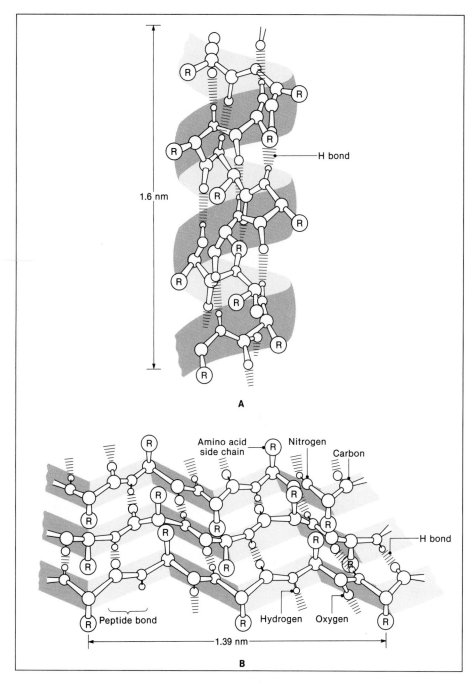

A

B

Figure 2.15
Two common types of secondary structure. (A) The α-helix is held in place by hydrogen bonds between the hydrogens (small dark circles) and the oxygens (large open circles). (B) A β-sheet.

In figure A: 1.6 nm, H bond, R

In figure B: Amino acid side chain, Nitrogen, Carbon, H bond, Peptide bond, Hydrogen, Oxygen, R, 1.39 nm

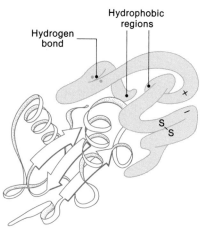

Hydrophobic regions

Hydrogen bond

Figure 2.16
Levels of structure in a hypothetical protein with two polypeptide subunits. The subunit on the left is shown in a "ribbon model," while the one on the right is shown in a "sausage model." The primary structure is the sequence of amino acids in the polypeptide strands (not represented). The secondary structure consists of the α-helices (spirals) and β-sheets (broad arrows) shown in the ribbon model. The tertiary structure is the overall shape of the protein, which is determined by hydrogen and disulfide bonds and by interactions between charged and hydrophobic regions (sausage model). The quaternary structure is the association of the two polypeptides into a single functional unit.

MECHANISM, MEASUREMENT, AND MICROSCOPY

Besides synthesizing organic molecules, organisms *appear* to violate the Second Law of Thermodynamics in another way: They concentrate the molecules they need inside themselves while releasing unwanted molecules into the environment. In this way they increase order inside themselves. There is no actual violation of the Second Law, however, because any increase in order inside an organism is more than matched by increased disorder in its environment, in the form of heat and metabolic wastes. Nevertheless, such increases in internal order require some explanation, because they seldom occur without life. Until a century ago, many biologists would have said that increasing internal order was a special attribute of life, like the ability to synthesize organic molecules. Nowadays, however, we assume that there are **mechanisms** that enable organisms to increase internal

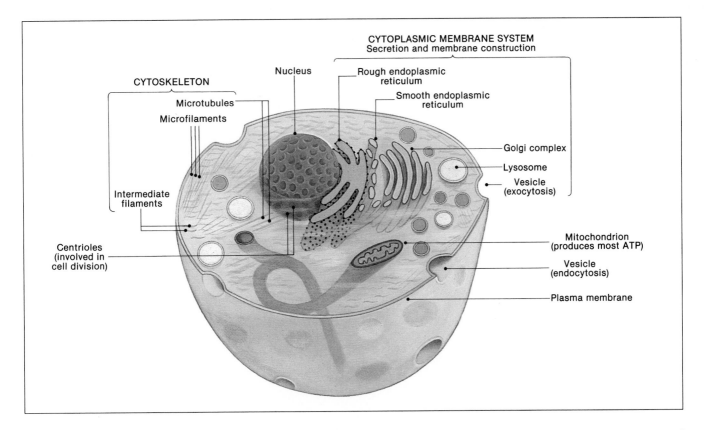

CYTOSKELETON

Microtubules
Microfilaments

Intermediate
filaments

Centrioles
(involved in
cell division)

Nucleus

CYTOPLASMIC MEMBRANE SYSTEM
Secretion and membrane construction

Rough endoplasmic
reticulum

Smooth endoplasmic
reticulum

Golgi complex

Lysosome

Vesicle
(exocytosis)

Mitochondrion
(produces most ATP)

Vesicle
(endocytosis)

Plasma membrane

Figure 2.17
A generalized animal cell, showing most of the organelles. The cell consists of the plasma membrane, the cytoplasm, and the nucleus. The plasma membrane surrounds every cell and controls the passage of material into and out of it. In the cytoplasm, the cytoplasmic membrane system, which includes rough and smooth endoplasmic reticulum and the Golgi complex, produces vesicles for new plasma membrane, lysosomes, and secretion. Other vesicles bring in certain materials, which are released into the cell after the vesicle is digested by lysosomes. Mitochondria produce most of the energy used by cells; centrioles are involved with the movement of genetic material during cell division, and ribosomes on the rough endoplasmic reticulum produce protein. These three organelles will be described in Chapters 3, 4, and 5, respectively. All these organelles are considered part of the cytoplasm. The nucleus, which is not considered part of the cytoplasm, is enclosed by a porous nuclear membrane that is connected with the cytoplasmic membrane system. The cell shape is maintained by tubules and filaments of the cytoskeleton.

order. Like materialism, the search for mechanisms is another characteristic of modern science.

The mechanisms responsible for life are largely associated with discrete structures called **organelles** ("little organs"). These organelles include the **plasma membrane** that encloses all cells, the nucleus that controls protein synthesis and heredity, and several others located within the **cytoplasm** (Figure 2.17). The organelles within the cytoplasm are bathed in a fluid called **cytosol.** Before attempting to understand the structure and functioning of organelles, it is important to understand how we can know anything at all about such structures. Most are so small, less than one-thousandth of a millimeter (1 micrometer), that it is hard even to imagine them. The first box (p. 36) is intended to help you grasp just how small they are. The second box (pp. 37–38) explains how microscopy is used to study the structure of cells.

THE PLASMA MEMBRANE

Structure.   The plasma membrane is largely responsible for keeping the low-entropy cytoplasm from mixing with the high-entropy environment of cells. Its importance is clear from the fact that damage to the plasma membrane usually kills the cell. Under the electron microscope the plasma membrane has a three-layered appearance, because the metals used to stain the membrane adhere only to the inner and outer surfaces. This simple appearance in the electron microscope belies a much more complicated structure that is still being revealed by ingenious use of electron microscopy and other techniques.

According to the **fluid mosaic model** that has been widely accepted since 1972, the plasma membrane consists of a liquid film of lipids with proteins floating within (Figure 2.21 on p. 39). In most animal cells the lipids account for about half the mass of the plasma membrane. As explained above, these phospholipids spontaneously form a bilayer that is the main structural component of the

Dimensions of Life

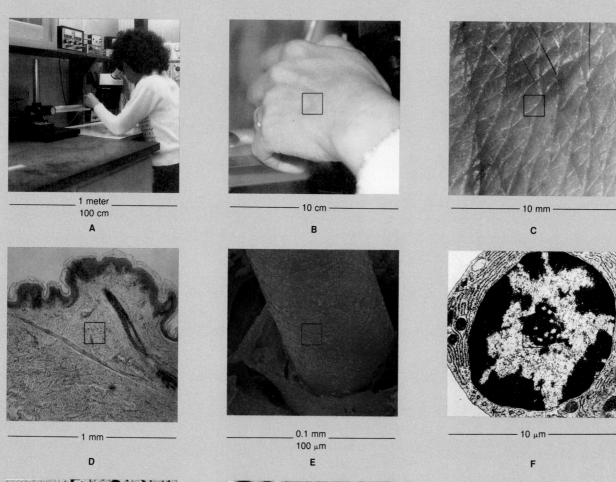

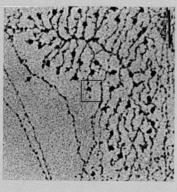

A
1 meter
100 cm

B
10 cm

C
10 mm

D
1 mm

E
0.1 mm
100 μm

F
10 μm

G
1 μm

H
0.1 μm
100 nm

Figure 2.18
Relative sizes of structures commonly encountered in zoology. Each figure is 10 times as wide as the following and covers 100 times the area. (A) Humans and other large animals are conveniently measured in units of meters. (B) The width of the hand is approximately 0.1 meter, or 10 centimeters (cm). (C) With a field of view 1 cm wide, hairs become visible. (D) Objects smaller than 1 millimeter (mm) are not easily seen with the unaided eye. This stained slice of human skin showing a hair follicle was photographed using light microscopy at a magnification of 100. (E) Scanning electron microscopy reveals three-dimensional structure at a higher magnification. (F) Most animal cells are on the order of 10 micrometers (μm) wide. Their internal structure is best viewed with transmission electron microscopy. (G) Transmission electron microscopy at high power even reveals the molecules responsible for heredity. (H) The fine structure of molecules is beyond the limits of transmission electron microscopy, although computers can enhance the image and enable the artist to represent it. At such small dimensions another unit of measure, the angstrom ($\mathring{A}$, = 10^{-10} meter = 0.1 nm) is also common.

Our understanding of cells has been intimately linked to developments in microscopy. Even the name "cell" derives from *cellulae*, the name Robert Hooke gave to the hollow spaces in cork that he observed under the microscope in 1663. Antony van Leeuwenhoek (LAY-win-huke) was probably the first to see living cells, using simple microscopes that he made with glass beads as lenses, starting in about 1670. This Dutch cloth merchant had intended to use his microscopes to examine the closeness of weave, but his first sight of a drop of pond water drew him into a new world inhabited by bacteria and "animalcules." Monumental as these observations were, their significance was not grasped for more than a century. Most biologists were more interested in the easily seen structures of organisms, the plant and animal fibers, and regarded cells as by-products.

It was not until the 19th century that improved techniques in microscopy allowed biologists to realize the significance of cells. First Carl Zeiss and others perfected the **compound microscope,** which incorporates several lenses to increase magnification and to correct the distortions inherent in single lenses. Just as significant was the discovery of dyes and techniques for preserving and slicing biological specimens. By the middle 1800s observations based on these techniques (as well as philosophical and even political reasons that we can scarcely comprehend today) convinced most biologists that cells were not merely by-products of living tissues but were essential for all life. In 1839 Theodor Schwann pronounced what we still regard as the cell theory: "that there exists one general principle for the formation of all organic produc-

tions, and that this principle is the formation of cells." By 1858 Rudolf Virchow was bold enough to propose the dictum *"Omnis cellula e cellula"*: all cells come from cells.

By the turn of the century it was firmly established that the essence of life is in cells. All that remained was to discover how the materials and mechanisms of cells accounted for life. As before, our understanding of the internal workings of cells has continued to depend on advances in microscopical techniques.

LIGHT MICROSCOPY

Although the compound microscope long ago reached its theoretical limit of magnification, there have nevertheless been improvements in the techniques of light microscopy. Light microscopy depends on light passing through different structures of a specimen in different ways so that the structures can be distinguished from the background. After the light has passed through the specimen it is magnified by lenses (left side of Figure 2.19).

Ordinary light passes through all parts of cells in about the same way, so most cells look virtually transparent and structureless (Figure 2.20A). To remedy this problem the light can be manipulated by polarizing it, changing its angle of incidence, splitting it into two rays and recombining them in different ways, or by other sophisticated optical techniques. Each technique brings out different features of a cell, often quite dramatically (Figure 2.20B–D).

Another approach to revealing cellular structure with light microscopy is to use dyes that stain structures

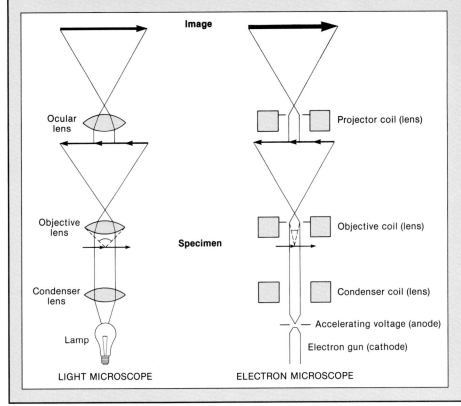

Figure 2.19
The pathway of light through a light microscope (left) compared with the pathway of electrons through a transmission electron microscope (right). The electron microscope is shown upside down for convenience.

(continued)

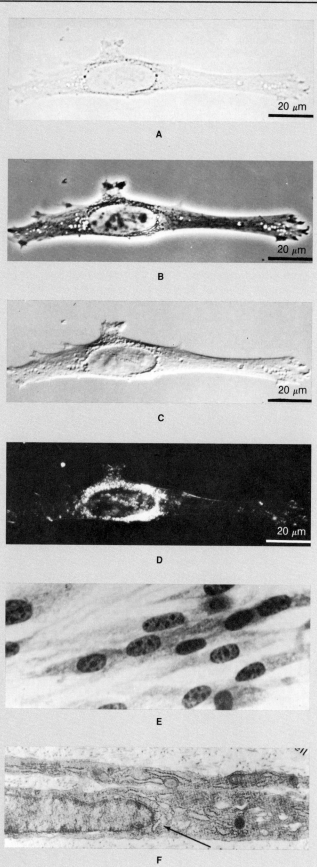

differentially. This approach ordinarily requires that the specimen be killed, **fixed** to coagulate proteins and hold the shape of cells, then preserved in alcohol or some other substance. The specimen is then embedded in paraffin and sliced into sections usually 10 to 20 micrometers (μm) thick and mounted on glass slides (Figure 2.20E).

ELECTRON MICROSCOPY

Although the techniques of light microscopy have become quite sophisticated, they cannot overcome a fundamental limitation. This limitation is the finite wavelength of light, which prevents any light microscope from distinguishing **(resolving)** an object smaller than about 1 μm. Many of the most interesting parts of cells are approximately this size and cannot be studied with the light microscope. Under certain circumstances, however, electrons also behave like waves, and they have much shorter wavelengths than light. This fact is the basis for the **electron microscope.** The most powerful electron microscopes have a **resolving power** 1000 times better than that of the light microscope. Electron microscopes can resolve objects as small as 1 nanometer (nm), which is about the size of many molecules.

Neither the cellular structures nor the stains used in light microscopy will absorb electrons, so heavy metals have to be used as stains in electron microscopy. In **transmission electron microscopy** the specimen first has to be killed, fixed, and embedded. The specimen is then sliced into sections less than 0.1 μm thick (ultrathin) to allow the electrons to pass through (right side of Figure 2.19). The section is then stained with heavy metals, taking advantage of the fact that the metal adheres to some molecules better than to others (Figure 2.20F). It is important to keep in mind that what one actually sees with the electron microscope is the metal stain, not the molecules in the specimen. With another type of electron microscopy, called **scanning electron microscopy (SEM)**, the specimen is usually not sliced. Instead, a metal stain is applied to the surface, allowing biologists to view the external structure of cells, tissues, and even entire animals with lifelike depth. [For examples, see the figure opening this Unit (page 16) and Figure 33.16A.]

Figure 2.20
Results of different techniques of microscopy. (A) Image of a living cell with ordinary light microscopy. The light passes through the unstained cell uniformly, so it appears almost transparent. (B) Phase-contrast microscopy brings out more details in the cell. (C) Differential-interference microscopy with Nomarski optics gives the cell a three-dimensional appearance. (D) In dark-field microscopy the cell is illuminated indirectly, so that only structures that scatter light are visible. (E) Here tissue containing several cells similar to those shown above have been killed, fixed, embedded in paraffin, sectioned, stained, and mounted on a slide. (F) Part of a similar cell (arrow) from tissue that has been sliced into ultrathin sections and stained for electron microscopy at low power.

Figure 2.21
The plasma membrane consists of a lipid bilayer with proteins embedded in it. Cholesterol helps regulate the fluidity of the membrane. The proteins are either thin strands with a hydrophobic α-helical structure or globular proteins. A hydrophobic part of the protein, generally an α-helix or a β-sheet, is embedded in the hydrophobic layer of the lipid bilayer, and hydrophilic parts of the protein are embedded in the outer layers. Other proteins are attached to the surface.

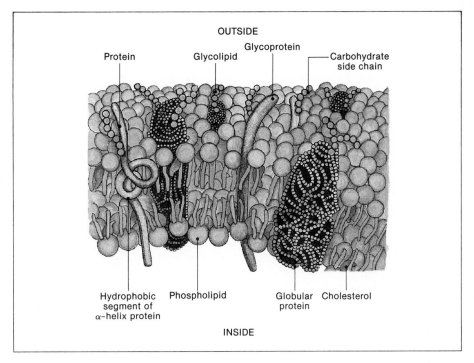

membrane. Each square micrometer of membrane contains approximately two million molecules of phospholipid. In most plasma membranes each phospholipid has a cholesterol molecule situated between the hydrophilic tails near the polar head group, which stiffens the tails, prevents the bilayer from becoming too fluid, and keeps the bilayer from hardening at low temperature. This is an important adaptation for animals that live in the cold. On the outer surface of the plasma membrane are **glycolipids:** lipids with carbohydrates attached. The function of these glycolipids is unknown but probably important.

Proteins account for the other half of the mass of plasma membrane. Most of these proteins have attached carbohydrates that protrude from the outer surface of the membrane only, making them **glycoproteins.** The glycoproteins are dissolved within the lipid bilayer and cannot flip over or pop out, because they are anchored by hydrophobic and hydrophilic interactions with phospholipids in the bilayer.

Diffusion. The ability of plasma membranes to prevent the passage of some molecules while allowing the passage of others is a good illustration of the application of scientific materialism and mechanism. Because of the lipid bilayer, only organic molecules that are soluble in lipids (hydrophobic) can diffuse through the membrane (Figure 2.22). Most molecules that are free to move in and around cells are hydrophilic, however, because they have to be dissolved in water. Therefore the lipid bilayer is impermeable to most of the organic molecules in organisms. Unless they are hydrophobic, organic molecules inside cells tend to stay in the cytoplasm, and those that are outside generally cannot enter unless they are brought in by some specific mechanism. In addition to blocking most organic molecules, the lipid bilayer also keeps out most inorganic molecules. There are only two kinds of inorganic molecules that can diffuse through the lipid bilayer: (1) small, nonpolar molecules, such as O_2; and (2) small, uncharged, polar molecules, including H_2O and CO_2.

In addition to H_2O, O_2, and CO_2, other natural molecules that can diffuse through the lipid bilayer are fats and steroids. There is also an increasing number

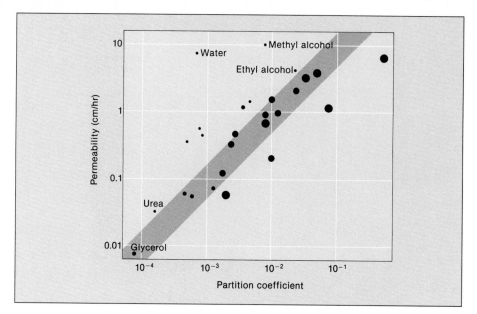

Figure 2.22
The relationship between the ability of a molecule to diffuse across a plasma membrane (its permeability) and its solubility in lipids. The solubility is measured as the partition coefficient, which is the maximum amount of a substance that will dissolve in a volume of olive oil, divided by the maximum amount that will dissolve in an equal volume of water. Generally, the higher the partition coefficient (the more hydrophobic the molecule), the higher its permeability. Permeability also depends somewhat on the relative diameter of the molecule, indicated by the size of the data points in the graph. Water and other small molecules are more permeant, and large molecules are less permeant.

of manufactured organic molecules that diffuse through plasma membranes, for good or ill. These include synthetic steroids, such as those in oral contraceptives. If they were not able to penetrate plasma membranes they could not be absorbed across the intestinal lining and would have to be administered by injection. Steroid hormones can also be absorbed through the skin. Some women have had unnecessary hysterectomies because steroids in some skin creams induced symptoms indistinguishable from uterine cancer. Of even more concern are industrial solvents, pesticides, and other manufactured substances that are designed to penetrate lipids. Because they are absorbed through skin, lungs, and intestines, and may also damage plasma membranes, many have a high potential for inducing cancers, mutations, and birth defects.

Some small hydrated ions (see Figure 2.4B) can diffuse through the plasma membrane without passing through the lipid bilayer, apparently because some of the glycoproteins form **channels.** Because of such channels, plasma membranes are generally permeable to H^+, K^+, and Cl^-. Usually they are less permeable to hydrated ions such as Na^+, which have larger water shells. For example, the plasma membranes of red blood cells are 50 times more permeable to K^+ and Cl^- than to Na^+. Most other ions have even larger water shells than Na^+ and are essentially impermeable. Because membranes are more permeable to some ions than to others, they are said to be **semipermeable.** This semipermeability gives rise to a voltage **(membrane potential)** across the plasma membrane, as will be explained in Chapter 7. This membrane potential is generally between 0.05 and 0.09 volt, negative inside the cell compared with outside. The negative membrane potential draws K^+ ions into the cell and repels Cl^- ions.

Under certain circumstances the permeability to sodium and other ions can be increased by opening **gated channels.** The signal opening the gate of such a channel can be a particular chemical or a change in membrane potential. For example, several different kinds of potassium channels are known, some voltage-dependent, some responsive to chemicals, and some responsive to the levels of calcium in the cytoplasm.

Diffusion through channels is usually limited to small ions; larger ions and even molecules such as glucose can diffuse across the plasma membrane by means of **facilitated diffusion,** in which a molecule diffuses down its concentration gradient by attaching to a carrier molecule that can move across the membrane.

Figure 2.23
Schematic representation of the operation of the Na⁺–K⁺ ATPase that is the sodium–potassium pump. In the upper left the binding of three Na⁺ ions (colored dots) triggers a change in the structure of the ATPase that moves the Na⁺ outside the cell, against diffusion and the negative membrane potential. Afterward, the binding of two K⁺ ions (black dots) restores the structure of the ATPase and transports the K⁺ into the cell.

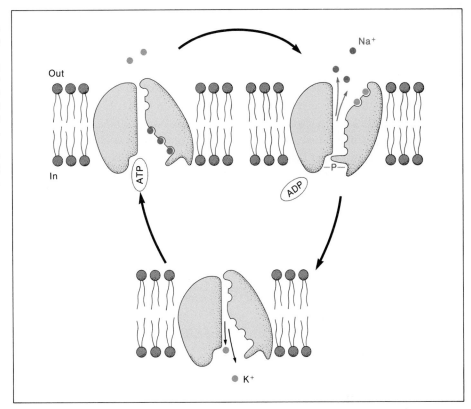

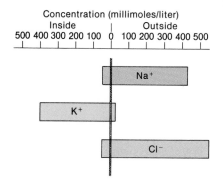

Figure 2.24
Typical concentrations of Na⁺, K⁺, and Cl⁻ inside and outside a cell, represented by the width of each bar.

Active Transport. The diffusion of water, ions, and other molecules, even if facilitated by a carrier molecule, is always from the side of the plasma membrane with the higher concentration to the side with the lower concentration. If diffusion were the only means by which a cell could bring in the substances it requires, it would be severely limited. There are other mechanisms, however, that enable a plasma membrane to transport materials in the direction opposite the one they would passively follow. These mechanisms are referred to as **active transport,** because they require that the cell use energy. The best-known example of a mechanism for active transport is the **sodium–potassium pump,** which pumps three sodium ions out of a cell while pumping two potassium ions into it. The sodium–potassium pump is also known as the **Na⁺–K⁺ ATPase,** because it obtains the energy to actively transport Na⁺ and K⁺ by breaking bonds in a molecule called ATP. It is not yet clear how the Na⁺–K⁺ ATPase works, but Figure 2.23 summarizes what it does.

Na⁺–K⁺ ATPases have been found in the plasma membranes of every organism examined, and by some accounts they use up to 40% of an organism's energy. Evidently they are important. One of the things sodium–potassium pumps do is to produce a lower concentration of Na⁺ inside the cell than outside. The pumps also bring some K⁺ into the cell, although the negative membrane potential is much more important in this regard and also maintains a low concentration of Cl⁻ inside the cell (Figure 2.24). Na⁺–K⁺ ATPases are also coupled to other kinds of pumps in the plasma membranes and are therefore indirectly responsible for other kinds of active transport. Some of these coupled pumps transport glucose, amino acids, and other nutrients from the intestine into the bloodstream, and from body fluids into cells, where they are used to produce energy and manufacture other organic molecules. In addition, there are also Ca^{2+} pumps, proton (H^+) pumps, and other kinds of pumps that operate independently of sodium–potassium pumps.

Exocytosis and Endocytosis. There are also mechanisms for getting materials out of or into cells without having them pass through the plasma membrane. These processes are called exocytosis and endocytosis. **Exocytosis** (Greek *exo-* out + *kytos* hollow vessel, i.e., a cell) occurs when material in the cytoplasm is packaged in a spherical membrane called a **vesicle**. The vesicle contents are then released externally when the vesicle fuses with the plasma membrane. During exocytosis the vesicle membrane is incorporated into the plasma membrane, which is how the cell builds new membrane. Besides the plasma membrane, several other organelles are involved in exocytosis, as will be described in more detail later. Among the important materials secreted by exocytosis are hormones, mucus, and substances that trigger muscle contraction or modify neural activity.

Endocytosis (Greek *endo-*, in) is essentially the reverse of exocytosis. Material

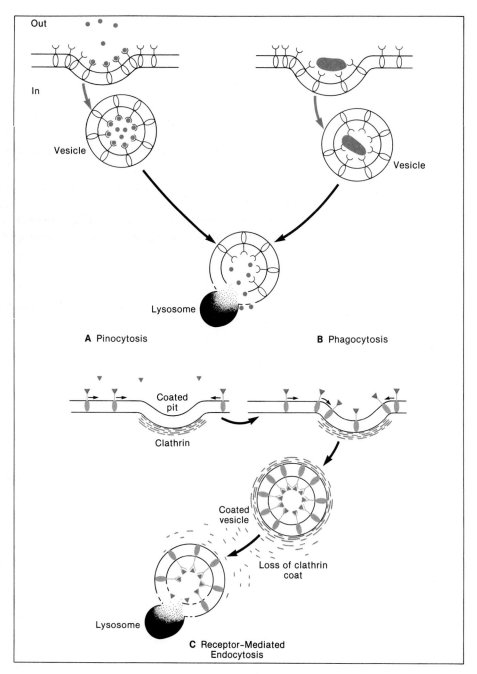

Figure 2.25
Three types of endocytosis, followed by digestion by a lysosome. (A) Pinocytosis, the uptake of water, along with dissolved materials and small particles in the water. (B) Phagocytosis, the uptake of solids. See also Figure 23.6. (C) Receptor-mediated endocytosis. Receptors carrying particular molecules migrate to pits coated on the inner surface with the protein clathrin. Vesicles form from these coated pits.

outside the cell binds to particular sites on the plasma membrane, which then invaginates, enclosing the material in a vesicle. The vesicle then pinches off the inner surface of the membrane and enters the cytoplasm. In the cytoplasm the vesicle contents are released when the vesicle is digested by an organelle called the **lysosome.** There are three types of endocytosis: pinocytosis, phagocytosis, and receptor-mediated endocytosis (Figure 2.25). **Pinocytosis** (Greek *pino*, to drink) refers to the uptake of water and dissolved material. **Phagocytosis** (Greek *phagein*, to eat) is the uptake of any undissolved material, such as food, viruses, bacteria, and cellular debris. **Receptor-mediated endocytosis** occurs when the receptors binding a particular substance migrate to a depression called a **coated pit,** where the endocytosis occurs. Figure 2.26 summarizes the mechanisms by

Figure 2.26
Summary diagram of the methods by which plasma membrane controls the entry and exit of material.

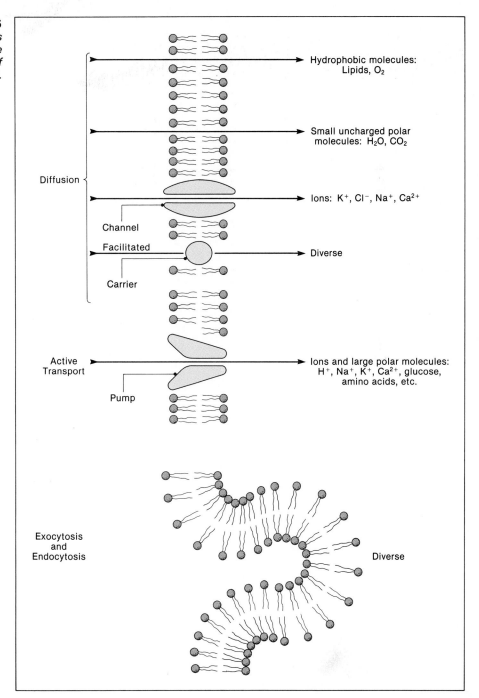

which the plasma membrane controls the movement of material into and out of cells.

WHY DO ANIMALS HAVE MORE THAN ONE CELL?

The importance of the plasma membrane in the transport of material into and out of the cell may partly explain a fundamental fact of life in animals and other multicellular organisms. These organisms consist of many small cells rather than one or a few large ones. The explanation has to do with the surface-to-volume ratio, a factor that recurs frequently throughout this text. As an object grows in size, its surface area increases more slowly than its volume. If a microscopic single-celled organism "decided" to become a larger organism by increasing the diameter of its cell tenfold, for example, its volume would increase a thousandfold (10^3), but its surface area would increase by only a hundredfold (10^2). Thus there would be a thousand times as much volume to supply nutrients to and eliminate wastes from, but only a hundred times the plasma membrane to do the job. For this and probably other reasons, single-celled organisms evolved into large organisms by keeping the size of their cells the same, but increasing their number.

The many cells that make up an animal do not function independently, but coordinate their activities, often by means of chemical messengers. Such chemical messengers are secreted by certain cells and often bind to specific **receptors** on the plasma membranes of other cells. In response to binding of the messenger the receptor molecule triggers a change on the inner surface of the membrane, causing a response in the cytoplasm. Among these chemical messengers are some hormones and synaptic transmitters that allow nerve cells to communicate with each other.

CELLULAR SHAPE, MOVEMENT, AND INTERCONNECTIONS

The Cytoskeleton. Although the plasma membrane is quite fluid, the cell is not merely a shapeless bag of cytoplasm. Cells have definite structure. This structure is partly due to a positive oncotic pressure resulting from a greater amount of protein and other molecules in the cytoplasm than outside the cell. In addition, there is an internal network of protein fibers that constitutes a **cytoskeleton.** The cytoskeleton consists of three major components: microtubules, microfilaments, and intermediate filaments. **Microtubules** are the thickest of the three (22 nm) and consist of subunits of **tubulin** and other proteins. Microtubules radiate from the nucleus of the cell to just under the plasma membrane. They are believed to act as a framework for the internal organization of cells and are also essential in cellular reproduction, as will be described in a later chapter. In nerve cells and perhaps other kinds of cells microtubules also serve as tracks along which vesicles and organelles move in both directions.

Microfilaments are the thinnest components of the cytoskeleton (6 nm) and are made of the protein **actin.** Microfilaments run just beneath the plasma membrane and appear to function in cell movements and as **stress fibers** that reinforce the plasma membrane (Figure 2.27). **Intermediate filaments** are between microtubules and microfilaments in diameter (7 to 11 nm) and consist of different proteins in different kinds of cells. The best known of these intermediate filaments is keratin, which was previously mentioned as the fibrous protein in horns, beaks, hooves, nails, and feathers.

Motility. Many of the proteins that make up the cytoskeleton are also involved in the movement of cells. Actin microfilaments, for example, are partly responsible for the contraction of muscle and some other kinds of cells. Actin is also believed

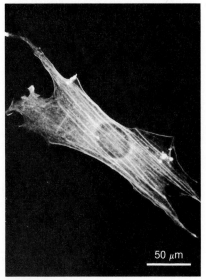

50 μm

Figure 2.27
Bundles of thin filaments revealed by fluorescent antibodies attached to the actin.

to be involved in a type of locomotion called **ameboid movement,** which is a kind of creeping movement of certain protozoans such as *Amoeba* and of white blood cells (see Figures 6.17 and 18.6). Microtubules are the main elements in **cilia** and **flagella,** which are hair-shaped structures on the outer surface of the plasma membrane (see Figure 6.18). Cilia are short and numerous and are responsible for locomotion in some protozoans and invertebrates and for moving fluids past tissues in many animals. Flagella are long and sparse and are responsible for the swimming of some protozoans and sperm and for the propulsion of fluid in sponges and some other invertebrates. All these types of movement are discussed in detail in Chapter 10. The formation of cilia and flagella is initiated by small, cylindrical organelles called **centrioles,** which are themselves made of microtubules (Figure 2.28).

Intercellular Connections. Most cells are connected to others in ways that affect their shape and restrict their movements. There are three kinds of **intercellular junctions:** desmosomes, tight junctions, and gap junctions (Figure 2.29). **Desmosomes** maintain the structural integrity of tissues by fastening together the plasma membranes of adjacent cells. **Tight junctions** block the passage of substances around cells by joining plasma membranes in a band that encircles each cell. **Gap junctions** form protein channels that allow the passage of monosaccharides, amino acids, and ions between neighboring cells.

CYTOPLASMIC MEMBRANES

The plasma membrane surrounding every cell is closely related to a system of membranes within the cytoplasm. This **cytoplasmic membrane system** is connected to the plasma membrane, exchanges phospholipids and glycoproteins with it, and prepares glycoproteins for exocytosis from the plasma membrane. The cytoplasmic membrane system comprises several distinct organelles, including

Figure 2.28
(A) Schematic diagram of two centrioles. (B) Electron micrograph showing two centrioles, one in cross section and one in longitudinal section. Diameter of each centriole is approximately 0.2 μm.

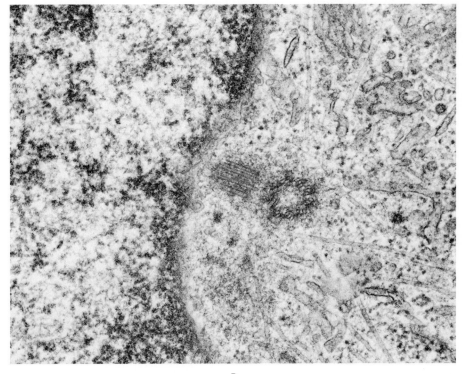

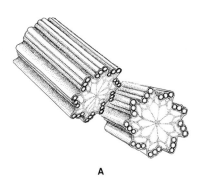

A

B

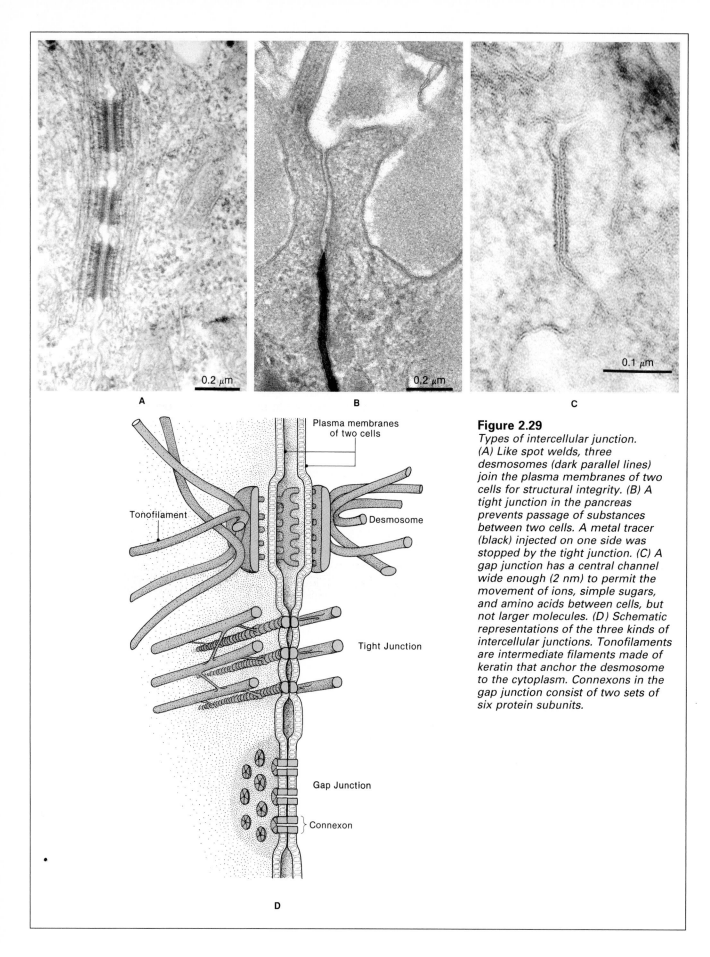

A

0.2 μm

B

0.2 μm

C

0.1 μm

Plasma membranes of two cells

Tonofilament

Desmosome

Tight Junction

Gap Junction

Connexon

D

Figure 2.29
*Types of intercellular junction.
(A) Like spot welds, three
desmosomes (dark parallel lines)
join the plasma membranes of two
cells for structural integrity. (B) A
tight junction in the pancreas
prevents passage of substances
between two cells. A metal tracer
(black) injected on one side was
stopped by the tight junction. (C) A
gap junction has a central channel
wide enough (2 nm) to permit the
movement of ions, simple sugars,
and amino acids between cells, but
not larger molecules. (D) Schematic
representations of the three kinds of
intercellular junctions. Tonofilaments
are intermediate filaments made of
keratin that anchor the desmosome
to the cytoplasm. Connexons in the
gap junction consist of two sets of
six protein subunits.*

endoplasmic reticulum, Golgi (GOAL-jee) complexes, and lysosomes. Briefly, proteins and glycoproteins are synthesized on endoplasmic reticulum, then packaged inside vesicles or as parts of the membranes enclosing the vesicles. The vesicles then enter Golgi complexes, which modify the glycoproteins in the vesicles, repackage the glycoproteins into new vesicles, and send them to their destinations. Some glycoproteins inside the vesicles are incorporated into lysosomes, which are organelles containing enzymes for intracellular digestion. Other glycoproteins are secreted during exocytosis (as in Figure 7.7B). Glycoproteins in the vesicle membrane become incorporated into the plasma membrane.

The first organelle of the cytoplasmic membrane system to be discussed is the **endoplasmic reticulum** (ER). Most ER is of a type called **rough endoplasmic reticulum,** because small organelles called **ribosomes** give it a grainy appearance in the electron microscope (Figure 2.30A and B). As will be described in Chapter 4, the ribosomes on rough ER initiate the synthesis of proteins according to genetic instructions from the nucleus. Proteins then complete their synthesis on chains of ribosomes in the cytosol and are used inside the cells. Other proteins have chains of sugars (oligosaccharides) attached to them, which convert them to glycoproteins and label them for further processing by a Golgi complex.

Glycoproteins from rough endoplasmic reticulum are packaged inside vesicles by a second type of endoplasmic reticulum. This type of ER is called **smooth endoplasmic reticulum,** because it lacks ribosomes. Smooth ER also differs from rough ER in forming a lacy network of tubules rather than sheets (Figure 2.30C and D). In many cells smooth ER is specialized for additional functions.

Figure 2.30
Rough endoplasmic reticulum (A, B) and smooth endoplasmic reticulum (C, D) shown by scanning electron microscopy (A, C) and transmission electron microscopy (B, D).

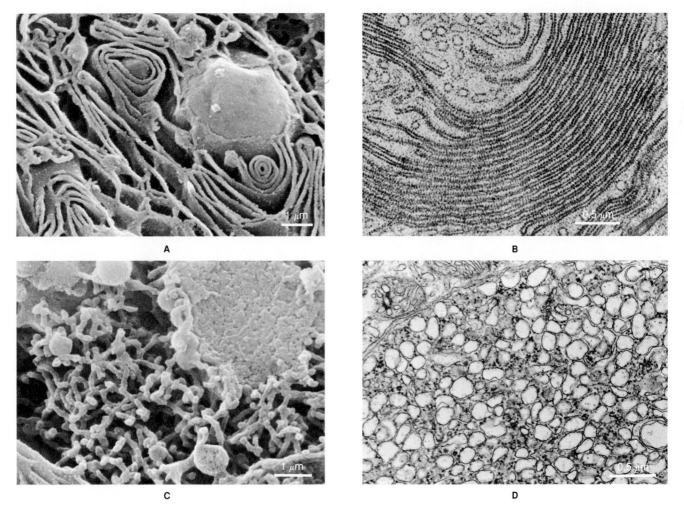

A

B

C

D

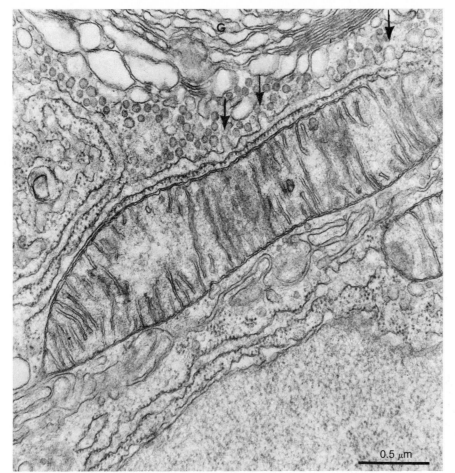

Figure 2.31
A Golgi complex (G) is shown in the upper portion of this electron micrograph. The vesicles of glycoprotein (color) were presumably on their way to the Golgi complex for further processing. Arrows point to vesicles forming on smooth ER. (Part of a mitochondrion, to be discussed later, is shown in the center, and part of the nucleus is at the bottom.)

0.5 μm

Vesicles containing the glycoproteins from endoplasmic reticulum are taken into **Golgi complexes** for further processing (Figure 2.31). A Golgi complex (= Golgi body or Golgi apparatus) consists of a stack of hollow, disc-shaped **cisternae.** Each cisterna absorbs vesicles, modifies the glycoproteins in them, repackages

Figure 2.32
Functioning of the cytoplasmic membrane system. Glycoproteins from rough endoplasmic reticulum are packaged inside vesicles or incorporated into the membranes of vesicles. The vesicles are then absorbed by a Golgi complex, which has three distinct groups of cisternae (compartments): cis, medial, and trans. Each cisterna absorbs a vesicle, modifies the glycoprotein in it according to its destination, repackages the glycoprotein in a new vesicle, and sends it on to the next cisterna. In the cis cisternae the glycoproteins destined to be incorporated into lysosomes (1) have phosphates attached to their carbohydrate chains (oligosaccharides). In the medial cisternae those glycoproteins are not changed, while those destined for secretion (2) are. Proteins incorporated into vesicle membranes progress through the Golgi complex and eventually become incorporated into plasma membrane (3). The trans cisternae repackage all the glycoproteins into vesicles and send them to their appropriate destinations.

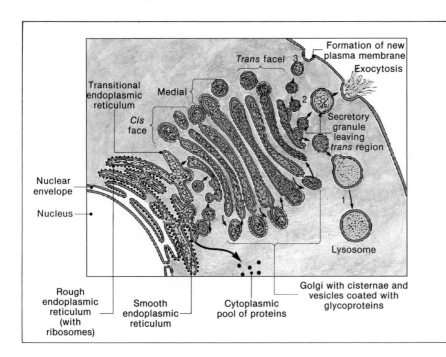

Transitional endoplasmic reticulum

Trans facel 3

Formation of new plasma membrane

Exocytosis

Medial

Cis face

2

Secretory granule leaving *trans* region

Nuclear envelope

Nucleus

1

Lysosome

Rough endoplasmic reticulum (with ribosomes)

Smooth endoplasmic reticulum

Cytoplasmic pool of proteins

Golgi with cisternae and vesicles coated with glycoproteins

the glycoproteins into new vesicles, then sends these vesicles on to the next cisterna. Each cisterna apparently modifies the glycoprotein in such a way that it is sent to its appropriate destination, whether to a lysosome, to a secretory vesicle, or to the plasma membrane (Figure 2.32).

Closely associated with the cytoplasmic membrane system is the **nuclear envelope,** which encloses the nucleus. The nuclear envelope consists of an inner and outer membrane penetrated by several thousand large pores (Figure 2.33). The pores, held open by a complex of proteins, permit the flow of genetic information from the nucleus to the ribosomes.

All the mechanisms described in this chapter require materials to form them and energy to operate them. Both the materials and the energy come from the nutrients in food. How those nutrients are processed is the subject of the next chapter.

Figure 2.33
The nuclear envelope. Pores in the nuclear envelope, at the lower left, are clearly visible. This scanning electron micrograph was prepared by freezing and fracturing the cell, eroding part of the exposed surface, then making a platinum replica (freeze-etch technique).

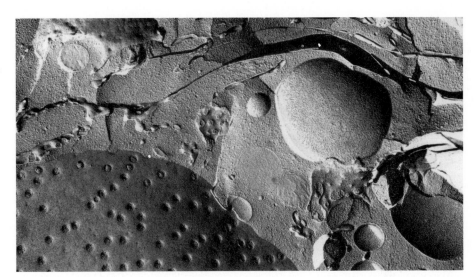

SUMMARY

Our understanding of life has progressed as far as it has because of the assumptions that life depends on the properties of matter that composes organisms and on physical and chemical mechanisms rather than on vital forces. These two assumptions, called materialism and mechanism, allow us to appreciate why water is essential to life. The molecular properties of water enable it to serve as a solvent for a wide variety of other essential materials. Among these are ions, such as H^+, and many other inorganic and organic substances. Another important property of water is its large capacity for thermal energy, which drives two essential processes: diffusion and osmosis.

Diffusion and osmosis tend to make the concentrations of molecules uniform, increasing entropy in accordance with the Second Law of Thermodynamics. There are numerous mechanisms, however, that maintain lower entropy inside cells than outside. Some of these mechanisms are associated with the plasma membrane, which allows the diffusion of some molecules but not of others, depending on the properties of the molecules. Generally, a molecule can diffuse through the lipid bilayer of the plasma membrane if it can dissolve in lipid, or if it is small and uncharged. Some small ions can diffuse through the plasma membrane through special channels made of protein. Other ions and molecules travel across the membrane by facilitated diffusion, in which they attach to a carrier molecule.

One kind of organic molecule that can diffuse through the lipid bilayer of plasma membrane is lipid itself, including triglyceride. Carbohydrates, such as sugars, are not soluble in lipid and therefore cannot diffuse through plasma membrane. Since they are important sources of energy for the cell, they have to be brought in by another process, called active transport. The mechanism of active transport depends on proteins in the plasma membrane.

Proteins are also unable to diffuse across plasma membrane. Many of them carry out essential functions inside the cells where they are produced. Among these proteins are enzymes, parts of the cytoskeleton that maintains the shape of the cell and enables it to move, and structures that form desmosomes, tight junctions, and gap junctions with other cells. Some proteins are secreted from the cell by exocytosis. The mechanisms of exocytosis are associated with the intracellular membrane system, which includes the endoplasmic reticulum and the Golgi complex. These organelles also produce new plasma membrane.

KEY TERMS

energy level
covalent bond
hydrogen bond
solvent
solute
concentration
ionic bond
acidity
diffusion
facilitated diffusion

osmosis
entropy
organic molecule
carbohydrate
lipid
hydrophobic
hydrophilic
phospholipid
peptide
plasma membrane

cytoplasm
channel
active transport
exocytosis
endocytosis
vesicle
microtubule
endoplasmic reticulum
Golgi complex

SELF-TEST

1. Describe three important properties of water.

2. Explain how the structure of water accounts for each of the properties in question 1.

3. Explain how each of the properties in question 1 is important to life.

4. Explain how diffusion and osmosis illustrate the Second Law of Thermodynamics.

5. For each of the following kinds of organic molecules, carbohydrates, lipids, and proteins, (a) describe the general structure, (b) name one example, (c) describe the general role in animals, and (d) explain how the molecular properties contribute to their performing that role.

6. Define the primary, secondary, tertiary, and quaternary structures of proteins. How is each kind of structure determined? Explain the importance of tertiary and quaternary structure in the functioning of proteins.

7. Sketch a portion of plasma membrane showing the main features of the fluid mosaic model.

8. Explain the role of phospholipids in the plasma membrane.

9. What kinds of molecules can diffuse through the lipid bilayer of the plasma membrane? How do ions and large polar molecules penetrate the plasma membrane?

10. Describe the roles of rough endoplasmic reticulum, smooth endoplasmic reticulum, and the Golgi complex in exocytosis.

READINGS

Allen, R. D. 1987. The microtubule as an intracellular engine. *Sci. Am.* 256(2):42–49 (Feb).

Atkins, P. W. 1987. *Molecules.* New York: Scientific American Library.

Boatman, E. S. et al. 1987. Today's microscopy. *BioScience* 37:384–394.

Bretscher, M. S. 1985. The molecules of the cell membrane. *Sci. Am.* 253(4):100–108 (Oct).

Bretscher, M. S. 1987. How animal cells move. *Sci. Am.* 257(6):72–90 (Dec).

Dautry-Varsat, A. and H. F. Lodish. 1984. How receptors bring proteins and particles into cells. *Sci. Am.* 250(5):52–58 (May).

De Duve, C. 1984. *A Guided Tour of the Living Cell*, 2 vols. New York: Scientific American Library.

Doolittle, R. F. 1985. Proteins. *Sci. Am.* 253(4):88–99 (Oct).

Ford, B. J. 1985. *Single Lens: The Story of the Simple Microscope.* New York: Harper & Row.

Karplus, M. and J. A. McCammon. 1986. The dynamics of proteins. *Sci. Am.* 254(4):42–51 (Apr).

Kessel, R. G. and R. H. Kardon. 1979. *Tissues and Organs: A Text-Atlas of Scanning Electron Microscopy.* San Francisco: W. H. Freeman.

Orci, L., J.-D. Vassalli, and A. Perrelet. 1988. The insulin factory. *Sci. Am.* 259(3):85–94 (Sept). (*Production and secretion of an important protein.*)

Rothman, J. E. 1985. The compartmental organization of the Golgi apparatus. *Sci. Am.* 253(3):74–89 (Sept).

Sharon, N. 1980. Carbohydrates. *Sci. Am.* 243(5):90–116 (Nov).

Staehelin, L. A. and B. E. Hull. 1978. Junctions between living cells. *Sci. Am.* 238(5):140–152 (May).

Unwin, N. and R. Henderson. 1984. The structure of proteins in biological membranes. *Sci. Am.* 250(2):78–94 (Feb).

Weber, K. and M. Osborn. 1985. The molecules of the cell matrix. *Sci. Am.* 253(4):110–120 (Oct).

Weinberg, R. A. 1985. The molecules of life. *Sci. Am.* 253(4):48–57 (Oct). (*Introduction to an issue devoted to this topic.*)

Figure 3.1
(A) A molecule of adenosine triphosphate (ATP). Adenosine diphosphate is identical except that it has one less phosphate. The crowding of negative charge in each phosphate gives the phosphate bond a moderate amount of energy. (B) ATP is synthesized when some of the energy released from a chemical reaction (A → B) is used to attach a phosphate group to ADP. This is referred to as a coupled reaction. The phosphate is either inorganic phosphate, or a phosphate group on the original molecule (A). When ATP drives a cellular process essentially the reverse kind of reaction occurs.

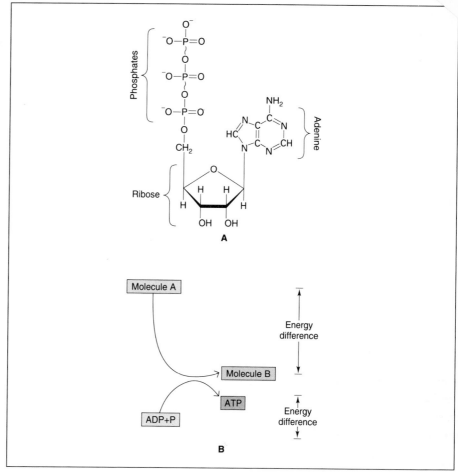

energy than can be released during the hydrolysis of the terminal phosphate bond of the ATP. Likewise the terminal phosphate bond in ATP must release more free energy during hydrolysis than is used by whatever mechanism uses the ATP. The amount of free energy released during the hydrolysis of the terminal phosphate bonds of a mole (505 grams) of ATP is generally calculated to be 7.3 kilocalories. (The exact amount depends on the conditions under which ATP is used.) This much energy is enough for about a minute of running, and it is about one-tenth the free energy provided by a slice of white bread.

ACTIVATION ENERGY

One of the major nutrients used by cells to make ATP is the simple sugar glucose (see Figure 2.9). By a long series of reactions that will be described in the next section, glucose breaks down bond by bond, releasing some of its free energy as heat and transferring some of the free energy to make ATP. Each day more than a kilogram of glucose breaks down in this way inside you. Yet you can walk into a chemistry stockroom and find jars of glucose sitting around not releasing energy as either heat or ATP. Although it contains a lot of free energy, glucose is stable. Otherwise it would break down in plants before animals consumed it, and it would break down in your bloodstream before your cells got it. Other nutrients are also stable. For that matter, the same is true for TNT. Even though these molecules have a lot of free energy to release, they do not do so until activated.

Like TNT, the breakdown of glucose and other nutrients must be activated by the addition of some form of energy. This added energy makes the molecule

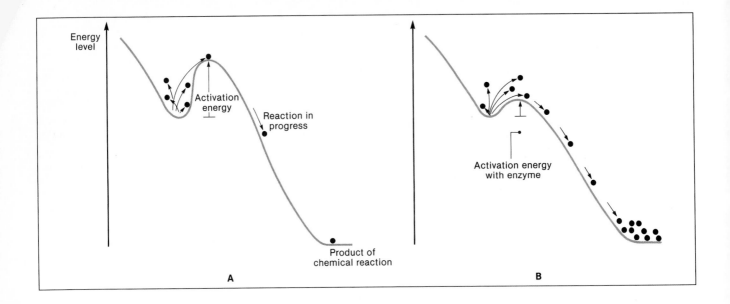

Figure 3.2
Activation energy and the role of enzymes. (A) Normally, the chemical breakdown of a nutrient molecule requires more activation energy than can be provided by thermal agitation. Therefore even though the chemical reaction is energetically favorable ("downhill"), it seldom occurs. (B) Enzymes lower the activation energy, accelerating the reaction.

unstable, thereby allowing it to react chemically. The amount of energy required to make a substance react chemically is called the **activation energy.** Activation energy can be visualized as a bump in a hillside, preventing a molecule from rolling down to a lower energy level (Figure 3.2A). In cells the only source of activation energy is usually thermal agitation, which, by itself, is seldom enough to make metabolic reactions proceed at an appreciable rate. One way to speed up chemical reactions in cells is to heat them. To some extent many animals, including ourselves, do this by maintaining our body temperatures above the environmental temperature. Obviously, however, there are limits to this approach, not the least of which is that heating the cells uses up most of the additional ATP produced.

ENZYMES

A far more practical mechanism for increasing a reaction rate is to reduce the activation energy by means of an **enzyme** (Figure 3.2B). An enzyme is an organic catalyst—a molecule that speeds up a chemical reaction, commonly by a factor of a million to a trillion. Enzymes are not changed by the reactions they catalyze, and they can therefore be used over and over again. With few exceptions, enzymes are proteins. The tertiary structure of the enzyme forms an **active site** in which the reacting molecule—the **substrate**—fits (Figure 3.3). The active site of an enzyme is generally highly specific for a particular substrate. An enzyme increases a reaction rate because the active site briefly holds the substrate in a position that favors the reaction. If the substrate is to be attached to another molecule, the enzyme holds the substrate and the molecule close to each other while they react. If the reaction involves the breaking of a bond in the substrate, the enzyme applies a strain to the bond.

Many enzymes require **cofactors** in order to function. Some cofactors are **coenzymes,** which are organic molecules that transfer a functional group to or from the substrate. ATP is a coenzyme that transfers a phosphate group, for example. Other cofactors are metal ions, such as magnesium, calcium, or zinc. In addition, many molecules have **allosteric effects** on enzymes. By binding to a particular allosteric site on an enzyme, they either increase or decrease the ability of the active site to bind to the substrate, thereby increasing or decreasing the rate of the reaction.

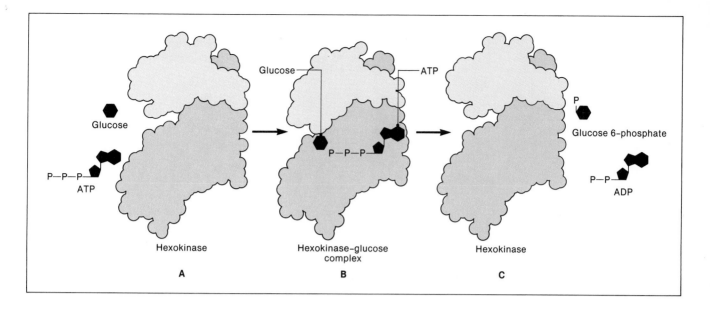

A

B

C

Glucose

ATP

Hexokinase

Glucose

ATP

P—P—P

Hexokinase–glucose
complex

P

Glucose 6-phosphate

P—P

ADP

Hexokinase

Figure 3.3

A model of the action of the enzyme hexokinase, which catalyzes the first reaction in the metabolism of glucose. (A) Glucose and ATP move to the active site. (B) The enzyme changes shape (a process called induced fit). The enzyme and its bound substrate are now called an enzyme–substrate complex. In this configuration the enzyme holds the glucose and an ATP molecule in such a way that a phosphate is transferred to the glucose. (C) After several microseconds the products of the reaction leave the enzyme, which is unaltered and ready to repeat the process.

Although enzymes work wonders in speeding up certain reactions, they cannot work miracles. They cannot force a reaction to proceed if the reaction is not spontaneous according to the Second Law of Thermodynamics. This means that a metabolic process won't proceed, even with an enzyme, unless it increases the total entropy. Nor can an enzyme make a metabolic process run "uphill," increasing the total free energy. Therefore, the only way to synthesize a molecule such as ATP is to couple the synthesis to a reaction such as the breakdown of glucose, which produces more entropy than ATP loses, and which loses more free energy than the ATP gains.

GLYCOLYSIS

The first stage in the breakdown of glucose to form ATP occurs in the cytosol. This stage, called glycolysis, involves the attachment of phosphate to sugar molecules, rearrangement of the sugar molecules so that energy in the sugars is concentrated in the phosphate bonds, then transfer of the phosphate bonds to ADP to form ATP. Although this sounds simple, it takes more than nine separate reactions (Figure 3.4). These reactions appear mystifying at first glance, but there is actually a good deal of logic in the sequence. The first four reactions accomplish three things:

1. They rearrange glucose into a fructose structure, which is more symmetric. The advantage of this symmetry is that the six-carbon sugar can be split into two identical three-carbon sugars (trioses), which can then be metabolized by the same set of enzymes, rather than have two separate pathways for each triose.
2. Glucose and fructose receive activation energy from two ATPs. This activation energy makes the fructose unstable, so it breaks down into the two trioses.
3. Each of the two trioses ends up with a phosphate group from one of the ATPs.

It might appear strange that the first steps in producing ATP actually use ATP. It is really no stranger, however, than investing money in a business in the expectation of making a profit later. In the remaining five reactions of glycolysis, in

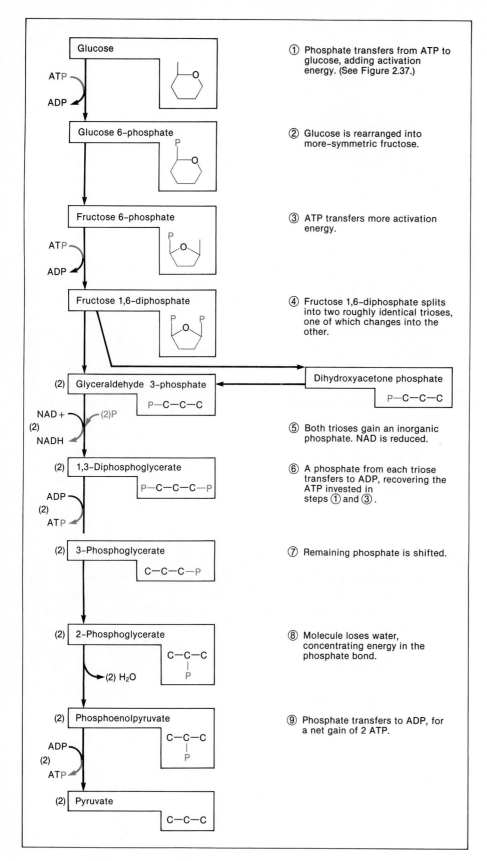

Figure 3.4
An outline of glycolysis, emphasizing the transfer of energy with phosphate (shown in color).

Glucose

ATP
ADP

Glucose 6-phosphate

Fructose 6-phosphate

ATP
ADP

Fructose 1,6-diphosphate

(2) Glyceraldehyde 3-phosphate   P—C—C—C

Dihydroxyacetone phosphate   P—C—C—C

NAD+
(2)   (2)P
NADH

(2) 1,3-Diphosphoglycerate   P—C—C—C—P

ADP
(2)
ATP

(2) 3-Phosphoglycerate   C—C—C—P

(2) 2-Phosphoglycerate   C—C—C | P

(2) H₂O

(2) Phosphoenolpyruvate   C—C—C | P

ADP
(2)
ATP

(2) Pyruvate   C—C—C

① Phosphate transfers from ATP to glucose, adding activation energy. (See Figure 2.37.)

② Glucose is rearranged into more-symmetric fructose.

③ ATP transfers more activation energy.

④ Fructose 1,6-diphosphate splits into two roughly identical trioses, one of which changes into the other.

⑤ Both trioses gain an inorganic phosphate. NAD is reduced.

⑥ A phosphate from each triose transfers to ADP, recovering the ATP invested in steps ① and ③.

⑦ Remaining phosphate is shifted.

⑧ Molecule loses water, concentrating energy in the phosphate bond.

⑨ Phosphate transfers to ADP, for a net gain of 2 ATP.

fact, the invested two ATPs are recovered, and a profit of two more ATPs is realized. The formation of two additional ATPs occurs because each of the two trioses acquires a second phosphate group (step 5 in Figure 3.4). In order for the triose and an inorganic phosphate to bind to each other in this reaction, they have to lose a hydrogen atom and an electron (e^-). H and e^- combine to form **hydride** ($H:^-$). This hydride is then transferred to a coenzyme that serves as an **electron carrier.**

The electron carrier in glycolysis is **nicotinamide adenine dinucleotide** (NAD+). NAD+ consists of nicotinamide (which is derived from the vitamin niacin), adenine, and two riboses joined by phosphates. NAD+ accepts the hydride to become NADH. In this process NAD+ is said to be **reduced** to NADH, while the triose is **oxidized.** NAD+ not only acquires a hydride during reduction, but it also gains a great deal of energy. As we shall see later, that energy can be used to form more ATP.

The end product of glycolysis, **pyruvate,** still has a considerable amount of free energy. Most cells in animals metabolize the pyruvate further and transfer that energy to ATP, as will be described in the next section. That process requires oxygen, however. If oxygen is not available the pyruvate is converted to a waste product and excreted from the cell. Since glycolysis by this method does not require oxygen, it is called **anaerobic glycolysis.** Another name for anaerobic glycolysis is **fermentation.** One familiar kind of fermentation involves the conversion of pyruvate into CO_2 and ethyl alcohol. This kind of fermentation is usually associated with yeasts, such as those used to brew beer. If deprived of oxygen, goldfish and certain small worms (nematodes, see Chapter 27) can also ferment pyruvate into CO_2 and alcohol (Crowe and Cooper 1971; Shoubridge and Hochachka 1980). Generally, however, animal cells with insufficient oxygen convert pyruvate to **lactate.** Lactate often forms in skeletal muscles during severe exercise, and it is largely responsible for the fatigue and soreness that result. During the production of lactate, pyruvate is reduced. This reduction is coupled to the oxidation of NADH, which means that the NAD+ is completely recyled.

Anaerobic glycolysis in most animal cells can be summarized by the following equation:

$$\text{glucose} + 2 \text{ phosphate} + 2 \text{ ADP} \rightarrow 2 \text{ lactate} + 2 \text{ ATP}$$

It is hard to imagine a more elegant way of producing ATP. An animal that produced all its energy by anaerobic glycolysis would not even have to breathe, since no O_2 is needed and no CO_2 is produced. Such an animal would be at a severe disadvantage, however, because of the small amount of ATP produced by this method. The amount of free energy that could be released from a mole of glucose is 686 kcal (kilocalories), but only 14.6 kcal of that energy finds its way into the two moles of ATP produced by glycolysis. This means that the efficiency of anaerobic glycolysis is just a little over 2% (14.6/686). A 2% return on investment is not impressive by any standard.

CELLULAR RESPIRATION

If oxygen is present in the cytoplasm, the pyruvate produced in glycolysis is not metabolized into lactate or other waste products. Instead, pyruvate undergoes further breakdown in which its free energy is eventually transferred to ATP as C—H bonds are broken and C is eliminated as CO_2. This complete catabolism of glucose under aerobic conditions is called **respiration,** or **cellular respiration** to avoid confusion with breathing. The use of the term respiration for both the aerobic breakdown of glucose and for breathing is not just an unfortunate coincidence. Cellular respiration requires O_2 and produces CO_2—the two main reasons for breathing.

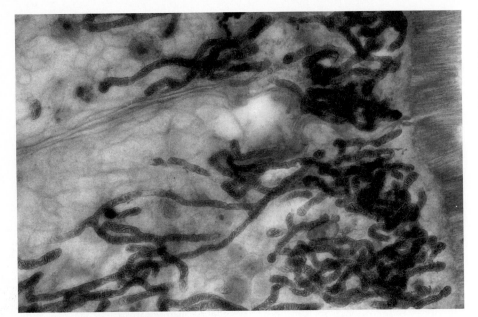

A

Figure 3.5
(A) Mitochondria in skin cells of a snail. Each cell may have from dozens to thousands of mitochondria, each typically 0.2 μm wide and several times as long. (B) Schematic diagram of a mitochondrion.

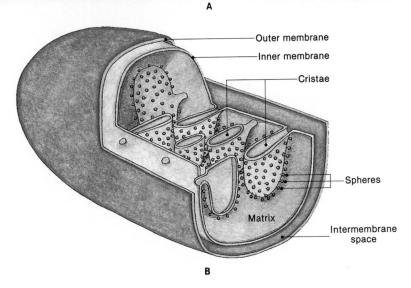

Outer membrane

Inner membrane

Cristae

Spheres

Matrix

Intermembrane space

B

The catabolism of pyruvate occurs in sausage-shaped organelles called **mitochondria** (singular, **mitochondrion**) (Figure 3.5). Mitochondria are bounded by two distinct membranes. The **outer membrane** is similar in composition to the plasma membrane and cytoplasmic membranes of the cell. The **inner membrane** is biochemically and functionally more like the plasma membranes of bacteria. This is one line of evidence that mitochondria originated as bacteria (see pp. 391–392). Also supporting this theory is the fact that mitochondria still retain a great deal of autonomy. They reproduce independently of the rest of the cell, and they have their own genetic material that controls the synthesis of some of the enzymes needed for cellular respiration.

Cellular respiration will be described in detail later, but for now it is important merely to see how it occurs in relation to mitochondrial structure (Figure 3.6). Cellular respiration involves two sets of processes: the Krebs citric acid cycle, which occurs in the central space, or **matrix** of the mitochondrion, and the respiratory chain, which occurs on the inner membrane, especially on infolding pockets called **cristae.** The **Krebs citric acid cycle,** also called the **tricarboxylic acid (TCA) cycle,** catabolizes pyruvate. Most of the free energy released by the catab-

Figure 3.6

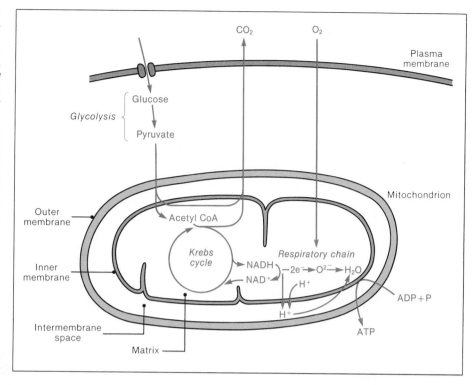

Figure 3.6
Outline of how most of the ATP is produced in cellular respiration. Pyruvate produced by glycolysis in the cytoplasm enters the mitochondrion, where it is converted to acetyl-CoA and CO$_2$. In the Krebs cycle the remaining carbons are lost as CO$_2$, and NADH is produced. Hydride (H:$^-$) from NADH enters the respiratory chain, where proton pumps use energy from the transport of e$^-$ to pump H$^+$ into the intermembrane space. Energy from the diffusion of H$^+$ back into the matrix generates ATP. The electrons are eventually accepted by oxygen atoms, which then combine with H$^+$ in the matrix to form H$_2$O.

olism of pyruvate is transferred with hydride to NAD+ which is reduced to NADH. Before it can transfer this energy to ATP the NADH must enter the **respiratory chain.** Some of the enzymes that catalyze the production of ATP in the respiratory chain are visible under the electron microscope as **spheres** that dot the cristae.

In the respiratory chain NADH is oxidized back to NAD+ when it loses its hydride. The hydride then splits into one proton (H$^+$) and two electrons. The energy associated with the electrons drives proton pumps. The proton pumps store the energy temporarily by transporting the H$^+$ from the hydride and other sources into the **intermembrane space** between the inner and outer membranes. The stored energy is released when H$^+$ diffuses back into the matrix. During this diffusion the energy drives a chemical reaction in which inorganic phosphate binds to ADP, forming ATP. Having expended their energy, the electrons attach to O$_2$, forming oxygen atoms that each have two electrons. Each oxygen atom then binds to two H$^+$ in the matrix, forming water as a waste product. This is the only function of oxygen in animal life: to serve as an **electron acceptor.** In fact, if O$_2$ were not used in this way it would build up to toxic levels that would oxidize organic molecules.

Krebs Cycle. Now let's consider the **Krebs citric acid cycle** in more detail. Every molecule of glucose produces two molecules of pyruvate. Each pyruvate starts one turn of the Krebs cycle. The Krebs cycle accomplishes four major things:

1. The three carbons in each pyruvate molecule are eliminated as CO$_2$. One carbon is eliminated as CO$_2$ immediately (step 1, Figure 3.7). The two carbons left from pyruvate bond to **coenzyme A** (a derivative of the vitamin pantothenic acid), forming **acetyl-CoA.** The function of coenzyme A is to reserve a bond on the acetyl-CoA for the attachment of a four-carbon molecule, oxaloacetate. This bonding forms the six-carbon molecule citrate. Citrate then undergoes a series of reactions in which two carbons are eliminated

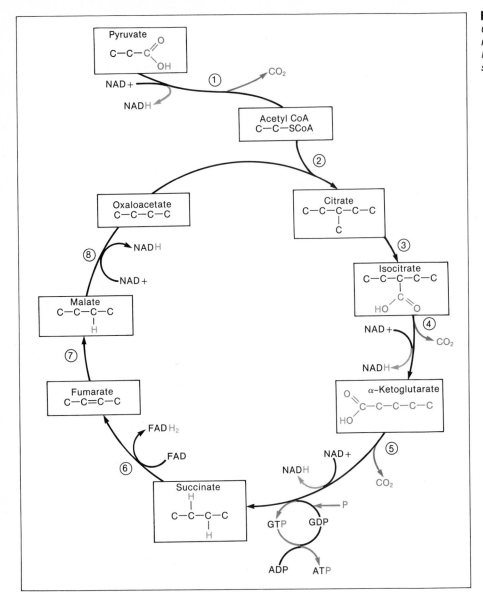

Figure 3.7
Outline of the Krebs cycle. The movement of carbons to CO_2, H to NADH, and phosphate to ATP are shown in color.

as CO_2 (steps 4 and 5), once more forming oxaloacetate. The cycle is completed when oxaloacetate attaches to another acetyl-CoA.

2. Free energy in the intermediates in the Krebs cycle is concentrated around hydrogen, which is then transferred as hydride to NAD+. In this way four molecules of NADH are produced for each molecule of pyruvate (steps 1, 4, 5, and 8 in Figure 3.7). Each of these NADH molecules has enough free energy associated with it to produce approximately three ATPs in the respiratory chain, thereby accounting for most of the ATP produced under aerobic conditions.

3. A molecule of **flavine adenine dinucleotide** (FAD) is reduced to $FADH_2$ (step 6). FAD is similar to NAD+ except that it has a flavine group (derived from the vitamin riboflavin) instead of a nicotinamide. Like NAD+, FAD is a coenzyme that gains free energy when it is reduced. This energy is sufficient to produce approximately two ATPs in the respiratory chain.

4. One molecule of ATP is produced (step 5). [Actually **guanosine triphosphate** (GTP), a molecule similar to ATP, is produced. GTP then transfers its terminal phosphate group to ADP.]

Figure 3.8

The respiratory chain. ① The NADH–dehydrogenase complex is named for an enzyme that catalyzes the oxidation of NADH to NAD+. Hydride from NADH dissociates into a proton (H+) and two electrons (e⁻). The proton, together with one from the matrix, is pumped into the intermembrane space. (Alternatively, the two hydrogen atoms can come from $FADH_2$.) ② The two electrons then transfer to coenzyme Q (Q). Coenzyme Q transfers the electrons to the b–c_1 complex. ③ The b–c_1 complex includes two cytochromes, b and c_1. (Cytochromes are proteins with an iron-bearing heme group that is readily reduced and oxidized. Their name comes from the fact that they give cells a slight color.) Energy transferred from the two electrons drives the b–c_1 complex to pump an undetermined number of H+ into the intermembrane space. ④ The two electrons then transfer to cytochrome c, which transfers them to the a–a_3 complex. ⑤ The a–a_3 complex, also known as cytochrome oxidase, includes cytochromes a and a_3. As the two e⁻ pass through, the complex pumps more H+ into the intermembrane space. The two e⁻ then transfer to an oxygen atom ($^1/_2O_2$). ⑥ The H+ pumped into the intermembrane space diffuses back into the matrix through the F_0F_1 ATPase, which is in a sphere on the cristae. This complex is essentially a reverse proton pump coupling the energy from H+ diffusion to the attachment of a phosphate group to ADP to make ATP. Each pair of protons combines with the oxygen atom to form water ($2H^+ + {}^1/_2O_2 \rightarrow H_2O$).

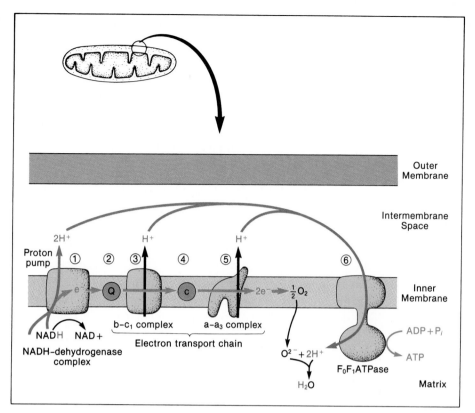

Table 3.1 A balance sheet for the production of ATP by aerobic catabolism of one mole of glucose.

The number of ATPs produced by one NADH from Krebs cycle is assumed to be three, and from glycolysis two. The number of ATPs from one $FADH_2$ is assumed to be two. Remember that the Krebs cycle occurs twice for each molecule of glucose.

Source of ATP	Moles of ATP
Glycolysis	2
2 NADH in glycolysis	4
Krebs cycle (2 GTP)	2
8 NADH in Krebs cycle	24
2 $FADH_2$ in Krebs cycle	4
Total	36

The Respiratory Chain. Although some ATP is produced in the Krebs cycle, most of the free energy in pyruvate ends up in NADH, with a little in $FADH_2$. This energy is transferred to ATP in the **respiratory chain.** The precise mechanism by which the respiratory chain produces ATP is still being worked out. The evidence currently available, however, supports the **chemiosmotic theory** first proposed by Peter Mitchell in 1961. The basic idea of Mitchell's chemiosmotic theory is that the respiratory chain powers **proton pumps** that transport hydrogen ions into the intermembrane space. The energy to pump the protons comes from electrons moving from higher to lower energy levels along an **electron transport chain.** This electron transport chain, as well as the proton pumps, are associated with three **enzyme complexes** bound to the inner membrane (Figure 3.8). The H+ then diffuses back into the matrix through another complex that is essentially a proton pump that runs in reverse. Instead of using ATP to transport H+, it couples the passive flow of H+ to the production of ATP.

The Efficiency of Cellular Respiration. The amount of ATP produced by cellular respiration depends on how much H+ is pumped into the intermembrane space, which varies with conditions in the cell. On average, however, one molecule of NADH from the Krebs cycle provides enough free energy to produce approximately three molecules of ATP. Under aerobic conditions there is also a net production of two molecules of NADH in glycolysis (step 5, Figure 3.4), which are used to make two molecules of $FADH_2$. These molecules of $FADH_2$, and the two from the Krebs cycle, transfer their H a little farther down the respiratory chain and therefore allow for the production of only about two ATPs each. Table 3.1 is an accounting of all the ATP produced. The "bottom line" is that the aerobic catabolism of one mole of glucose produces approximately 36 moles of ATP. (The exact amount may be as low as 25 moles or as high as 38 moles.)

The equation describing the complete aerobic breakdown of glucose is as follows:

$$C_6H_{12}O_6 + 6\ O_2 + 36\ ADP + 36\ \text{phosphate} \rightarrow 6\ CO_2 + 6\ H_2O + 36\ ATP$$

Comparison with the equation for glycolysis (see p. 57) shows that cellular respiration produces much more ATP but requires oxygen and produces carbon dioxide. Is the additional ATP from cellular respiration worth the trouble of having to breathe (not to mention the trouble of having to learn about it)? Assuming that each mole of ATP provides 7.3 kcal of free energy, the total yield is approximately 263 kcal. Since a mole of glucose can release 686 kcal of free energy, the efficiency of cellular respiration is approximately 263/686 = 38%. The remaining energy—62% of the energy in glucose—is lost as heat. This may seem wasteful, but in fact it is not bad compared with the most efficient artificial engine, the gas turbine, which has 40% efficiency. It is conservative compared with a well-tuned gasoline engine, which delivers only about 15% to 20% of the free energy in gasoline to the wheels. Cellular respiration is efficient, indeed, compared with the 2% efficiency of anaerobic glycolysis. If you are still not convinced that cellular respiration is worth the trouble, consider what happens when you stop breathing.

ALTERNATIVE METABOLIC PATHWAYS

Man does not live by glucose alone; nor does any other animal. Other carbohydrates, as well as fats and proteins, make up a significant portion of the diet. In addition, animals convert glucose into other carbohydrates and into fats and proteins. Many of these molecules are metabolized by alternative pathways that intersect the pathways of glycolysis and the Krebs cycle (Figure 3.9). For example, liver and muscle cells produce glycogen, a starchlike substance (see Figure 2.10B),

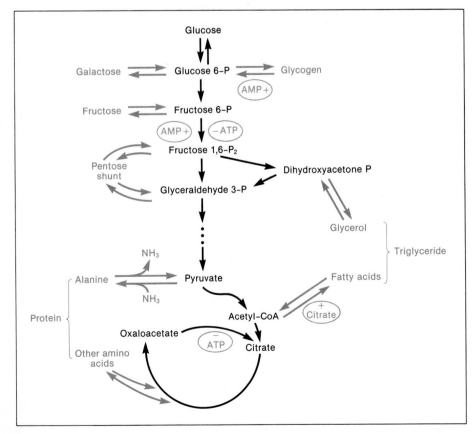

Figure 3.9
Alternative metabolic pathways. ATP, AMP, and citrate stimulate (+) or inhibit (−) reactions at the points shown.

whenever glucose is abundant. When glucose levels in the blood start to fall, the glycogen breaks down into glucose-6-phosphate, which is then catabolized in the cell or used to make glucose that is pumped into the blood. Glycogen can thus provide cells with a between-meal snack of glucose. Most adult humans store enough glycogen to provide about a day's worth of energy. Some other carbohydrates, especially ribose and other five-carbon sugars, do not fit so easily into the glycolytic pathway. A complicated series of reactions called the **pentose shunt** converts these pentoses from combinations of three- and six-carbon sugars.

Triglycerides (neutral fats) stored in adipose tissue or ingested with the diet are catabolized by converting them into glycerol and fatty acids. The glycerol is easily changed into dihydroxyacetone phosphate for use in glycolysis, and the fatty acids are (less easily) converted into acetyl-CoA for use in the Krebs cycle. If excess nutrients are available, triglycerides are synthesized by essentially reversing these processes. In other words, if you eat too much you get fat. Adult humans with normal body mass store enough fat to provide energy for a month.

The metabolism of proteins is complicated by the fact that 20 different amino acids are involved. In humans eight of these amino acids cannot, in fact, be either catabolized or synthesized. Some of the remaining 12 amino acids, however, are similar to intermediates in the Krebs cycle once they have lost their amine groups. The amino acid alanine, for example, is catabolized by deaminating it to form pyruvate, which then enters the Krebs cycle.

All these intersecting metabolic pathways (and these are only a few of them) make one wonder how the cell maintains any order at all. In general, each pathway is regulated by **feedback inhibition** from products of the pathway. For example, adenosine monophosphate (AMP), which accumulates from the breakdown of ATP, stimulates glycolysis and the breakdown of glycogen by its allosteric effect on certain enzymes (Figure 3.9). In contrast, ATP shuts off its own production by inhibiting glycolysis and the Krebs cycle. As the Krebs cycle slows down, citrate accumulates, which causes acetyl-CoA to be diverted into the synthesis of fatty acids. In addition to feedback inhibition, signals from outside the cell also affect metabolic pathways. The hormone insulin (see pp. 274–275), for example, stimulates the production of glycogen, thereby removing excess glucose from the blood. The hormones glucagon (see pp. 274–275) and adrenaline promote the breakdown of glycogen to provide additional glucose.

SUMMARY

Cells require energy to power their mechanisms, and they derive that energy from the chemical bonds in nutrients, especially glucose. Glucose is a stable molecule, requiring activation energy before it breaks down and releases the energy in its chemical bonds. Cells are able to extract that energy at normal body temperatures only because they produce enzymes that reduce the amount of activation energy needed.

The energy from chemical bonds in glucose is usually transferred to ATP, which can then transfer the energy to cellular mechanisms that require it. A few types of animal cell can produce ATP without oxygen in the series of reactions called glycolysis, in which glucose is converted into pyruvate. Only about 2% of the energy in glucose gets into ATP, however.

Most animal cells increase the efficiency by further metabolizing the pyruvate. Metabolism of pyruvate occurs in mitochondria, during the processes of cellular respiration. Cellular respiration consists of two major processes: the Krebs citric acid cycle and the respiratory chain. In the Krebs cycle the carbons in pyruvate are converted into CO_2, and most of the energy from the chemical bonds is transferred to NAD+ to form NADH. In the respiratory chain the NADH is converted back to NAD+, releasing protons and electrons. Energy from the electrons pumps protons against a concentration gradient. Energy from diffusion of the protons is then used to convert ADP and inorganic phosphate into ATP. Oxygen is required to accept the electrons after they have expended their energy. Cellular respiration is about 40% efficient.

Metabolism of most other nutrients share some of the same reactions as glycolysis and the Krebs cycle. The fatty acids in triglycerides, for example, are converted into acetyl-CoA, which is metabolized in the Krebs cycle.

KEY TERMS

metabolism
ATP
activation energy
enzyme

substrate
glycolysis
oxidation–reduction reaction
respiration

mitochondrion
Krebs citric acid cycle
respiratory chain

SELF-TEST

1. What are enzymes? How do they work? Give an example of a particular enzyme, and explain what it does.

2. Describe the structure and functioning of ATP.

3. The overall equation describing glycolysis in most animal cells is as follows:

glucose + 2 APD + 2 phosphate → 2 lactate + 2 ATP

Explain each term in the equation. ATP has more free energy than ADP; explain how it acquires that free energy. In what part of the cell does glycolysis occur?

4. Briefly describe the Krebs cycle. Briefly describe the respiratory chain. Diagram a mitochondrion showing where these two processes occur. What role does NADH play in the Krebs cycle and in the respiratory chain?

5. State the basic idea of Mitchell's chemiosmotic theory.

6. Cellular respiration is described by the following equation:

glucose + 6 O_2 + 36 ADP + 36 P → 6 CO_2 + 6 H_2O + 36 ATP

Which oxygen atoms on the right-hand side of the equation come from the O_2 on the left-hand side? Explain your answer. From where does the carbon in CO_2 come? By what series of reactions?

READINGS

RECOMMENDED READINGS

Dressler, D. and H. Potter. 1991. *Discovering Enzymes.* New York: Scientific American Library.

Hinkle, P. C. and R. E. McCarty. 1978. How cells make ATP. *Sci. Am.* 238(3):104–123 (Mar).

ADDITIONAL REFERENCES

Crowe, J. H. and A. F. Cooper, Jr. 1971. Cryptobiosis. *Sci. Am.* 225(6):30–36 (Dec).

Shoubridge, E. A. and P. W. Hochachka. 1980. Ethanol: novel end product of vertebrate anaerobic metabolism. *Science* 209:308–309.

Genetic Inheritance

Mother and young African elephant (Loxodonta africana).

CHAPTER OUTLINE

LEARNING OBJECTIVES

1. Are the traits acquired in the life of a parent passed on to the children?

2. Do inherited traits come from all parts of a parent?

3. What is the material basis of inheritance?

4. Do the effects of genes blend together, or are they segregated?

5. Does the inheritance of one gene affect whether another will be inherited, or are genes independently assorted?

6. Where are genes located?

7. How can their locations on chromosomes be determined?

8. How are genes duplicated so that each product of cell division receives an identical copy of them?

9. How is the amount of genetic material reduced to half in gametes so that the normal amount occurs when two gametes fuse during fertilization?

EARLY NOTIONS OF INHERITANCE

Anyone who has studied the materials and mechanisms of cells described in the previous chapters should be impressed, and perhaps mystified, by how well they perform their functions. How do cells produce materials and mechanisms that are so well adapted? Where do they get the information that tells them how to make an enzyme and how a microtubule is supposed to work? How is it that we can talk about *the* cell at all? Why are they and the organisms they compose all so similar, yet continually varying from one generation to the next?

The ancients had an explanation for these phenomena. It made such good sense that it is still often repeated by those who have not studied biology. According to this commonsense theory, cells and organisms essentially learn from their parents how to function, because the events that affect an organism in its lifetime are passed on to its offspring. We call this idea **the inheritance of acquired characters.** Inheritance of acquired characters has probably been common "knowledge" for thousands of years. *Genesis* 30, for example, records that Jacob produced spotted kids and lambs by placing spotted tree branches in front of the goats and sheep while they were conceiving.

Inheritance of acquired characters implies another commonsense theory: that every part of a parent's body contributes to the characteristics of the offspring. This would permit the visual system of Jacob's goats and sheep to influence their offspring. This idea is called **pangenesis,** from the Greek words for "all" and "origin." The earliest surviving statement of pangenesis occurs in writings of the medical students of Hippocrates from about 410 B.C., which state that "the semen comes from all parts of the body" (*On Airs, Waters, and Places* 14). In the 4th century B.C. Aristotle easily refuted both inheritance of acquired characters and pangenesis (*On the Generation of Animals* I. 17, 18). He noted, for example, that children are not born with the beards of their fathers, and that children often resemble their grandparents more than their parents.

Unfortunately, not even Aristotle could propose an explanation for heredity as seductive as inheritance of acquired characters and pangenesis. As late as 1809 Jean Baptiste Lamarck used inheritance of acquired characters (now called "Lamarckism") as a cornerstone of his theory of evolution. Even Charles Darwin accepted inheritance of acquired characters. In his book *The Variation of Animals and Plants under Domestication*, published in 1868, Darwin elaborated a theory of pangenesis that he was just as proud of as his theory of natural selection.

THE GENESIS OF GENETICS

Darwin did not know that just two years before publication of his book on pangenesis evidence against it and the inheritance of acquired characters had been published by an obscure monk, Gregor Mendel (Figure 4.1) Johann Mendel (1822–1884) was the son of poor peasants who scrimped to finance his education in physics and mathematics. At age 21 he became an Augustinian novice, taking the name Gregor. He was ordained a priest in 1847 and was appointed deputy teacher of high-school mathematics in 1849. After failing the biology portion of the teachers' qualifying examination, he was sent to the University of Vienna for two years to study zoology, plant physiology, chemistry, and physics. He applied again for the teachers' exam but had to withdraw because of sickness. At age 34 he entered the Altbrünn Monastery in what is now Brno, Czechoslovakia, and eventually became Abbot.

Mendel began his studies while engaged in one of the enterprises of the monastery, raising ornamental flowers. Mendel had noticed that whenever he allowed

Figure 4.1
Gregor Mendel.

pollen from one variety of plant to fertilize the flowers of a different variety of the same species, the resulting hybrids showed a remarkable consistency in bearing the traits of one variety or the other. Others, including Darwin, had already studied hybridization, but they looked at the overall features of hybrids. Since characters from both parents occurred in the hybrid offspring, they had concluded that the hereditary material blended together in a hopelessly complicated manner. Mendel, with the analytical approach of the physicist, decided to look at one character at a time. He chose garden peas, partly because their closed flowers resist random pollination by insects, and used true-breeding varieties, which consistently produce offspring with the same characteristics as the parents. In 1856 he began systematically cross-pollinating various true-breeding varieties of pea plants and studying the hybrids and the offspring of the hybrids. Mendel found, as one example, that if a plant with purple flowers were crossed with a plant with white flowers, then all the hybrid plants resulting from that cross had purple flowers. In Mendel's terminology, which we still use, the purple-flower character was **dominant** over the white-flower character, which was **recessive.**

If heredity were due to substances transmitted from the parents, what happened to the substance determining the recessive character? Did it simply disappear? To find out, Mendel allowed the hybrid flowers to become fertilized with their own pollen (a normal occurrence because of the closed flower structure), to see if any white flowers would reappear in the next generation. Of 929 plants produced from the self-fertilized hybrids, 705 bore purple flowers, and 224 had white flowers (Figure 4.2). The factors determining the recessive white-flower character had evidently survived within the purple-flowered hybrid. This in itself argued against the inheritance of acquired characters and pangenesis, since the white flowers in the second generation did not inherit the characters of their parents.

Figure 4.2
Mendel's experiment showing that a recessive character (white flowers) is not expressed in a hybrid but emerges in approximately one-fourth of the offspring of self-fertilized hybrids.

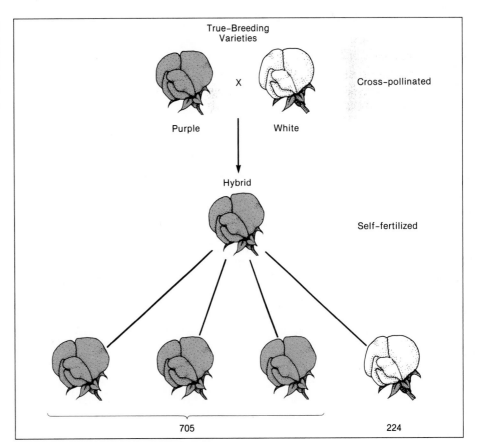

	Dominant character		Recessive character		Ratio
Form of seed	Round 5474		Wrinkled 1850		2.96:1
Color of albumen	Yellow 6022		Green 2001		3.01:1
Color of seed coat and color of flower	Grey–brown Purple 705		White White 224		3.15:1
Form of pod	Inflated 882		Constricted 299		2.95:1
Color of unripe pod	Green 428		Yellow 152		2.82:1
Position of flowers	Axial 651		Terminal 207		3.14:1
Length of stem	Long 787		Short 277		2.84:1
Totals	14,949		5010		2.98:1

Figure 4.3
Summary of Mendel's results from self-fertilized hybrids. True-breeding varieties of pea plants differing in the seven traits listed in the left column were cross-pollinated, producing hybrids with the dominant character. After the hybrids were self-fertilized, they produced offspring with the numbers of dominant and recessive characters shown in the boxes. The ratio of dominants to recessives for each character is given in the right column. The bottom row shows the total numbers and the ratio of dominant to recessive characters— 2.98:1, which Mendel rounded off to 3:1. (It appears that Mendel "cooked" his data to get a ratio so close to 3:1. But how did he know that the correct answer was 3:1?)

Mendel found similar results for six other characters (Figure 4.3). In every case about three-fourths of the offspring of self-fertilized hybrids had the dominant character, and about one-fourth had the recessive character. In other words, *the ratio of dominant to recessive characters in the offspring of hybrid parents is approximately 3:1.* When the plants with the recessive characters were self-fertilized, they too invariably produced offspring with the recessive characters. In other words, they were true-breeding. One-third of the plants with dominant characters were also true-breeding, producing only offspring with the dominant character. However, the other two-thirds of the plants with dominant characters produced both dominant and recessive offspring in the ratio of 3:1. Evidently they were hybrids like their parents.

Mendel proposed a simple mathematical model for these findings. He let *A* denote the dominant character, *a* the recessive character, and *Aa* the hybrid. He assumed that each **gamete** (ovum in a female flower or sperm cell in pollen) received either *A* or *a* with equal likelihood. In a self-fertilizing hybrid, therefore, an *A* ovum has an equal chance of being fertilized by *A* pollen or *a* pollen. Likewise an *a* ovum has an equal chance of being fertilized by *A* or *a* pollen. All the combinations, *AA*, *Aa*, *aA*, and *aa*, should therefore occur with equal frequency. *Aa* and *aA* are identical, so the ratio of *AA* to *Aa* to *aa* is 1:2:1. Since *AA* and *Aa* look alike, however, the ratio of those with the dominant character to those with the recessive character is 3:1. When those with the dominant characters self-fertilize, two out of the three are shown to be *Aa* hybrids.

SEGREGATION AND INDEPENDENT ASSORTMENT

This model works only if *A* and *a* do not blend, but remain **segregated** from each other during the formation of gametes. This principle of segregation has come to be known as **Mendel's First Law.** Mendel did not say so, but the simplest explanation for segregation is that each plant has a pair of "units" for each character— what we now call **genes**. The different kinds of genes for each character, whether dominant or recessive, we now call **alleles.** There may be many kinds of alleles for each gene, but only one allele is transmitted from a plant to each gamete.

The principle of segregation states that the allele determining a dominant character behaves independently of a recessive allele. But what about alleles for two different characters? For example, will round versus wrinkled seeds still occur in a 3:1 ratio whether or not the albumen (Figure 4.3) is yellow or green? If so, then three out of four of the offspring will be dominant in one trait, and three out of four of those plants will also be dominant in the other trait. Thus the proportion of offspring with both characters dominant should be $\frac{3}{4} \times \frac{3}{4} = \frac{9}{16}$. Similarly, the proportion with one character dominant and one recessive should be $\frac{3}{4} \times \frac{1}{4} = \frac{3}{16}$, and the proportion with both recessive characters should be $\frac{1}{4} \times \frac{1}{4} = \frac{1}{16}$. In other words, the ratio of offspring with the four combinations of characters should be 9:3:3:1.

To test this prediction Mendel allowed plants that were hybrid in two characters to self-pollinate. This procedure is now called a **dihybrid cross,** as opposed to the **monohybrid crosses** described previously. The two characters were round seeds versus wrinkled seeds, and yellow versus green albumen. Out of 556 seeds produced, the expected numbers were 313 ($556 \times \frac{9}{16}$) round with yellow albumen, 104 round with green albumen, 104 wrinkled with yellow albumen, and 35 wrinkled with green albumen. The actual numbers were 315, 108, 101, and 32. This close agreement (some say suspiciously close) supported the hypothesis that the different characters assort independently of each other. This principle of **independent assortment** is now known as **Mendel's Second Law.** As we shall see, however, it is a "law" that is widely disobeyed.

Mendel's research was published in 1866, but at that time few biologists could understand the mathematical reasoning or the far-reaching implications. It was not until 1900 that three researchers independently obtained some of the same results, then discovered Mendel's paper. By then the inheritance of acquired characters and pangenesis were virtually forgotten issues, so the fact that Mendel disproved them was unremarkable. However, other research (to be described later) had prepared the minds of a few biologists to see its revolutionary implications. In fact, the rediscoverers saw more in Mendel's work than perhaps even Mendel saw. They credited him with discovering that traits were inherited as particles, which they named genes and alleles. They also created a new vocabulary to describe various patterns of inheritance. Individuals that developed from zygotes with both alleles of the same type (both dominant or both recessive) were said to be **homozygous.** Individuals with two different alleles for the same gene were **heterozygous.** The actual genetic makeup of the individual was called the **genotype.** The appearance or other property of the organism, which depends partly on genotype and partly on environmental influences, was named the **phenotype.**

Using this terminology we would say that Mendel's true-breeding plants were either homozygous dominants or homozygous recessives, and that when they were crossed with each other they produced heterozygotes that developed into plants with the dominant phenotype. When the heterozygous plants self-fertilized they produced offspring with a genotype ratio of 1:2:1, and a phenotype ratio of 3:1. An equally important contribution of the early geneticists were statistical and mathematical methods to determine whether actual genotype and phenotype ratios are consistent with theoretical predictions. One of these methods is the **Punnett square,** which allows for the simple calculation of genotype and phenotype ratios (Figure 4.4).

CHROMOSOMAL INHERITANCE

During the 35 years that Mendel's work was being ignored, the mechanism of inheritance was being approached from a different direction by biologists using the improving techniques of microscopy to study cell structure. One of these

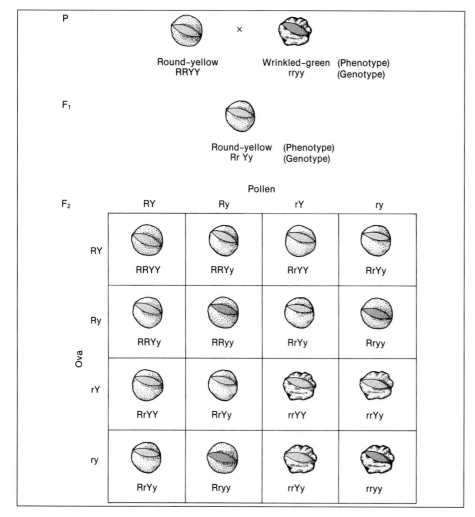

Figure 4.4
A Punnett square that predicts the ratios of the four phenotypes resulting from the dihybrid cross of pea plants heterozygous for form of seed and color of albumen, discussed previously in the text. Capital letters denote dominant alleles; lowercase letters denote recessive alleles. At top, in the parental (P) generation, a plant that is true-breeding (homozygous) for the dominant alleles for round seeds and yellow albumen is crossed with a plant that is true-breeding for the recessive alleles for wrinkled seeds and green albumen. Part of the seed coat is shown torn away to reveal the yellow or green albumen. The phenotypes are, respectively, round-yellow and wrinkled-green, and the genotypes are symbolized RRYY and rryy. The F_1 (first filial) generation has the dominant phenotype (round-yellow), and its genotype is heterozygous (RrYy). This hybrid produces pollen and ova with either the R or r allele, and either the Y or y allele. These recombine with equal probability in the F_2 generation produced by self-fertilized hybrids. Counting the number of squares with each of the four possible phenotypes gives the 9:3:3:1 ratio predicted from independent assortment.

pioneering cytologists was **August Weismann** (1834–1914). Weismann devoted his early career to disproving the inheritance of acquired characters and pangenesis. At first he took the straightforward approach of cutting off the tails of 40 generations of mice, demonstrating that tails in the 40th generation were just as long as those in the first. (Actually this was a superfluous experiment. Jews and Arabs had been circumcising boys for hundreds of generations with no apparent effect on foreskins.)

In the 1870s and 1880s Weismann took a more sophisticated approach with microscopical observations of the cnidarian *Hydra*. He observed that the cells that eventually produce gametes (sperm and eggs) are present in definite locations from the earliest stages of development. Thus gametes do not come from all over the body, and they have little opportunity to acquire characters during the life of the animal. Weismann therefore proposed that the **germ plasm** (the reproductive tissue that produces gametes) is separate and distinct from the **somatoplasm** (the body tissue), and that only the germ plasm contributes to the traits of the offspring. In essence, Weismann was proposing that what is really reproduced generation after generation is germ plasm, and that the body of an organism is merely a device that enables the germ plasm to reproduce.

Weismann and others continually refined the germ-plasm theory using the latest results from microscopical studies. One crucial line of study concerned the **nucleus.** Cytologists observed that when one cell divides into two, the nucleus also divides into two nuclei. In the early 1880s Oskar Hertwig and Hermann Fol also observed that the nucleus of a sea urchin's sperm unites with the nucleus of the ovum during fertilization. Both these observations suggested to Weismann and others that the germ plasm was located in the nuclei.

Later the germ plasm was localized in what we now call **chromosomes,** which are thread-shaped structures visible in nuclei just before cells divide. Generally, in body cells, chromosomes occur in pairs, so the cells are said to be **diploid** (Greek *diploos* double). The laboratory organisms commonly used in the 19th century have so many chromosomes that it was hard to follow their movements under the microscope. Edouard van Beneden and Theodor Boveri found, however, that the parasitic worm *Ascaris bivalens* has only two pairs of chromosomes. They were therefore able to see that only one member of each pair of chromosomes enters each newly formed ovum or sperm cell. The gametes are therefore **haploid** (Greek *haploos* single). When a sperm cell and an ovum unite during fertilization, the resulting **zygote** has its number of chromosomes restored to the normal diploid number. Weismann had previously predicted that something like this halving of the genetic material in gametes would have to occur to prevent the amount of hereditary material from doubling with each generation. He and others therefore concluded that the germ plasm was in the chromosomes.

Weismann's distinction between somatoplasm and germ plasm and his conclusion that the chromosomes are responsible for heredity suggest two different processes of heredity. First, there must be some way by which genetic information in chromosomes is passed from a zygote to the somatoplasm to determine the characteristics of the organism. Second, there must be a process by which the gamete is made haploid so that the zygote becomes diploid. The two processes are called, respectively, mitosis and meiosis.

MITOSIS

Cytologists in the 19th century found that each zygote divides into two **daughter cells,** each of which divides into two more daughter cells, and so on through

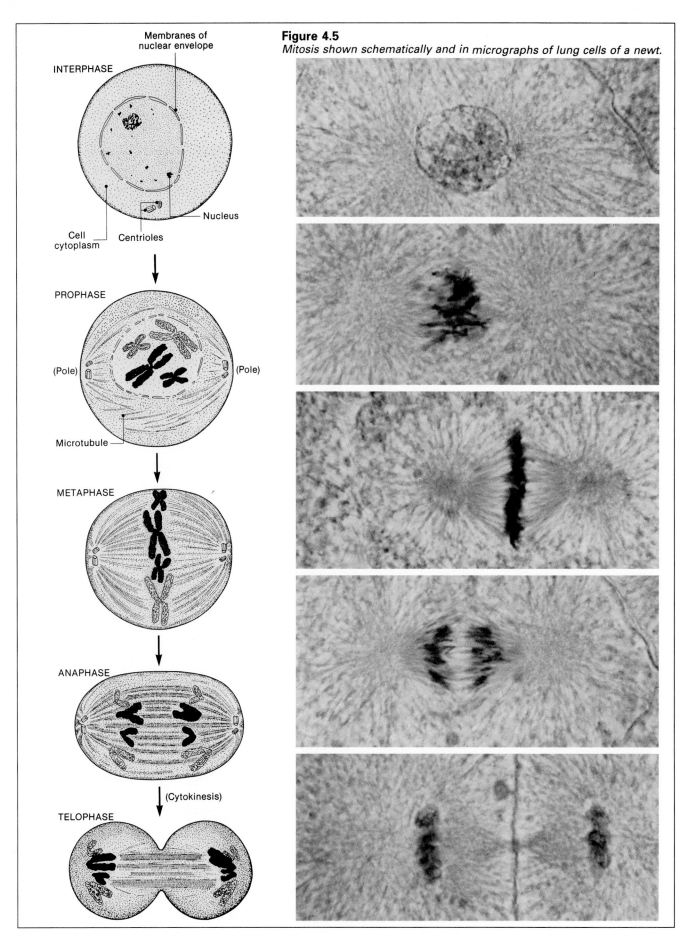

INTERPHASE

Membranes of
nuclear envelope

Cell
cytoplasm

Centrioles

Nucleus

PROPHASE

(Pole)

(Pole)

Microtubule

METAPHASE

ANAPHASE

(Cytokinesis)

TELOPHASE

Figure 4.5
Mitosis shown schematically and in micrographs of lung cells of a newt.

embryonic development and later growth. Just before each cell division the chromosomes become visible within the nuclei. Cytologists could then see that every cell has the same number of chromosomes as the original zygote. They could also see that the chromosomes move about in a quite remarkable pattern called **mitosis.** Mitosis accounts for the number of chromosomes remaining constant after each cell division.

Although mitosis is a continuous process, it is convenient to divide it into four phases (Figure 4.5). First is **prophase.** During prophase the **nuclear envelope** breaks up, and **chromatin,** the material that forms the chromosomes, condenses and becomes visible. Each chromosome consists of two sister **chromatids** attached to each other at a **centromere.** Also during prophase other changes occur that prepare the cell for division. In animal cells a pair of **centrioles** (see Figure 2.29) in the cytoplasm migrate to each of two opposite **poles.** These centrioles are composed of microtubules, and they apparently organize the assembly of other microtubules into the **spindle apparatus.** The spindle apparatus radiates from the two poles like iron filings around the poles of a magnet.

Spindle fibers attach to each chromosome at a **kinetochore.** The kinetochore is a protein located at the centromere. The spindle fibers guide the chromosomes to a disc-shaped area between the poles. When the chromosomes are lined up midway between the centrioles, the cell is said to be in **metaphase.** Chromosomes are most easily seen during metaphase (Figure 4.6). Metaphase is followed by **anaphase.** During anaphase the sister chromatids separate toward opposite poles, and each chromatid becomes a new chromosome. Arrival of the daughter chromosomes at each pole signals the onset of the last phase, **telophase.** During telophase the chromosomes begin to disperse into chromatin, and a new nuclear envelope starts to form around them. Mitosis concludes at this point, when two nuclei have formed.

Figure 4.6
Scanning electron micrograph of a metaphase chromosome. In metaphase the chromosome consists of tightly condensed strands of chromatin. The constriction shows the location of the centromere, to which the kinetochore (not shown) attaches. The ends are defined by special regions of chromatin called telomeres, which are not visible here.

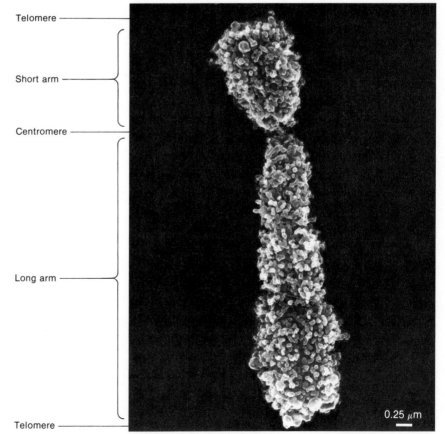

Telomere

Short arm

Centromere

Long arm

Telomere

0.25 μm

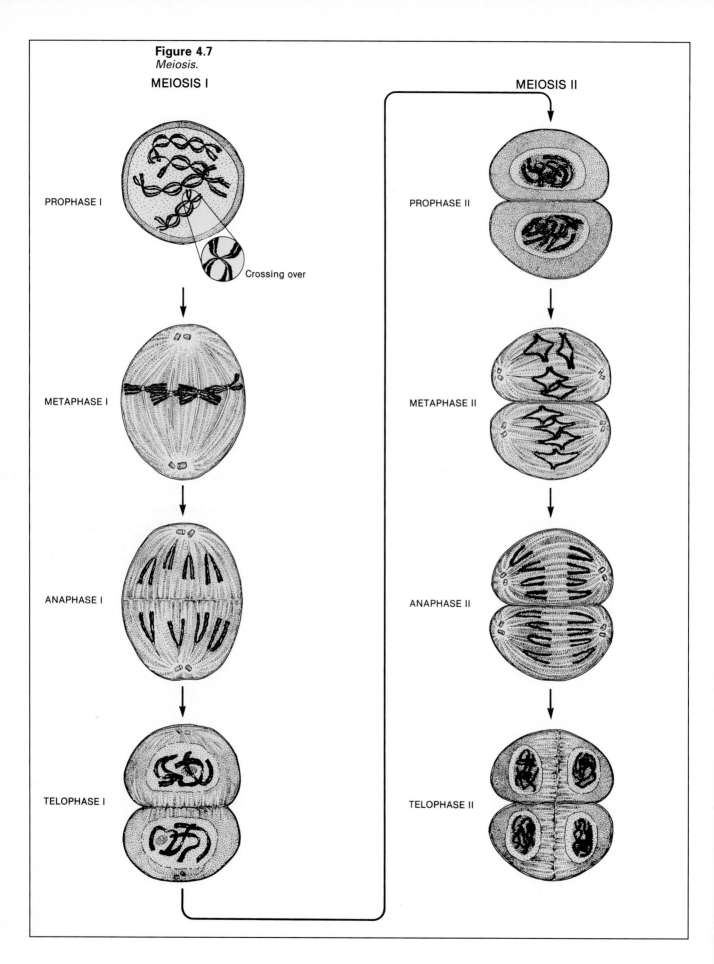

Figure 4.7
Meiosis.

MEIOSIS I

MEIOSIS II

PROPHASE I

Crossing over

METAPHASE I

ANAPHASE I

TELOPHASE I

PROPHASE II

METAPHASE II

ANAPHASE II

TELOPHASE II

Usually the cytoplasmic components of the cell also start to move apart, in a process called **cytokinesis.** The plasma membrane becomes constricted, dividing the cell into two diploid daughter cells, each with a nucleus containing a set of chromosomes identical to those of the parent cell. After telophase the cells enter the state called **interphase,** in which the chromatin directs the growth and other activities of the cell.

MEIOSIS

As the result of mitosis most cells in an animal inherit copies of the same chromosomes that were in the zygote. Gametes are the major exceptions. Mitosis will clearly not serve for the production of gametes, since they have to be haploid. Instead of mitosis, therefore, cells produce gametes through a process called **meiosis.** Meiosis begins in much the same way as mitosis, with each chromosome in a diploid cell appearing as two sister chromatids. The main difference is that there are two cell divisions, called **meiosis I** and **meiosis II.** During meiosis I the two chromatids do not separate, as in mitosis. Instead, each chromosome goes to one of two daughter cells. Each of these cells then divides again in meiosis II. This time the chromatids do separate, and each becomes a chromosome. Thus four haploid cells result from meiosis.

It is convenient to divide meiosis into several phases that are similar to the phases of mitosis (Figure 4.7). The first phase is **prophase I** (prophase of meiosis I), in which the chromosomes appear much like those in prophase of mitosis. In prophase I, however, each pair of **homologous chromosomes** (those that carry the same genes) line up side by side, and the four chromatids join. This association is called **synapsis,** and the interconnected chromatids are said to form a tetrad. The chromatids in each tetrad then intertwine in a process called **crossing over** (Figure 4.8). Although it could not be observed in the 19th century, genetic evidence to be described later shows that segments of chromatids from one homologous chromosome actually interchange with the chromatids of the other homologous chromosome during prophase I.

During prophase I the chromosomes attach to spindles and migrate to an area between the two cellular poles, as in prophase of mitosis. Homologous chromosomes are still joined as tetrads during **metaphase I.** Metaphase I is followed by **anaphase I,** in which the homologous chromosomes migrate toward opposite poles. Meiosis I is completed when the chromosomes arrive at the poles, during **telophase I.** At this point each chromosome still consists of two attached sister chromatids, although parts of each chromatid may have come from the homologous chromosome that is now at the opposite pole. After telophase I comes cytokinesis, followed by division into two daughter cells.

Each of the two daughter cells formed in meiosis I has one set of chromosomes, each consisting of two chromatids. They may then proceed directly to metaphase II, or they may enter a brief period of interphase, during which the chromatin disperses. In that case the chromosomes reappear in **prophase II,** the first phase of meiosis II. During prophase II each chromosome, consisting of two sister chromatids, attaches to spindles and moves to a position between the poles. This is followed by **metaphase II, anaphase II,** and **telophase II,** which resemble the corresponding phases of mitosis, but with a different result. In mitosis a chromatid from both homologous chromosomes goes to each daughter cell. In meiosis II, on the other hand, there are only two sister chromatids, and these separate to opposite poles and become haploid chromosomes. Cytokinesis and cell division then result in the production of haploid cells. The development of gametes from these cells is described in Chapter 6.

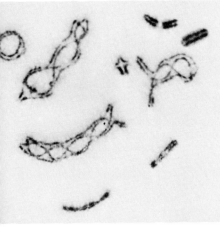

Figure 4.8
Synapsis and multiple crossing over during prophase I of meiosis in a grasshopper. Pairs of homologous chromosomes form a tetrad of intertwining chromatids.

THE LINKAGE OF GENES TO CHROMOSOMES

Almost immediately after Mendel's work was rediscovered in 1900 some geneticists realized that the movements of chromosomes during mitosis and meiosis were similar to the patterns by which genes are inherited by somatic cells and gametes. One of the first to make this observation was W. S. Sutton. In 1903, while a graduate student, Sutton pointed out that, like genes, chromosomes occur in pairs in somatic cells and are transmitted individually through gametes. During his brief scientific career before he had to go into medicine to make a living, Sutton used the lubber grasshopper *Brachystola* to show that chromosomes also remain distinct from each other, and that the inheritance of one chromosome in a homologous pair does not affect the inheritance of a chromosome in a different pair. In other words, chromosomes, like genes, obey the rules of segregation and independent assortment. Sutton's proposal that genes are parts of chromosomes seems obvious to us now, but it was widely opposed at the time.

One opponent of the idea was Thomas Hunt Morgan. The demonstration that genes are indeed parts of chromosomes was due mainly to T. H. Morgan himself, along with his students at Columbia University and millions of little flies named *Drosophila melanogaster* (Figure 4.9). When he began his work with drosophila in about 1909, Morgan was known mainly as an embryologist interested in evolution. At the time he doubted Mendel's work, as well as Darwin's theory of natural selection. He expected that his research would show that inheritance in animals is not due to genes, and that sudden changes (mutations) were far more important in evolution than was natural selection.

Morgan chose *Drosophila*, commonly called a fruit fly, allegedly because he could not get funding to work with mammals. Whatever the reason, drosophila was a good choice. They are inexpensive to raise in large numbers and have a short life cycle. Each female can lay several hundred eggs, providing a large sample of offspring for statistical analysis. Within two days each egg hatches into a maggot, which develops into a pupa and finally an adult within ten days. The adult becomes sexually mature within two days and can live for two more weeks. Another major advantage of drosophila is that it has only four pairs of chromosomes, which facilitates tracking them through mitosis and meiosis.

White-Eyed and Wild-Type Fruit Flies. Drosophila provided three main lines of evidence that convinced Morgan and almost every other biologist that genes do determine heredity, and that they are parts of chromosomes. The first line of evidence began in 1909 with a chance discovery by Calvin B. Bridges, who was then an undergraduate hired to wash the milk bottles in which the flies were raised. One day Bridges noticed a male fly with white eyes, in contrast to the normal red eyes of the "wild-type" drosophila. Rather than flick the mutant fly away, Bridges brought it to Morgan. Morgan then set out to study the inheritance of the mutation, expecting that it would disobey the Mendelian laws. Instead, the hybrids resulting from crossing the white-eyed male with a red-eyed female were all red-eyed, as would be expected if the white-eyed character were recessive. By this time the hybrids produced by the crossing of true-breeding parents were referred to as the **F₁ generation**. These F₁ hybrids were then interbred, producing the **F₂ generation**. The F₂ generation consisted of red-eyed and white-eyed flies in a 3:1 ratio, again as Mendel would have predicted. There was one major un-Mendelian feature, however. All the white-eyed flies were males; no females had white eyes. Somehow the inheritance of the white-eye character depended on the sex of the offspring.

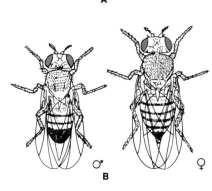

Figure 4.9
Major contributors to 20th century genetics. (A) T. H. Morgan in the Fly Room at Columbia University around 1917, surrounded by milk bottles in which fruit flies were raised. (B) Male and female Drosophila melanogaster: *length approximately 2 mm. Drosophila is commonly called a fruit fly, but it is more correctly called a vinegar fly or pomace fly (family Drosophilidae) to distinguish it from true fruit flies (family Tephritidae).*

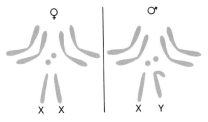

♀   ♂

X   X   X   Y

Figure 4.10
The four pairs of chromosomes in female and male Drosophila melanogaster. *The pairs at the top are autosomes; the pairs at the bottom (X and Y) are sex chromosomes.*

A simple explanation for the inheritance of the white-eye character was already at hand in the fact that some chromosomes differ, depending on gender. Of the eight chromosomes in *Drosophila*, six occur in three similar, homologous pairs, regardless of the sex of the fly. Such chromosomes are called **autosomes.** In males the two chromosomes of the remaining pair are dissimilar (Figure 4.10). One is designated X and the other Y. Females have two X chromosomes. This pair of chromosomes, either XY in males or XX in females, are called **sex chromosomes.** (A similar arrangement of sex chromosomes also occurs in humans and most other animals. In many animals, however, the male lacks a Y chromosome, and is designated XO. In moths, butterflies, birds, and many lizards and snakes, it is the females that have different sex chromosomes.) Every ovum of drosophila has one X chromosome from the female parent. Every sperm cell, however, has either the X chromosome or the Y chromosome from the male parent. An ovum fertilized by a sperm cell with the X develops into a female, and an ovum fertilized by a sperm cell with the Y develops into a male.

The explanation for the inheritance of the white-eye character by half the males and by no females in F_2 is that the recessive allele for the white-eye character is on the X chromosome. In other words, the allele is **sex linked.** The F_1 females were heterozygous for the allele, so half the F_2 males inherited the dominant allele and were therefore red-eyed like the wild type. The other F_2 males, however, inherited the X chromosome with the recessive white-eye allele. They had white eyes, since there was no allele on the Y chromosome to block expression of the recessive allele (Figure 4.11). This explanation was confirmed by crossing white-eyed F_2 males with the heterozygous F_1 females. Half the offspring, whether male or female, were white-eyed.

Figure 4.11
A Punnett square explaining the inheritance of a sex-linked allele for white eyes in the F_2 generation of drosophila. Above and to the side of the square are shown the two parents and their gametes. w^+ represents the dominant wild-type allele for red eyes, and w represents the recessive allele for white eyes.

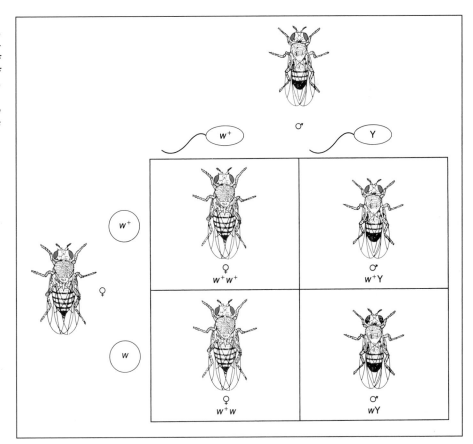

At first the fact that the white-eye allele was inherited in exactly the same pattern as the X chromosome did not convince Morgan that genes are parts of chromosomes. Bridges later became a graduate student in Morgan's lab, however, and found that X linkage of the white-eye allele could also explain eye color of abnormal flies that inherited an extra X or Y chromosome. This ability to explain this rare phenomenon convinced even Morgan. It is now established that each gene has a particular location, or **locus**, on a particular chromosome.

Linkage Groups. The evidence that genes are parts of chromosomes raised another question, the answer to which provided the third line of evidence that genes are on chromosomes. Since there are many more genes than chromosomes, many genes must have their loci (LOW-sy, plural of locus) on the same chro-

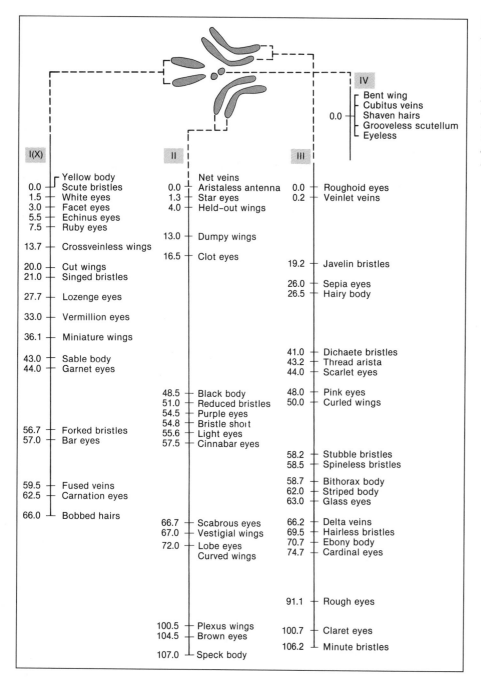

Figure 4.12
A linkage map showing the locations of alleles for some mutations in drosophila. The Y chromosome is not represented, although it is known to carry at least seven genes affecting male fertility, among others. The numbers represent map distances of the loci from one end of the chromosome, as inferred from the probability of recombination.

mosome. How, then could there be independent assortment? All the alleles on a chromosome should be inherited together. In fact, many geneticists had found that there *is* a tendency for certain alleles to be inherited together. The opportunity to study whether this tendency was due to genes occurring on the same chromosome arose after Bridges discovered numerous other mutations. Another of Morgan's students, A. H. Sturtevant, also spotted numerous mutants (in spite of being colorblind). Within a few years they had identified 85 mutations and had studied their patterns of inheritance. They found, for example, that if a fly carried both an allele for white eyes and an allele for hairy wings, its offspring tended to inherit both mutations or neither. Since the white-eye allele was known to be on the X chromosome, the logical inference was that the allele for hairy wings was also on the X chromosome. A mutant allele causing bent wings was inherited independently from those two mutations, however, suggesting that it was autosomal (due to an allele located on an autosome). By this kind of reasoning the 85 mutations were divided among different **linkage groups.** The number of linkage groups was four: the same as the number of pairs of chromosomes.

Two alleles in the same linkage group were not always inherited together, however. Evidently pairs of alleles on a single chromosome could become separated during crossing over in prophase I. In this way alleles on two homologous chromosomes **recombine** in different combinations. The greater the distance between the locus of one allele and the locus of another, the more likely this recombination. Sturtevant and others laboriously worked out the frequencies of recombination of numerous alleles. By this procedure they were able to map the relative distances between loci for the alleles. The ability to produce such **linkage maps** confirmed the physical identity of genes as parts of chromosomes (Figure 4.12).

That alleles on the same chromosome can become separated accounts for the fact that, in general, genes are independently assorted. Nevertheless, in some cases assortment is not independent. This naturally raises the question whether extraordinary luck or intercession by the saints guided Mendel in choosing the seven traits he studied. In pea plants there are only seven pairs of chromosomes, so it might appear that each of the seven genes must have been on a different chromosome. In fact, however, several of the genes for those traits are on the same chromosomes, but at loci far enough away that the linkage is weak (Novitski and Blixt 1978).

SUMMARY

Gregor Mendel's research with garden peas disproved the age-old assumptions of inheritance of acquired characters and pangenesis and provided the first clues for the existence of genes. Mendel found that the offspring of two true-breeding varieties all had the appearance of only one of the parents. When these hybrids were self-fertilized or cross-pollinated, however, one fourth of the offspring did not look like the hybrid parents, but like the other true-breeding variety. Mendel's interpretation, in modern terms, was that both true-breeding varieties were homozygous, one for a dominant allele and the other for a recessive. The hybrid offspring were heterozygous in genotype but had the phenotype of the dominant parent. The offspring of two hybrids, however, has a probability of one in four of getting a recessive allele from both parents. Such homozygous recessives therefore had the recessive phenotype. Likewise, one in four offspring of two hybrids was homozygous dominant, and half were heterozygous. Since these had the dominant phenotype the phenotypic ratio of dominant to recessive was 3:1.

This pattern of inheritance implies two principles: segregation and independent assortment. By segregation we mean that each allele behaves separately; that the effects of a dominant allele do not blend with the effects of a recessive allele. Independent assortment means that the probability of inheriting one allele is not affected by the inheritance of another. Mendel demonstrated the validity of independent assortment by means of a dihybrid cross, but we now know that assortment is not independent for two genes on the same chromosome.

Evidence that chromosomes are involved in inheritance was found in the late 19th century by cytologists unaware of Mendel's research. They found that the cells of most organisms are diploid; that is, they have two copies of each chromosome. Gametes, however, are haploid, with only one chromosome of each pair. During fertilization two gametes combine, producing a diploid cell that divides repeatedly to produce all the diploid cells of a new organism. The movement of chromosomes prior to such cell division is called mitosis, and it occurs in four phases: prophase, metaphase, anaphase, and telophase. The movement of chromosomes

that produce haploid gametes is called meiosis. It is divided into meiosis I and meiosis II, each of which has prophase, metaphase, anaphase, and telophase.

The unification of Mendel's work with that of the cytologists began early in this century, when it was observed that chromosomes obey the principles of segregation and independent assortment, just as genes do. The proof that genes are parts of chromosomes came first from studies of *Drosophila*, in which particular mutations, such as white eyes, were inherited along with particular chromosomes. By using pairs of mutations and calculating the probability of their being inherited together, the position of mutated genes could be mapped on the chromosomes.

KEY TERMS

dominant	heterozygous	chromatin
recessive	genotype	meiosis
gene	phenotype	crossing over
allele	chromosome	linkage
segregation	diploid	autosome
independent assortment	haploid	sex chromosome
homozygous	mitosis	sex linked

SELF-TEST

1. Explain how Mendel's research overturned the theories of pangenesis and inheritance of acquired characteristics.

2. Explain how a genotypic ratio of 1:2:1 in the F_2 generation of Mendel's studies gave a phenotypic ratio of 3:1.

3. "Mendel's First Law" proposes that a dominant allele remains segregated from its recessive counterpart. Describe the experimental evidence for segregation.

4. "Mendel's Second Law" proposes that different alleles assort independently. Describe the experimental evidence for independent assortment. Give one example of an exception to independent assortment.

5. Prior to 1900, what kind of evidence suggested that chromosomes were involved in heredity? Describe two kinds

of evidence that convinced Morgan and others that genes were parts of chromosomes.

6. Construct a Punnett square showing the genotypes and phenotypes of F_1 offspring of a white-eyed (homozygous recessive, sex-linked) female and a red-eyed (wild-type) male drosophila. Use Figure 3.11 as a model.

7. Explain the relationship of each of the following to each other: chromatin, chromatid, chromosome, centromere, kinetochore.

8. Explain the differences between mitosis and meiosis that make mitosis suitable for transmitting identical genetic information to all daughter cells, and meiosis suitable for transmitting different genetic information to gametes.

READINGS

RECOMMENDED READINGS

Crow, J. F. 1979. Genes that violate Mendel's rules. *Sci. Am.* 240(2):134–146 (Feb).

Moore, J. A. 1986. Science as a way of knowing—genetics. *Am. Zool.* 26:583–747. (*Comprehensive treatment of classical genetics in*

a historical context. See also other papers on genetics in the same issue.)

ADDITIONAL REFERENCES

Allen, G. E. 1983. T. H. Morgan and the influence of mechanistic materialism on the

development of the gene concept 1910–1940. *Am. Zool.* 23:829–843. (*See also other papers on Morgan in the same issue.*)

Novitski, E. and S. Blixt. 1978. Mendel, linkage, and synteny. *BioScience* 28:34–35.

Genetic Control

Albino eastern gray squirrel (Sciurus carolinensis).

CHAPTER OUTLINE

LEARNING OBJECTIVES

1. What is the molecular nature of a gene?
2. How is the genetic material replicated for cell division?
3. How is inherited information encoded in a gene?
4. Is this information used directly to form the product of a gene, or is it first transcribed into some other form?
5. How is genetic information translated into a gene product?
6. Can changes in gene number due to chromosomal aberrations produce genetic defects?
7. Can changes in the structure of a single gene produce inherited defects?
8. How have genetic changes led to evolution?
9. Can an understanding of genetics have practical applications?
10. Is it possible to alter the genetic composition of an organism, or to recombine the genetic material from two different species?

WHAT ARE GENES MADE OF?

It took Morgan and his students less than a decade to convince themselves and almost everyone else that heredity is due to genes that are parts of chromosomes. During the following two or three decades others showed that the existence of genes was consistent with Darwin's theory of natural selection (see p. 362). In the meantime there was naturally a good deal of curiosity about the materials and mechanisms of genes. By the early 1940s it was shown that in the red mold *Neurospora crassa* genes direct the production of enzymes. We now know that genes in other organisms also direct the production of enzymes, as well as other proteins. The gene is, in fact, loosely defined in the phrase "one gene, one polypeptide chain."

Also by the 1940s there was good evidence that the genes consist of deoxyribonucleic acid (DNA), at least in viruses and bacteria. Many geneticists ignored the evidence, however. They reasoned that the complex phenomena of heredity would require that genes consist of molecules that are equally complex. DNA, however, is rather simple. It consists only of a chain of sugars (deoxyribose) linked to each other by phosphate groups, with a nitrogen-containing structure called a **nitrogenous base** linked to each sugar. There are only four kinds of nitrogenous bases in DNA—a variety that appears too limited to account for the complexity of inheritance. Proteins, with their 20 naturally occurring amino acids, seemed much more likely candidates for genes.

One person who helped sustain interest in DNA was the physicist Erwin Schrödinger. In his book *What Is Life?* (1944), Schrödinger approached the question of heredity as a naïve physicist wondering what kind of substance could be stable enough at body temperature to preserve information for hundreds and thousands of years. His answer was not protein, which, as everyone knows, breaks down quickly. The most thermodynamically stable molecules were crystals. However, genes could not consist of perfectly regular crystals like those of table salt (see Figure 2.4A), because these would contain little information. Schrödinger therefore reasoned that genes must be crystals that vary in structure from one position to another. Schrödinger's little book enticed many physicists looking for challenges as great as relativity and quantum mechanics had been, as well as some looking for a subject less potentially destructive than nuclear physics. Naturally the physicists brought with them their tools of the trade, such as x-rays and radioactive tracers. Using these techniques among others, scientists soon established that Schrödinger's crystal was DNA.

WHAT DO GENES LOOK LIKE?

Once DNA was established as the material basis of heredity, many scientists, including the new breed of biophysicists inspired by Schrödinger, began a deliberate race to determine its molecular structure. The race was won in a remarkably short time by a young American postdoctoral student, James D. Watson, and his mentor, Francis Crick, at Cambridge University. Watson (1968) has given a personal account of this discovery in an entertaining best-seller that should be read for the insight it provides into scientific creativity.

Watson and Crick relied on two major clues. First was a discovery by E. Chargaff that there is a pattern to the amounts of the four nitrogenous bases—adenine, thymine, guanine, and cytosine—in DNA. The number of adenine bases in DNA is approximately equal to the number of thymines, and the number of guanines is similar to the number of cytosines. The second line of evidence was from analysis of the patterns into which x-rays are diffracted as they pass through crystals

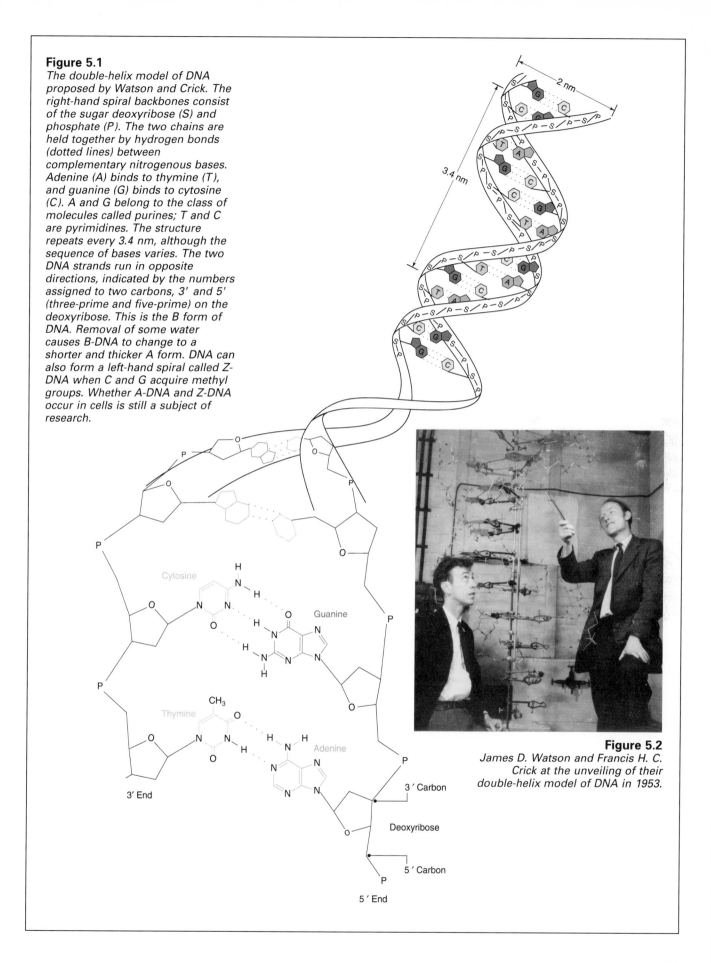

Figure 5.1
The double-helix model of DNA proposed by Watson and Crick. The right-hand spiral backbones consist of the sugar deoxyribose (S) and phosphate (P). The two chains are held together by hydrogen bonds (dotted lines) between complementary nitrogenous bases. Adenine (A) binds to thymine (T), and guanine (G) binds to cytosine (C). A and G belong to the class of molecules called purines; T and C are pyrimidines. The structure repeats every 3.4 nm, although the sequence of bases varies. The two DNA strands run in opposite directions, indicated by the numbers assigned to two carbons, 3' and 5' (three-prime and five-prime) on the deoxyribose. This is the B form of DNA. Removal of some water causes B-DNA to change to a shorter and thicker A form. DNA can also form a left-hand spiral called Z-DNA when C and G acquire methyl groups. Whether A-DNA and Z-DNA occur in cells is still a subject of research.

2 nm

3.4 nm

Cytosine

Guanine

Thymine

Adenine

3' End

CH₃

3' Carbon

Deoxyribose

5' Carbon

5' End

Figure 5.2
James D. Watson and Francis H. C. Crick at the unveiling of their double-helix model of DNA in 1953.

of DNA. The x-ray diffraction studies of DNA by Maurice Wilkins and Rosalind Franklin suggested that the sugar and phosphate backbones of DNA were in the form of a helix, or spiral.

In the early 1950s these findings were still imprecise and controversial, and most biologists would have advised waiting for further research. Rather than take this careful inductive approach, Watson and Crick used a kind of reasoning that had proved successful to physicists. They tried to guess a structure of DNA that would work. Using simple models with the sugar–phosphate backbones represented by wire and the bases cut out of sheet metal, they tried various arrangements to see if one was consistent with the meager evidence, and also intuitively compelling. One early model had three backbones twisted around each other and the bases sticking out. They abandoned this triple-helix model because it did not have the "ring of truth" that one expects from a correct theory of this importance. As Watson (1968) put it, the model "smelled bad." Finally, in 1953 Watson and Crick hit upon the idea of having two backbones outside, like rails of a spiral staircase, with the bases sticking inward like steps. To their delight this model worked only if adenine projected inward toward thymine, and guanine projected toward cytosine. The model therefore neatly explained Chargaff's ratios. Variations in the sequence of bases would account for the genetic information, while the uniform sugar–phosphate backbones would preserve the crystalline structure outside. In addition, the duplication of DNA during mitosis and meiosis could be explained by the separation and copying of both strands. The model explained so many properties of genes that the subsequent experimental proof was anticlimactic. In the words of Watson (1968), "A structure this pretty just had to exist" (Figures 5.1 and 5.2).

By the time of Watson and Crick the works of Mendel, Morgan, and others were classic. Only later did the relationship of DNA to genes and chromosomes become clear. We now know that each chromosome contains a single molecule of DNA, and that genes occur in series along this molecule. During mitosis the DNA is tightly condensed around proteins (see Figure 4.6). The degree of condensation is evident from the fact that the average human chromatid is only a few micrometers long but contains a molecule of DNA approximately 5 cm long. During interphase the chromatin partially decondenses (Figure 5.3). Functioning chromatin has a "beads-on-a-string" structure in which DNA is periodically looped around beadlike proteins called **histones.** The histones may serve as "spools" that keep the DNA from unraveling and may play some role in turning genes on and off. Each histone and the length of DNA associated with it make up the fundamental unit of chromatin structure, called the **nucleosome.**

The DNA in bacteria and in mitochondria differ in structure and functioning from the nuclear DNA of eukaryotes. Instead of strands of DNA organized around histones, mitochondrial DNA occurs as a single closed loop without associated proteins. In this and other ways it is like bacterial DNA, as might be expected from other evidence that mitochondria originated as prokaryotic endosymbionts (see pp. 391–392). The genes on mitochondrial DNA specify some proteins of the respiratory chain, as well as other molecules needed to produce those proteins. The mitochondrial DNA also duplicates itself when the mitochondria reproduce. One of the most interesting properties of mitochondrial genes is that they are generally inherited from the mother since the mitochondria of sperm remain outside the ovum during fertilization.

THE CELL CYCLE

During interphase genes carry out their essential functions of directing the synthesis of proteins and of duplicating themselves in preparation for mitosis. These different functions occur in a regular sequence corresponding to the cell cycle. The cell cycle starts with interphase in a new daughter cell at the end of a previous mitosis. It continues through the end of the following mitosis. Interphase is divided into three parts: two **gap phases,** G_1 and G_2, that are separated by the **S (synthetic) phase** (Figure 5.4). During G_1 genes actively direct the production of enzymes and other proteins, especially those needed for the daughter cell to grow to its final size. G_1 is followed by S phase, in which the cell synthesizes new DNA, histones, and other proteins needed to duplicate chromatids prior to mitosis. During G_2 the genes continue to direct the synthesis of proteins, especially those

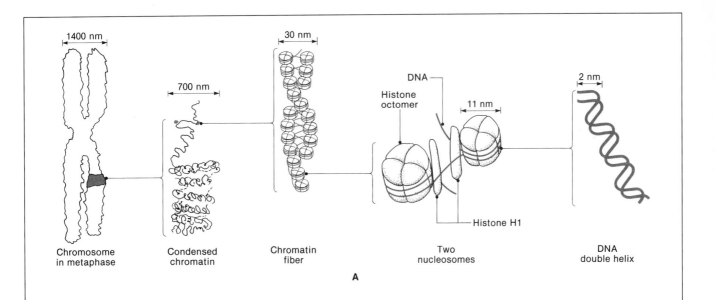

1400 nm

700 nm

30 nm

DNA

Histone octomer

11 nm

2 nm

Histone H1

Chromosome in metaphase

Condensed chromatin

Chromatin fiber

Two nucleosomes

DNA double helix

A

Figure 5.3

The organization of DNA in chromosomes. (A) In metaphase the DNA in chromosomes is highly condensed by proteins. During interphase the chromatin is partially unraveled but is still supercoiled into a chromatin fiber approximately 15 times as thick as a single DNA molecule. Portions of DNA that are functioning apparently unravel further into a series of nucleosomes. Each nucleosome consists of approximately 70 nm of DNA (approximately 200 base pairs), part of which is wrapped 1.8 times around a histone octomer consisting of four pairs of different histones. Another type of histone (H1) apparently keeps the DNA bound to the octomers. (Arrangement of H1 is conjectural.) (B) Removing the histones from a metaphase chromosome allows the DNA to unravel. (Arrow points to a loop of DNA.) Only about half the DNA of this human chromosome is shown. The dark structure is the protein scaffolding of the chromosome.

2 μm

B

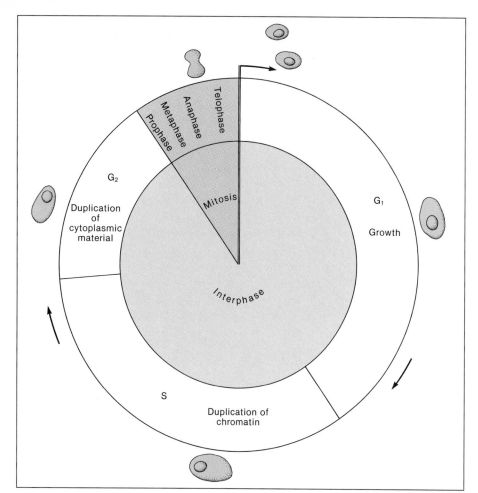

Figure 5.4
The cell cycle. In animals and other eukaryotes the cell cycle lasts from several hours to many years. The first gap phase, G_1, is the most variable in duration and depends on how small the daughter cell is compared to its final size. S phase, during which DNA and histones are synthesized, lasts 6 to 10 hours. G_2 lasts 3 to 5 hours, and mitosis takes 1 to 2 hours.

needed to make cytoplasmic materials for the two new daughter cells. In other words, during S phase the DNA undergoes replication. Also during S, and during G_1 and G_2, the DNA directs the production of proteins. This production involves the transcription of the information in the DNA into a messenger, and the translation of that messenger into a protein.

REPLICATION

During S phase the cell must construct a replica of its chromatin so that both daughter cells will receive a complete stock of genes identical to its own. Since the sequence of bases in DNA contains the genetic information, the problem of replicating the DNA is essentially one of synthesizing a new molecule of DNA with the same sequence of bases. One of the major achievements of the Watson–Crick model of DNA is that the solution to this problem falls out almost immediately. Because of the complementary base pairing of adenine to thymine, and of guanine to cytosine, it is only necessary to unzip the two strands of the double helix and use each one as a **template** for the construction of a complementary strand (Figure 5.5A). In this way each of the two old strands of DNA gets a new complementary strand. Since half the old DNA is conserved on each new molecule, the process is called **semiconservative replication.**

The base and the energy required to attach it to the new DNA strand comes from a triphosphate consisting of three phosphates attached to deoxyribose, to which is attached the base. An example of a triphosphate is deoxyadenosine tri-

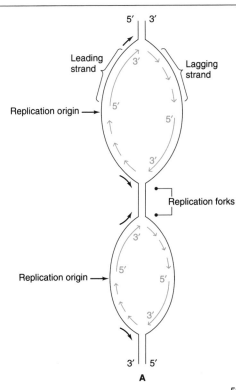

Figure 5.5

Schematic representation of semiconservative replication. No attempt is made to represent the helical structure of DNA or histones on the DNA. (A) Replication occurs at a rate of about three bases per minute at numerous replication origins. At each replication origin the two DNA strands are separated by an enzyme (DNA helicase), forming two replication forks that move away from each other. As the replication forks move apart, "unzipping" the double helix, each single strand of DNA serves as the template for the synthesis of a new strand. Nucleotides can be added to a growing new strand (color) only at the 3' end of DNA, so each new strand can elongate only in its 5'-to-3' direction indicated by arrows. Therefore only one strand, the leading strand, can grow steadily in the same direction as the movement of the replication fork. The other,

lagging strand, must wait for the replication fork to move a certain distance (100 to 200 bases), then grow in the opposite direction. An enzyme, DNA ligase, splices the sections of the lagging strand together. Existing histones remain with one new double helix. Additional histones are synthesized and added to the other. Replication stops when one replication fork hits another. (B) DNA at a replication fork. On the left a deoxyadenosine triphosphate (dATP) has just lost two phosphates, leaving an adenosine nucleotide that is about to join the leading strand opposite a thymidine. On the right deoxyguanosine triphosphate (dGTP) is about to add a guanosine to the lagging strand. Addition of new nucleotides is catalyzed by DNA polymerase. Note that the newly synthesized double helices will have identical base sequences.

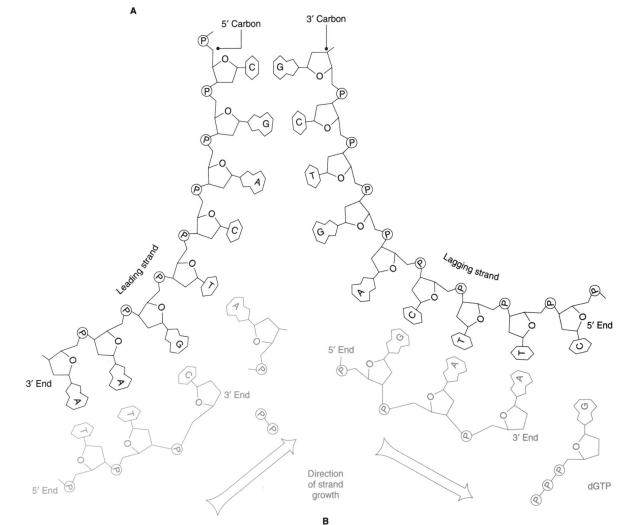

phosphate (dATP), which has adenine as its base. dATP is identical to adenosine triphosphate (ATP) except that its sugar is deoxyribose, which has one less oxygen atom than ribose (see Figure 2.35A). Like ATP, dATP transfers energy when the two terminal phosphates break off. The molecule that is left is the **nucleotide** adenosine, which consists of adenine, deoxyribose, and one phosphate. The energy released from dATP is used to bond the adenosine to the growing DNA strand, adding adenine to the sequence of bases (Figure 5.5B). In the same manner dGTP, dCTP, and dTTP provide the energy to attach the nucleotides guanosine, cytidine, and thymidine where they are called for by the template strand of the DNA. Enzymes called **DNA polymerases** catalyze the attachment of these nucleotides to the new DNA strand.

The enzymes involved in semiconservative replication also help ensure the correctness of the sequence of bases. After DNA polymerases catalyze the attachment of nucleotides during replication, it **proofreads** the sequence. Because of proofreading, the wrong nucleotide is inserted only about one time in a billion, or approximately three times during each mitosis of a human cell. DNA polymerases also help repair damaged DNA. Although the double helix is intrinsically stable, ultraviolet radiation from sunlight, ionizing radiation, and other hazards can alter the bases, turning genetic information into gibberish. DNA polymerases and other repair enzymes usually recognize such abnormal bases, cut them out, and substitute the correct nucleotide called for by the complementary DNA strand. **DNA ligase** then catalyzes the patching of the new nucleotides to the old strand.

TRANSCRIPTION

Throughout interphase genes direct the synthesis of all the proteins needed by the cell. The Watson–Crick model of DNA immediately suggests that the sequence of nucleotides in DNA determines the sequence of amino acids in proteins. In eukaryotes, however, DNA remains inside the nucleus while protein synthesis occurs in the cytoplasm. It follows, therefore, that the genetic control of the production of protein must occur in two stages. First, the base sequence in the DNA must be **transcribed** into a messenger. Next, the messenger carries the genetic information to the cytoplasm where it is **translated** into a sequence of amino acids in a polypeptide.

The messenger that is produced by transcription and decoded during translation is another kind of nucleic acid, called **ribonucleic acid** (RNA). RNA is similar to DNA except for the following differences:

1. RNA is normally single stranded.
2. The sugar in ribonucleic acid is ribose rather than deoxyribose.
3. In place of thymidine RNA has uridine (Figure 5.6).

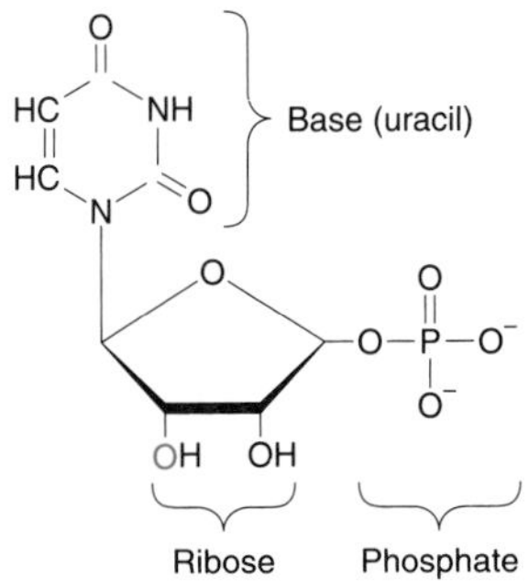

The expression of genetic information, whether for wrinkled peas, white eyes in drosophila, or skin color in humans, begins with the transcription of the genetic information of DNA into **messenger RNA** (mRNA). The process of transcription is similar in many respects to the replication of a complementary strand of DNA. A section of the DNA double helix unzips, and an **RNA polymerase** enzyme moves along one of the DNA strands attaching complementary nucleotides to the growing mRNA strand. Wherever a thymine (T) occurs on the DNA strand, the RNA polymerase uses a molecule of ATP to add an adenosine (A) nucleotide to the end of the mRNA strand. Similarly, wherever G or C occurs on the DNA, the RNA polymerase adds a cytosine or guanosine, respectively. A major difference

Figure 5.6
A nucleotide of RNA consists of a phosphate, attached to the sugar ribose, to which is attached a nitrogenous base. In this example the base is uracil, forming the nucleotide uridine. Three of the RNA bases—adenine, guanine, and cytosine—are identical to those found in DNA nucleotides. Uracil occurs in RNA instead of thymine. Deoxyribose in DNA differs from ribose in that it lacks an oxygen (color).

between transcription and replication is that when the RNA polymerase encounters an A on the DNA strand, it attaches the nucleotide with uracil (U) rather than thymine. Another major difference is that the RNA strand does not remain attached to the DNA strand but continually peels off as it is synthesized. The double helix zips up again about one turn past the point of transcription.

Transcription begins 20 to 30 bases away from a special sequence of DNA bases called the **promoter.** Promoters bind the RNA polymerase in one orientation only, thereby determining the direction along the DNA in which the polymerase will transcribe. Since transcription can occur in only one direction, the polymerase will transcribe one strand of the DNA and not the other (Figure 5.7). Transcription stops when the polymerase reaches a **termination sequence.** Typically, the distance between the promoter and the termination sequence is a few thousand bases, requiring several minutes to transcribe.

In eukaryotic cells the resulting RNA strand, called the **primary transcript** or **mRNA precursor,** is still not ready to direct the synthesis of a protein. First the primary transcript must undergo **processing.** The first step in mRNA processing

Figure 5.7
Transcription. Beginning 20 to 30 nucleotides "downstream" from the promoter sequence on DNA, RNA polymerase II unzips a turn of the double helix and synthesizes a complementary strand of RNA in the 5'-to-3' direction. Transcription stops at the terminator sequence. Note that the base sequence on the primary RNA transcript is identical to the base sequence on the nontranscribed DNA strand, except that uracil substitutes for thymine.
Later the primary transcript will undergo processing, during which a poly-A tail will be attached at the 3' end, a cap of methyl guanine will be attached at the 5' end, and introns will be deleted to form the mature mRNA.

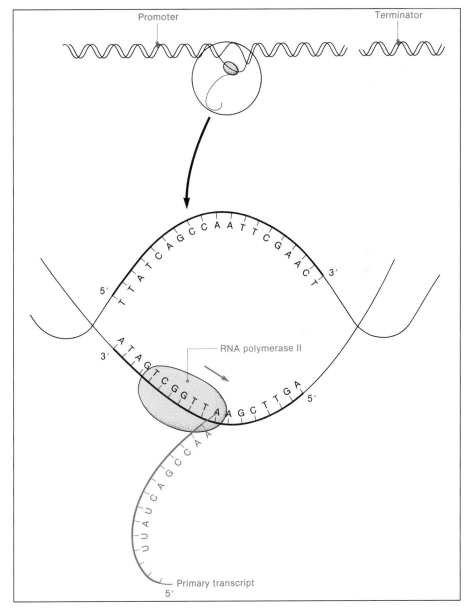

is the attachment of a **5′ cap** of methyl guanosine. Following this capping, the other end of the primary transcript is cut, and a **poly-A tail** consisting of 150 to 200 adenosines is attached. Then, in seeming defiance of the economy of nature, up to 90% of the laboriously transcribed RNA is cut out and discarded. The discarded bits of RNA (as well as the sequences of DNA that code for them) are called intervening sequences—**introns,** for short. The remaining portions of mRNA (as well as the sequences of DNA that code for them) are called **exons.** An average gene consists of 15 to 20 exons. There is further **editing,** by addition or deletion of bases here and there, before the mature mRNA is ready. This mature mRNA, consisting of a series of edited exons with a poly-A tail and a cap at the ends, is then exported through the pores in the nuclear envelope (see Figure 2.34) into the cytoplasm.

TRANSLATION

Messenger RNA directs the assembly of a polypeptide by specifying the amino acids that are to be attached as the polypeptide is produced. These instructions are contained in the sequence of bases in the mRNA, which was transcribed from the DNA of the gene coding for the polypeptide. The process by which the language of genes is interpreted in the language of proteins is called, appropriately, translation. The key to translation is the **genetic code.** With only four different bases in mRNA and 20 different amino acids in proteins, the genetic code clearly cannot simply specify one amino acid whenever a particular base occurs. Even two bases would be inadequate to specify a particular amino acid, since the number of combinations of four bases taken two at a time is only 16 (4^2). However, the number of combinations of three bases is 64 (4^3), which is more than adequate to specify 20 amino acids. Intensive research throughout the 1960s established that the genetic code is indeed based on triplets of bases. Each triplet on the mRNA, called a **codon,** specifies an amino acid to be added during the synthesis of a protein, or it signals the stop or start of translation (Table 5.1). Most codons have

TABLE 5.1   The genetic code.

Phenylalanine PHE	UU(U or C)
Leucine LEU	UU(A or G) and CU(U, C, A, or G)
Serine SER	UC(U, C, A, or G) and AG(U or C)
Tyrosine TYR	UA(U or C)
Cysteine CYS	UG(U or C)
Tryptophan TRP	UGG
Proline PRO	CC(U, C, A, or G)
Histidine HIS	CA(U or C)
Glutamine GLN	CA(A or G)
Arginine ARG	CG(U, C, A, or G) and AG(A or G)
Isoleucine ILE	AU(U, C, or A)
Methionine MET	AUG
Threonine THR	AC(U, C, A, or G)
Asparagine ASN	AA(U or C)
Lysine LYS	AA(A or G)
Valine VAL	GU(U, C, A, or G)
Alanine ALA	GC(U, C, A, or G)
Aspartic acid ASP	GA(U or C)
Glutamic acid GLU	GA(A or G)
Glycine GLY	GG(U, C, A, or G)
Stop	UAA, UAG, and UGA
Start	First AUG at 5′ end

The name and abbreviation of each amino acid is followed by a list of the codons that specify the amino acid in the synthesis of proteins. Parentheses enclose all the alternative bases in the third position of the codon that specify the same amino acid. The codons UAA, UAG, and UGA stop synthesis of the protein. The codon AUG codes for methionine, except for the first AUG near the 5′ end of the mRNA, which starts translation.

"synonyms" that code for the same amino acid. The genetic code is therefore said to be **degenerate.**

The genetic code is virtually universal among all organisms. For example, UUC is the codon for phenylalanine not only in humans but also in viruses, bacteria, fungi, plants, and other animals. This universality is one of the most compelling arguments for evolution, since the genetic code could otherwise be just as arbitrary as the dots and dashes of Morse code. Only a few minor "dialects" in the genetic

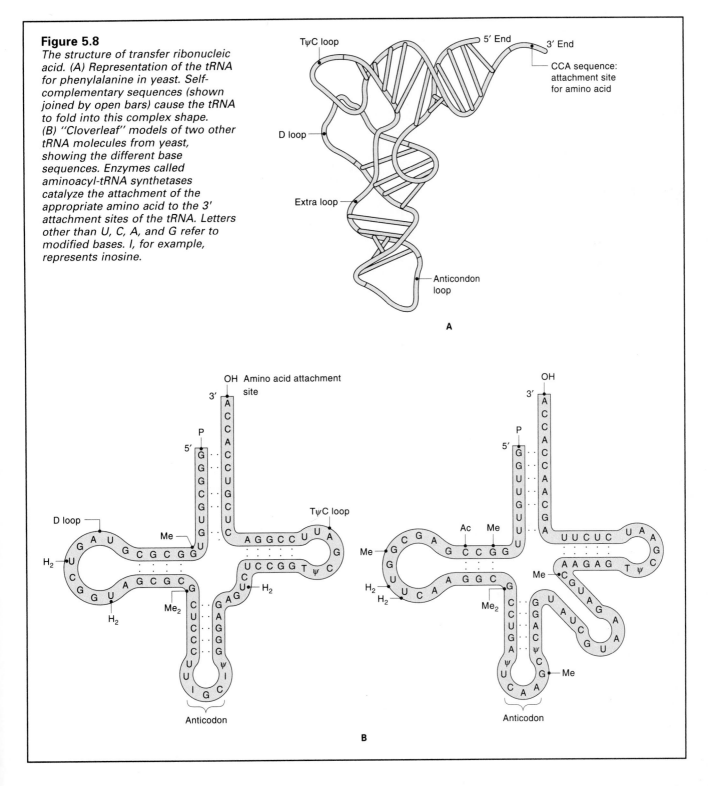

Figure 5.8
The structure of transfer ribonucleic acid. (A) Representation of the tRNA for phenylalanine in yeast. Self-complementary sequences (shown joined by open bars) cause the tRNA to fold into this complex shape. (B) "Cloverleaf" models of two other tRNA molecules from yeast, showing the different base sequences. Enzymes called aminoacyl-tRNA synthetases catalyze the attachment of the appropriate amino acid to the 3' attachment sites of the tRNA. Letters other than U, C, A, and G refer to modified bases. I, for example, represents inosine.

code have been discovered. For example, in certain protozoans called ciliates UAA and UAG are codons for glutamine rather than stop codons. Also, in the mitochondria of mammals AGA and AGG are stop codons rather than codons for arginine, and AUA is a codon for methionine rather than isoleucine.

Translation of the genetic code is mediated by a different kind of RNA, called **transfer RNA** (tRNA) (Figure 5.8). Transfer RNA is similar to mRNA except that many of the bases are altered. Each kind of tRNA can attach to a different amino acid. It also has a different combination of three bases, forming an **anticodon.** Each anticodon binds to a complementary codon on mRNA during translation. For example, the tRNA that carries the amino acid phenylalanine has the sequence AAG as its anticodon. AAG binds to the codon UUC on mRNA, causing the tRNA to transfer the phenylalamine to the protein being synthesized. Like most tRNAs, this one has some **wobble** to it, so it will also bind to the codon UUU. Wobble partly accounts for the degeneracy of the genetic code.

The actual synthesis of proteins takes place on organelles called **ribosomes.** Ribosomes are made within the cell nucleus in a dark-staining region called the **nucleolus.** They occur either free in the cytoplasm or attached to rough endoplasmic reticulum (see p. 47). Each ribosome consists of a large and a small subunit comprising several kinds of **ribosomal RNA** (rRNA) and more than a hundred different proteins (Figure 5.9). A ribosome begins translation at the start (AUG) codon nearest the 5′ cap, then moves along the mRNA in one direction (Figure 5.10). It moves three bases at a time, pausing at each codon until the tRNA matching the codon attaches. An enzyme then catalyzes the formation of a peptide bond between the tRNA's amino acid and the peptide chain being synthesized. This process continues until the ribosome reaches a stop codon.

Ribosomes spend an average of 50 milliseconds at each codon, attaching amino acids at a rate of about 20 per second. By the time the first ribosome has reached the stop codon, dozens of others may have begun transcribing the same mRNA, resulting in a chain of ribosomes linked to a single mRNA strand, forming a **polysome.** In this way a single mRNA can direct the production of many copies of a polypeptide simultaneously. Actual measurements of the total rate of protein synthesis are impressive. A silkworm caterpillar (*Bombyx mori*) makes 10^4 copies of the mRNA for fibroin, the protein in silk. Over a period of four days each of these mRNAs is translated into 100,000 fibroin molecules.

CHROMOSOMAL ABERRATIONS

Like all complex mechanisms, those of genetics sometimes malfunction. In humans some 4000 disorders are known to result from genetic malfunctions. Some of these are associated with **chromosomal aberrations:** changes in the number or structure of chromosomes. Some chromosomal aberrations result from **nondisjunction:** the failure of chromatids to separate during meiosis. Nondisjunction can change the number of chromosomes in a set, a condition known as **aneuploidy** (AN-you-PLOY-dee). In many species of plants, and in some animals, the entire set of chromosomes sometimes fails to disjoin, resulting in the gain of one or more entire sets of chromosomes. Strangely, this state, called **polyploidy,** is often not as harmful as inheriting an extra copy of just one chromosome. Apparently an extra balanced set of chromosomes is less harmful than an imbalance in the number of chromosomes within a set. In fact, plant breeders often induce polyploidy to improve crop species. Chromosomal aberrations also include changes in the structure of a chromosome. These are often due to the **translocation** of a fragment of a chromosome to the end of a different chromosome. The individual

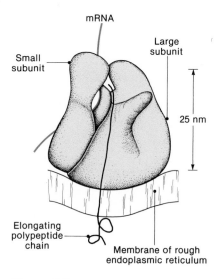

mRNA

Small
subunit

Large
subunit

25 nm

Elongating
polypeptide
chain

Membrane of rough
endoplasmic reticulum

Figure 5.9
Model of a ribosome bound to endoplasmic reticulum, showing the position of mRNA and the protein during translation.

Figure 5.10
Schematic representation of translation and protein synthesis. Before translation the two ribosomal subunits are separated. (A) Translation begins when an initiator tRNA carrying methionine, together with certain initiation factors, binds to a small subunit. (B) This initiation complex then binds to the start codon of mRNA. (C) A large ribosomal subunit then attaches to the small subunit in such a way that a P site of the large subunit is next to the initiator tRNA. The codon on the 3' side of the start codon is near the A site on the large subunit. This is where the appropriate aminoacyl-tRNA (tRNA with its attached amino acid) attaches. (D, E) After the aminoacyl-tRNA attaches, a peptide bond is formed between its amino acid and the methionine (MET). During this step the methionine detaches from the initiator tRNA, forming a dipeptide, and the initiator tRNA then drops off the P site. (F) The ribosome then moves three bases in the 5'-to-3' direction on the mRNA. The tRNA with the attached dipeptide (called a peptidyl-tRNA) is then at the P site of the ribosome. (G) Another aminoacyl-tRNA then attaches to the A site. A peptide bond joins its amino acid to the dipeptide, the tRNA at the P site drops off, the ribosome shifts again to the next codon, and so on. (H) This process continues until the A site of the ribosome reaches a stop codon. When that occurs a release factor attaches at the A site, terminating translation. The protein is released, and the two ribosomal subunits separate from the mRNA.

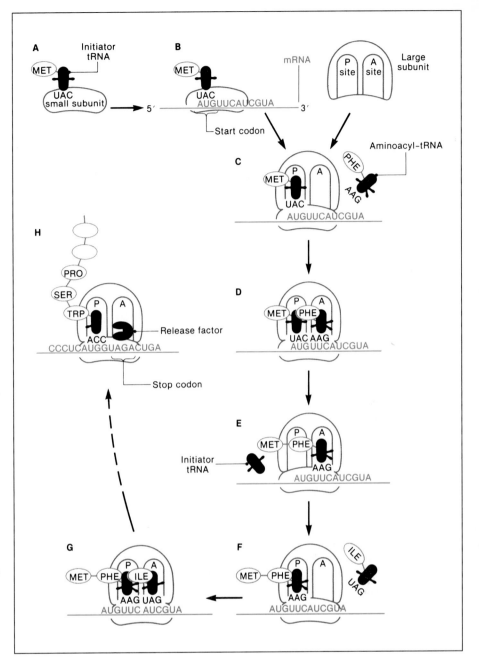

therefore receives an extra copy of the genes on the fragment or lacks a copy of those genes.

Aneuploidy, polyploidy, and translocation show up in the **karyotype,** which is the complement of chromosomes as they appear under the microscope. Karyotypes of human fetuses can be obtained by collecting fetal cells from the amniotic fluid through the process of amniocentesis.

Down Syndrome. Several kinds of abnormal karyotypes are frequent in humans. Probably the most familiar is Down syndrome, also called trisomy 21. Down syndrome is due to inheritance of an extra copy of chromosome 21 (Figure 5.11) or translocation of part of chromosome 21 to some other chromosome. In the United States, Down syndrome occurs in approximately 1 out of 700 live births. Most cases are thought to be due to some mishap of meiosis in ova, because

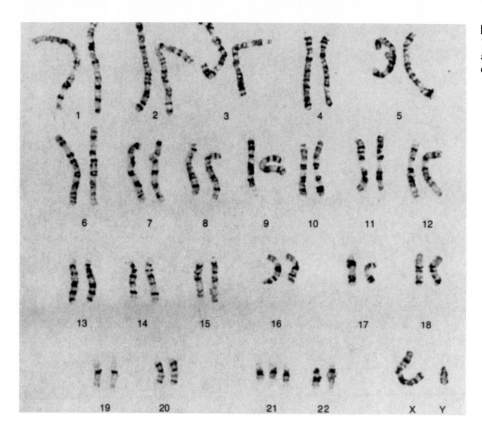

Figure 5.11
The karyotype of a male with Down syndrome. Note the extra chromosome numbered 21.

the chances of having a child with Down syndrome rises sharply if the mother is over age 35. Aneuploidy and translocation affect other chromosomes as well, but they are generally lethal to the fetus (Volpe 1987). Chromosome 21 is one of the smallest chromosomes, however, and presumably contains fewer genes.

About 20% of fetuses trisomic for chromosome 21 survive to birth, although with severe abnormalities. The major consequences of Down syndrome are severe mental retardation, defects in the heart and immune system, increased risk of leukemia and cataracts, and characteristic anatomical features such as flattened face, short stature, and unusual palm creases. With improved medical care and support, many victims of Down syndrome survive beyond age 35.

Aneuploidy of Sex Chromosomes. Other common chromosomal aberrations in humans involve aneuploidy of X or Y chromosomes. Trisomy XXY results in the development of a male, since the Y chromosome determines maleness in humans, but the additional X chromosome leads to testicular underdevelopment and feminization (Figure 5.12A). An individual with only the X chromosome (XO) develops as a female who is sterile and generally has impaired intelligence and other abnormalities (Figure 5.12B). The most common aneuploidy of the sex chromosomes, affecting an average of several males in ten thousand, is trisomy XYY. Individual XYY males are difficult to pick out of a crowd, but on average they are taller and slightly retarded. They are also more likely to be in prison than are men with normal karyotypes. This has led some to suppose that genes on the extra Y chromosome increase criminality. It is just as likely, however, that the intellectual impairment leads them to commit impulsive crimes and to make mistakes that get them convicted. This is a good illustration of how risky it is to jump to conclusions about genetic causes for behavior. (See pp. 410–413 for further discussion of the genetics of human behavior.)

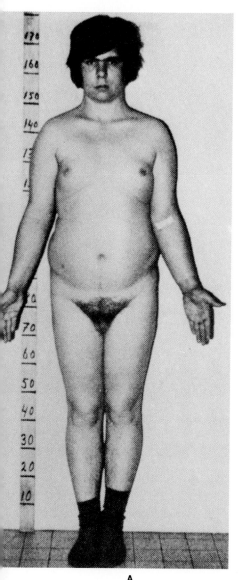

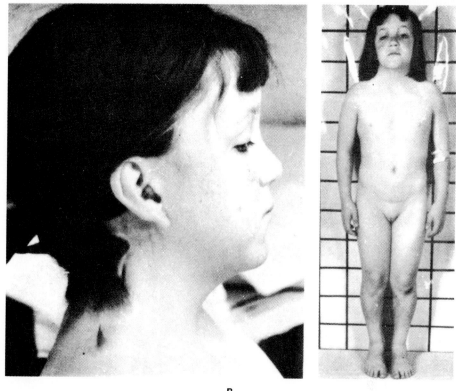

B

A

Figure 5.12
(A) Males with Klinefelter's syndrome, trisomy XXY, have undeveloped testicles and enlarged breasts. (B) Women with Turner's syndrome, XO, have underdeveloped sexual characteristics, short stature, hairline extending to the shoulders, and thick, webbed neck.

CHANGES IN CHROMOSOMAL STRUCTURE IN EVOLUTION AND DEVELOPMENT

Although devastating to individuals, chromosomal aberrations are often harmless or beneficial over the evolutionary scale of time. At some time in the evolution of humans, for example, two chromosomes fused, reducing our haploid number from the 24 found in living apes to our present 23. Apparently this reduction in chromosome number did little harm, presumably because there was no loss of genes. These and other changes in chromosome number have occurred, apparently randomly, throughout evolution (Table 5.2).

Chromosomes also undergo a variety of lesser changes in structure. Some of these result from **unequal crossing over.** Unequal crossing over occurs in prophase I of meiosis when two homologous chromosomes do not break in exactly the same place. One of the chromosomes therefore loses some genes or parts of genes to the other. Two other causes of chromosomal change are inversion and transposition. **Inversion** occurs during synapsis when a loop forms in a chromatid, then twists, breaks off, and reattaches backward. In **transposition** a segment of DNA called a **transposing element,** or sometimes a "jumping gene," moves to a different location on the same chromosome or another chromosome. Unlike other chromosomal changes, which are almost always random and harmful, many inversions and transpositions are examples of "programmed gene rearrangements" that occur in a systematic and useful manner (Borst and Greaves 1987). They allow for the duplication and deletion of genes at appropriate times in development, and they enable gene segments to recombine in a wide range of combinations. The latter function accounts for the ability to produce millions of different antibodies from just a few genes.

Geneticists now believe that changes in chromosome structure account for an

Table 5.2 Diploid chromosome numbers for animals representative of major taxa.

Even within classes of animals the number can vary widely, suggesting that chromosomal aberrations occur randomly. In many fishes, reptiles, and birds the chromosomes occur in two distinct sizes: macrochromosomes and microchromosomes.

SPONGES

Freshwater sponge *Spongilla lacustris*	10
Marine sponge *Leucosolenia ciliata*	26

CNIDARIANS

Jellyfish *Aurelia flavidula*	18
Hydra *Hydra vulgaris*	32

FLATWORMS

Schistosome *Schistosoma haematobium*	14
Planarian *Planaria torva*	16

NEMATODES (MALES)

Soil nematode *Rhabditis* sp.	13
Intestinal roundworm *Ascaris lumbricoides*	43

MOLLUSCS

Sea slug *Aplysia fasciata*	24
Land snail *Helix pomatia*	54

ANNELIDS

Leech *Dina lineata*	18
Earthworm *Lumbricus terrestris*	36

ARTHROPODS

Crayfish *Cambarus clarkii*	200
Mosquito *Culex pipiens*	6
House fly *Musca domestica*	12
Ant *Formica sanguinea* (females; males are haploid)	48
Ant *Myrmecia pilosula* (females; males are haploid)	2

FISHES

Northern pike *Esox lucius*	18
Gold fish *Carassius auratus*	100

AMPHIBIANS

Grass frog *Rana pipiens*	26
Clawed frog *Xenopus laevis*	36

BIRDS

House sparrow *Passer domesticus*	76
Pigeon *Columba livia*	80

MAMMALS

Chinese hamster *Cricetulus griseus*	22
Golden hamster *Mesocricetus auratus*	44
Cat *Felis catus*	76
Dog *Canis familiaris*	78
Donkey *Equus asinus*	63
Horse *Equus caballus*	64
Rhesus monkey *Macaca mulatta*	42
Gorilla *Gorilla gorilla*	48
Chimpanzee *Pan troglodytes*	48
Human *Homo sapiens*	46

Major source: Altman, P. L. and D. S. Dittmer, 1972. *Biology Data Book,* vol. 2. Bethesda, MD: FASEB. Data for *Myrmecia pilosula* from Crosland and Crozier 1986.

essential process in evolution: the acquisition of new genes. Translocation, aneuploidy, inversion, transposition, and unequal crossing over can give rise to duplicate genes that subsequently mutate and produce new products. Evidence that new genes arise by duplication and subsequent mutation lies in the existence of **protein families.** One example of a protein family is the three pigments in the human eye that are responsible for color vision. All three genes for these proteins are similar and close to each other on the X chromosome and are therefore likely to have arisen through unequal crossing over (Botstein 1986).

Instead of taking on a new role, a duplicated gene can also lose its promoter and become a **pseudogene.** Pseudogenes are never transcribed, but they are recognizable because their base sequences are similar to those of functioning genes. Pseudogenes and introns make up a large proportion of most organisms' **genomes** (the genetic material in a haploid set of chromosomes). In humans about 98% of the chromatin consists of such "junk DNA." Perhaps because so much of an animal's genome is junk that is apparently gained and lost at random, there is no consistent relationship between phylogeny and genome size (Figure 5.13).

POINT MUTATIONS

Most mutations do not affect the number or structure of chromosomes, but only a single DNA base pair. These are called **point mutations.** Point mutations occur with a frequency of about once in every billion base pairs, although **mutagens,** such as radiation and certain chemicals, increase the rate. Most point mutations are **neutral:** they have no effect on the gene product because they occur in introns or at the "wobbly" third base of a codon. Point mutations are more common in these harmless positions, apparently because they do not reduce the ability to reproduce (Lewontin 1986, pp. 816–817). Some point mutations have no effect on the functioning of a protein even if they do change the amino acid sequence.

Figure 5.13
The range of genome size in a variety of animals. The lack of relationship between DNA content and either phylogenetic position or physiological complexity is called the C-value paradox. Genome size, however, is well correlated with cell size.

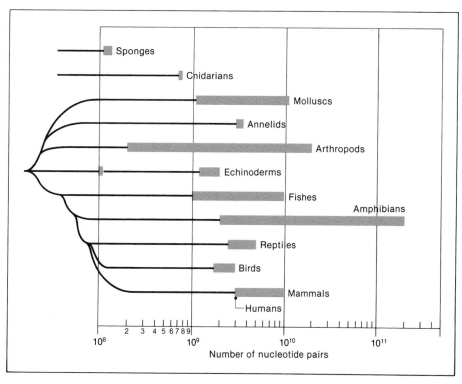

Bovine (cow) insulin differs from human insulin in three different amino acids out of 51, yet bovine insulin works quite well in human diabetics. Apparently the amino acid differences occur at sites on the insulin that are not crucial for its functioning.

At the other extreme are mutations that affect a gene for ribosomal or transfer RNA. Such **missense mutations** can stop all translation or affect every protein by causing the wrong amino acids to be inserted. Some point mutations also change a codon for an amino acid into a stop codon. These are called **nonsense mutations,** because they result in meaningless peptide fragments. There are also mutations that add or delete a base. These are called **frameshift mutations,** because they shift the **reading frame,** altering the message of every triplet following the site of the mutation.

Even harmful point mutations may have no immediate effect if they are recessive, as most are. It is likely that you have three, four, or five such recessive mutations. The normal homologous allele is usually dominant because it produces enough normal product to compensate for the mutant allele. Only if an individual inherits one mutant allele from each parent, and becomes homozygous recessive for the mutation, will it be expressed. The likelihood of both parents being carriers of a mutated form of the same allele is normally remote, however. The risks increase appreciably when there is **inbreeding,** because there is then a much greater risk that both parents have inherited the same mutation from the same ancestor.

Apparently the harmful effects of inbreeding have led to structural barriers against it in many plants and behavioral barriers in many animals. In many human cultures inbreeding was taboo long before the risks were understood. Inbreeding is a major concern in rearing captive animals and in managing small nature reserves. For that reason managers of zoos frequently exchange animals for breeding purposes. Inbreeding also threatens wild endangered species, which may succumb to internal genetic disorders even if they escape external threats.

Even without inbreeding, disorders due to certain point mutations do occur repeatedly in populations. Defects due to recessive point mutations linked to the X chromosome are especially common. One example of such a sex-linked trait is the inability to distinguish red and green, which affects 5% of males of northern European descent. A more serious X-linked mutation is Duchenne muscular dystrophy, which totally paralyzes about 1 in every 3300 boys. The cause of the disease is a mutation in the gene for a muscle protein called dystrophin, which is important for regulating calcium flow. One might think that a gene that restricted its own opportunities for inheritance so severely would soon disappear from a population. Because females have a normal gene on one of their X chromosomes, however, they can act as **carriers,** passing the mutation on for several generations before it occurs in a male and is selected against. Half the male children of carriers will inherit the disorder. Moreover, the gene for dystrophin is the largest ever discovered and presents a large target for spontaneous mutation. Approximately one-third of all cases are due to new mutations.

A similar set of circumstances is responsible for most cases of **hemophilia,** a disorder that impairs blood clotting in about one male out of every 10,000 (Figure 5.14). In the most common form of hemophilia a point mutation prevents an allele from producing normal factor VIII, one of the molecules essential for the clotting of blood. Until injections of factor VIII became widely available, hemophiliacs usually died of internal or external bleeding before they reached age 20. Factor VIII consists of 2332 amino acids, so, like the gene for dystrophin, its gene is a relatively large target for spontaneous mutation.

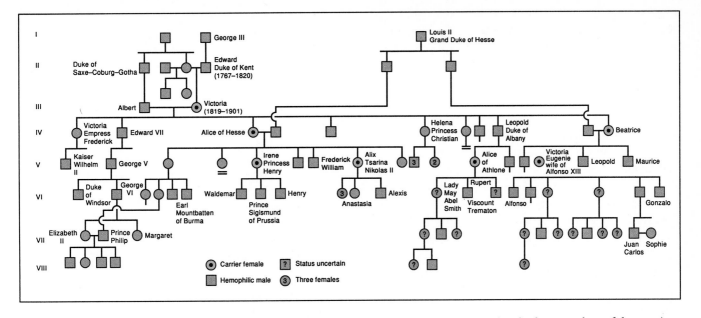

Figure 5.14

A pedigree chart showing the occurrence of hemophilia among the royal families of England, Germany, Russia, and Spain. One of the first steps in studying a genetic mutation is the recording of such pedigrees. This particular example is shown not because royalty are more susceptible to hemophilia than others, but because they record their pedigrees in unusual detail.

Sickle-Cell Anemia. Hemophilia is an example of a frequent harmful mutation that survives in large numbers because its effects are not immediately selected against. There are also genetic defects that survive in large numbers because selection favors them under certain conditions. The classic example is sickle-cell anemia. The allele for sickle-cell anemia is the mutated gene on chromosome 11 that codes for the beta chain of hemoglobin, the oxygen-carrying molecule that fills red blood cells. This point mutation consists of the substitution of an adenine for a uracil in one codon. This causes valine to be substituted for glutamic acid at the sixth position from the amino end of the beta chain. The change in hemoglobin structure causes the red blood cell to assume a sickle shape when the oxygen level in blood is low. Those who are homozygous for the allele suffer from sickle-cell anemia, and most die from clogged blood vessels. The allele is not recessive, so heterozygotes are also affected, though not as severely. Heterozygotes, who are said to have **sickle-cell trait,** live normal lives under most circumstances. Only when their blood is depleted of oxygen do the abnormal cells become sickle shaped, resulting in severe pain and tissue damage.

In spite of the clear disadvantages of the allele, it persists in large areas of Africa and southern Asia. Eight percent of Americans of African descent carry the allele. (A related disorder, **beta thalassemia,** affects some of Mediterranean descent.) Why hasn't the allele disappeared because of its adverse effects on the ability of its carriers to reproduce? The first clue to this mystery came from maps showing that the native range for sickle-cell anemia coincides with that for malaria. It happens that the sickling of red blood cells causes them to lose potassium, which inhibits the reproduction of the malaria parasite *Plasmodium* within the red blood cells (see pp. 478–481). The debilitating effects of sickle-cell trait are therefore compensated by resistance to malaria. Homozygous sickle-cell anemia is not compensated, of course, since it is just as debilitating as malaria. Apparently, however, the disadvantages of homozygosity are balanced against the advantages of heterozygosity. Loss of the allele due to the death of homozygotes is offset by the increased reproduction of the allele by heterozygotes.

GENETIC SCREENING

An understanding of genetics at the molecular level has led to a variety of techniques for analyzing the genes carried by individual organisms. Genetic screening requires only a small sample of blood, saliva, or other body fluids. Tiny amounts of DNA from such samples can be multiplied many times by **polymerase chain reaction** (PCR). In PCR the DNA is repeatedly cleaved into single strands, then treated with polymerase to double the amount of DNA. The DNA can then be mixed with radioactive **complementary DNA** (cDNA) probes. These are segments of DNA with a base sequence similar to that of a gene for which an individual is being screened. When mixed with the single-stranded DNA of the individual, the cDNA probe binds with the gene, revealing its presence by the radiation.

Another powerful tool for genetic screening uses **restriction enzyme** from bacteria. Restriction enzymes cut DNA at specific sequences of bases, usually where four to six bases on each strand have the same sequence in opposite directions. For example, the restriction enzyme EcoRI cleaves DNA at the sequence GAATTC, whose complementary strand reads the same in reverse. Because of point mutations at these sequences, a restriction enzyme will cut at one site on the DNA of one individual, but not in another. Thus the lengths of the fragments will vary from one individual to another. The different DNA fragment lengths are called **restriction fragment length polymorphisms,** or **RFLPs.** Numerous RFLPs can be isolated in this way, and it is virtually impossible for all of them to be identical in two individuals.

RFLPs make it possible to identify an individual organism and its relationship to others. This process is referred to as **DNA fingerprinting.** The best-known and most controversial applications of DNA fingerprinting are in criminal trials, in which guilt or innocence may be established from samples of semen or blood. DNA fingerprinting is becoming increasingly important to zoologists as well. It is used in wildlife management to ensure that animals bred to restore endangered species are not too closely related to each other. It is being used in zoos to prevent unwanted hybridization by being sure that animals selected for breeding belong to the same subspecies. DNA fingerprinting is also used in wildlife conservation to determine whether tusks, furs, or other animal products were obtained illegally from protected animals. Zoologists who study animal behavior also use DNA fingerprinting to determine the degree to which animals that are thought to be monogamous are faithful to their mates. (Apparently few are.)

RFLPs also serve as **markers** that indicate the presence of a harmful mutation. If a particular RFLP is found to occur consistently in individuals suffering from

---

Mapping the Human Genome

Using RFLPs and linkage mapping, researchers have located the genes for many disorders, including Huntington's disease, Duchenne muscular dystrophy, cystic fibrosis, familial Alzheimer's disease, and neurofibromatosis (so-called "Elephant man's disease) (Martin 1987; White and Lalouel 1988). Ultimately, it should be possible to map all of the 100,000 human genes, providing a linkage map similar to that for drosophila chromosomes (see Figure 4.12). Such a gene map could be the first step toward a brave new world in which humans directly intervene in altering their own genetic makeup and that of other organisms.

An undertaking that seems just as ambitious is already underway. This is the Human Genome Project, which aims to determine all three billion base pairs in the human genome. Researchers around the world are already engaged in the effort. There seems to be little objection on ethical grounds, but the estimated cost of $3 billion dollars over the 15 years of the project has raised some doubts. Although the total cost is only 1% of the U.S. military budget for one year, some biologists question its wisdom. They point out that 98% of the money would be spent on introns and other "junk" DNA, and they worry that such a large-scale project will drain resources from more worthwhile biological research.

a mutation, then the presence of the marker can be used to determine whether a person is likely to be carrying the mutant allele. One application is in identifying potential victims of Huntington's disease. Huntington's disease results from an autosomal dominant allele that causes the brain to degenerate, starting in middle age and continuing until death comes after decades of suffering. Without a genetic marker the disease cannot be detected until its symptoms begin, usually after victims have already had children. Each child of a Huntington's victim has a 50–50 chance of having inherited the allele. Using an RFLP marker to detect the presence or absence of the allele would alleviate their uncertainty and would help them to decide whether to have children.

GENETIC ENGINEERING

Those of us alive today may be among the last to grow up believing that genetic defects are as inescapable as an unlucky throw of the dice. We may be among the first to be able to load the genetic dice or take back the throw, by repairing such defects through genetic engineering. The main tool enabling genetic engineers to fulfill this goal is **recombinant DNA** technology. Recombinant DNA technology refers to the artificial transfer of a gene or genes from one organism into another. It is becoming a standard tool in research, wildlife management, and other areas of zoology. It is important, therefore, to understand what the technology involves.

Some bacteria naturally exchange genetic material from one species to another in the form of a circular strand of DNA called a **plasmid.** This natural recombination accounts for the spread of antibiotic resistance among different species of bacteria. *Agrobacterium tumefaciens*, the bacterium responsible for crown gall disease in plants, also transfers plasmids to the plants it infects. In certain cases genetic engineers can substitute desired genes for those on a plasmid, then allow bacteria to transfer the plasmid. Other techniques for introducing foreign genetic material include **electroporation,** which involves subjecting cells to electric shocks to make their membranes permeable to the DNA, and **biolistics,** in which foreign DNA is literally shot into cells on tiny metal projectiles. Probably the most common way to introduce foreign genetic material is to use viruses. When viruses infect cells they introduce their own genetic material, which then essentially hijacks the infected cell and forces it to produce more viruses. The genetic material of viruses can be DNA, or in the case of **retroviruses,** RNA. Retroviruses are particularly promising, because the RNA is first transcribed into a DNA copy, which then often becomes permanently incorporated into the DNA of the infected cell. Every human contains copies of such DNA sequences acquired from retroviruses. By removing certain genes from the viral genetic material and substituting a desired gene, genetic engineers can use viruses to introduce the desired genes into cells without commandeering the cells to make new viruses.

A long-range objective of recombinant DNA technology is to replace a mutated gene with a normal gene. More immediate goals, some of which have already been achieved, include the following five:

1. **Manufacture of drugs and other useful organic substances in large quantities by recombining the gene for those substances into bacteria.** Already human insulin to treat diabetes, growth hormone to prevent dwarfism, and factor VIII to treat hemophilia are being manufactured by this process.

2. **Recombination of genes directly into food crops and animals to achieve**

The Recombinant DNA Controversy

Like every new technology, recombinant DNA technology has generated controversy. Some of this controversy is over philosophical questions, such as whether humans should be playing God with their own genes or tampering with the wisdom of billions of years of evolution. These are certainly questions worth considering, but history suggests that they are not likely to deter genetic engineers any more than the question of whether humans were intended to fly deterred the Wright brothers.

Other controversies deal with potential dangers that cannot be so lightly dismissed. One can easily imagine a demented genetic engineer using recombinant DNA technology to "improve" the AIDS virus to make it as communicable as the common cold. Even a well-meaning genetic engineer might accidentally start a plague. One researcher was about to concoct a strain of intestinal bacteria that would digest cellulose and thereby provide an additional source of carbohydrate until it occurred to him that such a bacterium might cause intestinal gas or fatal diarrhea (Wade 1977). Another potentially unforeseen outcome might be oil-digesting bacteria that would infect automobiles and other machines. Now there are government regulations that are supposed to prevent this kind of thing. However, the history of nuclear energy and other government-regulated scientific activities provides little confidence in such regulations. In the short history of genetic engineering there have already been several cases in which scientists deliberately bypassed regulations (Crawford 1986; Schneider 1987; Sun 1986).

It is also unsettling when genetic engineers reveal their ignorance of ecology by claiming that releasing a genetically altered organism into the environment will have only the intended consequences without any side effects. In real life, as Garrett Hardin says, "You can never do just one thing." Introducing natural organisms can wreak havoc upon an ecosystem, so engineered organisms surely have the same potential. Even if an engineered organism has direct effects that are entirely beneficial, there are indirect effects that are difficult to predict. Who will profit, for example, from using the new technology to increase milk production when many farmers are already going out of business because of surpluses (DuPuis and Geisler 1988; Tangley 1986)? Will the new technology be affordable by small farmers, or will it accelerate the growth of huge agribusinesses? Will the technology ever relieve hunger in underdeveloped countries? If so, what effect will that have on overpopulation?

Again, however, experience suggests that these concerns will have little effect on the progress of recombinant DNA technology. Any technology that is profitable to someone will be done even if it is detrimental to others. Engineers continue to make cars, and consumers continue to buy them, even though 50,000 Americans die in them each year.

These are problems that few academic scientists or students have much contact with or control over. There is another class of problems, however, that emerges from the increasing entanglement of academic scientists and students with industry (Blumenthal et al. 1986a,b). These problems present numerous potential conflicts between the scientific motive and the profit motive. For example, should a scientist abandon the traditional free exchange of information in order to protect the trade secrets of a company sponsoring his or her research? Should a corporate sponsor have the power to block a publication or a student's thesis in order to protect its trade secrets? Should a teacher ignore a student's interests and direct that student toward research projects that will lead to profitable patents? Should a company contributing to the research in a university have any influence in the hiring and promotion of faculty or in the selection of students? These are problems that individuals and committees on campuses are having to deal with more frequently, and over which they can have some control.

the same effects as those just described. Genes for human growth hormone have already been recombined into fishes (Lewis 1988) and mice (Figure 5.15) experimentally. Genes that increase the production of particular amino acids in rice and other grains could increase the quality of protein available to large numbers of people.

3. **Production of organisms with unique abilities.** It might be useful, for example, to have bacteria that could digest oil spills, plastics, and other pollutants.

4. **Giving organisms the genetic ability to tolerate environmental stresses.** Plants and fishes can be given genes for the production of antifreeze compounds that will enable them to expand their ranges into colder regions. Another possibility is that threatened species could be given the ability to tolerate environmental damage.

Figure 5.15
Ordinary mouse (left) and a "super mouse" bearing recombined genes for human growth hormone. The ovum from which the "super mouse" developed had genes injected into it on circles of DNA called plasmids (pBR), which also incorporated a promoter from a mouse gene. Subsequently, the promoter was activated in a variety of tissues, which synthesized the growth hormone.

5. **Generating new research.** The research possibilities are unlimited in areas such as genetics, cell biology, and development. One rather bizarre possibility is that DNA from human mummies and extinct species such as the woolly mammoth might be recombined in living animals.

SUMMARY

The information that is passed from generation to generation is encoded in the genetic material, deoxyribonucleic acid. DNA consists of two helical chains of deoxyribose alternating with phosphate, with one of four nitrogenous bases attached to each deoxyribose. The sequence of the four bases—thymine, cytosine, adenine, and guanine—encodes the information. This sequence is copied when DNA is replicated for mitosis and meiosis, thus enabling the genetic information to pass to daughter cells and gametes.

The base sequence also controls the sequence of amino acids in proteins produced by a cell and thus controls much of the activity of cells and of the organism. The information is first transcribed to messenger RNA, which is similar to DNA except that it is single-stranded, has ribose instead of deoxyribose, and has uracil instead of thymine. After processing, the mRNA goes to ribosomes where the base sequence is translated into an amino acid sequence. The bases on the mRNA are divided into sets of three, each of which is a codon that specifies an amino acid. Each codon selects a particular transfer RNA that carries a particular amino acid. The tRNA then inserts its amino acid in the growing polypeptide chain as it is synthesized.

Normally, DNA is coiled tightly in chromosomes, and each gene has a certain location in a certain chromosome. Changes in the number and structure of chromosomes will therefore affect the number of genes, with consequences that are often detrimental, but sometimes normal and beneficial. These kinds of changes include aneuploidy, in which the number of chromosomes differs from that of the normal karyotype, and translocation, in which part of a chromosome changes location. Another kind of genetic change results from point mutation, in which one base in DNA is substituted for another. Such changes may have no effect on the protein coded by the gene or can lead to changes in the protein that may result in disease.

Genetic screening is now being used increasingly to detect mutations, as well as for other purposes. Ultimately, it should be possible to map all human genes, and even to determine the sequence of all bases in human DNA. This knowledge will increase the ability to change the genetic composition of humans and other organisms. Already genes can be transferred from one organism to another by recombinant DNA techniques, leading to improvements in food and other products.

KEY TERMS

deoxyribonucleic acid
double helix
nitrogenous base
nucleosome
cell cycle
replication
nucleotide
transcription

ribonucleic acid
translation
genetic code
codon
ribosome
transfer RNA
chromosomal aberration
aneuploidy

karyotype
point mutation
restriction enzyme
genome
recombinant DNA
plasmid

1. Describe some of the key discoveries leading to the conclusion that DNA, rather than protein, is the hereditary material.

2. Describe the major activities of genes during each part of the cell cycle: G_1, S, G_2, mitosis.

3. Name and describe the process by which genes are duplicated in mitosis and meiosis.

4. Antidiuretic hormone is a peptide with the amino acid sequence CYS-TYR-PHE-GLN-ASN-CYS-PRO-ARG-GLY. Using the genetic code (Table 5.1), write a sequence of mRNA bases that would code for this peptide. In addition to this sequence, what other bases would the processed mRNA have? Write the sequence of DNA bases from which your mRNA sequence would have been transcribed. In addition to that DNA sequence, what additional bases would have been associated with the gene?

5. The hormone oxytocin is identical to antidiuretic hormone except that ILE is found in the third position instead of PHE, and LEU is found in the next-to-last position in place of ARG. From Table 5.1, explain how each of these two substitutions could have arisen during evolution as a result of single point mutations.

6. Describe how transfer RNA translates the sequence or codons in mRNA to an amino acid sequence. What role do ribosomes play in translation?

7. Describe a chromosomal aberration and describe its consequences in humans. Describe a disease caused by a point mutation.

8. Explain what recombinant DNA is. How can it be produced artificially? Why?

_____ **READINGS** _____

RECOMMENDED READINGS

Arnheim, N., T. White, and W. E. Rainey. 1990. Application of PCR: organismal and population biology. *BioScience* 40:174–182.

Chambon, P. 1981. Split genes. *Sci. Am.* 244(5):60–71 (May).

Darnell, J. E. Jr. 1985. RNA. *Sci. Am.* 253(4):68–78 (Oct).

Dickerson, R. E. 1983. The DNA helix and how it is read. *Sci. Am.* 249(6):94–111 (Dec).

Felsenfeld, G. 1985. DNA. *Sci. Am.* 253(4):58–67 (Oct).

Grivell, L. A. 1983. Mitochondrial DNA. *Sci. Am.* 248(3):78–89 (Mar).

Hirsch, M. S. and J. C. Kaplan. 1987. Antiviral therapy. *Sci. Am.* 256(4):76–85 (Apr). (*How drugs that interfere with transcription and translation combat viruses.*)

Holliday, R. 1989. A different kind of inheritance. *Sci. Am.* 260(6):60–73 (June). (*How different genes may be activated in different cells.*)

Kornberg, R. D. and A. Klug. 1981. The nucleosome. *Sci. Am.* 244(2):52–64 (Feb).

Lawn, R. M. and G. A. Vehar. 1986. The molecular genetics of hemophilia. *Sci. Am.* 254(3):48–54 (Mar).

Murray, A. W. and M. W. Kirschner. 1991. What controls the cell cycle. *Sci. Am.* 264(3):56–63 (Mar.).

Nathans, J. 1989. The genes for color vision. *Sci. Am.* 260(2):42–49 (Feb).

Paterson, D. 1987. The causes of Down syndrome. *Sci. Am.* 257(2):52–60 (Aug).

Ross, J. 1989. The turnover of messenger RNA. *Sci. Am.* 260(4):48–55 (Apr).

Sapienza, C. 1990. Parental imprinting of genes. *Sci. Am.* 263(4):52–60 (Oct). (*Identical genes may have different effects, depending on which parent they come from.*)

Stahl, F. W. 1987. Genetic recombination. *Sci. Am.* 256(2):90–101 (Feb).

Steitz, J. A. 1988. "Snurps." *Sci. Am.* 258(6):56–63 (June). (*The processing of mRNA.*)

Todorov, I. N. 1990. How cells maintain stability. *Sci. Am.* 263(6):66–75 (Dec). (*How cells recover after an antibotic blocks translation.*)

Verma, I. M. 1990. Gene therapy. *Sci. Am.* 263(5):68–84 (Nov).

Watson, J. D. 1968. *The Double Helix: A Personal Account of the Discovery of the Structure of DNA.* New York: Signet.

Weintraub, H. M. 1990. Antisense RNA and DNA. *Sci. Am.* 262(1):40–46 (Jan).

White, R. and J.-M. Lalouel. 1988. Chromosome mapping with DNA markers. *Sci. Am.* 258(2):40–48 (Feb).

ADDITIONAL REFERENCES

Borst, P. and D. R. Greaves. 1987. Programmed gene rearrangements altering gene expression. *Science* 235:658–667.

Botstein, D. 1986. The molecular biology of vision. *Science* 232:142–143.

Blumenthal, D. et al. 1986a. Industrial support of university research in biotechnology. *Science* 231:242–246.

Blumenthal, D. et al. 1986b. University–industry research relationships in biotechnology: implications for the university. *Science* 232:1361–1366.

Crawford, M. 1986. Researcher reprimanded for pseudorabies test. *Science* 234:667–668.

Crosland, M. W. J. and R. H. Crozier. 1986. *Myrmecia pilosula,* an ant with only one pair of chromosomes. *Science* 231:1278.

DuPuis, E. M. and C. Geisler. 1988. Biotechnology and the small farm. *BioScience* 38:406–411.

Lewis, R. 1988. Fish: new focus for biotechnology. *BioScience* 38:225–227.

Lewontin, R. C. 1986. How important is genetics for an understanding of evolution? *Am. Zool.* 26:811–820.

Martin, J. B. 1987. Molecular genetics: applications to the clinical neurosciences. *Science* 238:765–772.

Palmiter, R. D. et al. 1983. Metallothionein–human GH fusion genes stimulate growth of mice. *Science* 222:809–814.

Schneider, K. 1987. Tearful scientist halts gene test. *The New York Times,* 4 September 1987.

Schrödinger, E. 1944 [1967 reprint]. *What Is Life?* New York: Cambridge University Press.

Sun, M. 1986. Biotech firm gets another black eye over experiment. *Science* 231:1242.

Tangley, L. 1986. Biotechnology on the farm. *BioScience* 36:590–593.

Volpe, E. P. 1987. Developmental biology and human concerns. *Am. Zool.* 27:697–714.

Wade, N. 1977. Dicing with nature: three narrow escapes. *Science* 195:378.

6

Development

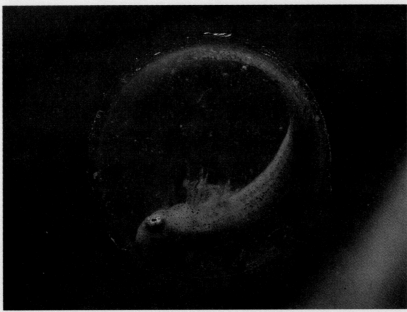

Spotted salamander embryo (Ambystoma maculata).

CHAPTER OUTLINE

LEARNING OBJECTIVES

1. Is each animal preformed within a gamete, or does it develop by gradually acquiring new structures?
2. How are gametes produced?
3. How do gametes of the same species come together so effectively during fertilization while resisting gametes of other species?
4. How does a zygote develop into many different kinds of cells?
5. How is the fundamental body plan of an animal established?
6. Is the fate of each cell determined from the beginning, or is it acquired gradually?
7. How does an embryo that develops in an egg or in a uterus obtain nutrients and oxygen?
8. What mechanisms guide cell movement during development?
9. Is there a relationship between normal development and the growth of tissues in cancer?
10. Is there a relationship between development and the processes of metamorphosis and regeneration?
11. Does the development of an individual tell us anything about the evolution of its species?

EARLY IDEAS OF DEVELOPMENT

It is quite likely that until the domestication of animals around 10,000 years ago, many people did not associate copulation with reproduction. Until people were able to observe that birth regularly followed mating in their animals, copulation and having babies were two unrelated things, each miraculous in its own way. Discovering that babies develop as a result of semen in the womb in no way lessened the awe and mystery. Even with our knowledge of cells and genes it is difficult for us not to feel the same way. Nevertheless, we know that underlying the mystery are materials and mechanisms that we can potentially understand scientifically.

One of the first to approach development with a scientific perspective was Aristotle in the 4th century B.C. In his book *On the Generation of Animals*, Aristotle says humans develop out of menstrual fluid that has been organized and vitalized by the semen. We are apt to laugh at such an idea now, but it took 2000 years to improve upon it. During the Renaissance, when the scientific approach was reinvented, many philosopher-scientists picked up the question of development just where Aristotle had left it. In the 17th century William Harvey studied development of the chick and concluded that all animals develop from an egg: *ex ovo omnia*. (Of course, Harvey did not believe all animals develop from the kind of things one might serve scrambled with English muffin. The term egg refers to any female gamete.)

Later Harvey looked for eggs in the uterus of female deer that had just mated. He did not find them but concluded that they must be there nevertheless. He also found no seminal fluid and decided that the semen must fertilize the egg by a kind of "contagion." He thought fertilization was due to a vapor in the same way that diseases were then thought to come from bad air. Harvey's emphasis on eggs led early microscopists to look for them in mammals, and in 1672 they found something that fit their expectations. (Actually, it was the follicle in which the egg develops. The mammalian egg was not identified until more than a century later.)

Several years after the egg was thought to have been discovered, Antony van Leeuwenhoek observed human sperm through his microscope but thought they were parasites. At about the same time, Nicolas Hartsoeker discovered sperm independently and realized their significance. The discoveries of eggs and sperm helped establish development as a product of materials and mechanisms, not of spirits or nebulous vapors. Nevertheless, they left unsolved the question of how a fully formed animal emerges from such a tiny beginning. One explanation advocated by many scientists, including Hartsoeker, is referred to as **preformationism.** Preformationism is the belief that each organism was already preformed within either the egg or the sperm (Figure 6.1). With our knowledge of cells and genetics it is not hard to come up with half a dozen arguments against preformationism now. Indeed, there were good arguments against it in its own time, one being that each unborn animal would have to contain its own preformed offspring, which would contain their own offspring, and so on for the entire future of the species.

EPIGENESIS

By the 19th century preformationism had been replaced by its alternative, epigenesis. Epigenesis refers to the gradual development of new structures that did not exist previously as such. In the modern view of epigenesis, an animal begins as a fertilized egg (zygote) that lacks the features that will eventually appear. When the zygote begins to develop it becomes an **embryo,** which gradually acquires the features characteristic of the animal. Although animals are extremely diverse, all

Figure 6.1
Hartsoeker's representation of a human infant preformed within the head of a sperm. The diamond shape is the fontanel between the cranial bones of the infant. Hartsoeker never claimed to actually see such sperm, but another preformationist claimed to see tiny chickens, horses, and mules in the semen from these species.

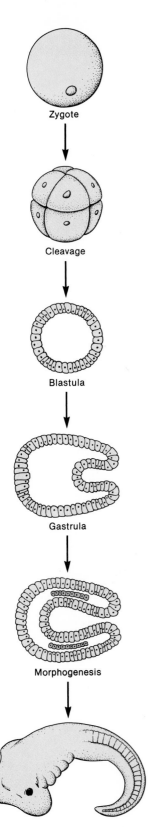

Zygote

Cleavage

Blastula

Gastrula

Morphogenesis

Figure 6.2
The modern epigenetic view. All animals develop through these embryonic stages.

develop through the same sequence of processes, called cleavage, gastrulation, and morphogenesis (Figure 6.2):

1. In **cleavage** the zygote divides repeatedly, resulting in a mass of similar cells, called a **blastula.**
2. The blastula then puckers inward during the process called **gastrulation.**
3. Gastrulation marks the beginning of **morphogenesis,** in which the embryo begins to take on a definite shape and cells start to become differentiated from each other.

GENETICS AND DEVELOPMENT

Before describing cleavage, gastrulation, and morphogenesis in detail, it is necessary to consider the role of genetics in development. One of the many mysteries of development is that not only the form of the organism but also the form and function of each of its cells are largely determined by the genetic information that was in the zygote from which it developed. With few exceptions, each cell in the adult animal has the same genetic information that was in the zygote. Why, then, aren't the cells of skin, muscle, liver, and all the other organs the same? Two answers have been proposed. In 1897 August Weismann (see p. 71) suggested that during development some genetic material was lost with each cell division, so that each daughter cell was genetically different from others. In 1934, however, Thomas Hunt Morgan (see pp. 76–78) proposed that every cell in an organism contains the same genetic information, but that different combinations of genes are activated in different cells.

Studies of mitosis (see pp. 71–73) support Morgan's theory. Generally, both daughter cells produced by mitosis receive the same genetic information, so differences between two daughter cells must be due to different combinations of genes being activated or turned off. Morgan's hypothesis was dramatically confirmed in the early 1960s by John B. Gurdon at Oxford. Gurdon destroyed the nuclei from eggs of African clawed frogs, *Xenopus laevis*, and replaced them with nuclei from the intestine of tadpoles. Some of these eggs went on to develop into normal frogs genetically identical to the "donor" tadpole. Therefore, even though most genes are inactive in the intestinal cells of a tadpole, they must have been present in order to provide all the information to make a complete frog.

The conclusion emerging from these studies is that development is a process of repeated cell division in which different combinations of genes become activated in different cells. These different gene combinations presumably account for differences in cellular functioning, in the formation of different tissues and organs, and in the final shaping of the organism. The next level of problem to be solved in genetics is to find out what signals activate and inactivate the different gene combinations. A few of these signals and the genes they affect are known. Generally, however, watching cells in a developing animal is like watching traffic from an airplane. There is some kind of orderliness in the overall activity, and individual movements seem to be purposeful, but it is hard to say what determines the destination of each one.

GAMETOGENESIS

The development of an organism from a fertilized egg begins long before fertilization, since the gametes themselves must first develop. The development of gametes, called gametogenesis, begins with **primordial germ cells.** The primordial germ cells that produce sperm are called **spermatogonia,** and the process of

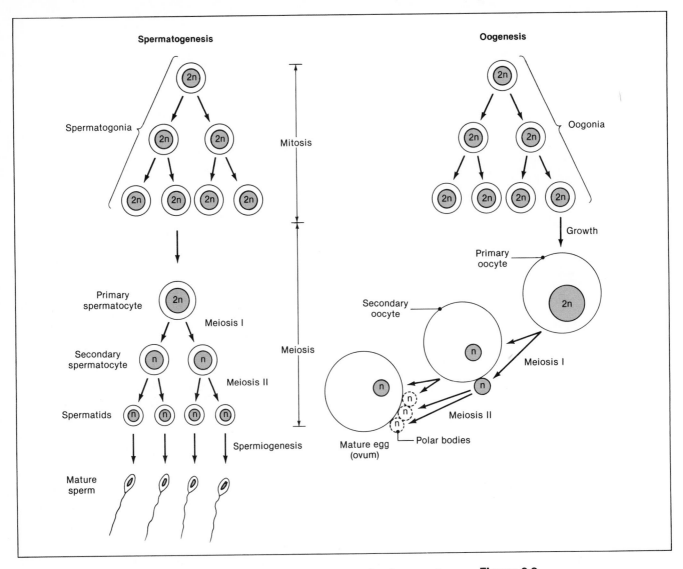

Figure 6.3
Spermatogenesis and oogenesis.

sperm production is called **spermatogenesis.** Spermatogonia first undergo repeated mitosis, producing genetically identical diploid **primary spermatocytes** in enormous numbers (billions in men) (Figure 6.3). Each primary spermatocyte then undergoes meiosis (see pp. 74–75). Meiosis I results in two haploid secondary spermatocytes. Because of crossing over during meiosis I, secondary spermatocytes are all genetically different from each other. Each of these undergoes meiosis II, resulting in a total of four haploid **spermatids.** The spermatids mature (during **spermiogenesis**) into functional sperm.

The earliest stages of egg production, during **oogenesis,** are similar to those of spermatogenesis. The germ cells, called **oogonia** (pronounced oh-ah-GO-nee-uh), first proliferate mitotically. A few of the oogonia then grow and develop into genetically identical **primary oocytes.** Primary oocytes undergo meiosis I, producing genetically diverse **secondary oocytes** and small, functionless **first polar bodies.** In meiosis II each secondary oocyte divides into a mature egg, or **ovum,** and a **second polar body.** Human females produce up to a million oogonia during their first trimester of life as a fetus. These oogonia begin meiosis but stop after prophase I. Fewer than 500 then resume meiosis, usually one at a time starting at puberty and continuing until menopause. In humans and some other species, meiosis II occurs only after fertilization. The first polar body also completes meiosis, bringing the total number of polar bodies to three.

FERTILIZATION

The overall function of fertilization is to combine two haploid gametes into a single zygote with the normal diploid number of chromosomes. This seemingly straightforward objective is complicated by the fact that the egg is usually exposed to numerous sperm simultaneously. In fact, in humans and many other species, numerous sperm are required to achieve fertilization, although only one sperm enters an ovum. The entry of more than one sperm **(polyspermy)** must be prevented or the number of chromosomes will be abnormal. In species in which fertilization occurs outside the body, the egg also generally has to resist fertilization by sperm from different species. Fertilization is therefore complicated by the conflicting needs of encouraging fertilization by sperm of the same species

Figure 6.4
(A) Sperm from various groups of animals. Some, such as those of crustaceans, are nonmotile. Sperm of round worms and arthropods propel themselves with ameboid motion, but most sperm use flagella. All drawn at the same scale. (B) Detail of a human sperm cell.

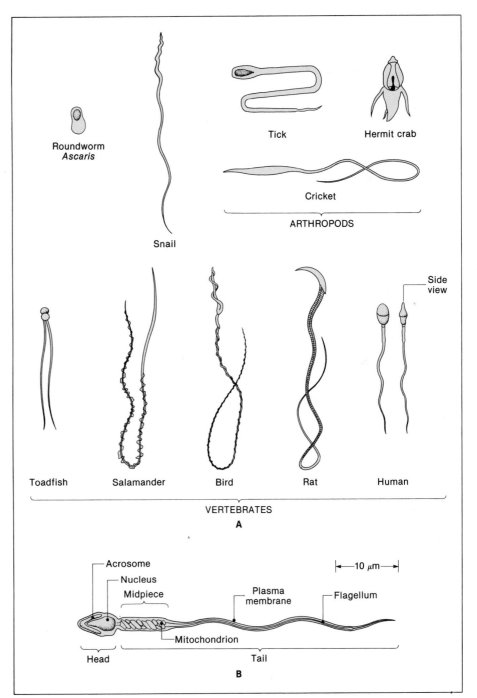

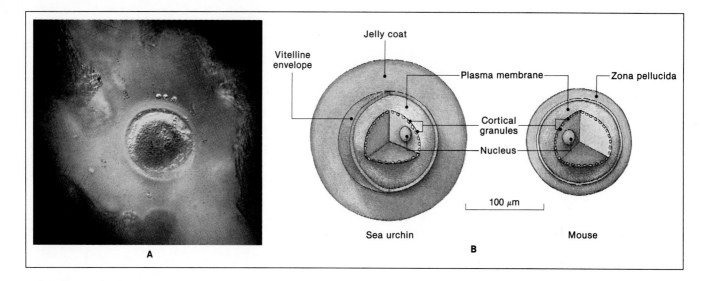

Figure 6.5
(A) A human oocyte in the oviduct. The halolike ring around the oocyte is the zona pellucida. (B) Schematic comparison of the ovum from a sea urchin and the oocyte from a mammal.

while preventing polyspermy and rejecting sperm of another species. There is an enormous variety of ways in which different groups of animals achieve these goals, but it has been possible to study fertilization in only a few groups. Echinoderms, especially sea urchins (see Figure 34.1C), are favorite subjects of fertilization studies. Recent advances in tissue culture have also made it possible to study fertilization in mammals. Fertilization will be described in each.

Before fertilization can occur, sperm must move into the vicinity of the egg. The sperm of most animals propel themselves actively, using long **flagella** (see p. 217; Figure 6.4). The flagellum of each sperm is powered by mitochondria densely packed in the **midpiece.** Anterior to the midpiece and separated by a neck is the **head** of the sperm, which contains the "payload" of tightly condensed chromatin in the nucleus. The egg is generally many times larger than the male gamete and incapable of moving under its own power (Figure 6.5). Most eggs are enclosed in protective substances that the sperm must penetrate. Sea urchin eggs are enclosed in a thick **jelly coat** and a **vitelline envelope.** Mammalian eggs are enclosed in a **zona pellucida.**

Depending on the species, fertilization occurs at different times in the development of the egg. In sea urchins meiosis is complete by the time of fertilization, and the egg has matured into an ovum. In mammals, however, the egg is arrested in meiosis I at the time of fertilization and is therefore properly referred to as an oocyte rather than an ovum. As soon as fertilization occurs, the developing zygote may be referred to as an embryo. (The term "embryo" is sometimes applied to any developmental stage before birth, especially in early stages before there is some recognizable form. In humans the term is generally applied only up to the eighth week after fertilization, after which the term "fetus" is used.)

Fertilization proceeds by several steps. First is a loose and nonspecific **attachment** of the sperm to the jelly coat (sea urchins) or zona pellucida (mammals). Next comes a stronger and specific **binding** between molecules on the sperm and receptor molecules on the vitelline envelope or in the zona pellucida. Generally, the receptor molecules bind only sperm of the same species. In sea urchins the receptor molecules on the vitelline envelope must match molecules called **bindins** on the sperm. In mammals the receptor is **ZP3,** one of the three kinds of glycoproteins that compose the zona pellucida.

The next step, called the **acrosome reaction,** consists of the rupture of the sperm's **acrosome** (= acrosomal vesicle; Figure 6.4B), which releases enzymes that digest a pathway through the jelly coat or zona pellucida. In sea urchins the acro-

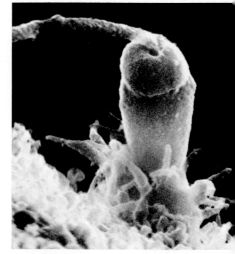

Figure 6.6
Scanning electron micrograph showing fertilization of a sea urchin egg. A group of microvilli on the egg's plasma membrane has formed a "fertilization cone" that draws the sperm cell inside. A group of sperm microtubules called the acrosomal process has penetrated the jelly coat and bound to the vitelline membrane.

Figure 6.7
The cortical reaction in sea urchins. (A) At fertilization only a few sperm have attached to the egg. (B) Sperm continue to attach, but the fast block prevents polyspermy. (C) Thirty seconds after fertilization the cortical reaction has cleared a zone around the point of fertilization (arrow). (D) Within three minutes the vitelline membrane is hardened, and sperm are dispersed.

some reaction is triggered by a carbohydrate in the jelly coat. In mammals it is triggered by ZP3. The acrosome reaction is followed by the fusion of the plasma membranes of both gametes. The sperm do not have to force their way into the egg. A group of tiny projections of the egg's plasma membrane, called **microvilli,** actually rise up and pull the sperm inside (Figure 6.6).

Fertilization triggers a series of reactions in the egg, some of which prevent a second fertilization. Within 3 seconds of fertilization in the sea urchin there is a rapid change in the voltage across the plasma membrane from about -60 milli-volts inside the egg to as much as $+20$ millivolts. Laurinda Jaffe, while a graduate student at UCLA, showed that this voltage change acts as a **fast block** that pre-

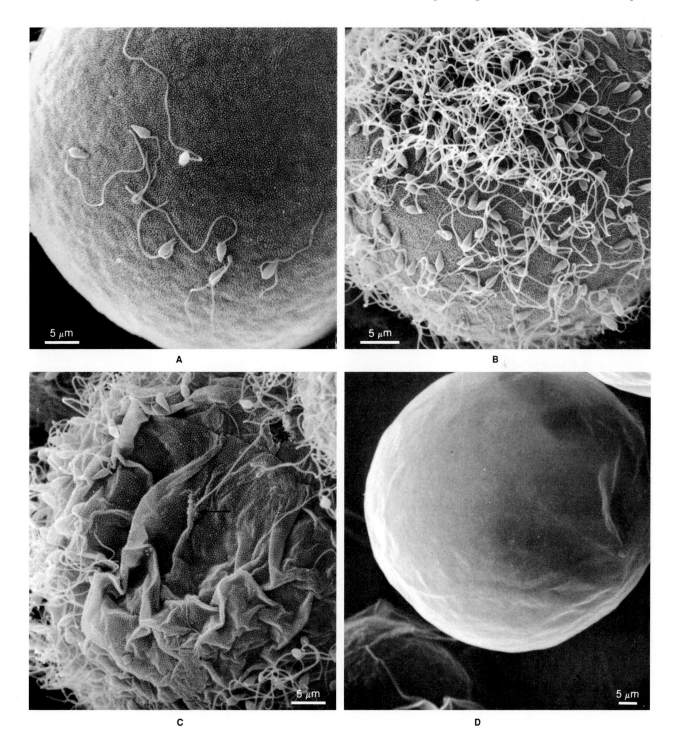

vents polyspermy. A similar process occurs in some other animals, but not in most mammals. Subsequent changes act as a stronger **slow block** to polyspermy. The first of these is the **cortical reaction,** which is the release of the contents of **cortical granules** just beneath the plasma membrane (Figure 6.5B). In sea urchins the cortical reaction alters the plasma membrane to prevent further fertilization (Figure 6.7). In mammals the cortical reaction blocks polyspermy by triggering a **zona reaction** in which the zona pellucida hardens and ZP3 changes so that it no longer binds sperm.

After the sperm cell enters the egg, all but the chromatin and one centriole detaches and disintegrates. The nuclear envelope of the sperm then disintegrates temporarily, allowing the chromatin to decondense. It then reaggregates, enclosing the **male pronucleus.** After the oocyte nucleus completes meiosis and becomes haploid, it becomes the **female pronucleus.** In sea urchins microtubules from the sperm centriole draw the male pronucleus directly toward the female pronucleus until the two pronuclei fuse into a diploid **zygote nucleus.** In mammals the sperm takes a more roundabout approach (Figure 6.8).

Fertilization also triggers **egg activation.** Egg activation refers to a sharp rise in the metabolic activities that are necessary for continued development. (Genetic activity in the zygote nucleus does not begin until later, however.) Activation does not depend on any molecular contribution from the sperm but is apparently due only to mechanical stimulation. In many cases a pin prick will trigger activation, and development will continue up to a certain stage. In most parthenogenetic species the eggs require neither fertilization nor activation.

CLEAVAGE

Egg activation initiates the series of cell divisions called cleavage. Unlike most mitotic divisions, those involved in cleavage are not followed by the G_1 phase during which cellular growth occurs (see p. 84). Instead, each mitosis is followed directly by S phase, in which DNA is replicated for the next mitosis. The total mass of cells therefore remains the same as in the zygote, and each cell, or **blastomere,** gets smaller with each division. (That is, the blastomeres become more like somatic cells in size.) Essential cellular activities depend on a massive amount of messenger RNA (mRNA) and other molecules already stored in the oocyte before fertilization. The main thing achieved during cleavage is multicellularity. Cleavage takes days in frogs and produces tens of thousands of blastomeres. It is much slower in mammals, with humans taking a week to produce a few hundred blastomeres (Figure 6.9). Cleavage results in the formation of a distinctive sphere of blastomeres called the **blastula** (Greek *blastos* a bud). In mammals the blastula is called a **blastocyst.** Generally, the blastula contains a fluid-filled cavity, the **blastocoel.**

Cleavage Patterns. The sizes and movements of blastomeres follow what is called the **cleavage pattern.** Cleavage patterns tend to be conserved among related groups of animals and are of considerable interest to zoologists because they are thought to provide clues to the evolutionary relationships among various groups of animals. (See Chapter 21 for further discussion.) One of the factors affecting cleavage pattern is the amount and distribution of reserve nutrients, especially proteins, in the egg. These reserve nutrients make up the **yolk.** Eggs that develop quickly into larval or adult forms that can feed themselves, or that are nourished by the mother by means of a placenta, generally have little yolk. Eggs

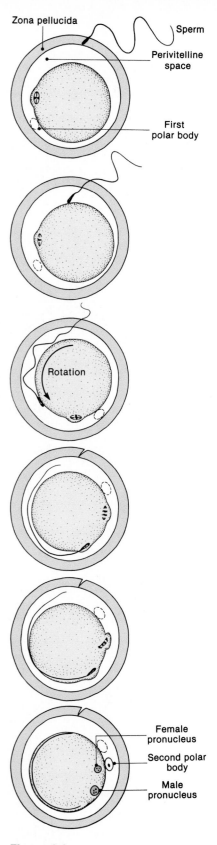

Figure 6.8
Fertilization in mammals takes more than a day. The sperm rotates the oocyte and enters sideways. The oocyte then completes meiosis II, forming the second polar body and female pronucleus.

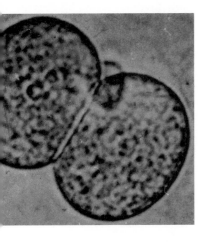
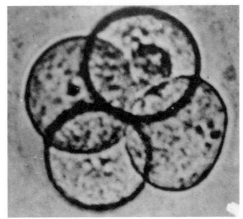
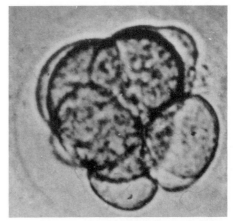

Figure 6.9

(A) Cleavage in a mammal (mouse), beginning with the two-cell stage following first cleavage, and ending with the blastocyst (the preferred term for blastula in mammals). (B) Cleavage in women occurs as the zygote is propelled down the oviduct by cilia. During the first day after fertilization the oocyte completes meiosis. Four cleavages occur over the next four days, resulting in a morula of approximately 16 cells. The morula then develops into a free blastocyst that enters the uterus. Approximately five to six days after fertilization the blastocyst implants into the uterine lining.

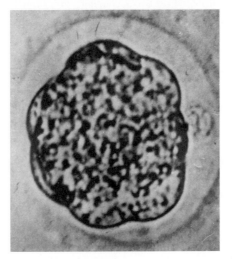
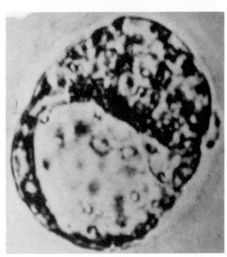

A

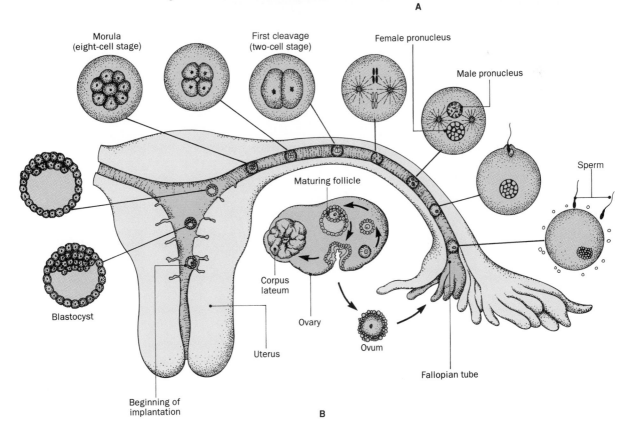

Morula (eight-cell stage)

First cleavage (two-cell stage)

Female pronucleus

Male pronucleus

Sperm

Maturing follicle

Corpus lateum

Ovary

Ovum

Fallopian tube

Uterus

Blastocyst

Beginning of implantation

B

with sparse yolk that is evenly distributed are said to be **isolecithal** (Figure 6.10). Animals such as reptiles, birds, and many fishes undergo prolonged development outside the mother's body with no external supply of nutrients. Their eggs therefore have large amounts of yolk, which is often concentrated in one place. Such eggs are said to be **telolecithal.** The eggs of cephalopod molluscs, such as squids and octopuses, and those of amphibians and some fishes are moderately telolecithal, or **mesolecithal.** In insects and most other arthropods the eggs are said to be **centrolecithal,** because the yolk is concentrated toward the center.

In isolecithal and mesolecithal eggs the entire egg can divide, and cleavage is therefore termed **holoblastic** (Greek *holo-* whole). Among animals with holoblastic cleavage, the two most common patterns of cleavage are spiral and radial. In **spiral cleavage** the first two cleavages are equal, resulting in four blastomeres of the same size (Figure 6.11). The furrows separating the four cells are said to be **meridional,** by analogy with the meridians on a globe. The third cleavage is unequal, dividing the four blastomeres into four **micromeres** and four larger **macromeres.** The micromeres spiral into the furrows between the macromeres, disrupting the meridional pattern of the cleavage furrows. This spiral pattern continues during subsequent cleavages.

In sponges, cnidarians, echinoderms, and the invertebrate chordate *Branchiostoma* (see Figure 35.7) there is **radial cleavage.** Radial cleavage also occurs in amphibians, which have meroblastic eggs. In radial cleavage daughter cells tend to remain radially aligned, and the furrows remain meridional (Figure 6.12). As in spiral cleavage, the third cleavage is unequal, resulting in micromeres and macromeres of unequal size. The micromeres occur at what is called the **animal pole** of the embryo. The macromeres, which contain more yolk, occur at the **vegetal pole.**

Besides spiral and radial, there are two other cleavage patterns that occur in animals with holoblastic cleavage. In humans and other mammals that develop with the aid of a placenta (placental mammals), there is a **rotational cleavage pattern.** After the second cleavage one pair of daughter cells lies at right angles to the other pair (see four-cell stage of Figure 6.9A, and Figure 6.13). Cleavage in mammals is also unusual in that the blastomeres do not divide simultaneously, so at any one time the cell number may not equal an integral power of two. Cephalopod molluscs (squids and octopuses), and the invertebrate chordates called tunicates (see Figure 34.6) have a **bilateral cleavage pattern.** The first cleavage furrow defines a plane of bilateral symmetry. Cleavage thereafter results in blastomeres on one side of the plane being mirror images of blastomeres on the other side.

Other animals have **meroblastic cleavage,** meaning that only part (Greek *meros* part) of the egg divides, because yolk is heavily concentrated in the other part. In the telolecithal eggs of fishes, reptiles, and birds, the dense yolk squeezes the dividing cells into a small disc, resulting in a **discoidal cleavage pattern.** In arthropods with centrolecithal eggs, cleavage occurs only on the outer surface, resulting in a **superficial cleavage pattern.**

Figure 6.10
Eggs classified according to the amount and distribution of yolk.

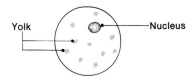

ISOLECITHAL
Sponges, cnidarians, echinoderms, many flatworms, nematodes, rotifers, most molluscs, many annelids, invertebrate chordates, placental mammals

MESOLECITHAL
Cephalopod molluscs, amphibians, some fishes

TELOLECITHAL
Many fishes, reptiles, birds

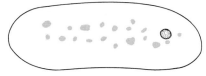

CENTROLECITHAL
Most arthropods

Figure 6.11
The spiral cleavage pattern. Arrows show pattern of daughter cells from each cleavage. In three dimensions the arrowheads would point slightly toward the reader.

Four-cell stage

Eight-cell stage

Micromere
Macromere

Sixteen-cell stage

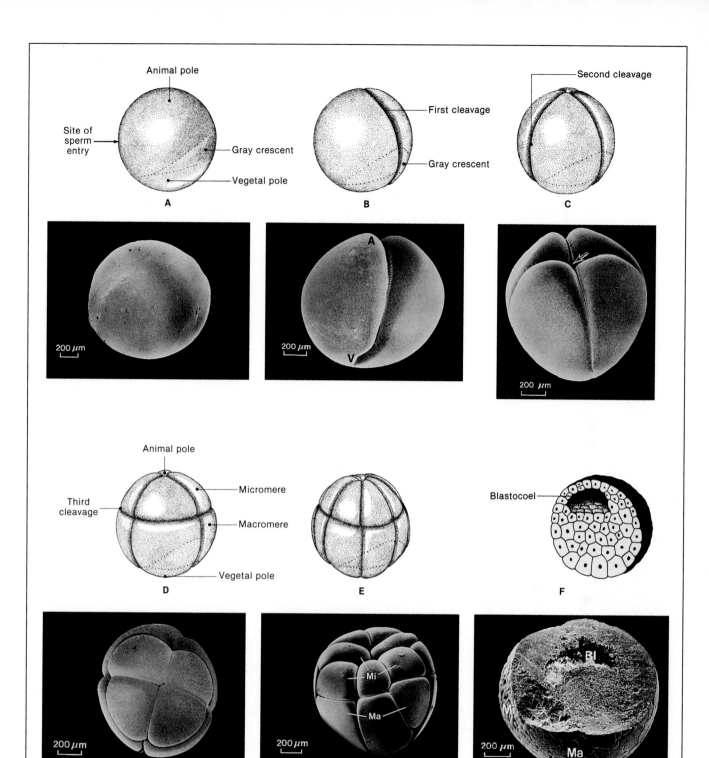

Figure 6.12
Schematic representations and scanning electron micrographs of radial cleavage in a frog egg. (A) Zygote about an hour after fertilization and before first cleavage. Pigment in the animal pole has migrated toward the site of sperm cell entry, producing a less-pigmented gray crescent. (B) First cleavage is from pole to pole, intersecting the site of fertilization and bisecting the gray crescent. (C) Second cleavage is meridional at right angles, forming a cross (arrow) at the animal pole. (D–E) Third through sixth cleavages produce micromeres (Mi) at the animal pole that tend to remain radially aligned with macromeres (Ma) at the vegetal pole. (F) Late blastula, with the blastocoel toward the animal pole.

(Parts A, and F halftones: *Richard B. Kessel and Harold W. Beams;* Part B halftone: *Reprinted from Beans, H. W. and Kessel, R. G., Cytokinesis: A comparative study of cytoplasmic division in animal cells,* American Scientist, *1976, 63(3), 279–290;* Parts D, and E halftones: *Reprinted from Kessel, R. G. and Shih, C. Y.,* Scanning Electron Microscopy in Biology, *Springer-Verlag, New York, 1976, pp. 338–341.)*

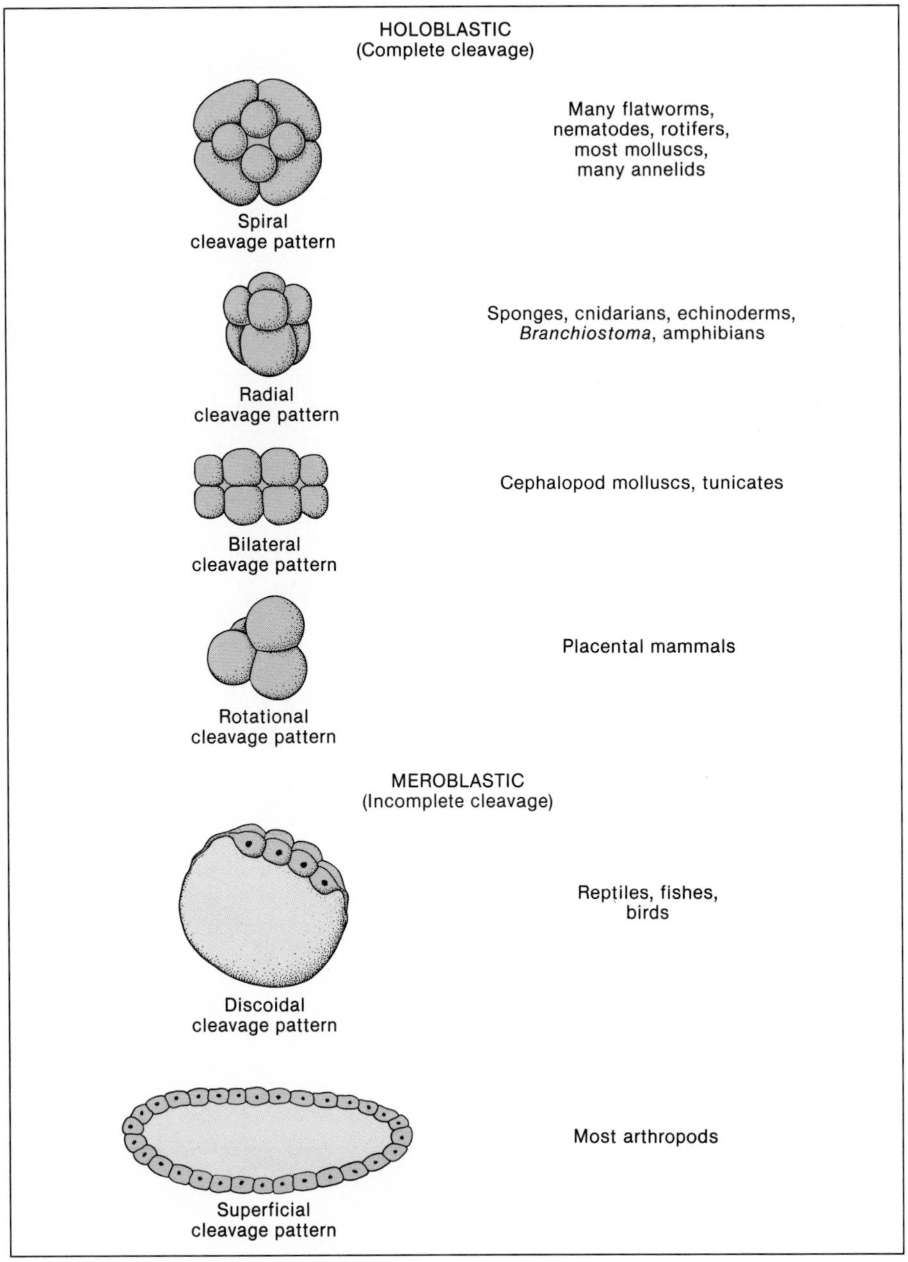

Figure 6.13
Pictorial summary of types of eggs and cleavage patterns.

HOLOBLASTIC
(Complete cleavage)

Spiral
cleavage pattern

Many flatworms, nematodes, rotifers, most molluscs, many annelids

Radial
cleavage pattern

Sponges, cnidarians, echinoderms, *Branchiostoma*, amphibians

Bilateral
cleavage pattern

Cephalopod molluscs, tunicates

Rotational
cleavage pattern

Placental mammals

MEROBLASTIC
(Incomplete cleavage)

Discoidal
cleavage pattern

Reptiles, fishes, birds

Superficial
cleavage pattern

Most arthropods

Determinate Versus Indeterminate Cleavage. In many species blastomeres begin early in cleavage to "decide" what they are going to be when they grow up. In particular, **germ cells** that will later develop into gametes are determined during cleavage, just as Weismann proposed (see p. 71). By destroying blastomeres or injecting dye into them, it is often possible to determine what contribution each one makes to the organism. One can then construct a **fate map** of the blastomeres. In some cases the embryos are transparent enough to allow the construction of fate maps simply by observing blastomeres as they divide and differentiate. One example is the roundworm *Caenorhabditis elegans* (pronounced SEE-no-RAB-dit-iss ELL-i-ganz; phylum Nematoda; see p. 549 for details). All 959 of its somatic cells, as well as its gametes, have been traced to just six blastomeres (Figure 6.14).

In *C. elegans* and many other invertebrates the fates of blastomeres are determined from the start and are not affected by position or interactions within the

Figure 6.14
(A) The adult nematode Caenorhabditis elegans (approximately 1.5 mm long). (B) An embryo of C. elegans approximately 90 minutes after fertilization, in the 24-cell stage. Already it is determined that the anterior of the worm will be to the left, and dorsal will be to the top as shown on the page. A special microscopical technique (Nomarski optics) makes the nuclei look like depressions (scale 10 μm). Letters identify stem cells. (C) The origin of the six stem cells and the fates of their descendants. The zygote P_0 divides into P_1 and stem cell AB. P_1 then divides into P_2 and EMSt, and so on. (Hypodermis covers the animal and secretes the outer cuticle.) (See von Ehrenstein and Schierenberg 1980 for details.)

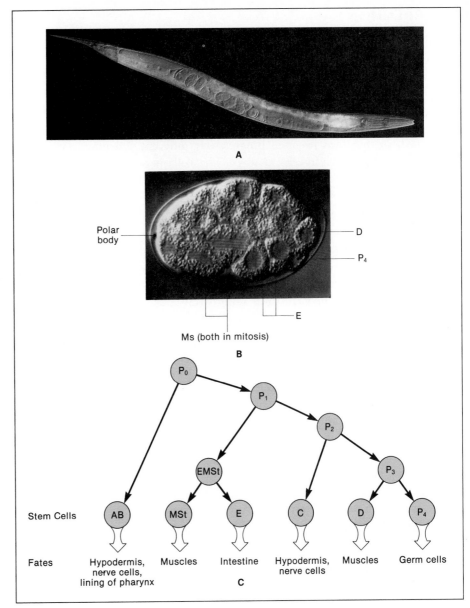

embryo. If two blastomeres are separated they will not develop into identical twins, but each will develop for a limited time into whatever part of the animal it would have become had the cells not been separated. This is called **determinate cleavage.** Another term for determinate cleavage is **mosaic development.** Like a mosaic, the entire organism is assembled from pieces (blastomeres) that are not interchangeable. Determinate cleavage probably results from different kinds and amounts of messenger RNA and other maternal factors going to different blastomeres as they divide.

In many other species, however, blastomeres have the capacity to modify their fate under abnormal conditions. For example, in echinoderms and chordates if two blastomeres are separated after the first cleavage, each blastomere can develop into a complete, normal animal. This sometimes happens in humans, producing identical twins. This type of cleavage is said to be **indeterminate.** Indeterminate cleavage is also called **regulative development,** because each blastomere seems to regulate its own development in response to position in the embryo and interactions with other blastomeres.

GASTRULATION

During cleavage the blastomeres have generally been following a pattern of division and movement controlled by maternal factors that were in the ovum. By the time the blastula forms this division and movement have slowed down considerably. Then suddenly something triggers the cells to take a more active role in epigenesis. Blastomeres begin to transcribe their own DNA instead of maternal mRNA inherited with the ovum. In each cell a different combination of genes is transcribed, so that each becomes progressively different and interacts with its neighbors in increasingly divergent ways. The cells begin to divide and move around dramatically. Out of a mass of seemingly identical blastomeres an embryo takes shape, and tissues and organs form. This process is called **morphogenesis.** The stage of development that marks the transition from cleavage to morphogenesis is called gastrulation.

In most animals gastrulation lays down the basic "tube-within-a-tube" body plan, with a gut running longitudinally through the body. This process is most clearly seen in echinoderms such as the sea urchin (Figure 6.15). In the sea urchin blastomeres at the vegetal pole push inward (invaginate), forming an opening called the **blastopore.** The blastopore leads into the primitive gut cavity, called the **archenteron.** Eventually, the archenteron completely penetrates the embryo and emerges at what will later be the mouth. Since this second opening, not the blastopore, becomes the mouth, echinoderms are said to be **deuterostomes** (Greek *deutero-* second + *stoma* mouth). Chordates, including humans and other vertebrates, are also deuterostomes. In most other animals the blastopore develops into the mouth. Since this is the first opening in the embryo, these animals are called **protostomes** (Greek *proto-* first). Many zoologists consider the distinction between protostomes and deuterostomes significant in classification, as will be discussed in Chapter 21.

In addition to establishing the basic body plan, gastrulation in most animals divides the cells into three **germ layers.** The outer germ layer is **ectoderm,** and cells descending from it will generally become epidermis, nervous tissue, and related cells. The inner germ layer, the **endoderm,** will usually become gut and

Figure 6.15
Gastrulation in a sea urchin. Parts of egg are shaded to show fates. (A–C) Radial cleavage produces micromeres at the vegetal pole. (D) In the blastula the micromeres migrate into the blastocoel and form diffuse connective tissue (primary mesenchyme). (E) Gastrulation. The vegetal plate invaginates, forming the blastopore and archenteron. Primary mesenchyme constructs spicules that compose the skeleton. The apical tuft of cilia will enable the embryo to swim. (F) Continued development of the archenteron forms the mouth and digestive tract. (G) Resulting echinopluteus larva, which swims and feeds on its own. See Figure 33.16 for external view.

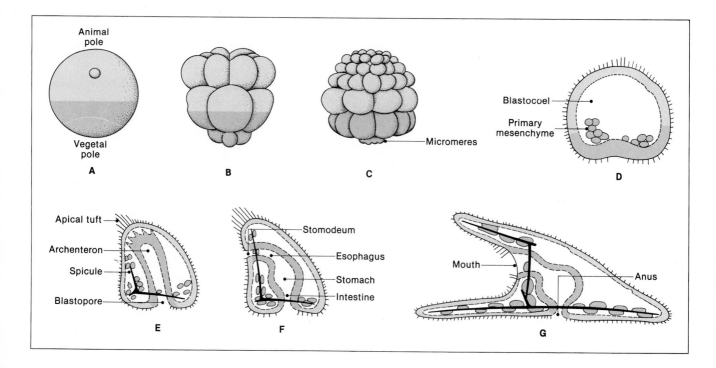

various internal glands and organs. A third germ layer, the **mesoderm,** develops from endoderm or (less often) ectoderm. Mesoderm usually gives rise to blood, bone, and connective tissues, as well as to muscle. Most animals have these three germ layers and are said to be **triploblastic.**

In most animals one of the most important creations of the mesoderm is a particular type of body cavity called the **coelom** (SEAL-um). Technically, a coelom is a body cavity that develops from mesoderm and is enclosed by a **peritoneum,** which is a membrane consisting of nucleated cells. The coelom generally arises in one of two different ways. In most protostomes and in advanced chordates the mesoderm splits into two layers, and the coelom forms between them (Figure 6.16A). These animals are said to be **schizocoelous** (Greek *schizo*- split). In many deuterostomes (echinoderms and primitive chordates) the mesoderm forms a pouch that pinches off the archenteron, enclosing the coelom within it (Figure 6.16B). These animals are termed **enterocoelous** (Greek *entero*- gut).

In vertebrates the formation of the gastrula is somewhat different from that in echinoderms. In amphibians, for example, the yolk prevents simple invagination of the vegetal pole. Instead, cells in the area of the gray crescent migrate into the blastocoel (Figure 6.17A). This movement produes a slitlike blastopore into which cells on the dorsal lip migrate and form the archenteron. Animals with even more yolk, such as fishes, reptiles, birds, and egg-laying mammals, begin morphogenesis without forming either a blastopore or an archenteron.

The gastrulas of vertebrates also form **extraembryonic membranes** that are

Figure 6.16
Representation of two methods of coelom formation from mesoderm. (A) Most protostomes and many chordates are schizocoelous: The coelom forms by the splitting of mesoderm. (B) Echinoderms and primitive chordates are enterocoelous: The coelom forms as an outpocketing of the lining of the archenteron.

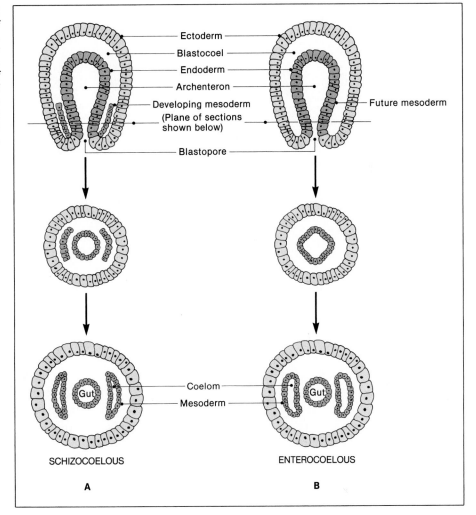

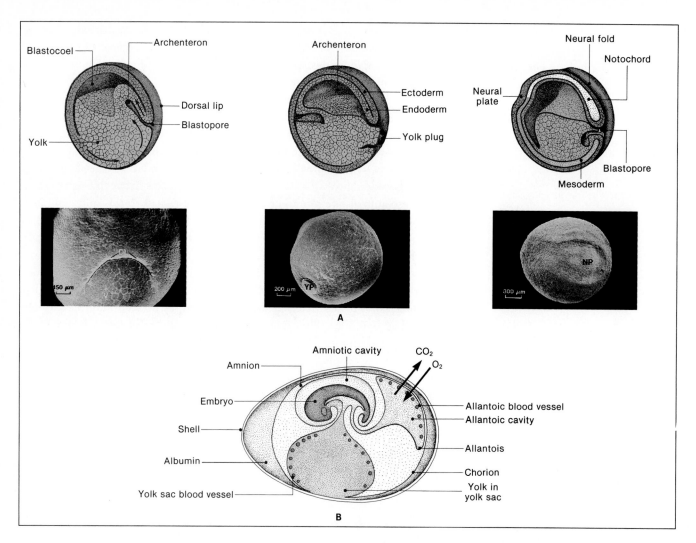

essential in maintaining a suitable environment for the embryo (Figure 6.17B). The largest of the extraembryonic membranes is the **yolk sac,** which is an extension of the primitive gut that encloses the yolk. Reptiles, birds, and mammals have an additional extraembryonic membrane, the **amnion,** that contains the amniotic fluid that moistens and cushions the embryo. The amnion is considered an evolutionary milestone, since it enabled these vertebrates—the **amniotes**—to develop outside water. A third kind of extraembryonic membrane of amniotes is the **allantois,** which stores metabolic wastes and helps exchange oxygen and carbon dioxide. Finally, amniotes have a **chorion** that also exchanges respiratory gases.

Even though humans and other **placental mammals** develop within the moist environment of the uterus, their gastrulation is much like that of vertebrates with yolky eggs. Presumably, this similarity reflects our evolution from reptiles. The cells in the blastocyst that will later form the fetus are restricted to an **inner cell mass,** as if they were squeezed there by yolk (Figure 6.18). Some of these cells separate into a **hypoblast.** This hypoblast forms a yolk sac like that in fishes, reptiles, and birds, even though in placental mammals there is no yolk for it to enclose. There is also an allantois that stores metabolic wastes, although it is vestigial in humans and many other placental mammals. Later it becomes the lining of the urinary bladder.

The remaining cells of the inner cell mass form the **epiblast.** The epiblast subdivides into two layers, one of which forms the amnion. The other layer is the **embryonic epiblast,** from which the embryo will develop. The blastocyst is sur-

Figure 6.17
(A) Gastrulation and subsequent development in the frog, continuing from Figure 6.12. (Left) Gastrulation, about 10 hours after fertilization. There are around 30,000 cells. Cells in the region of the gray crescent migrate inward, forming a slitlike blastopore that will become the anus. Cells migrating over the dorsal lip (DL) of the blastopore and into the blastocoel form the archenteron. V indicates vegetal pole. (Middle) Ectoderm and endoderm form. Migration of ectoderm over the vegetal pole covers the yolk except for a yolk plug (YP) at the blastopore. (Right) Ectoderm covers the entire gastrula. Some endoderm forms mesoderm. Gastrulation is complete after about eight hours. The number of cells has more than doubled. The notochord and neural folds are the future location of the brain and spinal cord. NP indicates neural plate. (B) The embryo of a bird, emphasizing the extraembryonic membranes. Compare Figure 39.18.

Figure 6.18
Changes in the human blastocyst leading up to gastrulation. (A) The blastocyst before implantation. Compare Figure 6.9. (B) The blastocyst during implantation by the trophoblast. Hypoblast cells from the inner cell mass have separated from the epiblast to form a yolk sac. (C) Approximately 10 days later, just before gastrulation. Trophoblast is proliferating. Amniotic ectoderm from the epiblast is forming an amnion enclosing the amniotic cavity.

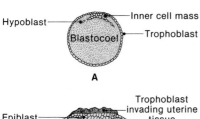

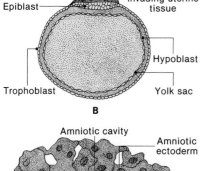

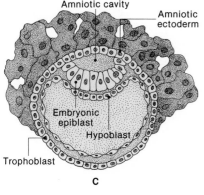

Figure 6.19
Gastrulation in the human embryo. (A) Dorsal view. Cells migrating into the primitive streak anteriorly will become the notochord; those migrating laterally will become endoderm and mesoderm. (B) Longitudinal section through the embryo at the time of gastrulation, about 17 days after fertilization. The embryonic disc lies between the amniotic cavity and the yolk sac, and all are attached by a connecting stalk to the developing placenta. The anterior of the embryonic disc is to the left.

rounded by a layer of blastomeres that forms a **trophoblast.** About five or six days after fertilization the trophoblast somehow works its way into the uterine lining during implantation. During the following week the trophoblast will develop into the chorion, which will extract nutrients from the mother. Later it will develop into the fetal part of the placenta.

Cells accumulate at the posterior end of the embryonic epiblast of amniotes, forming a **primitive streak** into which other cells migrate. Cells entering the anterior end of the primitive streak **(Hensen's node)** and continuing anteriorly form the **notochord.** The notochord is a rod of stiff tissue that is one of the hallmarks of the phylum Chordata, although it disappears before birth in mammals and most other vertebrates. Cells that migrate laterally into the primitive streak will become endoderm and mesoderm (Figure 6.19).

MORPHOGENESIS

Differentiation into Tissues. Gastrulation is the first part of morphogenesis, during which each embryonic cell makes a commitment that it and all its progeny will develop into a particular type of cell in the adult. This commitment is called

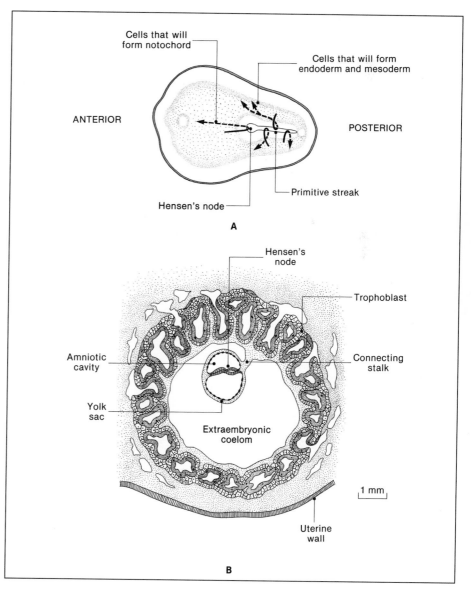

determination. Determination is usually not immediately apparent at the time it happens, but gradually cells become visibly different from each other in a process called **differentiation.** Groups of similar cells then form **tissues,** which are collections of cells that are structurally and functionally related to each other. There are four fundamental types of tissues: epithelium, connective tissue, nerve, and muscle. In mammals some 200 subtypes are recognized.

Epithelial tissue derives from ectoderm and endoderm. It serves as a covering for the body surface and for internal cavities and tubes and also forms the secretory portion of glands (Figure 6.20). Epithelial tissue can consist of one layer of cells

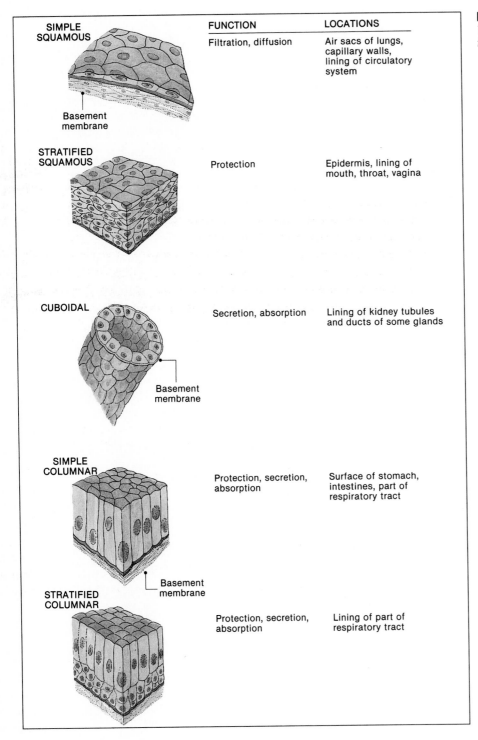

	FUNCTION	LOCATIONS
SIMPLE SQUAMOUS	Filtration, diffusion	Air sacs of lungs, capillary walls, lining of circulatory system
STRATIFIED SQUAMOUS	Protection	Epidermis, lining of mouth, throat, vagina
CUBOIDAL	Secretion, absorption	Lining of kidney tubules and ducts of some glands
SIMPLE COLUMNAR	Protection, secretion, absorption	Surface of stomach, intestines, part of respiratory tract
STRATIFIED COLUMNAR	Protection, secretion, absorption	Lining of part of respiratory tract

Figure 6.20
Types of epithelial tissue, and their functions and locations.

Figure 6.21
(A) Types of tissue formed from the three germ layers. (B) Assembly diagram for one human being.

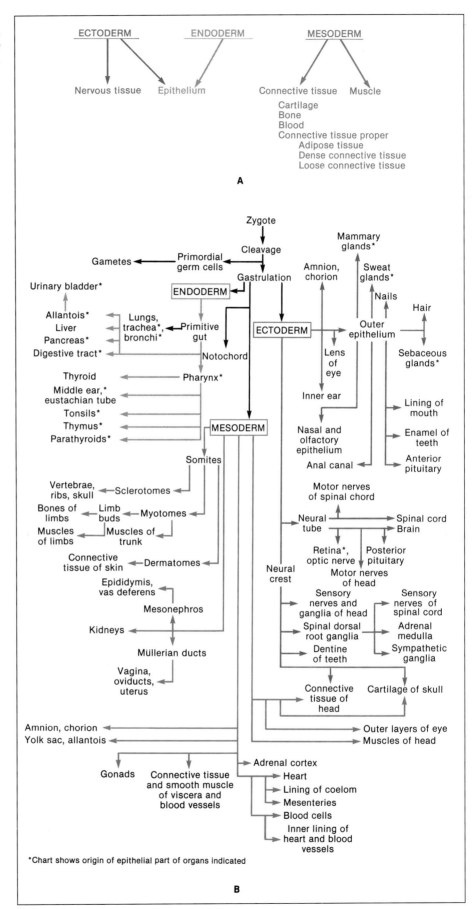

(simple epithelium) or two or more layers (stratified epithelium). Epithelium is also classified as squamous, cuboidal, or columnar, depending on whether the cells are flattened, cubical, or column shaped. One surface of epithelium is free, while the other is bound by a basement membrane (= basal lamina). The basement membrane consists mainly of protein fibers that provide mechanical support for the epithelium.

Connective tissue derives from mesoderm and includes cartilage and bone (see pp. 200–203), blood (see pp. 223–224), and connective tissue proper. The latter tissue includes adipose tissue (fat), which serves as an energy reserve and as thermal and mechanical insulation. Connective tissue proper also includes fibroblast cells that secrete collagen and other fibers that interconnect other kinds of tissues. Connective tissue proper is categorized either as dense (tendons and lig-

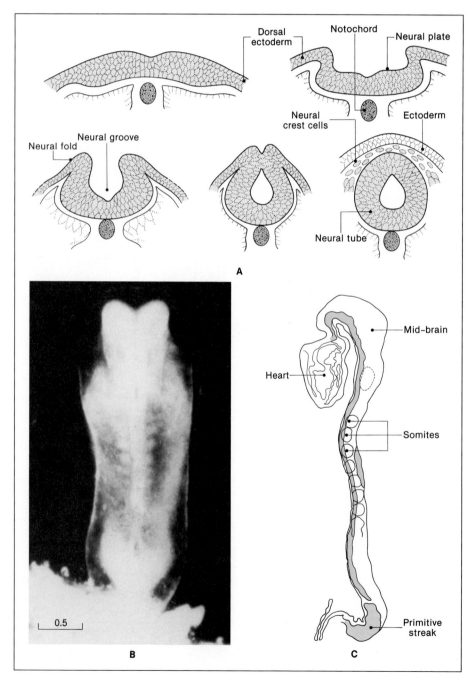

Figure 6.22
(A) Cross-sectional representation of neurulation. For an external view of the neural folds and neural plate in a frog see Figure 6.17. The notochord induces ectoderm on each side to rise up in neural folds that fuse together. This curls the neural plate between the folds, forming the neural tube that will become the brain and spinal cord. Between the neural tube and outer ectoderm is the neural crest, parts of which will migrate and form a variety of neural and other tissues.
(B) Dorsal view of a human embryo during neurulation, at about 21 days after fertilization (approximately four days after Figure 6.19). Neural folds rising over the head will form the brain. Neural folds posteriorly have already closed over the neural tube that will become the spinal cord. Note the amnion enclosing the embryo. (Scale in millimeters.)
(C) Drawing showing left side of the embryo in part B.

aments) or as loose. The remaining two types of tissue, neural and muscular, derive from ectoderm and mesoderm, respectively, and are described in later chapters. Figure 6.21 traces the lineage of various tissues in the human.

Neurulation.   During morphogenesis tissues somehow shape themselves into organs, a process referred to as **organogenesis.** The first organ to form in humans and many other vertebrates is the central nervous system, which includes the brain and spinal cord. The rudiments of the brain and spinal cord take shape as the **neural tube** forms along the notochord. This process is called **neurulation,** and the embryo at this stage is called a **neurula** (Figure 6.22A). Neurulation also forms **neural crest** cells that will migrate and give rise to other parts of the nervous system, as well as to dentine of the teeth, and other structures. (Refer to Figure 6.21.)

Continued Morphogenesis in the Human.   Soon after neurulation, the heart and major blood vessels start to form, and the heart starts to beat a few days later. At about the same time mesoderm on each side of the neural tube collects into dense structures called **somites.** The number of somites gradually increases, reaching the maximum of 40 at around one month of gestation in humans. These somites will differentiate into skeleton, muscles, and connective tissues of the skin (see Figure 6.22). Approximately 100 primordial germ cells for the next generation are already determined. They originate in the yolk sac but soon migrate to the hindgut.

By the fourth week after fertilization the human embryo has taken on a recognizable form (Figure 6.23A). The heart continues to beat as a large bulge outside the chest, and all or most of the somites have formed. The outline of the eye can be made out. On each side of the heart are limb buds that will become arms, and farther back are the limb buds that will become legs. Between the posterior limb buds is a tail—a relic of our evolution. **Pharyngeal slits** in the neck also attest that our ancestors had gills. The embryo is approximately 6 mm long at this stage, and the placenta is fully able to handle the increasing demand for nutrients and oxygen.

Figure 6.23
(A) Human embryo at about one month after fertilization. (Length approximately 6 mm.) (B) At about six weeks. (Length approximately 15 mm.)

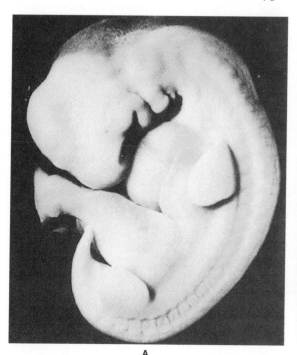

A

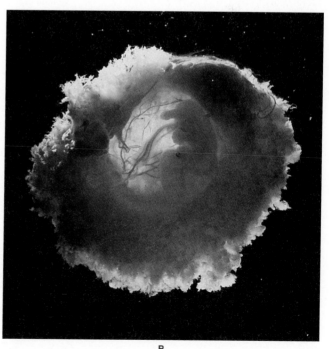

B

By about six weeks the embryo has grown to approximately 15 mm long. Outlines of fingers and toes are visible, although they are linked by webs of skin (Figure 6.23B). The pharyngeal slits are becoming incorporated into the neck and developing face. At about this time a gene on the Y chromosome of male embryos switches on, producing a **testis determining factor** (TDF). TDF causes the primordial gonads to develop into testes rather than ovaries. The testes then secrete testosterone and other substances that shape the development of male internal reproductive organs and genitals (Figure 6.24). **Anti-Müllerian duct factor** from the testes causes the Müllerian ducts to degenerate. Each of the two temporary **mesonephric kidneys** and the **Wolffian duct** that drains it then develop into male reproductive organs associated with one testis. In female embryos the absence of these secretions from the testes causes the degeneration of the two **mesonephric kidneys** and the Wolffian ducts. Müllerian ducts then continue to develop into the oviducts, uterus, and upper part of the vagina.

By the time these changes are under way at around the eighth week, the term "embryo" gives way to "fetus." The human fetus has reached essentially its final form (Figure 6.25). Although it is only about 29 mm long, it has approximately 90% of the 4500 named structures that will appear in the adult (O'Rahilly and Müller 1987). The fetus still has about 17 more weeks of development, however, before it has a chance of surviving outside the mother's body, even with support from the best technology available today. The external genitals are developed, although one cannot tell by looking whether they are male or female. The tail and the webs between fingers have degenerated by a process of **controlled cell death.** (Sometimes, however, the tail or webs persist even at birth.) The skeleton is present in the form of cartilage that will later be replaced by bone, a process that is not complete until after puberty.

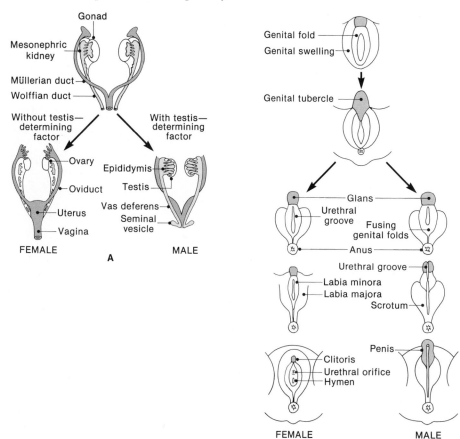

Figure 6.24
Changes in internal and external sex organs beginning at about eight weeks after fertilization. (A) TDF causes the gonad to develop into a testis. The testis then secretes anti-Müllerian duct factor, which causes the Müllerian ducts to degenerate, and testosterone, which causes the Wolffian ducts to develop. Each mesonephric kidney and Wolffian duct then becomes the epididymis, vas deferens, and seminal vesicle. (Compare Figure 16.4.) In females the mesonephric kidneys and Wolffian ducts degenerate, and the Müllerian ducts become the oviduct, uterus, and upper part of the vagina. (B) Changes in external genitalia. In females the genital tubercle becomes the clitoris, and the genital folds become the labia minora. In males the genital tubercle and genital folds of the embryo become the penis. The genital swelling becomes the labia majora or the scrotum.

The subsequent prenatal development of a fetus mainly involves growth of the organs and development of their ability to function. In the human fetus during the third month the nervous system and muscles begin to function. The fetus kicks, makes facial expressions, sucks, swallows, and jerks as if startled. These are spontaneous or reflex movements, rather than conscious actions. The parts of the brain responsible for consciousness, perception, and voluntary movement will not fully develop until months after birth. The lungs and kidneys are developed, although the placenta continues to do their work for them. By the end of the first trimester, about three months after fertilization, the fetus is about 6 cm long.

During the second trimester (fourth through sixth months) the fetus grows to a sitting height of approximately 35 cm. Bones replace much of the cartilage, and the fetus can kick hard enough to make its presence known to the mother. The heart can be heard beating. The skin is wrinkled and red, covered with downy hair, and protected by a cheesy material. The eyes open. Perhaps 10% of fetuses delivered around the end of the second trimester survive, but only if body temperature and breathing are maintained artificially. During the final trimester the fetus grows at a rate that, if sustained after birth, would make it weigh 200 pounds by its first birthday. A major change during this period is that the lungs and circulatory system prepare for the transition to air-breathing. Before birth the blood is diverted into arteries that bypass the lungs, but within minutes after birth ducts in major arteries close, forcing blood into the lungs.

MECHANISMS OF DEVELOPMENT

Anyone who saw an automobile or a watch assemble itself would quite rightly be amazed, yet most people regard a human assembling itself as natural and common. One thing that makes scientists different is that they never lose the capacity to be amazed by the commonplace. Another difference is that they want to understand causes. Advances in molecular and cellular biology in recent years have made it possible for developmental biologists not only to describe development but to understand some of its mechanisms. The space available is sufficient only to mention a few of those affecting pattern formation, morphogenesis, and growth.

Pattern Formation. Apparently, the basic pattern of an animal is established early in development by a series of genetically induced events. One aspect of pattern formation is the determination of **anterior–posterior** and **dorsal–ventral axes.** In *Drosophila* the anteroposterior and dorsoventral axes are already determined in the unfertilized egg by maternal factors that affect gene activities in blastomeres (Nüsslein-Volhard et al. 1987). In amphibians the dorsoventral and anteroposterior axes are defined by the distribution of yolk and the site of sperm entry, as shown in Figure 6.12. These axes in turn specify the site of gastrulation.

Another aspect of pattern formation that is important in annelids, arthropods, chordates, and a few other groups is **segmentation,** which is also called **metamerism** (see pp. 610–612). Segmented animals develop from a lengthwise series of segments or metameres. Early in development each segment is identical to the others. Subsequently, however, the activation of different combinations of genes in each segment causes the segment to form quite different structures. For example, the anterior segment of insect embryos will form antennae, while segments farther back will form legs. The antennae, legs, and other structures that have similar embryological origins are said to be homologous. In chordates segmentation is apparent in the somites, which give rise to homologous groups of muscles and bones. In vertebrates, for example, the vertebrae are homologous with each other.

Some insight into the genetic control of pattern formation in segmented animals is being provided by the study of certain mutations in *Drosophila*. One type of mutation affecting segmentation is the **segmentation mutation,** which affects the number or pattern of body segments. One example of a segmentation mutation is *fushi tarazu* (*ftz*), which in Japanese means "not enough segments." Fruit flies homozygous for the *ftz* allele die early in the larval stage with only half the normal number of segments. Other segmentation mutations result in the skipping of every odd- or every even-numbered segment, the development of an abdomen at both ends, or the elimination of groups of segments.

Another type of mutation affecting segmentation is the **homeotic mutation.** A homeotic mutation changes the identity of a segment, causing its appendages to develop into the homologous appendages of a different segment. For example, the homeotic mutation called *Antennapedia* results in fruit flies with tiny legs growing out of their heads where antennae would normally be. *Bithorax* mutations cause fruit flies to develop an extra pair of wings on the posterior thoracic segment, where peglike structures would normally occur (Figure 6.26). One of the intriguing discoveries about homeotic and segmentation genes in *Drosophila* is that most of them (30 according to a recent count) include an identical sequence of 180 base pairs. This sequence is called the **homeo box.** Even more exciting, the same

homeo box occurs in most animals, including not only insects but earthworms, sea urchins, frogs, chickens, mice, and humans.

Genes control segmentation and other aspects of pattern formation by directing the synthesis of molecules that activate or repress other genes. Such activation or repression is referred to as **induction.** Induction was dramatically demonstrated in 1924 by Hans Spemann and Hilde Mangold. Spemann and Mangold took the dorsal blastopore lip (Figure 6.17A) of one newt and grafted it into the gastrula of another newt in an area that was fated to become the belly. The transplanted dorsal blastopore lip continued to gastrulate and induced the host embryo to develop a second larva joined at the belly as a Siamese twin (Figure 6.27). No other part of the gastrula has this ability to induce development of a second larva. Spemann also demonstrated induction by removing the optic cup of a salamander embryo, which develops into the back of the eye, and inserting it into ectoderm that was fated to become the belly. The transplanted optic cup induced the ectoderm to develop the lens of an eyeball on the belly! Recent demonstrations of induction have been just as bizarre. If certain tissue from the jaw of a chick embryo develops in the presence of tissue from the jaw of a mouse embryo, the jaw of the chick develops teeth (Kollar and Fisher 1980). Teeth in chicks are normally as scarce as . . . well, hen's teeth, but the genes for teeth formation are evidently still present in birds as a legacy from their reptilian forebears. Induction is thought to be due to the release of substances called **morphogens.** Retinoids, which include vitamin A and retinoic acid, are most likely morphogens.

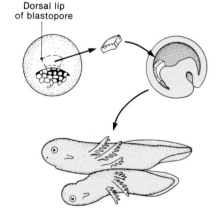

Dorsal lip
of blastopore

Figure 6.27
Spemann's and Mangold's demonstration of induction. The dorsal blastopore lip of the newt Triturus cristatus *was transplanted to the gastrula of* T. taeniatus *in the area fated to become the belly. The fact that* T. taeniatus *is pigmented, while* T. cristatus *is transparent, enabled Spemann and Mangold to determine whether later tissues came from the transplant or the host. The transplant induced the development of Siamese twin newts formed mainly of pigmented host tissue.*

Mechanisms of Morphogenesis. Once the pattern of an animal has been formed, the processes of morphogenesis direct particular cells to move to appropriate places in accordance with that pattern. These cellular movements are due to **extension, adhesion,** and **contraction** of cells. The migration of micromeres during gastrulation in sea urchins (Figure 6.15D) and the subsequent movement of the primary mesenchyme illustrate these processes. First the cells extend **pseudopods** with tips that adhere to appropriate cells on the inner surface of the blastocoel. Once the pseudopods are attached they contract, pulling the rest of the cell in. Cell extension and contraction are apparently due to fibrous proteins such as actin inside the cell (see p. 44). The adhesion of appropriate cells to each other is attributed to specific glycoproteins called **cell-adhesion molecules** (CAMs), which occur on the outer surface of the plasma membranes. CAMs may also be responsible for similarly differentiated cells adhering to each other to form tissues. Other glycoproteins called **substrate-adhesion molecules** (SAMs) also help guide cell migration and adhesion.

Perhaps the most impressive feat of morphogenesis—and of all development—is the organization of the human brain. Somehow during embryonic development each of the hundreds of billions of developing nerve cells in the brain selects a few other nerve cells with which to form connections. In fact, the brain continues to be "rewired" after birth and well beyond. Superfluous connections are deleted until the adolescent has only about half as many connections as when he was eight months old. Complex as this phenomenon is, it too appears to depend on the same three processes of extension, adhesion, and contraction. Each nerve cell extends a process called a **growth cone** from which thin extensions reach out like feelers (Figure 6.28). Once these extensions contact an appropriate target cell, they adhere and contract, pulling the growth cone and causing the nerve cell to grow toward the target. Of course, this simple account hardly begins to explain how a particular nerve cell "knows" which of the hundreds of billions of other nerve cells is an appropriate target. Human brains have a great deal of research to do before they understand their own development.

Mechanisms of Growth. Later embryonic development and early postnatal life are periods of rapid growth for most animals. This growth is not random but involves the stimulation of cell division in particular types of tissue by **growth factors.** The first growth factor was discovered in the 1950s by Rita Levi-Montalcini, Viktor Hamburger, and Stanley Cohen. They found that a protein in certain tumors, the venom of snakes, and the saliva of male mice was necessary for the growth and maintenance of particular nerve cells (those derived from the neural crest). They called the substance **nerve growth factor** (NGF). Afterward came the discovery of several other growth factors, such as **epidermal growth factor** (EGF). EGF stimulates division of epidermal, neural, and some other kinds of cells and also stimulates spermatogenesis. It also occurs in milk and stimulates continued development of the infant's digestive tract.

DEVELOPMENTAL DEFECTS AND CANCER

The multitude of events required for the development of a new organism provides ample opportunity for something to go wrong. In fact, approximately one-third of all human embryos miscarry, often without the mother knowing she was pregnant. Other developmental abnormalities are not fatal to the embryo but result in defects in the child. **Developmental defects** are of two types: **genetic defects** and **congenital defects.** Genetic defects are due to point mutations or chromosomal aberrations and result from absent or incorrect gene products during meiosis or development. Down syndrome (see p. 93) is just one of many such genetic defects. Congenital defects are not inherited but result from external factors, called **teratogens,** that interfere with a normal process of development. In humans virtually any substance that can be transferred from maternal to fetal blood through the placenta is a potential teratogen. The list of known and suspected teratogens includes viruses, such as the type that causes German measles, alcohol, and numerous drugs, including aspirin.

Tumors are also a type of abnormal development since they result from inappropriate cell division. When a tumor simply divides in place it is said to be **benign** and can generally be treated by surgical removal. In certain tumors, however, cells migrate to other parts of the body, in which case the tumor is referred to as **malignant** or **cancerous.** Cancers are also treated by removing the tumor,

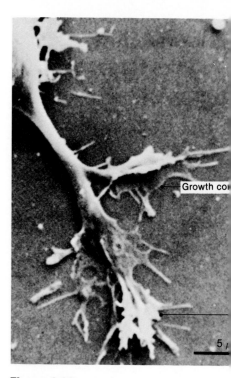

Figure 6.28
Scanning electron micrograph of a nerve cell in tissue culture searching for an appropriate target cell. The cell body containing the nucleus is at the upper left. The process running to the lower right divides and expands into two growth cones, from which thin filopodia extend.

but radiation or chemotherapy must also usually be employed to kill any cells that may have invaded other tissues. Cancers are classified into several groups, depending on the kinds of tissues they affect. **Carcinoma,** the most common type, begins in epithelium, especially in lungs, breasts, and intestines. **Sarcoma** arises from bone, cartilage, fat, muscle, or other tissues of mesodermal origin. **Lymphoma** and **leukemia** occur in solid tissues or circulating cells, respectively, that originate in lymph or blood.

Cancers generally result from activation of genes called **proto-oncogenes** and the inactivation of **suppressor genes.** Obviously it is not the normal function of these genes to induce cancer. The normal function of proto-oncogenes is to produce growth factors, receptor molecules for such factors, or other molecules involved in embryonic development. However, when proto-oncogenes become active later in life (i.e., when they become **oncogenes**), they induce cancers. Likewise, suppressor genes normally inhibit growth. The failure of these genes can induce cancer. Any mutagen, such as ionizing or solar radiation, or various chemicals **(carcinogens)** can cause cancer by activating oncogenes and inactivating suppressor genes. Generally, more than one gene must mutate, but there are dozens of each kind, and exposure to various mutagens over several decades can have a cumulative effect. Some oncogenes can also be acquired through infection by certain viruses. Oncogenic viruses have oncogenes built into their genomes, and these can become incorporated into the genomes of infected cells. One example of an oncogenic virus is the Epstein–Barr virus, which initiates Burkitt's lymphoma. Epstein–Barr and other oncogenic viruses are widespread, and there is no practical way to avoid exposure, such as by avoiding people with cancer.

METAMORPHOSIS AND REGENERATION

As if once were not impressive enough, many animals develop twice during their life cycles. They are born as **larvae,** then undergo **metamorphosis** into adults with completely different forms. Familiar examples of larvae include the tadpoles of frogs and the caterpillars of butterflies and moths. Animals that undergo such metamorphosis are said to have **indirect development.** Development directly into a form that resembles the adult is called **direct development.** In some cases larvae are essentially embryos that continue their development outside the egg, consuming food instead of yolk. The grubs and maggots of certain insects are examples. In most cases, however, the larva is a competent organism in its own right, adapted for essential functions that the adult cannot perform. In sponges, cnidarians, and many molluscs, for example, larvae are often responsible for dispersing the species to new habitats, swimming by means of cilia or drifting with currents. The adults, in contrast, are often sedentary and adapted mainly for reproduction.

Just as different parts of the egg, or different blastomeres, are fated to become different parts of the organism, different parts of the larva are fated to be reorganized into adult parts. This is clearly seen in *Drosophila.* Different parts of the maggot, called **imaginal discs,** are already set aside and determined to become various parts of the adult (Figure 6.29). Changes in the levels of hormones (especially juvenile hormone) trigger the development of imaginal discs into adult structures (see Figure 9.13). In frogs and other amphibians, increased levels of thyroid hormones trigger metamorphosis.

Like metamorphosis, **regeneration** is a kind of postnatal development of the adult form. Animals vary widely in their ability to regenerate. Certain flatworms can regenerate a new cerebral ganglion ("brain") if the original one has been removed surgically, but humans and other mammals are incapable of replacing even a single nerve cell that has died. Most insects and amphibians can regenerate

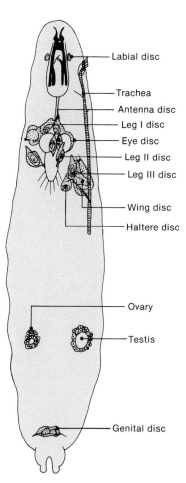

Figure 6.29
Imaginal discs in the larva of a drosophila. The term "imaginal" comes from the fact that the adult insect is called the "imago."

Labial disc
Trachea
Antenna disc
Leg I disc
Eye disc
Leg II disc
Leg III disc
Wing disc
Haltere disc
Ovary
Testis
Genital disc

a lost leg, but the best humans and other mammals can do is to regenerate much of the liver or a lost fingertip. Not only can many starfish regenerate a lost arm, but the lost arm can regenerate the rest of the starfish! One of the main reasons for studying regeneration, aside from the intrinsic interest and the potential for regenerating lost organs in humans, is that it may provide clues to developmental processes. In some respects regeneration and embryonic development are similar. Both require cells that are initially undifferentiated but that can differentiate into a variety of tissues. In regeneration an accumulation of such cells at the site of an amputation is called **blastema.** In salamanders and other tailed amphibians, which are the only vertebrates that can regenerate limbs as adults, blastema forms on the stump a week or so after amputation of a limb. Positional information then directs the organization of the blastema into a functional limb.

DEVELOPMENT AND EVOLUTION

As noted previously, one of the main motivations for the study of development is that it provides insights regarding taxonomic relationships among groups of animals. In fact, this was a primary concern of embryology even before animals were

Figure 6.30
An illustration of von Baer's discovery that the development of vertebrates proceeds from the general to the specific. The embryos in each column are from the same species, and those in each row are at comparable stages of development.

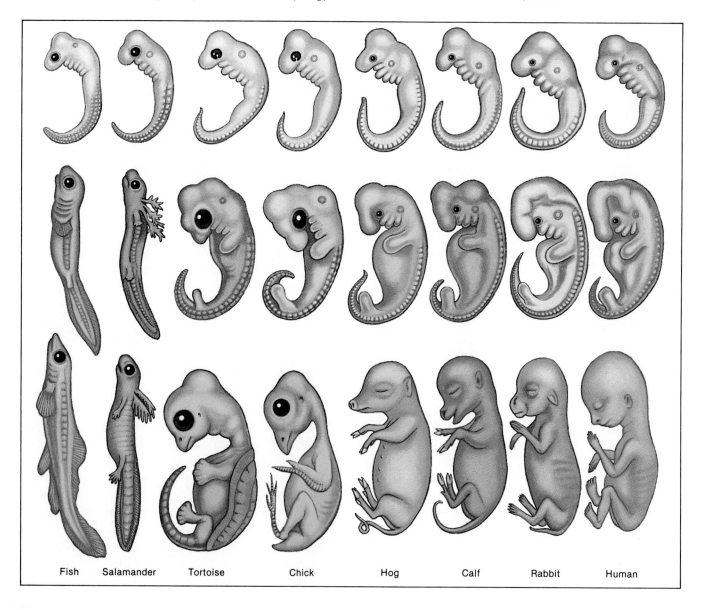

| Fish | Salamander | Tortoise | Chick | Hog | Calf | Rabbit | Human |

known to be related to each other by evolution. In the early 19th century the Estonian biologist Karl Ernst von Baer made a number of observations that suggest such a relationship, although he explicitly rejected any evolutionary implications. What is known as **von Baer's law** states that embryonic development in vertebrates goes from general forms common to all vertebrates to increasingly specialized forms characteristic of classes, orders, and lower taxonomic levels. Thus the early embryos of all vertebrates, whether fish, frog, hog, or human, all look alike (Figure 6.30). Later it is possible to tell the human embryo from the fish but not from the hog, and still later one can see a difference between the two mammals.

In 1866, inspired by the publication of Darwin's *Origin of Species*, Ernst Haeckel extended von Baer's law to include invertebrates and brought out its evolutionary implications. Unfortunately, Haeckel went too far with his **biogenetic law,** which states that "the history of the fetus is a recapitulation of the history of the race [species]." In other words, ontogeny (the development of the individual) is a recapitulation of phylogeny (the evolution of the species). The statement that ontogeny recapitulates phylogeny is patently false. Even though all vertebrates evolved from fishes, there is no stage of development in which a reptilian, avian, or mammalian embryo looks like adult fish, and certainly not like an invertebrate ancestor. Nevertheless, the biogenetic law was enthusiastically embraced by biologists for several decades. As is always the case when people discover they have been fooled, biologists were even more enthusiastic in rejecting the biogenetic law early in this century and rejected von Baer's law for good measure.

Von Baer's insight is now respectable once more (Gould 1977), although few biologists would call it a law. The explanation for von Baer's "law" is probably similar to the reason why no one tries to design an automobile from the ground up. It is much easier to start with a successful design and tinker with it, even if it means retaining obsolete features. Changing the basic design of the chassis would have serious and unforeseen consequences for the engine, transmission, and suspension that may lead to the car's failure in the marketplace. Early changes in development may likewise have consequences that lead to the failure of the organism in its natural environment. Thus human embryos develop pharyngeal slits, hair, and a tail perhaps because those that have the genetic mutations needed to omit those steps die from other consequences.

The study of developmental biology also provides other insights regarding evolution. One of these is the realization that evolution can result not only from mutations in the genes that are directly responsible for structures but also in the genes that regulate the activity of genes during development. Mutations such as those described above that affect segmentation can lead to radical changes in structure, some of which might be beneficial. It is not difficult to imagine, for example, that the *Bithorax* mutant could give rise to an entirely new group of four-winged flies.

Figure 6.31
Neoteny in salamanders. (A) Larvae of salamanders, such as the grotto salamander Typhlotriton spelaeus, *are generally aquatic and have external gills, which they lose after metamorphosis to adults. (B) In neotenic species, such as the Texas salamander* Eurycea neotenes, *the adult retains the external gills. Neoteny in some salamanders is due to the absence of a rise in thyroid hormones that normally triggers metamorphosis. Injecting thyroid hormones into larvae that are normally neotenic will cause them to lose their external gills.*

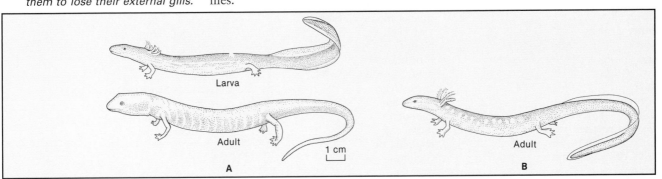

Larva

Adult

1 cm

A

Adult

B

Mutations in genes that affect development could also give rise to evolution by changing the timing of developmental events. Differences in developmental timing are, in fact, well known in some groups of animals and are referred to as **heterochrony.** Heterochrony causes certain organisms to develop certain features either earlier or later in life than they would ordinarily. One kind of heterochrony is **neoteny,** in which most features are retarded in development, so that the adult resembles a juvenile of a related species. Another kind is **progenesis,** which is the accelerated development of adult features in a juvenile. Either neoteny or progenesis can result in **paedomorphosis,** in which features usually characteristic of adults occur in an animal with a juvenile form. [The definitions of these and related terms have become confused, but see Gould (1977) for clarification.] Neoteny can clearly be seen in adult animals that look like the larvae of related species (Figures 6.31 and 36.18). It seems likely that neotenic species evolved as a result of mutations that affected the timing of developmental events. Some zoologists think humans evolved this way—that we are essentially neotenic apes.

SUMMARY

Animals undergo several stages of embryonic development, during which each cell becomes increasingly specialized as the number of genes expressed decreases. During gametogenesis, primordial germ cells undergo repeated mitosis, producing genetically identical diploid spermatocytes or primary oocytes. These then undergo meiosis, becoming genetically diverse haploid sperm or ova. During fertilization, a sperm cell propels itself toward an ovum, binds to receptor molecules specific for its species, then digests a pathway into the ovum. This triggers a cortical reaction that helps prevent polyspermy. Inside the ovum the haploid male pronucleus and the haploid female pronucleus combine to form a diploid zygote. The zygote then begins to divide repeatedly.

Repeated division, called cleavage, results in a blastula, which consists of blastomeres with a total mass no larger than that of the zygote. The cells follow specific cleavage patterns, depending on the amount and distribution of yolk and on the kind of animal. In some species development is determinate, with the fate of each blastomere already determined.

Beginning with gastrulation, the cells become increasingly specialized in gene expression, and they begin to move about.

This process, called morphogenesis, lays down the basic body pattern. During gastrulation, in most animals three germ layers are established: ectoderm, which will form nerve and skin; endoderm, which will become gut and other internal organs; and mesoderm, which will form blood, bone, connective tissue, and muscle. In mammals the formation of these germ layers is associated with the primitive streak, which occurs near the future site of the spinal cord and brain. As morphogenesis continues the cells continue to differentiate and to aggregate as different kinds of tissues. The mechanisms responsible for cell movements during morphogenesis are cellular extension, adhesion, and contraction.

Abnormal developmental processes can have severe consequences, including birth defects and cancer. Metamorphosis and regeneration in some animals appear to be related to normal patterns of development. Developmental patterns appear to have changed gradually during evolutionary history, providing a useful tool for inferring the evolutionary history of animals. Some changes, such as heterochrony, may explain some evolutionary changes.

KEY TERMS

germ cell
gametogenesis
sperm
egg
fertilization
embryo
cleavage
blastula
cleavage pattern
spiral cleavage pattern
radial cleavage pattern
determinate cleavage

indeterminate cleavage
gastrulation
blastopore
protostome
deuterostome
germ layer
ectoderm
endoderm
mesoderm
triploblastic
coelom
amnion

yolk
differentiation
tissue
segmentation
induction
growth factor
metamorphosis
regeneration
heterochrony

1. Explain the implications of the experiment by Gurdon, in which he transplanted the nuclei from intestinal cells of tadpoles into frog eggs.

2. Diagram the processes of spermatogenesis and oogenesis. Which stages are diploid, and which are haploid? What are polar bodies?

3. Eggs must encourage fertilization by sperm of the same species, while preventing polyspermy and fertilization by a different species. Explain how these conflicting requirements are achieved.

4. Briefly describe what occurs during cleavage, gastrulation, and morphogenesis.

5. How does the distribution of yolk affect cleavage?

6. Gastrulation is said to lay down the basic body plan of an animal. Explain this statement for a sea urchin, a frog, and a human.

7. Name the three germ layers. Name three structures in your body that originate from each germ layer.

8. Human embryonic development is apparently inefficient in many ways. For example, a tail is formed only to be eliminated, and there is a yolk sac but no yolk. How do most zoologists explain the occurrence of these structures?

9. Describe one example illustrating how advances in the understanding of cellular, genetic, and molecular biology have contributed to our understanding of the mechanism of development.

10. Cancer is said to be a defect of development. In what sense is this true?

11. Regeneration is also often compared with development. In what ways are they similar? How do they differ?

READINGS

RECOMMENDED READINGS

Beaconsfield, P., G. Birdwood, and R. Beaconsfield. 1980. The placenta. *Sci. Am.* 243(2):94–102 (Aug).

Beams, H. W. and R. G. Kessel. 1976. Cytokinesis: a comparative study of cytoplasmic division in animal cells. *Am. Sci.* 64:279–290.

De Robertis, E. M., G. Oliver, and C. V. E. Wright. 1990. Homeobox genes and the vertebrate body plan. *Sci. Am.* 263(1):46–52 (July).

Edelman, G. M. 1984. Cell-adhesion molecules: a molecular basis for animal form. *Sci. Am.* 250(4):118–129 (Apr).

Edelman, G. M. 1989. Topobiology. *Sci. Am.* 260(5):76–88 (May).

Epel, D. 1977. The program of fertilization. *Sci. Am.* 237(5):128–138 (Nov).

Gehring, W. J. 1985. The molecular basis of development. *Sci. Am.* 253(4):152–162 (Oct). (*The homeo box in drosophila.*)

Gerhart, J. et al. 1986. Amphibian early development. *BioScience* 36:541–549.

Goodman, C. S. and M. J. Bastiani. 1984. How embryonic nerve cells recognize one another. *Sci. Am.* 251(6):58–66 (Dec).

Hall, B. K. 1988. The embryonic development of bone. *Am. Sci.* 76:174–181.

Hynes, R. O. 1986. Fibronectins. *Sci. Am.* 254(6):42–51 (June).

Kline, D. 1991. Activation of the egg by the sperm. *Bioscience* 41:89–95.

Levi-Montalcini, R. and P. Calissano. 1979. The nerve-growth factor. *Sci. Am.* 240(6):68–77 (June).

Nilsson, L. 1977. *A Child Is Born*. New York: Delacorte Press. (*Extraordinary photographs of human development.*)

Sachs, L. 1986. Growth, differentiation and the reversal of malignancy. *Sci. Am.* 254(1):40–47 (Jan).

Stent, G. S. and D. A. Weisblat. 1982. The development of a simple nervous system. *Sci. Am.* 246(1):136–146 (Jan). (*The leech.*)

Wassarman, P. M. 1988. Fertilization in mammals. *Sci. Am.* 259(6):78–84 (Dec).

Weinberg, R. A. 1988. Finding the anti-oncogene. *Sci. Am.* 259(3):44–51 (Sept).

ADDITIONAL REFERENCES

American Zoologist 22:1–220 (1982) includes several papers on pattern formation, and *American Zoologist* 27:411–723 (1987) and 29:481–670 (1989) include several papers on development in general.

Bryant, S. V., D. M. Gardiner, and K. Muneoka. 1987. Limb development and regeneration. *Am. Zool.* 27:675–696.

England, M. A. 1983. *Color Atlas of Life before Birth*. Chicago: Yearbook Medical Publishers. (*Detailed color photographs of human development.*)

Foe, V. E. and G. M. Odell. 1989. Mitotic domains partition fly embryos, reflecting early cell biological consequences of determination in progress. *Am. Zool.* 29:617–652.

Gilbert, S. F. 1988. *Developmental Biology*, 2nd ed. Sunderland, MA: Sinauer Associates.

Gilbert, S. G. 1989. *Pictorial Human Embryology*. Seattle: University of Washington Press.

Gould, S. J. 1977. *Ontogeny and Phylogeny*. Cambridge, MA: Harvard University Press.

Kollar, E. J. and C. Fisher. 1980. Tooth induction in chick epithelium: expression of quiescent genes for enamel synthesis. *Science* 207:993–995.

Moore, J. A. 1987. Science as a way of knowing—developmental biology. *Am. Zool.* 27:415–573. (*Key concepts of development in historical context.*)

Mulnard, J. G. 1967. Analyse microcinématographique du développement de l'oeuf de souris du stade II au blastocyste. *Arch. Biol.* (*Liege*) 78:107–138.

Nüsslein-Volhard, C., H. G. Frohnhöfer, and R. Lehmann. 1987. Determination of anteroposterior polarity in *Drosophila*. *Science* 238:1675–1681.

O'Rahilly, R. and F. Müller. 1987. *Developmental Stages in Human Embryos*. Washington, DC: Carnegie Institution of Washington.

Tickle, C. 1981. Limb regeneration. *Am. Sci.* 69:639–646.

von Ehrenstein, G. and E. Schierenberg. 1980. Cell lineages and development of *Caenorhabditis elegans* and other nematodes. In: B. M. Zuckerman (Ed.), *Nematodes as Biological Models*, Vol. 1. New York: Academic Press, pp. 1–71.

Unit Two: Maintaining the Cellular Environment

These musk oxen live year-round in one of the most hostile environments of North America, yet their cells are constantly maintained in a nearly optimal internal environment. Body temperature stays around 38°C even when the air temperature is less than 40°C below freezing, thanks largely to qiviut (KEE-vee-ut), the fine undercoat of fur. After the introduction of firearms into the North, musk oxen were hunted almost to extinction, but now they support an important wool industry.

Musk oxen also maintain suitable levels of water, nutrients, and oxygen in their internal environments. Their cells always have sufficient water even though the external environment is as dry as the deserts of southwest America. The light snowfall does have the advantage of not covering the scattered tufts of grass, so they can provide their cells with sufficient nutrients. Their ability to convert grass into nutrients is impressive: an average male musk ox weighs almost as much as a Guernsey bull, but it requires only about one-sixth as much food.

Maintaining the correct internal levels of temperature, water, nutrients, and oxygen in any animal requires the coordinated functioning of excretory, digestive, and respiratory systems. These systems, in turn, require the neural, muscular, and hormonal systems to coordinate them. We shall examine these coordinating systems first in the following chapters. Next will be the circulatory system, which transports hormones, heat, water, nutrients, and oxygen to the internal environment of cells. Finally, we shall examine reproductive systems, which enable species to survive even after individuals inevitably succumb to the harshness of the external environment.

Overview

Nerve Cells

Squid (Loligo).

LEARNING OBJECTIVES

1. What kinds of signals enable parts of the body to communicate rapidly with each other?

2. How do different concentrations of ions produce a voltage across a membrane?

3. How do changes in the permeability to ions generate action potentials?

4. How do action potentials travel along a nerve cell membrane?

5. How do action potentials transmit messages to nerve and muscle cells?

6. How do animals learn and remember?

7. What causes mental disorders?

8. How do animals see, feel, and otherwise perceive their environments?

THE NATURE OF NEURAL SIGNALS

As you read these words you may imagine that they are somehow entering into your brain, speaking almost audibly to an inner self. A moment's thought tells you, however, that the words stop at the retinas of your eyes, and that there must be some way the retinas convert their images into a form that can be used within your dark and silent brain. Millions of your nerve cells are in fact translating these words into an electrical and chemical code. Understanding how nervous systems handle this task and others is the ultimate challenge of science, for science itself results from such processes.

Until a few centuries ago most people believed that understanding the brain was safely beyond the ability of humans, because our thoughts, sensations, and behavior were too miraculous to comprehend. This belief began to fade in the 17th century, thanks largely to the French philosopher-mathematician-scientist René Descartes (pronounced day-cart). Descartes argued that all animals were machines, and that humans were animals with souls. It followed, therefore, that humans—even their thoughts, sensations, and behavior—could be understood like any other machine. Descartes suggested, for example, that nerves activated muscles by squirting vapor into them. (Perhaps he had been inspired by the steam-powered, lifelike robots which were then the rage of Paris.) Descartes' explanation of how the brain controls behavior was wrong, of course, but it opened the way to studying the mechanisms of the brain without encroaching on religion.

More recently, understanding the mechanisms of the brain has seemed even closer, thanks to new techniques for understanding the electrical and chemical signals used by nerve cells. Nerve cells occur in almost every animal, and in every case they function by transmitting electrical signals along the plasma membranes that enclose them (Figure 7.1). These membrane signals carry information from sensory receptors into the central nervous system (the brain and spinal cord in vertebrates), and send commands out of the central nervous system to muscles and glands.

MEMBRANE POTENTIALS

Neural signals result from changes in the voltage across the plasma membranes, which is called the membrane potential. A membrane potential occurs because a plasma membrane is not equally permeable to all ions, and because ions do not have equal concentrations on each side of the membrane. The impermeant ions (those to which the membrane is not permeable) include proteins and amino acids. Since most of these molecules are negatively charged at the pH usually found in cells and body fluids, they are anions and are represented by A^-. There are more of these impermeant anions inside the cell than outside. A second important ion that is usually impermeant is sodium (Na^+), which has a higher concentration outside the cell than inside because of active transport by the sodium–potassium pumps (see p. 41).

In contrast to Na^+ and A^-, potassium (K^+) and chloride (Cl^-) ions ordinarily move freely through channels in membranes (see p. 40). You might expect that because of diffusion there would be equal concentrations of K^+ and Cl^- on each side of the membrane. A^- attracts K^+ into the cell, however, and Na^+ attracts Cl^- outside (Figure 7.2). The unequal distribution of K^+ and Cl^- satisfies a fundamental physical requirement: The net charge inside a conductor must be zero. Thus the positive charge (mainly K^+) inside a cell equals the negative charge (mainly A^-). In the interstitial fluid, the negative charge (mainly Cl^-) equals the positive charge (mainly Na^+).

Understanding the activities of nerve cells requires an understanding of some fundamental electrical concepts. **Voltage,** also called **potential** and **electromotive force,** is a force that tends to move electrons, ions, and other charged matter. **Charge** is a fundamental property, like mass, length, and time. Just as gravity tends to move mass, voltage tends to move charge. Unlike gravity and mass, however, voltages and charges come in pairs: positive and negative. Positive voltage attracts electrons and negatively charged ions (anions). Negative voltage attracts positively charged ions (cations).

The amount of charge moved by a voltage in a unit of time is the **current.** Increasing the voltage increases the current. The movement of charge due to a voltage can be opposed by electrical **resistance,** which is analogous to friction. The greater the electrical resistance of a substance, the less current there will be through it when a particular voltage is applied.

A current through a resistance will generate a voltage across the resistance, with the magnitude of the voltage proportional to both the current and the resistance. If the current consists of cations, such as Na^+ or K^+, moving through channels in the plasma membrane, the voltage produced across the membrane will be negative on the surface where the ions enter and positive on the surface where they exit. Voltage can also be produced by the separation of positive charges from negative charges by an object. The voltage produced is directly proportional to the amount of charge and inversely proportional to a property of the object called **capacitance.** The voltage across the capacitance is negative where the negative charges accumulate.

Figure 7.1
(A) Nerve cells in the brain. Numerous processes called dendrites spread from the cell bodies and receive signals from other nerve cells. A single long process called an axon runs from each cell body and branches out to contact the dendrites of other nerve cells. (B) The structure of a representative nerve cell.

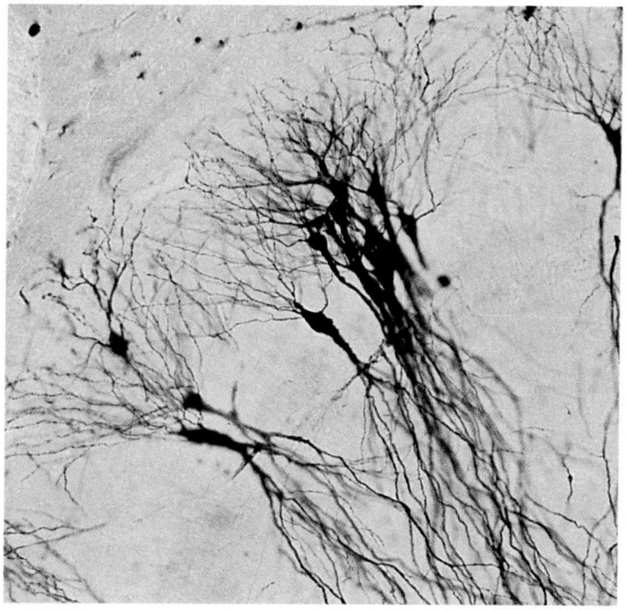

A

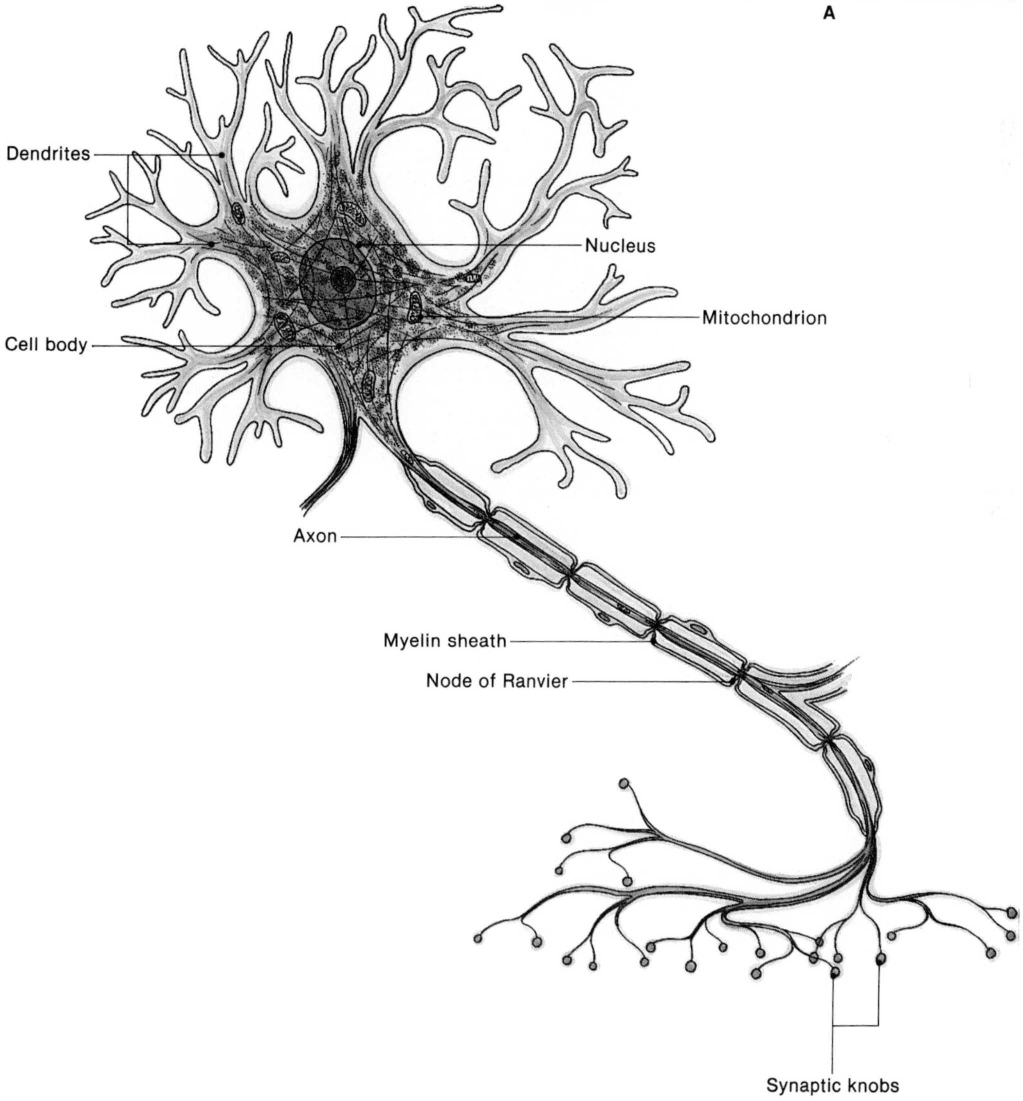

Dendrites

Nucleus

Mitochondrion

Cell body

Axon

Myelin sheath

Node of Ranvier

Synaptic knobs

B

Because they are permeant and in unequal concentrations on each side of the membrane, K^+ tends to diffuse out of the cell, and Cl^- tends to diffuse into the cell. The slight diffusion of K^+ out and Cl^- in produces a charge separation across the plasma membrane, with more negative charge on the inner surface of the membrane and more positive charge on the outer surface. The diffusion of these two ions continues until the voltage produced by the charge separation exactly counterbalances the tendency of the ions to diffuse further. The exact balance of the voltage and diffusion leads to an **equilibrium** in which there is no net diffusion. At equilibrium, the negative voltage inside the cell holds the K^+ in and keeps the Cl^- out. This voltage is called the **resting membrane potential.** The resting membrane potential is generally from 50 to 70 millivolts (0.05 to 0.07 volts), with the inside negative.

Neural signals result from changing the membrane permeability to one or more ions, chiefly potassium and sodium. Chloride is at equilibrium with the resting membrane potential, so increasing the permeability of the membrane to Cl^- (P_{Cl}) will not affect the rate of diffusion. It will, in fact, tend to keep the membrane potential at its resting value. In contrast, potassium is somewhat out of equilibrium, because the sodium-potassium pumps transport more K^+ into a cell than can be held there by the resting membrane potential alone. As a result, increasing the permeability of the membrane to K^+ (P_K) will allow K^+ to diffuse out more readily. The increased movement of K^+ will produce additional negative voltage across the membrane. The resulting increase in the polarization of the membrane is called **hyperpolarization.** Sodium concentrations are far out of equilibrium with the resting membrane potential, because of active transport. This means that an increase in the permeability to Na^+ (P_{Na}) above its normally low value will allow much more Na^+ to enter the cell, by both diffusion and attraction to the negative voltage. The voltage generated by the inward movement of Na^+ will be

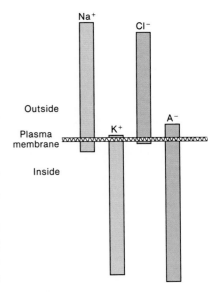

Figure 7.2
Relative concentrations of ions and charged molecules inside and outside a typical axon. A^- represents impermeant anions. Positively and negatively charged ions are shown in different colors. Note that the concentration of positively charged ions equals the concentration of negatively charged ions on each side of the membrane.

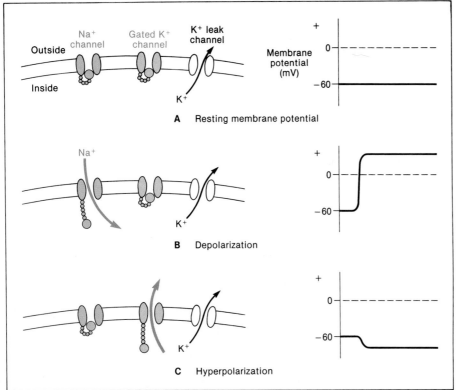

Figure 7.3
Schematic representation of how changes in permeability to Na^+ and K^+ change membrane potentials. Changes in ion permeability are due to opening and closing of ion channels, represented here according to the "ball-and-chain" model of ion channels. (A) At rest sodium channels are closed, so P_{Na} is negligible. P_K is much larger, allowing some of the K^+ that is pumped in to leak out. (B) Opening sodium channels allows a small amount of Na^+ to enter the cell and depolarizes the membrane. (C) Opening the gated potassium channels, which increases P_K, allows more K^+ to diffuse out and hyperpolarizes the membrane.

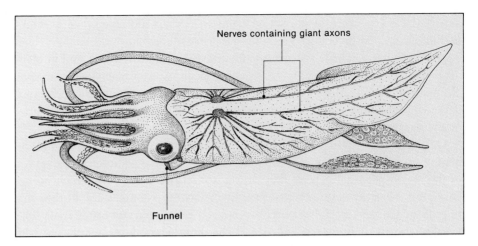

Figure 7.4
The squid Loligo pealei, *showing the location of giant axons. Giant axons trigger jet-propelled escape from predators by causing the body to contract and force water out the movable funnel.*

Nerves containing giant axons

Funnel

positive on the inside, and will reduce the magnitude of the membrane potential, or even reverse it to positive on the inside. Such a decrease or reversal of membrane potential is called **depolarization.** All electrical signals in nerve cells are depolarizations and hyperpolarizations resulting from changes in permeability (Figure 7.3).

ACTION POTENTIALS

The type of neuronal signal responsible for carrying information over long distances is the **action potential.** An action potential is a brief change in membrane potential that travels along the plasma membranes of certain nerve and muscle cells. Cells capable of producing action potentials are said to be **excitable cells.** In nerve cells, action potentials generally occur only on the plasma membranes surrounding the long processes called **axons** (Figure 7.1B). Our understanding of the action potential is due largely to studies on **giant axons** of the squid *Loligo* (Figure 7.4). A. L. Hodgkin, A. F. Huxley, and others who pioneered these studies just before and after World War II chose these axons because they are up to half a millimeter in diameter. The size permitted the insertion of fine wires into the axon to measure directly the voltage across the membrane. (Nowadays the same thing can be done with much smaller axons, by penetrating the membrane with **microelectrodes** having tip diameters of less than a micrometer.)

Hodgkin and Huxley found that when a giant axon is stimulated with a brief, depolarizing shock, an action potential occurs across the membrane. At the start of the action potential the membrane rapidly depolarizes. In fact, there is an overshoot, with the voltage inside the axon becoming positive. The membrane potential then returns to inside-negative and hyperpolarizes briefly. The entire action potential lasts only about 3 milliseconds (Figure 7.5). Hodgkin, Huxley, and others eventually learned that the depolarization, recovery, and hyperpolarization of the action potential are due to changes in the permeability of the membrane to sodium and potassium ions. These permeability changes occur in excitable cells because the channels for Na$^+$ and K$^+$ open and close in response to changes in the membrane potential.

This is how the changes in permeability produce an action potential: First the stimulus depolarizes the membrane slightly. This depolarization opens the Na$^+$ channels somewhat, increasing P_{Na}. The resulting sodium current depolarizes the membrane further. This depolarization opens the sodium channels further, which increases P_{Na} more, and so on. This sequence accounts for the rapid depolarization

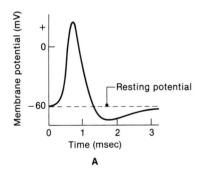

A

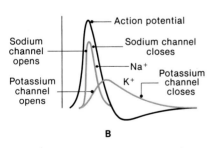

B

Figure 7.5
(A) An action potential from an isolated squid giant axon. (B) The changes in permeability to sodium and potassium that cause the action potential.

during the rising phase of the action potential. When the membrane potential becomes inside-positive, it triggers a separate change that closes the sodium channels. This helps terminate the action potential. In squid giant axons, and in many other axons, the depolarization also slowly opens voltage-dependent potassium channels, increasing P_K. The resulting K^+ current helps bring the membrane potential back to the resting level. In fact, it causes the membrane to hyperpolarize momentarily.

Once the action potential is triggered the changes in permeability and membrane potentials play themselves out automatically. The magnitude and duration of the action potential depend only on the concentration of ions, which do not change in a healthy animal. Consequently, the action potentials are all the same in each animal, and they either occur completely or they do not occur at all. This is called the **all or none law.** As long as the stimulus produces a sufficient triggering depolarization, an action potential will occur. In other words, there exists a **threshold** for an action potential. As long as a stimulus depolarizes the membrane beyond threshold, the size of the action potential will not increase with the size of the stimulus, any more than the bang from a stick of dynamite will increase if you light it with a blowtorch rather than a match. The threshold exists because a below-threshold stimulus increases P_{Na} too little to overcome the increase in P_K.

The energy that drives the action potential does not come from the stimulus, but from the high external concentration of sodium that was previously produced by the sodium–potassium pumps. The numbers of ions moving across the membrane during an action potential are relatively minuscule. For a squid giant axon with a radius of 0.2 mm and a length of 8 cm, only 10^{-12} mole of Na^+ enters per action potential. This is only 1/10,000 as much Na^+ as is already in the cell and requires only 3×10^{-13} mole of ATP to pump out.

Although the mechanisms for the action potential were discovered in the squid giant axon, the same principles have been found to apply to other kinds of excitable cells. There are some notable differences in detail, however. In many axons of mammals and other vertebrates, for example, voltage-dependent potassium channels are not involved in the recovery of the resting potential, so there is no hyperpolarization at the end (Waxman and Ritchie 1985).

Anesthetics and Neurotoxins. The above details enable neurophysiologists to understand the effects of many substances that have long been used in medicine. For example, benzocaine, xylocaine, and cocaine, which are **local anesthetics** when applied directly to a nerve, block action potentials by preventing P_{Na} from increasing. **General anesthetics,** on the other hand, depress neural activity in the brain by increasing P_K, which hyperpolarizes the membrane farther below threshold.

Many of the details of neuronal membranes have been worked out by a kind of chemical dissection, using chemicals produced by other animals. One especially useful **neurotoxin** is tetrodotoxin, which is produced by symbiotic bacteria in a poison gland of the puffer fish *Spheroides.* Tetrodotoxin (TTX) plugs the sodium channels, thereby preventing action potentials. Eating puffer fish without removing the ovaries, liver, and intestines often causes death by paralysis. (In spite of that, or perhaps because of it, puffer fish is a delicacy in Japan, where it is served as *fugu.* Even though chefs must be licensed to prepare *fugu,* approximately 100 people die each year from eating it.) Not surprisingly, a substance as useful as TTX has also evolved in other fishes, as well as in certain salamanders, newts, and the blue-ringed octopus *Hapalochlaena.* A similar substance, saxitoxin, is produced by the protozoans responsible for "red tides." Saxitoxin often kills fish that eat the protozoans and sometimes causes "paralytic shellfish poisoning" in people who eat molluscs that have fed on the protozoa.

Batrachotoxin, from the poison-dart frog *Phyllobates* of South America, exerts its fatal action in just the opposite way, by holding the sodium channels open. The toxin from the American scorpion *Centruroides* makes the sodium channels abnormally responsive to depolarization. Toxins in the venom of the North African scorpion *Leiurus* and from the sea anemone *Anemonia* apparently slow the closing of the sodium channels at the end of the action potential. Such toxins normally help the animals in predation and defense but are also useful to neurophysiologists.

CONDUCTION

Action potentials would be useless unless they went from one place to another in the nervous system. Axons conduct action potentials automatically, because an action potential at one place on a membrane is a large enough depolarization to excite adjacent areas of membrane. Thus one action potential triggers another farther along the membrane, and that triggers another, and so on. Consequently, an action potential starting at one end of an axon appears to travel to the other end. In fact, it is not the same action potential that is traveling, but a series of continually regenerating action potentials. Usually an action potential begins at the end of the axon near the cell body and travels in one direction. The hyperpolarization at the end of the action potential keeps the membrane from reexciting itself, so the action potential does not backtrack.

Most axons of invertebrates conduct at speeds of less than 1 meter per second (m/sec). At that slow rate it might take a large animal several seconds to find out that it is being eaten by a predator. The speed of conduction increases with the diameter of the axon, however, and many invertebrates have large-diameter **giant axons** that trigger rapid escape responses from potential predators. Besides the squid, the earthworm and other annelids, cockroaches and some other insects, and many fishes have giant axons. Squid giant axons have the largest diameter and the fastest conduction speed—approximately 25 m/sec.

MYELIN

Even greater conduction speeds occur in many vertebrate axons with small diameters, thanks to myelin. Myelin consists of sheaths of membrane that partially surround the axon. One type of myelin occurs in nerves outside the brain and spinal cord. This myelin in peripheral nerves is due to **Schwann cells** that line up along the axons during embryonic development, then spiral around the axons, wrapping them in layers of membrane. Spaces, called **nodes of Ranvier** (RAHN-vee-ay), remain between each deposit of myelin at intervals of about 1 mm (Figures 7.1B and 7.6). The myelin surrounding axons in the brain and spinal cord is similar in structure but is due to different cells, the **oligodendroglia.**

In both types of myelin, action potentials can occur only at the nodes of Ranvier, where the axon membrane is exposed. Instead of conducting through the plasma membrane, the voltage from an action potential at one node passes through the interstitial fluid to the next node, where it evokes an action potential. That action potential then excites the next node, and so on. This type of conduction is called **saltatory conduction,** from the Latin word *saltare*, meaning "to leap." The reason saltatory conduction is so much faster is that current passes more rapidly through the interstitial fluid than an action potential conducts along membrane. Consequently, myelinated axons conduct at speeds up to 100 m/sec, even though

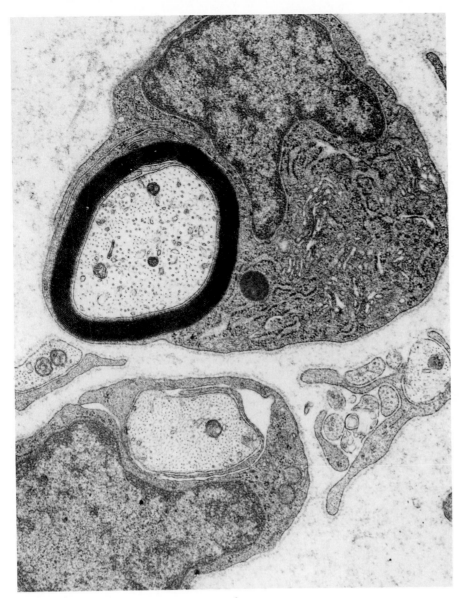

Figure 7.6
Electron micrographs of myelin. Compare Figure 7.1B. (A) Two Schwann cells from a nerve of the laboratory rat are shown in cross section forming myelin. The Schwann cell in the lower left has just begun encircling the axon with layers of membrane, while the Schwann cell in the upper right has already laid down many darkly staining layers of membrane. (B) A single node of Ranvier in a longitudinal section of an axon.

A

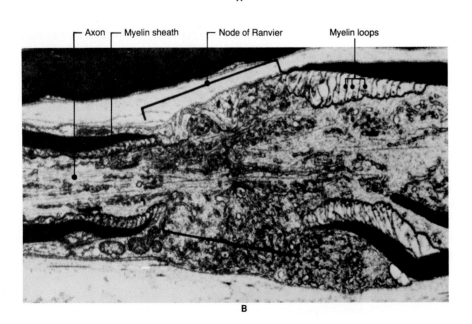

Axon   Myelin sheath   Node of Ranvier   Myelin loops

B

they generally have smaller diameters than unmyelinated axons. Another advantage of myelination is that it conserves energy. Na^+ does not have to be pumped back out all along the axon after each action potential, but only at the nodes.

SYNAPTIC CONNECTIONS

It would not matter how far or how fast action potentials conducted if there were nothing for them to do when they arrived at the end of an axon. What they have to do is to evoke some response in another cell, usually either a muscle cell or another neuron. The structures by which nerve cells evoke such responses are **synapses.** It would be reasonable to guess that synapses simply transmit action potentials from one cell to another, but that is correct only for a minority of synapses—the **electrical synapses.** Electrical synapses occur at gap junctions (see Figure 2.23C), often in rapidly conducting pathways such as the giant fibers that trigger escape in crayfish. The advantage of electrical synapses is that they conduct action potentials very rapidly. Their disadvantage is that rapid conduction of action potentials is all they can do.

Most synapses are **chemical synapses,** which can either evoke or inhibit action potentials in other nerve and muscle cells, to varying degrees. A chemical synapse works by releasing one or more substances called **synaptic transmitters** (= neurotransmitters) onto the membrane of another cell. Some of the approximately 60 known transmitters will be noted later. Synaptic transmitters are packaged within spherical **vesicles** and are released by exocytosis (see p. 42; Figure 7.7). Often the vesicles are located in bulb-shaped endings called **synaptic knobs,** which are often located near dendrites of other nerve cells (Figure 7.1B). The membrane of the synaptic knob is said to be **presynaptic,** and the opposite membrane is **postsynaptic.** Presynaptic and postsynaptic membranes are separated by a **synaptic cleft.**

Although the synaptic cleft is only 0.02 μm wide, action potentials cannot jump it. In fact, they simply vanish at the presynaptic membrane. Before they vanish, however, they cause some of the synaptic vesicles to fuse with the presynaptic membrane. Vesicles then rupture, releasing the synaptic transmitters within them. How action potentials trigger the release of transmitter is not fully understood. Often it requires an increase in calcium permeability in the presynaptic membrane, which allows Ca^{2+} to enter the synaptic knob. After the molecules of transmitter diffuse across the cleft, they bind to specific **receptor molecules** on the postsynaptic membrane.

The binding of the transmitter to receptor molecules changes the potential of the postsynaptic membrane by changing the permeability to one or more ions. The changes in potential are called **postsynaptic potentials.** If the transmitter increases the permeability to Na^+, the postsynaptic membrane will be depolarized. A depolarizing postsynaptic potential is called an **excitatory postsynaptic potential** (EPSP), because it tends to excite the target cell to produce action potentials. If the transmitter increases the permeability to K^+, the postsynaptic membrane will be hyperpolarized. Such a hyperpolarization is called an **inhibitory postsynaptic potential** (IPSP), because it inhibits the generation of action potentials. A transmitter that increases the permeability to Cl^- can also inhibit action potentials by keeping the resting membrane potential constant, resisting the effects of EPSPs.

EPSPs and IPSPs are examples of **graded potentials.** Unlike action potentials, graded potentials occur in various sizes, and they decrease in size as they conduct along membrane. Graded potentials are useless for transferring information over

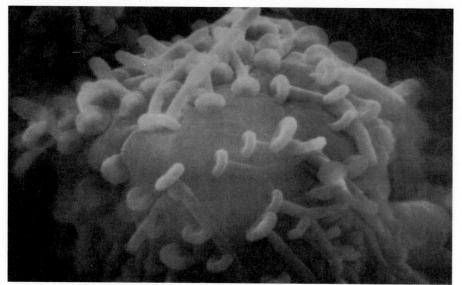

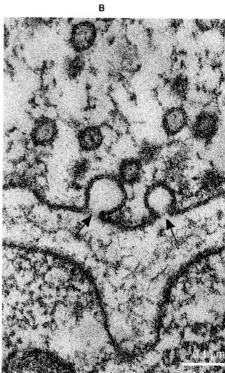

Figure 7.7
(A) Scanning electron micrograph of synaptic knobs in the sea hare Aplysia californica. (B) Transmission electron micrograph of the neuromuscular synapse of the common grass frog Rana pipiens. The nerve and muscle were quickly frozen just after stimulation and were thereby caught in the act of releasing the transmitter, acetylcholine. In this species each vesicle of 40 nm diameter contains approximately 10,000 molecules of acetylcholine. (C) Diagram of a generalized chemical synapse releasing transmitter.

B

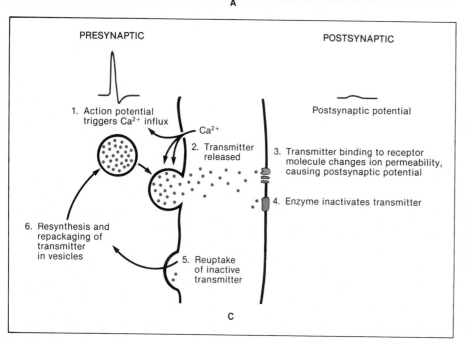

PRESYNAPTIC

POSTSYNAPTIC

Postsynaptic potential

1. Action potential triggers Ca²⁺ influx

Ca²⁺

2. Transmitter released

3. Transmitter binding to receptor molecule changes ion permeability, causing postsynaptic potential

4. Enzyme inactivates transmitter

6. Resynthesis and repackaging of transmitter in vesicles

5. Reuptake of inactive transmitter

C

long distances, but they are satisfactory in small nerve cells like those in the retina of the eye, and in transferring postsynaptic responses over a dendrite and cell body to the axon. If the EPSPs are not canceled by IPSPs, and if they are still above threshold by the time they reach the excitable portion of an axon, they can evoke action potentials.

In the meantime the transmitter must be inactivated and removed from the postsynaptic membrane and the cleft. The transmitter is usually inactivated by an enzyme, then transported back into the synaptic knob in a process called **reuptake.** The entire sequence of synaptic events occurs in about 1 msec—much less time than it takes to read about it.

Not even the Treasurer of the United States can imagine a hundred billion of anything, but if you could imagine a hundred billion nerve cells, and if you could imagine thousands of excitatory and inhibitory synapses on each one, then you would have some idea of the complexity of your brain. No computer will ever

Figure 7.8

Representation of a simple neural circuit theoretically capable of performing the following logical operation: "If A and not B then C." In this case the operation determines whether the individual will arise in the morning. Neuron A fires when the alarm clock rings and excites neuron C, which causes the individual to wake up. Neuron B fires if the individual realizes he has no morning classes; it inhibits C. It has been shown that such circuits with excitatory and inhibitory synapses are capable of handling any logical proposition that is either true or false.

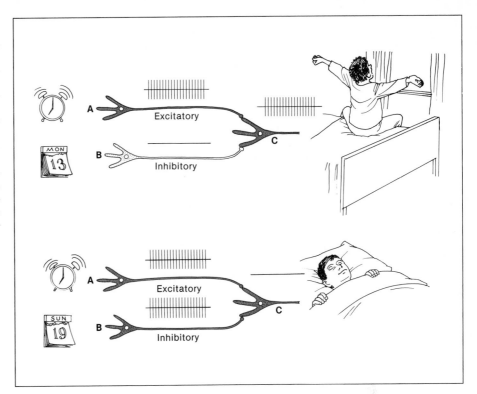

achieve such elaborate "wiring." The resulting power of nervous systems and the advantages of chemical over electrical synapses are suggested by the simple example in Figure 7.8.

LEARNING AND MEMORY

One of the most impressive things about the complex interactions of neurons is that they can be modified to allow an animal to learn how to respond to new situations. These modifications responsible for learning and memory are believed to occur at synapses, either strengthening or weakening the postsynaptic responses. Such changes could occur in a variety of ways: a neuron could grow or lose synaptic contacts, the number of vesicles released per action potential could change, the postsynaptic cell could change the number of receptor molecules, the rate of inactivation and reuptake of transmitter could change, or the permeability of either the presynaptic or postsynaptic membrane could change. Perhaps any combination of these mechanisms of learning could be found somewhere among animals. However, three frequently studied systems—the sea slugs *Aplysia* and *Hermissenda* and the part of the mammalian brain called the hippocampus—all rely on similar mechanisms (Barnes 1986). Learning in these systems apparently occurs by the following sequence of events (Figure 7.9):

1. A certain type of synapse increases the level of an intracellular **second messenger,** such as cyclic adenosine monophosphate (cAMP), and increases the intracellular concentration of Ca^{2+} (Figure 7.7).
2. The second messenger and Ca^{2+} cause a long-term activation of **kinase** enzymes. The kinases close certain ion channels, including at least one type of potassium channel. These enzymes also further increase the intracellular Ca^{2+} concentration.
3. The decrease in P_K increases the size of postsynaptic potentials.

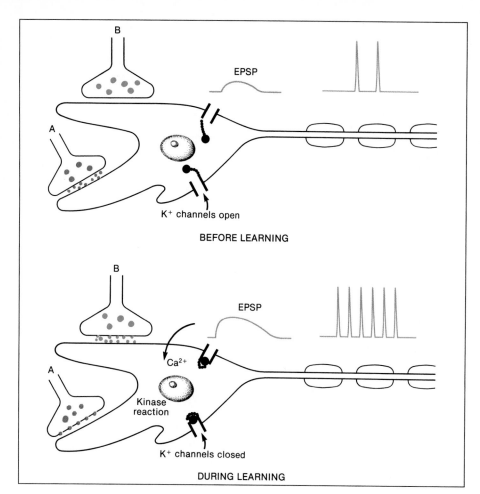

Figure 7.9
Changes in neuronal functioning during some kinds of learning, generalized from studies on molluscs and the mammalian hippocampus. Before learning, certain potassium channels are open, and the size of EPSPs evoked by a neuron (A) is small. During learning a second neuron (B) releases a transmitter that activates a second messenger system. The second messenger causes a long-term increase in kinase activity and intracellular Ca^{2+}, which close potassium channels. This increases the size of EPSPs in response to A, and therefore the response by the cell.

SYNAPTIC CHEMISTRY

The same features of chemical synapses that make them capable of learning also make them susceptible to chemical disturbances. Although most neurons are protected from substances in blood, some that enter the body through the intestine and lungs can affect them. Among these substances are many plant toxins, such as **tetrahydrocannabinol** (THC), the active ingredient in marijuana. THC appears to work in the brain by decreasing the production of cyclic AMP. Even certain nutrients in ordinary foods can affect synaptic functioning by altering amounts of synaptic transmitters. Other environmental factors, as well as genetically inherited traits, can also affect synaptic functioning. Many psychologists now believe that most behavioral disorders are essentially disorders of synaptic functioning, and most psychiatric treatment now depends heavily on modifying synaptic functioning with drugs. Table 7.1 describes some chemical transmitters, their modes of action, where they occur, and ways in which their actions are affected by drugs.

PRINCIPLES OF SENSORY RECEPTION

So far we have passed over the question of the origin of the stimuli that evoke action potentials. Obviously they do not normally come from the electrical devices Hodgkin, Huxley, and others use in their experiments. Often neural activity originates at sensory receptors, which convert energy from the environment into graded changes in membrane potential. These changes in potentials across the

Table 7.1 Synaptic transmitters and drugs that affect their functioning.

ACETYLCHOLINE (ACh). Excitatory. Released by parasympathetic nervous system (discussed later) and by synapses that control vertebrate skeletal muscle. Also common in the central nervous systems of both vertebrates and invertebrates. Alzheimer's disease, a progressive degeneration of the brain in many older people, results from death of neurons that secrete ACh.

Nicotine binds to ACh receptors and mimics the action of ACh, exciting neurons in low concentrations and blocking them at high concentrations. Nicotine was the first insecticide, but insects quickly learned to avoid the strong odor. Now its main use is to addict people to tobacco.

Curare, a substance from certain South American plants, blocks ACh receptors at synapses onto skeletal muscle, causing paralysis. The first use was as an arrow poison; now it is commonly used in surgery to keep muscles from contracting.

Diisopropylfluorophosphate (DFP) and related **organophosphorus compounds** destroy the enzyme (acetylcholinesterase) that normally inactivates ACh in the synaptic cleft. These chemicals were first developed as nerve gas for warfare against humans. They are now used in the war against insects, but they still kill and injure many farm workers.

NOREPINEPHRINE. Excitatory or inhibitory, depending on the site. A transmitter in the CNS of various animals; released by the sympathetic nervous system (discussed later). High levels of norepinephrine (and serotonin) occur in the brain during the manic phase of bipolar affective (manic–depressive) disorder, and low levels occur during the depressive phase.

Amphetamines stimulate the release of norepinephrine; prolonged use can cause psychosis by depleting the synaptic vesicles.

DOPAMINE. Excitatory or inhibitory. Secreted in CNS. High levels or high sensitivity to dopamine in the human brain is associated with psychosis. Low levels in one area of the brain occur in Parkinson's disease, characterized by tremors, loss of facial expression, and shuffling gait.

Antischizophrenic (= neuroleptic) drugs, such as Thorazine and Haldol, depress dopamine synapses. Prolonged use of these drugs induces symptoms like those of Parkinson's disease.

Cocaine inhibits the reuptake of dopamine (and norepinephrine) from cells that release it, making more available for other cells. Cocaine produces a euphoric mood in its users. A highly purified and addicting form, called "crack," can cause cardiac arrest by constricting blood vessels to the heart.

Amphetamines promote the release of dopamine (as well as norepinephrine). They also inhibit the storage of dopamine in vesicles.

L-DOPA, a metabolic precursor of dopamine, is used to treat Parkinson's disease.

SEROTONIN. Excitatory in CNS. Changes together with norepinephrine in bipolar affective (manic–depressive) disorder.

Hallucinogenic drugs often affect serotonin synapses. **Dimethyltryptamine** (DMT), for example, is chemically similar to serotonin and mimics its effects. **Lysergic acid diethylamide** (LSD) blocks receptors for serotonin.

Tryptophan, an amino acid, is a metabolic precursor to serotonin. Increasing the levels of tryptophan in the brain induces drowsiness.

ENKEPHALIN. Inhibitory, in CNS of vertebrates.

Opiates (heroin, morphine, codeine, etc.) bind to the enkephalin receptors and mimic its action, causing euphoria and blocking pain.

GAMMA AMINOBUTYRIC ACID (GABA). Inhibitory (increases P_{Cl}) in CNS of vertebrates and many invertebrates. Inhibitory transmitter on muscles of arthropods.

Valium, a tranquilizer, increases the action of GABA. **Ethyl alcohol** may exert its intoxicating effect partly in the same way as Valium, and partly by disrupting the structure of neuronal membranes.

151

membranes of sensory receptors are called **receptor potentials.** In many receptors the receptor potential is a depolarization that triggers action potentials directly in an axon that is part of the receptor cell. In other receptors the receptor potential activates a synapse that then generates action potentials. In either case, the action potentials conduct through **sensory** (= afferent) neurons into the central nervous system. This is the only known way that information about the outside real world gets into the nervous system. This book and everything else that is part of what we call reality are nothing more than action potentials, as far as your brain is concerned.

The fact that all sensations are based on action potentials presents a problem, since every action potential is identical to every other one in the nervous system. How can we distinguish among the countless different odors, sounds, colors, tastes, and other sensations using identical odorless, silent, colorless, and tasteless action potentials? The answer must be that there is a kind of sensory code based on the pattern of action potentials, like Morse code, which is based on patterns of dots and dashes. Patterns of action potentials can vary in only three ways: in frequency (number per second), in time of occurrence, and in the nerve cells in which they occur.

1. **The frequency of action potentials conveys information about the intensity of the stimulus.** The greater the stimulus intensity, the more action potentials evoked in a sensory neuron per second (Figure 7.10). The brain interprets the frequency of action potentials as intensity; for example, loudness or brightness. Some receptor cells are especially suited to providing this kind of information. They produce a constant frequency of action potentials as long as the stimulus intensity stays the same. Such receptors are said to be **tonic.**

2. **The time of occurrence of action potentials provides information about the time of occurrence of the stimulus.** This may seem obvious, but some receptors, called **phasic** receptors, are better suited than others for providing this kind of information. In contrast to tonic receptors, phasic receptors initiate action potentials only when there is a change in stimulus intensity. The beginning of a sound, for example, triggers a brief burst of action potentials from certain receptor cells in the ear, and this brief burst goes to the brain where it indicates precisely when the sound intensity increased. The system is so precise that you can tell which side a sound came from by the phasic response from one ear occurring a little sooner than the phasic response from the other ear. A phasic response to an increase in intensity is called an "on" response. Other phasic receptors produce "off" responses, triggering action potentials only when the intensity decreases. Other phasic receptors produce both "on" and "off" responses.

3. **Which neurons conduct the action potentials provides information about the modality, quality, and location of the stimulus.**

   Modality refers to the major category of sensation, whether sight, sound, smell, taste, or touch. Humans perceive any stimulus as belonging to the visual modality if the stimulus evokes action potentials in axons of the optic nerve or in the visual areas of the brain. Getting hit in the eye, for example, produces the visual sensation of "seeing stars."

   Quality refers to subcategories within each modality. Color and pitch are qualities within the modalities of vision and hearing, for example. Different receptors and axons respond to different colors and different pitches.

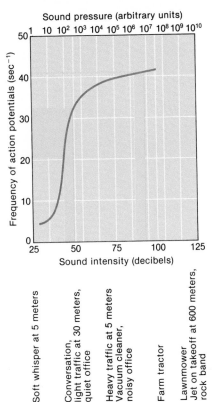

Figure 7.10
The frequency of action potentials in an auditory axon of a cat in response to brief tone bursts of different loudness. Sound intensity is given in the decibel (dB) scale, which is proportional to subjective loudness, and also as sound pressure, which reflects the actual force on the eardrum. The threshold for hearing is 0 dB. In a quiet room a person can barely detect a 1-dB sound. Every 10 dB represents a tenfold increase in sound pressure. The bars show that a sound that would produce a large frequency change at low background intensity would produce little change at high background levels. Thus, like most sensory systems, the auditory system combines extreme sensitivity with an enormous functional range of intensities.

Location of a stimulus is often determined by which receptor is stimulated, which determines which neurons are excited. If an itching sensation is detected by receptors on the elbow, then action potentials will occur in axons in the brain that respond to that elbow.

Somehow from this meager input of information, animals are able to perceive their environments. Perhaps, like us, other animals can use this neural code to create a rich image of the world. Until techniques are available to measure what another animal is thinking, however, we should resist assuming that the world appears to other species as it appears to us. In fact, we cannot even say that every human perceives stimuli in the same way. We have all learned to call light with a wavelength of 650 nanometers "red," but there is no way to tell whether the sensation of redness is the same for everyone.

STRETCH RECEPTOR OF THE CRAYFISH

The way in which receptor cells convert stimuli into receptor potentials can be explained for a few types of receptor. One of the most easily understood is the stretch receptor of crayfish and other crustaceans. These receptors evoke action potentials whenever a muscle stretches, thereby indicating the positions of legs and other parts of the animal's body. Crayfish stretch receptors are essentially neurons attached to the muscles (Figure 7.11). When the muscles are stretched, so are the receptors. The stretch increases the sodium permeability of the receptor membrane, perhaps by mechanically opening the sodium channels. The increase in P_{Na} produces a depolarizing receptor potential that can evoke action potentials. In the tonic stretch receptors, the greater the stretch, the larger the depolarization, and the higher the frequency of action potentials. Generally parallel to the tonic stretch receptor is a phasic stretch receptor that provides information about the time of occurrence of the stretch. As is common in receptors, the sensitivity can be reduced by inhibitory nerve cells conducting action potentials from the central nervous system.

Figure 7.11
(A) Two stretch receptors attached to parallel muscle fibers in a crayfish. One is a tonic receptor and the other is a phasic receptor. The inhibitory neuron to the receptors can block the receptor potential. (B) Receptor potentials recorded by microelectrodes impaling the cell bodies, and action potentials recorded from the axons. Although the stimulus to both receptors is the same, the phasic receptor responds only at the start of stimulation.

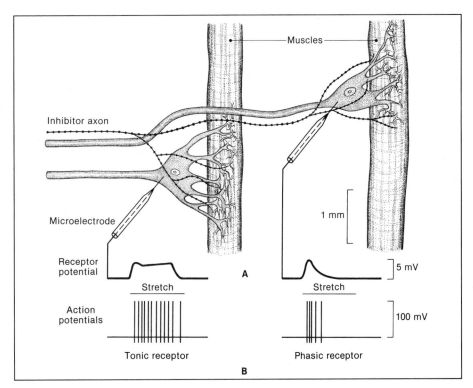

RETINAL RODS AND CONES

Among the most complex receptor cells are the photoreceptor cells of vertebrates. These are called rods or cones, depending on their shapes (Figure 7.12A). Rods have cylindrical **outer segments** that contain approximately 2000 disc-shaped membranes bearing light-absorbing pigments. In rods this pigment is **rhodopsin,** a yellow substance that absorbs a broad range of wavelengths. Rhodopsin combines a vitamin-A derivative called **retinal** with a protein called **opsin.** Cones are similar, except that their outer segments are cone shaped, and the pigment, **iodopsin,** has different opsins. There are three different opsins among the cones, and each absorbs the greatest amount of light at a slightly different wavelength from the other two. These three wavelengths are perceived by humans as being the colors red, blue, and green.

Rods also differ from cones in their maximal sensitivity. Rods are so sensitive that they can respond to individual photons, but they become blinded in bright daylight. Rods are therefore useful mainly at night and in heavy shadow. Cones are not sensitive enough to work in darkness, but the different iodopsins enable them to transmit information about color in bright light. Although there are only three kinds of iodopsins, any color can be perceived from the combination of cones it stimulates. For example, stimulation of both red-sensitive and blue-sensitive cones would indicate the color purple. Good vision in the dark requires a

Figure 7.12
(A) Structure of mammalian rod and cone. Note the presence of cilia, which characterize ciliary photoreceptors. The other retinal cells will be discussed later. (B) Mechanism of transduction in rods and cones. In the dark (upper part of figure), transducin and phosphodiesterase (PDE) are inactive, and cyclic GMP keeps the sodium channels open. In light (lower part), the retinal portion of rhodopsin (or iodopsin) changes from the cis to the trans form, which activates transducin, which then activates PDE. PDE breaks down the cGMP into GMP, allowing the sodium channels to close. The result is a hyperpolarizing receptor potential. Each opsin activates many molecules of transducin, which activates many molecules of PDE, which hydrolyzes many molecules of cGMP. This amplification makes rods responsive to even a single photon of light.

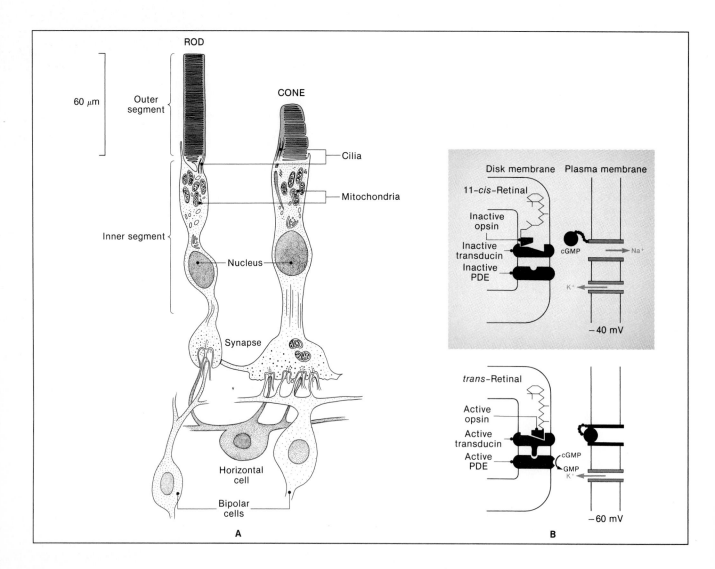

large number of rods, while color vision in daylight requires numerous cones. The retinas of most mammals, especially nocturnal ones, have primarily rods. The retinas of humans and other primates compromise by having only cones in the center of focus, called the **fovea,** and a mixture of rods and cones in peripheral areas. Humans have a total of about 100 million rods and 3 million cones.

The mechanism by which rods and cones convert light into electrical signals is still being worked out, but already the results are surprising. Unlike the crayfish stretch receptor, rods and cones depolarize when they are *not* stimulated—that is, in the dark. The reason is that their membranes have a high permeability to Na^+ in the dark. When stimulated by light the sodium permeability decreases, producing a hyperpolarizing receptor potential (Figure 7.12B). This receptor potential conducts to the **inner segment,** which contains a chemical synapse. This synapse leaks transmitter in the dark, but in response to the hyperpolarizing receptor potential it reduces the rate of transmitter release. This change in synaptic activity affects other retinal cells, which produce action potentials in the optic nerve.

PHOTORECEPTOR ORGANS

Receptor cells usually occur as parts of sensory organs that protect and tune the receptor cells to appropriate stimulation. The survey of animals in the final chapters describes many specialized receptor organs. For now we focus on organs that illustrate general principles, especially in humans and other vertebrates.

Photoreceptor organs include eyes and other structures that respond to electromagnetic radiation with wavelengths in or near the range that humans perceive as light. The exact range varies with species. Rattlesnakes, pythons, and boa constrictors have not only eyes but infrared (IR) receptors that allow them to "see" the warmth given off by prey. Insects, birds, and fishes have eyes that respond to ultraviolet (UV) wavelengths that are invisible to us.

There are two main types of photoreceptor cells, both of which may have evolved from the simple light-detecting **eyespots** of certain protozoans (see Chapter 22). The **rhabdomeric** type of photoreceptor cell occurs in most invertebrates, such as flatworms, molluscs, annelids, and arthropods. These photoreceptors are organized around either submicroscopic projections of cell membrane **(microvilli)** or around layers of membrane **(lamellae).** (See Chapter 31 for details in crustaceans and insects.) The **ciliary** type of photoreceptor develops from a modified cilium (see Figure 7.12A) and occurs in protozoans, in cnidarians such as jellyfish, and in vertebrates such as ourselves.

Photoreceptor organs range in complexity from simple light detectors to complex eyes capable of analyzing detailed images. The simplest photoreceptor organ is the **eyespot,** which may be little more than a nerve cell that contains pigment. Some animals have **simple eyes,** which are like eyespots except that they have a lens that focuses light onto the photoreceptor cell. If a simple eye contains several photoreceptor cells, it may be able to detect the position of light, depending on which cells the lens focuses it on. Only if there are numerous photoreceptor cells at the correct focusing distance from the lens will the eye be able to analyze an image. (See Figure 28.28 for examples of each type in molluscs.)

Vertebrate eyes have millions of ciliary photoreceptor cells (rods and cones) combined in each **retina** (Figure 7.13A). A lens focuses light onto the retina, and each rod or cone samples the light from a small area of the visual field. The lenses in fishes, amphibians, and snakes move back and forth to focus near or distant objects on the retina. In most reptiles, birds, and mammals the lens becomes rounder to focus light from nearby objects. In terrestrial animals the change in

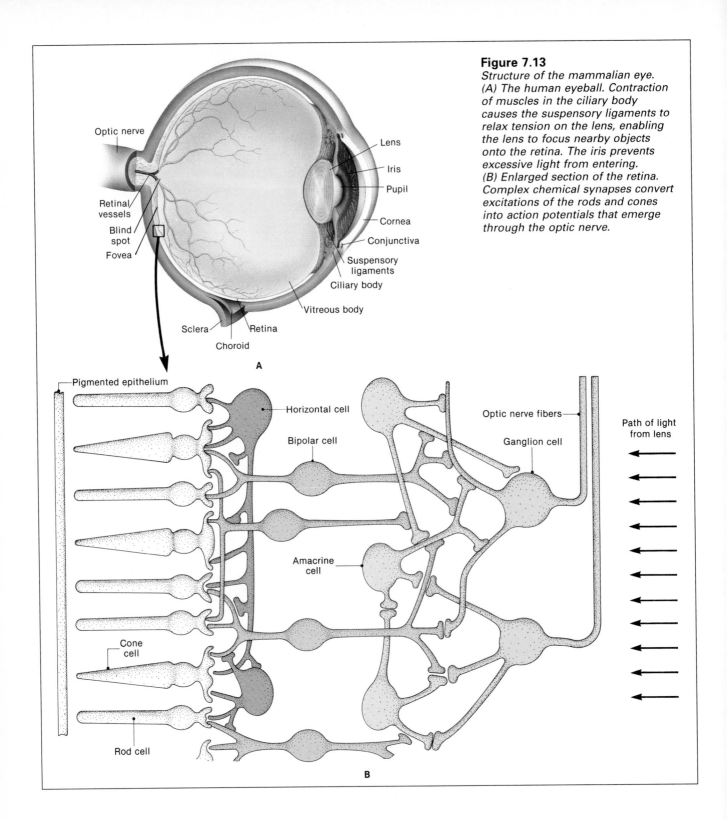

Figure 7.13
Structure of the mammalian eye. (A) The human eyeball. Contraction of muscles in the ciliary body causes the suspensory ligaments to relax tension on the lens, enabling the lens to focus nearby objects onto the retina. The iris prevents excessive light from entering. (B) Enlarged section of the retina. Complex chemical synapses convert excitations of the rods and cones into action potentials that emerge through the optic nerve.

refraction as light goes from air into tissue enables the clear covering of the eye, the **cornea,** to help the lens focus light. The diameter of the **pupil** controls the amount of light entering the eye.

The image on the retina is inverted, although that does not matter as far as the brain is concerned. (If you wore goggles that invert images, the world would look normal after a few days. Then, if you took the goggles off, everything would look upside down.) The light also seems to come in the wrong way, through the other retinal cells and the inner segments before reaching the outer segments. This does not affect vision, however, because the cells and inner segments are transparent. Any light not absorbed by the rhodopsin or iodopsins in the outer segments is stopped by a layer of **melanin** pigment. Many animals, however, have a reflecting layer, called the **tapetum,** that bounces the light back through the rods and gives them a second chance to absorb the light. This tapetum causes the characteristic eye shine of cats and many other nocturnal animals.

Although the vertebrate eye has many fine features, the image it forms on the retina is really quite poor. A camera store that sold human eyes would soon be out of business. The reason is that no simple lens—not even a living one—can form a sharp image over a wide area for all colors. That is why good camera lenses are so expensive; they combine many separate elements, each of which corrects defects in the others. Instead of using multiple lenses, however, vertebrate eyes correct the fuzzy image neuronally, through interactions of the retinal cells. In addition to the rods and cones, there are also **horizontal, bipolar, amacrine,** and **ganglion** cells. Axons of the ganglion cells carry all the visual information to the brain through the optic nerves. In the fovea, where vision is sharpest, the cells of the retina are interconnected in such a way that they enhance contrast. In areas of the retina nearest the nose, the cells are interconnected to detect movement "out of the corner of the eye." Thus a great deal of visual processing occurs even before action potentials reach the brain. Embryologically, the retina is, in fact, a part of the brain.

THERMORECEPTORS, NOCICEPTORS, AND MECHANORECEPTORS

Somesthetic (body sense) receptors respond to thermal and mechanical stimulation of the skin, to pain, and to the position of limbs. They include thermoreceptors that react to either heat or cold, nociceptors that produce the sensation of

Figure 7.14
A few of the dozen or more receptor types in mammalian skin. All are apparently mechanoreceptors; thermoreceptors and pain receptors have not been identified by structure. The functions tentatively assigned to each type are as follows (Iggo and Andres 1982): (1) Hair follicle receptor: bending of hair; (2) Pacinian corpuscle: most effective stimulus is high-frequency vibration, but normal response is to deep pressure; (3) Meissner's corpuscles: most effective stimulus is low-frequency "flutter"; (4) Merkel's corpuscles: tonic mechanoreceptor responsive to touch in small area; (5) Ruffini corpuscle: tonic mechanoreceptor responsive to stretch of skin (for many decades thought to be warm receptor); (6) Krause bulb: function unknown (long thought to be cold receptor); (7) free nerve endings: probably multifunctional (since these are the only receptors in the cornea of the eye.

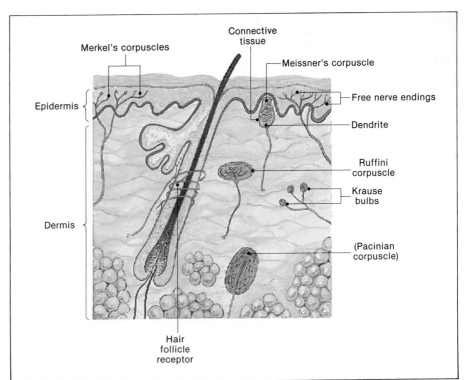

pain in response to substances released by damaged tissue, and various mechanoreceptors. Little is known about the structure and functioning of thermoreceptors and nociceptors. Mechanoreceptor cells are associated with a variety of organs that respond to many types of mechanical stimulation. These organs include stretch receptors like those in crayfish and vertebrates (see Figures 7.9 and 8.2), as well as receptors for touch, pressure, acceleration, and position with respect to gravity. Some mechanoreceptors appear to be little more than the ends of nerve cells, while others have elaborate accessory structures that tune the receptor cells to particular types of stimulation (Figure 7.14).

The most complex mechanoreceptor organs in vertebrates are the three inner-ear mechanisms for detection of rotation, orientation in a gravitational field, and sound. All three organs have **hair cells,** each with a bundle of about 50 hair-shaped projections of various lengths (Figure 7.15). Bending these "hairs" in a certain direction depolarizes the membrane and causes the release of transmitter from the synaptic end of the hair cell. The resulting EPSP then triggers action potentials. Bending the projections in the opposite direction hyperpolarizes the membranes and reduces the release of transmitter. Hair cells occur in the vestibular apparatus and the organs for hearing (Figure 7.16A).

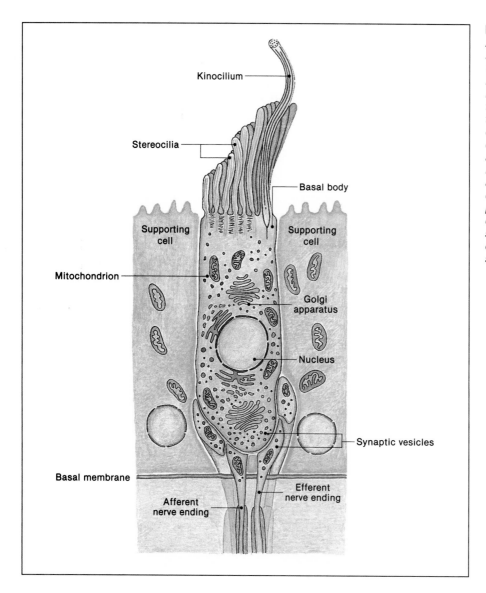

Kinocilium

Stereocilia

Basal body

Supporting cell

Supporting cell

Mitochondrion

Golgi apparatus

Nucleus

Synaptic vesicles

Basal membrane

Afferent nerve ending

Efferent nerve ending

Figure 7.15
An individual hair cell showing some of the 50 or so stereocilia (which are not true cilia) and one kinocilium. The role of the kinocilium is not clear; it is absent in the cochlea of mammals. Bending the stereocilia toward the kinocilium depolarizes the plasma membrane of the hair cell, and bending them away from the kinocilium hyperpolarizes it. Bending at right angles to the kinocilium has no effect. These changes in membrane potential regulate the release of excitatory transmitter from the synaptic portion of the cell. Note the efferent neuron that controls the sensitivity of the hair cell.

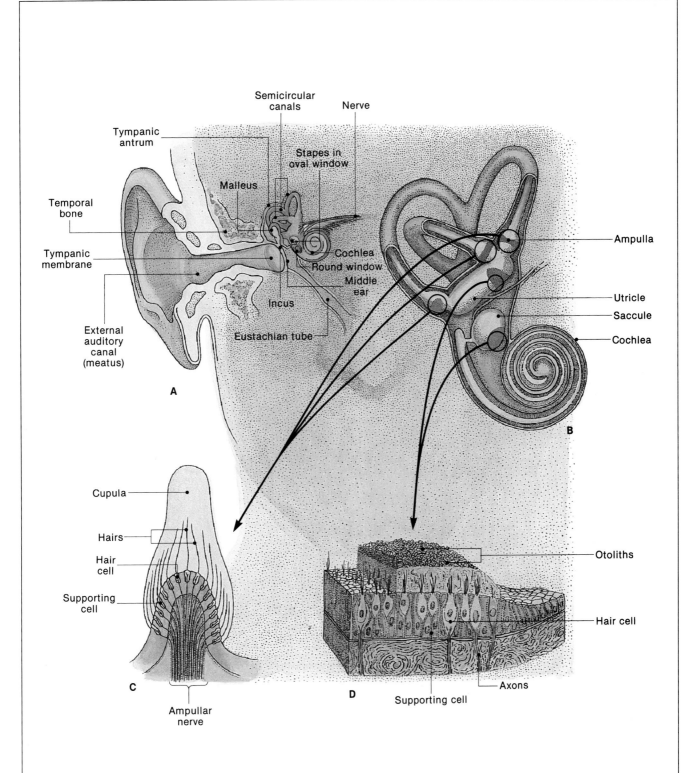

Figure 7.16

(A) The auditory organs and vestibular apparatus. (B) The inner ear of a human. The three semicircular canals detect rotation, the otolith organs (utricle and saccule) detect linear acceleration (gravity), and the cochlea detects sound. (C) Enlarged section of one ampulla of the semicircular canals. Hair cells are stimulated by the relative movement of endolymph against the cupula. (D) A portion of the utricle or saccule. The weight of otoconia stimulates hair cells.

The Vestibular Apparatus.   Information from the vestibular apparatus, combined in the brain with information from stretch receptors in the neck and eye muscles, helps maintain balance and distinguish movement of the body from movement of the surroundings. The vestibular apparatus includes the semicircular canals and the otolith organs. Each of the three **semicircular canals** of each inner ear responds to rotation in a different plane, such as turning the head, falling forward, or tilting sideways. During such events the fluid **endolymph** that fills the semicircular canals tends not to move while the surrounding tissue is moving. (The same effect occurs when you quickly turn a glass of water.) In a bulge called the **ampulla,** the motion of the endolymph relative to surrounding tissue pushes against a gelatinous flap called the **cupula,** into which the hair cells project. The resulting deflection of the hair cells triggers action potentials in neurons that go to the brain (Figure 7.16B, C).

The **otolith organs**—the **utricle** and **saccule**—are responsible for balance. They are named for otoliths (= otoconia), which are dense, gelatinous structures containing crystals of calcium carbonate (Figure 7.16D). Projections of hair cells are bent by the weight of the otoliths, generating action potentials that carry information about the direction of gravity.

Hearing.   Hearing is also due to a response by hair cells to mechanical stimulation. In this case the mechanical stimulation consists of waves of increasing and decreasing pressure rippling from a source of sound. Although most of us take hearing for granted, it is a rare ability in animals, occurring only in some arthropods and vertebrates. The **frequency** of the sound is the number of cycles of pressure change per second, measured in hertz (Hz). Humans perceive different sound frequencies as pitch. The difference between minimum and maximum pressures during a wave determines the intensity—or loudness—of the sound.

Mammals, birds, and some reptiles (crocodilians) hear by means of the **cochlea.** In humans each of the two snail-shaped cochleae is about the size of a house fly. The cochlea responds to vibration of the **middle ear bones,** which, in turn, respond to vibration of the **eardrum** (= tympanum) (Figure 7.16). Pressure changes in the endolymph inside the cochlea are conducted to the **basilar membrane** inside, which supports the **organ of Corti** (Figure 7.17). The mechanosensitive hair cells lie in the organ of Corti, with the hairs pointed out toward the **tectorial membrane.** As the organ of Corti vibrates up and down in response to sound, the tectorial membrane rubs against the hair cells. The hair cells then release synaptic transmitter that generates action potentials in axons of the auditory nerve. The louder the sound, the greater the stimulation of the hair cells, and the higher the frequency of action potentials (Figure 7.10). As usual, the intensity of stimulation is coded by the frequency of action potentials.

The pitch of sound appears to be coded according to which receptors along the cochlea are most intensely stimulated. Georg von Békésy, who began as a telephone engineer, devoted much of his life to figuring out how the cochlea determines pitch. After many extremely ingenious experiments on this tiny, delicate organ embedded within the skull, he concluded that for each pitch there is a particular portion of the basilar membrane that vibrates with the greatest amplitude. You can demonstrate the effect with a fishing rod or similar long, flexible pole. Shaking the rod rapidly produces the most vibration near your hand. Shaking it slowly causes the tip of the rod to vibrate most. In mammals the end of the basilar membrane nearest the middle ear bones vibrates maximally at high pitches, and the other end responds most to low pitches. Thus the hair cells near the middle ear bones send out the greatest frequency of action potentials in response

Figure 7.17
Cross section of one turn of the cochlea. Pressure bends the basilar membrane and the organ of Corti on it, causing the tectorial membrane to deflect the hair cells. The hair cells then evoke action potentials in axons of the auditory nerve. The greater the deflection of the hair cells, the higher the frequency of action potentials, and the greater the perceived loudness.

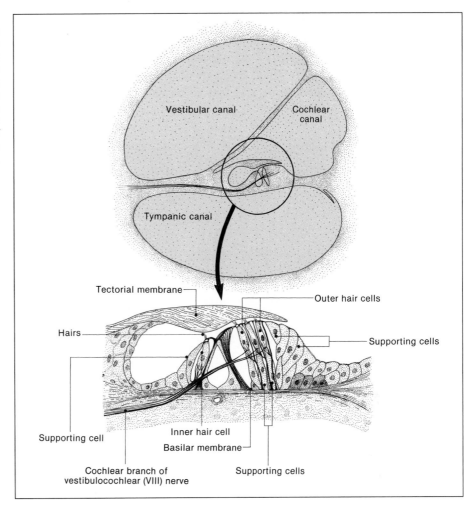

to high pitches, and the hair cells at the other end send out the greatest frequency of action potentials in response to low pitches. Intermediate pitches produce the maximal response from the hair cells at intermediate positions along the basilar membrane. Recent research indicates, however, that this is only part of the solution to pitch discrimination. In mammals and other terrestrial vertebrates, some tuning also occurs in the hair cells themselves (Hudspeth 1985; Rhode 1984; Zwislocki 1981).

CHEMORECEPTORS

Chemoreceptors in humans respond to taste (= gustation) and smell (= olfaction), but these two modalities do not begin to describe all the kinds of chemoreception that occur among animals. Fishes and many other aquatic animals, for example, have chemoreceptors on the body surface that respond to substances given off by prey, predators, and members of their own species. Terrestrial animals are also capable of a wide range of chemoreception that would be hard to categorize as either taste or olfaction. Many insects, in particular, have virtually an entire dictionary of chemicals by which they communicate.

In spite of their diversity, most chemoreceptors operate on similar principles. Generally, the outer surface of the chemoreceptor cell bears molecules that bind to particular substances, depending on the molecular shape and charge distribution of the substance. Like the interaction of a synaptic transmitter with the postsynaptic membrane receptor, this interaction triggers action potentials by changing membrane permeability. In order to detect a large variety of substances, several types of chemoreceptors are needed, each specific for a particular type of substance. Different animals have chemoreceptors that respond to different substances. Bees are indifferent to urine, and dogs do not go around sniffing flowers.

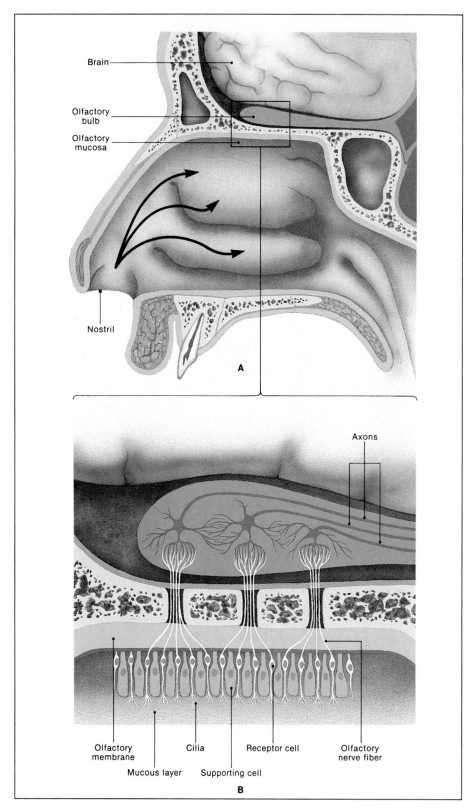

Figure 7.18
(A) Location of the olfactory epithelium and the olfactory bulb of the human brain. (B) Structure of the olfactory cells. The numerous cilia are embedded in a mucous layer; their function is unknown. Odorant molecules carried through the mucus by odorant-binding proteins interact with plasma membranes, which then generate action potentials that travel the short distance to the olfactory bulb.

Animals also differ in the sensitivity of their chemoreceptor organs. Most dogs, for example, have a sense of smell so keen that they trust their noses more than their eyes. (They ignore rabbit tracks in the snow.)

In mammals the detection of airborne chemicals is called **olfaction.** Chemoreceptors for olfaction lie in the **olfactory epithelium** on the roof of the nasal cavity. Short axons conduct action potentials from the olfactory epithelium to the two **olfactory bulbs** in the brain, where the information is analyzed (Figure 7.18). Neural pathways from the olfactory bulbs go to the deeper areas of the brain that are related to emotional and survival reactions, which may explain why odors often evoke strong reactions that are more emotional than cognitive. Most mammals, as well as amphibians and reptiles, also have another set of olfactory receptors in **Jacobson's organ** (= vomeronasal organ) in the roof of the mouth. In humans, Jacobson's organ occurs only in the fetus.

The other type of chemoreception is called **taste. Taste cells** are located in **taste buds** on the tongue, which are located on **taste papillae** (Figure 7.19). Human taste cells respond to four primary taste qualities—sour, salt, sweet, and bitter. How a substance tastes depends on its molecular properties. Acidic sub-

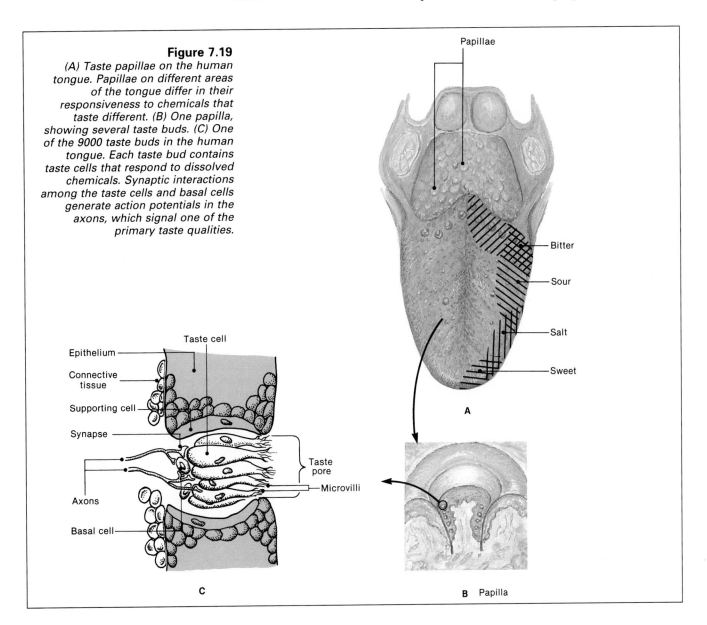

Figure 7.19
(A) Taste papillae on the human tongue. Papillae on different areas of the tongue differ in their responsiveness to chemicals that taste different. (B) One papilla, showing several taste buds. (C) One of the 9000 taste buds in the human tongue. Each taste bud contains taste cells that respond to dissolved chemicals. Synaptic interactions among the taste cells and basal cells generate action potentials in the axons, which signal one of the primary taste qualities.

stances release H^+, which blocks potassium channels on certain taste cells, giving rise to a sour taste. The salt taste appears to arise from the influx of Na^+ or K^+ through sodium channels. Sugars bind to molecular receptor molecules. Bitter-tasting substances (often plant toxins) work by a variety of mechanisms. The main function of taste may be to inform the animal that what it is about to eat contains needed calories or minerals (sweet or salt), or is acidic or toxic (sour or bitter). In this latter role taste buds serve a function analogous to that of human tasters who used to sample the food of kings and queens to be sure it had not been poisoned. Like tasters who served unpopular rulers, each taste cell survives for only a few days.

ELECTRORECEPTORS AND MAGNETORECEPTORS

The most recently discovered receptor type is the electroreceptor, which is found in all cartilaginous fishes (such as sharks and rays), many bony fishes, and some amphibians. These enable some fishes to navigate by means of electric fields generated by ocean currents, avoid solid objects that deflect electric fields, and communicate with members of their own species by means of weak electric signals. Electroreceptors are so sensitive in some fishes that they can detect the electric field produced by the muscles of prey buried in sand. Since electroreception occurs mainly in fishes, further discussion can wait until Chapter 35. Organs responsible for sensitivity to magnetic fields have not been discovered. The evidence for their existence is that some birds, insects, salamanders, and perhaps mammals use the Earth's magnetic field to navigate during homing and migration (see pp. 421–422).

SUMMARY

Nerve cells rapidly communicate information from sensory receptors to the central nervous system, and from the CNS to muscles and glands. This information is in the form of action potentials—reversals of membrane voltage that last only a few milliseconds. Action potentials are triggered by changes in the permeability to ions, mainly by an increase and then a decrease in sodium permeability. Various chemical substances, such as anesthetics and neurotoxins, can alter the generation and conduction of action potentials. Communication from one nerve cell to another is most often by release of chemical transmitters from synapses. The transmitters change the permeabilities of postsynaptic membranes, producing postsynaptic potentials that either excite or inhibit action potentials. Synaptic functioning can be altered by a variety of chemicals and during learning.

Information from sensory receptors is encoded as patterns of action potentials in afferent neurons. The intensity of stimulation is encoded as frequency of action potentials from tonic receptors, the time of stimulation is signaled by brief bursts of action potentials from phasic receptors, and the modality, quality, and location are encoded according to which neurons are conducting the action potentials. Sensory receptors vary in complexity. Some, such as the crayfish stretch receptor, produce a receptor potential that directly excites an afferent axon. In others, such as rods and cones of the vertebrate retina, the receptor potential first affects release of synaptic transmitter onto other nerve cells, which then interact in a complex way before exciting afferent cells.

Receptor cells are generally parts of receptor organs, which can be classified according to modality. Photoreceptors, such as eyes, respond to electromagnetic radiation that is visible to humans, and to infrared or ultraviolet. Somesthetic receptors include thermoreceptors that respond to temperature, nociceptors that respond to painful stimuli, and mechanoreceptors that respond to such stimuli as touch, pressure, and sound. Vertebrate mechanoreceptors include the vestibular apparatus, which senses head orientation and movement, and the cochlea, which is responsible for hearing. Chemoreceptors respond to chemical stimulation, and in humans include the cells responsible for olfaction and taste.

KEY TERMS

resting membrane potential	depolarization	axon
action potential	hyperpolarization	all or none law

threshold
conduction
myelin
synapse
synaptic transmitter

synaptic knob
postsynaptic potential
receptor potential
tonic
phasic

rhodopsin
vestibular apparatus
cochlea

SELF-TEST

1. Explain why there are unequal concentrations of the highly permeant ions K^+ and Cl^- on each side of the plasma membrane.

2. Explain how the unequal distribution of K^+ and Cl^- across a membrane gives rise to a voltage across the membrane.

3. Which ion is most important in initiating an action potential? Explain.

4. Why is the speed of axonal conduction important? Describe two means of increasing the speed of conduction.

5. All stages of neural functioning depend on changing membrane voltages by changing the permeability to one or more ions. Illustrate this statement by explaining excitatory and inhibitory synapses.

6. List all the ways you can imagine in which the function of chemical synapses could be modified by either learning or drugs.

7. Imagine a pin prick on the back of your left hand. Explain in your own words how your brain knows (a) the intensity of the stimulus, (b) when the stimulus is occurring, and (c) that the stimulus is cutaneous, that it is a sharp pain rather than pressure, and that it is located on the back of your left hand.

8. Explain how the crayfish stretch receptor works.

9. Explain how hair cells can respond differently, depending on whether they are in the semicircular canals, the otolith organs, or the cochlea.

READINGS

RECOMMENDED READINGS

Alkon, D. L. 1983. Learning in a marine snail. *Sci. Am.* 249(1):70–84 (July).

Alkon, D. L. 1989. Memory storage and neural systems. *Sci. Am.* 261(1):42–50 (July).

Bloom, F. E. 1981. Neuropeptides. *Sci. Am.* 245(4):148–168 (Oct).

Dunant, Y. and M. Israël. 1985. The release of acetylcholine. *Sci. Am.* 252(4):58–66 (Apr).

Gottlieb, D. I. 1988. GABAergic neurons. *Sci. Am.* 258(2):82–89 (Feb).

Hudspeth, A. J. 1983. The hair cells of the inner ear. *Sci. Am.* 248(1):54–64 (Jan).

Jacobs, B. L. 1987. How hallucinogenic drugs work. *Am. Sci.* 75:386–391.

Kety, S. S. 1979. Disorders of the human brain. *Sci. Am.* 241(3):202–214 (Sept).

Keynes, R. D. 1979. Ion channels in the nerve-cell membrane. *Sci. Am.* 240(3):126–135 (Mar).

Koretz, J. F. and G. H. Handelman. 1988. How the human eye focuses. *Sci. Am.* 259(1):92–99 (July).

Lester, H. A. 1977. The response to acetylcholine. *Sci. Am.* 236(2):106–118 (Feb).

Levine, J. S. and E. F. MacNichol Jr. 1982. Color vision in fishes. *Sci. Am.* 246(2):140–149 (Feb).

Llinàs, R. R. 1982. Calcium in synaptic transmission. *Sci. Am.* 247(4):56–65 (Oct).

Loeb, G. E. 1985. The functional replacement of the ear. *Sci. Am.* 252(2):104–111 (Feb).

Masland, R. H. 1986. The functional architecture of the retina. *Sci. Am.* 255(6):102–111 (Dec).

Morell, P. and W. T. Norton. 1980. Myelin. *Sci. Am.* 242(5):88–116 (May).

Newman, E. A. and P. H. Hartline. 1982. The infrared "vision" of snakes. *Sci. Am.* 246(3):116–127 (Mar).

Nilsson, D.-E. 1989. Vision optics and evolution. *BioScience* 39:298–307.

Parker, D. E. 1980. The vestibular apparatus. *Sci. Am.* 243(5):118–135 (Nov).

Poggio, T. and C. Koch. 1987. Synapses that compute motion. *Sci. Am.* 256(1):46–52 (May). (*Motion detectors in the retina.*)

Schnapf, J. L. and D. A. Baylor. 1987. How photoreceptor cells respond to light. *Sci. Am.* 256(4):40–47 (Apr).

Snyder, S. H. 1986. *Drugs and the Brain.* New York: Scientific American Library.

Stryer, L. 1987. The molecules of visual excitation. *Sci. Am.* 257(1):42–50 (July).

Van Dyke, C. and R. Byck. 1982. Cocaine. *Sci. Am.* 246(3):128–141 (March).

Wurtman, R. J. 1982. Nutrients that modify brain function. *Sci. Am.* 246(4):50–59 (Apr).

Wurtz, R. H., M. E. Goldberg, and D. L. Robinson. 1982. Brain mechanisms of visual attention. *Sci. Am.* 246(6):124–135 (June).

Zwislocki, J. J. 1981. Sound analysis in the ear: a history of discoveries. *Am. Sci.* 69:184–192.

ADDITIONAL REFERENCES

Barnes, D. M. 1986. Lessons from snails and other models. *Science* 231:1246–1249.

Hudspeth, A. J. 1985. The cellular basis of hearing: the biophysics of hair cells. *Science* 230:745–752.

Iggo, A. and K. H. Andres. 1982. Morphology of cutaneous receptors. *Annu. Rev. Neurosci.* 5:1–31.

Rhode, W. S. 1984. Cochlear mechanics. *Annu. Rev. Physiol.* 46:231–246.

Waxman, S. G. and J. M. Ritchie. 1985. Organization of ion channels in the myelinated nerve fiber. *Science* 228:1502–1507.

8

Nerve Systems

Short-eared owl (Asio flammeus).

CHAPTER OUTLINE

Neural Subsystems

Reflexes and the Spinal Cord
Servomechanisms and Muscle Control
Central Pattern Generators

Nervous Systems of Invertebrates

Autonomic Nervous System of Vertebrates

Internal Environment of the Brain and Spinal Cord

Sizes of Vertebrate Brains

Parts of the Vertebrate Brain

The Brainstem

Motor Functions of the Cerebrum and Cerebellum

Sensory Functions of the Cerebral Cortex

Lateralization
Topographical Organization
Feature Extraction

Language

Emotion

LEARNING OBJECTIVES

1. Can the functions of an entire nervous system be inferred from its parts?

2. What is the neural basis for a reflex?

3. Are there other kinds of fundamental behaviors and neural subsystems?

4. How do the nervous systems of different kinds of animals differ in organization and size?

5. How does the nervous system control activities of the heart, digestive system, and other organ systems?

6. How is the human brain organized?

7. How does the brain control movement?

8. How does the brain perceive sensory information?

9. How does the human brain produce and interpret language?

10. What is the neural basis of emotion?

NEURAL SUBSYSTEMS

No sensible person would claim to understand an entire culture after having met just one individual from that culture. Likewise, we should not presume that our consideration of individual cells in the previous chapter provides an understanding of an entire nervous system. Even when all the nerve cells and their connections are known, it is seldom possible to deduce the functions of the system. Scientists have worked out the complete "wiring" diagram of the 302 neurons and 8000 synapses in the nematode *Caenorhabditis elegans*, yet no one pretends to understand this little worm's behavior. Fortunately, however, nervous systems often appear to be collections of similar subsystems that perform similar functions in different animals. Among these functions are reflexes, servomechanisms, and central pattern generators.

Reflexes and the Spinal Cord. A reflex is a stereotyped response due to a relatively simple connection from receptors to muscles or glands. Examples of reflexes include the eye blink, the withdrawal of the hand from a painful stimulus, vomiting, and swallowing (Table 8.1). In the simplest reflexes there is only one synapse directly connecting afferent neurons to efferent neurons. These reflexes are called **monosynaptic reflexes.**

The best-known example of a monosynaptic reflex is the **withdrawal reflex,** which pulls a limb away from injury (Figure 8.1). The neural pathway begins with nociceptors (pain receptors), which excite afferent neurons with axons in a nerve to the spinal cord. Near the spinal cord the nerve splits into two branches: the **dorsal root** and the **ventral root.** In mammals almost all afferent axons enter the spinal cord via the dorsal root, and all efferent axons leave by the ventral root. The axons from nociceptors branch into the **gray matter** of the spinal cord, which contains synapses and cell bodies. There the axons have excitatory synapses onto motor neurons, which synapse with **flexor** muscles of the injured limb. These muscles pull the limb toward the body.

The effect of the withdrawal reflex is to pull the limb away from the stimulus that initiated the reflex. At the same time, other axons inhibit the motor neurons to the **extensor muscles** of the same limb, which would tend to extend the limb toward the source of injury. Still other axons send action potentials to the opposite side of the spinal cord to excite the extensor muscles and inhibit the flexor muscles, so that the other limb can support the body. All this happens before the brain knows about it. In fact, these reflexes occur even without a brain. Only later do

Table 8.1 Some reflexes of vertebrates.
Many reflexes normally occur only in immature animals before the brain has developed control over the spinal cord. In mature animals they may occur following injury to the brain or spinal cord. (As an exercise draw diagrams like Figure 8.1C that could account for some of the reflexes below.)

Withdrawal reflex. Withdraws limb from injury. See text for further description.

Grasping reflex. In infant primates the fingers and toes flex in response to pressure on the palm or sole of the foot. This presumably helps infants grasp limbs and the mother's hair.

Rooting and sucking reflexes. In infant mammals a touch on the cheek causes the infant to move its mouth toward the object. This rooting reflex allows the infant to find the mother's nipple. Contact on the lips initiates sucking.

Swallowing reflex. Pressure on the soft palate at the back of the mouth triggers contraction of the muscles involved in swallowing.

Vestibulo-ocular reflex. When the semicircular canals sense rotation of the head, the eyes rotate in the opposite direction. This occurs even in the dark. The function is to keep the gaze fixed on one spot even when the head is moving.

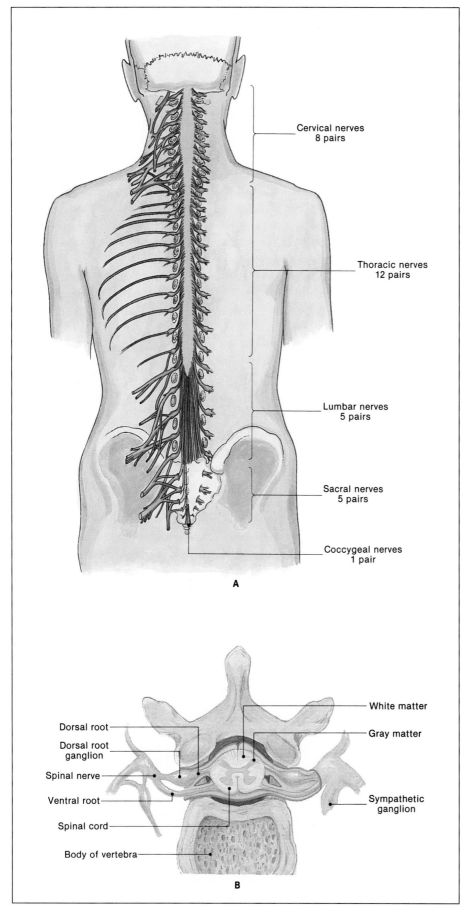

Cervical nerves
8 pairs

Thoracic nerves
12 pairs

Lumbar nerves
5 pairs

Sacral nerves
5 pairs

Coccygeal nerves
1 pair

A

Dorsal root

Dorsal root
ganglion

Spinal nerve

Ventral root

Spinal cord

Body of vertebra

White matter

Gray matter

Sympathetic
ganglion

B

Figure 8.1
(A) The position of the spinal cord in a human showing the many nerves branching to each side. (B) A cross section of the spinal cord. The lateral nerves split into the dorsal and ventral roots at the spinal cord. Almost all the axons of afferent neurons enter the dorsal roots and have their cell bodies in the dorsal root ganglia. Efferent neurons have axons that emerge through the ventral roots from cell bodies in the gray matter. This area is gray because it contains unmyelinated synapses and cell bodies. The white matter consists of myelinated axons linking the brain to the spinal cord, and various levels of the spinal cord to each other. (C) (facing page) A diagram representing part of the pathway for the withdrawal reflex in a human.

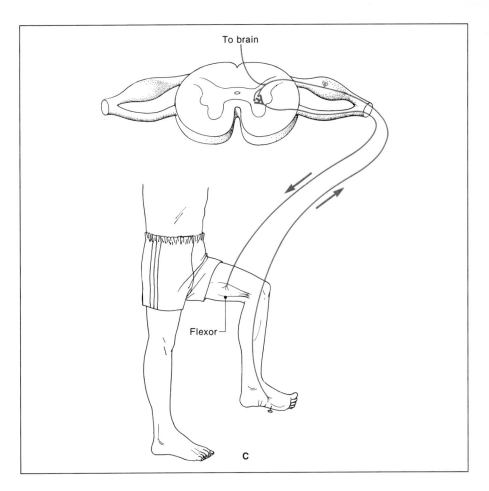

To brain

Flexor

c

action potentials reach the area of the brain that senses pain, after they ascend the **white matter** of the spinal cord. (White matter is white because of the myelin surrounding the axons.)

Servomechanisms and Muscle Control.   Once triggered, reflexes occur without further modification. Servomechanisms, in contrast, are neural subsystems that continually modify their output. As an example of a servomechanism, consider the fact that you can hold your hand out in a steady position without looking at it or thinking about it. The neural subsystem responsible for this ability begins with stretch receptors in your skeletal muscles. These **muscle spindles** respond to change in muscle length, and therefore to change in posture (Figure 8.2A). Each gram of muscle has from several to hundreds of spindle receptors. Each spindle has muscles of its own at each end. The receptor portion is in the middle. Stretch on the receptor portion triggers action potentials in afferent axons **(Ia afferents)** to the spinal cord. This stretch can be due either to contraction of the muscle in which the spindle occurs, or to contraction of the muscles within the spindle (due to voluntary activation through **gamma efferent neurons**).

Figure 8.2B shows how the spindle receptor organ enables you to hold out your hand steadily. In the gray matter of the spinal cord the Ia afferents from muscle spindles excite alpha motor neurons, which excite the muscles in which the spindles occur. If the hand starts to sag under the pull of gravity, the muscle spindles will be stretched. This will increase the frequency of action potentials in the alpha motor neurons, which will cause the muscle to contract with more force, restoring the hand to its original position. This mechanism maintains constant position even if someone unexpectedly drops a weight in your hand.

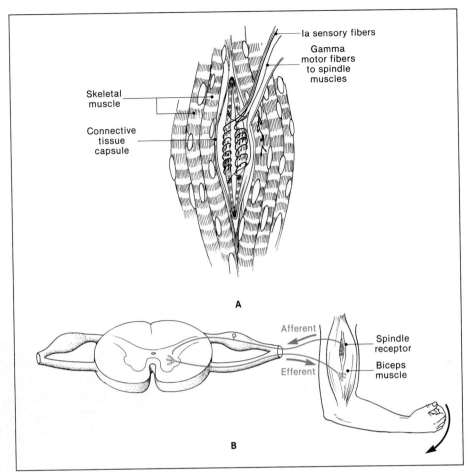

Figure 8.2
(A) The spindle receptor organ of a mammal. The receptor cells include the spiral endings of group Ia afferent axons. Stretch applied to these endings generates action potentials in the Ia afferents. At each end of the spindle are muscle fibers controlled by small motor neurons called gamma efferent fibers.
(B) How the muscle spindle receptor maintains constant limb position. If the biceps muscle starts to stretch because of the weight on the lower arm, the spindle receptor generates action potentials in the Ia afferent, which excites the alpha motor neuron to the biceps.

Central Pattern Generators. A third elementary subsystem is the central pattern generator, which is a network of neurons that produces action potentials without sensory input (unlike a reflex), and in a pattern that does not directly depend on sensory input (unlike a servomechanism). Central pattern generators are often responsible for rhythmic behaviors, such as breathing and walking. They include at least one nerve cell that produces action potentials spontaneously. Associated neurons may then shape the pattern of action potentials.

NERVOUS SYSTEMS OF INVERTEBRATES

There is a good correlation between the complexity of an animal's nervous system and the complexity of its behavior. Sponges, which do not have any nerve cells, spend their adult lives attached to substratum, hardly behaving at all. *Hydra,* jellyfishes, sea anemones, and other biradially symmetric animals (phyla Cnidaria and Ctenophora) are somewhat more active. They have nerve cells that form a diffuse **nerve net** in which stimulation at one point sends electrical activity radiating throughout the system (Figure 8.3). They have no collection of nerve cells that would justify the term "central nervous system."

Only in bilaterally symmetric animals does one see a central nervous system. Such organisms have a front and rear, and there is a tendency for sensory receptors and the nerve cells associated with them to be concentrated at the end that encounters the environment first. The collection of nerve cells in the head forms a **cerebral** (= cephalic) **ganglion.** Cerebral ganglia are often called "brains," but

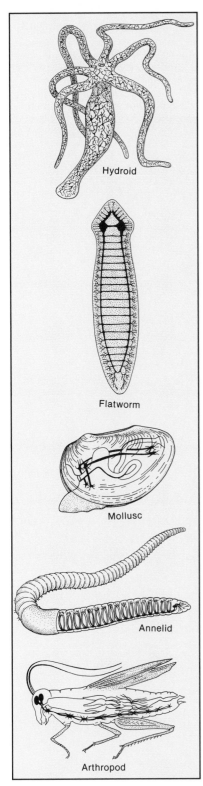

Figure 8.3
Organization of nervous systems in various invertebrates.

this term can be misleading in invertebrates, since the cerebral ganglion is seldom as dominant in invertebrates as the brain is in vertebrates.

Ganglia greatly increase the combinations of interactions of nerve cells and thus allow more complex and varied behaviors. Even the simplest bilaterally symmetric animals, such as flatworms (phylum Platyhelminthes), can perform complex behaviors such as learning to negotiate a maze. Snails and other molluscs have several ganglia interconnected in a loop and are model organisms for studying the mechanisms of learning. In annelids, such as earthworms, the central nervous system consists of a chain of ganglia. Each ganglion controls one segment of the animal and is linked to the next ganglion by a pair of nerves called **connectives.** Connectives are handy for research, since it is easy to record action potentials in the axons within them, and they can be cut to isolate ganglia from each other. Arthropods, such as insects and crustaceans, have nervous systems similar to those of annelids, except that most of the segmental ganglia become fused into a few larger ganglia. A typical insect has several abdominal ganglia, from one to three thoracic ganglia that control legs and wings, and cerebral ganglia that control the head and coordinate many functions of other ganglia.

AUTONOMIC NERVOUS SYSTEM OF VERTEBRATES

The central nervous system communicates with the rest of the body by means of axons in nerves that make up the **peripheral nervous system.** In vertebrates the peripheral nervous system includes the autonomic nervous system, which exercises involuntary control over digestion, circulation, breathing, copulation, and some other functions. The autonomic nervous system has two divisions, the sympathetic and the parasympathetic. Each division releases a different **neurosecretion** from synapselike nerve endings.

The **sympathetic division** tends to be active during stress ("fight or flight" situations). It secretes norepinephrine, which inhibits digestion and sexual arousal, and stimulates circulation and breathing. It also stimulates the release of the hormone epinephrine (= adrenaline) from the adrenal medulla (see p. 187). The sympathetic division originates in a chain of ganglia on each side of the middle portion of the spinal cord (Figure 8.4). The **parasympathetic division** tends to be active during rest. Its neurosecretion is acetylcholine, which stimulates digestion and sexual arousal and inhibits circulation and breathing. The parasympathetic division originates at both ends of the spinal cord. The effects of these two divisions on various organ systems will be detailed in later chapters.

INTERNAL ENVIRONMENT OF THE BRAIN AND SPINAL CORD

Unlike the peripheral nervous system, which is necessarily exposed to all kinds of dangers as it carries messages to and from various bodily outposts, the central nervous system is a well-guarded command center. Nerve cells in the brain and spinal cord are cushioned against mechanical and chemical shocks by cerebrospinal fluid and glial cells. **Cerebrospinal fluid** (CSF) acts as a liquid cushion for cells in the brain and spinal cord. It also circulates water, ions, nutrients, and other small, soluble molecules through large cavities in the brain, called **ventricles,** through the **central canal** in the spinal cord, and between the outer surface of the central nervous system and the surrounding **arachnoid membrane** (Figure 8.5). The **choroid plexus** helps regulate the composition of CSF by actively transporting ions and certain molecules from the blood. CSF eventually leaks into the blood through villi in the arachnoid membrane. The choroid plexus and arachnoid

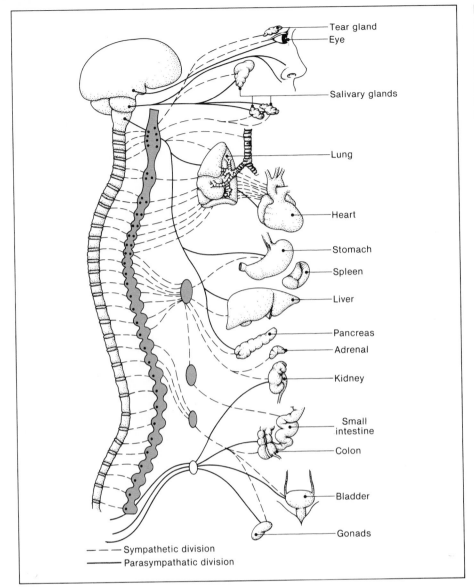

Figure 8.4
Schematic diagram of the human autonomic nervous system showing the organs it affects. The spinal cord is shown to the left of the chain of sympathetic ganglia. (See Figure 8.1B for an anatomically correct representation of the sympathetic ganglia.)

Labels in figure: Tear gland, Eye, Salivary glands, Lung, Heart, Stomach, Spleen, Liver, Pancreas, Adrenal, Kidney, Small intestine, Colon, Bladder, Gonads

– – – Sympathetic division
——— Parasympathatic division

membrane make a **blood–CSF barrier** that helps regulate the substances to which nerve cells are exposed.

Nonneural supporting cells called **glial cells** also help regulate the composition of CSF, in addition to serving a variety of other functions. The glial cells called **astrocytes** are associated with blood capillaries and actively transport glucose and other nutrients from the blood into the CSF. Astrocytes also prevent the entry of many other substances from blood into the CSF, thereby forming a **blood–brain barrier.** Other glial cells also help regulate ion concentrations surrounding nerve cells, metabolize synaptic transmitters, and may communicate with each other much as nerve cells do. During development other glial cells help axons reach their destinations, and form myelin sheaths around axons (see p. 145). Considering all they have to do, it is perhaps not surprising that glial cells outnumber nerve cells ten-to-one.

SIZES OF VERTEBRATE BRAINS

Humans are understandably proud of their brains and curious about how they compare with those of other vertebrates. In general, the mass of the brain increases

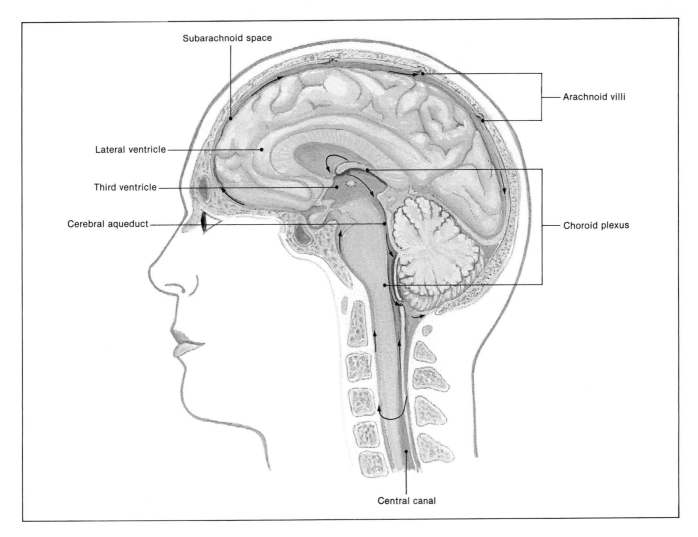

Subarachnoid space

Arachnoid villi

Lateral ventricle

Third ventricle

Choroid plexus

Cerebral aqueduct

Central canal

Figure 8.5
The formation and circulation of the cerebrospinal fluid. The choroid plexus transports ions and certain molecules from the blood into the cerebrospinal fluid and excludes blood cells and other materials. Certain glial cells associated with capillaries (not shown) transport glucose and other nutrients into the CSF. The resulting CSF circulates through the ventricles of the brain and the central canal of the spinal cord, supplying the interstitial fluid that surrounds nerve cells. The CSF eventually returns to the bloodstream through villi of the arachnoid membrane.

with the mass of the body, and birds and mammals have larger brains than fishes, amphibians, and reptiles of the same body mass (Figure 8.6). Naturally the largest brains occur in the huge mammals, such as whales and elephants. Together with porpoises, however, we get about seven times as much brain as we deserve based on body mass.

PARTS OF THE VERTEBRATE BRAIN

Just as important as total brain size are the relative sizes of different parts of the brain. The brains of all vertebrates develop from three embryonic divisions: the **forebrain** (= prosencephalon), the **midbrain** (= mesencephalon), and the **hindbrain** (= rhombencephalon). These three divisions begin development with similar proportions in all vertebrates, but the parts grow at different rates in different groups of vertebrates (Figure 8.7). In birds and mammals, the forebrain grows disproportionately large and gives rise to the **cerebrum.** The cerebrum is divided into two cerebral hemispheres. Its outer layer, the **cerebral cortex,** is responsible for perception and conscious behavior in mammals, as will be described later. In mammals, and especially in humans, the cerebrum grows so large that its surface seems to crumple, and it seems to spill over the sides. The hindbrain develops into the **cerebellum,** which coordinates learned movements. The cerebellum is large in birds relative to other vertebrates, presumably because of the greater demands on it for coordination in flight. All three divisions contribute to the

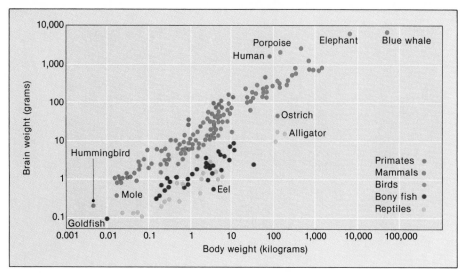

Figure 8.6
The relationship of brain mass to body mass in vertebrates. Birds and mammals generally have larger brains than fishes and reptiles of the same body size. The graph of brain mass to body mass for birds and mammals on this double-logarithmic plot is approximately linear, indicating that brain mass is generally proportional to (body mass)^p, where p is a number. The value of p here would be $^2/_3$, but graphs based on different data give a value of p equal to $^3/_4$.

brainstem, which is the part of the brain that would be left after removing the cerebrum and cerebellum.

THE BRAINSTEM

A major part of the brainstem is the **thalamus**—the upper part of the brainstem that is enclosed within the cerebrum. Much of the thalamus consists of **nuclei:**

Figure 8.7
Representative brains from several vertebrate classes drawn on a scale that emphasizes the relative size of different parts, rather than the actual size.

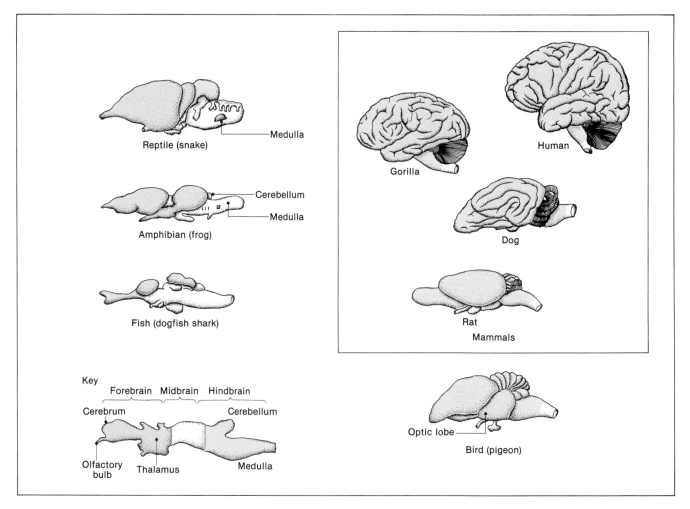

clusters of cell bodies and synapses that process information to and from the cerebrum. For example, the lateral geniculate nuclei process information from the retinas and forward it to the visual areas of the cerebrum. Just beneath the thalamus is the **hypothalamus** (Figure 8.8). As we shall see in the following chapters, the hypothalamus helps regulate feeding, water and ion balance, body temperature, and the secretion of hormones from the nearby pituitary gland.

Farther down the brainstem is the **tectum,** which coordinates reflex movements of the head and eyes toward a sound or a movement detected by the retina. Below it is the **pons** (Latin for bridge), whose horizontal axons bridge the two halves of the brain. The pons helps with orientation toward visual stimuli and also helps regulate breathing. A little farther down is the **medulla oblongata,** which regulates circulation and breathing, and which triggers the more-or-less vital functions of coughing, sneezing, swallowing, and hiccupping.

Running through the midbrain, pons, and medulla is the **reticular formation,** which maintains wakefulness by sending action potentials to all parts of the brain. The reticular formation is, in turn, activated by impulses from sensory areas of the brain. Sleep results when one part of the reticular formation turns off this activation. Incidentally, sleep is a necessity for most vertebrates, but it is not clear why. Some humans get along well without ever sleeping.

MOTOR FUNCTIONS OF THE CEREBRUM AND CEREBELLUM

Voluntary movements are somehow initiated in the motor areas of the cerebral cortex (Figure 8.9). As action potentials from the motor cortex descend toward the spinal cord, the pattern is modified by the **cerebellum** so that it is appropriate to circumstances. For example, the patterns of action potentials from the cerebrum that stimulate the muscles to maintain upright posture are modified by the cerebellum using information from joint and muscle receptors, from the retinas, and from the vestibular apparatus. The cerebellum also uses information from these receptors to coordinate skilled movements, such as throwing a ball. The cerebellum "computes" the weight of the ball and the distance of the target, then modifies the pattern of action potentials from the motor cortex accordingly as it descends

Figure 8.8
The human brain shown as if split down the middle. The corpus callosum is a band of axons connecting the two cerebral hemispheres.

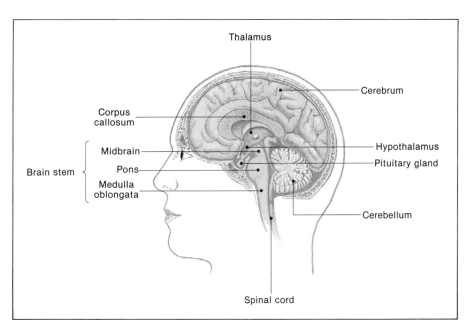

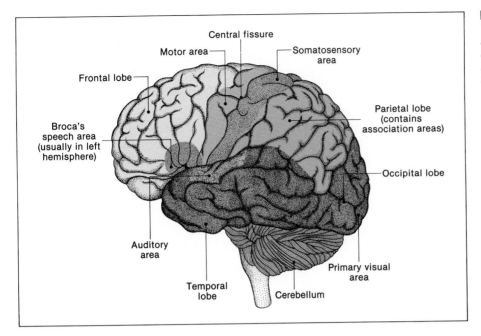

Figure 8.9
The location of the primary sensory and motor areas of the human left cerebral cortex. The major anatomical divisions and landmarks are indicated.

to the motor neurons. Damage to the cerebellum usually results in staggering, bizarre posture, and clumsiness. Another part of the cerebellum calculates one's position in the near future. Damage to this area will cause a person to continually bump into walls, even if the walls are clearly visible.

SENSORY FUNCTIONS OF THE CEREBRAL CORTEX

Only 350,000 efferent axons control all of a person's actions, but it takes two to three million afferent axons to provide enough information for the person to know how to act. Nearly half the sensory input enters through the optic nerves and nearly half through the spinal cord. The cerebrum is responsible for analyzing all the information brought in through the afferent neurons and for generating muscle contractions through the efferent neurons. Just as the motor area of the cerebral cortex has a specific location, so do the areas responsible for various kinds of sensory perception (Figure 8.9). Virtually all of the rear half of the cerebral cortex (the **occipital lobe**) is involved in visual perception. Perception of sound is localized in the **temporal lobes** near the ears. Just behind the motor cortex is the somatosensory cortex, which is responsible for perception of taste, pain, touch, and other bodily sensations. The **parietal lobe,** between the occipital lobe and the somatosensory area, is an association area, which coordinates information from all the sensory modalities.

Lateralization.   In spite of the diversity of tasks carried out by the cerebral cortex, it appears to be organized on similar principles throughout. One such principle is lateralization. Lateralization refers to the fact that each cerebral hemisphere is responsible for movements and sensations in the opposite side of the body. For example, the somatosensory cortex of the left hemisphere interprets information from somatic receptors of the right side of the body, and the visual cortex of the right cerebral hemisphere interprets visual stimuli seen on the left (right half of the retina). The **amytal test** demonstrates this lateralization dramatically. When the barbiturate amytal is injected into one carotid artery in the neck, the opposite half of the body immediately goes limp and numb.

Topographical Organization. A second principle is that adjacent areas of the cerebral cortex handle information from adjacent receptors. As shown in Figure 8.10A, for example, this topographical organization preserves the shape of an arrow as seen by the human eyes across the two areas of primary visual cortex. In the primary auditory cortex this principle gives a "tonotopic" organization: pitches that are close to each other excite areas of the auditory cortex that are close to each other. The somatosensory cortex is organized like a warped map of the body surface (Figure 8.10B).

Feature Extraction. The preceding paragraph might lead you to think that the sensory cortex is a kind of TV screen on which pictures of the world and the body are projected. If that were true, however, one would have to imagine a little person inside the brain watching the screen, with a little person in his brain watching a screen, and so on. In fact, the sensory cortex makes an abstract rather than a realistic picture of the world and the body. This abstract representation is due to a third principle of organization: Sensory areas of the cerebral cortex function as feature extractors. That is, instead of reproducing a replica of the stimulus, each area of the cerebral cortex detects a particular feature of the stimulus, such as color, form, or movement. After each section analyzes the feature it forwards the information to a higher level for integration.

Feature extraction is most easily understood for vision, thanks to studies pioneered by David Hubel and Torsten Wiesel using electrodes implanted into visual areas of the brains of cats, monkeys, and even human volunteers. Hubel, Wiesel, and many others have found that analysis of vision begins in the retina. Each ganglion cell of the retina responds only to stimulation in its **receptive field.** The receptive field of a neuron is the group of receptors that directly affects the neuron's electrical activity (Figure 8.10A). Each ganglion cell therefore responds only to stimulation in a small part of a visual stimulus. Receptive fields overlap, so each area of a visual stimulus is covered by several ganglion cells. Each ganglion cell is specialized for a different kind of feature extraction. Some ganglion cells respond only to red light and others to movement, for example.

Each ganglion cell transmits information about its receptive field in the form of action potentials that conduct through axons in the optic nerve to a **lateral geniculate nucleus.** After further processing in the lateral geniculate nucleus, the information goes to the **primary visual cortex.** Like a ganglion cell, each neuron in the lateral geniculate nucleus and in the primary visual cortex has a receptive field and responds only to particular features. For example, some cells in the primary visual cortex produce action potentials only if an edge in the receptive field is oriented within about 20° of a preferred angle (Figure 8.11). (Twenty degrees is equal to the movement the minute hand of a clock makes in 3 minutes and 20 seconds.)

Information from the primary visual cortex is then distributed through several hundred neural pathways that interconnect more than a dozen other visual areas. Each pathway transmits information about one particular feature, such as color, form, or movement. Finally, the information reaches two areas that are no longer selective for receptive field or for color. The first of these, in the parietal lobe, is responsible for perception of position and movement of objects. The second, in the temporal lobe, is responsible for recognition of objects. This separation of functions explains why damage to the parietal lobes can leave a person unable to determine the position of objects but still able to recognize them, and why damage to the temporal lobes can render a person incapable of recognizing familiar objects

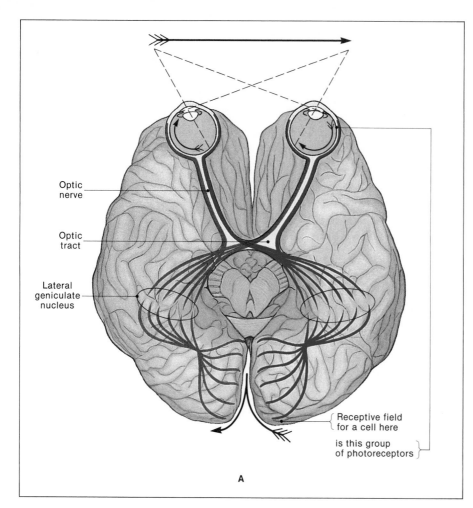

Figure 8.10

(A) An illustration of receptive fields and of the "retinotopic" organization of the primary visual cortex in the cat. Each neuron of the visual cortex responds only to a small group of receptors making up the receptive field. Adjacent areas of the visual cortex respond to adjacent areas of the visual field. The lateral geniculate nucleus is in the thalamus. (B) The somatotopic organization of the somatosensory and primary motor cortex. (C) (facing page) How the somatotopic organization of the brain is determined. Here the right cerebral cortex of an epilepsy patient is exposed under local anesthesia prior to surgery. The patient is facing right, and the temporal lobe, outlined by the large veins, is at the bottom of the picture. To map the motor cortex, different areas, marked by numbered pieces of paper, are electrically stimulated, and the resulting movements are recorded by the surgeons. A similar procedure is used to map sensory areas, except that the fully awake patient reports the sensations he feels as a result of the stimulation. In this patient stimulation of the following locations evoked the following results: (1) tingling in the left thumb; (2) tingling in left ring finger; (3) tingling in left middle finger; (4) flexing of left fingers and wrist; (8 and 13) complex memories.

Optic nerve

Optic tract

Lateral geniculate nucleus

Receptive field for a cell here

is this group of photoreceptors

A

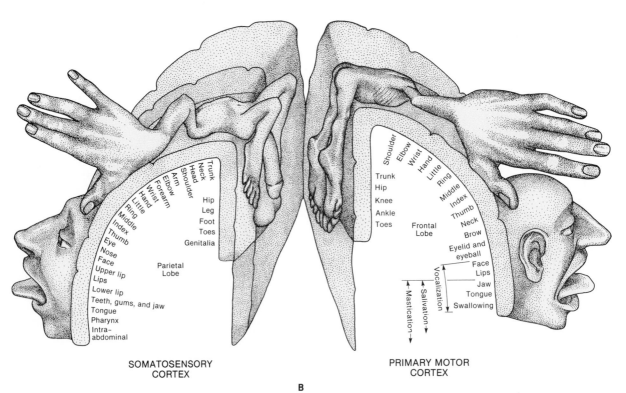

SOMATOSENSORY CORTEX

PRIMARY MOTOR CORTEX

B

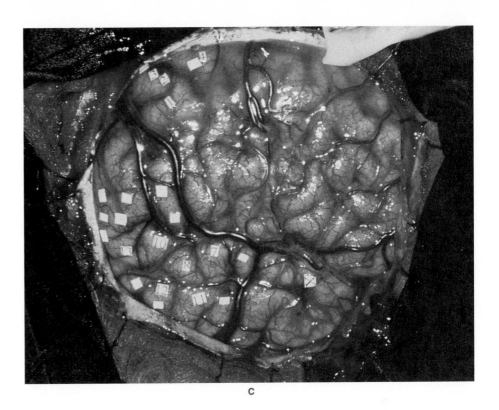

c

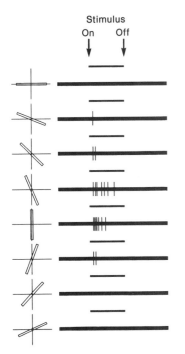

Stimulus
On   Off

Figure 8.11
The response of a single neuron in the primary visual cortex of a cat to a bar of light at various angles. For this simple cell the bar of light must be in the center of the receptive field and must be oriented at a certain angle (vertical in this example) in order to evoke a discharge of action potentials.

while still able to determine their location.

It is still largely a mystery how the cerebral cortex puts all this fragmented information back together into a coherent picture. Actually, it seems to make up much of the information, filling in details even when information from receptors is absent (Figure 8.12).

LANGUAGE

Language ability is largely restricted to **Broca's area** just anterior to the motor region for the lips and tongue (Figure 8.9). Broca's area is in the left cerebral hemisphere in 95% of right-handed people and 70% of left handers. Damage to Broca's area renders people incapable of speaking more than halting, "telegraphic" sentences, although they can still comprehend speech normally. Understanding of language occurs in **Wernicke's area,** which is between Broca's area and the primary visual and auditory areas of the left cerebrum. Damage to the pathways connecting the visual areas to Wernicke's area leaves people unable to read, even though they see normally and can understand speech. Damage between the auditory cortex and Wernicke's area leaves people unable to understand speech, without affecting the ability to read.

Normally the two halves of the brain exchange linguistic and other information through axons in the **cerebral commissure.** If the cerebral commissure is cut, however, as is sometimes done to control epilepsy, each half of the brain can think and behave independently of the other. Some of the consequences are startling. If a familiar object, like a hammer, is placed in the left hand of a blindfolded, commissurotomized ("split-brain") patient, he can use the object normally, just as anyone can. The patient, however, cannot name the object. If the object is placed in the right hand, he can name it. If two different scenes are flashed in the left and right visual fields, the patient can talk about the right visual field only,

Figure 8.12
A demonstration of the brain's ability to fill in missing information. With the left eye closed, focus the right eye on the circle and move the page closer or farther away until you find the distance where the X is no longer visible. At this point the image of the X is focused on the blind spot of the retina, where the optic nerves enter and receptors are lacking. Note that you do not see a black hole there, because the brain fills in the blind spot. It "assumes" that whatever is hidden in the blind spot looks just like the area surrounding it. Now advance the tip of a pencil slowly toward the X. Note that the tip seems to disappear as its image on the retina enters the blind spot. As you continue to advance the pencil, however, it suddenly appears whole as soon as the image of the tip passes out of the blind spot. Apparently the brain fills in the missing portion of the pencil by assuming that it is like the rest of the pencil.

although the right half of the brain can still perceive the left visual field. In one experiment a commissurotomized patient was shown a picture of a flower in the right visual field and a picture of a nude in the left. The patient grinned and said, "Wow! That's some flower!"

EMOTION

Not even the most elusive of all neural processes, those underlying emotions, have escaped the probing of neuroscientists. Particular emotions appear to be produced by particular areas of the brain, especially by a group of structures that encircle the brainstem and make up the **limbic system.** The limbic system includes part of the hypothalamus, the hippocampus, an almond-shaped structure called the **amygdala,** and others. The amygdala generates violent behavior, and some convicted criminals have voluntarily had their amygdalas removed as a means of controlling their violence. In the 1960s J. M. R. Delgado found that he could evoke or inhibit aggression by electrically stimulating the amygdala through electrodes implanted into the brains of human volunteers. Stimulating other areas of the brain evoked different emotions. Stimulating areas of the thalamus and temporal lobes often caused patients to report unexplainable feelings of anxiety. Some would get up and look behind doors and cabinets. Stimulating the frontal or temporal lobes often evoked pleasurable sensations that patients compared to orgasms. Such patients sometimes flirted with the interviewer. According to Delgado, the patients felt no less human while being electronically manipulated; they clearly understood the difference between the evoked reactions and their own free will.

SUMMARY

Nerve systems are complex in their organization and in the behaviors they produce, but parts—subsystems—can be understood. Simple connections between afferent and efferent nerve cells, such as occur in the vertebrate spinal cord, are responsible for reflex responses to stimulation. Another kind of subsystem, the servomechanism, continually monitors and corrects itself to control such behaviors as maintaining posture. Another subsystem is the central pattern generator, which produces patterns of action potentials underlying rhythmic behavior.

The nervous systems of different animals are extremely diverse in organization and complexity. Some, such as those of jellyfishes, are diffuse networks. Bilaterally symmetric animals generally have a ganglion at the anterior, and often several elsewhere. In vertebrates the peripheral nervous system includes an autonomic portion, with a sympathetic and para-

sympathetic division that has opposite effects on digestion, copulation, breathing, circulation, and other functions.

The vertebrate central nervous system comprises the brain and spinal cord. The brain consists of the cerebrum, cerebellum, and brainstem. The cerebellum coordinates learned movements. The outer layer of the cerebrum, the cerebral cortex, is divided among areas responsible for voluntary movements and for the perception of various types of sensation. The cortex of each cerebral hemisphere is responsible for movements and sensations of the opposite side of the body. It has a topographical organization, with adjacent areas of cortex responsible for adjacent areas of the body. Each nerve cell responds to selective features of its receptive field, such as color or pitch. In most people an area in the left cerebral cortex is responsible for language. Various areas of the brain generate emotions.

KEY TERMS

reflex
gray matter
white matter
dorsal root
ventral root
muscle spindle
servomechanism
central pattern generator
nerve net

ganglion
autonomic nervous system
sympathetic division
parasympathetic division
cerebrospinal fluid
choroid plexus
glial cell
cerebrum
cerebellum

brainstem
cortex
somatosensory cortex
lateralization
receptive field
Broca's area
cerebral commissure

SELF-TEST

1. Sketch a simple diagram that could account for the eye-blink reflex when something touches the cornea.

2. In addition to reflexes, other neural subsystems include servomechanisms and central pattern generators. Which of these would be responsible for the coordination of eye and hand movements? Which could account for the movements of a dog's leg when it scratches?

3. Describe some levels of neural organization found among invertebrates, proceeding from simplest to most complex.

4. Define the following: central nervous system, peripheral nervous system, autonomic nervous system. sympathetic division, parasympathetic division.

5. What types of function are handled by various parts of the brainstem?

6. Describe the major features of cerebral organization. That is, how are the functions of the cerebrum distributed anatomically?

7. What parts of the brain are involved in the coordination of movement, and how are they involved?

8. Give an example of feature extraction.

9. Trace the flow of information through your brain as you read this question, formulate an answer, and write it.

READINGS

RECOMMENDED READINGS

Aoki, C. and P. Siekevitz. 1988. Plasticity in brain development. *Sci. Am.* 259(6):56–64 (Dec).

Camhi, J. M. 1980. The escape system of the cockroach. *Sci. Am.* 243(6):158–172 (Dec).

Delgado, J. M. R. 1969. *Physical Control of the Mind.* New York: Harper & Row.

Gallistel, C. R. 1980. From muscles to motivation. *Am. Sci.* 68:398–408.

Glickstein, M. 1988. The discovery of the visual cortex. *Sci. Am.* 259(3):118–127 (Sept).

Goldstein, G. W. and A. L. Betz. 1986. The blood–brain barrier. *Sci. Am.* 255(3):74–83 (Sept).

Hubel, D. H. 1979. The brain. *Sci. Am.* 241(3):44–53 (Sept). (*Introduction to an entire issue on the brain.*)

Hubel, D. H. 1988. *Eye, Brain, and Vision.* New York: Scientific American Books.

Jerison, H. J. 1976. Paleoneurology and the evolution of mind. *Sci. Am.* 234(1):90–101 (Jan). (*The relation of brain size to intelligence among different species.*)

Kalil, R. E. 1989. Synapse formation in the developing brain. *Sci. Am.* 261(6):76–85 (Dec).

Kimelberg, H. K. and M. D. Norenberg. 1989. Astrocytes. *Sci. Am.* 260(4):66–76 (Apr). (*The varied functions of a glial cell.*)

Livingstone, M. S. 1988. Art, illusion and the visual system. *Sci. Am.* 258(1):78–85 (Jan).

Livingstone, M. and D. Hubel. 1988. Segregation of form, color, movement, and depth: anatomy, physiology, and perception. *Science* 240:740–749.

Mishkin, M. and T. Appenzeller. 1987. The anatomy of memory. *Sci. Am.* 256(6):80–89 (June).

Morrison, A. R. 1983. A window on the sleeping brain. *Sci. Am.* 248(4):94–102 (Apr).

Sacks, O. 1985. *The Man Who Mistook His Wife for a Hat.* New York: Summit Books. (*Fascinating and instructive account of behavioral consequences of neurological disorders.*)

Spector, R. and C. E. Johanson. 1989. The mammalian choroid plexus. *Sci. Am.* 261(5):68–74 (Nov).

ADDITIONAL REFERENCES

American Zoologist 30:403–714 (1990) has several excellent papers on neurobiology and behavior.

Maunsell, J. H. R. and W. T. Newsome. 1987. Visual processing in monkey extrastriate cortex. *Annu. Rev. Neurosci.* 10:363–401.

9

Hormones and Other Molecular Messengers

Hormone-induced sex differences in evening grosbeaks (Hesperiphona vespertina).

CHAPTER OUTLINE

LEARNING OBJECTIVES

1. What kinds of chemical messages help coordinate physiological functioning?

2. Where are the hormones of humans and other vertebrates produced?

3. How do hormones work?

4. What is the connection between the nervous system and the endocrine system?

5. Do invertebrates also have hormones?

CHEMICAL MESSENGERS

An organism has the same problem coordinating its cells that a large organization has in coordinating its members. Information should reach the intended target, and none other. There are two approaches to solving this problem. One is to use a common language but direct the information through a channel that goes only to the recipient. This method is used when one person calls another on the telephone and when the brain sends action potentials to particular muscle cells. The other approach is to broadcast the message widely but in a language that only the intended target can understand. This approach is used when spies transmit coded messages by radio. It is also the method used by hormones and other chemical messengers responsible for coordinating cellular activity in animals. Many cells intercept the message, but only target cells have the receptor molecules to decode it.

The best-known chemical messengers are hormones. By definition, **hormones are secreted by specific tissues, and they travel through the blood to evoke a response in distant cells.** The organs that secrete hormones are called **glands.** Since hormones travel within the circulatory system, they are also called **endocrine** secretions, to distinguish them from exocrine secretions such as sweat, tears, and digestive juices, which stay out of the bloodstream.

Besides hormones, there are other chemical messengers that resemble hormones in their chemical properties or modes of action but that, for one reason or another, do not fit the definition of a hormone. Among these "hormonelike" messengers are tissue factors that coordinate local cellular activities (Table 9.1). Like hormones, tissue factors are secreted by specific cells and interact with target cells by binding to molecular receptors. Unlike hormones, tissue factors act locally and do not travel through the bloodstream. Synaptic transmitters also work like hormones in that they are released from specific cells and bind to particular receptors on other cells. In fact, certain nerve cells, called **neurosecretory cells,** release hormones in the same way that nerve cells release transmitter. The similarities among hormones, tissue factors, and synaptic transmitters may reflect a common evolutionary origin (Figure 9.1).

Prostaglandins.   Among the most common hormonelike substances are the prostaglandins. These were first discovered in the 1930s by researchers who noted that something in human semen caused the smooth muscle of the uterus to con-

Table 9.1   Some hormonelike substances.

INVERTEBRATES

Neurosecretions in *Hydra* (phylum Cnidaria) and in the planarian *Dugesia* (phylum Platyhelminthes) stimulate growth, regeneration, and gonadal activity. Little is known of the structure of these substances.

VERTEBRATES

Vitamin D is a steroidlike substance produced by skin exposed to sunlight. Vitamin D is converted in the liver and kidney to a form required for the absorption of Ca^{2+} across the intestinal lining.
Kinins are peptides that coordinate local blood flow by relaxing arterial smooth muscle and increasing the flow out of blood capillaries. They also cause contraction of smooth muscle in the uterus and gut. Kinins are secreted in an inactive form and are activated by an enzyme called **kallikrein.**
Histamine, which is similar to the amino acid histidine, is released from **mast cells** of injured tissues, triggering swelling and other signs of inflammation (see p. 236).
Prostaglandins See the discussion under this heading and Figure 9.2.

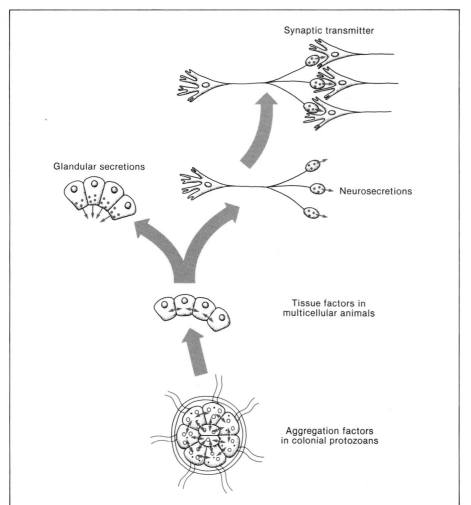

Figure 9.1
A comparison of various types of chemical coordination, showing their possible evolution. Secretions such as aggregation factors in colonial protozoans may have been the precursors in multicellular animals of the factors that coordinate development and activities of tissues. Certain of those tissues that secreted these factors may then have become specialized as endocrine glands, and others as neurosecretory cells. Some of the latter may then have evolved into nerve cells.

Synaptic transmitter

Glandular secretions

Neurosecretions

Tissue factors in multicellular animals

Aggregation factors in colonial protozoans

tract. Thinking that the substance came from the prostate gland, they called it prostaglandin. That particular prostaglandin actually came from the seminal vesicles, and other prostaglandins are secreted by a variety of other tissues. Prostaglandins break down so quickly that they have only local effects. They are sometimes referred to by the self-contradictory term "local hormone."

The basic prostaglandin structure has 20 carbons with a five-carbon ring (Figure 9.2). Prostaglandins are synthesized from fatty acids and retain the lipid solubility that allows them to spread through tissue. Besides affecting uterine smooth muscle, various prostaglandins affect blood pressure, blood clotting, inflammation, and numerous other processes. Aspirin appears to act as an anti-inflammatory drug by inhibiting the formation of one type of prostaglandin, PGE. Prostaglandins occur not only in mammals but also in other vertebrates and in invertebrates. In fact, the richest source of raw materials for the manufacture of prostaglandins is the sea whip coral *Plexaura homomalla* (phylum Cnidaria).

SOME MAJOR HORMONES OF VERTEBRATES

There are so many hormones that a single chapter can do little more than survey the best-known ones, which are naturally those of humans and other mammals. Fortunately, such a survey is not as limited in scope as it might seem, because the endocrine organs and hormones that occur in mammals also occur in most other

COO⁻

CH₃

HO

OH

PGE₁

Figure 9.2
A prostaglandin. Prostaglandins are synthesized from fatty acids in cell membranes.

Tools of Endocrinology

If you had been born a century ago you would not now be bothered with a chapter on hormones, for none had been discovered. The existence of hormones was first demonstrated by W. M. Bayliss and E. H. Starling in 1902, at a time when the most sophisticated apparatus available was a simple lever that scratched a mark on a smoked drum when a drop of fluid fell on the lever. Using this device, Bayliss and Starling found that a dog's pancreas increased its rate of secretion after acidic solution was injected into the intestine, even after nerves to the pancreas had been cut. This finding suggested that a chemical messenger from the intestine circulated to the pancreas and stimulated secretion. In order to test this hypothesis, they crushed some of the lining of the gut after it had been exposed to acid, filtered it, then injected the extract into the blood of another dog. Within about a minute this dog's pancreas increased its rate of secretion. Thus was discovered the first hormone, now called secretin.

For decades thereafter the standard techniques of endocrinology remained as simple as these. One approach was to remove a suspected endocrine organ, note the effects, then see if the effects were reversed by injecting extracts from the gland. The only problem with this approach is that hormones work in such minute quantities that it may take box-car loads of gland to get enough extract to test. Such techniques continue to bear fruit, however, especially in preliminary studies. The revolution in molecular biology has made possible an arsenal of more sophisticated techniques that have enabled endocrinologists to rapidly expand the list of known hormones. (Be grateful you don't have to study endocrinology a century from now!) Some of the techniques that have revolutionized endocrinology are described below.

1. **Tissue culture.** Small amounts of tissue kept alive in dishes (*in vitro*) can often be used to determine whether a suspected hormone acts on the tissue. Much smaller samples are needed, and conditions can be carefully controlled.

2. **Antibody labeling.** This technique is used to locate tissues that secrete a known hormone. Antibodies to a hormone are first obtained by injecting the hormone into an animal in which it does not normally occur. The animal then produces antibodies against the hormone. These antibodies are collected, coupled to radioactive isotopes, fluorescent molecules, or other markers and applied to slices of tissue from an animal that produces the hormone. The antibody will bind to any molecule of the hormone, revealing where it is secreted.

3. **Radioimmunoassay (RIA).** This technique is used to measure the **titer** (concentration) of minute amounts of hormone in solution. Antibodies to a hormone, plus a known amount of radioactively labeled hormone, are added to a sample containing an unknown titer of hormone. The unlabeled molecules already in the sample will compete with the labeled molecules for binding sites on the antibodies. The degree of competition will be proportional to the titer. After rinsing away all the hormone that did not find antibody to bind to, a measurement of radioactivity indicates how much labeled hormone did bind, and therefore how much unlabeled hormone there must have been competing with it in the sample.

4. **Protein sequencing and synthesis.** The sequence of amino acids in a protein hormone can be determined by breaking up the hormone with various digestive enzymes that cleave peptide bonds adjacent to particular amino acids. This procedure yields a variety of small, overlapping fragments from which the original sequence can be determined. Once the sequence is determined, unlimited quantities of the peptides can be synthesized for study. Sequencing and synthesis are now automated.

---

vertebrates, although they sometimes have different functions. For example, prolactin, which stimulates milk production in female mammals, also occurs in amphibians, which don't produce milk. In amphibians prolactin stimulates the regeneration of lost limbs, as well as some reproductive functions. Figure 9.3 and Table 9.2 summarize the major endocrine organs and best-known hormones of mammals and most other vertebrates. The hormones of invertebrates are quite different from those of vertebrates and will be surveyed later in this chapter.

WHAT HORMONES DO

The numerous hormones listed in the previous section obviously have diverse functions. They are similar, however, in that almost all act on many different cells,

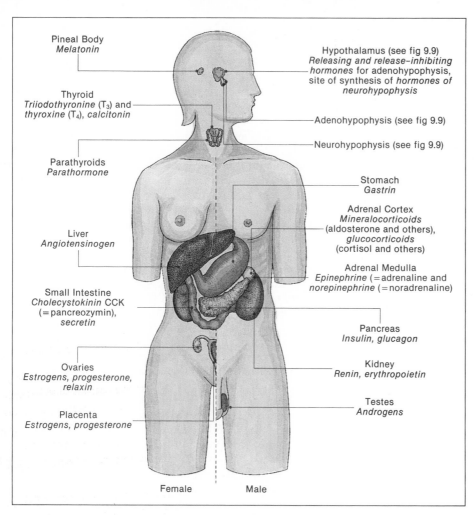

Table 9.2 **Major endocrine organs and their hormones in human and many other vertebrates.**

HYPOTHALAMUS

(See Figure 9.9)
Releasing and release-inhibiting hormones to the anterior pituitary.
Hormones of the posterior pituitary.

ANTERIOR PITUITARY

(See Figure 9.9)

POSTERIOR PITUITARY

(See Figure 9.9)

PINEAL GLAND (= PINEAL BODY)

Melatonin. Derivative of tryptophan. Secreted in humans and many other mammals with a daily rhythm. In many vertebrates the level of melatonin appears to help maintain rhythms, such as the seasonal rhythm of reproduction (see p. 420). (In amphibians, melatonin lightens skin color by inhibiting secretion of melanocyte-stimulating hormone MSH from the anterior pituitary.)

THYROID

Triiodothyronine T_3 and **Thyroxine** T_4. Derived from two tyrosine molecules with three or four iodine atoms attached (see Figure 9.6). T_4 is converted to T_3, the active form. T_3 is lipid soluble and diffuses into cells, where it binds to receptor molecules. T_3 and T_4 increase metabolic rates by stimulating mitochondrial activity. Also essential for development of the nervous system. (Thyroxine stimulates metamorphosis in many animals, such as jellyfish, fishes, and frogs.)
Calcitonin. A peptide. Reduces Ca^{2+} levels in blood by stimulating uptake by bone.

PARATHYROIDS

Parathyroid hormone. A peptide. Increases Ca^{2+} levels in the blood by stimulating removal from bone and the excretion of phosphate (PO_4^{3-}) by kidney. Stimulates conversion of vitamin D to its active form, which increases Ca^{2+} absorption by the intestine.

ADRENAL CORTEX

Mineralocorticoids (aldosterone and others). Steroids (see p. 188). Increase volume of body fluids by stimulating Na^+ reabsorption by kidney.
Glucocorticoids (cortisol and others). Steroids. Increase glucose levels in blood by stimulating conversion of fat and amino acids to glucose. Stimulates repair of tissue.
Estrogens and **Testosterone.** See below, under testes and ovaries.

ADRENAL MEDULLA (= CHROMAFFIN TISSUE)

Epinephrine (= adrenaline) and **Norepinephrine** (= noradrenaline). (See Figure 9.6.) Catecholamines, derived from tyrosine. Increase pumping of blood, blood pressure, and oxygen consumption. (The adrenal medulla is best considered a part of the sympathetic nervous system.)

HEART

Atrial natriuretic hormone (= cardionatrin, atriopeptin, or auriculin: discovered in 1984; there is not yet a consensus on a name). A peptide. Reduces blood volume and blood pressure.

KIDNEY

Renin (pronounced REE-nin to avoid confusion with the digestive enzyme rennin). A protein. Activates angiotensin II hormone (see under liver) in response to low blood flow in the kidneys.
Erythropoietin. A glycoprotein. Stimulates production of red blood cells in response to reduced availability of O_2.

LIVER

Angiotensinogen. A peptide. Precursor of angiotensin II, activated by renin in the blood. Angiotensin II increases blood pressure, stimulates thirst, and stimulates the secretion of aldosterone by the adrenal glands and antidiuretic hormone from the posterior pituitary.

PANCREAS

Insulin. A protein. Decreases concentrations of glucose in blood by stimulating uptake of glucose by most cells and by stimulating its conversion to glycogen by liver and muscle.
Glucagon. A protein. Increases concentration of glucose in blood by stimulating breakdown of glycogen in liver and muscle.

STOMACH

Gastrin. A protein. Secreted in response to peptides and proteins in stomach. Stimulates secretion of acid and protein-digesting enzyme by stomach.

SMALL INTESTINE

Cholecystokinin CCK (pronounced ko-lay-SIS-toe-KY-nin) (= pancreozymin PZ). A peptide. Secreted in response to fats and amino acids in small intestine. Stimulates release of digestive enzymes by pancreas and release of bile by gallbladder.
Secretin. A peptide. Secreted in response to food and acid in stomach and intestine. Stimulates release of bicarbonate by pancreas. Inhibits stomach motility.

TESTES

Androgens, such as testosterone. Steroids. Promote male behavior, morphology, and reproductive functions.

OVARIES

Estrogens, such as estradiol. Steroids. Promote female behavior, morphology, and reproductive functions.
Progesterone. A steroid. Promotes functioning of uterus and breasts.
Relaxin. A peptide. Relaxes pelvic ligaments and cervix to ease delivery of infant.

PLACENTA

Estrogens and **Progesterone.** See above under ovaries.
Human chorionic gonadotropin hCG. A protein. Essential for maintenance of pregnancy.

usually over periods ranging from minutes to perhaps years. In this respect the endocrine system contrasts with the nervous system, which generally acts on limited sites in the body, usually quite briefly. The nervous and endocrine systems are not entirely separate, however. Some nerve cells secrete hormones, and some stimulate endocrine glands to release their hormones. Many hormones, in turn, affect the functioning of the nervous system. Another generalization that can be made about hormones is that almost always they are secreted in response to an imbalance in one of the requirements for the lives of cells in the internal environment: nutrients, oxygen, water, ions, or temperature. The hormone helps correct the imbalance, thereby helping to maintain **homeostasis.** (The major exceptions to this generalization are the reproductive hormones, which do not help maintain homeostasis.)

Other chapters in this unit will present many examples of how hormones help regulate homeostasis, but just one example will be presented briefly here to illustrate the principles involved. The requirement for life that is regulated in this example is the blood sugar glucose, which supplies most of the energy requirements of cells. The hormone to be describe is **insulin,** although several others are also important in regulating blood glucose concentrations. As noted above, insulin comes from certain cells of the pancreas, and its main effect is to reduce the level of glucose in the blood by stimulating most cells to take in glucose and by stimulating the liver and muscles to store it as glycogen. Normally, however, insulin is secreted only when the blood glucose level rises above normal, as it does following a meal containing starches or other carbohydrates. The increasing level of glucose in the blood directly stimulates the cells of the pancreas to release insulin, which then circulates through the bloodstream, stimulating cells to take in the excess glucose. In this way the blood glucose level is prevented from becoming so high that it might cause glucose to be wasted in the urine or affect the functioning of the brain.

HOW HORMONES WORK

Steroid Hormones. It is convenient to divide hormones into two major classes—steroids and nonsteroids. Because of the chemical differences between the two classes, they work on their target cells in entirely different ways. Steroid hormones are secreted mainly by the gonads (androgens and estrogens) and the adrenal cortex (glucocorticoids and mineralocorticoids). They have a four-ring molecular structure derived from cholesterol (Figure 9.4), and they can usually dissolve in lipids. Thus they can diffuse directly across the lipid bilayer of plasma membranes and work within cells. Once inside a target cell the steroid binds to a specific receptor molecule. In the cell's nucleus the hormone–receptor complex alters the transcription of genes, thereby changing the rate of production of particular proteins (Figure 9.5). Only certain cells have the receptor molecule for a given steroid, so even though steroid hormones penetrate all cells, they normally act only on target cells.

Nonsteroid Hormones. Most nonsteroid hormones are proteins, glycoproteins, peptides, or catecholamines (epinephrine, norepinephrine, and dopamine). The molecular receptors for these hormones are glycoproteins bound to the plasma membrane of target cells. Nonsteroid hormones are generally not lipid soluble and are therefore usually unable to penetrate cell membranes. There are at least two exceptions, however. First, the small thyroid hormones, thyroxine and triiodothyronine, are lipid soluble and act on intracellular receptors in much the same

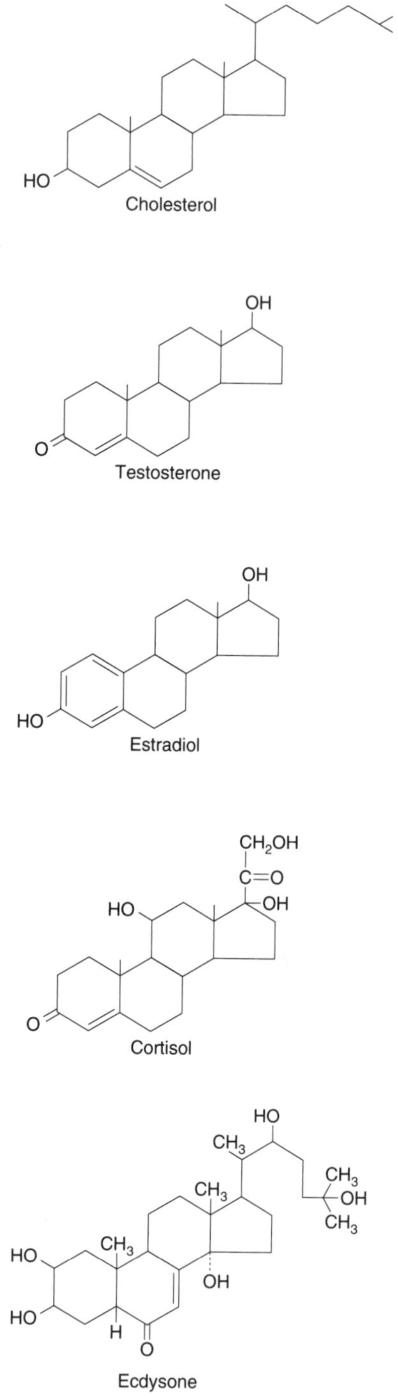

Figure 9.4
Cholesterol and some of the steroid hormones derived from it. All the steroids share a similar four-ring structure; the variety of their effects is due entirely to variations in side groups. All these hormones occur in vertebrates, except for ecdysone, which occurs in arthropods.

Figure 9.5

A representation of the action of steroid hormones. An essential feature of steroids is their lipid solubility, which allows them to pass through membranes and evoke their responses within target cells, which have receptors for them.

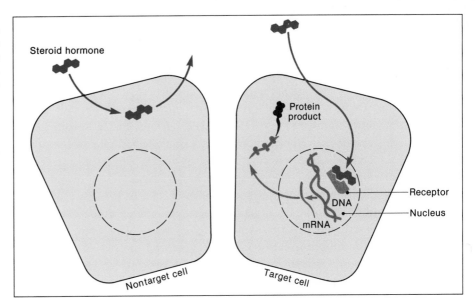

way that steroid hormones do (Figure 9.6). Second, after insulin binds to the receptor on the target cell surface, the receptor and bound insulin are both brought into the cell. The receptor then serves as an enzyme mediating the effect of insulin (Rosen 1987). Whether the insulin has any further effect in the cell is not known. Other nonsteroid hormone receptors may work in a similar fashion.

The effectiveness of a hormone depends on its concentration and also on the number of receptor molecules and how well they bind to the hormone. Changing the number and sensitivity of receptors is one way that endocrine systems regulate themselves. Prolonged exposure to a hormone may cause a cell to reduce the number of receptors for that hormone or for a different one. Progesterone, for example, normally decreases the effectiveness of estrogens by reducing the number of receptors for them. There can also be abnormal changes in receptor sensitivity due to genetic defects in the structure of the receptor or to antibodies that the immune system has, for some reason, begun producing against the receptors. Most cases of diabetes result from a reduction in the binding of insulin receptors to insulin, rather than an inability of the pancreas to produce insulin.

Second Messengers. How does the binding of a few molecules of nonsteroid hormones on the outer surface of a cell trigger a response in the cell? The answer is that second messenger molecules carry the signal from the hormone receptor

Figure 9.6

The structures of some of the nonsteroid hormones. The thyroid secretes thyroxine T_4 and triiodothyronine T_3. Dopamine from the hypothalamus, norepinephrine from the sympathetic nervous system, and epinephrine from the adrenal medulla are all catecholamines. See Figure 9.10 for other nonsteroids.

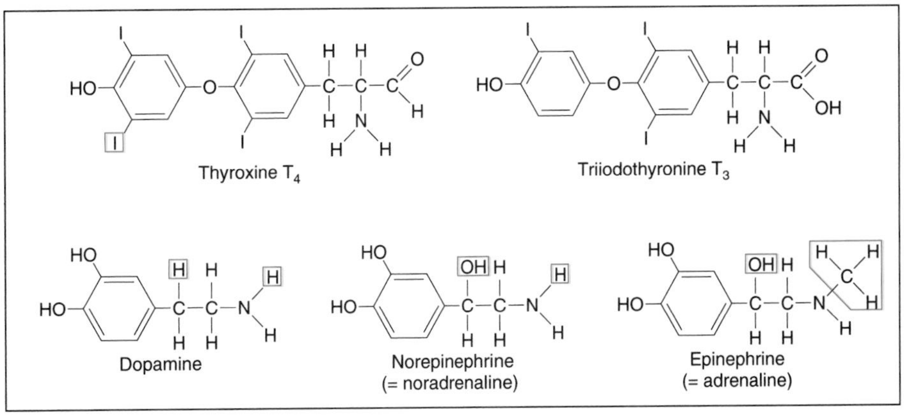

into the cell. Each second messenger molecule then triggers a number of reactions, thereby amplifying the response of the cell. The best-known second messenger is **cyclic adenosine monophosphate** (cAMP) (Figure 9.7). Cyclic AMP is like the ordinary adenosine monophosphate of RNA except that its phosphate is attached to the ribose at two places rather than at just one. Cyclic AMP is produced from ATP in response to the activation of a **G protein,** which occurs after the hormone binds to its receptor (Figure 9.8). Cyclic AMP then activates enzymes that catalyze the target cell's response to the hormone. For example, when glucagon binds to its receptor on liver cells, the resulting increase in cAMP triggers an increase in the activity of enzymes that convert glycogen into glucose. The following hormones also work through cAMP as second messengers: MSH, ACTH, TSH, ADH, LH, parathyroid hormone, norepinephrine, and epinephrine.

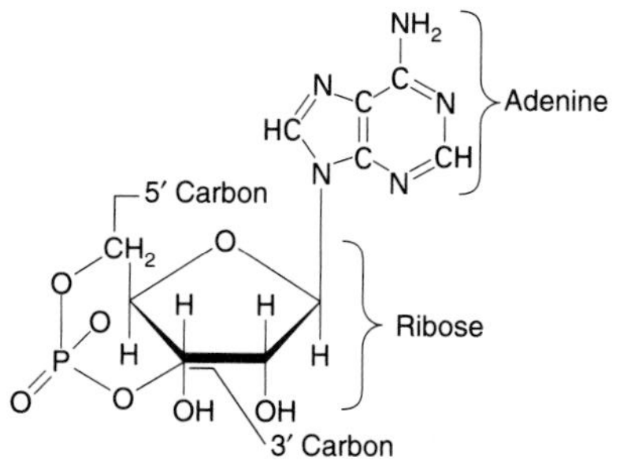

Cyclic AMP is only one of several known second messengers. Some cells have **cyclic guanosine monophosphate** (cGMP) as a second messenger, and others use Ca^{2+}. Others use a **polyphosphoinositide system** in which the binding of hormone to receptor causes a membrane phospholipid (phosphatidyl inositol; see Figure 2.12) to split. Each of the products, **inositol triphosphate** (IP_3) and **diacylglycerol** (DAG), then serves as a second messenger. All the second messenger systems function not only in response to hormones but also in response to neurotransmitters and tissue factors.

Figure 9.7
Cyclic adenosine monophosphate (cAMP).

HORMONES OF THE HYPOTHALAMUS AND PITUITARY

The pituitary (also called the **hypophysis**) is only the size of a pea in humans, but it is probably the most important endocrine gland, because it regulates the secretion of so many other hormones. The pituitary is therefore traditionally called "the master gland." In fact, however, the pituitary simply passes on instructions from the hypothalamus, so it might be more appropriate to call it the "overseer gland." The pituitary is actually two distinct types of tissue that come together during embryonic development (Figure 9.9). Each part takes its orders from the hypothalamus in a different language. The **anterior pituitary** develops from tissue

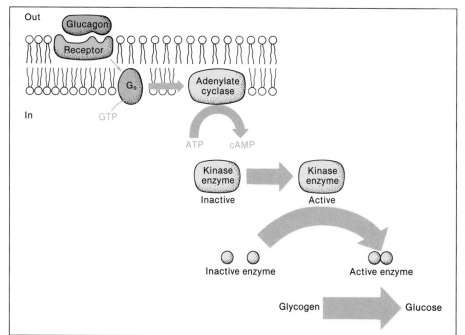

Figure 9.8
The action of glucagon mediated by cyclic adenosine monophosphate (cAMP). Glucagon is released when the amount of glucose in the bloodstream declines. The binding of glucagon to its receptor on the plasma membrane of a liver cell activates a G protein (G_s), which binds to guanosine triphosphate (GTP). The activated G protein stimulates the production of cAMP, a second messenger. cAMP then activates a kinase enzyme that attaches a phosphate group to an inactive form of another enzyme. The active enzyme then catalyzes the breakdown of glycogen and the release of glucose by the liver cell. The increasing widths of the arrows indicate that each step in the sequence amplifies the response.

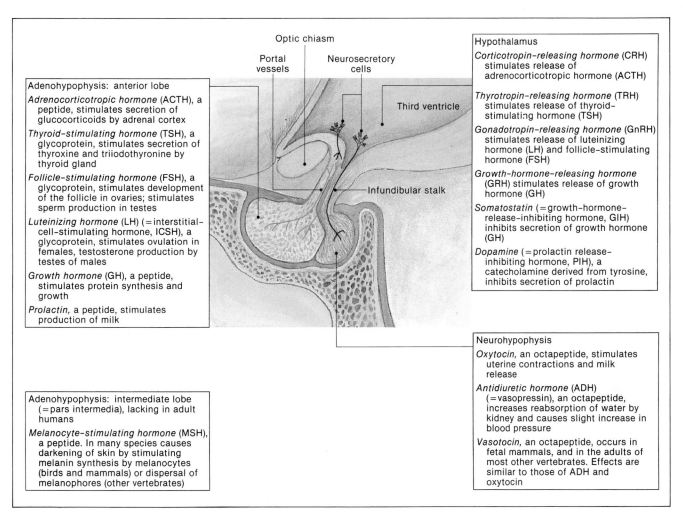

Adenohypophysis: anterior lobe

Adrenocorticotropic hormone (ACTH), a peptide, stimulates secretion of glucocorticoids by adrenal cortex

Thyroid-stimulating hormone (TSH), a glycoprotein, stimulates secretion of thyroxine and triiodothyronine by thyroid gland

Follicle-stimulating hormone (FSH), a glycoprotein, stimulates development of the follicle in ovaries; stimulates sperm production in testes

Luteinizing hormone (LH) (=interstitial-cell-stimulating hormone, ICSH), a glycoprotein, stimulates ovulation in females, testosterone production by testes of males

Growth hormone (GH), a peptide, stimulates protein synthesis and growth

Prolactin, a peptide, stimulates production of milk

Hypothalamus

Corticotropin-releasing hormone (CRH) stimulates release of adrenocorticotropic hormone (ACTH)

Thyrotropin-releasing hormone (TRH) stimulates release of thyroid-stimulating hormone (TSH)

Gonadotropin-releasing hormone (GnRH) stimulates release of luteinizing hormone (LH) and follicle-stimulating hormone (FSH)

Growth-hormone-releasing hormone (GRH) stimulates release of growth hormone (GH)

Somatostatin (=growth-hormone-release-inhibiting hormone, GIH) inhibits secretion of growth hormone (GH)

Dopamine (=prolactin release-inhibiting hormone, PIH), a catecholamine derived from tyrosine, inhibits secretion of prolactin

Labels on figure: Optic chiasm; Portal vessels; Neurosecretory cells; Third ventricle; Infundibular stalk

Neurohypophysis

Oxytocin, an octapeptide, stimulates uterine contractions and milk release

Antidiuretic hormone (ADH) (=vasopressin), an octapeptide, increases reabsorption of water by kidney and causes slight increase in blood pressure

Vasotocin, an octapeptide, occurs in fetal mammals, and in the adults of most other vertebrates. Effects are similar to those of ADH and oxytocin

Adenohypophysis: intermediate lobe (=pars intermedia), lacking in adult humans

Melanocyte-stimulating hormone (MSH), a peptide. In many species causes darkening of skin by stimulating melanin synthesis by melanocytes (birds and mammals) or dispersal of melanophores (other vertebrates)

Figure 9.9
Hormones of the hypothalamus and pituitary in humans.

in the roof of the mouth and secretes hormones in response to **releasing hormones** and **release-inhibiting hormones** from the hypothalamus. All the releasing and release-inhibiting hormones are peptides except one (dopamine). They reach the anterior pituitary by means of a short **portal vessel.**

The **posterior pituitary** (= neurohypophysis) develops as an extension of the hypothalamus, and it releases its hormones in response to action potentials. The hormones of the posterior pituitary are actually produced in cell bodies in the hypothalamus. They travel down axons and are released from large vesicles in much the same way that synaptic transmitters are released. The cells that secrete the hormones are therefore **neurosecretory cells.** The neurosecretions from the neurohypophysis are octapeptides (peptides with eight amino acids). At least ten are secreted by the posterior pituitaries of various vertebrates. Oxytocin and antidiuretic hormone (= vasopressin) are the two secreted by the human posterior pituitary (Figure 9.10).

Stress. As a part of the brain, the hypothalamus coordinates the activity of the endocrine system with that of the central nervous system. Many of its regulatory functions will be described in later chapters in this unit, but for now it will suffice to give an illustration of how the hypothalamus and pituitary integrate neural and hormonal responses. The stimulus in this example is stress, which physiologists define as any physiological or psychological challenge or trauma that tends to upset homeostasis. Pain, threat, or some other stressful stimulus triggers the hypothalamus to release **corticotropin releasing hormone** (CRH). CRH is a peptide that circulates to the anterior pituitary, where it stimulates the release of **adrenocorticotropic hormone** (ACTH).

ACTH is a peptide that circulates to the adrenal cortex, stimulating it to secrete a number of steroid hormones. The most important of these hormones in combatting stress are the **glucocorticoids,** especially **cortisol.** Glucocorticoids stimulate the conversion of amino acids into glucose as an energy source. They also have near-miraculous powers to promote healing. For a while after their discovery in the late 1940s they were used for everything from severe burns to diaper rash. Physicians became more conservative, however, after observing that prolonged use produces such side effects as loss of hair, brittleness of bones, sterility, and suppression of immune responses. Normally the endocrine system does not secrete glucocorticoids or other potentially harmful hormones for long periods. ACTH limits the secretion of glucocorticoids by inhibiting its own release by the anterior pituitary, and also by inhibiting the release of CRH from the hypothalamus.

SOME HORMONES OF INVERTEBRATES

Virtually every invertebrate examined has been found to produce hormones. The chemical identity and mode of action of invertebrate hormones are often quite different from those of vertebrates. Table 9.3 indicates some of the most important hormones of invertebrates and the functions they serve.

HORMONAL COORDINATION OF INSECT MOLTING

Rather than give a cursory glance at a large number of invertebrate hormones, it may be more rewarding to see how hormones help coordinate just one function—molting in insects. Insects have to molt in order to grow and to develop into adults (see pp. 635–637). For his pioneering studies on molting, Vincent B. Wigglesworth chose an animal that some might consider unsavory—the blood-sucking bug *Rhodnius.* This bug normally molts four times into a juvenile of increasing size, and then a fifth time into an adult. Wigglesworth found that if the last **instar** (juvenile stage) of *Rhodnius* were joined to an early instar, the last-instar juvenile molted into an extra juvenile stage rather than into an adult. Evidently a substance in the early instar, called **juvenile hormone** (JH), suppressed the development of adult features. Subsequent experiments revealed that JH is normally secreted in juveniles until the final molt, and that JH must be absent for development into an adult. Wigglesworth's simple but effective experiments left an impression on an entire generation of insect endocrinologists. Figure 9.11B shows a more literal way in which Wigglesworth left his mark.

It occurred to Carroll M. Williams of Harvard University that Wigglesworth's discovery could be used as the basis for a "third-generation insecticide." First-generation insecticides, using such natural toxins as nicotine, had failed because insects quickly learned to avoid the strong taste or odor of the substances. Second-generation insecticides, using synthetic chemicals such as DDT, had failed because the insects had become tolerant to them. These problems would not be expected to occur with a synthetic juvenile hormone or anti-juvenile hormone, however. Chemicals that disrupted the functioning of JH in insects would also be safer than previous insecticides, since they would not affect other animals.

It turned out that plants had come up with the same idea long before Williams. While working in Williams' lab, Karel Slàma found that the insects he had brought from Czechoslovakia would not develop normally. He eventually traced the problem to American paper towels that he had placed in the insect containers.

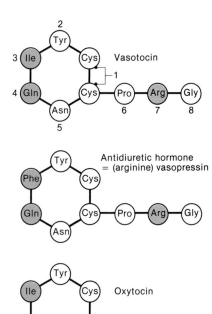

Figure 9.10
Three of the ten or more hormones secreted from the neurohypophyses of various vertebrates. Of these, nine octapeptides are identical except at the sites numbered 3, 4, and 7. Arginine vasotocin appears to be the most primitive of these octapeptides and may have been the one from which the others evolved. It is replaced by vasopressin in adult mammals. The vasopressin of most mammals is often called arginine vasopressin to distinguish it from lysine vasopressin in pigs, which differs at amino acid number 7.

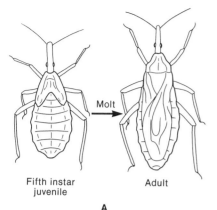

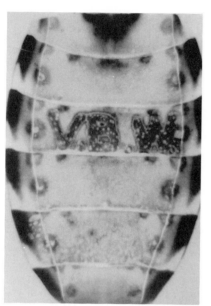

B

Figure 9.11
(A) The last juvenile instar and the adult of the blood-sucking bug Rhodnius prolixus. *The insect molts after a meal has sufficiently distended the abdomen. In the last molt the adult acquires wings, and the cuticle on the abdomen becomes paler and less mottled. (B) The initials of Vincent B. Wigglesworth inscribed into the abdominal cuticle of a* Rhodnius *adult. Juvenile hormone was used to write the initials on the juvenile prior to its last molt. The hormone caused the cuticle to retain the juvenile pattern.*

Table 9.3 The sources and effects of some hormones of invertebrates.

ANNELIDS

Polychaetes

Juvenile hormone inhibits gonads and stimulates growth and regeneration. A **gonadotropin** stimulates development of eggs.

Earthworms

A neurosecretion stimulates formation of gametes and sex characteristics and increases concentrations of sugar in the blood.

Leeches

A neurosecretion stimulates development of gametes and triggers color changes.

MOLLUSCS

Gastropods

Egg-laying hormone from bag cells in abdominal ganglion of marine slug *Aplysia*. A neurosecretion in the common land snail *Helix* stimulates spermatogenesis; another hormone stimulates egg development; hormones from ovary and testis stimulate accessory sex organs.

Cephalopods

Optic gland in the eye stalk of octopus, squid, and so on produces a hormone that stimulates egg development, proliferation of spermatogonia, and secondary sex characteristics. A secretion from this gland kills the female after she has brooded her young.

ARTHROPODS

Crustaceans

The Y organ in the head produces **ecdysone,** which triggers molting of old cuticle. The X organ-sinus gland (XOSG) system near the optic nerves releases **molt-inhibiting hormone** (MIH), which inhibits release of ecdysone. The XOSG system also releases other hormones that control pigments in the cuticle, increase glucose in the blood, and regulate other functions.

Insects

Corpora cardiaca (plural of corpus cardiacum; Figure 9.13) release hormones that regulate water balance, **adipokinetic hormone** that promotes the conversion of fat to blood sugar, and hormones that regulate heartbeat.
Diapause hormone arrests development of the eggs; **prothoracicotropin** (PTTH) stimulates the prothoracic glands to secrete the molting hormone **ecdysone**; and **eclosion hormone** triggers unique movements of the pupa that allow it to emerge as an adult (Figure 9.13).
Bursicon promotes hardening and tanning of the exoskeleton. Corpora allata (plural of corpus allatum; Figure 9.13) secrete **juvenile hormone** (JH), which prevents metamorphosis of a larva to the adult stage and is required for fertility in adults.
Prothoracic glands and ovarian follicles secrete **ecdysone,** which triggers molting (Figure 9.13).

ECHINODERMS

Hypodermal cell layer of radial nerve secrets **gonad-stimulating substance.**
Follicle cells of ovary secrete **1-methyl-adenine,** which indirectly stimulates maturation of gametes.

The wood of balsam fir that went into the paper contained a substance that mimics JH and prevents development into adults (Figure 9.12). Other plants, such as the common bedding plant *Ageratum*, produce an anti-juvenile hormone that destroys the secretory cells of JH. Juvenile insects that feed on these plants develop prematurely into small adults.

Juvenile hormone III

A

B

In another series of studies, Williams found that the isolated abdomens of the pupae of cecropia moths (*Hyalophora cecropia*) would molt into adult abdomens if the cerebral ganglia and the **prothoracic gland** were implanted into the abdomens. It has since been found that **prothoracicotropic hormone** (PTTH) from the cerebral ganglia normally stimulates the prothoracic gland to secrete a steroid called **ecdysone,** which triggers molting. Figure 9.13 summarizes how all these hormones normally function in insect development.

Figure 9.12
(A) Juvenile hormone (JH), normally secreted in the early stages of insect development, prevents the formation of adult features.
(B) Some plants, such as the balsam fir tree and the sweet basil herb, produce substances (juvocimenes) that mimic JH. Application of the JH analog from sweet basil to the last nymphal instar of the large milkweed bug Oncopeltus fasciatus promotes the development into additional juvenile stages. The milkweed bug on the extreme left is a normal last-instar juvenile. The second insect from the left was treated with a large dose of JH analog at a similar stage of development. Instead of developing into a normal adult like the bug on the far right, it developed into an extra juvenile instar. A smaller dose of the JH mimic resulted in the formation of the in-between insect (third from left). (See Bowers and Nishida 1980.)

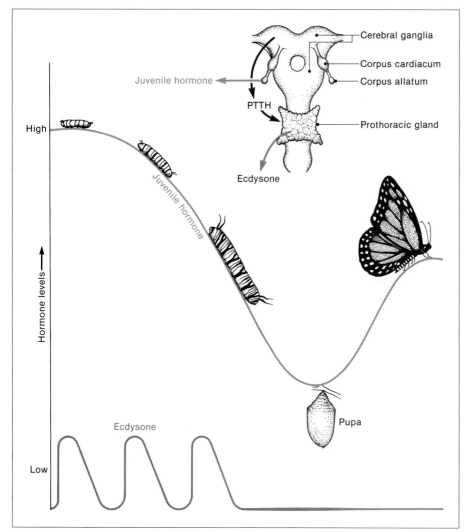

Figure 9.13
A summary of the hormonal control of development in a moth. High levels of juvenile hormone (JH) from the corpora allata during early juvenile stages keep the animal in the caterpillar stage. Ecdysone must be released from the prothoracic gland (in response to prothoracicotropin PTTH from the cerebral ganglion) during a critical period prior to each molt. The insect becomes a pupa when JH falls below a certain level. JH levels rise again in the adult and stimulate reproductive functions.

SUMMARY

Many physiological functions are coordinated by chemical messengers, including tissue factors and hormones. Tissue factors work locally in the tissues that secrete them, while hormones travel in blood to organs that are generally far from the glands that secrete them. In vertebrates the secretion of many hormones is influenced by the pituitary gland, which is in turn controlled by the hypothalamus. In mammals the pituitary gland has an anterior lobe, which secretes hormones in response to releasing hormones from the hypothalamus, and a posterior lobe, which releases neurosecretions.

Hormones are generally classified as either steroids or nonsteroids. Steroids enter cells, where they bind to a receptor molecule and induce gene activity. Nonsteroids generally bind to receptor molecules on the plasma membrane, which then activate a second messenger system that triggers the cellular response.

Different hormones are secreted by various invertebrates. In insects, for example, several hormones coordinate molting and development. Ecdysone triggers molting, and juvenile hormone prevents a juvenile from molting into an adult.

KEY TERMS

gland
endocrine
neurosecretion
steroid

second messenger
pituitary
releasing hormone
adrenocorticotropic hormone

juvenile hormone
ecdysone

SELF-TEST

1. What conditions must be satisfied before a chemical messenger is considered to be a hormone? Describe some chemical messengers that are not hormones. How do they differ from hormones?

2. List as many endocrine organs of mammals as you can, and compare your list with Figure 9.3.

3. Which of the endocrine organs in Figure 9.3 release their hormones by neurosecretion? Which ones secrete steroid hormones? Which ones affect at least one other endocrine gland?

4. Explain how the action of steroid hormones on their target cells differs from the action of nonsteroid hormones on their target cells.

5. Explain how the functioning of the anterior pituitary differs from that of the posterior pituitary.

6. In what way are the secretions of the posterior pituitary similar to each other?

7. Explain how mimics of juvenile hormone or anti-juvenile hormones might be used as insecticides. What advantages and disadvantages would such third- and fourth-generation insecticides have over conventional insecticides?

READINGS

RECOMMENDED READINGS

Baker, J. J. W. and G. E. Allen. 1970. *The Process of Biology: Primary Sources.* Reading, MA: Addison-Wesley. (*See excerpts of "The mechanism of pancreatic secretion" by Bayliss and Starling.*)

Berridge, M. J. 1985. The molecular basis for communication within the cell. *Sci. Am.* 253(4):142–152 (Oct).

Cantin, M. and J. Genest. 1986. The heart as an endocrine gland. *Sci. Am.* 254(2):76–80 (Feb).

Carmichael, S. W. and H. Winkler. 1985. The adrenal chromaffin cell. *Sci. Am.* 253(2):40–49 (Aug).

Evans, H. E. 1968. *Life on a Little-Known Planet.* New York: E. P. Dutton. (*Chapter 9 is an engaging history of research in insect endocrinology.*)

Rasmussen, H. 1989. The cycling of calcium as an intracellular messenger. *Sci. Am.* 261(4):66–73 (Oct).

Sapolsky, R. M. 1990. Stress in the wild. *Sci. Am.* 262(1):116–123 (Jan).

Scheller, R. H. and R. Axel. 1984. How genes control an innate behavior. *Sci. Am.* 250(3):54–62 (Mar). (*The egg-laying hormone of a sea slug.*)

Snyder, S. H. 1985. The molecular basis of communication between cells *Sci. Am.* 253(4):132–140 (Oct).

Uvnäs–Moberg, K. 1989. The gastrointestinal tract in growth and reproduction. *Sci. Am.* 261(1):78–83 (July).

ADDITIONAL REFERENCES

Bowers, W. S. and R. Nishida. 1980. Juvocimenes: potent juvenile hormone mimics from sweet basil. *Science* 209:1030–1032.

Feir, D. 1984. Inhibition of gland development in insects by a naturally occurring antiallatotropin (anti-hormone). In: C. L. Harris (Ed.), *Tested Studies for Laboratory Teaching: Proceedings of the Third Workshop/Conference of the Association for Biology Laboratory Education (ABLE).* Dubuque, IA: Kendall/Hunt, Chapter 6.

Kaltenbach, J. C. 1988. Endocrine aspects of homeostasis. *Am. Zool.* 28:761–773.

Laufer, H. and R. G. H. Downer (Eds.). 1988. *Endocrinology of Selected Invertebrate Types.* New York: Alan R. Liss.

Rosen, O. M. 1987. After insulin binds. *Science* 237:1452–1458.

10

Integument, Skeleton, and Muscle

Male purple finch (Carpodacus purpureus).

CHAPTER OUTLINE

LEARNING OBJECTIVES

1. How do animals protect their internal environments from hostile external environments?

2. How do animals support their bodies against their own weight and external forces?

3. How are the varied movements and forces of contraction produced by an animal?

4. What are the structural and biochemical mechanisms of muscle contraction?

5. From where does the energy for muscle contraction come?

6. How are some muscles specialized for power and others for efficiency?

7. Are the muscles of the heart and of the intestine different from those that move the body?

8. Why do animals with different body sizes move differently?

INTRODUCTION

Animals are multicellular organisms that cannot produce their own nutrients, so they have to move about through often inhospitable environments to obtain the necessities of life for their internal environments. A protective body covering, a supportive and protective skeleton, and mechanisms for locomotion are therefore important for most animals. The body covering, called an **integument,** separates this internal environment from the external environment. Animals that are fortunate enough to live in environments that are similar to their interstitial fluids may get by with a thin integument of a single cell layer. Other animals live in such hostile habitats that they require much thicker integuments.

Some animals have integuments so strong that they also serve as skeletons— an **exoskeleton,** to be precise. Other animals, such as ourselves, have **endoskeletons,** which are enclosed within the integument. Still other animals lack a solid skeleton but produce an internal hydrostatic pressure that serves as a **hydroskeleton.** All these skeletons prevent animals from collapsing under their own weight or from external forces.

Skeletons also provide support for the structures that either move the animal through its external environment or move the environment past the animal. Such movements are essential for an organism that has to eat, since it would soon run out of food if it did not go in search of it or did not bring new supplies in on water currents. Animal movements are generally produced by muscles, although many small, aquatic animals use cilia.

INTEGUMENT

Invertebrate Epidermis. The function of an integument is to protect the carefully maintained interstitial fluid of an animal from the external environment. As noted previously, the differences between the interstitial fluid and the environment are sometimes so small, especially in marine animals, that the integument need be no more than a single layer of cells, called an **epidermis.** The epidermis is easily permeable to nutrients, wastes, and respiratory gases, often enabling animals to do without special digestive, excretory, or respiratory organs to bring in or eliminate them. In larger invertebrates, especially if they are freshwater or terrestrial animals, the epidermis has additional adaptations for protection and support. Many, including flatworms, earthworms, and snails, have epidermal cells that secrete **mucus,** which is a watery solution of specialized glycoproteins (Figure 10.1A). The mucus protects against abrasion and in some animals also forms a track on which to glide during locomotion.

In many animals the epidermis secretes a protective shell. The shells of corals, molluscs, and many other invertebrates are made largely of calcium carbonate $CaCO_3$ (see Figure 28.4). In other animals, such as annelids and arthropods, the epidermis secretes a tough, water-resistant **cuticle.** (In such animals the epidermis is also called a **hypodermis.**) The cuticle of insects and other arthropods consists of **chitin** (KY-tin) and proteins in rigid plates linked by flexible membrane (see Figure 30.3). A disadvantage of shells and cuticle is that they require some provision to allow the animal to grow within them. Molluscs and some other animals with calcareous shells continually enlarge the shells as they grow. Arthropods periodically shed the old, outgrown cuticle in a process called **molting.**

Skin. The integument of vertebrates, known to all as skin, affords good protection and also grows as the animal does. Skin has two layers, the outer epidermis and the dermis beneath it (Figure 10.1B). The **epidermis** of vertebrates is quite

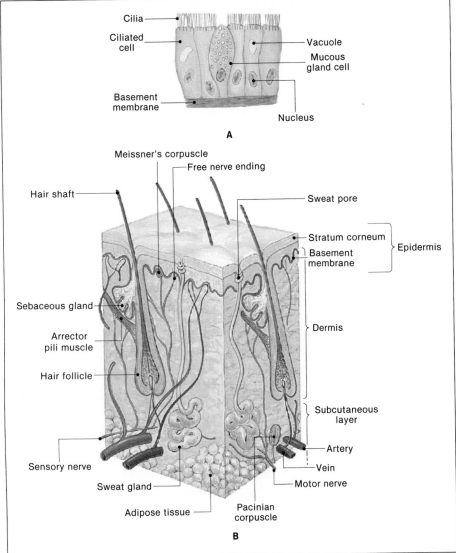

Cilia

Ciliated cell

Vacuole

Mucous gland cell

Basement membrane

Nucleus

A

Meissner's corpuscle

Free nerve ending

Hair shaft

Sweat pore

Stratum corneum

Basement membrane

Epidermis

Sebaceous gland

Dermis

Arrector pili muscle

Hair follicle

Subcutaneous layer

Artery

Vein

Sensory nerve

Motor nerve

Sweat gland

Adipose tissue

Pacinian corpuscle

B

Figure 10.1
Representative integuments of an invertebrate and a vertebrate. (A) The simple integument of a flatworm (phylum Platyhelminthes) consists of an epidermis supported by a basement membrane. Flatworms use a combination of mucus and cilia on the epidermal cells in locomotion. (B) The more complex integument of a human consists of a multilayered epidermis and dermis. Compare Figure 2.18D.

different from the epidermis of invertebrates in that it has several layers of cells. The innermost cells undergo frequent division, which pushes the older cells toward the surface, forming a **cornified** layer that is resistant to abrasion and is impermeable to water. Much of the cytoplasm in cells of the cornified layer is replaced by **keratin,** a complex of proteins held together by disulfide bonds. This process of **keratinization** eventually kills the cells, which end up forming a scaly layer on the skin surface. Billions of cornified cells from this layer are normally shed each day, many of them accumulating as house dust.

In many vertebrates the cornified layer becomes greatly modified in ways that are characteristic of the class to which the vertebrate belongs. The scales of reptiles, for example, are esentially thick plates of keratin from cornified cells. Feathers of birds are essentially reptilian scales adapted for flight and thermal insulation. In addition, the beaks and claws also consist largely of keratin from cornified cells. The hooves, claws, nails, and some horns of mammals also consist of keratin. The hair of mammals consists of keratin also, but it does not arise from cornified cells. Instead, other epidermal cells form a **follicle** that migrates beneath the epidermis and produces the hair.

Beneath the epidermis is the **dermis,** or true skin. Dermis consists mainly of connective tissue with large amounts of the fibrous protein **collagen.** Collagen accounts for the toughness and flexibility of skin. (Contrary to advertisements for many skin lotions, however, there is no way that collagen applied externally can penetrate the epidermis to restore the skin.) The dermis of mammals also includes the hair follicles, cutaneous sensory receptors (see Figure 7.14), and several types of gland cells that originate in the epidermis. The glands include **sebaceous glands,** which produce a fatty secretion **(sebum)** that is released through hair follicles (Figure 10.1B). In addition, there are two types of sweat glands. **Eccrine sweat glands** produce a watery solution that cools the body by evaporation. **Apocrine sweat glands** produce an odorous secretion that aids in courtship and other kinds of social communication—in most mammals, at least. In humans, the apocrine glands occur mainly in the arm pits and groin.

Coloration.   In addition to providing structural support and protection, many integuments also provide coloration. Colors may serve as camouflage or, for venomous, toxic, or bad-tasting animals, they may warn off potential predators. In addition, colors often serve in social communication, to help members of the same species identify individuals, their gender, reproductive status, social rank, and so on. (These topics are covered in Chapter 20.) Color is due either to **pigments** or to fine structures that refract or reflect light **(structural colors)** (Table 10.1). Pigments may be dispersed in epidermis, feathers, or hair, or they may be contained in special cells called **chromatophores.** Under neural and hormonal control, the pigments can condense or expand within the chromatophores, reducing or intensifying the color. This control over chromatophores accounts for the chameleon's famous ability to change appearance. Finally, parts of some animals are colored simply because their blood or whatever they have eaten shows through their tissues.

SKELETAL SUPPORT

Hydrostatic Skeletons.   Animals that are small or buoyant generally do not require a skeleton, because their own weight and movement do not generate significant mechanical stresses in their bodies. Larger animals, especially terrestrial

Table 10.1   Animal colors and their sources.

White is due to the reflection of all wavelengths of light. In insects it is often a structural color—**structural white**—due to dust-sized, transparent particles that scatter all wavelengths. A silvery white, like that of fish scales, is also produced by reflection from guanine, adenine, or some other purine. These pigments are often found in a type of chromatophore called an **iridophore.**

Iridescence. Interference between incident and reflected light causes some colors to be blocked while others are intensified, depending on the angle of viewing. The luster of pearls, peacock feathers, and some butterfly wings is a structural color due to overlapping translucent layers **(interference layers)**. The cuticles of some insects and the scales of some fishes are finely grooved and produce iridescent **diffraction colors**. Some birds' feathers contain bundles of thin tubes that produce diffraction colors.

Blue is generally a structural color called **Tyndall blue,** which is due to the scattering of light by particles about the size of the wavelength of light. Blue light is scattered most, producing a blue color in the same way that scattering of light by air produces a blue sky.

Green is generally produced by Tyndall blue filtered through one of the yellow pigments noted below. Some birds, however, have feathers with green pigments.

Yellow, orange, and red are generally due to carotenoids or pteridines, often in a type of chromatophore called a **xanthophore.**

Brown, gray, and black are due to melanins. Chromatophores containing melanins are called **melanophores.**

ones, require some kind of skeleton, even if it is not immediately recognizable as such. Many soft-bodied invertebrates have a **hydrostatic skeleton** (= hydroskeleton) that is not a distinctive structure but the effect of internal pressure of body fluids. In animals with hydrostatic skeletons there are typically two sets of muscles, one running longitudinally along the body and one running circularly around the body. Coordinated contractions of these muscles allow for locomotion, for example, the **peristaltic progression** of earthworms (Figure 10.2). Without the hydrostatic skeleton, the body would simply collapse when these muscles contract.

Exoskeletons.   Many other invertebrates have exoskeletons: rigid structures enclosing all or part of the body. As was previously noted, exoskeletons generally take the form of shells or cuticle. Usually the exoskeleton consists of hard plates hinged together by flexible connective tissue. Movement is generally produced by contraction of muscles attached to the inner surfaces of the plates. In clams, oysters, scallops, and other bivalve molluscs, for example, the two parts of the shells, called **valves,** are closed by contraction of muscle attached to both valves. An elastic ligament opens the valves when the muscles relax. In insects and other arthropods the muscles attach to the cuticular plates. Contraction of one set of muscles moves a joint in one direction, and contraction of another set of muscles (antagonist muscles) moves the joint in the opposite direction (Figure 10.3).

Endoskeletons.   Endoskeletons are enclosed by other body tissues. The endoskeletons of sponges consist of mineral **spicules** (see Figure 23.3) and fibers of **spongin** that keep the body from collapsing. Adult sponges remain attached to substratum (they are sessile), so they have no need for muscles attached to the endoskeleton. The endoskeletons of echinoderms such as starfishes and sea urchins are made up of small, calcareous plates called **ossicles** (see Figure 33.3).

The most familiar endoskeletons are, of course, your own and those of vertebrates you eat. Vertebrate skeletons are composed of **cartilage** and bone. Skeletons are entirely **cartilaginous** in the invertebrate chordate *Branchiostoma* (amphioxus; see Figure 34.7), in jawless vertebrates (hagfishes and lampreys; see Figure 35.1), and in sharks, rays, and related fishes (class Chondrichthyes; see Figure 35.2). In other vertebrates, cartilage forms the skeleton of the embryo, but

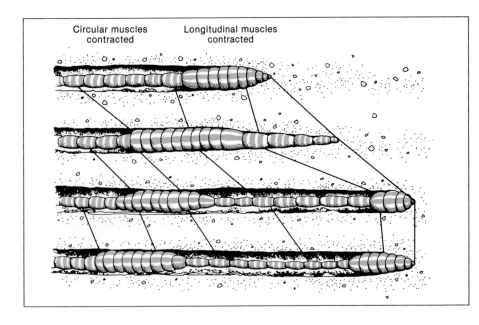

Circular muscles contracted   Longitudinal muscles contracted

Figure 10.2
How an animal that lacks bones can burrow through earth. The earthworm Lumbricus terrestris uses peristaltic progression, which is possible only because the hydrostatic skeleton keeps the body from collapsing when its muscles contract. Contraction of the longitudinal muscles thickens each anterior segment, allowing it to make stronger contact with the sides of the burrow. The circular muscles at the anterior end then contract, narrowing the segments and extending them forward (second part of the sequence). A wave of contraction of the circular muscles moves backward, alternating with contraction of the longitudinal muscles, resulting in forward movement. Progression is shown by the lines connecting every fifth segment.

Figure 10.3

Exoskeleton and muscle attachment in two arthropods. (A) The exoskeleton of the lobster Homarus americanus. *(B) Detail of the crusher claw, showing the attachment of the opener and closer muscles. (C) Attachment of flight muscles in insects. The arrangement on the left, which occurs in dragonflies, uses elevator and depressor muscles that attach to the wing. Elevators pivot the wings up, and depressors pull them down. The contraction of each muscle is synchronous with a burst of action potentials from the nervous system. The arrangement on the right is typical of flies, bees, beetles, and bugs. The muscles do not attach directly to the wings, but to the exoskeleton of the thorax. Both elevators and depressors are asynchronous muscles, which contract in response to being stretched, as will be described later. The elevators pull the top of the thorax down, causing the wings to spring up. This stretches the depressors, which run longitudinally through the thorax. When they contract they bow the thorax out, which depresses the wings. This in turn stretches the elevators, causing them to contract, and so on.*

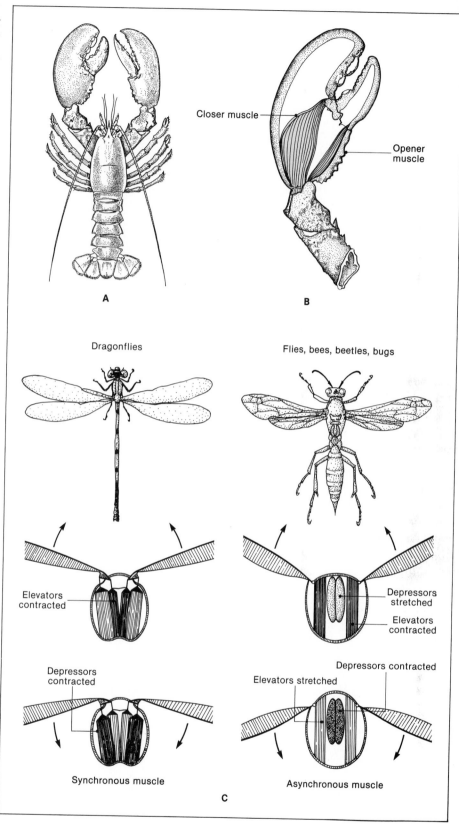

bone later replaces most of the cartilage. Cartilage does remain, however, as flexible supports in the nose, ears, and trachea and as cushions between joints (Figure 10.4A). The ability of cartilage to remain flexible, yet retain its shape and resist damage, lies in the fact that it consists largely of water. The water is held in place

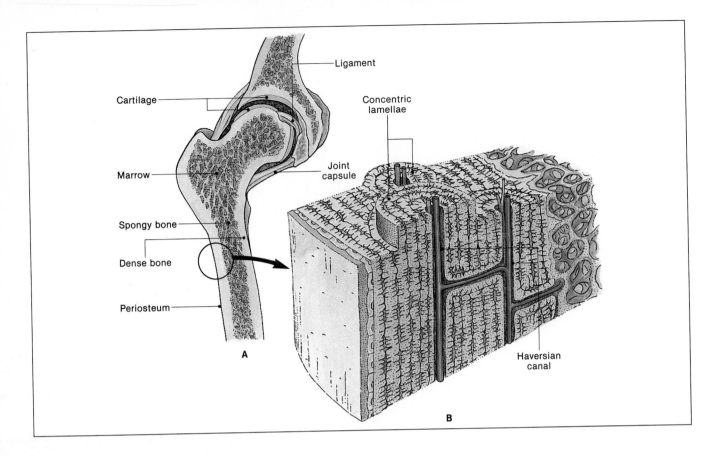

Ligament

Cartilage

Concentric
lamellae

Joint
capsule

Marrow

Spongy bone

Dense bone

Periosteum

A

Haversian
canal

B

by a network of collagen and complex molecules called **proteoglycans,** both of which are secreted by cells called **chondrocytes.**

Most of the vertebrate skeleton is composed of **bone.** One type of bone, called **dermal bone** (= membranous bone), is produced by layers of embryonic connective tissue rather than by replacement of cartilage. Dermal bone occurs in the skull and face of mammals and in such nonskeletal structures as fish scales, turtle shells, and antlers. Most bone is produced by cells called **osteoblasts** (Greek *osteon* bone + *blasti* a bud), which replace cartilage with calcium carbonate and calcium phosphate. During embryonic development in birds and mammals, osteoblasts deposit these minerals in concentric layers **(lamellae).** The centers of these layers remain as hollow channels, called Haversian canals, which contain blood vessels and nerves. The result is **dense bone** (= compact bone), which provides most of the structural support of the skeleton (Figure 10.4). In bones such as those of the arms and legs, dense bone surrounds a core of **spongy bone** (= trabecular or cancellous bone). The spaces in spongy bone are filled with marrow, which produces blood cells.

Contrary to appearances, dense bone is living, dynamic tissue. Osteoblasts continue to work throughout life, responding mainly to hormones such as thyrocalcitonin from the thyroid gland (see p. 186). Osteoblasts also respond to local mechanical stresses by depositing minerals where they are needed. Other cells, called **osteoclasts** (Greek *klasis*, a breaking), can break down the minerals of bone. This is usually in response to hormones, such as parathyroid hormone from the parathyroid glands (see p. 187), which signal a need for the calcium in the body fluids. The lack of mechanical stress, as when bedridden, and the absence of estrogens following menopause can also lead to excessive loss of minerals from bones.

Figure 10.5 shows the major bones of the human skeleton. Similar bones occur in other vertebrates, with sizes and shapes adapted to different functions. These

Figure 10.4
(A) Bone and cartilage at a human hip joint. The cartilage helps cushion the bones and reduces friction between them. Ligaments hold the bones together. The bones are surrounded by a sheath (the periosteum). The outer dense bone encloses the spongy bone. At the ends of long bones such as these, the spongy bone spaces are filled with red marrow, which produces red blood cells. (B) Details of dense and spongy bone.

Figure 10.5
Major bones of the human skeleton.

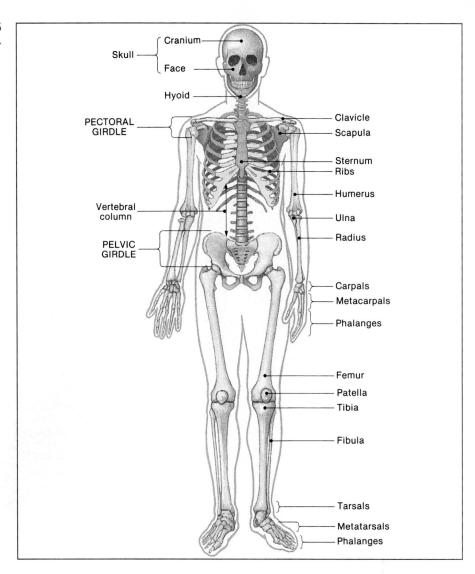

bones are divided into those of the **axial skeleton** and of the **appendicular skeleton.** The axial skeleton lies along the body axis and includes the bones of the **skull, vertebral column,** and **thoracic basket** (rib cage). The appendicular skeleton comprises the bones of the arms and legs, which attach to the **pectoral girdle** and the **pelvic girdle,** respectively.

MUSCULAR ANATOMY

In addition to supporting the body, the endoskeleton provides sites for the attachment of skeletal muscles. These attachments are a type of connective tissue called **tendon.** Generally, each end of the muscle attaches to two different bones that are separated by one or more joints. Contraction of a muscle then causes one bone to move with respect to the other. Muscles can only exert force while contracting, not while lengthening, so they have to be arranged in **antagonistic** pairs. For example, the biceps brachii in the front of the upper arm is a **flexor** muscle that helps bring the hand toward the body. The triceps brachii on the opposite side of the arm is an **extensor** that extends the hand away from the body. Figure 10.6 shows some of the most important skeletal muscles in the human. You should try to locate some of them on yourself and observe the movements they cause.

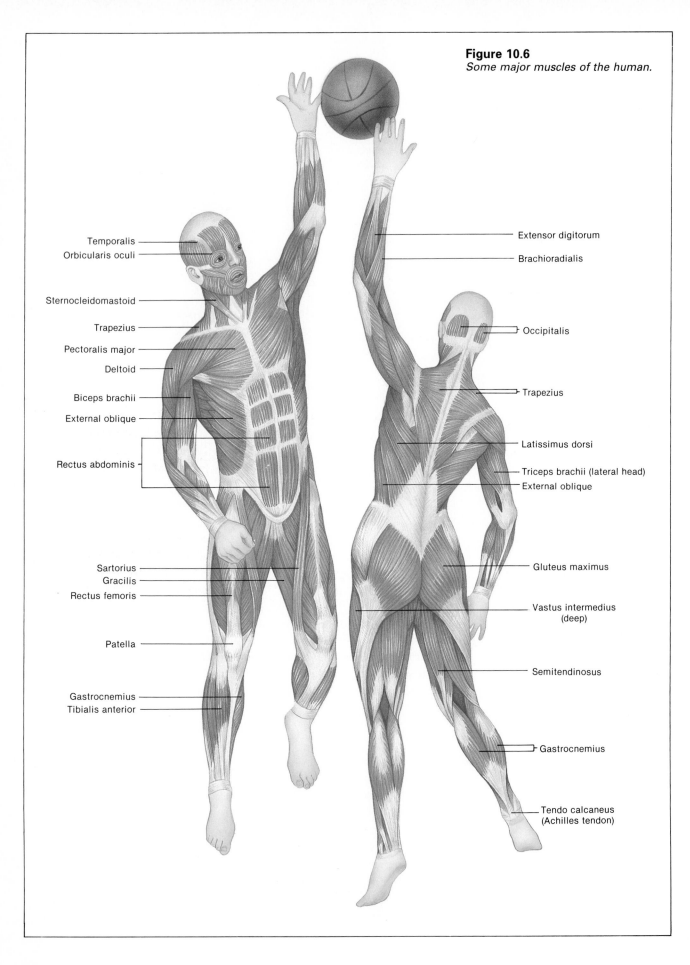

Figure 10.6
Some major muscles of the human.

Temporalis
Orbicularis oculi
Sternocleidomastoid
Trapezius
Pectoralis major
Deltoid
Biceps brachii
External oblique
Rectus abdominis
Sartorius
Gracilis
Rectus femoris
Patella
Gastrocnemius
Tibialis anterior

Extensor digitorum
Brachioradialis
Occipitalis
Trapezius
Latissimus dorsi
Triceps brachii (lateral head)
External oblique
Gluteus maximus
Vastus intermedius (deep)
Semitendinosus
Gastrocnemius
Tendo calcaneus (Achilles tendon)

The way in which a muscle is attached to the skeleton determines not only the type of motion produced when it contracts but also the speed and force with which the body part moves. This is so because the bones move like levers. Figure 10.7 illustrates this point for the human biceps brachii. Arthropods use the same lever principle, except that the muscles and skeleton are "inside out" (Figure 10.3B).

ACTIVATION OF SKELETAL MUSCLE

Galen in the 2nd century and Descartes in the 17th thought movement was due to the expansion of a neural vapor within muscles. Their concepts were probably influenced by the steam technology of their times, just as our concepts are influenced by computer technology. We now think of the muscle contractions that are responsible for all animal behavior as being controlled by patterns of electrical activity from the brain and spinal cord. This electrical activity, in the form of action potentials in motor nerve cells, interacts with skeletal muscle through **neuromuscular synapses.** These neuromuscular synapses, also known as neuromuscular junctions or endplates, work much like the chemical synapses within the central nervous system. The major differences between neuromuscular synapses and CNS synapses are structural. Neuromuscular synapses in vertebrates are usually elongated extensions of the axon, rather than knob-shaped structures. (Compare Figures 7.5 and 10.8.) Neuromuscular synapses also differ from synapses in the CNS in the way vesicles are arranged. In neuromuscular synapses the vesicles are not distributed randomly but are clustered just opposite **subjunctional folds** in the postsynaptic membrane of the muscle cell. The subjunctional folds contain the receptor molecules that bind the transmitter substance released by the neuromuscular synapse.

Another difference is that neuromuscular synapses do not release as large a variety of excitatory and inhibitory transmitters. In vertebrates, in fact, there are no inhibitory neuromuscular transmitters, and the only excitatory transmitter is acetylcholine (ACh). ACh depolarizes the postsynaptic membrane of the muscle cell, producing an **endplate potential** (EPP) that is similar to an excitatory postsynaptic potential. The EPPs travel to the excitable portion of the muscle mem-

Figure 10.7
The human elbow, showing the attachment of the biceps brachii muscle to the radius and scapula. The numbers show the mass (in kilograms) that can be lifted and the speed (cm/sec) of movement of the muscle and of the hand. Because of the lever action of the joint, the force and speed are not the same for the hand as for the muscle.

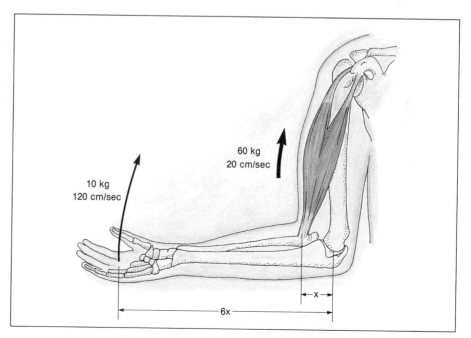

60 kg
20 cm/sec

10 kg
120 cm/sec

x

6x

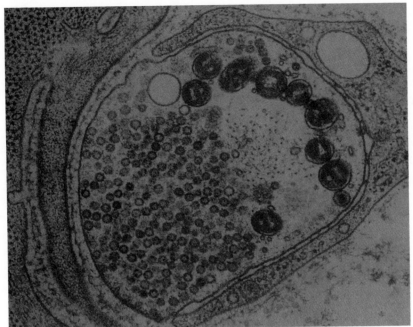

A

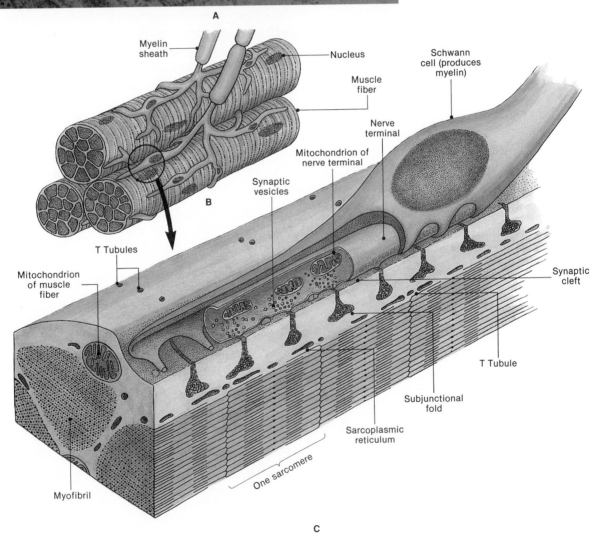

Figure 10.8
The structure of neuromuscular synapses in vertebrate skeletal muscle. (A) Electron micrograph of a cross section of the neuromuscular synapse of the leopard frog Rana pipiens. *Each of the vesicles is about 40 nm in diameter. Parts of the subjunctional folds can be seen in the postsynaptic membrane on the muscle. (B) Neuromuscular junctions. (C) An enlargement of the circled portion of (B), showing a longitudinal section of the neuromuscular junction.*

Myelin sheath

Nucleus

Muscle fiber

Schwann cell (produces myelin)

Nerve terminal

Mitochondrion of nerve terminal

Synaptic vesicles

B

T Tubules

Mitochondrion of muscle fiber

Synaptic cleft

T Tubule

Subjunctional fold

Sarcoplasmic reticulum

One sarcomere

Myofibril

C

brane, where they evoke action potentials. In vertebrates and a few invertebrates the action potentials arise from inward sodium currents like those in axons. In most invertebrates the action potentials result from inward calcium currents. In both cases the action potentials trigger muscle contraction after a sequence of events to be described in later sections of this chapter.

CONTROL OF MUSCLE CONTRACTION

As with the action potentials in axons, those in the membranes of muscles are all identical. How, then, can action potentials trigger so many different kinds of movement? As with action potentials from sensory receptors, the information directing bodily movements must be encoded in the pattern of action potentials to the muscles. One kind of muscle information is easily decoded: **the CNS controls the type of movement by activating different muscles.** Sending action potentials to a flexor muscle, for example, will cause a movement that is just the opposite of sending action potentials to the antagonistic extensor muscle.

A second principle of the motor code is revealed by the pattern of connections between motor neurons and skeletal muscle fibers. As Figure 10.9 shows, each branch of a peripheral nerve going to a particular muscle contains many motor axons, and each axon connects to different muscle cells. In vertebrates each muscle cell, also called a muscle fiber, receives synaptic input from just one motor axon, but each motor axon can have synapses onto several muscle fibers. Since an action potential in the motor axon conducts to all its branches and activates all the muscle fibers with which they connect, all those muscle fibers contract as a unit. A **motor unit** is therefore defined as one motor axon and all the muscle fibers it controls. The fewer the muscle fibers in a motor unit, the more precise the control over the muscle. The muscles controlling eye movements have as few as three muscle fibers per motor unit, while the large gastrocnemius muscle in the calf of the leg has 2000 fibers per motor unit.

The previous paragraph suggests the second principle of the motor code: **the**

Figure 10.9
Two small muscles innervated by branches of one nerve. Each nerve includes numerous axons, with each axon of a motor neuron controlling one motor unit. One motor unit—with three muscle fibers controlled by one axon—is shown in color.

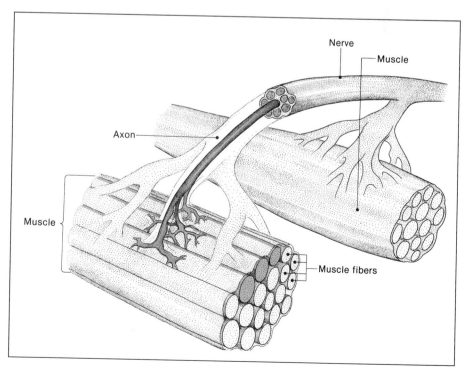

strength of muscle contraction increases with the number of motor units activated. This principle is illustrated in Figure 10.10, which shows the effect of electric shocks applied to a nerve going to a frog's gastrocnemius muscle. Increas-

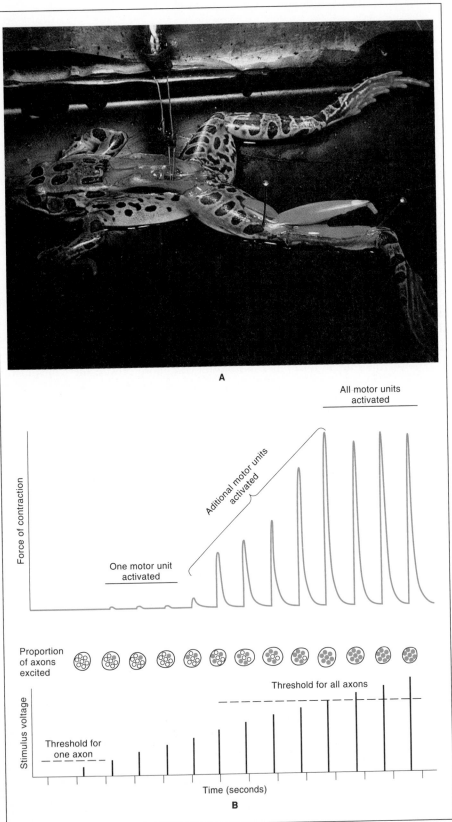

A

All motor units activated

Aditional motor units activated

One motor unit activated

Force of contraction

Proportion of axons excited

Threshold for all axons

Stimulus voltage

Threshold for one axon

Time (seconds)

B

Figure 10.10
How the force of contraction increases with increasing number of motor units activated. (A) The experimental preparation. A grass frog was pithed (its brain and spinal cord were destroyed by a probe to prevent voluntary movement and pain). Electrodes were then placed on the sciatic nerve near the spinal cord, and a force transducer was attached by a thread to the Achilles tendon on the gastrocnemius muscle. This muscle helps power the hop in the frog. (B) Brief shocks applied to the sciatic nerve evoked twitches in the gastrocnemius muscle, and the force of the twitches increased with the voltage of the shocks. Note, however, that although the voltage was increased gradually, the force of contraction increased suddenly. A representation of the sciatic nerve in cross section shows that the increasing voltage excited an increasing number of motor units in all-or-none fashion.

ing the voltage of the shocks to the nerve increases the number of axons in the nerve that are stimulated above threshold for action potentials. Therefore, increasing the stimulus voltage increases the number of motor units activated. The result is an increase in the force of the twitches. Presumably the central nervous system regulates the force of contraction by **recruiting** the appropriate number of motor units. Note in Figure 10.10 that the force of contraction in the frog's gastrocnemius muscle increases suddenly as the stimulus voltage is increased gradually. The reason for this is the all-or-none behavior of action potentials: A motor unit either responds or it does not.

Each brief shock to the nerve evokes a **twitch** in the muscle. A twitch is the response to a single action potential, and it is the briefest movement a muscle can produce. Brief as it is, however, a twitch lasts at least a hundred times as long as the action potential that evokes it. This means that many action potentials can occur during a single twitch, and the effect of all these action potentials can **summate** to produce a contraction of greater force. Thus, increasing the number of action potentials in a given period increases the force of contraction. This is the third principle of the motor code: **The strength of muscle contraction increases with the frequency of action potentials.** This principle is illustrated in Figure 10.11, which was obtained by increasing the frequency of shocks to the same nerve as in Figure 10.10.

To summarize, the type of movement is controlled by directing action potentials to different sets of muscles, and the force of movement is controlled by activating the appropriate number of motor units within each muscle with the appropriate frequency of action potentials.

Figure 10.11

The effect of stimulus frequency on the force of contraction. The procedure was the same as in Figure 10.10 except that instead of increasing the voltage of the shocks, the frequency was increased. When the frequency was high enough (about two shocks per second) each twitch began before the previous one was over, and the twitches began to summate. As the frequency reached about 15 shocks per second, individual twitches disappeared, and the contraction became steady. This steady contraction, called tetanus, is the normal mode of muscle contraction.

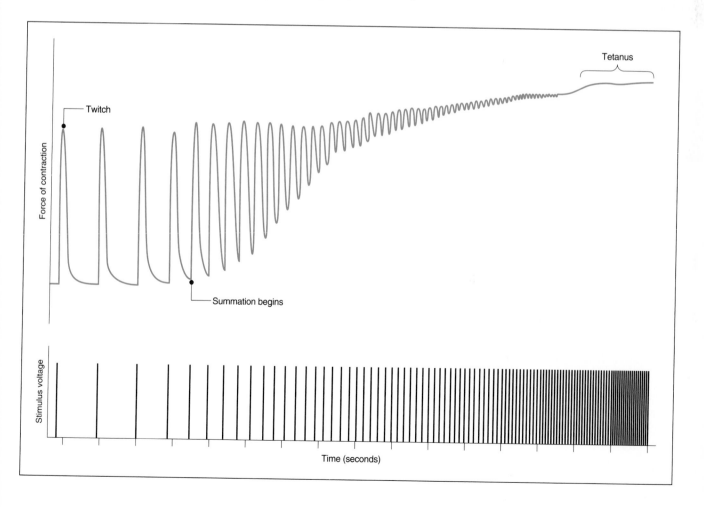

HOW MUSCLES CONTRACT

The Structure of Skeletal Muscle. Antony van Leeuwenhoek, the first microscopist, observed that meat, which is skeletal muscle, has an intriguing pattern of bands or striations running through its fibers (Figure 10.12A). These bands are now designated **A bands** and **I bands**. The A and I bands repeat from one end of a fiber to the other, with each A band enclosed by two I bands. Until a few decades ago the only contribution of these striations to the understanding of skeletal muscle was a new name: **striated muscle.** Striations were generally ignored earlier in this century because there was no place for them in the prevailing theory of contraction, which assumed that elastic filaments contracted within the fibers like rubber bands.

Refinements in light and electron microscopy in the early 1950s paved the way for a new theory in which the banding pattern of striated muscle did fit. A. F. Huxley (cited in Chapter 7 for his work on the squid giant axon) invented a new type of microscope that allowed him to see that the A bands of frog muscles did not shorten during contraction. This was contrary to the old theory of elastic filaments. Independently, H. E. Huxley (not directly related to A. F.) helped refine electron microscopy to reveal that muscle fibers have two types of filament in them, neither of which changes length. **Thick filaments** occur only in A bands. **Thin filaments** occur in I bands and extend into the A bands interspersed among the thick filaments. Both types of filament can be seen in Figure 10.8C, which shows a cross-section of an A band. The thin filaments connect to a **Z line,** which is actually a Z disc in three-dimensional life. Thus the sequence ZIAIZ repeats all along the length of the muscle fiber. Each repetition is called a **sarcomere** (Figure 10.12B). Presumably the function of an entire muscle can be understood from just one sarcomere.

In 1954 both Huxleys jointly proposed an alternative theory, which is now generally accepted. Their **sliding filament theory** proposes that the thin and thick filaments slide among each other, and that the force of muscle contraction comes from the interaction of the two types of filament. This theory is now usually presented as established fact—a situation that both Huxleys have lamented.

Biochemical Studies of Muscle. Many biochemical studies support the basic idea that filaments generate the force of contraction that slides thin filaments past thick filaments. Each thick filament consists of several hundred long protein strands with globular protein heads at each end. This protein is called **myosin.** The thin filaments consist of two twisted strands of globular proteins called **actin.** (This protein is found in microfilaments in all cells.) One line of support for the sliding filament theory comes from studies of purified actin and myosin mixed to form a gel-like suspension. If ATP is added to this suspension, the actin and myosin bind together to form a precipitate, but only if Ca^{2+} is also added. Apparently the force of contraction produced by the interaction of thick and thin filaments uses ATP as an energy source and Ca^{2+} as an activator.

In electron micrographs one can actually see connections, called **cross-bridges,** between actin and myosin. These cross-bridges are the heads of the thick filaments. Biochemical studies have shown that these heads have ATPase enzyme activity in the presence of actin. It appears, therefore, that the cross-bridges from myosin to actin are the sites where the energy from ATP is converted to the mechanical energy of contraction. How this energy conversion occurs is still uncertain. There is some evidence that the ATP causes the cross-bridges at both ends of a thick filament to pull the actin toward the middle of the sarcomere. Since this happens all along the length of a muscle fiber, the fiber shortens.

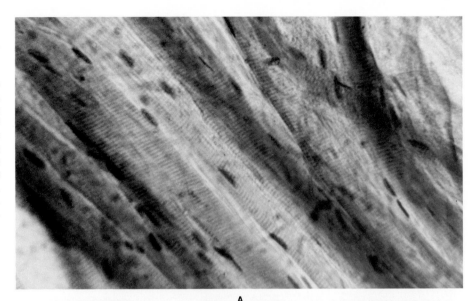

A

B

The role of Ca^{2+} as the activator of contraction is supported by studies using **aequorin,** a protein from the jellyfish *Aequorea aequoria* that emits light in the presence of Ca^{2+}. Muscle fibers that have been filled with aequorin emit a pulse of light just before they contract, suggesting that an increase in the concentration of Ca^{2+} in the sarcomeres triggers contraction. Other biochemical studies have shed light of another kind on the role of calcium ions. **Troponin,** a protein normally bound to actin, consists of three subunits that change their configuration in the presence of Ca^{2+}. Present evidence suggests that the burst of Ca^{2+} that triggers contraction does so by causing this change in troponin. The change in troponin configuration may then act on **tropomyosin,** another protein found on actin. It is thought that tropomyosin blocks the sites on actin where the myosin heads attach, but that the change in troponin configuration displaces the tropomyosin, allowing the cross-bridges to form. Once the myosin heads attach, their ATPase activity triggers contraction of the muscle (Figure 10.13).

As usual, answering these questions has raised others. From where does the

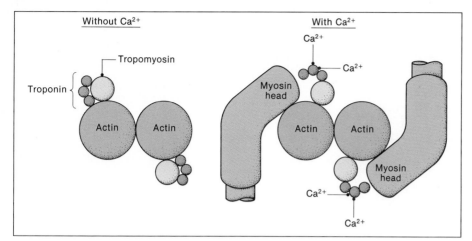

Figure 10.13
A model of the calcium-induced changes in troponin that allow cross-bridges to form between actin and myosin. The view here is as if we were sighting along the length of the double-stranded actin filaments. In the presence of Ca²⁺ the three subunits of troponin change configuration, allowing tropomyosin to slide into the groove between the actin strands. This uncovers the binding site for the myosin heads, which then attach to form the cross-bridges.

Ca^{2+} enter the sarcomere, and how do action potentials trigger its entrance? For invertebrate muscles the answer to this question appears to be that the influx of Ca^{2+} that initiates the action potential also triggers contraction. For vertebrates the answer is more complicated. It has long been known that a type of smooth endoplasmic reticulum (see p. 47) called **sarcoplasmic reticulum** (Figure 10.12) actively takes up Ca^{2+} from an external solution. It seems reasonable, therefore, that Ca^{2+} is stored in the sarcoplasmic reticulum when the muscle relaxes and is released when contraction is initiated. Apparently the calcium permeability of the membranes of sarcoplasmic reticulum increases in response to the action potential in the muscle fiber, releasing the Ca^{2+} that triggers contraction. Action potentials travel to the sarcoplasmic reticulum by way of **T tubules** (transverse tubules) that run from the plasma membrane into the fibers.

Summary of the Sliding Filament Theory. All this information and theory, accumulated over several decades in many laboratories by many techniques, can now be integrated into a coherent idea of the mechanism by which a muscle fiber contracts (Figure 10.14). Action potentials in motor axons release acetylcholine from a neuromuscular synapse, which evokes a depolarizing endplate potential (EPP) on the plasma membrane of the muscle fiber. This EPP triggers action potentials in the plasma membrane, which travel down the T tubules that encircle the myofibrils. The action potentials trigger an increase in calcium permeability in the membrane of sarcoplasmic reticulum, which allows the Ca^{2+} within the sarcoplasmic reticulum to leak out. The Ca^{2+} causes a change in the configuration of troponin on the actin molecules, which displaces tropomyosin from the binding site for the myosin heads. The myosin heads can then attach to the actin, forming cross-bridges. The myosin heads also convert ATP energy into mechanical energy, which causes the cross-bridges to pull on the actin filaments at each end.

Relaxation occurs when the action potentials cease, and the calcium permeability of the sarcoplasmic reticulum returns to its resting level. Calcium pumps then transport the Ca^{2+} from the sarcomeres into the sarcoplasmic reticulum, breaking the cross-bridges. It may seem incredible that something as weak as the bond between two protein filaments could account for the strength in your own muscles, let alone the strength of a professional weight-lifter. There are millions of cross-bridges in each muscle fiber, however, and there are thousands of muscle fibers in a muscle. A skeletal muscle 1 cm² in diameter has, in fact, enough cross-bridges to support a mass of up to 3 kg. Thicker muscles generate greater forces in direct proportion to their cross-sectional area.

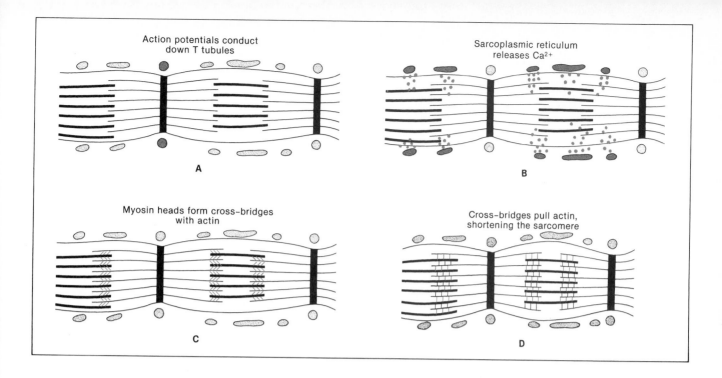

Figure 10.14
Activation of vertebrate skeletal muscle. (A) Action potentials in the motor axon evoke action potentials in the muscle fiber membrane. These action potentials conduct down the T tubules. (B) The action potentials trigger the release of Ca^{2+} from the sarcoplasmic reticulum. (C) Ca^{2+} causes a change in the structure of troponin, which allows cross-bridges to form. (D) The cross-bridges (myosin heads) swivel, pulling the actin and attached Z discs toward the center of the sarcomere.

ENERGETICS OF MUSCLE CONTRACTION

Sources of Energy.   The energy for muscle contraction comes from the conversion of ATP into ADP and inorganic phosphate. Approximately 70% of the available energy is used directly for contraction; the remainder is used in transporting Ca^{2+} back into the sarcoplasmic reticulum during relaxation. There is not enough ATP in a sarcomere to power more than a few twitches. If that small amount is depleted, the result is **cramping,** in which cross-bridges form but there is no contraction. The muscle simply becomes rigid. The same thing occurs after death, in **rigor mortis.** Cramping does not often occur, however, because molecules called **phosphagens** usually maintain a steady level of ATP. In vertebrates and some invertebrates the phosphagen is creatine phosphate; in most invertebrates it is arginine phosphate. Phosphagens work in one of two ways. In muscle fibers with many mitochondria, phosphagen makes it possible for ADP to be converted back to ATP without having to go back to the mitochondria. Instead, the ADP remains on the myosin, and the phosphagen brings it fresh phosphate. In muscles that do not rely on mitochondrial production of ATP, phosphagens serve as a reserve supply of phosphate that replenishes the ADP, as represented by the following equation:

$$\text{creatine-P} + \text{ADP} \longrightarrow \text{creatine} + \text{ATP}$$

Fatigue.   Muscles usually fatigue from some other cause before they run out of ATP. Fatigue—the reduction in force that a muscle can generate—is often due to the accumulation of by-products from the production and breakdown of ATP. Inorganic phosphate, which is released when ATP is used, is one cause of fatigue (Nosek et al. 1987). Lactic acid, from the anaerobic production of ATP (see pp. 55–57), also contributes to fatigue, by reducing the pH inside the sarcomeres. In addition, the lactic acid damages muscle tissue, causing soreness.

Power and Efficiency.   One expects that as the result of evolution animals come to use their muscles in such a way that they work optimally. For some muscles the optimal performance would occur when they do as much work as they can in the least time. For others optimal performance would occur when they do as much work as they can using the least energy. In other words, some muscles must produce peak power, while others must have peak efficiency. The speed of contraction determines both power and efficiency, but that speed is not necessarily the same for both. In fact, the optimal speed for power is generally about 1.5 times that for efficiency. Such considerations may explain why animals shift **gaits** at different speeds. Horses and dogs, for example, shift from a walk to a trot to a gallop as they move faster. Each gait may permit the muscles to shorten at about the optimum speed for either efficiency or power, regardless of the speed of locomotion. The gears of a bicycle serve a similar function, permitting the leg muscles to work at a speed that is optimal for either power or efficiency regardless of how fast the bicycle is moving.

VARIETIES OF SKELETAL MUSCLE

Numerous variations in the structure and function of muscle fibers occur among vertebrates, and even greater variety can be found among invertebrates. One type of variation is based on the fact that a muscle fiber cannot function at optimal efficiency and produce maximum power at the same time, for reasons described in the previous paragraph. Thus muscle fibers tend to be specialized for either efficiency or power. Those specialized for efficiency are called **slow-twitch** or **Type I fibers.** They produce ATP aerobically, and they contract relatively slowly. They have a reddish color due mainly to a generous supply of blood vessels and to the cytochromes in their numerous mitochondria. Type I fibers also have a large amount of the oxygen-binding pigment **myoglobin.** Muscles specialized for power are **fast-twitch** or **Type II white fibers.** They contract more rapidly but rely on the anaerobic breakdown of glycogen for their ATP. Type II fibers have a light color, few mitochondria, and little myoglobin. They are also generally thicker than Type I fibers.

Both types of fiber can occur within the same muscle, although many muscles have much more of one type of fiber than of the other. Slow-twitch fibers predominate in muscles that must contract for long periods to maintain an erect posture. A familiar example of slow-twitch fibers is the dark meat of chickens, which occurs in the legs and back. Fast-twitch fibers predominate in muscles that contract rapidly but not often. The light meat in the breast (flight muscles) of chickens consists largely of fast-twitch fibers. If required to work for long periods, fast-twitch fibers quickly fatigue and get sore. Intermediate types of muscle also occur. In ducks and most other birds that can fly for long distances, the breast muscles consist largely of an intermediate type of fiber—the **fast-contracting red fiber.** This type, also called the **Type II red fiber,** has many mitochondria but can contract rapidly.

Training may affect the proportions of muscle types in a muscle. Sprinters generally have a greater than average proportion of fast-twitch fibers in their muscles, while marathoners usually have a greater proportion of slow-twitch fibers, and middle-distance runners have about average proportions of fast- and slow-twitch fibers. These training effects may result from the fact that sustained stimulation of the motor nerve to a fast-twitch muscle will transform it into a slower muscle. Training also increases the strength of muscles in several ways. Muscle cells that are active can produce ATP more rapidly, because their blood supply, level of oxygen-binding myoglobin, and amounts of metabolic enzymes all increase. In addition, exercise stimulates muscle fibers to produce a larger number of filaments, which increases the diameter of the fibers. There may also be an increase in the number of muscle fibers (Jolesz and Sreter 1981).

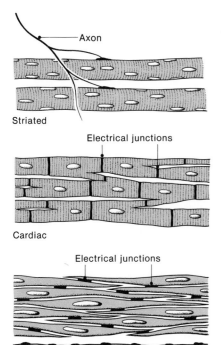

Figure 10.15
The three types of vertebrate muscle: skeletal, cardiac, and smooth. While skeletal muscle is multinucleate, cardiac and smooth muscle have only one nucleus per cell. Both skeletal and cardiac muscle are striated and fiber shaped, while smooth muscle is not striated (by definition) and is spindle shaped. Neither cardiac nor smooth muscle cells receive direct innervation through neuromuscular junctions, but smooth muscle can be controlled by nerve cells that release transmitters in their vicinity. Both cardiac and smooth muscle cells are electrically interconnected.

CARDIAC MUSCLE

Vertebrates have two other types of muscle besides skeletal muscle. Cardiac muscle, of which hearts are made, has the striations and parallel filaments of skeletal muscle but differs from it in many respects (Figure 10.15). The cells of cardiac muscle do not fuse into cylindrical fibers, but remain separate. Cardiac muscle cells are, however, linked to each other by **intercalated discs,** which work like electrical synapses. The cells branch out to form interconnecting networks that ensure that action potentials trigger contraction throughout the entire mass of cardiac muscle. Cardiac muscle also differs from skeletal muscle in not requiring excitation by the nervous system. Instead, some of the cardiac muscle cells generate action potentials spontaneously, and these action potentials conduct through the intercalated discs and trigger contraction in other cardiac muscle cells. Thus vertebrate hearts can go on contracting after they have been entirely removed from the body. Cardiac muscle also differs from skeletal muscle in having fewer T tubules (virtually none in most vertebrate classes), and in getting more of its Ca^{2+} influx across the plasma membrane rather than from the sarcoplasmic reticulum only. The relevance of this to heart function will be explained in Chapter 11.

SMOOTH MUSCLE

A third type of vertebrate muscle is smooth muscle, so called because it lacks the striations of skeletal and cardiac muscle (Figure 10.15). Smooth muscle is usually found surrounding hollow organs, such as blood vessels, the digestive tract, and the urinary bladder. Smooth muscle cells are spindle shaped. Their thick and thin filaments are not arranged in neat array like those of striated muscles, but in such a way that during contraction each cell twists like a cloth being wrung out (Warshaw et al. 1987). Like the cells of cardiac muscle, those of smooth muscle are electrically coupled to each other so that an action potential in one cell sends a wave of contraction throughout the muscle. One example of such a wave of contraction is the **peristalsis** that occurs in the esophagus during swallowing, from which the term peristaltic progression (see p. 200) is derived.

Action potentials in smooth muscle result from an influx of Ca^{2+} rather than Na^+. The same Ca^{2+} triggers the contraction. There is no troponin for the Ca^{2+} to bind to in smooth muscle. Instead, Ca^{2+} binds to **calmodulin,** a troponinlike protein that regulates calcium action in many cells. A Ca–calmodulin complex acts on the myosin heads to activate cross-bridge formation. Contraction is extremely slow in smooth muscle and can be excited by some nerve endings and inhibited by others. These nerve endings are not synapses, and they are not always required to trigger contraction. Smooth muscle of the intestine, for example, will contract in response to direct mechanical stimulation. The excitability of many smooth muscles is controlled by the autonomic nervous system, with the parasympathetic and sympathetic divisions having opposite effects.

TYPES OF INVERTEBRATE MUSCLE

A few differences among invertebrate muscles should be mentioned to indicate the numerous departures from the vertebrate patterns described above. The control of vertebrate striated muscle is not at all like that of most animals (Lehmann and Szent-Gyorgyi 1975). In the majority of invertebrates Ca^{2+} triggers contraction not only by its effect on troponin but also by an effect on myosin. Molluscs, in fact, have no troponin. Other differences occur in the pattern of innervation of skeletal muscle. In arthropods, for example, a typical muscle fiber is innervated by not just one but several excitatory neurons. One excitatory nerve fiber produces a large depolarization that triggers a fast contraction, and another excitatory nerve fiber produces a small depolarization and a slow contraction. There is also at least one inhibitory neuron that causes hyperpolarization. This multiple innervation seems to be a way of getting fine gradations of force with a limited number of muscle fibers.

Another variation in muscle function occurs in bees and wasps, flies, beetles, and bugs. The flight muscles of these animals do not contract in synchrony with the action potentials in their motor axons but often at a frequency higher than the maximum frequency of action potentials. These muscles are therefore called **asynchronous muscles** (= fibrillar muscles). Neural stimulation of these asynchronous muscles maintains a constant level of Ca^{2+} in the sarcomeres, but cross-bridges do not form until the muscles are quickly stretched. The movement of the wing upward therefore activates the muscles that produce the following down-stroke (Figure 10.3C). In a midge this cycle can happen 1000 times a second.

Another interesting variation in muscle function is found in the adductor muscles of bivalve molluscs such as mussels, oysters, scallops, and clams. Once the cross-bridges form in these **catch muscles,** they remain rigid with little further use of ATP. These smooth muscles give bivalves the ability to "clam up" against predators for days at little energetic expense. For their size they are the strongest muscles known.

MOVEMENT WITHOUT MUSCLE

Ameboid Movement.   Many protozoans, and many cells within animals, propel themselves by means other than muscle. One such mechanism is ameboid movement. As the name suggests, this type of movement occurs in *Amoeba* and some other protozoans (see Chapter 22). Humans and many other vertebrates have white blood cells that wander about and engulf foreign matter by ameboid movement. Under the light microscope it appears that ameboid motion results from a flow of **endoplasm,** a fluid form of cytoplasm, in the direction of overall motion (Figure 10.16). The endoplasm moves into projections of the cell called **pseudopodia** and then turns into a clearer, less-fluid **ectoplasm.** Researchers are still debating, after decades of work, whether this movement represents a push from the rear of the cell or a pull from the front, or both. The transformation from

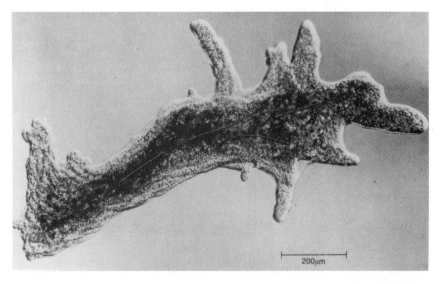

Figure 10.16
Light micrograph of the giant ameba Chaos carolinensis. *Blurring of the central portion of this protozoan resulted from the streaming of cytoplasm responsible for ameboid movement.*

Figure 10.17
(A) Electron micrograph of a flagellum from a sperm cell of the sea urchin Lytechinus. The "9 + 2" configuration of microtubules is the axoneme. (B) Cilia of the protozoan Opalina beat in waves. The distance between successive waves of cilia is about 8 μm. (C) Human sperm cells, each with a single flagellum about 40 μm long.

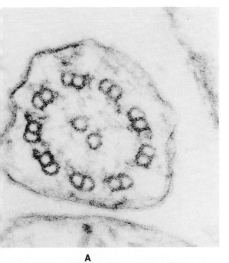

A

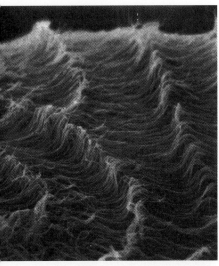

B

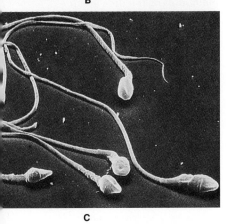

C

endoplasm in a **sol** form to ectoplasm in a **gel** form can occur even in isolated cytoplasm. Like muscle contraction, it requires the presence of actin, myosin, and Ca^{2+}.

Cilia and Flagella.   Another type of cellular movement without muscle uses cilia and flagella. Cilia and the flagella of eukaryotes are virtually identical in structure, but cilia are usually shorter (less than 15 μm long) and occur in large groups. Flagella are up to 2 mm long (in some insect sperm) and usually occur individually or in small groups. Both cilia and flagella have similar core structures, called **axonemes.** The axonemes usually have a distinctive "9 + 2" arrangement of nine microtubule doublets surrounding two microtubules (Figure 10.17A). Cilia and flagella attach to a **basal body** (= kinetosome) in the cytoplasm. How they are controlled is a mystery.

Cilia are responsible for the locomotion of many protozoans such as *Paramecium*, some small invertebrates such as flatworms and ribbon worms (phyla Platyhelminthes and Nemertea), some aquatic molluscs, and some invertebrate larvae. In some animals cilia circulate water during filter feeding and respiration. In mammals cilia lining the respiratory tract propel debris-trapping mucus away from the lungs, and cilia lining the oviduct propel the oocyte toward the uterus.

BODY SIZE AND LOCOMOTION

Allometry.   Among the favorite characters of fairy tales are humans and other animals who have grown or shrunk incredibly. Although such dwarfs and giants are usually portrayed as being ordinary in every way except size, their movements would in fact be far from ordinary. Consider what would happen if you suddenly grew to ten times your present height and retained the same proportions. Your weight would have increased in proportion to the cube of your height. However, the cross-sectional areas of your body and limbs would have increased only in proportion to the square of your height. Thus you would weigh a thousand (10^3) times as much, but your muscles and bones would be only a hundred (10^2) times stronger. If you now weigh 130 pounds, you would weigh a staggering (literally) 130,000 pounds, and all that weight would have to rest on legs adapted to support only 13,000 pounds. Your bones would probably be crushed, or at the very least you would be pinned to the floor by your own weight. Other problems would arise if you shrank to one-tenth your present height while retaining the same proportions. You would weigh 0.13 pound (about 2 ounces) but would have the strength appropriate for a person ten times as massive. You would probably smash yourself against the ceiling the first time you took a step.

The study of the differential effects of changing linear, area, and volume dimensions in organisms is referred to as **allometry** (Greek *allo* different + *metron* a measure) or **scaling.** Allometric considerations suggest that during evolution there have probably been many compromises among linear dimension, area, and volume. For example, it is probable that in the evolution of squirrels there were times when being larger would have helped protect them from predators. On the other hand, the increased weight would have hindered their escaping through the trees. Allometric constraints are evident in several mathematical regularities in animals of various sizes. One of the most studied of these regularities is in the energy an animal uses in locomotion. For animals ranging in size from mosquitoes to horses, the energy needed for an animal to transport a unit of its body mass for a given distance is proportional to the animal's mass raised to some power that depends on whether the animal swims, flies, or runs. That is, energy/(kg · km) is proportional to (mass)^k where k is a constant number for each type of locomotion. As explained in Figure 10.18, the value of k suggests that the energy per kilogram needed to swim or run a kilometer is proportional to one divided by the body length.

Swimming. This and other allometric relationships have led a growing number of zoologists to look for physical explanations. The case of swimming is most easily understood, since buoyancy allows us to ignore the effects of gravity. Swimming is possible when an animal can continue its forward movements between swimming strokes. This can happen only if the **inertial force** (the acceleration of the animal with respect to the water) exceeds the **viscous force**, which tends to stop the animal. The ratio of inertial to viscous forces is given by the **Reynolds number,** Re:

$$Re = \frac{(\text{length of body}) \times (\text{speed}) \times (\text{density of water})}{(\text{viscosity of water})}$$

$$= 100 \times (\text{length of body in cm}) \times (\text{speed in cm/sec})$$

All else being equal, therefore, the higher the Reynolds number the more easily an animal can swim. For similarly shaped animals, the longest and fastest ones will expend less energy in moving a kilogram of their body mass for one kilometer. The higher Reynolds numbers of larger animals may explain why larger fishes are more efficient as swimmers, as shown in Figure 10.18.

The blue whale (*Balaenoptera musculus*), with a body length of about 2500 cm and a cruising speed of 1000 cm/sec, has an Re of more than 10^8. It should be able to swim with extreme efficiency, although no one has yet figured out how to demonstrate this. Salmon, which have to be highly efficient to migrate upstream without eating, have an Re of about 10^6. Human swimmers, perhaps surprisingly, have an Re about equal to that of salmon. We are less efficient probably because we are not as streamlined. For much shorter and slower swimmers Re drops

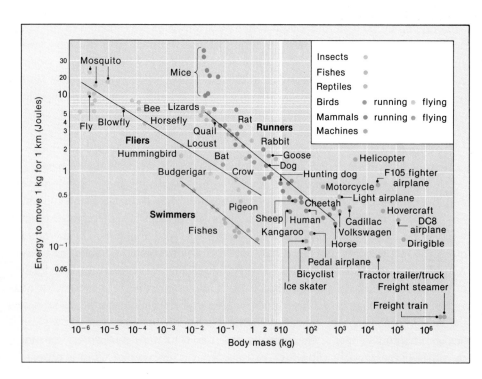

Figure 10.18
The energy required for an animal to move a kilogram of its body mass one kilometer, versus its total body mass. This is a double logarithmic plot, in which each unit on an axis represents a tenfold change. Plotting data in this way is equivalent to taking the logarithm of both sides of an equation. If one side of the equation is proportional to the other variable raised to a power, this method plots the data along a straight line with a slope that equals the power. This technique is common in allometry. (See Figures 8.6 and 13.10 for other examples.) The slopes for swimming and running are approximately $-\frac{1}{3}$. This shows that for swimming and running the energy required to move a kilogram of body mass one kilometer is proportional to one divided by the cube root of the body mass. Since the body mass is proportional to the cube of body length, the energy per kilogram per kilometer is proportional to one divided by body length. The slope for flying is approximately $-\frac{1}{4}$.

sharply. Viscous forces overcome inertial forces, and water starts to act like syrup or even tar. Swimming for such animals is inefficient, if not impossible. Diving beetles, with an Re of about 10^2, are at about the limit at which real swimming is possible. Invertebrate larvae and fish hatchlings with lower Reynolds numbers join the plankton, drifting passively with the current. Under a microscope many protozoans with even smaller values of Re often appear to move quite rapidly, but they cannot actually swim. A 0.02-cm paramecium moving through water at about 0.1 cm/sec (Re approximately 0.2) is essentially crawling through the water with its cilia.

Flying. These limitations imposed by body size also apply to flying if one corrects for the much lower density and viscosity of air. (For the same length and speed, Re for flying is about $\frac{1}{15}$ that for swimming.) The ability to glide through the air between wing strokes also depends on having a high Re. The California condor (*Gymnogyps californianus*) has an Re value on the order of 10^5 and can glide for extended periods—or could while it still survived in the wild. Even large dragonflies and butterflies have Re values large enough (more than 10^3) to permit periodic gliding. Smaller birds and insects have to flap continually to maintain forward motion between wing strokes.

Unlike swimming, flying is complicated by the fact that the animal must not only move forward but also generate **lift** to overcome gravity. Scientists do not fully understand how flying animals generate lift (see Chapter 32 for discussion of flight in insects and Chapter 38 for flight in birds). There is, however, one fact that is firmly grounded (so to speak) in numerous unsuccessful attempts at human flight: The surface area of wings must be large enough to generate enough lift to overcome the weight of the body. In other words, the **wing loading** (body mass divided by wing area) must not be too large. Body mass generally increases as the cube of body length, but wing area increases only as the square of body length, so there is an upper limit on body size for flapping flight.

Figure 10.19
Double logarithmic plot of the lift produced by flapping in more than 160 insects, birds, and bats versus the mass of their flight muscles. The data plot along a line with slope equal to 1.0, showing that lift is directly proportional to the mass of flight muscle.

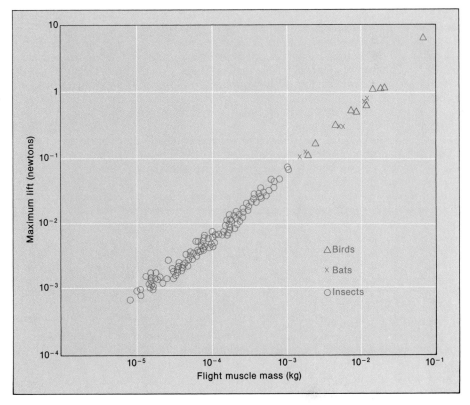

The generation of lift by flapping while in flight is hard to study, for obvious reasons. It is much easier to study how an animal takes off from a dead start by flapping. By adding weights until animals were no longer able to take off, Jim Marden (1987) has measured the maximum lift generated by more than 160 insects, bats, and birds of numerous species. He found that the maximum lift is directly proportional to the mass of the flight muscles (Figure 10.19). This result is rather surprising for two reasons. First, allometric considerations suggest that lift should be proportional to $(mass)^{2/3}$ rather than directly proportional to mass, since force is proportional to the cross-sectional area of muscle. Second, the result implies that the diverse flight muscles among insects, bats, and birds all generate the same lift per gram.

Marden found that the main difference between strong fliers and weak ones is the **flight muscle ratio** (FMR), defined as the mass of the flight muscles divided by the total body mass. Animals with FMRs less than 0.16 (flight muscles less than 16% of body mass) cannot take off at all from a standing start in calm. If they fly at all they must jump into the wind or get a running start. Marden also showed that differences in wingspan, area, and shape have little effect on the lift during takeoff, although they are undoubtedly important once the animal is airborne (Viscor and Fuster 1987).

SUMMARY

The internal environments of animals are protected by integuments that range in complexity from a thin epidermis to thick shells, cuticles, and skins. The skin of vertebrates consists of a protective epidermis and an underlying dermis. Vertebrate skin also includes many other structures, such as hairs, sensory organs, and glands, as well as pigments and structures that produce color.

The bodies of most animals are supported by skeletons. The skeleton may simply be internal fluid pressure—a hydroskeleton—or it may be a rigid structure. Many invertebrates have rigid exoskeletons that are parts of the integument, while other animals have endoskeletons. In vertebrates the skeleton consists mainly of bone.

Skeletal muscle attached to the skeleton enables animals to move. Different movements are evoked by activating muscles that attach to different bones and by varying the number of motor units activated and the frequency of action potentials. The connection between nerve cells and muscle cells is the neuromuscular synapse, which triggers changes in the voltages across membranes in muscle cells. These voltages trigger a release of calcium, which triggers interaction between two types of protein filament in the muscle. According to the sliding filament theory, muscle contraction is not due to shortening of filaments but to sliding of thin filaments past thick filaments. In response to calcium, thick filaments form cross-bridges with thin filaments, then pull on the thin filaments to shorten the muscle. The energy for contraction comes from ATP.

Skeletal muscle fibers are specialized for either efficiency or power. The former are slow-twitch fibers, and the latter are fast-twitch fibers. In addition to these two types of skeletal muscle cells, vertebrates also have cardiac muscle, which makes up the heart, and smooth muscle, which controls the movement of the digestive tract and the diameter of hollow and tubular organs. Numerous other kinds of muscles occur in invertebrates. Other mechanisms for movement are ameboid motion and cilia or flagella.

The strength of the skeleton and the forces required for movement depend to a large extent on body size. Length, surface area, and volume change by different amounts when body size changes. Swimming and flying are most efficient for long animals, but flying is impossible when body mass reaches a certain limit.

KEY TERMS

integument	flexor	sliding filament theory
epidermis	extensor	myosin
dermis	neuromuscular synapse	actin
hydrostatic skeleton	motor unit	cross-bridge
exoskeleton	striated muscle	sarcoplasmic reticulum
endoskeleton	thin and thick filaments	T tubules
cartilage	sarcomere	phosphagen

slow-twitch and fast-twitch fibers
cardiac muscle
smooth muscle

ameboid movement
cilium
flagellum

allometry

SELF-TEST

1. Describe how the epidermis of an invertebrate differs from that of a vertebrate.

2. What functions are served by the epidermis? What functions are performed by the dermis?

3. Give an example of an animal with each type of skeleton—hydrostatic, exoskeleton, and endoskeleton—and explain how the contractions of its muscles produce locomotion.

4. Explain three factors that affect the force you exert with one hand.

5. From memory make a sketch that shows a motor axon, a motor unit, a muscle fiber, and a sarcomere. Compare your sketch with Figures 10.8 and 10.9.

6. Describe the relationship among the following: actin, myosin, thin filaments, thick filaments, I band, A band.

7. State the essence of the sliding filament theory. Describe a competing theory and explain why it is not as widely accepted as the sliding filament theory.

8. Describe the functions of Ca^{2+} in movement and support. (Don't forget the role of Ca^{2+} in the neuromuscular junction.)

9. For each of the following, state one similarity and one difference between it and vertebrate skeletal muscle: insect asynchronous muscle; the catch muscle of molluscs; ameboid movement; the movement of cilia or flagella.

10. Examine the pictures of very large and very small mammals in Chapter 39. What do you notice about the thickness of the legs in relation to body size? How would you explain this relationship?

11. Fill in the following table:

	Slow-Twitch	Fast-Twitch
Speed of contraction (fast or slow)		
Usage (continuous or intermittent)		
Number of mitochondria (few or many)		
Type of metabolism (aerobic or anaerobic)		
Resistance to fatigue (little or great)		
Color (red or white)		

12. Fill in the following table:

	Skeletal Muscle	Cardiac Muscle	Smooth Muscle
Typical location in vertebrates			
Voluntary control?			
Striated?			
Each cell with only one nucleus?			
Cells interconnect electrically?			
Neuromuscular junctions present?			

READINGS

RECOMMENDED READINGS

Alexander, R. M. 1991. How dinosaurs ran. *Sci. Am.* 264(4):130–136 (Apr).

Caplan, A. I. 1984. Cartilage. *Sci. Am.* 251(4):84–94 (Oct).

Denny, M. W. 1990. Terrestrial *versus* aquatic biology: the medium and its message. *Am. Zool.* 30:111–121. (*Swimming versus flying.*)

Govind, C. K. 1989. Asymmetry in lobster claws. *Am. Sci.* 77:468–474.

McMahon, T. A. and J. T. Bonner. 1983. *On Size and Shape.* New York: Scientific American Books.

Smith, K. K. and W. M. Kier. 1989. Trunks, tongues, and tentacles: moving with skeletons of muscles. *Am. Sci.* 77:28–35.

Tucker, V. A. 1975. The energetic cost of moving about. *Am. Sci.* 63:413–419.

Webb, P. W. 1988. Simple physical principles and vertebrate aquatic locomotion. *Am. Zool.* 28:709–725.

ADDITIONAL REFERENCES

Holye, G. 1983. *Muscles and Their Neural Control.* New York: Wiley-Interscience. (*Unique source of information for various groups of animals.*)

Jolesz, F. and F. A. Sreter. 1981. Development, innervation and activity-pattern induced changes in skeletal muscle. *Annu. Rev. Physiol.* 43:531–552.

Lehmann, W. and A. G. Szent-Gyorgyi. 1975. Regulation of muscular contraction: distribution of actin control and myosin control in the animal kingdom. *J. Gen. Physiol.* 66:1–30.

Marden, J. H. 1987. Maximum lift production during takeoff in flying animals. *J. Exp. Biol.* 130:235–258.

Nosek, T. M., K. Y. Fender, and R. E. Godt. 1987. It is diprotonated inorganic phosphate that depresses force in skinned skeletal muscle fibers. *Science* 236:191–193.

Viscor, G. and J. F. Fuster. 1987. Relationship between morphological parameters in birds with different flying habits. *Comp. Biochem. Physiol.* 87A:231–249.

Warshaw, D. M., W. J. McBride, and S. S. Work. 1987. Corkscrew-like shortening in single smooth muscle cells. *Science* 236:1457–1459.

11

Circulation and Immunity

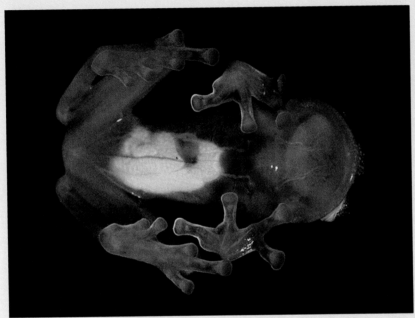

Glass frog (Centrolenella) with heart visible through skin.

CHAPTER OUTLINE

LEARNING OBJECTIVES

1. How does an animal produce and maintain a suitable environment for its cells?

2. What is the relationship of the interstitial fluid to blood?

3. How do the circulatory systems of various animals differ?

4. How are the hearts of different animals constructed?

5. How does the human heart work?

6. How does the heart "know" when to increase and decrease its activity?

7. What happens to blood as it circulates?

8. What makes blood clot at the appropriate time?

9. Why don't foreign cells take over the body fluids and tissues of an animal?

BODY FLUID

Animals are believed to have evolved from single-celled marine organisms that used the sea as a source of oxygen, minerals, nutrients, and thermal energy and as a dump for metabolic wastes. In some marine animals, such as sponges and cnidarians (sea anemones, jellyfishes, corals) the cells are still essentially bathed in seawater. In many marine animals, however, the body tissues are so large that most cells are far from the sea. Moreover, many animals live in habitats that would instantly kill an exposed cell. Nevertheless, the cells of all animals are still bathed in a kind of sea—a salty liquid called the **interstitial fluid.** This internal sea is created and maintained by the organ systems of the animals themselves, as will be described in later chapters of this unit.

Interstitial fluid bathes all cells but usually does not flow freely around them. Consequently, nutrients, oxygen, and other necessities must be constantly resupplied, and metabolic wastes must be carried off. Generally, the interstitial fluid draws upon a much larger reservoir of fluid that circulates more freely, transporting the necessities of life and removing the wastes. In many animals this reservoir is the fluid in a body cavity called the **pseudocoel** (SUE-doe-seal) or **coelom** (SEAL-um). This and other functions of these body cavities will be detailed later (see p. 544). In many animals the interstitial fluid is regenerated by **blood** rather than, or in addition to, the fluid in the pseudocoel or coelom. By definition, blood flows through a **circulatory** (= vascular) system that is separate, at least in part, from the body cavity.

In many invertebrates the blood is similar to the interstitial fluid. In other invertebrates, and in all vertebrates, blood contains specialized cells and molecules that aid in its functions. Removing the cells from vertebrate blood leaves the fluid portion, called **plasma.** Plasma often contains substances that cause **clotting,** which reduces blood loss in case of injury. If plasma is allowed to clot, and the clot is removed, the remaining fluid is **serum.** Blood may also contain **respiratory pigments,** which increase the ability of blood to transport oxygen, as will be described in the next chapter. The respiratory pigments may simply be suspended in the blood, or they may be contained in blood cells. In vertebrates the respiratory pigment **hemoglobin** is contained in **red blood cells** (erythrocytes). Still other blood cells and molecules combat viruses, bacteria, and parasites. In vertebrates the **white blood cells** (leukocytes) serve these functions. (See Figure 11.1 and the figure on p. 18.)

OPEN AND CLOSED CIRCULATORY SYSTEMS

Some of the enormous variety of circulatory systems will be described in the chapters dealing with various groups of animals. For now we can consider two major types of system—closed and open (Figure 11.2). In **closed circulatory systems** blood is always confined within the vascular system (arteries, capillaries, veins, and heart), and the blood is different from the interstitial fluid. Closed circulatory systems occur in vertebrates, such as ourselves, and in annelids such as earthworms (Figure 11.3A), cephalopod molluscs such as the octopus, and a few other groups (see Figure 21.10 for summary). Our own closed system will be examined more closely later in this chapter.

In **open circulatory systems** blood is not always confined within the vascular systems. Instead, the blood that leaves the heart simply empties into spaces that make up the **hemocoel** (HEEM-oh-seal; Figure 11.3B). The blood then moves through the hemocoel in a leisurely fashion, usually under low pressure. Open

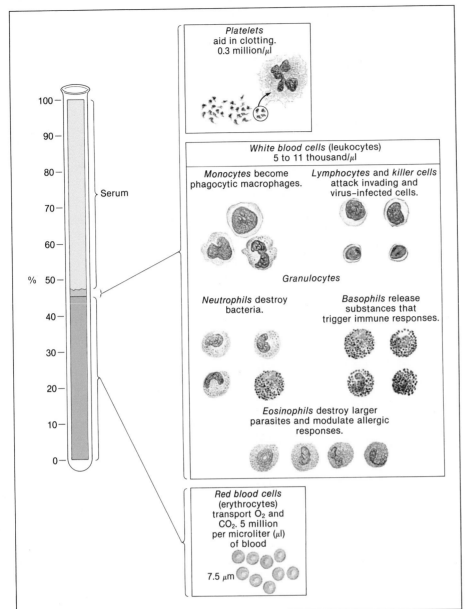

Platelets
aid in clotting.
0.3 million/μl

White blood cells (leukocytes)
5 to 11 thousand/μl

Monocytes become phagocytic macrophages.

Lymphocytes and *killer cells* attack invading and virus–infected cells.

Granulocytes

Neutrophils destroy bacteria.

Basophils release substances that trigger immune responses.

Eosinophils destroy larger parasites and modulate allergic responses.

Serum

Red blood cells (erythrocytes) transport O₂ and CO₂. 5 million per microliter (μl) of blood

7.5 μm

Figure 11.1
The composition of human blood. The figure at left shows blood from a normal person after removing the clotting proteins and centrifuging it to make the cells settle. A little over half the volume is the fluid serum. Approximately 45% (a little less in women) is red blood cells. Between serum and red blood cells is a "buffy coat" containing platelets and white blood cells. Platelets, red blood cells, and white blood cells make up the formed elements, all of which originate as stem cells in bone marrow. Platelets are fragments of cells called megakaryocytes. All the white blood cells appear colorless under the light microscope; special stains are used to reveal structural differences such as intracellular granules.

Figure 11.2
The relationships among the external environment, blood, and the interstitial fluid. (A) In closed circulatory systems blood is confined within a system of vessels. As it circulates through capillaries in the respiratory and digestive systems it absorbs oxygen and nutrients for delivery through other capillaries to metabolizing cells. The blood absorbs carbon dioxide and metabolic wastes from those cells and delivers them to the respiratory and excretory systems for elimination. (B) In open circulatory systems there is no clear distinction between blood and the interstitial fluid in some tissues, because the blood is not confined to the vessels of the circulatory system.

systems occur in most molluscs and arthropods, and in some invertebrate chordates. The molluscs with open circulatory systems are clams, snails, and other sluggish forms that don't demand much oxygen from their blood. Insects are able to sustain high levels of activity with open circulatory systems by having air tubes that bring air directly to cells, as will be described in the next chapter. An advan-

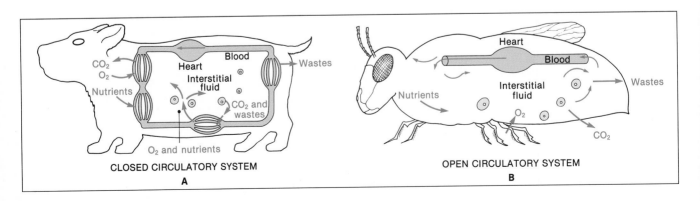

CLOSED CIRCULATORY SYSTEM
A

OPEN CIRCULATORY SYSTEM
B

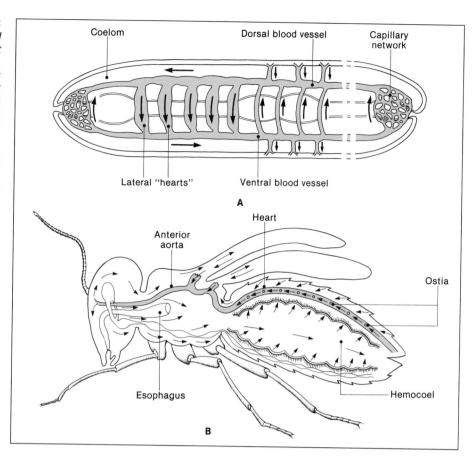

Figure 11.3
(A) Schematic diagram of the closed circulatory system of the giant earthworm Glossoscolex giganteus. Because of the large size of this Brazilian species (up to 0.5 kg), it has been possible to measure its blood pressure directly. In the ventral blood vessel the pressure varies between 35 and 70 mmHg. (B) Schematic diagram of the open circulatory system of a generalized insect. Blood is drawn into the heart through one-way openings called ostia and is then pumped through the anterior aorta. From there it percolates among the organs in the body cavity (hemocoel). In contrast to closed circulatory systems, open systems generally have low pressures. In the grasshopper Locusta, for example, blood pressure varies from about 2 to 6 mmHg.

tage of open circulatory systems is that blood pressure is usually low, so there is little danger of bleeding to death from an injury.

HEARTS

Blood is pumped by contractions of the heart, aided in some species by contractions of blood vessels. The term "heart" is applied to a wide variety of hollow structures, but almost all consist of muscle, and most have one-way valves that direct the flow of blood (Figures 11.3 and 11.4). It would be useful to be able to generalize about heart structures in various groups of animals, but there is no convenient basis for such a classification. Zoologists at one time thought that such a classification could be based on whether the stimulus for contraction originates in the nervous system or within the heart itself: that is, on whether the hearts are **neurogenic** or **myogenic,** respectively. Unfortunately, both kinds often occur within the same taxonomic group. In vertebrates, however, all hearts are myogenic. That is, the action potentials that trigger contraction of the cardiac muscle originate within the heart itself, rather than in neural tissue.

COMPARISON OF VERTEBRATE HEARTS

The hearts of different groups of vertebrates differ primarily in the number of chambers in them. The two major kinds of chambers are the **atrium** (= auricle) and the **ventricle.** Atria pump blood from veins into the ventricles, and the ventricles pump the blood into arteries. In addition, between the major vein and the atrium of fishes, amphibians, and most reptiles there is a **sinus venosus** that serves as a reservoir for the atrium. In birds and mammals the sinus venosus occurs only in the fetus. In the evolution from fishes to birds and mammals there has been a trend toward greater separation between the blood flow to respiratory organs and the blood flow to other body tissues. This separation required the subdivision of the heart into a larger number of chambers.

In fishes the sinus venosus, the atrium, and the ventricle are all in series, so there is no separation of oxygenated from deoxygenated blood (Figure 11.4A). Oxygen-depleted blood enters the sinus venosus from the major vein, the **vena cava.** The atrium and then the ventricle pump the blood through the major artery, the **aorta.** The blood then enters the capillaries in the gills, where oxygen is absorbed and carbon dioxide eliminated. The blood loses much of its pressure in the gills before it continues through arteries to the systemic circulation.

In amphibians two atria empty into a single ventricle. Deoxygenated blood from the body tissues enters the sinus venosus from the major vein, the vena cava, and then goes into the right atrium (Figure 11.4B). At the same time, blood that has been oxygenated in the lungs or skin passes through the **pulmonary vein** to the left atrium. The deoxygenated and oxygenated blood from the two atria enter the single ventricle, but structures within the ventricle reduce mixing (see Figure 36.8). Most of the oxygenated blood leaves the ventricle through the aorta, while the deoxygenated blood goes to the lungs and skin through the **pulmocutaneous artery.** Deoxygenated blood becomes oxygenated in the lungs if the amphibian is breathing air and in the skin if the animal is submerged.

The hearts of most reptiles are similar to those of amphibians except that the sinus venosus is fused with the right atrium, and the ventricle is more nearly divided into two chambers (Figure 11.4C). In crocodiles, alligators, and related reptiles (order Crocodilia) there is complete division of the ventricle into two chambers. Oxygenated blood therefore does not mix with deoxygenated blood while the crocodilian is breathing. If the animal holds its breath under water, however, a valve in the pulmonary artery shunts deoxygenated blood into the aorta, bypassing the lungs.

Like crocodilians, birds and mammals have four-chambered hearts, with complete separation between the **pulmonary circulation,** which pumps oxygen-depleted blood through the lungs, and the **systemic circulation,** which pumps the oxygenated blood to metabolizing tissues (Figure 11.4D). The complete separation of systemic and pulmonary circulation is necessary, for these animals have much higher levels of activity than reptiles. The occasional child born with a hole in the septum between the two ventricles will die within a few years unless the hole is surgically repaired. The separation of the circulations to the lungs and to the brain also allows birds and mammals to carry their heads high above their hearts. The left ventricle must produce a pressure high enough to get the blood up to the brain, but the right ventricle must produce a lower pressure to avoid flooding the lungs with fluid.

The existence of a separate pulmonary circulatory system raises an interesting question. Does it function in a mammalian embryo, where lungs are of little use and are, in fact, collapsed? The answer is no. High resistance to flow through the collapsed lungs forces the blood from the pulmonary artery into the aorta through a bypass called the **ductus arteriosus.** The newborn's first breath expands the lungs, which reduces the resistance to blood flow and allows the blood to flow through the pulmonary artery. Within hours the ductus arteriosus also closes.

STRUCTURE AND FUNCTION OF THE HUMAN HEART

More is known about the hearts of humans than about those of any other species, mainly because the prevalence of heart disease provides a great deal of incentive,

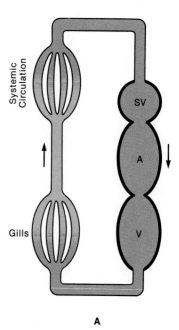

BONY FISHES

A

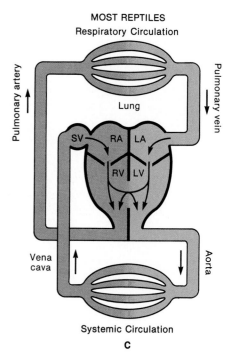

MOST REPTILES
Respiratory Circulation

Systemic Circulation

C

Figure 11.4
Schematic diagrams of the circulatory patterns of vertebrates. (A) In bony fishes three chambers pump in series: the sinus venosus (SV), the atrium (A), and the ventricle (V). There is no separation of respiratory and systemic circulation. (B) In amphibians the blood from the lungs and skin enters the left atrium (LA), and blood from the body enters through the sinus venosus and right atrium (RA). The blood from both atria

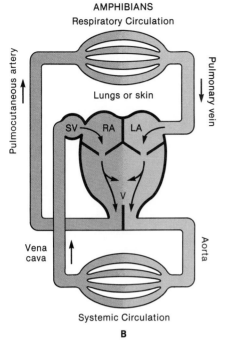

AMPHIBIANS
Respiratory Circulation

Pulmocutaneous artery

Pulmonary vein

Lungs or skin

SV   RA   LA

V

Vena
cava

Aorta

Systemic Circulation

B

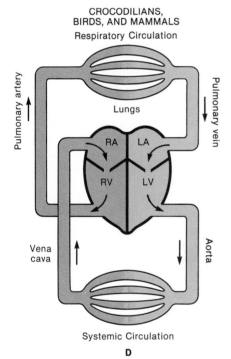

**CROCODILIANS,
BIRDS, AND MAMMALS**
Respiratory Circulation

Pulmonary artery

Pulmonary vein

Lungs

RA   LA

RV   LV

Vena
cava

Aorta

Systemic Circulation

D

empties into the one ventricle,
which then pumps it into the
respiratory and systemic
circulations. (C) In reptiles there is a
greater degree of anatomical
division of the ventricle into two
halves (RV and LV). (D) In birds and
mammals the ventricle is completely
divided, forming a four-chambered
heart, with the flow of blood
through the lungs completely
separated from the flow to other
tissues.

as well as material, for research. Figure 11.4D outlines the circulatory "plumbing" of humans. As in the four-chambered hearts of all birds and mammals, there is complete separation of systemic from pulmonary circulation, so it is useful to consider each side a separate heart. The "right heart" pumps deoxygenated blood from the body to the lungs, and the "left heart" pumps oxygenated blood from the lungs to the rest of the body. The two hearts are connected anatomically and beat in synchrony, but blood does not flow directly from one to the other.

Figure 11.5 gives a more anatomically correct view of the human heart. Deoxygenated blood enters the right atrium from the **superior and inferior venae cavae.** When the right atrium contracts it forces blood into the right ventricle through the **atrioventricular (AV) valve.** The right AV valve has three cusps of tissue and is therefore also called the **tricuspid valve.** Contraction of the atrium produces little pressure because its walls are so thin. The atrium functions mainly as a reservoir that ensures complete filling of the ventricle before it contracts.

There is normally a delay of 0.07 second between atrial and ventricular contraction, which allows time for the ventricle to fill. When the ventricle contracts the pressure quickly rises above the atrial pressure, forcing the AV valve shut. The only exit for the blood from the right ventricle is through another one-way valve into the pulmonary artery. This valve has flaps of tissue shaped like half-moons, so it is called a **semilunar valve.** When the right ventricle relaxes, the semilunar valve helps ensure that the blood does not flow back into it. Instead, the elastic arteries squeeze the blood through the narrow capillaries in the lungs.

After the blood has been oxygenated in the lungs it enters the left atrium through pulmonary veins. The left atrium contracts at the same time as the right atrium, emptying the oxygenated blood under low pressure into the left ventricle through the left atrioventricular valve. This AV valve has two cusps of tissue and is called the **bicuspid valve.** It must have reminded the ancient anatomists of a bishop's pointed hat, or miter, for it is also called the **mitral valve.** The left ventricle contracts at the same time as the right ventricle, forcing the bicuspid valve shut and sending the blood through a semilunar valve in the aorta. From the aorta the oxygenated blood enters the systemic arteries and capillaries. When the left ventricle relaxes, the aortic semilunar valve closes. The arterial blood pressure then slowly drops as the elastic arteries squeeze the blood into the smaller arteries and capillaries.

ELECTRICAL ACTIVITY OF THE HEART

The above description explains the mechanical events of the heart, but it does not tell us why the heart contracts. Contraction is, in fact, due to a flow of electrical potentials that is entirely separate from the flow of blood. Like all vertebrate hearts, the human heartbeat is myogenic, which means that the electrical stimulus for contraction originates in muscle cells of the heart itself. The stimulus for each contraction is an action potential that is produced spontaneously. Many muscle cells are capable of producing action potentials spontaneously, but those that produce them with the highest frequency act as the **pacemaker** for the entire heart. Ordinarily, a cluster of cells in the **sinoatrial (SA) node** on the back of the right atrium is the pacemaker. The action potentials arise because the voltage across the plasma membranes slowly decreases until it reaches threshold. The action potential then travels rapidly to adjacent cardiac muscle cells through **intercalated discs** (p. 215), triggering contractions and new action potentials as it goes.

An action potential from the SA node quickly spreads through the two artria. In mammals, however, atrial and ventricular muscle fibers are not electrically con-

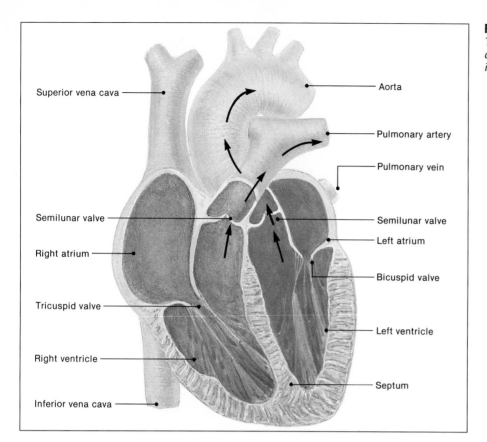

Figure 11.5
The human heart, showing the chambers and valves. Arrows indicate the direction of blood flow.

Superior vena cava

Aorta

Pulmonary artery

Pulmonary vein

Semilunar valve

Semilunar valve

Left atrium

Right atrium

Bicuspid valve

Tricuspid valve

Left ventricle

Right ventricle

Septum

Inferior vena cava

nected, so the action potentials must take a special route to excite the ventricles. This pathway consists of the **atrioventricular (AV) node,** two **bundles** of conducting fibers, and **Purkinje fibers** going to the ventricular walls. The AV node receives the action potentials from the atria and is also responsible for delaying them 0.07 second, allowing the ventricles to fill. After the delay, the AV node sends the action potentials through the right and left bundles, which conduct them through the Purkinje fibers. These fibers then activate all areas of the two ventricles simultaneously (Figure 11.6).

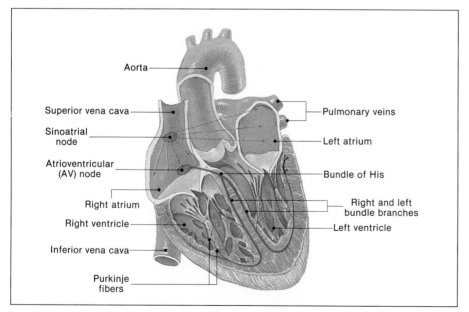

Figure 11.6
The structures of the human heart that trigger and coordinate its contraction. The sinoatrial (SA) node includes pacemaker cells that trigger action potentials (arrows), and therefore contraction, in the atria. The atrioventricular (AV) node picks up action potentials from the atria, delays them while the ventricles fill with blood, then sends them to the ventricles via the two bundles and the Purkinje fibers.

Aorta

Superior vena cava

Sinoatrial node

Atrioventricular (AV) node

Right atrium

Right ventricle

Inferior vena cava

Purkinje fibers

Pulmonary veins

Left atrium

Bundle of His

Right and left bundle branches

Left ventricle

Clinical Signs of Cardiac Functioning

The action potentials of cardiac cells spread throughout the body and can easily be recorded from the skin as the **electrocardiogram** (ECG, or EKG following the German spelling) (Figure 11.7). The first event in the ECG, the P wave, results from the combined action potentials when cells of the atria contract. Next comes the QRS complex, which results from the combined action potentials in the cells of the ventricles. Finally, the T wave results from repolarization of the membranes of the ventricles. Changes in the shape or interval between waves indicate a disturbance to electrical conduction in the heart. Such changes can be used to diagnose disorders such as heart attack or heart block (blockage of the bundles).

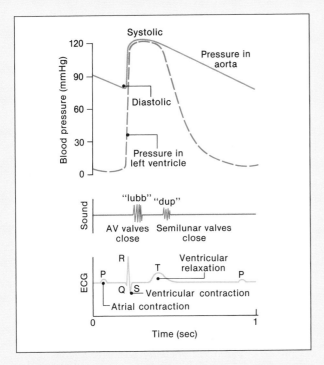

Figure 11.7
A graph of arterial blood pressure, sounds, and the electrocardiogram (ECG) of the normal human heart. The P wave results from the electrical activity of atrial contraction. The QRS complex results from electrical activity of ventricular contraction. Note that at the time of the QRS complex the blood pressure shoots up from its lowest (diastolic) value of about 80 mmHg to its maximum (systolic) value of about 120 mmHg. The rise in ventricular pressure causes the atrioventricular valves to shut, producing the "lubb" heart sound. The T wave of the ECG results from repolarization of membranes when the ventricles relax. At this time the arterial pressure starts to exceed the ventricular pressure, which forces the two semilunar valves shut, producing the "dup" heart sound. Although ventricular pressure remains low until the next contraction, the arterial pressure declines gradually, because the blood can only trickle through the fine capillaries.

The most familiar external signs of heart activity are the **heart sounds**, commonly designated "lubb" and "dup." The lubb sound occurs at about the time of ventricular contraction and is due to closing of the AV valves. The dup sound occurs during ventricular relaxation, as the semilunar valves close. The heart sounds are therefore useful in diagnosing disorders of the valves. One such disorder is a **heart murmur**, in which a blowing or roaring during the first or second heart sound indicates poor closing of an AV or semilunar valve, respectively. Even in this age of advanced medical technology, one of the most valuable talents of the physician is the ability to assess the functioning of the heart simply by listening to its sounds with a stethoscope.

Another time-honored part of a physical examination is the measurement of **arterial blood pressure.** The measurement of blood pressure uses the fact that the height to which a fluid rises in a tube is proportional to the pressure applied to the fluid at the base of the tube. The device used to measure blood pressure, called a **sphygmomanometer,** consists simply of a tube partly filled with mercury. The procedure for measuring blood pressure is to compress the brachial artery in the upper arm with an inflatable cuff attached to the sphygmomanometer. The physician then listens with a stethoscope for sounds of blood flowing past the cuff as the pressure is slowly released. A gauge on the cuff shows the pressure being applied to the artery. The blood will spurt audibly through the artery when the cuff pressure falls just below the highest pressure as the left ventricle contracts. At this point the pressure of the cuff will elevate the mercury column by a height proportional to the blood pressure, so the blood pressure can be read simply as that height in millimeters of mercury (mmHg, or torr). This peak arterial pressure is called the **systolic pressure.** As the cuff pressure continues to fall a final sound can be heard as it closes the artery only momentarily. This sound indicates when the minimum or **diastolic pressure** occurs, just before the ventricle contracts. Normal systolic pressure in humans is 120 mmHg, and normal diastolic pressure is 80 mmHg.

Systolic and diastolic pressures are usually indicated like a fraction, though they are never divided:

$$\frac{120}{80}$$

A systolic pressure of 120 mmHg is enough to raise a column of blood to a height of 1632 mm = 1.632 meters. For most people that is greater than the distance between the feet and the heart, so even when you stand on your head your heart pumps blood to your feet. Excessive systolic or diastolic blood pressure often indicates **hypertension** (high blood pressure), which can put an added strain on the heart and circulatory system, possibly leading to weakened heart muscle (heart failure), heart attack, or stroke.

Historical Aspects of Circulation

Nothing in physiology has quite the impact of blood and the beating heart. This and the fact that bleeding and the cessation of the heartbeat are so often associated with death probably account for ancient beliefs that life itself is located within the blood or the heart. According to Jewish belief that is literally as old as Moses, "the blood is the life." This is the basis for the kosher preparation of meats. Some religions still oppose blood transfusions because of the ancient association of life and blood.

One of the first to try to understand blood scientifically was Galen, a Roman physician of the second century. Galen based his ideas on what he could see in cadavers. Since the veins appear to the naked eye to be closed at their ends, he thought blood simply surged back and forth in them, carrying what he called "animal spirit." He knew that some blood also flowed from the right ventricle into the lungs, but since the diameter of the pulmonary artery in a cadaver is smaller than that of the vena cava, he concluded that not all the blood that entered the right ventricle from the vena cava went to the lungs. Some of it, he thought, must pass through invisible pores in the septum into the left ventricle, where it mixed with **pneuma** in the air from the lungs. Since the arteries of cadavers are usually empty of blood, Galen assumed that the mixture of pneuma and blood formed an invisible "vital spirit" that pulsed to and fro in the arteries.

Scientists in the Renaissance at first tended to trust Galen more than their own eyes. When anatomy students tried in vain to find pores between the two ventricles by probing with straws, many professors explained that the authority of Galen was superior to the efforts of students, or that they just weren't making hearts the way they did in Galen's day. In 1553 Michael Servetus wrote a book entitled *The Restoration of Christianity* in which he suggested that all the blood passed through the lungs and that none flowed directly between the ventricles. Servetus failed to convince many people, perhaps because Calvin burned him and all but three copies of his book for religious heresy.

In the 17th century, scientists began to realize the dangers of relying on the views of dead authorities, based on dead animals, and began experimenting on live animals. One of the first biologists to use the experimental approach was William Harvey, the physician to King Charles I. In 1628 Harvey published a book in which he argued that the heart was a pump and that there is a **circulation** of blood in one direction only. He based his argument partly on simple experiments that showed that valves prevent the blood from flowing through the veins away from the heart. You can perform one of Harvey's experiments in a few seconds using your own forearm, as shown in Figure 11.8. After doing this demonstration you may well wonder why it took 14 centuries for anyone to think of it. The only reasonable answer must be that the manipulation of nature by experimentation was not considered as reliable as pure reasoning.

The 17th century microscopists Malpighi and Leeuwenhoek soon confirmed Harvey's theory of circulation by directly observing connection between arteries and veins. Numerous other experiments have confirmed that the heart is "merely" a pump. Few people still revere it or the blood as the seat of life and spirit. Yet, even in an age when having "a change of heart" is no longer just a figure of speech, the drama of circulation still holds its mysteries for scientists and nonscientists alike.

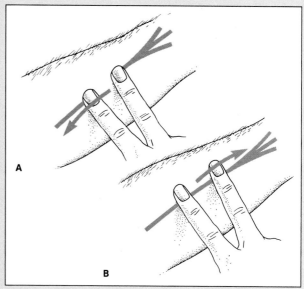

Figure 11.8
A simple experiment to demonstrate the one-way circulation of blood. (A) Press the fore and middle fingers over a prominent vein on the inner arm, then sweep the forefinger along the vein to squeeze the blood toward the heart. Then lift the forefinger while keeping the vein compressed with the middle finger. The cleared portion of the vein remains invisible because the blood has been squeezed out of it, and valves prevent blood entering from the direction of the heart. (B) Now reverse the experiment, gently sweeping the middle finger along the vein away from the heart, then lifting it. The vein remains visible because valves prevented the blood from being squeezed out in the direction away from the heart.

THE CONTROL OF CARDIAC OUTPUT

Blood carries almost all the oxygen required by vertebrates. Consequently, the cardiac output—the volume of blood pumped per minute—must be adjusted to match the demand for oxygen. The equation for calculating cardiac output shows how such adjustment is made. Cardiac output equals the rate of contraction multiplied by the volume of blood pumped during each contraction of the ventricle (the **stroke volume**):

$$\text{cardiac output} = \text{heart rate} \times \text{stroke volume}$$

Thus there are two and only two ways by which cardiac output can be adjusted to match the demand: by changing heart rate and by changing stroke volume. At rest the average human heart has a rate of 72 beats per minute and a stroke volume of 75 ml per beat, for a cardiac output of 5.4 liters per minute (about 60 barrels per day!). Even that prodigious amount is too little during exercise or unusual metabolic demand, so either rate, or stroke volume, or both must be increased.

Control of the cardiac output is mainly by neurosecretions from both divisions of the autonomic nervous system—the parasympathetic and sympathetic (see pp. 171–172). At rest the parasympathetic nervous system is more active and slows the resting rate of the human heart to about 72 beats/min on average. The parasympathetic fibers responsible for slowing the heart are in the **vagus nerve,** so this slowing is called **vagal inhibition.** The vagus nerve originates in the part of the brainstem called the medulla, which controls vagal inhibition. The vagus nerve slows the heart by releasing acetylcholine, which increases the voltage across membranes of pacemaker cells, making them take longer to reach threshold for each action potential. The medulla also activates vagal inhibition in response to increased pressure in the aorta and carotid arteries. This response probably helps prevent cardiac output from becoming too high. (Pressing firmly on the closed eyes also slightly reduces heart rate. The significance of this phenomenon is not clear.) The sympathetic nervous system increases cardiac output by releasing norepinephrine onto cardiac muscles. Norepinephrine increases heart rate. It also increases the force of cardiac-muscle contraction, which increases the stroke volume above its average resting level of 75 milliliters per beat. Circulating epinephrine from the adrenal medulla increases cardiac output in similar ways.

PERIPHERAL CIRCULATION

Arteries. The ceaseless beating of the heart is so dramatic that it tends to overshadow the silent streaming of blood through arteries, capillaries, and veins. Yet the functioning of these vessels is as important and complex as that of the heart. The arteries, which by definition conduct blood away from the heart, must have the resilience to inflate and deflate some 40 million times a year throughout your life. Figure 11.9A shows the elastic tissue that makes this possible. In the more peripheral arteries and their small branches, the **arterioles,** the middle layer of tissue consists largely of smooth muscle. Hormones and the autonomic nervous system control the tension of these muscles, thereby regulating local blood flow and overall blood pressure. Contraction of these arterial smooth muscles causes **vasoconstriction,** which reduces blood flow through the arteries. If vasoconstriction occurs in many arteries simultaneously, the blood that is circulating through them is squeezed into a smaller volume, which increases the pressure. Thus an increase in vasoconstriction, as well as an increase in cardiac output, increases blood pressure. Relaxation of the smooth muscles causes **vasodilation,** which has an effect opposite to vasoconstriction on blood flow and blood pressure.

Veins. Veins and their smaller branches, the **venules,** also have a layer of smooth muscles. However, contraction of these muscles, resulting in **venoconstriction,** has a different effect than vasoconstriction. On average the veins contain about 70% of the blood, and venoconstriction forces some of that reserve into circulation during exercise and stress. The pressure in veins remains low even during venoconstriction. Generally, venous blood pressure is about one-tenth as high as blood pressure in large arteries (Figure 11.9B). Venous blood pressure results mainly from hydrostatic pressure due to gravity. The effect of this hydrostatic pressure can be seen by watching how your veins swell as you lower your outstretched arm below the level of your heart.

With such low blood pressure, how do veins get the blood back up to the heart? The answer is the **venous pump,** which is not a separate organ itself but the result of one-way valves and massaging by neighboring skeletal muscles. When skeletal muscles contract they compress nearby veins, forcing the blood in the only direction the valves permit it to flow: toward the heart.

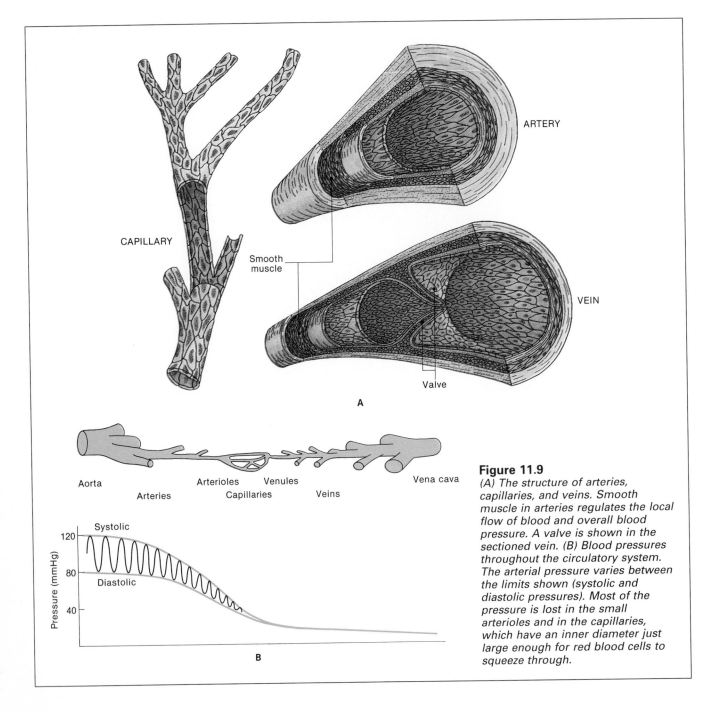

Figure 11.9
(A) The structure of arteries, capillaries, and veins. Smooth muscle in arteries regulates the local flow of blood and overall blood pressure. A valve is shown in the sectioned vein. (B) Blood pressures throughout the circulatory system. The arterial pressure varies between the limits shown (systolic and diastolic pressures). Most of the pressure is lost in the small arterioles and in the capillaries, which have an inner diameter just large enough for red blood cells to squeeze through.

Capillaries. The formation of interstitial fluid, which is what circulation is all about, is due to the capillaries. In contrast to arteries and veins, capillaries have extremely thin walls with gaps between their cells. Blood pressure within the capillaries forces plasma through the gaps, leaving the blood cells and large proteins behind. The resulting filtrate of blood is the interstitial fluid. Why doesn't all of the plasma get filtered out of the blood, drowning the tissues in interstitial fluid? One reason is that at any one time most capillaries are closed off by bands of smooth muscle at the arteriole end—the **precapillary sphincters.** Precapillary sphincters constrict when local conditions (low CO_2, normal temperature, and normal pH) indicate that there is adequate blood flow.

The Starling Effect. A second reason why excess interstitial fluid does not form is the Starling effect, named for E. H. Starling. The Starling effect refers to a balance between the blood pressure, which tends to force plasma out of the capillaries, and the **colloid osmotic pressure,** which tends to draw fluid back into the capillaries. The colloid osmotic pressure (also called oncotic pressure) is due to proteins (mainly serum **albumin**) that are too large to pass across the capillary walls, and that are therefore in higher concentration in the blood than in the interstitial fluid. At the arterial end the blood pressure in the capillary is greater than the colloid osmotic pressure, so fluid leaves the capillary. As blood squeezes through the narrow capillary, however, it loses much of its remaining pressure. At the venule end, therefore, the blood pressure is lower than the colloid osmotic pressure, so fluid is drawn back into the blood (Figure 11.10).

The Starling effect has many everyday consequences. One of these is **swelling** (= edema), which occurs when blood pressure is excessive or osmotic pressure is diminished. For example, standing for long periods without contracting the leg muscles (venous pumping) can cause blood to pool in the veins. This can increase pressure at the venous ends of capillaries to the point where it exceeds the osmotic pressure, thereby preventing the return of fluid from the tissues into the blood. Swelling also occurs in injured tissues, because they release chemicals that make capillaries leakier. The leaky capillaries allow proteins to enter the interstitial fluid, eliminating the osmotic pressure difference. Another manifestation of the Starling

Figure 11.10
A capillary network. The precapillary sphincters remain closed much of the time in tissues with low oxygen demand, such as skin. Even though the blood pressure at the arteriole end tends to force fluid out of the thin and highly permeable capillary walls, the colloid osmotic pressure of the blood at the venule end draws the fluid back in.

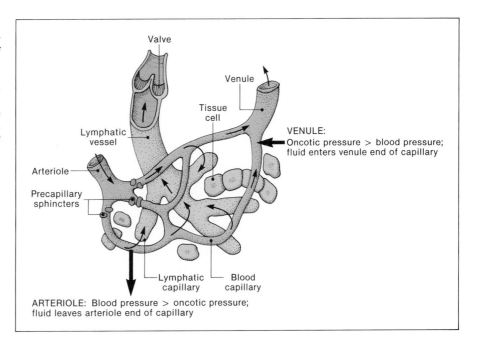

effect is sadly becoming more familiar: the swollen bellies of starving children. If there is too little protein to maintain the osmotic pressure of blood, fluid collects in the body cavity.

HEMOSTASIS

Since the blood of closed circulatory systems is pumped under high pressure, there is a constant danger that an injury to an artery will cause a fatal loss of blood. The pressure in large arteries is so high that nothing but an externally applied constriction will prevent death if the artery is cut. In veins and in smaller arteries, however, there are mechanisms that automatically limit blood loss. This limitation of blood loss is called **hemostasis.** (Note the difference between hemostasis and homeostasis.) One mechanism of hemostasis is the contraction of smooth muscles in injured arteries and veins. This contraction may be sufficient to close the cut ends of the blood vessel. In addition, blood **platelets** are attracted to the site of injury by the exposed cell contents. Platelets, which are actually fragments of certain cells **(megakaryocytes),** adhere to the injured site and to each other and form a **platelet plug** (Figure 11.11A).

For larger injuries a second mechanism of hemostasis comes into play: **clotting.** Clotting (coagulation) results from a chemical chain reaction that forms strands of a protein called **fibrin,** which entangle red blood cells at the site of injury. Approximately 16 proteins, ions, and other factors are involved in the series of reactions leading to clotting. In the final steps, **prothrombin** changes to **thrombin,** and thrombin converts the globular protein **fibrinogen** into the fibrous form called fibrin. A few hours after it forms, the clot begins to shrink, squeezing out serum. This shrinkage, called **clot retraction,** results from the contraction of the

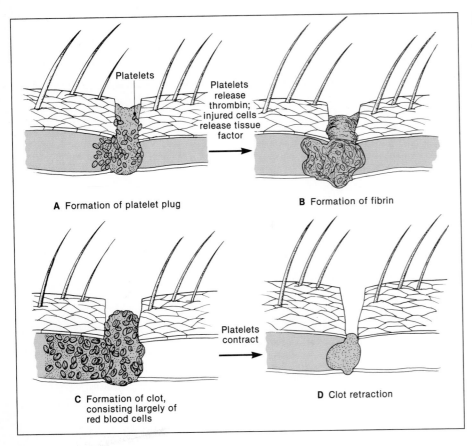

Platelets

Platelets release thrombin; injured cells release tissue factor

A Formation of platelet plug

B Formation of fibrin

Platelets contract

C Formation of clot, consisting largely of red blood cells

D Clot retraction

Figure 11.11
Representation of a cut in skin, showing some key steps in hemostasis. (A) Collagen from damaged cells causes blood platelets to adhere to the damaged vessel, forming a platelet plug. (B) Thrombin from the platelets and a tissue factor from injured cells then trigger the formation of fibrin strands from the protein fibrinogen in the blood. (C) Red blood cells become entangled in the fibrin and make up the bulk of the clot. (D) Finally the platelets contract, and clot retraction pulls the tissue together and forms a solid plug.

platelets. (Platelets contain more actin and myosin than any other cells except those of muscle.) Clot retraction pulls the injured tissue together and forms a more solid barrier to blood loss. Eventually, the clot is removed by an enzyme that breaks up the fibrin.

THE LYMPHATIC SYSTEM

From 1% to 4% of the interstitial fluid is not recycled back into capillaries by the Starling effect. In addition, some protein and cells escape from blood, and debris from dead cells also tends to accumulate in interstitial fluid. Why, then, aren't we all puffy from stuff that has collected in our tissues over the years? The answer is that there is a second circulatory system—the lymphatic system—that keeps the interstitial fluid free of such material. The fluid that circulates in the lymphatic system is called **lymph**. Lymph enters the lymphatic system through the closed but highly permeable ends of **lymphatic capillaries** (Figure 11.12). Lymph then flows into **lymphatic vessels.** In mammals these lymphatic vessels have one-way valves as in veins, so lymph is propelled by a mechanism similar to venous pumping. (Fishes, amphibians, and reptiles, however, have contractile vessels, called "lymph hearts," that pump lymph.) Lymphatic vessels all converge at one of two **thoracic ducts** near the collar bones, where the lymph empties into veins. If it

Figure 11.12
The lymphatic system and lymphoid tissue of humans. Lymph passes through lymph nodes, then empties from the lymphatic ducts into the subclavian veins. Lymph from the head and right arm drains into the lymphatic duct of the right subclavian vein. Lymph from the lower body and left arm drains into the lymphatic duct of the left subclavian vein.

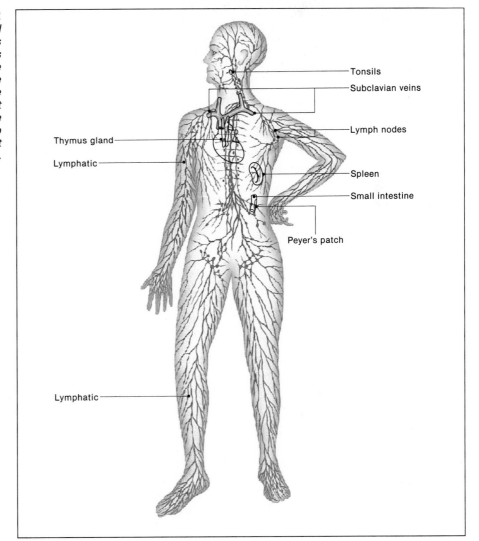

Tonsils
Subclavian veins
Lymph nodes
Thymus gland
Lymphatic
Spleen
Small intestine
Peyer's patch
Lymphatic

were not for the lymphatic system, we would all eventually show symptoms like those of elephantiasis, which occurs when the nematode worm *Wuchereria bancrofti* clogs the lymph vessels. Look at Figure 26.7 and be thankful if you have a healthy lymphatic system.

The lymphatic system also performs other essential functions. Some lymph vessels in the small intestine have small projections called **lacteals** that absorb digested fats (see Figure 13.9). The lymphatic system is also associated with **lymphoid organs,** including lymph nodes, tonsils, spleen, thymus, and Peyer's patches in the small intestine (Figure 11.12). Lymphoid organs produce infection-fighting cells called **lymphocytes.** In addition, each organ has specific functions. **Lymph nodes** contain phagocytic cells that engulf bacteria, white blood cells that have leaked into infected tissues, and other cellular debris. The lymph nodes also contain lymphocytes. Many lymph nodes are located in the groin, arm pits, and neck. These nodes, commonly referred to as "glands," are normally from 1 to 25 mm in diameter, but they become noticeably swollen when they have to deal with an infection. The tonsils, likewise, become enlarged during certain infections. The spleen serves as a reservoir for red blood cells and as a site where lymphocytes mature. The thymus is another major site where lymphocytes develop the ability to combat infections, as will be discussed later.

GENERAL DEFENSES AGAINST INVASION

An animal's interstitial fluid is a suitable habitat not only for its own cells but also for a variety of viruses, bacteria, fungi, protozoans, and parasitic animals. (Some of the parasitic protozoans and animals will be described in later chapters.) Probably all animals have mechanisms to resist such invaders, but these mechanisms are not well understood in invertebrates. For humans and other mammals, however, our understanding of the various defense mechanisms has increased at a revolutionary pace during the past two decades. One such mechanism is **lysozyme,** an enzyme that breaks down bacterial cell walls. Lysozyme occurs in urine, tears, saliva, and other secretions and protects the natural openings of the skin against bacterial infections.

The general defense against viruses includes a group of glycoproteins called **interferon.** Interferon is produced by infected cells that are already doomed from having their genetic machinery commandeered to produce new viruses. Before the infected cell succumbs, however, it manages to secrete interferon, which interferes with the ability of other infected cells to produce new virus. Interferons are effective against a broad array of viruses, including those that cause certain cancers, herpes, viral pneumonia, hepatitis, influenza, and the common cold.

A general defense against all kinds of invaders is **inflammation.** Inflammation is a reaction to any kind of damage to tissues. The signs of inflammation—warmth, pain, itching, reddening, and swelling—are familiar to anyone who has had a cold, a sunburn, or an allergic reaction. Inflammation is triggered by a variety of hormonelike substances. **Histamine,** for example, dilates arteries, thereby increasing the flow of blood to the area. This causes the reddening. Other substances, called **cytokines,** make capillaries and blood vessel walls leaky. This leakiness causes swelling, since blood proteins leak into the interstitial fluid, upsetting the balance between blood pressure and colloid osmotic pressure (the Starling effect). More important, the leakiness allows certain white blood cells to enter the tissue spaces. First come the granulocytes, which directly attack invading cells (Figure 11.1). These are soon followed by **macrophages,** whose name literally means "big eater." Macrophages engulf and digest many of the invaders and damaged cells.

IMMUNITY

Besides general defenses, there are specific defenses against particular viruses, foreign cells, or complex foreign molecules. These specific defenses are termed immunity. An immune response is often triggered during the early stages of inflammation. Certain molecules that occur in invaders but not in the invaded animal trigger the immune response. These molecules are called **antigens.** Each type of invader has its own specific antigens. Macrophages and certain other types of **antigen-presenting cells** incorporate antigens into their plasma membranes. Certain white blood cells have receptor molecules on their plasma membranes that bind to particular antigens. When a receptor on one of these white blood cells binds to antigen on an antigen-presenting cell, the white blood cell initiates an immune response.

The white blood cells largely responsible for the immune response are two types of lymphocytes: the *T* **cells** and the *B* **cells.** The formal name for *T* cells is **thymus-dependent lymphocytes,** because they mature in the thymus of the embryo. They act primarily against cancerous or virus-infected cells and parasites. *T* lymphocytes are responsible for the rejection of transplanted hearts and other organs. When a *T* cell's receptors come into contact with an antigen-presenting cell bearing a foreign **(nonself)** antigen, it divides into a **clone** of *T* cells bearing receptors for the same antigen. There are several kinds of *T* lymphocytes, but the ones directly responsible for killing cells are the **cytotoxic *T* cells.** For several days after exposure to a nonself antigen, cytotoxic *T* cells circulate through lymph and blood, binding to and destroying any cells bearing the antigen (Figure 11.13).

The other type of lymphocyte that reacts to antigens presented by antigen-presenting cells are the *B* **cells,** which mature in the bone marrow where they are produced. Upon exposure to an antigen-presenting cell bearing a particular antigen, a *B* lymphocyte with receptors for that antigen divides into clones of identical

Figure 11.13
Scanning electron micrograph of a cytotoxic T *cell (bottom) killing an invading cell. The* T *cell has secreted a protein called perforin that binds to the target cell's plasma membrane and forms numerous small holes. Here a large hole has formed in the attacked cell's plasma membrane. Natural killer cells and eosinophils work in a similar way. Shown 5700 times actual size.*

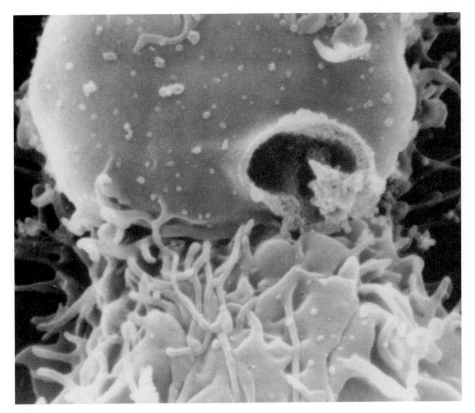

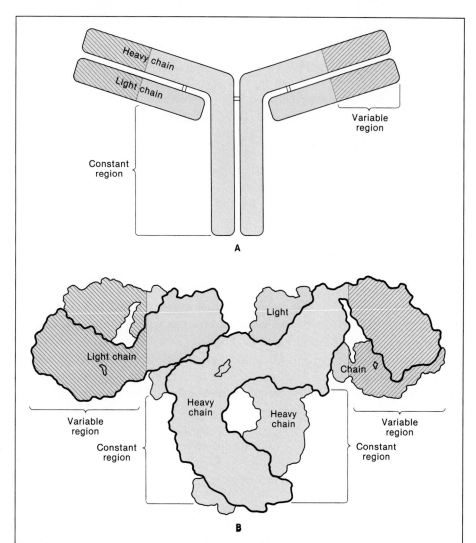

Figure 11.14
(A) A schematic representation of an antibody. (B) The actual outline of an antibody. Antibodies consist of two pairs of proteins: two identical heavy chains and two identical light chains. There are five types of heavy chain, giving rise to five types of immunoglobulin (IgG, IgA, IgM, IgE, IgD) with distinct functions. There are antibodies against virtually any potential antigen, because the variable regions can form more than 16 million kinds of attachment site that bind to antigens.

plasma cells. These plasma cells then release protein **antibodies** (= immuno-globulins) that bind to large molecules, such as toxins and the proteins on viruses and foreign cells. Each antibody has two variable regions, each with a structure that binds to a specific antigen (Figure 11.14). The binding of antibodies causes viruses, cells, or molecules to **agglutinate** in a harmless clump. By themselves, antibodies are incapable of destroying cells. Antibodies can, however, trigger an attack by **natural killer cells,** which resemble cytotoxic *T* cells. In addition, anti-bodies attract a system of proteins called **complement,** which attracts phagocytic cells and helps destroy bacteria.

To summarize (Figure 11.15), antigen-presenting cells alert *T* and/or *B* lym-phocytes of an invader. When a *T* cell is exposed to its particular antigen it repro-duces into a clone of cytotoxic *T* cells that bind to and destroy cells bearing the antigen. When a *B* cell is exposed to its antigen it rapidly reproduces into a clone of plasma cells that produce antibodies against the antigen. Antibodies immobilize the antigen and enable killer cells to destroy invading organisms. With the help of complement, antibodies also kill bacteria.

The reactions of *T* and *B* lymphocytes depend on a variety of factors. Helper lymphocytes stimulate the responses of both *T* and *B* cells. In addition, lympho-cytes, macrophages, and other cells involved in immunity release a variety of cyto-kines that trigger inflammation and coordinate the immune response.

Figure 11.15
An outline of some major interactions in the regulation of the immune response.

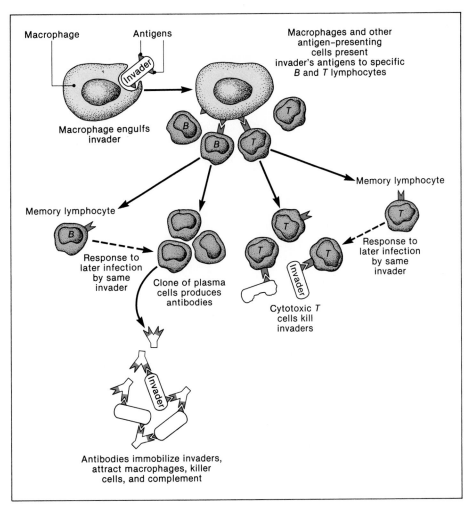

Macrophage   Antigens

Macrophages and other antigen–presenting cells present invader's antigens to specific *B* and *T* lymphocytes

Macrophage engulfs invader

Memory lymphocyte

Memory lymphocyte

Response to later infection by same invader

Response to later infection by same invader

Clone of plasma cells produces antibodies

Cytotoxic *T* cells kill invaders

Antibodies immobilize invaders, attract macrophages, killer cells, and complement

IMMUNOLOGICAL MEMORY

The immune response may take several days to build up the first time a particular antigen is encountered. If the same or a similar antigen is encountered later, however, the response can be quite rapid. This phenomenon is called immunological memory. After an invasion is successfully repelled by the *B* and *T* cells described previously, some of the cloned cells remain in reserve as **memory lymphocytes.** Memory lymphocytes are capable of recognizing the antigen and of triggering an immediate and massive production of cytotoxic *T* cells and/or plasma cells upon a second exposure to it. Immunological memory therefore protects against repeated infection by the same invader. In some cases, however, immunological memory can be a nuisance or even a hazard. It is the basis for **allergies,** which are due to immune reactions against antigens that are not harmful in themselves. In some cases an immune reaction can be so sudden and massive that it causes a sudden vasodilation that drastically lowers blood pressure. Such **anaphylactic shock** is a serious hazard to people who are especially sensitive to certain antigens, such as those in bee stings.

Immunological memory makes it possible to **vaccinate** people and animals against infections, by inoculating them with the antigens of infectious viruses or bacteria. The vaccine can be prepared from a related virus. Vaccination with the cowpox virus, for example, was used to induce immunological memory against the smallpox virus. Smallpox virus now survives only in laboratory cultures, and the only reminders of its existence are the round scars on the arms of people old enough to have been vaccinated against it. Vaccine can also be prepared from weakened virus, as in polio vaccine. People and other animals can also be immunized against toxins by several small injections of the toxin. Herpetologists and others who must handle poisonous snakes, for example, can be immunized in this way against venom. The venom can be injected by syringe in small doses, but some herpetologists simply wait for the snakes to make the injections. A person who has not been previously immunized against a toxin can be treated with **antiserum.** Antiserum is the serum from animals (often horses) that contains antibodies to a toxin or other antigen. It is prepared by inoculating the animal with the antigen to trigger antibody production.

Just as lymphocytes can acquire immunological memory of nonself molecules, they can also "learn" not to respond to **self** molecules, which are normally present in the body. Usually this "learning" occurs during embryonic development, as the *T* cells that would attack cells within the embryo and *B* cells that would produce antibodies against self molecules in the embryo are deleted or inactivated. Occasionally, however, lymphocytes attack self molecules such as receptors to synaptic transmitters or hormones. The result is an **autoimmune disease.** Examples of autoimmune diseases include myasthenia gravis (a muscle degeneracy), one form of diabetes mellitus, rheumatoid arthritis, and multiple sclerosis.

BLOOD TYPING

ABO Grouping.   For centuries physicians dreamed of transfusing blood from a healthy person into one who was bleeding to death. Early attempts, however, had variable success. Some patients were helped almost immediately, while others were killed just as quickly. Around the turn of the century, research on animals showed that the death resulted from the sudden agglutination (clumping) and destruction of the injected red blood cells. Karl Landsteiner (1868–1943) found that he could classify people into four groups—A, B, AB, and O—depending on whether their red blood cells survived mixing with the blood of others. Blood cells from people in group A would usually not agglutinate in blood from those in group A or AB but would agglutinate in blood from people in group B or O. Likewise, type B blood cells would not agglutinate in type B or AB blood but would agglutinate in type A or O. Type O blood cells would usually not agglutinate in blood of any type, but type AB blood cells would agglutinate in any type of blood other than AB.

We now know that the Landsteiner blood groups depend on whether A and/ or B antigens are present on red blood cells. If either of those antigens is not on the red blood cells of the recipient of the transfused blood, that molecule will be foreign, and he or she will have antibodies against it. These antibodies will agglutinate the transfused red blood cells, and the cytotoxic *T* cells will destroy them. To prevent this, blood typing is now routinely performed using antiserum for each antigen. For example, if the red blood cells in a drop of blood agglutinate when mixed with A antiserum, but not when mixed with B antiserum, the blood is type A. (What would be the blood type if the red cells agglutinate in both A and B antisera? Why wouldn't type O blood agglutinate in either serum?)

Rh Factor.   In 1940, ten years after he received the Nobel Prize for discovering the ABO grouping, Landsteiner, together with A. S. Wiener, discovered another important antigen: the Rh factor. The Rh factor, named for the rhesus monkey *Macaca mulatta* in which it was discovered, is either present (Rh+) or absent

AIDS

There is no better way to appreciate the importance of the immune system than to consider what happens when it is destroyed, as it is in victims of AIDS—acquired immune deficiency syndrome. AIDS is caused by a human immunodeficiency virus (HIV) that attacks *T* helper cells, thereby weakening the defense against other infections. Most cases of AIDS are diagnosed approximately five years after infection with HIV, and most of those with AIDS die within a few years, usually from infection by some invader that would otherwise have immediately been destroyed by the immune system. The ultimate killers are often protozoans that cause pneumonia and toxoplasmosis (see p. 478), tuberculosis, or a rare cancer of the blood vessels called Kaposi's sarcoma. HIV also damages brain cells.

HIV occurs mainly within macrophages and is apparently spread by direct transfer of blood or through sexual intercourse. The most common routes of infection are the following:

1. **Sexual intercourse, either anal or vaginal.** Anal intercourse among male homosexuals and bisexuals accounts for over half the cases in the United States. Since this mode of transmission is now well known, most homosexuals and bisexuals take precautions to avoid transmission. The most rapidly increasing mode of infection is now vaginal intercourse between heterosexuals. The danger of infection with HIV increases with the number of sex partners. The use of condoms by heterosexuals reduces the risk of infection approximately tenfold.
2. **Receipt (not donation) of blood or blood products by injection or transfusion.** Sharing of hypodermic syringes by intravenous drug users is now the major mode of HIV transmission, accounting for approximately one-fourth of new cases. Hemophiliacs were formerly at high risk, because they must take periodic injections of a blood-clotting factor that they lack. Now, however, the risk of acquiring AIDS from blood has all but been eliminated by tests that detect the presence of antibodies against HIV in donated blood. A positive test for the antibody merely indicates that the potential donor has been exposed to the virus, not that he or she is still carrying HIV or will develop AIDS.
3. **Transmission from an infected mother to her fetus through the placenta.**

AIDS was first identified in 1981 from a cluster of fewer than a hundred cases. Since then more than 125,000 cases of AIDS have been reported in the United States, and more than 60% of the victims have already died. At least one million Americans are believed to be infected with HIV. In spite of intensive publicity regarding the causes of AIDS, there are still more than 35,000 new victims each year. For those in a high-risk group, the chances of getting AIDS this year is similar to the average American's chance of getting cancer or dying of a heart attack. For them AIDS is understandably frightening. Many people outside the high-risk groups are also frightened, however, and some have reacted in ways reminiscent of the days of the Black Death. Otherwise decent and responsible people have called for the ostracism or tattooing of those who test positive for antibodies to HIV, even if they do not have the disease, in spite of a total lack of evidence that AIDS victims are a threat to ordinary people.

AIDS is, in fact, not a highly contagious disease. One heterosexual encounter with someone known to be infected with AIDS carries a risk of infection of 1 in 500, even if a condom is not used. There is no known case in which AIDS was transmitted from an infected person to a family member who was not a sexual partner or newborn infant, even if the family member engaged in normal affectionate contact and shared dishes or even toothbrushes. No nurse who has administered mouth-to-mouth resuscitation to an AIDS victim has acquired the disease. In one study only 4 out of 870 health-care workers who accidentally jabbed themselves with HIV-infected needles subsequently tested positive for the virus. Out of 104 workers who were splashed with blood from AIDS victims, none became infected. Heterosexuals who are not extremely promiscuous and who do not share syringes with others are therefore much more likely to die from overweight, smoking, or automobile accidents than from AIDS.

(Rh−) on red blood cells. The presence or absence of the Rh factor is indicated with a + or − following the ABO type. For example, a common blood type in the United States is O+. Seventeen other blood groupings are now known, but these result from weak antigens that can be ignored for purposes of transfusions. They are, however, often useful in tracing racial descent and in deciding paternity suits. Still, to be on the safe side, a drop of a potential donor's blood is tested in a patient's plasma prior to transfusion to be sure some antigen other than the A, B, or Rh factor will not cause the red blood cells to agglutinate.

Knowing the Rh factor is crucial in another way that clearly illustrates the importance of immunity. If an Rh− woman gives birth to an Rh+ baby, red blood cells from the baby can leak into the maternal circulation across the disintegrating placenta. These Rh+ red blood cells can then trigger the production of Rh antibodies by the mother, and these antibodies can enter the bloodstream of a later fetus. If this fetus is also Rh+, the antibodies from the mother will destroy the fetus' red blood cells. The result is called **erythroblastosis fetalis,** and it can kill the fetus or produce brain damage. Erythroblastosis fetalis is now rare in the United States, because Rh− mothers who have just given birth to Rh+ babies are injected with anti-Rh antibodies. These antibodies destroy any Rh+ red blood cells in her circulation before they can trigger immunity.

SUMMARY

Animals must create and maintain an internal environment—the interstitial fluid—in which their cells can live. Generally, the source of water, ions, nutrients, oxygen, and other essential components of the interstitial fluid is blood. Blood usually consists of cells and plasma and is pumped through a circulatory system by a heart. Arthropods and some other invertebrates have open circulatory systems, while other invertebrates and vertebrates have closed circulatory systems, in which the blood is always contained within the heart and blood vessels.

Hearts vary in structure and functioning. In fishes the heart consists of three chambers in series—the sinus venosus, the atrium, and the ventricle—and there is no separation of oxygenated and deoxygenated blood. Amphibians have two atria with some separation of oxygenated from deoxygenated blood. In some reptiles, and in birds and mammals, there are two atria and two ventricles. There is complete separation of the pulmonary circulation, which pumps deoxygenated blood to the lungs, from the systemic circulation, which pumps oxygenated blood from the lungs to the rest of the body.

Vertebrate hearts are myogenic: the action potentials that trigger contractions originate in pacemaker cells in the sinoatrial node of the heart. The action potentials conduct over the atria, then to the atrioventricular node, where they are delayed before going to conducting bundles. The bundles and Purkinje fibers then conduct the action potentials to the ventricles. Contraction of the atria therefore precedes contraction of the ventricles. One-way atrioventricular and semilunar valves prevent backflow of the blood. Cardiac output is regulated by effects of the autonomic nervous system and hormones on the heart rate and stroke volume.

Blood is pumped through arteries under high pressure. It then flows into capillaries and veins, losing pressure along the way. Because of the Starling effect, fluid tends to leave capillaries at the arterial end and to reenter at the venous end. Fluid and cellular debris that accumulate in intercellular spaces return to the bloodstream through the lymphatic system, which is also important in immune responses. Loss of blood is limited by several processes of hemostasis, especially clotting.

There are several general defenses against invading organisms, including inflammation. Certain processes of inflammation also trigger immune responses, which are directed against specific invaders. Upon exposure to specific antigens, *T* lymphocytes attack cells that bear that antigen. *B* cells become plasma cells that release antibodies against the antigen. The response of antibodies to A, B, and Rh antigens on red blood cells is the basis for blood typing.

KEY TERMS

interstitial fluid
blood
hemocoel
plasma
serum
red blood cell
white blood cell
pulmonary circulation
systemic circulation
atrium

ventricle
artery
vein
capillary
Starling effect
atrioventricular valve
semilunar valve
sinoatrial node
atrioventricular node
cardiac output

hemostasis
lymph
inflammation
immunity
T lymphocyte
B lymphocyte
antigen
antibody

SELF-TEST

1. Explain why most large animals require a circulatory system.

2. Explain how the following body fluids are related to each other, and how they differ: blood, plasma, serum, lymph, interstitial fluid.

3. What is the difference between open and closed circulatory systems? What are the advantages and disadvantages of each? Name one group of animals with open circulatory systems.

4. Briefly compare the hearts of various vertebrates. What is meant by the following statement: "In the evolution of vertebrates there has been a trend toward greater separation of pulmonary and systemic circulations"?

5. Diagram the human heart from memory and show the pathway of blood flow through it.

6. Explain what is meant by saying that the heart is myogenic. Briefly explain how the contraction of the heart is coordinated.

7. What role is served by nerves going to the human heart?

8. Explain how interstitial fluid is formed from blood.

9. What are the roles of the lymphatic system?

10. Briefly discuss heart transplants. What property of the heart permits it to continue beating after its removal from the body? How does the recipient's body attempt to reject the transplant?

11. Describe as completely as you can how the body would defend itself against the venom of a rattlesnake.

READINGS

RECOMMENDED READINGS

Ada G. L. and G. Nossal. 1987. The clonal-selection theory. *Sci. Am.* 257(2):62–69 (Aug).

Atkinson, M. A. and N. K. Maclaren. 1990. What causes diabetes? *Sci. Am.* 263(1):62–71 (July). (*Insulin-dependent diabetes is an autoimmune disease.*)

Baglioni, C. and T. W. Nilsen. 1981. The action of interferon at the molecular level. *Am. Sci.* 69:392–399.

Cohen, I. R. 1988. The self, the world and autoimmunity. *Sci. Am.* 258(4):52–60 (Apr).

Cooper, E. L. 1990. Immune diversity throughout the animal kingdom. *Bioscience* 40:720–722. (*Introduction to several papers on this topic.*)

Doolittle, R. F. 1981. Fibrinogen and fibrin. *Sci. Am.* 245(6):126–135 (Dec).

Gallo, R. C. and L. Montagnier. 1988. AIDS in 1988. *Sci. Am.* 259(4):41–48 (Oct). (*Introduction to an entire issue on AIDS.*)

Golde, D. W. and J. C. Gasson. 1988. Hormones that stimulate the growth of blood cells. *Sci. Am.* 259(1):62–70 (July).

Grey H. M., A. Sette, and S. Buus. 1989. How T cells see antigen. *Sci. Am.* 261(5):56–64 (Nov).

Laurence, J. 1985. The immune system in AIDS. *Sci. Am.* 253(6):84–93 (Dec).

Leder, P. 1982. The genetics of antibody diversity. *Sci. Am.* 246(5)102–114 (May).

Marrack, P. and J. Kappler. 1986. The *T* cell and its receptor. *Sci. Am.* 254(2):36–45 (Feb).

Nadel, E. R. 1985. Physiological adaptations to aerobic training. *Am. Sci.* 73:334–343.

Old, L. J. 1988. Tumor necrosis factor. *Sci. Am.* 258(5):59–75 (May).

Rennie, J. 1990. The body against itself. *Sci. Am.* 263(6):106–115 (Dec).

Robinson, T. F., S. M. Factor, and E. H. Sonnenblick. 1986. The heart as a suction pump. *Sci. Am.* 254(6):84–91 (June).

Rose, N. R. 1981. Autoimmune disease. *Sci. Am.* 244(2):80–103 (Feb).

Rosenberg, S. A. 1990. Adoptive immunotherapy for cancer. *Sci. Am.* 262(5):62–69 (May).

Tonegawa, S. 1985. The molecules of the immune system. *Sci. Am.* 253(4):122–131 (Oct).

Wakelin, D. 1984. *Immunity to Parasites: How Animals Control Parasitic Infections.* Baltimore: Edward Arnold.

Warren, J. V. 1974. The physiology of the giraffe. *Sci. Am.* 231(5):96–105 (Nov). (*How the giraffe solves the circulatory problems arising from its long neck.*)

Young, J. D.-E. and Z. A. Cohn. 1988. How killer cells kill. *Sci. Am.* 258(1)38–44 (Jan).

Zucker, M. B. 1980. The functioning of blood platelets. *Sci. Am.* 242(6):86–103 (June).

ADDITIONAL REFERENCES

American Zoologist 29:367–480 (1989) has several papers on immune responses to parasites in a variety of animals.

Snyder, J. A. and D. Fritsch. 1984. Two reliable and inexpensive lysozyme assays for teaching enzymology and microbiology. In C. L. Harris (Ed.), *Tested Studies for Laboratory Teaching: Proceedings of the Third Workshop/Conference of the Association for Biology Laboratory Education (ABLE).* Dubuque, IA: Kendall/Hunt, Chapter 11.

12

Oxygen

Aquatic spider (Argyroneta aquatica) *with air supply.*

CHAPTER OUTLINE

LEARNING OBJECTIVES

1. Why do gills, lungs, and other respiratory organs have to be so delicate?

2. How are respiratory organs adapted to increase the supply of oxygen and the elimination of carbon dioxide?

3. What is the significance of the color of blood?

4. Why do some animals have gills, some have lungs, and some have neither?

5. How does respiration in mammals differ from that of other kinds of vertebrates?

6. What causes breathing, and how does it change to keep pace with the demand for oxygen?

n the previous chapters of this unit we looked at the neural, muscular, and hormonal systems that coordinate the homeostatic mechanisms that regulate the interstitial fluid, and the circulatory system that creates it. In this and the following three chapters we shall consider how the interstitial fluid is provided with the necessities that make it a hospitable internal environment for animal cells.

Oxygen is one of the necessities for cells in animals, although it was not required in the single-celled organisms from which animals evolved. How did animals come to depend on oxygen? For about two billion years the world was practically devoid of O_2. In fact, O_2 was (and still is) toxic, since it oxidizes organic molecules. Eventually, however, some organisms evolved the ability to produce energy by photosynthesis and eliminated the waste product, O_2, into the seas and atmosphere. Because of this pollution, many organisms, including the ancestors of animals, had to change their way of living. They survived this early ecological crisis by developing the ability to detoxify oxygen by combining it with hydrogen to produce water. They then turned necessity into a virtue by evolving ways to

Figure 12.1
Some animals and the levels of oxygen in their habitats. Fishes and the hydra at the bottom of the figure represent animals that live in water, which has lower levels of oxygen. Near the middle of the figure the earthworm, snail, and frog represent animals that can breathe air but require moist habitats. Such animals generally evolved from aquatic forms. At the top of the figure are completely terrestrial animals, which generally evolved most recently.

extract ATP in the process. Since then animals have been not only tolerant of oxygen but dependent on it for the production of most of their ATP. This may explain why one of the major trends in the evolution of animals has been toward living in air, where it is easier to obtain oxygen than in water (Figure 12.1).

PRINCIPLES OF GAS DIFFUSION

No animal has yet evolved a mechanism for actively transporting oxygen, so O_2 has to enter body fluids simply by diffusion through moist, permeable membranes (see pp. 39–40). Carbon dioxide, the waste produced in the oxidation of organic nutrients, must also be eliminated by diffusion. The cellular membranes across which O_2 and CO_2 diffuse are either totally part of the integument of the animal or mainly confined to respiratory organs, such as gills and lungs. The direction and rate in which O_2 and CO_2 diffuse depend on their concentrations on each side of the membrane, as long as these gases are dissolved in water on both sides of the membrane. Thus O_2 diffuses from seawater through the membranes of a shark's gills into its blood plasma as long as the concentration of O_2 in the water is higher than that in the plasma.

For terrestrial animals the situation is different, because O_2 dissolves much more readily in air than it does in water. A terrestrial animal can easily suffocate in air that has a higher concentration of O_2 than the plasma has, because the O_2 is more soluble in air than in the plasma. Instead of considering concentration, therefore, we must think more generally in terms of **partial pressure.** Partial pressure is the contribution by one kind of molecule in a fluid to the total pressure of the fluid. For example, the partial pressure of O_2 (pO_2) in air is 21% of the atmospheric pressure, since 21% of a volume of air consists of O_2. At sea level the average atmospheric pressure is 760 torr (= mmHg; millimeters of mercury), so the pO_2 is 21% of 760 torr = 160 torr. The overall direction of diffusion is always from a high partial pressure to a low partial pressure. In effect, the partial pressures on each side of a membrane push O_2 in opposite directions, so the net rate of diffusion is proportional to the pO_2 on one side minus the pO_2 on the other side. The rate of diffusion therefore increases in direct proportion to the difference in partial pressure. If the partial pressures are equal, there is no net diffusion (Figure 12.2).

The rate of diffusion is also directly proportional to the area available for diffusion. Thus two lungs are twice as good as one. Another factor affecting the rate of diffusion is membrane thickness. The thinner the membrane, the higher the rate of diffusion through it. To summarize, the rate of diffusion across a membrane is proportional to the difference in partial pressures of the molecule on each side of the membrane, multiplied by the area of the membrane, divided by the thickness of the membrane. (The algebraic equivalent of this statement is known as Fick's equation.) These principles explain many of the features in the enormous variety of respiratory organs by which various animals obtain oxygen. No matter how diverse they are in structure, gills, lungs, and other such respiratory organs are all adapted to produce a large difference between the pO_2 in the environment and that in the body fluids, and to increase the area and minimize the thickness of tissue through which O_2 must diffuse.

One means of maintaining a large difference in pO_2 is by keeping the plasma and external fluids moving, for example, by circulating blood and by breathing. Keeping the external fluid moving prevents it from being depleted of O_2, and keeping the plasma moving prevents it from becoming saturated with O_2. Examples will be discussed in various animals groups later.

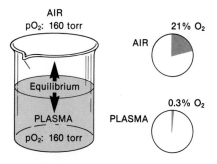

Figure 12.2
When considering the diffusion of gas from air to water or blood plasma, the partial pressure of the gas, rather than its concentration, determines the direction and rate of diffusion. Diffusion of O_2 from air to plasma stops when the partial pressures are the same, even though the amount of O_2 per milliliter of plasma is much less than the amount per milliliter of air.

Table 12.1 Types of respiratory pigment, listed in order of frequency of occurrence.

	Hemoglobin	Hemocyanin	Hemerythrin	Chlorocruorin
Animal Groups	Vertebrates; some flatworms, nematodes, annelids, crustaceans, insects	Some molluscs; horseshoe crabs, crustaceans, some spiders	Mainly sipunculids; priapulids, brachiopods; some nematodes, polychaete annelids	Some polychaete annelids
Metal	Iron	Copper	Iron	Iron
Location in Blood	Cells or plasma	Plasma	Cells	Plasma
Color: Oxygenated	Red	Blue	Violet	Green
Deoxygenated	Red-purple	Colorless	Colorless	Colorless

RESPIRATORY PIGMENTS

Respiratory pigments are a second adaptation for achieving large differences in pO_2. Respiratory pigments bind O_2, thereby removing it from blood plasma. This keeps the pO_2 in the plasma low, thereby favoring continued diffusion of O_2 into the plasma. In the tissues, respiratory pigments unload the O_2 into plasma. This maintains a high pO_2 that favors diffusion to the cells. Respiratory pigments also transport CO_2 from tissues to the respiratory organs, and they give the blood of animals their various colors, although that is seldom important biologically. Respiratory pigments occur either suspended in the plasma or contained within blood cells, depending on the species. They consist of a metal in an organic matrix; it is the metal that binds the oxygen. Several types of respiratory pigments occur in various animal groups (Table 12.1).

HEMOGLOBIN

Structure. By far the most studied respiratory pigment is hemoglobin (Hb), primarily because it occurs in so many animals, including humans. Hemoglobins consist of **hemes** and proteins. The hemes of all hemoglobins are identical and similar to the hemes in the cytochromes of mitochondria. In animals in which the

Figure 12.3
The structure of vertebrate hemoglobin (Hb). There are two β and two α protein chains (globins), each of which embraces a heme group (iron porphyrin). One O_2 binds to the iron atom in each of the heme groups.

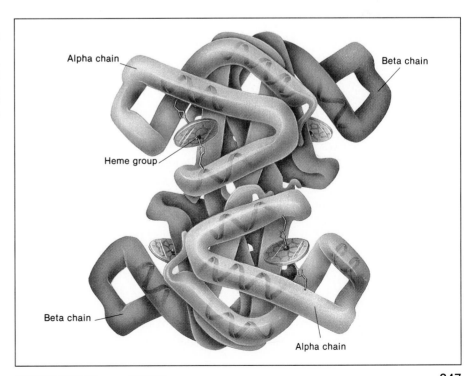

Alpha chain
Beta chain
Heme group
Beta chain
Alpha chain

hemoglobin occurs within cells, there are only a few heme units per hemoglobin (four in the hemoglobin of humans and other vertebrates). Each heme contains an iron atom that binds an O_2 molecule. The proteins vary widely, indicating that they evolved independently among different groups of animals. The protein is a **globin** with two α (alpha) and two β (beta) chains. Each chain bears one heme group (Figure 12.3). When O_2 binds to heme there is no exchange of electrons, so the iron is not oxidized by the oxygen. (In other words, hemoglobin does not rust.) Instead, the iron simply holds the O_2 in place electrostatically. A hemoglobin molecule bearing O_2 is said to be **oxygenated,** rather than oxidized. Oxygenated hemoglobin, abbreviated HbO, has the familiar red color of blood that has been exposed to air. Hemoglobin that is deoxygenated has a darker color.

Oxygen Binding and Release. Hemoglobin increases the amount of O_2 that can be carried by a given volume of blood seventyfold. Just as important, however, hemoglobin gives up its oxygen to the tissues that need it (Figure 12.4). How does hemoglobin "know" that it is supposed to bind O_2 in the respiratory capillaries and release it in the systemic capillaries? Of course it doesn't know in any conscious sense. The hemoglobin molecules simply respond to the conditions around them. When O_2 binds to one heme it increases the ability of other hemes in the same Hb molecule to bind to O_2. In other words, O_2 increases the **affinity** of Hb for O_2. Therefore, when blood is in the respiratory organs, where the pO_2 is high, Hb binds O_2 more strongly. When blood is in the systemic capillaries, where the pO_2 is low, Hb has a lower affinity for O_2 and therefore releases it to the tissues.

The effect of O_2 on the affinity of Hb is shown in the **oxygen dissociation curve** (= oxygen equilibrium curve, Figure 12.5). At high pO_2 virtually all the Hb is oxygenated (100% HbO). As the blood enters tissues and the pO_2 of the plasma falls, the Hb loses its affinity for O_2. Hb therefore gives up its oxygen. In the range of pO_2 normally found in tissues (20 to 40 mmHg) the oxygen dissociation curve is especially steep. Thus even a small drop in pO_2 in the tissue causes Hb to unload a lot of O_2 into the plasma. The steepness of the curve in the range of pO_2 found in tissues is not merely a stroke of good luck: The structure of the globin molecules that determines the shape of the oxygen dissociation curve has evidently undergone extensive evolution.

The globin molecules are also adapted to respond to CO_2 and other conditions in the plasma. CO_2, which is released by active tissues, reduces the affinity for O_2, causing Hb to give up even more of its O_2 to those tissues. This adaptation, called the **Bohr effect,** is represented in the oxygen dissociation curve by a shift of the curve to the right (Figure 12.6). High temperature and low pH, which also occur in active tissue, have similar effects on Hb. As a result, HbO unloads even more oxygen in the presence of highly active tissue than it would ordinarily.

Fetuses of humans and most other mammals produce a **fetal hemoglobin** that has a greater affinity for oxygen than does the Hb of their mothers. This enables the fetus to extract O_2 from the mother's blood, even though the pO_2 in the uterus is as low as on Mt. Everest. Skeletal muscle, especially the slow-twitch kind, also has a substance that has a higher affinity for O_2 than does Hb (Figure 12.6). This substance, **myoglobin,** is similar to the β chain of Hb. Since myoglobin binds O_2 more strongly than Hb does, it easily obtains the O_2 that is essential for the production of ATP.

CO_2 TRANSPORT

Red blood cells and hemoglobin also play important roles in the transport of carbon dioxide from tissues to the respiratory organs. CO_2 diffuses from cells into

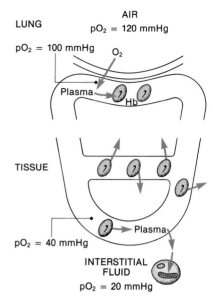

Figure 12.4
The diffusion of oxygen in mammalian lungs and in tissue. Hemoglobin can be viewed both as a means for increasing the oxygen capacity of blood and as a means of maintaining a large partial-pressure difference that favors the diffusion of O_2. In the lungs Hb removes O_2 from plasma, maintaining a lower pO_2 in plasma than in the air in the lungs. Oxygen therefore continues to diffuse into plasma. In other body tissues, Hb releases the O_2 into the plasma, maintaining a higher partial pressure in plasma than in interstitial fluid, which favors diffusion of O_2 into interstitial fluid.

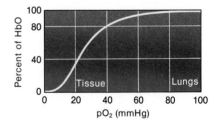

Figure 12.5
The oxygen dissociation (oxygen equilibrium) curve shows how the binding of O_2 by Hb depends on the partial pressure of O_2 (pO_2) in the surrounding plasma. In the lungs, where the pO_2 is high, Hb has a high affinity for O_2, and virtually all the Hb is in the oxygenated state (HbO). In the tissues, where pO_2 is about 30 to 40 mmHg, Hb loses its affinity for O_2 and gives it up to the plasma and interstitial fluid.

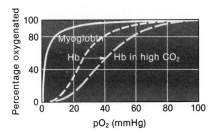

Figure 12.6
Carbon dioxide shifts the oxygen dissociation curve of Hb to the right, reducing the affinity for O_2 in tissues even if pO_2 remains the same. Thus Hb gives up more O_2 to tissues. Myoglobin (left curve) has a higher affinity for O_2 than does Hb, so it takes O_2 away from blood to be used by muscle mitochondria.

interstitial fluid and plasma, and from plasma into red blood cells. Some carbon dioxide reacts with water to form carbonic acid, which then dissociates into bicarbonate and hydrogen ion:

$$CO_2 + H_2O \rightarrow H_2CO_3 \rightarrow HCO_3^- + H^+$$

Ordinarily this reaction is too slow to carry off large amounts of CO_2, but red blood cells have an enzyme, **carbonic anhydrase,** that catalyzes the formation of the bicarbonate. Hemoglobin binds the H^+, accelerating the dissociation of the bicarbonate and preventing the acidification of blood. Some CO_2 is also transported by binding to the amine groups on the globin of hemoglobin, forming **carbamino compounds** (HbCO$_2$). These reactions reverse when the blood arrives in the respiratory organs.

INTEGUMENTARY EXCHANGE OF GASES

As noted above, the cellular membranes for diffusion of O_2 and CO_2 must be thin and have a large surface area. In very small animals the integument is so thin and its area is so large compared with the volume of the body that they do not require gills or other respiratory organs. Among these animals are roundworms (phylum Nematoda) and annelids (phylum Annelida), such as earthworms and leeches. Even some quite large animals have such thin tissues that oxygen has to diffuse only a short distance to reach any cell. These include jellyfishes and other cnidarians (phylum Cnidaria) and flatworms (phylum Platyhelminthes).

In many large animals with thick tissues, all or part of the integument is adapted for respiration and can properly be called a respiratory organ. Certain salamanders satisfy all their requirements for O_2 by integumentary respiration. In many others, such as land snails, eels, and frogs, integumentary exchange supplements gills or lungs. Frogs and most other amphibians have lungs, but even when breathing air they eliminate as much CO_2 through their skin as from their lungs. When submerged, of course, the integument is the only functioning respiratory organ in the frog. Unlike breathing through lungs, integumentary exchange requires no neural control. One can therefore destroy the nervous systems of a frog by "pithing" it with a probe, and the other organs will not deteriorate from anoxia. This is one reason why frogs are used in so many experiments.

Special adaptations for integumentary respiration include some or all of the following in various species:

1. Reduced thickness of the integument. For example, the thin skin of amphibians permits it to act as a respiratory organ under water.
2. Increased area. Some amphibians have folds of skin that provide additional area.
3. Increased vascularization, with capillaries lying close beneath the skin. This has the effect of increasing area, as well as quickly carrying off the oxygen so that pO_2 stays low in the blood.
4. Means of keeping the integument moist and permeable to O_2. Earthworms, amphibians, and many other animals with integumentary respiration continually secrete mucus for this function and generally have to stay in damp habitats.

GILLS

Larger and more active animals require more specialized respiratory organs. In aquatic species these organs are often gills. Gills have evolved independently in

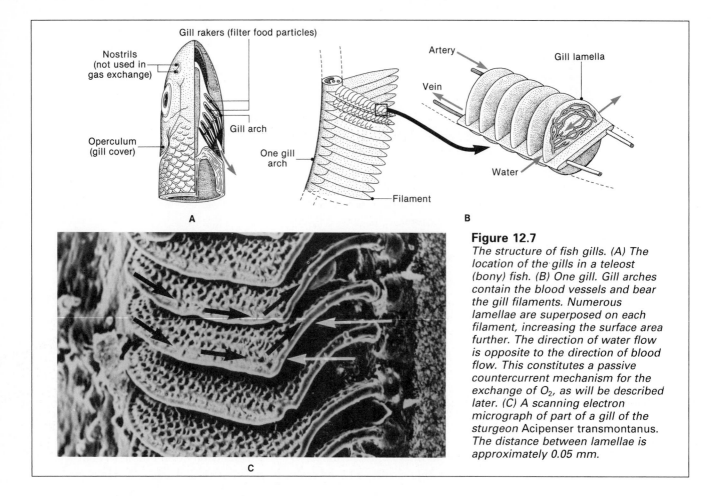

Figure 12.7
The structure of fish gills. (A) The location of the gills in a teleost (bony) fish. (B) One gill. Gill arches contain the blood vessels and bear the gill filaments. Numerous lamellae are superposed on each filament, increasing the surface area further. The direction of water flow is opposite to the direction of blood flow. This constitutes a passive countercurrent mechanism for the exchange of O_2, as will be described later. (C) A scanning electron micrograph of part of a gill of the sturgeon Acipenser transmontanus. The distance between lamellae is approximately 0.05 mm.

many groups, including molluscs, crustaceans, and fishes. They vary greatly in their structure, but all have large areas and thin membranes, usually with numerous sheets of tissue (Figure 12.7). Because of this structure, gills are usually delicate and require the buoyant support of water. Most animals with gills cannot breathe

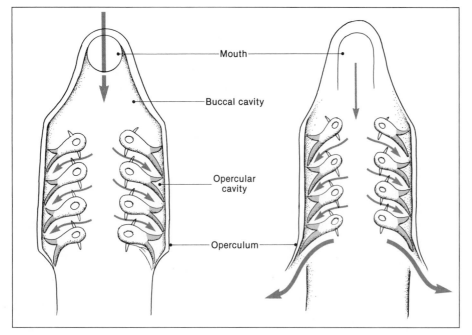

Figure 12.8
Buccal pumping in a fish. Water is first drawn into the buccal cavity (left). The mouth then closes, and the buccal cavity contracts, forcing water into the opercular cavity (right). After a short delay the operculum opens, emptying the opercular cavity.

air directly, since the gills collapse when out of water. Some crustaceans and fishes, however, can breathe air, using reinforced gills that do not collapse when out of water.

Animals with gills usually have some mechanism that keeps oxygenated water continually flowing past them so that the external pO_2 stays high. Bivalve molluscs, such as clams, maintain the water flow by means of cilia on the gill surfaces. Sedentary animals often pump water past the gills by body contractions. Fishes such as bass, catfish, and aquarium fishes that swim slowly in calm water take water into the mouth **(buccal cavity),** then force it past the gills in the **opercular chamber.** This process is called **buccal pumping** (Figure 12.8). Fishes that swim rapidly or live in swift water, such as tuna and trout, can maintain the flow of water simply by opening their mouths into the current. The circulation of blood through the gills of fishes is in the direction opposite to that of the water. Fish gills therefore take advantage of **countercurrent exchange,** which increases the amount of O_2 they extract from the water. (See box.)

Countercurrent Exchange

Countercurrent exchange is an elegant adaptation for extracting more oxygen from water, and it also occurs in numerous other organs adapted to maximize the exchange of molecules or heat. The following explanation of countercurrent exchange is for diffusion of molecules, but it applies as well for conduction of heat. The basic arrangement for a **passive countercurrent system** is that two fluids with different concentrations (or partial pressures, or temperature) flow past each other in opposite directions, separated by a permeable membrane. If the two fluids were flowing in the same direction side-by-side, the concentrations in both fluids would soon become equal, and diffusion would stop (Figure 12.9A). By having the fluids flow in opposite directions, however,

one fluid is always beside another fluid with a different concentration. Diffusion therefore continues, and more molecules are exchanged (Figure 12.9B).

Much higher concentrations can be produced in an **active countercurrent exchange system.** In an active system the fluid flows through a tube that doubles back so that the direction of flow reverses (Figure 12.9C). As fluid leaves the system, molecules are actively transported from it into the fluid that enters the system. This process repeats over and over, generating extremely large concentrations of the molecule. Examples of active countercurrent exchange in the kidney and in the swimbladder of fishes will be described later (see Chapters 14 and 35).

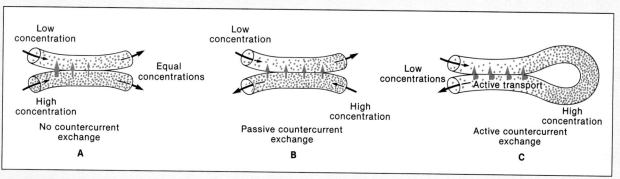

Figure 12.9
Countercurrent exchange. (A) A fluid with high concentration flowing past a fluid with low concentration in the same direction loses molecules by diffusion. Soon both solutions have equal concentrations. (B) If the fluids move in opposite directions, however, the fluid with the higher concentration is always adjacent to the fluid with the lower concentration. It therefore loses many more molecules to the other fluid by diffusion. This is a passive countercurrent exchange system, such as in fish gills (Figure 12.7). (C) In an active countercurrent exchange system molecules are continually recycled back into a fluid by active transport, resulting in a very large concentration.

TRACHEAE

The intricate surfaces of gills present such a high resistance to water flow that in some species 20% of the oxygen obtained by gills is used just to produce the energy to keep water flowing past them. Little wonder that many animals evolved mechanisms for breathing air, which not only contains more oxygen but also flows more easily. The invertebrates that are best adapted to air-breathing are spiders and insects. Spiders depend mainly on book lungs, which will be described later. In addition, hollow tubes called tracheae (TRAY-key-ee) contribute to respiration in spiders and are the main respiratory organs in insects. The principle of tracheal respiration is straightforward: tracheae simply provide a pathway from air to tissues (Figures 12.10 and 32.10). In some cases the smallest tracheae, called **tracheoles,** connect directly to the plasma membranes of individual cells. The outer end of a trachea, the **spiracle,** is guarded by a valve that opens when internal CO_2 levels are high, but that otherwise stays closed, reducing the loss of water from the body. The diffusion of air through the tracheae is often passive, but some insects can pump air through the tracheae by contracting their bodies.

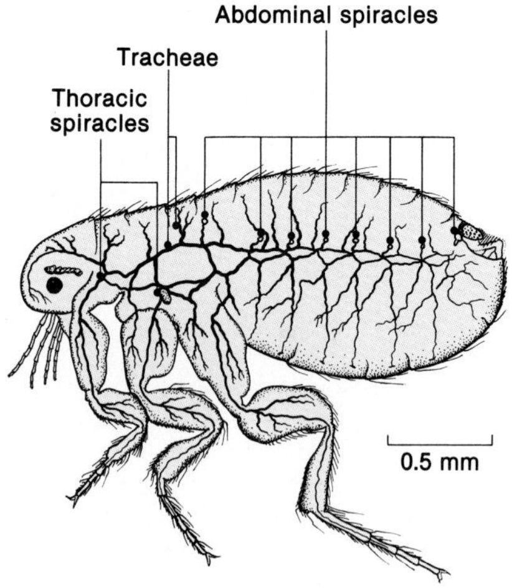

One advantage of tracheal respiration is that the circulation of blood does not have to be efficient, since the supply of oxygen does not depend on it. Insects therefore get by with an open circulatory system with very low blood pressure. On the other hand, tracheal respiration does limit the size of insects, since O_2 does not diffuse readily through long, narrow tubes. Most insects are therefore only a few millimeters wide, and the largest insects are only(!) about 20 cm wide.

Figure 12.10
The tracheal system of an insect (a flea). Air enters through the spiracles and diffuses through tracheae and tracheoles. For further detail see Figure 32.10.

LUNGS

The other major type of organ for breathing air is the lung, which usually consists of one or more internal air-filled cavities. Lungs evolved independently among three major groups: pulmonate molluscs, spiders, and vertebrates. Pulmonate molluscs (Latin *pulmo* lung) are the land snails and slugs and some freshwater species that evolved from them. The lungs of pulmonates are body cavities that take in air through an opening on one side. The lungs of spiders are called **book lungs** because they consist of flat pouches stacked like pages in a book (see Figure 30.12).

The lungs of vertebrates originated as simple sacs branching from the digestive tracts of certain fishes during a dry spell some 400 million years ago. Apparently the lungs originally served as air reservoirs and respiratory organs that allowed fishes to burrow into mud to escape drought. They also maintained buoyancy while fishes were in water. In most of the bony fishes the lungs became specialized as swim bladders for maintaining buoyancy, and they lost their ability to exchange respiratory gases. In some fishes, however, the lungs retained their respiratory abilities. Some of these fishes evolved into amphibians, and some of those evolved into reptiles. Reptiles then diverged into two other groups with lungs—birds and mammals.

RESPIRATION IN AMPHIBIANS AND REPTILES

Although the lungs of all vertebrates had the same evolutionary origins, they diverged greatly in structure. The lungs of amphibians and most reptiles are essentially sacs. Folds in the walls of the lungs, especially in reptiles, increase the surface area for gaseous exchange (Figures 12.11A and 37.13). All vertebrates with lungs have to breathe to maintain a high pO_2 and a low pCO_2 in the air within the lungs. The pattern of breathing varies widely. Amphibians draw fresh air into the buccal (mouth) cavity, exhale stale air from the lungs past the fresh air, then force

Figure 12.11
(A) The lungs of a frog. The lungs of reptiles are similar except that they usually have more extensive folding of the walls. (B) Schematic representation of breathing in the bullfrog Rana catesbeiana. *① With the nostrils open and the glottis closed, the buccal cavity expands, allowing atmospheric pressure to force air in (color). ② The frog then opens its glottis, and the elasticity of the lungs and contraction of the body force out the air already in the lungs. There is little mixing with air in the buccal cavity. ③ The frog then closes its nostrils and contracts the throat, forcing fresh air into the lungs. ④ With the glottis closed, frogs then flutter the throat to flush out the buccal cavity before beginning the cycle again.*

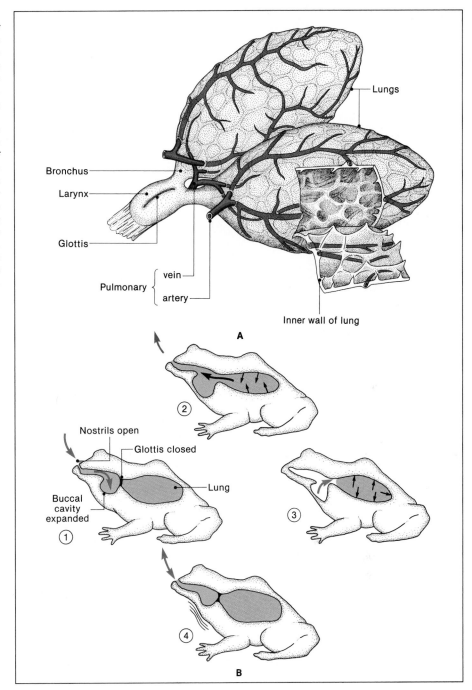

the fresh air from the buccal cavity into the lungs (Figure 12.11B). Reptiles, like mammals, inhale by expanding the body, which enlarges the lungs and allows air to enter. The major difference between breathing in reptiles and mammals is that reptiles hold their breaths for as long as several minutes after each inhalation, in contrast to the more regular inhalation and exhalation of mammals. Reptiles can hold their breaths for so long because their metabolic rates are so much lower than those of mammals.

RESPIRATION IN BIRDS

Unlike the lungs of other vertebrates, those of birds contain numerous parallel tubes called **parabronchi** (Figure 12.12A). Each parabronchus is divided length-

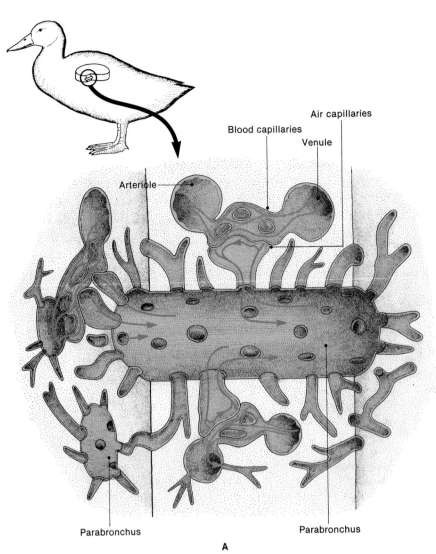

Arteriole

Blood capillaries

Air capillaries

Venule

Parabronchus   Parabronchus

A

Figure 12.12
The respiratory system of a bird.
(A) Detail of lung tissue, showing several parabronchi. Blood and air move in opposite directions (arrows), establishing a countercurrent exchange.
(B) Inhalation. The air sacs expand, drawing fresh air (red) mainly into the posterior air sacs, while the lungs contract, forcing air from the previous inhalation into the anterior air sacs. (C) Exhalation. Air sacs are compressed. Fresh air is pushed through the parabronchi of the expanding lungs, while old air in the anterior air sacs is exhaled.

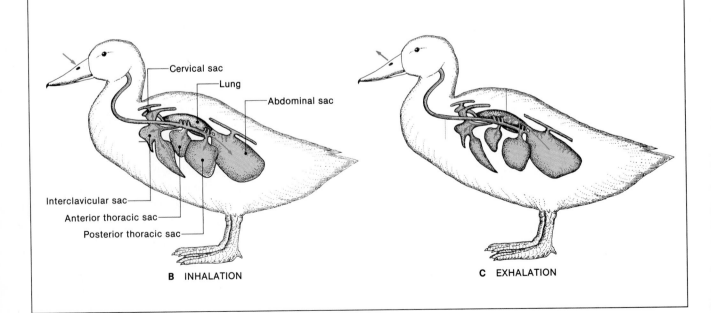

Cervical sac

Lung

Abdominal sac

Interclavicular sac
Anterior thoracic sac
Posterior thoracic sac

B INHALATION   **C** EXHALATION

wise into many **air capillaries,** which increase the surface area for exchange of gases with the blood capillaries surrounding the parabronchi. In contrast to the sacs in the lungs of amphibians, reptiles, and mammals, in which air stagnates, the parabronchi permit air to flow completely through the lungs in only one direction during both inhalation and exhalation. **Air sacs** play an essential role in this one-way air flow. During inhalation the posterior air sacs expand, drawing in fresh air (Figure 12.12B). The anterior air sacs also expand, and the lungs contract, so air flows out of the parabronchi into the anterior air sacs. During exhalation the air sacs are compressed while the lungs expand (Figure 12.12C). The fresh air in the posterior air sacs is therefore forced through the parabronchi, and the used air in the anterior air sacs is exhaled.

RESPIRATION IN MAMMALS

The lungs and associated structures in mammals are more like those of reptiles than of birds (Figure 12.13). Air inhaled through the nasal passages and/or mouth enters the **trachea** (windpipe), then passes into each lung through a **bronchus.** The trachea is built of various connective tissues, including C-shaped rings of cartilage that bend rather than break when compressed. The trachea lies in front of the esophagus in the neck, which means that food has to pass over the opening of the trachea (the **glottis**) when it is swallowed. This "design defect" presents a danger of choking while eating, but most of the time the **epiglottis** blocks the glottis during swallowing.

Figure 12.13
The lungs and air passages in a human. Air enters the lungs through the trachea and two bronchi. The larynx, or voice box, is part of the trachea.

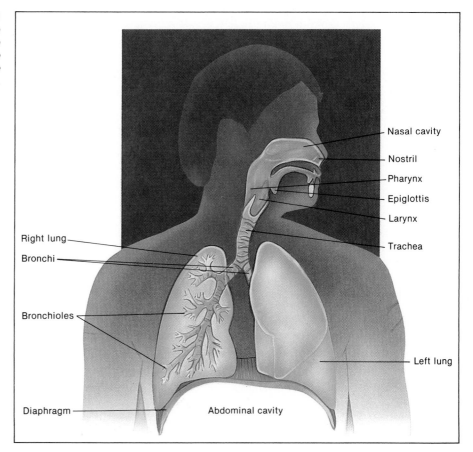

Labels: Nasal cavity, Nostril, Pharynx, Epiglottis, Larynx, Trachea, Right lung, Bronchi, Bronchioles, Left lung, Diaphragm, Abdominal cavity

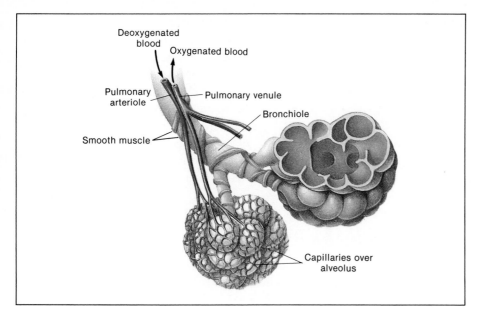

Figure 12.14
A bronchiole and several alveoli, with details of an alveolus shown. The diameter of each alveolus is approximately 0.1 mm.

Labels in figure:
Deoxygenated blood
Oxygenated blood
Pulmonary arteriole
Pulmonary venule
Bronchiole
Smooth muscle
Capillaries over alveolus

Each bronchus divides within the lung to form a network of numerous branches like an inverted tree. The smallest twigs in this bronchial tree are called **bronchioles.** Each bronchiole terminates with a cluster of hollow sacs called **alveoli** (plural of alveolus), where the actual exchange of oxygen and carbon dioxide occurs (Figure 12.14). Breathing ventilates the alveoli, bringing in fresh air that is rich in O_2 and removing the air that is laden with CO_2. As deoxygenated blood from the pulmonary artery flows through capillaries surrounding the alveoli, CO_2 diffuses out of the plasma while O_2 diffuses in. As one would expect from the requirements from diffusion, the alveolar walls are thin. Also, the presence of numerous small alveoli provides a much larger surface area than would a few large alveoli with the same volume.

It takes many additional adaptations to keep the alveoli functioning. Mucus lining the respiratory passages protects them from the drying effects of air and traps dust and bacteria. Cilia lining the bronchi and bronchioles continually propel the mucous layer toward the mouth where it may be swallowed or spit out (depending on how you were raised). The importance of bronchial cilia is evident in someone who has destroyed his with cigarette smoke. "Smoker's cough" is the only way such a person can bring up the mucus. The alveoli are also protected by antibiotic secretions and phagocytes that discourage bacteria, fungi, and other invaders, and by enzymes that break down any mucus that enters.

BREATHING IN MAMMALS

In mammals **inhalation** (= inspiration) is normally an active process, requiring contraction of skeletal muscles that enlarge the thoracic cavity. This enlargement reduces the pressure within the lungs by as much as 30 mmHg below atmospheric pressure, so air enters the alveoli. As the thorax expands, the lung is protected from abrasion by a surrounding double membrane called the **pleura.** The pleura also absorbs water or air that may become entrapped between the lungs and the chest wall.

Inhalation is triggered by neural impulses from the medulla of the brainstem to two sets of skeletal muscles: the diaphragm and the intercostal muscles. The

Figure 12.15

The mechanisms of inhalation and exhalation in humans. (A) Bursts of action potentials trigger inhalation by stimulating the diaphragm and intercostal muscles. The diaphragm flattens, and the ribs move upward and outward. Both actions increase thoracic volume and therefore reduce the pressure within the lungs. (B) During normal exhalation the ribs and diaphragm return to their relaxed positions, decreasing thoracic volume.

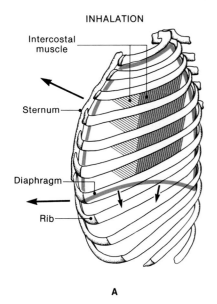

INHALATION

Intercostal
muscle

Sternum

Diaphragm

Rib

A

EXHALATION

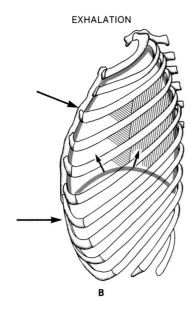

B

diaphragm is a sheet of skeletal muscle between the thoracic and abdominal cavities. When the diaphragm contracts it flattens out, increasing thoracic volume (Figure 12.15). The **intercostal muscles** lie in diagonal bands between the ribs (as you can see the next time you eat ribs). The contraction of these muscles causes the ribs to swivel up from their attachments on the vertebrae, and this enlarges the chest cavity. From their anatomical arrangement it is thought that the outer layer, consisting of the **external intercostal muscles,** is responsible for inhalation. During normal **exhalation** (= expiration) the diaphragm and intercostal muscles simply return to their relaxed positions. Air can also be exhaled forcefully, during coughing and speaking, for example. The **internal intercostal muscles** may be partly responsible for forced exhalation. Forced exhalation also involves contractions of the abdominal muscles, which press on the abdominal contents and force the diaphragm upward.

REGULATION OF BREATHING

Like the other essential requirements of cells, oxygen is maintained in appropriate levels in the interstitial fluid. This homeostasis is accomplished partly by changing the flow of air through bronchi and partly by changing the rate and depth of breathing. The flow of air depends on bronchial diameter, which is controlled by smooth muscles (Figure 12.14) that respond to neural and hormonal signs. During exercise or stress the sympathetic nervous system releases norepinephrine, which causes the smooth muscles to relax. The hormone epinephrine (adrenaline) from the adrenal medulla has a similar effect. Relaxation of the bronchial muscles allows the bronchi to dilate, letting air pass more easily into the alveoli. In the absence of stress or exercise, the parasympathetic nervous system releases acetylcholine onto these smooth muscles, which causes them to contract. The bronchial smooth muscles also contract during attacks of asthma. (In the days before low-nicotine cigarettes, asthma sufferers could get relief by smoking cigarettes, since nicotine blocks acetylcholine receptors. Now asthma sufferers use inhalers containing chemicals that mimic norepinephrine.)

The rate and depth of breathing are controlled by bursts of action potentials to the diaphragm and intercostal muscles from the medulla. The medulla contains a central pattern generator that automatically sends out bursts of action potentials that control the rate and depth of breathing. The rate of these bursts of action potentials (the number of bursts per minute) equals the rate of inhalation. The frequency of action potentials within each burst controls the force of muscle contraction, and therefore the depth of breathing.

Generally, both the rate and depth of breathing increase during exercise and stress, thereby ventilating the alveoli more completely and increasing pO_2. Increased levels of CO_2 in the cerebrospinal fluid directly cause the medulla to increase the rate and depth of breathing. In addition, the medulla increases rate and depth of breathing in response to action potentials from **carotid and aortic bodies** in the main arteries to the head (the carotid arteries) and in the aorta where it bends above the heart (the aortic arch). The carotid and aortic bodies are sensitive to pH, pO_2, and especially to pCO_2. Because the medulla responds to changes in pCO_2 more than to pO_2, CO_2 is added to oxygen tanks to "remind" the medulla that it is still necessary to inhale. Hyperventilation—breathing deeply and rapidly—can reduce the pCO_2 to the point that the medulla fails to trigger inhalation for several minutes. Some underwater swimmers take advantage of this phenomenon to inhibit the urge to breathe. Sometimes it works permanently.

SUMMARY

The adaptations by which various animals obtain oxygen reflect the principles of diffusion. Oxygen diffuses at a rate that is proportional to the difference between its external and internal partial pressures, proportional to the surface area for diffusion, and inversely proportional to the thickness of the diffusion surface. Animals therefore must maintain a sufficient difference in pO_2 and have a thin, broad surface across which O_2 can diffuse. Many animals keep the external and internal fluids moving to maintain large differences in pO_2. Respiratory pigments such as hemoglobin serve the same function, by removing O_2 from blood plasma in respiratory tissues where pO_2 is high and releasing the O_2 in other tissues where pO_2 is low.

In small animals and those with thin tissues, the integument is thin enough and large enough in relation to body volume that no special respiratory organs are needed to obtain sufficient O_2. In many larger animals the integument is specially adapted as a respiratory organ. The adaptations include reduced thickness, increased area, increased blood circulation, and maintenance of a moist surface. The same adaptations occur in the gills of aquatic animals and the lungs of terrestrial animals. The lungs of amphibians, reptiles, and mammals are essentially sacs subdivided to various degrees. The lungs of birds consist of numerous parabronchi through which air flows into and out of air sacs. Lungs are ventilated during breathing, the pattern of which varies in different groups of animals. The lungs of mammals contain numerous alveoli. Breathing results from contractions of intercostal muscles and the diaphragm and is controlled by the medulla.

KEY TERMS

partial pressure
respiratory pigment
hemoglobin
integumentary respiration
gill

buccal pumping
countercurrent exchange
trachea
lung
parabronchus

bronchus
alveolus
diaphragm
intercostal muscle

SELF-TEST

1. Why must partial pressure rather than concentration be used in figuring the direction and rate of diffusion of oxygen in terrestrial animals?

2. What respiratory pigment occurs in vertebrates? Describe its structure and function. How is its function modified in response to O_2 and CO_2?

3. Write a brief biography of a hemoglobin molecule as it makes a complete journey through the circulatory system.

4. Name four types of respiratory organ. For each type give the common name of an animal with that organ. Why is that type of respiratory organ found in that animal?

5. Explain why respiratory organs are so delicate.

6. How do the lungs of birds differ in structure and function from those of amphibians, reptiles, and mammals?

7. In mammals, which muscles are involved in breathing? What causes them to contract? How is the rate and depth of breathing adjusted to meet the requirements for O_2?

READINGS

RECOMMENDED READINGS

Feder, M. E. and W. W. Burggren. 1985. Skin breathing in vertebrates. *Sci. Am.* 253(5):126–142 (Nov).

Kanwisher, J. W. and S. H. Ridgway. 1983. The physiological ecology of whales and porpoises. *Sci. Am.* 248(6):111–120 (June).

Schmidt-Nielsen, K. 1971. How birds breathe. *Sci. Am.* 225(6):72–79 (Dec).

Sebel, P. et al. 1985. *Respiration: The Breath of Life.* New York: Torstar. (*A lavish presentation of the physiology and sociology of breathing.*)

ADDITIONAL REFERENCES

Graham, J. B. 1988. Ecological and evolutionary aspects of integumentary respiration: body size, diffusion, and the invertebrata. *Am. Zool.* 28:1031–1045.

Liem, K. F. 1988. Form and function of lungs: the evolution of air breathing mechanisms. *Am. Zool.* 28:739–759.

13

Nutrients

Deer mouse (Peromyscus maniculatus) *eating blackberries.*

LEARNING OBJECTIVES

1. What components of food make it essential for animals?

2. How are the feeding mechanisms of different animals adapted for different kinds of foods?

3. What similarities and differences occur among the digestive systems of different animals?

4. What processes are involved in digestion?

5. How do the nutrients from foods get from the digestive system to the cells of the body?

6. What mechanisms control the level of nutrients available to cells, the rate at which an animal uses nutrients, and the rate at which those nutrients are replaced by feeding?

t is a common theory of children that the food they eat simply falls into their hollow legs, and when the legs get full they have to go to the bathroom. Only later do children realize that this theory leaves a number of questions unanswered. Why does the food that goes in look so different from what comes out? Why does one feel the need to eat? And what is the good of it all, anyway? The objective of this chapter is to answer those questions and others. A concise answer is that the interstitial fluid must contain nutrients if it is to serve as a suitable internal environment for cells. These nutrients are released from food and absorbed into the body during the process of digestion.

INORGANIC NUTRIENTS

Like all molecules, nutrients are either inorganic or organic. The inorganic nutrients include ions and metals such as sodium and potassium, which produce membrane potentials of neurons and other cells, calcium, which is an intracellular messenger and major component of bone, and iron, in hemoglobin and cytochromes (Table 13.1). Many other elements must be present in animal diets in small amounts. The requirement for some of these **trace elements** is so small that it is often difficult to determine the consequences of a deficiency, since they cannot be totally excluded from experimental diets. One of the trace elements is **copper,** which mammals require in extremely small amounts for hemoglobin synthesis. Copper also occurs in cytochrome oxidase in mitochondria, and in the respiratory pigment hemocyanin in some invertebrates. Other trace elements are **cobalt** (in vitamin B_{12}), **iodide** (in the hormones thyroxine and triiodothyronine), **arsenic, chromium, fluoride, manganese, molybdenum, nickel, selenium, silicon, tin, vanadium, zinc,** and probably others. While essential in trace amounts, most of these elements are toxic in large concentrations.

ORGANIC NUTRIENTS

The organic nutrients include carbohydrates, fats, proteins, and vitamins. Cells need organic nutrients for two main reasons: for energy and for materials to make new cells. Organic nutrients consist mainly of six elements, easily remembered by the acronym "CHNOPS." By definition, **carbon** (C) is an element of all organic molecules, representing 9.5% of all the atoms in organic nutrients. The other

Table 13.1   Some important minerals required in the human diet.

Sodium (Na^+) and **Chloride** (Cl^-). These are the most common ions in the interstitial fluid and in blood. Their net concentrations are largely responsible for maintaining the osmotic pressure in cells. They are also important in producing resting and action potentials across cell membranes. They must be present in the diet in rather high levels (several grams per day for humans, primarily as NaCl).

Potassium (K^+). This is mainly an intracellular ion, important in maintaining the membrane voltage. Several milligrams are required each day.

Calcium (Ca^{2+}) and **Magnesium** (Mg^{2+}). These divalent cations are among the most important in coordinating cellular functions. Ca^{2+} is a second messenger that triggers the release of synaptic transmitter, hormones, enzymes, and other secretions. It also triggers the contraction of muscles. It is important in calcium phosphate, the major mineral in bone, and in calcium carbonate, the major component of shells. Mg^{2+} is required as a cofactor for enzymes such as ATPases.

Iron (Fe^{2+}). This metal occurs in hemoglobin, myoglobin, and the cytochromes. Relatively small amounts are needed in the diet because Fe^{2+} is recycled. When the spleen breaks down dead red blood cells it transfers the iron to the liver, where it is stored on the protein ferritin until bone marrow needs it to produce more hemoglobin.

elements and their proportions in organic nutrients are **hydrogen** (63%), **nitrogen** (1.4%), **oxygen** (25.5%), **phosphorus** (< 1%), and **sulfur** (< 1%).

Carbohydrates. Carbohydrates serve primarily as immediate energy sources during cellular metabolism (see pp. 55–63). Metabolism of 1 gram of carbohydrate produces about 4 kilocalories of energy (4 kcal/g), most of it heat, and little less than half ATP. [One kcal is enough energy to raise the temperature of 1 kilogram of water 1 degree Celsius. It is the same as the more familiar Calorie (capitalized) used in nontechnical literature such as diet books.] An adult human ordinarily requires between 2000 and 3000 kcal (Calories) per day, most of which comes from carbohydrates. These carbohydrates include starches and simple sugars, such as glucose, sucrose, and lactose (Table 13.2). Carbohydrates not immediately needed are stored in cells as the starchlike substance glycogen.

Lipids. Lipids include several kinds of molecules, including fats, phospholipids, and cholesterol. Fats are **triglycerides** (= triacylglycerols), which consist of a three-carbon glycerol backbone to which are linked three fatty acids (see Figure 2.11). (Fats that are liquid at body temperature are called oils.) Fats are stored mainly in **adipose tissues** as long-term energy reserves, providing about 9 kcal/g. The average American has about 150,000 kcal of fat reserve (about 17 kg), which is enough to sustain life for up to 75 days. Animals can synthesize most fatty acids, but for most species there are some **essential fatty acids** that cannot be synthesized and that must therefore be present in the diet. For humans there are three essential fatty acids (arachidonic, linoleic, and linolenic). Phospholipid is similar to triglyceride except that a phosphate-linked group substitutes for one fatty acid (see Figure 2.12). Phospholipids make up much of the substance of cell membranes. Cholesterol is also a major component of biological membranes, and the raw material for the synthesis of steroid hormones (see Figure 10.4).

Proteins. Proteins serve as enzymes, hormones, and antibodies. Proteins also form collagen, actin, myosin, and other structural components of cells. Although many proteins are more-or-less permanent structures, others are continually broken down and resynthesized. Proteins are the energy source of last resort, supplying 4 kcal/g, for a total of up to 24,000 kcal in a starving human.

Proteins consist of precise sequences of amino acids (see Figure 2.13). Most of the 20 amino acids in protein can be synthesized by all animals, but there are also some **essential amino acids** that must be obtained from the diet. For most animals, including humans, there are nine essential amino acids (cysteine, isoleucine, leucine, lysine, methionine, phenylalanine, threonine, tryptophan, and valine). The

Table 13.2 Some important carbohydrates.

Glucose is the major blood sugar of vertebrates and most other animals (see Figure 2.9).

Sucrose—common table sugar—combines two simple sugars and is therefore a **disaccharide** (see Figure 2.10A). It consists of glucose linked to the fruit sugar, fructose.

Lactose is a disaccharide consisting of glucose linked to the similar sugar, galactose. Lactose occurs in milk, so it is mainly a nutrient for infant mammals.

Starch is the most common source of carbohydrate for herbivores (plant-eating animals) and most omnivores (animals such as humans that eat both plants and animals). Starch consists of many molecules of glucose, so it is a **polysaccharide**.

Glycogen is a starchlike polymer of glucose that is synthesized as a between-meal energy reserve (see Figure 2.10B). In humans the skeletal muscles store about 480 kcal of glycogen, and the liver stores about 280 kcal.

lack of just one of these amino acids is as damaging to growth and health as the lack of all of them, for the synthesis of a protein halts when mRNA calls for an amino acid that is not available. Unlike fatty acids, amino acids cannot be stored for use later, so proper growth and functioning require that all the essential amino acids be taken in frequently and at about the same time. The "quality" of a protein is a measure of how closely the proportions of essential amino acids match the proportions required by the animal. For each animal the food with the highest quality of protein is obviously another animal of the same species. Most animal proteins are similar enough, however, that carnivores and omnivores can get enough of the essential amino acids without resorting to cannibalism. Herbivores, however, may find it difficult to get the essential amino acids from plants. Many herbivores compensate by eating a variety of plants simultaneously, with each plant food making up for the deficiencies of others. Human vegetarians have culturally evolved the same strategy. Mexicans developed a liking for beans with rice. Beans make up for the deficiency of lysine in the rice, and the rice makes up for the deficiencies in methionine and cysteine in beans. Similarly, Jamaicans have a taste for rice and peas, and Indians mix wheat and pulses (seeds of legumes).

Vitamins. Vitamins are organic molecules that are required in the diet but that are not essential fatty acids or amino acids. Vitamins serve various functions (Table 13.3), but unlike other organic nutrients they are not broken down, so they provide no energy. (This fact surprises most Americans, who have been misled by generations of advertising to believe that vitamins produce energy.) The same vitamins are not required by all species. For example, ascorbic acid (vitamin C) is not a vitamin for amphibians, reptiles, and many birds and mammals, which can synthesize it. It is a vitamin, however, for invertebrates, fishes, some birds, and primates (including humans). There are two groups of vitamins: the lipid-soluble vitamins that can be stored in body fat, and the water-soluble vitamins that cannot be stored and that must therefore be ingested frequently. While some of each vitamin is required, too much can be harmful. This is especially so for fat-soluble vitamins, which accumulate in the body.

DISSOLVED NUTRIENTS

The first cells probably lived in their food. They simply absorbed nutrients from the "organic soup" across their cell membranes. Some animals still obtain at least a portion of their nutrients by absorption of **dissolved organic material** (DOM) across the body surface. Animals that depend mainly on DOM require a large surface area relative to body volume. They are therefore either long and thin or they are quite small. Such animals often do not even have digestive organs, since the nutrients are either already broken down to a form suitable for metabolism, or they can be broken down by intracellular digestion using **lysosomes** (see pp. 42–43). Many aquatic invertebrates, such as sponges, various larvae, and small worms, obtain at least part of their nutrition in this way. Dissolved organic material amounts to only about 1 milligram per liter of seawater, so quite a lot of water must pass over such an animal to provide significant nutrition. **Endoparasites,** which live within the gut or tissues of other animals, also commonly absorb nutrients across the body surface. Among these are tapeworms and some flukes (phylum Platyhelminthes; see Chapter 25).

Many other animals acquire some of their nutrients from **endosymbionts** such as algae or bacteria that live within the animal. Some corals, sponges, and clams derive a large portion of their nutrients from photosynthesis by endosymbiotic algae **(zooxanthellae).** Some sponges benefit from photosynthetic cyanobacteria.

Table 13.3 Vitamins required by humans.

FAT-SOLUBLE

A (carotene) is a precursor for retinal, a part of the visual pigments, and is required for the maintenance of epithelial tissue. Deficiency causes night blindness and retardation of growth; excess causes neurological disturbances and arthritislike symptoms. Sources include green leafy and orange-yellow vegetables, milk, egg yolks, liver, and fish-liver oil.

D (ergocalciferol D_2 and cholecalciferol D_3). Vitamin D_3 is synthesized by skin exposed to the ultraviolet rays of sunlight, so it does not normally fit the definition of a vitamin. The liver and kidneys convert the two forms of vitamin D into 1,25-dihydroxyvitamin D, which is required for absorption of Ca^{2+} by the gut and for bone growth. Deficiency causes bone malformations (rickets) in children; excess causes abnormally high levels of calcium, which cause neurological and kidney disorders. Dietary sources include egg yolks, butter, liver, and fish-liver oils.

E (tocopherols) serves as an antioxidant that preserves membrane structure. Vitamin E is so abundant, and so little is needed, that deficiency symptoms are never observed. Sources include vegetable oils, wheat germ, leafy vegetables, egg yolk, margarine, and legumes.

K (phylloquinone) is required for the synthesis of prothrombin, which is required for blood clotting. In many mammals sufficient vitamin K is produced by the bacteria in the large intestine. Other sources include leafy vegetables, vegetable oils, and pork liver.

WATER-SOLUBLE

B_1 (thiamine) is part of an enzyme required for the production of ATP by mitochondria. Deficiency usually occurs only in people subsisting on polished rice, or in alcoholics, malnourished pregnant women, or others with inadequate intake of the vitamin or inability to absorb it. Mild deficiency causes neurological disturbances, such as tiredness, irritability, insomnia, and pain. Severe deficiency causes beriberi, with severe neurological and circulatory disruptions that may be fatal. Sources of thiamine include dried yeast, whole grains, meat, nuts, legumes, and potatoes.

B_2 (riboflavin) is part of FAD, which is involved in the production of ATP by mitochondria. Deficiencies, which are rarely seen in humans, appear mainly as skin disorders. Sources include milk, cheese, liver, meat, eggs, and enriched cereals.

Niacin (nicotinic acid) is part of NAD and NADP, which help produce ATP in the mitochondria. (Nicotinic acid is not chemically related to nicotine.) The main symptom of deficiency is pellagra, which damages the nervous system, skin, and digestive tract. Sources include dried yeast, liver, meat, fish, legumes, and enriched cereals.

Pantothenic acid is part of coenzyme A in the Krebs cycle in mitochondria. It is present in so many foods that deficiency symptoms, mainly neurological, occur only in animals on experimental diets.

B_6 (pyridoxine and related molecules) are coenzymes for metabolism of amino acids and fat. Deficiencies are rare, resulting in neurological and skin disorders. Sources include dried yeast, liver, whole-grain cereals, fish, and legumes.

Folacin (folic acid) is a coenzyme for metabolism of amino acids and nucleic acids and in the maturation of red blood cells. Deficiency results in anemia. Sources include green leafy vegetables, meat, liver, and dried yeast.

B_{12} (cyanocobalamin) is required for formation of red blood cells and the metabolism of nucleic acid. Unlike most vitamins, which are synthesized by plants, vitamin B_{12} is produced only by certain bacteria. Humans obtain it mainly in meat, liver, eggs, and milk. Absorption depends on an "intrinsic factor" from the stomach. Failure of this factor results in pernicious anemia. Deficiencies can also result from extreme vegetarianism, or as the result of a fish tapeworm (see p. 536) or excessive population of *Escherichia coli* bacteria in the gut, both of which absorb large amounts of the vitamin.

Biotin is a coenzyme in fat and glycogen synthesis and in amino acid metabolism. Deficiency is rare in humans, producing skin irritation. Sources include meat, liver, and egg yolks.

C (ascorbic acid) is required for collagen synthesis and in the maintenance of the intracellular matrix of bone and cartilage cells. The name "ascorbic" refers to the ability to prevent scurvy, a degeneration of the skin and gums. Sources include citrus fruits, tomatoes, potatoes, cabbage, and green peppers.

The beard worm *Riftia pachyptilia* (phylum Pogonophora) obtains nutrients from sulfur-metabolizing bacteria that it cultivates within a special chamber. Some mammals, perhaps including humans, obtain significant amounts of vitamins B_{12} and K from bacteria in the large intestine.

FEEDING

To the extent that animals can acquire nutrients by absorption across the body surface or from endosymbionts, they are relieved from the necessity of feeding. Most animals, however, must find, ingest, and digest liquid or solid food to create an internal environment with suitable amounts of nutrients. Except for reproduction, probably no other activity requires as much of an animal's attention and ingenuity as nutrition. One can understand much of an animal's structure and behavior simply by knowing how it feeds. There are numerous modes of feeding that depend on whether the food is liquid or solid and, if solid, the size of the food. A sampling of feeding mechanisms follows.

Suspension Feeding. When food occurs as large numbers of small particles, suspension feeding is common. In suspension feeding, food particles suspended in liquid are trapped by some anatomical structure, secretion, or device constructed by the suspension-feeding animal. Each food particle can be as small as 1 micro-

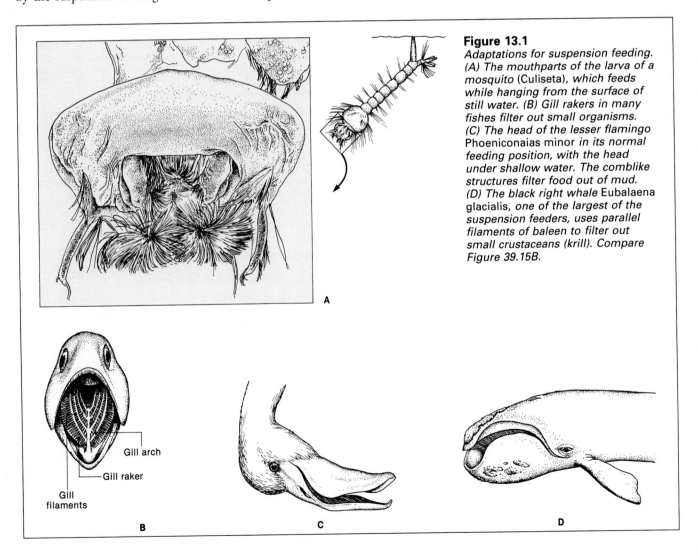

Figure 13.1
Adaptations for suspension feeding. (A) The mouthparts of the larva of a mosquito (Culiseta), which feeds while hanging from the surface of still water. (B) Gill rakers in many fishes filter out small organisms. (C) The head of the lesser flamingo Phoeniconaias minor in its normal feeding position, with the head under shallow water. The comblike structures filter food out of mud. (D) The black right whale Eubalaena glacialis, one of the largest of the suspension feeders, uses parallel filaments of baleen to filter out small crustaceans (krill). Compare Figure 39.15B.

Gill arch

Gill raker

Gill filaments

A

B

C

D

meter (the fine detritus consumed by the small crustacean *Daphnia*) or as large as several centimeters (the crustacean krill consumed by baleen whales). One type of suspension feeding is filter feeding, in which a filtering structure acts like a sieve, trapping anything above a certain size. Other suspension feeders use mucus, electric charge, and other surface properties to collect food. Obviously the ratio of food to nonfood particles must be high for suspension feeding to be efficient, so the presence of suspension feeders is often a good indication that water has not been polluted by inorganic debris. Figure 13.1 illustrates some of the varieties of suspension feeding, and others will be described in Unit Four.

Biting and Swallowing. Methods other than suspension feeding must be employed for relatively large food masses that occur in isolated lumps. There are three general ways of obtaining large foods, depending on the size of the food and whether it is a living animal. (1) If each food particle is smaller than the feeding animal and is sedentary (single cells, small plant parts, or small carrion) the feeder can ingest it in one or a few bites. Numerous examples will come to mind, such as birds eating berries. (2) If the food is larger, the animal can chew, burrow, or scrape its way through it. Carrion beetles, termites, and many endoparasites feed in this manner. (3) If the food is another animal capable of fleeing, the feeding animal(s) must first capture it and then either swallow it whole or bite off parts. A few of the adaptations for capture and ingestion are illustrated in Figure 13.2, and many others will be described in Unit Four.

Figure 13.2
Mechanisms for the capture and ingestion of food. (A) Hydra (phylum Cnidaria) has nematocysts that discharge when prey contact the trigger (cnidocil). The barb and thread poison the prey, immobilizing it for capture by the tentacles. The prey is then taken through the mouth into the gastrovascular cavity where enzymes from gland cells digest it. Nutrients are absorbed into cells by endocytosis. (B) The dentition of mammals. Incisors are used for biting, canines for tearing meat, and premolars and molars for shearing (lion) or crushing (squirrel). The numbers and sizes of each kind of tooth vary with the diet of the mammal. Carnivores have more pronounced canines than do herbivores.

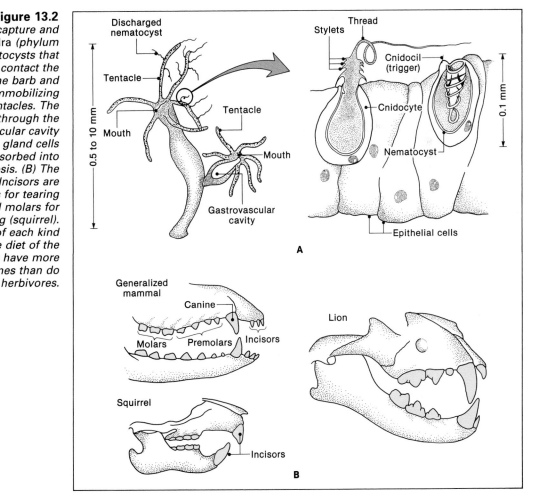

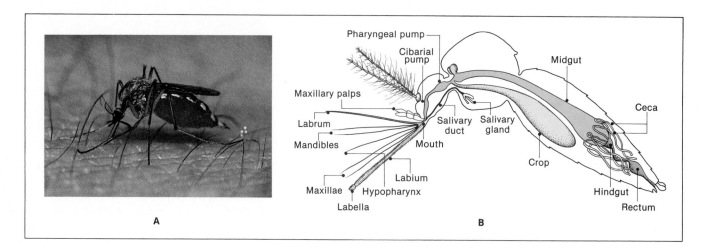

Pharyngeal pump
Cibarial pump
Midgut
Maxillary palps
Ceca
Labrum
Salivary duct
Salivary gland
Mandibles
Mouth
Crop
Labium
Maxillae
Hypopharynx
Hindgut
Labella
Rectum

A

B

Piercing and Sucking. A different method is needed for fluid food, such as the sap of plants or the body fluids of other animals. Animals that feed on fluid often have piercing and sucking mouthparts. For example, aphids can pierce the tough cuticle of plants to get to the sap, and mosquitoes can penetrate the epidermis of mammals to tap the bloodstream (Figure 13.3).

INTRACELLULAR VERSUS EXTRACELLULAR DIGESTION

A few animals can absorb small bits of food directly into cells by **endocytosis** (see pp. 42–43). Endocytosis encompasses both **phagocytosis** (cellular uptake of solids) and **pinocytosis** (uptake of liquids). Food is enclosed within a vacuole within the cell, then broken down by enzymes in the lysosome into molecules that can be metabolized (Figure 13.4A). This mechanism of intracellular digestion provides all or some of the nutrients in protozoans, sponges, cnidarians, flatworms, rotifers, bivalve molluscs, and primitive chordates.

Digestion of larger masses of food must be extracellular (Figure 13.4B). Extracellular digestion occurs not only outside cells, but anatomically outside the body, for the inside of the digestive tract (the **lumen**) is really an extension of the external environment. A starfish, in fact, does not swallow food but everts its stomach over it. Theoretically, humans could do the same if their dinner companions did not object.

DIGESTIVE SYSTEMS

Extracellular digestion requires digestive systems, which vary widely among different groups of animals. The simplest system is merely a cavity in which digestive enzymes convert food into nutrients that can be absorbed. The **gastrovascular**

Figure 13.3
(A) A female mosquito (family Culicidae) piercing human skin and sucking blood. Note the change from the suspension feeding larva (Figure 13.1A). (B) Side view of the mosquito with the mouthparts separated. During piercing of skin the sheathlike labium folds out of the way. Blood is drawn through the labrum into the midgut, where it is digested. All insects have the same mouthparts, but they are greatly modified for different kinds of foods (see Figure 32.14).

Figure 13.4
A comparison of intracellular (A) and extracellular digestion (B). In intracellular digestion digestive enzymes from lysosomes combine with food in a vacuole. In extracellular digestion the enzymes are released not only outside the cells but morphologically outside the body. The nutrients are then absorbed into the body fluids.

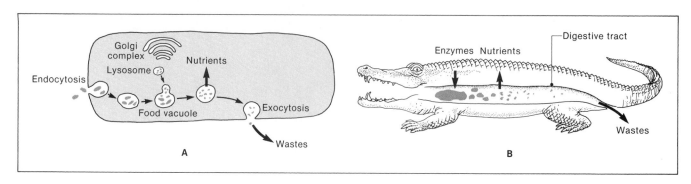

Golgi complex
Nutrients
Lysosome
Endocytosis
Food vacuole
Exocytosis
Wastes

A

Enzymes Nutrients
Digestive tract
Wastes

B

cavity of cnidarians such as *Hydra* is an example of such a simple system (Figure 13.2A). More complex systems have additional organs for the secretion of enzymes and the absorption of nutrients (Figure 13.5). The gastrovascular cavity is an example of an **incomplete digestive system.** This does not mean that it is unfinished, but that there is only one opening that serves as both mouth and anus. Most animals (see Figure 21.10 for survey) have **complete digestive systems.** The advantage of having an anus separate from the mouth probably does not have to be explained.

Ruminants. Among the most elaborate digestive systems are those of ruminants such as cattle, sheep, and deer (order Artiodactyla). These mammals have a large fore-stomach—the **rumen**—in which the initial breakdown of plant material occurs (Figure 13.5D). The partially digested material (the cud) is then regurgi-

Figure 13.5
Various types of digestive organs. (See also Figure 13.2A.) (A) Flatworms (phylum Platyhelminthes) typically have a branched gut with few specialized digestive organs. There is a mouth but no anus, so this is an incomplete digestive system. Other digestive systems in this figure are all complete, with a separate mouth and anus. (B) Insects have highly developed digestive systems. (Compare Figure 13.3B.) In the omnivorous cockroach the salivary gland secretes an enzyme (amylase) that digests starch. The crop serves primarily as a food reservoir, holding up to two months' supply. The proventriculus acts as a gizzard, breaking up solids. Digestion and absorption occur primarily in the midgut and ceca, which secrete a variety of enzymes. The rectum reabsorbs water from the feces. The Malpighian tubules are not part of the digestive tract but excrete metabolic wastes and regulate ion concentrations. (C) The digestive tract of the pigeon Columba livia. *The crop stores food and, in this particular species, secretes a milklike fluid for the young. (D)* Bos, *a device for converting grass into milk. The digestive system, especially the four-chambered stomach, occupies most of the abdominal cavity. Grass is fermented by microbes in the rumen, then regurgitated for a second chewing before being swallowed again. Following this rumination the grass enters the reticulum.*

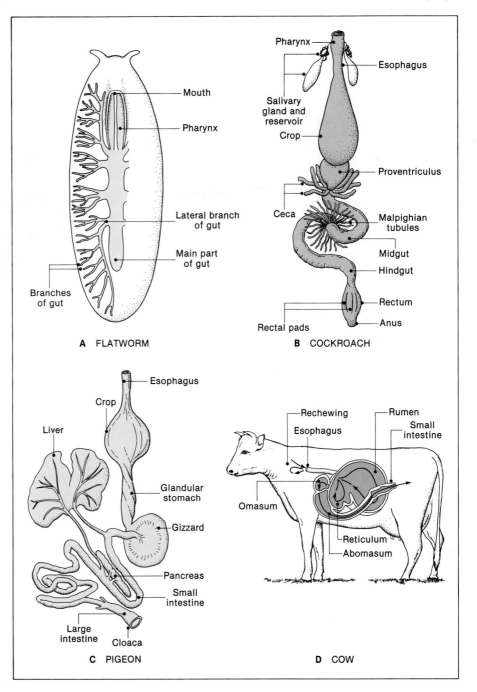

A FLATWORM

B COCKROACH

C PIGEON

D COW

tated for chewing, reswallowing, and further digestion in other parts of the digestive tract. This process of regurgitation and chewing is called **rumination.** (Regurgitation by camels and llamas is also called rumination, but these animals are not considered to be ruminants, because they do not have rumens.) The rumen serves as a huge culture flask in which bacteria and protozoa partially digest cellulose and starch. The microorganisms obtain their energy by fermenting the sugars released, but the anaerobic environment of the digestive tract prevents them from completely oxidizing the sugars to CO_2 and H_2O. The fermentation products are therefore available to the host animal as nutrients.

An arrangement similar to that of ruminants occurs in a South American bird called the hoatzin (pronounced WAT-sin; *Opisthocomus hoazin;* Grajal et al. 1989). Many rodents and rabbits also have a large **cecum** (SEE-come) at the junction of the small and large intestines that houses microbial fermentation. These animals recover the released nutrients by eating their own feces. The guts of termites and some cockroaches contain bacteria and protozoa that digest wood, a food that is unavailable to virtually all other animals.

ABSORPTION OF NUTRIENTS

The function of digestion is to break food down into nutrient molecules that can be absorbed into blood and interstitial fluid. Absorption is therefore certainly as important as feeding and digestion. The mechanisms by which nutrients are absorbed across the cells lining the digestive tract are similar to those by which other molecules enter cells—namely, diffusion and active transport (see pp. 39–41). **Diffusion** is always from high to low concentration, usually from the lumen into the blood. The nutrients absorbed by diffusion are small ions that can enter through channels in cell membranes, and hydrophobic molecules (lipids), which can pass through the phospholipid bilayer. Therefore, the following nutrients are probably the ones absorbed by diffusion:

Potassium and chloride ions

Triglycerides (after digestion to monoglycerides, diglycerides, and fatty acids)

Steroids (cholesterol and steroid hormones)

Thyroxine and triiodothyronine

Lipid-soluble vitamins A, D, E, and K

Other lipid-soluble molecules, such as alcohol and certain toxins

All other nutrients are probably absorbed by some form of **active transport.** Sodium and calcium are probably absorbed into the bloodstream by pumps in the membranes of cells in the intestinal lining. There are also pumps for glucose, amino acids, and some vitamins.

DIGESTION IN HUMANS

Chewing. Humans are omnivores. That does not mean that we literally eat everything, but it does mean that our digestive systems are capable of handling a wide variety of foods from both plants and animals. Our 32 **teeth,** for example, have a variety of shapes adapted for different foods (Figure 13.6). As you can verify with your tongue, each half of each jaw has two chisel-shaped **incisors** in front that are adapted for biting, one **canine** adapted for tearing meat, and two **premolars** and three **molars** adapted for grinding vegetable material and shearing

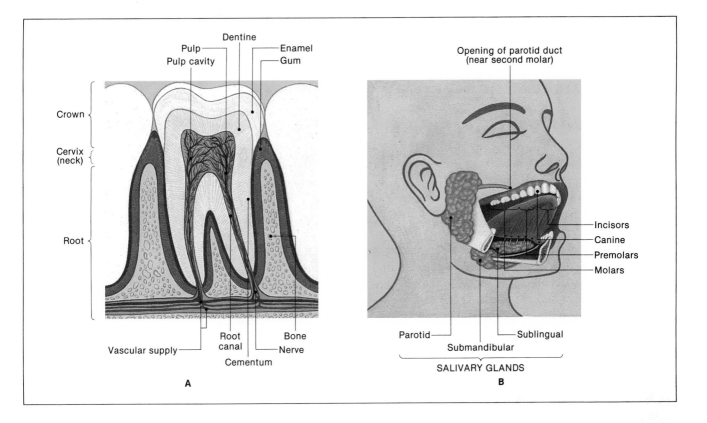

Figure 13.6
Digestive structures in the human mouth. (A) A longitudinal section of a molar. (B) Salivary glands and teeth.

meat. (Chances are you will have the hindmost of the molars, the "wisdom teeth," removed.) As food changes consistency during chewing, the tongue manipulates it to the appropriate teeth.

The tongue and teeth also mix the food with **saliva** from the salivary glands. Normally, human salivary glands secrete about a liter per day, but most of it is water that gets recycled by swallowing. Saliva also contains mucus, which lubricates the chewed food, and an **amylase** enzyme that helps digest starch into the disaccharide **maltose.**

Swallowing. After food has been chewed to a certain consistency, it forms into a **bolus** that the tongue pushes to the back of the mouth. Pressure of the bolus against the soft palate closes off the nasal passages and pushes the epiglottis down to block the glottis (the opening of the trachea). The bolus also stimulates touch receptors at the opening of the pharynx, at the back of the mouth. These receptors activate a swallowing center in the brainstem, which triggers contractions of numerous skeletal muscles in the pharynx. These muscles force the bolus into the tubular **esophagus** that connects to the stomach. Smooth muscles in the esophagus produce waves of contractions, called **peristalsis,** which propel the bolus to the stomach.

The Stomach. The stomach is a storage area that eliminates the need to eat continuously. It is separated from the esophagus and the small intestine by the **cardiac and pyloric sphincters.** In mammals the stomach secretes **hydrochloric acid** (HCl), which helps destroy contaminants, and **pepsin,** an enzyme that helps digest protein. The reduction in pH by the HCl inactivates the salivary enzymes but activates the pepsin. Contractions of the smooth muscle in the stomach churn the food and secretions into a watery, acidic mixture called **chyme** (pronounced kyme).

You might wonder how cells that secrete HCl and protein-digesting enzymes avoid digesting themselves. For HCl the answer is that the acid is secreted in two different processes. **Oxyntic cells** (= parietal cells) secrete H^+ and, by a separate process, secrete Cl^-, so the strong acid HCl is never in the cell (Figure 13.7). The pepsin is secreted in the inactive form called **pepsinogen.** Pepsinogen gets activated in the stomach when HCl and other pepsin molecules remove a portion of the protein that blocks the active site on the pepsinogen (Figure 13.7). This explains how the secretory cells keep from destroying themselves, but what keeps the acid and protease from digesting the stomach? Normally, the stomach lining, called the **mucosa,** is protected by a layer of mucus, as is the entire lumen of the digestive tract. Even with the protection of mucus, however, the acid and digestive enzymes destroy some ten billion mucosal cells a day.

The Small Intestine and Associated Digestive Organs. The stomach is important, but people can live without one, because most digestion occurs in the small intestine. When the chyme enters the first section of small intestine (the **duodenum**) it mixes with enzymes and other secretions from several sources. The main source of digestive enzymes is the **pancreas,** which is located just below the stomach in humans. Pancreatic enzymes enter the small intestine through a duct. (They are therefore **exocrine** secretions of the pancreas, not to be confused with the *endo*crine secretions, insulin and glucagon, which go into the bloodstream.) The pancreatic enzymes include lipase, several proteases, and an amylase similar to salivary amylase. As with pepsin, the pancreatic proteases are secreted in inactive form, so they do not digest the exocrine cells that secrete them. The pancreatic enzymes cannot function in an acidic environment, but the pancreas also releases bicarbonate (HCO_3^-), which neutralizes the acidity of the chyme. **Goblet cells** in the mucosa secrete mucus, which protects the intestinal lining

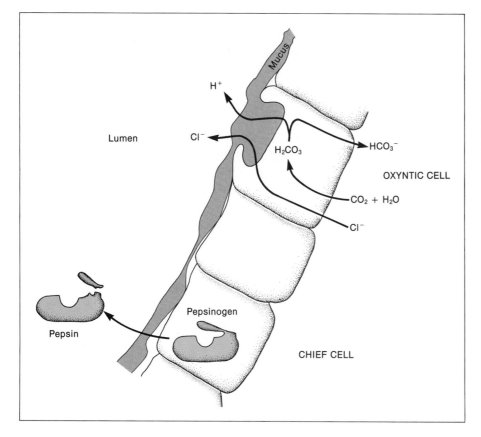

Figure 13.7
Hydrogen ion comes from the weak carbonic acid in oxyntic (= parietal) cells and is actively transported into the stomach by proton (H^+) pumps. Pepsin is secreted as inactive pepsinogen by the chief cells, and the active site of the enzyme is not uncovered until the pepsinogen is converted to pepsin. A layer of mucus protects the stomach lining from the acid and enzyme.

William Beaumont

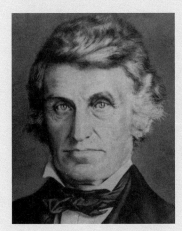

William Beaumont
Born: 21 November 1785 in Lebanon, Connecticut
Died: 25 April 1853 in St. Louis, Missouri
Education: Studied medicine by reading and apprenticeship Army surgeon; Pioneer in the study of stomach physiology

Considering its importance, it is surprising how recent our understanding of the stomach is. Long after the heart was known to be a pump, the human stomach was described by some authorities as a mill, a fermenting vat, or a stew pan. William Beaumont provided the first indication that it was none of these. Beaumont got his medical training the old fashioned way, by reading and apprenticing to a physician. He was licensed in 1812 and immediately joined the Army in Plattsburgh, New York, for an ill-fated invasion of Canada at the start of the War of 1812. Beaumont was also present at the Battle of Plattsburgh, which turned out much better for the United States. According to local accounts, Beaumont's interest in digestive functions was apparent even then. It is said that he had the intestines of dead soldiers strung about his office. After the war Beaumont was released from the Army and settled in Plattsburgh, where he practiced medicine and sold pharmaceuticals, groceries, tobacco, and liquor.

In 1819 Beaumont accepted a commission in the Army and was posted at Fort Mackinac (MAC-kin-aw) in the remote Upper Peninsula of Michigan. At that time this was the northwest frontier of the United States, and Beaumont was the only physician within 300 miles. Beaumont's name might have passed into oblivion had it not been for an accident in 1822. A 19-year-old French-Canadian fur trader, Alexis St. Martin, was accidentally hit in the left side by a shotgun blast at close range. Beaumont was summoned and quickly pronounced the case hopeless. Fortunately, Beaumont was mistaken, for over the next several years St. Martin slowly recovered. St. Martin's survival was largely due to Beaumont, who took

him into his home and personally cared for him after the Army and his employers cut off financial support.

The healing was not complete, however. A hole remained in St. Martin's side and stomach, providing Beaumont with a view of the inner workings of the stomach. In 1825 Beaumont began to take advantage of the situation to study the role of the stomach in digestion. At first the grateful St. Martin cooperated, but a few months later he ran away to Canada, having recovered sufficiently to canoe his family all the way back to Montréal from Prairie du Chien, Wisconsin! Four years later Beaumont found St. Martin and convinced him to resume experimentation. Further attempts to run away were thwarted by having St. Martin enlist in the Army. It is said, however, that when Beaumont looked in on St. Martin's stomach he often found it inflamed with alcohol. Supposedly, St. Martin took the most direct means to achieve that end, emptying the bottle directly into the opening in his side. Experiments continued until 1833, when Beaumont published his findings. St. Martin then managed to return to Canada and refused all further efforts of Beaumont and other scientists to continue the experiments. After St. Martin died in 1883 at age 80, his widow, afraid of grave-robbing scientists, allowed the body to decay before burying it at twice the normal depth in an unmarked grave.

Beaumont's work quickly attracted the attention of physiologists. European researchers, especially, were taken with the achievements of the "Frontier Doctor." While many of Beaumont's findings had already been suggested from studies of regurgitated gastric juice and partly digested food, Beaumont confirmed the following points by direct observation:

1. Gastric digestion results from a secreted fluid.
2. This secretion occurs at numerous points on the stomach lining.
3. Eating stimulates gastric secretion, and emotion influences secretion.
4. The fluid is acid as soon as it is secreted. (It had been thought that the acidity was due to decomposition.)
5. The stomach empties rather quickly.
6. Foods differ in their digestibility, and alcohol and other substances affect digestion.
7. Gastric digestion is due not only to the action of acid but to some other substance. (Enzymes were unknown at the time.)

from digestion. (Figure 13.8 shows the arrangement of the major digestive organs, and Table 13.4 summarizes the digestive secretions.)

Bile. Lipids are usually the most difficult nutrients to digest, because they form large globules rather than dissolving in the watery chyme. The digestive system solves this problem the same way dishwashing detergents do: by **emulsifying** the large fat globules into smaller ones. Small droplets present a greater surface area

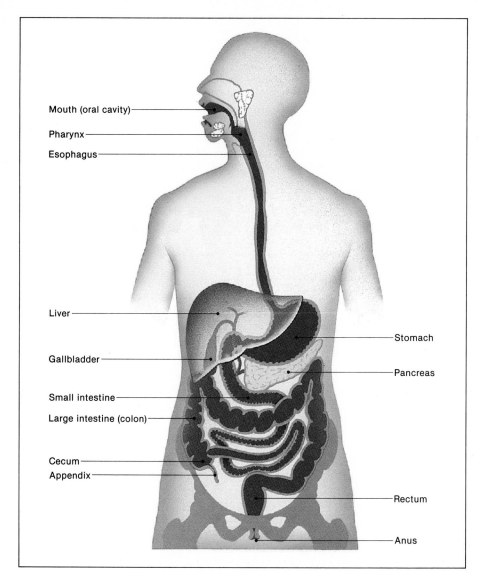

Figure 13.8
The human digestive tract consists of the mouth, pharynx, esophagus, stomach, small intestine, large intestine, and rectum. The part below the esophagus is often called the gastrointestinal (GI) tract. Food does not enter the liver, gallbladder, or pancreas. Rather, these organs secrete bile salts, enzymes, and bicarbonate into the small intestine, where they aid digestion.

Mouth (oral cavity)
Pharynx
Esophagus
Liver
Gallbladder
Small intestine
Large intestine (colon)
Cecum
Appendix
Stomach
Pancreas
Rectum
Anus

for lipases to work on. Some emulsification results from the movements of the stomach, and further emulsification results from substances in bile: a fluid produced by the liver and stored in the gallbladder. Bile contains **bile salts** and phospholipids, which can mix with both lipids and water. This property allows them to break up the large fat droplets. Bile also contains substances not directly related to digestion, including cholesterol and bilirubin. Bilirubin is a breakdown product of hemoglobin and gives feces their characteristic color. (Note that "feces" is always plural.) One symptom of a disorder of the liver or gallbladder, such as jaundice or gallstones, is gray feces and yellowish skin, due to the retention of bilirubin.

The Large Intestine. By the time material has passed through the three or so meters of the human small intestine, virtually all the absorbable nutrients have been removed, leaving a brown watery fluid containing indigestible material such as cellulose. This material, called **chyle** (kyle), enters the large intestine, where it first passes a pouch called the **cecum.** In humans the cecum is small and ends with an **appendix,** which is of no known value except to surgeons. The rest of the large intestine reabsorbs water into the body, contains bacteria, and stores feces until it is convenient to eliminate them. The reabsorption of water is due

Table 13.4 **Enzymes and other digestive secretions in humans (and most other vertebrates).**

MOUTH

Salivary amylase breaks down starch to the disaccharide maltose by hydrolyzing every other bond between the glucose subunits.

STOMACH

Pepsin, released as inactive pepsinogen by chief cells, hydrolyzes particular peptide bonds, breaking up proteins into peptide fragments.
HCl, released by **oxyntic** (parietal) cells, serves mainly to kill bacteria and activate pepsin.

LIVER

Bile salts and **phospholipids** in bile emulsify large fat droplets into small ones, which present more surface area for lipases to work on in the small intestine. Bile is stored in the gallbladder.

PANCREAS

Pancreatic amylase breaks down starch, like salivary amylase.

Other **carbohydrases** digest different carbohydrates. For example, **sucrase** (= invertase) digests sucrose, and **lactase** digests lactose (milk sugar). (Lactase occurs in infants, but not always in adults. Therefore many adults cannot digest most dairy products.)

Several proteases, the most important of which are **trypsin, chymotrypsin,** and **carboxypeptidase,** hydrolyze different peptide bonds. Carboxypeptidases hydrolyze the last peptide bond at the carboxyl end of a peptide. Like pepsin, these proteases are released in inactive form (trypsinogen, chymotrypsinogen, procarboxypeptidase).

Pancreatic lipase hydrolyzes the ester bonds linking fatty acids in triglycerides, thereby converting triglycerides to diglycerides, monoglycerides, and fatty acids.

Bicarbonate (HCO_3^-) neutralizes acid from the stomach.

---

to sodium pumps, which set up an osmotic gradient that draws water out of the feces and into the bloodstream.

The bacteria in the large intestine constitute a complex and poorly understood community called the "intestinal flora." These bacteria ferment cellulose into gases, produce vitamins B_{12} and K, and contribute from a fourth to a half of the bulk of feces. *Escherichia coli,* the research bacterium familiarly known as *E. coli,* is the best-known member of the intestinal flora, although it represents less than 1% of the cells in this community. Feces are eliminated through the anus by the process of defecation, which combines voluntary and reflex contractions of skeletal and smooth muscles.

Control of Digestion.   Just thinking about food can stimulate digestive activities such as stomach movements (often audibly and at the most solemn moments). The autonomic nervous system plays a major role in regulating digestive movements and secretions (see pp. 171–172). Generally, the **parasympathetic branch** stimulates digestive activities (salivation and movements and secretions of the stomach and intestines), and the **sympathetic branch** inhibits them. Since the sympathetic branch is active during stress, a dry mouth and indigestion are symptoms of stress. In addition, there are networks of nerve cells within the digestive tract that coordinate its movements. These **myenteric** and **submucosal plexuses** have been referred to as a "little brain," because they seem to function independently of the central nervous system.

Hormones also help coordinate digestive functions. **Gastrin** stimulates stomach motility and the secretion of pepsin and HCl by chief and oxyntic cells. Gastrin is secreted from the lining of the stomach, but like all hormones it goes into the bloodstream to reach its target cells. After circulating back to the stomach lining, it triggers the secretion of enzyme and acid into the stomach lumen. The stimulus for gastrin secretion is the presence of peptides in the stomach.

The lining of the first portion of the small intestine secretes two families of hormones that help coordinate digestion. One family includes **cholecystokinin** (CCK), which is also called **pancreozymin** (PZ). The two names of CCK–PZ indicate their two roles in coordinating digestion: (1) stimulating the gallbladder to contract and release bile and (2) stimulating the pancreas to release digestive enzymes. Chole-cysto-kinin means gall-bladder-mover, and pancreo-zymin alludes to the pancreatic enzymes. The second family of hormones includes **secretin,** which stimulates the secretion of bicarbonate by the pancreas and the synthesis of bile by the liver. The stimulus for secretion of CCK–PZ and secretin is the presence of chyme in the small intestine. Thus these two families of hormones are involved in negative feedback loops. They trigger the release of substances to digest the material that was the original stimulus for hormone secretion. At the same time secretin and other intestinal hormones inhibit stomach motility and secretion, giving the small intestine a chance to catch up with the digestion of the chyme already released.

Absorption of Nutrients in Humans.   Nutrient molecules released from food by digestion are absorbed across the mucosal lining of the small intestine. Both diffusion and active transport are involved, as described previously, and both rely on the large surface area of the mucosa. The large surface area is due to extensive folding and to **villi,** which are fingerlike projections that give the mucosa a velvety texture. Individual cells on each villus have further submicroscopic projections called **microvilli,** which comprise a **brush border** that provides even more surface area (Figure 13.9). If all the villi and microvilli were flattened out, the intestinal mucosa could serve as a tennis court, though a rather slippery one.

The large surface area allows virtually complete diffusion of the fatty acids and mono- and diglycerides that result from the digestion of triglycerides. In addition, steroids, vitamins A, D, E, and K, thyroxine and triiodothyronine, alcohol, and other lipid-soluble molecules easily diffuse across the mucosa. These substances do not enter the bloodstream directly, but first go into **lacteals** within the villi. The lacteals, named for the milky appearance of the fat-laden lymph within them, are projections of the lymphatic system (see Figure 11.12). The nutrients in the lacteals circulate in the lymph until they enter the bloodstream at a vein beneath the left collar bone.

Simple sugars, amino acids, water-soluble vitamins, most ions, and most other nutrients are absorbed by active transport. Pumps or carriers in the plasma membranes transport the nutrients across the outer cells of the villi. The nutrients then apparently diffuse into capillaries within the villi. The blood from the small intestine flows first through a portal vein to the liver. The liver then "decides" whether to store the glucose as glycogen, whether to metabolize the amino acids for energy or to use them for protein synthesis, and so on.

REGULATION OF BLOOD-GLUCOSE CONCENTRATION

The levels of a nutrient in the blood and interstitial fluid naturally tend to vary with the amount of the nutrient absorbed with the diet. The resulting fluctuations could raise havoc among many cells. This is especially true in the brain, because the activity of nerve cells varies in direct proportion to levels of glucose in the cerebrospinal fluid. Normally such fluctuations are limited by the hormones insulin and glucagon, which are secreted into the bloodstream by endocrine cells located in the islets of Langerhans of the pancreas. Excessive levels of blood glu-

Figure 13.9
The structure of the mucosal lining of the small intestine.

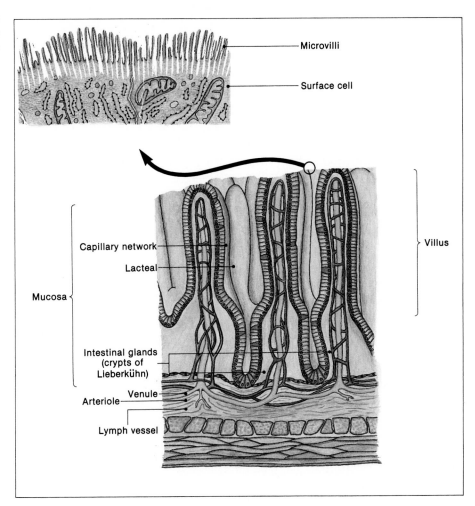

Microvilli

Surface cell

Villus

Capillary network

Lacteal

Mucosa

Intestinal glands
(crypts of
Lieberkühn)

Venule

Arteriole

Lymph vessel

cose (hyperglycemia) trigger the secretion of **insulin** by the β (beta) cells of the pancreatic islets. Insulin then circulates through the blood and stimulates cells, especially in muscle and adipose tissue, to take up more glucose. Muscle and liver cells then convert the glucose into the storage form glycogen. This negative-feedback mechanism prevents glucose levels from getting too high immediately following the ingestion of carbohydrates. When glucose levels fall below normal (hypoglycemia), the α (alpha) cells of the pancreatic islets secrete **glucagon.** Glucagon stimulates liver cells to convert their glycogen reserves into glucose, thereby restoring blood-glucose concentrations.

Diabetes mellitus is a common disorder that illustrates the importance of this homeostatic regulation of blood-glucose levels. One type of diabetes, **Type I,** or **insulin-dependent diabetes,** results from the inability of the pancreas to secrete insulin. The cause of Type I diabetes is destruction of the β cells by a person's own immune system. The more common form of diabetes mellitus, afflicting approximately 5% of American adults, is called **Type II,** or **insulin-resistant diabetes.** It develops gradually, primarily in overweight people, especially those with "pear-shaped" obesity, rather than with body fat distributed higher on the torso. The cause of Type II diabetes is the inability of receptor molecules on cell membranes to respond to insulin. Cells therefore do not take up the glucose. The excess glucose in the blood spills over into the urine, osmotically bringing water with it and increasing urine volume.

METABOLIC RATES

Trying to follow the consumption of all the various nutrients throughout an animal would be so difficult that few scientists would attempt it. It is relatively easy, however, to measure an animal's overall rate of energy exchange, defined as the metabolic rate (MR). There are several ways of measuring MR, the most straightforward being **direct calorimetry.** Since all the energy released by the metabolism of nutrients is ultimately converted into heat (either during metabolism or while using the ATP it produces), one can determine the rate of energy conversion by measuring the heat produced by an animal over a period of time. In order to measure MR one could therefore submerge an animal in a known volume of water and measure the rise in water temperature. This procedure is obviously inconvenient for terrestrial animals. A more convenient method is to measure the oxygen consumption. This method, called **indirect calorimetry,** works because a given nutrient requires a certain amount of O_2 for complete oxidation. MR is therefore usually measured as the volume of O_2 consumed in a given time, and this volume can be converted to kilocalories. For a typical human adult the metabolic rate is about 2000 kcal (Calories) per day. That is, a human produces enough heat in a day to heat 2000 liters of water by 1°C, or 1 liter of water by 2000°C! Exercise, growth, increased thyroid activity, and exposure to cold sharply raise the MR above this level.

The First Law of Thermodynamics states that energy is neither created nor destroyed, so all the energy released as heat, stored in the body, and lost as waste must have come from food. That is,

$$\text{caloric intake} = \text{MR} \times \text{time} + \text{stored energy} + \text{waste}$$

Usually the energy lost in waste is negligible. If the caloric intake is less than the amount expended (MR × time), then the stored energy must be negative. That is, if an animal uses more metabolic energy than it takes in with food, it will lose weight. Calories absorbed in excess of the amount used will be stored, increasing body weight. If caloric intake equals caloric expenditure, weight will be stable. Therefore it takes at least 2000 kcal/day to maintain constant body weight in an average adult. About five slices of apple pie provide that much energy (though not all the essential nutrients). Increased activity will increase caloric requirements. Walking briskly for 2 hours, for example, will require another piece of pie.

Body size also affects metabolic rate. This is not surprising, since a large animal has more metabolizing tissue than a smaller one. It is surprising, however, that the metabolic rate does not increase in direct proportion to body weight. An animal that weighs twice as much as another usually has a metabolic rate that is less than twice as high. This means that an average gram of mouse tissue has a higher metabolic rate than an average gram of elephant tissue (Figure 13.10). No one quite understands the reason for this, but it implies that large animals are more efficient in their metabolism. A 2.5-gram shrew must, in fact, eat almost constantly while awake, while most large animals have time for other things.

REGULATION OF FOOD INTAKE

Most animals maintain constant weights throughout their adult lives, as long as sufficient food is available. They must therefore have mechanisms that regulate food intake. In house flies a simple stretch receptor in the abdomen indicates when enough food has been ingested (see p. 408). The mechanisms in most other mammals, including humans, are more complicated and poorly understood in spite of considerable research. The **hypothalamus** appears to be involved in the regulation of caloric intake, but how is still controversial. Electrical stimulation of the **lateral**

Figure 13.10
The "mouse to elephant" graph of metabolic rate per gram of body mass versus the body mass. Both axes are plotted as powers of 10 (logarithmic). The relationship is linear with a slope of $-\frac{1}{4}$, which indicates that the metabolic rate (MR) divided by body mass is proportional to the body mass to the $-\frac{1}{4}$ power: $MR = aW^{-1/4}$, where a is a constant. Thus the larger the animal the lower the metabolic rate in an average gram of tissue in a resting mammal. The MR of an average gram of tissue in a mouse is approximately 115 calories per gram per day, while that in an elephant is only about 15 calories per gram per day. Similar graphs occur for nonmammals, although the exponent is not always -0.25. It ranges from -0.30 for freshwater fishes to -0.14 for turtles.

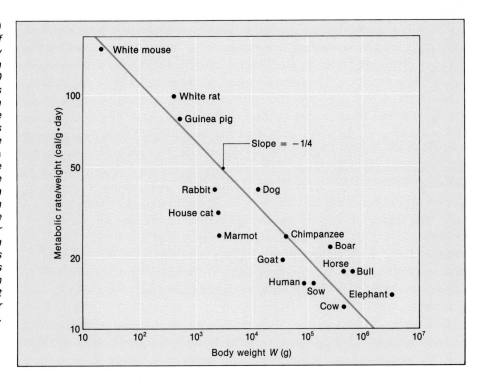

hypothalamus in some animals causes them to increase their food intake, and destruction of lateral areas of the hypothalamus makes them reduce food intake. Consequently, the lateral hypothalamus is often considered to contain a "feeding center." Some physiologists point out, however, that stimulating and destroying these areas affect the overall activity of the animals, and eating may simply be one of the outlets for that activity. Electrical stimulation of the **ventromedial hypothalamus** reduces food intake, and destroying it has the opposite effect, so this area is often considered a "satiety center." The concept of the satiety center is now also in doubt, because the stimulation and destruction can easily affect nearby nerve tracts.

Whether or not such neural centers exist, it is likely that hunger and satiety are evoked by some kind of signal associated with caloric intake. Candidates for a satiety signal include stretch receptors in the stomach, increased blood-glucose levels, by-products of lipid synthesis, and cholecystokinin.

1. Stretch receptors do signal when the stomach is full, but most animals stop eating long before these receptors are activated.
2. Glucose is a logical choice as a "satiety signal," and glucose-responding nerve cells do occur in the hypothalamus. High levels of glucose could stimulate satiety, and low glucose levels could stimulate appetite. However, glucose levels fluctuate too much to keep body weight constant year after year.
3. A more logical choice for long-term regulation would be some by-product of lipid metabolism. The synthesis of body fat could increase the concentration of some metabolic intermediate, which would then produce the sensation of satiety. If body fat were being broken down a different metabolite could stimulate feeding. There is little experimental evidence supporting this mechanism.
4. Several studies indicate that cholecystokinin (CCK) secreted by the small intestine during digestion suppresses appetite. CCK released by the gut does not enter the brain, however, but may activate neural pathways locally. CCK is also a neurotransmitter in the brain, where it may serve as a satiety signal.

There is ample evidence all around us that in societies where food is plentiful and appetizing the mechanisms that regulate food intake either do not function properly or are overridden by other factors affecting the balance between caloric intake and expenditure. The consequence is usually excessive body weight, which is called obesity if it is greater than 10% to 20% above normal for body height. Obesity can contribute to heart failure, diabetes, and other health-threatening disorders. The First Law of Thermodynamics implies that there is only one cause of excess body weight: taking in more calories than are expended. One cause of excess body weight may therefore be inadequate expenditure of calories. In many mammals excess calories are consumed by **brown adipose tissue** (= brown fat). Brown adipose tissue accounts for up to 6% of body weight in infants and mammals well adapted to cold. In the average human adult brown adipose tissue is only about 1% of the body weight, and it is concentrated in the neck, upper back, chest, and around the kidneys. Brown fat has a large number of mitochondria that do not produce much ATP, but break down fats into heat, thereby burning off excess calories. Many humans may be obese because this brown adipose tissue is inactive.

The other possible cause of excessive body weight is excessive caloric intake. Overeating may result from a variety of behavioral or physiological causes. Perhaps it is due to the simple fact that food tastes better now than it has throughout almost all of our evolution. Most animals, and most humans for that matter, have a bland, monotonous diet lacking the great variety of sugars and other flavorings we in the luckier parts of the world enjoy. For the less fortunate, avoiding overeating may simply be a matter of getting tired of the food. If rats are allowed to eat tasty human foods, they also overeat by 70% to 80% and quickly become obese (Le Magnen 1983). Conversely, humans with nothing but one flavor of pizza to eat three times a day would probably never become obese.

Pharmaceutical firms are naturally interested in understanding and controlling the mechanisms that regulate caloric intake. Considering the billions of dollars spent annually on diets and weight-control pills that don't work, one can only imagine how much money could be made from a pill that does work. The only thing better than a pill to curb the appetite would be one that would allow you to eat all you wanted without gaining weight. Hundreds of such pills have, in fact, been claimed already. For example, a nationally circulated weekly television guide carried an ad for pills that would "burn fat" and let you "melt away pounds without diets or exercise." The pills were, in fact, nothing but benzocaine to anesthetize the taste buds before eating. They came with a little booklet of diets and exercises.

SUMMARY

Nutrients are either inorganic ions and metals or organic molecules. Among the latter are carbohydrates, the chief sources of energy, fats, the long-term stores of energy, proteins, and vitamins. Some animals obtain nutrients by absorbing them in dissolved form, but most get them by feeding. Major modes of feeding include suspension feeding, biting and swallowing, and piercing and sucking.

Once food is ingested it must be digested to break it into molecules small enough to be absorbed into the body fluid. Some animals rely mainly on intracellular digestion, but most use extracellular digestion, in which digestive enzymes are secreted into the lumen of a digestive tract. Lipid-soluble nutrients such as fats and certain vitamins then diffuse across

the lining of the digestive tract into the body fluids. Most other nutrients, such as sugars and amino acids, must be actively transported into body fluids.

In humans the digestive enzymes include salivary amylase, which digests starch, pepsin, which partially digests proteins in the stomach, and various carbohydrases, proteases, and lipases that are secreted by the pancreas into the small intestine. Other important digestive secretions include hydrochloric acid from the stomach, bicarbonate from the pancreas, and bile from the liver. Nutrients are absorbed mainly across the lining of the small intestine, which has a large surface area by virtue of its folds, villi, and microvilli. Lipid-soluble molecules diffuse into lacteals in the villi, and carbohydrates and amino acids are pumped into capillaries in the villi.

Digestion is controlled by the autonomic nervous system and by hormones such as gastrin, cholecystokinin, and secretin. The concentration of glucose in the blood is regulated by the hormones insulin and glucagon. The rate at which nutrients are used—the metabolic rate—depends on the overall level of activity of the animal and also on body size. The rate of food intake is thought to be controlled by areas of the brain associated with the hypothalamus. Satiety is initiated by signals, including perhaps receptors in the stomach, glucose levels, a by-product of lipid metabolism, or cholecystokinin.

KEY TERMS

vitamin	chyme	cholecystokinin
suspension feeding	pancreas	secretin
intracellular digestion	bile	villi
extracellular digestion	trypsin	microvilli
ruminant	chymotrypsin	lacteal
amylase	carboxypeptidase	α and β cells
stomach	lipase	diabetes mellitus
pepsin	gastrin	calorimetry

SELF-TEST

1. List as many categories of nutrients as you can (e.g., carbohydrates), and state the general function of each.

2. Describe the feeding methods that are appropriate for the following types of food: food particles in large numbers but much smaller than the feeding animal; large food masses; fluid food.

3. Digestion is said to occur outside the body in mammals. Explain what that means.

4. Describe the overall process of extracellular digestion.

5. Which kinds of nutrients can be absorbed by diffusion, and which ones must be absorbed by active transport?

6. Some animals feed on plants that contain chemicals that are potentially toxic, carnivores ingest their prey's hormones, and humans ingest laxatives and other drugs that would be harmful if they entered the bloodstream. Many such molecules are safe to eat, however, because they are inactivated during digestion. What kinds of molecules would be inactivated by digestion? What kinds of molecules would not be destroyed during digestion? Of these, what kinds could get into the bloodstream?

7. Summarize the digestive functions of the following: mouth, stomach, small intestine, pancreas, liver, large intestine.

8. Explain the differences between digestive enzymes and digestive hormones. Give an example of each, and explain how it functions.

9. Diagram a villus from memory and indicate the routes followed by fatty acids and by amino acids during absorption.

10. Explain why glucose concentrations in the blood remain fairly constant even though the ingestion of glucose is periodic.

11. Define metabolic rate, and describe two general methods for measuring it.

12. Explain what is meant by a satiety signal. What are some of the possible candidates for satiety signals? How could a knowledge of satiety signals be exploited for weight control?

READINGS

RECOMMENDED READINGS

Dethier, V. G. 1962. *To Know a Fly.* San Francisco: Holden-Day. (*An entertaining account of Dethier's research on feeding in flies.*)

Merritt, R. W. and J. B. Wallace. 1981. Filter-feeding insects. *Sci. Am.* 244(4):132–144 (Apr).

Mertz, W. 1981. The essential trace elements. *Science* 213:1332–1338.

Moog, F. 1981. The lining of the small intestine. *Sci. Am.* 245(5):154–176 (Nov).

Owen, J. 1980. *Feeding Strategy.* Chicago: University of Chicago Press.

Sanderson, S. L. and R. Wassersug. 1990. Suspension-feeding vertebrates. *Sci. Am.* 262(3):96–101 (Mar).

ADDITIONAL REFERENCES

Cohen, I. B. (Ed.). 1980. *The Career of William Beaumont and the Reception of His Discovery.* New York: Arno Press. (*Includes two of Beaumont's early notebooks, an account of his work, and an evaluation of his influence.*)

Grajal, A. et al. 1989. Foregut fermentation in the hoatzin, a neotropical leaf-eating bird. *Science* 245:1236–1238.

Le Magnen, J. 1983. Body energy balance and food intake: a neuroendocrine regulatory mechanism. *Physiol. Rev.* 63:314–386.

Myer, J. S. 1939. *Life and Letters of Dr. William Beaumont.* St. Louis: Mosby.

14

Water and Ions

White-tailed deer (Odocoileus virginianus).

LEARNING OBJECTIVES

1. Why does all life on Earth depend on water?

2. What happens when animals become dehydrated?

3. Why are body fluids salty?

4. How do terrestrial and freshwater animals maintain ionic concentrations in their body fluids?

5. What kinds of organs are used in regulating the composition of body fluids?

6. How does the kidney work?

7. How are the functions of the kidney regulated?

8. How do desert and marine vertebrates conserve water and eliminate excess salt?

OSMOLARITY   TONICITY

Hyperosmotic — Hypertonic

Iso-osmotic — No shrinking or swelling — Isotonic

Increasing concentration

Hypo-osmotic — Hypotonic

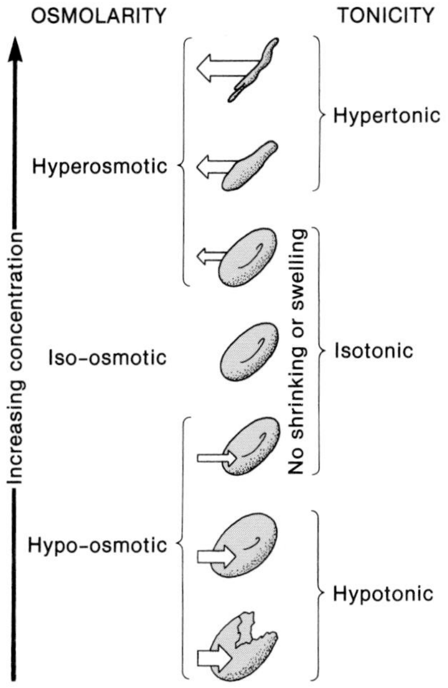

Figure 14.1
The effects of varying concentrations on a red blood cell. The direction and thickness of the arrows indicate the net direction and rate of water movement. The left side of the figure represents the effects of concentration on osmolarity. A red blood cell (or any other cell or animal) in an iso-osmotic fluid neither gains nor loses water due to osmotic pressure. Increasing the concentration makes the solution hyperosmotic, and the cell tends to lose water. Decreasing the concentration below the iso-osmotic level makes the solution hypo-osmotic, and the cell tends to gain water. Tonicity refers to the effect of a concentration on shrinking or swelling. Whatever its concentration, a solution is isotonic to a particular cell or animal if there is no shrinking or swelling. If the solution makes the cell or animal shrink, it is hypertonic. If it makes the cell or animal swell, it is hypotonic.

For virtually all of their history cells have evolved in an environment of seawater. Not surprisingly, therefore, the cells of animals still require an environment similar to seawater, even when the animals themselves live in fresh water or on land. The maintenance of an essentially marine internal environment is not difficult for most animals that live in the sea, but for animals that live in fresh water or on land the task presents problems. First, the tendency of water to move osmotically from low to high solute concentrations threatens to flood the tissues of freshwater animals and to dehydrate the tissues of terrestrial animals. Second, the tendency of ions to diffuse out of or into the body threatens the ionic balance across cell membranes, which is required for their proper functioning. This chapter examines how homeostatic systems cope with these threats.

THE IMPORTANCE OF WATER AND IONS

Water has many unique properties that make it not only useful but necessary for life (see pp. 20–22). One of these essential properties is its ability to dissolve a wide range of substances, including ions. A few animals, such as some nematodes, tardigrades, and rotifers, can survive almost total drying in an inactive state called **cryptobiosis,** but most animals die if only partially dehydrated. Humans, for example, are about 75% water, and a loss of only 15% to 20% of that water is fatal. Even a loss of 10% (just 5 kg of body weight) is life-threatening.

Ions are as essential for life as water (see p. 23). In many animals the concentrations of ions required in the interstitial fluid do not match the ions available in the food or external environment. For example, herbivores ingest large amounts of potassium from the cellular fluid of plants, yet the potassium concentration in their interstitial fluids must remain low to avoid depolarizing the membranes of neurons and other cells. In such cases homeostatic systems must regulate the concentration of the ion.

THE IMPORTANCE OF CONCENTRATIONS

Not only the concentrations of individual ions but also the total ionic concentration of the interstitial fluid must be regulated. In addition, non-ionic solutes contribute to the total concentration of the interstitial fluid. This total concentration is called the **osmolarity.** Incorrect osmolarity can affect protein structure and function and can cause water to move into or out of cells or organisms by osmosis. Such osmotic movement of water into or out of a cell can lead to swelling or shrinking of cells and tissues, which can be fatal. A solution with an osmotic concentration higher than that of a cell or organism will tend to draw water out of the cell or organism by osmosis. Such a solution with a high osmotic concentration is said to be **hyperosmotic** (Figure 14.1). A solution that has an osmotic concentration that is lower than that of a cell or organism, such that the cell or organism tends to take in water from the solution, is said to be **hypo-osmotic.** A solution in which there is neither a net loss nor a net gain of water by the cell or organism is termed **iso-osmotic.** In most cases an animal must have an interstitial fluid that is iso-osmotic with respect to its cells. Either the animal must live in an environment that naturally maintains an iso-osmotic interstitial fluid, or the animal must have mechanisms to regulate the osmotic concentration of that fluid.

Although all animals have interstitial fluids that resemble seawater, there are wide variations in osmolarity and in the concentrations of individual ions (Table 14.1). There is some order in these variations, however, that depends on the ionic concentrations in the environment. For example, a fish that lives in seawater (SW)

Table 14.1   The concentrations of ions in seawater, fresh water, and in the body fluids of animals representative of different habitats.

Numbers below each ion are concentrations in millimolar (mM).

	Na	K	Ca	Mg	Cl
"Average" Seawater	470	10	10	54	548
Marine Invertebrate Starfish *Asterias*	428	10	12	49	487
Marine Vertebrate Sculpin *Myoxocephalus*	194	4	3	2	177
Terrestrial Invertebrate Cockroach *Periplaneta*	161	8	4	6	144
Terrestrial Vertebrate Human *Homo*	142	4	5	2	104
Freshwater Invertebrate Crayfish *Cambarus*	146	4	8	4	139
Freshwater Vertebrate Salmon *Oncorhynchus*	147	9	3	1	117
Fresh Water (average North American river)	0.39	0.04	0.52	0.21	0.23

is likely to have a higher sodium concentration in its interstitial fluid than a fish that lives in fresh water (FW). In general, the relationship of ion concentrations should go as follows:

SW ≥ marine animal > terrestrial animal ≥ freshwater animal > FW

The basis for this relationship is that less energy is required to maintain the difference between the internal and external concentrations if that difference is small. A major exception to this generalization is that marine vertebrates, such as bony fishes (e.g., sculpin), reptiles (sea snakes and turtles), and mammals (whales and seals) generally maintain internal concentrations more like those of freshwater animals. The explanation is that these and other vertebrates evolved from freshwater fishes.

OSMOCONFORMERS

Many marine animals do not have to regulate their ionic concentrations, because seawater is unvarying and similar in ionic composition to their interstitial fluids. For small marine animals it would also take too much energy to maintain a concentration difference across their large body surfaces. Many marine animals, especially small ones, therefore lack mechanisms that would regulate internal ionic concentrations if they happened to be placed in an environment that differed from seawater. Such animals are said to be osmoconformers.

Many osmoconformers die if placed in water that is more concentrated or less concentrated than seawater. The reason is that their cells either lose or gain water osmotically, and the shrinking or swelling destroys membranes and organelles. Some osmoconformers, however, can survive being placed in dilute or concentrated solutions, even though their interstitial fluids lose or absorb water. Although the osmolarity of the interstitial fluid changes, their cells do not shrink or swell because the intracellular osmolarity also changes. When the interstitial fluid loses water, the cells keep from shrinking by increasing the levels of free amino acids in the cytosol. When the interstitial fluid absorbs water, the cells keep from swelling by decreasing the free amino acids in the cytosol. The champion survivor of

Speaking of Concentrations

The total **concentration** of dissolved material depends on the number of individual particles in a given volume of solution. Even in a few milliliters of fresh water the number of dissolved particles is unimaginably large (billions or trillions). Rather than enumerate particles, therefore, it is more convenient to express concentrations in moles per liter. A mole equals 6.022×10^{23} particles. A concentration of X moles/liter can also be expressed as an X molar concentration. Millimolar (mM) concentrations are more common in living systems.

As long as the concentration is small enough that particles do not interact with each other, the osmotic effect—the **osmolarity**—will be directly proportional to the sum of the molar concentrations of all the dissolved particles. Thus a 1 mM solution of NaCl will have about twice the osmolarity of a 1 mM solution of sucrose, since the NaCl will dissociate into 1 millimole of Na^+ plus 1 millimole of Cl^-. $CaCl_2$ will have three times the osmolarity of an equal concentration of sucrose. At the concentrations normally found in body fluids, many salts do not completely dissociate into ions, so the osmolarity is somewhat lower than would be predicted simply from adding concentrations. Osmolarity is measured in osmoles/liter, or osmolar (Osm). Average seawater is 1 osmolar. Human interstitial fluid has an osmolarity of approximately 300 mOsm.

Osmolarity is one of several properties, called **colligative properties,** that depend on the concentration of particles and not on the nature of those particles. Therefore osmolarity can be measured by measuring any of those other colligative properties. The **freezing point depression**—the number of degrees below 0°C at which a solution freezes—is one such property. Many **osmometers** determine the osmolarity of a solution by measuring the temperature at which it freezes. While pure water freezes at 0°C under standard conditions, average sea-

water, for example, freezes at -1.86°C. Seawater therefore has a freezing point depression of 1.86°C. The measurement in degrees Celsius can be converted to osmoles, but often it is not. One often sees osmolarities expressed in degrees Celsius.

Salinity is another way of expressing osmotic concentration, especially in marine biology. Salinity is essentially the number of grams of solids remaining after the water has evaporated from 1 kilogram of solution. Salinity therefore indicates the number of parts per thousand of dissolved material. Since percentage (parts per hundred) is indicated by the symbol %, salinity is indicated by the symbol $\permil$. The dissolved solids in average seawater make up 3.5% of its weight, so average seawater has a salinity of $35\permil$.

The scale in Figure 14.2 may prove useful in converting among the various units of measurement. As a point of reference for comparison to other solutions it would not be a bad idea to remember that average seawater has an osmolarity of 1000 mOsm, a freezing point depression of 1.86°C, and a salinity of $35\permil$.

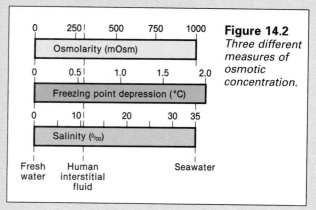

Figure 14.2
Three different measures of osmotic concentration.

---

osmotic stress, the sea slug *Elysia chlorotica*, does not vary its intracellular levels of amino acids but uses a different, unidentified intracellular substance to maintain osmotic balance. This mollusc can survive gradual changes in concentration ranging from 0.024 to 2.4 times that of seawater (Pierce et al. 1983). Although the cells of osmoconformers do not shrink or swell, the entire animals do, as the interstitial fluid loses or absorbs water. Such animals are theoretically "living osmometers," because one could use them to measure the osmolarity of a solution by weighing them before and after they were placed in the solution.

OSMOREGULATORS

Many marine animals maintain internal levels of ions that are lower than those of seawater. Such animals are osmoregulators. Freshwater and terrestrial animals are also osmoregulators, since they have to maintain ionic concentrations in the interstitial fluid that are higher than those in their environments. The most adept

It is doubtful whether much progress in physiology would have been possible without the ability to keep tissues functioning outside the body. Early physiologists were handicapped because dehydration destroyed isolated organs. Adding water destroyed tissues even more quickly than dehydration. Eventually, someone found out that the correct amount of sodium chloride in the water would allow many organs to continue functioning for longer periods. These NaCl solutions, called **physiological salines,** are useful because they provide the correct osmotic pressure for cells, and because they provide the two most prevalent ions (Na^+ and Cl^-) found in most extracellular fluids. Even so, few tissues can continue functioning for long in simple salines.

In 1883 a combination of error, insight, and research provided the solution to this problem—in both senses of the word "solution." Sidney Ringer was conducting research on isolated frog hearts at London University College Hospital when he noticed that the hearts were continuing to beat in the physiological saline for much longer than the usual half hour. At first he thought this

was just one of many seasonal variations in frog physiology, and he actually published this incorrect explanation. Then Ringer found out that a laboratory technician had erroneously made up the physiological saline using ordinary tap water instead of laboratory distilled water. Ringer quickly realized that ions in tap water were required for the normal functioning of frog hearts, and by trial and error he found the optimum concentrations of calcium and potassium salts. The result was an improved solution that is still known as **Ringer's solution.** It consists of 6 grams of NaCl, 0.22 gram of KCl, and 0.29 gram of $CaCl_2$ per liter of solution. Bicarbonate (0.2 gram of $NaHCO_3$) is added to buffer the pH. Ringer's solution works well for most amphibians, but for other animals many other physiological solutions, known collectively as Ringer's solutions, have been developed. For example, Locke's, Tyrode's, and Krebs' solutions are for various vertebrate tissues. The major ions in Ringer's solutions are Na^+, K^+, Ca^{2+}, Mg^{2+}, H^+ (pH), and Cl^-. Often glucose is added as an energy source.

---

osmoregulators live in environments with the greatest variability in ionic composition, such as **estuaries** where fresh water empties into salt water. The blue crab *Callinectes sapidus*, which inhabits estuaries, can maintain a fairly constant internal osmotic concentration no matter how dilute the external water is. Animals that can tolerate wide ranges of osmolarity are said to be **euryhaline** (Greek *eurys* broad + *halin* of salt). Those that can tolerate only a limited range are said to be **stenohaline** (Greek *stenos* short) (Figure 14.3).

MECHANISMS OF OSMOREGULATION

There are only a few things an osmoregulator can do to prevent or compensate for the loss or gain of water due to osmosis, or the loss or gain of ions by diffusion.

1. **The osmoregulator can actively transport ions in the direction opposite from that of diffusion.** The mechanism for such active transport is generally an **ionic pump** located in the plasma membranes of certain cells (see p. 41). Such pumps enable animals living in a hyperosmotic environment to pump out excess ions to compensate for diffusion and intake with food. Ionic pumps can also be used to set up a concentration difference that will transport water osmotically. Using this arrangement, an animal in a hyperosmotic environment can pump in ions to bring in water, then pump the ions back out. Conversely, animals living in hypo-osmotic environments can pump ions in to compensate for loss, or use ion pumps to eliminate excess water.

2. **The osmoregulator can vary its permeability to water or to ions.** This approach allows for control of the amounts of water and ions entering or leaving the body.

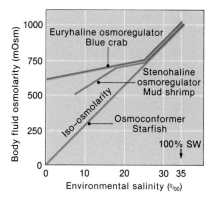

Figure 14.3
Changes in internal osmolarity versus external osmolarity for three representative animals: the osmoconforming starfish Asterias, *the euryhaline osmoregulating blue crab* Callinectes sapidus, *and the stenohaline osmoregulating mud shrimp* Upogebia pugettensis. *The X and Y axes are labeled in different but equivalent units. The internal osmolarity (Y axis) for the starfish is the same at all points as the external osmolarity (X axis).*

OSMOREGULATORY ORGANS

Different organisms employ these principles in a variety of ways. Freshwater protozoans and sponges osmoregulate by means of **contractile vacuoles** (= water expulsion vesicles) in each cell (Figure 14.4). The vacuole slowly swells with the osmotically absorbed water until it reaches a certain size, then quickly expels its contents. Contractile vacuoles are surrounded by mitochondria, but it is not known how their energy is harnessed to actively transport water into the vacuoles and out of the cytoplasm.

Most animals have kidneys or analogous organs for osmoregulation (summarized in Figure 21.10). Osmoregulatory functions in many invertebrates will be described in the chapters dealing with them. The kidneys of vertebrates will be described more thoroughly later in this chapter. All these organs generally work in two stages: filtration and active transport. First blood or other body fluid is filtered into the osmoregulatory organ, leaving cells, proteins, and other large molecules behind. The filtrate thus consists mainly of water, ions, and small dissolved molecules such as glucose and amino acids. Pumps also actively transport excess ions and molecules into the filtrate. This active transport from body fluid to filtrate is called **secretion.** Ions and molecules needed by the animal are pumped out of the filtrate and back into the body fluid. Active transport from filtrate into body fluid is called **reabsorption.**

NEPHRIDIA

These principles of osmoregulation operate in the osmoregulatory organs of most invertebrates, which are often tubes called nephridia. Unlike the kidneys of humans and other vertebrates, which form a filtrate of blood, nephridia form a filtrate of some other body fluid, such as coelomic fluid. Also unlike vertebrate kidneys, the pressure that drives the body fluid into the nephridium is not blood pressure but usually the activity of either cilia or flagella. After the filtrate is produced, the nephridium secretes excess ions into it and reabsorbs needed ions, nutrients, or water back into the body fluid. The resulting fluid containing excess ions, metabolic wastes, and water is then excreted out of the body through a **nephridiopore** at the external end of the nephridium.

Protonephridia. The internal ends of nephridia are either closed or open. Nephridia with closed internal ends are called protonephridia, and those with open

Figure 14.4
A freshwater amoeba, showing a contractile vacuole.

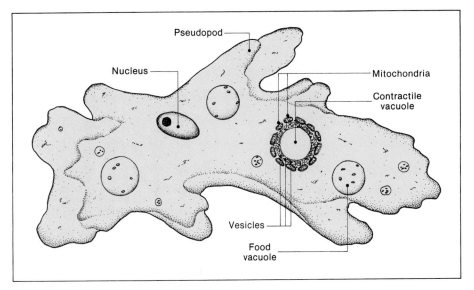

internal ends are called metanephridia. Protonephridia usually have branched tubes. They occur mainly in animals that lack coeloms (acoelomates and some pseudocoelomates) and also in the primitive chordate *Branchiostoma* (amphioxus). A striking feature of many protonephridia are **flame cells** located at the closed internal ends of the tubes (Figure 14.5). Flame cells are named for the flickering tuft of cilia that projects into the nephridial tube. The cilia may be responsible for producing the pressure that filters the body fluids into the nephridia.

Metanephridia. Metanephridia (often called simply nephridia) occur only in invertebrates with true coeloms. Metanephridia are unbranched tubes that open not only at the nephridiopore but also at the internal end, called the **nephrostome** (Figure 14.6).

OSMOREGULATORY ORGANS OF ARTHROPODS

Antennal Glands. Several different osmoregulatory mechanisms occur among arthropods, depending on whether they are aquatic or terrestrial. Some crustaceans have antennal glands (= green glands) in the head that conserve potassium and calcium and excrete excess sulfate and magnesium. In freshwater crustaceans, such as crayfish, the antennal glands also eliminate excess water. A filtrate of blood enters an **end sac,** then usually a **labyrinth,** and finally a coiled **tubule.** Reabsorption and secretion occur in all three structures, forming urine that is stored in a bladder until it is emptied through a pore at the base of the antenna (Figure 14.7).

Malpighian Tubules. Most insects and spiders are terrestrial arthropods whose main problem is to conserve water while eliminating metabolic wastes and excess ions. These functions are served by a combination of Malpighian tubules and the rectum. From two to several hundred Malpighian tubules are attached to the digestive tract between the midgut and hindgut and wave about within the abdom-

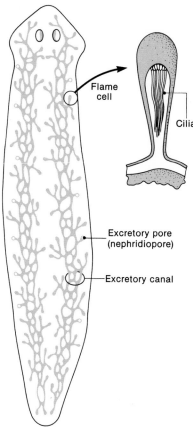

Figure 14.5
The protonephridia of a planarian (phylum Platyhelminthes). Protonephridia often have flame cells at the internal ends (inset). Protonephridia of some animals have solenocytes, which differ from flame cells in having only one flagellum.

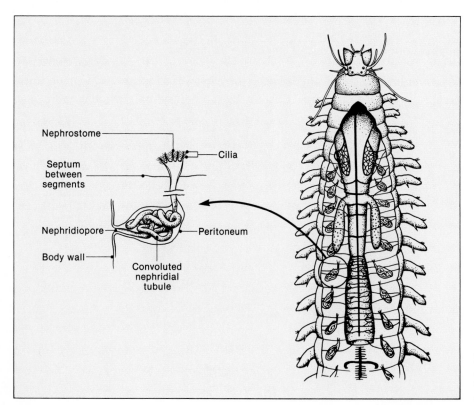

Figure 14.6
Metanephridia of the marine annelid worm Nereis. *The tubule is unbranched and open at both ends.*

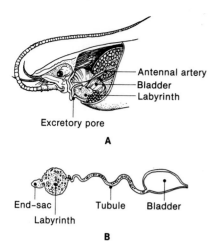

Antennal artery
Bladder
Labyrinth

Excretory pore

A

End-sac
Labyrinth
Tubule
Bladder

B

Figure 14.7
The antennal gland (green gland) of the crayfish Astacus pallipes, *in its natural (A) and extended (B) configurations.*

inal cavity (see Figures 13.5B and 32.9). These tubules transport ions, especially potassium, out of the blood and into the gut for elimination with the feces. Body fluids are filtered into the Malpighian tubules osmotically. As the fluid passes through the gut into the rectum, water is reabsorbed by **rectal pads.**

EXCRETION OF NITROGEN WASTES

Organs that regulate ion concentrations in interstitial fluids are often called excretory organs, because in addition to regulating ion concentrations they excrete certain metabolic wastes. The most important of these wastes is the nitrogen from amino acids. There are three major forms of nitrogenous waste: ammonia, urea, and uric acid (Figure 14.8). **Ammonia** is extremely toxic (as you could guess from inhaling its fumes), but it is easily diluted in water. It is the major nitrogenous waste of aquatic animals, since they can simply excrete it into the environment as fast as it is produced. Many terrestrial animals, including all mammals, most amphibians, and some turtles, produce **urea** as their major nitrogenous waste. Urea denatures proteins but is less toxic than ammonia, so it can be tolerated in the bloodstream until eliminated by the kidneys. Oviparous terrestrial animals, including most reptiles and birds, eliminate nitrogenous waste as **uric acid.** Unlike ammonia and urea, which are soluble and would accumulate to toxic levels in an egg, uric acid is insoluble. The insolubility of uric acid also gives the adult reptile or bird the advantage of not requiring much water to eliminate it.

STRUCTURE AND FUNCTION OF HUMAN KIDNEYS

Unlike nephridia, the kidneys of vertebrates form a filtrate of blood rather than of another body fluid, and the filtration pressure is due mainly to blood pressure. Several kinds of kidneys occur in vertebrates, as will be described later. The functioning of the human kidney is representative of the kidneys of adult reptiles, birds, and other mammals. The kidney of the adult human is a fist-sized organ that is . . . well, kidney-shaped. (Mammals are the only animals in which the

Figure 14.8
The most common nitrogenous wastes of various groups of vertebrates.

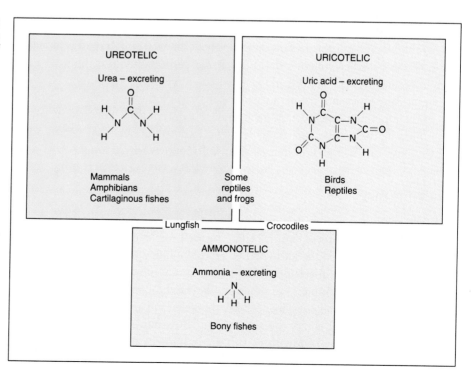

UREOTELIC

Urea – excreting

Mammals
Amphibians
Cartilaginous fishes

URICOTELIC

Uric acid – excreting

Birds
Reptiles

Some reptiles and frogs

Lungfish _____ Crocodiles

AMMONOTELIC

Ammonia – excreting

Bony fishes

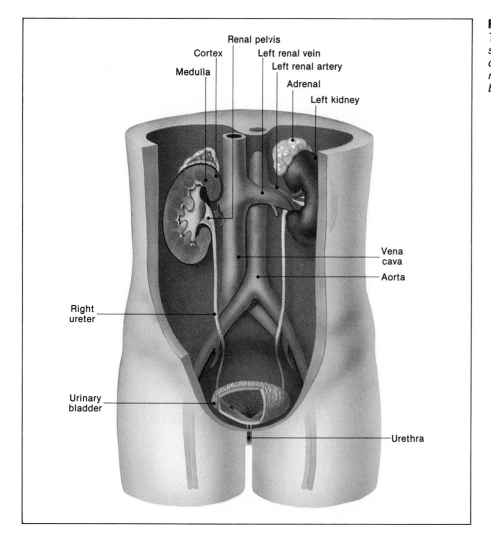

Renal pelvis
Cortex
Left renal vein
Medulla
Left renal artery
Adrenal
Left kidney

Vena cava
Aorta

Right ureter

Urinary bladder

Urethra

Figure 14.9
The structure of the human renal system. The ureters transport urine, containing excess ions and metabolic wastes removed from the blood, into the urinary bladder.

kidney has a definite shape.) The two kidneys are located in the lower back on each side of the aorta and vena cava (Figure 14.9). At any given time about one-fifth of the blood from the aorta is entering the kidneys through the **renal arteries.** A slightly smaller volume exits through the **renal veins** into the vena cava. The difference between arterial and venous blood volumes equals the 1.5 liters of urine that daily flows into the **urinary bladder** through the **ureters.** This description of the plumbing suggests that the overall function of the kidneys is to remove water, bloodborne wastes, and excess ions. To understand how they perform this task we must look at them on microscopic and submicroscopic levels.

THE NEPHRON

Each human kidney contains about a million subunits called nephrons that do the actual work of the kidneys (Figure 14.10). In each nephron arterial blood enters a tuft of capillaries called the **glomerulus.** Here the blood pressure drives the first stage of renal function, **filtration.** (Filtration in glomeruli and other capillaries is often called **ultrafiltration.**) As in other capillaries, the fluid portion of the blood is forced across the walls, leaving behind the cells and proteins. The resulting filtrate (or ultrafiltrate) enters the surrounding **Bowman's capsule,** which conducts it into the **proximal convoluted tubule.** In humans approximately 180 liters per day make it this far. Fortunately, not all of the filtrate goes into urine. In the

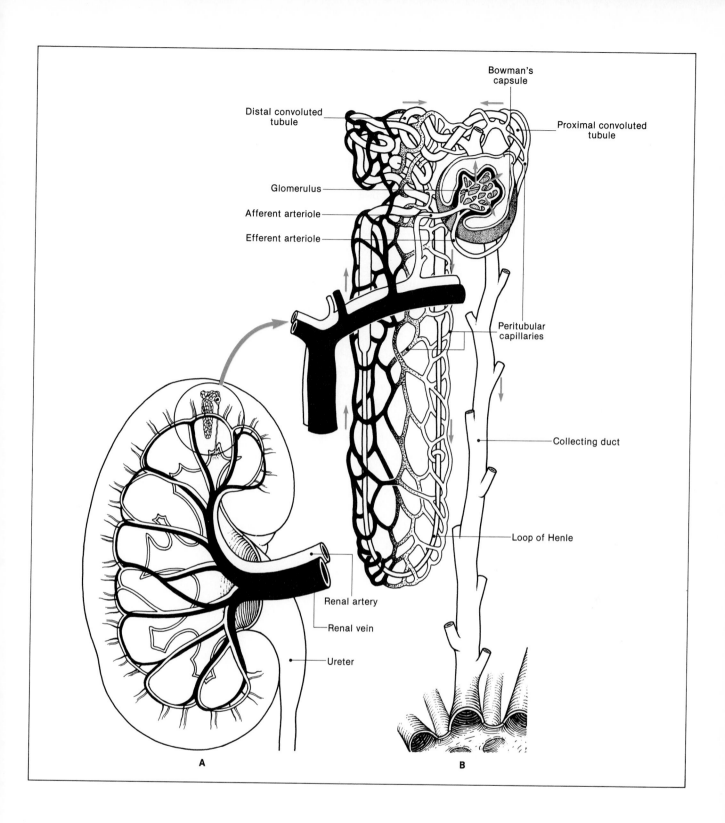

Bowman's capsule

Distal convoluted tubule

Proximal convoluted tubule

Glomerulus

Afferent arteriole

Efferent arteriole

Peritubular capillaries

Collecting duct

Loop of Henle

Renal artery

Renal vein

Ureter

A

B

Figure 14.10
(A) A sectioned human kidney. The location of one nephron is indicated. (B) A nephron. Arrows indicate the flow of fluid in the nephron. The total length of an outstretched nephron is from 2.0 to 4.4 cm.

proximal tubule much of the inorganic ions and almost all of the glucose and amino acids are pumped out of the filtrate. These ions and molecules then diffuse back into the blood in the **peritubular capillary,** which surrounds each nephron. At the same time most of the water in the filtrate is reabsorbed osmotically. The lining of the proximal tubule has microvilli similar to those of the small intestine, and the reabsorption of the nutrients in the tubule may be due to sodium-coupled pumps like those in the gut.

The tubular fluid next passes through a hairpin turn called the **loop of Henle.** The functioning of Henle's loop is a complicated subject best left for later. For now it is enough to note that the loop of Henle aids in the reabsorption of water. In the **distal convoluted tubule** there is further active transport of inorganic ions. Depending on whether the concentration in the blood is too high or too low, there will be secretion or reabsorption of H^+, K^+, or other ions. Herbivores, for example, are apt to have an excess of potassium in the bloodstream, so pumps in the distal tubule secrete K^+ out of the peritubular capillaries into the distal tubule. H^+ is either reabsorbed or secreted to adjust the pH of blood. Urea, the nitrogenous waste that always comes to mind when we think of urine, has no special mechanism for secretion, probably because urea would destroy any protein that would serve as its pump. By the time the tubular fluid enters the **collecting duct** it is essentially dilute urine, called pre-urine. Humans produce about 9 liters of pre-urine each day. Under normal circumstances most of the water in the pre-urine is reabsorbed, leaving concentrated, industrial-strength urine.

HENLE'S LOOP

The ability to produce concentrated urine depends on the loops of Henle, as was first suspected from the fact that vertebrates that have to conserve water generally have more and longer loops of Henle. Fishes, amphibians, and reptiles lack loops of Henle, and they are incapable of producing urine that is hyperosmotic to plasma. Birds have short loops of Henle and can produce urine that is slightly hyperosmotic. In mammals the loops of Henle occur in both short and long forms. The short forms are contained entirely within the outer portion, the **cortex,** of the kidney. The long forms extend into the inner portion, the **medulla.** The greater the proportion of long to short Henle's loops, the more concentrated the urine and the better the mammal can survive water stress. In the mountain beaver *Aplodontia rufa,* which lives in wet areas of the American Northwest, there are no long loops of Henle. This rodent's urine is only 2.4 times as concentrated as its blood plasma. In contrast, the kangaroo rat's loops of Henle are all long, and its urine is from 10 to 14 times as concentrated as its plasma. Approximately 14% of the loops of Henle in humans are long, and our urine averages about 4 times the osmolarity of plasma. Depending on circumstances, human urine can be from one-fifth to five times as concentrated as plasma.

It took many years for physiologists to understand how the loops function. It is now known that the loop of Henle works as an elaborate version of an active countercurrent-exchange mechanism (see p. 251). Active transport of Na^+ and Cl^- and passive movement of urea set up a large osmotic gradient that increases from cortex to medulla. As pre-urine passes through the collecting ducts it is exposed to an increasing osmotic concentration that draws out most of the water.

The system is driven by active transport at the ascending thick portion of the loop, at the place marked ① in Figure 14.11. Pumps in this portion of the loop transport both Cl^- and Na^+ from the tubular fluid into the surrounding space. The increased concentrations of Na^+ and Cl^- then draw water out of the descending loop ② and out of the distal tubule. The net effect of this loss of both ions and water from the fluid in the loop is that urea becomes highly concentrated in the fluid. However, the tubule is impermeable to urea except in later parts of the collecting duct ③. Urea therefore diffuses out of the collecting duct only in the inner medulla. The increased concentration of urea surrounding this portion of the collecting duct draws water out of fluid in the loop of Henle. The loss of water produces extremely high Na^+ and Cl^- concentrations at the bend. Some

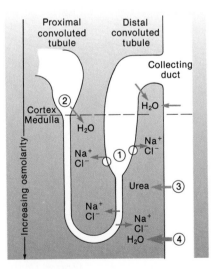

Figure 14.11
A loop of Henle showing the active transport of ions at point ①, and the passive movements of ions, urea, and water that produce an osmotic gradient that reabsorbs water from the collecting duct at point ④. See text for explanation.

of these ions therefore diffuse out, adding to what was actively transported at ①. The result of all this is a gradient of ions and especially of urea that increases from cortex to medulla. As pre-urine flows through this gradient in the collecting duct, the increasing concentrations draw out more and more water ④. This water then enters the blood in the peritubular capillaries.

HORMONAL CONTROL OF KIDNEY FUNCTION

Antidiuretic Hormone.   The amount of water reabsorbed depends on how permeable the collecting ducts are to water. This permeability varies, depending on how much water is needed in the body fluids, and is controlled by antidiuretic hormone (ADH), also called vasopressin. The posterior pituitary secretes ADH whenever the hypothalamus detects high osmolarity in the cerebrospinal fluid. ADH helps dilute body fluids by making the collecting ducts more permeable, thereby reducing the amount of water lost in the urine. Since a diuretic is a substance that increases urine production, ADH is an antidiuretic. ADH also helps dilute body fluids by promoting thirst, and it also increases blood pressure.

The importance of ADH is clear from the disorder **diabetes insipidus.** Like the more common form of diabetes, diabetes mellitus, diabetes insipidus is characterized by excessive urine production and thirst. It results from either a reduced ability to secrete ADH or a lack of responsiveness in the collecting ducts. The result is that untreated individuals excrete up to 10 times the normal volume of urine, up to 15 liters per day. Before synthetic ADH became available, victims of diabetes insipidus spent much of their lives in the bathroom. Alcohol also temporarily inhibits ADH secretion, which explains the excessive urination and subsequent thirst that typically accompany drinking.

Renin–Angiotensin.   Two hormones, renin (REE-nin) and angiotensin, together help ensure that fluid is filtered by the glomerulus into the tubules at a rate sufficient for the tubules to function. This rate of filtration, called the **glomerular filtration rate** (GFR), is crucial since reabsorption and secretion can occur only if there is filtrate. When the blood pressure and therefore GFR are low, the **juxtaglomerular apparatus** (JGA) near the glomerulus secretes renin into the bloodstream. The JGA also secretes renin if the level of NaCl in the tubule is too low or in response to stimulation by the sympathetic nervous system. Renin is a hormone and also an enzyme that converts **angiotensinogen** from the liver into angiotensin I. An enzyme in the lungs converts angiotensin I to angiotensin II. Angiotensin II has several effects, all of which help counteract the low blood pressure: (1) it increases blood pressure by causing vasoconstriction; (2) it stimulates the secretion of ADH; (3) it stimulates thirst; and (4) it stimulates secretion of aldosterone.

Aldosterone and Atrial Natriuretic Hormone.   Aldosterone is a steroid hormone (mineralocorticoid) secreted by the adrenal cortex. Aldosterone stimulates the Na–K pumps in the loops of Henle and the collecting ducts, increasing the reabsorption of Na^+ and therefore the amount of water reabsorbed into the blood. The overall effect is to increase the blood volume and pressure.

Aldosterone generally works for only a matter of weeks. Eventually, the increase in blood volume causes the heart to release a peptide called atrial natriuretic hormone. Atrial natriuretic hormone counteracts the effects of aldosterone by inhibiting its secretion, inhibiting sodium reabsorption, inhibiting the renin–angiotensin system, and causing dilation of arteries. Since high blood pressure (**hypertension**) is often the result of excessive Na$^+$ in body fluids, there is a great deal of interest in the possible therapeutic effects of atrial natriuretic hormone.

Parathyroid Hormone. Finally, parathyroid hormone stimulates the reabsorption of Ca^{2+} by the loop of Henle and the distal convoluted tubule. The net effect is to increase the level of Ca^{2+} in the body fluids. Parathyroid hormone also promotes the release of Ca^{2+} from bones.

TYPES OF KIDNEYS IN VERTEBRATES

The structure and functioning of kidneys differ among different groups of vertebrates, and similar differences occur within each vertebrate during embryonic development. There are three kinds of vertebrate kidney: the **pronephros,** the **mesonephros,** and the **metanephros.** The pronephros, which lies anteriorly in the abdomen, appears only briefly in many vertebrates, and not at all in the human embryo. In some vertebrates, however, the pronephros is the first osmoregulatory and excretory organ of the embryo, and it remains as the functioning kidney of tadpoles and other amphibian larvae. During embryonic development, or during metamorphosis in amphibians, the pronephros is replaced by the mesonephros, which is elongated and posterior to the pronephros. The mesonephros is the functioning embryonic kidney of humans and many other vertebrates, and it remains as the permanent kidney of adult fishes and amphibians (see Figures 35.10, 35.21, and 36.7). In reptiles, birds, and mammals the mesonephros gives way during embryonic development to the metanephros, which becomes the functional kidney of the adult.

ADAPTATIONS OF VERTEBRATE KIDNEYS FOR DIFFERENT HABITATS

Bony Fishes. Whether mesonephric or metanephric, the kidneys of adult vertebrates all work on principles similar to those described for humans, but with adaptations suitable for their own environments (Figure 14.12). In freshwater fishes the nephrons of the mesonephric kidneys have glomeruli that produce large amounts of filtrate. Their nephrons, however, lack loops of Henle, and they are therefore incapable of producing urine more concentrated than blood plasma. Of course, they would not need to, since these fishes are not normally in any danger of dehydration. Their problem is to eliminate excess water and to absorb and retain ions.

For marine bony fishes the problem is just the opposite. Like all vertebrates, they evolved from freshwater fishes, and they retain low internal ionic concentrations (Table 14.1). Their osmotic problem is to conserve water and eliminate ions. The nephrons of their kidneys lack glomeruli, which prevents water loss by filtration. Only a small amount of urine forms in the tubules and eliminates Mg^{2+} and SO$_4^{2-}$. More important are the gills, which actively transport Cl$^-$ and Na$^+$ out of the body fluids.

Cartilaginous Fishes. Sharks and rays (cartilaginous fishes) are also marine vertebrates that evolved from freshwater ancestors, and they too have ion concentrations lower than those in seawater. Unlike marine bony fishes, however, their body fluids are iso-osmotic or even hyperosmotic to seawater. The high concentration in their body fluids is due to **osmolytes.** One osmolyte in cartilaginous

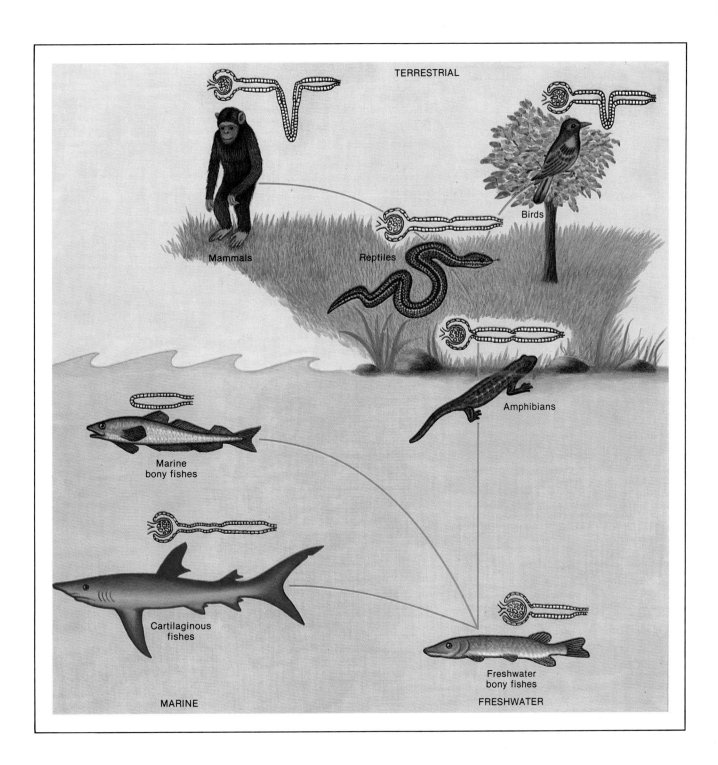

TERRESTRIAL

Mammals

Reptiles

Birds

Amphibians

Marine
bony fishes

Cartilaginous
fishes

Freshwater
bony fishes

MARINE

FRESHWATER

Figure 14.12
Variations in nephron structure among various vertebrates. Lines represent evolutionary relationships.

fishes is the nitrogenous waste urea. The level of urea in the blood of a shark or ray is many times the concentration that would kill us and most other animals. Sharks and rays not only tolerate the urea, they require it for muscle function. The tolerance to urea was a deep mystery for many years, until it was found that sharks and rays have a second group of nitrogenous osmolytes, **methylamines,** that counteract the detrimental effects of urea. Because of the osmolytes water loss is not a problem. The kidneys of cartilaginous fishes have glomeruli that produce copious urine. There is still a tendency for sodium, chloride, and other ions to accumulate in the body fluids by diffusion, however. These ions are eliminated by the **rectal gland** (see Figure 35.10), which excretes a concentrated solution into the posterior end of the digestive tract.

Amphibians and Reptiles. Most adult amphibians are adapted to living on land, but their permeable skin makes them highly dependent on a moist environment. Like freshwater fishes, the nephrons of their mesonephric kidneys have glomeruli, and they are incapable of producing urine more concentrated than blood plasma. In many amphibians, however, the urinary bladder can reabsorb water from the urine. The metanephric kidneys of reptiles also have glomeruli and are incapable of forming hyperosmotic urine. Reptiles are less dependent than amphibians on a moist environment, however, because they have impermeable skin. As noted previously, the metanephric kidneys of birds and mammals are more adapted to terrestrial life, since most have nephrons with loops of Henle that produce concentrated urine.

Mammals. Marine mammals have kidneys that require little water to eliminate the salts ingested with their food. It takes only 650 milliliters of water for a whale's kidneys to eliminate the salt in 1 liter of seawater, providing a net gain of 350 ml of pure water. In contrast, the less-efficient kidneys of humans require 1.35 liters of water to eliminate the salt from 1 liter of seawater, for a net *loss* of 350 ml. This explains why shipwrecked sailors survive longer if they drink nothing than if they give in to the urge to drink seawater.

Mammals that live in the desert also osmoregulate primarily by means of efficient kidneys that can excrete concentrated urine. The kangaroo rat *Dipodomys merriami* of the American Southwest has such efficient kidneys that it never has to drink water. As long as the kangaroo rat remains in its burrow during the daytime to avoid excessive evaporation, the water produced in mitochondria (see p. 61) makes up for losses (Figure 14.13). Extraordinary as this seems, the common house mouse *Mus musculus* is almost as good at doing without water, unless it gets into our protein-rich food and needs extra water to eliminate the urea.

Figure 14.13
The daily water budget for the kangaroo rat Dipodomys merriami, *which drinks no water. A daily water budget for humans is shown for comparison. About half our water is contained in food, and a little less than that comes from drinking. Sixty percent of the water taken in is lost in the urine, and most of the rest by evaporation.*

KANGAROO RAT

Gain (ml)		Loss (ml)	
Present in food	6.0	Urine	13.5
Drinking water	0.0	Evaporation	43.9
Metabolic product	54.0	Feces	2.6
	60.0		60.0

HUMAN

Gain (ml)		Loss (ml)	
Present in food	1200	Urine	1500
Drinking water	1000	Evaporation	950
Metabolic product	350	Feces	100
	2550		2550

SALT GLANDS

Reptiles, birds, and mammals that live in or near the sea face special osmoregulatory challenges. Although their kidneys can produce urine more concentrated than blood plasma, they are incapable of eliminating all the salt taken in with food and water. Most marine reptiles and birds therefore depend on **salt glands** that secrete a highly concentrated solution. Salt glands have probably evolved independently in various groups of marine reptiles. In marine lizards the salt gland empties through the nostril. In sea turtles the salt gland is located in the orbits of the eye, and tears carry away the hyperosmotic secretion. In sea snakes the salt gland is located beneath the tongue. Saltwater crocodiles such as *Crocodylus acutus* (see Figure 37.6A), which lives around the tip of Florida, have salt glands in their tongues (Taplin and Grigg 1981). Almost any bird fed salty foods will develop enlarged nasal glands above the eyes that can serve as salt glands.

SALT INTAKE

All animals require an adequate intake of salt, especially NaCl. For marine animals this obviously presents no problem. Most terrestrial and freshwater carnivores also get sufficient salt, because their prey are likely to have optimal ion concentrations in their tissues. Herbivores inhabiting sea shores or former sea floors also get sufficient salts from eating plants that have absorbed the ions. Many freshwater and terrestrial herbivores, however, have to supplement the dietary intake of ions. Many freshwater animals obtain ions by actively absorbing them from water through the gills, intestine, and skin. Terrestrial herbivores generally have to find deposits of salt, especially NaCl. These animals usually have a tremendous hunger for salt, and any source is a valuable commodity. Porcupines will consume leather gloves, tool handles, and even the seats of outdoor toilets to get the salt left by human sweat. Many groundhogs risk being hit by cars to eat salt that has been spread on highways to melt snow.

In human cultures in which meat, milk, or blood are the main foods, supplementary salt is seldom needed. In other cultures, however, salt is as essential to people as to most other terrestrial animals. In the United States, for example, the ability to eat just one salted peanut is considered proof of superior willpower. Roman soldiers gladly accepted salt as payment for their services. (The word "salary" is derived from the Latin word *salarius*, meaning "of salt.") Ancient rulers waged wars for control of salt deposits, and Jesus praised his disciples as "the salt of the earth." Now that salt is plentiful we may be getting too much of a good thing. The average American consumes 4 to 6 grams of NaCl per day, which is 5 to 10 times the recommended intake. Increased sodium concentration in body fluid increases the production of renin, which increases blood pressure in many people. Since hypertension leads to heart failure and other circulatory diseases, this salt-hunger may be as risky for humans as for groundhogs on the side of the road.

SUMMARY

The concentrations of several ions are crucial for the proper functioning of cells. In addition, the total concentration of ions and other molecules in the body fluids has a profound osmotic effect on cells. Some marine animals, which live in environments where ion concentrations are constantly suitable, do not regulate concentrations in their body fluids:

They are osmoconformers. Other animals, especially marine vertebrates and freshwater and terrestrial animals, must be osmoregulators, maintaining internal concentrations different from external concentrations. Osmoregulation utilizes at least one of the following two operations: (1) active transport of ions, possibly coupled to osmotic transport of water;

(2) changing the permeability to water or ions. These operations are generally carried out in two stages: (1) filtration of blood or other body fluids and (2) reabsorption of needed substances back into the blood and secretion of excess substances out of the blood into the urine. A variety of organs carry out these operations. Freshwater protozoans and sponges use contractile vacuoles. Many invertebrates have nephridia. Some crustaceans have antennal glands, and insects have malpighian tubules. All vertebrates have kidneys.

In mammals the nephrons in the kidneys form a filtrate of blood in the glomerulus and surrounding Bowman's capsule, and the tubule carries out reabsorption and secretion. The filtrate enters the proximal convoluted tubule, where most of the water and almost all the glucose and amino acids are reabsorbed. It then passes through the loop of Henle to the distal convoluted tubule, where reabsorption and secretion of ions occurs. The pre-urine then enters the collecting duct. Active transport of Na^+ and Cl^- in the loop of Henle sets up a concentration gradient that draws urea out of the collecting duct. The resulting concentration gradient of urea and ions draws most of the water out of the pre-urine, forming concentrated urine. Active transport and permeability are adjusted by several hormones. The structure of the kidney varies in different groups of vertebrates. In addition, various vertebrates have special adaptations, such as salt glands, for living in marine and desert habitats.

KEY TERMS

osmolarity
osmoconformer
osmoregulator
contractile vacuole
filtration
secretion
reabsorption

nephridium
antennal gland
Malpighian tubule
nitrogenous waste
urea
nephron
glomerulus

Bowman's capsule
proximal convoluted tubule
distal convoluted tubule
loop of Henle
collecting duct
salt gland

SELF-TEST

1. Define osmolarity. Why is it important?

2. How does the osmolarity of body fluids generally compare with the osmolarity of the environment for the following groups of animals: marine invertebrates; marine vertebrates; freshwater vertebrates and invertebrates? (Use the terms "iso-osmotic," "hypo-osmotic," and "hyperosmotic" in your answers.) What problems in osmoregulation must animals in each group solve?

3. Graph the changes in internal osmolarity versus external osmolarity for an osmoconformer, a stenohaline osmoregulator, and a euryhaline osmoregulator. In labeling the axes of your graph, what units of measure can you use?

4. Describe the following types of osmoregulatory organs: contractile vacuoles, nephridial organs, antennal glands,

Malpighian tubules, kidneys. What kinds of animals have each type of organ?

5. Sharks, marine bony fishes, sea snakes, sea birds, and whales have body fluids with lower ionic concentrations than in seawater. How do members of each of these groups maintain this state?

6. Diagram a nephron and label its parts. Indicate where filtration occurs. What molecules are reabsorbed, and where? Which ions are secreted, and where?

7. Give a general explanation for how the loops of Henle enable the kangaroo rat to survive without drinking water.

8. List four hormones involved in regulating kidney function. Briefly state the function of each hormone.

READINGS

RECOMMENDED READINGS

Beauchamp, G. K. 1987. The human preference for excess salt. *Am. Sci.* 75:27–33.

Cantin, M. and J. Genest. 1986. The heart as an endocrine gland. *Sci. Am.* 254(2):76–80 (Feb).

Crowe, J. H. and A. F. Cooper, Jr. 1971. Cryptobiosis. *Sci. Am.* 225(6):30–36 (Dec).

Dantzler, W. H. 1982. Renal adaptations of

desert vertebrates. *BioScience* 32:108–113.

Fertig, D. S. and V. W. Edmonds. 1969. The physiology of the house mouse. *Sci. Am.* 221(4):103–110 (Oct). (*The mouse's ability to conserve water is one key to its success.*)

Schmidt-Nielsen, K. 1981. Countercurrent systems in animals. *Sci. Am.* 244(5):118–128 (May). (*Among other functions, these help conserve water in the kidneys and in the noses of many mammals.*)

ADDITIONAL REFERENCES

Pierce, S. K., M. K. Warren, and H. H. West. 1983. Non-amino acid mediated volume regulation in an extreme osmoconformer. *Physiol. Zool.* 56:445–454.

Taplin, L. E. and G. C. Grigg. 1981. Salt glands in the tongue of the estuarine crocodile *Crocodylus porosus. Science* 212:1045–1047.

15

Temperature

Japanese snow monkey (Macaca fuscata) *in hot spring.*

CHAPTER OUTLINE

LEARNING OBJECTIVES

1. Why are the activities of an animal so dependent on body temperature?

2. Why are some animals "cold-blooded" and others "warm-blooded"?

3. How do animals such as insects and frogs compensate for cold and avoid freezing?

4. How do birds and mammals maintain constant body temperatures that are different from environmental temperatures?

5. What are the benefits and costs of maintaining a constant body temperature?

6. What mechanisms are involved in keeping body temperature constant?

WHY TEMPERATURE MATTERS

Few people would doubt that the subjects of the previous three chapters—oxygen, nutrients, water, and ions—are necessities that animals actively seek out and regulate within themselves. A suitable temperature is also such a necessity, as animals themselves demonstrate by their behavior (Figure 15.1). Many animals seek favorable environmental temperatures and have elaborate mechanisms for regulating the temperatures of their internal environments. This chapter describes why temperature is so important and the ways in which animals adapt to temperature.

The temperature range for animal life extends from below −50°C (−58°F) for some Arctic insects, up to 50°C (122°F) for a certain fish (*Barbus thermalis*) in hot springs of Sri Lanka. This upper limit is well above the temperature of 45°C that is painful to humans. Most animals live within a narrower range of environmental temperatures, usually between 0°C and 35°C. Life is restricted to particular environmental temperatures because body temperature tends to become like the environmental temperature, and body temperature has a profound effect on the rates of all physiological processes. The reason is that almost all physiological processes depend on chemical reactions, and these reactions require **activation energy** (see pp. 53–54). The source of this activation energy is usually the thermal agitation of molecules. An increase of 10°C in body temperature typically doubles or triples the rate of a biological process (Figure 15.2). A change in body temperature can therefore seriously affect the rates of activities. Low body temperature can kill an animal by reducing its rates of activity, even if the temperature is above freezing. High temperatures can be just as lethal by destroying enzymes and other proteins, even if the temperatures are far below the temperatures for boiling or combustion.

ECTOTHERMS AND ENDOTHERMS

Small animals, especially aquatic ones, have body temperatures that are essentially the same as their environmental temperatures, since a small animal has a large heat-conducting surface relative to its body mass, and water conducts heat readily.

Heat is energy due to molecular motion. **Temperature** is a measure of the average of this energy per molecule. Heat is commonly measured in calories or kilocalories (cal or kcal), and temperature is measured in degrees. It takes 1 calorie of heat to increase the temperature of 1 gram of water by 1 degree Celsius. Since animals consist mainly of water, it also takes about 1 calorie to warm 1 gram of tissue by 1 degree. Thus for a person who weighs 50 kg it takes 50,000 cal (50 kcal) to raise the body temperature by 1°C, from the normal of 37°C to 38°C. This heat can be absorbed from the environment or produced as a by-product of metabolism.

Figure 15.1
Painted turtles Chrysemys picta *crowd onto a log to bask in the sun after spending a winter buried beneath a frozen pond. The turtles expose themselves to predators and go to considerable trouble to climb onto the logs.*

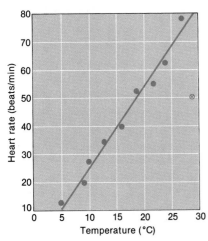

Figure 15.2
The rate of contraction of the isolated heart of a grass frog Rana pipiens *in Ringer's solution at various temperatures. Irreversible damage, indicated by the fall in heart rate (the circled X), occurs at a little above 27°C. Physiologists express the dependence of a physiological rate on body temperature by the Q_{10}, which is the rate of a process at one temperature, divided by the rate at a temperature 10°C lower. In this example, the Q_{10} over the temperature range 10°C to 20°C was approximately $^{52}/_{25} = 2.08$. (Data from Harris 1976.)*

Such animals, in which the body temperature is determined primarily by the environment rather than by internally produced heat, are referred to as ectotherms (Greek *ecto-* outside). Ectotherms are generally also **poikilotherms,** meaning that their body temperatures are variable (Greek *poikilo-* varied). This need not be the case, however. Ectotherms that live in the constant temperature of a cave will have a constant body temperature, and lizards, though ectothermic, can maintain a constant body temperature by moving to cooler or warmer places.

Many animals, especially large, terrestrial ones, have body temperatures that depend more on their own heat production than on environmental temperature. Such animals are called endotherms (Greek *endo-* within). Endothermy enables some animals, primarily birds and mammals, to maintain fairly constant body temperatures in spite of varying environmental temperatures. These animals are therefore homeotherms (Greek *homeo-* same). Not all endotherms are homeotherms, however. Bluefin, skipjack, and yellowfin tuna (genus *Thunnus*) maintain body temperatures above environmental temperatures by conserving body heat, although the body temperature is not constant.

ADAPTATIONS OF POIKILOTHERMS TO SEVERE COLD

Torpor and Diapause. When any animal's body temperature drops so low that physiological activities slow down drastically, the animal is said to be in a state of torpor. An animal in torpor may appear to be dead, but its normal activities return as soon as the body is warmed. During spring and fall an ectotherm may become torpid each night, then recover during the day. Torpor can also last throughout the winter, as with amphibians, reptiles, and some insects. Torpor is simply a natural response to the reduction in enzyme activities and requires no special adaptation. The advantage of seasonal torpor is that it allows an ectotherm to survive winters without feeding, since its metabolic rate is so low. A disadvantage of torpor is that a sudden rise in temperature will immediately arouse the animal, perhaps at a time when there is no food available. This hazard is avoided by an adaptation called diapause, which occurs in some ectotherms, such as the house fly *Musca domestica*. Diapause is like torpor except that some physiological mechanism prevents arousal during brief warm periods. Torpor and diapause are often called "hibernation." It is advisable, however, to reserve the term hibernation for a quite different process that occurs only in one bird and a few mammals, as will be described later.

Protection Against Freezing. While torpor and diapause allow animals to remain alive at low temperatures, they will not protect an animal from destruction of body tissues by ice crystals. One common protection in ectotherms exposed to cold is lowering the freezing point by increasing the concentration of solute in body fluids. For freshwater animals, such as fishes and frogs, the normal concentrations of body fluids are sufficient to prevent freezing as long as they are in liquid water, since the water freezes at a higher temperature than the body fluids. Some animals that are exposed to freezing temperatures, especially terrestrial ectotherms, lower their freezing points by increasing the concentrations of certain molecules, such as glucose and glycerol.

Besides lowering the freezing point by increasing concentration, glycerol is also an **antifreeze.** Antifreezes lower the freezing point by changing the arrangement of water molecules. Glycerol is a major antifreeze in insects and some frogs (and at one time in automobile radiators). The larva of the wasp *Bracon* has a 5 molar concentration of glycerol that enables it to survive −45°C winters in Canada and

Siberia. Another type of antifreeze in insects and in certain spiders and mites is **thermal hysteresis protein** (THP). THPs are somewhat mysterious in that they lower the freezing point without affecting the temperature at which body fluid melts once it is frozen. Some Antarctic fishes, such as *Trematomus*, produce **antifreeze peptides** and **antifreeze glycopeptides.**

Some insects, Antarctic fishes, frogs, and turtles actually live at temperatures below their freezing points in a state of **supercooling.** Supercooled fluids are below the freezing temperature, but they do not crystallize because there is no **nucleator** on which crystals can get started. One disadvantage of supercooling is that a supercooled animal that contacts a nucleator freezes instantly. A school of Antarctic fish becomes flash-frozen if one of them bumps into a piece of ice.

Some ectotherms actually survive with up to 65% of their body water frozen. So far the animals known to survive partial freezing include barnacles and molluscs exposed at low tide, some frogs, painted turtle hatchings (*Chrysemys picta*), and some insect larvae. Instead of avoiding freezing, these animals actually produce nucleators that encourage the freezing of water in extracellular spaces. At the same time, however, they increase the levels of antifreeze compounds that keep the ice crystals small. As the extracellular water freezes, the concentration of ions in the liquid water naturally increases, and this increased concentration draws water out of the cells osmotically. The intracellular concentration therefore increases, lowering the freezing point. This reduced freezing point protects the cells from damage due to freezing, but the cells would be damaged from shrinkage if it were not for certain **cryoprotectant** molecules, such as the disaccharide trehalose and the amino acid proline.

ACCLIMATION OF ECTOTHERMS TO LOW TEMPERATURE

Acclimation is a physiological adjustment to an environmental change that occurs over a period of days or longer. Acclimation to low temperature is quite common in ectotherms that live more than a few months in temperate regions (Figure 15.3). The effect of low temperature is to reduce the rates of biochemical reactions, but this effect can be compensated for during acclimation by changing enzymes that catalyze the reactions. One approach is simply to increase the amount of enzyme. A second approach is to reduce the amount of thermal activation energy required by producing a different form of an enzyme. The different enzyme, called an **isoenzyme** (= isozyme), requires different genes to direct its synthesis. One example of this method of acclimation occurs in the brains of trout and involves acetylcholinesterase, the enzyme that breaks down the synaptic transmitter acetylcholine. In the winter and spring, trout brains produce an isoenzyme of acetylcholinesterase that works best at low temperature. In summer and autumn the gene for that isoenzyme is turned off, and another gene produces an isoenzyme of acetylcholinesterase that works best and resists denaturation at higher temperatures.

COMPENSATION FOR ENVIRONMENTAL TEMPERATURE BY CHANGING BODY TEMPERATURE

Behavioral Mechanisms. Although poikilotherms cannot maintain a constant body temperature, they often have behavioral mechanisms that enable them to alter it. Being animals, they can seek out more favorable **microclimates,** areas

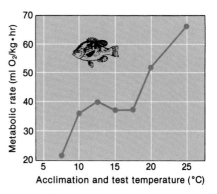

Figure 15.3
Compensation to environmental temperature by the common pumpkinseed sunfish Lepomis gibbosus. *After allowing the fish to acclimate to each of the temperatures indicated by a datum point, its metabolic rate (rate of oxygen consumption) was measured at that temperature. Over the range of 10°C to 18°C the fish had about the same metabolic rate. (Data by J. L. Roberts.)*

where the local temperature differs from the general air or water temperature. Numerous studies on trout, bluegill sunfish, and other fishes have demonstrated that they select the temperature at which their growth and reproduction are best. Another behavioral mechanism, **basking** in the infrared radiation from the sun, can elevate the body temperature above the air temperature (Figure 15.1). Insects and reptiles can often be found on cool, sunny mornings basking with their bodies oriented so as to absorb the most solar radiation.

Partial Endothermy.   Many poikilotherms can compensate for low environmental temperature by temporarily producing and conserving more body heat to raise the body temperature. Since the body temperature during those periods depends mainly on the heat produced by the animal, this phenomenon is called partial endothermy. The rise in body temperature can be due to basking, as mentioned previously, or to **thermogenesis.** Thermogenesis is any kind of physiological activity whose major effect is the production of heat. Partial endotherms generally also have adaptations for conserving the heat.

Bernd Heinrich of the University of Vermont has studied many examples of partial endothermy among insects. His best-known study is on the bumble bee *Bombus.* Heinrich grabbed bumble bees while they were foraging for nectar and inserted tiny thermocouples to measure the temperatures in various parts of their bodies. He found that even when the air temperature was only 5°C, the temperature of the flight muscles of these "cold-blooded" animals was at least 36°C, and sometimes as high as 45°C. (The body temperatures of humans and other "warm-blooded" mammals is only around 37°C.) At the same time, the abdomen of the bumble bees was often 20% cooler than the thorax. Heinrich and Ann E. Kammer found that bumble bees first shiver to warm up their flight muscles to the minimum temperature required for hovering, then maintain that temperature by conserving heat while flying. Bumble bees conserve heat from the flight muscles by countercurrent heat exchange (see p. 251), which prevents the heat from circulating into the abdomen along with the blood (Figure 15.4). This allows bumble bees to keep their flight muscles at working temperatures as they forage among widely separated flowers. If flowers are close together on a plant, however, bumble bees allow the flight muscles to cool, and they simply walk from flower to flower.

THERMOGENESIS

Partial and complete endotherms raise their body temperatures above environmental temperatures by thermogenesis. Usually thermogenesis is divided into two categories: shivering and nonshivering thermogenesis. **Shivering** is a rapid contraction of muscles, which converts the energy from ATP into heat rather than useful mechanical work. Shivering, as well as increased muscle tension, are important mechanisms of thermogenesis during the first minutes of exposure to cold.

Nonshivering thermogenesis refers to metabolic processes with low efficiencies in which most of the released energy goes into heat. In mammals nonshivering thermogenesis occurs mainly in **brown adipose tissue** (= brown fat), which takes its name from the large amount of cytochromes in the mitochondria. Brown adipose tissue constitutes up to 6% of a mammal's body weight, especially in infants, small mammals, and mammals acclimated to cold. Brown adipose tissue is unusual in having mitochondria in which the electron transport chain becomes uncoupled from the production of ATP. Instead of producing ATP, the energy from the metabolism of fatty acids produces heat.

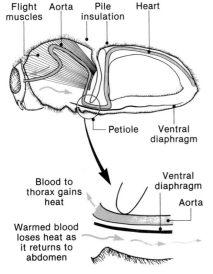

Flight muscles · Aorta · Pile insulation · Heart · Petiole · Ventral diaphragm

Blood to thorax gains heat · Ventral diaphragm · Aorta · Warmed blood loses heat as it returns to abdomen

Figure 15.4
Countercurrent heat conservation in the bumble bee Bombus *sp. As blood passes anteriorly through the aorta it picks up heat (represented by darker red) from the flight muscles. The warmed blood then passes posteriorly through the narrow "waist" (petiole) between the thorax and abdomen, where it transfers much of the heat back to the blood in the aorta. On warm days the bumble bee raises the ventral diaphragm so that more heat flows into the abdomen, where it can be lost from the ventral surface. Heat from the ventral surface also allows the queen bumble bee to incubate her eggs, by shivering while she presses her abdomen against them.*

Bernd Heinrich

Bernd Heinrich
Born: 19 April 1940, in Germany
Education: University of Maine, B.A., M.S.; UCLA, Ph.D. (Zoology)
Professor of Zoology, University of Vermont. Ultramarathon runner: U.S. record for distance in 24 hours (157 miles); U.S. record for 100 km on road; 100 mile record on track (12 hr 27 min).
Author of: *Bumblebee Economics,* a popular account of thermoregulation and energetics in bumblebees; *In a Patch of Fireweed,* an autobiographical account of his life and research; *One Man's Owl; Ravens in Winter;* and numerous research articles.

Excerpts from an interview a few days after Heinrich won the Lake Waramaug Ultramarathon (100 km) and a few days before he set the 100-mile track record.

You start your book In a Patch of Fireweed *with your early childhood in postwar Germany, in the first few years of your life, when your main concern seems to have been foraging [in the Hahnheide Forest] to get enough to eat. I wonder if that affected your later research. . . . I don't want to get too psychoanalytical. . . .*

It sounds that way, but no, because I went into biology at first just being interested in all kinds of mechanisms of living things. I got really interested in cell biology at University of Maine. I was also very much interested in field biology, but at that time all the big advances were made in molecular biology, so I thought the really significant stuff was going to be in molecular biology. Eventually as I got more secure, I suppose, I decided that, "Well, I'll do what's more fun for me."

So you switched from cell biology into field studies for the fun of it. You use that word in your book over and over again: "fun."

Yes. I just enjoy it.

The switch from molecular biology to field biology was quite a decision for a young graduate student to make.

Oh it was, it was. A really tough decision, and I agonized over it for a long, long time. I said, "Well, I've got a long life ahead, and if I'm going to be 30, 40, 50 years, or whatever on a lab bench, and if I don't really enjoy it, I'm not going to make any progress." I got much satisfaction out of it [cell biology]. I was really enthusiastic. I got several papers. There was intellectual stimulation, but I was getting results very slowly, and I felt there was an awful lot of competition, and you have to have access to tremendous resources when you get into cellular biology. You're tied to huge machines that cost thousands and thousands of dollars, so you're tied into getting money. You're tied into being at a place where you can get that, so I felt that I'd lose a lot of freedom. I liked the intellectual challenge, but there were too many costs associated with it.

It seems to me that your running is almost as great a passion as your science. I mean who would run 157 miles in 24 hours? Is there any connection between that and your science? Do you ever get any insights from running that you couldn't have gotten from reading and from experiments on animals?

I doubt it.

That's a real disappointment. I thought you were going to say something about bioenergetics. But maybe something your running and your research do have in common is a high threshold for pain.

That is true. I talk about its being fun all the time, but that might be a little bit misleading, because it's in fun that you exert yourself the most and are willing to suffer the most. But you get satisfaction back from it. So when I look back on the 24-hour run, I had to suffer pretty hard for 24 hours, and actually for a couple of months beforehand getting in training. But I still look at it in terms of fun, because now for years later I'll be having that satisfaction. And the same with the work with the bees. I might have to really extend myself at times and really go through a lot of pain and effort. It's sort of like a long-distance run to go through a research project. You have to work on it, because if you slow down and stop, you essentially have destroyed the project. But then afterwards you get all the benefits.

There are some aspects of it that are more painful than others I imagine. What parts of research do you find the least fun?

Well, for example, right now I was doing some stuff on heat transfer in carpenter bees, and its not much fun to have to get this animal set up with these little thermocouples—and you have to have a whole bunch of them

302

inside—they are extremely fine things like hairs. They have to be in exactly the right place, and they pull out if you look at them sideways. It's frustrating to get them in just the right place and to get it set up when the animal with its legs flailing is destroying it all the time. You have to do it over and over again. It really takes a lot of persistence sometimes.

There seems to be a growing interest in the kinds of studies you're doing. A sort of . . . I don't know whether to call it a counterrevolution. Maybe that's just my impression, but do you sense that people are coming back to the classical biology studies of the kind you're doing?

Well, there seems to be an awful lot of work that way. It seems like in the 50s and 60s there was sort of a downgrading of this type of biology, but I think in the meantime studies of behavior, ecology, et cetera have matured more and have gained a lot more interest. If you look at the journals, the outpouring of work is just phenomenal since then, so I think there is a lot of work being done.

I'm wondering how much of your present research and things like the Bumblebee Economics *book depend on your formal education, and how much depended on your self-education. Is there any way you could divide that up? If you had to go back to college again would you have taken the same courses and curriculum?*

Well, I think it's always pretty hard to say exactly what you're going to need and precisely how a course you might take is going to feed into what you're going to do. I'm sure that a broad range of biology courses really helped me a lot in getting a feel for what the different types of things were in biology. In that way I could choose better, and relate better by having the whole picture. So it's a matter of giving me more confidence in doing what I'm doing. Basically most of it is probably learned on my own, but if I wasn't sure about what I had to learn then I might spend most of my time reading. Certainly there are many potential places for input from courses directly to, let's say *Bumblebee Economics.* But in that particular case I think it would mostly be in my graduate education, in actually working in the lab and actually seeing what other graduate students were doing, and in going out in the field and observing for myself and taking the measurements.

At one point in In a Patch of Fireweed *you describe a feeling of being trained to fill a niche in the job market, and your dissatisfaction with that.*

Yeah, that's right. I went to college essentially knowing I wanted to be a farmer or cabinet-maker or something practical. I wanted to work with my hands, and wanted to be outside and be my own boss. I had absolutely no concept at all that education was correlated with a job. I thought you went to college to learn something, simply for the sake of knowledge and for no other reason whatsoever. That it's a means to earning a living never even occurred to me.

I imagine there are quite a few students who will read your book on Bumblebee Economics *and kind of resent the fact that you've taken the problem from them. That problem is no longer available for them to work on. Do you ever wonder yourself whether you're going to run out of projects—use them all up?*

I sometimes think about it. When I first finished the study with the moth thermoregulation [for the Ph.D.] I said, "Well, here now I've found *the* solution to how insects thermoregulate—at least the ones which are important. I've skimmed the cream." I said, "Now there's no place to go but down, and there's just little details to pick up." And then I'd pick up bumblebees, and I'd see something entirely new that I didn't even imagine, and I'd work on them a little bit more, and I'd see something else that I didn't imagine. I keep thinking I'm going to run out, but in actuality I just get overwhelmed with new problems.

Do you deliberately go out looking for projects?

No, I don't do it deliberately, ever. I like being outdoors, so I might spend a day out in the woods just going out looking at birds and watching things. Just being outdoors. It's like . . . if you really like a woman then you see a lot of beautiful things.

CONTROL OF HEAT EXCHANGE BY ENDOTHERMS

Endothermy implies that the body temperature can differ from the environmental temperature. That, in turn, implies that the endotherm can control the tendency of heat to be lost from the body in the cold, or gained from a warmer environment. Physicists know of only four ways by which heat is exchanged—conduction, convection, radiation, and vaporization—so these are the only processes by which an endotherm can control heat exchange.

Conduction. Conduction is the transfer of thermal agitation of molecules in one body to the molecules in another body with which it is in contact. Heat loss by conduction is proportional to the surface area of the animal (S), the **thermal conductivity** of the surface (C), and the difference between the core body temperature (T_b) and the ambient (environmental) temperature (T_a): $SC(T_b - T_a)$. Since S, C, T_b, and T_a are the only variables in this equation that the animal controls, endotherms can control conduction only by changing the exposed surface area, changing the conductivity of the surface, changing the body temperature, or seeking an area with a different temperature. Some animals do change the core body temperature, as will be described later, and most do seek out more favorable environmental temperatures.

Animals also control conduction by changing the exposed surface area S and the conductivity C. Humans exposed to cold reduce S by folding their arms, thereby preventing heat loss from the inner surface of the arms and part of the torso. Dogs stretch out to increase their exposed surface area when they are warm. The importance of surface area to homeothermy is suggested by the large size of birds and mammals compared with the size of most poikilotherms. Even the smallest mammal, Savi's pygmy shrew (*Suncus etruscus*) of the Mediterranean, has a mass of 2 grams, which is much larger than the mass of most poikilotherms. The smallest bird, the bee hummingbird (*Mellisuga helenae*) of the Caribbean, has a mass of 3 grams. Because homeotherms are larger, they have less surface area per gram of tissue through which heat can conduct (and also radiate).

Conductivity C in bumblebees and some other partial endotherms is reduced by a coat of hairlike **pile** that traps warmed air. This reduces the thermal conductivity C and thereby reduces heat loss due to conduction. In terrestrial homeotherms hair and feathers serve a similar function. Figure 15.5 shows the importance of hair in thermoregulation by mice. The effectiveness of hair can be increased by **piloerection,** due to contractions of smooth muscles attached to hair (the arrector pili muscles in Figure 10.1B). Birds have a similar mechanism to fluff the feathers. Humans have lost most of their fur, so piloerection now produces only ineffectual "goose pumps." Many mammals and birds also have a layer of low-conductance fat beneath the skin. Seals, whales, and other marine mammals have the most impressive layers of such **subcutaneous fat** in the form of blubber.

Convection. Convection is like conduction except that heat is conducting to a moving fluid that continually carries off the heat. As air or water is warmed by contact with an animal, it becomes less dense and rises, carrying heat with it. This warm air or water is then replaced by cooler air or water, which absorbs more heat from the animal. The resulting currents are called natural convection. There can also be forced convection due to wind and water currents. The rate of heat

Figure 15.5
The metabolic rates of mice at various temperatures, before and after destroying the thermoregulatory effect of the fur by wetting. (A) The experimental apparatus. A laboratory mouse is placed in a small bottle with a packet of soda lime to absorb CO_2 and water vapor. Any change in the volume of air in the bottle results from the consumption of oxygen. Movement of a bubble in a graduated pipette indicates the rate of O_2 consumption. A thermometer measures air temperature in the bottle, which can be varied by covering the bottle with ice. (B) Results compiled by students in the author's animal physiology course at SUNY–Plattsburgh from 36 mice before and after wetting. The vertical bars show the variation in metabolic rates (standard error of the mean). Wetting the fur of the mouse shifts the curve upward and to the right, showing that higher metabolic rates occur at correspondingly higher environmental temperatures because of the loss of insulation, and because of vaporization of the water. At the lowest temperatures the metabolic rates declined, perhaps from hypothermia.

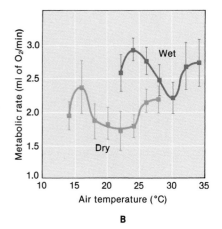

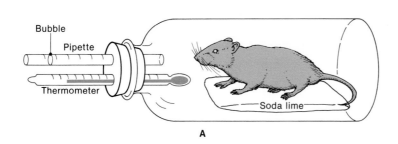

loss depends on the surface area and on the air or water speed. Changing the position of the body can increase or decrease the speed of natural or induced convection currents, thereby changing the rate of heat loss. Pile, hair, and feathers that slow the movement of convection currents can also aid endotherms in adjusting heat loss by convection.

Radiation. Radiation is the loss of thermal energy by the emission of electromagnetic radiation in the infrared range. As with conduction and convection, the rate of heat loss by radiation is proportional to the surface area. If the difference between the effective environmental temperature (T_a) and the surface temperature of the animal (T_s) is small, the rate of heat loss by radiation is also roughly proportional to the difference between T_a and T_s. The rate of radiant heat loss is therefore proportional to $S(T_s - T_a)$. This means that an animal can control the rate of heat loss by changing its exposed surface area as described previously, by changing its skin temperature (which is not necessarily the same as core body temperature T_b), or by seeking an area with a different effective temperature T_a.

Figure 15.6
(A) In most fishes, such as the freshwater drum Aplodinotus grunniens, *the major arteries (red) and veins radiate directly to the body surface, where much heat is lost to the water. (B) In bluefin* Thunnus thynnus *and other tunas a rete mirabile ("wonderful net") of intertwined veins and arteries recycles much of the heat from venous blood to arterial blood by countercurrent exchange.*

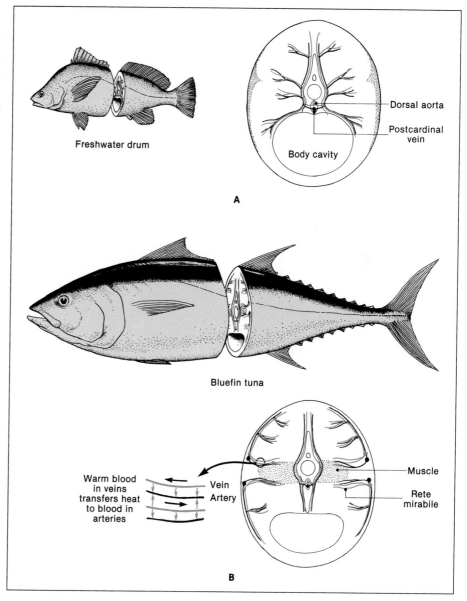

T_a is not necessarily the temperature you would measure with a thermometer. The air temperature on a clear night may be above freezing, but an animal will be radiating heat to deep space, which has an effective temperature far below freezing. On a clear night an animal is therefore much better off under a tree, rock, or other kind of shelter that will reflect infrared. On a clear day, of course, animals can gain radiant heat by basking in the infrared radiation from the sun. Seeking shelter and basking can therefore be viewed as two behavioral means of controlling radiation by changing T_a.

Some partial endotherms also reduce heat loss due to radiation by reducing the surface temperature T_s. Tuna and mackerel sharks, for example, reduce surface temperature by means of a **rete mirabile** ("wonderful net") of intertwined veins and arteries that serves as a countercurrent exchange mechanism for heat (see p. 251; Figure 15.6). Homeotherms also alter the skin temperature, by vasodilation and vasoconstriction. **Vasoconstriction**—the constriction of arteries—reduces the flow of blood to the skin, so that in an environment with low T_a the skin temperature will be less than the core body temperature. **Vasodilation** increases the flow of blood—and therefore heat—to the skin, so that more heat is radiated from the body. The effects of vasodilation and vasoconstriction are easily seen in pink or pale skin.

Homeotherms also use vasoconstriction and vasodilation to regulate countercurrent exchange of heat from venous blood warmed by muscle activity to cooler arterial blood (Figure 15.7). Because of vasoconstriction and countercurrent heat exchange, the feet, noses, and other extremities of sled dogs, ducks, and other

Figure 15.7
Countercurrent heat exchange in the human arm. (A) In cold weather the superficial veins of the arm constrict, reducing heat loss from the surface. The deep veins dilate, so heat from the venous blood recycles to the nearby arteries. (B) In warm temperatures the surface veins dilate and eliminate body heat, and the deep veins constrict.

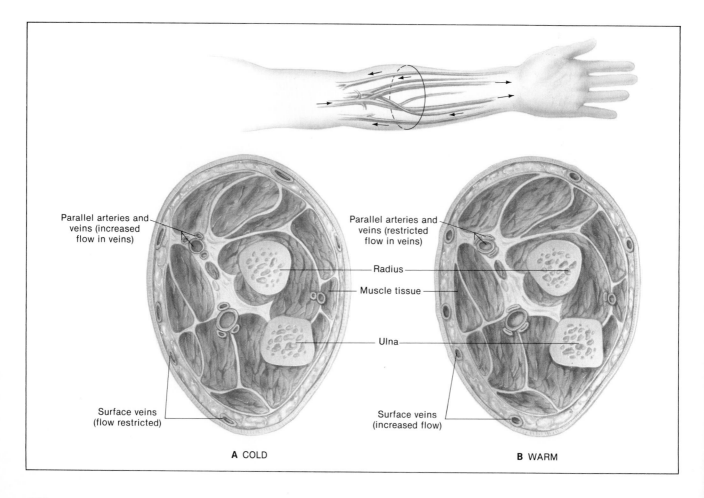

Parallel arteries and veins (increased flow in veins)

Parallel arteries and veins (restricted flow in veins)

Radius

Muscle tissue

Ulna

Surface veins (flow restricted)

Surface veins (increased flow)

A COLD

B WARM

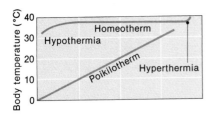

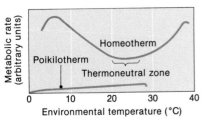

Figure 15.8
The idealized body temperatures and metabolic rates of a homeotherm and a poikilotherm, both acclimated to 22°C. The upper graph shows that for the poikilotherm the body temperature equals the environmental temperature. The homeotherm maintains a constant body temperature over a wide range of environmental temperatures—about 10°C to 35°C in this example. The lower graph shows that for the poikilotherm the metabolic rate increases continuously with body temperature until a few degrees above the acclimation temperature, where enzymes denature and the animal dies. The metabolic rate of the homeotherm varies with a U-shaped relationship to environmental temperatures over the range in which the body temperature can be regulated. The minimum metabolic rate often occurs at the acclimation temperature and a few degrees above or below, in the thermoneutral zone. At extremely low temperatures the animal is not able to produce or conserve enough heat to thermoregulate, and the body temperature drops (hypothermia). This will cause enzyme activities to slow down, which will make it even more difficult to thermoregulate. Death is inevitable unless heat is applied from an external source. At extremely high temperatures the heat produced as a by-product of thermoregulation may actually exceed the amount of heat lost. When that happens the body temperature starts to rise (hyperthermia), the enzymes speed up their rates, and even more heat is produced. In a matter of seconds the body temperature can rise sufficiently to denature proteins, resulting in death.

animals exposed to cold for long periods often have temperatures barely above freezing. One reason the cells are able to function at such temperatures is that lipids in their cell membranes remain oily at temperatures that would congeal the lipids in other cells.

Vaporization. Vaporization removes heat that is needed to convert water from a fluid to a gaseous phase. The rate of heat loss by vaporization is directly proportional to the amount of water lost, which depends on how fast the animal can evaporate water. Ultimately, that depends on how much water the animal can afford to lose in order to keep cool. In general, birds and hairy mammals evaporate water from the lungs, mouth, and nasal passages by **panting.** Many hairy mammals, such as mice, also lick themselves to vaporize water (and to reduce the insulation of fur). Mammals with sparse hair vaporize water by **sweating** (perspiring). A human can sweat as much as 10 liters per day. Even at room temperature (20°C) there is some **insensible perspiration.**

The rate of vaporization depends on the relative humidity of the air. The more water vapor already in the air, the more difficult it is to vaporize more. Humans who have adapted to tropical climates where the relative humidity is high do not waste water in futile sweating. It would also be useless for aquatic animals to sweat, since the water obviously cannot evaporate. In relatively dry climates where there is plenty of drinking water, however, vaporization is of major importance in endothermy. Of all the channels for heat loss, vaporization is the only one that can cool the body below the environmental temperature.

HOMEOTHERMY

Among living animals, only birds and mammals add body temperature to their repertoire of homeostatic mechanisms. These homeotherms maintain body temperature within a few degrees of a set point for extended periods. Mammals usually have core body temperatures of around 37°C over a wide range of environmental temperature, and birds maintain a body temperature of about 41°C. Birds and mammals can therefore remain active at environmental temperatures in which they would freeze if they were poikilotherms, and they do not need to adjust enzyme amounts or produce alternative isoenzymes to compensate for changes in body temperature. These are obvious advantages over poikilotherms. Homeotherms pay dearly for these advantages, however.

The Cost of Homeothermy. In a homeotherm the metabolic rate varies as a U-shaped curve over the range of temperatures at which it successfully thermoregulates (Figures 15.5B and 15.8). The minimum metabolic rate for a homeotherm often occurs at the acclimation temperature and a few degrees above or below. This is the **thermoneutral zone,** in which homeothermy can be maintained by changing body posture and other methods that require little additional energy. At environmental temperatures above or below the thermoneutral zone a homeotherm must expend additional energy by shivering, panting, and other heating or cooling mechanisms described earlier. In a poikilotherm, on the other hand, the metabolic rate increases continuously with body temperature until the lethal temperature is reached. Even at the highest temperature it can tolerate, however, the metabolic rate of a poikilotherm is never as high as that for a homeotherm of the same size. According to Robert Bakker (1975), "The total energy budget per year of a population of endothermic birds or mammals is from 10 to 30 times higher than the energy budget of an ectothermic population of the same size and adult body weight." This means that birds and mammals must obtain at least ten times the amount of food they would need if they were poikilotherms.

Homeothermy would not be nearly as expensive if the set points for body temperature were not so high. Maintaining a body temperature of 37°C when the environmental temperature averages, say, 25°C, is like setting your room thermostat to 98°F and opening the windows to keep cool. Why have the "thermostats" of birds and mammals evolved to be so high? Perhaps the high body temperature is necessary to prevent overheating when the environment gets unusually warm. If our normal body temperature were 25°C, we would save many calories that would not have to be replaced by eating, but whenever the environmental temperature rose about 25°C it would be difficult to keep from overheating. The only way to avoid overheating would be vaporization of sweat, which might lead to dehydration. By keeping the body temperature near the maximum encountered in the environment, we reduce the threat of overheating and of dehydration. Another possible reason for high body temperature is that we cool much faster the hotter we are compared with the environment. During exercise, therefore, it is easier to eliminate excess heat if the body temperature is high to begin with (Heinrich 1977).

Homeostatic Control.   If you destroy a thermostat, the furnace or air conditioner will respond erratically, if at all. If you apply ice or a lighted candle to a thermostat, the system will react as if the temperature of the entire room had changed. Similar effects occur if portions of the **hypothalamus** of the brain are damaged, cooled, or heated. Destroying one area of a dog's hypothalamus will make it unresponsive to heat. In a hot room the dog will not pant and would die of hyperthermia. Heating an area of the dog's hypothalamus will trigger panting even when the body and environmental temperatures are normal. Cooling a different area of the hypothalamus will trigger shivering and piloerection. These experiments suggest that homeothermy in mammals is coordinated in the hypothalamus.

The hypothalamus apparently gets its information about body temperature from two sets of **thermoreceptors.** Peripheral thermoreceptors in the skin provide an early warning of environmental temperature, and central thermoreceptors in the spinal cord and in the hypothalamus itself sense the core body temperature. Information from these thermoreceptors is compared with the **set point** (37°C), which is somewhat programmed into the hypothalamus. The hypothalamus then triggers the appropriate behavior and responses of the autonomic nervous system, which controls piloerection, vasodilation, vasoconstriction, and sweating (Table 15.1).

Table 15.1   Behavioral and autonomic responses of mammals to cold and heat.

Cold	Heat
BEHAVIORAL	
Huddling	Stretching out
Exercise	Inactivity
AUTONOMIC	
Piloerection	Sweating or panting
Vasoconstriction	Vasodilation

RESETTING THE THERMOSTAT

Heterothermy.   The term *homeo*therm must be taken rather loosely, for no bird or mammal maintains exactly the same core body temperature throughout its life. Body temperature usually varies over a cycle of about 24 hours, even if environmental temperature is constant. Human body temperatures are generally a degree or so lower at 2 A.M. than at 2 P.M. (This is an example of a **circadian rhythm;** see Chapter 20.) In some species the daily variation in body temperature is even greater, and these animals are referred to as heterotherms. The camel (*Camelus*) reduces panting if deprived of water and lets its daytime temperature rise to 41°C. At night the camel's hypothalamus lets body temperature drop to 35°C, thereby reducing conduction, convection, and the radiative heat loss to the clear desert sky. In other words, camels have automatic set-back thermostats analogous to those now available for homes.

Some hummingbirds conserve energy by allowing the body temperature to drop to around 14°C each night. Many small mammals also let their body temperatures drop for several hours during cold weather. During these periods of reduced body temperature the animal is inactive, in a state of **daily torpor.** In addition to daily cycles of body temperature, some homeotherms also have seasonal variations. For example, squirrels allow the body temperature to drop several degrees while they sleep through periods of severe cold. This "winter-sleep" reduces the amount of heat lost by conduction and radiation.

Hibernation.   Winter-sleep, as well as the torpor and diapause of poikilotherms, is sometimes called hibernation. It would be best, however, to reserve the term for a quite different phenomenon that occurs only in one species of bird, the common poorwill (*Phalaenoptilus nuttallii*) and a few species of mammals, including bats, ground squirrels, and woodchucks. Hibernation differs from the torpor and diapause of poikilotherms in that the hibernating animal is still a homeotherm, but with a set point usually between 2 and 3°C. In the autumn, as if they knew what lay ahead, mammals that hibernate put on large amounts of fat, which will later act as insulation and nutrient reserves. They then den up and start making "test drops" of the body temperature as they become dormant. Eventually, they enter a period of reduced body temperature that lasts from hours to weeks. Because of the reduced body temperature, enzyme activity is slowed, and the hibernator is dormant. Hibernators often arouse briefly during the winter, especially in response to disturbance or life-threatening cold. Arousal is perhaps the trickiest part of hibernation, because the body temperature must be raised using enzymes that are working at far below the normal body temperature. The rapid warming may be aided by thermogenesis in the large amounts of brown adipose tissue that most hibernators have.

Fever.   Homeotherms and even some poikilotherms, such as lizards, can turn up the thermostat in response to bacterial infections. The resulting fever has been shown to inhibit reproduction in many disease-causing bacteria. Fever is triggered by proteins called **pyrogens** (Greek *pyro*-fire), especially interleukin-1 from macrophages (see p. 236). Pyrogens apparently turn up the set point in the hypothalamus, making it "think" the body temperature is too low. It therefore triggers vasoconstriction, shivering, and heat-conserving behaviors until the body temperature equals the new set point. After the infection is over the set point falls back to 37°C. The hypothalamus then senses that the body temperature is too high, so it triggers panting or sweating, vasodilation, and cooling behaviors until the body temperature returns to normal. People with infections often go through cycles of chill and fever, with the temperature rising and falling as if two roommates could not agree on a setting for the thermostat.

ACCLIMATION IN HOMEOTHERMS

It is a familiar fact that one can get used to different seasons and climates. In New England the first 40°F day in autumn can seem unbearably cold, but in spring the same temperature can induce euphoria. In mammals a major mechanism for acclimation to cold is increased **thermogenesis.** Another major mechanism is **delayed responsiveness.** Animals acclimated to cold start shivering at lower temperatures than nonacclimated animals, and those acclimated to heat begin sweating at higher temperatures. Acclimated animals also show **increased tolerance for discomfort.** Many mammals also grow extra hair and fat in the winter, but this is not an example of acclimation since it occurs even if they are kept warm.

In order to study the ability of humans to acclimate, P. F. Scholander (1958) somehow persuaded eight Norwegians to camp out for six weeks, sleeping at below-freezing temperatures in one-blanket sleeping bags. (Perhaps the "volunteers" were his students.) At the end of a six-week acclimation period Scholander measured the metabolic rates and foot temperatures of his subjects as they slept and compared the measurements with those of Norwegians who had never before attempted to sleep under such conditions. As one would expect, the unacclimated subjects never managed to get to sleep, and their metabolic rates were higher than their day-time resting rates. Also the metabolic rates were extremely irregular. Looking at the jagged line on the graph of metabolic rate, one can almost see the subjects stomping and swearing throughout the night (Figure 15.9). In spite of that activity they were not very successful homeotherms. The temperature of their feet dropped to as low as 18°C. In contrast, the acclimated Norwegians kept their feet at a comfortable sleeping temperature by maintaining a steady, high metabolic rate.

Scholander then went to Australia to make comparable measurements on aborigines, who preferred to sleep naked at subfreezing temperatures. For his studies, however, Scholander had the Australians use sleeping bags like those of the Nor-

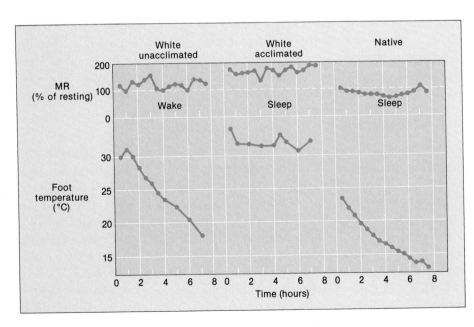

Figure 15.9
The metabolic rates and foot temperatures of three men sleeping or trying to sleep in the cold. The unacclimated Caucasian represented in the left part of the figure had an irregular, elevated metabolic rate, could not keep his feet warm, and could not sleep. After acclimation a Caucasian (middle part) could sleep and keep his feet warm by maintaining a high metabolic rate. The Australian native could also sleep, even without additional metabolic expenditure to keep his feet warm.

wegians. In order to ascertain that the conditions were comparable, Scholander himself made an unsuccessful attempt to sleep next to them. "We ourselves provided the control material. It turned out, that while we shivered and thrashed about all night, wishing that the sun would rise, the Australians lay motionless, snoring their way through the night, usually on a subbasal metabolic rate, cooling off even more than we did on the surface." That is, the Australian aborigines managed to sleep without the elevation in metabolic rate of the Norwegians, by abandoning total homeothermy. They allowed their foot temperatures to drop below 15°C and simply ignored the discomfort.

It is interesting that the Norwegians' and the Australians' hypothalami chose different but appropriate strategies for acclimation. Assuming that Scholander was not so heartless as to starve as well as freeze his Norwegian volunteers, it would make sense for their hypothalami to maintain homeothermy of the feet regardless of the metabolic cost. On the other hand, the Australian aborigines and their ancestors had probably experienced frequent periods of inadequate caloric intake. It would make more sense for their hypothalami to conserve energy by not attempting to keep their feet warm. We should not be too surprised that the hypothalamus can learn to balance the need for homeothermy against the need to conserve energy. The hypothalamus is, after all, a part of the brain.

SUMMARY

The rates of all the activities of an animal depend on the rates of biochemical reactions, which depend on body temperature. In ectotherms the body temperature depends on the environmental temperature. Most ectotherms are therefore poikilotherms—with varying body temperatures. Ectotherms can, however, compensate for low environmental temperatures by producing more enzymes or a different isoenzyme. Many avoid the destructive effects of freezing by such means as increasing the concentration of solute in body fluids, by antifreezes, or by supercooling. Some poikilotherms are partial endotherms: They have periods in which the body temperature is determined by their production and conservation of heat.

In complete endotherms the body temperature is almost always determined by the production and conservation of body heat, and most complete endotherms are homeotherms. The production of heat is either shivering or nonshivering thermogenesis. Heat is conserved or eliminated by changing the area, temperature, and conductivity of the body surface, which affects conduction, convection, and radiation of heat. Vaporization of water by panting or sweating is another means by which endotherms eliminate body heat.

Homeothermy allows most physiological functions to occur at a constant rate regardless of environmental temperature. The major exception of this rule is the metabolic rate, which is much higher even at the acclimation temperature and is even higher at colder or hotter temperatures. Some homeotherms conserve energy at extreme temperatures by allowing the body temperature to deviate from normal.

KEY TERMS

heat
temperature
ectotherm
endotherm
poikilotherm
homeotherm
torpor

acclimation
partial endothermy
thermogenesis
brown adipose tissue
conduction
convection
radiation

vaporization
piloerection
vasoconstriction
vasodilation
thermoneutral zone
hibernation
pyrogen

SELF-TEST

1. Explain how temperature affects an animal's activities.

2. What two properties of enzymes can be modified to compensate for a reduction in body temperature?

3. Define the following: poikilotherm, homeotherm, ectotherm, endotherm. In general, what kinds of animals are found in each category? Name some specific examples.

4. Describe three ways in which poikilotherms avoid freezing.

5. What are the four processes by which heat is exchanged between an animal and its environment? For each of the four, give one method in which an endotherm can modify the rate of heat loss by that process.

6. In the experiment illustrated in Figure 15.7, students typically observe the following responses in dry mice as the temperature falls: the mouse's fur fluffs out, the mouse curls up, the ears and tail become pale, and the mouse shivers. As the temperature rises they observe that the mouse licks its fur, and its ears and tail become red. Explain the significance of each of these observations. Which of these actions are behavioral, and which ones are under autonomic control?

READINGS

RECOMMENDED READINGS

Bakker, R. T. 1975. Dinosaur renaissance. *Sci. Am.* 232(4):58–78 (Apr).

Carey, F. G. 1973. Fishes with warm bodies. *Sci. Am.* 228(2):36–44 (Feb).

Coutant, C. C. 1986. Thermal niches of striped bass. *Sci. Am.* 255(2):98–104 (Aug).

Eastman, J. T. and A. L. DeVries. 1986. Antarctic fishes. *Sci. Am.* 255(5):106–114 (Nov).

French, A. R. 1988. The patterns of mammalian hibernation. *Am. Sci.* 76:568–575.

Heinrich, B. 1973. The energetics of the bumblebee. *Sci. Am.* 228(4):96–102 (Apr).

Heinrich, B. 1979. *Bumblebee Economics.* Cambridge, MA: Harvard University Press.

Heinrich, B. 1981. The regulation of temperature in the honeybee swarm. *Sci. Am.* 244(6):146–160 (June).

Heinrich, B. 1984. *In a Patch of Fireweed.* Cambridge, MA: Harvard University Press.

Heinrich, B. 1987. Thermoregulation in winter moths. *Sci. Am.* 256(3):104–111 (Mar).

Heinrich, B. and G. A. Bartholomew. 1972. Temperature regulation in flying moths. *Sci. Am.* 226(6):70–77 (June).

Heinrich, B. and G. A. Bartholomew. 1979. The ecology of the African dung beetle. *Sci. Am.* 241(5):146–156 (Nov).

Jordan, W. 1984. The bee complex. *Science 84*:58–65 (May). (*On Bernd Heinrich.*)

Kanwisher, J. W. and S. H. Ridgway. 1983. The physiological ecology of whales and porpoises. *Sci. Am.* 248(6):110–120 (June).

Lee, R. E. Jr. 1989. Insect cold-hardiness: to freeze or not to freeze. *BioScience* 39:308–313.

Schmidt-Nielsen, K. 1981. Countercurrent systems in animals. *Sci. Am.* 244(5):118–128 (May).

Southwick, E. E. and G. Heldmaier. 1987. Temperature cold in honey bee colonies. *BioScience* 37:395–399.

Storey, K. B. and V. M. Storey. 1990. Frozen and alive. *Sci. Am.* 263(6):92–97 (Dec).

ADDITIONAL REFERENCES

Cossins, A. R. and K. Bowler. 1987. *Temperature Biology of Animals.* London: Chapman and Hall.

Harris, C. L. 1976. Temperature vs. rate of the isolated frog heart: three hypotheses. *Physiol. Teacher* 5(3):1–3.

Heinrich, B. 1977. Why have some animals evolved to regulate a high body temperature? *Am. Naturalist* 111:623–640.

Schmid, W. D. 1988. Supercooling and freezing in winter dormant animals. In: R. W. Peifer (Ed.), *Tested Studies for Laboratory Teaching: Proceedings of the Ninth Workshop/Conference of the Association for Biology Laboratory Education* (*ABLE*). Dubuque, IA: Kendall/Hunt, Chapter 11.

Scholander, P. F. 1958. Studies of man exposed to cold. *Fed. Proc.* 17:1054–1057.

16

Reproduction

Ringed snake hatching (Natrix natrix).

LEARNING OBJECTIVES

1. What good is sex?

2. Do all animals have sex?

3. Do all animals have two sexes?

4. What processes are involved in human reproduction?

5. How are reproductive organs stimulated to develop and function at the appropriate times?

6. What causes each sex to look and behave in a manner conducive to reproduction?

7. How does a fetus obtain nutrients and other support from the mother?

8. How is milk produced?

ost animals decline and die soon after they lose the ability to reproduce. For the female octopus, in fact, there is literally no life after sex. A "suicide gland" (the optic gland) does her in soon after she ovulates. Reproduction is also the end of life in another sense of the word: Reproduction is the end for which physiology is the means. No matter how impressive an animal's nervous system, however mighty its muscles and intricate its hormonal and circulatory systems, they are merely the means for creating an internal environment in which the reproductive cells can function. It is therefore appropriate that this unit on physiology end with reproduction.

Not only physiology but also virtually every other aspect of an animal's functioning is geared in one way or another to perpetuating the species. The developmental processes by which gametes are produced, fertilized, and develop into new individuals were described in Chapter 6. Chapter 20 describes the behavior that enables animals to breed and to nurture their offspring. This chapter concentrates on the physiological mechanisms specifically involved in producing gametes and in breeding.

THE BIRDS AND THE BEES

Asexual Reproduction. "The birds and the bees" are traditionally regarded as models of human reproduction, but they and most other animals differ from humans and from each other in the ways by which they reproduce. Practically the only thing all animals have in common in their reproduction is that they *can* reproduce sexually, although some frequently choose not to do so. Sponges, flatworms, ribbonworms (phylum Nemertea), and some echinoderms such as starfishes can reproduce by **fragmentation.** (See Figure 21.10 for a taxonomic summary of this and other modes of reproduction.) In fragmentation an individual divides into two or more parts, each of which develops into a new individual. Another mode of asexual reproduction is **budding,** in which a new individual develops from a bud of cells in another individual. In *Hydra* and related cnidarians, as well as many sponges, new individuals develop by **external budding** of cells on the body surface (see Figure 24.14). Many freshwater sponges and ectoprocts produce **internal buds** that enable the species to survive harsh winters and other environmental stresses in a dormant state (see Figure 23.7).

A problem with asexual reproduction is that the new individual is genetically identical to the individual from which it developed. Thus an entire population could accumulate genetic defects, or become so genetically homogeneous that it would be wiped out by an environmental change. In sexual reproduction, on the other hand, each new individual generally combines genes from two individuals. This genetic recombination increases the likelihood that in some individuals adaptations will evolve that will enable the species to survive environmental changes. This may well be the only benefit of sex, from an evolutionary point of view. Sex also has its drawbacks, however. One problem with sex is that the formation of gametes by meiosis produces about as many males as females, yet only a few males are needed to fertilize many females. The surplus males merely use food and space that would better serve females and their offspring. Another problem with sexual reproduction is that finding mates of the correct species and opposite gender requires a great deal of energy and time and exposes animals to predators. Some animals have modified forms of sexual reproduction that avoid some of these disadvantages.

Figure 16.1
Sequential hermaphroditism in the Fairy basslet, Anthias *sp. The males at the bottom were once females like those above.*

Hermaphroditism. In many species of animals, reproduction is sexual, but each individual functions as both male and female. Such species are said to be **hermaphroditic,** after the mythological character Hermaphroditus, who had his body fused with that of a nymph. Hermaphroditic species are also called **monoecious** (mon-EE-shus), from the Greek for "one house." Like asexual reproduction, hermaphroditism avoids a surplus of males and the problem of finding mates. Hermaphroditic individuals serve as both sexes either simultaneously or sequentially. **Simultaneous hermaphroditism,** in which each individual has both types of reproductive organs at the same time, is common in sponges, corals, and other animals that cannot go searching for mates. Other simultaneous hermaphrodites include flatworms, snails (see Figure 28.19), earthworms (see Figures 29.17 and 29.18), and certain fishes. Some simultaneous hermaphrodites are capable of fertilizing themselves, thereby producing offspring that are genetically identical to themselves. Usually, however, two or more simultaneous hermaphrodites exchange gametes.

In **sequential hermaphroditism** each individual is first one gender and then the other (Figure 16.1). Sequential hermaphroditism in which females change into males is called **protogyny** (Greek *proto-* first + *gyne* female). If the change is from male to female, it is called **protandry** (Greek *andros* male). Sequential hermaphroditism avoids genetic homogeneity while ensuring a supply of both sexes. In some species the sex change occurs if there are too few of one gender. In schools of the sea bass *Anthias squamipinnis,* for example, removing a certain number of males causes an equal number of females to change to males. These protogynous fishes then have a greater chance of breeding. Female saddleback wrasse *Thalassoma duperrey* change to males if they see that their school consists of a large proportion of small (usually female) fish. In species in which there are harems of females dominated by a single male, the largest female may improve its chances of mating by changing into a male.

Hermaphroditism probably evolved in species in which each individual was permanently one sex or the other, as in most animals. These species are said to be **dioecious** (die-EE-shus; from the Greek for "two houses"). Dioecious species avoid genetic homogeneity, but they may have problems in ensuring an appropriate number of each gender and in allowing mates to find each other. Being monoecious solves those problems. Another solution is to be dioecious, but to keep mates attached to each other. The female anglerfish *Ceratias hobolli* keeps her tiny mate permanently on the outside of her body. *Enteroxenus,* a molluscan parasite of sea cucumbers, keeps her mate inside her body. *Enteroxenus* was thought to be hermaphroditic for many years until it was discovered that her "testis" was actually the degenerate male.

Parthenogenesis. Another solution to the mate-supply problem in dioecious species is for the female to reproduce without having her eggs fertilized. This arrangement is called parthenogenesis (Greek *parthenos* virgin + *genesis* birth). Parthenogenesis is fairly common among sedentary and sparsely distributed animals, such as rotifers and certain insects. In some insects parthenogenesis is a mechanism that produces different types of individuals adapted for different functions. In honey bees (*Apis mellifera*) male drones are produced parthenogenetically from unfertilized eggs. In aphids parthenogenetic and sexual generations alternate and are quite different from each other. Among vertebrates parthenogenesis occurs in several species of fish, some desert lizards, and domestic turkeys.

In animals except butterflies, moths, birds, and some reptiles, the females have two identical sex chromosomes (see p. 77), so the parthenogenetic females produce only daughters. In such species the males are genetically superfluous and may even disappear, resulting in **unisexual species.** In some unisexual species the females occasionally practice **hybridogenesis,** in which males of related species fertilize the parthenogenetic females. In whiptail lizards parthenogenetic females take turns acting like males, thereby inducing ovulation in each other (see Figure 37.15B). In some species in which parthenogenesis is common, however, a genuine male of the same species is required for reproduction, even if its genes are not. In some fishes, for example, the eggs must be fertilized by sperm, but the male's chromosomes are not used. This phenomenon is called **gynogenesis.**

SEXUAL VARIATIONS

External Fertilization. In spite of the disadvantages, the majority of animals reproduce using two separate sexes. Evidently sex has some advantages that outweigh all the disadvantages, but there is considerable controversy about what they may be (see p. 383). Pleasure is usually not one of the advantages. In many dioecious species individuals simply release sperm or ova into the environment without even seeing each other (Figure 16.2). The sperm may be chemically attracted to ova of the same species, or they may simply swim or drift until they encounter them. Fertilization then usually occurs outside the body and is called external fertilization. External fertilization exposes gametes to damage from the environment, especially in fresh water or air. Exposed gametes are also likely to become food for other species. In many external fertilizers the risk is reduced by releasing sperm and eggs at the same time. Corals inhabiting extensive areas of reef, for example, simultaneously release their gametes during the same hour in **mass spawnings.** Depending on species, the means of synchronizing the release of gametes may be a chemical **(pheromone)** or a visual, tactile, or auditory signal during **courtship.** Courtship will be described in more detail later (see pp. 428–430).

Internal Fertilization. Internal fertilization is less hazardous for gametes and prevails among flatworms, earthworms, molluscs, insects, reptiles, birds, and mammals. Internal fertilization often requires some type of courtship behavior to ensure that the object of an animal's intentions belongs to the correct gender and species and is capable of producing gametes. Among predators, courtship is also important to prevent a mate from becoming a meal. Courtship climaxes with sperm transfer, usually through **copulation.** The sperm may be transferred in a fluid **(semen)** or in a packet called a **spermatophore.** The copulatory and sperm-transfer organ is usually a **penis,** although other appendages may be adapted to that purpose. Octopuses use an arm (Figure 16.3), spiders use a pedipalp (see Figure 30.14A), and sharks use special claspers (see Figure 35.7). Some animals achieve internal fertilization without copulation. In scorpions, some insects, and salamanders, the male releases a spermatophore that the female then takes into her oviduct.

Oviparity. In almost all external fertilizers the young develop outside the females. Even in species with internal fertilization most of the embryonic development occurs outside the female, usually in an **egg** that is protected by a shell or some other structure. Such species are said to be oviparous (Latin *ovum* egg + *parere* to bring forth). Oviparity is common in many invertebrates and in birds, most fishes, amphibians, reptiles, and primitive mammals (monotremes) such as the duck-billed platypus. Following release of the egg from the mother **(oviposition),** the developing embryo obtains its nourishment from **yolk.** Although the

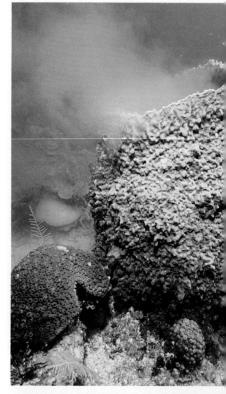

A

B

Figure 16.2
Two examples of external fertilization. (A) A basket sponge releases sperm into seawater, causing divers to report "smoking sponges." (B) Two chorus frogs Pseudacris *sp. in amplexus, the breeding embrace. The smaller male behind releases sperm as the female releases the white eggs.*

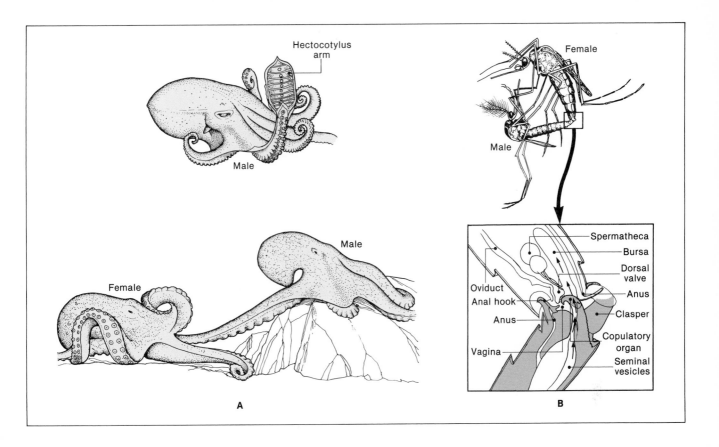

Figure 16.3

Two examples of intromittent organs for the introduction of sperm during internal fertilization. (See the next figure for another example.) (A) The male octopus has one arm specialized as a "copulatory arm" with which he inserts spermatophores into the female. After mating, the end of the copulatory arm detaches, rendering him incapable of reproducing again. Some eminent scientists of the 19th century mistook the detached end of the arm for a parasite in the female. They named the "parasite" Hectocotylus octopodis, *and the copulatory arm is still often called the hectocotylus arm. (B) Internal fertilization in the yellow-fever mosquito* Aedes aegypti *involves an intricate interlocking of genitals. The male's clasper and anal hook stabilize and widen the female's vagina. The copulatory organ is then engaged, widening the opening through which semen enters the bursa. All this takes about 20 seconds. Afterward, the sperm, about a thousand of them, swim out of the bursa and into the spermathecae to fertilize ova as they are deposited.*

parents cannot provide oxygen, nutrients, or other physiological support directly to the embryo inside the egg, many oviparous species do provide other kinds of support. Many **brood** their eggs, guarding or warming them during development. Some species of fish even brood eggs within the mouth or gill cavity. The rare (perhaps extinct) Australian frog *Rheobatrachus silus* broods eggs within its stomach (see Figure 36.16).

Ovoviviparity. Some female insects retain the eggs within their bodies during embryonic development, so the young hatch while still in the oviduct. This phenomenon is called ovoviviparity (Latin *vivus* living). Ovoviviparous mothers do not provide any nourishment or oxygen to the embryo, but simply serve as brooding chambers. Often ovoviviparity is said to occur in some fishes, amphibians, and reptiles as well as insects. In these vertebrates, however, the egg shell is essentially absent, and the female provides at least some nutrient or oxygen through specialized membranes (a **placenta**) in the oviduct (Smith 1986).

Viviparity. Viviparity occurs in all placental mammals (i.e., mammals except monotremes) and perhaps the "ovoviviparous" vertebrates noted in the previous paragraph. Females of viviparous species give birth to living young after nurturing them within their bodies during embryonic development. Some of the adaptations for viviparity will be discussed in the next section and in the chapter on mammals (Chapter 39).

REPRODUCTIVE FUNCTIONING OF MALE MAMMALS

All mammals are internal fertilizers, and all but the monotremes are viviparous. Mammals are accordingly equipped with quite elaborate organs for fertilization and for bearing young. The functioning and coordination of these organs are similar in most mammals and can be represented by humans. Figure 16.4 shows the reproductive organs in the adult human male. The external genitalia include both the beginning and the end of the male's direct role in reproduction. The **seminiferous tubules** in the **testes** produce sperm in the process called spermatogenesis (see p. 108). The sperm mature in the **epididymis** on each testis, and they are stored there until ejaculation. A tube called the **vas deferens** transports sperm to the **urethra** of the **penis** during ejaculation. The sperm are propelled by contractions of smooth muscles in the epididymides (ep-ee-DID-ee-MY-deez, plural of epididymis) and the **vasa deferentia** (plural of vas deferens).

During ejaculation a man releases about 300,000,000 sperm in 3 milliliters of semen. Semen consists of secretions from three sets of glands:

1. Two **bulbourethral glands** (= Cowper's glands) secrete a clear, mucus-containing solution for lubrication.
2. Two **seminal vesicles** produce a viscous fluid containing fructose as an energy source for the sperm. The seminal vesicles also secrete hormonelike prostaglandins (see pp. 183–184).
3. The **prostate gland** produces a milky fluid that maintains a suitably alkaline environment for the sperm.

Intromission of the normally flaccid penis is made possible by **erection,** assisted in most mammals by a bone. In humans erection is due to three cylinders of erectile tissue that fill with blood. The circulatory changes responsible for erection are controlled by the autonomic nervous system. The parasympathetic nervous system stimulates erection in response to pleasurable stimulation of mechanoreceptors in the penis, as well as olfactory, visual, and other types of stimulation. The sympathetic nervous system inhibits erection in response to fear or aggression. The sympathetic nervous system is, however, responsible for ejaculation.

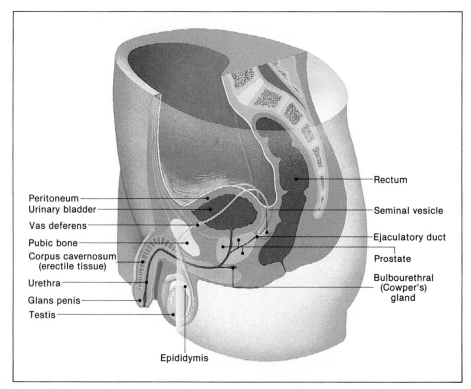

Figure 16.4
The reproductive organs of a human male.

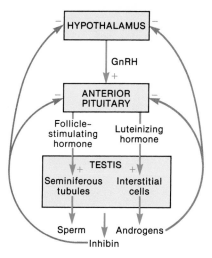

Figure 16.5
Stimulatory (+) and inhibitory (−) interactions among the male reproductive hormones.

HORMONAL CONTROL OF REPRODUCTION IN MALES

Reproductive physiology in the male is also controlled by hormones, especially two **gonadotropic hormones** from the anterior pituitary. One such gonadotropin is **follicle-stimulating hormone** (FSH). FSH stimulates production of sperm in the **seminiferous tubules** of the testes. The second gonadotropin from the anterior pituitary is **luteinizing hormone** (LH). LH, also called **interstitial-cell-stimulating hormone** (ICSH) in males, stimulates the interstitial cells (Leydig cells) of the testes to produce **androgens**. Androgens, such as **testosterone,** are steroid hormones that stimulate development and functioning of the reproductive organs. Androgens also promote **secondary sex characteristics.** These are traits that are useful but not essential for reproduction. In men they include hair on the face and body, deep voice, and behavioral characteristics such as aggression (Rubin et al. 1981). Androgens are also **anabolic steroids,** which means that they promote growth of muscle tissue.

Androgens also have a negative effect on their own release, by inhibiting the secretion of LH, and possibly FSH. Androgens may act directly on the anterior pituitary, or they may act through the hypothalamus by inhibiting secretion of **gonadotropin releasing hormone** (GnRH), the peptide that stimulates release of both FSH and LH (see p. 319). This inhibition tends to keep the reproductive state of men from fluctuating as much as in females, although androgens do surge during sexual arousal. The testes also produce **inhibin,** which inhibits GnRH and FSH secretion. These hormonal interactions are illustrated in Figure 16.5.

REPRODUCTIVE FUNCTIONING OF FEMALE MAMMALS

The reproductive organs of a woman are illustrated in Figure 16.6. The **vagina** is a receptacle for the penis, and it, as well as the clitoris and labia, has mechanoreceptors that produce the sensation of pleasure in women. Oogenesis, the production of eggs, occurs in the two **ovaries** (see p. 108). In mammals the egg is called an **oocyte** (OH-ah-site), and release of the oocyte is called **ovulation.** After ovulation the oocyte enters one of the two **oviducts** (= Fallopian tubes) and is propelled toward the **uterus** (= womb) by cilia. Fertilization, if it occurs, usually happens in the oviduct, and embryonic development begins before the fertilized egg (= **zygote**) arrives in the uterus (see Figure 6.9). In women and other primates the uterus has a single chamber that is adapted for the development of one offspring at a time. Most other mammals have a Y-shaped **bicornate uterus,** with two horns in which litters of offspring develop.

Girls are born with about a million primary oocytes, of which about 400,000 survive past puberty. A few hundred of these develop, usually one at a time, within **follicles** in the ovaries (Figure 16.7). Ovulation usually occurs approximately every 28 days for three decades, first in one ovary and then the other. During ovulation the oocyte is ejected into the abdominal cavity, but it is usually picked up by the ciliated **fimbriae** at the end of the oviduct. The follicle remains in the ovary, where it forms a **corpus luteum** (Latin for "yellow body").

HORMONAL CONTROL OF REPRODUCTION IN FEMALES

Coordination of reproductive physiology and behavior in the female mammal is orchestrated by an impressive array of mechanisms, many of them hormonal. As in males, the anterior pituitary releases two gonadotropins, follicle-stimulating hormone (FSH) and luteinizing hormone (LH). FSH in women stimulates devel-

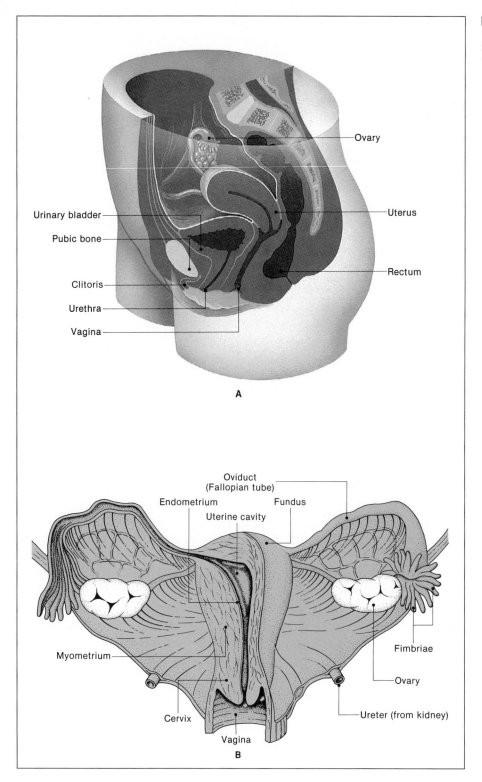

opment of a follicle. LH triggers ovulation and stimulates the development of the corpus luteum. The secretions of both FSH and LH are due to gonadotropin releasing hormone (GnRH) from the hypothalamus. The follicle and the corpus luteum are also endocrine organs involved in regulating the ovarian cycle. The follicle produces steroid hormones called **estrogens,** including estradiol. Estrogens stimulate the development of the uterine lining—the **endometrium**—preparing it for implantation (see pp. 120–121). Estrogens also promote secondary

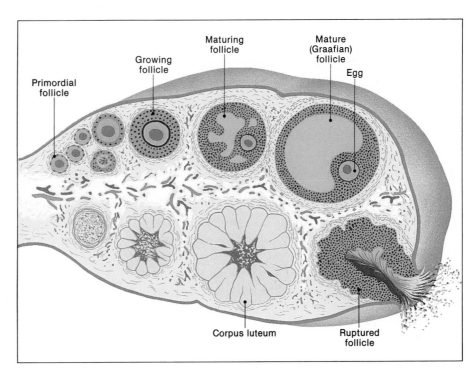

Figure 16.7
An ovary, showing stages in the maturation of a follicle (clockwise from upper left).

Primordial follicle

Growing follicle

Maturing follicle

Mature (Graafian) follicle

Egg

Corpus luteum

Ruptured follicle

sexual characteristics, such as absence of hair on the face and body, high-pitched voice, more subcutaneous fat, and appropriate reproductive behaviors. (Male rats that have been castrated and injected with estrogens build nests, something no self-respecting male rat would ordinarily do.) The corpus luteum also produces estrogens, as well as another steroid hormone, **progesterone.** Like estrogens, progesterone stimulates the endometrium in preparation for implantation.

In addition, estrogens and progesterone have feedback effects on the hypothalamus and anterior pituitary. Estrogens from the ripening follicle and from the corpus luteum inhibit FSH secretion, preventing the development of other oocytes. Estrogens also stimulate LH secretion, triggering ovulation. Progesterone from the corpus luteum then inhibits LH secretion. The ovary also produces inhibin, which inhibits production of FSH. Besides acting directly on the anterior pituitary, estrogens and inhibin may inhibit FSH secretion by inhibiting release of gonadotropin releasing hormone from the hypothalamus. These interactions are illustrated in Figure 16.8.

ESTROUS AND MENSTRUAL CYCLES

In many female mammals the hypothalamus releases GnRH according to a regular cycle. By controlling the release of FSH and LH, this cycle also controls the development and ovulation of the oocyte. In most mammals the females become sexually receptive at the time of ovulation and are then said to be "in heat" or, more politely, in **estrus.** In these mammals the reproductive cycle is called the **estrous cycle.** (Note the difference in spelling of the noun "estrus" and the adjective "estrous.") In humans and gorillas, the females are sexually receptive without regard to whether there is an oocyte ready to be fertilized. In these species sex serves not only for the conception of offspring but also to keep both parents together during the long gestation and upbringing of the child. In these and other primates the decline in levels of estrogens and progesterone as the corpus luteum degenerates causes some of the endometrium to slough off, forming a **menstrual**

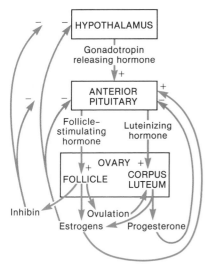

HYPOTHALAMUS

Gonadotropin releasing hormone

ANTERIOR PITUITARY

Follicle-stimulating hormone

Luteinizing hormone

OVARY
FOLLICLE
CORPUS LUTEUM

Inhibin

Ovulation

Estrogens

Progesterone

Figure 16.8
Stimulatory (+) and inhibitory (−) interactions among the female reproductive hormones.

discharge. In primates, therefore, the cycle is called the **menstrual cycle** rather than the estrous cycle.

The changes in hormone levels and in the ovary and endometrium are represented in Figure 16.9. The first day of menstruation is arbitrarily taken as day one of the cycle. At this time estrogen and progesterone levels are low, so secretion of FSH is not inhibited. As FSH levels rise a new follicle ripens, and an oocyte develops within it. The ripening follicle then produces estrogens, which trigger a surge in LH secretion. The LH surge triggers ovulation and formation of a corpus luteum. The corpus luteum continues to produce estrogens and also progesterone, which terminates LH secretion. Both estrogens and progesterone stimulate development of the endometrium. If conception does not occur the corpus luteum degenerates, levels of estrogens and progesterones drop, and a new cycle begins.

Many female mammals do not have estrous or menstrual cycles but depend on external stimuli to trigger ovulation. The external stimulation may be seasonal, so that ovulation and estrus occur only at a particular time of the year. This ensures that young will be born when temperature, food supply, and other conditions are more favorable to their survival. In some cats, rabbits, and other mammals the

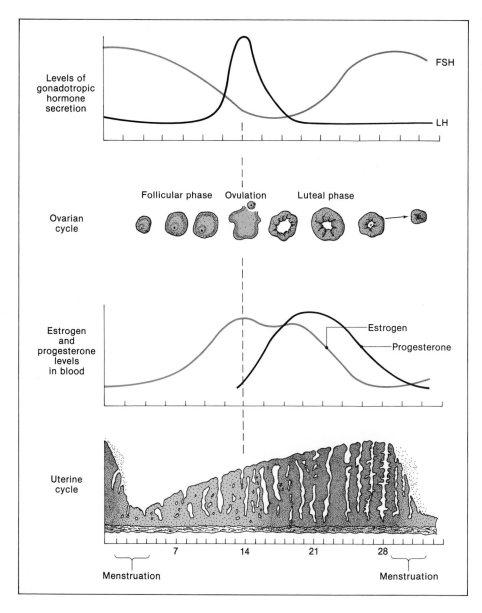

Figure 16.9
Changes in hormone levels, the follicle, and the endometrium during a menstrual cycle. The graph begins five days before the onset of a menstrual period (menses).

Birth Control

SURGICAL METHODS

Birth control is becoming increasingly important in the management not only of human populations but also of wild and captive animals. One approach to birth control is **castration:** the surgical removal of ovaries or testes. Since castration also removes the sources of sex steroids, it has not enjoyed much success lately among humans. It is most often used by farmers and pet owners to prevent breeding and to make animals more docile.

A less radical surgical procedure is **vasectomy:** cutting and tying the vasa deferentia to prevent sperm from being ejaculated. Vasectomy may now be the most common surgery in the United States and is probably the most common form of contraception in married men. It is performed through the scrotum under local anesthetic during a short visit to the surgeon's office. Vasectomy is also becoming increasingly common in the management of animal populations in zoos. In some parts of the United States, beaver populations are controlled by vasectomy. The analogous procedure in females is **tubal ligation:** tying the oviducts. Blocking the oviducts by tubal ligation is more complicated than vasectomy, since it requires entry into the abdominal cavity.

Another surgical approach to birth control is **abortion,** in which the fetus is removed from the uterus by scraping or sucking it from the endometrium, or by injecting hypertonic salt solution. Abortion is generally more dangerous to the woman than is contraception, although it is less risky than pregnancy itself.

HORMONAL INTERFERENCE

In Europe a drug called **RU 486** has gained wide acceptance as a safe and effective means of inducing abortion. RU 486 works by blocking the receptor molecules for progesterone in the uterus. It is usually used in conjunction with prostaglandin (see pp. 183–184), which induces uterine contractions. In the United States hormonal interference is used only to prevent conception. The popularity and social impact of chemical contraception can be judged by the fact that of all the pills available in the United States, only one is instantly recognized as *the* pill. *The* pill usually consists of a combination of synthetic estrogens and progesterone that inhibits FSH and/or LH production. If used according to directions, the pill is completely effective in preventing conception. The advantage of using steroids as oral contraceptives is that they can be swallowed and absorbed without being destroyed by digestive enzymes.

MECHANICAL AND OTHER

Several other techniques are widely used, even though they are less effective in humans and generally impractical in other animals. One approach to contraception is spermicidal jelly or foam that kills sperm. These are about 80% effective when used alone, and more so when used with a diaphragm: a flexible device that fits over the cervix (neck) of the uterus and prevents entry by sperm. The condom also mechanically prevents sperm from fertilizing eggs. One advantage of the condom over all other forms of contraception is that it reduces the spread of AIDS and other sexually transmitted diseases. However, because of slipping and leaking, condoms fail up to 14% of the time.

In China and some other developing countries the main method of birth control is the **intrauterine device** (IUD). IUDs come in a variety of forms and are inserted into the uterus. Like any foreign object in the uterus, an IUD blocks implantation. IUDs are approximately 98% effective. Because of some risk of uterine damage, however, physicians seldom prescribe them in the United States, and many manufacturers have stopped selling them to avoid lawsuits. [The threat of litigation is one reason why there is little effort to develop new contraceptives in the United States, according to Djerassi (1989).]

One family-planning method that does not rely on technology is the **rhythm method,** in which intercourse is avoided for three days before and after ovulation. (Eggs and sperm can each survive for only about three days in the female.) The rhythm method reduces the likelihood of conception *if* the time of ovulation can be accurately determined. Often it cannot. Out of 100 women using the rhythm method exclusively, between 13 and 24 will get pregnant within a year. The rhythm method is therefore used mainly because it is the only one permitted by the Catholic Church.

females ovulate only when stimulated during copulation. This **reflex ovulation** avoids wasting eggs and time.

PREGNANCY

If conception occurs, the reproductive system of the female shifts out of the menstrual cycle and prepares for pregnancy. This radical change in the physiology of the mother-to-be is triggered by the **blastocyst**: the cluster of cells that develops from the fertilized ovum (see pp. 112–113). The outer layer of cells of the blastocyst, called the **trophoblast** (see pp. 120–121), forms a **chorion**. One of the functions of the chorion is to secrete a protein hormone called **human chorionic gonadotropin** (hCG). hCG (or simply **chorionic gonadotropin** in nonhumans) enters into the circulatory system of the mother and prevents the corpus luteum from degenerating. Consequently, instead of the levels of estrogens and progesterone declining, the levels of these steroid hormones increase. Estrogens and progesterone continue to stimulate development of the endometrium, and menstruation does not occur. hCG is therefore the basis for the traditional first hint of pregnancy, the "missed period." hCG is also the basis for modern tests for pregnancy, which use antibodies to detect hCG in the urine of a pregnant woman.

Besides secreting hCG, the chorion also forms fingerlike processes called **chorionic villi,** which embed themselves into the endometrium around five or six days after fertilization. During the next week these chorionic villi develop into the fetal portion of the **placenta** (Figure 16.10). The main function of the placenta is to establish an exchange of nutrients, metabolic wastes, respiratory gases, and other materials between the fetus and the blood flowing in the mother's endometrium. Nutrients, oxygen, and other materials in the blood of the mother diffuse or are transported across the membranes of the villi and into the circulatory system of the fetus. Carbon dioxide and other metabolic wastes from the fetus move in the opposite direction into the maternal circulation.

At around the third month after conception the level of hCG declines, and the placenta takes over most of the production of estrogens and progesterone. In addition to stimulating uterine development, these hormones promote development of the mammary glands, preparing them for **lactation**—the production of milk. At the same time progesterone inhibits lactation until after birth. Estrogens and progesterone, as well as other placental hormones, also suppress the *T* cells in the mother's immune system, preventing the rejection of the foreign embryonic tissue developing within her.

Figure 16.10
(A) A human fetus at about the eighth week of gestation (compare Figure 6.27), showing the placenta. (B) Detail of the placenta showing some of the chorionic villi surrounded by pockets of circulating maternal blood. Blood flows away from and toward the fetus through vessels in the umbilical cord.

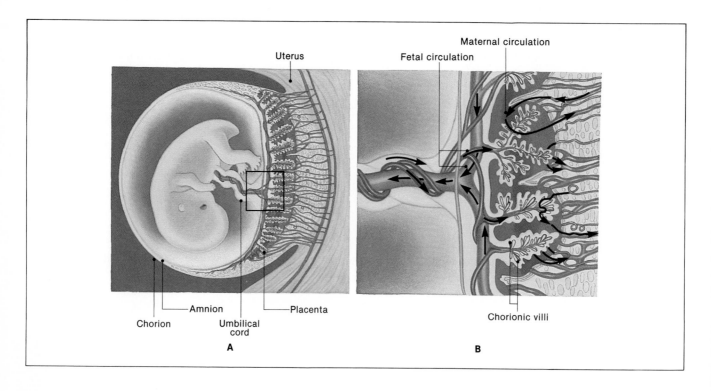

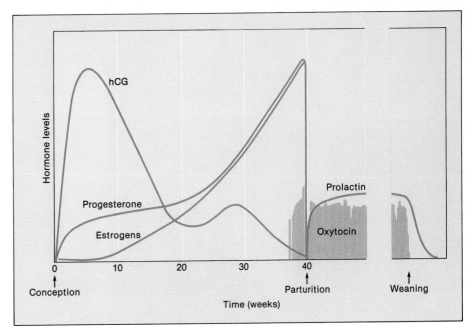

Toward the end of the gestation period uterine contractions begin, gradually becoming stronger and more frequent. These contractions are due to increased secretion of **oxytocin** by the posterior pituitary (see p. 191) and increased numbers of oxytocin receptors on smooth muscle (**myometrium**) of the uterus. The uterine contractions culminate in **parturition**—the delivery of the child (Figure 16.11).

LACTATION

The mother's levels of estrogens and progesterone drop sharply as soon as the placenta is delivered after the birth of the baby. This allows **prolactin** from the anterior pituitary to initiate lactation. Prolactin continues to be secreted as long as suckling occurs regularly, because nerve impulses from the nipple prevent the hypothalamus from secreting a hormone (dopamine) that inhibits prolactin secretion. These nerve impulses also trigger the release of oxytocin, which causes the actual discharge of the milk. This discharge, which goes by the unromantic name **milk let-down,** is due to contractions of smooth muscles around the sacs (**alveoli**) that contain the milk (Figure 16.12).

As long as the nipple is stimulated regularly, lactation can continue for several years. If suckling occurs several times every hour, as it does in some cultures, the release of FSH and LH are blocked, and the mother remains infertile. Lactation thus serves as a natural means of spacing births. As the young are **weaned,** however, the level of prolactin drops, lactation ceases, and the mother resumes ovulation.

In addition to its nutritional and family-planning benefits, human milk protects infants from one of their greatest threats, intestinal infections. Peyer's patches in the small intestine of the mother (see Figure 11.12) contain plasma cells that produce antibodies to bacteria ingested by the mother. These plasma cells migrate to the breasts and enter the milk, thereby providing infants with a source of antibodies to bacteria that they too are likely to encounter. Ironically, in developing countries, where children would benefit most from improved resistance to disease, mothers are giving up nursing in favor of bottle-feeding, so that they can be like

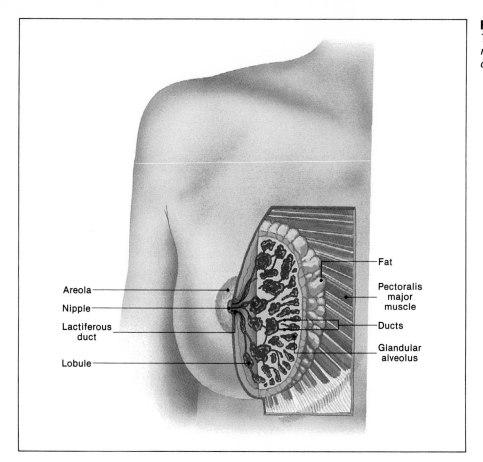

Figure 16.12
The structure of the human mammary gland. Milk is secreted by cells of the alveoli.

Areola

Nipple

Lactiferous
duct

Lobule

Fat

Pectoralis
major
muscle

Ducts

Glandular
alveolus

women in developed countries. Meanwhile, increasing numbers of women in developed countries are realizing the benefits of nursing.

SUMMARY

All species of animals are capable of sexual reproduction, which has the advantage of increasing genetic diversity, but the disadvantage of producing superfluous males and requiring sometimes complicated matings. Some animals avoid these problems by reproducing asexually, by fragmentation or budding. Others are simultaneous or sequential hermaphrodites, combining the functions of both sexes in one animal. Other animals are parthenogenetic: the female produces offspring without the eggs being fertilized. Most animals, however, are dioecious and reproduce only sexually. Fertilization may occur externally or internally, and the females may be oviparous, ovoviviparous, or viviparous. All mammals have internal fertilization, and all but a few are viviparous.

The testes of male mammals produce sperm, which is ejaculated with semen. Reproductive functions are coordinated largely by hormones, especially two gonadotropins, FSH and LH, from the anterior pituitary. Androgens from the testes stimulate development and functioning of the reproductive organs, as well as secondary sexual characteristics. In females, follicles in the ovaries produce oocytes, which are released during ovulation into an oviduct. FSH and LH are released in a cyclic fashion corresponding to the estrous or menstrual cycle. The follicle produces estrogens; then after ovulation it develops into a corpus luteum, which produces both estrogens and progesterone. These hormones stimulate reproductive functions as well as secondary sexual characteristics.

Following conception, chorionic gonadotropin from the blastocyst prevents the corpus luteum from degenerating, and the blastocyst becomes implanted in the endometrium. A placenta then forms and exchanges oxygen, nutrients, and other necessities from the mother's blood to the fetus. Estrogens and progesterone, first from the corpus luteum and then from the placenta, maintain the endometrium. These hormones and also prolactin prepare the breasts for lactation. Oxytocin triggers uterine contractions that lead to parturition. After birth, prolactin stimulates lactation, and oxytocin causes milk let-down.

KEY TERMS

fragmentation
budding
hermaphrodite
monoecious
dioecious
parthenogenesis
external fertilization
internal fertilization
oviparity
ovoviviparity
viviparity
testis

vas deferens
seminal vesicle
prostate gland
vagina
ovary
oviduct
uterus
ovulation
follicle
corpus luteum
follicle-stimulating hormone
luteinizing hormone

androgen
estrogens
progesterone
estrus
estrous cycle
menstrual cycle
chorionic gonadotropin
prolactin
oxytocin
lactation

SELF-TEST

1. What are the disadvantages of sexual reproduction? How do fragmentation and budding avoid those disadvantages? How do hermaphroditism and parthenogenesis avoid those disadvantages?

2. What are the advantages of internal fertilization?

3. Define the following terms: oviparous, ovoviviparous, viviparous. In what kinds of animals does each of them occur? What are the advantages and disadvantages of each?

4. Describe the career of a human sperm cell from the time it leaves the testis until it fertilizes an ovum.

5. Diagram a sequence of neural and hormonal events that could account for the reflex ovulation of a cat or rabbit after copulation.

6. From your understanding of hormonal effects, sketch a graph showing changes in hormones, the follicle, and the uterus throughout a complete menstrual cycle. Compare your graph with Figure 16.9.

7. In your graph for the preceding question, show how the levels of hormones would change after ovulation if the egg is fertilized. What triggers these changes in the mother?

8. What are the roles of prolactin and oxytocin during pregnancy and lactation?

READINGS

RECOMMENDED READINGS

Beaconsfield, P., G. Birdwood, and R. Beaconsfield. 1980. The placenta. *Sci. Am.* 243(2):94–102 (Aug).

Bronson, F. H. 1984. The adaptability of the house mouse. *Sci. Am.* 250(3):116–125 (Mar).

Cole, C. J. 1984. Unisexual lizards. *Sci. Am.* 250(1):94–100 (Jan).

Eberhard, W. G. 1990. Animal genitalia and female choice. *Am. Sci.* 78:134–141.

Frisch, R. E. 1988. Fatness and fertility. *Sci. Am.* 258(3):88–95 (Mar).

Gould, S. J. 1983. *Hen's Teeth and Horse's Toes.* New York: W. W. Norton. (*See Chapter 1 on reproductive curiosities of fishes.*)

Kirkpatrick, J. F. and J. W. Turner Jr. 1985. Chemical fertility control and wildlife management. *BioScience* 35:485–491.

Short, R. V. 1984. Breast feeding. *Sci. Am.* 250(4):35–41 (Apr).

Ulmann, A., G. Teutsch, and D. Philbert. 1990. RU 486. *Sci. Am.* 262(6):42–48 (June).

Warner, R. R. 1984. Mating behavior and hermaphroditism in coral reef fishes. *Am. Sci.* 72:128–136.

ADDITIONAL REFERENCES

Djerassi, C. 1989. The bitter pill. *Science* 245:356–361.

Eberhard, W. G. 1985. *Sexual Selection and Animal Genitalia.* Cambridge, MA: Harvard University Press. (*On hypotheses regarding the different shapes of genitalia.*)

Rubin, R. T., J. M. Reinisch, and R. F. Haskett. 1981. Postnatal gonadal steroid effects on human behavior. *Science* 211:1318–1324.

Smith, H. M. 1986. Ovoviviparity: spurious for ectotherms. *BioScience* 36:292.

Tyler, M. J. et al. 1983. Inhibition of gastric acid secretion in the gastric brooding frog, *Rheobatrachus silus. Science* 220:609–610.

Unit Three: Interactions of Animals with Their Environments and with Each Other

Overview

The topics of the previous two units are essential for a complete understanding of how animals function. Understanding how cells and physiological systems work, however, cannot completely explain such interactions as cleaning symbiosis. Would anyone who completely understands cells and physiology have predicted that the cleaner wrasse (Lambroides dimidiatus) would feed on parasites and dead tissues on other fish, and that the oriental sweetlips (Plectorhynchus orientalis) would let it? Because of the incompleteness of cellular and physiological explanations, many zoologists specialize in other approaches to understanding such interactions of animals with each other and their environments.

A zoologist specializing in ecology would wonder what contributions cleaning symbiosis makes to both cleaner and client and might try to measure the contributions in terms of energy. A zoologist interested in evolution might wonder how such an interaction ever evolved and might look for antecedents in related species or in the fossil record. A zoologist specializing in behavior would be more interested in how the cleaner and client recognize each other's intentions and would try to find out by carefully observing the behavior under different circumstances.

The chapters in this unit describe some of the questions, assumptions, and techniques of zoologists with each of these interests.

17

Ecology

Oxpeckers (Buphagus sp.) eating parasites on black rhinoceros (Diceros bicornis).

CHAPTER OUTLINE

LEARNING OBJECTIVES

1. What kinds of factors make a tropical rainforest differ from a desert, or a marsh differ from a coral reef?

2. What determines the abundance of a species?

3. Why are certain species found with each other?

4. Why do the species in an area sometimes change?

5. How do carbon, oxygen, water, and nitrogen circulate among organisms and among soil, air, and water?

6. Why are there usually fewer animals than plants, and fewer meat eaters than plant eaters?

7. Why do the numbers of organisms in an area often remain constant?

nimals get the requirements for life—oxygen, nutrients, water, ions, suitable temperature—from the physical environment and from other organisms. Every animal therefore depends on at least some parts of the physical environment, and on at least one other organism. In fact, it is probably near the truth to say that every animal interacts with everything else, at least indirectly. The study of the interactions of organisms with each other and their environment is called **ecology**, a word derived from the Greek work *oikos*, meaning "house." The word evokes an image of all living and nonliving things thrown into a house—our planet—and left to work out their relationships. "Ecology" also calls to mind "economics": the study of the production, distribution, and consumption of goods. For living things, goods are the inorganic and organic—the abiotic and biotic—requirements for life.

ABIOTIC INTERACTIONS

It is impractical to study the interaction of everything on earth with everything else, so ecologists generally focus on one particular ecological system at a time. An ecological system, or **ecosystem**, is an association of all the organisms in an area, together with their physical environment. This definition of ecosystem is rather imprecise, but one does not have to be a trained ecologist to see that there are differences between such ecosystems as grassland and open ocean. Such differences arise partly from differences in **abiotic** factors, such as oxygen, water, ions, and temperature. It is a familiar fact that different species live in different ecosystems. Which species can live in a particular ecosystem, and which ones cannot, are ultimately determined by the **limits of tolerance** of the species. The limits of tolerance are the ranges of temperatures, oxygen levels, ionic concentrations, and other abiotic factors in which a particular organism can survive. The upper and lower extremes of the limit of tolerance are **lethal limits,** where there is either too much or too little of an abiotic factor to permit the organism to live (Figure 17.1).

Even within the limits of tolerance, conditions may not be optimal for particular organisms, and they may show signs of stress such as hypothermia, gasping, or dehydration. Although organisms may be able to survive such stressful conditions, they are less likely to be found there, because they generally do not reproduce well in those conditions. Animals simply go elsewhere if they can. Organisms are more likely to be found where abiotic factors are within their **preferred range,** and especially near the **optimum.** Theoretically, the preferred habitat for a given species should simply be one that combines the preferred ranges for each abiotic factor. In reality, however, one physical factor can affect the preferred range of another. For example, the optimal temperature for a fish might be 20°C in well-aerated water, but in a real environment the fish might suffocate in such warm water, because warm water holds less oxygen, and high temperature increases oxygen demand.

A further complication is that the levels of all these physical factors usually change over time. There are annual cycles of temperature and precipitation, for example, and these cycles differ in various parts of the world. Temperate zones have large seasonal variations in temperature and small variations in precipitation. The most stressful time of the year is generally the winter, when many animals die from cold and lack of food. In contrast, the temperature in tropical regions remains fairly constant through the year, but in many areas a season of heavy rains alternates with a dry season. The dry season is the most stressful to animals, when vegetation is scarce and watering places dry up.

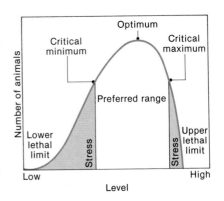

Figure 17.1
A hypothetical tolerance curve, showing the expected numbers of animals at various levels of an abiotic factor: oxygen, water, ions, or temperature.

BIOTIC FACTORS

Even if an animal could find a habitat in which all the abiotic factors are in its preferred range throughout the year, that animal could not survive there without some other organism to eat. For an animal that ate plants, for example, not only would there have to be adequate oxygen, water, ions, and temperature for itself, but also suitable sunlight, wind or water velocity, and substrate stability for the plants it eats. These and other dependencies of organisms on each other give rise to numerous biotic factors. Biotic factors are those associated with other organisms. They are primarily associated with the exchange of energy and nutrients from one organism to another.

Animals have to eat because they, unlike plants, are incapable of using raw solar energy and most inorganic matter directly. The energy and matter must first be **fixed**—incorporated into organic molecules by **autotrophs.** The term "autotroph" comes from the Greek words for "self" and "food," and it refers primarily to photosynthesizers—certain bacteria, algae, and green plants. Autotrophs are called **primary producers,** and their fixation of matter and energy in forms that can be used by animals and other organisms is called **primary production.** Primary production supports not only the autotrophs themselves, but also the animals and other organisms that feed on them and on each other. Such nonautotrophic organisms are called **heterotrophs.** Heterotrophs are also called **consumers,** because they obtain energy by consuming organic material synthesized by other organisms.

Ecologists are very much interested in primary production, because it sets the limit on how many organisms can live in a given ecosystem. Primary production varies in different areas, and also at different times, since photosynthesis shuts down when light, temperature, moisture, and certain minerals are inadequate. Much of the primary production of summer days may be used by producers themselves to sustain life during the night and winter. Whatever is left over is the **net primary production.** The annual net primary production is the yearly increase in organic matter or energy within an area. As Figure 17.2 shows, ecosystems vary greatly in annual net primary production.

ECOSYSTEM TYPES

Different combinations of abiotic and biotic factors divide the planet into distinctive types of ecosystem. For example, there are six continent-sized **faunal regions** or **faunal realms** (see Figure 19.11). Geological and evolutionary history have been most important in defining these faunal regions, so further discussion is more appropriate in a later chapter on evolution. Within each faunal region there are different combinations of abiotic factors that lead to distinctive associations of plant and animal species. These ecosystem types include prairie, tropical rainforest, lake, desert, and so on. They are easily recognized except where one grades into another. Often there is no sharp boundary between grassland and desert, or wetland and pond, for example. Terrestrial ecosystem types are called **biomes.** The two most important abiotic factors defining a biome are temperature and moisture (Figures 17.3 and 17.4). These two factors determine which kinds of plants grow in a biome, and the plants largely determine the kinds of animals there. Aquatic ecosystems are mainly divided into freshwater or marine, depending on ionic concentration. There are several types of freshwater and marine ecosystems, as listed in Figure 17.2.

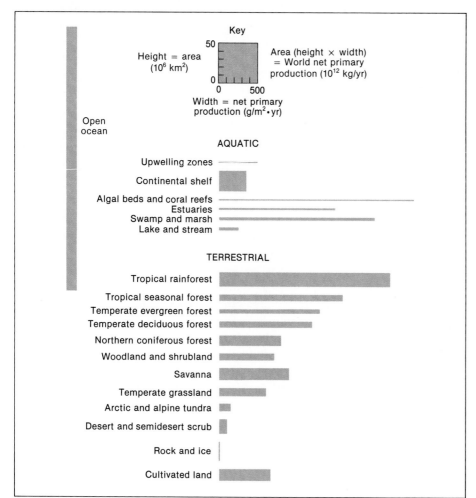

Figure 17.2
Primary production in various aquatic and terrestrial ecosystems. The height of each rectangle represents the total area of the ecosystem type on Earth. The width of each rectangle represents the average annual net primary production per unit area, estimated by comparing the dry weight of organic material at the beginning and end of one year. The area of each rectangle therefore represents the world net primary production for that type of ecosystem.

BIOMES

Tundra. On the polar ice caps no plant grows. The few animals, such as polar bears in the Arctic and penguins in the Antarctic, obtain food from the seas and are best considered members of an aquatic ecosystem rather than a biome. The northernmost biome is the belt of **arctic tundra,** where extreme cold and extended darkness are the two major facts of life. There is little precipitation, but so little water evaporates that the upper layer of soil is saturated in summer. Beneath this soggy layer is permanently frozen **permafrost.** Trees cannot take root under such conditions, so the vegetation consists of short grasses and other nonwoody plants. Similar plants are often found on mountain tops above the tree line, in **alpine tundra,** where the soil is too shallow to anchor trees against the wind. The only animals that survive year-round in Arctic tundra are the eggs and larvae of insects and a few species of mammals. The mammals include the arctic fox, snowshoe hare, lemming and other rodents, caribou, polar bear, and musk-ox. One of the few birds that live year-round in the arctic tundra is the ptarmigan, but many migratory birds arrive in the summer. In summer the populations of mosquitoes and other insects often reach plague proportions.

Northern Coniferous Forest. Vast forests of cone-bearing evergreen trees, such as spruces and fir, range south of the arctic tundra, in the **taiga** (pronounced TYE-guh) of North America and Eurasia. Similar coniferous forests grow at higher elevations farther south. These forests often seem dark and forbidding,

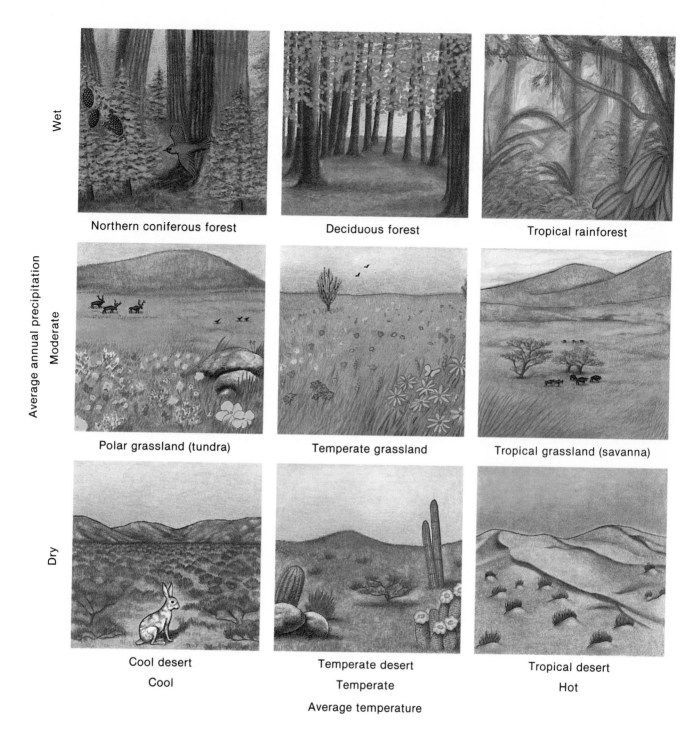

Average annual precipitation

Wet

Moderate

Dry

Northern coniferous forest

Deciduous forest

Tropical rainforest

Polar grassland (tundra)

Temperate grassland

Tropical grassland (savanna)

Cool desert

Temperate desert

Tropical desert

Cool

Temperate

Hot

Average temperature

Figure 17.3
Temperature and rainfall determine which plants and animals can survive in a biome.

because the needle-shaped leaves of conifers do not transmit sunlight. Little vegetation grows on the forest floor, not only because of the darkness but also because the soil is acidic and poor in nutrients. Most animals, including squirrels, birds, and hordes of insects, therefore forage high in the trees. Broad-leaved shrubs and trees do grow along streams, lakes, and other clearings, however. Deer, bear, beaver, moose, and hare can often be found feeding there.

Temperate Deciduous Forest.   In moist temperate regions, such as the eastern United States, conifers give way to broad-leaved trees that lose their leaves in autumn. Even after the leaves reemerge in spring, deciduous forests appear lighter than coniferous forests, because the leaves transmit more light to the forest floor.

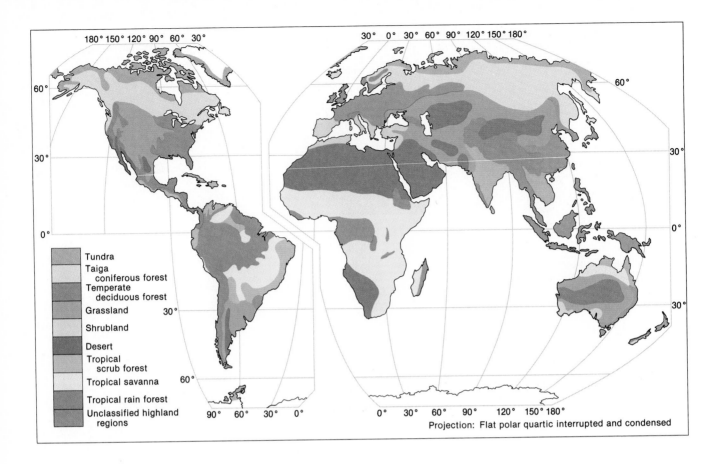

The fallen leaves of autumn form a nutrient-rich **humus** on the forest floor. In contrast to coniferous forests, therefore, the floors of deciduous forests are often crowded with flowering plants, shrubs, and small trees. Animals are diverse and abundant at all levels. Among the most conspicuous are deer, black bears, squirrels, chipmunks, rabbits, and birds of all kinds.

Figure 17.4
The major biomes of the world.

Temperate Grassland (Prairies).   In central North America and in parts of Africa, South America, and Asia, groundwater is too scarce to support trees. Grasses and other nonwoody plants are dominant plants, and grazers such as antelopes and bison are (or were) the dominant animals. Natural prairies are now quite rare, since the deep, fertile soil proved too tempting as farmland, and the absence of forest cover made the animals easy targets for hunters. Smaller vertebrates, such as prairie dogs, rabbits, foxes, coyotes, and prairie chickens, still thrive on the remaining grasslands.

Tropical Grassland (Savanna).   In the savannas of Africa and South America the average annual rainfall is abundant, but periodic dry seasons and fires limit trees to isolated clumps. These trees, such as acacia and baobabs in Africa, are centers of intense animal activity. Ants nest within thorns on the acacias and pay their rent by protecting the trees from other insects. Only giraffes browse among the branches. The baobab trees provide nest sites for birds, at least until they are knocked over and eaten by elephants. In Africa the grasses support the largest diversity of hoofed animals on earth, including zebras, giraffes, and many species of antelope, such as wildebeest, topi, and gazelle. They graze peacefully but watchfully in view of lions, hyenas, and other predators that wait to take advantage of the weak or lame.

Desert. Deserts occur wherever evaporation exceeds rainfall. In the **hot deserts** of the Sahara and the American Southwest the high evaporation is due to high daytime temperatures. **Cool deserts** occur in areas of extremely low precipitation, such as in Washington and Oregon, where the Cascade Mountain range intercepts moisture from the prevailing winds. Desert plants and animals must be adapted to quickly absorb and hold whatever moisture is available. Many desert animals obtain much of their water from metabolism, and their excretory organs are adapted to conserve water. In hot deserts they are often nocturnal and remain underground during the heat of the day. Among these are numerous reptiles, insects, scorpions, and rodents. Grasslands can become deserts if they are stripped of vegetation, which shades the soil and holds its moisture. **Desertification** is occurring in much of Africa because of overgrazing by cattle.

Tropical Rainforest. Tropical rainforests occur in equatorial regions with constant high temperature and moisture. This ecosystem type will be described in greater detail near the end of this chapter. A related biome, the **semi-evergreen, seasonal tropical forest,** occurs in equatorial areas subject to alternating dry and rainy seasons.

FRESHWATER ECOSYSTEM TYPES

Ponds and Lakes. Bodies of water with little or no current are called **lentic** ecosystems. Whether a lentic body of water is called a pond or a lake is often a matter of local custom, but in general ponds are more shallow than lakes. Natural ponds and lakes occur most often within moist biomes, especially northern coniferous forests. Beavers frequently create ponds, and Ice Age glaciers have hollowed out deep basins that now hold lakes. Lakes can be divided into three zones, depending on how much sunlight penetrates the water to support photosynthesis. The **littoral zone** includes shallower water, usually around the lake margin, where sunlight penetrates to the bottom. Rooted plants, such as pond lilies, grow there, and a variety of animals take advantage of the plants for food and shelter. These animals include leeches, snails, and numerous insects that may spend all or part of their lives under water. Small fishes and amphibians are often numerous in the littoral zone, and larger vertebrates, such as moose and raccoon, come from the forests to feed on them.

Enclosed by the littoral zone is open water where sunlight does not penetrate deeply enough to support rooted plants. In the upper portions of the open water is the **limnetic zone,** where sunlight is sufficient to support photosynthesis by algae. Algae and other drifting organisms are called **plankton.** Plankton include not only photosynthesizers (**phytoplankton**), but also tiny animals (**zooplankton**). Plankton are eaten by fishes and other small animals, and they, in turn, are eaten by larger fishes of the limnetic zone. Beneath the limnetic zone, in the **profundal zone,** no plant or animal occurs except for fishes escaping the heat of summer. Often the bottoms of lakes are covered in muck, where anaerobic bacteria break down organic material that drifts down from the productive limnetic zone.

Lakes in temperate areas often show **thermal stratification** because the density of water varies with temperature (Figure 17.5). Water is most dense at 4°C (39°F), so both warmer and colder water will float above 4°C water. In winter, therefore, 4°C water settles at the bottom, with colder water above it, and ice on top. In summer, warmer water remains above the colder, denser water on the bottom. This thermal stratification also leads to **oxygen stratification,** partly because of stagnation and partly because there is no photosynthesis on the bottom. Water at

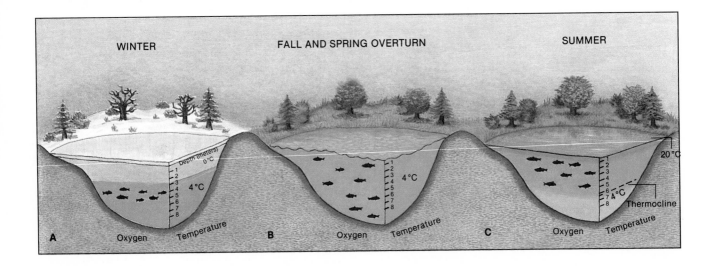

the bottom therefore lacks oxygen, while water at the surface is continually replenished from the air. In the spring and fall there is generally a circulation of water, which redistributes the temperature and oxygen in lakes. The **spring overturn** occurs when the surface water warms to 4°C, and wind or some other disturbance mixes it with the bottom water at the same temperature. **Fall overturn** occurs when the surface water cools to 4°C and sinks below the warmer, less dense water on the bottom. Spring and fall overturns often bring up bottom sediments, including nutrients that can be used by photosynthesizers.

Rivers and Streams. Rivers, streams, and other moving fresh water are **lotic** ecosystems. The differences between lotic and lentic ecosystems are mainly due to the force of moving water. In a slow-moving river one might find the same species as in a lake. In a swift brook, however, one would find animals such as trout and specialized insect larvae, which are able to withstand the current.

Marshes, Bogs, and Other Wetlands. Wetlands occur wherever water covers the land for at least part of the year. They can result from periodic flooding from rivers or lakes, or from poor drainage. The periodicity of flooding largely determines the species composition. A wooded riverbank that is briefly flooded each spring may differ little from nearby upland areas. A marsh or bog, in contrast, will support only shallow-rooted plants such as cattails and shrubs. Few trees can take root in the soggy, oxygen-depleted soil. Marshes are among the most active ecosystems. They are essential nesting and feeding sites for ducks, geese, and many other birds. Wetlands also support many invertebrates, amphibians, reptiles, and small mammals. Wetlands that do not provide good footing are avoided by large mammals, including humans.

MARINE ECOSYSTEM TYPES

Estuaries. Estuaries occur at river mouths and bays where fresh water mixes with seawater. Most estuaries are surrounded by tidal marshes that are flooded at high tides. Because the salinity of the water varies with the tide and with the distance from the source of fresh water, estuaries provide a complex variety of habitats. They are also highly productive, because rivers continually supply nutrients. Mammals generally inhabit only the edges of estuaries, but many birds can

Figure 17.5
Seasonal changes in the distribution of temperature and oxygen in a temperate-zone pond, and the effect on the distribution of fish. (A) In winter the water at the surface is coldest, and denser water at 4°C settles to the bottom. Oxygen cannot diffuse through snow-covered ice into the water and has the highest concentration at middle levels. (B) In spring the surface water warms to 4°C, and wind causes an overturn that eliminates both thermal and oxygen stratification. A similar overturn occurs in the fall when surface waters cool to 4°C. (C) In summer, warm, oxygenated water near the surface may be sharply separated from cold, oxygen-poor water by a layer called the thermocline.

take advantage of the entire surface area. The water hosts a variety of fishes, crustaceans, and molluscs capable of tolerating the changing salinity (see Chapter 14).

Continental-Shelf Waters (Inshore Waters). The greatest diversity of marine animals occurs near shore, in waters above the gently sloping continental shelves. In fact, inshore waters have the highest species diversity of any type of ecosystem. Even the casual beachcomber is impressed by the variety of mollusc shells and other signs of animal life brought in by the waves. Most flounder, cod, salmon, tuna, and other commercial fishes are netted in continental-shelf waters. One reason for the high productivity is the continual supply of nutrients from land and from sediments stirred up by wave action. Inshore waters are so turbid with sediments that sunlight seldom penetrates more than 30 meters. The **intertidal zone** and seashore are interesting parts of the continental-shelf ecosystem because of the diversity of seabirds, crustaceans, and other animals there. Tide pools, where water collects at low tides, often host a fascinating variety of animals (Figure 17.6).

Open Ocean. As the continental shelf drops off sharply to depths of 7 kilometers or more, one enters the open-ocean (**pelagic**) ecosystem. Here the water is clearer, and sunlight penetrates to 100 or 200 meters. In spite of the increased potential for photosynthesis by phytoplankton, however, the absence of nutrients makes open ocean one of the least productive ecosystems per square kilometer. Since two-thirds of the Earth's surface is open ocean, however, it is the largest source of primary productivity (Figure 17.2). Open oceans will be described later in this chapter.

Upwelling Zones. Upwelling regions occur where local winds or currents force nutrient-rich cold water from the ocean bottom to the surface. There photosynthesizing phytoplankton incorporate the nutrients into organic molecules, which then become available to the zooplankton that consume the phytoplankton, and to larger animals that consume plankton. Upwelling zones are about four times as productive as open-ocean areas. The importance of upwelling regions becomes evident every few years in a major upwelling region off the coast of Peru and Ecuador. An invasion of warm Pacific water coinciding with weak trade winds produces a thermal stratification that prevents the cold water from rising to the surface for 6 to 18 months. When extreme, such an event is called El Niño—the boy—because it often arrives around Christmas. (The opposite condition—cold surface temperature and strong trade winds—is termed La Niña.) El Niño of 1982–1983 reduced production to one-twentieth its normal value, resulting in a sharp decline in fish populations and the closing of many fisheries. All the newborn Galápagos fur seals (*Arctocephalus galapagoensis*) died during that year, and no seabirds bred.

INTERACTIONS AMONG ORGANISMS

Since organisms depend on each other for food and other biotic factors, they inevitably interact with each other. These interactions can be classified into several categories, the major ones being competition, predation, and symbiosis. **Competition** occurs whenever organisms (not necessarily animals) in the same habitat depend on the same resource. The more similar organisms are to each other the

Figure 17.6
Some of the inhabitants of a Pacific tide pool.

1 Bushy red algae, *Endocladia*
2 Sea lettuce, green algae, *Ulva*
3 Rockweed, brown algae, *Fucus*
4 Indescent red algae, *Iridea*
5 Encrusting green algae, *Codium*
6 Bladderlike red algae, *Halosaccion*
7 Kelp, brown algae, *Laminaria*
8 Western gull, *Larus*
9 Intrepid marine biologist, *Homo*
10 California mussels, *Mytilus*
11 Acorn barnacles, *Balanus*
12 Red barnacles, *Tetraclita*
13 Goose barnacles, *Mitella*
14 Fixed snails, *Aletes*
15 Periwinkles, *Littorina*
16 Black turban snails, *Tegula*
17 Lined chiton, *Tonicella*
18 Shield limpets, *Collisella pelta*
19 Ribbed limpet, *Collisella scabra*
20 Volcano shell limpet, *Fissurella*
21 Black abalone, *Haliotis*
22 Nudibranch, *Diaulula*
23 Solitary coral, *Balanophyllia*
24 Giant green anemones, *Anthopleura*
25 Coralline algae, *Corallina*
26 Red encrusting sponges, *Plocamia*
27 Brittle star, *Amphiodia*
28 Common starfish, *Pisaster*
29 Purple sea urchins, *Strongylocentrotus*
30 Purple shore crab, *Hemigrapsus*
31 Isopod or pill bug, *Ligidium*
32 Transparent shrimp, *Spirontocans*
33 Hermit crab, *Pagurus*, in turban snail shell
34 Tidepool sculpin, *Clinocottus*

more likely they are to depend on the same resources, so competition is likely to be most intense between members of the same species and of decreasing intensity between members of different species, genera, families, and so on. The concept of competition has deep ecological and evolutionary implications, as will be discussed later in this chapter and in the next.

Predation refers to the capture and eating of one animal, the **prey,** by another animal, the **predator.** Unlike competing animals, predators and prey tend to be quite different from each other. Not only do they not compete with each other for food, but the adaptations required for each to survive are quite different. Like competition, predation has profound ecological and evolutionary consequences.

Symbiosis includes any close association of organisms belonging to different species (not necessarily animals). In everyday speech the term symbiosis often refers to a mutually beneficial association, which biologists call **mutualism.** Besides mutualism, however, biologists refer to two other kinds of associations as symbiotic. One of these is **parasitism,** in which one organism (the **parasite**) benefits from living on or in another living organism (the **host**), which is harmed. The second is **commensalism,** in which one organism (the **commensal**) benefits from associating with a host, which is neither helped nor harmed. Although Table 17.1 summarizes these interactions clearly, it is often difficult to distinguish predation from parasitism, or mutualism from commensalism.

THE CONCEPT OF NICHE

The interactions of an organism with others, together with all the other biotic and abiotic factors that determine the organism's place in an ecosystem, are summarized in the term "niche." A niche can be thought of as a job that can be filled only by those with the exact qualifications. Just as there are often many applicants for one job, there are often two or more species that compete for the same niche. It is generally assumed that where there is competition for a niche, only one species will successfully compete for it. This idea is called **the principle of com-**

Table 17.1 Types of interactions among organisms.

The symbols +, −, and 0 indicate whether the individual benefits, suffers, or is unaffected by the interaction.

Interaction	Effect		Example
Competition	Competitor −	Competitor −	Coyote/fox
Predation	Predator +	Prey −	Coyote/rabbit
Symbiosis Mutualism	Mutualist +	Mutualist +	Cleaner fish/grouper
Parasitism	Parasite +	Host −	Tapeworm/human
Commensalism	Commensal +	Host 0	Some bacteria/human

petitive exclusion. It is also called "Gause's principle," after G. F. Gause who published a famous study of it in 1934. Gause found that *Paramecium aurelia* and *P. caudatum* thrived when cultured separately in a particular medium, but when the two protozoans were cultured together, *P. aurelia* suppressed the population of *P. caudatum*. Similar results have been demonstrated with flour beetles (*Tribolium*) and with *Drosophila* in the laboratory. As we shall see in a moment, however, there is some doubt whether the principle of competitive exclusion applies in nature.

COMMUNITIES

Within each ecosystem are numerous areas where particular species associate with each other. These distinctive associations, called communities, can be found repeatedly in many different areas. For example, the assortment of bacteria, fungi, and insects in rotting logs is so typical that ecologists refer to them simply as the "rotting-log community." Many ecologists explain the persistence of such communities as the consequence of the principle of competitive exclusion. According to this view, a community will necessarily consist only of those organisms that can successfully compete for the niches in a particular habitat. Since the 1980s this static view of communities has been tested and often found inadequate. Although some communities do, indeed, appear to fit this model, others are apparently much more dynamic. In many communities predation and chance events such as storms keep populations so low that there is little competition.

Another reason for reexamining the static view of community structure is that the principle of competitive exclusion appears to have little applicability in nature. In spite of Gause, vast herds of wildebeest, zebra, topi, and Thomson's gazelles graze peacefully together on the grasses of the Serengeti. They apparently avoid competition by feeding on different parts of the grasses. Zebra tend to feed first, taking the tall grass stems. Topi and wildebeest then follow, taking the leaves and lower portions of the grasses that were exposed by the zebra. Finally, Thomson's gazelles forage primarily on new grass shoots, as well as broad-leaved plants and their fruits (Figure 17.7). There are numerous other examples in which animals that would otherwise compete with each other become specialized for different resources. (See p. 376 for example.) It might be argued that these animals are not really occupying the same niche and are therefore not violating the principle of

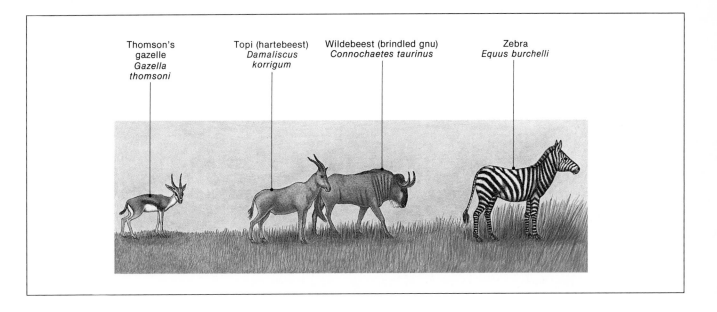

Figure 17.7
Four species of grazing ungulates in Africa. At the close of the wet season, each species specializes on different parts of plants, thereby avoiding competition.

competitive exclusion. If every niche has to be defined so narrowly and differently in every case, however, then neither the concept of niche nor the principle of competitive exclusion is of much value in formulating general predictions or explanations.

SUCCESSION

Communities often undergo regular sequences of changes, usually following some natural or artificial disturbance. Such regular changes are called succession. If succession begins with a complete absence of life, including soil organisms, it is called **primary succession.** On freshly exposed rock, for example, primary succession often begins with lichens (mutualistic associations of fungi and algae). The lichens trap wind-blown soil and also slowly decompose the rock, creating a substratum on which mosses can grow. The mosses may eventually shade out the lichens. Dead mosses accumulate in a layer of organic humus, which will eventually allow grasses and other shallow-rooted plants to grow and replace the mosses. These too will contribute to the humus, which will eventually become deep enough to support larger shrubs and then trees. All these stages of primary succession are often visible on different parts of the same rock.

Successional change that begins with some life already present is called **secondary succession.** Agriculture is the most common initiator of secondary succession in North America. In the eastern United States the most familiar example of secondary succession is **old-field succession** on abandoned farms. After crop lands and pastures are abandoned and no longer mowed or grazed, they revert to tall grasses, then to weeds and shrubs. Weeds and shrubs require a great deal of sun, but they themselves often create so much shade that their own seedlings cannot develop in the area. Eventually, some species of trees may rise above the weeds and shrubs, shading them out. The seeds of shade-tolerant trees may then germinate beneath these trees and eventually grow up and shade them out. Succession ends when the area is covered in shade-tolerant trees whose seedlings can grow up in the shade of their parents. Such stable plant communities are called **climax** stages of succession. Oak and hickory dominate climax forests in the mid-Atlantic states. Farther north, beech, maple, and birch are the dominant species in deciduous climax forests.

Three explanations of succession have been proposed (Brown 1984):

1. Species in early stages of succession alter their habitat to their own detriment, but to the advantage of the species in later stages of succession. This seems most consistent with primary succession.
2. Early species inhibit the growth and survival of others. Only after the early species die can later stages of succession occur. This theory seems most consistent with what occurs during secondary succession.
3. Early species neither inhibit later ones nor prepare a place for them. They become established first simply because they reproduce more quickly than later species. Later species become established only because they can compete successfully with other species. This may occur with both primary or secondary succession, although there is little evidence for it.

Faunal Succession.   As the plants in a community change during succession, so must the animals that depend on the plants for food and habitat. Such changes in animal species are called faunal succession (Figure 17.8). In addition to passively following plant succession, animals also contribute to it (Naiman 1988). Grazing animals retard old-field succession by cropping the shoots of weeds and trees. Insects can have a similar effect (Brown 1984). The animal with perhaps the most obvious impact on succession is the beaver *Castor canadensis*, whose dams convert wooded valleys into ponds. The ponds then gradually fill with organic matter, forming meadows that eventually succeed to level woods.

NUTRIENT CYCLES

As noted previously, animals cannot use carbon, nitrogen, and many other elements until they have been fixed into organic molecules by autotrophs. **Carbon fixation** occurs primarily in green plants and algae, when they convert carbon dioxide and water into oxygen and glucose through photosynthesis. During pho-

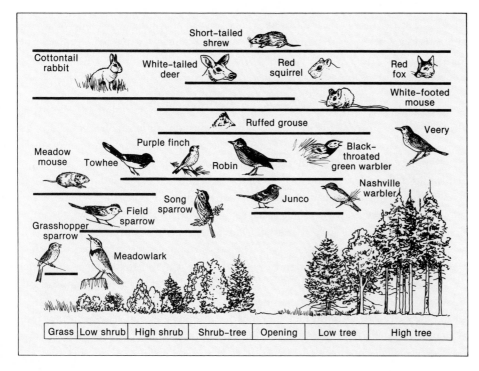

Figure 17.8
Faunal succession in the northeastern United States. The bars beneath each species show its range of habitats. As fields gradually succeed to climax forest (left to right), the animal community also succeeds.

tosynthesis, red and blue light supplies the energy that links the carbon atoms from CO_2 into chemical bonds that make glucose ($C_6H_{12}O_6$). Cells then break down the glucose and extract the energy in a form they can use. The breakdown of glucose through cellular respiration releases the CO_2 back into the environment. Cellular respiration is therefore essentially the opposite of photosynthesis.

$$6\ CO_2 + 6\ H_2O + \begin{array}{c} \text{solar} \\ \text{energy} \\ \text{ATP} \end{array} \overset{\text{photosynthesis}}{\underset{\text{respiration}}{\rightleftarrows}} C_6H_{12}O_6 + 6\ O_2$$

The Carbon Cycle. Photosynthesis and cellular respiration are complementary to each other, cycling carbon between an inorganic state in CO_2 and an organic state in glucose. The alternation between CO_2 and glucose forms a major loop of the carbon cycle (Figure 17.9). Most carbon drops out of this loop for long periods in a **sedimentary cycle.** In the sedimentary cycle much carbon exists as fossil fuels—peat, coal, and oil—until it is liberated as CO_2 during combustion. Calcium carbonate ($CaCO_3$) and other minerals in mollusc shells, coral, and other marine animals also enter the sedimentary cycle. For example, carbon in the shells of molluscs that lived 400 million years ago is still locked in the sedimentary cycle as limestone. Only through the slow process of weathering, or the even slower process of volcanic action, is this carbon eventually released to participate once more in photosynthesis and respiration. Because of the intertwining of biological and geological states, this and other nutrient cycles are also called **biogeochemical cycles.**

Decomposers, which are usually bacteria or fungi, are essential links in the biogeochemical cycles of carbon and other nutrients. Without the decomposition of complex organic molecules, much of the carbon and other nutrients in them would not recycle after organisms died. Animals that feed on excrement, dead organisms, and other nonliving organic matter—the **detritus feeders**—are also important, for they provide decomposers with access to dead plants and animals. Detritus feeders include numerous small animals of the soil (Figure 17.10). Most people would rather not think about the activities of dung beetles, carrion beetles,

Figure 17.9
The carbon cycle begins with photosynthesis (on the left) and is completed in part by cellular respiration in plants, animals, and other organisms (right). During the cycle, carbon alternates between CO_2 dissolved in air and water (top) and in organic molecules in organisms (bottom). This portion of the cycle accounts for only a small fraction of the carbon in the carbon cycle. Almost 2000 times as much carbon is in a slower sedimentary cycle (center) as fossil fuels, animal shells, and other biogenic minerals. Weathering and combustion eventually complete that cycle as well.

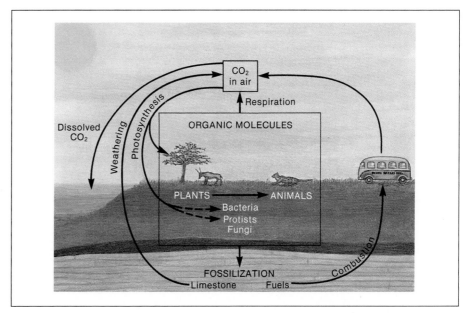

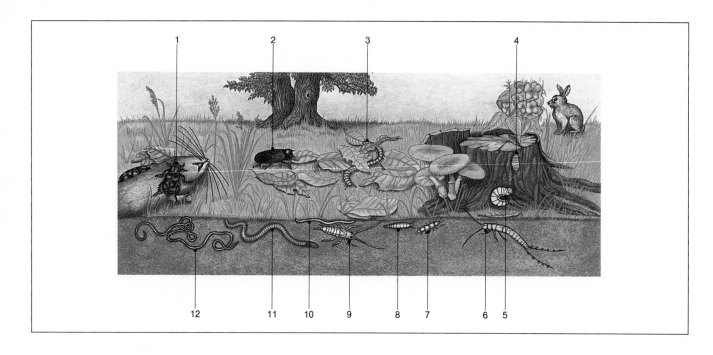

Figure 17.10
Many animals help complete nutrient cycles by feeding on detritus. The following animals are represented here: (1) *carrion beetle;* (2) *dung beetle;* (3) *millipede;* (4) *pill bug (a terrestrial crustacean);* (5) *beetle larva;* (6) *bristletail;* (7) *pauropod (belonging to an insectlike group of arthropods);* (8) *proturan;* (9) *springtail;* (10) *potworm;* (11) *earthworm;* (12) *nematode.*

and other detritus feeders, but without them life would be even more unthinkable. Fields and forests would be nightmarish landscapes of fallen and undecayed trees, dead animals, and excrement.

Oxygen and Water Cycles. Coupled to the carbon cycle is an oxygen cycle. Oxygen is produced during photosynthesis and is consumed during respiration and the combustion of fossil fuels. Animals are as dependent on photosynthesizers for their O_2 as for their organic carbon and energy. The oxygen cycle is also coupled to another important biogeochemical cycle: the water cycle (= **hydrologic cycle**). Almost all the oxygen available to animals and other organisms comes from water broken down during photosynthesis. The water cycle is completed during cellular respiration, when O_2 and H^+ combine to form H_2O.

The Nitrogen Cycle. Because nitrogen is an essential element in proteins and many other organic molecules, the nitrogen cycle is also important (Figure 17.12). N_2 makes up 78% of the atmosphere, but before it can be used in the synthesis of organic molecules it must be converted into water-soluble nitrate (NO_3^-) or ammonia (NH_3). This conversion is called **nitrogen fixation.** Some nitrogen is fixed as nitrate by lightning and by combustion in gasoline engines, and some is introduced as ammonia in synthetic fertilizers. The major source of fixed nitrogen, however, is nitrate from nitrogen-fixing bacteria. Many nitrogen-fixing bacteria are free-living in soil and water, but bacteria in the genus *Rhizobium* live within root nodules of legumes (peas, alfalfa, soybeans, etc.) and provide nitrates directly to their hosts. Nitrogen is usually the least-abundant nutrient in soils, so plant growth is often limited by how many nitrogen-fixing bacteria live in an area.

Another biological source of fixed nitrogen for plants are nitrogenous wastes, such as urea, and the proteins of dead organisms. These are converted to ammonia and nitrate by decomposers, including **ammonifying bacteria.** Ammonifying bacteria convert nitrogen-containing molecules into ammonia, much of which is used directly by plants. **Nitrite bacteria** then convert some of the ammonia into nitrite (NO_2^-), which may then be converted into nitrate (NO_2^-) by **nitrate bacteria.** The nitrate can then be used by plants. Much of the nitrite and nitrate, however,

The Greenhouse Effect

The inputs and outputs of the carbon cycle resemble the incomes and expenditures of a government, a household, or a corporation, so ecologists speak of the "carbon budget." The carbon budget would be balanced if CO_2 production by respiration and combustion equaled CO_2 uptake by photosynthesis and absorption in oceans. Like so many other budgets, however, the carbon budget is unbalanced. Since 1958 the amount of CO_2 in the atmosphere has increased at least 12%, as measured in the relatively clean air of Hawaii (Figure 17.11). This increase is attributed mainly to burning of fossil fuels, clearing of forests, and reduction in the amount absorbed by the oceans. There is a lively controversy about which of these factors is most important.

The debate over CO_2 is not merely of academic interest, because CO_2, along with H_2O and many synthetic gases, reflects back to Earth much of the heat that would otherwise be radiated into space. Atmospheric CO_2 therefore has a greenhouse effect, increasing temperature on the Earth's surface. The level of CO_2 is expected to double by the year 2050, and the increased greenhouse effect is expected to raise average global temperature. Computer models give different estimates of how large the increase could be, but estimates range around 2.5°C. By the time we know for certain, it will be too late to do anything about it. The increased CO_2 and warming could have a variety of effects. Crop yields might increase by 60% to 80% worldwide, although many areas that are now productive could become deserts. The increased temperature could also melt the polar ice caps, flooding coastal areas, including many large cities.

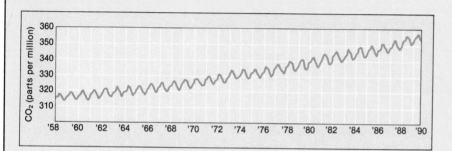

Figure 17.11
Levels of CO_2 in parts per million (ppm) measured at Mauna Loa on the island of Hawaii since 1958. CO_2 declines each summer, because of photosynthesis, and rises during each winter. This annual cycle is superimposed on a rising trend.

Figure 17.12
The nitrogen cycle.

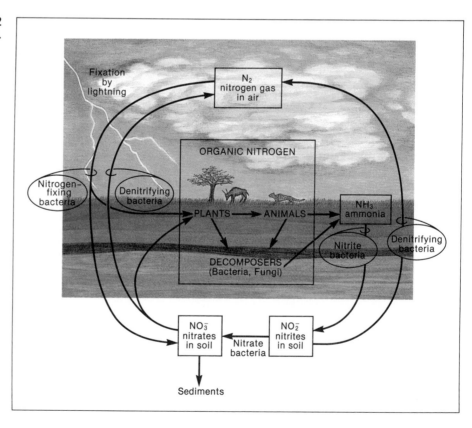

347

are removed from biological utility when **denitrifying bacteria** convert them back into gaseous nitrogen (N_2). Denitrifying bacteria are the main reason why modern agriculture is so dependent on artificial fertilizers to replenish the nitrogen in soils. If denitrifying bacteria were destroyed, food production could presumably be greatly expanded in countries that cannot afford artificial fertilizers. So little is known about these bacteria, however, that destroying them would be foolhardy (Payne 1983).

ENERGY

Most biological energy is associated with chemical bonds, so the flow of energy through an ecosystem follows the same path as carbon during photosynthesis and cellular respiration. This connection between organic molecules and energy explains why primary production can be expressed in units of energy as well as of mass. The movements of carbon and of energy differ in one major respect, however. Carbon cycles between CO_2 and organic molecules, but energy flows in one direction—from sunlight to dissipated heat. Therefore, *during every transfer of energy—such as in photosynthesis and respiration—some energy is lost as unrecoverable heat.* Eugene P. Odum expresses it succinctly: "Matter circulates; energy dissipates." The dissipation of energy is a corollary of the Second Law of Thermodynamics, and it explains why life on Earth requires the Sun to continuously replenish the lost energy. To put it another way, life is not a perpetual motion machine.

Trophic Levels. The dissipation of energy as heat explains why the amount of energy available to an animal depends on how far along the animal is in the sequence of consumption. The position of an animal in this sequence is called its **trophic level.** The lowest trophic level is that of **primary consumer**—an animal that eats plants. Primary consumers include **herbivores,** which eat plants only, and **omnivores,** which eat both plants and animals. Some of the energy ingested by a primary consumer is lost as indigestible plant fiber in feces. The rest is **assimilated** into the animal. Part of the assimilatd energy is lost as dissipated heat or is used immediately by the primary consumer. Whatever energy is left is available to produce new tissues, through fat storage, growth, and reproduction.

In the long run, only this new biomass produced by part of the assimilated energy is available as an energy source at the next trophic level, that of the **secondary consumer.** A secondary consumer is an animal that eats a primary consumer. Secondary consumers can be omnivores or, if they eat only other animals, **carnivores.** Once again, the secondary consumer loses some of the ingested energy (hair, bones, and other indigestible wastes) in feces. The secondary consumer loses some of the remaining assimilated energy as heat, and whatever is left is either used or stored in new biomass. Again, only this biomass can benefit an animal at the next trophic level that eats the secondary consumer. The process of energy dissipation continues in this way from trophic level to trophic level up to the **top carnivore.**

Ecological Growth Efficiency. The assimilated energy (energy available for use and new biomass) divided by the ingested energy is defined as the ecological growth efficiency. The ecological growth efficiency varies with the organisms involved but is often around 10%. One study on Isle Royale in Lake Ontario, for example, showed that for every 100 kcal of plant energy that moose consumed, only about 1 kcal was eventually assimilated by wolves that consumed moose. The

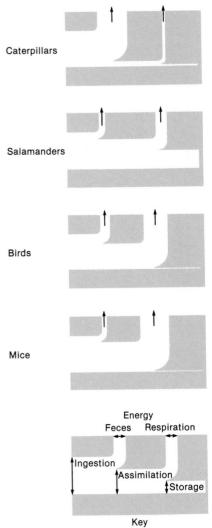

Figure 17.13
Energy inputs and outputs for four kinds of animals. Ingested energy is either assimilated or lost as feces (see key). Of the assimilated energy, part is consumed in mitochondrial respiration to satisfy the energy needs of the animal, and the remainder is stored as new tissue. Caterpillars lose much of the ingested energy as feces. Birds and mammals use much of their energy in cellular respiration to keep warm. Salamanders, which do not maintain a high body temperature, store the most energy as tissue. (Based on a study from Hubbard Brook Experimental Forest in New Hampshire.)

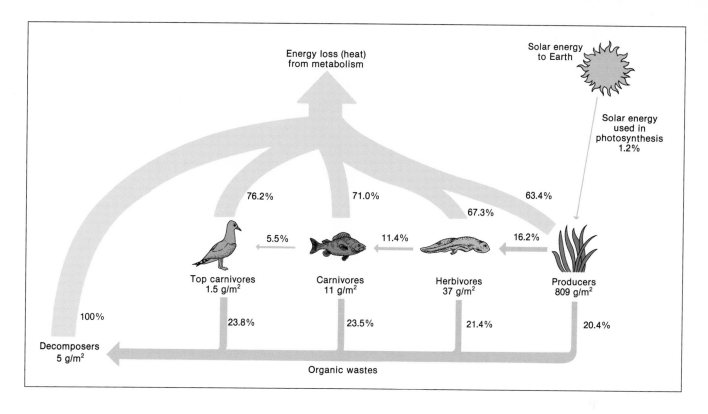

Energy loss (heat)
from metabolism

Solar energy
to Earth

Solar energy
used in
photosynthesis
1.2%

76.2%   71.0%   63.4%

67.3%

5.5%   11.4%   16.2%

Top carnivores   Carnivores   Herbivores   Producers
1.5 g/m²   11 g/m²   37 g/m²   809 g/m²

100%   23.8%   23.5%   21.4%   20.4%

Decomposers
5 g/m²

Organic wastes

Figure 17.14
Annual energy flow through an ecosystem, based on a study in Silver Springs, Florida. All the incident solar energy is eventually radiated back to space as heat (colored arrow pointing up), but 1.2% of it is first used by producers in photosynthesis (arrow from Sun). This energy is then available to producers and to consumers. Numbers beneath the producers and the consumers at three trophic levels represent the biomass in grams per square meter. During metabolism, producers convert 63.4% of the energy available from photosynthesis directly into heat (colored arrow). Of the energy made available in photosynthesis 20.4% goes into organic wastes and is eventually converted to heat during metabolism by decomposers (arrow down and to the left), and 16.2% of the energy is consumed by herbivores (arrow to left). Arrows from the herbivore and other trophic levels likewise represent the proportions of the energy input that go into heat, organic wastes, and to the next trophic level. (Data from Odum 1957.)

proportion of the assimilated energy that enters new tissue also varies. In birds and mammals much of the assimilated energy is used to keep the body warm and is therefore not available for new biomass (Figure 17.13).

Ecosystem Analysis. A few ecologists have actually undertaken the enormous task of accounting for all the energy transfers within a particular ecosystem. One classic example of such an ecosystem analysis was conducted at Silver Springs, Florida, by Howard T. Odum in the 1950s (Figure 17.14).

ECOLOGICAL PYRAMIDS

Pyramids of Energy and Biomass. Ecosystem analyses invariably show that there is less available energy at higher trophic levels than at lower ones. This relationship can be expressed by a pyramid of energy, one of the many useful concepts contributed by Charles Elton in his 1927 text on animal ecology. The pyramid of energy represents the amount of energy at each trophic level by the area of a section of the pyramid (Figure 17.15). Since higher trophic levels must get all their energy from lower trophic levels, and since much of the energy is lost with each energy transfer, upper levels have less area than lower levels. Hence the pyramid shape. Since the energy is stored in an organism's tissues, the pyramid of energy also implies a **pyramid of biomass.** In the pyramid of biomass the area of each level of the pyramid represents the "standing crop": the total mass of living organisms at that level at a given time.

The Pyramid of Numbers. If the average size of individual organisms is the same at each trophic level, then the pyramids of energy and biomass also imply a pyramid of numbers of individual organisms at each trophic level. Usually there are indeed fewer carnivores than herbivores in an area, and fewer herbivores than individual food plants. However, the exact numerical relationship between individuals in each trophic level depends on their relative body sizes. For example, even though ants are at a higher trophic level than an acacia tree, thousands of ants can live on one tree because each ant is so small. If the body sizes of the organisms are known, the number of animals at each trophic level can be accurately calculated from the amount of energy available to them. Using this principle, Schneider and Wallis (1973) were able to calculate the maximum number of Loch Ness monsters. They found that the energy available in Loch Ness could sustain 157 small Nessies (average body mass 100 kg) or 10 large ones (average mass 1500 kg).

FOOD CHAINS AND WEBS

The transfer of energy from organism to organism can be represented as a food chain: a linear sequence in which one species is food for the next species in the sequence. Each food chain generally consists of producers plus two trophic levels in terrestrial ecosystems, or three trophic levels in aquatic ecosystems. Usually several food chains intertwine to form a **food web**. Figure 17.16 shows some of the members of a food web in Antarctica. Even this simplified food web includes eight different food chains leading from the photosynthetic diatoms to the top carnivore, the killer whale *Orcinus orca*. This figure shows that many species occupy more than one trophic level.

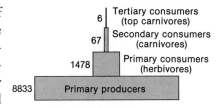

Figure 17.15
A pyramid of energy for the Silver Springs ecosystem shown in Figure 17.14. The area of each level of the pyramid represents the amount of energy (kcal/m²) in each trophic level.

Figure 17.16
A simplified representation of an Antarctic food web. See Laws (1985) for a more nearly complete representation and discussion of this food web.

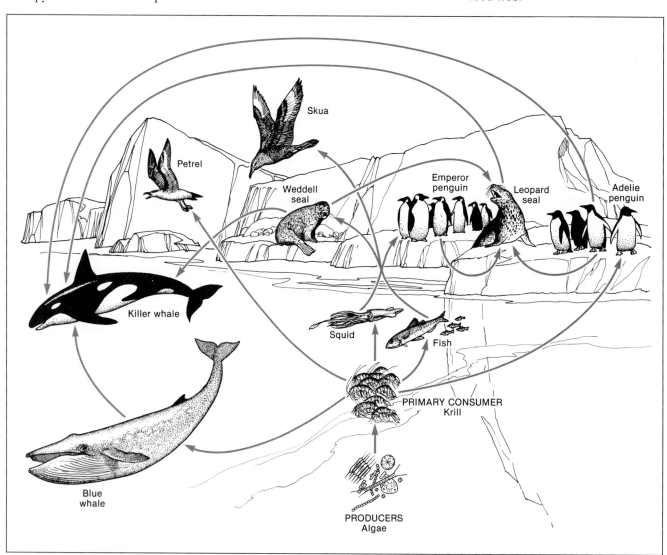

DIVERSITY AND STABILITY

It is commonly assumed that the more complex a food web, the greater the stability of populations in a community or ecosystem, since a consumer will not be totally dependent on just one food source. The population of killer whales should be fairly stable, for example, since they can switch to preying on leopard seals if penguins become scarce. Although the relationship of complexity to stability seems self-evident, some ecologists now doubt that it exists. Others believe that instead of complexity leading to stability, it is stability that leads to complexity. According to them, the stability of populations provides time for a diversity of species to become established in a community.

There are situations in which increasing diversity leads to instability rather than to stability. If an exotic species enters an ecosystem where it has no competitors, predators, or disease organisms, its population can soar while other species are extinguished. Humans are the most common agents for introducing exotic species, with consequences ranging from amusement to disaster. During the filming of Tarzan movies in Silver Springs, Florida, during the 1930s, a few rhesus monkeys *Macaca mulatta* escaped. These natives of India did well in Silver Springs, and they became a favorite tourist attraction. Now, however, some of them must be removed from the area because they have become numerous and bold enough to chase humans away. Less amusing consequences followed the accidental introduction of gypsy moths *Lymantria dispar* into a suburb of Boston in 1869. A naturalist had imported the gypsy moth from Europe in hopes of hybridizing it with the silkworm moth *Bombyx mori* to get the silk-making ability of *Bombyx* without its finicky taste for mulberry leaves. A few of the gypsy moths escaped, however, and now each summer throughout the eastern United States their descendants turn thousands of acres of foliage into a deluge of feces.

Stability also depends on other factors, such as area. Destroying half a forest, for example, does not necessarily result in merely a smaller forest. For reasons not fully understood, the number of species in the reduced area often declines, changing the character of the forest. As a rule of thumb, reducing area by one-tenth reduces species diversity by one-half. In one study, three tracts of 1, 10, and 100 hectares were isolated from the surrounding rainforest by clearing a wide perimeter (Lewin 1984). (A hectare is 10,000 square meters: approximately 2.5 acres.) In the 10-hectare area, species of birds that follow army ants and prey upon their fleeing victims disappeared within a few years, because there were not enough ant colonies to sustain them. Peccaries (piglike animals) abandoned even the 100-hectare tracts, and with them went three species of frogs that breed in the peccaries' mud-wallows.

POPULATION

The stability of populations in natural ecosystems is remarkable considering that most breeding pairs of animals can produce many offspring in a year. Many insects, for example, produce hundreds of fertilized eggs per year, a pair of birds commonly produces more than two eggs per year, and many mammals produce large litters of offspring. Even a couple of members of our own slow-breeding species could generate more than 10,000 new humans in a century if each couple had 10 children. Theoretically, the population of every species could increase at a rate proportional to the number of individuals at a given time. If the population of rabbits increased tenfold each year, then 10 rabbits one year could become 10^2 rabbits in the second year, 10^3 the next, 10^4 the next, and so on. In other words, **populations have the potential to increase exponentially.**

Exponential population growth does, in fact, occur when populations are allowed to reproduce unchecked by crowding, competition, disease, or predation. These conditions commonly occur during the first few years after an exotic species is introduced (Figure 17.17A). Exponential population growth cannot continue indefinitely, however. There are only 10^{72} electrons in the entire universe, so a species with a tenfold rate of population growth would use up all the electrons in less than 70 generations. Long before that, of course, the species would have killed off its food species. Once a species exhausts its food sources, its population necessarily "crashes" (Figure 17.17B).

Exponential growth and population crashes are necessarily short-term phenomena. Over the long term most natural populations fluctuate less extremely at levels below the maximum number that the ecosystem can support. This maximum sustainable population level is called the **carrying capacity** (Figure 17.18). A population may rise toward or above the carrying capacity temporarily, but then the depletion of food, the spread of communicable diseases and parasites, or simply the stress of overcrowding increases the death rate and decreases the birth rate. The population then falls below the carrying capacity, reducing disease and stress and allowing food souces to recover. The population may then start to increase once more. See box on page 355.

Population Cycles. Fluctuations in population around the carrying capacity often occur in regular patterns called population cycles. For example, the sheep population shown in Figure 17.18 peaked in about 1854, 1880, and 1910, suggesting a population cycle with a period of between 26 and 30 years. Such cycles are believed to result from the delayed effects of overpopulation. For example, it may take several years for food resources to be depleted, or for the populations of parasites and predators to increase.

A population cycle in one species can induce a cycle in another species in the same ecosystem. Figure 17.19 shows how the population cycle of the snowshoe hare *Lepus americanus* in Canada induced a cycle in the population of one of its predators, the lynx *Felis lynx*. Note that the lynx cycle lagged the hare cycle, because it took a year or so for the rise and fall of the hare population to influence the fertility and death rates of lynx.

A SURVEY OF SOME ECOSYSTEMS

So far we have been looking at the principles of ecology in a quite artificial manner, as if these principles operated independently of each other. This analytical

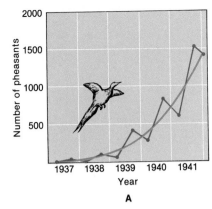

A

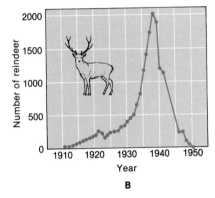

B

Figure 17.17
(A) Exponential growth in the population of ring-necked pheasants Phasianus colchicus *following introduction on Protection Island, Washington. The smooth curve averages the fluctuations due to spring hatching and winter kill. The population approximately doubled each year following introduction. (B) After 18 reindeer* Rangifer rangifer *were introduced onto the Pribilof Islands, Alaska, their population grew exponentially for 30 years. Then, presumably because of overgrazing, the population began to crash. A decade later there were only eight survivors.*

Figure 17.18
The population of domestic sheep Ovis aries *in Tasmania. The data points show populations averaged over several years. The dashed line indicates the presumed carrying capacity (K).*

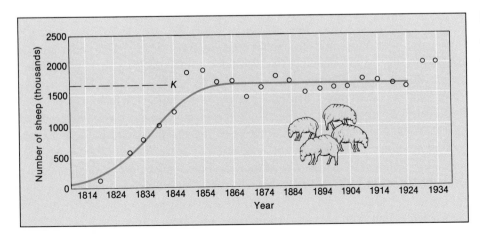

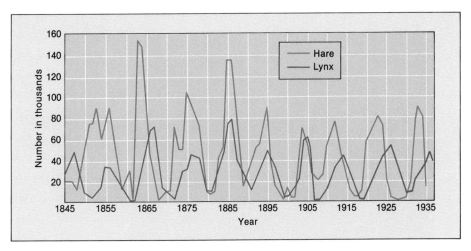

Figure 17.19
Populations of lynx Felis lynx *and of snowshoe hare* Lepus americanus, *as inferred from the number of pelts sold to the Hudson's Bay Company in Canada. Both species have population cycles with periods of about ten years, and the lynx cycle tends to lag behind the hare cycle. (Data from MacLulich 1937.) This is perhaps the most published graph in biology. For many years it was thought to show repeated overpredation of hares by lynx, followed by collapse of the lynx population and recovery of the hare population. It is now known, however, that the cycle in the hare population occurs even without lynx.*

approach is necessary for human comprehension, but it does not provide an appreciation of the complexity that makes ecology so difficult and so fascinating. In the space available here we can only examine a few ecosystem types to see how they "work." These examples are chosen because they include such a large number of animal species that they are frequently mentioned in later chapters.

Tropical Rainforests. Tropical rainforests now cover approximately 3% of the planet, but they account for more than 20% of its net primary production, and more than half of its biomass. Perhaps more than half the world's species live in tropical rainforests: the figure is uncertain because the majority of species have not even been classified. The diversity of plants is so great that one must often search for hours to find two individuals of the same species. In an 800-km² area of Central American rainforest 500 to 600 resident species of birds have been counted. Five hundred species of insects were caught in 2000 sweeps of a net near the ground, and many more species live in the canopy. Terry L. Erwin (1983) of the National Museum of Natural History exploded insecticide bombs in the canopy and found that the majority of insects brought down were previously unknown. Among these E. O. Wilson of Harvard identified 43 species of ants living on one tree in Peru—as many as live throughout Canada or the British Isles. With so many different species, no single one is numerous enough to be called dominant.

Such a diverse ecosystem might be expected to be stable, yet approximately one species per day is becoming extinct in tropical rainforests, and 4000 to 6000 have already been lost. This is about 10,000 times the natural rate of extinction. The cause of this instability is an exotic species—*Homo sapiens.* Humans are clearing tropical rainforests for lumber and pasture at a rate of between 100,000 and 200,000 km² per year, an area at least as large as Virginia. The study of tropical rainforests is therefore a race against time to discover what we can about the functioning of this ecosystem. The race is being run not only for pure science but also for practical benefits to humans. Pharmaceutical firms, for example, have found it more efficient to isolate new drugs from tropical plants than to synthesize them from scratch. About half the available pharmaceuticals originated from tropical plants, and more than a thousand anticancer drugs have been discovered in this way. Preservation of rainforests is also crucial to the survival of many American songbirds that overwinter there.

The immense diversity of tropical rainforests is due largely to **vertical stratification** of light, humidity, temperature, and wind speed, which gives each layer

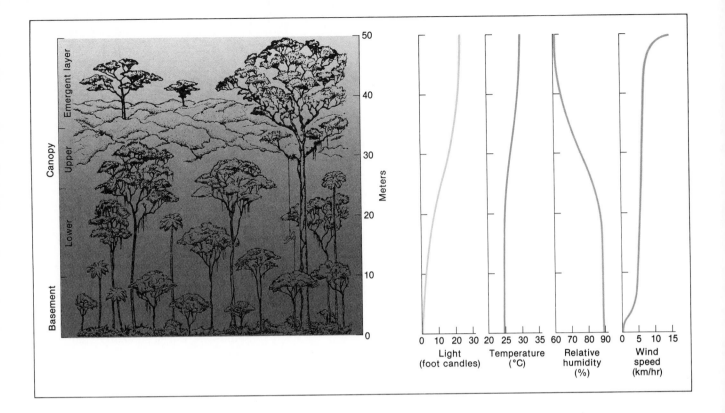

Figure 17.20
Vertical stratification in a tropical rainforest. From the canopy to the forest floor, light, temperature, and wind velocity all decrease, while humidity increases. Most primary production occurs in the canopy, which can be studied by ecologists with climbing equipment and exceptional courage (see figure of climber).

of forest its own microclimate (Figure 17.20). Light is only about 4% as bright on the ground as in the canopy. Consequently, most of the primary production occurs in the canopy, and there is little vegetation near the ground. ("Jungles" of thick undergrowth usually occur only along rivers or in abandoned clearings.) Vertical stratification of vegetation also leads to vertical stratification of animals. Many species spend their entire lives in the canopy without ever descending to the ground. These species include insects, frogs, lizards, birds, bats, and monkeys that feed on leaves, fruit, and nectar. The male birds are often brightly colored and engage in showy courtship displays that make them visible to potential mates in the thick foliage. Many mammals and insects in the canopy contribute to the ever-present chorus of courtship and other sounds.

Many tree-climbing species range between the canopy and the ground. These include frogs, many of which breed in water that collects in the leaves of **epiphytes**—plants that grow on the limbs and trunks of trees. Some snakes can also climb tree trunks in pursuit of prey. Most of all there are the ants and termites that range up and down the trees. These two groups alone account for one-third of the animal biomass in the rainforest. Biting ants defend many trees against both epiphytes and tree-climbing animals. These overly protective ants once prevented ecologists from studying the forest canopy, but now special climbing gear and platforms have solved that problem. Now ecologists have nothing to worry about except falling 40 or 50 meters.

Ants and termites are as important on the ground as in the trees. As detritus feeders they aid bacteria and fungi in decomposing fallen leaves, fruits, and dead organisms. So rapid is decomposition in the warm, moist tropics that almost no organic matter accumulates in the soil. The soil is also poor in minerals, because of leaching by the heavy rain and rapid absorption by the shallow roots of trees. The only type of agriculture that is practical in such soil is the "slash and burn" system. Burning the vegetation adds some minerals to the bare clay, but after a

r Selection, K Selection, and Life-History Traits

If certain assumptions are made, exponential population growth, followed by a leveling off near the carrying capacity, can be modeled mathematically by a **logistic equation:**

$$\frac{dN}{dt} = rN\frac{(K - N)}{K}$$

where dN/dt is the rate of population growth (number of new individuals per year); r is the natural rate of population growth if predation, starvation, and so on do not limit it; N is the population at a given time; K is the carrying capacity.

Note that when N is much smaller than K, as when a species is newly introduced into an ecosystem, $dN/dt = rN$ approximately. This equation describes an exponential curve. When N equals K, dN/dt is zero: the population stops increasing.

$(K - N)/K$ represents the unused proportion of the carrying capacity. For example, if the carrying capacity is 100, but the population is only 90, then 10% of the carrying capacity is still available for population growth. In words, this logistic equation says that the rate of population growth is proportional to the natural rate of growth, times the population, times the proportion of the carrying capacity still available.

This logistic equation suggests that high reproductive success can result from two kinds of adaptations: those that increase the reproductive rate r, and those that maintain the population as close as possible to the carrying capacity K. Many ecologists once thought that organisms tended to have either one kind of adaptation or the other, but not both simultaneously. Organisms with adaptations that increased r were said to be r-selected, and those with adaptations that maximized the population near K were called K-selected. Examples of r-selected animals would be gypsy moths, most flies, and many small rodents. Many birds and mammals, including humans, were considered to be K-selected.

Most ecologists now generally consider this to be an oversimplification based on unrealistic assumptions. Many species appear to be r-selected at one stage of life and K-selected at another. Moreover, the genetic traits that affect r and K can change within species. Rather than categorize each species as being r-selected or K-selected, ecologists now tend to focus on **life-history traits,** studying the role of individual adaptations to see how they affect population. In one such study David A. Reznick, Heather Bryga, and John A. Endler (1990) used guppies *Poecilia reticulata* from a river in Trinidad in which the adults, but not the juveniles, are preyed upon by a cichlid fish (*Crenicichla alta*). They transferred the guppies to a tributary in which there were neither guppies nor cichlid predators, but where there were killifish (*Rivulus hartii*) that prey on small juvenile guppies. As they had predicted, the change from predation on large guppies to small ones resulted in significant evolutionary change in the life history of the guppies. Within 11 years (30 to 60 generations) the guppies began to produce larger but fewer offspring per brood.

---

few plantings these become depleted, and the field must be left to succeed to jungle.

The Open Ocean. Like the tropical rainforests, oceans are vertically stratified. Near the surface is a "canopy" of **diatoms** and other photosynthesizers. Primary consumers, mainly zooplankton, swim up to the surface at night to feed on the phytoplankton. At dawn they reverse the **vertical migration,** thereby avoiding predation. Most zooplankters (the term for individual species of zooplankton) are only a few millimeters long, but they make up for their small sizes by being the most numerous animals on Earth. Among the most important are two groups of crustaceans, **krill** and **copepods** (numbers 21 and 35 in Figure 17.21; see pp. 668 and 670). As noted previously (Figure 17.16), krill are an essential link in food chains of polar regions. In temperate and tropical oceans 2000 to 3000 species of copepods, especially of the genus *Calanus,* dominate the zooplankton.

In contrast to the plankton are the animals that make up **nekton:** fishes and other large animals that swim actively rather than drift with currents. Nekton occur all the way down to the ocean floor, an average of 4 km below the surface. Nutrients for animals of the deep rain down as detritus from the productive regions above. Most of the animals that live in the perpetual darkness of the ocean

Figure 17.21

Plankton and nekton of the open ocean. The fishes shown at the bottom are members of the benthos—inhabitants of the ocean floor.

1 Dolphins, *Delphinus*
2 Tropicbirds, *Phaethon*
3 Paper nautilus, *Argonauta*
4 Anchovies, *Engraulis*
5 Mackerel, *Pneumatophorus*, and sardines, *Sardinops*
6 Squid, *Onykia*
7 *Sargassum*
8 Sargassum fish
9 Pilotfish
10 White-tipped shark
11 Pompano, *Palometa*
12 Ocean sunfish, *Mola mola*
13 Squids, *Loligo*
14 Rabbitfish, *Chimaera*
15 Eel larva, leptocephalus
16 Deep sea fish
17 Deep sea angler, *Melanocetus*
18 Lanternfish, *Diaphus*
19 Hatchetfish, *Polyipnus*
20 "Widemouth," *Malacosteus*
21 Krill, *Nematoscelis*
22 Arrowworm, *Sagitta*
23 Amphipod, *Hyperoche*
24 Sole larva, *Solea*
25 Sunfish larva, *Mola mola*
26 Mullet larva, *Mullus*
27 Sea butterfly, *Clione*
28 Copepods, *Calanus*
29 Assorted fish eggs
30 Stomatopod larva
31 Hydromedusa, *Hybocodon*
32 Hydromedusa, *Bougainvillia*
33 Salp (pelagic tunicate), *Doliolum*
34 Brittle star larva
35 Copepod, *Calocalanus*
36 Cladoceran, *Podon*
37 Foraminifer, *Hastigerina*
38 Luminescent dinoflagellates, *Noctiluca*
39 Dinoflagellates, *Ceratium*
40 Diatom, *Coscinodiscus*
41 Diatoms, *Chaetoceras*
42 Diatoms, *Ceraulus*
43 Diatom, *Fragilaria*
44 Diatom, *Melosira*
45 Dinoflagellate, *Dinophysis*
46 Diatoms, *Biddulphia regia*
47 Diatoms, *B. arctica*
48 Dinoflagellate, *Gonyaulax*
49 Diatom, *Thalassiosira*
50 Diatom, *Eucampia*
51 Diatom, *B. vesiculosa*

depths have light-emitting organs by which they lure prey and communicate (see Figure 36.33). In some cases the light is generated by mutualistic bacteria.

Coral Reefs. Coral reefs are the largest things on Earth created by organisms other than humans. (See Figure 24.14.) They are made primarily of calcium carbonate secreted by algae and several kinds of animals, especially coral polyps (phylum Cnidaria; see Figures 24.8 and 24.9), and certain sponges. Each small, cylindrical coral polyp secretes an external shell of $CaCO_3$. The polyps can live only on the outer surface of the reef, so each generation of polyp secretes a new layer of skeleton. Thin as they are, the layers accumulate year after year to depths of hundreds of meters. These organisms require clean, warm water and much sunlight, so almost all coral reefs are located near the surface on eastern continental shores or on submerged volcanoes in the tropics, where the water temperature stays above 20°C.

Charles Darwin identified three kinds of coral reefs:

1. **Fringing reefs,** which grow in narrow bands parallel to coasts in shallow water. This is the common type of reef along the Florida Keys and the Caribbean (Figure 17.22).
2. **Barrier reefs,** which form broader bands separated from coasts by lagoons. The most spectacular example is the Great Barrier Reef off the northeast coast of Australia, which is 2000 km long and 145 km wide. Actually the Great Barrier Reef is so large that it also includes fringing reefs within it.
3. **Atolls,** which are rings of coral that enclose lagoons thousands of meters across and only tens of meters deep. Atolls are formed on sinking volcanoes, as Charles Darwin first proposed during his voyage on the *Beagle* and as drilling confirmed in the 1950s.

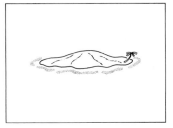

Fringing reef

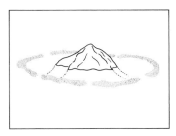

Barrier reef

Atoll

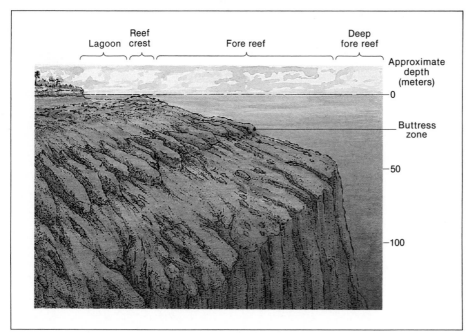

Figure 17.22
Structure of a fringing reef. The reef crest is a shallow band of growing coral that approximately parallels the coast. Often the reef crest encloses a shallow lagoon. Water in the lagoon is calm because the surf breaks against the reef crest. Most of the wave action is absorbed by the buttress zone in the sloping fore reef. Channels in the fore reef allow debris from eroding coral to fall into the deep fore reef instead of clogging the coral polyps. Few corals live in the deep fore reef, where light drops from 25% to 5% of the intensity at the surface.

The reason reef-building corals require sunlight is that the polyps obtain many of their nutrients from photosynthesizing protists (**zooxanthellae**) within their cells. The rate at which corals build reefs is closely coupled to the photosynthetic activity of these **endosymbiotic** (intracellular) protists. Coral polyps are also secondary consumers. At night their tentacles emerge from skeletons and prey upon zooplankton, stunning them with poisonous barbs (**nematocysts**). The prey are mainly small crustaceans and worms that graze on algae. Polyps also absorb small amounts of **dissolved organic material** from excrement and dead organisms. Coral reefs have the highest primary production of any ecosystem (Figure 17.2). This high productivity is rather paradoxical, because the clear waters above coral reefs are among the poorest in nutrients. The high productivity supports an enormous biomass and diversity of animals. Among the Caribbean reefs the common fishes alone number some 300 species.

The variety of habitats present in coral reefs partly accounts for the great diversity of animals that live among them. The most complete description of this complex habitat has been provided by the late Thomas F. Goreau and his family and colleagues for the fringing reef at Discovery Bay, Jamaica (Figure 17.22). Near shore the reef protects shallow lagoons where sea urchins and sea cucumbers (phylum Echinodermata) graze upon algae. The summit of the reef, called the **reef crest, coral rampart,** or **reef flat,** is exposed at low tide. It is dominated by large rounded and branching corals that form a network in which more delicate corals and algae live. In this highly productive zone space seems to be the most limited resource. Many species of coral use their stinging nematocysts against potential competitors, and sponges secrete chemicals that inhibit crowding. Invertebrates and fishes also compete for places to hide during the day. Among these animals are many that tend to erode the reef. Parrot fish chew up the coral skeletons with their beaklike mouths, the boring sponge *Cliona* takes out small flakes, and sea worms bore through it.

On the seaward side of the crest is the **fore reef** or **reef front.** Near the top of the fore reef, corals form a wall deeply scarred with channels that provide numerous and varied habitats for animals. More important, these channels protect the reef by dissipating wave action and by carrying away the fine, white sand of decomposed corals that would otherwise smother the coral polyps. The reef front then drops off almost vertically for approximately 100 meters into the darkness. As the depth increases, the light-requiring algae and symbiotic corals give way to sponges and other animals that feed on the detritus raining down from the productive zone above.

SUMMARY

The type of ecosystem, whether tropical rainforest, desert, marsh, coral reef, or other, is determined by such abiotic factors as moisture, temperature, and ionic concentration. In addition, the organisms in an ecosystem depend on biotic factors for food and other necessities. Animals in an ecosystem inevitably interact with each other through competition, predation, and symbiosis. The animals that succeed in such interactions make up part of a community. Many ecologists refer to the role of each species in a community as its niche and believe that competition for niches is the main factor that determines community structure. Others believe that predation and chance are more important than competition. Community structure may change over time in a process called succession.

Interactions of abiotic and biotic factors in ecosystems lead to nutrient cycles, in which carbon, oxygen, water, and nitrogen are exchanged among various organisms and with the air, soil, and water. Energy tends to be transferred along with carbon, except that instead of cycling, energy tends to dissipate. Consequently, there is less energy available at each trophic level, from producer, to primary consumer, to secondary consumer, and so on. This results in ecological pyramids of energy, biomass, and numbers, with less of each at higher trophic levels.

The transfer of energy from one organism to another can be represented as a food chain or, more realistically, as a food web. Increased diversity in a food web may be associated with greater stability in populations, either because complexity leads to stability, or because stability allows diversity to become established. In natural ecosystems populations are often stable, even though most organisms have the intrinsic capability of reproducing at an exponential rate. When the population reaches the carrying capacity, deaths from hunger, disease, and predation tend to equal the rate of birth. Often there are population cycles in which the population of a species fluctuates periodically around the carrying capacity.

Three of the most important ecosystems to zoologists are tropical rainforest, open ocean, and coral reef. Because of vertical stratification of light, humidity, and other abiotic factors, tropical rainforests provide a variety of habitats. Consequently, there is tremendous diversity; perhaps half of all animal species live there. Open ocean also supports a large number of species of zooplankton, as well as larger animals. Coral reefs, which are created by small animals called polyps, are also host to a tremendous species diversity, due to primary production by endosymbiotic zooxanthellae and the variety of habitats at various depths.

KEY TERMS

ecology	stratification	climax community
ecosystem	predation	biogeochemical cycle
limit of tolerance	symbiosis	decomposer
abiotic	mutualism	detritus feeder
biotic	parasitism	trophic level
primary production	commensalism	herbivore
producer	community	omnivore
consumer	niche	carnivore
autotroph	competition	ecological pyramid
heterotroph	competitive exclusion	food web
biome	succession	carrying capacity

SELF-TEST

1. In your own words explain what ecology is, and why the study of ecology is essential for a complete understanding of animals.

2. Using the concept of limits of tolerance, explain why different species of animals are found in different biomes. Name four biomes, and for each one name an animal species that can live there and one that cannot. For the latter species, explain why it is excluded.

3. What are some of the reasons for the differences in primary productivity shown in Figure 17.2? For example, why is primary productivity higher in a temperate deciduous forest than in a northern coniferous forest?

4. Name four types of aquatic ecosystem. For each one name an animal that lives in it.

5. Define "secondary succession" and describe an example of it, preferably from your own experience.

6. Figure 17.9 shows some terrestrial plants and animals involved in the carbon cycle. What kinds of marine organisms would play corresponding roles in each part of the carbon cycle?

7. Diagram the oxygen cycle. (*Hint:* Use Figure 17.9 as a model.)

8. Explain the meaning of the following phrase: "Matter circulates; energy dissipates."

9. Suppose that one area of the surface of Mars had been found to harbor 100 trillion cyanobacteria capable of photosynthesis, that the total biomass of these Martian cyanobacteria was 1 gram, and that they had an annual net primary production of 500 kcal. What can you say with certainty about the numbers, biomass, or stored energy of the animals (if any) living within this ecosystem? What can you say with less certainty but with some confidence?

10. What are krill and copepods? Explain why they are so important to marine ecosystems.

11. Explain in your own words why the stability of a community may increase with its diversity of species. Under what circumstances can increasing diversity reduce stability? Give an example.

12. Tropical rainforests and open oceans are both vertically stratified. Explain what this means. What effect does vertical stratification have on the diversity and distribution of various organisms at various depths in both ecosystems? (Where are the primary producers located? At what trophic level are the animals on the bottom? In each case, what kinds of animals are the major primary consumers?)

READINGS

RECOMMENDED READINGS

Bergerud, A. T. 1983. Prey switching in a simple ecosystem. *Sci. Am.* 249(6):130–141 (Dec).

Brill, W. J. 1977. Biological nitrogen fixation. *Sci. Am.* 236(3):68–81 (Mar).

Brownlee, S. 1984. Bizarre beasts of the abyss. *Discover* 71–74 (July).

Cronon, W. 1983. *Changes in the Land.* New York: Hill and Wang. (*A fascinating account of how Indians and colonists altered the ecology of New England.*)

Goreau, T. F., N. I. Goreau, and T. J. Goreau. 1979. Corals and coral reefs. *Sci. Am.* 241(2):124–136 (Aug).

Houghton, R. A. and G. M. Woodwell. 1989. Global climatic change. *Sci. Am.* 260(4):36–44 (Oct).

Jackson, J. B. C. and T. P. Hughes. 1985. Adaptive strategies of coral-reef invertebrates. *Am. Sci.* 73:265–274.

Laws, R. M. 1985. The ecology of the southern ocean. *Am. Sci.* 73:26–40.

May, R. M. and J. Seger. 1986. Ideas in ecology. *Am. Sci.* 74:256–267.

Perry, D. R. 1984. The canopy of the tropical rain forest. *Sci. Am.* 251(5):138–147 (Nov).

Post, W. M. et al. 1990. The global carbon cycle. *Am. Sci.* 78:310–326.

Ramage, C. S. 1986. El Niño. *Sci. Am.* 254(6):77–82 (June).

Schneider, S. H. 1989. The changing climate. *Sci. Am.* 261(3):70–79 (Sept).

Sebens, K. P. 1985. The ecology of the rocky subtidal zone. *Am. Sci.* 73:548–557.

White, R. M. 1990. The great climate debate. *Sci. Am.* 263(1):36–43 (July).

ADDITIONAL REFERENCES

Barber, R. T. and F. P. Chavez. 1983. Biological consequences of El Niño. *Science* 222:1203–1210.

Brown, V. K. 1984. Secondary succession: insect–plant relationships. *BioScience* 34:710–716.

Connor, E. F. and D. Simberloff. 1986. Competition, scientific method, and null models in ecology. *Am. Sci.* 74:155–162.

Davidson, J. 1938. On the growth of the sheep population in Tasmania. *Trans. R. Soc. S. Australia* 62:342–346.

Einarsen, A. S. 1945. Some factors affecting ring-necked pheasant population density. *Murrelet* 26:39–44.

Erwin, T. L. 1983. Beetles and other insects of tropical forest canopies at Manaus, Brazil, sampled by insecticidal fogging. In: S. L. Sutton and T. C. Whitmore (Eds.). *Tropical Rain Forests: Ecology and Management.* Edinburgh: Blackwell, pp. 59–75.

Lewin, R. 1983. Predators and hurricanes change ecology. *Science* 221:737–740.

Lewin, R. 1984. Parks: how big is big enough? *Science* 225:611–612.

MacLulich, D. A. 1937. Fluctuations in the numbers of the varying hare (*Lepus americanus*). University of Toronto Studies, Biol. Ser. No. 43.

Menard, H. W. 1986. *Islands.* New York: Scientific American Books. (*See Chapter 7 on coral reefs.*)

Naiman, R. J. 1988. Animal influences on ecosystem dynamics. *BioScience* 38:750–752. (*Introduction to several papers on beaver, moose, prairie dogs, and other animals.*)

Nicholson, A. J. 1958. The self-adjustment of populations to change. *Cold Spring Harbor Symp. Quant. Biol.* 22:153–173.

Odum, H. T. 1957. Trophic structure and productivity of Silver Springs, Florida. *Ecol. Monogr.* 27:55–112.

Payne, W. J. 1983. Bacterial denitrification: asset or defect? *BioScience* 33:319–325.

Reznick, D. A., H. Bryga, and J. A. Endler. 1990. Experimentally induced life-history evolution in a natural population. *Nature* 346:357–359.

Scheffer, V. B. 1951. The rise and fall of a reindeer herd. *Sci. Monthly* 73:356–362.

Schneider, W. and P. Wallis. 1973. An alternative method of calculating the population density of monsters in Loch Ness. *Limnol. Oceanogr.* 18:343.

Walker, J. C. G. 1984. How life affects the atmosphere. *BioScience* 34:486–491.

Walsh, J. J. 1984. The role of ocean biota in accelerated ecological cycles: a temporal view. *BioScience* 34:499–507.

18

Causes of Evolution

Male great frigate bird (Fregata minor) *displaying on the Galápagos Islands.*

CHAPTER OUTLINE

LEARNING OBJECTIVES

1. How did so many different species come to be?

2. How did the current ideas of evolution and natural selection arise?

3. What is the relationship of genetic mutation to evolution?

4. What kinds of experimental studies have convinced biologists of evolution?

5. How could phenomena as elaborate as mimicry and camouflage have evolved?

6. How do scientists determine the evolutionary history of a species?

7. Does evolution lead to progress? Do only the fittest survive? Do all features of a species contribute to fitness? Is evolution for the good of the species?

8. Is natural selection the only mechanism of evolution?

9. By what process does a new species evolve from a preexisting species?

10. What role does sex play in evolution? Why did sex evolve?

THE MAIN IDEAS OF EVOLUTION

Fossils speak as eloquently of past evolution as the ruins of temples and monuments speak of past civilizations. The anatomy, physiology, and genetics of living organisms likewise bear the imprint of their evolutionary heritage as indelibly as living cultures bear the imprint of their cultural heritage. There are as few biologists who doubt that evolution occurred as there are historians who doubt that ancient Rome existed and influenced our own culture. This is not to say that all biologists agree on every aspect of evolution. Even the definition of evolution has been argued. In the past evolution usually referred to the transformation of one species into another. Now, however, most biologists define evolution as *the change in the genetic makeup of an interbreeding group of organisms.* Such changes may or may not accumulate to the point where a taxonomist would recognize the emergence of a new species.

Many theories have been proposed to explain how and why evolution occurs, but the most widely accepted is the **synthetic** or **neo-Darwinian theory.** As both names for the theory suggest, it synthesizes Charles Darwin's theory of natural selection with modern genetics. The synthetic theory is based on three elementary facts of life:

1. Within natural **populations** (groups of interbreeding organisms) there are differences in the genetic makeup of individuals. These differences arise from genetic **mutations** and from **recombinations** of these mutations during sexual reproduction (see Chapter 4). Some individuals therefore inherit combinations of genes that make them better adapted than other individuals.
2. Many more individuals can be produced than can survive. Members of a population therefore compete with each other to survive and reproduce, and the number of offspring tends to be higher for those that are better adapted for the competition.
3. Thus there tends to be an increase in the frequency of those genes that contribute to such adaptations.

Natural selection means that evolution occurs because nature, in effect, selects which individuals will pass on their genes to the next generation. The resulting change in gene frequency within a species is evolution. If the change in gene frequency is within a single species, it is referred to as **microevolution.** The synthetic theory proposes that natural selection over long periods also leads to **macroevolution:** the origin of new species, genera, and higher taxa.

The synthetic theory itself is like a living organism. Because of competition from other theories, it has had to evolve to keep pace with a changing theoretical and experimental "environment." This chapter is an account of that evolution.

THE EVOLUTION OF DARWINISM

Like most of the really important ideas in science, evolution is simple now that some extraordinarily gifted people have discovered it for us. What surprises us now is that until two and a half centuries ago apparently no scientist realized that species could slowly change. One reason for the failure to appreciate evolution is that it is usually not apparent within the lifetime of a human, or even within the history of civilization. There are also historical reasons why it took so long to appreciate evolution. One such stumbling block was the overwhelming influence of Aristotle in Western thought. (See Harris 1981 for references and further details on the historical and philosophical portions of this chapter.) In the 4th century B.C. Aristotle created an all-embracing view of science and metaphysics that emphasized design and purpose. It would have made no sense to him for a species that had been designed for one purpose to change into another species designed for an entirely different purpose. In the 13th century Thomas Aquinas incorporated Aristotelian philosophy into Christian theology, and Aristotle's ideas became woven into the fabric of Christian culture. So ingrained are the concepts of design and purpose that even evolutionists often resort to the word "design" when describing adaptations, and they have to fight the temptation to think of adaptations as having a purpose rather than simply a function.

Early Evidence and Theories of Evolution. The idea that species could slowly change due to natural processes first arose in France in the middle of the 18th century, during a period called the Enlightenment. This period just prior to the French Revolution was a time of general discontent with all established authority, including the Church, the King, and Aristotle. Philosopher-scientists of the Enlightenment argued that human progress is possible under natural law (not under the laws of the Church and the King) and seized upon any evidence that humans and other species can naturally progress from lower to higher forms. The two earliest philosopher-scientists to clearly state this evolutionary view were **Pierre-Louis Moreau de Maupertuis** and **Denis Diderot.** Maupertuis (pronounced mo-per-TWEE) proposed in 1745 that each organism was formed from particles coming from all parts of its parents (**pangenesis**), and that new types of organisms arose when unique combinations of such particles were maintained for several generations. Diderot (deed-ROW) suggested in 1754 that entire species could be born, develop, and die just like individuals. Although Maupertuis and Diderot were politically motivated, they based their arguments on scientific evidence such as the following:

1. **Comparative anatomy.** It had recently been found that in spite of their outward differences, the bones in horse hooves and human hands were similar in number and position. This suggested that hooves and hands were not specially designed to serve the needs of each species but had come from a shared origin (Figure 18.1). Such structures that have a shared evolutionary origin are now said to be **homologous.** Structures such as the wings of bats and of butterflies are, in contrast, **analogous:** in spite of their superficial similarities, they evolved from different origins. The most striking examples of homologous structures are **vestigial organs,** which serve no function but resemble a functional organ in another organism, and presumably in ancestral species. Examples of vestigial organs include pelvic bones in some whales and snakes, and in humans the reduced tail bones and muscles, and much of the body hair.
2. **Comparative embryology.** Anatomical similarities are even clearer in the early embryos of related species, as Karl von Baer first noted for vertebrates (see pp. 132–133).
3. **Fossils.** Fossils had been known for many centuries but were generally treated as rare curiosities rather than as relics of extinct species. With increased digging for ores and fuels during the Industrial Revolution, however, fossils could no longer be ignored. Evolution was directly suggested by the fact that fossils were similar to, yet different from, living species. Fossils also raised the question of whether God would have designed species only to let them become extinct (Figure 18.2).

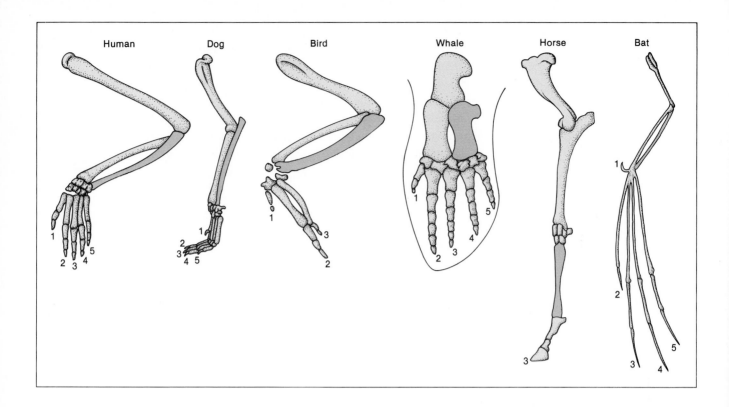

Human   Dog   Bird   Whale   Horse   Bat

Figure 18.1
Homologous structure of the limbs of vertebrates. Even such diverse structures as the bat's wing, a horse's leg, and a human arm have similar numbers and sequences of bones. The number of bones can vary, however, if two or more of them fuse together during evolution, as in the horse.

4. **The age of the Earth.** According to biblical chronology the world was created only within the last 6000 years—too recently for significant evolution to have occurred. Studies of the rate of cooling of iron spheres indicated, however, that if the Earth had begun as a molten mass, it must be at least hundreds of thousands of years old to have cooled to its present temperature. If the Earth were indeed so old, changes in species seemed more plausible. (In fact, we now known that the Earth is approximately 4.5 *billion* years old. The 18th-century experimenters did not know that heat is continually generated in the Earth's core by nuclear reactions.)

5. **Artificial selection.** Entirely new kinds of plants and animals had been produced in relatively short times by selective breeding. It seemed likely that what humans could do in a short time, nature could certainly do in the much longer time available (Figure 18.3).

In order to avoid the censors, Maupertuis and Diderot had to write briefly and obscurely. In 1809, however, the French Revolution gave **Jean Pierre Antoine de Monet de Lamarck** the freedom to write a long book on the transformation of species and its causes. Lamarck was a gifted scientist who revolutionized the classification of animals by distinguishing between vertebrates and invertebrates. He was also one of the first to use the word *biologie*, thereby recognizing the study of life as a distinct science. Lamarck believed there were two major causes of the transformation of species: a natural tendency for each species to progress toward a "higher" form, and the inheritance of acquired characters. The latter principle means that bodily changes that occur during the life of an organism can be inherited by its offspring. The concept of inheritance of acquired characters is nowadays referred to as **Lamarckism,** but it was popular thousands of years before Lamarck (see p. 66). Lamarck argued that the struggle of individual animals to survive causes their nerves to secrete a fluid that enlarges the organs involved in the

Figure 18.2
(A) One of the six known fossils of Archaeopteryx. The feathered wings clearly indicate that Archaeopteryx was a bird, but unlike any existing bird, it had teeth and vertebrae in its tail. Some fossils of Archaeopteryx, including the one most recently discovered, were catalogued as dinosaurs for many years.
(B) Archaeopteryx as it might have appeared in life, as a crow-sized predator.

struggle, and that the effects of this nervous fluid are passed on to succeeding generations. In this way, for example, giraffes evolved long necks through their attempts, generation after generation, to browse on higher and higher leaves. Lamarck never explained how organisms without nerves would evolve.

Lamarck's theory was quickly abandoned, mainly because of criticism by Baron Georges Léopold Chrétien Frédérick Dagobert Cuvier, one of the biggest names in science. Cuvier (pronounced KEW-vee-ay) was a pioneer in paleontology (the study of fossils) and in comparative anatomy. Although he rejected the idea of transformation of species, he certainly could not ignore the fossil evidence that different species had occupied the Earth in different ages in the past. His explanation for the changes in species was that life had been extinguished periodically by catastrophes, then replaced by different species. This theory is called **catastrophism.** When the first dinosaur skeletons were discovered by quarrymen digging for plaster of Paris to repair the ravages of the revolution, Cuvier must have regarded them as absolute proof of his theory. What else but a catastrophe could have extinguished such huge beasts?

Other Evolutionists Before Charles Darwin. While Lamarck was creating his theory, **Erasmus Darwin,** the most popular English poet and physician of his time, and grandfather of Charles Darwin, independently proposed a similar theory in a poem (1789) and in a medical book (1794). Dr. Darwin's theory was attacked and then forgotten in England. In 1813 **William Charles Wells,** who had grown up in Charleston, South Carolina, and gone to England during the American Revolution, argued that the black and white races of humankind had diverged because those with darker skins survived better in the tropics. Wells did not suggest that one species could change into another, but by today's definition of evolution we would have to call him an evolutionist. Moreover, he grasped the

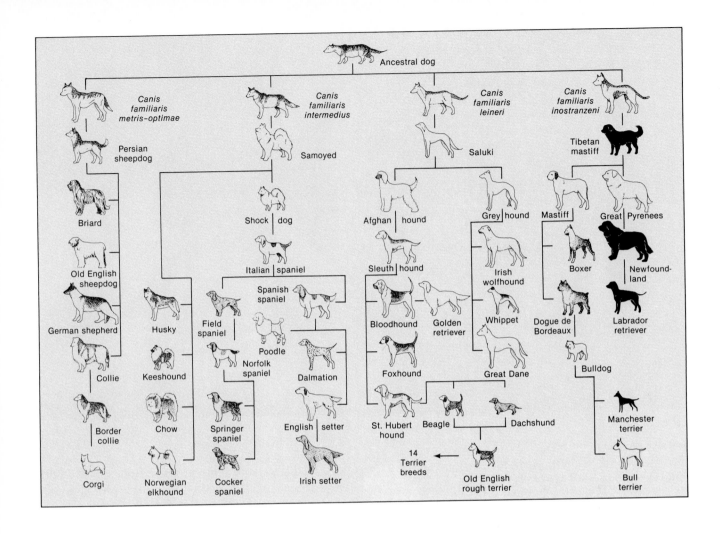

Figure 18.3
One of several pedigrees that have been suggested for the major breeds of dogs. All breeds are classified within the species Canis familiaris *because they are known to have diverged from common ancestors (possibly the Asiatic wolf) within recent times. The anatomical differences among many of the breeds are far greater, however, than the differences between many pairs of different but related species, such as wolves* Canis lupus *and coyotes* C. latrans. *A taxonomist with no knowledge of the history of dogs would probably classify the Saint Bernard and the chihuahua in different species.*

principle of natural selection. The one who came closest to anticipating Darwin's theory of natural selection was **Patrick Matthew**, a little-known Scotsman, who wrote his ideas in an appendix to a book on ship-building timber published in 1831. Unfortunately for Matthew, wooden ships were becoming obsolete, so his appendix went unnoticed. Another problem for acceptance of evolution was that **Charles Lyell** criticized it in his *Principles of Geology* in the 1830s.

FOUNDATIONS OF DARWINISM

Ironically, it was Lyell who laid the foundation on which Charles Darwin erected his arguments that evolution had occurred by natural selection. Lyell's contribution was to replace catastrophism with uniformitarianism. **Uniformitarianism** is the concept that geological changes in the past occurred in the same way as those occurring now. Instead of sudden worldwide catastrophes, geologists after Lyell started to believe that geological changes were mainly local and gradual.

Between 1836 and 1856 at least 19 others suggested that evolution had occurred, but none provided convincing evidence or adequate explanations for evolution. With few exceptions their writings attracted little attention until the publication of Darwin's *Origin of Species* in 1859. Even when Charles Darwin's theory was outlined at the Linnaean Society along with Alfred Russel Wallace's (see box) few people paid much attention. The president of the Linnaean Society summarized that year of 1858 as one in which nothing revolutionary had occurred

Creation "Science"

At the same time that Enlightenment philosophers were starting to doubt a recent creation of species, biblical scholars began to doubt the literal truth of the two differing accounts of creation in *Genesis*. Since then it has become increasingly rare to find either theologians or scientists who seriously believe that all species were created individually during a short period within the past few thousand years. Curiously, however, since 1970 there have arisen vocal groups of "creation scientists" who insist on just such a view. Even more curious, many school boards, some state legislatures, and even a President of the United States have taken the claims of creation scientists seriously enough to advocate teaching creationism in science courses.

Two groups largely responsible for this situation are the Institute for Creation Research and the Creation Research Society. Although they claim to represent science, their activities are not scientific in the usual sense. Most of their research is done in libraries, where they scour the scientific literature for disagreements about *how* evolution occurred and then misrepresent them as disagreements about *whether* evolution occurred. Contrary to their public claims, these groups admit in private that their motives are religious. Typical is a December 1989 letter from the Institute for Creation Research, which includes the following: "We are thankful that the Lord has, through the gifts of His people, kept the ICR witness active now for over 19 years. Please continue to pray and give . . . to keep the message of true creation going out through the coming year and—Lord willing—until Christ returns. Sincerely, in His Name, (signed) Henry M. Morris, Ph.D. President."

Further evidence of the real nature of creation "science" is the oath that all members of the Creation Research Society must sign, which includes the following: (1) "The Bible is the word of God, and . . . all of its assertions are historically and scientifically true; (2) All basic types of living things, including man, were made by direct creative acts of God during Creation Week as described by Genesis; (3) The great Flood described by Genesis . . . was an historical event. . . ; and (4) Finally, we are an organization of Christian men of science. . . ."

Creation scientists have succeeded mainly because of skilled debaters who effectively present their case to a public that is poorly educated in the theories and methods of science. Creationists have also taken advantage of the fact that at first most evolutionists were too busy with current scientific controversies to spend much time debating 18th-century controversies. Lately, however, evolutionists have been better prepared for debates, and many have taken the trouble to answer all of the creationist arguments. (See end of chapter for selected references.) More important in the long run, many science educators (such as Moore 1984) have begun to emphasize the necessity of teaching science in a more realistic manner—as a continuing effort to create testable theories about phenomena of the knowable world, rather than as a mere collection of immutable facts. Only in this way can we hope to avoid future situations in which people cannot tell the difference between dogma and science.

---

in science! It was Darwin's *Origin of Species*, published a year later, that changed the way biologists—and, indeed, people in general—think about life. It is therefore Charles Darwin who is rightly given most of the credit for establishing "Darwinism" as a field of biology.

THE ORIGIN OF THE SYNTHETIC THEORY

By the turn of the century *The Origin of Species* had convinced almost all biologists that evolution occurred. It was not so successful in convincing them that natural selection was the cause. In fact, Darwin himself often stated that natural selection was only one of several possible causes. One reason for the reluctance to accept natural selection was that many theories of inheritance, including Darwin's own theory (a form of pangenesis), allowed for the possibility that the environment or the organism itself directly altered the heredity of the organism. The work of Gregor Mendel (see pp. 66–70) disproved this idea, but it was unknown to Darwin and to most others until 1900. In fact, Mendel's work was first taken as an argument against natural selection, because Mendelians emphasized the importance of mutations rather than natural selection. They believed that evolution resulted from large genetic changes occurring within one generation, rather than from selection of small genetic variations occurring over many generations. During the first two decades of this century, therefore, many of the leading biologists were confidently telling their students that Darwinism was extinct.

The Convergent Evolution of Darwin and Wallace

Charles Robert Darwin

Born: 12 February 1809 in Shrewsbury, England
Died: 19 April 1882 in Down, England
Education: Undergraduate medical student for 1½ years at Edinburgh University. B.A. (in divinity) Christ's College, Cambridge University.
Naturalist. Author of *The Voyage of the Beagle, The Origin of Species, The Descent of Man, and Selection in Relation to Sex, The Expression of the Emotions in Man and Animals,* and numerous other books and articles.

Alfred Russel Wallace

Born: 8 January 1823
Died: 7 November 1913
Education: Six years of elementary school
Collector and naturalist. Author of *Darwinism* and several books of travel and natural history.

Figure 18.4
Charles Darwin at age 45, four years before receiving Wallace's manuscript.

Figure 18.5
Alfred Russel Wallace, probably sometime after 1862, when he returned to England from the Malay Archipelago.

Figure 18.6
Charles Lyell, who warned Darwin that Alfred Russel Wallace might also be working on the theory of natural selection. Lyell had previously played a crucial role in evolution by convincing most geologists that changes in the Earth had come about gradually over long periods of time.

Late in the spring of 1858 Charles Darwin (Figure 18.4) sat in his large home at Down, 20 miles outside London, reading the morning mail before returning to his study for another hour and a half of work. His eyes gazed wearily beneath his overhanging brows. He was not yet 50, but poor health and the labor of writing several books on geology and biology had left their marks. Now Darwin was struggling to complete a book summarizing evidence gathered for 20 years, which suggested that species had evolved through a process he called natural selection.

Darwin picked up an envelope and recognized it as being from young Alfred Russel Wallace (Figure 18.5), the gifted naturalist who was making his living in the Malay Archipelago by selling specimens he collected in the tropical forests. He eyed it warily. Less than two years before, Darwin's close friend Charles Lyell (Figure 18.6) had warned him that Wallace seemed on the verge of proposing a theory much like his own. Heeding Lyell's warning, Darwin had begun to write a brief summary for early publication, but the "brief summary" had already grown into a large book, and it was still growing. Darwin opened Wallace's envelope and found a cover letter asking Darwin's opinion on an enclosed manuscript. Darwin scanned the manuscript and was stunned. In a few pages it summarized the theory he had labored over for two decades.

It may have crossed Darwin's mind to destroy Wallace's paper and pretend he had never received it. Mail took several weeks between England and the Malay Archipelago, so Darwin could easily have put together his own short paper before Wallace discovered that his manuscript was missing. Instead Darwin wrote to Lyell: "Your words have come true with a vengeance. . . .

Please return me the [manuscript] which he does not say he wishes me to publish, but I shall of course, at once write and offer to send to any journal. So all my originality, whatever it may amount to, will be smashed, though my book, if it will ever have any value, will not be deteriorated; as all the labour consists in the application of the theory." A week later Darwin and Lyell decided it would not be dishonorable if some of Darwin's previous sketches of the theory were read before the Linnaean Society at the same time they presented Wallace's manuscript. In the meantime Darwin set to work condensing the big book he had started into an "abstract" that we now know as *The Origin of Species.*

The convergence of Charles Darwin and Alfred Russel Wallace on the same theory of evolution seems a remarkable coincidence, especially considering the differences in their backgrounds. Charles Darwin's father was a prominent physician and son of Erasmus Darwin, and his mother was of the famous Wedgwood family of potters. As a member of the landed gentry, Charles Darwin had few practical worries. In fact, he gave up studying medicine partly because he realized he would inherit a fortune anyway. His father convinced young Darwin to study theology as an alternative to becoming an idle sporting man. Charles Darwin agreed partly because ministering to the parishioners of some rural church would not interfere with his main interest, collecting beetles. Darwin was religious in a polite sort of way and had no objection to any of the doctrines of the Church of England, including the belief that God had created each species individually.

Wallace, on the other hand, was brought up without religion, or much of anything else. He had to begin earning a living after only six years of schooling. Through his own efforts he acquired enough training to become a teacher. Many hours in libraries, museums, and the field later made him a leading authority on natural history. These intellectual pursuits were by necessity mingled with the need to earn a living.

From the perspective of science the differences between Darwin and Wallace are less important than the similarities in their backgrounds. Chief among the similarities were: (1) collecting, which provided an appreciation of the variations within species; (2) traveling, which provided an appreciation of biogeography (the distribution of species); and (3) the reading of the *Essay on Population* by Thomas Malthus, which provided an appreciation of the struggle for survival within species.

COLLECTING

As a boy Darwin collected shells, pebbles, eggs, and other natural miscellany, but at Cambridge in the late 1820s he focused on beetles. The intensity of Darwin's passion for beetle collecting can be judged from the following episode described in his *Autobiography*.

> One day, on tearing off some old bark, I saw two rare beetles and seized one in each hand; then I saw a third and new kind, which I could not bear to lose, so that I popped the one which I held in my right hand into my mouth.

In trying to classify his treasures, Darwin could hardly have missed the differences among individuals within the same species (Figure 18.7). Variability is important in natural selection, for if there are no inherited differences among individuals, then there is nothing to select.

Wallace became interested in natural history at age 13 during his first job helping his brother, a surveyor. A decade later Wallace met Henry Walter Bates, who interested him in insects. Together they went to South America, planning to finance the voyage by selling rare

Figure 18.7
Variation within a single species of beetle, the Asiatic lady beetle Harmonia axyridis. *Each variant is geographically isolated from other populations, yet all belong to the same species.*

insects they collected. Collecting was to be Wallace's major source of income for the next 14 years. As with Darwin, the variations he observed within species played an important role in his conception of natural selection.

TRAVELS

Soon after graduating in 1831, Darwin received a letter from a former teacher offering a position as naturalist on a round-the-world surveying voyage aboard the *Beagle* (Figures 18.8 and 18.9). As the letter noted, Darwin was not a "finished naturalist," but he was "amply qualified for collecting, observing, and noting anything worthy to be noted in Natural History." (In addition, Darwin was

The Convergent Evolution of Darwin and Wallace *(continued)*

Figure 18.8
The Beagle *in the Straits of Magellan, between South America and Antarctica. Darwin was one of 74 people crammed on board this vessel, which was 90 feet long and 24½ feet wide. In the foreground are natives of Tierra del Fuego, whose wildness made a profound impression on Darwin.*

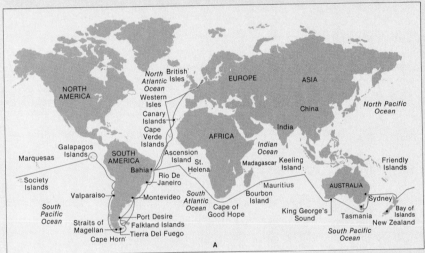

A

B

Figure 18.9
(A) The route of Darwin's voyage around the world. The Beagle *left England on 27 December 1831 (after a two-month delay by weather) and returned on 2 October 1836. (B) The Galápagos Islands. These islands arose from submerged volcanoes within the past three to four million years. Although located on the equator, the climate is cooled by the Humboldt current from the south.*

370

of the correct social class to serve as a companion to the 26-year-old captain. (The previous captain of the *Beagle* had committed suicide, probably from loneliness.) So at age 22 Darwin bade farewell to his family and the woman to whom he was practically engaged and set off on a hazardous voyage that was supposed to last three years, but which lasted five.

As a going-away present Darwin was given a copy of the first volume of *Principles of Geology*, which had just been published by a little-known lawyer named Lyell. Darwin thereby became the first naturalist to see the New World through uniformitarian eyes. With this perspective he made several important contributions to geology, including the correct explanation for the origin of coral atolls. Lyell's book also focused Darwin's attention on the fossil record and the geographical distribution of species. Observations in these two areas eventually convinced Darwin of evolution. This conversion did not occur until March 1837, however, at least five months after Darwin's return to England (Sulloway 1982b). Throughout the voyage Darwin believed the theory he had been taught by those who compromised between *Genesis* and the fossil record. This theory was similar to Cuvier's: species had been periodically extinguished by catastrophes, then replaced by newly created species.

The following observations eventually made Darwin doubt that theory:

1. In South America Darwin found the fossils of several animal species that were different from, but related to, the species still living in the area. Such discoveries had been made before. Catastrophists explained them as the result of God's having used similar species to replace the ones just destroyed. Darwin, however, realized that the evolution of the fossil species into the present species was more consistent with uniformitarianism.

2. Darwin found that similar animals on the Galápagos Islands differed from one island to the next, even though the islands were physically similar and close to each other. It seemed unreasonable that a creator would have made different species for each island. At first Darwin assumed that all the similar animals were simply varieties within the same species, but on returning to England he learned from the experts that his "varieties" were in fact distinct species. This made him wonder why it was so difficult to tell the difference between species and varieties if species had been created individually.

Wallace's first voyage to the Amazon between 1848 and 1852 ended in disaster when the ship with many of his specimens burned on the return trip. Undaunted, Wallace scraped together enough money for an expedition to the Malay Archipelago between 1854 and 1862. There his discoveries in biogeography paralleled those of Darwin. In 1855 he published a paper in which he too observed that the fossils of extinct species are found in areas where similar species now live. (This was the paper that prompted Lyell to warn Darwin of the possibility of being "scooped" by Wallace.) Wallace also discovered that the marsupials and other unique groups on Australia and neighboring islands were separated from the Oriental groups on other islands of the Malay Archipelago by a sharp boundary, now called Wallace's Line. As with the Galápagos species, the simplest explanation was that the species on both sides of Wallace's Line evolved in different directions after they became isolated from each other.

MALTHUS AND THE STRUGGLE FOR SURVIVAL

Although Darwin realized several months after the *Beagle* voyage that species had evolved, he merely hinted at his ideas in public. He was too busy arranging for his collections to be analyzed by experts, writing an account of the voyage (which subsequently became *The Voyage of the Beagle*), and making important contributions in geology. He must also have been distracted by his first cousin, Emma Wedgwood, for he married her early in 1839. Darwin also began to suffer from heart palpitations, vomiting, and other disorders that did not abate until *The Origin of Species* was finally published. Even if Darwin had had the time and good health to publish his idea of evolution right away, he must have known that doing so would have jeopardized his standing as a geologist. Lyell had demolished "Lamarckism," and most biologists, including Darwin, had a low opinion of other theories of evolution. Darwin knew that he needed more evidence to support his claim that evolution had occurred, and also a plausible explanation for how it occurred.

For clues to how nature altered species, Darwin began to study artificial selection, by which humans had altered numerous domesticated species. His major problem was to figure out how nature, lacking the consciousness of the breeder, could select which individuals would reproduce. Darwin read farm journals, quizzed breeders, and became a pigeon breeder himself, in hopes of learning the secret, but in the end the clue came from an essay on social theory, which Darwin read "for amusement" late in 1838. It was the *Essay on Population*, in which the Reverend Thomas Malthus pointed out that the ability of humans to reproduce far exceeds the capacity of the land to feed the resulting population. Malthus concluded that the human "struggle for survival" through famine, war, and social injustice was therefore natural and unavoidable. (Malthus mentioned the alternative of birth control only to dismiss it as a "vice" worse than starvation, war, and disease.)

Malthus intended his essay as an argument against French Enlightenment dreams of human progress, but

Darwin saw it as a mechanism for biological change. If more individuals are produced than can survive, then competition among members of each species is inevitable. Rather than stifle change, as Malthus thought, competition could cause evolutionary change. The best-adapted individuals would be the best competitors and would therefore contribute their heritable adaptations in greater proportion to later generations. The insight triggered by Darwin's chance reading of Malthus overthrew Artistotle in a single stroke. Adaptations need not result from design—they could result from natural selection of individuals in randomly varying populations. Darwin might have published his theory in 1838, only to have it forgotten. Instead, he continued to collect evidence and did other research, including an eight-year study of the anatomy and taxonomy of barnacles. As we have seen, it was Wallace's manuscript proposing the same theory that finally forced Darwin to publish.

By a remarkable coincidence Wallace's insight was also triggered by Malthus' *Essay on Population*, which he had read 12 years before in England. While lying down in his forest shack to recover from an attack of malaria, Wallace happened to remember Malthus' essay and related it to the problem of evolution that he had been thinking about. In the next two days he wrote out the main ideas of the theory of evolution by natural selection, unaware that the man he intended to send it to had already devoted 20 years to the same theory.

Contributions of Theoretical Population Genetics. The neo-Darwinian or synthetic theory of evolution emerged gradually as a combination of Darwin's theory of natural selection and modern genetics. On the theoretical front, G. H. Hardy and W. Weinberg showed independently in 1908 that one argument against natural selection was mathematically unfounded. It had been assumed that a new **allele** (a different form of a gene) arising in a single individual would quickly vanish by **blending** into the rest of the population. If that happened, then evolution by natural selection would hardly have a chance to get started. What is now called the **Hardy–Weinberg Law** shows, however, that alleles do not get diluted out of an **equilibrium population.** An equilibrium population is one that satisfies the following conditions:

1. The size of the population is practically infinite, so random variations in the reproduction of genes can be ignored.
2. Individuals mate at random. That is, the presence of a particular allele does not affect which individuals mate with each other.
3. The alleles do not increase or decrease the ability of organisms to reproduce. (In other words, natural selection does not occur in the population.)
4. There is no gain or loss of new alleles from another source. That is, there is no immigration or emigration.
5. There is no new mutation.

Obviously there has never been a population that satisfies all these conditions. The Hardy–Weinberg Law is nevertheless valuable, because it demonstrates that alleles do not automatically disappear by blending, even in infinitely large populations. Moreover, the Hardy–Weinberg Law serves in two ways as the starting point for all **theoretical population genetics.** First, since an equilibrium population is by definition a nonevolving one, the Hardy–Weinberg Law provides a baseline for measuring rates of evolution. Second, the Hardy–Weinberg Law suggests that any departure from the five equilibrium conditions listed above can cause a change in allele frequencies. Since evolution is now defined as a change in allele frequencies in a population, the previous sentence means that anything that upsets the conditions for equilibrium listed above can be a mechanism of evolution. This includes the random effects of small population size (genetic drift), nonrandom mating (sexual selection), natural selection, emigration or immigration, and mutation.

The second theoretical support for the synthetic theory came in the 1920s and 1930s when **R. A. Fisher, J. B. S. Haldane,** and **Sewall Wright** showed by three different mathematical proofs that natural selection is a more effective mechanism of evolution than is mutation. In order for random mutations to be the agent of evolution, they would have to occur so frequently that harmful mutations would undo any previous evolution that had occurred. Mutations are, of course, necessary for evolution, since they produce genetic variations. These variations in **genotype** then contribute to the variations in **phenotype** (the expressed characteristics of an organism) on which selection acts. By selecting one phenotype over another, nature indirectly selects one genotype over another, thereby altering allele frequencies.

Experimental Contributions. In addition to theoretical population genetics, numerous field studies during the 1930s and 1940s produced results that agreed more with the predictions of "selectionists" than with those of "mutationists." Among the major contributors to these studies were the following.

Theodosius Dobzhansky was a *Drosophila* geneticist who took his knowledge into the field. He discovered, among many other phenomena, that in natural populations there is a great deal of genetic variation. Consequently, evolution does not have to wait for just the right mutation to occur at just the right time.

Ernst Mayr is, among many other distinctions, a taxonomist who clarified the concept of species. Most biologists had thought of species as fixed types, rigidly defined by their appearance. This "typological" concept made it difficult to understand how one species could evolve into another. In contrast, Mayr defined species as "groups of actually or potentially interbreeding natural populations, which are reproductively isolated from other such groups." This **biological species concept** greatly clarifies thinking about living organisms that reproduce sexually, although it does not apply to asexual species or to fossils.

Bernhard Rensch drew attention to the way populations of a given species tend to vary with environmental conditions in an adaptive way, as predicted from natural selection. For example, a mammal in a cold climate tends to have shorter ears, tail, and feet than a member of the same species in a warmer climate (Allen's Rule). Rensch pointed out that this phenomenon could be explained by natural selection, since shorter appendages lose heat less readily.

George Gaylord Simpson was a paleontologist who showed that the fossil record was consistent with gradual evolution of adaptations to environmental conditions, as predicted by natural selection (Figure 18.10).

SOME RECENT STUDIES OF EVOLUTION

Competition in Darwin's Finches. According to the synthetic theory, without competition within a population there can be no selection. Competition therefore plays a crucial role in evolution, but not necessarily the obvious role pictured in Alfred, Lord Tennyson's phrase, "Nature, red in tooth and claw." In most animals (and certainly in plants and other organisms) competition is usually a quiet, invisible struggle to avoid predators and obtain energy, mates, and the other necessities for survival and reproduction. The struggle is so quiet and invisible that most

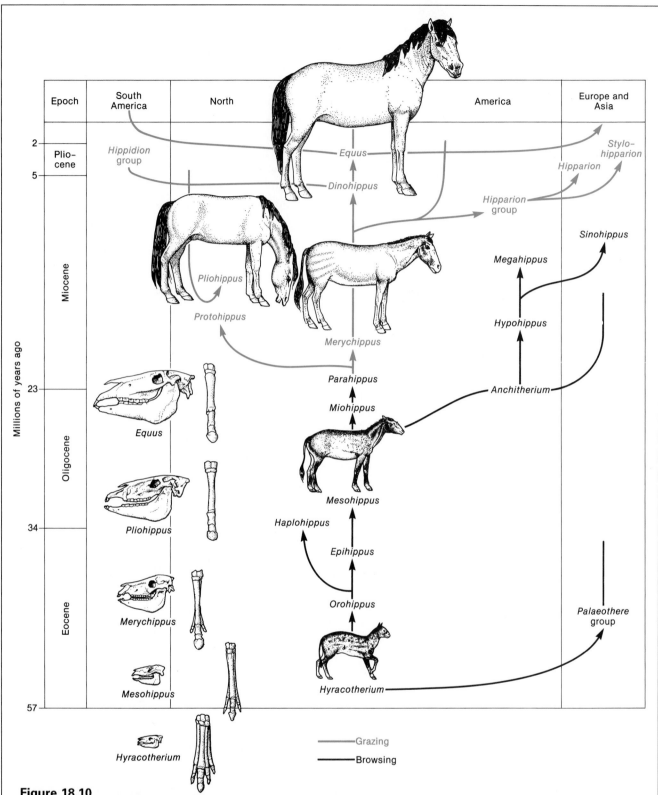

Figure 18.10
The evolution of horses, from the rabbit-sized Hyracotherium *that browsed on foliage to the modern* Equus *that grazes on grasses. The left vertical axis shows the time scale, and the top horizontal axis shows geographical distribution.*

The lower left shows fossils of the skulls and front feet on which George Gaylord Simpson based these reconstructions. The transition from browsing to grazing paralleled the decreased humidity in North America and the change from

tropical forest to prairie and savanna. During the last ice age horses crossed into Asia, then became extinct in America until the Spanish reintroduced them. This phylogeny is based on MacFadden (1985).

Three types of mimicry in two species: the monarch butterfly Danaus plexippus *and the viceroy butterfly* Limenitis archippus. *Most monarch butterflies become unpalatable (left) from feeding on toxic species of milkweed. Viceroys are also unpalatable, at least in some areas (Ritland and Brower 1991). Their similarity to monarchs reinforces the lesson to potential predators—orange and black butterflies are not good to eat. Such mimicry of two or more generally unpalatable species by each other is called Müllerian mimicry. Viceroys were formerly considered to be palatable but to be avoided by predators because of their resemblance to unpalatable monarchs. This type of mimicry would be Batesian mimicry. Some monarchs are also palatable and may be protected by automimicry of unpalatable members of their own species.*

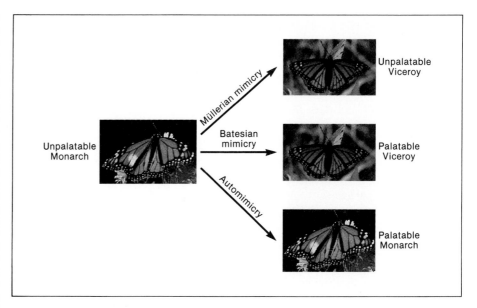

atable organism or object are more likely to leave offspring, and that after many generations the offspring will closely mimic the model.

The explanation from natural selection has experimental support in some cases, such as those in which the model is unpalatable and "advertises" the fact with **warning coloration.** Predators quickly learn to avoid prey with warning coloration and may then subsequently avoid mimics that are similarly colored (see Figure 37.5C,D). If the mimic is palatable, the situation is called **Batesian mimicry,** after its discoverer, Henry Walter Bates (Wallace's companion in South America). The viceroy butterfly *Limenitis archippus* was for many years assumed to be a palatable Batesian mimic of the monarch butterfly, *Danaus plexippus,* which generally becomes unpalatable from feeding on milkweeds. It now appears, however, that viceroys are also often unpalatable (Figure 18.13). These two unpalatable species may therefore reinforce each other's protection from predation. Such mimicry of one or more unpalatable species by another is called **Müllerian mimicry,** named for Fritz Müller.

Camouflage in Moths. Camouflage may be thought of as mimicry of the surroundings. (See Figures 32.18, 35.21B, and 36.6B for examples.) One of the most celebrated examples of camouflage occurs in the peppered moth *Biston betularia* in England. Like many other insects, many peppered moths have genes that produce the dark pigment melanin. If such darkening is more common in areas where soot covers the vegetation on which the insects rest during the day, it is called **industrial melanism.** Prior to the mid-19th century, industrial melanism was rare in peppered moths. Almost all had speckled light-gray wings that were hard to see against lichen-covered tree trunks. In 1849 near Manchester, where industries burned large amounts of coal, the first **melanic** (dark-winged) moth was reported. As pollution increased in England, so did the proportion of melanic moths. By 1886 up to 98% of peppered moths in some industrial areas were dark-winged.

Some biologists suggested that industrial melanism was due to the inheritance of acquired characters or to the direct induction of mutations by pollution. In the 1930s E. B. Ford, one of the early contributors to the synthetic theory, suggested that natural selection was the cause. He hypothesized that air pollution destroyed lichens and exposed the dark tree trunks on which the moths rested during the day. The camouflage of the light-winged moths was therefore destroyed, while that of the dark-winged moths was improved (Figure 18.14). Predatory birds then acted as the agents of natural selection, eating a greater proportion of the easily seen light-winged moths.

H. B. D. Kettlewell of Oxford University tested Ford's hypothesis by tagging hundreds of light and dark moths, releasing them in polluted and nonpolluted areas, and estimating the survivability of each form from the numbers recaptured. The results agreed with Ford's hypothesis. Near the industrial city of Birmingham, Kettlewell recaptured 55% of the dark moths and only 25% of the light moths. Near unpolluted Dorset, Kettlewell recaptured only 4.7% of the dark forms and 13.7% of the light forms.

As Kettlewell has pointed out, there are numerous factors that could account for these results, including the fact that the moths tend to migrate to areas where they are well camouflaged. The small proportion of light-winged moths recaptured in Birmingham and of dark-winged moths recaptured in Dorset could therefore have been due to migration out of each area. In order to see whether predation really was a factor, Kettlewell posed dead moths of both varieties on light or darkened tree trunks. He found that, indeed, birds did consume more light-winged moths on dark trunks, and more melanic moths on lichen-covered trunks. Governments have conducted their own "experiments" by tightening controls on air pollution. As expected, the proportion of melanic moths has declined in the past few decades as the air has become cleaner.

A

Biochemical Studies of Evolution.

In the past, evolutionists could compare only the anatomical differences between organisms as a means of inferring their evolutionary history. Since the 1960s, however, it has been possible to compare directly the structures of genes and proteins from different organisms. The various techniques of these studies are known collectively as **molecular phylogenetics.** The first technique of molecular phylogenetics to contribute significantly to evolutionary theory was **electrophoresis.** In this procedure proteins are placed into a gel, and an applied voltage makes the proteins move at different rates, depending on the net charges, sizes, and shapes of the proteins. These properties depend on the sequence of amino acids in each protein. By comparing the rate of movement of proteins from two different species, one can infer how closely related the species are. Another way of comparing the structures of proteins is by **immunological cross-reactivity.** If two species are related to each other, then antibodies against the proteins produced by one species should also bind (cross-react) with similar proteins in the other species.

A more precise method of comparing proteins is by determining the sequence of amino acids in them—a procedure called **sequencing.** The protein is first broken into small fragments using digestive enzymes. The amino acids in each fragment are then identified, and by comparing overlapping fragments the sequence can be deduced—if the protein is small enough and the resources of the investigators are large enough.

Because of redundancy in the genetic code, some point mutations in the DNA do not cause amino acid substitutions. In addition, proteins are generally more difficult to handle than the nucleic acids (DNA and RNA) that code for them. For these reasons **nucleic acid sequencing**—determining the sequence of bases in DNA or RNA—is becoming increasingly common as a tool of molecular phylogenetics. The choice of which nucleic acid to use depends on the taxonomic

B

Figure 18.14
Camouflage in light- and dark-winged peppered moths Biston betularia. *The dark wings are due to increased deposits of melanin, which result from a single dominant allele. (A) One dark moth and one light moth resting on the lichen-covered bark of a tree. (B) Dark and light moths resting on the bark of a tree in an area in which industrial pollution has destroyed the lichen.*

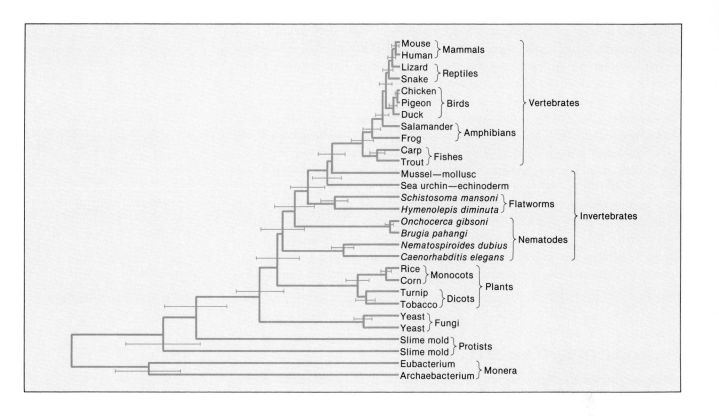

Figure 18.15
The phylogeny of 19 animals and other organisms, as inferred from comparisons of a ribosomal RNA. The number of nucleotide substitutions between two organisms, and therefore the evolutionary distance between them, is proportional to the total length along the horizontal lines connecting the two. Thus the farther left the vertical line connecting two horizontal lines, the earlier the inferred evolutionary divergence of the organisms joined by the two lines. Each light horizontal line represents the range of possible error of measurement. Several alternative patterns are possible within these ranges of error. This figure suggests, among other patterns, that flatworms diverged from other invertebrates before mussels and sea urchins did. Compare Figures 1.4 and 21.9.

scale at which comparisons are made. For comparisons among lower taxa (orders, families, genera, species) one must use a nucleic acid that accumulates mutations rapidly, such as mitochondrial DNA. If one is interested in comparing different kingdoms or phyla, one must use a highly conserved nucleic acid (one that mutates slowly), such as one of the RNAs in ribosomes.

Another technique, **DNA hybridization,** directly compares the DNA from two different organisms. The DNA is first heated to separate its two strands. When single strands of DNA from two related organisms are mixed, some of the strands from one organism "hybridize" with strands from the other organism. The more alike the DNA from the two organisms, the tighter will be the bonding between the bases in the hybrid DNA. The tightness of the bonding, and therefore the degree of relatedness of the species, can be determined by measuring the temperature required to "melt" the two strands apart again.

These techniques, as well as others, suggest the patterns, or **phylogenies,** by which species evolved from shared ancestors. Like phylogenies based on anatomy, molecular phylogenies are subject to errors and differences in interpretation. The main advantage of molecular phylogenetics is that each difference in a molecule is an independent datum, so each molecule contains a large amount of data. Computerized statistical tests are used to determine the phylogeny most consistent with the mass of data. Figure 18.15 shows the phylogeny deduced by comparing the base sequences in an RNA from the ribosomes of 11 vertebrates, 8 invertebrates, and some plants, fungi, protists, and bacteria. It shows that, as expected, the ribosomal RNA of a human is more like that of another mammal (the mouse) than that of a vertebrate in any other class (bird, reptile, amphibian, or fish). It also shows that human ribosomal RNA differs from that of any of the invertebrates more than it differs from any of the vertebrates. This result, together with independent research using different ribosomal RNAs (see Figure 22.9), supports the traditional classification based on comparative anatomy.

Research in molecular phylogenetics also provides an argument that comes as close to proof of evolution as anyone could hope for. This argument is based on the unlikelihood that exactly the same errors will occur twice independently. Compilers of street guides and trivia games routinely use this principle by making up erroneous addresses and "facts" to trap plagiarists. A plagiarist can always argue that the correct facts in his or her work came from the same source that another author used; it is more difficult to explain to a jury how the same fictitious address or fact appeared. Like plagiarized errors, similar **pseudogenes** occur in species that evolutionists have long argued were related. Pseudogenes are sections of DNA that are nonfunctional because they are neither transcribed into RNA nor translated into proteins. Several pseudogenes in humans also occur in apes, but not in other mammals (Max, 1990). One of these pseudogenes was apparently acquired from a virus. It is unlikely that the same useless pieces of DNA arose independently in humans and in several species of apes, but not in other species. The most likely explanation is that humans and apes all evolved from the same species that had those pseudogenes.

THE LOGIC OF NATURAL SELECTION

There are three kinds of arguments for natural selection as a mechanism of evolution: It is intuitively obvious, it is mathematically feasible, and it has been observed in fast-breeding organisms, such as antibiotic-resistant bacteria and pesticide-resistant insects. Each of these arguments shows that if there is genetic variability in a population, and if conditions favor the reproduction of some variants over others, then the allele frequencies for the favorable variants will increase at the expense of the less favored. Almost no evolutionist doubts, therefore, that natural selection is one mechanism—perhaps the only one—for the evolution of adaptations. This synthetic theory of evolution is so widely accepted that it could be called the orthodox view. Like most orthodoxies, it has given rise to several fallacies of logic that should be critically examined.

Evolution and Progress.   One of the most prevalent fallacies of evolution is the notion that it is progressive—that organisms can be ranked on a scale according to how far they have progressed toward the most highly evolved species (ourselves, naturally). According to this way of thinking, natural selection is slowly urging all the "lower" species upward on the scale of perfection atop which we perch. Chimpanzees have almost made it; cockroaches have a long way to go. In Darwin's time many British even fancied themselves on the topmost rung, with Mongoloid, Negroid, and other "less evolved" races on lower rungs. Biologists now laugh at such nonsense, but the fallacy of progressive evolution still lingers in nonscientific references to "lower" and "higher" species. When biologists use the terms "lower" and "higher" they are merely indicating that a taxonomic group arose earlier or later in evolution, not that the group is inferior or superior.

Survival of the Fittest.   The social theorist Herbert Spencer coined the phrase "survival of the fittest," and Alfred Russel Wallace persuaded Darwin that it would be a good summary of their theory of natural selection. "Survival of the fittest" is an unfortunate statement of a scientific theory, however, because it is not experimentally testable. Fitness is defined as the ability to produce viable offspring, so "survival of the fittest" means "survival of the populations that are best able to survive." Any "experiment" designed to test whether the fittest populations survive can obviously have only a foregone conclusion.

The Panglossian Paradigm. Many biologists assume, often without being aware of it, that every feature of an organism is optimally adapted for its function. Stephen Jay Gould and Richard C. Lewontin (1979) refer to this notion as the "Panglossian paradigm," after Dr. Pangloss, a character in a novel by Voltaire, who kept insisting through earthquakes and other catastrophes that this is the best of all possible worlds. The idea is useful when it stimulates research to discover how a presumed adaptation functions. Often, however, instead of creating a testable hypothesis, one succumbs to the temptation to make up a plausible story of how the adaptation evolved. Before breakfast any good biologist can come up with six explanations for why bats fly, and half a dozen equally good explanations for why mice do not. Such "explanations" are often embarrassingly similar to the *Just So Stories* in which Rudyard Kipling explained to children how the whale got its throat, how the camel got its hump, how the rhinocerous got its skin, and so on.

"For the Good of the Species." A common fallacy in thinking about natural selection is that it is capable of producing whatever adaptation benefits a species. Natural selection is not, however, a conscious force that cares about the species. The problem with "for the good of the species" thinking is that it often obscures more interesting ideas of evolution. For example, it is tempting to accept the following explanation uncritically: "Male deer charge each other with their potentially lethal antlers during contests for mating rights, but for the good of the species they have evolved behavioral mechanisms that make them avoid inflicting injury." It is certainly useful to the species for its members not to destroy each other, but a moment's thought reveals that this is not an adequate explanation for the evolution of this behavior. There could have been natural selection for this behavior only if it allowed the *individuals* displaying the behavior to leave more of their alleles in the next generation. Yet failing to kill an opponent, thereby perhaps losing the privilege of reproducing, would *decrease* the likelihood of a buck transmitting its genes into the next generation. More likely explanations (suggested in Chapter 20) are much more demanding, but more interesting, than "for the good of the species."

OTHER CAUSES OF EVOLUTION

Genetic Drift. According to the Hardy–Weinberg Law, equilibrium—the absence of evolutionary change—can occur in an infinitely large population. In a small population, however, the allele frequency may not be representative of the entire species simply because of sampling error. (Another example of sampling error occurs if you grab marbles out of a bag containing 500 black ones and 500 white ones. You are much more likely to get all marbles of one color if you grab only three than if you grab 300.) Such sampling errors in small populations can give rise to genetic drift, which is a random change in allele frequency. Such evolutionary changes in small populations are not likely to be as adaptive as those resulting from natural selection, but they may be just as significant in evolution.

Evolutionists have given considerably more emphasis to genetic drift as a mechanism of evolution since various techniques have revealed that there is more genetic variation in populations than was once thought possible. Most evolutionists once believed that natural selection rigorously eliminated all but the best-adapted organisms. They were surprised, therefore, when electrophoresis revealed a large **polymorphism** (variety) in the structures of most proteins. In *Drosophila*, 53% of proteins examined electrophoretically were found to be polymorphic within a single population, and the average fly was found to be heterozygous in 15% of its genes. The degree of polymorphism tends to be less for vertebrates, but even in humans more than 25 out of approximately 100 enzymes studied were found to be polymorphic (Lewontin 1982, chapter 3). The ABO and other blood groupings (see pp. 240–242) are examples of this polymorphism.

The Neutral Theory of Evolution.   Some evolutionists doubt that so much variability could occur if natural selection were the major factor in evolution. Instead, they offer the neutral theory of evolution, which proposes that most mutations have no effect on fitness, either because the mutations produce no change in the phenotype, or they do not affect the ability of the organism to reproduce. Selection is "blind" to such neutral mutations, so they arise unchecked and become established in populations through genetic drift. Neutralists, as advocates of the theory are called, contend that if evolution is defined as changes in allele frequencies, then evolution is due more to neutral mutations than to natural selection. Many adherents to the synthetic theory counter that since neutral mutations are by definition nonadaptive, they cannot account for the important changes that occur during evolution. This controversy is not likely to be resolved in the near future. In the meantime, it is safe to say that natural selection is the major cause of **adaptive** evolution, but that it may not be the cause of all evolutionary changes.

SPECIATION

According to the biological species concept, a new species begins when a group of organisms can no longer interbreed with the population from which it originated. How does this come about? One possible mechanism is that part of the population accumulates genetic changes that make it unable to reproduce with others in the population, even though they continue to live in the same areas. This pattern, called **sympatric speciation** (Greek *sym* together + *patris* native land), has been documented in some species. One mechanism that can lead to sympatric speciation, especially in plants, is polyploidy (see p. 92), a sudden change in chromosome number. Sympatric speciation appears to be less common than **allopatric speciation** (Greek *allo* other) in which a subpopulation first becomes physically separated from its parent population, then evolves differences that prevent the subpopulation from interbreeding with the parent population.

Physical separation leading to allopatric speciation can occur in two ways. First, in an extremely large population, individuals on the fringes may not be able to mix well with the majority of the population, and the gene frequencies of this subpopulation may be different from the population as a whole. The subpopulation can then give rise to a genetically different population. This result is called the **founder effect,** because the descendants of the subpopulation will resemble the founders just as the inhabitants of a small town often resemble the founders of the town. The second mechanism of physical separation is called **vicariance.** Vicariance can result from a geological event, such as the rise of a mountain range, from the death of individuals living in the middle of the population, or from the straying of part of the population from the rest. The different species of Darwin's finches are thought to have arisen by vicariance, as a subpopulation wandered or was blown from the mainland to a Galápagos island, and from there to other islands.

SEX AND EVOLUTION

Sexual Selection. In an equilibrium (nonevolving) population, mating is random. Since animals have sense organs and the ability to move, however, they are capable of some choice in mates. This results in sexual selection. Sexual selection is a mechanism of evolution that differs from others in that it depends on nothing more than one organism taking a fancy to another of the opposite sex. Many evolutionists believe that sexual selection can be just as arbitrary as this and can lead to the evolution of traits that would otherwise be counteradaptive. The elaborate tail of the peacock *Pavo cristatus*, for example, serves no function other than to make him attractive to females. In fact, it endangers the males by attracting predators and by interfering with other functions.

Some evolutionists point out, however, that sexually selected traits often depend on the overall fitness of the animal. A peacock that is infested with parasites is not likely to have a handsome tail. The fact that it can survive in spite of such a tail therefore advertises that it must be fit enough to avoid parasites. The peacock's tail would then be an example of "truth in advertising." In cases where sexually selected traits honestly represent overall fitness, sexual selection can be considered merely a special case of natural selection.

Is Sex Worth the Trouble? According to the synthetic theory, natural selection operates as if it were the purpose of each organism to get as many of its alleles as possible into the next generation. Yet most animals and many other organisms seemingly defy natural selection by wasting half their alleles in sexual reproduction. Not only does sexual reproduction make use of only half of each organism's alleles to make one offspring, but the evolution of sex has led to further disadvantages, including (1) loss of energy required to produce gametes, (2) time and energy lost in acquiring mates, (3) vulnerability during reproduction, (4) elaborate mechanisms needed to ensure that gametes are not wasted on a member of the wrong species, (5) cumbersome mechanisms of meiosis and fertilization, and (6) the existence of males that may contribute nothing but sperm. The synthetic theory leads us to believe that there is some advantage to sex that compensates for all the disadvantages, yet biologists are as uncertain as most 12-year-olds about whether sex is a good idea.

Of the several theories proposed to explain why sex evolved, probably the most widely accepted is that sexual reproduction increases genetic variability. First, the two pairs of chromosomes common in sexual species allow for **recombination:** the shuffling of alleles between homologous pairs of chromosomes. Second, sexual reproduction can increase genetic variability through **outcrossing:** the mixing of alleles from two different individuals. The advantage of increased genetic variability could be twofold. First, organisms with many variable offspring are likely to have at least *some* that are well adapted. Second, sexual species are less likely to become extinct, because the increased variability allows them to adapt more quickly to environmental change.

There are several problems with these arguments. The first advantage applies only to species with high fertility. In mammals and others with relatively few offspring, high genetic variability due to sexual reproduction might cause all the offspring to be defective rather than result in a few being better adapted. Second, there is no evidence that asexual organisms are less able to adapt than sexual ones. Bernstein and his colleagues (1985) have taken a totally different approach to explaining the evolution of sex. They suggest that sex provides a genetically defective individual with a good copy to use in repairing its DNA or in masking the mutation.

SUMMARY

That species evolved from preexisting species was first appreciated in the 18th century, but convincing evidence and a plausible explanation for it were not available until Charles Darwin and Alfred Russel Wallace proposed their theory of natural selection in 1858. Subsequently, natural selection was integrated with genetics in the synthetic theory of evolution. Evolution is now defined as any change in gene frequency in a population. Small changes in gene frequency can accumulate to the point that organisms can no longer interbreed in the population from which they originated. They then become members of a new species, according to the biological species concept. Speciation is generally allopatric, with a subpopulation becoming physically separated by geological or other means (vicariance), or by being isolated on the fringe of the parent population (founder effect).

Some common notions of evolution are not valid. Evolution and natural selection do not result in progress, and they do not necessarily happen just because they would be good for the species. Not all features that evolve increase fitness, and the idea of survival of the fittest is untestable.

Evolution is supported by many experimental studies, such as those demonstrating predictable changes resulting from competition in Darwin's finches, studies of mimicry of the monarch butterfly, and industrial melanism in peppered moths. Molecular biological studies also demonstrate that patterns of differences in proteins and amino acids are consistent with phylogenies deduced by anatomical comparisons. Numerous experiments also support natural selection, although there is also evidence for other mechanisms of evolution, such as genetic drift. Sexual selection may also be a form of natural selection. Although sex is thought to contribute to evolution by increasing genetic variety, how or why it evolved is a mystery.

KEY TERMS

synthetic (neo-Darwinian) theory
natural selection
adaptation
population
homologous
analogous
uniformitarianism

Hardy–Weinberg Law
biological species concept
Darwin's finches
mimicry
industrial melanism
molecular phylogenetics
genetic drift

neutral theory of evolution
sympatric
allopatric
founder effect
vicariance
sexual selection
recombination

SELF-TEST

1. State a modern definition of evolution.

2. Define the following terms: population, mutation, species, allele, genotype, phenotype, homologous, natural selection.

3. Describe three kinds of evidence that evolution has occurred.

4. Briefly describe the theory of evolution proposed by Charles Darwin and Alfred Russel Wallace. How does this theory differ from Lamarck's?

5. Briefly explain the synthetic theory of evolution in your own words. How does it differ from Darwin's original theory of evolution?

6. Describe some of the main contributions leading to acceptance of the synthetic theory of evolution.

7. What is the relevance of the Hardy–Weinberg Law to evolution?

8. Many creationists have charged that evolution is mathematically impossible because it depends entirely on chance. Does evolution depend only on chance? Which aspects of evolution do depend on chance, and which ones do not?

9. Creationists also charge that evolution cannot be observed, and it is therefore not a fit subject for science. Is this a valid argument? Give a specific example of evolution research that satisfies your definition of science.

10. Prepare an outline for an essay on the relationship of geology to evolution.

11. Prepare an outline for an essay on the relationship of genetics (both population genetics and molecular genetics) to evolution.

12. Define genetic drift. How can it cause evolution?

13. Describe the most likely way by which a new species arises.

READINGS

RECOMMENDED READINGS

Ayala, F. J. 1978. The mechanisms of evolution. *Sci. Am.* 239(3):56–69 (Sept).

Bishop, J. A. and L. M. Cook. 1975. Moths, melanism and clean air. *Sci. Am.* 232(1):90–99 (Jan).

Brower, L. P. 1969. Ecological chemistry. *Sci. Am.* 220(2):22–29 (Feb). (*On mimicry in butterflies.*)

Darwin, C. (N. Barlow, Ed.) 1958. *The Autobiography of Charles Darwin*. New York: W. W. Norton.

Darwin, C. 1859. *The Origin of Species*. Cambridge, MA: Harvard University Press. (*Reprint of the first edition. Also in numerous other editions. Still worth the effort.*)

Darwin, C. 1962. *The Voyage of the Beagle*. Garden City, NY: Doubleday. (*One of the all time great travel books. Widely available in paperback.*)

Gilbert, L. E. 1982. The coevolution of a butterfly and a vine. *Sci. Am.* 247(2):110–121 (Aug).

Gould, S. J. 1977. *Ever Since Darwin*; 1980. *The Panda's Thumb*; 1983. *Hen's Teeth and Horse's Toes*; 1985. *The Flamingo's Smile*. 1991. *Bully for Brontosaurus*. New York: W. W. Norton. (*Essays, mostly related to evolution, collected from Gould's "This View of Life" column in Natural History Magazine.*)

Harris, C. L. 1981. *Evolution: Genesis and Revelations*. Albany, NY: SUNY Press. (*The history and philosophy of evolutionism, with readings from the original contributions to ideas of the origin of species.*)

Herbert, S. 1986. Darwin as a geologist. *Sci. Am.* 254(5):116–123 (May).

Kettlewell, H. B. D. 1959. Darwin's missing evidence. *Sci. Am.* 200(3):48–53 (Mar).

Lewin, R. 1982. *The Thread of Life*. Washington, DC: Smithsonian Books. (*A beautifully illustrated introduction to evolution.*)

Mossman, D. J. and W. A. S. Sarjeant. 1983. The footprints of extinct animals. *Sci. Am.* 248(1):74–85 (Jan).

Owen, D. 1980. *Camouflage and Mimicry*. Chicago: University of Chicago Press.

Pietsch, T. W. and D. B. Grobecker. 1990. Frogfishes. *Sci. Am.* 262(6):96–103 (June). (*The anglerfish and related mimics.*)

Stebbins, G. L. and F. J. Ayala. 1985. The evolution of Darwinism. *Sci. Am.* 253(1):72–82 (July).

Wilson, A. C. 1985. The molecular basis of evolution. *Sci. Am.* 253(4):164–173 (Oct).

ADDITIONAL REFERENCES

Bernstein, H. et al. 1985. Genetic damage, mutation, and the evolution of sex. *Science* 229:1277–1281.

Berra, T. M. 1990. *Evolution and the Myth of Creationism*. Stanford, CA: Stanford University Press.

Berry, R. J. (Ed.). 1985. *Evolution in the Galápagos*. New York: Academic Press. Reprinted from *Biol. J. Linnaean Soc.* 21 (1 and 2).

Boag, P. T. and P. R. Grant. 1981. Intense natural selection in a population of Darwin's finches (Geospizinae) in the Galápagos. *Science* 214:82–85.

Clarke, B. 1975. The causes of biological diversity. *Sci. Am.* 233(2):50–60 (Aug).

Cook, L. M., G. S. Mani, and M. E. Varley. 1986. Postindustrial melanism in the peppered moth. *Science* 231:611–613.

Eldredge, N. 1982. *The Monkey Business: A Scientist Looks at Creationism*. New York: Washington Square Press.

Futuyma, D. 1982. *Science on Trial: The Case for Evolution*. New York: Pantheon.

Godfrey, L. R. 1983. *Scientists Confront Creationism*. New York: W. W. Norton.

Gould, S. J. and R. C. Lewontin. 1979. The spandrels of San Marco and the Panglossian paradigm: a critique of the adaptionist programme. *Proc. R. Soc. (London)* B205:581–598.

Grant, P. R. 1981. Speciation and the adaptive radiation of Darwin's finches. *Am. Sci.* 69:653–663.

Grant, P. R. 1986. *Ecology and Evolution of Darwin's Finches*. Princeton, NJ: Princeton University Press.

Kitcher, P. 1982. *Abusing Science: The Case Against Creationism*. Cambridge, MA: MIT Press.

Lewontin, R. 1982. *Human Diversity*. New York: Scientific American Library.

MacFadden, B. J. 1985. Patterns of phylogeny and rates of evolution in fossil horses: hipparions from the Miocene and Pliocene of North America. *Paleobiology* 11:245–257.

Max, E. E. 1990. (letter). *Creation/Evolution* 10(1):45–49.

Mayr, E. and W. B. Provine (Eds.). 1980. *The Evolutionary Synthesis*. Cambridge, MA: Harvard University Press. (*Accounts of the origin of the synthetic theory of evolution by many of its founders.*)

McCosker, J. E. 1977. Fright posture of the plesiopid fish *Calloplesiops altivelis*: an example of Batesian mimicry. *Science* 197:400–401.

Moore, J. A. 1984. Science as a way of knowing—evolutionary biology. *Am. Zool.* 24:467–534. (*A concise summary of evolutionary biology, with sound advice on teaching it.*)

Newell, N. D. 1982. *Creation and Evolution: Myth or Reality*. New York: Columbia University Press.

Qu, L.-H., M. Nicoloso, and J.-P. Bachellerie. 1988. Phylogenetic calibration of the 5′ terminal domain of large rRNA achieved by determining twenty eucaryotic sequences. *J. Mol. Evol.* 28:113–124.

Ritland, D. B. and L. P. Brower. 1991. The viceroy butterfly is not a batesian mimic. *Nature* 350:497–498.

Ruse, M. 1982. *Darwinism Defended: A Guide to the Evolution Controversies*. Reading, MA: Addison-Wesley.

Schluter, D., T. D. Price, and P. R. Grant. 1985. Ecological character displacement in Darwin's finches. *Science* 227:1056–1059.

Sulloway, F. J. 1982a. Darwin and his finches: the evolution of a legend. *J. Hist. Biol.* 15:1–53.

Sulloway, F. J. 1982b. Darwin's conversion: the *Beagle* voyage and its aftermath. *J. Hist. Biol.* 15:325–396.

19

History of Evolution

Chimpanzee (Pan troglodytes).

CHAPTER OUTLINE

LEARNING OBJECTIVES

1. What was Earth like before there was life?

2. When and how did life get started?

3. How did animals originate?

4. What were the earliest animals like?

5. Do organisms change slowly, rapidly, or at differing rates?

6. Have the kinds of organisms steadily increased, or have there been mass extinctions?

7. How have geological changes, such as continental drift and ice ages, affected life?

8. How do humans differ from apes, and how did those differences evolve?

9. What were the ancestors of humans like?

BEFORE LIFE ON EARTH

In the previous chapter we considered general principles of evolution that could apply to life on any planet. Now it is time to consider how these principles have shaped the course of evolution on the one planet whose life we vaguely know. We must begin the history of life on Earth at the beginning, for although the origin of life was different from its evolution, the particular way life originated has affected all subsequent evolution.

The Prebiotic Atmosphere. At present there is considerable uncertainty about the conditions that gave rise to life on Earth (Patrusky 1984). Most geophysicists believe that the Earth and the other inner planets of our solar system solidified from the dust and gas that remained after a nebula gave birth to our sun. Heat from collisions and radiation then softened our infant planet, allowing molten iron to sink and form the core. At the same time lighter material rose to the surface and formed a crust that remained relatively cool except at volcanoes. A variety of measures (see box) indicate that these events occurred some 4.5 billion years ago (Table 19.1). If you are familiar with the federal budget, a few billions may not impress you. However, 4.5 billion years is so long that you could have lingered over each word of this book for a hundred centuries and still have finished reading it long ago. The newly formed crust, unprotected by an atmosphere, was continually bombarded with icy comets and stony asteroids and probably looked something like the moon. Unlike the moon, however, the Earth had enough gravitational attraction to hold an atmosphere, which was created from water vapor, nitrogen, carbon dioxide, and other gases spewing from volcanoes. With an atmosphere came weather, probably in the form of severe thunderstorms. As the crust cooled oceans formed, and minerals leached from the rocks made the oceans salty.

For biologists one of the most important questions about the prebiotic Earth is the composition of its atmosphere, for it is generally thought that atmospheric gases reacting in shallow pools formed the first organic molecules from which life arose. Unfortunately, various geochemical models give different results for the early atmosphere. The present consensus, however, is that the prebiotic atmosphere consisted mainly of molecular nitrogen (N_2), water, and carbon dioxide. In short, it was much like our present atmosphere except that it contained much more CO_2 than now, and it lacked oxygen. The absence of oxygen is important, because the organic molecules needed to synthesize the first life would have been oxidized immediately if O_2 had been present. Oxygen is one reason why organic molecules do not now occur on Earth unless they are synthesized within the protected environment within cells.

Prebiotic Synthesis of Organic Molecules. How did the first organic molecules come about in the absence of organisms? Charles Darwin speculated in 1871 that organic molecules might have been synthesized in "some warm little pond," and most of the research on the prebiotic synthesis of organic molecules has elaborated on this theme. In the 1920s J. B. S. Haldane and A. I. Oparin independently suggested mechanisms by which simple molecules containing carbon and nitrogen might have reacted in water to form a "primordial soup" of organic molecules. In the early 1950s Stanley L. Miller and Harold C. Urey tested this idea by discharging a spark in a synthetic atmosphere containing the gases then thought to have been present in the prebiotic atmosphere (Figure 19.2). After a week, several amino acids common to life (as well as some that are not) had accumulated

Table 19.1 Milestones in the history of Earth.
Dates, in millions of years before present, are based on Harland et al. (1990).

Date	Eon	Major Events
	Phanerozoic	Ice ages. Evolution of humans. Birds, mammals, flowering plants.
500		Dinosaurs. First vertebrates. Sudden appearance of most animal phyla.
		Oldest fossils of multicellular animals.
1000		
1500	Proterozoic	First eukaryotes: single-celled algae.
2000		Oxygen-rich atmosphere. Aerobic bacteria.
2500		
3000		Oldest fossils—bacteria.
3500	Archean	Production of O_2 by photosynthetic bacteria. Absence of O_2. Origin of life. Earliest datable terrestrial rocks.
4000	Priscoan	Oceans form. "Primordial soup." Atmosphere of N_2, H_2O, CO_2. Intense volcanic activity.
4500		No atmosphere. Formation of Earth's crust. Origin of solar system.

in the apparatus. Many variations on this experiment, using different gases and energy sources, have demonstrated that not only amino acids but also nucleic acid bases and other biologically important molecules could easily have been formed before there was life. In fact, it is hard to see how the formation of a primordial soup could have been avoided.

The Origin of Genetic Control. Although the experiments of Miller and others indicate that organic molecules could have been synthesized prebiotically, there was still a major hurdle before they could form life. Somehow a mechanism had to arise by which the sequence of nucleic acid bases translated into the sequence of amino acids in proteins. At the same time some method had to arise to replicate the nucleic acids for reproduction. Since the replication of DNA now requires

Measuring the Ages of Rocks

At various times and places sand, mud, shells, and other sediments have slowly accumulated in layer upon layer for millions of years. Preserved in many of these strata, like photos in the pages of an album, are the fossils of organisms that lived when the deposits were formed. The oldest fossils are naturally in the deepest strata (Figure 19.1). By correlating the fossils in strata from various parts of the Earth, geologists during the past century were able to divide the history of the Earth into eons, eras, periods, epochs, and ages, much as taxonomists classified organisms into kingdoms, phyla, classes, and so on.

Of course there are no fossils to bear witness to the world before there was life. Moreover, the fossils themselves cannot tell us how old they are. Until a few decades ago geologists could only guess how old fossils and sediments were by estimating how long it took them to accumulate. **Radioactive dating** now allows the accurate measurement of strata. There are several methods of radioactive dating, all of which rely on the fact that radioactive isotopes decay into stable products at a rate that is not affected by temperature, pressure, or other variables. By comparing the amount of product with the amount of radioactive isotope in a rock, one can determine how long the isotope has been decaying, and therefore how old the rock is.

For example, half the atoms of potassium 40 (the isotope of potassium with a total of 40 protons and neutrons in its nucleus) decay into either argon 40 or calcium 40 every 1.28 billion years. That is, potassium 40 has a half-life of 1.28 billion years. Unlike calcium, argon is a gas at normal temperatures and does not react chemically; therefore all the argon found inside a rock must have gotten there by decay of potassium 40 within the rock. The greater the ratio of argon 40 to potassium 40 inside a rock, the older the rock must be. In order to be useful for dating, both the radioactive isotope and its decay product must be present in measurable amounts. Carbon 14, with a half-life of only 5730 years, is useful mainly for dating organic materials less than 60,000 years old.

A

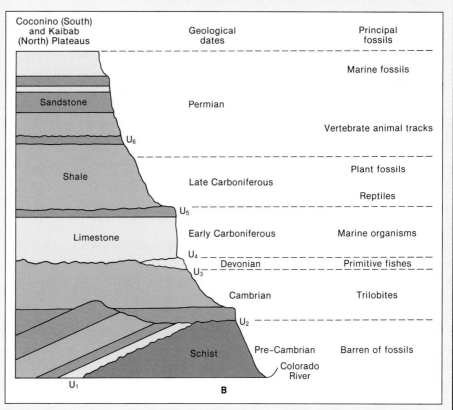

Figure 19.1

(A) The north rim of the Grand Canyon at its greatest depth, more than one mile. Erosion by the Colorado River has revealed distinctive strata formed by sediments of different materials deposited at different times. (B) The geological periods corresponding to the strata. The Grand Canyon exposes one of the most complete series of strata known, with most of the Paleozoic era represented. There are a few unconformities (U_1 through U_6) due to erosion or to the absence of sedimentation when the land rose above sea level. Unconformity U_3 represents a gap of approximately 100 million years.

Coconino (South) and Kaibab (North) Plateaus	Geological dates	Principal fossils
		Marine fossils
Sandstone	Permian	
		Vertebrate animal tracks
	U_6	
Shale	Late Carboniferous	Plant fossils
		Reptiles
	U_5	
Limestone	Early Carboniferous	Marine organisms
	U_4	
	U_3 Devonian	Primitive fishes
	Cambrian	Trilobites
	U_2	
Schist	Pre-Cambrian	Barren of fossils
	Colorado River	
U_1	**B**	

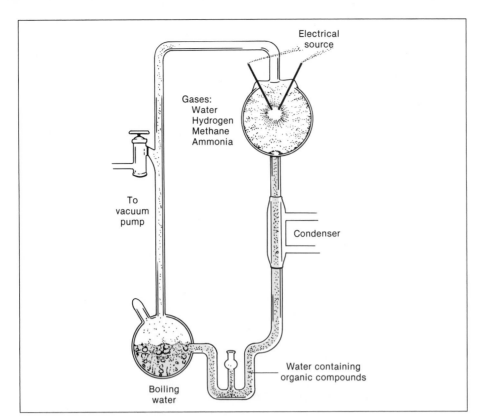

Figure 19.2
The apparatus used by Stanley L. Miller in 1953, which demonstrated how organic molecules could have been synthesized on the prebiotic Earth. The energy from a spark discharge (simulating lightning) drives the synthesis of amino acids from methane, ammonia, water, and hydrogen (the gases thought in 1953 to have been present in the prebiotic atmosphere).

proteinaceous enzymes, and protein synthesis requires DNA, we have a classic chicken-and-egg problem. One possible solution to this problem would be for a single molecule, perhaps RNA, to have served the functions of both genetic material and proteins. This idea has seemed more reasonable lately, since it has been discovered that some RNA can catalyze reactions on itself. After life had gotten a start by using RNA as both gene and enzyme, the machinery could have evolved by which the RNA directed the synthesis of both DNA for genes and proteins for enzymes. Eventually, RNA could have been relegated to its present role of translating the information from DNA to make protein. The idea is still controversial, however.

EARLY EVOLUTION: THE PROTEROZOIC EON

The First Organisms.   The problems of genetic control and replication appear to have been solved in a relatively short time. Fossils of microorganisms appear in rocks that are only a few hundred million years older than the formation of oceans. Microfossils more than three billion years old have been found in what is now Africa, Australia, and Canada. (These geographical references would have meant nothing three billion years ago. The continents have drifted so much it is hard to say where in the world Africa, Australia, and Canada were then.) The size of the fossils (a few micrometers in diameter), their lack of internal structure, and the appearance of reproduction by binary fission suggest that they were prokaryotes—organisms like existing bacteria that lacked nuclei and other membrane-bound organelles (Figure 19.3).

The chemistry of the rocks indicates that the atmosphere still lacked oxygen, so these early bacteria are presumed to have been anaerobic. Around two billion

Figure 19.6
Some animals of the early Cambrian period. The habitat was a tropical reef hundreds of feet below sea level, which now forms the Burgess Shale 8000 feet up in the Rocky Mountains of British Columbia. Most Burgess shale animals, including those numbered 2, 3, and 5 through 7, represent extinct phyla that were previously unknown. The animals were trapped by underwater avalanches of sediment. In the foreground such an avalanche has exposed a specimen of Ottoia *(phylum Priapulida)* ① *in its burrow. Members of two species of* Anomalocaris ② *loom overhead. These fearsome predators were the largest animals in this habitat. The mouth was once thought to be an animal itself.* Dinomischus ③ *was apparently sessile, like the sponges on which* Aysheaia ④ *(phylum Onychophora) fed. The aptly named* Hallucigenia ⑤ *may have walked on seven pairs of stiltlike legs, somehow feeding by means of the eight tubes on its back. Another strange feeding device occurred on the five-eyed* Opabinia ⑥. Wiwaxia ⑦ *had toothed jaws with which it fed as it crept along. Others shown here were arthropods, but not crustaceans.* Sarotocercus ⑧ *may have swum upside down.* Leanchoilia, Yohoia, Sidneyia, Habelia, *and the elegant* Marrella *(⑨ through ⑬) were adapted for walking on the bottom.*

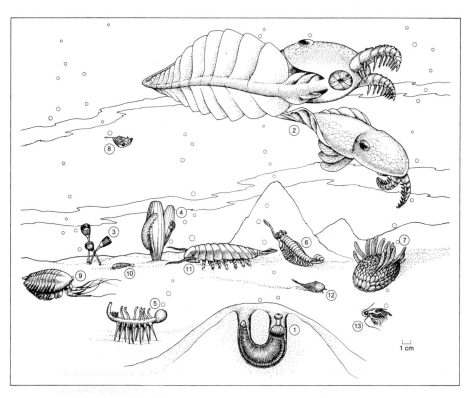

alive today not so much because they are fitter, but because their ancestors were merely lucky (Gould 1989). Our own phylum Chordata is one of the lucky survivors from the Cambrian (see Figure 34.16B). Less than 60 million years after the dawn of the Cambrian, the first vertebrates—jawless fishes—appeared. Within another hundred million years, the Silurian period, some plants and animals had declared their independence from the oceans in which life arose (Table 19.2).

The Pace of Evolution. The sudden increase in animal diversity, called the Cambrian explosion, has long been a problem for biologists and geologists, because they have come to accept uniformitarianism and a gradual pace of evolution—**gradualism**—without question. Uniformitarianism and gradualism are so ingrained that many scientists are reluctant to accept that the Cambrian explosion

Figure 19.7
Cornuproetus sculptus, one of the 4000 known species of trilobites that flourished during the Paleozoic era. This individual died during the Devonian period about 400 million years ago, in seas that covered what is now Czechoslovakia. Length approximately 1 cm. See pp. 638–640 for further description of trilobites.

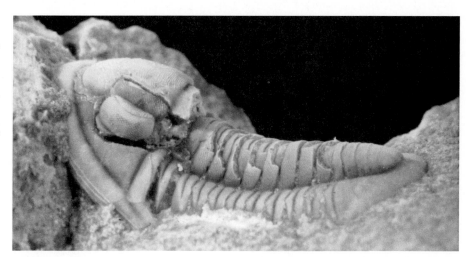

resulted from a sudden increase in diversity. Many prefer to believe that the diversity did arise gradually but appears sudden only because the fossils of the Proterozoic are so scarce. Such a scarcity could have occurred because pre-Cambrian animals lacked the shells and skeletons that fossilized so nicely in later strata. It is also possible that uplifting of the Earth's crust or changes in sea level prevented steady sedimentation in shallow water, where most fossilization occurs. Another explanation may be that the pre-Cambrian fossil record was wiped out, as happened to approximately 100 million years of Ordovician and Silurian strata in the north wall of the Grand Canyon (Figure 19.1). It does not appear that any of these are explanations for the Cambrian explosion, however.

There are also numerous instances in the fossil record in which fossils appear not to have changed during all the time that a stratum was being deposited, but the fossils in the stratum immediately above or below it are quite different. Because of the acceptance of uniformitarianism and gradualism, such **gaps** in the fossil record have usually been attributed to the kinds of geological disturbances described in the previous paragraph. From time to time some biologists and geologists pointed out that such gaps might actually result from a sudden quickening in the pace of evolution. This idea did not really sink in, however, until 1972, when Niles Eldredge of the American Museum of Natural History and Stephen Jay Gould of Harvard University gave it the name **punctuated equilibrium.** The idea of punctuated equilibrium is that evolution is generally characterized by long periods of equilibrium punctuated by brief periods of change. (To a paleontologist, of course, "brief" means within tens of thousands of years.) There are, in fact, some fossil sequences that show such sudden punctuations that cannot be explained by geological disturbances. There are also many undisturbed fossil sequences that show gradual evolution (Levinton 1988). In fact, even in a single series of fossils, some features changed rapidly while others changed gradually.

THE MESOZOIC ERA

Dominance by marine invertebrates in the Paleozoic era gave way to dominance by vertebrates, especially reptiles, during the Mesozoic era, from approximately 245 to 65 million years ago. The idea that reptiles were dominant is based more on size than on numbers. Invertebrates have always outnumbered vertebrates, but it is hard to consider an animal such as *Apatosaurus,* which weighed 33 tons, anything but dominant (see Figure 37.9).

Mass Extinctions. Another reason for the relative dominance of vertebrates over invertebrates in the Mesozoic era is that a **mass extinction** wiped out 19 of every 20 species of marine invertebrates during the Permian period just before the Mesozoic era. The Mesozoic era also ended with a mass extinction, between the Cretaceous and the Tertiary periods. During the Cretaceous/Tertiary (K/T) mass extinction approximately 70% of all species vanished.

The K/T mass extinction has aroused considerable interest because of a theory that it was caused by a collision of one or more meteorites with the Earth. This theory originated with the discovery of a thin, worldwide band of iridium in deposits dating 65 million years ago, at about the time of the K/T mass extinction. Iridium is rare on Earth, but common in asteroids, which are the usual sources of meteorites. The amount of iridium is consistent with impact by an asteroid some 10 km in diameter and 10^{15} kg in mass, or by several smaller asteroids at about the same time. The energy of the impact would have dwarfed the capabilities of all existing nuclear weapons. However, many paleontologists believe that volcanic activity, rather than asteroid impact, produced the iridium layer and caused the mass extinction. Either asteroids or volcanoes could have sent up a dust cloud that could have darkened the skies for several months, started a global fire storm, poisoned the air and water with acid and metals, lowered temperatures, and shut down photosynthesis.

Continental Drift. Another unpredictable influence on the diversity and distribution of animals was continental drift. Contrary to the almost universal belief prior to the 1960s, continents are not firmly anchored to Earth's core but are plates floating on a layer of molten rock. India, for example, is now plowing into Asia and forcing up the Himalayas, and South America and Africa are getting farther apart. According to theories of **plate tectonics,** molten lava pushing up from **mid-ocean ridges** causes the sea floor to spread at a speed of several centimeters per year. When the spreading sea floor encounters the margin of a continental plate it either pushes it away or sinks beneath it, pushing up mountains in the process (Figure 19.8).

Continental drift and mountain building have affected the evolution of animals in two major ways: by changing dispersal patterns and by altering climate. Continental drift affected climate in several ways:

1. As continents drifted toward the poles or toward the equator, their climates became colder or warmer.
2. The uplift of mountains changed wind patterns, and therefore the distribution of rainfall. Changes in ocean currents changed sea and air temperatures, as well as atmospheric moisture.
3. As islands aggregated into large land masses, their temperatures became more diverse and variable, and their humidities declined. The greater diversity of climates on large continents gave rise to a greater diversity of organisms. On the other hand, the ability of animals to disperse over large continents may have increased competition with previously isolated species, which may have become extinct as a result.

Figure 19.8
How plate tectonics builds mountains and moves continents. On the west coast of the Americas the Pacific plate beneath the floor of the ocean pushes under the lighter rocks of the continents. The resulting heat forms volcanoes and mountain ranges inland. On the east coast the American plate is being pushed away from the African and Eurasian plates by sea-floor spreading at the mid-Atlantic ridge.

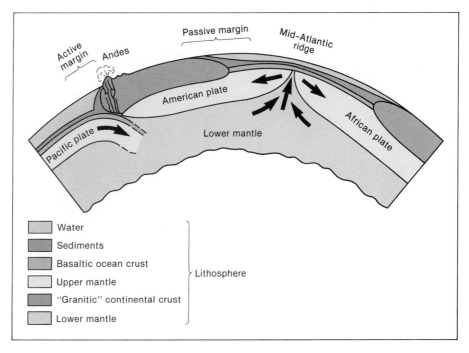

Continental drift has probably been occurring since the Earth's crust formed, but the patterns of drift are clearest in the Mesozoic era. Early in the Mesozoic era, until approximately 200 million years ago, all the present continents were aggregated into one land mass, called **Pangea** (Figure 19.9). Continents that are now widely separated probably shared many of the same species when they were in contact as Pangea. For example, fossils of *Mesosaurus*, a small, fish-eating reptile, are found in both Brazil and South Africa. By 180 million years ago, in the Jurassic period, Pangea split into two large masses, **Gondwana** and **Laurasia.** Gondwana consisted of what is now South America, Africa, Antarctica, Australia, and India. Laurasia included present-day North America, Europe, Asia, and Greenland. By the end of the Mesozoic era, South America had split from Gondwana, and Africa and India were joining Asia.

Continental drift helps to explain the present distribution of animals into six **faunal regions** (= faunal realms) (Figure 19.10). Consider the marsupials, for example. Marsupials, such as kangaroos and opossums, are native to Australia and South America, but not to Asia. This distribution suggests that they evolved some time after the breakup of Pangea, but before Gondwana split into Australia, South America, Africa, India, and Antarctica. If so, marsupials would also have lived on what is now Antarctica, at a time when the climate was much warmer there than it is now. This prediction has been confirmed by the discovery of a fossilized marsupial jaw in Antarctica (Woodburne and Zinsmeister 1982). Until it was colonized by Europeans, Australia remained isolated from Asia at Wallace's Line, and with few exceptions that island continent retained a uniquely marsupial assemblage of mammals. Other faunal regions are less distinctive because land bridges have periodically allowed certain species to gradually expand their ranges into new continents. For example, during the Cenozoic era Central America became a land bridge that allowed marsupials to invade North America. A land bridge to Europe

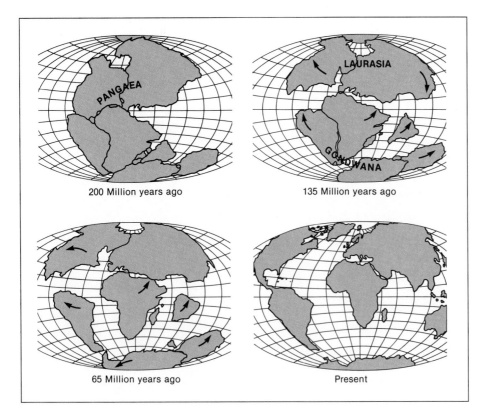

200 Million years ago

135 Million years ago

65 Million years ago

Present

Figure 19.9
Continental drift. Approximately 200 million years ago the continents were aggregated into one land mass, Pangea. By 135 million years ago Pangea was divided into Laurasia and Gondwana. Sixty-five million years ago South America and India were completely separate island continents. Subsequently, North America separated from Eurasia, a land bridge emerged between North and South America, and Australia separated from Antarctica.

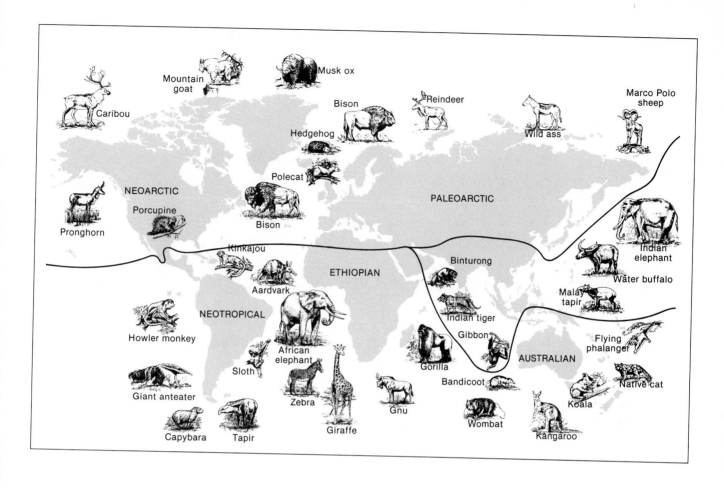

subsequently allowed placental mammals to enter North America, wiping out these marsupials except for a few, such as the Virginia opossum (*Didelphis virginiana*). The placental mammals then continued into South America, where they continued to outcompete most of the marsupials.

THE CENOZOIC ERA

The big reptiles declined around the end of the Mesozoic era, opening opportunities for the birds and mammals. Prior to the start of the present Cenozoic era, some 65 million years ago, the few mammals had been scarcely larger than rats. Most of them remained small throughout the early Cenozoic era, during the Tertiary period. Later, during the Quaternary period, however, many mammals evolved truly impressive proportions (Figure 19.11). As will be described later, a medium-sized, but no less interesting species, *Homo sapiens*, also evolved during the Quaternary period.

Ice Ages. One of the factors that probably contributed to the success of birds and mammals during the Cenozoic era was **homeothermy**—the ability to maintain a constant body temperature. Homeothermy seems to have evolved just in time, for the Cenozoic era has witnessed one of only seven **ice eras** that have occurred on Earth. An ice era is a period lasting about 50 million years during which ice ages occur. The average global temperature often drops only several degrees during an ice age, but that is sufficient to allow large glaciers to expand over the land. The most recent ice age lasted from about 120,000 until about 10,000 years ago. Its ice sheet was up to 3 km (2 miles!) thick and covered all of

Canada, the Great Lakes, and the northeastern United States as far south as New York (Figure 19.12). (Long Island was, in fact, formed from debris scraped up by the ice.) Three "little ice ages" have occurred in the past 10,000 years. The last one, between the 15th and 19th centuries, produced crop failures, famine, and wandering bands of marauders.

Ice ages have affected the evolution of animals not only through direct thermal effects but also by reducing and modifying habitats. It is obvious that the ice sheet deprived many terrestrial animals of habitat, but marine organisms were similarly affected. So much water was locked up in ice that the seas dropped 90 meters below their present level, exposing coral reefs and shallow sea margins. The lowering of the seas also exposed land bridges that allowed the migration of animals between faunal regions. Between 30,000 and 10,000 years ago the Bering Strait was a major route for the spread of mammals. Horses migrated from America to Asia, and other large animals spread in the reverse direction, with human hunters in pursuit.

Figure 19.11
Some large mammals of the Quaternary period (within the past two million years). (A) The Irish elk Megaloceros *had antlers with a spread of three meters. (B) The sabertooth cat* Smilodon *lurked around present-day Los Angeles and also spread to South America.*

Figure 19.12
During the height of the last ice age 18,000 years ago, ice sheets (white) covered much of the northern land masses. The lowered sea level changed the shapes of continents and exposed land bridges between them.

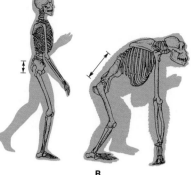
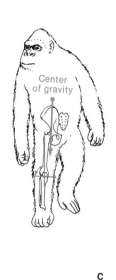
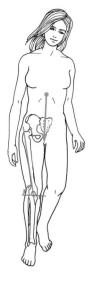

Figure 19.13
(A) The chimpanzee Pan troglodytes, *like most apes, can climb well but is likely to be found knuckle-walking on the ground. It can walk upright only briefly. (B) In chimpanzees there is a greater distance between the attachment of the vertebral column and the sockets of the femurs in the pelvis, making the animal unstable when upright. Note also that the vertebral column in the neck of the chimpanzee joins the skull at an angle, while that of the human is vertical and directly beneath the skull. (C) In chimpanzees the angle of the femur at the knee is approximately 90°. Therefore, when the chimpanzee lifts one foot during bipedal walking it must shift its body to keep the center of gravity over the other foot. Note also that in the ape the first digit of the foot is angled like a thumb rather than a big toe.*

HUMAN EVOLUTION

Differences Between Humans and Apes.

One of the most interesting events in the Cenozoic era—at least to us—was the origin of our own species. Contrary to the view widely held more than a century ago, humans did not suddenly appear on Earth but gradually distinguished themselves from apes through a series of evolutionary innovations, including the following:

1. **Bipedal locomotion,** rather than **brachiation** (swinging by the arms) or the **knuckle-walking** of chimpanzees and gorillas (Figure 19.13).
2. **Increased brain size,** characterized by a brain mass that is about 75% larger in humans than it would be in chimpanzees or gorillas if they were the same size as humans.
3. **Language,** which has communication abilities far exceeding those of the vocalizations and facial expressions of chimps and gorillas.
4. **Tool making and tool use,** which occur only sporadically in chimps and gorillas.

Until the 1970s most paleoanthropologists assumed that the evolution of all these differences would have taken at least 15 million years. In the 1980s, however, accumulating evidence from molecular phylogenetics indicated that the human line of evolution diverged from that of apes much more recently (Figure 19.14). Assuming a constant rate of mutation, the ancestors of humans probably diverged from those of chimps and gorillas between five and eight million years ago. This date has now been accepted by most paleoanthropologists as consistent with fossil evidence.

The study of human evolution has been revolutionized not only by molecular phylogenetics but also by a different way of thinking. Prior to the 1970s human evolution was commonly seen as an epic struggle of the ape to acquire a large brain and language and to master tools and fire in order to become a man. The term "man" is appropriate here because the role of females in evolution was generally ignored. During the Vietnam War this scenario took a backward turn, and the image of man as Killer Ape became fashionable. To some extent popular culture still influences paleoanthropology, as can be seen in some recent evolutionary scenarios that give the female star billing in human evolution and relegate the male to supporting cast.

Paleoanthropologists are now more likely to approach human evolution more

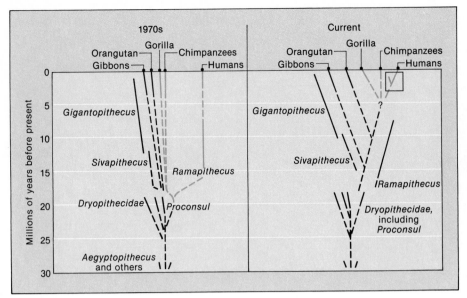

Figure 19.14
The time scale of hominid evolution widely accepted in the 1970s compared with the current view. Humans were thought to have diverged from all other apes from the common ancestor Proconsul *at least 15 million years ago.* Ramapithecus *was considered to be the first hominid. Because it is now known that humans, chimps, and gorillas shared an ancestor as recently as eight million years ago,* Ramapithecus *could not have been ancestral to humans, and it is uncertain whether* Proconsul *was. The question mark indicates that molecular phylogeny is not precise enough to resolve the order of divergence of humans, chimps, and gorillas. Most evidence indicates that humans are more closely related to chimps than to gorillas. Dashed lines represent inferred relationships. Details within the box in the upper right are shown in Figure 19.16.*

scientifically, by formulating testable hypotheses. Instead of asking how apes managed to improve and become more like us, they investigate the ecological conditions that made it advantageous to be more like a human than like an ape. For example, under what conditions is it energetically more advantageous to walk upright than to knuckle-walk; what ecological conditions favor eating meat rather than only fruits and leaves; and when is hunting more advantageous than scavenging? To answer such questions, paleoanthropologists have to study not only the fossils of our ancestors but also the climate, plants, and animals on which they depended. It is much more difficult and tedious work than making up stories of heroes, but it is much more scientifically rewarding.

Viewed in this way, the divergence of humans from apes is not a matter of humans leaving the apes behind in the evolutionary race, but simply a matter of each group making different uses of the available resources. As the forests of Africa gave way to savanna, the ancestors of chimps and gorillas remained in the trees, while our own ancestors went to the grasslands. Although our current understanding of human evolution is now on surer footing, there is still much disagreement. The number of fossil species is relatively small, so the discovery of a new one could easily upset present thinking. What follows is therefore a tentative description of some of the major events in our recent evolution.

Australopithecine Ancestors. Following the divergence of the human line from that of chimps and gorillas, the oldest known humanlike primate is *Australopithecus afarensis* (Figures 19.15 and 19.16). Fossils of *A. afarensis*, including those of the female named "Lucy," date to around 3.2 million years ago in eastern Africa. In many respects, such as dentition, this species is intermediate between apes and humans. What distinguishes *A. afarensis* as not just another fossil ape is that it was fully bipedal. In fact, even older (3.5-million-year) fossilized tracks of a bipedal hominid have been found (see Figure 20.5C). The long arms of *A. afarensis* suggest, however, that it had only recently left the trees and might still have climbed them to escape predators and to nest at night. The most surprising thing about *A. afarensis* was that its brain was so small. Up until its discovery most anthropologists had assumed that a large brain came before the ability to use tools,

Figure 19.15
(A) The mortal remains of "Lucy," the first specimen of A. afarensis discovered by Donald Johanson in 1974. She stood approximately a meter tall and would have weighed 60 pounds. The small pelvis and inward turn of the femur in this and other specimens indicates bipedality. The arms were relatively long, however, suggesting that brachiation was still used.
(B) Artist's reconstruction of Lucy. The amount of hair, shape of ears, and facial expression are speculative.

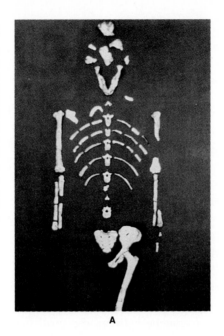
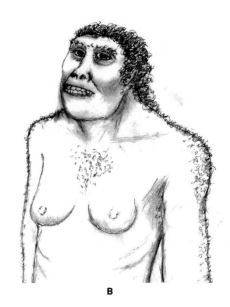

A

B

which then forced the evolution of upright posture to free the hands. *Australopithecus afarensis* shows that bipedal locomotion came before a large brain.

Australopithecus afarensis may have been an ancestor of larger and more numerous hominids: *Australopithecus africanus*. Raymond Dart discovered the first spec-

Figure 19.16
Major events in the evolution of humans.

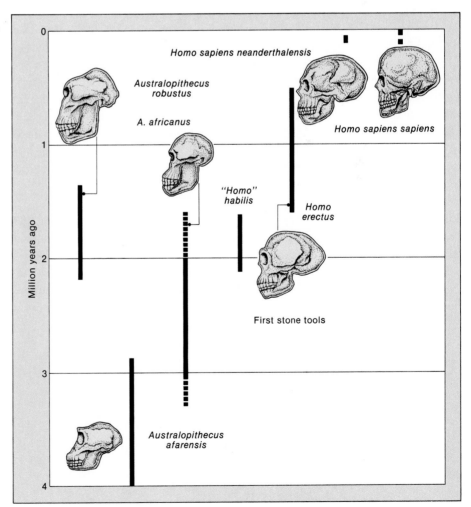

imen of *A. africanus*, the famous Taung Child's skull, in South Africa in 1924, long before anthropologists were ready to accept such a small-brain primate as anything but an ape. Far from being stupid, as its small brain might suggest, *A. africanus* may have been the first to make stone tools.

The First Humans. *Australopithecus africanus* became extinct some two million years ago and was replaced by several other species. One was a larger australopithecine, *A. robustus* (= *Paranthropus robustus*), which was about the size of a small modern man. Another was the first species recognized as human, *Homo habilis*—literally, "handy man." *Homo habilis* gets its name from the fact that its fossils are often found with large numbers of sharp stone flakes and other tools. (There is evidence that *A. robustus* also used tools.) *Homo habilis* males were approximately 1.3 meters tall and weighed 40 kg. Females were only half as large. Such differences in the sexes generally occur when a dominant male controls a harem of females, as in gorillas. *Homo habilis* was smaller than *A. robustus* and lived at the same time and places in southern and eastern Africa, and perhaps elsewhere. In some cases the skeletons of the two species have been found in the same deposits and caves, inspiring all kinds of tales about the struggle between the small but skillful human versus the large, clumsy australopithecine. They may not have competed with each other, however. Analysis of tooth wear suggests that *H. habilis* scavenged for meat while *A. africanus* ate fruits.

Both *A. robustus* and *H. habilis* became extinct, but descendants of *H. habilis* apparently evolved into *Homo erectus* (Figure 19.17A). (The name was given before it was realized that erect posture was nothing new.) *Homo erectus* was almost as large as modern humans, and the females were closer in size to the males. The reduced difference in size suggests a change in social organization to one of more cooperation between the sexes. The brain size enlarged dramatically, sometimes reaching 1100 cm³, which is within the normal range of modern human brains. Perhaps not surprisingly the tools produced by *H. erectus* were more sophisticated than those of its ancestors and included hand axes. Starting around a million years ago, *H. erectus* ranged from Africa into Europe and Asia, leaving fossils now named Peking Man and Java Man. They also left deposits of bones and tools that suggest that some groups of *H. erectus* may have set up camps periodically until they had hunted out the local game, then packed up and moved on.

Homo sapiens. *Homo erectus* became extinct all over the world some 400,000 years ago, but not before giving rise to our own species. *Homo sapiens* also spread throughout the Old World and diverged into various subspecies. One subspecies was *Homo sapiens neanderthalensis*—Neanderthal Man—who lived in Europe, the Near East, and Asia beginning about 100,000 years ago. In the popular mind the image of Neanderthal Man is that of a stooped and stupid person. This perception

Figure 19.17
The skulls of (A) Homo erectus, (B) a Neanderthal person, and (C) a Cro-Magnon (modern) person.

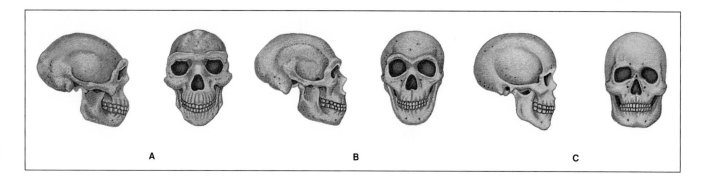

A   B   C

arose largely because the first study was of an arthritic skeleton and was done at a time when scientists were looking for an "ape-man." Neanderthals were undoubtedly of formidable appearance, though. Heavy ridges on the bones where muscles attached indicate that Neanderthals were powerfully built. The prominent brows and receding forehead and chin would have given them a brutish look (Figure 19.17B). Nevertheless, they may well have been quite nice folks. Their brains were actually larger on average than our own. They buried their dead and had the vocal apparatus and Broca's area of the brain required for speaking.

Neanderthal people disappeared some 35,000 years ago, perhaps killed off, outcompeted, or absorbed into groups of fully modern humans, *Homo sapiens sapiens* (Figure 19.17C). These early modern humans are called Cro-Magnon people, after a place in France where the oldest remains were found in a cave. They were ice age people as capable of creating cave art as of spearing a mammoth. Perhaps driven by climate changes and the pursuit of game, they spread throughout the world, even sailing to Australia and crossing the Bering land bridge to America. They were the parents of us all.

SUMMARY

Life is believed to have begun with bacteria-like organisms almost four billion years ago, at a time when the atmosphere lacked oxygen. Eventually, however, some of these microorganisms altered the atmosphere by producing oxygen as a by-product of photosynthesis. Some organisms acquired the ability to use the oxygen to make ATP, and these are believed to have become endosymbionts and then mitochondria within the ancestors of eukaryotic cells. Eukaryotic cells may also have acquired chloroplasts, flagella, and cilia by endosymbiosis. Some of these eukaryotic organisms became multicellular and evolved into animals before the Cambrian period some 600 million years ago.

Almost all the major groups of animals alive today, as well as many that became extinct, were present during the Cambrian period. The Cambrian and other periods of the Paleozoic era were dominated by marine invertebrates, but there were some terrestrial arthropods and reptiles. The most significant animals of the Mesozoic era were dinosaurs and other reptiles. There were also mammals, but they did not become dominant until the present Cenozoic era. Many dinosaurs and many other groups of animals became extinct at around the start of the Cenozoic era, during one of several mass extinctions.

Geological events have profoundly influenced evolution. The movement of continents during the Mesozoic era separated various groups of mammals from each other, dividing them into faunal realms. Ice ages during the Cenozoic further altered the distribution of animals by lowering sea levels and creating land bridges. Continental drift and ice ages also directly affected evolution by their effects on climate.

Humans evolved from a primate that was also ancestral to gorillas and chimpanzees. The main differences acquired during human evolution were bipedal locomotion, the ability to make and use tools, increased brain size, and language. The oldest known species that may have been ancestral to humans was *Australopithecus afarensis*, which was already bipedal more than three million years ago. The oldest ancestor considered to be human was *Homo habilis*, which made tools more than two million years ago. *Homo erectus*, which lived after *H. habilis*, had a brain almost as large as that of modern humans. Our own species, *H. sapiens*, dates back to 100,000 years ago and is represented by Neanderthal people. They may have been capable of speech. Modern humans, *H. sapiens sapiens*, originated in the form of Cro-Magnon people.

KEY TERMS

Proterozoic eon
endosymbiont
Paleozoic era
Mesozoic era
Cenozoic era
Cambrian period

mass extinction
continental drift
plate tectonics
faunal region
australopithecine
Australopithecus afarensis

Homo habilis
Homo erectus
Neanderthal
Cro-Magnon

SELF-TEST

1. List the main events of biological relevance that occurred in the first four billion years of the Earth's existence. Estimate how long ago each event occurred.

2. Briefly summarize the current state of knowledge regarding the origin of life.

3. Describe the Serial Endosymbiotic Theory of the origin of eukaryotes.

4. Briefly describe life in the Paleozoic era.

5. What effects did geological events, especially asteroid collisions and continental drift, have on evolution during the Mesozoic era?

6. How would you account for the success of mammals during the Cenozoic era?

7. What are the main distinctions between humans and apes? Briefly outline the sequence by which each of these distinctions could have evolved. What is the nature of the evidence for this sequence?

READINGS

RECOMMENDED READINGS

Alvarez, W., F. Asaro, and V. E. Courtillot. 1990. What caused the mass extinction? *Sci. Am.* 263(4):76–92 (Oct).

Briggs, D. E. G. 1991. Extraordinary fossils. *Am. Sci.* 79:130–141.

Buffetaut, E. and R. Ingavat. 1985. The mesozoic vertebrates of Thailand. *Sci. Am.* 253(2):80–87 (Aug).

Calder, N. 1983. *Timescale.* New York: Viking. (*A useful guide to changes throughout Earth's history.*)

Cartmill, M., D. Pilbeam, and G. Isaac. 1986. One hundred years of paleoanthropology. *Am. Sci.* 74:410–420.

Cech, T. R. 1986. RNA as an enzyme. *Sci. Am.* 255(5):64–75 (Nov).

Chorlton, W. 1983. *Planet Earth: Ice Ages.* Alexandria, VA: Time-Life Books. (*Beautifully illustrated.*)

Cloud, P. 1983. The biosphere. *Sci. Am.* 249(3):176–189 (Sept). (*Mainly on Proterozoic evolution.*)

Covey, C. 1984. The Earth's orbit and the ice ages. *Sci. Am.* 250(2):58–66 (Feb).

Eigen, M. et al. 1981. The origin of genetic information. *Sci. Am.* 244(4):88–118 (Apr).

Glen, W. 1990. What killed the dinosaurs? *Am. Sci.* 78:354–370.

Groves, D. I., J. S. R. Dunlop, and R. Buck. 1981. An early habitat of life. *Sci. Am.* 245(4):64–73 (May).

Horgan, J. 1991. In the beginning. . . . *Sci. Am.* 264(2):116–125 (Feb).

Leakey, R. E. 1981. *The Making of Mankind.* New York: E. P. Dutton.

Lewin, R. 1984. *Human Evolution: An Illustrated Introduction.* New York: W. H. Freeman.

Lovejoy, C. O. 1988. Evolution of human walking. *Sci. Am.* 259(5):118–125 (Nov).

Marshall, L. G. 1988. Land mammals and the great American interchange. *Am. Sci.* 76:380–388.

McMenamin, M. A. S. 1987. The emergence of animals. *Sci. Am.* 256(4):94–102 (Apr).

Pilbeam, D. 1984. The descent of hominoids and hominids. *Sci. Am.* 250(3):84–96 (Mar).

Rukang, W. and L. Shenglong. 1983. Peking Man. *Sci. Am.* 248(6):86–94 (June).

Russell, D. A. 1982. The mass extinctions of the late Mesozoic. *Sci. Am.* 246(1):58–65 (Jan).

Siever, R. 1983. The dynamic earth. *Sci. Am.* 249(3):46–55 (Sept). (*Introduction to an entire issue on the evolution of Earth and its life.*)

Simpson, G. G. 1983. *Fossils and the History of Life.* New York: Scientific American Library. (*An attractive summary of the significance of fossils by one of the founders of modern paleontology.*)

Stanley, S. M. 1984. Mass extinctions in the oceans. *Sci. Am.* 250(6):64–72 (June).

Stanley, S. M. 1987. *Extinction.* New York: Scientific American Library.

Vidal, G. 1984. The oldest eukaryotic cells. *Sci. Am.* 250(2):48–57 (Feb).

Walker, A. and M. Teaford. 1989. The hunt for *Proconsul. Sci. Am.* 260(1):76–82 (Jan).

ADDITIONAL REFERENCES

Briggs, D. E. G. and P. R. Crowther (Eds.). 1990. *Palaeobiology: A Synthesis.* Boston: Blackwell Scientific Publications. (*Concise reviews of all aspects of evolution.*)

Ciochon, R. L. and J. G. Fleagle. 1987. *Primate Evolution and Human Origins.* New York: Aldine de Gruyter. (*A collection of major research papers.*)

Cloud, P. and M. F. Glaessner. 1982. The Ediacarian period and system: metazoa inherit the earth. *Science* 217:783–792.

Conway Morris, S. and H. B. Whittington. 1979. The animals of the Burgess shale. *Sci. Am.* 241(1):122–133 (July).

Conway Morris, S. and H. B. Whittington. 1985. Fossils of the Burgess shale. Geological Survey of Canada, Misc. Rep. 43.

Glaessner, M. F. 1984. *The Dawn of Animal Life.* New York: Cambridge University Press.

Gould, S. J. 1989. *Wonderful Life: The Burgess Shale and the Nature of History.* New York: W. W. Norton.

Harland, W. B. et al. 1990. *A Geologic Time Scale 1990.* New York: Cambridge University Press.

Johanson, D. and M. Edey. 1981. *Lucy: The Beginnings of Humankind.* New York: Warner Books.

Levinton, J. 1988. *Genetics, Paleontology, and Macroevolution.* New York: Cambridge University Press.

Patrusky, B. 1984. Before there was biology. *Mosaic* 15(6):10–17.

Simon, E. L. 1989. Human origins. *Science* 245:1343–1350.

Tattersall, I., E. Delsen, and J. V. Couvering (Eds.). 1988. *Encyclopedia of Human Evolution and Prehistory.* New York: Garland.

Woodburne, M. O. and W. J. Zinsmeister. 1982. Fossil land mammal from Antarctica. *Science* 218:284–286.

20

Behavior

Three-spined sticklebacks (Gasterosteus aculeatus) *in nest.*

CHAPTER OUTLINE

LEARNING OBJECTIVES

1. What are the basic processes of animal behavior?

2. Are behaviors innate, learned, or both?

3. Are there genes for specific behaviors?

4. Why are some animals solitary and others social?

5. How do animals without language communicate?

6. How do animals tell the time of day or the seasons?

7. How do migrating animals find their way?

8. What biological functions are served by courtship?

9. Why do animals differ in their mating and parenting behaviors?

THE SCIENTIFIC STUDY OF ANIMAL BEHAVIOR

Many people who are impressed when a computer generates a complex and beautiful pattern on its screen will not stop to watch a spider constructing a web. Perhaps they take the spider for granted, imagining they could do the same thing if they could make the silk. For them the spider merely builds its web in anticipation of catching a meal, just like the fisherman repairing his nets. It takes a scientist to turn a "simple" phenomenon like the spider's web into a difficult problem. Scientists realize that the mechanisms and motivations of the spider are not those humans would have under similar circumstances, and that our own behaviors are not even as simple as we apparently think they are.

The error of ascribing to animals the thoughts, emotions, and motivations of humans is now rejected by most behavioral scientists as **anthropomorphic** (from the Greek words meaning "man" and "form"). The banishing of anthropomorphism was due largely to the behaviorist school of psychology, which argued that one could only observe animal behavior, not try to infer the mental states underlying it. Later behaviorists, like B. F. Skinner, went even further, arguing that mental states, such as free will, did not occur even in humans. The behaviorist approach to animal behavior has produced more experimentally testable theories than the anthropomorphic approach, but most of these theories have been based on studies of the albino laboratory rat under extremely artificial conditions. Behaviorists have therefore been accused of replacing the anthropomorphic view of animals with a "ratomorphic" view of humans. Many zoologists think that the behavior of rats under laboratory conditions is not only a poor model for the behavior of humans, but also a poor model for the natural behavior of rats.

NEURAL EVENTS IN BEHAVIOR

Reflexes. Zoologists are more likely than psychologists to be interested in the behavior of animals in their natural contexts and for their own sake, rather than primarily as models for human behavior. As biologists they are more likely to accept that neural, hormonal, and other physiological mechanisms cause behaviors. This assumption has, in fact, been verified for some of the simplest behaviors. One such simple type of behavior for which the mechanism is completely known is the reflex. As explained in Chapter 8 (see pp. 167–169), a reflex is a stereotyped response to a stimulus that is controlled by simple chains of nerve cells that link muscular or glandular output rather directly to sensory input. Reflexes can also account for some complex behaviors, such as the fly's uncanny ability to arrive just when and where dinner is being served and to select the tastiest dishes. Vincent Dethier (pronounced de-TEER) has studied these abilities in the black blow fly *Phormia regina*, which normally lives on nectar, sap, and decaying plants or animals. The fly locates its food by flying upwind when the chemoreceptors on its antennae detect a suitable odor. The fly then lands on the food and tastes it with chemosensory hairs on its feet. Dethier (1976) found that there are four chemoreceptor cells in each chemosensory hair: two sugar receptors and two salt receptors. There is a mechanoreceptor as well. The fly feeds only if the sugar and salt receptors signal that the food is sweet enough and not too salty. If the salt receptors in the feet are not too active, action potentials from the sugar receptors travel to the head ganglia and trigger the reflex lowering of the fly's proboscis (Figure 20.1). The ingestion of food through the proboscis is due to another reflex. Sugar receptors on the proboscis activate muscles that spread the mouthparts and pump food into the crop. Feeding then continues for a minute or so until the

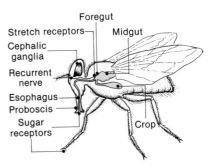

Figure 20.1
The feeding system of the blow fly Phormia regina. Sugar receptors on the feet trigger the reflex lowering of the proboscis. Sugar receptors on the proboscis then trigger the reflex ingestion of food. After a certain amount of food has been ingested, stretch receptors in the foregut and body wall activate the retraction of the proboscis by means of action potentials in the recurrent nerve.

sugar receptors become less responsive. Feeding also ceases if stretch receptors in the foregut or body wall signal that the fly has eaten enough. If the nerve carrying the information from these stretch receptors is cut, a fly will eat until it bursts.

Taxes.   Another kind of elementary behavior is a **taxis**, in which an animal orients its movement with respect to the direction of a stimulus. One example of a taxis is the flight of the fly upwind, in the direction of the source of an odor. This would be an example of a **chemotaxis**. A taxis toward or away from light is a positive or negative **phototaxis**. Taxes (pronounced TAX-eez) may result from neural subsystems called **servomechanisms** (see pp. 169–170). Unlike a reflex mechanism, which tends to produce an all-or-none behavior, a servomechanism continuously corrects its output using sensory input. Thus if a fly strays out of the path of the odor molecules, the response of chemoreceptors on its antennae will weaken. It will then change the direction of flight until the odor is again intercepted.

Kineses.   Another example of an elemental behavior is a **kinesis**, which is a change in the rate of movement rather than the direction of movement. Many animals display **chemokinesis** when they encounter molecules indicating the presence of food. The increased rate of movement, though not directed toward the source of food, increases the chances that they will encounter it by chance. Other animals become restless in the presence of light (**photokinesis**), heat (**thermokinesis**), or other adverse conditions, making it more likely that they will escape from those conditions.

Other Innate Behaviors.   Reflexes, taxes, and kineses tend to produce stereotyped behaviors that are characteristic of the species. These neural subsystems are "wired" into the nervous system during development, presumably according to genetic instructions. There are also more complex behaviors that appear to be inborn. For example, the female brown-headed cowbird *Molothrus ater* may never have heard the song of a male brown-headed cowbird before, since cowbirds are raised in the nests of other species. Yet the first time a mature female hears a male's song she responds with a stereotyped posture that signals sexual receptivity (West et al. 1981; Figure 20.2). Such apparently inborn behaviors are said to be **instinctive** or **innate**.

Learning.   Innate behaviors can be thought of as lying at one end of a spectrum, with completely learned behaviors at the other end. In between the extremes are innate behaviors that are modified by learning and behaviors that can be learned only if there is an innate ability. Most experimental psychologists recognize two major kinds of learning: classical conditioning and operant conditioning. The Russian physiologist Ivan Pavlov (1849–1936) was the first to study **classical conditioning,** which is sometimes called Pavlovian or Type I conditioning. Classical conditioning occurs when an animal learns to respond to an irrelevant stimulus, the **conditioned stimulus**, in the same way that it normally responds to a relevant stimulus, the **unconditioned stimulus**. The most famous example of classical conditioning was Pavlov's dogs, which began to salivate whenever a bell signaled feeding time. The American psychologist B. F. Skinner was the first to study **operant conditioning,** which is also called Skinnerian, instrumental, or Type II conditioning. Operant conditioning occurs when an animal gradually develops a behavior after the rewarding (reinforcement) of closer and closer approximations

Figure 20.2
The copulatory posture of the female cowbird Molothrus ater. *Even though the female has been raised in the nest of another species and may never before have heard the courtship song of a male cowbird, it responds with this posture the first time it hears the song as an adult.*

to that behavior. In a typical demonstration, a rat learns to press a button if it is rewarded with food every time it hits the button by chance.

Theories of classical and operant conditioning have had little influence on studies of behavior in nature, perhaps because these theories were developed under artificial laboratory conditions. However, both types of learning probably do occur in nature. Animals probably become either tame or shy of humans by associating our presence with benefit or danger (classical conditioning). Honey bees probably learn by operant conditioning when to arrive at flowers that open at a certain time of day.

Animals that are incapable of learning certain tasks are not necessarily stupid. Like humans, other animals may be quite adept at learning some tasks, but not others. Rats, for example, can learn complex mazes to avoid getting an electric shock. They cannot, however, learn to rear up on their hind legs to avoid shock, even though they are physically capable of standing on their hind legs. In many cases it would make little sense for members of a species to have the capacity to learn a task, because that task never occurs in their normal habitats.

GENES AND BEHAVIOR

Since innate behaviors do not have to be learned, they must be assumed to result from neural mechnisms that are determined genetically. In addition, there are patterns of behavior that are characteristic of each species and that appear to be genetically controlled. House cats, for example, behave like their lone, stealthy hunting ancestors, while dogs are more noisy and social, like their pack-hunting ancestors. These species differences persist even though domesticated cats and dogs have been raised under similar circumstances for thousands of years.

For a few behaviors the evidence for a genetic basis is as strong as Mendel's evidence for genetic factors in pea plants. By using an approach similar to Mendel's, Dilger (1962) demonstrated the role of genes in nest-building by lovebirds. One species of lovebird, *Agapornis personata*, carries its nest material (bark or shredded paper) to the nest site in its beak. Another species, *A. roseicollis*, carries its nest material tucked in its feathers. Hybrids of the two species attempt to carry nest material by both methods. They repeatedly tuck the material into their feathers, then grasp it in their beaks, then tuck it into their feathers again. Finally, they fly off with the material either in their beaks or in their feathers. Evidently the hybrids inherit alleles for both types of behavior.

Drosophila geneticists early in this century noticed that some of their mutant flies behaved differently. Since then *Drosophila* **behavioral mutants** have provided much material for studying the genetics of behavior. Some *Drosophila* mutants, such as those named *dunce* and *amnesiac*, cannot remember whether to turn right or left in a simple Y maze. Biochemical analysis shows that the normal allele of the *dunce* gene codes for an enzyme that breaks down cyclic AMP. Mutations of this gene result in abnormally high levels of cAMP that somehow interfere with learning.

Genetic mosaics of *Drosophila* have also provided insights into the role of genes in behavior. Genetic mosaics are individuals with genetically different parts resulting from a mutation during development. Some of the most interesting genetic mosaics combine both male and female parts. These **gynandromorphs** (from the Greek words meaning "female," "male," and "form") are produced when a mutation in a female embryo causes one X chromosome of one cell to become lost during mitosis. The cells that descend from that mutant cell are genetically male. Which part of the adult becomes male depends on where and at what stage of development the X chromosome was lost. These sex mosaics display some confused courtship behaviors. Normally a male *Drosophila* initiates courtship of a female by **orienting** toward her (Figure 20.3). If she has already mated she sticks out her genitalia, which discourages(!) him. If the female does not signal that she has mated, the male taps her abdomen with a foreleg. She may then avoid him, but the male will **follow** and eventually **extend** and vibrate one wing in a courtship **song**. The male also **licks** the female's genitalia, after which he usually attempts copulation. The female may eventually spread her wings and extend her genitalia to be mated.

Orienting, following, wing-extension, and licking occur only if a particular cluster of nerve cells in the dorsal part of the head ganglion is genetically male. This is true even if the rest of the gynandromorph is female. Even though wing-extension may occur in gynandromorphs with male head ganglia, singing (wing vibration) occurs only if particular cells in the thoracic ganglion are genetically male. The greater the number of male cells in the thoracic ganglia, the longer the gynandromorph persists in its attempts to mate. The greater the number of female cells in the abdomen, the more appealing a gynandromorph is to males. Gynan-

Figure 20.3
Behavior of the male Drosophila melanogaster *during courtship (left side, top to bottom). The smaller male begins by orienting toward the female. If she has not recently mated, he will tap her on the abdomen and extend and vibrate a wing. He will then lick the female's genitals and attempt copulation. Arrows point to the parts of the central nervous system that must be genetically male for each behavior to occur.*

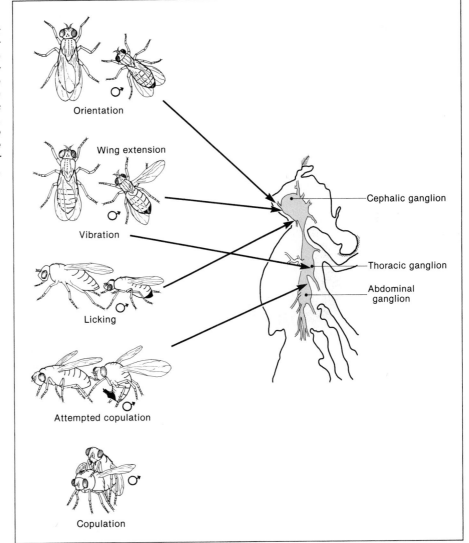

Orientation

Wing extension

Vibration

Licking

Attempted copulation

Copulation

Cephalic ganglion

Thoracic ganglion

Abdominal ganglion

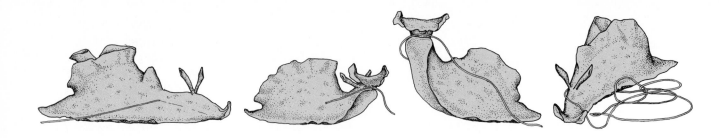

Egg laying in the marine snail Aplysia.

dromorphs with male anteriors and female posteriors will court females at the same time they are being courted by males.

In the mollusc *Aplysia*, Scheller and his co-workers have actually isolated the genes that produce peptides responsible for egg-laying behavior (Scheller and Axel 1984; Scheller et al. 1984). In this behavior the animal extrudes a string of egg cases from the genital pore on its side, grasps the string in its mouth, waves its head to pull it out, then attaches the string to an object (Figure 20.4). Egg laying normally occurs only after the animals copulate, which these hermaphrodites often do in groups forming long chains or circles. Unmated *Aplysia* will also perform the entire sequence of egg-laying behavior if they are injected with three peptides produced by a particular group of genes. All three peptides act as synaptic transmitters on particular nerve cells. The injection of each one of the three peptides evokes a different part of the egg-laying behavior.

THE EVOLUTION OF BEHAVIOR

Behavioral Fossils. If some behaviors are genetically inherited, and if they can affect the ability to reproduce, then it follows that those behaviors can evolve. Behavior does not fossilize, however, so studies of the evolution of behavior depend a great deal on imaginative hypotheses. In a few cases inferences are rather straightforward. Some animals of the distant past left fossil evidence of their behavior (Figure 20.5). Sometimes animals died so suddenly that their fossils provide a snapshot of their normal behavior. For example, fossils of young dinosaurs too large to have been recently hatched have been found still within nests. They suggest that adult dinosaurs bestowed a great deal of care upon the young. As another example, we can infer that *Archaeopteryx* flew from the fact that its fossilized feathers are asymmetric, unlike those of penguins and other birds whose symmetrical feathers serve only for insulation (see Figures 18.2 and 20.5B). In 1978 colleagues of Mary Leakey made a striking discovery about the locomotor behavior of humanlike (hominid) primates. While playing a game of elephant-dung tossing, they literally stumbled upon the footprints of hominids who had walked upright more than 3.5 million years ago (Figure 20.5C).

The Comparative Method. In some cases evolution provides the only apparent justification for a behavior. Consider, for example, the female balloon fly *Hilara sartor*, which mates only with males carrying empty silken balloons (Kessel 1955). The females have absolutely no use for such balloons, and one would expect that they would soon be as wise as a woman whose suitor kept giving her empty candy boxes. A possible explanation of the female balloon fly's seeming gullibility comes from a comparison with the courtship behavior of other flies in the same family (Empididae). This comparative method suggests the sequence through which the behavior may have evolved. The first stage may be represented by certain species

Genes for Behavior?

Many who do not hesitate to speak of the inheritance of wrinkling in peas or the inheritance of color-blindness in humans avoid speaking of the inheritance of human behaviors. Yet the phrase "gene for behavior" has essentially the same meaning as "gene for wrinkling in peas." It means simply that there is a gene that directs the synthesis of a protein (an enzyme, synaptic transmitter, or hormone) that affects the development or functioning of the nervous system in such a way that a particular behavior results. For most behaviors, there are likely to be many genes that contribute.

The reluctance to speak about genes for behavior no doubt arises from a fear that it will encourage the stupid and evil perversion of science that occurred in Nazi Germany and elsewhere. Even in the United States in fairly recent times some judges, physicians, and psychologists evidently thought insanity, idiocy, criminality, and poverty were genetic and prescribed forced sterilization as a remedy. Such absurdities also occurred long before any knowledge of genetics, however. Bringing back the good old days before Mendel probably would not eliminate them. It seems more likely, in fact, that knowledge is a better antidote than ignorance. Thanks to genetics, we know that it is not practical to eliminate genetic traits by genocide or forced sterilization. One reason is that in humans even inherited behavioral traits are highly variable because of environmental differences. Many of those with a gene for a particular behavior would not display the behavior, and many of those without the gene would display the behavior.

Although it seems likely that many human behaviors are influenced by genetics, it is difficult to say which behaviors are so influenced and to what extent. This will probably remain the case as long as it is impractical and unethical to perform the kinds of controlled-breeding studies on humans that can be done in peas and fruit flies. There is strong evidence, however, that some human behaviors are affected more or less directly by genetic traits. Schizophrenia, for example, is ten times more likely to occur in a son or daughter of a schizophrenic than in a biologically unrelated stepson or stepdaughter. The gene responsible for some forms of schizophrenia has been localized on chromosome 5. Schizophrenia is, of course, not a behavior in itself, but it certainly affects behavior.

The strongest evidence for genetic influences on human behavior comes from studies of monozygotic twins reared apart (MZAs). Because monozygotic twins are genetically identical, any unusual similarities in the behaviors of such twins reared separately may be due to genetics. Such similarities do, indeed, occur. One striking case involved monozygotic brothers who had been separated shortly after birth, with one raised by his father as a Jew in the Caribbean, and the other by his grandmother as a Catholic Nazi in Germany. When the two were reunited after 47 years, they naturally discovered many differences between themselves, but there were also many odd similarities. Both wore similar glasses and moustaches, liked spicy foods and sweet liqueurs, thought it was funny to sneeze in public, and saved rubber bands on their wrists. It is not clear what to make of such curiosities. Is there a gene for saving rubber bands on the wrist, or is that just one way in which particular gene combinations happen to be expressed? Could it be that any two people chosen at random would show an equal number of coincidental similarities?

A more scientifically rigorous approach uses personality tests to determine the degree to which behavior differs between MZAs compared with monozygotic twins reared together (MZTs). If the differences in scores of MZAs is similar to the differences in scores of MZTs, then those differences must be due more to heredity than to environment. Bouchard and his colleagues (1990) at the University of Minnesota have conducted such studies using more than 100 MZAs. They have found that for several personality traits the differences in scores are due more to genetics than to environment (Holden 1987). One of these traits is the Intelligence Quotient (IQ). It should be noted, however, that this research has been done mainly on middle-class Caucasians, so there is no basis for inferences about other groups. It does not follow, for example, that those with low IQ scores cannot benefit from education, or that differences in average IQ scores of different races are due to genes rather than to environment. It should also be remembered that although IQ is an extremely accurate measure of the ability to take IQ tests, it may not measure intelligence itself, whatever that is.

of the empidid family in which the male presents a female with dead prey during courtship. Three factors may have favored the evolution of this behavior.

1. Males with demonstrated prey-catching abilities would be more likely to pass on to succeeding generations genes contributing to that ability.
2. Males that ensured that their mates were well fed were more likely to get their genes into the next generation.
3. Males that provided the females with something other than themselves to eat were more likely to avoid becoming prey of the female during copulation.

A

B

C

The next stages of the sequence may be represented by species in which the male wraps the prey in silk before presenting it to the female. Possibly this gift-wrapping gives the male more time to copulate without becoming prey to the female, thereby assuring more nearly complete fertilization. In other species of empidid flies the male first sucks all the juices out of the wrapped fly, depriving the female and the eggs he fertilizes of nutrients, but still giving evidence of prey-catching ability and buying time for copulation. Still other species of empidids are nectar-feeding, so there is no need for the male to prove his predatory skill or to evade being eaten by the female. Yet the males of these species find insect fragments, wrap them up, and present them to the females. At this stage the behavior appears to be merely an evolutionary relic of a once-functional behavior. *Hilara sartor* represents the final step in the sequence, presenting only the wrapping without even a hint of prey.

COMMUNICATION

Auditory Communication.   The interaction of male and female balloon flies, like so many other behaviors, depends on communication. There are several channels for communication, each with particular advantages and disadvantages. Auditory communication, such as the courtship songs of birds, has the advantage of allowing animals to convey large amounts of information over long distances. In a typical area during the breeding season, for example, male birds of dozens of species sing their distinctive songs that attract females. Each song must encode at least enough information to identify the species. A disadvantage of this and other forms of auditory communication is that the sound may also attract potential predators.

Figure 20.5
Behavioral fossils. (A) Tracks of trilobites(?) from 500-million-year-old rocks on what is now the shore of Lake Champlain. The ripple marks indicate that these animals lived in shallow water. Scale is in centimeters. (B) Fossil feather of Archaeopteryx, a primitive bird that was in many ways like a dinosaur. Archaeopteryx was once thought to be incapable of flight, but the off-center placement of the feather shaft makes sense only in birds that fly. (C) Fossil footprints made by a small and a large hominid who walked through volcanic ash 3.5 million years ago. (Tracks on the right were made by the extinct three-toed horse Hipparion, *probably at a slightly different time.)*

In some species the song appears to be innate, with learning playing little or no role. One example would be the male cowbird, which may grow up without ever hearing the song of another male. In male white-crowned sparrows, *Zonotrichia leucophrys*, learning is more important. If males are isolated from the songs of adult males during a **sensitive period** lasting from the second week to a few months after hatching, they sing only a rudimentary song after they mature. Even if such a male hears a normal male song after the sensitive period, it can never learn to duplicate the song. On the other hand, if the sparrow hears a normal song during the sensitive period and is later isolated, it is able to sing normally even though it has not attempted to sing in the intervening months. Somehow during the sensitive period these and other birds not only learn to sing the song of their species, but they also avoid learning to sing like other species around them.

Figure 20.6
Visual communication between a dominant wolf (right foreground) and a subordinate (left foreground) regarding a carcass. Note how each element of the communication is opposite in each animal. (1) The dominant wolf bares its teeth as if to bite, while the subordinate grimaces. (2) The dominant sticks out its tail, while the subordinate tucks its tail between its legs. (3) The dominant's ears are forward, while the subordinate's are flattened. (4) The dominant's hackles (the hair on the back) are raised, while the subordinate's are not. (5) The dominant narrows its eyes, while the subordinate shows the whites of its eyes.

Visual Communication. Among mammals, facial expressions and body postures are the favored channels of visual communication. Charles Darwin argued that such gestures evolved as ritualized symbols for behaviors. For example, when wolves expose their teeth in a snarl, they are hinting at a behavior that may be forthcoming (Figure 20.6). Arthropods also use visual communication, but their rigid exoskeletons do not permit facial expressions. Fireflies more than compensate for this deficiency with their remarkable **bioluminescent organs.** Flying males solicit females by emitting pulses of light in a pattern that is specific for their species (Figure 20.7). Females reply with another species-specific pattern. The male then approaches the female, they exchange several more flashes, then copulate. The ability to communicate implies the ability to deceive, and many animals have mastered this ability almost as well as humans. When the female firefly *Photuris* sees the courtship flashes of a male of another species she can respond by flashing like the female of that species. The male is lured to the female *Photuris* but becomes her meat rather than her mate.

Figure 20.7
The patterns of light flashing by males of nine different species of fireflies. The females do not fly but answer the males by flashing from the ground (not shown).

Chemical Communication. Many animals communicate by chemical signals. (See E. O. Wilson's *Sociobiology*, 1975, pp. 232–233 for a survey.) Unlike visual and auditory signals, chemical messages can persist for some time after the messenger has left an area. They also have the advantage that they will be detected only by those with receptors that respond to the chemical, so they are less likely to attract predators. Some chemical messages have a hormonelike ability to induce specific behavioral responses in recipients in the same species. Such chemical messages are called **pheromones.** The best-known pheromones are insect sex attractants, many of which have been isolated and chemically analyzed. The first such pheromone to be studied was bombykol, which is produced in minute amounts by glands near the anus of the female silk moth *Bombyx mori.* The glands from half a million females had to be processed to yield 12 mg of pheromone. (One lab worker was reportedly overheard complaining, "The end is always in sight, but the work is never done.") A single molecule of bombykol is enough to evoke an action potential from the antenna of a male silk moth, and several hundred molecules are enough to make the male fly upwind, toward the female.

Other functions served by pheromones include alarming members of the same species, causing members of a group to aggregate, identifying members of a group, marking territorial boundaries, marking food trails, and indicating social rank. Many of these functions are performed by various pheromones in social insects (termites, bees, and ants) (Figure 20.8). The potential for deception has also been exploited in this channel of communication. The larvae of certain beetles induce ants to feed and care for them by secreting a pheromone that mimics that of the ant larvae.

Whether humans produce pheromones has long been an intriguing question. Many male mammals in zoos become particularly attentive to female attendants during their menstrual periods, so it is possible that human males might also respond in quite subtle ways. Human females might also respond. In 1971 Martha McClintock reported that the menstrual periods of women in college dormitories tended to become synchronized.

Tactile Communication: The Dance Language of Honey Bees.

One of the most remarkable examples of communication occurs within the darkness of the hives of honey bees and depends mainly on touch and vibration. Beekeepers have known for thousands of years that once one honey bee discovers a source of food, many honey bees soon converge upon it. Aristotle (*History of Animals*, Book IX, Chapter 40) thought that the first honey bee led the others to the food source, but the real explanation turns out to be much more interesting. When a worker honey bee, the **scout,** first discovers a rich source of food, she returns to the hive and performs a kind of dance to which her sister workers pay keen attention. Evidently these workers, called **recruits,** learn from the scout's dance what and where the food is, for within minutes they fly out and begin foraging on it.

In 1945, after more than 20 years of research, Karl von Frisch managed to decipher the vocabulary of the dance language. First the recruits easily discover the identity of the food by the odors clinging to the dancing scout. If the food is located close by (between 5 and 85 meters, depending on the race of bee), the scout performs a **round dance** in a broad figure-eight. The round dance apparently contains no information about the exact distance or direction of the food. The other workers make a special sound that signals the scout to stop dancing so that they can sample the scent. (If the scout doesn't stop dancing, it is stung to death.) The recruits then simply fly out in the vicinity of the hive until they locate the food by its odor.

For distant food sources the scout performs a **waggle dance,** running in a straight line while waggling its abdomen some 14 times per second and periodi-

Figure 20.8

Like many social insects, the imported fire ant Solenopsis saevissima produces a variety of pheromones. Some pheromones are not released by a specific gland. For example, a high concentration of CO_2 produced by a crowd of ants will attract stray ants. Ants also pick up odors that are unique to each nest, and they attack ants in the nest that have a different nest odor. Secretions from the body surface also prompt ants to groom each other. Numerous glands have also been identified as sources of various pheromones. A gland in the head of an attacked Solenopsis worker produces an alarm substance that produces excited behavior in other workers. At the same time Dufour's gland emits a substance that attracts the workers. When workers are foraging, this or another chemical is deposited by the stinger as a trail substance that other workers follow to the food source. The stinger is used not only to deposit trail substance but also for attack and defense, which has made this aggressive species a nuisance since its accidental importation into the southern United States from South America.

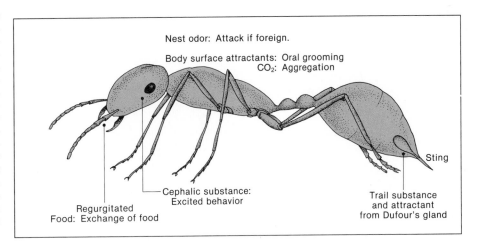

Nest odor: Attack if foreign.

Body surface attractants: Oral grooming
CO_2: Aggregation

Sting

Cephalic substance:
Excited behavior

Trail substance
and attractant
from Dufour's gland

Regurgitated
Food: Exchange of food

cally buzzing with her wings. She then circles back and repeats the performance. Evidently the waggle dance communicates information about the direction of the food, for instead of searching randomly within an area, the recruits fly more or less directly to the source. Evidently recruits also learn the distance of the source, for they take off with only enough fuel to fly the correct distance and back. Most of the recruits land within 20% of the exact distance and within 25° of the correct direction.

Von Frisch found that the scout communicates the distance and direction essentially by reenacting the trip out to the food source. If the food is very distant or requires flying into the wind, then the duration of the waggle portion of the dance increases. In one type of honey bee each second of the waggle portion of the dance translates as approximately 1 kilometer to the food source. If the waggle dance is performed on a horizontal surface, the waggle portion of the dance is aimed in the direction of the food (Figure 20.9). On the vertical combs that honey bees usually build, however, the scout makes a remarkable translation in its dance language. It performs the waggle portion of the dance at an angle with respect to vertical that equals the angle between the food source and the direction of the Sun. For example, if the food source is 20° to the right of the Sun, the scout orients its waggle dance 20° to the right of vertical.

Understandably, there have been doubts that any insect could perform such calculations of distance and angles and then translate them into an abstract dance language that could be understood by other bees. Some doubts arose from the fact that recruits do not necessarily have to rely on the information of the dance. Apparently honey bees have a kind of olfactory "map" of the areas around their hives and can often tell where to go simply from the odor brought back by the scout. Nevertheless, several studies have shown that honey bees can use the information in the waggle dance. One such study was done by James L. Gould while he was a graduate student at Rockefeller University. Gould's research relied on two curious phenomena. First, if a light is shone over a vertical comb, scouts will orient their dance with respect to the light, rather than to the vertical. Second, bees perceive bright light by means of three simple eyes called **ocelli,** while they perceive flowers and other shapes by means of the compound eyes. In one experiment Gould shone a bright light at an angle of 35° to the right of vertical inside the hive while scouts waggle-danced to communicate the location of an artificial flower. Scouts and recruits that did not have their ocelli covered performed and interpreted the dances using the light as a reference, and the majority of recruits arrived at the correct location. If the scouts had their ocelli covered with paint, however, they could not see the light and danced with reference to vertical. The

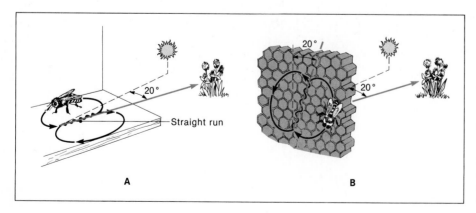

Figure 20.9
The waggle dance of a scout honey bee indicates the direction of a food source with reference to the sun. (A) When the scout dances on a horizontal surface the straight run points to the food source. (B) Usually the scout dances on the vertical combs within the hive, and the angle of the straight run with reference to gravity equals the direction of the food source with respect to the sun.

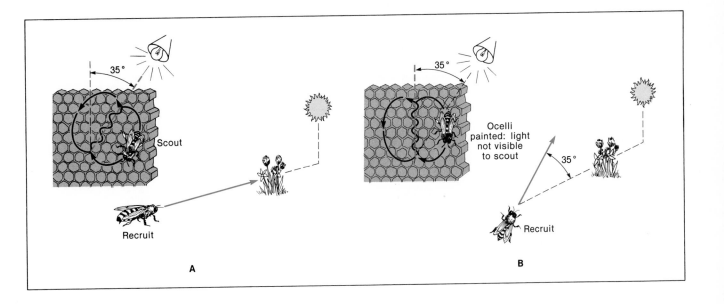

A   **B**

Figure 20.10
J. L. Gould's experiment demonstrating that the waggle dance does convey directional information. (A) A bright light inside the hive causes the scout to dance with reference to it rather than with reference to the vertical. If the food source is in the direction of the sun (angle = 0°), the scout will orient its dance toward the light, at 35° from vertical in this example. (B) Bees see the light with their simple eyes, the ocelli. If the ocelli of the scout is painted over so that it cannot see the light, it dances with reference to the vertical. If the ocelli of the recruits are not painted over, however, they interpret the dance with reference to the light and fly off in the wrong direction.

recruits that did not have their ocelli painted interpreted the dance with respect to the light. Consequently, most of these recruits arrived 35° to the left of the correct location (Figure 20.10).

BIOLOGICAL CLOCKS

Suppose a honey bee is allowed to observe a waggle dance that indicates that a source of nectar is in the direction of the Sun, but before she can begin foraging some curious zoologist imprisons her in a dark box for several hours. When the bee is released will it fly toward the Sun, which is no longer in the same direction as the food source? Or will the bee somehow know how far the Sun has moved during its captivity and fly in the correct direction? The remarkable answer is that the bee generally forages in the correct direction. Apparently bees can, in effect, tell the time of day and can compensate for the movement of the Sun. This time compensation suggests that bees have **endogenous clocks.** The only way to confuse a bee in this type of experiment is to stop its endogenous clock by cold or by anesthesia.

Circadian Rhythms. Numerous species in addition to bees have endogenous clocks. They are used not only for navigation but also to regulate feeding, body temperature, sleeping, and other activities. These activities often follow a daily rhythm that is synchronized with the rhythms of Earth—the cycles of day and night, heat and cold. If such rhythms have a periodicity of approximately one day, they are called circadian rhythms (Latin *circa* around, *dies* day). Many endogenous clocks maintain a circadian rhythm even without external cues, although the clocks tend to run a little fast or slow. In the spring of 1972 a French cave explorer volunteered to live alone in constant light and temperature in a Texas cave so that his endogenous rhythms could be studied. He went to sleep when he felt sleepy and ate when he felt hungry. Generally, he maintained a normal sequence of daily activities, but his self-imposed days usually lasted a little longer than 24 hours. By the time the experiment ended on August 10, the volunteer thought it was only mid-July.

Clock Works. In some birds and other vertebrates the **pineal gland** near the brainstem appears to be an important component of a circadian clock. The pineal gland (which is actually the vestige of a primitive eye) responds directly to light, which easily passes through the thin skulls of birds. Light suppresses the gland's production of the hormone melatonin. Consequently, there is normally a 24-hour rhythm of melatonin synthesis, with melatonin production ending at sunrise. Sue Binkley and others have found that the rhythm of the pineal gland's activity is due to the enzyme *N*-acetyltransferase, which catalyzes the synthesis of melatonin in the dark. Binkley discovered that a circadian rhythm of *N*-acetyltransferase activity persisted in chicks even if she kept them in the dark. Furthermore, even when the pineal gland was totally removed from the chick's brain and kept in the dark, its circadian rhythm of melatonin synthesis persisted for several days. It appears, therefore, that if the pineal gland in some species of birds is not *the* endogenous clock, it at least contains the "pendulum" for one endogenous clock. Melatonin may be the hands of that clock, telling other organs what time it is.

In mammals the pineal gland also plays a role in circadian rhythms, but it is not directly responsive to light and does not have the central role it has in birds. Instead, the rhythm of melatonin synthesis by the pineal glands of mammals appears to be driven by the **suprachiasmatic nuclei** (SCN). The SCN are two clusters of nerve cells lying just above the optic chiasm, where the optic nerves cross. The SCN stimulate melatonin synthesis by the pineal gland except when the retinas are exposed to bright light. Then a special neural pathway from the retinas inhibits the SCN, thereby inhibiting melatonin synthesis. Removing the two SCN from mammals (but not from birds) permanently abolishes some of their daily rhythms. The rhythms can be restored by implanting the SCN from another animal into their brains.

Lunar Cycles. Many animals display rhythms of activity that are cued by the position of the Moon. First there may be a rhythm with a periodicity of 24.8 hours (**circalunidian**) that is cued by the position of the Moon in its orbit around the Earth. In addition, many shore animals have a rhythm that depends on the 12.4-hour cycle of tides (**circatidal**), which depends on the position of the Moon. There may be an additional rhythm of 14.75 days, which equals the interval between quarter moons, when there is the greatest difference between sea level at high and low tide. The California grunion *Leuresthes tenuis* times its reproduction according to this cycle (Figure 20.11). At certain times of the year, shortly after high tide when the Moon is either full or new, these 18-cm, silver-blue fish come onto the beaches in huge numbers to fertilize and deposit their eggs in the sand. The next maximum high tide washes over the eggs, triggering hatching, uncovering the fry, and sweeping them out to sea.

Circannual Rhythms. Although annual rhythms are difficult to study because it takes so much longer to collect sufficient data, they have been demonstrated. Annual rhythms ensure that animals reproduce, migrate, grow long hair, or hibernate in the appropriate seasons. Many of these rhythms are triggered by environmental cues, such as day length and temperature. Seasonal variations in the length and brightness of daylight affect the daily rhythm of melatonin secretion, which may serve as a cue for annual reproductive cycles (Tamarkin et al. 1985). Although many annual cycles are apparently driven by external cues, there is also evidence for endogenous circannual clocks. (Or should they be called calendars?) The golden-mantled ground squirrel *Clitellus lateralis* enters hibernation during the autumn even with constant room temperature and day length.

Figure 20.11
Spawning grunion Leuresthes tenuis.

NAVIGATION

Biological Compasses: The Sun. How do sparrows and other migrating animals know which direction is north or south? What is their compass? Like the early attempts to identify *the* endogenous clock, early attempts to identify *the* compass now seem to have been oversimplified. In many birds there is not one compass, but several. When one compass cannot be used, the bird uses one or more of the others as backup. In homing pigeons *Columba livia*, which have been bred and trained for centuries to return to the home loft from vast distances, vision is important in navigation. It is not necessary, however, that they have a good view of the land over which they are flying. Pigeons fitted with frosted contact lenses home as well as pigeons with good vision. All that is required is that they be able to perceive the direction of the Sun. This has been demonstrated by keeping pigeons under artificial lighting, which was gradually shifted so that their circadian clocks became 6 hours out of phase with natural daylight. When the pigeons were released on a clear day in familiar territory, they attempted to home 90° in the wrong direction. Evidently they ignored the familiar landmarks and navigated by using the Sun's position, which was 90° from the position their endogenous clocks calculated. On a cloudy day the pigeons homed correctly, since they use familiar landmarks or other cues when they cannot see the Sun.

Stars. Birds that migrate at night have neither landmarks nor the Sun to guide them. Many, however, can use the stars as a compass. Apparently they learn which way is north by observing the stars as they rotate around the North Star. Afterward, they can orient toward the north simply by viewing some of these stars. If birds are raised in a planetarium in which the star field rotates around another star, they will later migrate in a direction that would be appropriate if that star were in the north. Birds apparently learn the star pattern during the weeks just prior to their first fall migration. If they can see the stars before or after that sensitive period, but not during it, they will never learn to migrate by means of the star pattern.

Magnetic Fields. If the night sky is cloudy, birds use other directional cues. After decades of skepticism, it is now clear that one of those cues is Earth's magnetic field. Evidence for the existence of magnetic compasses in birds is due largely to the late William T. Keeton of Cornell University, who found that attaching magnets near the heads of homing pigeons disrupted their navigation on cloudy days.

Migration of Monarch Butterflies. No migration is more remarkable for distance covered per unit of body weight than that of the monarch butterfly, *Danaus plexippus*. With four orange and black wings as flimsy as paper, these intrepid insects fly hundreds of kilometers. Fred Urquhart (1976, 1987) of the Scarborough Campus of the University of Toronto began studying monarch migration in 1937. He later enlisted the aid of his wife, Nora Urquhart, and a total of more than 4000 associate members of the Insect Migration Association. These associates, mostly amateurs of all ages and walks of life, have helped the Urquharts attach hundreds of thousands of small, numbered tags to the wings of living monarchs (Figure 20.12).

From tagging studies the Urquharts discovered that monarchs fly during the daytime at speeds of 5 to 20 km/hr and cover tens of kilometers per day. (The record for one day's flight is 425 km, between Waterford, PA and Franklin, WV

Figure 20.12
(A) A monarch butterfly Danaus plexippus just after receiving a numbered tag from a member of the Insect Migration Association. Recovery of such tagged monarchs permits tracking of the migration of the monarch. (B) Professor Fred Urquhart in the Sierra Madre Mountains of Mexico surrounded by over-wintering monarchs.

in 1983.) Many of the monarchs tagged in the western United States in late autumn are recovered among approximately 40 remaining wintering grounds along the California coast. (Several have been destroyed by developers in recent years.) Tagged monarchs from eastern North America turn up at one of 11 known sites in Mexico. At the wintering grounds monarchs roost in dense clusters. In spring the monarchs mate, then the females set out on their northeasterly journeys, depositing fertilized eggs on milkweed plants along the way. The eggs hatch into yellow, black, and white-striped caterpillars (see Figure 32.15) that feed on the milkweed plants, then metamorphose into adults that carry on the migration and reproduction. In late autumn the survivors reverse direction toward the wintering grounds. No single monarch makes the journey round-trip. As many as five generations can elapse between the spring and fall migrations.

In spite of the help of thousands of associates, it took the Urquharts four decades to locate the first Mexican wintering site. Finally, in 1975, one of the Mexican sites was discovered at an altitude of 3000 meters in the Sierra Madre Mountains, about 400 km northeast of Mexico City. The Urquharts were rewarded for their perseverance by finding themselves in a black and orange forest of over-wintering monarchs crowded so thickly in the trees that limbs broke under their weight (Figure 20.12B). After several decades these patient and intelligent scientists had located a spot that each of the monarchs had found in less than a month.

ETHOLOGY

So far we have considered the biological bases of behavior, gradually building in complexity to the impressive and poorly understood behavior of navigation. Later in this chapter we shall look at some categories of behavior, such as courtship and parental behavior, for which the basic biological processes are even more complex and poorly known. Getting a start at understanding such behaviors often begins with certain assumptions. Different zoologists begin with different assumptions and therefore different approaches to studying behavior.

One of the earliest approaches to the biological study of behavior was ethology, founded in the 1950s mainly by Konrad Lorenz and Niko Tinbergen. Like many scientific disciplines, ethology is difficult to define except by saying that it is what people who call themselves ethologists do. Such a definition is of little use, however, especially since the term ethologist is often applied to any biologist who studies behavior. Perhaps ethology can best be defined by contrasting it with psychological studies of animal behavior. Unlike most psychologists, ethologists insist on observing behaviors in the natural environment. They also are more interested in stereotyped behaviors that are characteristic of a species, and they assume that such behaviors are largely inherited and have evolved under the influence of natural selection. Ethologists call such stereotyped behaviors **fixed action patterns** (FAPs). They call the stimuli that evoke FAPs **sign stimuli.** When a sign stimulus is from a member of the same species and the FAP is a social behavior, the sign stimulus is called a **releaser.**

Courtship of the Three-Spined Stickleback. Some of the best-known studies in ethology have been on courtship, aggression, and territoriality, all of which are involved in the classic study by Tinbergen of a common freshwater fish called the three-spined stickleback (*Gasterosteus aculeatus*). Like many fishes, sticklebacks spend much of their lives in schools. During the breeding season, however, the male establishes a territory from which it excludes other males. Within its territory

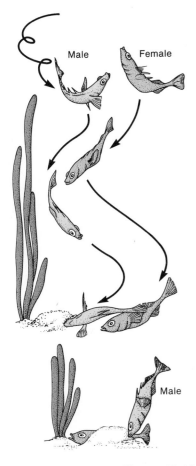

Figure 20.13
Courtship and mating in the three-spined stickleback Gasterosteus aculeatus. *The male zigzags toward a female whose belly is swollen with eggs, then he guides her to the nest. The male leads the female through the tunnel-shaped nest, but she is unable to exit because of her swollen belly. Below the male is shown prodding the female to lay her eggs. Once this occurs she is able to leave. The male then enters the nest and fertilizes the eggs.*

the male excavates a shallow depression in the sand, then builds a tunnel-shaped nest of algae glued together with a secretion from the kidneys. Immediately after it completes the nest, the male stickleback develops brilliant colors. The eyes turn bright blue, the back goes from dull brown to shiny blue, and the belly becomes red. If one brightly colored male strays into the territory of another, the resident male attempts to drive out the intruder by darting toward it with mouth open and spines erect on its back. If the trespasser does not immediately flee, the resident escalates the threat by making jerky movements while "standing on its head." The red belly is the main releaser of this territorial aggression. In fact, Tinbergen found that a male stickleback reacted more aggressively toward an artificial fish with a red underside than toward a real male stickleback without a red belly. Once Tinbergen even saw a male stickleback threaten a red mail truck passing outside a window near the aquarium!

While the males are preparing their nests and changing colors, the females develop a silvery gloss, their bellies swell with ripe eggs, and they begin to swim about in schools. If a school enters a male's territory, he swims toward the females just as he would toward an intruding male. Upon seeing the swollen bellies of the females, however, the male repeatedly turns away, then darts toward the school again. This is called a zigzag dance (Figure 20.13). A model of a fish with a swollen belly is an equally effective releaser of the zigzag dance by the male. The swollen, nonred belly is what determines whether the male zigzags amorously or darts aggressively. Even the zigzag frightens some of the females away, especially if their eggs are not fully ripe.

A female that is not intimidated turns toward the zigzagging male, which then goes straight to the nest. She follows, and the male leads her into the tunnellike nest, where her swollen belly entraps her. The male then prods the base of her tail to stimulate spawning. After the eggs have been deposited, the female easily escapes from the tunnel. Since she has lost her figure, the male chases her away. He then enters the nest and fertilizes the eggs. His job does not end there, however. As the fertilized eggs develop during the following seven or eight days, the male spends up to three-fourths of his time "fanning" them with his tail to meet their growing requirements for oxygen. The male's parental behavior continues even after the fry hatch, for he protects them for several weeks until both he and they are ready to join schools.

SOCIOBIOLOGY

Another productive approach to understanding animal behavior is sociobiology. Like ethologists, sociobiologists study behavior in the natural environment, and they assume that many behaviors are largely dependent on genetics and have evolved due to natural selection. As their title suggests, however, sociobiologists generally confine their interests to social behaviors, and they are less interested in the mechanisms of such behaviors than they are in the contributions of those behaviors to evolutionary fitness. Unfortunately, many people believe that sociobiology is nothing more than the naive assumption that all human behavior is genetically determined. This misconception is due largely to politically motivated attacks on E. O. Wilson and his 1975 book *Sociobiology: The New Synthesis*. In the first 546 pages of the book Wilson surveyed the biology of social behavior in nonhumans, and in the final 29 pages he attempted to extrapolate to explanations of how such human behaviors as homosexuality, entrepreneurship, and religion could have evolved. Attacks on those 29 pages were due largely to biologists in the United States who followed the Marxist doctrine that human social behavior is determined by the economic system, not by biology. Some others attacked sociobiology out of fear that establishing a biological basis for human behavior would encourage or excuse sexism, racism, and genocide. (See Harris 1981, Chapter 9, and Lumsden and Wilson 1983, pp. 36–45 for discussions of the political motivations of some of sociobiology's critics.) Of course, as with any new and interesting area, there were also criticisms by biologists motivated purely by science.

Altruism in Honey Bees. The real contributions of sociobiology have often been overlooked. Among these is one possible explanation for altruism, which has plagued evolutionists since Darwin. Darwin's theory of natural selection stressed competition and the struggle for survival, yet he and everyone else knew that many animals cooperate with other members of their species. For example, it is well known that in many species parents will sacrifice their lives to ensure the survival of their offspring. This kind of altruism presents no difficulties for Darwinism, since it directly favors an increase in the number of individuals bearing the genes that encourage parental altruism. Altruism toward those that are not offspring, however, was difficult for Darwinians to explain. Why, for example, do lionesses nurse the cubs of other females, thereby perhaps denying nutrients to themselves and their own offspring? And why do worker honey bees sacrifice their labor and even their lives to defend the hive, even though they usually do not reproduce?

In *The Origin of Species*, in the chapter entitled "Difficulties of the Theory," Darwin attempted to explain such altruism by arguing that it was for the good of the species. Today, however, evolutionists are skeptical of "for the good of the species" reasoning (see p. 381). The problem is that in the millions of years that there have been lionesses, there should have been at least one mutation that would lead a female to "cheat" the system and nurse only her own cubs. Such genes for "cheating" would have been favored during evolution, because the offspring of cheaters would have been more likely to survive. The problem with altruism in worker honey bees is that there is no way that a gene for altruism could favor the survival of workers' offspring, since they have no offspring.

The key to understanding altruism was first glimpsed by J. B. S. Haldane. So the story goes, Haldane was once asked if he would lay down his life for a brother. After a moment's calculation he replied that it would take two brothers or eight cousins. What he meant was that as far as evolution is concerned, it wouldn't matter if he transmitted his own genes to the next generation, or if an equal number of identical genes got transmitted by siblings (brothers or sisters) or by cousins. Statistically, the same number of genes would be transmitted to the next generation if he made it possible for two siblings or eight cousins to reproduce in his place. (This assumes that each individual would have the same average number of children.) All that is required is that the number of relatives reproducing in one's place equal the reciprocal of their relatedness. (**Relatedness** is the probability that two individuals have inherited the same gene from the same ancestor.)

W. D. Hamilton gave mathematical rigor to this insight in 1964. He argued that an organism's fitness—its potential for transmitting genes into the next generation—includes not merely its individual fitness but also a kinship component. The kinship component equals the sum for all genetic kin of the individual's contribution to their fitness, multiplied by their relatedness. This kinship component added to individual fitness equals the **inclusive fitness** of an organism. Therefore, even though each sterile worker honey bee has an individual fitness of zero, its inclusive fitness is quite large, because it includes its contribution to the fitness of the queen. By ensuring the fitness of the queen, honey bee workers improve the chances that the queen will transmit genes like theirs to the next generation. Of course honey bee workers do not mathematically calculate their inclusive fitness and the advantages of altruism, any more than most humans calculate how many siblings or cousins are worth dying for. It is simply that the genes that lead the workers to devote their lives to the reproductive success of the queen were selected because they had a greater chance of being reproduced.

An argument like that for honey bees might also account for the lioness' nursing of cubs of other lionesses. Generally, the females in a pride of lions remain together, so lionesses are likely to be sisters, mothers and daughters, and aunts and nieces. Therefore the cubs that the lioness nurses are likely to be related to her, even if they are not her own offspring. Aiding their survival therefore fosters the success of her own genes. Similar sociobiological explanations may apply to other types of altruism where groups of related individuals remain together.

Kin Selection. Even in groups of animals that are not all related, altruism can also occur if individuals can recognize and aid kin or discriminate against nonkin. One of the contributions of sociobiology is that it has prompted zoologists to look for evidence of **kin recognition** and kin selection. As a result, many surprising examples of kin recognition and selection have been discovered. Female Belding's ground squirrels *Spermophilus beldingi* of the American Northwest, for example, evidently distinguish between their full sisters and half sisters, for they are more likely to fight with the latter (Figure 20.14).

Reciprocity. This line of argument helps explain some of the controversy over human sociobiology. Since people of different races are easily recognized as being unrelated, it could be argued that racism has a genetic basis that evolved as a means of favoring one's own kin by discriminating against unrelated competitors. If one insists on deriving moral lessons from nature, however, then there are also species that show that aiding nonkin can be beneficial to all. This kind of altruism, in which unrelated individuals aid each other, is called reciprocity. An impressive example of reciprocity occurs in the vampire bat *Desmodus rotundus*, which feeds on the blood of mammals (including humans) in Central and South America. A vampire bat that has recently fed will regurgitate blood to an unrelated vampire bat that would otherwise starve to death. Reciprocity can evolve when the benefit to the recipient is greater than the cost to the donor, and when every individual has a good chance of being either recipient or donor.

BEHAVIORAL ECOLOGY

Mating in Dung Flies. Behavioral ecology, one of the most recent approaches to animal behavior, shares with ethology and sociobiology the insistence on studying behavior in nature and the assumption that behavior is largely inherited and has evolved due to natural selection. Unlike ethologists and sociobiologists, however, behavioral ecologists are most interested in the ecological costs and benefits associated with particular behaviors. Behavioral ecologists often begin with the hypothesis that behaviors are **optimal** for efficiency of energy use. A study by Geoff A. Parker on reproduction in yellow dung flies (*Scatophaga stercoraria*) illustrates the approach. Female dung flies visit fresh cow pats looking for a place to lay eggs. Males also congregate there, seeking females with which to mate. It takes about 100 minutes for a male to fertilize all of one female's eggs, so one might think that the optimal duration of copulation would be 100 minutes. This would indeed be the case if each male copulated with just one female. After copulating

Figure 20.14
Belding's ground squirrel
Spermophilus belding.

with one female and guarding her while she lays her eggs, however, the male flies off to find another female. Parker found that guarding and searching took an average of 2.5 hours. The male is therefore faced with a dilemma. Should he copulate for the full 100 minutes to fertilize all the eggs of each female, or should he stop copulating sooner so that he can get an early start on finding the next mate? Somewhere between 0 minutes and 100 minutes there is an optimal duration of copulation that will result in the male's leaving the maximum number of offspring. Behavioral ecology predicts that the actual duration of copulation will be near that optimum. By actually measuring the proportion of eggs fertilized for various durations of copulation, Parker determined that the optimal duration was 41 minutes. The actual average duration of copulation was 36 minutes.

The 5-minute difference between the theoretical optimum and the actual duration of copulation is probably within the normal range of variation that one expects in biology. There are many possible explanations for why the duration was not exactly optimal. The dung flies Parker studied could have evolved under circumstances in which the optimum was formerly 36 minutes. On the other hand, perhaps some other theory would provide a better explanation. As with all theories, the value of assuming optimality lies in its ability to help biologists generate testable hypotheses, not in proposing final answers.

LIVING IN GROUPS

Having looked at the biological elements of behaviors and different approaches to studying them, we can now consider some categories of behaviors that are perhaps the most interesting, but the least understood. One of these categories is group living. The most spectacular example of group living were the vast herds of buffalo (*Bison bison*) that extended across the American prairies as far as the eye could see. A less obvious example were the packs of wolves (*Canis lupus*) that lurked along the fringes of those herds. The advantages of group living for both wolves and bison are immediately evident. No single wolf would dare attack a bison, and no single bison could survive attack by a pack of wolves. A pack of wolves could, however, take on a sick bison that had strayed from the herd. The naturalist-artist George Catlin observed that "when the herds are travelling, it often happens that an aged or wounded [bison], lingers at a distance behind, and when fairly out of sight of the herd, is set upon by these voracious hunters [wolves], which often gather to the number of fifty or more, and are sure at last to torture him to death, and use him up at a meal."

Zoologists are a long way from being able to explain in fundamental neural or genetic terms why bison and wolves live in groups, but at least we can begin to understand the advantages that may have led to the evolution of group living. For wolves and other predators, the advantage is that there are more individuals looking for opportune prey. For bison and other prey species, living in groups provides a defense against attack, and the large number of eyes make it more likely that at least one individual will spot a predator and warn the others. Another advantage of group living in prey species is the **saturation effect.** Even a large pack of wolves would be unable to eat more than one buffalo at a time, so the rest of the herd would escape.

TERRITORIALITY

Many animals share a **home range** with members of their own species. In winter, for example, many birds flock together in a home range. Following the spring migration to their new habitats, however, most male birds undergo a Jekyll-to-Hyde transformation and establish a **territory** that they vigorously defend against other males of their species. The essential difference between a home range and a territory is that the territory is, by definition, defended against encroachment by others of the same species. Male birds advertise possession of their territories by songs and visual displays. An intruding male's song or plumage is the releaser for **territorial aggression** by the resident. Resident males will often attack a tape recorder playing another male's song, or a tuft of feathers the same color as the male's breeding plumage. The only way for a male bird to lurk about within the territory of another is to keep quiet and out of sight. Such "lurkers" seldom breed, however, for the songs and display are also required to attract females.

Like birds, many fishes establish breeding territories that are occupied by only one male and his mate (Figure 20.15). In other species a territory may be occupied by two or more males and their mates. In such group territories the defense is against encroachment by other groups. Wolves are the classic illustration of a species that defends a group territory. The average wolf pack is an extended family of from five to eight individuals with a territory of a few hundred square kilometers. An **alpha male** and an **alpha female** lead the pack. They define the pack's territory by releasing a pheromone during a characteristic **raised-leg urination** about every 450 meters as they patrol its perimeter. To wolves from neighboring territories these olfactory boundary markers are, in one sense at least, nothing to sniff at. Packs of wolves have been seen abandoning a deer chase rather than cross into another pack's territory.

AGGRESSION

Aggression often plays a major role in maintaining order among members sharing a territory. This kind of aggression should not be confused with defense or predation, which are sometimes referred to as aggression (Figure 20.16). Aggression

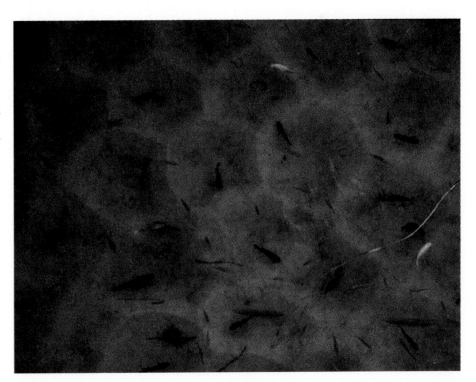

Figure 20.15
Territories of the mouthbrooder fish Tilapia mossambica. *Each large, dark fish is a male that has scooped out a depression in which to mate and brood the eggs. Note the hexagonal shape of most of the territories; hexagons naturally occur when each organism tries to maximize its own space on a flat surface.*

or threats of aggression are often used in establishing a **dominance hierarchy,** often called a pecking order. The dominance hierarchy determines which individuals have priority in mating, feeding, or other activities. Generally, the largest and fittest individuals dominate a hierarchy. In wolf packs the alpha male and the alpha female are dominant. The alpha male dominates a hierarchy of subordinate males, and the alpha female dominates the subordinate females. Probably only the alpha pair breeds within most packs. The alpha male directs the hunting, brings food to the female and the pups, and helps tend the pups. The alpha female determines the location of the den where she will bear her pups and sometimes directs the hunting.

Subordinate wolves sometimes challenge the alpha male or female, but unless the alpha is old or wounded the insurrection is usually put down without bloodshed. Visual communication is the most important means by which alpha wolves demonstrate and maintain their dominance (Figure 20.6). The above description should not be taken as the rule for all wolf packs, and certainly not for all species. Wolves are extremely adaptable animals, and the pattern of dominance can change under different circumstances.

In some species subordinate animals suffer considerably because of continual harassment from dominant individuals. Subordinate male baboons (*Papio*), for example, are often plagued by parasites and infected wounds because the glucocorticoid hormones secreted in response to social stress inhibit immune responses. In other species subordinate animals appear to suffer little stress as long as each individual knows its place in the hierarchy. When two roosters (*Gallus gallus*) are first introduced to each other, both their heart rates increase from a resting level of about 200 beats per minute to approximately 350 b.p.m. Within a minute, however, one of the roosters establishes its dominance by crowing or threatening the other, and the heart rates of both roosters quickly decline to resting levels (Harris and Siegel 1967). Evidently the stress to roosters of not knowing their place on the pecking order is greater than the stress of being subordinate.

COURTSHIP

Functions of Courtship.   Not every species of animal courts. Sponges, for example, simply emit sperm in vast clouds and let the currents carry them to eggs of the same species. Other animals, such as some insects and snakes, mate with little or no courtship. For many species, however, courtship is essential. There are many obvious disadvantages to courting, but also many benefits:

1. As noted before, in predaceous animals courtship keeps one mate from attacking the other.
2. Courtship ensures that an animal will not waste its gametes on a member of a different species. Probably only females of the same species would be seduced by the zigzag dance of a male three-spined stickleback, for example.
3. Courtship also ensures that the partner-to-be is physiologically capable of reproduction. For example, female sticklebacks whose eggs are not ripe are likely to flee a courting male.
4. In many species courtship appears to synchronize reproductive activities. For example, a female whooping crane *Grus americana* does not ovulate until a male has danced about her, jumping in the air and flapping his wings. (Whooping cranes are now so scarce that human volunteers perform the dance to get the females to produce the eggs needed to save this endangered

A

B

C

Figure 20.16
Three types of behavior in the rattlesnake Crotalus. *All these behaviors are sometimes called aggression, but they are quite different as far as the rattlesnake is concerned. (A) During dominance displays, rattlesnakes neck-wrestle until one snake is pinned by the other. The defeated snake then assumes a submissive posture and slinks away unharmed. They do not use the poison, which is as toxic to rattlesnakes as to other animals. Nor do they use the rattlers; snakes are deaf to airborne sounds. (B) In defense the rattlesnake uses its rattlers as a warning and bites if necessary. (C) In predation the snake silently stalks its prey, then seizes and swallows it whole.*

species. The eggs are then artificially inseminated with whooping crane semen, however.)

5. Some aspects of courtship appear to prolong copulation, helping to ensure that as many eggs as possible are fertilized. For example, the empty silk balloon that male empidid flies present to females may allow the male more time to copulate.

6. Courtship is also a mechanism of natural selection, which prevents individuals with undesirable traits from passing them on. In some empidid flies, for example, males that are genetically incapable of capturing prey are not likely to mate. The restricting of courtship to the dominant members of a hierarchy or to those with territory also helps ensure that individuals too weak to maintain dominance and territory will not pass genes into the next generation.

Leks. In certain species of insects, amphibians, birds, and mammals, males gather and display for females. These gathering places are called leks (Figure 20.17). Because mass courtship at a lek is so intense and easily observed, leks have been studied extensively. Nevertheless, there remains some disagreement over the function of a lek. It is usually assumed that the males select the site of the lek and that the females are drawn to the lek, where they exercise their preference in males. It is also possible, however, that the site of the lek is one that is normally frequented by females. This is called the "hotspot" theory. Bruce Beehler and Mercedes Foster (1988) of the Smithsonian Institution have suggested a third interpretation of leks, which they call the "hotshot" theory. They propose that before arriving at a lek the males compete among themselves, and that the winners (the hotshots) do most of the mating regardless of female preference.

Mating Systems. The successful males in a lek may copulate with many females but form no bond with any of them. Such **promiscuity** also occurs among many other species, such as insects. At first glance promiscuity appears to be the way to produce the most offspring, especially for males. In fact, however, the offspring in a promiscuous species may find it hard to survive, because their own male parents are competing with them for resources. Often more offspring can be produced if males and females remain mates through courtship, and both contribute to the offspring as parents. In many species one male and one female remain together as mates at least for one breeding period. This arrangement, or mating system, is called **monogamy** (Greek *mono-* single, *gamos* marriage).

Although monogamy seems most natural to people brought up in Western cultures, **polygamy** is the norm in many animals, and perhaps in most human cultures before they became influenced by the West. There are two kinds of polygamy. **Polygyny** (pronounced puh-LIJ-uh-nee; Greek *poly* many, *gyne* lady) is a mating system in which one male has more than one female mate. Polygyny is most common in environments in which resources are distributed in patches. Females that depend on the resources of the patch to raise their young have no choice but to mate with the male that controls the patch, even if they have to share the male with other females. In human cultures, polygyny also appears to be most frequent in patchy environments, such as deserts.

In **polyandry** (Greek *andro-* male) one female has more than one male mate. Polyandry generally occurs when the female is larger than the male or for some other reason controls resources on which the male depends. Honey bees, for example, are polyandric. The drones, which lack stingers and cannot forage, are

Figure 20.17
A male sage grouse Centrocercus urophasianus *struts at a Wyoming lek, using its showy display of plumage and the loud booming sound made by forcing air out of its esophageal sac.*

totally dependent on the hive, which in turn depends on the queen (see pp. 699–701). Polyandry is also common among the ploverlike birds (order Charadriiformes), in which the female is larger than the male. In phalaropes (genus *Phalaropus*) of the subarctic, the females have showier plumage and establish leks that attract males. The males perform many of the parental duties. In the related American jacana *Jacana spinosa*, which lives on lily-covered lakes in Central America, the female dominates a number of smaller males in her territory (Figure 20.18). The males are largely responsible for incubating her many clutches of eggs.

Reproductive Investment. It is evident that males and females of many species make a considerable investment of time and energy in securing a suitable mate through courtship. The female can be certain that her courtship efforts will not be wasted, since her eggs are certainly bearing her genes. Males, on the other hand, run the risk that their reproductive investment will be diluted or nullified by other males. In many species there are mechanisms that reduce this risk. Internal fertilization is one such mechanism that avoids fertilization by extraneous sperm. Even with internal fertilization, however, a male may find his reproductive efforts undone by a later male. Some male insects avoid cuckoldry by releasing an antiaphrodisiac pheromone onto just-mated females. Male wolves and other canids often cannot disengage their penes for as long as a half hour after ejaculation. This ensures that the male's sperm will have a chance to fertilize ova before another male comes along. In snakes the male secretes a **copulatory plug** that mechanically prevents a later mating of the female. Male spiny-headed worms (phylum Acanthocephala) cement the females' genitals shut.

In spite of such stratagems zoologists have found that in virtually every species studied the offspring in a brood are fathered by more than one male. This is true even for monogamous species. In the black-capped chickadee *Parus atricapillus*, for example, females will sneak away from the nest to mate with males that are higher ranked than their mates. Using the technique of DNA fingerprinting (see p. 100), zoologists have found that up to 70% of the offspring in birds' nests are not fathered by the male that helps raise them. The only "faithful" couples found so far occur in the California mouse *Peromyscus californicus*.

PARENTAL BEHAVIOR

Another way of protecting a reproductive investment is to help ensure the survival of eggs and offspring through parental behavior. Generally, parental behavior demands an enormous investment of time and energy. Sociobiologists have created a considerable body of **parental investment theory** to account for the circumstances in which parental behavior is a good investment. Parental behavior is clearly not advantageous when an animal cannot be sure that at least some of the eggs or offspring it is tending are bearing its own genes. It is not surprising, therefore, that parental behavior is uncommon in species with external fertilization. The exceptions occur mainly when the parent keeps track of the eggs from the moment of fertilization, as is the case with the male three-spined stickleback. The male banded jawfish *Opisthognathus macrognathus* can also be fairly sure of the eggs he has fertilized, for he quickly takes them into his mouth, where they develop. In other **mouthbrooder** fishes, such as those in the genus *Tilapia*, it is the female that incubates the eggs in her mouth. In the African mouthbrooder *Haplochromis burtoni* and others, fertilization actually occurs within the mouth (Figure 20.19).

Figure 20.18
American jacana Jacana spinosa.

Figure 20.19
When the female African mouthbrooder Haplochromis burtoni *(left) tries to take into her mouth the egglike spots on the male's anal fin, she will also take in some sperm that will fertilize the eggs that are already in her mouth.*

Figure 20.20
A reed warbler Acrocephalus scirpaceus *with a cuckoo* Cuculus carorus *nestling. It is easy for humans to see that something is amiss when the young dwarfs the parent, but the gaping mouth in the nest is an irresistible releaser to this surrogate parent. The absence of young warblers in the nest suggests that the cuckoo, with its larger mouth, outcompeted the young warblers for parental attention.*

Figure 20.21
Like many birds, goslings are imprinted to the first active object they see or hear during a sensitive period shortly after hatching. Usually that imprinting stimulus is the mother, but eminent ethologists will also do. This particular imprinting stimulus is Konrad Lorenz, who was one of the first ethologists to study imprinting.

Cooperative Breeding. Some species defy the logic of the preceding paragraph by including many individuals that do not have their own offspring but help raise the offspring of others. Such cooperative breeding is best known in more than 220 species of birds, such as the acorn woodpecker *Melanerpes formicivorus* of Central America and the American Southwest, the green woodhoopoe *Phoeniculus purpureus* of Africa, and the Florida scrub jay *Aphelocoma coerulescens*. These birds often live in flocks within territories that can support only one breeding pair each, either because food or nest sites are scarce, or because of competition from neighboring flocks. Presumably, cooperative breeding evolved because it benefited the nonreproducing helpers. But how? This may be an example of either kin selection or reciprocity.

Parental Behavior as Fixed Action Pattern. The extent to which parents will sacrifice for their offspring is familiar to anyone who has watched a robin tirelessly bringing worms to a nest full of gaping mouths. Humans are apt to interpret such selflessness as a manifestation of love, but many ethological studies suggest that in birds the feeding of young is a fixed action pattern. In many species of birds, any gaping mouth will evoke feeding by parents, whether the mouth belongs to their own young or not. Cuckoos and other **brood parasites** exploit this FAP to their own advantage (Figure 20.20). A further sign that parental behavior in birds is automatic can be seen when the chicks of gulls stray outside the nest. The parents apparently do not recognize a chick outside the nest, for they stand idly by while they starve to death or are eaten by predators.

Imprinting. Young birds and some young mammals react to parents with a social bond called imprinting. Imprinting is often manifest as a strong tendency in the young to follow and be with the parent. Imprinting also looks to humans like love, but it is also an automatic response. Young birds can become imprinted to any object they see (or hear, or smell) during a sensitive period shortly after hatching. The imprinting stimulus is usually the mother, but it can be a milk bottle, an electric train, or some other object (Figure 20.21). Birds imprinted to these stimuli grow up preferring them as companions and mates.

Figure 20.22
An infant rhesus monkey Macaca mulatta *clings to a surrogate mother in a study pioneered by Harry F. Harlow. If deprived of even this minimal substitute for parental behavior, monkeys fail to develop appropriate social behaviors.*

Love. Humans and other primates do not become imprinted, but strong social bonds do form between parents and offspring. For want of a more precise term, we can call this bond in primates love. Love in primates appears to be as essential a biological adaptation as mother's milk. Even if infant monkeys are provided with all their physiological needs, they develop pitifully abnormal behaviors unless they are mothered (Figure 20.22).

Reductionism

Many people will object to the direction this chapter is leading us—to the idea that even parental love can be reduced to an automatic behavior or physiological state, and presumably to the actions of genes and nerve cells. Critics of such ideas often brand them with the label reductionism, ignoring the fact that it is the role of all science to try to reduce complex and poorly understood phenomena to more elementary and better-understood phenomena. Critics of reductionism apparently hope that behavior—especially human behavior—will remain hidden from the cold light of science.

From a scientific perspective a more valid criticism of reductionism may be that it is not the most efficient approach to understanding behavior. An example from physics will illustrate why this may be. Theoretically, the temperature and pressure of a volume of gas could be calculated from the velocities of each molecule in the gas, using the laws of momentum and energy. In practice, however, such an approach is hopelessly difficult if the number of molecules is greater than two. The principles of thermodynamics were, in fact, derived from experience with large volumes, and no one has ever reduced them to the laws of physics that apply to interactions of individual molecules. It seems just as unlikely that anyone will ever reduce a complex animal behavior to the actions of genes or nerve cells. The complete reduction of behavior in a human seems even more remote. Any brain capable of understanding itself will probably be so complicated that it can never be understood. Nevertheless, while reductionism may never totally succeed, it has been the most fruitful approach so far in attempts to understand behavior.

SUMMARY

Zoologists study behavior from several approaches, all of which start with the premise that behavior is a product of evolution by natural selection. The oldest zoological approach is ethology, which emphasizes animal behavior in a natural context, as the result of neural mechanisms such as releasers and fixed action patterns. Sociobiology deals with social interactions such as altruism and the benefits that may have led to their having evolved. Behavioral ecologists start with the hypothesis that behavior is optimal and attempt to test that hypothesis by measuring the time and energy expenditures involved.

Although behaviors are generally complex, many can be understood in terms of more elementary processes, such as reflexes. A behavior can be innate, or learned, or a combination of the two. The two major kinds of learning are classical and operant conditioning. Innate behaviors are genetically based, as has been demonstrated in studies of nest building by lovebirds and courtship in *Drosophila*.

Living in groups can be advantageous to both predators and prey. It requires that individuals within the group be able to communicate through various channels. These channels may be auditory, as in bird songs, visual, as with fireflies, chemical, as with pheromones, or tactile, as in the waggle dance of the honey bee.

Many animals can determine the time of day or the season by means of endogenous clocks. These clocks regulate activities such as sleeping and waking during the day, and reproduction at different times of the year. They also influence the time and direction of migration. Many migrating animals can use the Sun or stars for navigation and can compensate for their movement by means of their endogenous clocks.

Many individual animals and some groups defend a territory. Within a group territory order is often maintained by a dominance hierarchy imposed by aggression. In some species aggression is closely related to courtship. Courtship ensures that a potential mate is of the appropriate species and sex, is fertile, and is not likely to turn predatory. It may also synchronize the physiological processes of mating and can serve as a means of sexual selection. Various animals have different mating systems, such as promiscuity, polygamy, and monogamy. Monogamy is often associated with intense parental behavior by both parents.

KEY TERMS

reflex
taxis
kinesis
classical conditioning
operant conditioning
pheromone
waggle dance
endogenous clock
circadian rhythm
ethology

fixed action pattern
sign stimulus
releaser
sociobiology
altruism
inclusive fitness
kin selection
reciprocity
behavioral ecology
home range

territory
aggression
dominance hierarchy
lek
mating system
monogamy
polygamy
polygyny
polyandry
imprinting

SELF-TEST

1. Name three people who have contributed to the scientific study of behavior, and briefly describe their contribution.

2. Describe a reflex and give an example of one.

3. Explain how classical conditioning differs from operant conditioning. Give an example from your own experience of each type of learning.

4. Describe one kind of experimental evidence that some behaviors can be genetically inherited. Exactly what do we mean when we say that a behavior is genetically inherited?

5. What is the evidence that many animals have endogenous clocks?

6. Birds can use several alternative methods of determining direction during migration. Describe two of those methods. What kind of evidence suggests that those methods are available?

7. Compare the following approaches to the scientific study of animal behavior: ethology, sociobiology, and behavioral ecology. What premises do these three approaches share? How do they differ? Give an example of a study characteristic of each approach.

8. What is a fixed action pattern? What evidence is there that fixed action patterns have an inherited, neural basis?

9. What are the advantages to predators and to prey of living in groups?

10. Describe a specific example of the role of communication in each of the following: predation, defense, courtship, parental behavior.

11. The term "aggression" is sometimes applied to several distinct types of behavior: territorial aggression, dominance, predation, and defense. Explain the differences among these, and give an example of each.

12. Explain three advantages of courtship. Support each advantage by giving an example.

13. Describe the circumstances under which parental behavior is a sound investment. What type of return do the parents get for their investment?

READINGS

RECOMMENDED READINGS

Alkon, D. L. 1983. Learning in a marine snail. *Sci. Am.* 249(1):70–84 (July).

Batra, S. W. T. 1984. Solitary bees. *Sci. Am.* 250(2):120–127 (Feb).

Bekoff, M. 1984. Social play behavior. *BioScience* 34:228–233.

Bekoff, M. and M. C. Wells. 1980. The social ecology of coyotes. *Sci. Am.* 242(4):130–148 (Apr).

Binkley, S. 1979. A timekeeping enzyme in the pineal gland. *Sci. Am.* 240(4):66–71 (Apr).

Blaustein, A. R. and R. K. O'Hara. 1986. Kin recognition in tadpoles. *Sci. Am.* 254(1):108–116 (Jan).

Bloom, F. E., A. Lazerson, and L. Hofstadter. 1985. *Brain, Mind, and Behavior.* New York: W. H. Freeman. (*See Chapter 5 on biological rhythms in humans.*)

Bonner, J. T. 1983. Chemical signals of social amoebae. *Sci. Am.* 248(4):114–120 (Apr).

Borgia, G. 1986. Sexual selection in bowerbirds. *Sci. Am.* 254(6):92–100 (June).

Clutton-Brock, T. H. 1985. Reproductive success in red deer. *Sci. Am.* 252(2):86–92 (Feb).

Crews, D. and W. R. Garstka. 1982. The ecological physiology of a garter snake. *Sci. Am.* 247(5):158–168 (Nov).

Dilger, W. C. 1962. The behavior of lovebirds. *Sci. Am.* 206(1):88–98 (Jan).

Dyer, F. C. and J. L. Gould. 1983. Honey bee navigation. *Am. Sci.* 71:587–597.

Emlen, S. T. 1975. The stellar-orientation system of a migratory bird. *Sci. Am.* 233(2):102–111 (Aug).

Fenton, M. B. and J. H. Fullard. 1981. Moth hearing and the feeding strategies of bats. *Am. Sci.* 69:266–275.

Forster, L. 1982. Vision and prey-catching strategies in jumping spiders. *Am. Sci.* 70:165–175.

Gould, J. L. and C. G. Gould. 1989. *Sexual Selection.* New York: Scientific American Library.

Gould, J. L. and P. Marler. 1987. Learning by instinct. *Sci. Am.* 256(1):74–85 (Jan).

Gwinner, E. 1986. Internal rhythms in bird migration. *Sci. Am.* 245(4):84–92 (Apr).

Halliday, T. 1980. *Sexual Strategy.* Chicago: University of Chicago Press.

Halliday, T. R. and P. J. B. Slater (Eds.). 1983. *Animal Behaviour,* 3 volumes. Salt Lake City, UT: W. H. Freeman. (*A beautifully illustrated presentation of reproductive behaviors.*)

Hölldobler, B. and E. O. Wilson. 1983. The evolution of communal nest-weaving in ants. *Am. Sci.* 71:490–499.

Holmes, W. G. and P. W. Sherman. 1983. Kin recognition in animals. *Am. Sci.* 71:46–55.

Horner, J. R. 1984. The nesting behavior of dinosaurs. *Sci. Am.* 250(4):130–137 (Apr).

Johnson, C. H. and J. W. Hastings. 1986. The elusive mechanism of the circadian clock. *Am. Sci.* 74:29–36.

Ligon, J. D. and S. H. Ligon. 1982. The cooperative breeding behavior of the green woodhoopoe. *Sci. Am.* 247(1):126–134 (July).

Lloyd, J. E. 1981. Mimicry in the sexual signals of fireflies. *Sci. Am.* 245(1):138–145 (July).

Partridge, B. L. 1982. The structure and function of fish schools. *Sci. Am.* 246(6):114–123 (June).

Prestwich, G. D. 1983. The chemical defenses of termites. *Sci. Am.* 249(2):78–87 (Aug).

Ryker, L. C. 1984. Acoustic and chemical signals in the life cycle of a beetle. *Sci. Am.* 250(6):112–123 (June).

Scheller, R. H. and R. Axel. 1984. How genes control an innate behavior. *Sci. Am.* 250(3):54–62 (Mar).

Seeley, T. D. 1983. The ecology of temperate and tropical honeybee societies. *Am. Sci.* 71:264–272.

Shettleworth, S. J. 1983. Memory in food-hoarding birds. *Sci. Am.* 248(3):102–110 (Mar).

Stacey, P. B. and W. D. Koenig. 1984. Cooperative breeding in the acorn woodpecker. *Sci. Am.* 251(2):114–121 (Aug).

Sulzman, F. M. 1983. Primate circadian rhythms. *BioScience* 33:445–450.

Sweeney, B. M. and others. 1983. Biological clocks. *BioScience* 33:424–450.

Thornhill, R. 1980. Sexual selection in the black-tipped hangingfly. *Sci. Am.* 242(6):162–172 (June).

Topoff, H. 1990. Slave-making ants. *Am. Sci.* 78:520–528.

Turek, F. W. 1983. Neurobiology of circadian rhythms in mammals. *BioScience* 33:439–444.

Urquhart, F. A. 1976. Found at last: the monarch's winter home. *Natl. Geogr.* 150:160–173 (Aug).

Waterman, T. H. 1989. *Animal Navigation.* New York: Scientific American Books.

West, M. J., A. P. King, and D. H. Eastzer. 1981. The cowbird: reflections on development from an unlikely source. *Am. Sci.* 69:56–66. (*Genetics and learning interact in the recognition by female cowbirds of the male's songs.*)

Wilkinson, G. S. 1990. Food sharing in vampire bats. *Sci. Am.* 262(2):76–82 (Feb).

Winfree, A. T. 1987. *The Timing of Biological Clocks.* New York: Scientific American Library.

ADDITIONAL REFERENCES

Beehler, B. M. and M. S. Foster. 1988. Hotshots, hotspots, and female preference in the organization of lek mating systems. *Am. Nat.* 131:203–219.

BioScience 33:545–582 (1983). (*The October issue is devoted to behavioral endocrinology.*)

Bouchard, T. J. Jr. et al. 1990. Sources of human psychological differences: The Minnesota study of twins reared apart. *Science* 250:223–228.

Dethier, V. G. 1976. *The Hungry Fly: A Physiological Study of the Behavior Associated with Feeding.* Cambridge, MA: Harvard University Press.

Hailman, J. P. 1985. Ethology, zoosemiotic and sociobiology. *Am. Zool.* 25:695–705.

Harris, C. L. 1981. *Evolution: Genesis and Revelations.* Albany, NY: SUNY Press. (*See Chapter 9 on sociobiology.*)

Harris, C. L. and P. B. Siegel. 1967. An implantable telemeter for determining body temperature and heart rate. *J. Appl. Physiol.* 22:846–849.

Holden, C. 1987. The genetics of personality. *Science* 237:598–601.

Holmes, W. G. and P. W. Sherman. 1983. Kin recognition in animals. *Am. Sci.* 71:46–55.

Kessel, E. L. 1955. Mating activities of balloon flies. *Systematic Zool.* 4:97–104.

Larson, J. R. and D. M. Meyer. 1984. Animal behavior experiments using arthropods. In: C. L. Harris (Ed.), *Tested Studies for Laboratory Teaching: Proceedings of the Third Workshop/Conference of the Association for Biology Laboratory Education (ABLE).* Dubuque, IA: Kendall/Hunt, Chapter 7.

Lumsden, C. J. and E. O. Wilson. 1983. *Promethean Fire: Reflections on the Origin of Mind.* Cambridge, MA: Harvard University Press.

McClintock, M. K. 1971. Menstrual synchrony and suppression. *Nature* 229:244–245.

Quinn, W. G. and R. J. Greenspan. 1984. Learning and courtship in *Drosophila:* two stories with mutants. *Annu. Rev. Neurosci.* 7:67–93.

Scheller, R. H. et al. 1984. Neuropeptides: mediators of behavior in *Aplysia*. *Science* 225:1300–1308.

Tamarkin, L., C. J. Baird, and O. F. X. Almeida. 1985. Melatonin: a coordinating signal for mammalian reproduction? *Science* 227:714–720.

Turek, F. W. 1985. Circadian neural rhythms in mammals. *Annu. Rev. Physiol.* 47:49–64.

Urquhart, F. A. 1987. *The Monarch Butterfly: International Traveller*. Chicago: Nelson-Hall.

Wilson, E. O. 1975. *Sociobiology*. Cambridge, MA: Belknap Press of Harvard University Press. (*Not only the definitive text on sociobiology, but an excellent source of information on social behaviors in all kinds of animals.*)

Unit Four:
The Diversity of Animals

The organisms pictured here represent a small sample of the enormous diversity in the animal kingdom. The solid structure of the coral reef was itself created by animals called polyps, which are classified in the phylum Cnidaria. Polyps are minute but so numerous that the calcium carbonate they secrete around themselves over the decades accumulates to form structures that dwarf any creation of humankind. The many nooks and crannies within the reef are habitats for a wide variety of animals. Corals are so still and so simple in organization that they were not even recognized as animals until relatively recently. Another cnidarian, the sea anemone, is also relatively simple and quiet. In contrast, the fishes shown here (phylum Chordata, subphylum Vertebrata) move rapidly and have a complex organization that makes them immediately recognizable as animals.

The other vertebrates shown here are among the most interesting of all animals. By various contrivances this one species has managed to adapt itself to almost every type of environment on earth. It is also the only species that systematically studies every other species of animal. The growing impact of our species makes such studies imperative for the survival of all animals, including ourselves.

Overview

21

Systematics

Frog (Hyla *sp.*) *with dragonfly* (Plathemis *sp.*).

CHAPTER OUTLINE

The Importance of Systematics

What Is a Species?

Taxa

Nomenclature

Evolutionary Systematics

Character Analysis

Molecular Phylogenetics

Numerical Taxonomy

Cladistics

The Fallacious Chain of Being

Body Plans

Kinds of Animals

LEARNING OBJECTIVES

1. What good is it to know the classification of animals?

2. Who decides on the scientific names of animals?

3. What approaches do biologists use to classify organisms?

4. How many major kinds of animals are there, and how are they distinguished from each other?

THE IMPORTANCE OF SYSTEMATICS

More than a million living species of animals have been named, and some zoologists believe that at least 30 times that many remain to be discovered (Wilson 1988). Although 13,000 new species of animals are named and described each year, most animals will remain forever unknown. Many are so minute and elusive that only the greatest good fortune would lead to their being noticed by someone who could appreciate their significance. Many other species are becoming extinct faster than they can be discovered, especially in tropical rainforests, where perhaps half the world's species are endangered by humans. Even in the United States the majority of species have not been described. Biology, medicine, and agriculture are constantly handicapped because researchers cannot identify the organisms they are working on.

In spite of our ignorance of most species, it is still possible to know a great deal about animals, because we can generalize from the species that have been studied to those that are related. Such generalization is possible only because the discipline of systematics, or biosystematics, allows us to classify them into related groups. From studies of a few vertebrates, or a few mammals, or a few fishes, we know what features to expect in other vertebrates or mammals or fishes. Without systematics, zoology texts would be enormous, disorganized collections of facts about thousands of apparently unrelated species.

WHAT IS A SPECIES?

It is of obvious importance to biologists to know what it is they are referring to when they use the term "species." Most adhere to the **biological species concept** (see p. 373), which defines a species as a group of organisms in which individuals are capable of interbreeding within the group, but not with individuals in other

Figure 21.1
Willow and alder flycatchers look virtually identical to us and were once considered the same species. They do not interbreed, however, so according to the biological species concept they are classified as separate species. Reproductive isolation is due to differences in habitat and song. The willow flycatcher is more likely to be found in old orchards and willows, and the male's song sounds like "fitz-bew." The alder flycatcher is more likely to be found in alder thickets near water, and the male's song sounds like "fee-bee-o."

Figure 21.2

In spite of differences in appearance, the Baltimore oriole and Bullock's oriole are now considered to be merely races within one species, the northern oriole.

such groups. Obviously the criterion that each species be reproductively isolated from every other species is of little use for organisms that reproduce asexually or for fossils. In addition, it is often impractical to study whether groups are reproductively isolated from each other. For that reason biologists often have to consider a group of organisms to be a species if they are sufficiently different in some way from all other such groups of organisms. This criterion is often valid, because sharp differences often indicate that a group does not interbreed with another group. In fact, such differences may be what keeps the groups reproductively isolated.

Often the definition of species, based on what appear to be significant differences, conflicts with the definition based on reproductive isolation. For example, the alder flycatcher and the willow flycatcher are virtually indistinguishable to human eyes and were once lumped together in one species. They were split into two species after it was discovered that slight differences in song and habitat make the males of one species unacceptable to the females of the other species (Figure 21.1). The opposite situation also occurs. Based on appearance, the Baltimore oriole of the eastern United States and Bullock's oriole of the West were once classified as different species. After they were caught interbreeding where their ranges overlap, they were lumped together as the northern oriole, much to the dismay of amateur ornithologists, who had to drop a species from their life lists (Figure 21.2).

TAXA

The goal of being able to generalize about various organisms is facilitated by grouping related species together, so that one can speak of "mammals," "vertebrates," "animals," or other such groups. The branch of systematics that deals with assigning species to these general groups is called **taxonomy.** Whenever a scientist first describes a new species, he or she suggests its relationship to other organisms by assigning the species to **taxa** (plural of taxon). These taxa must be specified in at least six categories. In zoology these categories are **kingdom, phylum** (plural phyla), **class, order, family,** and **genus** (plural genera). These levels are hierarchical, which means that the kingdom includes several phyla, each phylum includes several classes, and so on. A classification of our own species, for example, would be as follows:

kingdom Animalia

  phylum Chordata

    class Mammalia

      order Primates

        family Hominidae

          genus Homo

            species *Homo sapiens*

In addition, each of these required levels may be augmented by taxa with prefixes and suffixes such as super-, sub-, and infra-. For example, we classify ourselves in the subphylum Vertebrata, the suborder Haplorhini, infraorder Catarrhini, and subfamily Homininae. Unlike species, which can (sometimes) be determined by reproductive isolation, these taxa have no biological justification. They are created merely for the convenience of biologists.

NOMENCLATURE

Another essential area of systematics, called nomenclature, deals with the naming of species. The rules of nomenclature result in every species having a scientific name consisting of two words, such as the weird italicized words you have seen sprinkled throughout this book. These names often seem strange to those who are not fluent in Latin, and most zoologists know only a few of them. Nevertheless, they are a vast improvement over the names once used. In the Middle Ages organisms were given a descriptive name in Latin (the universal language of Europe), and as increasing numbers of organisms were discovered it became necessary to add more and more descriptive words to the names to distinguish similar species from each other. By the 18th century the name of the honey bee sounded like a Gregorian chant: *Apis pubescens, thorace subgriseo, abdominae fusco, pedibus posticis glabris utrinque margine ciliatis*—the fuzzy bee with the grayish thorax, dusky-brown abdomen, hairless hind feet bordered with small hairs on both sides.

In the 18th century Carolus Linnaeus (pronounced lin-NEE-us) simplified the lives of biologists considerably by assigning to each species a name consisting of only two words. We still use his **binominal system.** (The system was formerly called the "binomial system," and many zoologists still refer to it by that name.) The honey bee then became simply *Apis mellifera:* the bee that brings honey. The Linnaean (lin-NEE-an) system was quickly adopted throughout Europe. Unfortunately, however, there was no international body to ensure that species names introduced in one country would be accepted in other countries. English-speaking biologists often had one Latin name for a new species, and French-speaking biologists had another Latin name for the same species. They might as well have used their native tongues. In the early 1900s, however, an international organization was established to ensure that scientific names are used consistently around the world, no two animals have the same name, and the names do not change arbitrarily. These goals of **universality, uniqueness,** and **stability** are safeguarded by the International Commission on Zoological Nomenclature. Generally, the Commission selects the name that was proposed by the first person to publish a clear description of the species after 1758, when the 10th and last edition of Linnaeus' *Systema Naturae* was published. The Commission's *International Code of Zoological Nomenclature* (Ride et al. 1985) includes guidelines for assigning names to species and to higher taxa. The Commission explicitly states, however, that the guidelines must not restrict "the freedom of taxonomic thought or action." Some of the guidelines that are most helpful in becoming familiar with nomenclature are summarized below. Additional comments are in brackets.

1. The name of a species consists of two words (binomen) and that of a subspecies of three words (trinomen). In each case the first word is the name of the genus, the second word is the species epithet, and the third word, when used, is the name of the subspecies group. [Examples of binomens and trinomens: *Homo sapiens; Homo sapiens sapiens.* A subgenus may be given in parentheses after the generic name but is seldom used. The generic name may be used alone. In order to refer to a particular species when only its generic name is known, the genus is followed by "sp.": for example, *Rana* sp. To indicate more than one species in a genus, use "spp.": *Rana* spp.]

2. The name must be either Latin or latinized. [The name should be written in a different typeface, usually italics, from the body of text in which it occurs. If different typefaces are not available, the name is underlined.]

3. The name of the genus and subgenus must be a noun in the nominative singular.

4. The species epithet must be an adjective agreeing in gender (masculine, feminine, or neuter) with the generic name [*Homo sapiens*, wise man]; or a noun in the nominative singular (like the genus) [*Felis catus*, cat cat]; or a noun in the genitive case [*Charadrius wilsonia*, Wilson's plover]; or an adjective derived from the name of an organism with which the animal is associated [*Eurosta solidaginis*, a fruit fly that develops in the goldenrod *Solidago*]. [Exceptions to these rules are sometimes allowed. Classification of one species of the beetle *Agra* was so troublesome that the name *Agra vation* was accepted.]

5. Genus and subgenus names are always capitalized; species epithets never are, even if derived from a proper name. [This differs from the rule for plants, in which the species epithet may be capitalized if derived from a proper name. A generic name may also be used as the common name, in which case it is neither capitalized nor italicized: for example, drosophila.]

6. The name of the author who first described the species, or an abbreviation of his or her name, may be appended to the species names, not italicized. If the species has been transferred to a new genus, the name or abbreviation is enclosed in parentheses. [Example: *Periplaneta americana* (L.), the American cockroach, originally described by Linnaeus (L.) in another genus.]

7. The names of taxa above the genus rank are always one word, capitalized and plural. [Thus it is correct to say "The Hominidae are . . . ," even though many taxonomists recognize only one living species in the family.]

8. For higher taxa the Commission has not standardized terminology except to some extent at the family level. At this level superfamily, family, subfamily, and tribe, in descending order, are recognized. The name of the family group is formed by adding the appropriate suffix to the generic name of the "type genus" chosen by the author as being representative of the new family group. For superfamily the suffix is -OIDEA; [hence Hominoidea (pro-

nounced ha-min-oy-DEE-uh) is formed by changing the suffix of *Homines*, which is the plural of *Homo*]. For family the suffix is -IDAE [hence Hominidae, pronounced ha-MIN-i-dee]. For subfamily the suffix is -INAE (pronounced -EYE-nee), and for tribe it is -INI (-EYE-nigh).

EVOLUTIONARY SYSTEMATICS

Prior to the acceptance of evolution, taxonomy and nomenclature were used mainly to keep track of the many new species discovered by explorers and colonists. Generally, organisms were grouped together on the basis of their similarity or sometimes merely on the basis of their medicinal value. In *The Origin of Species*, however, Charles Darwin predicted that an evolutionary point of view would provide a natural basis for taxonomy, and, indeed, most taxonomists do now try to make classification of each group reflect its evolutionary history, or **phylogeny.** Until a few decades ago there was no name for this approach to systematics, because there was no alternative. Recently, however, it has come to be known by such names as traditional systematics and evolutionary systematics. The results of evolutionary systematics are published in verbal descriptions of taxonomic groups and also in the form of **dendrograms** or **phylogenetic trees,** such as Figures 18.11 and 21.4A.

CHARACTER ANALYSIS

Evolutionary systematists base their taxonomies on an analysis of similarities and differences in anatomy, molecular biology, behavior, and other traits, called **characters.** Simply put, a character is anything that can assist in classification. Not all similarities and differences are useful. Evolutionary systematists must first perform a character analysis to determine the nature and potential usefulness of characters. Character analysis involves the following:

1. **Determining whether the character is homologous or analogous. Homologous** characters, like the bones in birds' wings and human arms (Figure 18.1), are so similar in structure that they were most likely inherited from the same ancestral species. If, on the other hand, two groups have evolved the same character independently, the character is **analogous.** For example, the wings of bats and butterflies are analogous characters, since they clearly differ in structure. Only homologous characters provide any information about phylogeny.
2. **Determining whether a character is primitive or derived. Primitive** characters are those that are retained from earlier in an evolutionary lineage. For example, the four limbs of reptiles, birds, and mammals constitute a primitive character inherited from their amphibian ancestors. In this context the term "primitive" does not mean crude or poorly adapted. **Derived** characters, on the other hand, are later evolutionary modifications. The modification of the forelimbs of bats into wings is a derived character. Determining whether a character is primitive or derived helps establish whether a group is ancestral to or descended from another group.
3. **Estimating the degree of evolutionary difference between the characters in two groups.** The degree of evolutionary difference, along with the fossil record, provides information on the times of evolution. The degree of evolution also helps determine taxonomic levels; that is, it helps determine

whether two groups will be classified in the same genus, or family, or order, and so on.

4. Finally, character analysis involves **determining how much weight to give to different sets of characters.** Characters based on developmental patterns might suggest a totally different phylogeny from characters based on behavior, for example, so the evolutionary systematist must decide how much importance to assign to each type of character.

Often character analysis suggests several different ways in which a group of animals can be classified. Choosing among the alternatives often depends on what could be called the personality or style of the individual systematist. Knowledge, experience, and judgment combine in a complex way to shape this scientific personality or style. One systematist may be a "splitter" who tends to break up large groups of animals, while another systematist, for equally valid reasons, may be a "lumper" who tends to combine several groups into one. Because of such individual differences, you will often find different classifications for the same group of organisms. One is free to choose the classification that appears most consistent with phylogeny and is most convenient.

MOLECULAR PHYLOGENETICS

Since the 1950s an increasing number of taxonomists have used homologous molecules in different groups of animals as sources of characters (see pp. 378–380). For example, the sequences of bases in a ribosomal RNA from humans, lizards, chickens, sea urchins, and so on have been used to infer the phylogeny of these groups of animals (see Figure 18.15). This approach, called molecular phylogenetics, has the advantage that each molecule provides a large number of characters. For example, each base in a nucleic acid, or each amino acid in a protein, can be considered a separate character. Thus the number of similarities or differences in the base sequence or amino acid sequence can be used as a measure of the evolutionary difference between two groups. Systematists then use computers to select the most statistically probable phylogeny out of the vast number of possibilities.

The advantage of molecular phylogenetics has been demonstrated for ten strains of laboratory mice by comparing the phylogeny inferred from chromosomal differences with known breeding records (Fitch and Atchley 1987). The molecular phylogeny was exactly correct, even when five different statistical methods of analyzing the data were used. In contrast, phylogenies based on morphology (e.g., lower jaw structure) or life-history traits (litter size, body mass at different ages, etc.) gave different phylogenies, none of which was correct.

Although molecular phylogenetics provides taxonomists with a powerful tool, it is still essential that morphological and other characters be studied. After all, what one sees in the field are shells, bones, and skins—not molecules. Unfortunately, zoos, museums, and other traditional sources of information on such characters have often been neglected in the rush of enthusiasm for the new molecular tools.

NUMERICAL TAXONOMY

During the 1960s some taxonomists became dissatisfied with the subjectivity involved in character analysis and sought more objective approaches to classification. One attempt was numerical taxonomy, also called **phenetics.** Pheneticists avoid subjective aspects of character analysis simply by not doing any character analysis. They include as many characters as possible, whether homologous or analogous, and give them all equal weight. The assumption is that with a large number of characters, the effects of incorrect ones will cancel each other. Each character in each species is then graded numerically, and all the numbers are entered into a computer. The computer then groups the organisms together according to their similarity for all the different characters. There is no attempt to infer phylogeny from the results. Numerical taxonomy has been largely abandoned, but its statistical and computer methods are still widely used in molecular phylogenetics.

CLADISTICS

Cladistics, also called **phylogenetic systematics,** is an attempt at objective classification that is gaining acceptance. Cladists, also called phylogeneticists, avoid many of the problems of character analysis by limiting the analysis to determining whether characters are homologous or analogous. They do not weight characters, and they do not attempt to judge differences in the degree of evolution. They also have a more objective approach to determining whether a character is primitive or derived by using an **outgroup,** which is considered not to be closely related to the groups being classified. If a character occurs not only in the groups being classified but also in the outgroup, it is assumed to be primitive. In the special jargon of cladistics, primitive homologous characters are called **symplesiomorphies.** Homologous characters that do not occur in the outgroup, but are instead derived and shared by two or more groups under study, are called **synapomorphies.** Synapomorphies—derived homologies—are the only characters that are useful in establishing that two or more groups are related to each other phylogenetically.

Phylogeneticists next represent the groups as the end points of a pattern of branched lines, called a **cladogram** (Figure 21.3). A cladogram has the following properties:

Figure 21.3
A simple, hypothetical cladogram. A, B, and C represent three different groups of insects being classified by comparison with an outgroup. All four groups share a similar body shape, which is therefore a symplesiomorphy (primitive homologous character). The three groups A, B, and C share derived, homologous characters 1 and 2 that are not found in the outgroup and are therefore synapomorphies. Groups B and C also share synapomorphies 3 and 4 that are not found in A. They are therefore assumed to be more closely related to each other than either is to A. Together, B and C form a clade that is a sister group of A.

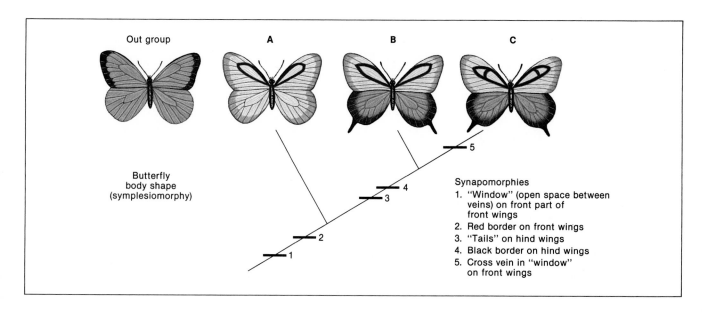

Out group   A   B   C

Butterfly body shape (symplesiomorphy)

Synapomorphies
1. "Window" (open space between veins) on front part of front wings
2. Red border on front wings
3. "Tails" on hind wings
4. Black border on hind wings
5. Cross vein in "window" on front wings

1. It begins with a single **node:** a point from which two or more lines branch. One or more of the lines then lead to other nodes, from which two or more other lines may branch, and so on.
2. A branch and all the branches from it form a **clade.** Clades that diverge from the same node are "sister" clades and have the same taxonomic rank.
3. Lines can only diverge from a node; they are not allowed to converge. Thus a clade includes only branches that can be traced back to only one node. Clades are always **monophyletic**, in other words. In contrast, evolutionary systematics may combine two or more lineages into a taxon that is **polyphyletic**.

The cladogram is set up in such a way that groups share the maximum number of synapomorphies. Actually computers usually do this work, because the number of possibilities can be enormous. With only ten species there are more than 280 million possible cladograms! There is no attempt to make the cladogram show times of evolutionary divergence, and, in fact, some cladists deny that the cladogram represents anything about evolution. Most, however, believe that the branching pattern of a cladogram does represent phylogeny.

Many zoologists who are more interested in the patterns than in the times of evolution have applauded phylogenetic systematics because cladograms represent hypotheses that can be more rigorously tested than most other hypotheses in systematics. Many, however, have been dismayed by some of the implications of cladistics. For example, because crocodilians (crocodiles, alligators, etc.) and birds diverged more recently than did turtles, snakes, and lizards, phylogeneticists group crocodilians in the same clade with birds, rather than with lizards, snakes, and turtles (Figure 21.4). If librarians were cladists they would either shelve books on crocodiles in the ornithology section or shelve books on birds in the herpetology section. Another problem is that many phylogeneticists consider only the end points of a cladogram to represent taxonomic groups. Thus a statement such as "birds evolved from reptiles" is nonsense to many cladists, even though birds evolved from dinosaurs, which are generally considered to have been reptiles. Either the group "reptiles" does not exist taxonomically, or birds should be considered reptiles. Likewise, either fishes do not exist as a taxonomic group, or amphibians, reptiles, birds, and mammals should be called fishes.

Another criticism of phylogenetic systematics is that it ignores all the information from the fossil record and from molecular phylogenetics regarding the times of evolutionary divergence. In addition, phylogenetic systematics takes no account of evolutionary change within a species; it considers only divergence. Finally, it appears that cladistics is no more immune to error than is evolutionary systematics. Neither approach can be any more reliable than the evidence on which it is based (Figure 21.5).

THE FALLACIOUS CHAIN OF BEING

Regardless of what method of taxonomy is used, it is essential that organisms be classified in some kind of system that allows us to generalize about them. Most zoologists recognize between 30 and 35 phyla of animals, which approaches the maximum that any individual can feel familiar with. The task of understanding animals is even more formidable if we wish to distinguish among them at the level of class, order, or lower taxonomic levels. Most people in Western cultures find it an irresistible temptation to deal with the vast diversity of life by ranking organisms on a linear hierarchy, from "lowest" to "highest." This is one of the legacies

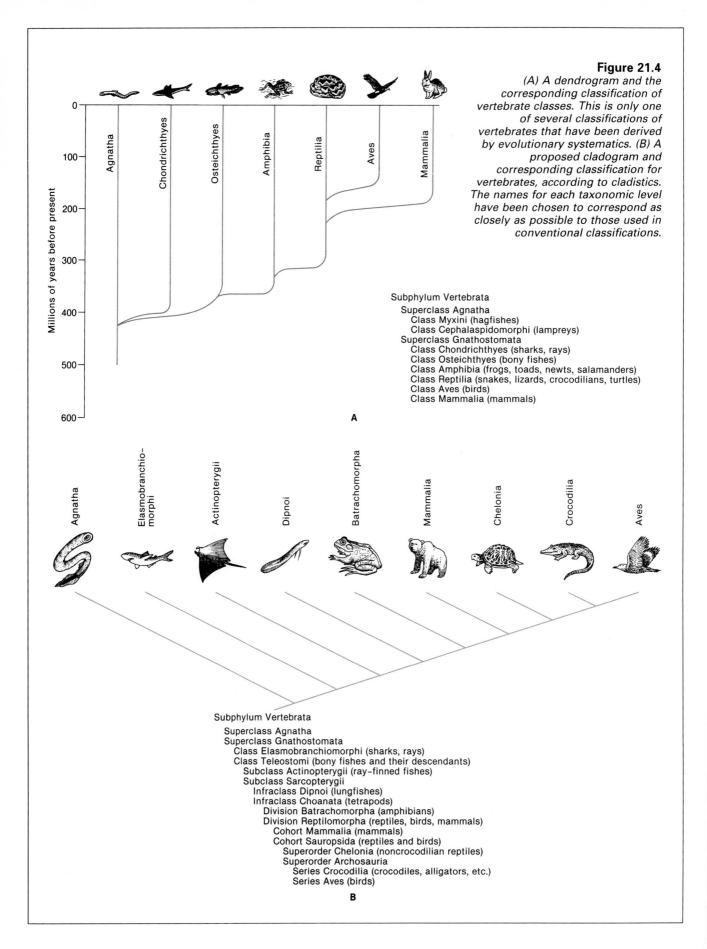

Figure 21.4
(A) A dendrogram and the corresponding classification of vertebrate classes. This is only one of several classifications of vertebrates that have been derived by evolutionary systematics. (B) A proposed cladogram and corresponding classification for vertebrates, according to cladistics. The names for each taxonomic level have been chosen to correspond as closely as possible to those used in conventional classifications.

Subphylum Vertebrata
    Superclass Agnatha
        Class Myxini (hagfishes)
        Class Cephalaspidomorphi (lampreys)
    Superclass Gnathostomata
        Class Chondrichthyes (sharks, rays)
        Class Osteichthyes (bony fishes)
        Class Amphibia (frogs, toads, newts, salamanders)
        Class Reptilia (snakes, lizards, crocodilians, turtles)
        Class Aves (birds)
        Class Mammalia (mammals)

A

Subphylum Vertebrata
  Superclass Agnatha
  Superclass Gnathostomata
    Class Elasmobranchiomorphi (sharks, rays)
    Class Teleostomi (bony fishes and their descendants)
      Subclass Actinopterygii (ray-finned fishes)
      Subclass Sarcopterygii
        Infraclass Dipnoi (lungfishes)
        Infraclass Choanata (tetrapods)
          Division Batrachomorpha (amphibians)
          Division Reptilomorpha (reptiles, birds, mammals)
            Cohort Mammalia (mammals)
            Cohort Sauropsida (reptiles and birds)
              Superorder Chelonia (noncrocodilian reptiles)
              Superorder Archosauria
                Series Crocodilia (crocodiles, alligators, etc.)
              Series Aves (birds)

B

of Aristotle, who believed that everything, from minerals, to plants, to animals, to humans, and then the heavenly spheres, could be arranged on a Chain of Being. To most people, the lowest animals are presumed to be the simplest, the least evolved, and the least successful. Often they are also considered to be the least worthy of existence.

This Aristotelian approach is inconsistent with phylogenies, which resemble bushes more than chains. It also leads to inevitable contradictions. For one thing, many parasitic animals are simple only because they have handed over many of their physiological functions to their hosts, while free-living members of the same group may be quite complex. There are also contradictions between lowliness and success, if we measure success by diversity and numbers. Roundworms (phylum Nematoda), many of which are "lowly," simplified parasites, may have more species than any other phylum, although taxonomists have gotten around to describ-

Figure 21.5
(A) A cladogram derived from comparison of teeth in carnivorous marsupials. (B) A dendrogram based on radioimmunoassay (RIA) of blood proteins. The dendrogram reveals several differences from the cladogram, particularly in the relationship of the Tasmanian wolf. RIA also allows for an estimate of evolutionary distance, which supplements the fossil record in showing the times of evolution in the dendrogram.

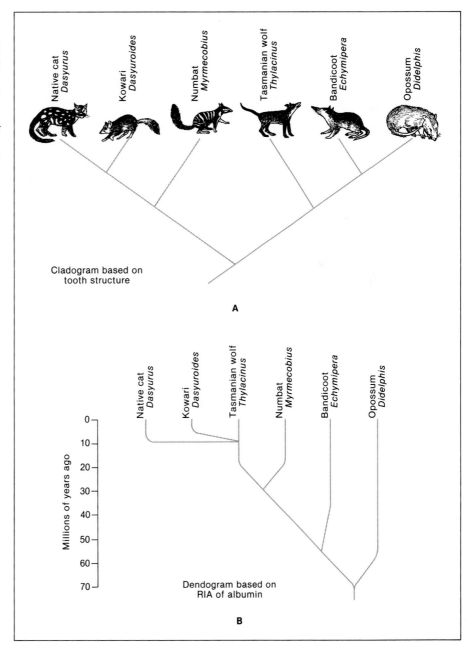

ing only 12,000 of them (Figure 21.6). There is also an enormous number of individual nematodes: 100 billion in a square meter of soil, for example. By comparison, chordates and most other "higher" taxa have failed miserably. "Lowliness," simplicity, and success are clearly of little value in organizing the animal phyla.

BODY PLANS

For the past century most zoologists have attempted to group animals at the phylum level according to their body plans. A body plan, often referred to by the German term *Bauplan*, is the overall structural and functional organization of an animal. In contrast to individual organs, body plans are established early in embryonic development and appear to be conserved during evolution.

One of the features of a body plan is its type of **symmetry,** if any. Some animals, including many sponges, are **asymmetric** (Figure 21.7). That is, there is no way that one can cut one of these animals and get two similar halves. Other animals, such as sea anemones, are **radially or biradially symmetric.** Radially symmetric animals have approximately cylindrical bodies that can be equally divided like a cake by cutting through the central axis at any angle. **Biradially symmetric** animals are essentially cylindrical in shape, but with some paired features distributed on opposite sides of the body. A birthday cake with two candles on opposite sides is biradially symmetric. Animals that are radially or biradially symmetric tend to be sessile (attached to substratum). Most animals that move about actively are **bilaterally symmetric:** there is only one way in which they can be cut down the midline into two approximately equal halves. When using type of symmetry in classification one must be careful to determine whether the type is primitive or derived. For example, the asymmetry of adult snails is considered to be a derived trait, for two reasons: (1) Other molluscs, and presumably the ancestors of snails, are bilaterally symmetric; (2) snails develop from bilaterally symmetric larvae.

Another feature of the body plans of many animals that is often assumed to provide clues to their kinship is **segmentation.** Segmented animals (annelid worms, arthropods, chordates, and perhaps others) develop from a series of similar masses of embryonic tissue (see p. 128).

Another useful feature is whether there is a fluid-filled, completely enclosed body cavity. Animals that are primitively asymmetric, radially symmetric, or biradially symmetric do not have such a body cavity. Nor do certain bilaterally symmetric animals, called **acoelomates** (ay-SEAL-oh-maytz). Some bilaterally

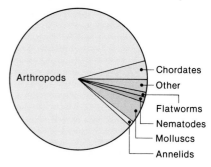

Figure 21.6
The proportion of the million named animals in various phyla. Only Chordata, Arthropoda, Mollusca, Annelida, Nematoda, and Platyhelminthes have more than 10,000 species each. One of the delightful implications of this chart is that anyone familiar with these phyla is automatically familiar with the vast majority of animals.

Figure 21.7
Types of symmetry in animals. The asymmetry of snails and the radial symmetry of echinoderms are derived, in the sense that these animals develop from larvae that are bilaterally symmetric. Biradially symmetric animals are radially symmetric in most features but have some paired structures on opposite sides of the body.

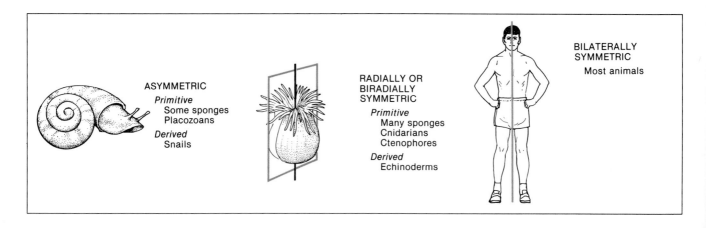

ASYMMETRIC
Primitive
Some sponges
Placozoans
Derived
Snails

RADIALLY OR BIRADIALLY SYMMETRIC
Primitive
Many sponges
Cnidarians
Ctenophores
Derived
Echinoderms

BILATERALLY SYMMETRIC
Most animals

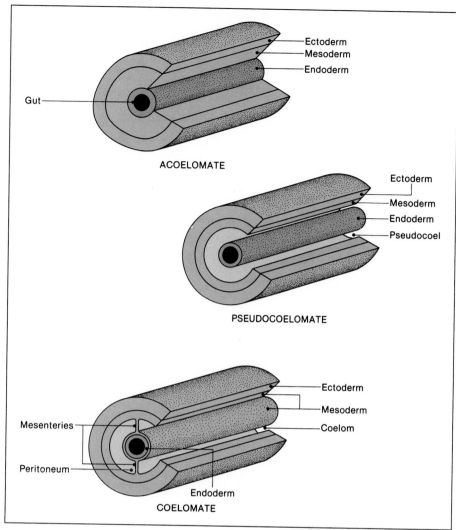

symmetric animals have a body cavity called a **pseudocoel** (SOO-doe-seal; = pseudocoelom) and are called **pseudocoelomates.** The pseudocoel is often a blastocoel that persists into adulthood and is often located between the endoderm and the mesoderm (see pp. 118–119; Figure 21.8). Other bilaterally symmetric animals, called **coelomates,** have a body cavity called a **coelom** (SEAL-um). A coelom is enclosed within mesoderm and lined with a cellular **peritoneal membrane** (= peritoneum). The functions of these body cavities are discussed in Chapters 26 and 27.

Many zoologists classify coelomates as either **protostomes** or **deuterostomes.** In animals in the division Protostomia the mouth (*stoma*) originates from the **blastopore,** which is the first (*proto*) opening to appear in the embryo (see p. 118). In animals in the division Deuterostomia the mouth originates from a second (*deutero*) opening, and the blastopore usually becomes the anus. Many protostomes and deuterostomes also differ in the pattern of cell movement during cleavage, in whether the fate of each cell is determined during cleavage (determinate versus indeterminate), and in whether the coelom originates from a split or from a pouch in the mesoderm (schizocoelous versus enterocoelous).

The validity of using these differences in body plan as a basis for classifying animals has often been doubted. (See Willmer 1990 for a thorough discussion.)

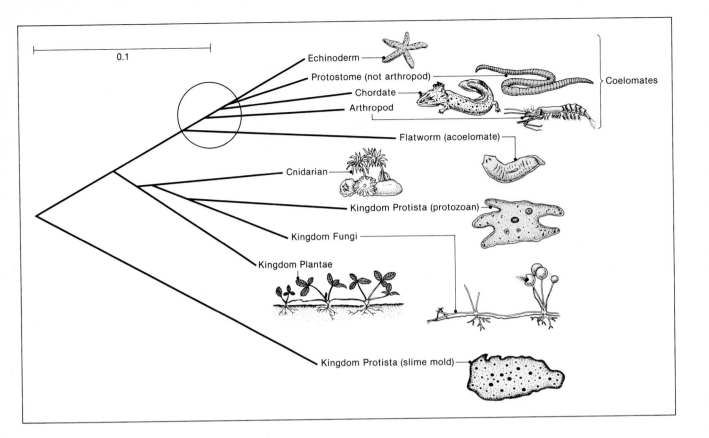

Figure 21.9
Phylogenetic tree inferred from comparisons of base sequences in RNA from the 18S subunit of ribosomes from representative animals of the major groups shown. Distance from one group to another along the line segments represents evolutionary distance, as determined by number of differences in RNA bases. (Scale bar represents 0.1 base difference per position.) According to this analysis, chordates, echinoderms, arthropods, and protostomes other than arthropods represent four groups of coelomates distinct from each other and from the acoelomates and cnidarians. The method is not sufficiently precise to determine the actual order of branching within the circle. Compare with Figure 18.15.

One problem, for example, is that pseudocoels and coeloms appear to have evolved independently in various phyla. To the extent that they are analogous, they are useless in inferring phylogeny. As another example, there are many exceptions to the protostomate/deuterostomate dichotomy. Recent evidence based on comparisons of base sequences in a ribosomal RNA also suggests that coelomates should be divided into not just two major divisions but into at least four: (1) worms, molluscs, and other protostomate animals other than arthropods; (2) arthropods (spiders, crustaceans, insects, etc.); (3) echinoderms (starfish, sea urchins, etc.); and (4) chordates, including vertebrates (Figure 21.9; Table 21.1).

Table 21.1   Major differences among the four major groups of coelomates identified by Field et al. (1988).

Each column shows whether the group is protostomate or deuterostomate, the main cleavage pattern, and whether the coelom originates enterocoelously or schizocoelously. For a description of cleavage patterns see pp. 112–116.
See Lake (1990) for a different interpretation.

Protostomes Besides Arthropods	Arthropods	Echinoderms	Chordates
Protostomes	Protostomes	Deuterostomes	Deuterostomes
Spiral	Superficial	Radial	Radial, discoidal, or rotational
Schizocoelous	?	Enterocoelous	Enterocoelous or schizocoelous

KINDS OF ANIMALS

Although such terms as pseudocoelomate, coelomate, protostome, and deuterostome probably do not reflect phylogeny, they are still useful as descriptions of body plans. They will probably continue to be used until someone determines a more valid way of organizing the animal phyla. The 32 phyla recognized in this text are listed in Table 21.2, arranged according to the body plans. Figure 21.10 is a pictorial summary of the phyla, intended to give a concise summary of the numbers of species, habitats, sizes, and physiological complexity. See also the inside front cover.

Table 21.2   The 32 phyla of animals recognized in this book.

Asymmetric, Radially, or Biradially Symmetric

Porifera (pore-IF-er-uh)   sponges
Placozoa (plak-oh-ZO-uh)
Mesozoa (mess-oh-ZO-uh)
Cnidaria (nigh-DARE-ee-uh)   hydra, jellyfish, sea anemones, etc.
Ctenophora (ten-OFF-or-uh)   comb jellies

Bilaterally Symmetric

Acoelomates

Platyhelminthes (plat-ee-hell-MINTH-eez)   flatworms
Gnathostomulida (NATH-oh-stom-YOU-lid-uh)   jaw worms
Gastrotricha (gas-tro-TRICK-uh)

Pseudocoelomates

Nematoda (NEE-ma-TOAD-uh)   roundworms
Nematomorpha (nee-MAT-oh-MORE-fuh)   horsehair worms
Kinorhyncha (KIN-oh-RING-kuh)
Loricifera (lore-i-SIFF-er-uh)
Priapulida (PRE-ah-PEW-lid-uh)
Rotifera (row-TIFF-er-uh)   rotifers
Acanthocephala (a-KANTH-oh-SEF-full-uh)
Entoprocta (IN-toe-PROK-tuh)

Coelomates

Ectoprocta (ek-toe-PROK-tuh)
Phoronida (for-OH-nid-uh)
Brachiopoda (brack-ee-OP-uh-duh)
Nemertea (NEM-ur-TEE-uh)
Sipuncula (sy-PUNK-you-luh)   peanut worms
Echiura (ek-ee-YOUR-uh)   spoon worms
Tardigrada (tar-di-GRADE-uh)   water bears
Onychophora (on-i-KOFF-or-uh)   velvet worms
Pogonophora (PA-gone-OFF-or-uh)   beard worms
Mollusca (ma-LUS-kuh)   snails, octopus, squid, shellfish, etc.
Annelida (an-EL-lid-uh)   segmented worms, leeches, etc.
Arthropoda (are-THROP-uh-duh)   insects, spiders, crustacea, etc.
Echinodermata (e-KINE-oh-DER-mah-tuh)   starfish, sea urchins, etc.
Chaetognatha (key-TOG-nah-thuh)   arrowworms
Hemichordata (HIM-i-core-DAY-tuh)   acorn worms, etc.
Chordata (core-DAY-tuh)   fishes, amphibians, reptiles, birds, mammals

Figure 21.10
A summation of animal phyla. Numbers of species mainly from Brusca and Brusca (1990) and Wilson (1988).

KEY:

SIZE

<1 mm 1 mm to 1 cm >1 cm

HABITAT

Terrestrial
Freshwater (T / F / P / M) Parasitic within other organisms
Marine
Dot indicates few species in a habitat

CIRCULATORY SYSTEM

- Blank — No heart or vessels
- Open system (blood not restricted to heart and vessels)
- Closed system

RESPIRATORY SYSTEM

- Blank — Through integument only
- G — Gills
- T — Tracheal tubes
- L — Lungs

OSMOREGULATORY SYSTEM

- C — Contractile vacuoles (Porifera only)
- P — Protonephridia (closed tubes)
- M — Metanephridia (open tubes)
- MT — Malpighian tubules (Onychophora and insects)
- K — Kidneys (Chordata only)

DIGESTIVE SYSTEM

- Incomplete (no anus)
- Complete

NERVOUS SYSTEM

- ✓ Present

REPRODUCTION

- ○ Asexual budding or fission
- ♀ Parthenogenesis
- ⚥ Hermaphroditic (monoecious)
- ⚤ Separate males and females (dioecious)

Phylum		Size	# Named species	Habitat	Physiology: Circulatory system	Respiratory system	Osmoregulatory system	Digestive system	Nervous system	Reproduction
PORIFERA	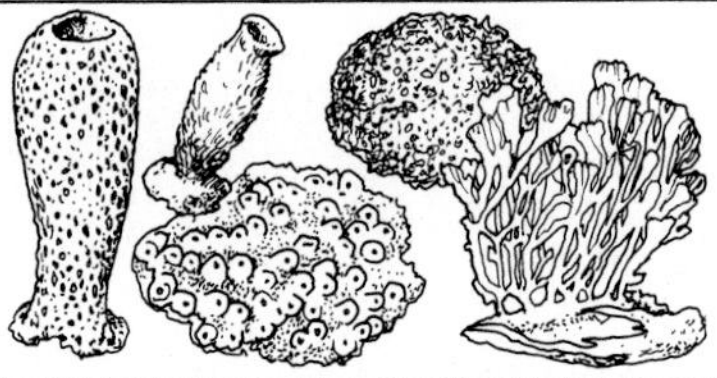	>1 cm	5000	F•, M			C			○ ⚥
PLACOZOA	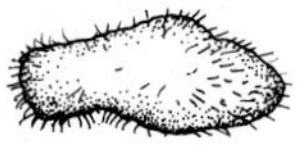	1 mm to 1 cm	1	M						○ ⚤?
MESOZOA	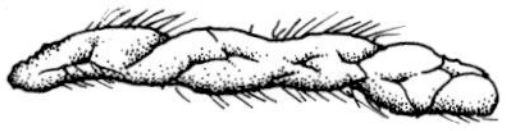	<1 mm; 1 mm to 1 cm	100	P						○ ⚥ ⚤
CNIDARIA	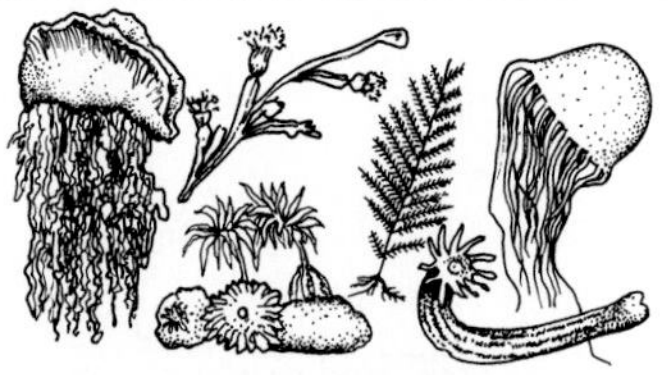	<1 mm; 1 mm to 1 cm; >1 cm	9000	F•, M				Incomplete	✓	○ ⚥ ⚤
CTENOPHORA	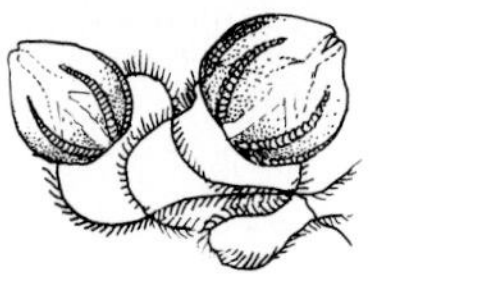	>1 cm	100	M				Incomplete	✓	⚥

ACOELOMATES									
						Physiology			
Phylum	Size	# Named species	Habitat	Circulatory system	Respiratory system	Osmoregulatory system	Digestive system	Nervous system	Reproduction
PLATYHELMINTHES		12,200	F • P / M			P	or None	✓	
GNATHOSTOMULIDA		80	M			P		✓	
GASTROTRICHA		340	F T P / M			None or P		✓	

PSEUDOCOELOMATES									
NEMATODA		12,000	F T P / M					✓	
NEMATOMORPHA		230	F P / M					✓	
KINORHYNCHA		150	M			P		✓	
LORICIFERA		10	M			P		✓	
PRIAPULIDA		17	M			P		✓	
ROTIFERA		1800	F • •			P		✓	
ACANTHOCEPHALA		700	P			None or P	None	✓	
ENTOPROCTA		150	• M			P		✓	

Phylum		Size	# Named species	Habitat	Physiology					
					Circulatory system	Respiratory system	Osmoregulatory system	Digestive system	Nervous system	Reproduction
ECTOPROCTA		⊐	4500	(M)				●	✓	♂ ♀ / ⚥ / ○
PHORONIDA		🔍 ●	15	(M)	♥		M	●	✓	♂ ♀ / ⚥ / ○
BRACHIOPODA		🔍 ●	335	(M)	♥		M	●	✓	⚥
NEMERTEA		⊐ 🔍 ●	900	(M)	♥		P	●	✓	○ / ⚥
SIPUNCULA		🔍 ●	250	(M)			M	●	✓	⚥
ECHIURA		🔍 ●	135	(M)	♥	G	M	●	✓	⚥
TARDIGRADA		⊐	400	(F/M)			MT?	●	✓	⚥
ONYCHOPHORA		●	80	(T)	♥	T	M	●	✓	⚥
POGONOPHORA		●	135	(M)	♥			M	✓	⚥
MOLLUSCA		⊐ 🔍 ●	50,000	(F/T/M)	♥ or ♥♥	G (L)	M	●	✓	⚥ / ♂♀

	COELOMATES Protostomes								
Phylum	Size	# Named species	Habitat	Circulatory system	Respiratory system	Osmoregulatory system	Digestive system	Nervous system	Reproduction
ANNELIDA		12,000	(F T M)	♥		P or M		✓	
ARTHROPODA		874,000	(F T P M)	♥	T L G	MT and other		✓	
	Deuterostomes								
ECHINODERMATA		6100	(M)	♥				✓	
CHAETOGNATHA		100	(M)					✓	
HEMICHORDATA		85	(M)	♥				✓	
CHORDATA		44,000	(F T M)	♥	G L	K (P)		✓	

SUMMARY

The function of zoological systematics is to group animals according to phylogeny so that one can generalize about different groups rather than about each of the million or so species. The most common approach to systematics is evolutionary systematics, which uses comparative anatomy, embryology, fossils, molecular phylogeny, reproductive isolation, and various other techniques to determine patterns and times of evolution. The first step in evolutionary systematics is character analysis, by which the systematist selects homologous characters, determines whether they are primitive or derived, estimates their degree of evolution, and determines how important they are.

Alternative approaches to systematics, such as numerical taxonomy and cladistics, attempt to eliminate some of the subjective elements of character analysis. Cladistics, an increasingly popular alternative to evolutionary systematics, uses synapomorphies (derived homologies) to relate groups to each other in cladograms. Cladograms may represent phylogenies but not the times of evolutionary divergence.

Another important function of systematics is nomenclature: the establishment of universal, unique, and stable names for each species. Each species is assigned a two-word name in accordance with the binominal system.

Between 30 and 35 phyla of animals are now generally recognized. These phyla are often grouped according to similarities in body plans, such as whether they are asymmetric, radially symmetric, biradially symmetric, or bilaterally symmetric; whether they are segmented or not; and whether they are acoelomates, pseudocoelomates, or coelomates. Phyla may also be distinguished by whether they are protostomes or deuterostomes. These differences in body plans probably do not always represent phylogeny, however.

KEY TERMS

species
nomenclature
binominal system
taxon
kingdom
phylum
class
order
family
genus
character

homologous
analogous
primitive
derived
molecular phylogenetics
evolutionary systematics
dendrogram
phylogeny
cladistics
synapomorphy
cladogram

body plan
asymmetric
radially symmetric
biradially symmetric
bilaterally symmetric
segmentation
acoelomate
pseudocoelomate
coelomate
prostostome
deuterostome

SELF-TEST

1. Define "character." What kinds of features of animals can be characters? What kinds of considerations enter into character analysis?

2. List the seven required taxonomic levels in descending sequence. What, if any, biological basis is there for each level?

3. What are the objectives of evolutionary systematics?

4. Describe the main differences between evolutionary systematics and cladistics.

5. Most people conceive of animals as ranked linearly from lowest to highest. Explain why this is wrong. Suggest a more useful approach to classifying animals.

6. Briefly explain the differences among the following body plan characters:

   asymmetric/radially symmetric/biradially symmetric/bilaterally symmetric

   acoelomate/pseudocoelomate/coelomate

   protostome/deuterostome

7. For each character listed in the previous question, name an animal with that character.

READINGS

RECOMMENDED READINGS

Blunt, W. 1971. *The Compleat Naturalist: A Life of Linnaeus.* New York: Viking.

Duellman, W. E. 1985. Systematic zoology: slicing the Gordian Knot with Ockham's Razor. *Am. Zool.* 25:751–762.

Goto, H. E. 1982. *Animal Taxonomy.* London: Edward Arnold.

Gould, S. J. 1980. The telltale wishbone. In: *The Panda's Thumb.* New York: W. W. Norton, pp. 267–277. (*Some implications of cladistics.*)

Gould, S. J. 1983. What, if anything, is a zebra? In *Hen's Teeth and Horse's Toes.* New York: W. W. Norton, pp. 355–365. (*Further implications of cladistics.*)

Jeffrey, C. 1973. *Biological Nomenclature.* London: Edward Arnold.

Patterson, C. 1980. Cladistics. *Biologist* 27:234–240.

Stafleu, F. A. 1971. *Linnaeus and the Linnaeans: The Spreading of Their Ideas in Systematic Botany, 1735–1789.* Utrecht: A. Oosthoek's Uitgeversmaatschappij.

Tangley, L. 1985. A national biological survey. *BioScience* 35:686–690.

ADDITIONAL REFERENCES

Ayers, D. M. 1972. *Bioscientific Terminology: Words from Latin and Greek Stems.* Tucson: University of Arizona Press.

Borror, D. J. 1960. *Dictionary of Word Roots and Combining Forms.* Palo Alto, CA: Mayfield.

Conway Morris, S., J. D. George, R. Gibson, and H. M. Platt (Eds.). 1985. *The Origins and Relationships of Lower Invertebrates.* New York: Oxford University Press.

Eldredge, N. and J. Cracraft. 1980. *Phylogenetic Patterns and the Evolutionary Process.* New York: Columbia University Press.

Field, K. G. et al. 1988. Molecular phylogeny of the animal kingdom. *Science* 239:748–753.

Fitch, W. M. and W. R. Atchley. 1987. Divergence in inbred strains of mice: a comparison of three different types of data. In: C. Patterson (Ed.), *Molecules and Morphology in Evolution: Conflict or Compromise?* New York: Cambridge University Press, pp. 203–216.

House, M. R. (Ed.). 1979. *The Origin of Major Invertebrate Groups.* New York: Academic Press.

Hyman, L. H. 1940–1967. *The Invertebrates.* (6 vols.). New York: McGraw-Hill. [*This monumental work by Libbie Hyman (1888–1969) is the starting point for invertebrate taxonomy.*]

Lake, J. A. 1990. Origin of the Metazoa. *Proc. Natl. Acad. Sci. USA* 87:763–766.

Lowenstein, J. M. 1985. Molecular approaches to the identification of species. *Am. Sci.* 73:541–547.

Mayr, E. 1981. Biological classification: toward a synthesis of opposing methodologies. *Science* 214:510–516.

Ride, W. D. L. et al. (Eds.). 1985. *International Code of Zoological Nomenclature,* 3rd ed. Berkeley: University of California Press.

Willmer, P. G. 1990. *Invertebrate Relationships: Patterns in Animal Evolution.* New York: Cambridge University Press.

Wilson, E. O. 1988. The current state of biological diversity. In: E. O. Wilson (Ed.), *BioDiversity.* Washington, DC: National Academy Press, Chap. 1.

GENERAL REFERENCES

Alexander, R. M. 1990. *Animals.* New York: Cambridge University Press.

Barnes, R. D. 1987. *Invertebrate Zoology,* 5th ed. Philadelphia: Saunders.

Barnes, R. S. K., P. Calow, and P. J. W. Olive. 1988. *The Invertebrates: A New Synthesis.* Palo Alto, CA: Blackwell Scientific Publications.

Brusca, R. C. and G. J. Brusca. 1990. *Invertebrates.* Sunderland, MA: Sinauer Associates.

Buchsbaum, R. 1987. *Animals Without Backbones,* 3rd ed. Chicago: University of Chicago Press.

Dorit, R., W. F. Walker, Jr., and R. D. Barnes, 1991. *Zoology.* Philadelphia: Saunders College Publishing.

Fingerman, M. 1981. *Animal Diversity,* 3rd ed. Philadelphia: Saunders.

Grzimek, B. (Ed.). 1968. *Grzimek's Animal Life Encyclopedia* (13 vols.). New York: Van Nostrand Reinhold. (*Available in relatively inexpensive paperback edition. Somewhat dated.*)

Hickman, C. P. Jr., L. S. Roberts, and F. M. Hickman. 1988. *Integrated Principles of Zoology,* 8th ed. St. Louis, MO: Times Mirror/ Mosby.

Laverack, M. S. and J. Dando. 1987. *Lecture Notes on Invertebrate Zoology,* 3rd ed. Palo Alto, CA: Blackwell Scientific Publications.

Lutz, P. E. 1986. *Invertebrate Zoology.* Reading, MA: Addison-Wesley.

Maeglitsch, P. A. and F. R. Schram. 1991. *Invertebrate Zoology,* 3rd ed. New York: Oxford University Press.

Margulis, L. and K. V. Schwartz. 1987. *Five Kingdoms: An Illustrated Guide to the Phyla of Life on Earth,* 2nd ed. San Francisco: W. H. Freeman.

Miller, S. A. and J. P. Harley. 1991. *Zoology.* Dubuque, IA: Wm. C. Brown.

Mitchell, L., J. Mutchmor, and W. Dolphin. 1988. *Zoology.* Menlo Park, CA: Benjamin-Cummings.

Orr, R. T. 1982. *Vertebrate Biology,* 5th ed. Philadelphia: Saunders.

Parker, S. P. (Ed.). 1982. *Synopsis and Classification of Living Organisms.* New York: McGraw-Hill.

Pearse, V., J. Pearse, R. Buchsbaum, and M. Buchsbaum. 1986. *Living Invertebrates.* Palo Alto, CA: Blackwell Scientific Publications.

Pennak, R. W. 1989. *Fresh-Water Invertebrates of the United States,* 3rd ed. New York: Wiley.

Pierce, S. K., T. K. Maugel, and L. Reid. 1987. *Illustrated Invertebrate Anatomy: A Laboratory Guide.* New York: Oxford University Press.

22

Prelude to Animals: Protozoa

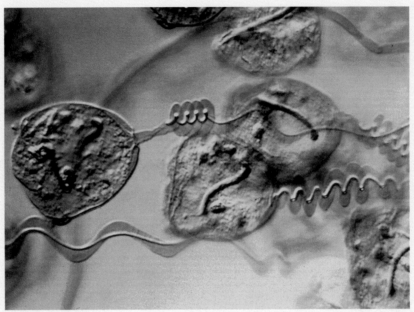

Vorticella.

CHAPTER OUTLINE

LEARNING OBJECTIVES

1. How do organisms with just one cell manage to live?
2. What kinds of protozoans are there, and how are they classified?
3. How do protozoa feed themselves?
4. How can protozoa behave without nerve cells?
5. How do protozoa reproduce?
6. How might animals have evolved from protozoa?
7. Why are protozoa important to animals?
8. What kinds of diseases can protozoa cause?
9. How does the protozoan that causes malaria manage to live in two animals as different as the mosquito and a human?

There is a little-known planet where millions of bizarre organisms pursue an ancient struggle for survival. Formless beasts flow like jelly, surrounding and engulfing their living prey. They in turn are sucked by whirling fans into gaping jaws. Others consume their victim's blood from within. Still others drift placidly in glassy shells, sparkling in the sunlight. The planet is our own—or a part of it as seen through a microscope. The organisms are protozoans (Figure 22.1).

When Antony van Leeuwenhoek first observed protozoa in 1674 he called them "animalcules"—little animals—because all organisms were supposed to be animals if they moved and plants if they didn't. Protozoa are no longer considered to be animals, but the proper study of zoology still includes protozoa. Animals and protozoans probably had the same ancestors, so we need to know about them in

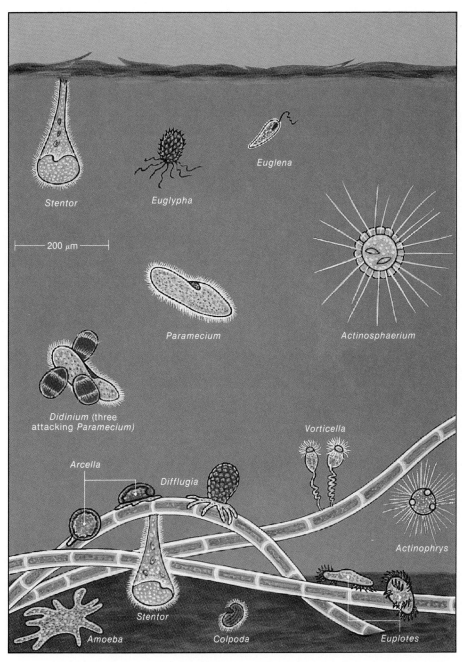

Figure 22.1
Common protozoa of a freshwater pond. Many will be discussed in this chapter. All are drawn to the same scale. The length of Paramecium *is about half the diameter of the period at the end of this sentence.*

order to understand how animal cells came to be as they are. Also, because of their kinship to animals and their ability to live independently of other cells, protozoans make excellent models for the study of many phenomena that occur in animals. Another reason for studying protozoa is that many of them have an enormous impact on animals, including ourselves.

THE UNICELLULAR WAY OF LIFE

The impressive thing about a protozoan, and the thing that is hardest to believe while watching one under a microscope, is that just one cell performs all the tasks essential to life. Some tasks that are essential for animals, however, are not necessary in a unicellular organism. Protozoa do not need neural and hormonal systems, since there is seldom a need for intercellular coordination. Respiratory, excretory, and circulatory systems are also unnecessary, because the small size of the protozoan creates an enormous ratio of surface to volume, which allows oxygen, carbon dioxide, and wastes to diffuse readily across the plasma membrane. About all that remains for a protozoan is to move to a favorable habitat, to feed, and to reproduce (Table 22.1).

Moving. Many protozoans are sessile, or they move passively with water currents or in animal hosts. Other protozoans, notably *Amoeba*, move by forming flexible extensions called **pseudopodia,** then flowing into them (see p. 216). Many protozoans move by cilia or flagella (see p. 217), which are often embedded in a protein coat called the **pellicle.** The pellicle provides a semirigid structure that transmits the force of the cilia or flagella to the entire body of the protozoan.

Feeding. Protozoans in the mature, feeding stage of life **(trophozoites)** obtain food in a variety of ways. Many eat organic detritus, bacteria, algae, other protozoa, and even small animals such as rotifers. Others live within the bodies of animals as endosymbionts, either absorbing nutrients from body fluids or intercepting nutrients in the digestive tract. A few protozoans produce their own nutrients through photosynthesis, although generally photosynthesizing protists are considered to be algae rather than protozoa.

Table 22.1   Characteristics of protozoa.

Protozoa PRO-toe-ZO-uh (Greek *proto* first + *zoon* animal).

Morphology: Unicellular eukaryotes. Various symmetries: none, radial, bilateral, spherical. Either one nucleus, many similar nuclei, or nuclei of two different sizes. Usually lacking both endo- and exoskeletons; some species secrete protective shells. Cells of colonial forms show little, if any, differentiation.

Physiology: No organs; organelles carry out most functions. Nutrition of various types: ingesting other cells, absorbing organic material, or photosynthesis.

Locomotion: Eukaryotic flagella, cilia, pseudopodia, or cytoplasmic streaming without pseudopodia. Some forms sessile.

Reproduction and Development: Reproduction either **asexual** by **fission,** or **sexual** by fusion of two individuals or gametes. No embryonic development.

Habitat, Size, and Diversity: Free-living in fresh water, seawater, or moist soil; or parasitic within other organisms. Mostly **microscopic** (5 to 250 μm), but some flagellates as small as 1 μm, and some foraminiferans up to 5 cm. Approximately **35,000 living species** described, with an average of more than one new species per day being identified. Approximately the same number of extinct species known from fossilized shells.

Reproducing. Protozoa reproduce by almost every method conceivable. The usual method is **binary fission,** with one protozoan dividing into two. Other forms of fission include **budding,** in which a new cell forms as a small growth from another, and **multiple fission,** in which one cell divides into numerous others. In some protozoans reproduction involves the merging of nuclei from two individuals (**conjugation**), or the merging of nuclei within a single individual (**autogamy**). These processes are said to be sexual. In contrast with more familiar forms of sexual reproduction, however, they do not involve gametes, and they do not produce a completely new individual.

Some protozoans do form a new individual from the fusion (**syngamy**) of gametes. These protozoan sexual practices approach those of animals and suggest possible solutions to the puzzle of how and why sex evolved. Unlike animals, however, not even these protozoans develop from embryos, and there is no visible difference between the sexes. The life cycles of many species involve a stage called a **cyst** or spore that is capable of surviving environmental stress. Cyst production is especially common in the parasitic species, which must survive the harsh environment when passing from one host to another.

A BEWILDERING DIVERSITY

Protozoans are usually classified in the kingdom Protista, which includes all eukaryotes that are neither fungi, plants, nor animals. As might be expected from such a definition, the kingdom is not a natural group. Recent results in molecular phylogenetics indicate that many protists are as different from each other as plants are from animals. In general, the members of kingdom Protista are considered algae if they photosynthesize and protozoans if they do not. The distinction is not always clear, however. *Euglena*, but not its close relatives, photosynthesize. Botanists therefore consider *Euglena* and its relatives plants, and zoologists consider them animals (Figure 22.2).

Until recently protozoa were divided into four major groups, based on their modes of locomotion. Protozoa with flagella made up the group Flagellata (= Mastigophora). Those that move by means of **pseudopodia** were classified in the group Sarcodina (= Rhizopoda). Those with cilia were included in the group Ciliata (= Ciliophora). Finally, protozoans without any of these means of locomotion, and which reproduce by spores (= cysts), constituted the group Sporozoa. Since the 1960s, however, the number of known species (living and fossil) has

Figure 22.2
(A) Scanning electron micrograph of Euglena. *Beating by the flagellum pulls the organism through water. Note the outer covering, called the pellicle, which consists of spiral strands of proteins. The pellicle is flexible enough to permit* Euglena *to change shape and squirm through tight places (euglenoid movement). Length approximately 60 µm. (B) Transmission electron micrograph of* Euglena. *The chloroplast enables* Euglena *to photosynthesize and gives it a green color. The stigma (eyespot) directs its movements toward light. The stigma is not a photoreceptor but a pigment that controls the amount of light hitting the base of the flagellum, and it therefore controls the direction of swimming.*

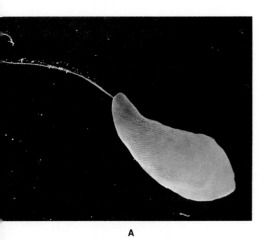

A

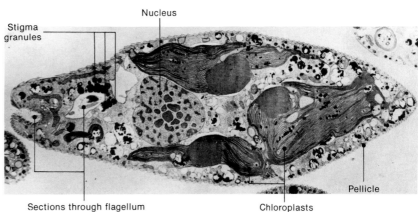

Stigma granules

Nucleus

Sections through flagellum

Chloroplasts

Pellicle

B

grown considerably, and numerous other discoveries, especially by electron microscopy, rendered this classification obsolete. A Committee on Systematics and Evolution of the Society of Protozoologists, chaired by N. D. Levine (1980), considered these discoveries in producing a revised classification of protozoa. This classification combined the flagellates and sarcodines in one phylum and erected several new phyla. The terms "ciliate," "flagellate," "sarcodine," and "sporozoan" therefore no longer correspond to the major groups of protozoans, although they are still commonly used.

In less than a decade discoveries in molecular phylogenetics suggested that revisions of the 1980 classification are in order:

1. Contrary to Levine et al., flagellates are not closely related to sarcodines or to any other group of protozoans (Baroin et al. 1988; Sogin et al. 1986).
2. Several protozoans now considered to be flagellates should be in other phyla, if not in other kingdoms. For example, dinoflagellates are more closely related to ciliates than to flagellates, and *Giardia* is as closely related to bacteria as it is to flagellates.
3. Ciliates should be distributed among several phyla, since they are genetically as diverse as the entire animal kingdom (Sogin and Elwood 1986).
4. Cellular slime molds, considered by Levine et al. to be sarcodines, are not closely related to sarcodines or other protozoans (Field et al. 1988; Qu et al. 1988).

In keeping with points 1 and 4, I have restored flagellates to their own phylum Mastigophora and deleted slime molds from the following classification, but I have not presumed to erect the other phyla that may be needed if points 2 and 3 are confirmed.

MOVING AND FEEDING

The Trouble with Being Small.   We can scarcely imagine the problems protozoans have in moving, because the forces they encounter are so different from those we face. Like other large animals, we must contend mainly against gravity and inertia. Even the viscosity of water (its resistance to flow) does not keep us from diving and swimming. The situation is quite different for protozoans. Because of their buoyancy and small mass they are not affected by gravity and inertia but are greatly affected by the viscosity of water. As a result, protozoans do not so much swim as crawl through water. Perhaps the only way for us to really understand protozoans would be to dive into a pool of molasses. Considering the problem of viscosity, protozoans move surprisingly rapidly, especially when you are trying to keep one in view under a microscope. Paramecia have been clocked at 1 millimeter per second, which is about 10,000 body-lengths per hour. For us that would be the equivalent of swimming more than 10 miles an hour.

Cilia.   Paramecia and many other protozoans move by means of cilia, which are short, numerous, hair-shaped organelles (see p. 217). Often the cilia are arranged in rows, called **kineties.** Like the cilia of vertebrates, which propel mucus along the bronchial tubes and ova through the oviducts, the cilia of protozoans generate a force parallel to the membrane surface. The beating of cilia all over the plasma membrane is somehow coordinated, as can be seen from the **metachronal waves** in Figures 9.18B and 22.3.

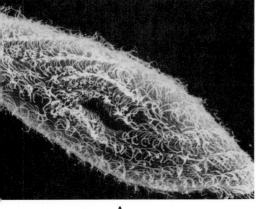

A

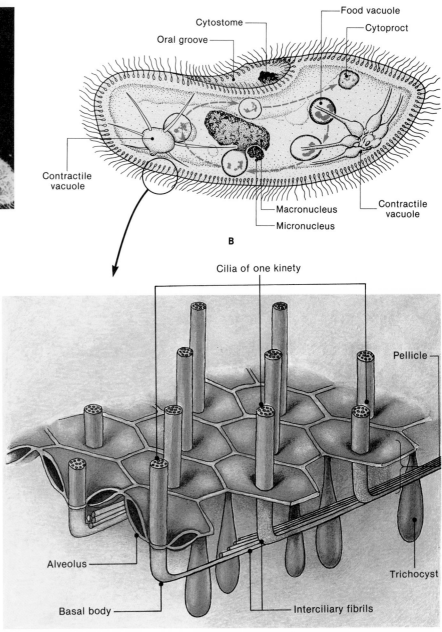

B

C

Figure 22.3
(A) SEM of Paramecium, *the most familiar ciliate (approximate length 200 μm). Approximately 10,000 cilia are evenly distributed in rows over the plasma membrane. The cilia are coordinated, as is apparent from the metachronal waves shown in the scanning electron micrograph. (See also Figure 10.17B.) The opening is the cytostome (cell mouth).*
(B) Schematic diagram of the organization of Paramecium. *Cilia bring food into the cytostome, where it enters a food vacuole for digestion. (See Figure 22.8B.) While food is being digested in the vacuole it migrates in a loop (shown by arrows) to the cytoproct where wastes are eliminated. The star-shaped organelles are contractile vacuoles (water expulsion vesicles) that eliminate water, an essential task in this freshwater genus (see p. 285). (C) The pellicle of* Paramecium *consists of double-walled chambers (alveoli) formed by plasma membrane and arranged in rows called kineties. Each alveolus is penetrated by a cilium, which originates from a basal body (= kinetosome). Kinetosomes in each kinety are linked together by interciliary fibrils (= kinetodesmata). These fibrils have been called "nervelike," but it is unlikely that they conduct action potentials. Trichocysts, which are not part of the pellicle, contain filaments that are ejected when* Paramecium *is attacked by a predator.*

In addition to propelling protozoans, cilia also bring food into the **cytostome** (literally the "cell mouth"). In many ciliates, such as *Paremecium, Vorticella,* and *Stentor,* cilia continually draw in water from which food is filtered into the cytostome. Once ingested, the food enters a **food vacuole** that merges with **lysosomes** containing digestive enzymes. Nutrients released by digestion are then transported across the vacuole membrane into the cytoplasm. In some species the cilia combine near the cytostome as specialized **buccal ciliature.** Buccal ciliature may be in the form of long, fin-shaped **undulating membranes,** or smaller, platelike **membranelles.** Cilia can also fuse into **cirri,** which some ciliates use for "walking" (Figure 22.4).

Some ciliates do not have cilia in the mature stage. They are sessile and obviously feed by some mechanism other than cilia. Suctorians, for example, lie in wait for other protozoans to pass, then snatch them by means of tentacles (Figure

(text continues on page 468)

Major Groups of Protozoans

Genera mentioned elsewhere in this chapter are noted.

Phylum Mastigophora mass-ti-GOF-for-uh (Greek *mastix* whip + *phora* bearing). At least one flagellum in mature organisms. One type of nucleus. Asexual reproduction by binary fission; rare syngamy (fusion of haploid gametes).

Subphylum Phytomastigophorea FIGHT-oh-mass-ti-go-FOR-ee-uh (Greek *phyton* plant). Either with chloroplasts, or similar to species with chloroplasts. *Euglena, Gonyaulax* (Figures 22.1, 22.2).

Subphylum Zoomastigophorea ZO-oh-mass-ti-go-FOR-ee-uh (Greek *zoo-* animal). Without chloroplasts, and not similar to species with chloroplasts. Many parasitic. *Codosiga, Giardia, Leishmania, Trichomonas, Trichonympha, Trypanosoma* (Figures 22.5, 22.14).

Phylum Sarcomastigophora SAR-ko-mass-ti-GOF-for-uh (Greek *sarx* flesh). With pseudopodia. One type of nucleus, with few exceptions. Sexual reproduction in some groups.

Subphylum Opalinata OH-pal-in-AH-tuh (Latin *opalus* opal). Numerous cilia (actually shortened flagella) in oblique rows over entire body surface, but not classified with ciliates because there is only one type of nucleus. No cytostome (mouth). Binary fission as well as syngamy with flagellated gametes. All parasitic in the rectums of amphibians. *Opalina* (Figure 22.5).

Subphylum Sarcodina sar-ko-DINE-uh (Greek *dinos* whirling). Locomotion by pseudopodia or protoplasmic flow without pseudopodia; flagella rare. Some with external or internal skeleton. Reproduction by fission or by flagellated gametes; rarely by ameboid gametes.

Superclass Rhizopoda rye-ZOP-ah-duh (Greek *rhiza* root + *podos* foot). Four of eight classes are listed below.

Class Lobosea low-BOSE-ee-uh (Greek *lobos* lobe). Lobe-shaped pseudopodia (lobopodia), sometimes narrowing to thread-shaped. *Amoeba, Arcella, Difflugia, Entamoeba* (Figures 10.16, 14.4, 22.1, 22.6, 22.7).

Class Filosea fill-OH-see-uh (Latin *filum* thread). Thread-shaped pseudopodia (filopodia). *Euglypha* (Figures 22.1, 22.7).

Class Granuloreticulosea GRAN-you-low-re-TICK-you-LOW-see-uh (Latin *granulum* little grain + *reticulum* small net). Reticulopodia: thin pseudopodia, usually interconnecting to form a net. In foraminiferans (Latin *foramen* opening + *ferre* to carry) reticulopodia extend through openings in an external skeleton. Sex-

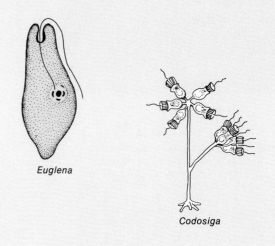

Euglena

Codosiga

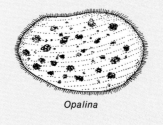

Opalina

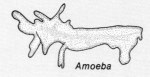

Amoeba

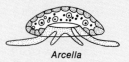

Arcella

ually reproducing generation alternates with asexual generation. Two sizes of nuclei in developmental stages of some species. *Globigerina* (Figures 22.7, 22.12).

Superclass Actinopoda ACT-in-OP-ah-duh (Greek *actinos* rays). Axopodia: pseudopodia reinforced by microtubules. Usually spherical and planktonic, often with external skeleton. Marine forms are called radiolarians. *Actinophrys, Actinosphaerium* (Figures 22.1, 22.7).

Phylum Labyrinthomorpha LAB-i-rinth-oh-MORE-fuh (Greek *labyrinthos* maze + *morphe* form). Cells spindle-shaped or spherical, moving within network of ectoplasm. Parasites of marine and estuarine algae.

Phylum Apicomplexa A-pi-kom-PLEX-uh (Latin *apex* + *complex* combination). With submicroscopical apical complex of variously shaped organelles that enhance entry into host cells and may protect it from expulsion by the cell. No cilia. Sexual reproduction in some species. All parasitic.

    Class Perkinsea per-KIN-see-uh. Small group of oyster parasites.

    Class Sporozoea SPORE-oh-ZO-ee-uh (Greek *spora* a sowing of seed). Locomotion in mature organisms by gliding, body flexion, or undulation of ridges. Pseudopodia, when present, used only for feeding. *Babesia, Eimeria, Plasmodium, Toxoplasma.*

Phylum Microspora my-CROSS-spore-uh (Greek *micros* small + *sporos* seed). Spore-forming parasites inside cells in nearly all major animal groups. Spore has coiled filament through which cytoplasm enters host cell. Without mitochondria.

Phylum Ascetospora as-see-TOSS-spore-uh (Greek *asketos* curiously made—referring to the strange and complex spore structure). Spore-forming parasites.

Phylum Myxozoa mix-oh-ZO-uh (Greek *myx* slime). Spore-forming parasites, especially in annelid worms and fishes. Considered by some to be animals. *Myxosoma cerebralis* (= *Triactinomyxon gyrosalmo*).

Phylum Ciliophora sill-ee-OFF-or-uh (Latin *cilium* eyelash). Cilia, or at least the membrane structures of cilia, present in at least one stage. Two types of nucleus in almost all species. Sexual reproduction of several types. Contractile vacuole usually present. Most species free-living. Classification within this phylum is controversial. *Blepharisma, Colpoda, Didinium, Euplotes, Heliophrya, Ichthyophthirius, Paramecium, Stentor, Stylonychia, Tetrahymena, Tokophrya, Vorticella* (Figures 22.1, 22.3, 22.4).

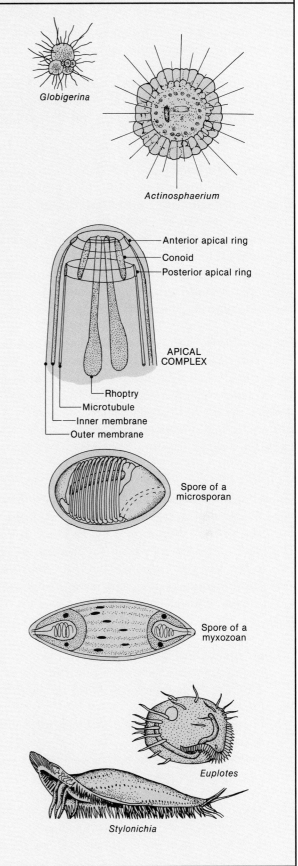

Globigerina

Actinosphaerium

Anterior apical ring
Conoid
Posterior apical ring

APICAL COMPLEX

Rhoptry
Microtubule
Inner membrane
Outer membrane

Spore of a microsporan

Spore of a myxozoan

Euplotes

Stylonichia

22.4). The tentacles usually paralyze prey within seconds, then begin drawing in their cytoplasm. Feeding by suctorians probably does not involve suction, in spite of their name (Rudzinska 1973). *Paramecium* often escapes suctorians, either by twisting free or by discharging numerous hair-shaped **trichocysts** from the pellicle. It is usually assumed that the trichocysts of ciliates repel an attack by predators in the same way that the arrows of Indians repelled the cavalry. However, numerous other functions have been suggested for trichocysts. Spoon and his colleagues (1976) have observed that the discharge of trichocysts loosens the cilia, perhaps making predators lose their grip. This would explain why trichocysts are ineffective against *Didinium*, which grabs *Paramecium* bodily (Figure 22.1).

Flagella. The beating of a flagellum generates a force parallel to the long axis of the flagellum. In some species the force generated by the flagellum is increased by an attached **undulating membrane** (Figure 22.5). In most flagellates, such as *Euglena*, this force pulls the protozoan through water, rather than pushing it, as occurs with sperm. Flagellates move at about 0.2 mm/sec, or about one-fifth as fast as ciliates. In *Codosiga* and some other flagellates the base of the flagellum is surrounded by a sievelike collar of microvilli that filter out particles of food. This mechanism is similar to that in the **choanocytes** (collar cells) that assist in suspension feeding in sponges and some other animals (see Figure 23.1C).

Pseudopodia. Pseudopodia occur mainly in the sarcodines (subphylum Sarcodina), but some flagellates can also form them, and some sarcodines can form flagella. Pseudopodia are mobile extensions that seem to result from the flowing of cytoplasm within them (see Figure 10.17). The pseudopodia bend downward, elevating the rest of the body from the substratum. Thus, when viewed from the side, a sarcodine appears to creep about "on tiptoes." In a race against a typical ciliate and a typical flagellate, a sarcodine would come in a distant third, creeping along at about 0.005 to 0.02 mm/sec. Like cilia and flagella, pseudopodia are used not only for locomotion but also for feeding. The process, called **phagocytosis,** has been studied extensively in *Amoeba* in an attempt to understand how vertebrate white blood cells ingest bacteria and other invaders. During phagocytosis, pseudopodia surround a food particle and enclose it in a **food vacuole** (Figure 22.6). This process can apparently occur anywhere on the plasma membrane. (Ameboid cells do not have cytostomes like those in paramecia.) Once the food vacuole is formed by phagocytosis, lysosomes merge with it and digest the food.

Four types of pseudopodia can be distinguished on the basis of shape (Figure 22.7).

1. **Lobopodia** are lobe-shaped and occur in *Amoeba, Arcella,* and others in class Lobosea. Some amoebas move as if the whole cell were a pseudopodium. This sluglike locomotion is called the **limax** form, in honor of the slug *Limax.*
2. **Filopodia** are long and thin, and they occur in *Euglypha* and other members of class Filosea.
3. **Reticulopodia** are similar to filopodia except that they interconnect to form a net that traps food. Reticulopodia occur in foraminiferans such as *Globigerina* (class Granuloreticulosea).
4. **Axopodia** are reinforced by regular arrays of microtubules, and they occur in members of the superclass Actinopoda, such as *Actinosphaerium.* Upon contact with food the microtubules shorten, pulling the axopods and food toward the body, which engulfs it by phagocytosis.

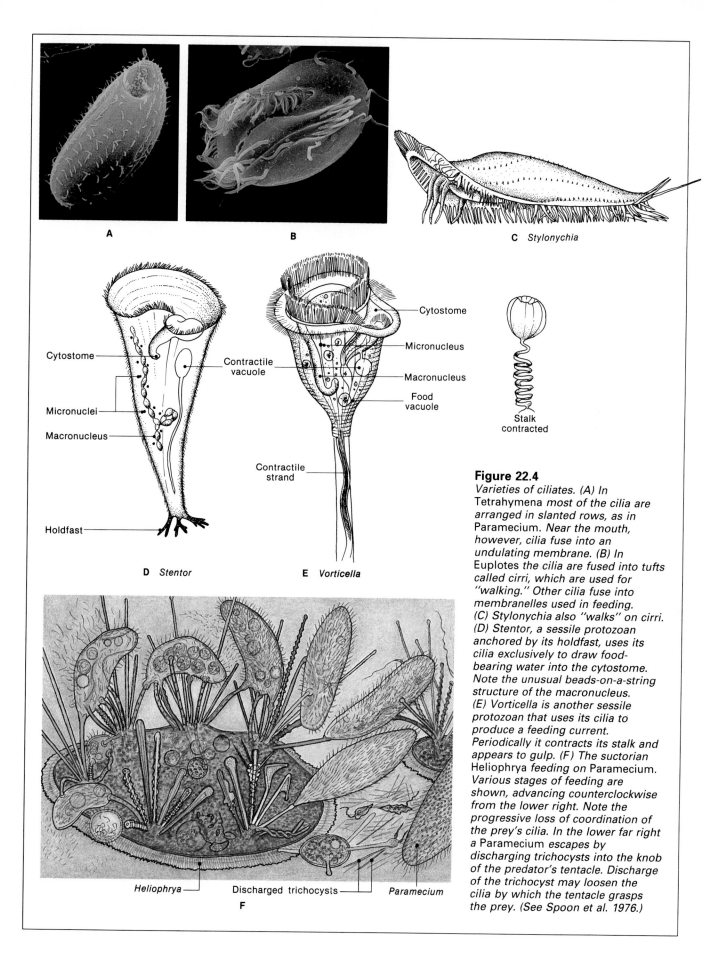

A

B

C   *Stylonychia*

Cytostome ——

Micronuclei —

Macronucleus —

Holdfast ——

D   *Stentor*

——— Contractile
vacuole

Cytostome

—— Micronucleus

—— Macronucleus

—— Food
vacuole

Contractile
strand ———

E   *Vorticella*

Stalk
contracted

Figure 22.4
*Varieties of ciliates. (A) In
Tetrahymena most of the cilia are
arranged in slanted rows, as in
Paramecium. Near the mouth,
however, cilia fuse into an
undulating membrane. (B) In
Euplotes the cilia are fused into tufts
called cirri, which are used for
"walking." Other cilia fuse into
membranelles used in feeding.
(C) Stylonychia also "walks" on cirri.
(D) Stentor, a sessile protozoan
anchored by its holdfast, uses its
cilia exclusively to draw food-
bearing water into the cytostome.
Note the unusual beads-on-a-string
structure of the macronucleus.
(E) Vorticella is another sessile
protozoan that uses its cilia to
produce a feeding current.
Periodically it contracts its stalk and
appears to gulp. (F) The suctorian
Heliophrya feeding on Paramecium.
Various stages of feeding are
shown, advancing counterclockwise
from the lower right. Note the
progressive loss of coordination of
the prey's cilia. In the lower far right
a Paramecium escapes by
discharging trichocysts into the knob
of the predator's tentacle. Discharge
of the trichocyst may loosen the
cilia by which the tentacle grasps
the prey. (See Spoon et al. 1976.)*

Heliophrya ——   Discharged trichocysts ——   *Paramecium*

F

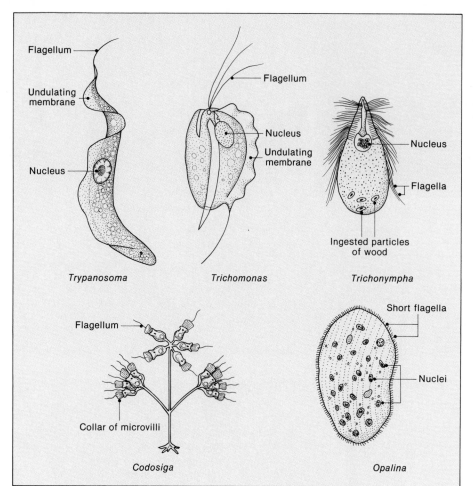

Figure 22.5
Some representative flagellates.
Trypanosoma, *with its single flagellum, is a parasite of animals, including humans.* Trichomonas vaginalis *infects the reproductive tracts of humans. Other species of* Trichomonas, *as well as* Trichonympha, *are symbiotic in the guts of wood-eating insects.* Codosiga *is colonial and sessile. The flagellum of each individual draws water through the sievelike collar, which filters out food. Because it has cilia,* Opalina *was long considered a ciliate. However, its many nuclei are all of one size, unlike those of ciliates.* Euglena *(Figures 22.1 and 22.2) is also a flagellate.*

BEHAVIOR

***Phototaxis in* Euglena.** A protozoan observed under a microscope seems to move about with as much purpose in life as we ourselves have. One must keep reminding oneself that the protozoan has no brain, or even a single nerve cell. Even without a microscope you can observe some impressive protozoan behavior. If you find a shaded, stagnant pond whose surface is green with *Euglena* and shine a bright light at one side, soon all the green will be at that end. The euglenoids swim to the bright side (**phototaxis**), leaving the other flagellates and ciliates swimming throughout the water and the ameboids on the bottom. The response of *Euglena* to light is due to the light sensitivity of the base of the flagellum. When *Euglena* is swimming directly toward or away from light, the base of the flagellum is continuously illuminated. When *Euglena* is swimming at an angle with respect to light, however, pigments in the **stigma** (= "eyespot"; Figure 22.2) shade the base of the flagellum periodically as the body rotates. Each time a shadow of the stigma falls on the base of the flagellum, the long axis of the flagellum turns a few degrees in the direction of the stigma and therefore toward the light. Thus the flagellum pulls the body of *Euglena* toward the light.

***Behavior of* Paramecium.** One of the most studied protozoan behaviors is the **avoiding reaction** of *Paramecium*. When *Paramecium* bumps into an object it does not persist in hammering at it, like a fly against a window. Instead, the slipper-shaped organism backs up a little, turns on its heel, as it were, and swims

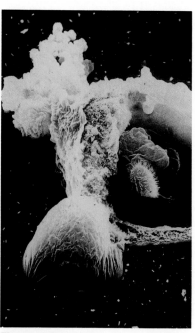

Figure 22.6
A SEM of an Amoeba *engulfing a ciliate.*

Figure 22.7
Four varieties of pseudopodia in the subphylum Sarcodina. Like Amoeba, Arcella *has the lobopodium form of pseudopodia. The lobopodia are restricted to the portion of membrane not covered by a shell (test). See also* Difflugia *in Figure 22.1.* Euglypha *has long, thin filopodia.* Globigerina *has reticulopodia, which are similar to filopodia except that they interconnect in a network. The reticulopodia of foraminifera project through openings in the test.* Actinosphaerium *has axopodia. Inset: Electron micrograph of an axopod of* Actinosphaerium *in cross section.*

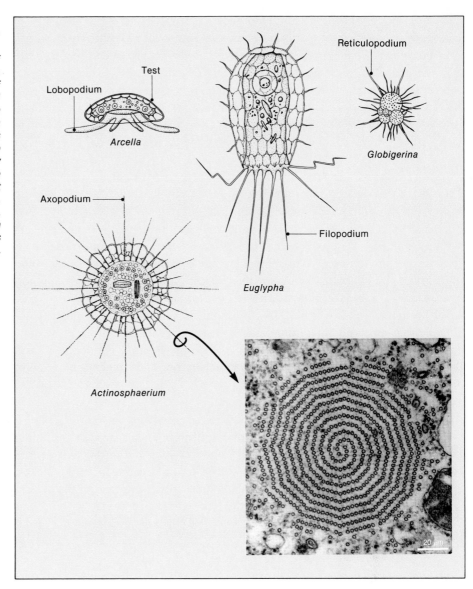

off to the side (Figure 22.8A). The intriguing question about the avoiding reaction is how the *Paramecium* "perceives" and responds to contact without sensory organs or a nervous system. The cilia of *Paramecium* and other ciliates are interconnected by fibrils (Figure 22.3) that have been called "nervelike" and "coordinating," but it is unlikely that these fibrils conduct action potentials, or that they are involved in ciliary coordination (Kung and Saimi 1982; Naitoh and Eckert 1969a,b). Coordination is more likely due to changes in the intracellular Ca^{2+} concentration triggered by changes in the voltage across the plasma membrane. Contact at the anterior end of *Paramecium* triggers a depolarizing action potential during which calcium flows into the cell. The Ca^{2+} causes the cilia to reverse their direction of beating, backing the paramecium out of trouble. After the action potential, Ca^{2+} is pumped out, and the cilia resume their normal beating. Similar mechanisms are responsible for **chemokinesis:** changes in the speed of movement in response to chemicals (Van Houten 1979). The cilia increase their rate of movement in repellent substances and slow down in attractive ones (Figure 22.8C).

As in animals, the behavior of *Paramecium* is under some degree of genetic control. Mutants called "pawns" have smaller action potentials and less influx of

Ca^{2+} in response to contact. Therefore these "pawns," like those in the game of chess, can only go forward. Another mutant of *Paramecium* is called "paranoiac," because its action potentials last so long that it has an exaggerated avoiding reaction (Cronkite 1979). There have been numerous attempts to demonstrate that protozoans can learn, in hopes that they would provide useful models for research. Unfortunately, protozoans modify their behaviors in only the most rudimentary ways (Applewhite 1979).

REPRODUCTION

Fission. Reproduction in protozoans is almost always asexual and usually by fission. Fission is preceded by mitosis, which differs in many protozoans from that of animal cells (see pp. 71–75). In some protozoans the spindle apparatus occurs within the nucleus, centrioles are not present, or the nuclear envelope may remain intact throughout mitosis. Mitosis must get quite complicated in *Amoeba* and some of the other sarcodines that have more than 500 chromosomes. The most common type of fission is **binary fission,** with one protozoan dividing into two roughly equal "daughter cells" (Figure 22.9). Binary fission in members of the phylum Ciliophora is especially interesting because there are two kinds of nuclei. In general, the larger type, the **macronucleus,** directs metabolism, development, and the physical traits of the cell. The **micronucleus** transmits genetic information during reproduction.

Multiple Fission and Budding. Another means of asexual reproduction is multiple fission, in which numerous daughter cells form at the same time. Multiple fission is especially common in the subphylum Sarcodina and in the class Sporozoea and will be described later in the discussion of malaria. In many ciliates, especially suctorians, a daughter cell forms from a small outgrowth of the parent, in the process called budding. Usually the bud forms outside the "parent," but *Tokophrya* and some other suctorians "give birth" to young that develop from internal buds.

Conjugation. There are several forms of reproduction that involve the recombination of genetic material from two individuals and are therefore said to be sexual. Conjugation in ciliophora is one of these. In *Paramecium* conjugation begins with two ciliates attaching to each other by their cytostomes. The micronucleus of each conjugant then undergoes meiosis, producing four haploid **pronuclei** (Figure 22.10). The conjugants give each other one pronucleus, which fuses with one already present to form a diploid micronucleus. The two conjugants then separate, and the new micronucleus in each one divides several times to form new micronuclei. The old macronucleus degenerates, but one of the new micronuclei enlarges and becomes a new one. Finally each individual divides twice, with everything except the micronuclei duplicating each time. The result is four cells with one micro- and one macronucleus each.

Not all pairs of ciliates in the same species will conjugate with each other. They must also belong to the same **variety.** Even belonging to the same variety does not always ensure that conjugation can occur. Two individuals must also belong to different **mating types** in order to conjugate, just as two animals must belong to different **sexes** in order to copulate. In some paramecia there are not merely two mating types but four or eight. (It broadens the mind to consider what human society might be like if animals had evolved from conjugating ciliates.)

Actinophrys and some other protozoans sometimes undergo a process called

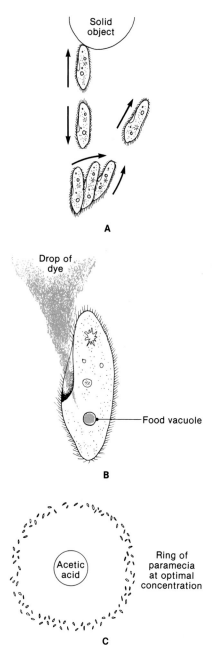

Figure 22.8
Behavior in Paramecium. *(A) The avoiding response of* Paramecium *when it swims into an obstacle. (B)* Paramecium *samples the environment ahead by drawing in a stream of water, as can be demonstrated by injecting a drop of dye in front of it. (C) A drop of acetic acid in water diffuses outwardly in a decreasing concentration. By means of chemokinesis, paramecia aggregate in a zone around the drop, avoiding the high acidity in the center, while taking advantage of the nutrient.*

Figure 22.9

Binary fission (A) in the flagellate Euglena, (B) in the ciliate Paramecium, and (C) in the sarcodine Arcella. Fission is lengthwise in flagellates and transverse in ciliates. In Arcella the shell does not divide, but new shell material is extruded along with the new daughter cell.

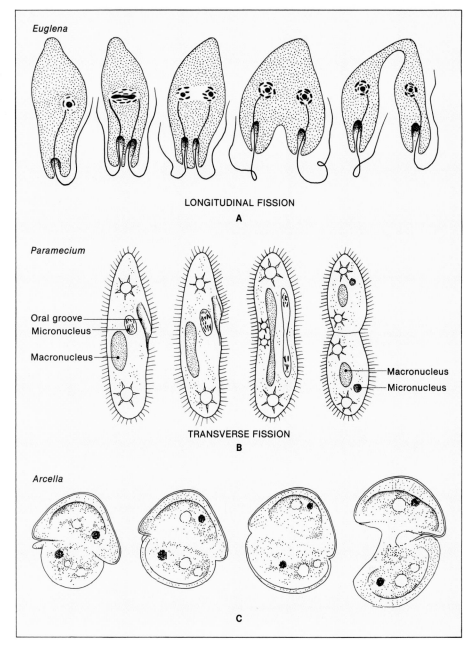

Euglena

LONGITUDINAL FISSION
A

Paramecium

Oral groove
Micronucleus

Macronucleus

Macronucleus
Micronucleus

TRANSVERSE FISSION
B

Arcella

C

autogamy, which is similar to conjugation except that it occurs entirely within a single individual. The haploid pronuclei that result from meiosis fuse without an exchange of genetic material between individuals.

Syngamy. In some protozoans (but not ciliates) reproduction is similar to that in animals, with one gamete fertilizing another and forming a zygote. This process is called syngamy in both protozoans and animals. Meiosis always occurs, preventing the chromosome number from doubling. As in humans it is completed in the zygote *after* fertilization, rather than during the formation of gametes. Even in sexual protozoans there is no development from an embryo, as occurs in animals.

Encystment. Many protozoans undergo changes in morphology associated with

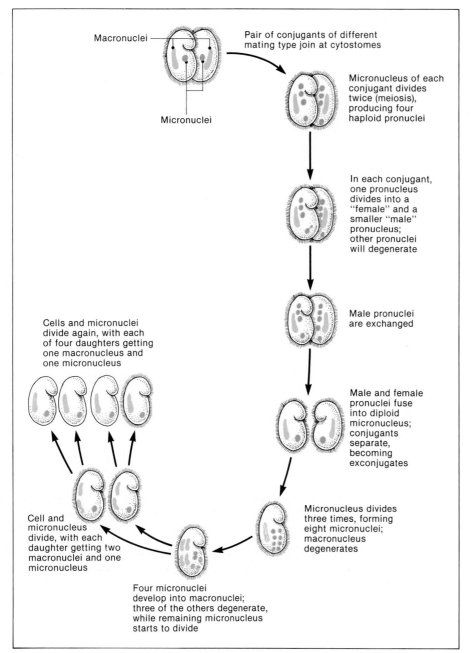

Macronuclei

Micronuclei

Pair of conjugants of different mating type join at cytostomes

Micronucleus of each conjugant divides twice (meiosis), producing four haploid pronuclei

In each conjugant, one pronucleus divides into a "female" and a smaller "male" pronucleus; other pronuclei will degenerate

Male pronuclei are exchanged

Male and female pronuclei fuse into diploid micronucleus; conjugants separate, becoming exconjugates

Micronucleus divides three times, forming eight micronuclei; macronucleus degenerates

Four micronuclei develop into macronuclei; three of the others degenerate, while remaining micronucleus starts to divide

Cell and micronucleus divide, with each daughter getting two macronuclei and one micronucleus

Cells and micronuclei divide again, with each of four daughters getting one macronucleus and one micronucleus

Figure 22.10
Conjugation in Paramecium caudatum. *Conjugation produces four daughter cells, all of which are likely to be genetically different.*

reproduction. In some species the zygote surrounds itself with a protective coating that forms a **cyst.** Several generations of binary fission may then occur within the cyst. Protozoans may also encyst prior to dispersal or in response to environmental stress. Such cysts are often called **spores.** Like the seeds of plants, they allow the protozoan to survive in a dormant state until environmental conditions are favorable. During encystment the protozoan loses any cilia or flagella, pumps out water, becomes spherical, and secretes a protective coating around itself. In this form protozoans typically survive winters, dry seasons, and other severe conditions. *Colpoda* cysts have survived in the laboratory for many hours at temperatures of 180°C below freezing, and for several hours in boiling water. Excystment occurs upon return to favorable conditions. Several species of protozoans were found cheerfully swimming about in museum soil samples that had been moistened for the first time in 49 years!

Figure 22.11
Asexual and sexual reproduction in Volvox. The hollow spherical colony is approximately 1 mm in diameter and contains up to 50,000 individuals. A colony reproduces asexually by forming buds inside that develop into "daughter" colonies. Specialized cells can also form gametes, usually in autumn. In some colonies the sexes are separate, as shown here. When a male colony encounters a female colony it binds to it, and sperm (microgametes) penetrate and fertilize the eggs (macrogametes). The zygote may then form a cyst that will develop into a new colony during the following spring.

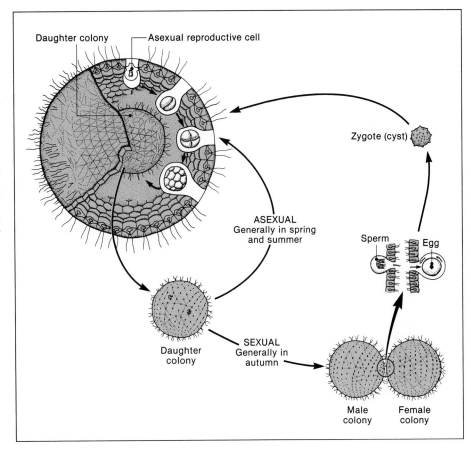

SPECULATIONS ON THE ORIGIN OF ANIMALS

Because of the similarity of certain protozoans to certain animals, it is attractive to assume that animals (metazoans) evolved from protozoans. Such evolution could have occurred when the cells of a colonial protozoan became differentiated, forming a multicellular animal. A version of this theory, called the **colonial flagellate** theory, was first proposed in 1874 by Ernst Haeckel, who argued from the following similarities of animals to colonial flagellates:

1. Some flagellates form hollow colonies that resemble the hollow blastula stage of animal embryos.
2. Syngamy with gametes of different sizes occurs in both colonial flagellates and in animals.
3. Among flagellates one can find cells that resemble flagellated, ciliated, ameboid, and numerous other types of cell that occur in animals.

Haeckel picked as a model of the ancestral colonial flagellate the photosynthesizing protist *Volvox*, which is common in stagnant water. *Volvox* colonies are hollow and consist of up to 50,000 cells. Each colony usually reproduces asexually from internal buds that form daughter colonies (Figure 22.11). In autumn, however, daughter colonies produce sperm and eggs. The fertilized eggs then develop into new colonies.

Haeckel's theory is still accepted by many zoologists, but there are serious problems with it. First, the main reason Haeckel chose *Volvox* is that its colonies resemble the hollow blastula stages of animal embryos. This fit into his now-discredited recapitulationist thinking, which held that the development of an animal goes through the same stages as the evolution of the species (see p. 133). A more serious objection is that *Volvox* and all other protists that have hollow colonies and reproduce and develop like *Volvox* are algae rather than protozoans and are more closely related to plants than to animals (Rausch et al. 1989).

An alternative theory, called the **syncytial ciliate** theory, holds that a multinucleate protozoan—probably a ciliate—became multicellular as each of its nuclei became partitioned from the others by internal membranes (Hanson 1977). The result was a multicellular organism that evolved into something like a turbellarian flatworm (phylum Platyhelminthes, class Turbellaria), which in turn evolved into the majority of other animals. One criticism of this theory is that the proposed process resembles embryonic development only in insects, not in flatworms or other animals where one would expect it. Another problem is that no known ciliate reproduces in the way that animals and *Volvox* do—by syngamy.

INTERACTIONS WITH HUMANS AND OTHER ANIMALS

Foraminiferans. Most people are not aware how much of their everyday surroundings are due to protozoa. Much of the chalk we write with and the limestone we walk upon was created by **foraminiferans** (phylum Sarcomastigophora, class Granuloreticulosea), which once made up a large portion of the plankton of many oceans. As foraminiferans died, their shells, up to 10 cm in diameter, sank and formed deep sediments on the ocean floor (Figure 22.12). The shells changed into chalk deposits, such as the famous White Cliffs of Dover in England, and limestone, such as that found in the quarries of Indiana and in the Egyptian Pyramids. One esteemed scientist early in this century was so impressed with foraminiferans that he thought all rocks, including meteorites, had been created by them (Gould 1980).

Indirectly, foraminiferans must also be credited with many of our petroleum products. The different shells produced by species of foraminiferans that lived in different ages have formed "indicator fossils" that petroleum geologists use in searching for oil. Because of foraminiferans, one is as likely to find protozoologists in the petroleum industry and in geology departments as in zoology departments.

Almost all the foraminiferans were wiped out in the mass extinction at the end of the Mesozoic era 65 million years ago (see pp. 396–397), but the survivors, especially *Globigerina*, continue to create a steady rain of shells to the ocean floor. "Globigerina ooze" is now accumulating at a rate of about 60 cm per century and covers approximately half the ocean floor. Pressure at depths greater than 4 km dissolves the calcium-based shells of these foraminiferans but not the silicon- or strontium-based shells of the radiolarians (phylum Sarcomastigophora, superclass Actinopoda). Instead of globigerina ooze, one therefore finds "radiolarian ooze" at the bottom of the deepest oceans. Radiolarian ooze has made its own contribution to civilization, having formed the flint that went into early tools.

Ecology. The immense numbers of protozoa imply that they must have an enormous day-to-day impact on ecology. Protozoa form essential links in food chains by consuming other protozoans and bacteria and then being consumed by a variety of small animals. Their decomposing of organic matter is important in breaking down pollutants, and their consumption of bacteria is essential for the proper functioning of sewage-treatment facilities. It may well be that on the whole protozoa prevent more diseases than they cause.

Figure 22.12
A sample of foraminiferan tests.

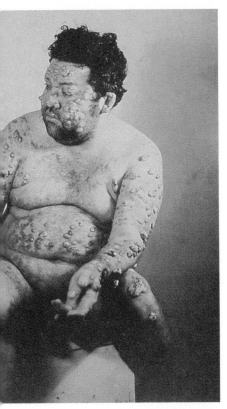

Figure 22.13
A victim of leishmaniasis. Some deformities are even more ghastly.

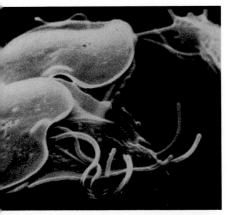

Figure 22.14
Giardia lamblia trophozoites attach to intestinal cells by means of the adhesive discs. Although these discs are sometimes called "suckers," it is doubtful that they adhere by suction. Each trophozoite is approximately 15 μm wide.

Symbioses. Many protozoans are symbiotic with animals, as harmless commensals, beneficial mutualists, or parasites. These symbioses are often highly specialized. Some species of suctorians, for example, have never been found anywhere except on the backs of turtles, and members of subphylum Opalinata are almost entirely restricted to the rectums of amphibians.

Some dinoflagellates called **zooxanthellae** (class Phytomastigophora, order Dinoflagellida) are mutualistic in reef-building corals, sea anemones, and clams (see Figure 24.10A). The photosynthetic contribution of these zooxanthellae is essential to the formation of coral reefs, so many other species are indirectly dependent on them as well. High temperature and other stresses can make corals evict their zooxanthellae. The consequent "bleaching" is a matter of periodic concern to ecologists (p. 517).

Table 22.2   Some important protozoan diseases of humans.

Leishmaniasis, caused by *Leishmania,* which is spread from other mammals by blood-sucking sand flies. Some forms cause temporary disfigurement (Figure 22.13), while others attack macrophages, producing an AIDS-like immune deficiency.

Trypanosomiasis (African sleeping sickness), caused by *Trypanosoma brucei rhodesiense* and *T. b. gambiense* (Figure 22.5). Transmitted throughout central Africa by the tsetse fly *Glossina.* Hundreds of thousands of victims, with 25,000 new cases per year. For the first two years the symptoms of sleeping sickness are fever, lethargy, and malaise (just not feeling well). Gradually the face swells and the lethargy gets worse as the trypanosomes attack the central nervous system. Eventually the desire to sleep overpowers the desire to eat, and the patient becomes malnourished. Finally the victim enters a coma from which he never awakens. Unlike most parasites that invade host cells to escape the immune system, *Trypanosoma* lives in the bloodstream exposed to antibodies. It survives by frequently changing its antigens (**variable surface glycoproteins**).

Chagas' disease, caused by *Trypanosoma cruzi.* Affects 12 million people in Central and South America. First invades macrophages, then attacks other tissues, especially nerve and muscle. Damage to the heart can cause death within a few years or after many years. Trypanosomes from the feces of the "kissing bug" *Triatoma* invade wounds or mucous membranes. (It was once thought that Charles Darwin's mysterious illness was Chagas' disease, but his symptoms are now known to have been quite different.)

Trichomoniasis, a common infection of the genital tract caused by *Trichomonas vaginalis* (Figure 22.5). Sometimes produces inflammation of the vagina but is often symptomless.

Giardiasis, caused by *Giardia lamblia* (Figure 22.14). Leeuwenhoek, curious about the cause of his own diarrhea, was the first to observe *Giardia.* It is common in the human intestine but does not always produce symptoms because cells of the gut lining are replaced every three or four days. In severe infections the entire lining of the intestine may be covered with the trophozoites. Symptoms then include abdominal pain, gas, and "explosive" diarrhea. Once considered an affliction of travelers to undeveloped countries, it is now common throughout the United States. Since the 1970s, careless defecation by increasing numbers of backpackers has brought *Giardia* cysts to even the remotest streams. Many animals are reservoirs for *Giardia,* continually reinfecting water. *Giardia* is unusual among eukaryotes in that it lacks mitochondria, endoplasmic reticulum, and Golgi bodies. These might have been lost as an adjunct of parasitism, but molecular studies suggest that *Giardia* is intermediate between prokaryotes and eukaryotes.

Amebic dysentery, caused by *Entamoeba histolytica.* Unlike the several other species of amoebas in the large intestine, *E. histolytica* sometimes attacks the tissues of the host. The infection may then spread to the liver or other organs, and damage to the colon may allow lethal bacteria to invade the abdominal cavity. More commonly the symptoms are diarrhea, vomiting, cramps, and malaise, which may be fatal to children. Approximately 10% of humans harbor the parasite, with about one-fourth of them showing symptoms. Transmission of cysts is through food and water, with flies and cockroaches often acting as vectors between exposed feces and food. The use of human feces as fertilizer is a common route of infection in much of the world.

Less beneficial are the dinoflagellates whose populations bloom to the point of creating "red tides" (which may actually be red, yellow, brown, colorless, or luminescent). *Saxitoxin*, a nerve blocker secreted by such dinoflagellates as *Gonyaulax*, can accumulate in clams and other molluscs. It does not appear to harm the molluscs but produces **toxic shellfish poisoning** in humans who eat them.

Several mutualistic species of *Trichomonas* and *Trichonympha* (Figure 22.5) inhabit the guts of termites and wood-eating cockroaches. Their combined weight may equal one-third the weight of the host. These flagellates, or perhaps the mutualistic bacteria within them, digest plant material that the animal host is incapable of breaking down. The host then "harvests" the excess population of protozoans. Since insects shed the inner lining of the gut during each molt, they lose their protozoans periodically. Many insects are then obliged to start a new culture by licking the anus of another insect. Some protozoans form cysts when the host starts producing the molting hormone ecdysone. They are then excreted before the host molts and find a new host by being consumed in the feces. Ruminant mammals such as cattle contain in their stomachs large numbers of ciliates that help digest the cellulose walls of plants. Unlike many termites and cockroaches, cattle fare just as well without their ciliates, so the symbiosis is probably commensal rather than mutualistic.

Table 22.3 Some important protozoan infections of animals.

Trypanosomiasis, caused by several species of *Trypanosoma*. Biology and transmission similar to that for human trypanosomiasis, described previously. Infects native ruminants of Central Africa and is usually fatal to cattle introduced from Europe.

Toxoplasmosis of cats, humans, and many other mammals, caused by *Toxoplasma gondii*. Cysts are transmitted through feces and undercooked meat. By means of the apical complex the trophozoite invades a cell, where it remains within a vesicle, unmolested by the immune system. Usually *Toxoplasma* lives in cells lining the gut, which are replaced rapidly enough that symptoms seldom occur. Severe cases can, however, kill kittens within a few weeks. Cats are especially liable to infection. Humans with immune deficiencies such as AIDS are also severely affected. Pregnant mammals, including women, can pass the infection to fetuses, with the resulting **congenital toxoplasmosis** causing miscarriages, stillbirths, and birth defects. Pregnant women should therefore avoid contact with cats.

Eimeria infection (coccidiosis) of numerous vertebrates, including turtles, chickens, ducks, sheep, and rabbits. Each host is infected by a particular species of *Eimeria*. Consequences include bloody diarrhea, sloughing of gut lining, and often death. *Eimeria tenella* causes expensive epidemics in the poultry industry.

Babesiosis, a disease of cattle, dogs, and other mammals, caused by *Babesia*. Like the malaria parasite, *Babesia* reproduces within red blood cells, causing fever. Death often occurs within a week in cattle newly introduced into an infested area. Economically, the most serious species is *B. bigemina,* which causes Texas cattle fever, also called red-water fever because of the bloody urine. This disease is not restricted to Texas but infects domestic as well as wild ruminants in all warm parts of the world. The principal vector is a tick (*Boophilus annulatus*) that transmits the protozoan to its offspring through the egg.

Whirling disease of salmonid fishes caused by *Myxosoma cerebralis* (= *Triactinomyxon gyrosalmo*). This protozoan attacks the cartilage of introduced species of salmonids, such as rainbow trout (*Salmo gairdnerii*). Damage to the organs of balance cause the fish to swim in circles. The protozoan responsible for whirling disease, *Myxosoma cerebralis,* is actually a different form of the protozoan *Triactinomyxon,* which was previously known to parasitize tubifex worms (phylum Annelida, class Oligochaeta) (Wolf and Markiw 1984). The appearance of the parasite differs so markedly in its two hosts that taxonomists had put the two forms into different classes.

Ick (= white-spot disease) in freshwater fishes, caused by *Ichthyophthirius multifiliis*. Attacks the skin, producing white spots. Can also produce fatal damage to gills. A serious pest for aquarium keepers.

Diseases.   Most people in undeveloped parts of the world are constantly aware of the ecological role of protozoa, because of the parasitic forms that give them and their animals diseases. Tables 22.2 (p. 477) and 22.3 (p. 478) summarize some of these diseases. A detailed treatment of them is possible only in a course on parasitology, but the later discussion of malaria will give some idea of the biological, economic, political, and sociological complexities of protozoan infections in general.

THE BIOLOGY OF MALARIA

The Disease.   Most people in developed countries think of malaria as a conquered disease, if they think of it at all. Yet there are 600 million people in the world with malaria, and the number is rapidly increasing. Each year approximately two million people, mostly children, die from the direct effects of malaria. Millions of others are so weakened that they cannot support themselves. Malaria therefore ranks as one of the most serious diseases of humankind. It is also one of the most difficult diseases to control, having defeated all the scientific breakthroughs that have regularly promised to eliminate it. At present, malaria is locked in battle against the latest techniques of biological engineering, and the outcome is uncertain. The difficulty of eliminating malaria is partly due to social factors (see box), but also to the intricate biology of the causative agent, the sporozoan *Plasmodium.*

The first symptoms of malaria—irregular chills and fever, malaise, and headache—are easily confused with influenza. These soon disappear, only to be followed by anemia and by debilitating chills and fever every few days. As the fever develops, shivering may literally rattle the bed. Fever occurs at two- or three-day intervals and is referred to as **tertian** or **quartan,** respectively. (These terms are

Figure 22.15
The life cycle of Plasmodium. *The stages that occur in red blood cells are superposed on a graph of the body temperature of a person with tertian fever.*

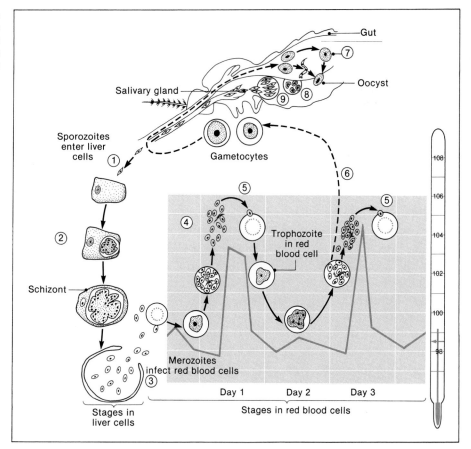

from the Latin for "three" and "four," following the Roman custom of starting a count from one rather than zero.) There are more than 50 species of *Plasmodium* that attack a variety of birds and mammals, but the most serious for humans is *P. falciparium*, which causes malignant tertian fever. *Plasmodium falciparum* is often fatal because it makes red blood cells attach to the inner surface of blood vessels, blocking circulation to the brain, kidney, heart, or other organs. *Plasmodium falciparum* causes about half the cases of malaria and almost all the deaths.

Until the turn of the century it was thought that malaria was caused by something in warm, humid air. (*Mal'aria* is Italian for "bad air.") The "something" turned out to be mosquitoes of the genus *Anopheles* (see Figure 13.3). Only a few of the 390 known species of *Anopheles* transmit the disease to humans. These are the species that occur in the same habitat as humans and that prefer human blood. Like many parasites, *Plasmodium* must complete different stages of its life cycle in two different hosts, both of which are highly specialized.

***Life Cycle of* Plasmodium.** The asexual stages of the life cycle occur in humans (steps 1 through 6 below), and the sexual stages occur in *Anopheles* (steps 7 through 9). Also like many other parasites, *Plasmodium* is highly selective about which tissues it attacks in each host. The life cycle, summarized in Figure 22.15, is as follows:

1. Humans acquire the *Plasmodium* parasite from *Anopheles* females that have previously fed on the blood of a malaria victim. (Male mosquitoes do not feed on blood.) While drawing the victim's blood, the mosquito ejects saliva containing several hundred *Plasmodium* in the **sporozoite** stage.

2. Within an hour each sporozoite that survives the immune system enters a liver cell. During the next 6 to 12 days, depending on the species, the sporozoite develops into a large, multinucleate **schizont,** which splits into 5000 to 10,000 individual cells by multiple fission (**schizogony**). The schizont eventually ruptures the liver cell, releasing thousands of **merozoites** into the blood. (*Plasmodium vivax* and *P. ovale* also produce **hypnozoites** in the liver. Hypnozoites remain dormant for long periods and cause relapses.)

3. Within seconds each merozoite invades a red blood cell and forms an ameboid **trophozoite.** The trophozoite feeds on the hemoglobin, turning it into an insoluble waste called **hemozoin.** The combination of erythrocyte destruction and the inability to recover the iron in hemoglobin produces anemia in the victim.

4. Each trophozoite undergoes schizogony once more, splitting into between 10 and 20 new merozoites that are released from what is left of the red blood cell. Each merozoite infects another red blood cell.

5. The release of the merozoites and/or their metabolic wastes triggers fever in the victim. At first the release is spread over time, producing a mild fever, but later it becomes synchronized with the daily rhythm in the host's body temperature, causing bed-rattling fever every 48 to 72 hours. (The synchronized release of merozoites may be an adaptation that swamps the victim's immune system with more invaders than it can handle at one time.)

6. After a few cycles some of the trophozoites develop inside blood cells into male **microgametocytes** and female **macrogametocytes,** rather than merozoites. If another female *Anopheles* attacks the victim, it can ingest some of these gametocytes. As they are being sucked in, the increase in pH resulting from the elimination of CO_2 from the surrounding blood triggers the gametocytes to develop into gametes. Several flagella sprout out of each male

gametocyte in a process called **exflagellation.**

7. In the gut of the mosquito, male gametes fertilize female gametes. The zygote then burrows into the gut lining, forming an **oocyst** (OH-oh-sist).

8. Thousands of sporozoites develop within the oocyst. This process, by which a zygote divides into many cells, is called **sporogony.** (Compare schizogony in 2 above.) Sporozoites are not spores. Since *Plasmodium* is always inside one or the other host, it never needs a spore stage to resist environmental stress. *Plasmodium* is therefore a sporozoan without spores.

9. The sporozoites migrate through the body cavity of the mosquito and enter the salivary glands to await the next human victim. The mosquito stage of the *Plasmodium* life cycle takes from 7 to 18 days, or longer if the temperature is low.

Natural Defenses. Adult humans mount an **immune response** against *Plasmodium* just as they do against any other foreign cell. Generally, however, the immune system is ineffective in preventing malaria. White blood cells have less than an hour to attack sporozoites, and it takes only one surviving sporozoite to

The Social Dimension of Malaria Research

All that we know about the biology of malaria was accumulated bit by bit by knowledgeable scientists motivated by the challenge of a difficult problem and by the wish to make a lasting and favorable impression on humanity. Students should know, however, that knowledge, curiosity, and good intentions do not ensure the smooth progress of science. In any kind of research with potentially important applications, social forces almost always intervene. The history of malaria research provides numerous examples of why scientists need to be aware of the political, economic, and cultural environment in which they work.

Malaria was described by Egyptians as early as 1500 B.C., so *Plasmodium* and *Anopheles* have had many centuries to adapt to the ways of *Homo*. One of the characteristics of humanity that *Plasmodium* has made good use of is the desire of some people to dominate others. Slave traders and Crusaders probably brought malaria into Europe from Africa and the Middle East. The Spanish Conquistadores then apparently brought it to America. Until the germ theory of disease was established, many Europeans probably thought malaria was God's punishment and racked their brains (and each other) trying to figure out what sin they had committed.

After Robert Koch and Louis Pasteur established that microbes cause diseases and that those diseases can be cured and prevented, the search began for a biological cause of malaria. Just as colonialism had been a major factor in spreading the disease, it became a major motive for stopping it. Malaria was a nuisance in European countries and the United States, but in tropical colonies it was an absolute plague, decimating legions of white soldiers, merchants, and administrators. It is no coincidence that

much of the research on malaria was by physicians attached to colonial armies.

In 1880 Louis Alphonse Laveran, a physician with the French army in Algeria, observed exflagellation in blood from malaria patients and concluded that a protozoan caused malaria. Acceptance of Laveran's discovery was slow in coming, for at least two reasons. First, other pathologists failed to confirm Laveran's findings of flagella thrashing vigorously about in the blood, because they habitually killed cells before looking at them. Second, Koch and Pasteur had emphasized that bacteria cause disease; the claim by an unknown military physician that protozoans can cause infection did not count for much in comparison. Eventually, however, most physicians were convinced.

The next step in the discovery of malaria's cause was due to Surgeon-Major Ronald Ross, a physician with the British army in India. Ross was passionately fond of poetry and mathematics and studied medicine only to satisfy his father. In India, however, he eventually became interested in malaria, which he studied mainly in his spare time in spite of superior officers who stupidly mismanaged his time. Eventually Ross discovered that only a few species of *Anopheles*, out of the swarms of other mosquitoes, transmitted the protozoan. He then knew that he had to look within those species to observe the development of *Plasmodium*, stained with the blood of its human victim. (Ross' superiors rewarded this crucial breakthrough in malaria research by assigning him to work on leishmaniasis as well!)

As a result of these discoveries, control of malaria did come to some areas, but slowly and more as a side effect of economics than of science. In the United States many

(continued)

breeding areas for *Anopheles* were drained, not so much to eliminate malaria as to increase the value of real estate. *Anopheles* was also inhibited by better housing and window screens and by mechanization, which allowed farm communities to spread out. Another economic factor was the cultivation of the tree *Cinchona ledgeriana*, from which the antimalaria drug quinine was derived. Quinine had been introduced to Europeans by South American natives in the 17th century, but it was too expensive for most people until *Cinchona* plantations were established in Southeast Asia. Quinine somehow disrupts the formation of merozoites in red blood cells and was the only treatment for malaria until the 1930s.

Americans had largely ceased to worry about malaria until Japan occupied the cinchona plantations during World War II. Scientists were then put to work developing alternatives to quinine, such as chloroquine. At about the same time DDT was developed, and the eradication of mosquitoes from large areas became feasible. After the war the World Health Organization began a massive campaign to exterminate *Anopheles*, and therefore malaria, with DDT. Parts of the world did not have the political or cultural organization for such an effort, but in other places the campaign was an overwhelming success. By the mid-1960s, after spending $6 billion (a lot of money in those days), malaria had been eliminated from 80% of its former range, and the number of cases dropped from 350 million in 1948 to 100 million in 1965. Most governments were so confident that malaria was doomed that they saw no reason to set up further research efforts, and young scientists avoided choosing malaria as a research topic.

It soon became clear, however, that *Anopheles*, like many other insects, was developing resistance to DDT, and that *Plasmodium* was becoming resistant to the antimalaria drugs. In the Vietnam War probably more soldiers were put out of action by chloroquine-resistant malaria than by bullets. Since that war the number of malaria victims has skyrocketed. Developed countries without large numbers of troops at risk were unconcerned until recently. The total spent in the United States is a paltry $26 million per year. Malaria, along with amebic dysentery, trypanosomiasis, and hookworm, has therefore become one of the Great Neglected Diseases of Mankind.

Recent advances in molecular biology have once again raised the hope that malaria can be defeated. The focus of the latest research is development of vaccines to induce the production of antibodies against sporozoites before the first infection occurs (Miller et al. 1986). A major hurdle has been to produce enough of the circumsporozoite protein to serve as an antigen to induce antibody production. This problem may be solved by the use of recombinant DNA techniques, using a cloned gene for the CS protein (see pp. 101–103). In 1984 two groups succeeded in cloning the gene for CS protein. Production of the vaccine was delayed, however, because of yet another nonscientific complication. The World Health Organization, which helped sponsor the research, insisted that all the research it aids be freely available to the public, while Genentech, which also supported the research and would manufacture the vaccine, was understandably reluctant to share trade secrets with potential rivals. Finally, the vaccine was produced, but initial trials on human volunteers have been disappointing.

cause the disease. Antibodies are produced too late to combat the first invasion of sporozoites, although they do respond to subsequent infections by attaching to the **circumsporozoite (CS) protein** that coats the sporozoites. Antibodies against CS protein are only partly effective, however, because the sporozoite simply sheds its CS coat and synthesizes a new one. Moreover, each CS protein has multiple copies of the antibody-attaching site, forcing the victim to use up copious amounts of antibody.

Although the immune system is largely ineffective in preventing malaria, it can reduce the symptoms of some forms of the disease to the point where victims can lead normal lives. The two major exceptions are, tragically, pregnant women and children under five. Pregnancy weakens the immune system, so malaria often strikes pregnant women, resulting in miscarriages, stillbirths, and prematurity. The immune systems of children under five are not yet mature enough to protect them.

Another type of natural defense against malaria is **sickle-cell trait,** often found in those whose ancestors came from malaria-infested parts of Africa. A single base substitution in the gene for the alpha chain of hemoglobin causes the red blood cell to become sickle-shaped in low oxygen. The change in shape causes potassium to leak out of the red blood cell, and the low K^+ concentration kills the trophozoites. Presumably sickle-cell trait evolved because the advantage of resistance to malaria outweighed the disadvantages of sickle-cell anemia. **Thalassemia,** a genetic variant common around the Mediterranean and in Southeast Asia, also confers resistance to malaria. In thalassemic red blood cells, the membrane is damaged by hydrogen peroxide released by trophozoites. This allows K^+ to escape, killing the parasite.

SUMMARY

Although most consist of only one cell, protozoans can nevertheless carry out all the functions of life. The large ratio of surface to volume eliminates the need for respiratory and circulatory systems, leaving locomotion, feeding, and reproduction as the main tasks. Many perform intricate behaviors, such as phototaxis and chemotaxis, in spite of the absence of a nervous system. Protozoans were formerly classified into four groups: the sarcodines, the ciliates, the flagellates, and the parasitic sporozoans. This classification is no longer accepted, although the informal names for those groups are still useful.

Most protozoans reproduce asexually by fission. Some, however, recombine genetic material from two individuals and are therefore said to be sexual. Conjugation is a common form of sexual reproduction in ciliates. It is generally assumed that animals evolved from protozoans. The colonial flagellate hypothesis is that animals evolved from an organism like *Volvox.* Alternatively, the syncytial ciliate hypothesis proposes that animals evolved from a ciliate in which the nuclei became separated by internal membranes. There is strong evidence against both hypotheses.

Protozoans interact with humans and other animals in a variety of ways. Foraminiferans were responsible for massive stone and chalk deposits and serve as indicator fossils for petroleum geologists. Protozoans are also important links in food chains and as decomposers. They also form important symbioses with some animals, such as reef-building corals. Many protozoans cause diseases in humans and other animals. One of the most serious is malaria.

KEY TERMS

pseudopodium
sarcodine
ciliate
flagellate
trophozoite
binary fission
budding
conjugation
autogamy

syngamy
spore
cytostome
food vacuole
lysosome
contractile vacuole
pellicle
trichocyst
phagocytosis

micronucleus
macronucleus
cyst
zooxanthella
sporozoite
merozoite
gametocyte

SELF-TEST

1. Based on its appearance, characterize each of the protozoans in Figure 22.1 as flagellate, ciliate, sarcodine. (Why aren't sporozoans represented here?)

2. In a pool of stagnant water, where would you expect to find the following types of Protozoa, and why? Photosynthesizing flagellates, other flagellates, amoebas.

3. Describe a process of asexual reproduction in a particular protozoan. Describe a process of sexual reproduction in a particular protozoan.

4. Sketch a *Paramecium* from memory, being sure to include the following organelles: cilia, contractile vacuole, food vacuole, cytostome, cytoproct, macronucleus, micronucleus. State the function of each of these organelles.

5. Describe the colonial flagellate theory of the origin of animals and discuss its validity.

6. What are foraminiferans and radiolarians? In what ways are they similar? How do they differ from each other? Why are geologists so interested in them?

7. Describe three diseases of humans or other animals caused by protozoans. Explain how the protozoan produces the disease symptoms, and how it is transmitted.

8. Starting with the sporozoite, arrange in correct sequence the following stages of *Plasmodium:* sporozoite, gametocyte, oocyst, schizont, trophozoite, merozoite, gamete. State where each stage occurs—in which host and in which organ.

READINGS

RECOMMENDED READINGS

Desowitz, R. S. 1981. *New Guinea Tapeworms and Jewish Grandmothers: Tales of Parasites and People.* New York: W. W. Norton. (*Informative and entertaining.*)

Donelson, J. E. and M. J. Turner. 1985. How the trypanosome changes its coat. *Sci. Am.* 252(2):44–51 (Feb).

Friedman, M. J. and W. Trager. 1981. The biochemistry of resistance to malaria. *Sci. Am.* 244(3):154–164 (Mar). (*How sickle-cell trait and thalassemia confer resistance to malaria.*)

Godson, G. N. 1985. Molecular approaches to malaria vaccines. *Sci. Am.* 252(5):52–59 (May).

Harrison, G. 1978. *Mosquitoes, Malaria and Man: A History of the Hostilities Since 1880.* New York: E. P. Dutton. (*An entertaining and instructive history of attempts to control malaria.*)

Jahn, T. L., E. C. Bovee, and F. F. Jahn. 1979. *How to Know the Protozoa*, 2nd ed. Dubuque, IA: Wm. C. Brown. (*A guide to identification.*)

Kabnick, K. S. and D. A. Peattie. 1991. *Giardia:* A missing link between prokaryotes and eukaryotes. *Am. Sci.* 79:34–43.

Kolata, G. 1984. Scrutinizing sleeping sickness. *Science* 226:956–959.

Lewin, R. 1982. Nairobi laboratory fights more than disease. *Science* 216:500–503.

McKelvey, J. J. Jr. 1973. *Man Against Tsetse.* Ithaca, NY: Cornell University Press. (*An account of attempts to control sleeping sickness.*)

Wakelin, D. 1984. *Immunity to Parasites: How Animals Control Parasitic Infections.* Baltimore: Edward Arnold.

See also relevant selections of General References listed at the end of Chapter 21.

ADDITIONAL REFERENCES

Anderson, O. R. 1983. *Radiolaria.* New York: Springer-Verlag.

Applewhite, P. B. 1979. Learning in protozoa. In: M. Levandowsky and S. H. Hutner (Eds.). *Biochemistry and Physiology of Protozoa*, 2nd ed. New York: Academic Press, Vol. 1, Chap. 11.

Baroin, A. et al. 1988. Partial phylogeny of the unicellular eukaryotes based on rapid sequencing of a portion of 28S ribosomal RNA. *Proc. Natl. Acad. Sci. USA* 85:3474–3478.

Cronkite, D. L. 1979. The genetics of swimming and mating behavior in *Paramecium.* In: M. Levandowsky and S. H. Hutner (Eds.), *Biochemistry and Physiology of Protozoa*, 2nd ed. New York: Academic Press, Vol. 2, Chap. 6.

Fenchel, T. 1986. *Ecology of Protozoa: The Biology of Free-Living Phagotrophic Protists.* Madison, WI: Science Tech/New York: Springer-Verlag.

Field, K. G. et al. 1988. Molecular phylogeny of the animal kingdom. *Science* 239:748–753.

Gould, S. J. 1980. Crazy old Randolph Kirkpatrick. In: *The Panda's Thumb.* New York: W. W. Norton, Chap. 22.

Hanson, E. D. 1977. *The Origin and Early Evolution of Animals.* Middletown, CT: Wesleyan University Press.

Jeon, K. W. (Ed.). 1961. *The Biology of Amoeba.* New York: Academic Press.

Klayman, D. L. 1985. *Qinghaosu* (Artemisinin): an antimalarial drug from China. *Science* 228:1049–1055.

Kung, C. and Y. Saimi. 1982. The physiological basis of taxes in *Paramecium. Annu. Rev. Physiol.* 44:519–534.

Laybourn-Parry, J. 1984. *A Functional Biology of the Free-Living Protozoa.* Berkeley: University of California Press.

Lee, J. J., S. H. Hutner, and E. C. Bovee (Eds.). 1985. *The Illustrated Guide to the Protozoa.* Lawrence, KA: Society of Protozoologists.

Levine, N. D. et al. 1980. A newly revised classification of the protozoa. *J. Protozool.* 27:37–58.

Marshall, E. 1990. Malaria research—what next? *Science* 247:399–402.

Miller, L. H. et al. 1986. Research toward malaria vaccines. *Science* 234:1349–1356.

Naitoh, Y. and R. Eckert. 1969a. Ionic mechanisms controlling behavioral responses of paramecium to mechanical stimulation. *Science* 164:963–965.

Naitoh, Y. and R. Eckert. 1969b. Ciliary orientation: controlled by cell membrane or by intracellular fibrils? *Science* 166:1633–1635.

Qu, L.-H., M. Nicoloso, and J.-P. Bachellerie. 1988. Phylogenetic calibration of the 5′ terminal domain of large rRNA achieved by determining twenty eucaryotic sequences. *J. Mol. Evol.* 28:113–124.

Rausch, H., N. Larsen, and R. Schmitt. 1989. Phylogenetic relationships of the green alga *Volvox carteri* deduced from small-subunit ribosomal RNA comparisons. *J. Mol. Evol.* 29:255–265.

Rudzinska, M. A. 1973. Do suctoria really feed by suction? *BioScience* 23:87–94.

Sogin, M. L. and H. J. Elwood. 1986. Primary structure of the *Paramecium tetraurelia* small-subunit ribosomal RNA coding region: phylogenetic relationships within the ciliophora. *J. Mol. Evol.* 23:53–60.

Sogin, M. L., H. J. Elwood, and J. H. Gunderson. 1986. Evolutionary diversity of eukaryotic small-subunit rRNA genes. *Proc. Natl. Acad. Sci. USA* 83:1383–1387.

Spoon, D. M. et al. 1976. Observations on the behavior and feeding mechanisms of the suctorian *Heliophrya erhardi* (Rieder) Matthes preying on *Paramecium. Trans. Am. Microsc. Soc.* 95:443–462.

Van Houten, J. 1979. Membrane potential changes during chemokinesis in *Paramecium. Science* 204:1100–1103.

Wolf, K. and M. E. Markiw. 1984. Biology contravenes taxonomy in the Myxozoa: new discoveries show alternation of invertebrate and vertebrate hosts. *Science* 225:1449–1452.

23

Sponges, Placozoa, and Mesozoa

Red vase sponge (Mycale *sp.*).

CHAPTER OUTLINE

LEARNING OBJECTIVES

1. Are sponges and other such simple organisms animals?

2. How do sponges obtain food and the other necessities of life while fixed to substratum?

3. What is the significance of the "spongy" structure of sponges?

4. Do sponges have organs? How are their cellular activities coordinated?

5. Do sponges have sex?

ARE SPONGES AND OTHER SIMPLE ORGANISMS ANIMALS?

The boundaries separating the kingdoms of life are as uncertain as those that separated the kingdoms of humans before there were accurate surveys and maps. Some of the colonial protozoans seem on the verge of invading the animal kingdom, while the animals discussed in this chapter seem to owe their allegiance to the kingdom Protista. Systematists avoid boundary disputes between protozoans and animals by defining animals as organisms that:

1. Are multicellular, with more than one type of cell;
2. Are heterotrophic (cannot produce their own food);
3. Reproduce sexually (at least sometimes) from a zygote formed from two different haploid gametes;
4. Go through a blastula stage during embryonic development.

The organisms described in this chapter, Porifera, Placozoa, and Mesozoa, satisfy these criteria, but in many other respects they are quite different from other animals. The major difference is that the functions of the cells are not coordinated by neural, hormonal, or other mechanisms. It is sometimes said that they have a cellular grade of organization rather than an organ-system grade of organization.

———— SPONGES: PHYLUM PORIFERA ————

INTRODUCTION

The structure and functioning of sponges are so different from those of other animals that sponges are often considered to have evolved separately and in parallel to most other metazoans. Together with the one species in the phylum Placozoa, they are often referred to as Parazoa (Greek *para* beside + *zoon* animal). It was not always clear that sponges were animals at all. Aristotle was inclined to think they were, but in the 16th and 17th centuries many people thought sponges were dried sea foam or shelters constructed by worms. It was not until 1766 that Aristotle's belief was confirmed by the discovery that sponges expel water. They were thought to be related to corals (phylum Cnidaria) at first, until one of Darwin's teachers, R. E. Grant, proposed for them the separate phylum Porifera.

As we saw in the previous chapter, protozoans make quite a good living as single cells. About 550 million years ago, however, another way of life evolved in the sponges. Instead of propelling their bodies through water and extracting food from it, they propel the water through their stationary bodies. The most significant features of sponges are all related to this basic plan (Table 23.1). These features are the canal system through which water circulates in the body, the cells that pump the water (the **choanocytes**), and the endoskeleton that keeps the body from collapsing.

In the simplest sponges the canal system consists of a central chamber, the **spongocoel** (SPUN-jo-seal), which is lined with choanocytes (= collar cells) (Figure 23.1). Choanocytes resemble *Codosiga* and some other flagellated protozoans in form and function (see Figure 22.5). The beating of their flagella draws in a continuous stream of water that brings in food and oxygen and removes wastes. Surrounding each flagellum is a fringe of microvilli, the collar, which traps the food that the choanocyte then phagocytizes. The choanocyte either uses the food itself or transfers it to **amebocytes.** Amebocytes distribute food to cells in the **mesohyl,** which is the part of the sponge not exposed to water. The life-giving

Table 23.1 Characteristics of sponges.

Phylum Porifera pore-IF-er-uh (Latin *porus* pore + *ferre* to bear).

Morphology: Multicellular animals without tissues or organs. Asymmetric or with radial symmetry. Numerous microscopic **ostia** by which water enters the **canal system** through the body. One or few **oscula** from which water exits. Water propelled by **choanocytes. Endoskeleton** of protein fibers and/or mineral spicules. No germ layers in embryo.

Physiology: A cellular grade of organization: physiological functions performed by individual cells. Osmoregulation in freshwater species by contractile vacuoles. No nervous, muscular, or hormonal systems.

Locomotion: Adults **sessile.** Larvae swim by means of flagella.

Reproduction: Asexual reproduction by buds, and **sexual** reproduction. Most species hermaphroditic. Fertilization external.

Development: Radial cleavage pattern. Development indirect (with a juvenile stage, called a larva, that does not resemble the adult).

Habitat, Size, and Diversity: All species aquatic; most marine. Distribution worldwide, including Arctic and Antarctic. Sizes range from 1 mm to 2 meters. Dull white or gray, or with red, orange, yellow, or purple pigments. Some freshwater species green from symbiotic algae. Approximately **5000 living species** described.

water enters the sponge through numerous microscopic **ostia** (the pores that give poriferans their name) and exits by one large **osculum** or several oscula. A unique skeleton of protein fibers and mineral **spicules** (little spikes) keeps the whole thing from collapsing.

CANAL SYSTEMS

As noted above, the basic body plan of sponges is a system of canals through which water brings in food and oxygen. Because there is no other circulatory system, each cell must be close to the water, and because water often has little organic material in it, quite a lot of it has to be pumped through. According to one estimate, a marine sponge has to pump 30,000 grams of water to add 1 gram of cells to its mass. As a sponge grows, however, the volume of cells that must be supplied increases as the cube of body length, while the area available for choanocytes increases only as the square of body length. This fundamental principle therefore limits the size of sponges. Sponges with the simplest body plans have an **asconoid** (bag-shaped) canal system (Greek *askos* bag). In asconoid systems the choanocytes occur only on the lining of the spongocoel (Figures 23.1A and 23.2A). Because of the limited surface area for choanocytes, asconoid sponges never grow larger than 10 centimeters.

Other sponges can grow larger because they have more elaborate canal systems that provide more area for choanocytes. In the **syconoid** canal system (Figure 23.2B) the choanocytes do not line the spongocoel but are found within numerous **radial canals.** The spongocoel of syconoid sponges serves merely as an exhaust leading to the osculum. Still more area is available for choanocytes in the majority of sponges, which have **leuconoid** canal systems. In most leuconoids there is no spongocoel at all. The choanocytes are located within numerous **flagellated chambers,** which draw in water from **incurrent canals** and expel it by an **excurrent canal** directly to the osculum. One 7-cm leuconoid sponge was estimated to have several million flagellated chambers. The large number of choanocytes enables some leuconoid sponges, such as the loggerhead, *Spheciospongia vesperia*, to reach a height of 2 meters and to pump approximately 2 liters of water every minute of every day.

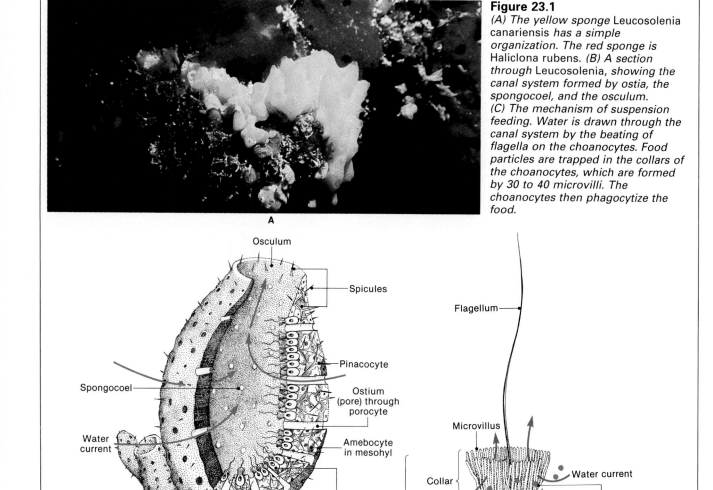

Figure 23.1
(A) The yellow sponge Leucosolenia canariensis *has a simple organization. The red sponge is* Haliclona rubens. *(B) A section through* Leucosolenia, *showing the canal system formed by ostia, the spongocoel, and the osculum. (C) The mechanism of suspension feeding. Water is drawn through the canal system by the beating of flagella on the choanocytes. Food particles are trapped in the collars of the choanocytes, which are formed by 30 to 40 microvilli. The choanocytes then phagocytize the food.*

THE ENDOSKELETON

The canal system will function only if the body can be prevented from collapsing. That is the function of the endoskeleton, which consists of a network of protein fibers and (usually) mineral spicules. The fibers are made of collagen, including a unique form called **spongin.** The spicules are either **calcareous** or **siliceous** and range in length from 10 μm to 40 cm, depending on the species. Calcareous spicules are composed mainly of calcium carbonate (in the form of calcite or aragonite); siliceous spicules are made of a silicon-containing mineral. In addition to helping support the body, the spicules also deter predators.

Figure 23.2
The types of canal system found in sponges. Darkened areas show the locations of choanocytes. (A) In asconoid sponges (Greek askos sac) the choanocytes are restricted to the spongocoel. (B) In syconoid sponges, named for the genus Sycon (see Figure 23.4A), the choanocytes occupy the sides of radial canals. The left and right halves of this figure show simple and complex versions of the syconoid arrangement. In the simple syconoid each ostium leads into only one radial canal. In the complex system each ostium conducts water into more than one radial canal. (C) Leuconoid sponges (named for Leuconia, now Leucandra) have flagellated chambers distributed throughout.

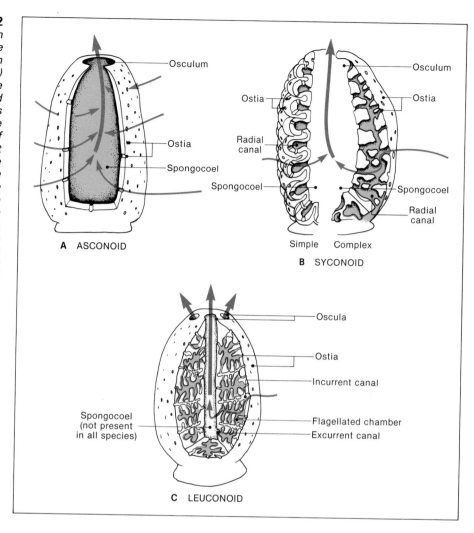

Some siliceous spicules are constructed on a strand of protein that serves as a kind of scaffolding. Cells called **sclerocytes** secrete the silica onto the strand, and the silica then crystallizes into a spicule with the required shape. Some calcareous spicules are constructed without such a scaffolding. In order to construct a three-rayed spicule, three sclerocytes congregate at the construction site and each divides into a **founder cell** and a **thickener cell** (Figure 23.3). Each founder cell moves

Figure 23.3
Formation of a spicule. (A) Three sclerocytes assemble. (B) They develop into three thickener cells and three founder cells; the latter lay down a sliver of calcite. (C) The founder cells move apart, elongating the calcite rays. (D) The thickener cells add to the spicule thickness.

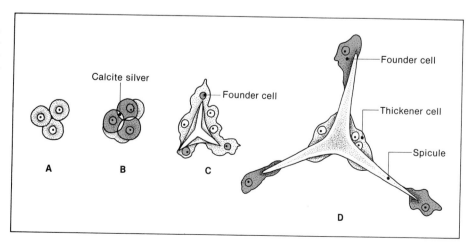

away, laying down a ray of calcite, and the thickener cells then move along these rays adding more $CaCO_3$. When the spicule is completed, all six cells slide off the rays and disintegrate. Virtually nothing is known about the factors that guide sclerocytes in constructing the spicules appropriate for their species.

CLASSIFICATION

The spongy nature of poriferans makes it difficult to classify them on the basis of their shapes. Taxonomists have therefore had to rely largely on their only solid and regular parts, the spicules. Three classes of sponges are now generally recognized, based largely on the composition and shape of spicules.

Class Calcarea (= class Calcispongiae) is characterized by calcite spicules with three or four rays (Figure 23.4). Calcarea are commonly called calcareous, limy, or chalky sponges. Their bodies are bristly with spicules, and they are usually dull-colored and less than 4 cm long. Calcareous sponges are all marine and more common in shallow than in deep water.

Sponges in **class Hexactinellida** (= class Hyalospongiae), commonly known as the glass sponges, are named for the six-rayed siliceous spicules. Often their spicules are cemented together into a roughly cylindrical or conical skeleton 10 to 30 cm tall, as in the beautiful Venus' flower basket *Euplectella aspergillum*, whose delicate skeleton graced many a Victorian mantel (Figure 23.5). This class is one of the least understood, owing to their inaccessibility in deep oceans. They are never found in depths less than 23 meters, and most live between 200 and 2000 meters. Some have been dredged up from depths greater than 6 km. Structural peculiarities include the lack of mesohyl and epidermis (pinacoderm), both of which are replaced by syncytia (fused cells). The linings of the flagellated chambers are also syncytial. Because of this syncytial rather than cellular grade of organization, some consider the Hexactinellida to belong to their own subphylum (Reiswig and Mackie 1983) or even to a separate phylum Symplasma (Bergquist 1985).

Species in **class Demospongiae** also have siliceous spicules (or none at all), but the spicules are not six-rayed (Figure 23.6). This class includes 90% of all species, mainly because taxonomists tend to include within it any sponge they cannot fit into the other classes. Demosponges can be found at all depths of the ocean down to 9 km, although each species has its preferred depth. All 150 of the freshwater species of Porifera belong to this class. Also included in this class are the bath sponges (*Spongia* and *Hippospongia*) that (obviously) do not have spicules.

Classes Calcarea and Demospongiae also include the **coralline sponges,** which are so remarkable that for about two decades they were given their own class (Sclerospongiae) (Vacelet 1985; Wood 1990). The skeletons of coralline sponges consist of a large base of calcium carbonate, in addition to siliceous spicules (Figure 23.7). Cells grow in a thin layer within and over the base. Also distinctive are the bulging excurrent canals that form star-shaped patterns (astrorhizae) radiating from each osculum. Fossil coralline sponges were long known, but all were thought to have become extinct until the 1960s. Some living species had been misclassified as corals. Many others had remained undiscovered on steep, dark reef banks where they were missed by dredging equipment (Hartman and Goreau 1970).

A

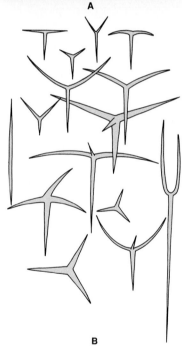

B

Figure 23.4
(A) Sycon, a member of class Calcarea. (B) Three- and four-rayed spicules from various calcareous sponges.

A CELLULAR GRADE OF ORGANIZATION

Sponges consist not only of choanocytes and sclerocytes but several other cell types as well (Table 23.2). Except for sharing food and reaggregation (see below),

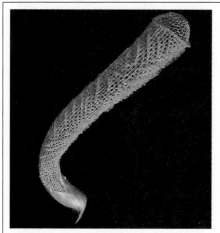

Figure 23.5
(A) The skeleton of Venus' flower
basket *Euplectella aspergillum, a
hexactinellid. Note the fusion of
spicules and the spun-glass
appearance at the end.
Approximately 30 cm long.
(B) Spicules of hexactinellids are
generally six-rayed.*

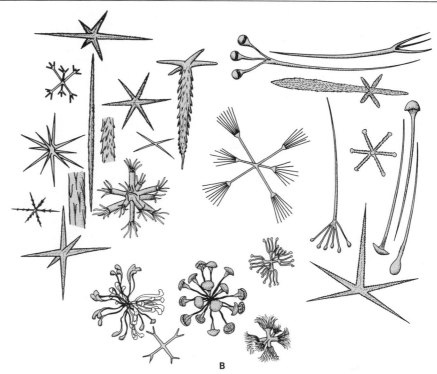

B

these cells generally act as if they were unconcerned with one another. They do
not form tissues, but simply cell layers. The cell layers are separated by a gelatinous mesohyl (often called mesenchyme or mesoglea) (see p. 492) that contains
cells of several types. There is no coordination of cellular activities by hormones
or nerves. In at least one species of glass sponge, however, rapidly conducting,
all-or-none electrical signals do turn off all the choanocytes simultaneously. This
electrical conduction is attributed to a network of syncytial amebocytes, called
neuroid cells (Mackie 1990). Neuroid cells are not nerve cells, and nerve cells
did not evolve from them.

Reaggregation. One of the most remarkable displays of a cellular grade of
organization occurs when cells of sponges are separated from each other. Not

Extant Classes of Phylum Porifera

Genera mentioned elsewhere in this chapter are noted.

Class Calcarea kal-sa-REE-uh (Latin *calcis* limestone).
Skeleton composed of spicules of calcium carbonate,
needle-shaped or with three or four rays. Includes species with asconoid, syconoid, and leuconoid canal systems. All marine. *Leucandra, Leucosolenia, Sycon*
(Figures 23.1, 23.4).

Class Hexactinellida hex-ACT-in-ELL-id-uh (Greek *hex* six
+ *aktis* ray). Siliceous six-rayed spicules, often fused into
networks. Syconoid or leuconoid. No pinacocytes, poro-

cytes, or mesohyl. Choanocytes fused (syncytial). All
marine. *Euplectella* (Figure 23.5).

Class Demospongiae dim-oh-SPONGE-ee-ee (Greek
demos people + *spongia* sponge). Skeleton of siliceous
spicules and/or spongin fibers, or (rarely) neither. Spicules are not six-rayed as in the hexactinellids. All
leuconoid. *Ceratoporella, Cliona, Cryptotethya, Halichondria, Haliclona, Hippospongia, Microciona, Mycale,
Spheciospongia, Spongia, Spongilla, Verongia* (Figures
23.6, 23.7, 23.8, 23.9).

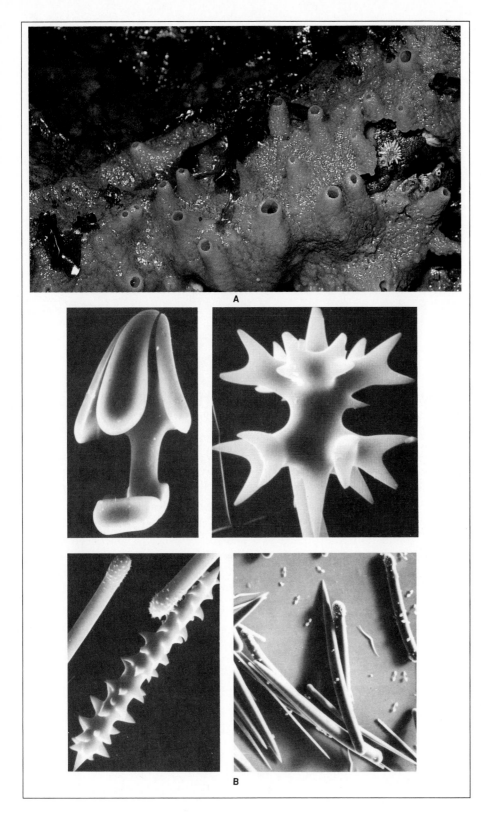

Figure 23.6
(A) Crumb-o'-bread sponge
Halichondria panicea, *so called because dried specimens resemble moldy bread. This demosponge encrusts rocks and seaweed and can often be seen at low tide.*
(B) Scanning electron micrographs of spicules from various demosponges.

A

B

''Mesohyl,'' ''mesoglea,'' ''mesenchyme'' are three of the confusing terms applied to the gelatinous material between inner and outer cells in a variety of invertebrates. Some of these terms are defined below. This text uses the term most common for each group of animals.

Mesenchyme: Primitive middle layer derived at least in part from ectoderm. Consists of cells and cell products in a jellylike mesoglea.

Mesoglea: Jellylike portion of mesenchyme. Synonymous with mesenchyme that contains no cellular material, as in many cnidarians and ctenophores.

Collenchyme: Mesenchyme with little cellular material, as in some cnidarians and ctenophores.

Parenchyme: Mesenchyme in which cells are densely packed, especially in acoelomates. Often spelled parenchyma.

only can they survive by themselves for a considerable time, but in some species they can actually reassemble into a working sponge. This phenomenon, called reaggregation, was discovered by H. V. Wilson around the turn of the century. It was long forgotten until some scientists began to use it as a model system for the study of transplant rejection. Wilson forced a redbeard sponge *Microciona pro-*

Table 23.2 Cells of sponges.

Cells of Inner or Outer Layers

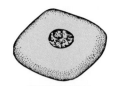

PINACOCYTE

Ostium — POROCYTE

Choanocytes (= collar cells; Figure 23.1):
1. Propel water through the sponge with their flagella.
2. Trap food particles with the collar, phagocytize them, and transfer them to amebocytes.
3. Give rise to cells that form sperm and ova in some species.

Pinacocytes line the outer surface (= epidermis, pinacoderm) and the nonflagellated channels in most sponges. (In the class Hexactinellida the epidermis is replaced by a syncytium—a multinucleated mass of cytoplasm resulting from the merging of many cells.)

Porocytes are tubular pinacocytes that surround each pore (except in hexactinellids). They can regulate water flow by contracting.

Cells in Mesohyl

AMEBOCYTE

Spongin

SPONGOCYTE

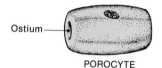

Myocytes contract around oscula and channels, thereby regulating water flow (in some demosponges).

Amebocytes
  Archaeocytes:
  1. Undergo differentiation into any other cell type.
  2. Transport food vacuoles from choanocytes into mesohyl.
  3. Form sperm and ova in some species.

Sclerocytes (= scleroblasts) form spicules. (Founder and thickener cells; Figure 23.3.)

Collencytes, lophocytes, and **spongocytes** form fibers of the protein collagen. The spongocytes secrete a form of collagen called **spongin,** which is unique to sponges.

Figure 23.7
(A) A sclerosponge Ceratoporella nicholsoni *photographed in its natural habitat, a reef off Jamaica. The star-shaped astrorhizal pattern is apparent around each osculum. (B) A schematic section of a sclerosponge showing the siliceous spicules and basal calcareous mass.*

A

Osculum

Astrorhizae

Excurrent canal

Incurrent canal

Flagellated chamber

Silicious spicule

Calcareous skeleton

B

lifera through a cloth to separate its cells. When allowed to stand in seawater for a few weeks, the cells reaggregated into a functioning sponge again. Wilson also mixed cells from the red *M. prolifera* with cells from the purple *Haliclona oculata*, and within a few weeks they somehow sorted themselves out and reaggregated into one red and one purple sponge again. Some sponges not only refuse to aggregate with cells of another species, but they also will not reaggregate with cells of another individual of the same species (Hildemann et al. 1979). Generally, however, sponges of the same species readily merge with each other. Reaggregated cells link together by means of a complex of proteins and carbohydrates called **aggregation factor.**

REPRODUCTION

Asexual Reproduction. Sponges can reproduce asexually by **budding.** In some species, mainly the hexactinellids, archaeocytes form external buds, usually on long stalks. The buds then drift away and form new sponges. Freshwater sponges, and some marine species, such as the boring sponge *Cliona lampa*, can reproduce from internal buds, called **gemmules.** Gemmules are small (0.3 mm), spherical structures capable of surviving environmental stresses, such as would be faced in a pond in winter (Figure 23.8). Before winter some unknown stimulus, perhaps reduced light or temperature, triggers gemmule formation. Archaeocytes group together, and a tough coat of spongin and spicules surrounds them. In spring the archaeocytes in the gemmules develop into choanocytes and all the other cell types required to start a new sponge.

Sexual Reproduction. Like all animals, sponges are also capable of reproducing sexually. Since they cannot move about to find mates they depend on the vicissitudes of water currents to ensure that sperm find an ovum of the correct species. Most individuals are hermaphroditic (monoecious), but each one usually produces ova and sperm at different times, thereby avoiding self-fertilization. The sexes look alike in diocious species. By examining the gametes produced by each individual, it has been found that females usually outnumber males. Sperm are often released in such prodigious numbers that divers report "smoking sponges" (see Figure 16.2A). The sperm drift downstream and are taken into the ostia of other sponges, where the choanocytes phagocytize them as if they were food. If the sperm are of the same species, however, the choanocytes do not pass them on to amebocytes for distribution as food. Instead, the choanocytes lose their collars and flagella, and they themselves carry the sperm to egg cells.

DEVELOPMENT

Development in sponges is indirect, meaning that there is a larval form that does not resemble the mature sponge. The larvae have flagella by which they move about, which they must do to disperse to new habitats. Some species are oviparous, with the eggs being released soon after fertilization. Most sponges, however, are viviparous, with some larval development occurring in the mesohyl of the parent. Often **nurse cells** attend the developing larva. In one study a sponge gave birth to an average of four to five larvae every minute for three or four days. Most demosponges, and probably some species in other classes, produce larvae called **parenchymulae.** The parenchymula larva is essentially a solid blastula (stereoblastula) covered with flagella (Figure 23.9A). Using the flagella, the parenchy-

A

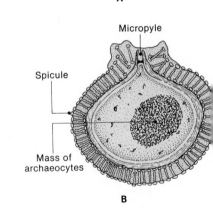

B

Figure 23.8
(A) The freshwater sponge Spongilla *in autumn, filled with gemmules. This sponge is frequently encountered on submerged vegetation in sunlit ponds and streams. Symbiotic algae give it a green color in spring and autumn. Scale in mm. (B) Diagram of a sectioned gemmule of* Spongilla. *In spring the archaeocytes will spill out of the micropyle and develop into new sponges.*

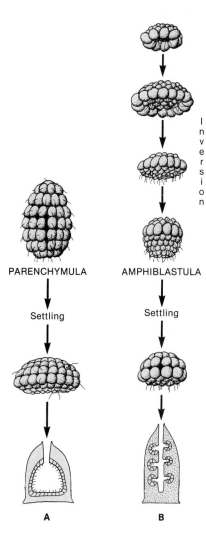

Figure 23.9
Development of two types of sponge larva. (A) A parenchymula larva. It was formerly believed that the external flagellated layer inverted inward and produced the choanocyte chamber after settling, but this does not appear to be the case (Bergquist and Glasgow 1986). (B) An amphiblastula larva.

PARENCHYMULA

AMPHIBLASTULA

Inversion

Settling

Settling

A

B

mulae swim about for up to two days. If they are lucky enough not to become food for some other organism and to find suitable substratum, they **settle** and begin developing into a sponge.

In most calcareous sponges and some demosponges there is a different type of larva, the **amphiblastula** (Figure 23.9B). The amphiblastula larva develops from a hollow blastula (coeloblastula) in which flagella are pointed inward within the blastocoel. The larva then undergoes **inversion,** which pushes the flagellated cells to one end (the animal pole) of the outer surface, allowing the larva to swim. The amphiblastula is then released from the parent's mesohyl and soon settles. After settling, the flagellated cells of the amphiblastula turn inward once more and become choanocytes.

INTERACTIONS WITH HUMANS AND OTHER ANIMALS

Considering that sponges lack teeth and claws and are not even capable of fleeing, they are remarkably free of predators. Apparently few animals relish a mouth full of spongin and spicules. However, a few bony fishes and the hawksbill turtle (*Eretmochelys imbricata*) feed exclusively on sponges (Meylan 1988). Some molluscs also prey on marine sponges, and the larvae of spongilla flies (order Neuroptera, family Sysiridae) feed on freshwater sponges. Sponges are generally free of bacterial infections, thanks to antibacterial secretions. Many coral-reef sponges, however, live mutualistically with cyanobacteria inside them—something no other animal does. The sponge provides a sunlit habitat for the cyanobacteria, and the cyanobacteria produce nutrients and oxygen (Wilkinson 1983).

The greatest problem faced by sponges is not predation or infection, but finding suitable substratum for attachment. Their chief competitors in this contest are the corals. Many sponges avoid being smothered by coral by secreting substances that kill corals or inhibit their growth (Sullivan et al. 1983). Boring sponges, such as *Cliona lampa*, create their own habitat by chemically etching holes in corals and shells. Their work can be found on just about any shell on a beach. Boring sponges are so active around Bermuda and some other places that they have changed the reef structure. Some sponges are provided a substratum by the decorator crab (*Dromia*), which attaches sponges to itself for camouflage and protection.

Numerous animals live inside the channels of sponges, taking advantage of the protection and continuous flow of water. One specimen of the loggerhead sponge *Spheciospongia vesparia* from Florida was found to harbor 16,000 snapping shrimp (*Synalpheus brooksi*)—impressive hospitality even for a sponge that can grow 2 meters tall. Rachel Carson in *The Edge of the Sea* comments on the considerable racket produced by so many snapping claws. Venus' flower basket *Euplectella* (see Figure 23.5) often contains a male and a female shrimp (*Spongicola*) that have set up housekeeping within the protective skeleton of the sponge and then grown too large to escape. Dried specimens were once popular gifts to newlyweds in Japan, symbolizing the wish that the couple remain together until death.

Humans have valued bath sponges (*Spongia* or *Hippospongia*) since the Bronze Age, nearly 4000 years ago. What is so valuable is not the entire sponge, but the skeleton that remains after the sponge is killed, dried, and kneaded, leaving the spicule-free spongin skeleton with its useful ability to hold up to 35 times its weight in liquid. As the demand for sponges has grown, divers have had to dive as deep as 30 meters for them. Until recently their only diving equipment was a heavy rock. Many divers off Tunisia still use only a lead weight and a rope. Kalymnos, Greece, was a major port for sponge divers for thousands of years until a blight wiped out their sponges, and other Mediterranean countries closed their territorial waters to foreign divers. Many of the Greek divers have now gone to Tarpon Springs, Florida, whose sponge industry has recently recovered from killer fungi and red tides (population explosions of dinoflagellates).

A wild sponge takes about five years to reach marketable size (12.5 cm) and sells for about four dollars. Many people are willing to pay that price for a bath sponge that is so much better than the synthetic kind. For many applications, such as metal polishing, there is no good substitute for a real sponge. Small pieces of elephant ear sponge *Spongia officinalis lamella* are also preferred for pottery making. Sponges may someday prove even more useful as a source of medicines. Cytosine arabinoside, from the Caribbean sponge *Cryptotethya crypta*, is a major anticancer drug that blocks DNA synthesis in tumors (Ruggieri 1976; Tucker 1985).

PHYLUM PLACOZOA

Phylum Placozoa is represented so far by only one confirmed species, *Trichoplax adhaerens* (Figure 23.10). This organism has been known since 1883 but was mistaken for the larva of a cnidarian or for a mesozoan until 1971. Until the late 1980s the only place *Trichoplax* had been found was on the walls of marine aquaria. Vicki B. Pearse has since found it swimming freely at various depths throughout the Pacific. It appears to be somehow protected from predation. It has a few thousand cells and is barely visible to the unaided eye (up to 3 mm). *Trichoplax* reproduces asexually by binary fission and budding, and also sexually. There are no organs and little differentiation of cells, as shown by the fact that a complete animal can regenerate from any excised portion.

Trichoplax moves in any direction, so there is no front, rear, or sides, and there is no symmetry. There are, however, differences between the dorsal and ventral surfaces. The dorsal surface consists of flat ciliated cells and shiny spherical cells ("bright spheres"), and the ventral surface has columnar ciliated cells and gland cells. Between the dorsal and ventral cells is a cavity containing fluid and fibrous cells. The fibrous cells contract the body in a way that makes *Trichoplax* resemble an amoeba. Locomotion is due, however, to the ventral cilia. The animal probably feeds by covering algae or other food, secreting digestive enzymes, and absorbing the nutrient molecules (Table 23.3).

Figure 23.10
Trichoplax adhaerens, *with a section cut away to show the two cell layers sandwiching a middle layer of fluid and fibrous cells (mesenchyme). Diameter approximately 3 mm.*

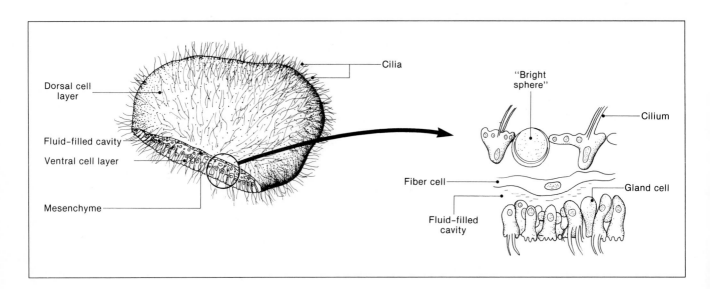

Table 23.3 Characteristics of Placozoa (based on one species).

Phylum Placozoa plak-oh-Z0-uh (Greek *plax* plate).

Morphology: Multicellular animals without tissues or organs. Outer layer of ciliated cells. Inner cavity contains fluid and fibrous cells. Asymmetric.

Physiology: Cellular grade of organization.

Locomotion: Swimming and creeping on ventral cilia.

Reproduction: Sexually, as well as by binary fission and budding.

Development: Unknown.

Habitat, Size, and Diversity: Marine. Natural history unknown. Size up to 3 mm. **One species** described.

PHYLUM MESOZOA

The Mesozoa are mysterious, small, parasitic, worm-shaped (vermiform) animals generally consisting of 20 to 30 cells. The phylum acquired its name more than a century ago from the belief that it was intermediate (Greek *mesos* middle) between protozoa and metazoa. This conjecture has found support in a molecular phylogeny that suggests a closer relationship to some protozoans than to most metazoans (Hori and Osawa 1987). At present the phylum is generally divided into two classes: Rhombozoa (or Dicyemida) and Orthonectida. Many zoologists, however, think these should be two different phyla (Table 23.4).

We are less ignorant about *Dicyama* than about any other mesozoan. *Dicyema* is a vermiform animal less than 7 mm long usually found in the kidney of octopus and other cephalopod molluscs. Virtually nothing is known about what the dicyemid finds so attractive about cephalopod urine, or why the cephalopod tolerates dicyemids in its kidneys. *Dicyema* most often reproduces asexually within the cephalopod kidney by releasing small vermiform larvae. The adults that produce vermiform larvae are called **nematogens** (Greek *nema* thread + *genes* born) (Figure 23.11). When the population grows to the point that virtually the entire inner surface of the kidney is covered, some factor triggers the production of sexually reproducing **rhombogens.** The larvae released by rhombogens are shaped like a toy top and covered with cilia. They are called **infusoriform larvae,** because they resemble ciliated protozoa (infusoria). After being released when the cephalopod urinates, infusoriform larvae sink to the bottom by virtue of two dense cells. Infusoriform larvae are not immediately infective to another cephalopod at this

Table 23.4 Characteristics of Mesozoa.

Phylum Mesozoa mess-oh-Z0-uh (Greek *mesos* middle).

Morphology: Two layers of cells (not homologous to germ layers of metazoan embryos). Asymmetric. No tissues.

Physiology: Lacking circulatory, respiratory, osmoregulatory, digestive, and nervous systems.

Locomotion: Cilia.

Reproduction: Asexual and sexual.

Development: Indirect.

Habitat, Size, and Diversity: Endoparasites of marine invertebrates. Sizes range from less than 1 to 7 mm. **Approximately 100 known species.**

Genera mentioned elsewhere in this chapter are noted.

Class Rhombozoa rom-bow-ZO-uh (Greek *rhombos* a top). Axial cell (through long axis of body) is surrounded by a single layer of cells and encloses reproductive cells that produce two types of larva. Symbiotic (presumably parasitic) in the kidneys of bottom-dwelling cephalopod molluscs (squids, cuttlefishes, octopuses).

> **Order Dicyemida** die-sigh-YEM-id-uh (Greek *di-* two + *kyema* embryo). Adult with separate, ciliated somatic cells; with head (calotte) region. *Dicyema* (Figure 23.11).

> **Order Heterocyemida** HET-er-oh-sigh-YEM-id-uh (Greek *hetero* different). Adult with syncytial, non-ciliated somatic cells; no head region.

Class Orthonectida OR-tho-NECK-tid-uh (Greek *orthos* straight + *nektos* swimming). Male and female adults free-swimming. Parasitic; form a multinucleate plasmodium less than 1 mm. Parasites in planarians, nemertines, polychaete annelids, echinoderms, and bivalve molluscs. *Rhopalura* (Figure 23.12).

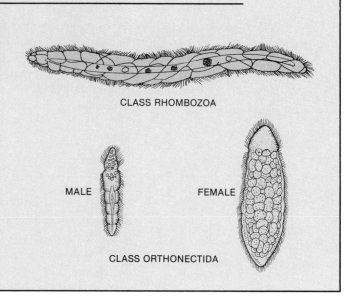

CLASS RHOMBOZOA

MALE

FEMALE

CLASS ORTHONECTIDA

stage but must undergo some unknown steps of maturation, perhaps in a secondary host. Both the vermiform and infusoriform larvae are produced within an **axial cell** that runs up the center of the adult. Within this cell are numerous other

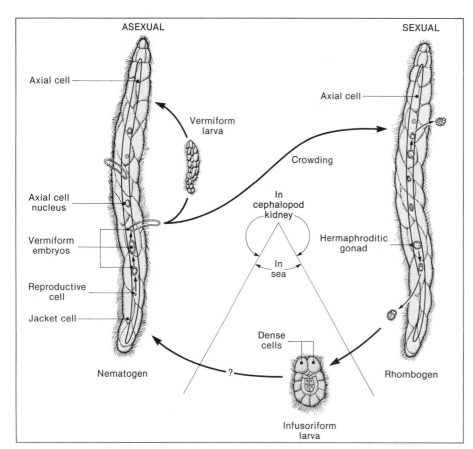

Figure 23.11
The life cycle of the dicyemid Dicyema truncatum. *Generally, the nematogen forms reproduce asexually in the kidney of a cephalopod mollusc. When the lining of the kidney becomes too crowded, hermaphroditic rhombogen forms are produced, and these reproduce sexually. The resulting infusoriform larvae are then released into water.*

Figure 23.12
Life cycle of the orthonectid
Rhopalura ophiocomae. *Length of
the adult female is approximately
0.25 mm.*

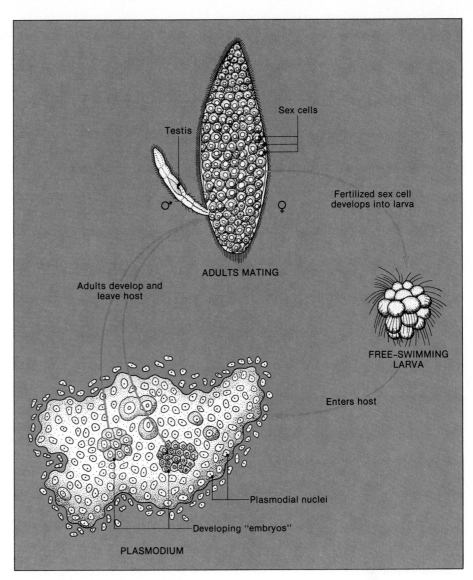

cells(!) that produce vermiform larvae in nematogens. In rhombogens the axial cell contains hermaphroditic gonads that produce self-fertilizing gametes.

Adults in class Orthonectida are free-swimming and bisexual (Figure 23.12). The dimorphic males and females are essentially containers of gametes. After internal fertilization the female gives birth to ciliated larvae. The larvae swim to a host and enter, then lose their cilia and begin parasitizing a variety of tissues. The orthonectid larva then undergoes repeated mitosis without cell division, resulting in a syncytial **plasmodium.** Some of the nuclei of the plasmodium then develop into male or female free-swimming adults, which leave the host and begin the next cycle.

SUMMARY

The three phyla discussed in this chapter are animals, since they are multicellular, heterotrophic, and sexual and have a blastula stage of development. Unlike most animals, however, there is a cellular grade of organization, in which there are no organs and no neural or hormonal coordination. These animals are also asymmetric.

Members of phylum Porifera are characterized by a canal system through which choanocytes pump water and extract food. The body is supported by an endoskeleton of collagenous fibers and spicules. Canal systems are classified as asconoid, syconoid, or leuconoid, in order of increasing complexity and increasing surface area for choanocytes. Four classes of sponges are distinguished, mainly on the basis of the spicules. Most species belong to class Demospongiae, which has siliceous spicules (if any) that are not six-rayed. Sponges reproduce asexually by budding, or sexually by releasing gametes. Larvae are flagellated and responsible for dispersal to new habitats.

Phylum Placozoa has just one known species, *Trichoplax adhaerens*, which is a flattened mass of cells that swims and crawls with cilia. Phylum Mesozoa includes small, worm-shaped parasites with only 20 to 30 cells.

KEY TERMS

spongocoel
choanocyte
ostium
osculum
spicule

cellular grade of organization
asconoid
syconoid
leuconoid
radial canal

flagellated chamber
archaeocyte
mesohyl
reaggregation
gemmule

SELF-TEST

1. Explain why choanocytes are essential to the life of sponges. How do ostia, oscula, and skeleton relate to the functioning of choanocytes?

2. Describe the three types of sponge canal systems, using the terms asconoid, syconoid, and leuconoid. What problem is solved by the development of the syconoid and leuconoid types?

3. Explain what is meant by the phrase "cellular grade of organization." Give one example illustrating how the cells of sponges behave independently. Give an example of how the cells of sponges interact.

4. What are the main criteria by which the three classes of sponge are distinguished? Name the three classes and describe the type of skeleton each has.

5. Describe how sponges reproduce sexually. What is a gemmule?

6. Both Porifera and Placozoa are often classified in the subkingdom Parazoa. What similarities do these phyla share that would support this classification? What are some of the important ways in which Porifera and Placozoa differ?

7. What is the significance of the term "Mesozoa"?

READINGS

RECOMMENDED READINGS

de Laubenfels, M. W. 1953. *A Guide to the Sponges of Eastern North America.* Coral Gables, FL: University of Miami Press.

Lapan, E. A. and H. J. Morowitz. 1972. The mesozoa. *Sci. Am.* 227(6):94–101 (Dec).

Wood, R. 1990. Reef-building sponges. *Am. Sci.* 78:224–235.

See also relevant selections in General References listed at the end of Chapter 21.

ADDITIONAL REFERENCES

Bergquist, P. R. 1978. *Sponges.* Berkeley: University of California Press. (*A detailed but readable account of research.*)

Bergquist, P. R. 1985. Poriferan relationships. In: S. Conway Morris et al. (Eds.), *The Origins and Relationships of Lower Invertebrates.* New York: Oxford University Press, pp. 14–27.

Bergquist, P. R. and K. Glasgow. 1986. Developmental potential of ciliated cells of ceractinomorph sponge larvae. *Exp. Biol.* 45:111–122.

Fry, W. G. (Ed.). 1970. *The Biology of the Porifera.* Zoological Society of London Symposium 25. New York: Academic Press.

Hartman, W. O. and T. F. Goreau. 1970. Jamaican coralline sponges: their morphology, ecology and fossil relatives. In: W. G. Fry (Ed.), *The Biology of the Porifera.* Zoological Society of London Symposium 25. New York: Academic Press, pp. 205–243.

Hildemann, W. H., I. S. Johnson, and P. L. Jokiel, 1979. Immunocompetence in the lowest metazoan phylum: transplantation immunity in sponges. *Science* 204:420–422.

Hori, H. and S. Osawa. 1987. Origin and evolution of organisms as deduced from 5S ribosomal RNA sequences. *Mol. Biol. Evol.* 4:445–472.

Mackie, G. O. 1990. The elementary nervous system revisited. *Am. Zool.* 30:907–920.

Meylan, A. 1988. Spongivory in hawksbill turtles: a diet of glass. *Science* 239:393–395.

Reiswig, H. M. and G. O. Mackie. 1983. Studies on hexactinellid sponges. III. The taxonomic status of Hexactinellida within the Porifera. *Philos. Trans. R. Soc. (London) Ser. B* 301:419–428.

Ruggieri, G. D. 1976. Drugs from the sea. *Science* 194:491–497.

Simpson, T. L. 1984. *The Cell Biology of Sponges.* New York: Springer-Verlag. (*Far more comprehensive than the title suggests.*)

Sullivan, B., D. J. Faulkner, and L. Webb. 1983. Siphonodictidine, a metabolite of the burrowing sponge *Siphonodictyon* sp. that inhibits coral growth. *Science* 221:1175–1176.

Tucker, J. B. 1985. Drugs from the sea spark renewed interest. *BioScience* 35:541–545.

Vacelet, J. 1985. Coralline sponges and the evolution of Porifera. In: S. Conway Morris et al. (Eds.), *The Origins and Relationships of Lower Invertebrates.* New York: Oxford University Press, pp. 1–13.

Wilkinson, C. R. 1983. Net primary productivity in coral reef sponges. *Science* 219:410–412.

24

Cnidaria and Ctenophora

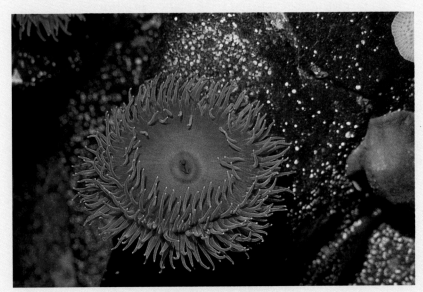

Green sea anemone.

CHAPTER OUTLINE

LEARNING OBJECTIVES

1. How are freshwater polyps, jellyfish, sea anemones, and corals related to each other?

2. How are these odd animals related to others?

3. What is the significance of the jellylike structure of jellyfish and related animals?

4. How do jellyfish, sea anemones, and similar animals get oxygen, nutrients, ions, and other necessities to their cells without a circulatory system or other organs?

5. How do they control their movements?

6. How does a jellyfish or a Portuguese man-of-war sting?

7. What are the jellylike globules called sea gooseberries that are often found on beaches?

INTRODUCTION

As we saw in the previous chapter, sponges manage to grow to large size by bringing the necessities of life to their cells on water currents. A different solution to life's problems has evolved in the phylum Cnidaria, which includes jellyfish, freshwater polyps, corals, and sea anemones, and in the phylum Ctenophora (Figure 24.1). Both cnidarians and ctenophores consist of two or three layers of cells organized around a central chamber, so no cell is far from the source of oxygen, nutrients, water, and ions. The cells of each layer are similar in type and function and therefore form tissues. This tissue grade of organization distinguishes them from the phyla considered in the previous chapter. The fact that the tissues are not organized into complex organs distinguishes them from the animals to be considered in later chapters. Another feature of cnidarians and ctenophores is that most have **biradial symmetry.** That is, the body tends to be radially symmetric, but some paired structures are positioned on opposite sides (see p. 450).

Because of these similarities cnidarians and ctenophores were once combined in the phylum Coelenterata. They are still often combined under the term "radiata," because of the assumption that radial symmetry was primitive and the bilateral tendencies evolved later. There was also a long-held belief that "coelenterates" were near the base of the evolutionary tree from which other animals branched out. The consensus now, however, is that the similarities of the two phyla to each other result from evolutionary convergence, and that they are not closely related to each other or to other animals. Nevertheless, both cnidarians and ctenophores satisfy all the requirements for being classified as animals, and it is convenient to consider them in the same chapter.

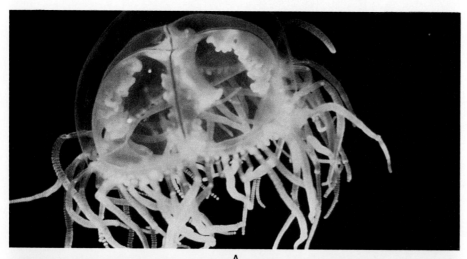

A

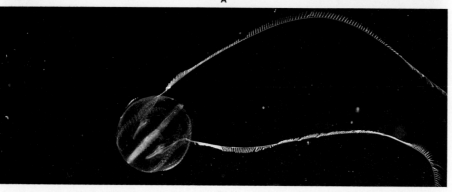

B

Figure 24.1
Members of the phyla Cnidaria and Ctenophora have thin tissues oriented radially or biradially around a hollow cavity. (A) The jellyfish Gonionemus (phylum Cnidaria). The four white structures beneath the transparent bell are gonads. Diameter of bell 2 cm. (B) The comb jelly commonly called sea gooseberries, Pleurobrachia pileus *(phylum Ctenophora). Diameter approximately 2 cm.*

CNIDARIANS

GENERAL ORGANIZATION

Gastrovascular Cavity. The gastrovascular cavity (= coelenteron) is central to the life of a cnidarian (Table 24.1; Figure 24.2). As its name implies, the gastrovascular cavity serves as both digestive and circulatory system, and most cnidaria spend much of their lives putting food into it. Cnidarians usually eat zooplankton, such as copepods (Crustacea), and large individuals can even ingest fish. The main tools for obtaining food are unique subcellular structures called **cnidae** (NY-dee; singular cnida), which spear, poison, and lasso prey. Most cnidae are of the stinging variety, called **nematocysts. Tentacles** pull prey toward the mouth, which draws it into the gastrovascular cavity. The mouth is the only opening into the gastrovascular cavity, so it also serves as an anus. (In other words, the digestive tract is incomplete.)

Tissue Layers. The lining of the gastrovascular cavity is a tissue layer called the **gastrodermis.** Gastrodermal cells secrete enzymes that partially digest prey in the gastrovascular cavity, then they absorb food particles for further digestion. Digestion is therefore both extracellular and intracellular. Flagella on certain gastrodermal cells draw water into the gastrovascular cavity, thereby producing a pressure that maintains a hydrostatic skeleton. The flagella also apparently mix food and enzymes. The external covering of the body is a tissue layer called the **epidermis.** Cells of the epidermis apparently obtain nutrients from the gastrodermis by diffusion. Epidermal cells may also absorb dissolved organic material directly. In many cnidarians, such as corals, the epidermal cells also harbor photosynthetic protists (**zooxanthellae**) that help produce food.

The epidermis and gastrodermis were formerly considered to be the only two tissue layers in cnidarians (and ctenophores). Cnidarians were therefore considered to be **diploblastic.** There is an increasing tendency now to regard as a third tissue

Table 24.1 Characteristics of cnidarians.

Phylum Cnidaria nigh-DARE-ee-uh (Greek *knide* nettle [referring to the stinging cnidae]).

Morphology: Biradial symmetry (essentially radial in hydroids). Either bell-shaped medusae or cylindrical polyps. With hollow tentacles. Three (or two) thin layers of tissue around a central gastrovascular cavity. No coelom. Hydrostatic skeleton, or elastic skeleton of mesoglea. Some also secrete chitinous sheaths or calcareous exoskeletons.

Physiology: Diffuse **nerve net;** no central nervous system. Cnidae used in predation and defense. Mostly carnivorous, but some contribution from dissolved organic material and photosynthesizing endosymbionts. Few organs; individual cells perform most physiological functions.

Locomotion: Planula larvae swim by means of cilia. Medusae swim by slow undulations due to epitheliomuscle and nutritive muscle cells. Polyps usually sessile. Some cnidae used in locomotion.

Reproduction: Asexual budding, or sexual. Usually dioecious. Fertilization usually external.

Development: Radial cleavage pattern. Cleavage indeterminate. Ciliated planula larvae.

Habitat, Size, and Diversity: All **aquatic;** almost all marine. Sizes of adult individuals range from less than 1 mm to 70 meters long, including tentacles. Approximately **9000 living species** described.

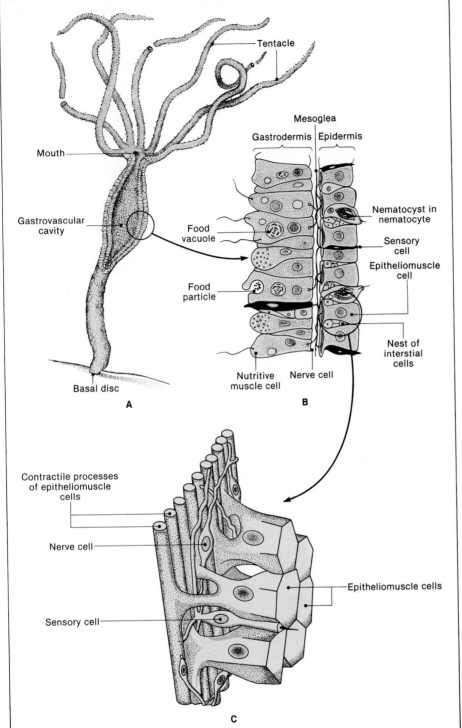

Figure 24.2
(A) Schematic longitudinal section of Hydra. (B) Representation of a section through the body wall of Hydra, showing the epidermis and the gastrodermis lining the gastrovascular cavity. The majority of epidermal and gastrodermal cells are contractile cells (epitheliomuscle and nutritive muscle cells). The latter absorb food in vacuoles and bear pairs of flagella. The interstitial cells differentiate into other cell types. Nematocytes produce the stinging structures called nematocysts. (C) Portion of the epidermis showing epitheliomuscle cells.

layer the **mesoglea** (= mesenchyme; see p. 493), which lies between the gastrodermis and the epidermis. Cnidarians (and ctenophores) would therefore be **triploblastic,** like most animals. (The three tissue layers of cnidarians are probably not homologous with those of other animals, however.) The mesoglea (Greek *mesos* middle, *gloia* glue) is a translucent, gelatinous material that is often invaded by neural and other kinds of cells. In some cnidarians, such as *Hydra*, it is merely

Figure 24.3
In general, Cnidaria alternate
between a sexual, swimming
medusa (left) and an asexual, sessile
polyp. Although they look quite
different externally, this schematic
diagram shows that both the
medusa and the polyp are organized
similarly.

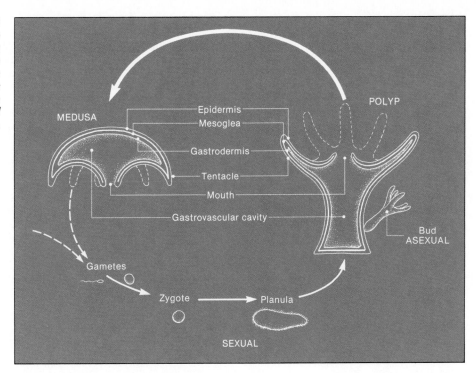

Figure 24.4
A Hydra *polyp with a bud forming
on one side and an ovary opposite.
In some species of* Hydra *budding
occurs only in the presence of a
secretion from bacteria or prey
(Rahat and Dimentman 1982). High
levels of CO_2 in crowded colonies
trigger the formation of gonads.*

an adhesive layer that holds the epidermis and gastrodermis together. In others, especially jellyfishes, the mesoglea constitutes most of the body. The mesoglea performs several functions:

1. It is an elastic skeleton, returning the animal to its normal shape after muscular contraction.
2. It maintains **neutral buoyancy,** because it contains less sulfate (SO_4^{2-}) than seawater and is therefore less dense.
3. It may be a reservoir of nutrients.
4. It increases body size without increasing the number of cells requiring oxygen and nutrients.

LIFE CYCLES

In many species of Cnidaria, each individual during its life changes from a **polyp** to a **medusa,** or vice versa (Figure 24.3). This is often considered to be the primitive state of cnidarians, but in most species only the polyp or the medusa stage occurs. Polyps are usually approximately cylindrical and attached to substratum, as in sea anemones, corals, and hydras. Polyps produce new polyps or medusae (if they occur) by asexual budding. If the medusa form is absent, as in hydras and corals, polyps reproduce new polyps either sexually or by budding (Figure 24.4).

Medusae are shaped like bells or umbrellas and can swim with a kind of jet propulsion by contracting the body. The familiar jellyfishes (Figure 24.1A) are in the medusa stage. Medusae usually have separate sexes, although they are difficult to tell apart. Fertilization usually occurs externally, and the zygote develops into a **planula** larva that swims by means of cilia. In most species the planula **settles** and forms a polyp. The polyp then buds asexually, with each bud developing into either a new medusa or a new polyp. The life cycle of the colonial cnidarian *Obelia* is similar to that sketched in Figure 24.3 and is commonly used as a typical example (Figure 24.5).

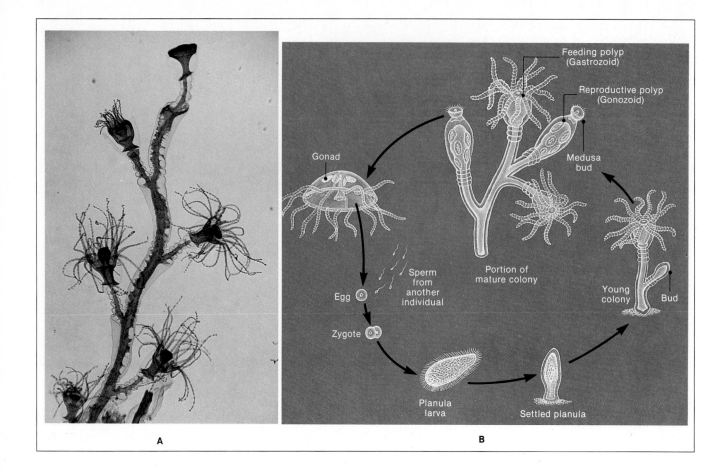

A

B

A SURVEY OF THE MAJOR GROUPS

Class Hydrozoa. The number of classes of Cnidaria ranges from three to six in recent taxonomies, but four is a common number. *Obelia* is a representative hydrozoan, since it includes both medusa and polyp states. The medusae of *Obelia* and other hydrozoans are distinguished from true jellyfishes (class Scyphozoa) by a shelflike **velum** around the inner margin of the bell, which concentrates the force of water expelled when the bell contracts. Most hydrozoans are marine, but the class also includes all freshwater cnidarians, such as *Hydra*. The class also includes fire corals and the Portuguese man-of-war. Perhaps the Portuguese man-of-war *Physalia* (Figure 24.6) is the most familiar hydrozoan since it is so hard to ignore. If its bright colors do not attract notice, contact with its stinging cnidae certainly will. *Physalia* superficially resembles a jellyfish, but instead of a medusa, it is a colony of polyps **(zooids).** Each polyp is adapted as a float, as a tentacle for predation, as a digestive organ, or as a reproductive organ.

Class Scyphozoa. Class Scyphozoa includes most of the cnidarians commonly called jellyfishes. (The term jellyfish is also a general term for a medusa in any class of Cnidaria.) The polyps of scyphozoans are reduced or absent, but the medusae are usually prominent, owing to the large, jellylike mesoglea. The largest, *Cyanea*, has a bell up to 2 meters in diameter and tentacles up to 70 meters long. The common jellyfish of the seashore, *Aurelia aurita*, is more typical in size, with a diameter of about 10 cm and short tentacles (Figure 24.7). This jellyfish is distributed from the equator to both poles. It is a major predator on the larvae of herring (*Clupea harengus*; Möller 1984).

Figure 24.5
(A) The colonial hydrozoan Obelia. *Width of each member of colony is approximately 1 cm. (B) The life cycle of* Obelia. *Note the similarity to Figure 24.3.*

Figure 24.6

The Portuguese man-of-war Physalia *resembles a jellyfish in appearance, as well as in the effect of its nematocysts. It is not a single medusa with dangling tentacles, however, but a floating colony of many polyps (zooids). The fishing polyps (dactylozoids) can extend to 13 meters. Prey—usually small fishes—are ingested by different polyps (gastrozoids, near the dactylozoids). Other polyps (gonozoids, on left) are responsible for sexual reproduction.*

The structure of *Aurelia* is representative of the class. There is a prominent four-part, radial symmetry. Four **oral arms** bearing cnidae lead into the mouth and gastrovascular cavity, and the gonad is shaped like a four-leaf clover (Figure 24.7A). The umbrella is scalloped in eight sectors (four or sixteen in other scyphozoans). On the margin of the umbrella where two sectors meet there is a sensory organ called the **rhopalium.** The rhopalium includes a statocyst organ for balance, as will be discussed later. In some scyphozoans the rhopalia also bear simple eyes. Four **radial canals** branch out from the center like the ribs of an umbrella and join the **ring canal** that runs around the margin.

Unlike some jellyfishes, *Aurelia* has a polyp stage (Figure 24.7C). *Aurelia* is also unusual in having internal fertilization. The males emit sperm from their mouths, and females draw it through their mouths into the gastrovascular cavity where

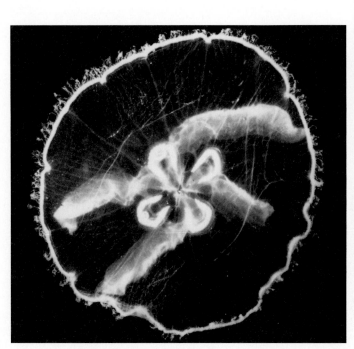

A

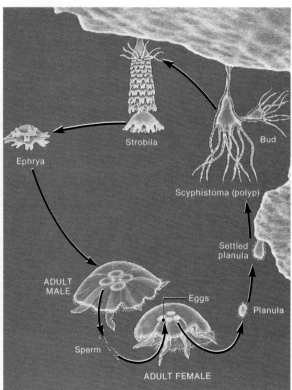

C

Figure 24.7

(A) The common jellyfish Aurelia aurita. *The clover-leaf pattern on the umbrella is the ''gonad'' (testis or ovary). Diameter of bell approximately 40 cm. (B) Structure of* Aurelia. *(C) The life cycle of* Aurelia. *Unlike some jellyfishes,* Aurelia *has a polyp stage, called the scyphistoma. The scyphistoma develops asexually into a strobila, which produces immature medusae, called ephyrae. Each ephyra then matures into an adult medusa.*

B

Major Extant Groups in Phylum Cnidaria

Genera mentioned elsewhere in this chapter are noted.

Class Hydrozoa hide-ro-ZO-uh (Greek *Hydra* a many-headed water serpent in Greek mythology + *zoon* animal). Asexual polyps and/or sexual medusae. Medusae, when present, have a velum (Latin for veil)—an inward fold of the edge of the bell. Gastrovascular cavity not divided by septa. Hydroids, fire coral, Portuguese man-of-war. *Chlorohydra, Hydra, Obelia, Physalia* (Figures 24.4, 24.5, 24.6).

Class Scyphozoa sigh-foe-ZO-uh (Greek *skyphos* cup). Free-swimming medusae with four-part radial symmetry. Polyp reduced or absent. Prominent gelatinous mesoglea. Margin of bell with rhopalia. No velum. Jellyfish. *Aequorea, Aurelia, Cyanea, Gonionemus* (Figures 24.1A, 24.7).

Class Cubozoa cube-oh-ZO-uh (Greek *kubos* cube). Bell cubical and bent inward at margin to form a velarium (not a true velum). Tentacles suspended from four flat pedalia at corners of bell. Cubomedusans. *Carybdea, Chironex* (Figure 24.8).

Class Anthozoa an-tho-ZO-uh (Greek *anthos* flower). No medusae; all sessile polyps. Pharynx leads into gastrovascular cavity, which is divided by eight or more septa (mesenteries) bearing cnidae. Mesoglea contains cells.

Subclass Octocorallia (= Alcyonaria) OK-toe-core-AL-ee-uh (Greek *octo* eight + *korallion* pebble). Eightfold biradial symmetry: eight tentacles; gastrovascular cavity divided by eight septa. Endoskeleton of calcareous spicules. All colonial. Includes soft corals, sea whips, sea fans, sea pens, and sea pansies. *Corallium, Gorgonia, Heliopora, Plexaura* (Figure 24.9).

Subclass Hexacorallia (= Zoantharia) HEKS-a-core-AL-ee-uh (Greek *hex* six). Often with sixfold biradial symmetry: gastrovascular cavity divided by six septa. Includes sea anemones, stony (true) corals, tube corals, black (thorny) corals. *Adamsia, Antipathes, Calliactis, Metridium, Palythoa, Platygyra, Pocillopora, Stomphia* (Figures 24.10, 24.11, 24.13).

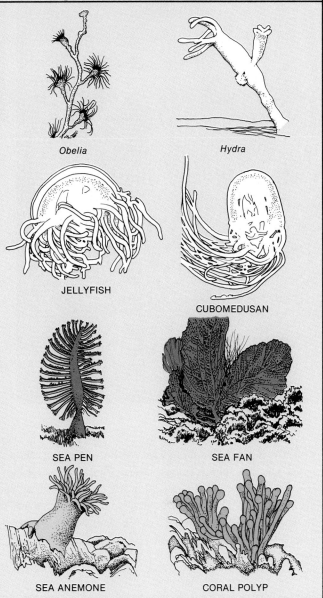

Obelia   *Hydra*

JELLYFISH   CUBOMEDUSAN

SEA PEN   SEA FAN

SEA ANEMONE   CORAL POLYP

fertilization occurs. Development of the ciliated planula larvae also occurs internally, within the oral arms of the female. The planula larva then drops off, settles on the sea floor, loses its cilia, and develops into a polyp.

Class Cubozoa. The 16 or so species of cubozoans or cubomedusans resemble true jellyfishes and are commonly called "box jellyfishes" because of their cubical umbrellas. Cubomedusans differ from jellyfishes, however, in having four **pedalia** with which they swim rapidly. Some species come into frequent and sometimes fatal contact with people, because they are attracted to prey in shallow water and by pollution and lights (Figure 24.8).

Class Anthozoa. Members of class Anthozoa have no medusae and differ so much in other respects from other Cnidaria that some taxonomists propose ele-

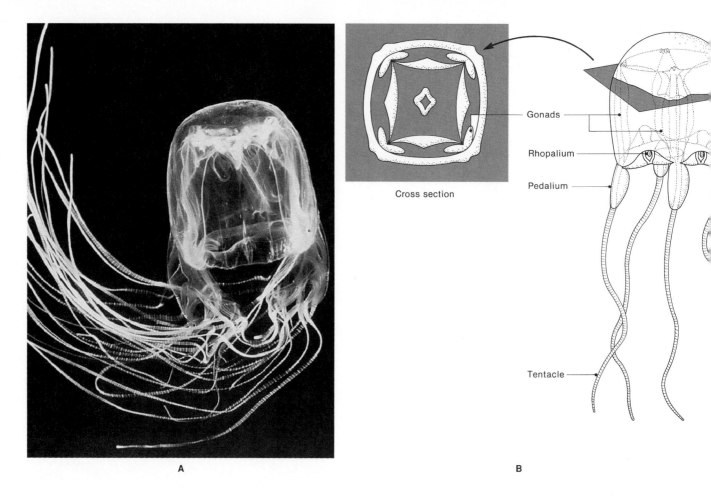

Cross section

Gonads

Rhopalium

Pedalium

Tentacle

A   B

vating Anthozoa to a subphylum. Anthozoans are generally divided into two sub-classes. Subclass Octocorallia includes soft corals, sea whips, sea fans, sea pens, sea pansies, and others (Figure 24.9). All are colonial polyps with eight tentacles. Subclass Hexacorallia is more diverse and includes sea anemones, stony (true) corals, tube anemones, and black (thorny) corals (Figure 24.10). Tube anemones and black corals are so different from the others that many taxonomists put them into their own subclasses (Ceriantheria and Antipatheria).

Figure 24.8
Since 1980 more than 70 swimmers in Australia have stumbled ashore in severe pain, then dropped dead within minutes. (A) The cause—nematocysts of the sea wasp Chironex fleckeri, *a cubozoan. Width of bell approximately 25 cm. (B) Structure of another cubozoan,* Carybdea marsupialis.

Figure 24.9
Octocoral anthozoans. (A) Colony of the red gorgonian Gorgonia adamsi. *(B) Closeup showing several extended polyps of a different species.*

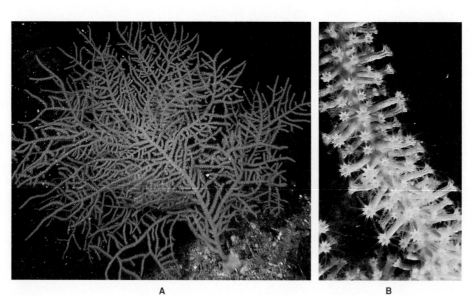

A   B

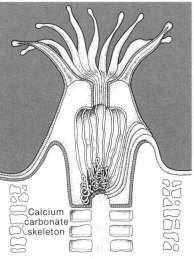

A

B

C

Figure 24.10
Hexacoral anthozoans. (A) Polyps of the stony coral Pocillopora sp. *The green color is due to symbiotic photosynthetic protists called zooxanthellae. (B) Schematic diagram of a polyp within its calcium carbonate skeleton. (C) A brain coral* Platygyra sinensis *spawning. Spheres containing sperm or eggs are being released from the mouths of polyps during mass spawning. The tentacles retracted when illuminated by the photographer's light.*

The most prominent individuals in class Anthozoa are the sea anemones (Figure 24.11). They are much more massive than the polyps of hydrozoans, ranging in diameter from less than 5 mm to more than a meter. Many form colorful beds of "flower animals" in warm, shallow waters. The hydrostatic skeleton that helps support such a large body is generated by the action of cilia at one or both ends of the long slit-shaped mouth, in a groove called the **siphonoglyph.** The siphonoglyph draws a steady current into the pharynx and gastrovascular cavity, and cilia elsewhere in the pharynx pump out the water. Support is also provided by **septa** (= mesenteries) that run vertically through the body. Six pairs of **complete septa** connect the pharynx to the outer wall, dividing the upper part of the gastrovascular cavity radially into six chambers. Pairs of **incomplete septa** partially divide these

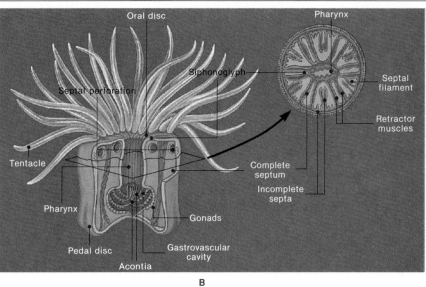

A

B

Figure 24.11
The sea anemone Metridium.
(A) Living specimen. Height 45 cm.
(B) Diagram of sectioned anemone.

chambers. In addition to reinforcing the body, the septa also increase the surface area and support the gonads.

On the edges of incomplete septa, below the gonads, are **filaments** with cnidae and gland cells that help subdue and digest prey. In some species the filaments continue into **acontia** that bear cnidae and can be extended through pores or the mouth to augment the tentacles in subduing prey. The numerous tentacles are arrayed in two rings on the **oral disc** surrounding the mouth.

CNIDAE

Functions. Cnidarians are the only animals that produce intracellular structures called cnidae, which are used in predation, defense, and locomotion. There are more than 20 types of cnidae. Some attach to solid objects by means of a sticky substance and are used for locomotion. Others entangle prey before capture by the tentacles. The most common type of cnida, the nematocyst, punches holes into prey or predators by means of **stylets,** then launches a hollow **thread** into the wound (see Figure 13.2A). The thread is about 0.5 mm long and serves as a syringe that injects a poison or paralytic toxin. The nematocyst is triggered by contact with a structure called the **cnidocil.** Usually chemoreceptors must first have detected the presence of a suitable victim, thereby preventing the waste of nematocysts if the cnidocil contacts an inorganic object.

Those who have swum into a jellyfish or Portuguese man-of-war will appreciate how effective these nematocysts can be. The toxin can sting for hours and produce inflammation. Attempting to wipe off or wash off the tentacles merely triggers more nematocysts. The best first aid is to dehydrate the nematocysts with powder or alcohol. (Beer is worth trying.) In severe cases the toxin destroys tissue and can produce a fatal cardiac arrest. The stings of the sea wasp *Chironex fleckeri* and some others in the class Cubozoa can kill a person within minutes or scar survivors for life. Not all animals are bothered by nematocysts. The sea slug *Glaucus* not only eats the Portuguese man-of-war but somehow manages to get the undischarged nematocysts into its skin for its own protection. The flatworm *Microstomum* uses the nematocysts from its prey, *Hydra*, in a similar way. Such nematocysts, taken from cnidarians by their predators, are called **kleptocnidae** (Greek *kleptein* to steal).

Mechanism of Discharge. The small size (0.1 mm) of the nematocyst and the speed of firing (within 3 msec) have made it difficult to learn what force ejects the thread, but various plausible candidates have been suggested. The **osmotic theory** holds that water enters the nematocyst capsule osmotically, swelling the capsule and ejecting the thread. The **tension theory** suggests that the thread is fired by tension built up springlike during synthesis. The **contractile theory** proposes that contracting structures surrounding the nematocyst squeeze out the thread. Holstein and Tardent (1984) in Zurich have managed to obtain evidence bearing on the mechanism by filming the discharge with ultrahigh-speed microcinematography. They found that in *Hydra attenuata* nematocysts are discharged in four steps, shown in Figure 24.12:

A. First the nematocyst swells by about 10%.
B. The nematocyst's cover flips open, and three stylets emerge and punch a hole in the prey or predator. (This phase takes less than 10 μsec, so the acceleration of the stylets is approximately 40,000 *G*!)
C. The stylets back out of the hole within about 150 μsec. This means that the initial force that opens the cover must have been expended.

Figure 24.12

Discharge of a nematocyst. (A) The capsule swells. (B) The operculum opens, and the stylets emerge and puncture the prey or predator. (C) Stylets withdraw. (D) Thread emerges and enters hole in prey or predator. (Based on ultrahigh-speed films of nematocysts of Hydra attenuata *by Holstein and Tardent 1984.)*

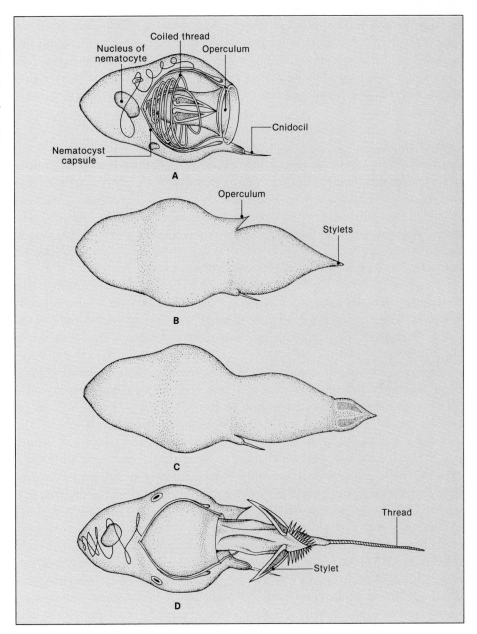

D. The thread shoots into the hole in the prey or predator, turning inside-out as it extends.

Holstein and Tardent suggest that the initial swelling of the capsule is due to osmosis (osmotic theory), and that the discharge of the thread is due to stored mechanical energy from the capsule wall (contractile theory).

NERVE CELLS AND MOVEMENT

Nerve Nets. Compared with nematocyst discharge, most behaviors of cnidarians are slow, simple, and nonspecific. This is to be expected from the diffuse nervous system, the general absence of complex receptor organs, and the poorly differentiated contractile cells. The nerve cells do not connect specific receptors with specific muscles via well-defined neural pathways, but form a nerve net that sends a wave of excitation radiating from the area of stimulation (see Figure 8.3).

In many cnidarians nerve cells can produce two types of action potential: sodium spikes that trigger fast tentacle retraction and other escape movements, and calcium spikes that trigger slower feeding movements. In jellyfish the epithelial cells also conduct electrically and mediate very slow movements. Many of the synapses have transmitter vesicles on both sides of the cleft, which enables them to transmit in two directions.

Contractile Cells. In many Cnidaria most of the cells of the epidermis and gastrodermis are capable of contracting. These **epitheliomuscle cells** and **nutritive muscle cells** (Figure 24.2) are not considered to be true muscle cells because they are not derived from the middle tissue layer, as in other animals. The contractile portion of epitheliomuscle cells runs lengthwise along the body and tentacles, while the contractile portion of nutritive muscle cells runs circularly around the gastrovascular cavity. These contractile cells do not attach to a hard skeleton but work against the elastic mesoglea and hydroskeleton.

Receptors. Judging from their behavior, Cnidaria must have chemoreceptors and touch receptors on the epidermis, although they are not well defined structurally. Perhaps the best-known evidence for chemoreceptors is the mouth-opening and tentacle movements of *Hydra* in the presence of prey. The substance that triggers this behavior is glutathione, a tripeptide that leaks from wounded animals. Many jellyfishes and cubomedusans also have combined receptor organs called rhopalia distributed around the margin of the bell (Figure 24.7B). Included in each rhopalium is a balance organ, the **statocyst,** and often several ocelli. The statocyst works like those found in other aquatic invertebrates. Dense structures rest on mechanoreceptor hairs, deflecting some of them, depending on the orientation of the organ with respect to gravity. Activation of a statocyst inhibits contraction on the same side of the medusa, so contractions on the opposite side tend to keep the animal oriented correctly.

Cubozoa have relatively elaborate ocelli, each with a lens and sensory layer reminiscent of those in the eyes of cephalopod molluscs and vertebrates. The ocelli connect directly to the nerve net without going through a ganglion, so it is doubtful that these animals can analyze much visual information. Nor is it clear what they would do with the information. The ocelli might detect luminescent prey or, since Cubozoa copulate, help in finding mates. Apparently ocelli are not the only means of detecting light. Even without ocelli *Chlorohydra* can move into sunlight, somersaulting end over end to where its photosynthesizing symbionts can function.

BEHAVIOR

Many cnidarians are sessile or slow-swimming, and they tend to behave as individuals. The large shoals of jellyfishes that sometimes wash onto beaches are not due to social schooling or to a gang assault on human swimmers, but simply to the vagaries of wind and water. Perhaps the closest thing to social behavior is **mass spawning,** in which all corals of many different species in a reef release their gametes at the same hour. The most complex behaviors are displayed by sea anemones. Some sea anemones, such as *Stomphia* and *Calliactis,* detach and swim away in the presence of chemicals released by certain predatory starfishes or sea slugs (Figure 24.13). Another interesting behavior of anemones is the withdrawal of their bodies during heavy wave action. Many sea anemones also go through periodic cycles of withdrawal and extension that are unrelated to wave intensity.

Figure 24.13
The sea anemone Stomphia coccinea *escaping from a predatory sea star* (Dermasterias imbricata). *The anemone first retracts its tentacles and releases its hold on the substratum (first three frames). By contracting its body it then moves away (last three frames). Fifteen minutes later it resettles. The substance that triggers the response has been isolated from the sea star and named imbricatine (Elliott et al. 1989).*

The function of this behavior is not known; it is simply referred to as the "mood" of the anemone.

REGENERATION IN *HYDRA*

Antony van Leeuwenhoek first described hydras in 1702, but these retiring organisms were unable to compete for attention among all the other marvels being revealed by the microscope. In the 1740s, however, Abraham Trembley, a Swiss naturalist working in The Netherlands, made hydras superstars among organisms by showing (he thought) that they bridge the gap between plants and animals. This created such a sensation because it seemed to confirm the assertion of philosophers that all organisms are linked from lowest to highest on a Chain of Being.

Trembley first noticed hydras as green, flower-shaped structures attached to plants in ditch water. Since they were green and apparently rooted, he first thought they were plants. He found, however, that they contracted when disturbed, and éven moved into sunlit parts of their containers. This puzzled Trembley, because only animals were supposed to move. Trembley decided to test whether his organisms were plants or animals by cutting them in two; only plants were supposed to be able to survive this procedure. To his surprise these organisms not only survived, but the flowerlike head regenerated a new body, and the decapitated body sprouted a new head. This phenomenon later inspired Linnaeus to name the organism *Hydra*, after the sea serpent that grew two heads for every one that was chopped off. Since then the genus that Trembley studied has been renamed *Chlorohydra*: green hydra. Trembley concluded that hydras must be plants, but then changed his mind for the last time after observing that they captured prey and that the green color was due to symbiotic algae.

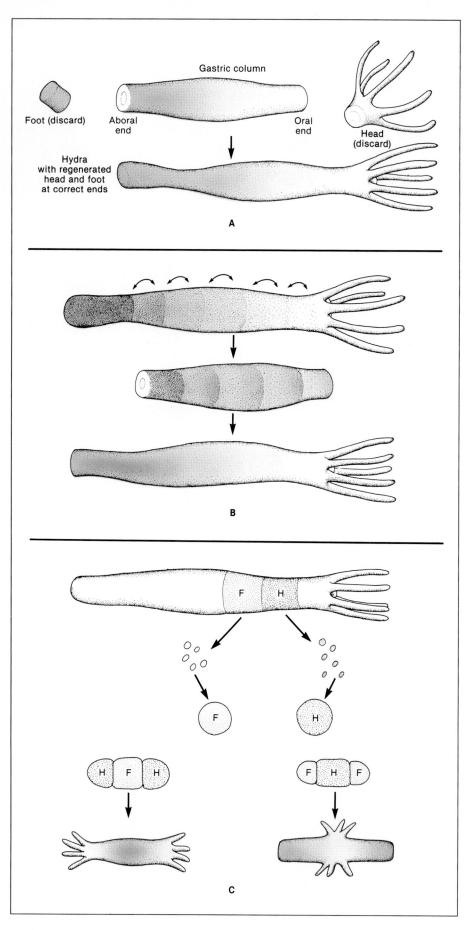

Figure 24.14
(A) Polarity in the regeneration of a new head and foot from the gastric column of Hydra. (B) Polarity is preserved even if the gastric column is cut into many sections, each of which is reversed. This indicates that polarity is due to a chemical gradient. (C) Polarity is preserved even in tissue clumps that have reaggregated from isolated cells.

Gastric column

Foot (discard)

Aboral end

Oral end

Head (discard)

Hydra with regenerated head and foot at correct ends

A

B

F   H

F

H

H F H

F H F

C

Trembley also discovered that the head from one hydra could be transplanted onto the body of another. (This experiment in 1742 was the first permanent transplant of animal tissue.) Since then more than 2000 published reports have followed Trembley's work on regeneration in hydras. Many such studies are now aimed at determining how embryonic cells "know" where they belong in the body and what they are supposed to become when they get there. Hydras are useful models in such studies of morphogenesis for two reasons. First, like some sponges, hydras are able to reaggregate into functioning animals after being separated into individual cells. Second, like some vertebrates, they display **polarity** during regeneration. If the head (mouth + tentacles) and the foot (basal disc + basal stalk) are removed from a polyp, the remaining **gastric column** always regenerates a new head at the oral end and a new foot at the other end (Figure 24.14A). Any section removed from the gastric column also regenerates a head and foot at the correct ends. Polarity can be reversed by grafting the head and foot onto the wrong ends of the gastric column and leaving them for a day.

There are at least two possible explanations for polarity: (1) either each section of gastric column includes oriented cells that mark the polarity; or (2) there is a chemical gradient due to different concentrations of substances at the oral and aboral ends. If the first theory were correct, then cutting the gastric columns into several sections and reversing each section while maintaining them in the same sequence should cause a reversal of polarity (Figure 24.14B). This does not occur, however. There are, in fact, gradients of four different peptides involved in polarity (Spencer 1989). The regeneration of a head is due to a high concentration of **head activator.** Once the head is regenerated it produces a **head inhibitor** that prevents another head from regenerating. **Foot activator** and **foot inhibitor** have similar effects at the opposite end.

Polarity is not disrupted even if cells are isolated and allowed to reaggregate. Gierer (1974) and his co-workers took cells from the oral end of the gastric column and allowed them to reaggregate into one clump (H) and similarly allowed cells from an adjacent region toward the foot to aggregate into another clump (F). If they put three clumps together in the sequence HFH, then tentacles formed at both ends. If three clumps were put together in the sequence FHF, tentacles formed in the middle.

INTERACTIONS WITH HUMANS AND OTHER ANIMALS

Coral Reefs. Of all the interactions of cnidarians with other animals, none is more impressive than those of a coral reef. Reef-building (**hermatypic**) corals are undoubtedly among the most ecologically important animals (see pp. 355–358). Coral reefs provide substratum for a multitude of other sessile marine animals and resting and hiding places for numerous mobile ones. Moreover, their sheer bulk (Figure 24.15) and locations buffer coastal organisms from the effects of storms. Reef-building by corals is largely dependent on the amino acid, glycerol, glucose, and other products leaked by mutualistic protists (**zooxanthellae**) sheltered within their polyps (see p. 476 and Figure 24.10A). Certain environmental stresses cause coral polyps to expel their zooxanthellae, a phenomenon called **bleaching.** Widespread episodes of bleaching in the Caribbean, believed to have been due to high water temperature, have caused a great deal of concern (Roberts 1990).

Among the few predators of cnidarians, the most significant are echinoderms. The crown-of-thorns starfish (*Acanthaster planci*) is one of the most voracious destroyers of corals (see Figure 33.18). During the 1960s and 1970s these starfish

appeared to endanger even the enormous Great Barrier Reef of Australia, but since then the threat has subsided as mysteriously as it began.

Symbioses. Zooxanthellae are only one of many kinds of organisms involved in symbiosis with cnidarians. The mutualistic green alga of *Chlorohydra* is another example, already noted. While the green hydra benefits from nutrients supplied by the algae, the algae benefit from the hydra's ability to move into sunlight. Another example of mutualism is the association between the Portuguese man-of-war and the fish *Nomeus*. *Nomeus* swims among the tentacles of the Portuguese man-of-war unharmed and safe from predators. In return, *Nomeus* probably serves as bait, luring predatory fishes into the tentacles of the Portuguese man-of-war. A similar mutualism occurs between certain species of sea anemones and damsel fishes (also called anemone fishes and clown fishes, family Pomacentridae). Each species of anemone secretes a specific substance that attracts a particular species of damselfish (Murata et al. 1986). Fishes in that species swim among the anemone's tentacles, feeding on scraps from the prey captured by the anemone, which they themselves may have lured (Figure 24.16). Many explanations have been proposed for why the nematocysts do not sting the damselfish, but none is completely satisfactory.

Perhaps the most bizarre example of mutualism in sea anemones is that between *Adamsia palliata* and the hermit crab *Eupagurus prideauxi*. A young *Adamsia* attaches to the mollusc shell that is occupied by a young hermit crab, then rides about with its mouth just above and behind that of the crab, feeding on scraps. The anemone gradually surrounds and absorbs the shell, replacing it with its own body and growing at the same rate as the crab. The crab not only is defended and camouflaged, but also, unlike other hermit crabs, it does not have to find a new shell when it outgrows the old one. *Adamsia* without a crab soon dies, and *Eupagurus* without the sea anemone is easy prey.

Humans. Only occasionally do humans prey on cnidarians for food. In the Orient some jellyfishes, as well as the highly dangerous *Chironex*, are eaten. They are probably not an essential item of diet, considering that they are 98% water. The most important interaction of humans with cnidarians is our impact on coral

Figure 24.16
The skunk clown fish Amphiprion akallopsisus.

reefs. Pollution from seaside development and oil spills clogs the feeding systems of coral polyps, shades their zooxanthellae, and promotes the growth of algae and other competitors. In the Philippines fishermen have destroyed 95% of the coral reefs by illegally using cyanide and dynamite to catch fish for food and for sale to unscrupulous pet dealers. Boats increasingly damage coral reefs in shallow waters, and collectors destroy large blocks of reef to sell as coffee-table curios.

Precious corals, especially red precious coral *Corallium rubrum*, have been prized for thousands of years for jewelry. Divers with modern equipment have made them rare in places. In parts of the Mediterranean the tiniest fragment of precious coral is quickly seized. Black corals (*Antipathes*) and blue corals (*Heliopora*) are also used for jewelry. Coral jewelry is not only attractive but was once believed to protect against the "evil eye." The incidence of "evil eye" has declined drastically in recent times, so precious corals are no longer prescribed much.

Other cnidarians may soon find a place in medicine, however (Ruggieri 1976; Tucker 1985). Palytoxin, from the Hawaiian soft coral *Palythoa toxica*, is one of the most potent toxins known and was once used as a spear poison. It is now used as an antitumor drug. After eight years of effort, organic chemists have succeeded in synthesizing palytoxin, which is one of the most complex organic molecules known, with 129 carbon atoms. Some other corals, such as *Plexaura*, produce both antitumor and antimicrobial substances. Surgeons are also experimenting with coral as a substitute for bone. Aequorin, from the jellyfish *Aequorea aequorea*, has been used to study intracellular processes in both normal and abnormal cells, because it emits light in proportion to calcium concentration.

CTENOPHORES

GENERAL FEATURES

Members of phylum Ctenophora, commonly called "comb jellies," are delicate marine animals sometimes encountered swimming or drifting harmlessly near the surface (Figures 24.1B and 24.17). Often they are found along beaches as translucent, gelatinous lumps called "sea walnuts" or "sea gooseberries." At night they may make their presence known by brilliant flashes of light. The use of submersible research vessels in recent years has shown that ctenophores are much more numerous in deep water than was previously suspected. The evolutionary origin of the ctenophores is unclear, partly because of a virtually complete lack of fossils. Like cnidarians, ctenophores have thin tissues oriented around a gastrovascular cavity (Table 24.2). Unlike cnidarians, however,

1. Ctenophores have no alternation of polyp and medusa generations.
2. They do not produce cnidae but capture food by adhesive cells (colloblasts).
3. Most swim by means of rows of fused cilia called comb plates.
4. Cleavage is determinate, as in most other invertebrates. (That is, removal of a cell during cleavage results in an incomplete embryo.)
5. Their tentacles (when present) are solid and consist only of epidermis.
6. They have true muscle cells derived from mesenchyme.

Ctenophores take their name from the eight rows of **comb plates** (Greek *kteis* comb) that most species use in swimming (Figure 24.18). Each comb plate is a strip of fused cilia. The combs move in sequence from the anus toward the mouth, making luminescent species flash like neon signs. The **statocyst** near the anal

Figure 24.17
Beroe cucumis *(class Nuda) preys on other ctenophores. It is about 1 cm in diameter. (Beroe is often spelled* Beroë, *in violation of the rules of nomenclature, to indicate that the* e *is pronounced and not part of the diphthong* oe.*)*

Table 24.2 Characteristics of ctenophores.

Phylum Ctenophora ten-OFF-or-uh (Greek *kteis* a comb + *phora* bearing).
Morphology: Biradial symmetry: opposite tentacles (if present) and internal canals on a radially symmetric body. Tentacles (if present) consist only of solid epidermis. Some species compressed into a flattened shape. **Monomorphic** (no alternating polyp and medusa). Always solitary and mobile. Thin tissues around a central gastrovascular cavity. No coelom.
Physiology: Few organs. **Statocyst** and **nerve net** control locomotion and feeding. Adhesive **colloblasts** trap food. **Hydrostatic skeleton** only.
Locomotion: By eight rows of **comb plates** (fused cilia), or by undulations of the body using true muscle cells.
Reproduction: Sexual reproduction only; most hermaphroditic.
Development: Cleavage determinate. Development indirect.
Habitat, Size, and Diversity: All **marine**. Size of adult from about 1 cm to 1.5 m. Approximately **100 living species** described.

opening coordinates the activity of the comb plates. This coordination allows the animal to swim in the preferred direction, usually horizontally with its mouth in front. The operation of this statocyst is quite elegant. The supporting cilia on

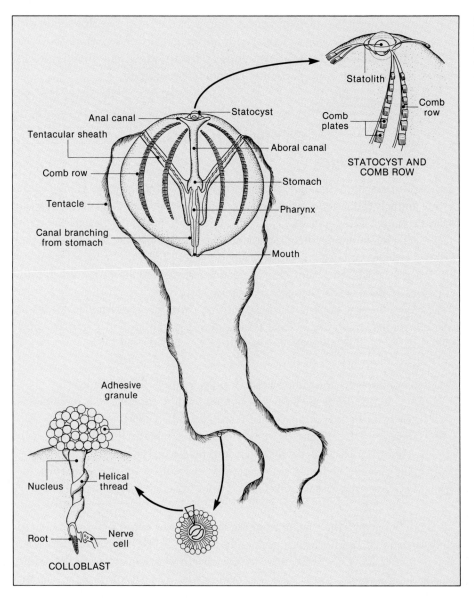

Figure 24.18
Pleurobrachia. *(Compare Figure 24.1B.)* Insets show details of colloblast and of statocyst and comb rows.

which the **statolith** rests normally beat with an intrinsic rhythm, and these cilia impose the same rhythm on all the cilia in the comb plates that radiate from the statocyst. When the statolith presses against a particular group of supporting cilia, because of gravity or acceleration, it changes their rate of beating, thereby changing the rate of beating in the cilia of particular comb plates.

FEEDING

Ctenophores are all predatory, although nutrition in some species is suplemented by photosynthetic algae. In ctenophores with tentacles, the **colloblasts** (glue cells) are the major organs of predation. [One species, *Haeckelia rubra*, has nematocysts that are apparently kleptocnidae acquired from cnidarian prey (Carré and Carré 1980; Mills and Miller 1984).] The colloblasts are not ejected, like nematocyst threads, but adhere to small organisms as the tentacles gracefully sweep the water. The tentacles then pass across the mouth, as if the ctenophore were licking its fingers. In the pharynx, food particles adhere to a mucous layer, which tufts of fused cilia (**membranelles**) propel into the stomach. Species with reduced or no tentacles feed by swimming with their mouths agape, trapping plankton on the mucous surfaces of the enlarged **oral lobes.** Some species simply swallow large prey whole. *Beroe* (Figure 24.17), which has no tentacles, aggressively pursues and swallows only other ctenophores, including those larger than itself.

Digestion begins in the pharynx and continues in the stomach. The gastrovascular cavities of ctenophores branch into canals that distribute the partially digested food for intracellular digestion. The ctenophore gastrovascular cavity differs from that of cnidarians in having two **anal canals** through which only a little of the wastes are excreted. (The digestive tract is considered to be incomplete.)

Figure 24.19
Two bilaterally symmetric ctenophores. (A) Cestum. Total length is approximately 1.5 meters. (B) Coeloplana differs from Cestum in the direction of flattening. Length approximately 1 cm.

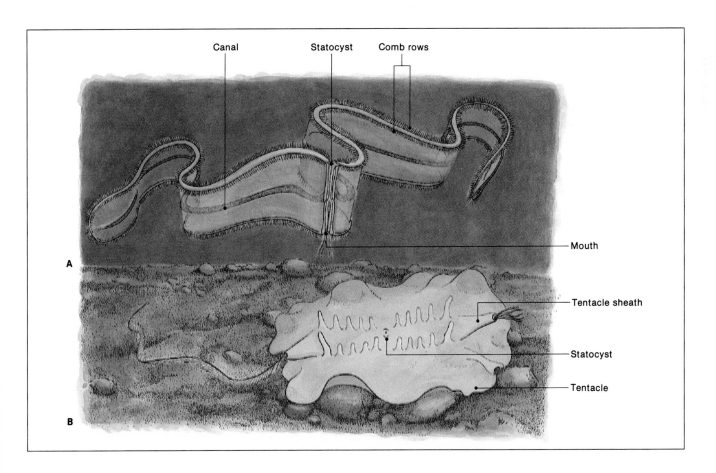

REPRODUCTION

Ctenophores are generally hermaphroditic and can reproduce only sexually, rather than by budding. Gametes generally exit through the mouth. Development occurs externally and produces a miniature of the adult, called a cydippid larva.

CLASSIFICATION

Two classes of ctenophores, Tentaculata and Nuda, are distinguished by the presence or absence of tentacles. The 50 species of *Beroe* (Figure 24.17) belong to class Nuda. *Pleurobrachia*, whose name means "side arms," is often considered representative of the tentaculates (Figures 24.1B and 24.19). The body of *Pleurobrachia* is from 1.5 to 2.0 cm in diameter, but the tentacles can extend from their protective **tentacle sheaths** to a length of approximately 15 cm. Most of the body is composed of mesoglea (often called **mesenchyme** or **collenchyme** in ctenophores; see p. 493). Within the class Tentaculata there is a wide divergence of shapes, with two genera having lost almost all semblance of biradial symmetry (Figure 24.19). *Cestum*, commonly called Venus' girdle, is flattened along the oral–aboral axis, while *Coeloplana* is compressed with the oral and aboral portions close to each other on opposite sides. *Coeloplana* lacks comb rows; it creeps about on substratum.

SUMMARY

Members of the phyla Cnidaria and Ctenophora generally have biradial symmetry and three (or two) thin layers of tissues enclosing a gastrovascular cavity. This body plan allows for a large size without a circulatory system or other complex organs to supply the necessities of life to all the cells. In spite of these similarities, major differences indicate that cnidarians and ctenophores are not closely related to each other. One difference is that cnidarians generally have a life cycle in which a sessile, asexual polyp alternates with a swimming, sexual medusa. No such alternation of generations occurs in ctenophores. Second, cnidarians prey using nematocysts, while ctenophores use colloblasts. Third, the tentacles of cnidarians include all tissue layers—gastrodermis, epidermis, and mesoglea (if present)—while the tentacles of ctenophores are solid and consist only of epidermis. Fourth, cnidarians move by means of nutritive muscle cells and epitheliomuscle cells, while ctenophores have true muscle cells. Another difference is that ctenophores swim by means of comb plates that are unique to the phylum.

Four classes of cnidarians are generally recognized: Hydrozoa, Scyphozoa, Cubozoa, and Anthozoa. Hydrozoans include freshwater polyps, such as *Hydra*, the Portuguese man-of-war, and fire corals. Scyphozoans are the true jellyfish, in which the medusa is the most prominent or only stage. Cubozoans are similar to jellyfish but are box-shaped. Anthozoans are extremely diverse polyps, including true corals, sea fans, sea whips, sea pansies, and sea anemones.

Cnidarians are virtually all carnivorous, using their cnidae to spear, poison, or entangle prey. Nematocysts first punch holes in prey, then shoot threads into the holes. Discharge of nematocysts is triggered by a combination of chemical and mechanical stimulation, without the involvement of the nervous system. The nervous system consists of a diffuse nerve net that causes slow movements of the body and tentacles during feeding, or fast movement during escape. There are few other organs. Parts of hydras can regenerate a new head or foot at the correct ends, because of chemical gradients that establish polarity.

KEY TERMS

biradial symmetry	polyp	cnida
gastrovascular cavity	medusa	nematocyst
gastrodermis	planula	mass spawning
epidermis	nerve net	polarity
mesoglea	radial canal	comb plate
epitheliomuscle cell	siphonoglyph	statocyst
nutritive muscle cell	septum	colloblast

SELF-TEST

1. Describe the ways in which Cnidaria and Ctenophora resemble each other. What differences account for their being classified in different phyla?

2. Explain how the cnidarians and ctenophores are able to maintain such large bodies without a circulating body fluid.

3. Some cnidarians alternate between polyp and medusa generations. What are the characteristics of each form? What advantages might the alternation of forms confer on a species?

4. Describe the life cycle of *Obelia*.

5. Describe two examples of mutualism involving members of class Anthozoa.

6. At the beach you find a biradially symmetric, gelatinous animal with tentacles. How can you tell from its appearance whether it is a scyphozoan, a cubozoan, or a tentaculate? Which of these would be safe to touch?

7. What are the differences between cnidae and colloblasts?

8. Describe the organization and functioning of the nervous systems of cnidarians and ctenophores.

9. Describe one experiment supporting the chemical-gradient theory of polarity in *Hydra*.

READINGS

RECOMMENDED READINGS

Faulkner, D. and R. Chesher. 1979. *Living Corals.* New York: Clarkson N. Potter. (*Beautifully illustrated.*)

Gierer, A. 1974. Hydra as a model for the development of biological form. *Sci. Am.* 231(6):44–54 (Dec).

Goreau, T. F., N. I. Goreau, and T. J. Goreau. 1979. Corals and coral reefs. *Sci. Am.* 241(2):124–136 (Aug).

Gould, S. J. 1985. A most ingenious paradox. In: *The Flamingo's Smile.* New York: W. W. Norton, Chap. 5. (*On the question of individuality in the Portuguese-man-of-war.*)

Jackson, J. B. C. and T. P. Hughes. 1985. Adaptive strategies of coral-reef invertebrates. *Am. Sci.* 73:265–274.

Koehl, M. A. R. 1982. The interaction of moving water and sessile organisms. *Sci. Am.* 247(6):124–134 (Dec). (*How anemones adapt to ocean currents.*)

Lenhoff, H. M. and S. G. Lenhoff. 1988. Trembley's polyps. *Sci. Am.* 258(4):108–113 (Apr).

See also relevant selections in General References listed at the end of Chapter 21.

ADDITIONAL REFERENCES

Carré, C. and D. Carré. 1980. Les cnidocysts du cténophore *Euchlora rubra* (Kölliker 1853). *Cah. Biol. mar.* 21:221–226.

Elliott, J. K. et al. 1989. Induction of swimming in *Stomphia* (Anthozoa: Actiniaria) by imbricatine, a metabolite of the asteroid *Dermasterias imbricata.* *Biol. Bull.* 176:73–78.

Holstein, T. and P. Tardent. 1984. An ultrahigh-speed analysis of exocytosis: nematocyst discharge. *Science* 223:830–833.

Mills, C. E. and R. L. Miller. 1984. Ingestion of a medusa (*Aegina citrea*) by the nematocyst-containing ctenophore *Haeckelia rubra* (formerly *Euchlora rubra*): phylogenetic implications. *Mar. Biol.* 78:215–221.

Möller, H. 1984. Reduction of a larval herring population by jellyfish predator. *Science* 224:621–622.

Murata, M. et al. 1986. Characterization of compounds that induce symbiosis between sea anemone and anemone fish. *Science* 234:585–587.

Rahat, M. and Ch. Dimentman. 1982. Cultivation of bacteria-free *Hydra viridis:* missing budding factor in nonsymbiotic hydra. *Science* 216:67–68.

Roberts, L. 1990. Warm waters, bleached corals. *Science* 250:213.

Ruggieri, G. D. 1976. Drugs from the sea. *Science* 194:491–497.

Spencer, A. N. 1989. Neuropeptides in the Cnidaria. *Am. Zool.* 29:1213–1225.

Tucker, J. B. 1985. Drugs from the sea spark renewed interest. *BioScience* 35:541–545.

25

Flatworms and Other Acoelomates

Marine flatworm (Pseudoceros zebra) on red alga.

CHAPTER OUTLINE

LEARNING OBJECTIVES

1. How do acoelomates function without a body cavity?

2. What insights might acoelomates provide about the earliest bilaterally symmetric animals?

3. What kind of animal is a tapeworm? What kind of animal causes schistosomiasis?

4. How is the life cycle of a parasitic flatworm adapted to infect not just one but two or more hosts?

INTRODUCTION

If the evolution of animals can be compared to the growth of a tree, then the animals we have studied so far appear to have grown out of the roots without reaching any great height. In this chapter we begin to study the limbs of the tree, starting with three groups generally assumed to be on the lowest branches. These are the acoelomate phyla **Platyhelminthes, Gnathostomulida,** and **Gastrotricha,** commonly known as flatworms, jaw worms, and gastrotrichs. These three phyla are considered together for two main reasons: (1) Unlike animals to be considered in later chapters, they lack a body cavity (pseudocoel or coelom; see pp. 450–451). Thus they are called acoelomates (a-SEAL-oh-maytz). Instead of a body cavity the tissue between ectoderm and endoderm—the mesenchyme—consists of densely packed mesodermal cells and is called **parenchyme.** (2) Acoelomates have similar solutions to the problem of getting the necessities of life to all their cells. The main solution is that all are flattened dorsoventrally, which provides a large ratio of surface to volume and ensures that no cell is far from a source of oxygen, nutrients, water, and ions, even though there is no circulatory system or body cavity.

Acoelomates also share the following similarities:

1. All species have ciliated epidermal cells.
2. All three phyla include species with protonephridia (osmoregulatory tubules open only at the external end).
3. All have a central nervous system consisting of cerebral ganglia and longitudinal nerve cords.

Acoelomates also share the distinction of being elongated, legless invertebrates known classically as **worms,** or helminths. Thus, unlike the asymmetric, radial, or biradial animals described in the previous two chapters, acoelomates are **bilaterally symmetric.** This allies them with almost all other animals, which are often called the **bilateria.** This similarity is one of the arguments for acoelomates having originated low on the main trunk of animal evolution. This conclusion is also supported by molecular phylogenetics, at least in the case of flatworms (see Figures 18.16 and 21.9). Another feature that acoelomates share with other bilateria, but not with sponges, cnidarians, or others described previously, is that there is no doubt that their tissues are derived from three germ layers: ectoderm, mesoderm, and endoderm. That is, they are **triploblastic.** In addition, their tissues are organized into discrete structures that perform specific functions. In other words, they have an **organ-system grade of organization,** rather than a cellular or tissue level of organization. Individual cells do not have to take care of all their own needs but are provided with an interstitial fluid that is maintained by organs.

FLATWORMS: PHYLUM PLATYHELMINTHES

CHARACTERISTICS AND MAJOR KINDS OF FLATWORMS

Flatworms vary considerably, mainly because some are free-living while others are parasitic within other animals. The major characteristics that most have in common are summarized above and in Table 25.1.

Table 25.1 Characteristics of flatworms.

Phylum Platyhelminthes plat-ee-hell-MINTH-eez (Greek *platys* flat + *helminthos* worm).

Morphology: No body cavity: **parenchyme** consisting of mesodermal cells separates ectodermal and endodermal tissues. **Bilateral symmetry.** Body usually flattened dorsoventrally.

Physiology: Central nervous system, typically with a pair of anterior ganglia and up to four pairs of longitudinal nerve cords joined by branches. Also a diffuse nerve plexus beneath the epidermis. Mainly carnivorous (predatory, scavenging, or parasitic); some with mutualistic algae in epidermis. Probable absorption of dissolved organic material across the epidermis, especially in endoparasites. Pharynx and often highly branched digestive cavity. Digestive tract (if present) incomplete (without a separate anus). Digestion partly intracellular. Osmoregulation by **protonephridia.** No circulatory, respiratory, or skeletal system.

Locomotion: Creeping on substratum, swimming, or passive within host. Duogland adhesive system for temporary anchorage to substratum.

Reproduction: Usually **monoecious,** but self-fertilization usually avoided. Asexual reproduction by transverse fission in some forms.

Development: Spiral cleavage pattern (when not aberrant due to heavy yolk). Determinate. Blastopore becomes mouth. Development direct, or indirect with one or more larval stages, including a trochophore in some species.

Habitat, Size, and Diversity: Almost all **aquatic** or **parasitic.** Sizes of adults from less than 1 mm to more than 10 meters long. Approximately **12,200 living species** described.

Classification within the phylum Platyhelminthes depends largely on the degree to which the group is free-living or parasitic. Members of **class Turbellaria,** which are almost all free-living, tend to have the best-developed organ systems, since they must search for and secure food and tolerate a variety of external environments (Figure 25.1). Turbellarians can creep about on cilia in a film of secreted mucus. Turbellarians can also move by alternately contracting longitudinal and circular muscles. On encountering suitable prey—usually a small animal—turbel-

---

Major Extant Groups of Phylum Platyhelminthes

Genera mentioned elsewhere in this chapter are noted.

Class Turbellaria TUR-bell-AIR-ee-uh (Latin *turbellae* a turbulence). Usually free-living, with ventral mouth, a pharynx that extends through the mouth, and a branched or lobed intestine. Locomotion by creeping on cilia in a layer of secreted mucus. Epidermis contains rodlike rhabdites that may secrete mucus that repels predators. Mostly hermaphroditic; some also reproduce by transverse fission. *Bipalium, Dugesia, Mesostoma, Pseudoceros* (Figures 25.1, 25.3).

Class Monogenea MON-oh-gen-EE-uh (Greek *monos* single + *genea* generation). Flukes with one juvenile form. Mainly leaf-shaped ectoparasites of fish. Attachment to host by hooks, clamps, and/or suckers on posterior. Gut well developed. Adult covered by nonciliated, syncytial tegument. Juveniles ciliated and free-swimming. Monoecious. *Gyrodactylus, Oculotrema* (Figure 25.6).

Class Trematoda TRIM-a-TOAD-uh (Greek *trematodes* with holes [referring to suckers]). Flukes with four or more larval forms. Adults mainly round or leaf-shaped endoparasites of vertebrates. Adult attaches to primary host's tissues by suckers. Gut with two branches. Adult covered with nonciliated tegument. Mostly monoecious. Some larval stages reproduce asexually. *Opisthorchis, Fasciola, Schistosoma* (Figures 25.7, 25.8).

Class Cestoda (= Cestoidea) ses-TOW-duh (Greek *kestos* belt). Adults elongated, flattened endoparasites in intestines of vertebrates. Usually one or more larval stages and intermediate hosts. Adult attaches to host's intestine by hooks and/or suckers. No mouth or digestive organ. Adult covered with nonciliated tegument. Tapeworms. *Diphyllobothrium, Dipylidium, Echinococcus, Taenia, Taeniarhynchus, Vampirolepis* (Figure 25.9).

A   B

Figure 25.1

(A) An unidentified marine flatworm. Many marine flatworms swim gracefully and are brightly colored. (B) Bipalium kewense, a terrestrial turbellarian native to the tropics of Asia. Having been inadvertently imported with ornamental plants, it is now common in gardens in the southern United States. It preys on earthworms and has become a serious pest in some worm farms. It is up to 25 cm.

larians cover it with their bodies and begin feeding through a ventral mouth.

The parasitic flatworms called **flukes** were formerly included in a single class Trematoda. Now, however, the **monogenetic flukes** are placed in their own **class Monogenea.** Monogenetic flukes reproduce only as adults (Greek *mono-* single + *genesis* origin). The eggs develop directly into nonparasitic juveniles that resemble the adults. Most adults are external parasites on the skin or gills of fish, to which they cling by suckers, hooks, or clamps. **Digenetic flukes,** which reproduce both as larvae and adults (Greek *di-* two), remain in **class Trematoda.** These trematodes have at least four larval stages, usually with two that parasitize a mollusc. The adult is usually an internal parasite of a vertebrate, in whose tissues it attaches by means of suckers.

Tapeworms, class Cestoda, are parasites of the vertebrate digestive tract, to which they attach by suckers and/or hooks. Tapeworms do not have a digestive system of their own. In fact, only the reproductive system is well developed. Development is indirect, and most have larval forms that parasitize a different intermediate host species.

TURBELLARIANS

All free-living flatworms are assigned to class Turbellaria, although the class also includes some species that are symbiotic in marine organisms. Since free-living forms cannot depend on a host to provide a homeostatically regulated environment, most turbellarians are fully equipped with their own organs. These organ systems enable them to inhabit a wide range of habitats: marine, fresh water, and terrestrial (Figure 25.1).

Feeding and Digestion.   In most turbellarians the most prominent organ is the **digestive cavity.** It is often highly branched into the thickest layer of tissue, the parenchyme. The digestive cavity partly substitutes for the absence of a circulatory system and body cavity. Most turbellarians have a pharynx that can be everted through the mouth. Digestive enzymes are secreted through the pharynx into the food, and the partially digested food can then be drawn through the pharynx into the digestive cavity. Partly digested food is then phagocytized into parenchymal cells for intracellular digestion (Figure 25.2).

Turbellarians are primarily predators and scavengers. Typically, they glide about with head slightly raised until they are over small prey such as nematodes, rotifers, copepods, and insects. The turbellarian then grips the prey and immo-

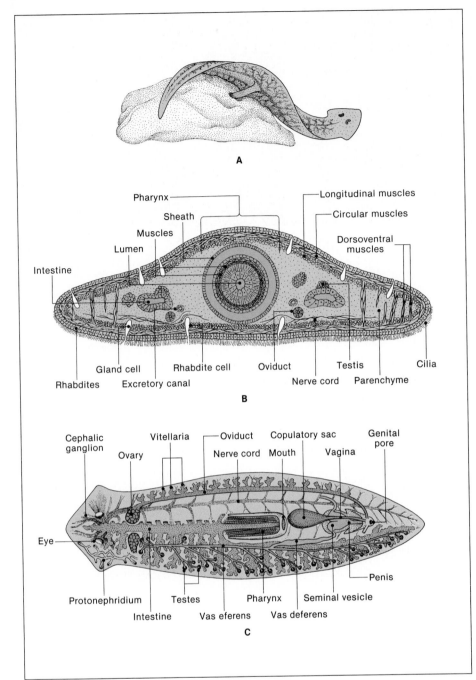

Figure 25.2
A free-living flatworm, the planarian Dugesia, from 0.5 to 2.5 cm long. (A) Feeding on a mass of food, using the pharynx that has been everted through the mouth. (B) Cross section of the body through the pharynx. (C) Schematic diagram of the organ systems. The upper part of the diagram shows the nervous system and the female reproductive organs. The medial and left parts show the digestive tract and male reproductive system. Part of the protonephridial system that courses throughout the parenchyme is shown in the left anterior portion. See also Figure 25.3.

bilizes it with mucus, which is secreted by the epidermis and by rod-shaped structures called **rhabdites.** Rhabdites are produced in the parenchyme and produce mucus when released from the epidermis. Mere contact with the mucus of some turbellarians paralyzes and kills prey (Figure 25.3). The mucus also appears to be toxic to potential predators. This may explain why many marine turbellarians display beautiful warning coloration. Many turbellarians also secrete a kind of glue that may help keep them anchored while struggling with prey, as well as during locomotion. This adhesive comes from a **viscid gland** cell on the ventral surface and goes onto microvilli of **anchor cells** for temporary anchorage. A **releasing gland** later secretes a substance that breaks the bond. The combination of viscid and releasing glands is called a **duogland adhesive system.**

Figure 25.3
The turbellarian Mesostoma *(left)
with its pharynx around a mosquito
larva. Turbellarians may significantly
reduce mosquito populations (Case
and Washino 1979).*

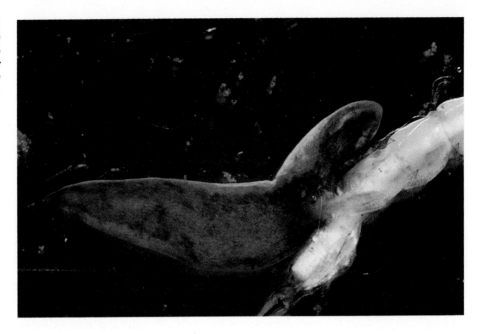

Excretion and Osmoregulation.

Excretion and Osmoregulation.   Ammonia and other metabolic wastes are eliminated partly through the epidermis and partly through **protonephridia.** Protonephridia are also responsible for osmoregulation, which is especially important in freshwater turbellarians, which would swell up without them. Protonephridia are, by definition, excretory tubules that are closed internally and open to the outside. The internal endings of the tubules terminate in **flame cells,** so called because each cell has one tuft of cilia inside that flickers like a flame (see Figure 14.5). In some flatworms each cell at the end of the protonephridia has numerous ciliated **flame bulbs.** The cilia of flame cells and flame bulbs are believed to pull fluid through the membranes of the flame cell, producing a filtrate of water and small molecules, but leaving cells and large molecules within the interstitial fluid. As the filtrate moves along the protonephridial tubules, active transport or diffusion may eliminate wastes and reabsorb nutrients into the parenchyme.

Nervous System.   Turbellarians are the first animals encountered so far in this unit that have central nervous systems. Since bilateral symmetry is associated with movement of an animal in one direction, it is natural for sensory receptors to be concentrated at the anterior end, which encounters the environment first, and for nerve cells to be concentrated at the same end. As in most bilaterally symmetric invertebrates, concentrations of nerve cells occur in two **cephalic ganglia** (= cerebral ganglia). Cephalic ganglia are sometimes called "brains," but unlike the brains of vertebrates, they have little control of the worm's overall activities. Most behavior is coordinated in the longitudinal nerve cords (**connectives**) and their ladderlike branches (see Figure 8.3). In addition, there is a **subepidermal nerve plexus** that functions independently of the cephalic ganglia, in much the same way as the cnidarian nerve net. The cephalic ganglia serve primarily as centers for the integration of sensory information from head receptors, including the two **eyespots** (ocelli). The dark areas of these eyespots, which resemble the irises of human eyes, are actually pigment cups that block the light from the medial direction (Figure 25.4). Without this pigment the animal could not determine the direction of the light, since the ocelli have no lenses that could form an image. The epidermis is equipped with a variety of other receptors, especially in the

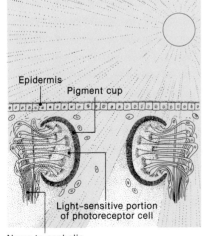

Epidermis

Pigment cup

Light-sensitive portion
of photoreceptor cell

Nerve to cephalic
ganglion

Figure 25.4
Ocelli of Dugesia. *The pigment cups
block light from the opposite side,
enabling the flatworm to determine
the direction of light.*

lateral lobes (**auricles**) on the head. Behavioral studies indicate that such receptors enable flatworms to react appropriately to a variety of chemical and mechanical stimuli. Some turbellarians also have a **statocyst** on top of the head that detects the direction of gravity.

Reproduction and Development. In turbellarians and most other flatworms, reproduction is sexual, except for occasional transverse fission. Like most flatworms, turbellarians are hermaphroditic and apparently favor copulation over self-fertilization. Copulation increases genetic diversity and may avoid the waste of gametes that occurs with external fertilization. All flatworms have a penis, and most also have a vagina. Those without a vagina achieve internal fertilization by **hypodermic impregnation,** with each member of a pair inserting its penis through the epidermis of the other. Many flatworms have numerous testes, distributed along each side, and two ovaries (Figure 25.2C). Sperm leave each testis through a **vas efferens,** and these vasa efferentia collect in a **vas deferens** (= sperm duct) on each side. The two vasa deferentia carry the sperm into the **penis bulb** where they are stored until copulation.

During copulation the penis bulb forces the sperm into the **copulatory sac** of the partner, where they remain until ovulation. During ovulation the ova travel down an **oviduct** on each side and are joined along the way by yolk cells from many **vitellaria** (= yolk glands). A few ova and thousands of yolk cells are then packaged into a cocoon that may be attached to substratum. The embryo usually develops directly into young, which resemble their parents except that they lack reproductive organs. Some marine species first hatch into **trochophore** larvae, which swim by means of a ring of cilia. (See Figures 28.7A and 29.5A for trochophores.) Trochophores are also a common larval form in molluscs, annelids, and several other phyla of coelomates.

Laboratory Studies of Planarians. The freshwater turbellarians called planarians are often used in laboratory studies. *Dugesia*, with its appealing cross-eyed look, is especially popular (Figure 25.2). It is easily collected from beneath leaves, rocks, and other objects in ponds and streams. A bit of raw meat will soon attract a crowd of them, and cultures are easily maintained for study in the laboratory. *Dugesia* is often used in studies of **regeneration.** Like *Hydra* (see pp. 515–517) they demonstrate polarity. That is, a section removed from the middle of a planarian will grow a new head and tail at the correct ends. If the head is split lengthwise and the two halves kept apart, each half will regenerate the missing parts to form a two-headed worm. This can be repeated indefinitely to produce planaria with ten or more heads. Flatworms also readily accept grafts of heads and other body parts, even from other species.

In most animals regeneration serves the obvious function of restoring lost limbs. In flatworms it serves at least two other functions. First, starved flatworms often metabolize some of their own body parts, so regeneration allows them to recover from prolonged food shortages. Second, regeneration enbles them to reproduce by transverse fission. In warm water, well-fed turbellarians often appear to pinch themselves in two, and each half regenerates into a new and complete individual. In some species transverse fission is incomplete; the flatworm produces additional body parts, but actual separation is delayed. The result is a chain of attached **zooids** that superficially resembles a segmented worm.

During the 1960s many psychologists used *Dugesia* as a simple model to study **learning.** They even had their own journal with the whimsical title *Worm Runner's Digest*. Numerous reports in this journal claimed that *Dugesia* could not only learn various types of mazes and conditioned responses but could acquire such learning by cannibalizing trained planaria! Other laboratories were unable to repeat the experiments, however, and they are now seldom mentioned. One problem with this type of research using planaria is that they "run" the mazes (actually they glide through them) on a layer of mucus secreted from the ventral surface. Care must be taken to ensure that the planarian is actually learning to run the maze correctly, rather than simply following an old mucous trail.

MONOGENETIC FLUKES

The main differences between turbellarians and flukes are adaptations of the latter to parasitism. One such adaptation is the **tegument,** a body covering that protects flukes from host defenses and from the external environment. The tegument was formerly considered to be cuticle—a protective layer of dead cells or secretions, such as occurs in insects. Electron microscopy has revealed, however, that the tegument actually consists of inactive portions of a syncytium whose active portions are submerged within the parenchyme (Figure 25.5). Other adaptations of flukes—and of parasites in general—compensate for the low probability that a particular parasite will find a proper host. There are three general kinds of adaptations that compensate for the poor odds:

1. Large numbers of offspring may be produced.
2. The reproductive or other behaviors of the parasite may coincide with times when the host is most vulnerable.
3. The parasite may modify the behavior of the host in such a way that the survival and reproduction of the parasite are enhanced.

Figure 25.5
The tegument of Fasciola hepatica, *the "liver rot" trematode of sheep and cattle.*

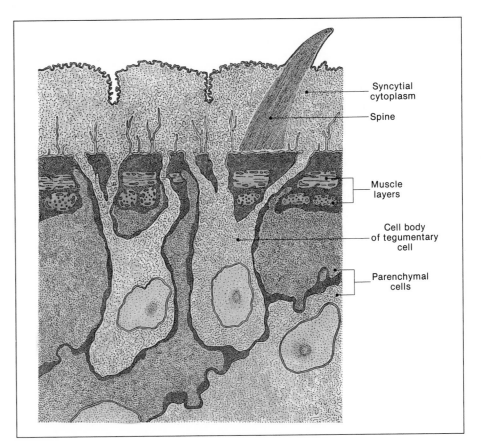

Syncytial cytoplasm

Spine

Muscle layers

Cell body of tegumentary cell

Parenchymal cells

Examples of all these kinds of adaptations will appear in this and the following two chapters.

Flukes are usually flattened, oval, and a few centimeters long. All flukes used to be placed in class Trematoda, but most monogenetic forms—those that reproduce only as adults—are now assigned their own class Monogenea. They usually have direct development with a single juvenile form called an **oncomiracidium** (ON-ko-MERE-a-SID-ee-um). Most oncomiracidia spend some time swimming by means of their cilia before they find a suitable host. Enormous numbers of oncomiracidia are produced, thereby compensating for the low probability that one will find a host. Once an oncomiracidium attaches to a host, it develops into an adult fluke, which remains attached by means of a posterior **opisthaptor** equipped with various hooks, clamps, and suckers.

Most adults of class Monogenea are ectoparasites on the skin or gills of fishes, but some parasitize other vertebrates that spend a lot of time in water. *Oculotrema hippopotami*, for example, lives on the eyes of hippos. Like many parasites, monogeneans are highly selective with regard to host species and even host tissues. A species may be found on one gill arch of a fish, but never on a different gill arch, or near the base of a gill filament, but never near the end. As is often the case with parasites, flukes and their hosts have both evolved mechanisms that ensure that the parasites do not endanger the lives of the hosts. This is not due to the kindness of the parasite but simply reflects the fact that the parasite's own success depends on survival of the host. Only when transmission to hosts becomes especially easy, as in crowded fish farms, do monogenetic flukes produce disease symptoms. The parasites may then become so numerous that the host is literally smothered by flukes covering the gills or by scar tissue caused by excessive damage. *Gyrodactylus*, for example, is a serious pest of trout, bluegills, and goldfish in ponds (Figure 25.6). One of the keys to the success of *Gyrodactylus* is that the young become pregnant before they are even born!

TREMATODES

Other flukes belong to class Trematoda, and most of these belong to subclass Digenea (die-GEN-ee-uh). The Digenea differ from the Monogenea in several ways:

1. Development is indirect.
2. Not only adults, but also larvae, reproduce.
3. Larvae are endoparasitic in one or more **intermediate hosts,** usually including a mollusc.
4. The adults are endoparasitic in various tissues. Most classes of vertebrates are **final** (= primary or definitive) hosts to one or more species of trematode.

It must be difficult for an endoparasite to survive in one unwilling host, yet digenetic flukes can survive in two or more different hosts. The adult is adapted to live and sexually reproduce in the tissues of a vertebrate, and there are at least four larval stages, two of which usually live in a mollusc. These larval stages are:

1. The **miracidium,** which hatches from the egg and usually infects a mollusc as intermediate host.
2. The **mother sporocyst,** which develops from the miracidium and asexually produces

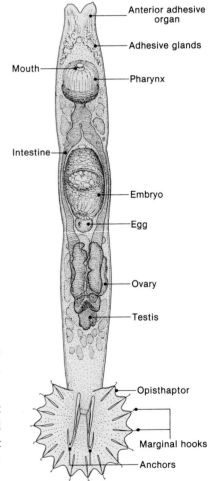

Figure 25.6
Gyrodactylus cylindriformis *showing the adhesive glands and opisthaptor by which it clings to the gills of fish. Length approximately 0.4 mm.*

3. either numerous **daughter sporocysts** or from one to three generations of **rediae,** depending on species. The major difference between the sporocyst and the redia is that the sporocyst lacks a mouth and digestive tract and obtains nutrients by absorption across the tegument, while the redia actively consumes tissues of the intermediate host.

4. Each daughter sporocyst or redia develops into a **cercaria,** which leaves the intermediate host and swims freely to infect a final host or a second intermediate host, depending on species.

Flukes that produce rediae also have at least one other larval state, the **metacercaria,** which follows the cercaria stage.

Life Cycle of the Liver Fluke.   The liver fluke *Opisthorchis sinensis* (formerly *Clonorchis sinensis*) is an example of a fluke with more than four larval stages. The liver fluke may reach several centimeters in length, and inhabits the bile duct, which connects the gallbladder to the liver. This fluke is a serious problem for people in the Orient, where custom and a lack of cooking fuel encourages the eating of raw fish. The life cycle of the liver fluke begins when fertilized eggs produced by the hermaphroditic adults get into the feces of the primary host (Figure 25.7). If the egg is fortunate enough to land in water and be ingested by a snail, a miracidium larva hatches out. The snail then becomes the first inter-

Figure 25.7
The life cycle of the human liver fluke Opisthorchis sinensis. *This parasite is common in the Orient, where fish is eaten raw and human feces are used as fertilizer. The adult flukes inhabit and damage the bile duct of the liver. Severe infections can cause fatal damage to the liver.*

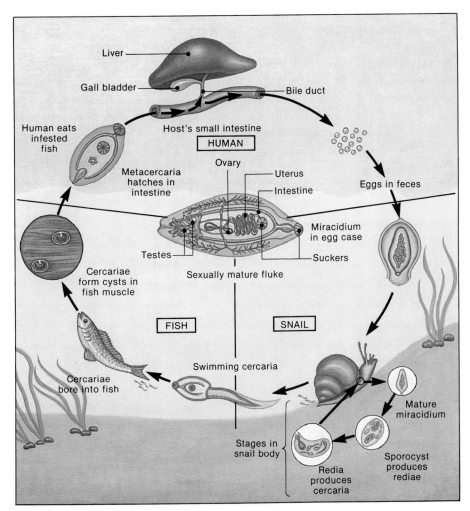

mediate host. Inside the snail's tissues the miracidium develops into a sporocyst, which then asexually produces thousands of rediae larvae. Each redia asexually produces up to 50 cercariae, which emerge from the snail's epidermis. Cercariae swim freely by means of their tails, then bore into the muscles or under the scales of a fish, which becomes the second intermediate host. There each cercaria becomes encysted as a metacercaria. The life cycle is completed when the final host, another vertebrate, eats the fish containing the metacercariae.

Schistosomiasis (= bilharziasis).   The blood fluke *Schistosoma* (SHISS-toe-SO-muh; Greek *schistos* split + *soma* body) is another serious cause of disease, affecting approximately 200 million people (Figure 25.8). Schistosomiasis can be caused by three species of *Schistosoma*, each with a different intermediate snail host, geographic distribution, and symptoms. *Schistosoma mansoni* is common in Africa, South America, and the Caribbean. *Schistosoma haematobium* is common in Africa and the Middle East. *Schistosoma japonicum* occurs in the Far East.

Schistosoma differs from the liver fluke in that it skips the redia and metacercaria stages. Miracidia produce sporocysts within snails, and the sporocysts produce more sporocysts. The second generation sporocysts then produce cercariae that penetrate the skin of humans, without a second intermediate host. Penetration of the skin by cercariae (Figure 25.8B) triggers an immune response that results in itching and a rash. (These immediate symptoms are similar to "swimmer's itch," which results when a species of *Schistosoma* that is common in North American lakes enters the skin. This species cannot use humans as a host, however.) These immediate symptoms are followed during the next 4 to 10 weeks by fever, abdominal pain, throat irritation, diarrhea, and enlargement of the liver and spleen.

In the meantime female schistosomes, sheltered within the groove of the males' bodies, begin releasing large numbers of eggs. Over the next several years these eggs lodge in blood vessels, triggering inflammation and growth of tissue around them. Blockage of blood flow damages organs, especially the liver, spleen, and lungs. Some eggs emerge through damaged vessels in the intestine or urinary bladder and are excreted in either the feces or urine, often accompanied by blood. Eventually the victim stops passing eggs, but affected organs may already have been damaged permanently, and often fatally.

Figure 25.8
(A) Male and female Schistosoma mansoni *copulating, as they are often found in their hosts. Note the gynecophoric canal of the male, which embraces the female and gives the genus its name. During copulation the female receives carbohydrates from the male.*
(B) Scanning electron micrograph of a cercaria of S. mansoni *entering the skin of a mouse. The cercaria secretes a protein-digesting enzyme that enables penetration.*

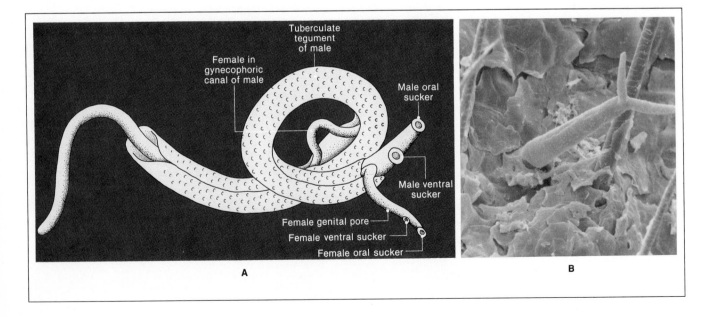

A

B

Schistosoma and Society

Schistosoma and *Homo* have probably had a long relationship. The bloody urine characteristic of *S. haematobium* is noted in ancient Egyptian papyri, and the eggs of this species have been found mummified along with their hosts of 3000 years ago. Conditions for *Schistosoma* are probably more favorable now than they have ever been, however. The combination of high population density, poor sanitation, and irrigated fields where snails thrive seems made-to-order for the disease. Poor sanitation and the need to live close to both drinking water and farm fields ensure that eggs will be excreted into the water, where host snails abound. The cercariae that develop in the snails readily find a human host. The victim can be a fisherman, a peasant working in a rice paddy, a woman washing clothes, or a child playing in the water.

To those in developed countries it would appear to be a simple matter to eliminate schistosomiasis. All one need do is interrupt the life cycle of *Schistosoma* long enough. This could be done in three ways:

1. Destroy the flukes in their human victims. There are now several drugs that will rid human victims of *Schistosoma*, and new drugs and vaccines are being developed.
2. Destroy the snail hosts. There are poisons capable of ridding streams of snails without endangering vertebrates. In the People's Republic of China, since 1955 workers have systematically dug up the banks of streams and irrigation ditches to destroy the snails.
3. Improve sanitation and provide alternatives to fishing, planting, swimming, and washing in contaminated water.

In most countries none of these solutions is affordable.

In Gambia, for example, the amount of money available per year to treat *all* health problems is less than 50 cents per person. This is not enough for drugs or poisons, and certainly not enough for sewage systems, water treatment, or swimming pools. In some countries people are so debilitated by schistosomiasis and other diseases that they are incapable of earning even 50 cents for health care.

In principle, an improvement in the economy of impoverished countries should help eliminate the conditions that favor parasites. In practice, however, attempts to improve the economies of less-fortunate countries sometimes do more harm than good, because of ignorance of the biological situation. One example is the Aswan High Dam in Egypt, which was largely financed and planned by Britain, the Soviet Union, and the United States. This dam was intended to provide flood control, irrigation, and power, but it has also provided an enlarged habitat for snails. As predicted, the resulting increase in schistosomiasis deprived many Egyptians of the hoped-for benefits from the dam. Before the dam was completed in 1971, approximately 5% of people in the 500-mile valley between Aswan and Cairo were victims of schistosomiasis. Four years later the prevalence was approximately 35% in people living below the dam, and 76% in people living above the dam.

Hope for the eradication of schistosomiasis has been raised by the development of an inexpensive chemical (niclosamide) that can be applied to the skin to block penetration by the cercariae (Cherfas 1989). The U.S. Army has begun tests of the antipenetrant, but there does not appear to be much interest except to protect the armies of developed nations when they invade undeveloped countries. The focus has now shifted to development of a vaccine (Cherfas 1991).

TAPEWORMS

Of all Platyhelminthes, those of class Cestoda are most dedicated to a life of parasitism (Figure 25.9). They have no choice, because they cannot synthesize certain essential nutrients, and they lack a digestive tract. The adults get their food already digested in the intestine of a vertebrate host. The adaptations of cestodes to parasitism reveal a terrible kind of beauty. The tegument of the tapeworm must be permeable to nutrients, yet resist damage by digestive enzymes and alkaline secretions of the host. The teguments of tapeworms are adapted for these conflicting requirements in two ways. First, they are covered with **microtriches** (singular = microthrix), which resemble the microvilli on the vertebrate intestinal lining. Like the microvilli, the microtriches increase the surface area for the absorption of nutrients. Second, many tapeworms secrete substances that inhibit digestive enzymes. Tapeworms also reduce the pH to a level at which they, but not the digestive enzymes, can function (Uglem and Just 1983). The reduction in pH may be due to acidic products of anaerobic glycolysis, which is the worm's major source of ATP in the oxygen-deficient environment of the gut. Since glycolysis is so inefficient, tapeworms must consume a lot of the host's food. Hence the phrase, "he must have a worm," is applied to someone with an excessive appetite.

In most cestodes (those in the subclass Eucestoda) a **scolex** (= holdfast) with a combination of suckers and hooks keeps the tapeworm from being swept away by traffic in the gut (Figure 25.9B). Most drugs that are effective against cestodes work by making the gut so inhospitable or the worm so sick that the scolex lets go. Behind the scolex a **germinative zone** asexually produces a steady stream of subunits called **proglottids.** In some species an individual may produce a dozen proglottids daily, for as long as 20 years. Hundreds of proglottids, collectively referred to as a **strobila,** may form a chain more than 10 meters long in some species. The proglottids are united by the nerve cords, protonephridia, and tegument, but reproductively each proglottid is an individual (Figure 25.9C). Any two proglottids, either on the same or on different tapeworms, are capable of exchanging sperm. Inside the uterus of the proglottid, fertilized ova develop into larvae, called **oncospheres,** which remain enclosed within shelled eggs. Eggs then either leave the proglottid, or the oldest proglottids at the end of the strobila break off with up to 100,000 eggs inside. **Gravid** (egg-bearing) proglottids often crawl out of the anus. Proglottids of *Dipylidium caninum*, glistening white and several millimeters square, can sometimes be seen crawling on an infected dog, cat, or child. Proglottids can also be found crawling about in the bed or clothing of an infected person. Once the eggs of a tapeworm are released they must usually be ingested by an intermediate host in order to hatch. Humans are seldom intermediate hosts of *D. caninum*, since most of us avoid eating food contaminated by feces. However, the unsanitary habits of dogs, and humans kissing them, sometimes leads to people becoming intermediate hosts (Figure 25.10). From the worm's point of view humans make poor intermediate hosts, since we seldom get eaten by another vertebrate that could serve as a final host.

Usually the intermediate host is a vertebrate prey of the final host, or an ectoparasite, such as a flea or louse, that is commonly eaten by the final host. After the intermediate host ingests the eggs, its digestive secretions hatch them, and an oncosphere emerges from each. The oncospheres then bore through the intestinal

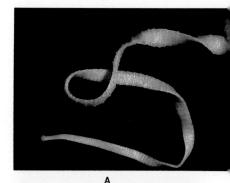

A

B

C

Figure 25.9
(A) Part of the fish tapeworm Diphyllobothrium latum. *The scolex is at the upper right. (B) Scanning electron micrograph of the scolex of a tapeworm, showing the circlet of hooks and two of the suckers. (C) A mature proglottid of a tapeworm filled by gonads of both sexes. The function of Mehlis' gland is unknown. Width 5 mm.*

Table 25.2   Some important tapeworms of humans and other hosts.

Species	Intermediate Hosts	Final Hosts
Fish tapeworm, *Diphyllobothrium latum*	Copepod, fish	Human, other fish-eating mammals
Dog tapeworm, *Dipylidium caninum*	Flea	Dog, cat, human
Beef tapeworm, *Taeniarhynchus saginatus*	Cattle	Human
Pork tapeworm, *Taenia solium*	Pig	Human
Dwarf tapeworm, *Vampirolepis nana*	Flour beetle (optional)	Mouse, human
Unilocular hydatid, *Echinococcus granulosus*	Moose, sheep, human	Wolf, coyote, dog

Figure 25.10
Unilocular hydatid cysts of Echinococcus granulosus *in the brain of a 34-year-old woman. The day before her death, she was admitted to the hospital with epileptic seizures. Hydatid cysts can also form in other organs, sometimes accumulating several liters of fluid each. Humans become infected from unhygienic association with dogs. Herbivores such as moose and sheep are more likely than humans to be intermediate hosts, because they pick up the eggs from the feces of canines. The canines then complete the life cycle when they prey upon the moose or sheep.*

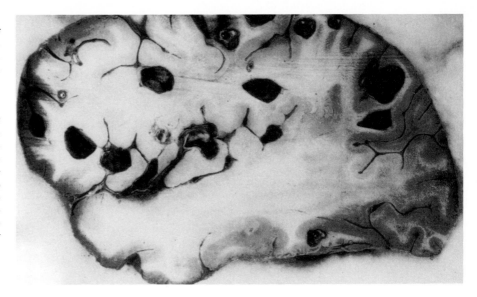

wall and into the blood or lymphatic circulatory system of the intermediate host. The blood or lymph then takes them to the skeletal muscle, heart, or some other organ. The oncospheres can survive in a variety of tissues, since they quickly secrete a protective cyst around themselves. They eventually develop into a variety

Figure 25.11
The life cycle of the beef tapeworm Taeniarhynchus saginatus.

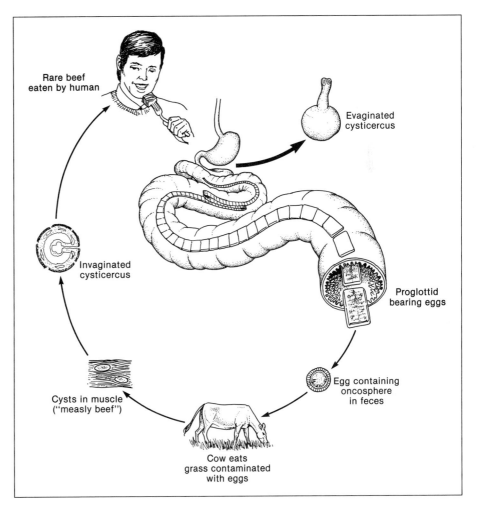

of forms, depending on species. For example, in the beef tapeworm and the pork tapeworm (*Taeniarhynchus saginatus* and *Taenia solium*) the cyst develops into a juvenile called a **bladder worm** or **cysticercus** (SIS-tee-SIR-kus). Each of the cysticerci (SIS-tee-SIR-sy) is essentially an inside-out scolex that everts after the infected tissue of the intermediate host is eaten by the final host. The scolex then attaches to the lining of the intestine by means of suckers and hooks.

Table 25.2 lists some common tapeworms and their hosts. Since the intestine of one mammal is much like that of another, the species that serves as final host depends mainly on whether its diet includes the intermediate host. Figure 25.11 illustrates the life cycle for the beef tapeworm *Taeniarhynchus saginatus* (formerly *Taenia saginata*). This worm often infected humans in the United States until meat inspection became widespread. Even now infections can occur. Only 80% of domestic beef is inspected, one out of four samples with cysts are missed during inspection, and beef is often served so rare that cysticerci are not killed.

JAW WORMS:
PHYLUM GNATHOSTOMULIDA

Gnathostomulids (Figure 25.12) were discovered in 1956 but were considered to be tiny turbellarians until 1969. Among the numerous characteristics that they share with turbellarians in addition to the flattened shape and acoelomate condition are the lack of a circulatory system, hermaphroditism, protonephridia, ciliated epithelium, and a pharynx (Table 25.3). However, Gnathostomulida differ from Platyhelminthes in the following ways:

1. Instead of flame bulbs in the protonehridia, there are **solenocytes,** which have flagella rather than tufts of cilia. (In Gnathostomulida solenocytes are also called **cyrtocytes**—curved cells.)
2. The epithelium differs from that of turbellarians (and all other animals except gastrotrichs) in having only one cilium per cell.
3. Instead of extending the pharynx out of the mouth when feeding, gnathostomulids scrape food up with a comb-shaped **basal plate** and a pair of hard **jaws.**

Table 25.3 Characteristics of gnathostomulids.

Phylum Gnathostomulida NATH-oh-stom-YOU-lid-uh (Greek *gnathos* jaw + *stoma* mouth).

Morphology: Bilaterally symmetric. Dorsoventrally flattened. Parenchyme poorly developed. Ventral mouth with comb-shaped **basal plate** and two **jaws** made of hard organic material.

Physiology: Central nervous system consists of cephalic ganglia and longitudinal nerve cords. Protonephridia with solenocytes (= cyrtocytes). No circulatory, respiratory, or skeletal system. Digestive tract usually incomplete, but some have functional anal pore.

Locomotion: Cilia; one per epidermal cell.

Reproduction: Monoecious. Fertilization internal.

Development: Development direct.

Habitat, Size, and Diversity: Found on marine plants or in sediments worldwide. Sizes of adults from 0.3 to 1 mm. Approximately **80 living species** described.

Figure 25.12
Structure of the gnathostomulid
Problognathia minima.

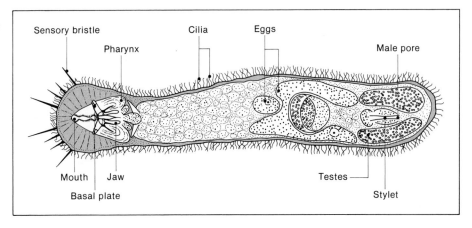

Jaw worms are less than 1 mm long and feed on bacteria, protists, and fungi. None is known to be parasitic. They are usually found on marine algae and plants or in shallow sediments. Many marine forms are found in the black, anaerobic layer produced by sulfur-metabolizing bacteria. Many species live among sand grains, where they share their **interstitial** habitat with other microscopic animals, such as gastrotrichs and nematodes. Six thousand gnathostomulids were found in a 1-liter sample of such sediment. In spite of their vast numbers, probably fewer than one-tenth of the living gnathostomulid species are known. Because they are transparent and small, they are easily overlooked (Table 25.3). Also, it requires special techniques to free them from sand without damage. Jaws of gnathostomulids were once thought to have formed **conodonts:** common fossils 200 to 500 million years old. It is now believed that conodonts were formed by primitive eel-like chordates (Aldridge et al. 1986; Gould 1985).

PHYLUM GASTROTRICHA

Gastrotrichs (Figure 25.13) are usually considered to be pseudocoelomates, because a small space can be seen between the ectoderm and mesoderm in specimens prepared for microscopy. It has long been suspected that gastrotrichs are acoelomates, however, because the space is largely filled with mesodermal cells, and their bodies are organized much like those of flatworms. W. D. Hummon (1982 and personal communication) has shown, in fact, that the "body cavity" is an artifact from fixing the tissue for microscopy. Gastrotrichs should therefore be considered acoelomates. Like turbellarians, gastrotrichs glide on cilia, have duo-gland adhesive systems for temporary attachment, and are hermaphroditic. Freshwater species have protonephridia with solenocytes. As in gnathostomulids, there is only one cilium per epidermal cell (Table 25.4).

Gastrotrichs are all free-living and use their ventral cilia to draw organic debris and protists into the mouth. Marine forms occur on coral and in subtidal or intertidal sediments. Freshwater species occur in bogs and marshes—seldom in clear streams. With a hand lens one can often find them creeping about on lily pads and duckweed. They are easily mistaken for ciliated protozoa. Some are parthenogenic, but most are sequentially hermaphroditic (protandric). That is, each individual functions first as a male, then as a female. Fertilization is internal,

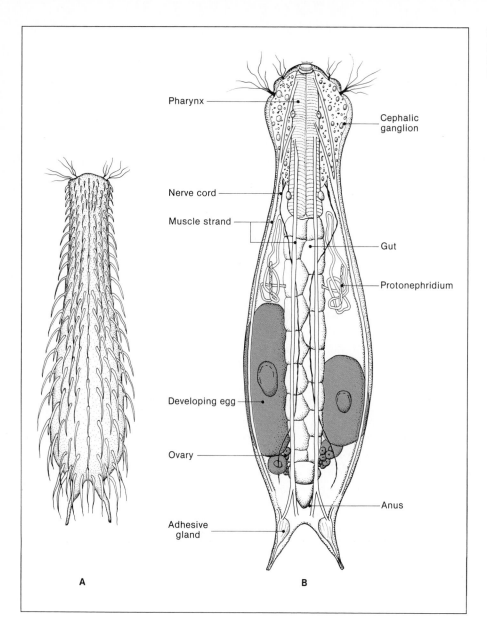

Figure 25.13
(A) External view of the freshwater gastrotrich Chaetonotus *sp. (B) Internal structure.*

Pharynx

Cephalic ganglion

Nerve cord

Muscle strand

Gut

Protonephridium

Developing egg

Ovary

Anus

Adhesive gland

A

B

Table 25.4   Characteristics of gastrotrichs.

Phylum Gastrotricha GAS-tro-TRICK-uh (Greek *gaster* belly + *trichos* hairs).

Morphology: Bilaterally symmetric. Dorsoventrally flattened.

Physiology: Feeding by means of ventral cilia; one cilium per epidermal cell. Protonephridia with solenocytes bearing one or two flagella. No circulatory, respiratory, or skeletal system. Central nervous system consists of cephalic ganglia and longitudinal nerve cords. Digestive tract complete.

Locomotion: By muscular creeping on ventral cilia. With duogland adhesive system for temporary attachment.

Reproduction: Monoecious. Fertilization internal.

Development: Development direct.

Habitat, Size, and Diversity: Living in sediments and on plants; about half the species are marine and half freshwater. Sizes of adults from 0.1 to 4 mm. Approximately **340 living species** described.

with an acting male transferring a spermatophore (sperm packet) to an acting female. Gastrotrichs produce the largest ova in relation to adult body size of any animal. Two types of egg can be formed by freshwater gastrotrichs. One has a thick shell and must undergo drying, freezing, or heating before it will develop. The other has a thin shell that hatches immediately. Development is direct in both cases.

SUMMARY

Flatworms, jaw worms, and gastrotrichs (phyla Platyhelminthes, Gnathostomulida, and Gastrotricha) lack body cavities. Access of oxygen, nutrients, water, and ions to cells is provided by a flattened body, which gives a large surface area per volume. In the larger flatworms a highly branched digestive cavity also aids in the distribution of nutrients. There are also protonephridia and other organ systems that help maintain an interstitial fluid. All are bilaterally symmetric. Other features shared by the three phyla include locomotion by creeping on ciliated cells, and the presence of a central nervous system with cephalic ganglia.

Flatworms include turbellarians, monogenetic flukes, digenetic flukes (trematodes), and parasitic tapeworms (cestodes). Turbellarians generally prey on smaller organisms by everting the pharynx and secreting digestive enzymes. Monogenetic flukes are usually ectoparasites on the gills of fish, to which they attach by the opisthaptor. Trematodes parasitize vertebrates as adults, and molluscs as larvae. Fertilized eggs leave the vertebrate primary host in the feces or urine, then hatch into miracidia larvae. The miracidia then enter a mollusc intermediate host, in which they develop into a sporocyst and then into daughter sporocysts or rediae. They then form cercariae that leave the mollusc and enter a new final host either directly or after further development. Liver flukes and schistosomes are two trematodes that cause diseases in humans.

Tapeworms inhabit the intestines of vertebrates, to which they attach by hooks and suckers on the scolex. They produce a chain of proglottids, each of which sexually produces eggs that contain oncospheres. The oncospheres are ingested by an intermediate host, usually an insect or vertebrate, in which they encyst as cysticerci. The life cycle is completed when a new final host eats the intermediate host.

KEY TERMS

planaria
fluke
tapeworm
bilateral symmetry
parenchyme
monogenetic
digenetic
duogland adhesive system

protonephridium
flame cell
tegument
opisthaptor
trochophore larva
oncomiracidium
miracidium
sporocyst

redia
cercaria
metacercaria
scolex
proglottid
oncosphere
cysticercus
solenocyte

SELF-TEST

1. Explain the term "acoelomate."

2. Acoelomates are said to have an organ-system grade of organization. Illustrate what this means by describing an organ system of an acoelomate.

3. Name the four classes of Platyhelminthes, and describe the major features that distinguish them from each other.

4. Explain how the differences among turbellarians, flukes, and tapeworms are correlated with their degrees of parasitism.

5. Explain how irrigation and poor sanitation favor schistosomiasis.

6. Describe the life cycle of any tapeworm, using the terms oncosphere, proglottid, and cysticercus.

7. Gnathostomulids and gastrotrichs are both small acoelomates. List the major similarities and differences between them.

RECOMMENDED READINGS

Desowitz, R. S. 1981. *New Guinea Tapeworms and Jewish Grandmothers: Tales of Parasites and People*. New York: W. W. Norton (*Informative and entertaining.*)

See also relevant selections in General References listed at the end of Chapter 21.

ADDITIONAL REFERENCES

Aldridge, R. J. et al. 1986. The affinities of conodonts: new evidence from the Carboniferous of Edinburgh, Scotland, UK. *Lethaia* 19:279–291.

Boaden, P. J. S. 1985. Why is a gastrotrich? In: S Conway Morris et al. (Eds.), *The Origins and Relationships of Lower Invertebrates*. New York: Oxford University Press, pp. 249–260.

Case, T. J. and R. K. Washino. 1979. Flatworm control of mosquito larvae in rice fields. *Science* 206:1412–1414.

Cherfas, J. 1989. New weapon in the war against schistosomiasis. *Science* 246:1242–1243.

Cherfas, J. 1991. New hope for vaccine against schistosomiasis. *Science* 251:630–631.

Gould, S. J. 1985. Reducing riddles. In: *The Flamingo's Smile*. New York: W. W. Norton, Chap. 16. (*On conodonts.*)

Hummon, W. D. 1982. Gastrotricha. In: S. P. Parker (Ed.), *Synopsis and Classification of Living Organisms*, Vol. 1, New York: McGraw-Hill, pp. 857–863.

Sterrer, W., M. Mainitz, and R. M. Rieger. 1985. Gnathostomulida: enigmatic as ever. In: S. Conway Morris et al. (Eds.), *The Origins and Relationships of Lower Invertebrates*. New York: Oxford University Press, pp. 181–199.

Uglem, G. L. and J. J. Just. 1983. Trypsin inhibition by tapeworms: antienzyme secretion or pH adjustment? *Science* 220:79–81.

26

Nematodes and Other Pseudocoelomates

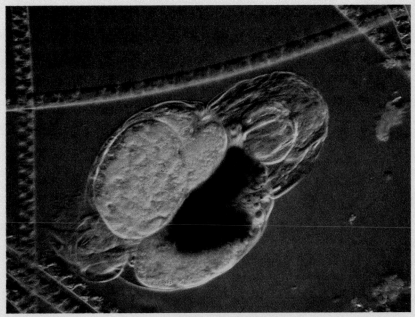

A *rotifer* (Notomata cerberus).

LEARNING OBJECTIVES

1. What is the significance of a body cavity?

2. What features account for the enormous numbers and diversity of nematode worms?

3. What kind of animals are intestinal roundworms and pinworms? What causes elephantiasis?

4. How are these animals adapted for their parasitic way of life?

INTRODUCTION

The eight phyla considered in this chapter—**Nematoda, Nematomorpha, Kinorhyncha, Loricifera, Priapulida, Rotifera, Acanthocephala,** and **Entoprocta**—are extremely diverse. The reason they are often considered together is that, unlike the acoelomates considered in the previous chapter, the digestive tract is not merely hollowed out of a more-or-less solid body. Instead, the digestive tract lies within a **body cavity**, producing a "tube-within-a-tube" organization. This body cavity encloses other organs as well and is filled with fluid that may contain unattached cells of various kinds. There are two major kinds of body cavities: the **coelom** and the **pseudocoel.** Most animals, including ourselves, have a coelom (SEAL-um), which by definition forms within mesoderm and is surrounded by a cellular membrane called the **peritoneum.** Nematodes and other pseudocoelomates have a pseudocoel (SOO-doe-seal; also called pseudocoelom): a cavity that lacks a mesodermal peritoneal membrane (see Figure 21.8).

It was formerly believed that all pseudocoels are simply blastocoels that persist after embryological development (see p. 112). It now appears that pseudocoels can arise in several ways in different phyla, and even in different groups within a phylum (Maggenti 1976). Thus the pseudocoelomates are not a natural taxonomic group. Despite the prefix "pseudo-," the pseudocoel is just as real as the coelom, and just as functional as a coelom. Even some coelomates (molluscs, arthropods, and three other groups described in the next chapter) have what is technically a pseudocoel, although it is called a hemocoel in those animals. The functions of pseudocoels (and of coeloms) include the following:

1. The pseudocoel (or coelom) provides room for the digestive tract to grow, enabling the surface area for the absorption of nutrients to grow at the same pace as body volume. It also allows the digestive tract to move within the body, enabling food to be propelled through it.
2. The body cavity (pseudocoel or coelom) may also provide greater flexibility in the development of other organs so that they can grow as needed. This may be especially important for gonads, allowing them to enlarge seasonally and to produce large eggs.
3. The body cavity may circulate nutrients, oxygen, water, and ions, substituting for a blood vascular system.
4. The body cavity may hold metabolic wastes and excess water or ions. Evidence for this function comes from the fact that many osmoregulatory and excretory organs filter the fluid in the body cavity.
5. The body cavity may function as a hydrostatic skeleton, providing a semirigid body structure against which muscles can contract.

Five of the pseudocoelomate phyla—Nematoda, Nematomorpha, Rotifera, Kinorhyncha, and Priapulida—as well as the acoelomate phylum Gastrotricha (often considered pseudocoelomate), are sometimes combined in the single phylum Aschelminthes (pronounced ask-hel-MIN-theez). The phylum Aschelminthes is now widely regarded as obsolete, however. Further revisions can be expected, for we know little about the relationship of the pseudocoelomates to each other and to other groups. Our ignorance about their origin, and therefore about the early evolution of animals, is due to the lack of a fossil record for these soft-bodied animals, and to relative neglect of them by zoologists. Except for the economically important parasitic species, even the huge phylum Nematoda has received inadequate attention.

ROUNDWORMS: PHYLUM NEMATODA

The best-known phylum of pseudocoelomates is Nematoda. Like flatworms, they appear to have arisen early in the evolution of bilaterally symmetric animals (see Figure 18.6; Hori and Osawa 1987). With so much time to diversify they now rival insects as the most numerous and diverse animals on Earth. Only about 12,000 species have been described, but there are probably millions awaiting discovery and description. Virtually every sample of soil, water, animals, or vegetation in an unstudied area will yield new species. The number of individual roundworms is as impressive as the number of species. A square meter of bottom sediment off the coast of The Netherlands was found to harbor more than 4 million nematodes; in one rotting apple 90,000 nematodes of several species were counted; and 10^{11} nematodes have been estimated to live in an acre of farmland. As N. A. Cobb remarked in 1915, "If all the matter in the universe except the nematodes were swept away, our world would still be dimly recognizable [as] a thin film of nematodes. . . ."

ORGANIZATION OF THE BODY

The tube-within-a-tube construction is apparent on dissecting a large nematode, such as the intestinal parasite *Ascaris*, which may reach a length of 30 cm (Figure 26.1). The inner tube is the complete digestive tract, which includes the mouth, pharynx, intestine, rectum, and anus. (Table 26.1) Together with the reproductive organs, the digestive tract is suspended within the pseudocoel. (These tissues shrink in alcohol and other preservatives, so the pseudocoel is not really as spacious as it appears in Figure 26.1.) The outer tube consists of the muscular body wall that is enclosed by a syncytial epidermis and a tough **cuticle** consisting of several layers of collagen. The cuticle protects parasitic forms from digestive enzymes, and free-living forms from abrasions and other hazards. Many zoologists attempting to kill and preserve nematodes in alcohol have been surprised to find them still thrashing about hours later.

Table 26.1   Characteristics of nematodes.

Phylum Nematoda NEE-ma-TOAD-uh (Greek *nema* thread).

Morphology: Vermiform (worm-shaped): long, round, bilaterally symmetric, without appendages, tapered at both ends. With a **pseudocoel. Cuticle** consisting of the protein collagen. Outer layer often decorated with grooves or structures shaped like hairs, rods, or bumps. Epidermis syncytial.

Physiology: Central nervous system consists of nerve ring around pharynx, ventral and caudal ganglia, and several nerve cords. Complete, unbranched digestive tract. No circulatory or respiratory system. No protonephridia. Some species have **gland cells** or **renette cells** that are assumed to be osmoregulatory because they are located in the pseudocoel and have ducts to the outside.

Locomotion: Longitudinal muscles contract against cuticle and hydrostatic pressure in pseudocoel, producing dorsoventral waves. No appendages. No cilia or flagella except for cilia in sensory receptors.

Reproduction: Usually **dioecious** and dimorphic. Sperm ameboid.

Development: Spiral cleavage pattern, often not evident. **Eutelic:** number and position of somatic cells constant in all adults of same species. Four juvenile stages, often called larvae, resemble the adult.

Habitat, Size, and Diversity: Aquatic (in fresh water or seawater), terrestrial (in moist soil), or **endoparasitic.** Lengths of adults from a few millimeters to approximately 1 meter. Approximately **12,000 living species** described.

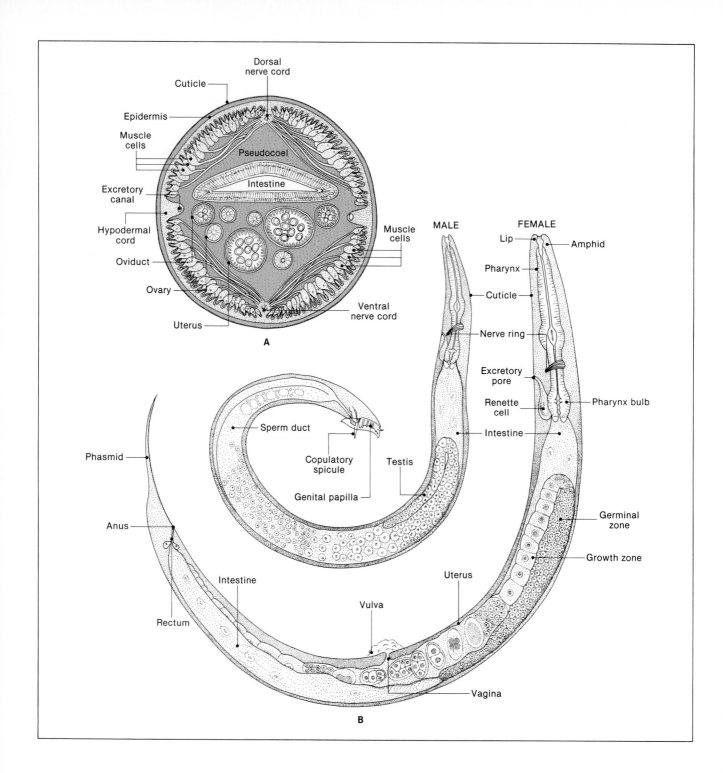

Dorsal
nerve cord

Cuticle

Epidermis

Muscle
cells

Pseudocoel

Intestine

Excretory
canal

Hypodermal
cord

Oviduct

Ovary

Uterus

Muscle
cells

Ventral
nerve cord

A

MALE

FEMALE

Lip

Amphid

Pharynx

Cuticle

Nerve ring

Excretory
pore

Renette
cell

Pharynx bulb

Intestine

Sperm duct

Copulatory
spicule

Genital papilla

Testis

Germinal
zone

Growth zone

Phasmid

Anus

Rectum

Intestine

Uterus

Vulva

Vagina

B

LOCOMOTION

Unlike most worms, which have muscles arranged both lengthwise along the body and circularly around the body, nematodes have only the longitudinal muscles. They do not have circular muscles that would maintain an internal pressure and keep the body from collapsing when the longitudinal muscles contract. Instead, high pressure in the pseudocoel serves as a **hydroskeleton,** and the tough cuticle serves as an **exoskeleton.** The alternating contraction of the dorsal and ventral longitudinal muscles produce dorsoventral waving of the body that propels the worm slowly among soil particles or host cells. The dorsal and ventral muscles

Figure 26.1
(A) Cross section of the parasitic nematode Ascaris *showing the tube-within-a-tube organization. Note that muscle fibers branch to nerve rather than vice versa. (B) Structure of the female and male free-living, freshwater nematode Rhabditis. Ameboid sperm are shown maturing within the testis of the smaller male. Eggs and embryos are shown developing within the uterus of the female.*

Classification of Nematoda

The classification of Nematoda is satisfactory only for lower taxa. Two classes are tentatively recognized.

Class Adenophorea (= Aphasmidia) ad-DEAN-oh-FOR-ee-uh (Greek *aden* gland + *phoros* bearing). Mainly free-living in marine sediments. No phasmids (posterior sensory organs). Amphids (anterior sensory organs) large, variously shaped. Excretory canals rudimentary or lacking. *Trichinella* (Figure 26.8).

Class Secernentea (= Phasmidia) sess-ur-NIN-tee-uh (Latin *secernens* secreting). Mainly parasitic, or microbivorous in soil. Rarely aquatic. Phasmids present. Amphids pore-shaped. Excretory system with lateral canals. *Ascaris, Caenorhabditis, Dracunculus, Enterobius, Necator, Onchocerca, Rhabditis, Toxocara, Wuchereria* (Figures 26.1, 26.5, 26.6B).

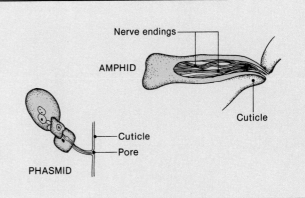

send out branches to the nervous system, rather than the other way around. (This pattern was once thought to be unique to nematodes, but it has been found also in flatworms, gastrotrichs, and other groups.) The muscle branches are interconnected by electrical synapses, so all the dorsal muscles contract in synchrony, as do all the ventral muscles.

NERVOUS SYSTEM

The dorsal and ventral muscles are controlled by a dorsal and a ventral nerve cord, respectively. The rest of the central nervous system consists of a nerve ring around the pharynx, with several ganglia connected to it. There are various sensory organs, including the relatively complex **amphids** recessed on each side of the head. Parasitic species generally have **phasmids**, which are similar to amphids but located near the tail. These organs apparently include receptors for chemicals released by hosts and prey and for sex attractants.

REPRODUCTION AND DEVELOPMENT

Reproduction in nematodes is conventional in many respects but unusual in others. Most species have separate, dimorphic sexes, and fertilization is internal. In some species the female apparently crawls along the curl of the male's tail to effect copulation. Because the hydrostatic pressure in the pseudocoel is quite large, however, a male probably could not insert a penis into the female. Instead of a penis, therefore, males have one or two **copulatory spicules** that hold the female's **vulva** open and allow sperm to enter the vagina. The sperm are unusual in lacking acrosomes and in being ameboid rather than flagellated (see Figure 6.3). Evidently they work well enough: Some females lay hundreds of thousands of shelled eggs every day. Embryogenesis continues within the eggs. Out of the embryonated egg hatches the first of four juvenile stages, L1 through L4, which are often called larvae even though they resemble the adults (Figure 26.2). The juveniles are covered with cuticle, which they must shed (**molt**) before developing into the next, larger stage.

In many species adverse environmental conditions cause development to stop during a particular juvenile stage, producing a juvenile that is capable of surviving for months or years until conditions improve. In many species this resistant form

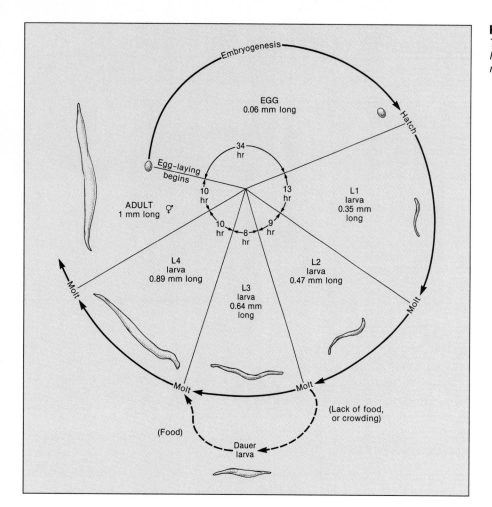

Figure 26.2
The 3¹/₂-day life cycle of a hermaphrodite of the free-living nematode Caenorhabditis elegans.

is the third juvenile state, which is then called a **dauer larva** (German *dauer* enduring). A "crowding pheromone" or the absence of food triggers formation of dauer larvae in *Caenorhabditis elegans*, and secretions from food organisms trigger resumption of the normal life cycle (Golden and Riddle 1982).

CRYPTOBIOSIS

The phenomenal success of roundworms must depend in part on their ability to survive and succeed in adverse conditions. Anywhere there is a film of water, there are probably nematodes. In fact, in many places where there is no water there are nematodes, although in a state of arrested activity called cryptobiosis. Cryptobiosis also occurs in several other groups of animals, including rotifers (Crowe and Cooper 1971; Rensberger 1980). Leeuwenhoek first observed it in rotifers in 1702. The discovery of cryptobiosis in nematodes in 1743 created a sensation in France. Soon all the best people were entertaining their guests by bringing nematodes back from the dead with a drop of water. The phenomenon also ignited a century-long debate between science and the church, provoked by scientists who insisted on calling the phenomenon "resurrection." Most scientists believed that the nematodes actually returned to life after being dead. Clerics, however, insisted that life after death was a miracle that God did not lightly bestow upon worms. They argued that the nematodes were not actually dead, but merely in a state of extremely low metabolism. For once the clerics were right.

Caenorhabditis elegans: The New Drosophila?

Nematodes have contributed to some of the most important concepts in biology. The discovery that fertilization involves the fusion of two pronuclei in the egg was first made in nematodes by O. Bütschli. E. van Beneden used nematodes to show that the chromosome number is halved during meiosis and restored during fertilization. Some of the first studies of sex chromosomes and of cleavage were also made using nematodes. Around the turn of the century the study of development in *Ascaris* eggs led to important discoveries about how cells "decide" what they are going to be when they grow up. These studies had to use fixed, whole mounts of eggs, however. Since the cells migrate and the embryo moves during development, it was impossible with those methods to keep track of development beyond the 102-cell stage of embryogenesis. After a lapse of more than half a century, at least a dozen laboratories around the world have once again resumed the study of development in a nematode, and once again a humble roundworm is yielding fundamental biological discoveries.

The modern era of nematode development began in 1963 with a proposal by Sidney Brenner of the Medical Research Council's Laboratory of Molecular Biology in Cambridge, England. Brenner perceived that molecular biology had progressed to a point where the genetic control of development could be completely understood if one could find a suitable species to study. A suitable species would have to have a short life cycle so that complete development could be followed in a reasonable time. The cells would also have to be few so that every one could be traced. The animal would have to be small and easily cultivated in large numbers for controlled breeding experiments. The species would also have to have a suitable genetic system. All these advantages and more were provided by a free-living nematode *Caenorhabditis elegans* (SEE-no-rab-DYE-tiss ELL-i-ganz).

1. The life cycle of *C. elegans* takes only 3.5 days (Figure 26.2). The adult has only 959 somatic cells (excluding gonads), and the development of each cell is identical in every individual (see Figure 6.16). That is, the animal has the property of **eutely** (YEW-tell-ee).
2. The length of the adults is only about 1 mm. They are easily cultured on agar plates provided with a lawn of *Escherichia coli* bacteria on which to feed. Each hermaphroditic parent produces an average of 300 offspring.
3. There are only about 80 million pairs of DNA nucleotides comprising 2000 to 4000 genes. These are distributed on five pairs of autosomes, plus one X chromosome in the male and two X chromosomes in the hermaphrodite. (The male and hermaphrodite resemble, respectively, the male and female *Rhabditis* shown in Figure 26.1.) The hermaphrodites are protandric, which means they first produce sperm, then ova. The hermaphrodite stores the sperm and can use them to fertilize its own eggs. If the hermaphrodite mates with a male, however, that male's sperm are used first. Whether the sperm are acquired by copulation or self-fertilization, half the offspring are males, and half are hermaphrodites.
4. Self-fertilizing hermaphrodites are advantageous for research, because they tend to have homozygous offspring in which recessive mutations are easily detected. Another advantage of self-fertilization is that even if mutations prevent the hermaphrodites from mating, they can still produce offspring.
5. An additional advantage of *C. elegans* is that they survive freezing in liquid nitrogen ($-196°C!$), so mutants can be stored indefinitely for later study.
6. Finally, the eggs, juveniles, and adults are all transparent. Therefore it is not necessary to kill, fix, and stain them for microscopy. Cells can be observed dividing and migrating in living eggs and juveniles using Nomarski optics.

In spite of doubts from many biologists that the project was feasible, or even worthwhile, Brenner published his first paper on the genetics of *C. elegans* only ten years after his initial proposal. Within another decade, by 1983, the complete lineage of the 959 somatic cells had been worked out by Einhard Schierenberg at the Max-Planck Institute for Experimental Medicine in Göttingen, Germany, H. R. Horvitz at the Massachusetts Institute of Technology, John Sulston at the MRC Laboratory of Molecular Biology, and their many colleagues. Judith Kimball was mainly responsible for tracing the lineages of the 55 cells that make up the male gonads and the 143 cells of the hermaphrodite's gonads. John White and his colleagues pieced together a thousand electron micrographs to trace the formation of all 302 nerve cells and every one of their 8000 synapses.

Some of the conclusions from this study were completely unexpected. First, commonly accepted generalizations about the fate of the three germ layers are not absolute (see pp. 121–125). Some muscle cells do not come from mesoderm, and some nerve cells do not come from ectoderm, for example. Second, bilateral symmetry in the adult does not result from bilaterally symmetric development in the embryo. Homologous cells on each side of the adult may come about in quite different ways. Third, development is uneconomical: 20% of neurons die soon after they develop, apparently when only one daughter cell formed by mitosis is needed. Finally, the development of this simple worm cannot be summarized in a few simple statements but must be described cell-by-cell. The organism results from numerous biological interactions among individual cells, none of which has any "conception" of what the final structure will be like.

Figure 26.3
Scanning electron micrograph of a nematode trapped by a predatory fungus (Arthrobotrys anchonia). Three cells in the fungal loops expand on contact, trapping the prey. The fungus then enters the worm and digests it. At least 150 other species of fungi also prey on nematodes.

INTERACTIONS WITH HUMANS AND OTHER ANIMALS

Nematodes are generally divided between free-living and parasitic forms. Free-living species feed on all manner of organic debris, bacteria, fungi, algae, protozoans, and small animals such as other nematodes. Of course they are also food for other organisms (Figure 26.3). Their ecological importance must be enormous. In agriculture we know how important nematodes are. They cause billions of dollars per year in damage to livestock and crops. Some of those billions of dollars are spent on nematocides. However, these poisons also kill beneficial nematodes that prey on insect pests, consume bacteria and fungi, decompose biological toxins and wastes, and aerate soil. In fact, nematodes that infect insects and other agricultural pests are now sold in dehydrated form (just add water!) as an alternative to expensive chemicals. The parasitic forms are equally important to humans in much of the world, where virtually everyone is host to one or more forms. Some of the most important parasites are described in the following sections.

LARGE INTESTINAL ROUNDWORMS

Ascaris is one of the largest nematode parasites, and it is therefore a common subject for nematode research. In *Ascaris lumbricoides*, the species that infects humans, adult females are up to 30 cm long, and the males are about 20 cm long. They occur in the intestines of approximately one person in six around the world, mainly where sanitation is poor and feces are used as fertilizer. Even in parts of the United States ascariasis is not uncommon. The major cause of infection is fecally contaminated food. The infective second-stage (L2) juvenile can survive up to seven years in its egg, so even food that is free of feces can transmit the infection. Once in the small intestine, digestive secretions hatch the egg, and the L2 juvenile emerges. Curiously, even though the L2 juvenile is in the intestine it does not directly develop into an intestinal parasite. It first must pass through the

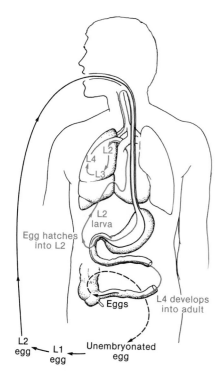

Figure 26.4
The life cycle of the large intestinal roundworm Ascaris lumbricoides.

Labels in figure: Egg hatches into L2; L2 larva; L4; L2; L3; Eggs; L4 develops into adult; L2 egg; L1 egg; Unembryonated egg

mucosa into the lymph or blood veins, then circulate through the right heart and into a lung (Figure 26.4). The L2 juvenile then has to break into an alveolus (air sac), where it remains for 10 days while molting twice into an L4 juvenile. The L4 then ascends the trachea and finally gets back into the digestive tract by being swallowed. This roundabout journey makes sense only if the species evolved from an ancestor that bored through skin and had to pass through the lung to get into the gut. The L4 arrives in the lower small intestine about two months after having been swallowed the first time as an egg. The juvenile then molts into a fertile adult and begins feeding on the intestinal contents. If another worm of the opposite gender is available, mating occurs. Each female then begins releasing up to 200,000 eggs per day into the feces to complete the cycle.

Infection by *Ascaris* produces a variety of symptoms, depending on the site and stage of infection, and on the number of worms. When the L2 juveniles penetrate the intestine they usually do not cause much damage, presumably because there is such a rapid turnover of the intestinal lining. However, L2 juveniles that go astray into the wrong organ can cause local inflammation. A great deal of damage, including a severe form of pneumonia, can be produced when large numbers of L2 juveniles break through the alveoli. Adults in the small intestine usually cause only minor symptoms as long as there are few of them and they remain in the intestine. Sometimes females, perhaps following the urge to pass through the curled tail of a male for mating, wander into the appendix, bile duct, or pancreatic duct, where they can cause serious blockage. They may also get into the lungs and other organs by wandering up the esophagus. They can also exit through the mouth or anus. While not physically damaging, the psychological effect of a foot-long worm coming out the anus or mouth should not be underestimated.

Heavy infections by adults can block the intestine or penetrate its lining, with fatal consequences. *Ascaris* can also cause malnutrition and underdevelopment, by robbing the human host of its food and by releasing toxic metabolic wastes. The resulting poor health and increased demand for food make it more difficult for a victim to work his or her way out of the poverty that is often associated with the unsanitary conditions causing infection. The life cycle of the parasite is entwined in the economic cycle of the host.

The common roundworms of puppies and kittens, *Toxocara canis* and *T. cati*, are similar in appearance and life cycle to *Ascaris*, but smaller. All puppies and 20% of kittens are infected with *Toxocara* until they are "wormed." People, especially children, can also become infected with *Toxocara*, even though *Toxocara* cannot complete its life cycle in a human. The juveniles, however, wander about through a variety of organs, producing an inflammation.

HOOKWORMS

Hookworms are also parasitic nematodes of the intestine, but they are quite different from *Ascaris* in their life cycle and mode of parasitism. *Necator americanus* is the major hookworm of humans, infecting hundreds of millions of people in warmer regions of the world, including several million in the southeastern United States. In *Necator* adult females are approximately 1 cm long, and males are about 7 mm long. The life cycle of *Necator* begins when an egg is released with feces onto the soil. Within a week the free-living L1 juvenile hatches out and molts twice into an infective L3 juvenile, which can survive for several weeks in warm, moist soil if protected from sunlight. In order to complete the life cycle, however, it must burrow into the skin of a human, often a barefoot child. The juvenile then circulates with the blood into a lung, where it breaks through into the trachea and is swallowed. On arriving in the upper portion of the small intestine, L3 molts twice into an adult. The adult then begins moving about on the intestinal lining, where it finally attaches to villi with its "teeth" and begins feeding on blood (Figure 26.5). Each worm consumes 0.03 ml of blood per day, and infection by many worms can use up to 200 ml of blood per day. The anemia that results often weakens the victim significantly.

PINWORMS

One of the most common nematodes in humans is the pinworm *Enterobius ver-micularis*. It seldom constitutes a health problem, however, because it apparently feeds on wastes and bacteria in the cecum and appendix of the large intestine, rather than on the host's nutrients or tissues. Usually the most troublesome symptom of infection is embarrassment. In the United States perhaps one out of three children and one out of six adults are victims. It is highly contagious and therefore tends to be shared by every member of a family. *Enterobius* thrives in spite of excellent sanitation because it does not depend on fecal contamination of food or soil, but on the spread of eggs or juveniles directly from the anus.

The life cycle of *Enterobius* usually begins at night. After copulating in the large intestine, the males die, and the females make a one-way journey to the anus. Each centimeter-long female then lays as many as 16,000 eggs on the anus. Four to seven hours later the L1 juveniles hatch out. If bathing does not occur, the juveniles can migrate back into the anus. A juvenile must go all the way up to the small intestine to complete its development before returning to the large intestine as an adult. The L1 juveniles can also get into the small intestine in eggs transferred directly from the anus to the mouth by the host. Transmission by this route is more likely because the females cause anal itching when they lay eggs. The tiny eggs can also be picked up from bed clothes and furniture and can even be inhaled. These eggs provide the most convenient method of diagnosis, which is done by dabbing the anus with cellophane tape and inspecting it with a microscope.

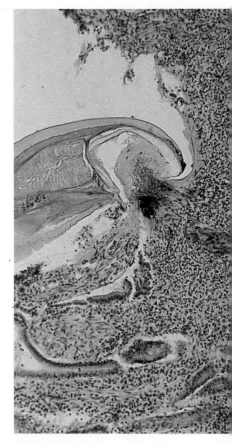

Figure 26.5
Microscopic section showing the attachment of a hookworm to the intestinal lining. A "tooth" deep in the mouth of the worm penetrates the mucosa and makes it bleed.

GUINEA WORMS

Of all animals *Dracunculus medinensis* may well be the least lovable. The female of this species grows to be a meter long and spends its adult life eating its way under human skin. The species is ovoviviparous, which means that the female retains the eggs in her body during early development. Eventually pressure, irritation, or aging of the female causes L1 juveniles to erupt from the mother. The human host reacts by forming a blister in the skin, and when the blister breaks, millions of juveniles spill out. Since the blisters often occur on the leg or foot, and since victims often try to find relief from the burning pain by bathing, the juveniles often emerge directly into water. This is essential for completion of the life cycle, for the L1 juvenile must be ingested by a particular freshwater crustacean within about a week in order to complete its development. This crustacean is the copepod *Cyclops*, a frequent inhabitant of drinking water in parts of Africa, India, and the Middle East (Figure 26.6A). The juvenile invades the body cavity of this intermediate host, then molts twice within approximately two weeks into an infective L3 juvenile.

The L3 juvenile in the copepod enters the final host, *Homo*, with drinking water. For the next three weeks the juvenile burrows through a variety of host tissues until it reaches the subcutaneous connective tissue in the pelvic region.

Then the third and fourth molts occur, forming an adult approximately 43 days after the contaminated water was drunk. By the third month of infection mating occurs. Males then die, but the female continues to eat and grow beneath the skin.

A number of drugs are effective against guinea worms, but for economic reasons the treatment that is still widely used is to wind out the female a little at a time on a stick (Figure 26.6B). This must be done over a period of several weeks. If the worm breaks it retreats, introducing bacteria beneath the skin. The timing of the cycle is often such that 30% to 40% of adults and children are incapacitated at the time of planting or harvesting. To most of us it would appear that instead of treatment, the potential victims should try prevention, either by keeping *Cyclops* or people out of the drinking water. It is not easy, however, to get this message into remote villages, or to change habits that have persisted for thousands of years. In spite of the difficulties the World Health Organization has targeted the guinea worm for extinction by 1995. This goal is possible because *Dracunculus* cannot complete its life cycle in any animal except a human.

FILARIAL WORMS

There are eight species of filarial worms parasitic in human tissues, and almost all have blood-sucking insects as intermediate hosts. *Wuchereria bancrofti* is the best-known filarial worm, because of the gross disfigurement of the legs and genitals that it can cause (Figure 26.7). This condition is commonly called "elephantiasis." **Filariasis** is the appropriate name, however, since the disease is caused by filarial worms, not by elephants. Filariasis occurs when lymphatic vessels are blocked by female and male worms, which are 10 and 4 cm long, respectively. The disease can occur anywhere in the world that is warm enough for the nematodes to develop within insect intermediate hosts. There are now approximately 400 million victims in tropical regions around the world, and some experts believe it is the world's fastest spreading disease.

Adult filarial worms feed on lymph and mate within the lymph vessels, and the females give birth to juveniles called **microfilariae.** The microfilariae pass through the thoracic lymph duct into the bloodstream, where they can be ingested by a blood-sucking insect. The insect is the intermediate host and the vector for the transmission to humans. Most microfilariae are produced during sleep, so the intermediate hosts are often nocturnal mosquitoes. Inside the insect the microfilariae penetrate the gut and molt twice into infective L3 juveniles. When the insect feeds on a human again, these juveniles emerge from the insect's mouth and enter the wound in the human's skin. Once in the lymphatics, the L3 molts twice into an adult.

Another filarial worm, *Onchocerca volvulus*, has a similar life cycle except that its intermediate hosts are black flies (*Simulium* spp.). *Onchocerca* microfilariae infect the skin and cornea of the eyes, causing **onchocerciasis.** In the skin onchocerciasis causes an itching that sometimes drives the victim to suicide. The worms are so dense that positive diagnosis is made simply by snipping off a small piece of skin and seeing the worms in it. Infections of the cornea cause blindness. Onchocerciasis is often called **river blindness,** because it is more prevalent around rivers, in which black fly juveniles live. In some villages in Africa and South America, almost everyone over 50 is blind from onchocerciasis, and some of the most productive land is uninhabitable. River blindness may soon be history, however. The Onchocerciasis Control Program of the World Health Organization is eliminating black fly juveniles from rivers, and the pharmaceutical firm Merck & Com-

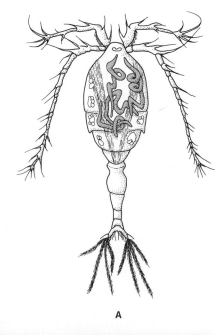

A

B

Figure 26.6
(A) Guinea worm juveniles in the copepod Cyclops, *the intermediate host.* Cyclops *is about the size of the head of a pin. (B) The traditional way of removing* Dracunculus, *by winding it a little each day onto a stick.*

pany is distributing without cost the drug Ivermectin, which kills the microfilariae before they can cause blindness.

TRICHINOSIS

Trichinella was discovered in 1835 by a first-year medical student, James Paget, in London. Paget noticed that calcified particles in the cadaver he was dissecting kept dulling his scalpels. Other students and faculty had previously dismissed such particles as bone fragments, but under the microscope Paget saw that they were actually encysted worms. Later they were named *Trichinella spiralis*, and still later they were found to cause the disease now called trichinosis or trichinellosis. Humans and other animals usually get trichinosis by eating cysts in raw or under-cooked meat. In the small intestine the cysts are digested away, freeing the juveniles to molt into adults. The adults then burrow into the mucosa lining the gut, where they mate. Human victims generally suffer nausea and abdominal pain at this time. During the next 4 to 16 weeks each female may give birth to 1500 juveniles before she dies. Some juveniles enter the digestive tract and spread through feces to a new host. Other juveniles enter the blood circulation and eventually arrive in muscle. There the juveniles burrow into muscle fibers, which then serve as "nurse cells." Protected from the immune system by these nurse cells, the juveniles curl up and become encysted until ingested by a carnivore or omnivore. In the encysted state juveniles can live for 10 months, absorbing nutrients and growing to a diameter of 1 mm (Figure 26.8).

In humans the cysts of *Trichinella* merely produce vague muscle aches often mistaken for flu. The adults and juveniles can, however, produce severe pain, inflammation, fever, and fatal damage as they burrow through tissues. Humans can avoid trichinosis by not consuming the meat of any carnivore or omnivore, especially pork, without cooking it thoroughly. These precautions would appear to be simple in wealthier countries, yet approximately 1 in 25 Americans has *Trichinella* worms in his muscles (Kolata 1985). Most cases are discovered only during autopsy. The incidence was much higher until feeding pigs table scraps

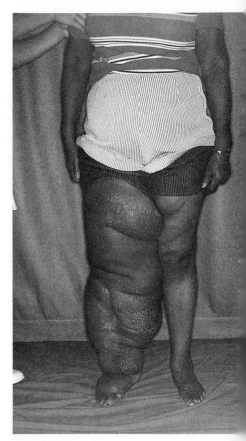

Figure 26.7
A victim of elephantiasis.

Figure 26.8
Juvenile Trichinella spiralis *encysted in muscle.*

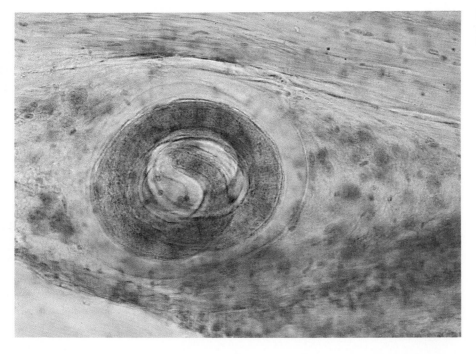

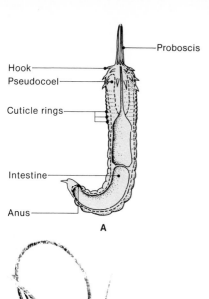

Proboscis
Hook
Pseudocoel
Cuticle rings
Intestine
Anus

A

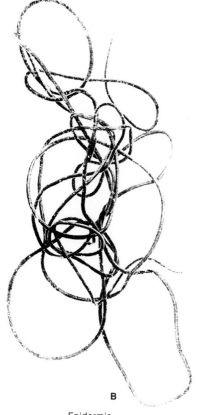

B

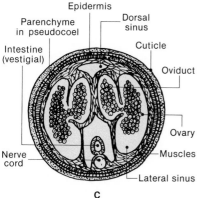

Epidermis
Parenchyme in pseudocoel
Dorsal sinus
Intestine (vestigial)
Cuticle
Oviduct
Ovary
Nerve cord
Muscles
Lateral sinus

C

Figure 26.9
(A) Larva of a nematomorph with proboscis extended. Length approximately 250 μm. (B) An adult Gordius. (C) Cross section of a female Gordius.

including pork was made illegal. Even now, however, approximately one in every thousand pigs in the United States gets trichinosis by eating rats or other pigs' tails. Pork and wild game should be cooked until it is no longer pink, or, in the case of pork sausages, until all the meat reaches 170°F.

HORSEHAIR WORMS: PHYLUM NEMATOMORPHA

Adult nematomorphs are long and thin and can often be found in puddles, contorted into knots. They are commonly called horsehair worms, because they were so common in watering troughs that many people thought they came from horses' hair. They are also known as gordian worms, after King Gordius, who tied an intricate knot and declared that whoever untied it would rule the world. Zoologists once considered them to be nematodes because of the external similarity of the adults, and because they also have a cuticle, longitudinal muscles only, and a nerve ring around the pharynx (Table 26.2). However, the larvae of nematomorphs, which are parasitic in arthropods, are quite different from juvenile nematodes (Figure 26.9). The larvae in fact resemble priapulid worms and kinorhynchs. One of the many proposals for classifying Nematomorpha would combine them into a new phylum with Priapulida and Kinorhyncha (Malakhov 1980).

Nectonema agile, the one species of marine nematomorph that lives in seas bordering the United States, has larvae that parasitize crabs. The free-living adults can sometimes be found along the beach on moonless summer nights. Other nematomorphs spend their brief adult lives reproducing in fresh water. They do not—in fact, cannot—feed, because the digestive tract is degenerate. The anus serves only as an exit for gametes. The males crawl and swim actively until they find a female, deposit sperm near her anus, then die. The ova are fertilized as they emerge, often in strands more than 2 meters long. Several weeks later the eggs hatch into larvae that parasitize insects. How the larvae get into an insect host is unknown. They may be ingested with water, or they may drill their way in with the retractable proboscis. The larvae absorb nutrients across the cuticle and develop into adults inside the host. They emerge only when the insect returns to water, which the nematomorph larvae somehow induce it to do. If the adult develops in autumn it overwinters as a cyst.

Table 26.2 Characteristics of horsehair worms.

Phylum Nematomorpha (= Gordiacea) NEE-mat-oh-MORE-fuh (Greek *nema* thread + *morphe* form).

Morphology: Vermiform. Pseudocoel usually filled with parenchyme.

Physiology: Nerve ring around pharynx. Digestive tract degenerate. Juveniles, endoparasitic in arthropods, absorb nutrients across **cuticle**. No respiratory, circulatory, or osmoregulatory system.

Locomotion: Longitudinal muscles produce undulations of body.

Reproduction: Dioecious. Fertilization external.

Development: Blastopore becomes mouth. Larvae resemble adult Priapulida and Kinorhyncha.

Habitat, Size, and Diversity: Adults free-living, mainly in fresh water. **Juveniles parasitic** in arthropods. Lengths of adults up to 70 cm. Approximately **230 living species** described.

PHYLUM KINORHYNCHA

Kinorhynchs are tiny, free-living worms that burrow through marine sediments, feeding on diatoms and organic material. The organ for burrowing is an extrusible proboscis and circlet of spines (**scalids**) similar to those of nematomorph larvae. The kinorhynch burrows by forcing its head into the sediment, anchoring the spines, then pulling the body forward by retracting the head. Unlike nematodes and nematomorphs, kinorhynchs have not only longitudinal but also circular and diagonal muscles. There are six juvenile stages, each followed by a molting of the chitinous cuticle. The cuticle of the trunk is divided into 11 overlapping joints, but the body is generally not considered to be segmented (Figure 26.10, Table 26.3).

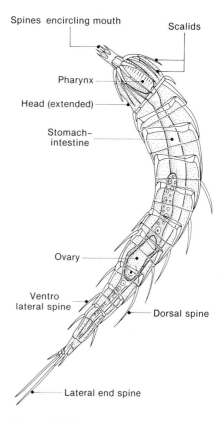

Figure 26.10
An adult kinorhynch Echinoderes *with head extended. Total length approximately 0.2 mm.*

Table 26.3   Characteristics of kinorhynchs.

Phylum Kinorhyncha (= Echinodera) KYN-oh-RING-kuh (Greek *kinein* to move + *rhynchos* beak).

Morphology: Retractable head with scalids. Vermiform. Body flattened ventrally and domed dorsally. Body divided into proboscis, head, neck, and trunk.

Physiology: Nerve ring around pharynx. Digestive tract complete in adults. Pair of **protonephridia** with solenocytes (as in the acoelomates Gnathostomulida and Gastrotricha). No respiratory or circulatory system.

Locomotion: Burrowing with **retractile head.**

Reproduction: Dioecious. Sexes similar. Fertilization never observed, but presumed to be internal.

Development: Direct.

Habitat, Size, and Diversity: Free-living in marine sediments worldwide, to depths of 6 km. Length of adults less than 2 mm. Approximately **100 living species** descibed.

PHYLUM LORICIFERA

The newest phylum in the animal kingdom was erected in 1983 by Reinhardt Kristensen of the University of Copenhagen. Kristensen came across Loricifera in sand and gravel from Denmark, Greenland, and the Coral Sea beginning in 1975, usually while looking for gnathostomulids and other interstitial animals (**meiofauna**). Some of the delicate specimens of Loricifera were lost in microscopic preparation, however, and others were larvae that could not be used as a basis for a new phylum. In 1982 Kristensen was spending his last day on a project in France when he received a large sample of bottom gravel. Since he did not have time to extract the small animals by the standard method, he decided to bathe them in fresh water to osmotically loosen them from the gravel. He then preserved all the meiofauna in formalin for later study. To his delight, this sample yielded a definitive series of larvae and adults that he named Loricifera.

More than 50 species of loriciferans have been found, but their numerous scalids make them difficult to describe. So far only 10 species have formally been described and named. The first one formally defined was *Nanaloricus mysticus* (Latin *nano* dwarf + *lorica* corset; *mysticus* secret). Its retractable head and thorax ringed by scalids suggest a kinship with kinorhynchs (Figure 26.11). Also like kinorhynchs, they have mouths on cones with stylets (Table 26.4). The **lorica** that girdles the abdomens of adults and larvae resembles that of some rotifers. The adults are sedentary, but the larvae swim by means of two propeller-like **toes.**

Table 26.4 Characteristics of loriciferans.

Phylum Loricifera lore-i-SIFF-er-uh (Latin *loricus* corset + *fero* to bear).

Morphology: Head, neck, and thorax retract into abdomen. **Lorica** of six cuticular plates covers abdomen.

Physiology: Nervous system consists of a cephalic ganglion, a ring of ganglia around the pharynx, and at least one abdominal ganglion. Pair of protonephridia. Digestive tract complete.

Locomotion: Adults believed to be sedentary; larvae propel themselves by two caudal toes.

Reproduction: Dioecious; sexes differ. Fertilization never observed, but presumed to be internal.

Development: Indirect, with perhaps three to five larval stages.

Habitat, Size, and Diversity: Free-living in marine sediments worldwide, to depths of several kilometers. Length of adults up to 383 μm. **Ten species** formally described.

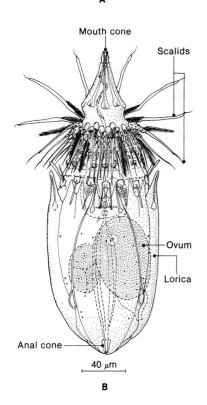

─ Toe

20 μm

A

Mouth cone

Scalids

─ Ovum

Lorica

Anal cone ─

40 μm

B

Figure 26.11
(A) A larva (Higgins-larva) of Nanaloricus mysticus, *the first loriciferan discovered. (B) Adult female* N. mysticus. *Length of the female is approximately 230* μm.

PHYLUM PRIAPULIDA

Priapulida are often considered to be coelomates, but recent evidence indicates that the lining of the body cavity is not cellular, and therefore not a peritoneum (Land and Nørrevang 1985; Malakhov 1980). Priapulids live partly buried in marine sediments. They feed by grabbing passing polychaete worms (phylum Annelida) with a kinorhynch-like proboscis. The body of the adult resembles a nematomorph larva, except that most priapulids have one or two unique caudal appendages (Figure 26.12). Like loriciferans and some rotifers, the larvae have a lorica of cuticular plates. Like nematodes, priapulids have a nerve ring around the pharynx, and like kinorhynchs they have protonephridia with solenocytes (Table 26.5).

Table 26.5 Characteristics of priapulids.

Phylum Priapulida PRE-ah-PEW-lid-uh (Greek *priapos* penis).

Morphology: Arguably pseudocoelomate. Vermiform. Bilateral symmetry, tending toward radial symmetry. Cuticle with numerous projections in rows that give the appearance of segmentation. Yellow or brown.

Physiology: Predatory by means of eversible proboscis. Nerve ring around pharynx. Digestive tract complete in adults. **Protonephridia** with solenocytes. No respiratory or circulatory system (although cells in body cavity of some species contain the respiratory pigment hemerythrin).

Locomotion: Burrowing by contraction of longitudinal and circular muscles.

Reproduction: Dioecious. Fertilization external. One species lacks males and is presumably parthenogenetic.

Development: Indirect. Larvae enclosed in cuticular lorica that is molted between stages.

Habitat, Size, and Diversity: Free-living in marine sediments worldwide, to depths of several kilometers. Lengths of adults 2 mm to 8 cm. **Seventeen living species** described.

PHYLUM ROTIFERA

Except for nematodes, rotifers make up the largest and most diverse group once included in the phylum Aschelminthes. Molecular phylogenetics suggests that they are more closely related to coelomates than to nematodes (Hori and Osawa 1987), and they may have become pseudocoelomates by losing the peritoneum. Rotifers can easily be found with a hand lens in virtually any freshwater habitat: rivers, lakes, bogs, sandy beaches, ditches, and even moisture that collects in mosses and tree bark. They are so numerous that they constitute a major link in aquatic food webs. A few species contribute to marine plankton, and some are symbiotic on or in other animals. Some species are colonial. Like many nematodes, rotifers are capable of withstanding almost complete desiccation in a state of cryptobiosis (see p. 548) and can also survive freezing in liquid nitrogen.

Most species attach to a substratum by means of a **foot** with one or more **toes** or swim by means of a ciliated **corona.** The constantly whirling cilia make rotifers, also called wheel animals, fascinating animals to watch under the microscope. Their transparent bodies provide a window to their inner workings (Figures 26.13 and 26.14). Behind the mouth can be seen the uniquely modified pharynx, called the **mastax,** with its rapidly grinding jaws (**trophi**). Also unique in many species is the **retrocerebral sac,** which is thought to be a sensory organ. The digestive tract is complete with an anus. In females the anus empties into a single receptacle called the **cloaca** (klo-A-kuh) (Latin *cloaca* sewer). By definition, a cloaca is also the terminus of the oviduct and of the excretory system. In freshwater rotifers the excretory system is quite important and consists of a pair of protonephridia that empty into a bladder (Table 26.6).

Chances are good that any rotifer you see will be a female that reproduces parthenogenetically. The only two species that have a conventional sex life are otherwise unconventional; they are parasitic on the gills of a particular marine crustacean. Rotifers in the class Monogononta seem to lie between these extremes, producing males only when they are needed. During most of the summer the females of *Asplanchna* and other species of monogononts are parthenogenetic, producing only diploid, thin-shelled eggs that hatch into females only. Such females are said to be **amictic** (Greek *a* not + *miktos* mixed). Late in autumn, however, unknown stimuli cause the eggs to develop into females that are said to be **mictic.** Mictic females lay haploid eggs that hatch into degenerate males. The males then hypodermically impregnate the mictic females (including perhaps their own mothers). The females then lay fertilized diploid eggs with thick shells. These "winter eggs" are actually encysted females. The thick shells enable "winter eggs" to survive drying, cold, and other adverse conditions. They can also be distributed widely by wind or by adhering to birds and other animals. Upon reaching favorable, moist conditions, these eggs hatch into the next generation of parthenogenetic (amictic) females.

Table 26.6   Characteristics of rotifers.

Phylum Rotifera row-TIFF-er-uh (Latin *rota* wheel + *ferre* to bear).

Morphology: With anterior ciliated **corona** and posterior **toes.** Cuticle, forming a **lorica** in some species, but skeletal support due mainly to intracellular structures. Many tissues **syncytial.** Transparent; often taking on color of prey.

Physiology: Nervous system consisting of dorsal ganglion and longitudinal nerves. Digestive tract complete in females; usually degenerate in males. **Protonephridia** with flame cells. No respiratory or circulatory system.

Locomotion: Free-swimming forms use corona. Sessile forms attach by a foot.

Reproduction: Dioecious. Most species entirely or partly parthenogenetic. Many species mainly parthenogenetic, but periodically reproducing sexually.

Development: Direct. Spiral cleavage. **Eutelic.**

Habitat, Size, and Diversity: Mainly free-living **predators** in fresh water. Lengths of adult females 50 μm to 2 mm. Approximately **1800 living species** described.

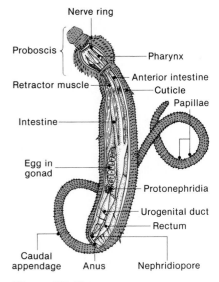

Figure 26.12
The priapulid Tubiluchus corallicola, *approximately 1 mm long.*

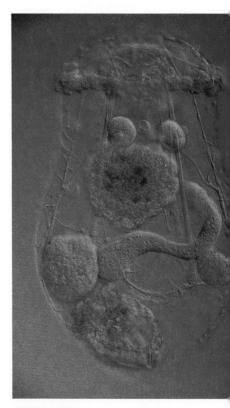

Figure 26.13
A female Asplanchna sieboldi, *one of the largest rotifers (up to 2 mm long). Near the middle is the stomach with two round digestive glands on top. The curved organ is the vitelline (yolk) gland, with an ovum attached on the right. On the lower left are two embryos within the uterus.*

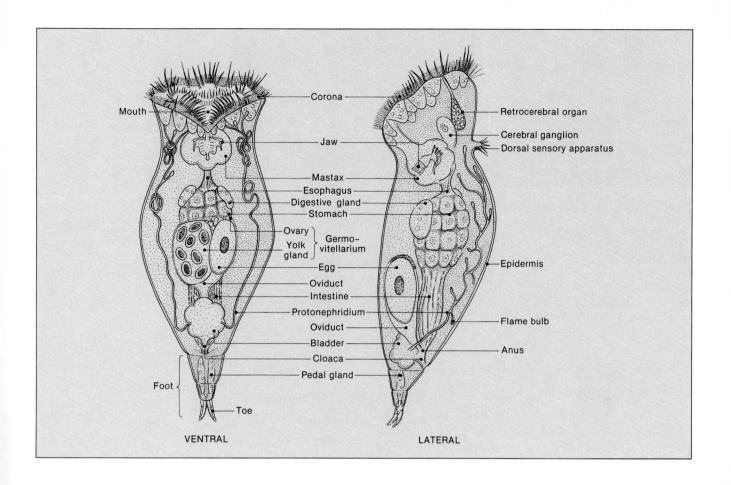

Figure 26.14
Ventral and lateral views of a female rotifer Epiphanes. *Length approximately 0.5 mm.*

Labels in figure:
Mouth, Corona, Jaw, Mastax, Esophagus, Digestive gland, Stomach, Ovary, Yolk gland, Germo-vitellarium, Egg, Oviduct, Intestine, Protonephridium, Oviduct, Bladder, Cloaca, Pedal gland, Foot, Toe, VENTRAL, Retrocerebral organ, Cerebral ganglion, Dorsal sensory apparatus, Epidermis, Flame bulb, Anus, LATERAL

SPINY-HEADED WORMS: PHYLUM ACANTHOCEPHALA

Acanthocephalans are seldom seen because they are entirely parasitic. Juveniles are parasitic in the body cavities of insects or crustaceans, the intermediate hosts (Figure 26.15). The adults live in the small intestines of fishes, birds, and mammals, attached by the spiny proboscis that gives the phylum its name. Many acanthocephalans modify arthropod behavior in ways that make them more susceptible to predation by vertebrate final hosts. Janice Moore (1984) at the University of New Mexico found that when the pill bug *Armadillidium vulgare* (actually a terrestrial crustacean) is infected by the acanthocephalan *Plagiorhynchus cylindraceus*, it is less likely to be in dark, damp places. In other words, it is more likely to be out in the open where it is easy prey for birds.

After copulation within the final host, females release eggs containing embryos, called **acanthors,** into the host's feces. If an appropriate species of insect or crustacean ingests the egg while eating feces, the egg hatches, and the acanthor bores through its gut. Once in the intermediate host's body cavity, the acanthor develops into another embryonic form, the **acanthella.** The acanthella eventually develops into an encysted **cystacanth.** If the intermediate host is then eaten by the appropriate species of final host, the cystacanth develops into an adult. Since humans do not intentionally eat arthropods that eat feces, they are seldom infected by acanthocephalans. Sometimes the cystacanth passes through the digestive tracts of several species before it reaches the intestine of a suitable final host. Up until this time the proboscis has been withdrawn into its receptacle. On entering the small

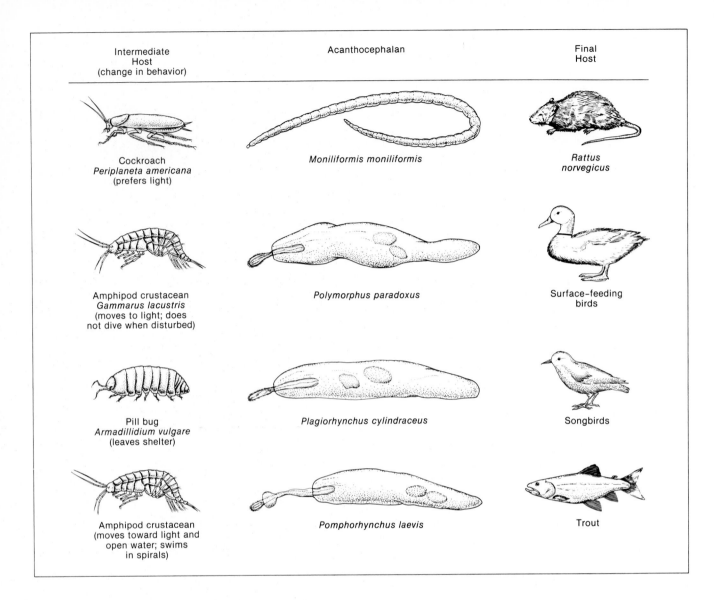

Intermediate Host (change in behavior)	Acanthocephalan	Final Host
Cockroach *Periplaneta americana* (prefers light)	*Moniliformis moniliformis*	*Rattus norvegicus*
Amphipod crustacean *Gammarus lacustris* (moves to light; does not dive when disturbed)	*Polymorphus paradoxus*	Surface-feeding birds
Pill bug *Armadillidium vulgare* (leaves shelter)	*Plagiorhynchus cylindraceus*	Songbirds
Amphipod crustacean (moves toward light and open water; swims in spirals)	*Pomphorhynchus laevis*	Trout

intestine of an appropriate vertebrate, however, the proboscis everts (Figure 26.16B). The recurved spines then penetrate the intestinal wall. A heavy infection

Figure 26.15
Some acanthocephalans and their intermediate and final hosts. As noted in parentheses, many acanthocephalans alter the behavior of the intermediate hosts in ways that make them more likely to be preyed upon by the final hosts.

Table 26.7 Characteristics of spiny-headed worms.

Phylum Acanthocephala a-KANTH-oh-SEF-full-uh (Greek *akantho* thorn + *kephale* head).

Morphology: With spiny, retractable proboscis. Body of adult usually flattened in life, but often cylindrical following preparation for microscopy. Body covered by syncytial tegument.

Physiology: Parasitic; no digestive system. Nutrients absorbed across tegument. Nervous system with ganglion in proboscis receptacle. Protonephridia, when present, with flame cells. No respiratory system. No circulatory system; lacunae within the tegument may distribute nutrients.

Locomotion: Passive in host.

Reproduction: Dioecious. Fertilization internal.

Development: Development indirect. Cells of nervous system and gut eutelic.

Habitat, Size, and Diversity: Seldom seen because they are internal parasites. Distributed worldwide. Lengths of adults 1 mm to 70 cm. Approximately **700 living species** described.

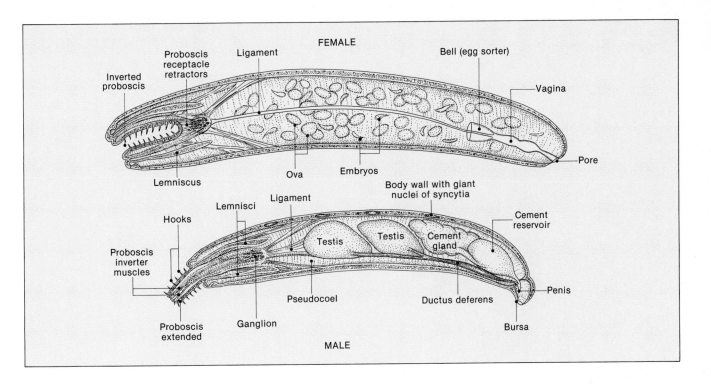

Proboscis
receptacle
retractors
Ligament
Bell (egg sorter)
Inverted
proboscis
Vagina
Lemniscus
Ova
Embryos
Pore
Body wall with giant
nuclei of syncytia
Lemnisci
Ligament
Cement
reservoir
Hooks
Testis
Testis
Cement
gland
Proboscis
inverter
muscles
Pseudocoel
Ductus deferens
Penis
Proboscis
extended
Ganglion
Bursa
MALE

Figure 26.16

Structure of a male and female acanthocephalan. The function of the lemnisci is unknown.

can cause massive and painful destruction of the mucosa. Acanthocephalans have no mouth or digestive tract but absorb the host's food directly across the tegument (Table 26.7).

In the vertebrate's intestine cystacanths soon develop into adults and start to reproduce. The male impregnates the female by means of a penis, then seals the female's vagina with a secretion from his **cement gland.** The primary function of the cement gland is presumably to prevent copulation by a subsequent male, but males have also been known to cement the genitalia of other males, which keeps them from copulating (Abele and Gilchrist 1977). The cement dissipates from the female in time for her to begin releasing eggs. Another curious feature of acanthocephalan reproduction is that the female has an "egg sorter" that releases the mature eggs but withholds the immature ones for further development (Figure 26.16).

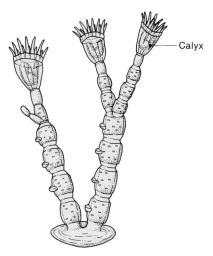

Calyx

Figure 26.17

Part of a colony of the freshwater entoproct Urnatella. *The calyx dies each autumn, but a new one is regenerated from the stalk in the following spring. Height of each zooid approximately 3 mm.*

PHYLUM ENTOPROCTA

Entoprocts are an obscure group with uncertain relationships to the other animals. Some zoologists consider them to be acoelomates, because the body cavity is filled with gelatinous material. In several ways they resemble hydras (Figure 26.17). Like *Hydra*, each individual **zooid** (zo-oyd) in a colony has a mouth surrounded by a ring of tentacles (Table 26.8). Each zooid also has the capacity to regenerate and to reproduce by budding and is mainly sessile, although some can move by somersaulting. Instead of a gastrovascular cavity, however, each zooid has a complete, U-shaped digestive tract, with the mouth and anus both enclosed by tentacles (Figure 26.18). Also the functioning of the tentacles differs greatly from that of *Hydra*. Instead of capturing prey with nematocysts and pulling them to the mouth, the tentacles of entoprocts have cilia that draw water past the tentacles, trapping food in a layer of mucus. Entoprocta superficially resemble Ectoprocta (next chapter), and the two phyla have often been confused. Ectoprocts are coelomates,

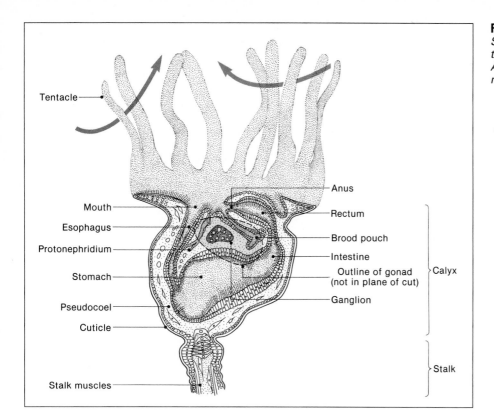

Figure 26.18
Structure of the calyx and part of the stalk of the entoproct Barentsia. *Arrows show the direction of water movement.*

however, and their anal opening lies outside the ring of tentacles. Nearly all entoprocts are marine, but one freshwater genus, *Urnatella*, is oddly distributed in lakes in the eastern and midwestern United States, India, and a few other isolated places.

Table 26.8   Characteristics of entoprocts.

Phylum Entoprocta (= Kamptozoa) IN-toe-PROK-tuh (Greek *entos* within + *proktos* anus).

Morphology: Bilateral symmetry, tending toward radial symmetry. Solitary or forming colonial "animal mats." Each individual or zooid with a **calyx** (body) atop a **stalk** with adhesive base. Exoskeleton present.

Physiology: Digestive system complete and U-shaped, with both anus and mouth within circle of from 6 to 36 ciliated **tentacles**. Cilia draw water past tentacles, trapping food particles on mucus. Nervous system with ganglia. **Protonephridia** with flame cells. No circulatory or respiratory system.

Locomotion: Adults sessile; larvae creep by means of cilia.

Reproduction: Mainly monoecious. Fertilization internal. Reproduction also by budding in colonial species.

Development: Indirect. Spiral, determinate cleavage. Blastopore becomes anus.

Habitat, Size, and Diversity: Almost all **marine.** Lengths of zooids 1 to 5 mm; colonies greater than 1 cm. Approximately **150 living species** described.

SUMMARY

Nematodes, nematomorphs, rotifers, kinorhynchs, loriciferans, priapulids, acanthocephalans, and entoprocts generally have little in common except a pseudocoel: a body cavity without a cellular lining. The digestive tract, if present, lies within the pseudocoel, giving a tube-within-a-tube organization. The pseudocoel provides room for the digestive sys-

tem and other organs to enlarge, may substitute for a circulatory system, and functions as a hydroskeleton.

The most numerous and best-known pseudocoelomates are roundworms (nematodes). They move about within soil or organisms by contracting the longitudinal muscles against the hydroskeleton of the pseudocoel and the stiffness of the cuticle. Many are free-living, but others are parasitic. One of the most important roundworms is the free-living *Caenorhabditis elegans*, which is used extensively in studies of developmental genetics. Among the important parasitic forms in humans are large intestinal roundworms, hookworms, filarial worms, Guinea worms, pinworms, and trichinella worms.

Other pseudocoelomates are diverse in appearance and life-styles, and of uncertain relationship to each other and to other animals.

KEY TERMS

pseudocoel	copulatory spicule	mictic
cuticle	eutely	amictic
cryptobiosis	filarial worm	

SELF-TEST

1. Explain what a pseudocoel is. How does it differ from a coelom? Give three possible functions of the pseudocoel.

2. There are many uncertainties in the classification of pseudocoelomates. Give two examples of uncertain phyla, and explain the source of the uncertainty.

3. Nematodes are said to be successful. Explain the basis for this statement.

4. *Caenorhabditis elegans* has made major contributions to our

understanding of development. Briefly describe the nature of these contributions. State four reasons why this species is so suitable for this research.

5. Describe the life cycles of two species of parasitic nematodes.

6. Among the eight phyla in this chapter there is a great diversity of feeding mechanisms. Describe four of these mechanisms and name one phylum in which the mechanism occurs.

READINGS

RECOMMENDED READINGS

Crowe, J. H. and A. F. Cooper, Jr. 1971. Cryptobiosis. *Sci. Am.* 225(6):30–36 (Dec).

Gilbert, J. J. 1980. Developmental polymorphism in the rotifer *Asplanchna sieboldi. Am. Sci.* 68:636–646. (*How vitamin E alters the morphology of the female.*)

Moore, J. 1984. Parasites that change the behavior of their hosts. *Sci. Am.* 250(5):108–115 (May).

Rensberger, B. 1980. Life in limbo. *Science 80* (Nov):36–43. (*On cryptobiosis.*)

See also relevant selections in General References listed at the end of Chapter 21.

ADDITIONAL REFERENCES

Abele, L. G. and S. Gilchrist. 1977. Homosexual rape and sexual selection in acanthocephalan worms. *Science* 197:81–83.

Clément, P. 1985. The relationships of rotifers. In: S. Conway Morris et al. (Eds.), *The Origins and Relationships of Lower Invertebrates*. New York: Oxford University Press, pp. 224–247.

Cobb, N. A. 1915. Nematodes and their relationships. In: *U.S. Department of Agriculture Yearbook*. Washington DC: U.S. Government Printing Office, pp. 457–490.

Cox, F. E. G. (Ed.). 1982. *Modern Parasitology*. Boston: Blackwell Scientific Publications.

Croll, N. A. (Ed.). 1976. *The Organization of Nematodes*. New York: Academic Press.

Crompton, D. W. T. and B. B. Nickol (Eds.). 1985. *Biology of the Acanthocephala*. New York: Cambridge University Press.

Golden, J. W. and D. L. Riddle. 1982. A pheromone influences larval development in the nematode *Caenorhabditis elegans. Science* 218:578–580.

Hori, H. and S. Osawa. 1987. Origin and evolution of organisms as deduced from 5S ribosomal RNA sequences. *Mol. Biol. Evol.* 4:445–472.

Kenyon, C. 1988. The nematode *Caenorhabditis elegans. Science* 240:1448–1453.

Kolata, G. 1985. Testing for trichinosis. *Science* 227:621, 624.

Kristensen, R. M. 1983. Loricifera, a new phylum with Aschelminthes characters from the meiobenthos. *Z. zool. Syst. Evol.* 21:163–180.

Land, J. van der and A. Nørrevang. 1985. Affinities and intraphyletic relationships of the Priapulida. In: S. Conway Morris et al. (Eds.), *The Origins and Relationships of Lower Invertebrates*. New York: Oxford University Press, pp. 261–273.

Levine, N. D. 1980. *Nematode Parasites of Domestic Animals and of Man*, 2nd ed. Minneapolis: Burgess.

Lewin, R. 1984a. Why is development so illogical? *Science* 224:1327–1329. (*Research on Caenorhabditis elegans.*)

Lewin, R. 1984b. The continuing tale of a small worm. *Science* 225:153–156.

Lorenzen, S. 1985. Phylogenetic aspects of pseudocoelomate evolution. In: S. Conway

Morris et al. (Eds.), *The Origins and Relationships of Lower Invertebrates*. New York: Oxford University Press, pp. 210–223.

Maggenti, A. R. 1976. Taxonomic position of Nematoda among the pseudocoelomate bilateria. In: N. A. Croll (Ed.), *The Organization of Nematodes*. New York: Academic Press, pp. 1–10.

Malakhov, V. V. 1980. Cephalorhyncha, a new phylum uniting Priapulida, Kinorhyncha, Gordiacea and a system of Aschelminthes worms. *Zool. Zh.* 59:485–499. (*In Russian, with English abstract.*)

Marx, J. L. 1984. *Caenorhabditis elegans:* getting to know you. *Science* 225:40–42.

Nicholas, W. L. 1984. *The Biology of Free-Living Nematodes*, 2nd ed. New York: Oxford University Press.

Poinar, G. O. 1983. *The Natural History of Nematodes*. Englewood Cliffs, NJ: Prentice-Hall.

Roberts, L. 1990. The worm project. *Science* 248:1310–1313.

Walsh, J. 1986. River blindness: a gamble pays off. *Science* 232:922–925.

Wharton, D. A. 1986. *A Functional Biology of Nematodes*. Baltimore: Johns Hopkins University Press.

Wood, W. A. et al. (Eds.). 1988. *The Nematode* Caenorhabditis elegans. Cold Spring Harbor, NY: Cold Spring Harbor Laboratory.

Zuckerman, B. M. (Ed.). 1980. *Nematodes as Biological Models*, 2 vols. New York: Academic Press. (*An excellent summary of recent research.*)

27

Miscellaneous Coelomate Protostomes

An onychophoran (Peripatus).

CHAPTER OUTLINE

Introduction

PHYLUM ECTOPROCTA

PHYLUM PHORONIDA

PHYLUM BRACHIOPODA

PHYLUM NEMERTEA

PEANUT WORMS: PHYLUM SIPUNCULA

SPOON WORMS: PHYLUM ECHIURA

WATER BEARS: PHYLUM TARDIGRADA

VELVET WORMS: PHYLUM ONYCHOPHORA

BEARD WORMS: PHYLUM POGONOPHORA

LEARNING OBJECTIVES

1. What is the significance of the coelom?

2. What are protostomes?

3. How are the nine minor phyla of coelomate proto-stomes different from other animals and from each other?

4. What are some of the ways by which these diverse groups obtain food?

INTRODUCTION

The animals to be discussed in the rest of this book are all generally considered to be coelomates. That is, all have a body cavity—the **coelom**—that develops within mesoderm and is surrounded by a cellular membrane called the **peritoneum** (see Figure 21.8). Often the peritoneal membrane forms **mesentery** membrane, which holds the digestive tract and other organs suspended within the coelom. The coelom performs the same functions as a pseudocoel (see p. 544), and it is sometimes difficult to tell them apart.

The coelomates are traditionally divided into two groups, depending on the fate of the blastopore (see p. 118). In chordates, echinoderms, and a few other phyla the blastopore becomes the anus, so they are considered to be deuterostomes. In arthropods, annelids, molluscs, and most of the animals discussed in this chapter the blastopore becomes the mouth, so they are considered to be protostomes. Many protostomes also share other similarities in development. For example, they are generally **schizocoelous,** which means that the coelom originates as a split in the mesoderm. In addition, the pattern of cleavage tends to be **spiral** and **determinate** (see pp. 451–452). There are many exceptions to these patterns, however, and some zoologists doubt their value in phylogenetics (Willmer 1990). Molecular phylogenetics also suggests that instead of two major groups, coelomates should be divided into at least four groups: protostomes except arthropods, arthropods, echinoderms, and chordates (see Figure 21.9). The coelomates discussed in this chapter—the phyla **Ectoprocta, Phoronida, Brachiopoda, Nemertea, Sipuncula, Echiura, Tardigrada, Onychophora,** and **Pogonophora**—are all nonarthropod protostomes.

Figure 27.1
Part of a colony of Bugula neritina *ectoprocts.* Bugula *is common on rocks and pilings along the Atlantic coast and is easily confused with algae.*

PHYLUM ECTOPROCTA

The Ectoprocta were once joined with the Entoprocta in phylum Bryozoa (Greek *bryon* moss), because many members of both phyla resemble each other in forming mosslike colonies (Figure 27.1). Many zoologists still prefer the name Bryozoa for the phylum Ectoprocta and refer to ectoprocts as bryozoans or "moss animals." Unlike entoprocts, however, ectoprocts are coelomates, and the anus of each zooid lies outside its ring of tentacles. The latter difference explains the term ectoproct (Greek *ektos* outside + *proktos* anus). The tentacles constitute a **lophophore,** which is defined as a ring of tentacles enclosing the mouth but not the anus. Each tentacle contains a branch of the coelom. The fact that the anus lies outside the lophophore is important not only to taxonomists but to the ectoprocts themselves. This arrangement allows the cilia on the tentacles to maintain a stream of food-bearing water directly toward the mouth without drawing in the animal's own feces. (Compare Figures 26.18 and 27.2.)

Because the phyla Phoronida and Brachiopoda also have lophophores, Ectoprocta are usually thought to be more closely related to them than to Entoprocta, even though they do not look like each other. The three phyla—Ectoprocta, Phoronida, and Brachiopoda—are referred to as **lophophorates.** Classification of the lophophorates is confused by the fact that the pattern of development in many of their species is more like that of deuterostomes than protostomes. In all three phyla cleavage is radial, as in many deuterostomes. Also in many ectoprocts and brachiopods the mouth does not develop from the blastopore. But then, neither does the anus. In fact, many animals in these two phyla do not form blastopores in the embryo. In phoronids both the mouth and the anus develop from the

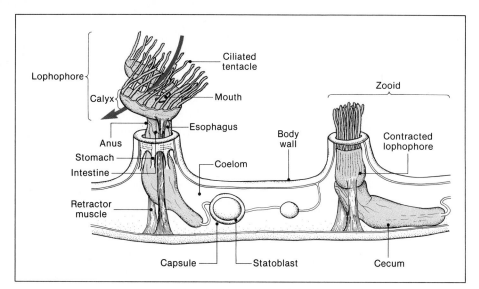

Figure 27.2
The ectoproct Plumatella, *which occurs on the undersides of rocks in fresh water. Arrow shows direction of water current. The statoblast is a resistant structure that asexually regenerates a new colony after the old one dies in winter.*

blastopore. Many zoologists resolve these uncertainties by placing the lophophorates in a group separate from both the protostomes and the deuterostomes, while others consider them deuterostomes. (See Willmer 1990 for discussion.) As noted before however, comparisons of molecular structures indicate that ectoprocts and brachiopods, and presumably phoronids also, are related to protostomes such as molluscs and annelids (Field et al. 1988; Hori and Osawa 1987).

Ectoprocts are colonial, often consisting of millions of zooids with their coeloms interconnected. Each zooid consists of a fleshy **polypide** living within a secreted chamber called the **zoecium** (zo-EE-she-um; from the Greek for "animal house"). The polypide includes the lophophore, digestive tract, and the nervous system and muscles. The zoecia may be cubic, conical, oval, or cylindrical, depending on the species. The colonies may be mossy, globular (Figure 27.3), brushy, or crusty. Many colonies are protected by an external covering that includes various proportions of gelatin, **chitin** (the substance in insect exoskeletons), calcium carbonate, and sometimes sand (Table 27.1).

Ectoprocts can also defend themselves by retracting their delicate lophophores

Table 27.1   Characteristics of ectoprocts.

Phylum Ectoprocta (= Bryozoa) ek-toe-PROK-tuh (Greek *ektos* outside + *proktos* anus).

Morphology: Each zooid is bilaterally symmetric. **Lophophorate:** circular or crescent-shaped lophophore encloses mouth but not anus. **Exoskeleton** of protein, chitin, CaCO₃, and/or sand.

Physiology: Ciliated tentacles of lophophore draw water toward mouth and past tentacles. Lophophore retracted by muscles; extended by muscles, deformation of exoskeleton, or coelomic pressure. Digestive system complete, U-shaped. Nervous system with nerve ring around esophagus, and ganglion between mouth and anus. No excretory, respiratory, or circulatory system.

Locomotion: Adults usually **sessile** and colonial. Some freshwater colonies creep.

Reproduction: Monoecious. Fertilization internal. Colonies grow by asexual budding.

Development: Protostomate, although cleavage tends to be radial, and the mouth does not develop from the blastopore. Neither schizocoelous nor enterocoelous; coelom forms during metamorphosis of larva.

Habitat, Size, and Diversity: Most species marine. Found worldwide. Length of each zooid less than 1 mm; colonies greater than 1 cm. Approximately **4500 living species** described.

Figure 27.3
A large ectoproct colony Pectinatella magnifica *found in Lake Champlain. Diameter approximately one meter.*

Genera mentioned elsewhere in this chapter are noted.

Class Phylactolaemata fill-ACT-toe-LEE-mah-tuh (Greek *phylax* guard + *laemos* throat). All freshwater species. Large cylindrical zooids with crescent-shaped lophophores. All zooids similar (monomorphic). Produce overwintering statoblasts. *Pectinatella*, *Plumatella* (Figure 27.3).

Class Stenolaemata STEE-no-LEE-mah-tuh (Greek *stenos* narrow). Marine. Tubular zoecia in calcareous shells. Circular lophophore. Some polymorphism.

Class Gymnolaemata GYMN-no-LEE-mah-tuh (Greek *gymnos* naked). Mostly marine. Circular lophophore. Polymorphism common: avicularia, etc. present. *Bugula* (Figure 27.1).

into the zoecia, using some of the fastest muscles known (Thorpe et al. 1975). At the same time, the disturbed zooids alert others in the colony by means of nerve impulses. Still others sprout thorny growths when preyed upon (Harvell 1984). In some ectoprocts, such as *Bugula*, some zooids are specialized as **avicularia,** which defend the colony by means of sharp, snapping beaks like those of birds. Other ectoproct colonies have zooids equipped with a long bristle that sweeps off sediments and the larvae of sponges, coral, and other animals that compete for space.

Since adult zooids are confined within zoecia, their sex lives are somewhat constrained. Colonies generally grow by asexual budding of the zooids, and new colonies are started from eggs produced and fertilized in the same hermaphroditic zooid. A zooid broods the eggs within the coelom or in a special zoecium (**ovicell**). The larvae are mobile and thus able to disperse the species to new habitats. Each larva selects a suitable substratum, then starts a new colony by reproducing asexually. Many freshwater ectoprocts can also reproduce from **statoblasts,** which have tough capsules that can survive winter.

Evidently this kind of life has been successful; ectoproct fossils have been dated as far back as the early Ordovician, half a billion years ago. Petroleum explorers value ectoprocts as "indicator fossils." Someday living ectoprocts may also be valued as an ally in the battle against cancer. Seventeen antitumor substances have been isolated from *Bugula neritina* alone (Tucker 1985). Now, however, *Bugula* and some other marine ectoprocts are known mainly for their ability to foul up ship bottoms and pilings.

PHYLUM PHORONIDA

Phoronids are a small phylum of lophophorates that live in shallow seas, either singly in sediments or in tangled groups on pilings and rocks. Each phoronid is confined in a leathery or chitinous tube that it secretes. Many are colored bright orange, pink, green, or yellow. A group of them with extended lophophores can make the sea floor resemble a flower bed (Figure 27.4A). The lophophore has up to 50 ciliated tentacles in two spirals. Cilia on the tentacles propel water downward, and food particles are trapped in mucus in a food groove between the spirals. Cilia then propel the food-bearing mucus to the mouth (Figure 27.4B).

Phoronida have some interesting physiological innovations, probably related to their half-buried existence (Table 27.2). First, they have **closed circulatory systems,** with blood entirely confined within vessels. There is no heart; blood is

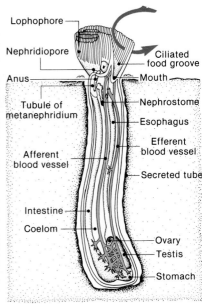

Figure 27.4
(A) A "garden" of phoronids,
Phoronis vancouverensis.
(B) Longitudinal section of P.
australis. *The tentacles are shown
cut. In life they are about as long as
the trunk, approximately 5 mm.
Arrow shows direction of water
current.*

pumped by contracting vessels. Like vertebrates, they have red blood cells containing the oxygen-transporting pigment hemoglobin. Second, adult phoronids have a pair of **nephridia,** which are excretory tubules open both to the outside of the body and at the coelom. Such nephridia are called **metanephridia** to distinguish them from the protonephridia of some acoelomates and pseudocoelomates, which are closed on the internal end by flame cells or solenocytes. Larval phoronids also have protonephridia. Most phoronids are hermaphroditic, but fertilization is generally external and by sperm from another individual. The eggs are released through the metanephridia, which therefore serve as **gonoducts.** This is a common arrangement in coelomates with metanephridia. The eggs exit among the tentacles and are brooded there in some species. The larvae are free-swimming members of the planktonic community and eventually settle on the appropriate substratum before metamorphosis into an adult.

Table 27.2   Characteristics of phoronids.

Phylum Phoronida for-OH-nid-uh (Phoronis: surname of the goddess Io in Greek mythology).

Morphology: Lophophorate: spiral lophophore with 20 to 50 tentacles enclosing mouth but not anus. Adults **vermiform,** inhabiting a tube secreted by the epidermis. Not colonial.

Physiology: Digestive system complete, U-shaped. Probably some contribution to nutrition by absorption of dissolved organic material. Nervous system diffuse, mainly in epidermis. Adults have pair of **metanephridia,** which also serve as **gonoducts.** Closed circulatory system with red blood cells; no heart.

Locomotion: None in adult.

Reproduction: Mainly monoecious. Fertilization external. Reproduction also by transverse fission or budding in some species.

Development: Protostomate even though some are enterocoelous, and cleavage is radial in some species. Development indirect; larvae free-swimming.

Habitat, Size, and Diversity: Marine, to depths of 400 meters. Distributed worldwide. Length from 1 mm to 50 cm. Approximately **15 living species** described.

PHYLUM BRACHIOPODA

Brachiopods are easily confused with bivalve molluscs such as clams and mussels and were once included in phylum Mollusca. Unlike molluscs, however, the **valves** (shells) of most brachiopods are unequal in size and lie on the ventral and dorsal sides of the animal rather than on the left and right. (Many brachiopods have one valve that resembles a Roman oil lamp, and they are therefore called lamp shells.) Also unlike molluscs, which have fleshy bodies, brachiopods have large, curled lophophores as their most prominent organs (Figure 27.5). Another difference between brachiopods and molluscs is that brachiopods are usually attached to substratum by a **pedicel,** ventral side up (Table 27.3). The pedicel serves as an anchor against the turbulence of shallow water. Muscles attached to the pedicel can also contract to shake sediment off the brachiopod (Richardson 1986).

Brachiopods have apparently fallen on hard times in the past 250 million years. They are among the most common fossils from the Paleozoic, but there are only about 1% as many species alive now. The cause of their decline may have been competition from bivalve molluscs (Thayer 1985). Some brachiopods have survived quite well, however. *Lingula* has remained virtually unchanged for more than half a billion years. It may well be the oldest "living fossil."

Table 27.3 Characteristics of brachiopods.

Phylum Brachiopoda brack-ee-OP-oh-duh (Greek *brachion* arm + *podos* foot).

Morphology: Lophophorate: lophophore with two circular or spiral arms within the shell. Shell of two **calcareous valves,** usually unequal in size.

Physiology: Digestive system either complete or without anus. Nervous system with several ganglia connected to a nerve ring around intestine. With **metanephridia** that function as gonoducts as well as excretory organs. Open circulatory system with contractile **heart.**

Locomotion: Adults usually **sessile,** attached by pedicel. A few species use the pedicel for locomotion. Larvae free-swimming.

Reproduction: Dioecious. Fertilization usually external.

Development: Molecular phylogenetics suggests kinship with protostomes, but mouth does not develop from blastopore. Cleavage also radial as in deuterostomes, and some species are enterocoelous. Development indirect.

Habitat, Size, and Diversity: Marine: at all depths from intertidal to abyssal. Length from 1 mm to more than 9 cm. Approximately **335 living species** described.

Major Groups of Phylum Brachiopoda

Genera mentioned elsewhere in this chapter are noted.

Class Inarticulata in-are-TICK-you-LOT-uh (Latin *in* without + *articulus* joint). Valves without hinge; held together by muscles only. Valves similar to each other, composed of calcium phosphate and chitin. Lophophore without skeletal support. Digestive system complete. Schizocoelous. *Lingula.*

Class Articulata. Valves hinged, unequal, composed of calcium carbonate. Lophophore with skeletal support. No anus. Enterocoelous. *Magellania* (Figure 27.5A).

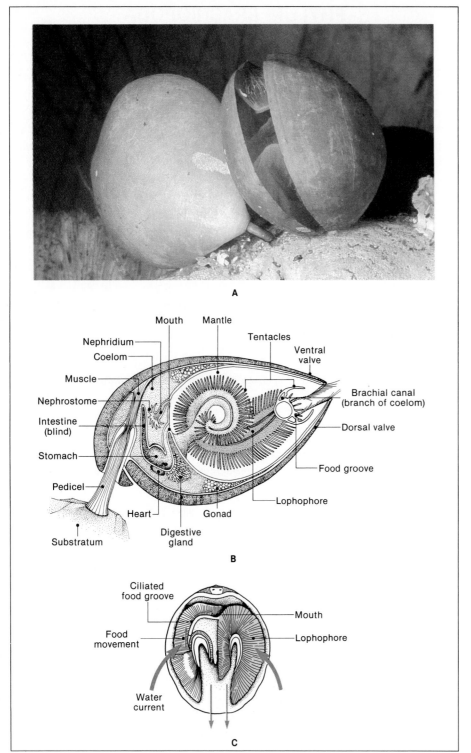

Figure 27.5
(A) Unidentified brachiopods, one with gape open showing lophophore, and another showing pedicel. (B) Side view of the brachiopod Magellania. This is a typical position, with the ventral valve uppermost. (C) Vertical view into one valve of a brachiopod, showing the lophophores and the direction of water and food movement.

A

Mouth Mantle
Nephridium Tentacles
Coelom Ventral valve
Muscle
Nephrostome Brachial canal (branch of coelom)
Intestine (blind)
Stomach Dorsal valve
Pedicel Food groove
Substratum Lophophore
Heart Gonad
Digestive gland

B

Ciliated food groove
Mouth
Food movement Lophophore
Water current

C

PHYLUM NEMERTEA

Members of phylum Nemertea (= Rhynchocoela, Nemertina, Nemertinea; Figure 27.6) are generally considered to be acoelomates, because they do not have a body cavity as such, and because they resemble Platyhelminthes in several ways. They move over substratum by muscle contractions, or by means of cilia and a mucous

Figure 27.6
The nemertean Lineus *sp. on coral.*

Figure 27.7
The anatomy of the nemertean
Amphiporus pulcher. *(A)*
Longitudinal section. (B) Cross
section.

secretion, and they have nervous systems and protonephridia like those of flatworms. Nemerteans differ from flatworms, however, in being mostly dioecious and in having a complete digestive tract and a vascular system for the circulation of blood. There is no heart: blood is propelled by contractions of large vessels, first in one direction, then the other (Figure 27.7). The main reason for consid-

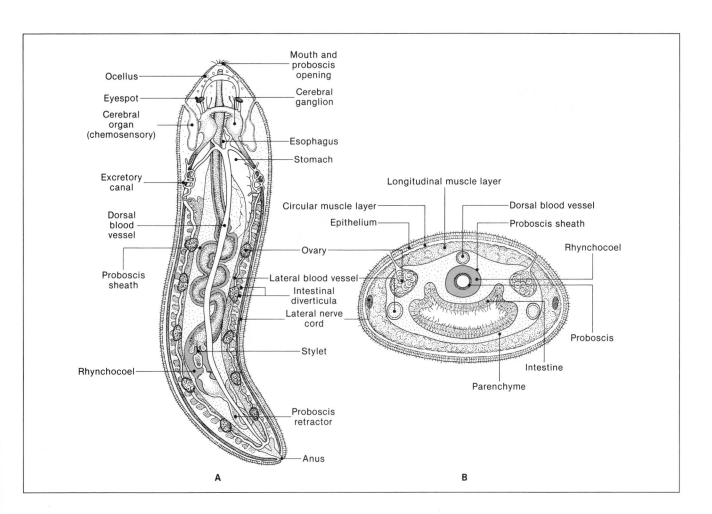

ering nemerteans to be coelomates is the **rhynchocoel** (RING-ko-SEAL), a chamber that houses the proboscis (Greek *rhynchos* snout + *koilos* cavity). The rhynchocoel forms within mesoderm and is lined with a cellular membrane, so it is technically a coelom (Turbeville and Ruppert 1985). Molecular phylogeny also suggests that nemerteans are more closely related to lophophorates, molluscs, and other coelomates than to flatworms (Hori and Osawa 1987).

Most nemerteans live among seaweed and gravel, where they scavenge and prey mainly upon annelid worms and other marine organisms, including diatoms, flatworms, nematodes, molluscs, crustaceans, and small fishes. Their main tool of predation, as well as locomotion, is the distinctive **proboscis,** which resides in the rhynchocoel when it is not being used. By contracting the body, the nemertean shoots out its proboscis, often with great accuracy and for a distance up to three times its body length. The proboscis turns inside out as it extends. It coils around the target, trapping it in sticky and sometimes toxic mucus. In some species nail-shaped **stylets** also stab the prey repeatedly. The proboscis is not connected to the digestive tract. Contraction of a muscle pulls prey to the mouth, then withdraws the proboscis back into the rhynchocoel.

Nemerteans have impressive powers of regeneration. If a proboscis is damaged or hopelessly entangled, the nemertine simply sheds it and grows a new one. This regenerative ability also enables some nemerteans to reproduce asexually by **fragmentation.** An adult can spontaneously break up, and each fragment can develop into a new individual. Nemerteans also reproduce sexually, with external fertilization and usually with separate sexes. Cleavage is spiral and determinate, as in most protostomes. Development is generally direct (Table 27.4).

Nemertean worms are common along the sea coast, and many are brightly colored and large. The long, round ones are often called boot lace worms, and the long, thin ones are often called ribbon worms. The longest animal in existence is a nemertean, *Lineus longissimus*. One specimen was almost 60 meters long. It is difficult to measure the body length accurately, because many nemertines can stretch to ten times their resting length. In spite of their size and appearance, nemertines are not often seen since they do not attack humans, and they burrow

Table 27.4 Characteristics of nemerteans.

Phylum Nemertea (= Rhynchocoela) NIM-ur-TEE-uh (Greek *Nemertes* a sea nymph, unerring).

Morphology: Adults with an eversible **proboscis** normally sheathed within the **rhynchocoel,** which is technically a coelom. Most of the body filled with parenchyme.

Physiology: Some organ systems similar to those of Platyhelminthes: protonephridia with flame bulbs; no respiratory or skeletal system; a central nervous system consisting of cephalic ganglia and longitudinal nerve cords. Up to several hundred planarian-type eyespots in some species. A **closed circulatory system,** but no heart. **Complete digestive tract;** digestion partly intracellular. Mainly carnivorous. Probable absorption of dissolved organic material.

Locomotion: By muscular contraction in adults. Proboscis sometimes used for gripping and burrowing.

Reproduction: Usually **dioecious.** Fertilization external. Asexual reproduction by fragmentation.

Development: Spiral, determinate cleavage. Rhynchocoel is schizocoelous. Development generally direct.

Habitat, Size, and Diversity: Almost all marine. Some symbiotic in mantle cavity of molluscs, or in gills or egg masses of crabs. Adults range from less than 1 mm to almost 60 meters long. Approximately **900 living species** described.

into sand or mud during the day and at low tide. Fishermen who know how to find them, however, often use nemerteans for bait. Several genera, such as *Prostoma*, are common in freshwater ponds. Several species have made the transition from marine to terrestrial life and are sometimes accidentally imported into greenhouses. A few are symbiotic. *Carcinonemertes errans* lives on the egg masses of crabs, including the commercially important Dungeness crab (*Cancer magister*).

PEANUT WORMS: PHYLUM SIPUNCULA

Like lophophorates, sipunculans also have ciliated tentacles that enclose the mouth but not the anus. These tentacles are not part of a lophophore, however, because the space within each tentacle is not part of the coelom. Instead, the tentacles are part of the **introvert:** the retractable anterior portion of the worm (Figure 27.8). The introvert is extended by pressure in the coelom due to contraction of the body wall, and the tentacles are extended by fluid pressure in the **compensation sac.** Sipunculans use their tentacles to feed on detritus on the ocean floor, where they live in burrows, mollusc shells, crevices of coral reefs, or among roots of plants. Some can burrow into coral and may cause considerable damage to reefs. Although many sipunculans are colorful and large (up to a meter long), they are seldom seen because of their reclusiveness, scarcity, and inactivity. When disturbed they slowly pull in the introvert. When thus contracted, some sipunculans resemble peanuts: hence the common name peanut worm.

Figure 27.8
(A) The sipunculan Dendrostomum pyroides *with tentacles everted. (B) Longitudinal section with introvert extended.*

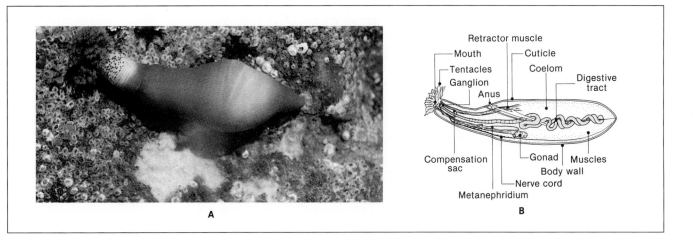

Table 27.5   Characteristics of sipunculans.

Phylum Sipuncula sy-PUNK-you-luh (Latin *siphunculus* a small pipe).

Morphology: Retractable **introvert** with tentacles.

Physiology: Digestive system complete and U-shaped. Nervous system with ganglion and connectives around esophagus. Pair of **metanephridia,** which function as gonoducts. No circulatory or respiratory system, but some species have hemerythrin in coelomic fluid.

Locomotion: Usually none in adults.

Reproduction: All but one species **dioecious.** Fertilization external. Some reproduction by transverse fission.

Development: Development either direct or with one or two larval stages, including a trochophore.

Habitat, Size, and Diversity: Marine; worldwide, especially in tropics. Adults usually **sessile,** often burrowed in sediment or coral. At all depths from intertidal to abyssal. One species, *Golfingia procera,* parasitic in the annelid *Aphrodite.* Length from several centimeters to almost a meter. Approximately **250 living species** described.

Most sipunculans have separate sexes, although they cannot be distinguished by external appearance. Gametes exit from the gonads in the coelom through the metanephridia, which are therefore gonoducts. Fertilization occurs externally. Reproduction is synchronized by some substance in sperm that induces females to release their ova. Development in some species is direct, with the egg immediately forming a worm. In other species the egg first develops into a **trochophore larva,** which has a ring of cilia around its body for locomotion (Table 27.5). Similar larvae also occur in molluscs and annelids (see Figures 29.7A and 30.5A). This similarity, among others, leads many zoologists to believe that sipunculans are related to both annelids and molluscs (Hori and Osawa 1987; Rice 1985).

_____ SPOON WORMS: PHYLUM ECHIURA _____

Echiurans are also burrowing marine animals, but unlike some phyla discussed previously in this chapter, they do not have tentacles. Instead, echiurans burrow and feed with a ciliated, mucus-coated, nonretractable proboscis, which has a **gutter** that brings detritus into the mouth (Figure 27.9). The proboscis is extremely elastic; it can extend more than 150 cm in an echiuran with a body only 40 cm long. Often the proboscis is spoon-shaped, giving the phylum its common name of "spoon worms." Echiurans are often fairly large, and those that are not transparent may be colored red, yellow, brown, or gray. However, they are seldom seen except by clam diggers, because they spend much of their lives within burrows, or in crevices or shells abandoned by other species. Although fossils of spoon worms are not found, remnants of their burrows appear in 450 million-year-old rocks.

Spoon worms are dioecious and release their gametes through the pore of the metanephridium (gonoduct) for external fertilization (Table 27.6). The larva is a trochophore, suggesting to some taxonomists that echiurans, like sipunculans, are related to molluscs and annelids. This finding is also consistent with molecular phylogeny (Hori and Osawa 1987). In most echiurans the sexes are similar, but the genus *Bonellia* shows one of the most bizarre examples of sexual dimorphism of any animal. The females are about 2 meters long, including the proboscis, and resemble other echiurans. The males, however, are only about 1 to 3 mm long and live inside the females. Whether a larva develops into a female or male

Figure 27.9
(A) The echiurid Tatjanellia grandis feeding with trunk buried in sand. (B) The female Bonellia tasmanica from the coast of southeastern Australia. This species is green from a pigment in the algae it eats. The genus is unusual in its sexual dimorphism. The tiny male is located within the nephridium of a female. Total length approximately 10 cm.

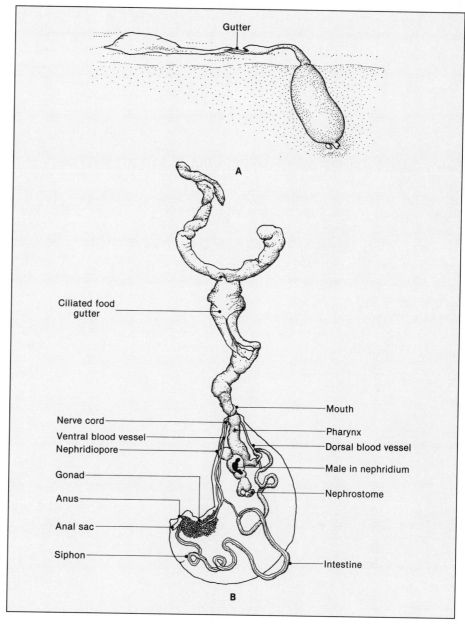

Table 27.6   Characteristics of spoon worms.

Phylum Echiura ek-ee-YOUR-uh (Greek *echis* snake + *oura* tail).

Morphology: Ciliated, mucus-coated proboscis, which is highly extensible but cannot be retracted into the body. **Vermiform.**

Physiology: Digestive system complete. Nervous system without ganglia. **Metanephridia** function mainly as gonoducts. Excretion of metabolic wastes due mainly to **anal sac.** Closed circulatory system; no heart. No respiratory system, but proboscis and anus may function like gills. Cells in coelom carry hemoglobin.

Locomotion: Adults **sessile,** often burrowed in sediment or living in abandoned shells.

Reproduction: Dioecious. Fertilization external in most species.

Development: Indirect, with free-swimming **trochophore** larva.

Habitat, Size, and Diversity: Marine; worldwide, especially on the bottoms of warm oceans. At all depths from intertidal to abyssal. Length of body from a few millimeters to 40 cm. Proboscis up to 1.5 meters long when extended. Approximately **135 living species** described.

depends on whether it settles near the proboscis of a female. A substance from the female's proboscis causes a larva to migrate into her nephridium and to develop into a male. Inside the nephridium the degenerate male functions essentially as a testis, fertilizing the female's eggs, which develop within the nephridium. The nephridium has no excretory function in this genus but serves only as a uterus.

___ WATER BEARS: PHYLUM TARDIGRADA ___

The name "water bears" was given to the tardigrades by an 18th-century zoologist who was apparently reminded of bears as he watched them lumber along on his microscope slide (Figure 27.10). Tardigrades are somewhat smaller than bears— less than a millimeter long—and they usually live in pond sediments, in the water film in soil or plants, or in marine sediments. They feed on plants, nematodes, rotifers, and other animals (including other tardigrades), by piercing them with two **stylets** that protrude from the mouth, then sucking out their juices. Some consider tardigrades to be pseudocoelomates related to rotifers, because the main body cavity is a hemocoel, which is technically a pseudocoel (Table 27.7). Others consider the membrane-lined cavity around the gonad to be a coelom (**gonocoel**).

Figure 27.10
Scanning electron micrograph of the tardigrade Echiniscus.

Table 27.7 Characteristics of water bears.

Phylum Tardigrada tar-di-GRADE-uh (Latin *tardus* slow + *gradus* step).

Morphology: Coelom reduced to cavity around gonad. Main body cavity a hemocoel. Body not divided into head, thorax, abdomen. Proteinaceous **cuticle** divided, but opinion differs on whether body is segmented. Four pairs of **legs** with four to eight claws each.

Physiology: Nervous system with cephalic ganglion around pharynx, and four ventral ganglia. Usually with pair of eyespots. Digestive tract complete. No circulatory or respiratory system. Branches from gut considered by some to be excretory organs: Malpighian tubules like those of insects.

Locomotion: Slow creeping on legs.

Reproduction: Dioecious. Sexes similar. Some species parthenogenetic.

Development: Enterocoelous. Development direct, with periodic molting of cuticle. **Eutelic.**

Habitat, Size, and Diversity: Aquatic, mainly in freshwater ponds or in water film in soil or on plants. Some interstitial in marine sands. Some parasitic on crustaceans, mussels, and sea cucumbers. Worldwide, and at elevations ranging from the Himalayas to the ocean abyss. Body length from 0.1 to 1.7 mm. More than **400 living species** described.

Some regard them as relatives of arthropods because of the hemocoel, the periodically molted cuticle, the resemblance to mites, and branches from the gut that resemble Malpighian tubules (Figure 27.11).

Tardigrades are dioecious, but in some species males have never been identified, and the females are presumed to be parthenogenetic. Apparently females release eggs only when about to molt. In some species the copulating males inject sperm between the old and new cuticle, and the fertilized eggs are shed with the old cuticle, in which they develop. In other species copulation does not occur; males deposit sperm on the molted cuticle containing the eggs. Tardigrade eggs hatch into miniatures of the adults. The juveniles grow by increasing the size of cells, rather than by increasing the number of cells through mitosis. Like nematodes and some other pseudocoelomates, tardigrades have a constant number of cells (**eutely**).

Ordinarily, tardigrades live for only a few months, but if they are gradually dried out they can live for decades or possibly centuries in a state of **cryptobiosis** (Crowe and Cooper 1971; Nelson 1975; Rensberger 1980). During cryptobiosis the body, called a **tun,** is barrel-shaped, inactive, and light enough to be dispersed

Figure 27.11
Longitudinal section of a generalized tardigrade. Note the bands of muscles that move the legs.

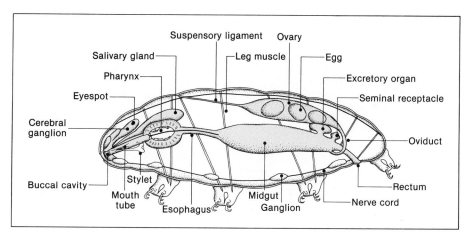

by wind. In this condition it can survive temperatures near absolute zero (− 270°C) and as high as 150°C, as well as ionizing radiation a thousand times more intense than the lethal dose for humans. On being returned to water at normal temperature they resume their normal activities within a few hours.

VELVET WORMS: PHYLUM ONYCHOPHORA

Velvet worms, also known as walking worms, are seldom seen because they avoid light and live in rainforests. Even when seen they are likely to be mistaken for shiny green, blue-black, orange, or whitish caterpillars as they creep over the forest floor on their 14 to 43 pairs of clawed appendages (Figure 27.12). The resemblance to arthropods also includes a chitinous cuticle, dorsal tubular heart, open circulatory system, hemocoel as the major body cavity, and breathing tubes (**tracheae**) (Table 27.8). Some zoologists also consider the body to be segmented, as in arthropods. Onychophorans differ from arthropods in having a cuticle that does not harden, appendages that are not jointed, no thoracic or abdominal ganglia, and openings of the tracheae that cannot close to limit the loss of body moisture. The latter deficiency restricts onychophorans to moist environments, so they are active only at night and during rain. The mode of eating also differs from that of any arthropod. Onychophorans entrap small prey by squirting them with jets of slime from two **oral papillae** at distances up to 30 cm. They then regurgitate digestive juices onto the prey.

While many zoologists see a resemblance to arthropods, others relate onychophora to annelids, pointing out that the cuticle molts in patches, and a metanephridium empties near each leg. Onychophoran locomotion is also more like that of earthworms and other annelids than of insects, in spite of appearances. The appendages are not truly legs that can be manipulated individually, but **lobopodia** that move passively as a peristaltic wave of contraction passes along the body (Figure 27.13). Many zoologists consider onychophorans to be intermediate

Figure 27.12
The onychophoran Peripatoides *in Queensland, Australia. The head is to the left front. Note the antennae.*

Table 27.8 Characteristics of velvet worms.

Phylum Onychophora on-i-KOFF-or-uh (Greek *onyx* claw + *phora* bearing).

Morphology: Coelom greatly reduced; main body cavity a hemocoel. Wormlike, but with 14 to 43 pairs of legs (lobopodia).

Physiology: Free-living, feeding on arthropods and molluscs. **Slime glands** eject sticky secretion from oral papillae that traps prey and potential predators. Digestive enzymes released onto prey, and nutrients sucked up. Digestive tract complete. Nervous system with cephalic ganglion only. Two longitudinal nerve cords with ladderlike connections. Metanephridium with pore near each leg. Dorsal, tubular **heart** and open circulatory system. Hydrostatic skeleton. **Tracheal** tubes for breathing.

Locomotion: Peristaltic contraction aided by lobopodia.

Reproduction: Dioecious. Internal or external fertilization. Mostly viviparous, but some ovoviviparous or oviparous.

Development: Direct.

Habitat, Size, and Diversity: Terrestrial, but in moist habitats only. Length from 1.4 to 15 cm. Approximately **80 living species** described.

---

between annelids and arthropods, but most consider them to be sufficiently different to constitute their own phylum. (See Boudreaux 1979, chapter 3, for discussion.)

Onychophorans are apparently ancient animals. Fossils half a billion years old in the Burgess shale of British Columbia have been interpreted as onychophoran. (See number 4 in Figure 19.6.) On the other hand, their present distribution in India, Africa, South America, Australia, and islands formerly connected to these continents suggests that they originated on Gondwana during the Mesozoic, between 180 and 65 million years ago. (See Figure 19.9.)

BEARD WORMS: PHYLUM POGONOPHORA

Beard worms are among the most mysterious animals alive. They were not even discovered until this century, when fragments were dredged up from the ocean depths. They were once considered to be deuterostomate, because their coeloms seemed to be divided by septa into three chambers, as in some other deuterostomes. Then in 1963 it was discovered that this conclusion was based on samples

Figure 27.13
Longitudinal section of a female Peripatus sp., dorsal view. Note the presence of a uterus in this viviparous species. For simplicity the tubular heart along the dorsal midline is not shown, nor are the tracheal tubes or nephridia near each leg. Length approximately 7 cm.

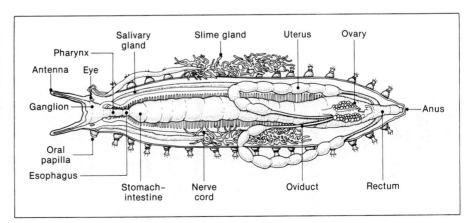

that had lost their segmented tails (**opisthosomes**), which have several coelomic compartments. Some zoologists now think beard worms belong in the same phylum with other segmented worms (Annelida), but the majority think they are only close relatives of annelids.

Pogonophorans live within upright chitinous tubes that they secrete around themselves, with from one to many thousand tentacle-like **branchiae** protruding from the upper end of each tube. The branchiae are the "beard" of beard worms. Without their tubes many beard worms would look like threads that are badly frayed at one end (Figure 27.14). The branchiae are often called tentacles, but

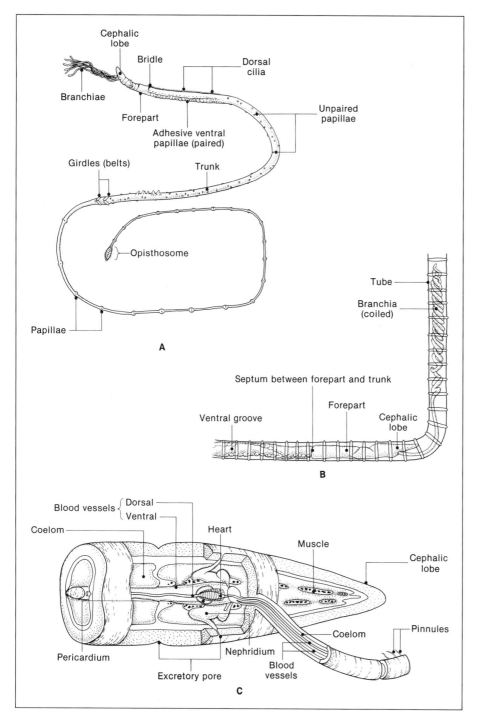

A

B

C

Figure 27.14
(A) A pogonophoran removed from its tube. The body has four divisions: cephalic lobe, forepart, trunk, and opisthosome. The thickness of the body is greatly exaggerated in this schematic figure. (B) Upper portion of the pogonophoran Siboglinum *retracted in its tube. (C) Detail of cephalic lobe of* Siboglinum *with its single branchia.*

Table 27.9 Characteristics of beard worms.

Phylum Pogonophora PA-gone-OFF-or-uh (Greek *pogon* beard + *phora* bearing).

Morphology: Several body cavities, at least one of which is a coelom. Segmented. **Vermiform.** From one to hundreds of thousands of **branchiae.** Living within a **chitinous tube.**

Physiology: No digestive tract in adults. Dissolved nutrients absorbed by branchiae or from mutualistic chemoautotrophic bacteria. Closed circulatory system with heart; hemoglobin dissolved in blood. No special respiratory organ. Nervous system without ganglia. Metanephridia.

Locomotion: None in adults.

Reproduction: Dioecious. Males release water-borne spermatophores that may be caught by females' branchiae. Fertilization presumably external, in tube of female.

Development: Considered a protostome, although cleavage is more nearly radial than spiral. No blastopore, mouth, or anus. Possible trochophore larva in at least one species.

Habitat, Size, and Diversity: Marine, on ocean floors from 100 meters to 10 km deep. Many in sulfur-rich waters, especially near hydrothermal vents. Length up to 3 meters. Approximately **135 living species** described.

Figure 27.15
Vestimentiferan tube worms Riftia pachyptila *at the Galápagos hydrothermal vent. Other members of the hydrothermal vent community shown here are crabs and mussels.*

they absorb dissolved material and are not used to capture solid food. Pogonophorans have no mouths, intestines, or digestive gland (Table 27.9). In their typical cylindrical arrangement the branchiae resemble an intestine, complete with microvilli that increase the surface area.

Some pogonophorans apparently do not even have to absorb dissolved organic material but can live on mutualistic sulfur-metabolizing bacteria. These **vestimentiferan** tube worms have been discovered in several areas rich in sulfur, as well as in other habitats. One such vestimentiferan species, *Riftia pachyptila*, was discovered near hydrothermal vents off the Galápagos Islands by a crew of the submersible research vessle *Alvin*. *Riftia pachyptila* is unusually thick—2 to 3 cm compared with less than a millimeter for most beard worms. *Riftia* is up to 3 meters long and lives in its glistening white tube that stands up near hydrothermal vents. Its more than 200,000 branchiae stick out like a rolled tongue, bright red with the hemoglobin in the blood (Figure 27.15) (Jones 1981b).

Riftia obtains much of its food from **chemoautotrophic bacteria** that live within a special chamber, the **trophosome,** in its body (Childress et al. 1987). Apparently the beard worm harvests the bacteria or their metabolic products. This type of bacteria synthesizes carbohydrate [CH$_2$O] using the energy released from the oxidation of hydrogen sulfide, according to the following reaction:

$$CO_2 + H_2S + O_2 + H_2O \rightarrow H_2SO_4 + [CH_2O]$$

The hemoglobin dissolved in the blood is crucial for this process. It not only carries oxygen, like other hemoglobins, but also binds the hydrogen sulfide, which would otherwise kill the animal. The bacteria have both H$_2$S and the O$_2$ to oxidize it brought directly to them by the hemoglobin.

Meredith Jones (1985a) of the Smithsonian Institution has proposed that the vestimentiferans be assigned to their own phylum Vestimentifera. Discoveries regarding this group are accumulating so rapidly, however, that this proposal could soon require modification. For now it seems better to follow the earlier and more conservative suggestion of Jones (1981a) and consider the vestimentiferans to be a subphylum of Pogonophora.

SUMMARY

Animals classified in the nine phyla in this chapter all have a coelom, which is a chamber formed within mesoderm and lined with a cellular peritoneum. These phyla are all considered to be protostomate, which means that in most the blastopore becomes the mouth, the coelom forms by schizocoely, and cleavage is spiral and determinate. There is considerable uncertainty, however, about their relationships to each other and to other phyla. These phyla also have in common the fact that they are now represented by few species compared with other coelomate protostomes—molluscs, annelids, and arthropods.

Three of the phyla, Ectoprocta, Phoronida, and Brachiopoda, are united by having lophophores, which are circles of tentacles enclosing the mouth but not the anus. Nemerteans are often considered to be acoelomates related to flatworms, but they have a coelom into which the proboscis retracts. Onychophorans and tardigrades share features similar to those of arthropods, and pogonophorans appear to be related to annelids.

KEY TERMS

coelom	determinate cleavage	rhynchocoel
protostome	lophophore	trochophore
schizocoelous	gonoduct	trachea
spiral cleavage	valve	branchia

SELF-TEST

1. Explain what a coelom is and why it is important.

2. What is a lophophore? How does it function? What is its taxonomic significance?

3. Some phyla in this chapter have tentacles but are not lophophorates. Name the phyla. Why are they not considered to be lophophorates?

4. Name two phyla discussed in this chapter that do not feed by tentacles. Describe how they do feed.

5. For each of the nine phyla give a characteristic or combination of characteristics not found in any of the other eight phyla.

READINGS

RECOMMENDED READINGS

Childress, J. J., H. Felbeck, and G. N. Somero. 1987. Symbiosis in the deep sea. *Sci. Am.* 256(5):114–120 (May). (*On beard worms and others.*)

Crowe, J. H. and A. F. Cooper, Jr. 1971. Cryptobiosis. *Sci. Am.* 225(6):30–36 (Dec).

Nelson, D. R. 1975. The hundred-year hibernation of the water bear. *Nat. Hist.* 84(7):62–65.

Rensberger, B. 1980. Life in limbo. *Science* 80(Nov):36–43.

Richardson, J. R. 1986. Brachiopods. *Sci. Am.* 255(3):100–106 (Sept).

Tucker, J. B. 1985. Drugs from the sea spark renewed interest. *BioScience* 35:541–545.

See also relevant selections in General References at the end of Chapter 21.

ADDITIONAL REFERENCES

American Zoologist 17:3–147 (1977) includes several papers on the biology of lophophorates. *American Zoologist* 25:1–151 (1985) includes papers on nemerteans.

Boudreaux, H. B. 1979. *Arthropod Phylogeny: With Special Reference to Insects.* New York: Wiley-Interscience. (*Onychophora are discussed in Chapter 3.*)

Conway Morris, S. et al. (Eds.). 1985. *The Origins and Relationships of Lower Invertebrates.* New York: Oxford University Press.

Field, K. G. et al. 1988. Molecular phylogeny of the animal kingdom. *Science* 239:748–753.

Harvell, C. D. 1984. Predator-induced defense in a marine bryozoan. *Science* 224:1357–1359.

Hori, H. and S. Osawa. 1987. Origin and evolution of organisms as deduced from 5S ribosomal RNA sequences. *Mol. Biol. Evol.* 4:445–472.

Jones, M. L. 1981a. *Riftia pachyptila*, new genus, new species, the vestimentiferan worms from the Galápagos Rift geothermal vents (Pogonophora). *Proc. Biol. Soc. Wash.* 93:1295–1313.

Jones, M. L. 1981b. *Riftia pachyptila* Jones: observations on the vestimentiferan worm from the Galápagos Rift. *Science* 213:333–336.

Jones, M. L. 1985a. On the Vestimentifera, new phylum: six new species, and other taxa, from hydrothermal vents and elsewhere. *Biol. Soc. Wash. Bull.* 6:117–158.

Jones, M. L. 1985b. Vestimentiferan pogonophores: their biology and affinities. In: S. Conway Morris et al. (Eds.), *The Origins and Relationships of Lower Invertebrates.* New York: Oxford University Press, pp. 327–342.

Rice, M. E. 1985. Sipuncula: developmental evidence for phylogenetic inference. In: S. Conway Morris et al. (Eds.), *The Origins and Relationships of Lower Invertebrates.* New York: Oxford University Press, pp. 274–296.

Ross, J. R. P. (Ed.). 1987. *Bryozoa: Present and Past.* Bellingham, WA: Western Washington University Press.

Thayer, C. W. 1985. Brachiopods versus mussels: competition, predation, and palatability. *Science* 228:1527–1528.

Thorpe, J. P., G. A. B. Shelton, and M. S. Laverack, 1975. Colonial nervous control of lophophore retraction in cheilostome Bryozoa. *Science* 189:60–61.

Turbeville, J. M. and E. E. Ruppert. 1985. Comparative ultrastructure and the evolution of nemertines. *Am. Zool.* 25:53–71.

Willmer, P. 1990. *Invertebrate Relationships.* New York: Cambridge University Press.

Woollacott, R. M. and R. L. Zimmer (Eds.). 1977. *The Biology of Bryozoa.* New York: Academic Press.

28

Molluscs

Mexican dancer nudibranch (Coryphella iodinea).

CHAPTER OUTLINE

LEARNING OBJECTIVES

1. What features of molluscs make them so diverse, common, and successful?

2. How are snails, clams, octopus, and other molluscs similar, and how are they different?

3. What are the structure and function of the mollusc shell? How do molluscs make their shells?

4. How do molluscs confined within a shell manage to reproduce?

5. How do snails, clams, and other shelled molluscs move?

6. How do squid swim?

THE IMPORTANCE OF MOLLUSCS

Before the arrival of Europeans in what is now northern California, a Yurok man could get a bride from a wealthy family for 10 strings of the tusk-shaped shells of the mollusc *Dentalium pretiosum* (Figure 28.1A). The fine for adultery was five strings of tusk shells. European aristocrats preferred to use gold for such transactions and save the seashells for aesthetic enjoyment.

Until the 18th century Europeans collected shells only for their beauty, with little thought to the molluscs (also spelled mollusks) that made them. Lamarck

Figure 28.1
A variety of molluscs. (A) Shell of the scaphopod Dentalium elephantinum *among other mollusc shells. (B) The chiton* Mopalia muscosa. *(C) The common land snail* Helix. *Note the two sets of head appendages, with eyes on the upper, longer pair. (D) The razor clam* Tagelus plebeius, *which normally stays buried in marine sediment, pumping water through its long siphons. (E) In some molluscs, such as the giant octopus,* Octopus dofleini, *the shell is reduced or absent.*

(later to be reviled for his theory of evolution) was one of the first to realize that if dead shells are so interesting and beautiful, then the living molluscs must be even more so. During the French Revolution he convinced the government to employ in the Museum of Natural History a specialist in shells and molluscs, as well as experts on other groups of animals. Thus in France and elsewhere, even before the terms "zoologist" and "scientist" had been invented, there were people whose chief occupation was collecting and studying molluscs and their shells. We would now call them malacologists and conchologists. These were perhaps the first people actually paid for studying animals—something others had done for mere pleasure.

We know much more about the contributions of molluscs to the origins of zoology than we know about the origins of molluscs themselves. Their body plans are so different from those of other animals that some zoologists do not consider them coelomates, and some suggest that they are not directly related to other protostomes (Willmer 1990). Evidence from molecular phylogenetics, however, supports the majority view that molluscs are closely related to annelids and other coelomate protostomes (Field et al. 1988; Hori and Osawa 1987). The fossil record of molluscs goes back more than half a billion years, to the Cambrian period. By then they had already evolved all the major groups now recognized. At least 35,000 extinct species are known from their fossilized shells. The grandest of these were the ammonoids (Figure 28.2), whose spiral shells had diameters up to 2 meters. They were quite common throughout the Paleozoic and Mesozoic eras, then became extinct rather suddenly. Possibly they were among the many victims of the Cretaceous/Tertiary mass extinction (see pp. 396–397), or perhaps predators such as marine reptiles and crabs finally developed the ability to crush the ammonoids' shells. In spite of such catastrophes, approximately 50,000 species of molluscs survive, making the phylum Mollusca one of the most diverse. Another measure of molluscan success is the wide range of habitats molluscs have invaded. Most are marine organisms, but many live in fresh water and moist terrestrial habitats, and a few are parasitic. About the only thing no mollusc can do is fly.

Figure 28.2
Fossils of ammonites (Dactylioceras) from about 200 million years ago.

GENERAL ORGANIZATION

The great diversity of molluscs makes it difficult to discuss them as a single group. Many zoologists attempt to remedy this difficulty by describing **the hypothetical ancestral mollusc.** Unfortunately, this animal has never existed except in textbooks, and specialists in molluscs have generally abandoned it. Instead of describing such a hypothetical ancestor, it is just as well to describe the general features shared by most molluscs and discuss the variations later.

The body of a mollusc can generally be divided into the shell and the fleshy, living part. The fleshy parts of a mollusc can be further divided into the **foot** and the **visceral mass** (= visceral hump) (Figure 28.3). The foot is generally adapted in a variety of ways for locomotion. The visceral mass includes the organs for digestion, circulation, reproduction, and respiration.

The visceral mass also includes two external flaps of tissue called the **mantle** (= pallium). The mantle secretes the shell and encloses a **mantle cavity.** The mantle cavity performs many of the functions that the coelom performs in other animals. The fluid in the mantle cavity, which in aquatic molluscs is continually replaced with water from the outside, carries away excess water, ions, and wastes and helps circulate nutrients and oxygen. One final structure that is unique to molluscs and is found in all groups except bivalves and a few others is the **radula** (RAD-jul-uh). In most forms the radula is a rasping organ near the mouth that is used in scraping up algae and other food.

SHELLS

The shell is absent in one class of molluscs (Aplacophora) and is either absent or vestigial in octopus, squid, and cuttlefish (class Cephalopoda). In most molluscs, however, the shell is obvious and important. It generally supports the soft body and protects it from predation. Little else about the shell is obvious. We still do not know the function, if any, of many of the ornate shapes and colors that make shells so attractive and valuable. Nor do we understand the developmental pro-

Figure 28.3
Representations of six classes of molluscs showing the shell and visceral mass. The latter is indicated by the digestive tract and includes the foot (color) and mantle. The position of the radula is also indicated.

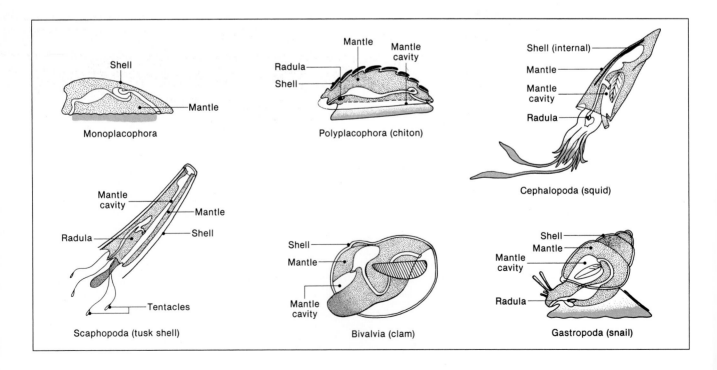

Figure 28.4

(A) Valves of the freshwater mussel Margaritifera margaritifera *showing the internal nacre (top) and outer periostracum. Part of the periostracum has eroded from the oldest part of the shell (the umbo), exposing the chalky prismatic layer. Attachment sites (scars) for the muscles that open and close the shells are visible on the inner surface. The teeth prevent the shells from slipping sideways. The annual growth rings indicate that this 8-cm specimen was approximately six years old. (B) Schematic representation of the mechanism that opens and closes the shell. The elastic ligament at the umbo opens the shell. The closer muscles are a type known as catch muscles (see p. 216) and are among the most powerful known. (C) Cross section of the edge of the shell and mantle.*

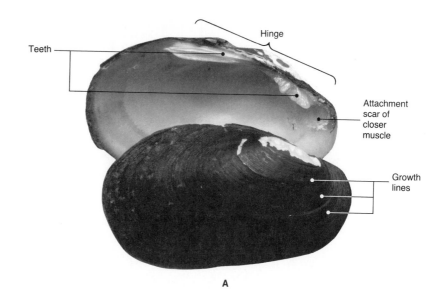

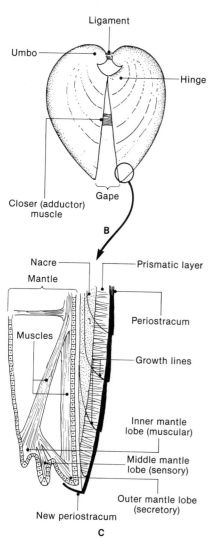

cesses that direct molluscs in the designing of their shells. We do know, however, that the shells are made by the mantle in three parts: the inner **nacreous layer,** the middle **prismatic layer,** and the outer **periostracum** (Figure 28.4). The mantle forms the nacreous layer, or **nacre,** by continuously secreting a solution of calcium carbonate. As the $CaCO_3$ precipitates it forms thin layers of the mineral aragonite. Since this process continues through life, the nacreous layer thickens as the mollusc ages.

Many molluscs, such as oysters, also secrete nacre around particles of foreign tissue or debris that become lodged between the mantle and the shell. After several years hundreds or thousands of layers of nacre will have accumulated, forming a pearl. Another name for the nacreous layer of the shell is therefore mother-of-pearl. Mother-of-pearl is often iridescent, because light at certain wavelengths (colors) is canceled by its reflection through the partially transparent nacreous layers, while other wavelengths are reinforced by their reflections. Humans value this iridescence in mother-of-pearl, but, of course, it is invisible and therefore useless in the living mollusc.

The prismatic layer of the shell is chalky white and consists of prism-shaped crystals of $CaCO_3$ (aragonite or calcite) in a protein matrix. This layer is formed by the leading edge of the mantle and therefore expands as the mollusc grows. The periostracum is also formed by the mantle edge in a continuous sheet over the outer surface of the prismatic layer. It is the only part of the shell that contributes colors visible from outside. The periostracum is made of the smooth, hard, proteinaceous substance **conchin** (KON-kin; formerly called conchiolin) that protects the prismatic layer from abrasion and dissolution by acids. Protection from naturally occurring acids is especially important in freshwater and terrestrial species.

THE FOOT

As the name implies, the foot is usually responsible for locomotion. In many species it secretes a layer of mucus, like the familiar slime trails left by snails and slugs. The mollusc glides upon this mucus track by waves of ciliary or muscle contraction. This form of locomotion also occurs in many flatworms (see Chapter 25), which leads some zoologists to think flatworms were the direct ancestors of molluscs. Some freshwater snails can also move upside down beneath the water surface (see Figure 2.3A). In bivalves (clams, oysters, etc.) the foot is used for locomotion in another way, by rooting into sand or mud and pulling the animal along. In squid, octopus, and other cephalopods, the foot is modified into a funnel for jet propulsion. In these and most other molluscs the head is closely associated with the foot.

THE RADULA

In most molluscs the radula bears teeth on a membrane that is bent around a cartilaginous support (the **odontophore**) (Figure 28.5). Ordinarily the radula is kept within a **radula sac** beneath the mouth. By contracting certain muscles to the odontophore and radula membrane, the radula can be extended and made to grind off food particles, much as a belt sander grinds off bits of wood. The radula also serves as a conveyor belt, carrying food particles into the mouth. The number of teeth ranges from a few to hundreds of thousands, depending on the species. In some species the radula is used to scrape algae off rocks or to bore through the shells of other molluscs. This is rather hard on the teeth, which are made of **chitin,** the same substance in the cuticle of insects. To compensate for wear, the teeth are continuously regenerated from their bases.

RESPIRATION

In most molluscs the respiratory organs are **gills** (Table 28.1). As discussed in Chapter 12, gills are adapted to exchange O_2 and CO_2 in water by having a large surface area and thin membranes. In molluscs the gill also has unique adaptations that justify giving it the special name **ctenidium** (Greek *kteis* a comb). Ctenidia consist of sets of **filaments** (= lamellae) covered with cilia (Figure 28.6). As the cilia propel water across the surface of the filaments, oxygen diffuses across the membrane into the blood, and carbon dioxide diffuses out. In some molluscs, such as clams and other bivalves, the cilia also sort out particulate matter, sending food particles on a string of mucus to the mouth. After passing the gills, the water stream usually goes past the anus and the outlets of the kidneys, carrying off wastes. In many molluscs water enters and leaves through **incurrent and excurrent siphons.** Before it reaches the gills the incoming stream of water is sampled by a sense organ, the **osphradium,** which may detect silt, food, or predators.

Some molluscs do not have gills but exchange respiratory gases directly across the mantle surface. Snails have adapted to terrestrial life by losing the gills and having part of the mantle modified into a **lung** for air-breathing. Some of these **pulmonate** snails have reinvaded aquatic habitats but still retain the lung. Often they can be seen in ponds rising to the surface for air.

Figure 28.5
(A) Side view of a generalized radula. When the mollusc is not feeding, the radula is retracted by the odontophore retractor muscles. During feeding the odontophore protractor muscles move the odontophore forward, pressing the radula against substratum. The radula protractor and retractor muscles then slide the radula back and forth. (B) Scanning electron micrograph of the radula teeth of the snail Cupedora. *Each tooth is approximately 35 μm long.*

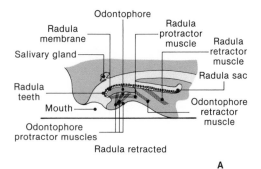
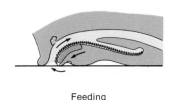
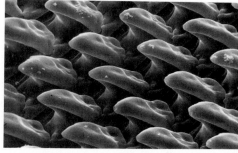

A   B

Table 28.1 Characteristics of molluscs.

Phylum Mollusca ma-LUS-kuh (Latin *mollis* soft).

Morphology: Usually considered coelomate, although coelom reduced to a pericardium. Bilateral symmetry, but many forms secondarily asymmetric. Usually with **shell** formed by **mantle. Mantle cavity** performs some of the functions of a coelom. Major body cavity is the **hemocoel.**

Physiology: Gaseous exchange usually by **gills** (ctenidia) in mantle cavity, or mantle may be modified into lung. Many with unique **radula** for scraping up food particles. Digestive tract complete, but much digestion intracellular. Nervous system with several pairs of ganglia linked by nerves. Circulatory system open or closed; with heart. Osmoregulation and excretion by metanephridia.

Locomotion: Muscular **foot,** which may be highly modified for functions other than locomotion. **Cilia** and **mucus** often important in locomotion and other organ functions.

Reproduction: Mostly dioecious, but some monoecious. Copulation common even in monoecious species, but fertilization is more often external.

Development: Spiral cleavage pattern, except bilateral in cephalopods. Protostomate. Development usually indirect, with **trochophore larva,** and a veliger larva in some species.

Habitat, Size, and Diversity: Primarily marine, but many freshwater and terrestrial species, and a few parasitic. Adult body length from less than 1 mm to more than 20 meters (giant squid *Architeuthis,* including tentacles). Approximately **50,000 living species** described.

Figure 28.6

(A) Gills (ctenidia) in a freshwater clam. Cilia on the gills draw water into the filaments through pores. Water then passes through tubes into the suprabranchial chamber, and out through the excurrent aperture. O₂ and CO₂ are exchanged with blood circulating through the filaments. (B) As water passes through the pores in the filaments, particles of food, sand, and debris are filtered out. Food collects on a mucous string, which is then propelled by cilia toward the mouth, and from there to the stomach and intestine within the foot. (See Figure 28.21.) Sand and debris settle out.

CIRCULATION

The circulatory system consists of a heart and blood vessels. In most species the heart has three chambers: two atria and one ventricle. The heart, along with the gonads and kidneys, is enclosed within a chamber called the **pericardium,** which is the greatly reduced coelom. In most forms the copper-containing pigment **hemocyanin** transports oxygen in the blood. In squid, octopus, and other cephalopods, which are much more active than other molluscs, the circulatory system is closed, but in most molluscs the circulatory system is open. That is, the blood (= **hemolymph**) is not confined to the heart and vessels but percolates under low pressure through irregular channels and sinuses in the tissues. These channels and sinuses make up a **hemocoel.** The hemocoel is the major body cavity in these molluscs and is larger and more important by far than the coelom.

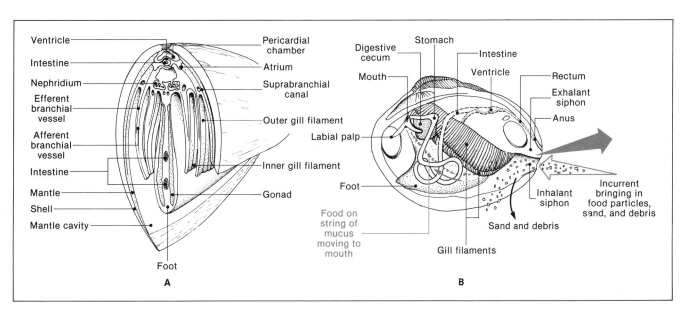

NERVOUS SYSTEM

The central nervous system of molluscs typically consists of a ring of ganglia (see Figure 8.3). A pair of **pedal ganglia** usually control the foot, **cerebral ganglia** integrate sensory information, and other ganglia control the functions of other parts of the body. The nervous system of many molluscs also secretes hormones that regulate such functions as egg laying and growth (see Table 9.3). Most molluscs move at the proverbial snail's pace, and they are generally regarded as rather dull, but this may be because few zoologists or psychologists have had the patience to test their intellectual acuity. Some molluscs, especially octopus, squid, and other cephalopods, are active predators with keen visual and tactile perception and have the ability to learn complex discrimination tasks.

OSMOREGULATION AND EXCRETION

Molluscs have one or more pairs of **nephridia.** The nephridia remove excess water, ions, and metabolic wastes from the coelomic fluid and transport them into the mantle cavity for excretion. The mollusc nephridium is of a type called a **metanephridium,** since its tubule has both an external pore (**nephridiopore**) and an internal opening (**nephrostome**). The nephridia of molluscs are often called kidneys, although they function quite differently from the kidneys of vertebrates, since they filter coelomic fluid rather than blood. Also, unlike the kidneys of vertebrates, the nephridia of many molluscs serve as channels (gonoducts) that transport gametes from the gonads into the mantle cavity.

REPRODUCTION AND DEVELOPMENT

Most molluscs are either one sex or the other, although one would never know it by looking at them. Snails and some others are hermaphroditic. **Development** is direct in a few groups, especially freshwater snails and some bivalves. In most marine molluscs there is a **trochophore larva** (Figure 28.7A), as in some flatworms, annelids (see Figure 29.5), and other protostomes. Trochophores are characterized by a ring of cilia that helps compensate for the adult mollusc's limited ability to travel to new habitats. In some groups of molluscs, and in no other phylum, another larval form, the **veliger,** occurs after the trochophore (Figure 28.7B). The veliger also has cilia for motility and, unlike the trochophore, also has the beginnings of the shell and most of the adult organs.

CLASSIFICATION

Most taxonomists now recognize seven or eight classes of molluscs, based mainly on differences in the foot and shell. Usually these differences are quite apparent, making it easy to identify on sight the class to which a mollusc belongs. The first three or four classes to be described are interesting but minor in terms of the number of species. (See Purchon 1977, appendix A, for more-complete discussions of these groups.)

Aplacophorans. Fewer than 300 species of wormlike, shell-less molluscs comprise **class Aplacophora** (Greek *a* without + *plax* plate). Most aplacophorans are about 2.5 cm long and have calcareous scales and spicules instead of shells. Most creep on deep ocean floors and are often found entwined in colonial hydroids or soft corals (phylum Cnidaria) on which they feed. The foot is modified into a

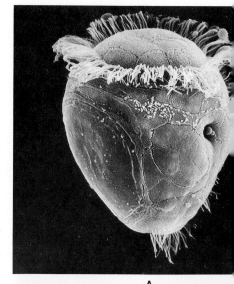

A

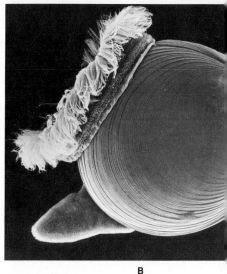

B

Figure 28.7
(A) Scanning electron micrograph of the trochophore larva of the shipworm Teredo navalis. *Trochophores are defined by the circle of cilia (prototroch). See Figure 29.5A for anatomical details of a trochophore, and Figure 28.20A for an adult* Teredo. *Diameter approximately 25 μm. (B) SEM of the veliger larva of a different shipworm,* Lyrodus pedicellatus. *The foot (lower left) and valves are evident. Locomotion is due to the ring of cilia. Diameter of the valves approximately 175 μm.*

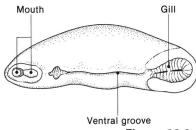

Figure 28.8
An aplacophoran, Neomenia
carinata.

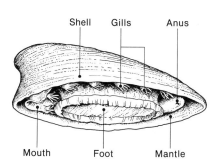

Figure 28.9
A monoplacophoran, Neopilina. *The
longest dimension is up to 3.7 cm.*

groove up the ventral midline (Figure 28.8). Many taxonomists recognize as a separate class (Caudofoveata) about 70 species that spend most of their lives burrowed vertically in the sea floor, feeding on microbes and detritus with the posterior end uppermost.

Monoplacophorans.   **Class Monoplacophora** was known from fossils but had been considered extinct until the 1950s. Then while examining some preserved specimens that had been collected deep off the west coast of Mexico a few years before, Henning Lemche of Copenhagen came across one labeled as a limpet. Turning it over, he saw that the animal had five pairs of gills and was therefore not only not a limpet, but also not a member of any other class of molluscs thought to be extant. Thus the first recent monoplacophoran, *Neopilina galatheae,* was discovered. Since then about a dozen living species of Monoplacophora have been collected from around the world at depths of 2 to 7 km. As the name implies, monoplacophorans have a one-piece shell (Figure 28.9). The stomach contents suggest that *Neopilina* feeds mainly on radiolarians (protozoa in the group Actinopoda). Lemche (1957) suggests that *Neopilina* lives on its back, but a similar monoplacophoran, *Vema hyalina,* has been photographed creeping across the ocean floor on its foot. One of the most interesting features of Monoplacophora is the serial repetition of organs, with five or six pairs of gills, six pairs of nephridia, eight pairs of pedal-retractor muscles, and so on. Some zoologists consider this to be evidence that molluscs are related to segmented animals such as annelids.

Chitons.   Members of **class Polyplacophora** are commonly referred to as **chitons,** because their dorsal shells of eight overlapping plates resemble Greek tunics with that name (Figure 28.1A). No other mollusc has such a shell. The shells of chitons are also different in other ways. Chiton plates have only two layers; an outer **tegmentum** made of conchiolin and calcium carbonate, and an inner **articulamentum** that is entirely calcareous. The plates are embedded in a part of the mantle called the **girdle.** Often the girdle is defended with spines, scales, and

Figure 28.10
*(A) Ventral view of a chiton.
(B) Internal structure of a chiton.*

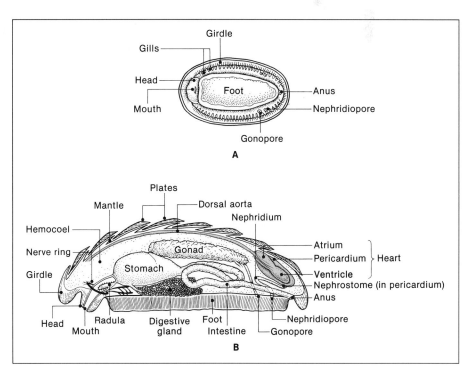

bristles. The plates also bear numerous smaller projections, each of which is a sense organ called an **esthete.** The esthetes may have simple tactile or visual receptors, or in some species, well-developed eyes, complete with lenses. They allow the chiton to detect predators and wave action even though its head is shielded under the shell. The mantle cavity is a groove between the foot and the edge of the mantle. There are from 6 to 88 gills (Figure 28.10).

The 600 or so species of chitons are up to 40 cm long and are fairly common in the intertidal zone and in deeper waters. Often they can be found in shallow water, creeping along on the broad ventral foot and scraping up algae and invertebrates with the radula. Getting a look at the foot and the radula can be difficult, however. When disturbed, the foot and mantle edge create suction against the substratum, making it hard for waves and predators, as well as curious zoologists, to dislodge the animal.

Tusk Shells. **Tusk shells,** such as the species mentioned at the start of this chapter, belong to **class Scaphopoda** (Figure 28.1B). They are also called tooth shells. The shell is open at both ends. Scaphopods spend most of their lives with the posterior end sticking up from the sand or mud in water as deep as 6 km. Cilia in the mantle cavity draw water in through the buried anterior end, and sudden contractions of the foot periodically squirt the water out the posterior opening of the shell (Figure 28.11). Cilia and mucus on the foot and tentacles (= **captacula**) trap protozoa and detritus and transfer them to the mouth and radula. There is no gill; oxygen from the water diffuses directly into the mantle. The coelom, heart, and blood vessels, if present, are so reduced that many anatomists doubt their existence.

Major Groups: Gastropods, Bivalves, and Cephalopods. The next three classes, Gastropoda, Bivalvia, and Cephalopoda, comprise the majority of molluscs (Figure 28.1C, D, and E). Each of these classes will be discussed later in a separate section. **Class Gastropoda** includes forms with a one-piece shell, such as snails, limpets, conchs, and whelks, as well as slugs and other relatives, in which the shell is greatly reduced. **Class Bivalvia** (= Pelecypoda) includes oysters, mussels, clams, and other molluscs in which the shell consists of two **valves. Class Cephalopoda** includes squids, octopuses, nautiluses, and other molluscs with tentacles.

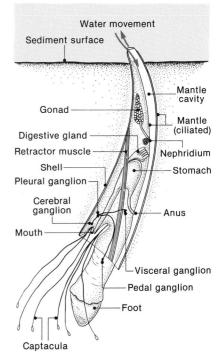

Figure 28.11
Structure of the scaphopod Dentalium.

Major Extant Groups in Phylum Mollusca

Genera mentioned elsewhere in this chapter are noted.

Class Aplacophora AYE-plak-OFF-for-uh (Greek *a* without + *plakos* plate + *phoros* bearing). Vermiform, up to 30 cm long. No shell but with calcareous scales and spicules. *Neomenia* (Figure 28.8).

APLACOPHORAN

Class Monoplacophora MON-oh-plak-OF-for-uh (Greek *monos* single). With single cap- or cone-shaped dorsal shell. Nephridia, gills, and other organs serially repeated. Benthic. Dioecious; fertilization internal. *Neopilina, Vema* (Figure 28.9).

MONOPLACOPHORAN

Major Extant Groups in Phylum Mollusca *(continued)*

Class Polyplacophora POL-ee-plak-OF-for-uh (Greek *poly* many). Shell of eight overlapping plates. Flattened. Head reduced; foot broad and flat. Benthic. Dioecious. Chitons. *Mopalia* (Figure 28.1B).

Class Scaphopoda skaf-FOP-oh-duh (Greek *skaphe* trough + *podos* foot). Tooth- or tusk-shaped shell open at both ends. Burrowing in marine sediment, head down. With tentacles. No gill; perhaps no circulatory system, except for blood sinuses. Dioecious; fertilization external. Tusk shells. *Dentalium* (Figure 28.1A).

Class Gastropoda gas-TROP-oh-duh (Greek *gaster* belly). Shell of one piece, often spirally coiled; absent in slugs. Body undergoes torsion during development. Secondarily asymmetric. Both trochophore and veliger larvae may occur, or development may be direct.

    Subclass Prosobranchia pro-so-BRANK-ee-uh (Greek *pros* forward + *branchia* gills). Gills anterior to heart as result of torsion. One pair of tentacles. Mainly dioecious. Limpets, abalone, periwinkles, conchs, cowries, slipper shells, whelks, cone shells. *Busycon, Conus, Cypraea, Diodora, Haliotis, Lambis Littorina, Murex, Thais, Trunculariopsis* (Figure 28.12).

    Subclass Opisthobranchia oh-PISTH-oh-BRANK-ee-uh (Greek *opisthe* behind). Gill, if present, posterior, as result of detorsion. One to four pairs of tentacles. Shell, gill, and/or mantle cavity reduced or absent. Marine. Monoecious. Bubble shells, sea hares, pteropods (sea butterflies), sea slugs (order Nudibranchia). *Aplysia, Coryphella, Glaucus, Hermissenda* (Figures 20.4, 28.13).

    Subclass Pulmonata pull-mon-ATE-uh (Latin *pulmo* lung). Mantle cavity modified into lung. Degree of detorsion variable. Shell, if present, a spiral. Monoecious. Development direct. Terrestrial and freshwater snails and slugs. *Ariolimax, Cupedora, Helix, Limax* (Figure 28.1C).

Class Bivalvia (= Pelecypoda) bi-VALVE-ee-uh (Latin *bi* two + *valva* folding door). Shell of two lateral valves. Foot usually wedge-shaped. No radula. Gill large and used for suspension feeding. Usually dioecious, with external fertilization. With both trochophore and veliger larvae. Benthic, in either fresh or salt water. Clams, mussels, oysters, scallops, cockles, shipworms. *Bankia, Chlamys, Dreissena, Lyrodus, Margaritifera, Meleagrina, Mercenaria, Mytilus, Ostrea, Pecten, Tagelus, Teredo, Tridacna* (Figure 28.1D, 28.20, 28.22).

Class Cephalopoda SEF-a-LOP-oh-duh (Greek *kephale* head). Head well developed, with prominent arms or tentacles, eyes, and jaws. Siphon for jet propulsion. Dioecious. Development direct. All marine.

    Subclass Nautiloidea naw-til-OY-dee-uh (Greek *nautilos* sailor). External coiled shell, divided into chambers. Many (80 to 90) tentacles. One genus, *Nautilus* (Figure 28.24).

    Subclass Coleoidea col-ee-OY-dee-uh (Greek *koleon* sheath). Shell reduced, internal, or absent. Eight or ten arms with suckers. Squid, cuttlefish, octopus. *Loligo, Octopus, Sepia, Todarodes* (Figures 28.1E, 28.26).

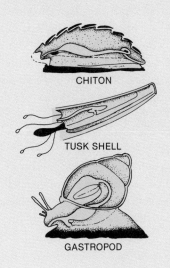

CHITON

TUSK SHELL

GASTROPOD

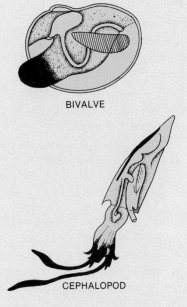

BIVALVE

CEPHALOPOD

GASTROPODS

More than one-third of all mollusc species are gastropods. They are so diverse that it is convenient to divide the class into three subclasses. Members of subclass Prosobranchia produce many of the shells familiar on beaches—the periwinkles, conchs, cowries, whelks and cone shells. Another prosobranch is the abalone (AB-a-LOW-nee), which, if you live near the Pacific, may be familiar on your dinner plate (Figure 28.12). The subclass Opisthobranchia includes sea hares such as *Aplysia* and *Hermissenda*, which are famous for their contributions to research on the mechanisms of behavior (see p. 149) (see Figures 20.4 and 28.13). Subclass Pulmonata includes the air-breathing snails and slugs, familiar scourges of the garden (see Figures 28.1C and 28.14).

Figure 28.15 is a schematic representation of the internal structure of gastropods. The dominant features are the coiled shell and the foot. The shell encloses the visceral mass and continues to enlarge at the anterior edge as the animal grows. When threatened, many gastropods retreat into the mantle cavity and seal the entrance with a tough **operculum** on the back of the foot. This is a bit of a bother, because the leading edge of the mantle must be repositioned precisely in order to resume shell growth.

Torsion. One of the peculiarities of gastropod organization evident in Figure 28.15 is the way the digestive tract doubles back on itself, with the mouth and anus at the same end. This is due to torsion, which is one of the few features found in all living gastropods, and in no other group. Torsion begins in the veliger larva (except in snails and others that do not have distinct larval stages) as the rear of the visceral mass makes two 90° twists to the right, bringing the mantle cavity behind and above the head (Figure 28.16). Depending on the species, torsion can occur within a few minutes or over a much longer period. The immediate cause of torsion is a muscle on the right side that is stronger than the corresponding muscle on the left. The evolutionary cause is not as well understood. Numerous advantages of torsion have been proposed, but each has weaknesses, and none has been tested experimentally. One obvious advantage is that torsion enables the gastropod to protect its delicate head and foot by pulling them into the mantle cavity. The same advantage could be provided, however, simply by making the shell a little larger.

Having the anus hanging over the gills is a definite hygienic concern. Many gastropods avoid fouling their gills by tying up the feces in strings of mucus. Others with coiled shells avoid fouling by having just one gill and eliminating feces toward the side without a gill. Primitive gastropods with two gills, such as the abalone *Haliotis* and the keyhole limpet *Diodora*, void the feces through one or more holes in the shell. Gastropods in class Opisthobranchia avoid the danger of fouling by **detorsion,** in which the gut is straightened out to some degree. The sea hare *Aplysia* shows the largest degree of detorsion, with the anus almost completely at the rear of the animal.

Coiling. In addition to torsion, many gastropods also undergo the separate process of coiling in the development of the visceral mass and shell. All living gastropods, including those without shells, show signs of having descended from species with coiled shells. Coiling is one way that a shell can grow along with the animal without becoming so long that it impedes locomotion. (The other way is to have roughly circular shells, as in bivalves.) Without coiling, gastropods would have shells like scaphopods and would probably have to be sedentary like them.

Figure 28.12
Prosobranchia. (A) The yellow periwinkle Littorina obtusata. *(B) Overturned shell of the spider conch* Lambis. *(C) The Chinese cowry* Cypraea chinensis. *The pink mantle is partly separated, exposing the shell. (D) Shell of the lightning whelk* Busycon contrarium *with egg cases. The elongated portion of the shell is the siphonal canal, which encloses the siphon. Whelks use the sharp edge of the shell to pry open bivalves. This species is unusual among gastropods in that it is left-handed (sinistral). That is, the shell coils counterclockwise away from the apex, which is at the opposite end from the siphonal canal. (E) The abalone* Haliotis *showing the series of holes for the escape of respiratory gases. As the abalone grows it closes off older holes and develops new ones.*

A

B

C

D

E

Figure 28.13
The sea slug Hermissenda, *a popular subject for the study of the neural mechanisms of learning.*

The simplest form of coiling is one in which all whorls are in the same plane (planospiral). In other words, the shell will lie flat on either side. Such shells are satisfactory for some aquatic snails (see Figure 2.3A), but they are too bulky for gastropods that have to support the mass of the shell. A more compact form that is less likely to be tipped over by gravity results when each whorl develops a little to one side of the preceding whorl (conispiral). This pattern leaves the older and smaller whorls—those forming the **apex** and **spire**—sticking out to one side.

Figure 28.14
The banana slug Ariolimax columbianus *of the American Northwest, showing the large respiratory pore (pneumostome). This mollusc is the mascot of the University of California at Santa Cruz. Approximately 20 cm long.*

(Whether the apex is on the right or left is determined early in development by the same gene that determines the direction of the spiral movement of cells during cleavage.) This arrangement leaves the shell off balance (Figure 28.17B). This problem is solved by a shell that is partially turned to one side, with the spire tilted over the back (Figure 28.17C). The evolution of this adaptation must have

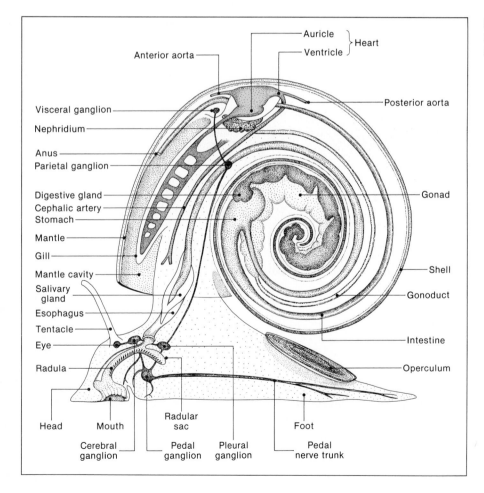

Figure 28.15
Internal structure of a generalized gastropod.

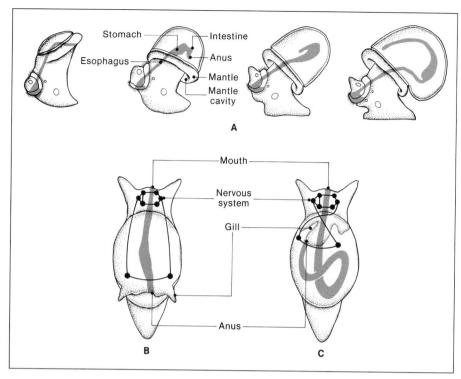

Figure 28.16
(A) Torsion in the veliger larva of a gastropod. The digestive tract is shown in color to indicate movement of the visceral mass. (B) An adult gastropod as it would appear without torsion. The nervous system and digestive tract are indicated. (C) A torted adult. The result of torsion is that the mantle cavity enclosing the gills and anus lies above and behind the head, and the nervous system is twisted into a figure eight.

Stomach — Intestine
Esophagus — Anus
Mantle
Mantle cavity

A

Mouth
Nervous system
Gill
Anus

B   C

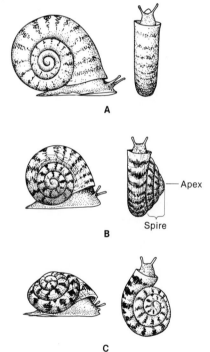

A

B
Apex
Spire

C

Figure 28.17
Side and top views of coiled gastropod shells. (A) A planospiral coil is appropriate for a buoyant mollusc, but its high profile would be easily upset by gravity or water currents. (B) A more compact conispiral shell results from offsetting each new turn in the coil to one side. The weight is now unbalanced, however. (C) Balance is restored by turning and tilting the shell so that its weight is over the foot.

presented an additional problem: the weight of the shell would have rested on one of the delicate gills, nephridia, and auricles of the heart. In modern gastropods this final problem is solved by omitting those organs on the side that would be beneath the shell. As a result all gastropods, even those that have subsequently lost the shell, have only one gill (if any), one nephridium, and one auricle.

Defense and Predation. The shell is a constant refuge against predation, drying, and other threats, but it restricts habitat to places where calcium is abundant. Snails, for example, are plentiful only in fresh water or soils near limestone, which consists of fossilized shells (including perhaps those of the snails' ancestors). Where calcium is lacking one often finds slugs but few snails. Slugs make up for the lack of a protective shell by being active only when it is damp, and by secreting mucus that is sticky and noxious. (One zoologist has reported that slug mucus numbs the tongue.)

Although there is plenty of calcium in the oceans, sea slugs and sea hares also lack shells. Like land slugs, most of these opisthobranchs also defend themselves with toxic mucus, which many of them obtain from sponges and other prey. Opisthobranchs often advertise their toxicity with showy **warning coloration** that probably deters many would-be predators. John Steinbeck in *Cannery Row* refers to "orange and speckled nudibranchs [that] slide gracefully over the rocks, their skirts waving like the dresses of Spanish dancers." Other shell-less gastropods protect themselves in more novel ways. *Glaucus* shoots predators with nematocysts that it gets from its own prey, the Portuguese man-of-war (see p. 507). *Aplysia* distracts predators by secreting a cloud of **ink** when disturbed. Most gastropods graze and browse placidly, but a few are as innovative in predation as they are in defense. Cone shells (*Conus*) have modified radula teeth that they use like harpoons to inject neurotoxins into passing fish (Barinaga 1990; Olivera et al. 1985). (The toxin can also be lethal to a human who steps on *Conus*.) *Conus purpurascens* enhances its hunting abilities by **aggressive mimicry**, using its red, worm-shaped proboscis to lure prey (Figure 28.18).

Reproduction.   Reproduction is another area in which gastropods reveal their inventiveness. Some slugs will court for several days before copulating. A pair of slugs *Limax maximus* (Figure 28.19), which may be more than 10 cm long, sometimes display their commitment to each other by dangling from a tree limb by a 40 cm strand of mucus while they exchange sperm packets. One is tempted to applaud such exuberance, but it may be only a way to avoid predation during this vulnerable time. No less strange is the courtship of certain snails, such as the common *Helix*. A pair of these snails will display their mutual interest by stabbing each other in the foot with a **dart**. The calcareous or chitinous dart, sometimes called a "love arrow," is a centimeter long in some snails. Fortunately, unlike Cupid's arrows, these darts are not aimed for the heart. In the 18th century the French philosopher-scientist Maupertuis speculated that the function of the dart was to sexually excite these lethargic animals. This hypothesis is still often repeated, but there does not appear to have been any attempt to test it experimentally.

Locomotion.   Some gastropods, such as sea hares, swim by undulatory movement of the body, and a few, such as conchs, drag themselves along with a proboscis. In the majority of gastropods, however, the foot is responsible for locomotion. Snails, for example, move by ripplelike contractions of the foot on a mucous trail. This is an inefficient form of locomotion, since the speed is limited by the rate of mucous secretion, and the secretion alone uses up to one-fourth of the gastropod's total energy budget. The efficiency is not as low as it might be, however, because of a peculiar property of the mucus. It sets up like glue beneath stationary portions of the foot, but turns liquid under a part of the foot that starts to move. In some limpets the mucus also recovers some energy by trapping and stimulating the growth of algae, which the limpets eat as they retrace their mucous trails during tidal migrations (Conner and Quinn 1984).

BIVALVES

Members of class Bivalvia include the familiar clams, mussels, oysters, and other animals with shells consisting of two valves (Figure 28.20). Unlike the valves of brachiopods (see pp. 569–570), those of bivalves are positioned laterally and are symmetrical to each other. Normally the valves are held open by an elastic ligament near the hinge, but when threatened the bivalve can "clam up" using its **adductor muscles** (Figure 28.4B). (The adductor muscle is what you eat when you have bay scallops.) Another distinctive feature of bivalves is the wedge-shaped foot; hence the former name of the class, Pelecypoda (Greek *pelekys* axe + *podos* foot).

Locomotion.   Most bivalves that move about do so by inserting the foot into soft substratum, inflating it with blood, when using it as an anchor to pull the body along. Other bivalves use the foot to burrow into sand. The razor clam *Tagelus* (Figure 28.1D) uses its foot to burrow deeply, then extends two long incurrent and excurrent siphons up through the sediment to maintain the essential flow of water past the ctenidia. Other bivalves use the valves for locomotion or burrowing. The scallop *Pecten* claps the two valves together, forcing jets of water out around the hinge and propelling the scallop in the direction of the gape. The

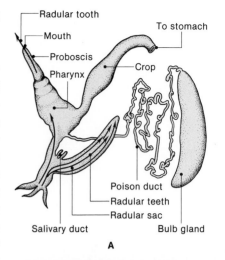

Figure 28.18
(A) The feeding mechanism of the marine snail Conus. *The radula is highly modified into a poison "harpoon" gun. (B) Aggressive mimicry by* Conus purpurascens *of the eastern Pacific. The sequence begins when the cone shell senses a passing fish and sticks out its wormlike proboscis. The fish nibbles at the lure, is stung in the mouth, and ingested.*

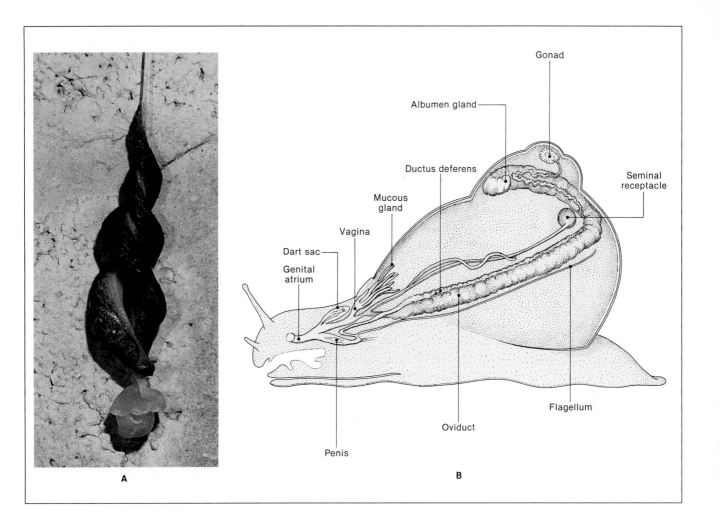

Gonad

Albumen gland

Ductus deferens

Seminal
receptacle

Mucous
gland

Vagina

Dart sac

Genital
atrium

Flagellum

Penis

Oviduct

A

B

ship worms *Teredo* and *Bankia* use their tiny valves to bore into sea grasses and
wooden pilings or boat hulls. Many other bivalves are sessile as adults. The oyster
Ostrea cements its left valve to substratum, and sea mussels such as *Mytilus* attach
themselves by **byssal threads** secreted from the rudimentary foot. These byssal
threads contain a substance that rivals epoxy as an adhesive and has the additional
advantage of working under water. Researchers are studying ways to use the adhesive in dentistry or medicine.

Feeding and Digestion.   Bivalves are unusual among molluscs in that they lack
radulae. Most use the gills (ctenidia) for filter feeding. As cilia draw water across
the surface of the ctenidia they also bring in particles of food and inorganic debris
(Figure 28.5). The particles cannot pass through the pores into the gills and therefore become trapped in a continuously secreted layer of mucus. Cilia separate the
debris from the food, which they propel to the mouth in a mucous string. The
mucous string is reeled into the stomach by another unique feature in bivalve
digestion, the **crystalline style** (Figure 28.21). The style is kept rotating by cilia
lining the style sac. The crystalline style also releases digestive enzymes, such as
amylase, as it grinds against the **gastric shield** on the stomach lining. Nutrient
molecules released during digestion in the stomach enter the **digestive glands,**
where further digestion occurs intracellularly. Larger indigestible particles go into
the intestine and are eliminated.

One important variation in this pattern of feeding should be mentioned. Six

A   B

species of **giant clams** and one closely related cockle do not move about and feed in detritus-rich sediments but stay anchored in nutrient-poor coral reefs with the valves gaping upward. Some of these giant clams (*Tridacna gigas*) are a meter long and have a mass of 250 kg. This prodigious size naturally raises the question of what they eat. Contrary to old sailors' tales, giant clams probably do not subsist on human flesh, since their digestive systems are no better equipped to handle meat than are those of other clams. Much of their nutrition comes from photosynthetic protists called zooxanthellae (see p. 476). These protists are housed in the mantle just inside the edges of the valve and produce glycerol and alanine whenever the host opens its shell in the tropical sunlight (Figure 28.22). Often the mantle of giant clams has bright pigments that block the wavelengths of sunlight that would damage the clam's tissues.

Reproduction and Development.   Bivalves appear to have used up their share of evolutionary innovation on the ctenidia and foot, since the other organs are fairly simple. There is not even a concentration of receptors and neural tissue to indicate where a head would be. This simplicity also extends to reproduction. As a rule, bivalves show little of the courtship or other social behaviors displayed by gastropods and some other molluscs. Most of them simply discharge gametes into the water, where fertilization occurs. There is usually no parental behavior toward the trochophore or veliger larvae. The majority of gametes and larvae become food for other species, but each female releases tens of millions of eggs per season.

Various behavioral adaptations also increase the odds of successful reproduction. Some female bivalves inhale sperm, thereby effecting internal fertilization. The female Pacific Coast oyster *Ostrea lurida* also broods her larvae within the mantle cavity until they become **spats**—veligers ready to attach to substratum. Some clams have special veliger larvae, called **glochidia,** that attach to the gills or skin of fish (Figure 28.23). The glochidia encyst in the fish tissue, where they are nourished and protected until able to live independently. The female clam *Lampsilis ventricosa* lures fish that serve as adoptive parents for her glochidia. She

Figure 28.20
(A) Two shipworms Teredo *in wood. Note the highly modified valves around the suckerlike foot. Shipworms do considerable damage by digesting the cellulose in wooden structures.* Teredo, *a native of the Caribbean, nearly sank Columbus. (B) The scallop* Chlamys *showing numerous eyes bordering the gape. Bivalves show little cephalization; even the eyes are not concentrated in the head.*

Figure 28.21
Digestive system of a bivalve. (A) External view of the stomach and digestive ceca, in which extracellular and intracellular digestion, respectively, occur. See Figure 28.6 for the location of the stomach in the animal. (B) Cutaway view of the stomach showing the crystalline style reeling in the mucous string bearing food particles. The crystalline style releases digestive enzymes as it grinds against the gastric shield. The ciliated sorting area separates small food particles from larger, indigestible bits. Arrows indicate the movement of small particles to the digestive ceca, and of larger particles to the intestine.

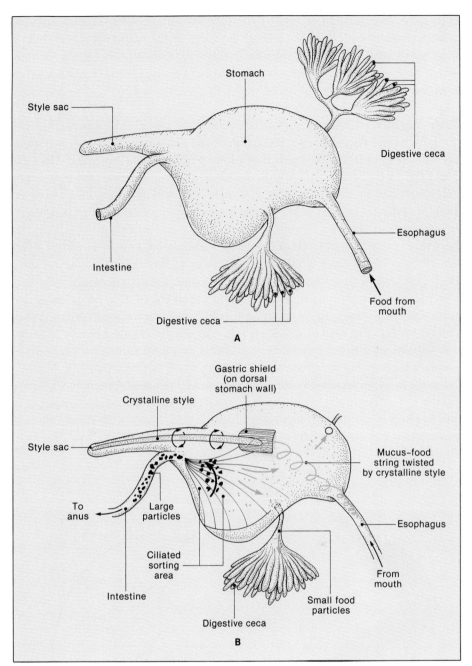

has a foot shaped remarkably like a small fish, even down to an eyespot. When a larger fish approaches the lure for a meal, the clam releases her glochidia, which then hitch a ride in the fish's gills.

CEPHALOPODS

Cephalopods are named for the foot, which is integrated into the head in the form of arms, tentacles, and/or a siphon. Water is expelled through the siphon for locomotion, while tentacles and arms are used mainly in predation. Cephalopods are divided into two quite different subclasses that differ in the number and form of the tentacles. Subclass Coleoidea comprises squids, cuttlefishes, and octopuses, all of which have eight arms, with suckers lining the inner surfaces. Squids and cuttlefishes have in addition two longer tentacles with cup-shaped suckers at the ends. Chitinous teeth or hooks around the edges of the suckers augment suction in grasping prey. Subclass Nautiloidea includes six species in the genus *Nautilus*, which have more than 90 sticky tentacles without suckers, each retractable in its own sheath.

Body Support.   Unlike most molluscs, octopuses have no shell, and their bodies are bulbous and flaccid when relaxed. Squids and cuttlefishes have only a vestigial shell inside the mantle. In squids it is a long **pen** (= gladius) running down the back, and in cuttlefishes it is the flat **cuttlebone** often seen in bird cages. The cuttlebone has air-filled chambers that make the cuttlefish buoyant. The giant squid *Architeuthis*, which has the largest mass of any invertebrate (up to 450 kg), is also buoyant. In fact, it weighs less than 5 grams in its natural habitat. Its buoyancy is due to high levels of low-density ammonium ions in its tissues. (This was discovered by Clyde Roper and two others as they were cooking up some *Architeuthis* steaks to celebrate completion of a doctoral examination.)

Unlike most cephalopods, *Nautilus* has a massive shell that holds gases that make the animal buoyant. As the living part of *Nautilus* grows, it enlarges its shell and moves into the newest, widest area. It seals off the old chamber by secreting a calcareous septum behind it (Figure 28.24). The animal cannot simply leave an empty chamber, however, because it lives at depths of more than a hundred meters, where the pressure would crush it. Instead, a **cameral fluid** that is ionically similar to nautilus blood (as well as to seawater) fills the abandoned chamber until the new septum is formed. Gradually, cameral fluid is removed osmotically and replaced by a gas—essentially air with the oxygen removed. The low density of

Figure 28.22
The giant clam Tridacna *on a coral reef. The mantle edges contain symbiotic zooxanthellae.*

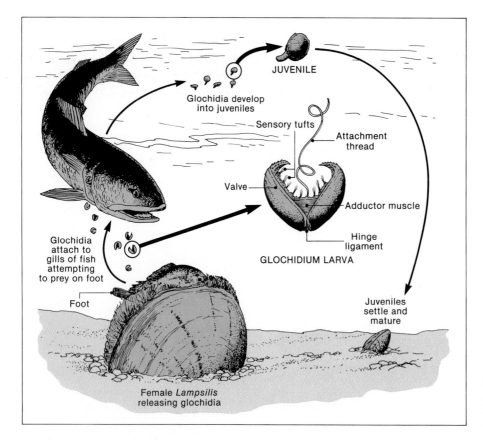

JUVENILE

Glochidia develop
into juveniles

Sensory tufts

Attachment
thread

Valve

Adductor muscle

Hinge
ligament

GLOCHIDIUM LARVA

Glochidia
attach to
gills of fish
attempting
to prey on foot

Foot

Juveniles
settle and
mature

Female *Lampsilis*
releasing glochidia

Figure 28.23
Life cycle of the freshwater clam Lampsilis ventricosa, *including details of the glochidium larva.*

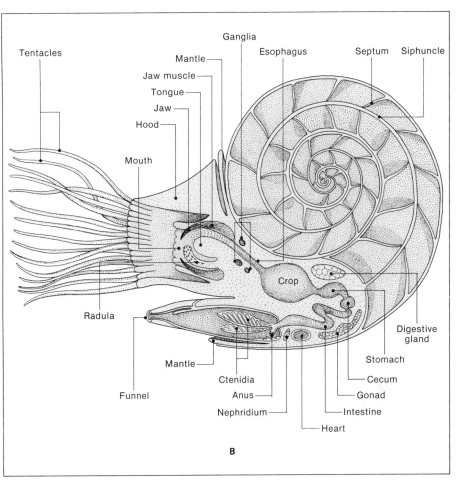

Figure 28.24
(A) Nautilus *feeding. (B) Internal
structure of* Nautilus.

the gas compensates for the added mass of shell and body, thereby achieving buoyancy.

Nautilus replaces the cameral fluid with gas even at depths greater than 500 meters, where the pressure is more than 50 atmospheres. This is equivalent to blowing up a balloon with a 25-ton boulder on it! This feat is performed by the **siphuncle,** the strand of tissue that runs through all the chambers. The siphuncle actively transports ions out of the cameral fluid, creating a local osmotic gradient that draws the water out of the chambers. The gas then infiltrates the chambers, presumably by diffusion (Ward et al. 1980; Ward 1987).

Locomotion. Octopuses usually prowl the ocean floors at a typical molluscan pace, but squids and cuttlefishes are swift predators of the open oceans. Both have streamlined bodies and fins. They, as well as octopuses, can also swim rapidly by expelling water from the mantle cavity through the funnel. This mode of swimming is only about one-fourth as efficient as swimming by salmon. It is good enough, however, to enable the squid *Todarodes pacificus* to swim continuously for six weeks at a speed of 300 cm/sec during 2000-km migrations (Gosline and DeMont 1985). Jet propulsion is especially rapid in the escape response, which is triggered by the famous **giant axons** of squids (see pp. 143–145).

Contraction of the mantle cavity is due to bands of radial and circular muscles that are enclosed in a collagenous **tunic** (Figure 28.25 inset). The tunic maintains the overall shape of the mantle and keeps it from expanding lengthwise when the

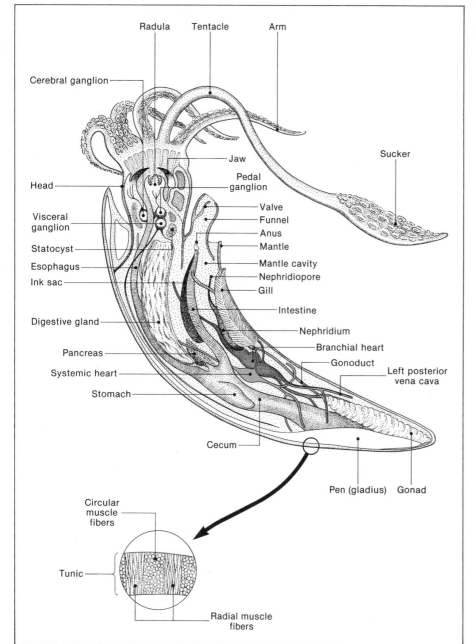

Figure 28.25
Internal structure of the squid Loligo. The inset shows a section of the mantle, with radial and circular bands of fibers enclosed by collagenous sheaths called the tunic. Radial fibers contract lengthwise in the plane of the paper, thickening as they do so, and thereby expanding the diameter of the body. Circular muscles compress the body when they contract.

muscles contract. When the radial muscles contract they thicken, thereby expanding the diameter of the body. This draws water past the head into the mantle cavity. During the escape response the circular muscles contract, squeezing the water out through the siphon. Escape from predators is often aided by another distinctive cephalopod trait, the release of a blob of mucous **ink,** which may startle or confuse the predator.

Internal Organization. Unlike most molluscs, cephalopods have a closed circulatory system capable of sustaining high metabolic levels (Figure 28.25). In many cephalopods there is not only a **systemic heart** that pumps oxygenated blood from gills to body tissues but also an accessory **branchial heart** that returns deoxygenated blood from the body tissues to the gills. The digestive system begins with a pair of chitinous jaws shaped like those of a parrot. There is a radula, but it is used only in swallowing the chunks of prey bitten off by the jaws. Digestive secretions are produced by **salivary glands,** a **liver,** and a **pancreas,** which are, of course, not homologous with vertebrate organs of the same names. The anus is near the gills, but strong contractions of the mantle prevent fouling.

Behavior. Cephalopods display a variety of complex behaviors. Octopuses can quickly learn to approach or avoid various visual, chemical, and tactile stimuli, and they retain memory for long periods. Some of the most interesting behaviors of coleoids derive from their ability to change their appearances. This feat is accomplished by means of pigment-containing **chromatophore** cells, which expand when muscles surrounding them contract. Depending on the species, yellow, orange, red, blue, and black chromatophores may be present in three layers, allowing the coleoid to assume a variety of patterns or to flash from one color to the next. Coleoids change their appearances during courtship and other social behaviors, presumably as a form of communication (Figure 28.26). The chromatophores give squids, cuttlefishes, and octopuses an uncanny talent for camouflage. Some squids and cuttlefishes also have **photophores:** luminescent organs they can turn on and off. Photophores may be used for communication but are also known to serve in camouflage. By using the photophores on the ventral surface to match the amount of overhead illumination, squids make themselves invisible to prey and predators below them (Young and Roper 1976).

As would be expected from such behaviors, cephalopods have good vision. The eyes of coleoids show striking similarities to those of vertebrates. (Compare Figures 7.13A and 28.27D.) Like our own eyes, they have a focusing lens, an iris that regulates the amount of light, and a retina that converts the image into action potentials. That these are examples of evolutionary convergence, rather than products of shared ancestry, is shown by developmental differences and the fact that the photoreceptors of the mollusc retina aim toward the lens. Figure 28.27 suggests a possible sequence by which the octopus eye could have evolved.

Figure 28.26
Two cuttlefish Sepia *copulating. Note the differences in appearance due to the chromatophores.*

Reproduction. Cephalopods have separate sexes, and they copulate. Male coleoids do not use a penis as the intromittent organ but have the bottom left arm modified for that function. This arm plucks a sperm packet (**spermatophore**) from the male's mantle cavity and inserts it into the female's mantle cavity (see Figure 16.3A). The arm then detaches in the female and is never regenerated, so each male can mate only once in its life. Some 19th-century zoologists thought the detached arms found in females were parasites, which they named *Hectocotylus octopodis*. The copulatory arm of the male is still often called the **hectocotylus arm.** Like males, each female is limited to one mating, because a secretion of the **optic gland** kills her while it promotes development of the fertilized eggs. If the optic gland is surgically removed the female survives mating but is sterile. Most coleoids live for only one or two years, but nautiloids, which do not self-destruct in mating, may live for more than 20 years.

INTERACTIONS WITH HUMANS AND OTHER ANIMALS

Molluscs are important to humans as well as to other animals as food. According to one study, sperm whales alone consume as large a mass of squid as humans do of all species of fish combined. Squid is, of course, also an item in the human diet. Fishing boats from Korea and other countries often use drift nets up to 30 miles long to catch squid. These nets also catch and kill thousands of sea birds, turtles, and other nonfood animals. Because of pressure from environmental groups, Japan has dropped this practice. Other molluscan foods include clams, oysters, mussels, abalone, snails (escargot), and bay scallops (not sea scallops, which are plugs of skate flesh). Giant clams (*Tridacna*) have long been a staple in the diet of Pacific Islanders, and more than 15 groups of scientists are now studying ways to domesticate them (Heslinga and Fitt 1987).

The consumption of molluscs goes back centuries. In the Mediterranean area fishermen for thousands of years have been dropping pots into the water to take advantage of the tendency of octopuses to crawl into dark refuges. Archaeologists frequently find ancient mounds of discarded shells (middens) indicating that bivalves have long been important in the human diet. Oysters were first cultivated for food in the first century B.C. by a Roman. Humans have also found a way to use oysters to increase the food supply indirectly: The crushed shells attract microorganisms that kill nematodes that are agricultural pests.

Molluscs also nourish humans culturally. Rare and beautiful shells have been prized throughout history, and many are still extremely valuable to collectors. In some cultures mollusc shells served for money. The *wampum* of East Coast Indians was made from the purple or white shells of the edible clam, or quahog, *Mercenaria mercenaria*. Until colonists taught Indians the mercenary uses of *wampum*, it was merely a ceremonial item. Shells and other parts of molluscs have long been used in bodily adornment. Pearls immediately come to mind. Many bivalves form pearls when a foreign object lodges between the shell and the mantle, but the pearls of most species lack the desired luster. Pearl oysters are still sometimes gathered by divers who go as deep as 40 meters, equipped only with a clamp over their noses. Now, however, most pearls are cultured in Japan. They are produced by the oyster *Meleagrina*, usually after a piece of nacre from a different species has been inserted beneath their shells. Demand for the clam shell used as a seed for cultured pearls has created an economic boom in the upper Mississippi region and has also endangered the species. Bivalve shells also provide the material from which cameos are carved. Finally, the ink of *Sepia* is still used as a source of the artists' ink called sepia. During the first 2000 years B.C. three species of marine snails, *Murex bran-*

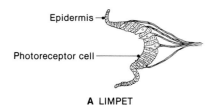

A LIMPET

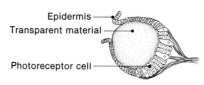

B ABALONE

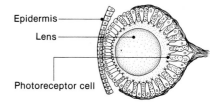

C SNAIL

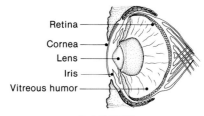

D OCTOPUS

Figure 28.27
Photoreceptor organs of molluscs, suggesting a possible sequence by which the octopus eye could have evolved. (A) The eyespot of a limpet. (B) The abalone eye, which contains a transparent substance that may have been the precursor of a lens. (C) The eye of a pulmonate. (D) The eye of an octopus.

daris, *Thais haemastoma*, and *Truncrulariopsis trunculus*, made their contribution to decoration as the sources of a permanent purple dye for cloth. The dye was laboriously collected from hypobranchial glands in the mantles of vast numbers of snails, then processed by secret recipes. In Rome all but a select few were forbidden to make or use the "royal purple," even if they could afford to do so. All of the Hebrews of the Exodus, however, sewed a single band of the purple into their garments to remind them of the commandments (*Numbers* 15:38). Jews no longer follow the practice, because the art of making the authentic purple has been lost.

Royalty also found less benign uses for molluscs. Agrippina, the mother of Emperor Nero, is said to have poisoned her son's rivals with secretions from sea slugs. Other molluscs that have detrimental impacts on humans include snails, which are essential intermediate hosts for the fluke *Schistosoma* (see pp. 534–535). Shipworms cause extensive damage to wooden piling and boat hulls. In some harbors in the United States shipworms are becoming increasingly troublesome, as improved water quality allows more of them to survive. Another mollusc, the zebra mussel *Dreissena polymorpha*, has recently begun to make an even bigger nuisance of itself in the Great Lakes, clogging municipal water intakes and outcompeting other bivalves and sport fishes. The mussel is believed to have been introduced in 1986 when a ship legally emptied the ballast water it had carried across the Atlantic. Although scientists in 1981 had warned of just such a threat from dumping ballast water, Congress has still not outlawed the practice. The zebra mussel is expected to incur costs of $4 billion during the next decade.

Humans are undoubtedly more harmful to molluscs than vice versa. Pollution from sewage, toxic wastes, and agricultural runoff remains high in Chesapeake Bay and many other areas. Because mussels and oysters are filter feeders, they tend to accumulate pollutants. At approximately 200 sites along the U.S. coast they are collected and analyzed as a means of monitoring water pollution. Many oysters have also been hit hard by a mysterious leukemia-like cancer and by the protozoan parasites MSX and Dermo (*Haplosporidium nelsoni* and *Perkinsus marinus*).

SUMMARY

Molluscs are extremely diverse, but most have a similar body plan. The living tissue is usually divided into a foot and a visceral mass. The foot is adapted in a variety of ways for locomotion. The visceral mass includes the mantle, which secretes the shell and forms a mantle cavity enclosing the respiratory organs. Another unique feature of most molluscs is the radula, which is used to scrape up food or bore into shells. Although molluscs have a small coelom enclosing the heart, the main body cavity is the hemocoel through which the open circulatory system pumps blood. Classification is based largely on the form of the shell. There are three major groups. Gastropods include snails, conchs, whelks, and others with a one-piece, coiled shell, as well as slugs and related forms in which the shell is reduced or absent. Mussels, clams, oysters, and others with a shell of two valves are classified as bivalves. In cephalopods—octopus, squid, cuttlefish, and nautiloids—the foot is modified into tentacles.

Many gastropods, such as snails and slugs, creep about on a mucous secretion. In some the mucus is toxic. The larvae of gastropods undergo a process of torsion by which the rear portion of the body bends forward over the anterior end. In most gastropods a separate process of coiling makes the shell more compact.

Bivalves generally use the foot as an anchor during locomotion, although some swim by clapping the two valves together. Most feed by filtering food particles out of the water current that flows over the gills (ctenidia). The particles are trapped in a mucous strand that is reeled in by the crystalline style. Food is digested intracellularly in the digestive gland with the aid of enzymes from the crystalline style.

Cephalopods are generally more active than other molluscs and can swim rapidly by compressing the body wall to produce a jet of water through the funnel. Squids, cuttlefishes, and octopuses have eight tentacles with suckers, and either a reduced shell or none at all. Nautiloids have numerous sticky tentacles and an elaborate spiral shell that maintains buoyancy.

KEY TERMS

foot
visceral mass
mantle
mantle cavity
radula
ctenidium
crystalline style

nacre
prismatic layer
periostracum
hemocoel
pulmonate
trochophore larva
veliger

torsion
adductor muscle
byssal threads
cameral fluid
siphuncle
hectocotylus arm

SELF-TEST

1. Identify the class to which each mollusc in Figure 28.1 belongs.

2. For each of the classes listed in the previous question, describe the foot and explain its function.

3. Mucus plays several roles in the lives of molluscs. Describe three such roles and give an example of a mollusc in which mucus serves that function.

4. Define the mantle and describe its functions.

5. What are the functions of the shell in molluscs? Explain how the shell grows to keep pace with the growth of the mollusc. How do some molluscs adjust the shapes of their shells to keep pace with body size?

6. Describe the radula and the gill of a mollusc. What roles do they play?

7. Describe the process of torsion. In what kinds of molluscs does it occur?

READINGS

RECOMMENDED READINGS

Burch, J. B. 1962. *How to Know the Eastern Land Snails.* Dubuque, IA: Wm. C. Brown.

Gosline, J. M. and M. E. DeMont. 1985. Jet-propelled swimming in squids. *Sci. Am.* 252(1):96–103 (Jan).

Gould, S. J. 1985. *The Flamingo's Smile.* New York: W. W. Norton. (*Chapters 3, 9, 11, and 16 discuss various aspects of molluscs, including Gould's own research on snails.*)

Hoffmann, R. 1990. Blue as the sea. *Am. Sci.* 78:308–309. (*On royal purple.*)

Lindner, G. 1979. *Field Guide to Seashells of the World.* New York: Van Nostrand Reinhold.

Linsley, R. M. 1978. Shell form and evolution of gastropods. *Am. Sci.* 66:432–441.

Morse, A. N. C. 1991. Ho do planktonic larvae know where to settle? *Am. Sci.* 79:154–167. (*Mainly on abalone.*)

Rehder, H. A. 1981. *The Audubon Society Field Guide to North American Seashells.* New York: Alfred A. Knopf.

Roper, C. F. E. and K. J. Boss. 1982. The giant squid. *Sci. Am.* 246(4):96–105 (Apr).

Safer, J. F. and F. M. Gill. 1982. *Spirals from the Sea: An Anthropological Look at Shells.* New York: Clarkson N. Potter.

Scheller, R. H. and R. Axel. 1984. How genes control an innate behavior. *Sci. Am.* 250(3):54–62 (Mar). (*The egg-laying hormone of the marine snail* Aplysia.)

Solem, A. 1974. *The Shell Makers: Introducing Mollusks.* New York: Wiley.

Ward, P. 1983. The extinction of the ammonites. *Sci. Am.* 249(4):136–147 (Oct).

Ward, P. D. 1988. *In Search of Nautilus: Three Centuries of Scientific Adventures in the Deep Pacific to Capture a Prehistoric-Living-Fossil.* New York: Simon & Schuster.

Ward, P., L. Greenwald, and O. E. Greenwald. 1980. The buoyancy of the chambered nautilus. *Sci. Am.* 243(4):190–203 (Oct).

Yonge, C. M. 1975. Giant clams. *Sci. Am.* 232(4):96–105 (Apr).

See also relevant selections in General References at the end of Chapter 21.

ADDITIONAL REFERENCES

Barinaga, M. 1990. Science digests the secrets of voracious killer snails. *Science* 249:250–251.

Conner, V. M. and J. F. Quinn. 1984. Stimulation of food species growth by limpet mucus. *Science* 225:843–844.

Dance, S. P. 1966. *Shell Collecting: An Illustrated History.* London: Faber & Faber.

Eyster, L. S. and M. P. Morse. 1984. Early shell formation during molluscan embryogenesis, with new studies on the surf clam, *Spisula solidissima. Am. Zool.* 24:871–882.

Field, K. G. et al. 1988. Molecular phylogeny of the animal kingdom. *Science* 239:748–753.

Heslinga, G. A. and W. K. Fitt. 1987. The domestication of reef-dwelling clams. *Bio-Science* 37:332–339.

Hori, H. and S. Osawa. 1987. Origin and evolution of organisms as deduced from 5S ribosomal RNA sequences. *Mol. Biol. Evol.* 4:445–472.

Lemche, H. 1957. A new living deep-sea mollusk of the Cambro-Devonian class Monoplacophora. *Nature* 179:413–416.

Martin, A. W. and I. Deyrup-Olsen. 1984. Control of surface exudation by slugs. In: C. L. Harris (Ed.), *Tested Studies in Laboratory Teaching: Proceedings of the Fourth Workshop/ Conference of the Association for Biology Laboratory Education (ABLE).* Dubuque, IA: Kendall/Hunt, Chap. 3.

Nixon, M. and J. B. Messenger (Eds.). 1977. *The Biology of Cephalopods.* New York: Academic Press.

Olivera, B. M. et al. 1985. Peptide neurotoxins from fish-hunting cone snails. *Science* 230:1338–1343.

Purchon, R. D. 1977. *The Biology of the Mollusca,* 2nd ed. New York: Pergamon.

Ward, P. D. 1987. *The Natural History of Nautilus.* Boston: Allen & Unwin.

Wells, M. J. 1978. *Octopus: Physiology and Behavior of an Advanced Invertebrate.* New York: Wiley.

Willmer, P. 1990. *Invertebrate Relationships.* New York: Cambridge University Press, Chap. 10.

Young, R. E. and C. F. E. Roper. 1976. Bioluminescent countershading in midwater animals: evidence from living squid. *Science* 191:1046–1048.

29

Annelids

Ornate worm (Amphitrite ornata).

CHAPTER OUTLINE

LEARNING OBJECTIVES

1. Why are earthworms, leeches, and polychaetes all classified as annelids when they look so different?

2. What is the significance of segmentation in annelids?

3. How do polychaete worms feed, especially the sedentary ones?

4. How do earthworms burrow?

5. How do they mate while living in burrows?

6. How do leeches locate prey?

7. How do leeches prevent blood from clotting while feeding? What keeps the blood from spoiling while in their guts?

THE SIGNIFICANCE OF SEGMENTATION

A year before he died in 1882, Charles Darwin charmed thousands with a book on *The Formation of Vegetable Mould through the Action of Worms, with Observations on their Habits*. What apparently captivated Darwin's readers was the revelation of how the unseen labors of the humble earthworm contributed to the fertility of the land, and thereby to their own existence. Earthworms belong to the phylum Annelida, along with other oligochaetes and with two other major groups. One group, the polychaetes, are not quite as familiar as earthworms, and the other, leeches, are not quite as charming.

Oligochaetes, polychaetes, and leeches look very different from one another, but they share certain similarities that lead taxonomists to group them in the same phylum (Table 29.1). One of these similarities is that their bodies are divided into **segments** (Figure 29.1A). Annelids are therefore commonly called segmented worms. The segments, also called **somites** or **metameres,** are not produced by repeated division and subdivision of the body, as might be expected. Instead, they are generated sequentially from the rear of the embryo, the way a machine turns out sausage links. (See p. 128 for more on segmentation in development.) True segmentation (= metamerism) occurs unequivocally in only three other phyla: Pogonophora, which some taxonomists regard as annelids, or at least close relatives (see pp. 579–581); Arthropoda (next three chapters); and our own phylum Chordata. The essential features of segmentation are that the coelom and the muscles of the body wall show serial repetition, at least in the embryo. That is, the coelom and muscles are homologous from one metamere to the next. In addition, the nervous system, excretory organs, and other organs may also be serially repeated.

Table 29.1   Characteristics of annelids.

Phylum Annelida an-EL-lid-uh (Latin *annellus* little ring).

Morphology: Body **segmented**. Integument covered with thin, nonchitinous cuticle secreted by epidermis. Often bearing **setae** made of chitin. Coelom divided into a cavity on each side of each segment, except in leeches, which have one large coelomic cavity filled with spongy connective tissue.

Physiology: Gaseous exchange usually through integument; sometimes through specialized respiratory structures. Digestive tract complete. Central nervous system a ladderlike array of ganglia on a pair of ventral nerves (which may be fused into a single cord). Cephalic ganglion dorsal to esophagus. Circulatory system closed, with ventral and dorsal longitudinal vessels in all but leeches. Respiratory pigments often present in blood cells or plasma, and/or in coelomic fluid. Many segments with pairs of nephridia for osmoregulation. A layer of circular muscle surrounds bundles of longitudinal muscles; diagonal and dorsoventral muscles also present in some groups.

Locomotion: Creeping or swimming with parapodia (many polychaetes), peristaltic burrowing (earthworms), or undulatory swimming (leeches). Coelomic pressure produces hydrostatic skeleton; important for support as well as locomotion.

Reproduction: Polychaetes mainly dioecious; external fertilizers. Oligochaetes and leeches hermaphroditic; internal fertilizers. Asexual budding or fragmentation in some polychaetes.

Development: Cleavage pattern often spiral. **Trochophore larvae** in most polychaetes; development direct in oligochaetes and leeches.

Habitat, Size, and Diversity: More than half the species are marine polychaetes, but there are many freshwater and terrestrial species, especially oligochaetes and leeches. Distributed worldwide except on Antarctica and Madagascar. Body length from less than 1 mm to approximately 3 meters. Approximately **12,000 living species** described.

Figure 29.1
(A) The clam worm Nereis virens, *a polychaete annelid. The segmentation of the body is evident. (B) A few of the 200 or more somites of Nereis, emphasizing segmentation. The cross section at the left is at one septum at the front of a segment.*

A

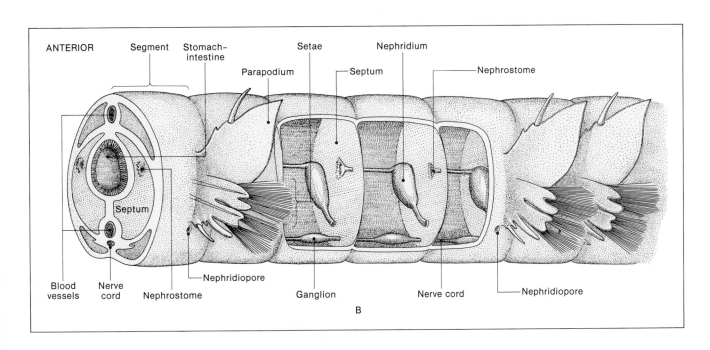

B

In arthropods and chordates most of the signs of segmentation are obscured before adulthood. In these phyla segments fuse and differentiate into distinctive body regions, such as the head, thorax, and abdomen. Each of these regions is called a **tagma** (plural tagmata). Annelids show little if any of this **tagmosis** (= tagmatization). Earthworms, for example, are essentially alike from front to rear. In most annelids segmentation remains apparent throughout life, with each segment evident externally as a little ring (*annellus* in Latin). Segmentation also remains evident inside most annelids. Each somite usually retains throughout life its own set of muscles, its own coelomic cavity, its own ganglion on the ventral nerve cord, its own nephridia, and many other organs (Figure 29.1B).

Zoologists have proposed several hypotheses to explain the adaptive advantages of segmentation. One of the most plausible hypotheses is that the segmentation of the coelom and muscles was an adaptation for burrowing, as explained in Chapter 10 (see p. 200). Another possibility is that segmentation is simply a developmentally convenient way of achieving large body size. Rather than increase the size of every organ in the body, the embryo simply repeats parts. Segmentation has undoubtedly evolved independently in chordates and perhaps arthropods, so there may have been different advantages in each of these groups.

ORIGINS AND CLASSIFICATION

The origins of most phyla are uncertain because of a paucity of evidence. The origin of annelids is obscured by too many clues, many of them contradictory. The segmentation and body shape of annelids suggest a kinship to Pogonophora, while the nonchitinous cuticle that molts in patches and the position of nephridiopores near each appendage are like those of Onychophora. The trochophore larvae of some Annelida are like those of some members of the phyla Echiura, Sipuncula, and Mollusca. Except for fossilized worm tracks and burrows, there are few paleontological clues to assist in unraveling annelid evolution. Many taxonomists once considered a group of small, simple annelids called Archiannelida to be relics of the early evolution of annelids. Now, however, archiannelids are regarded as being polychaetes that have become simplified as an adaptation to living within sediment.

Within the phylum classification is a little clearer. Most zoologists recognize three classes: polychaetes (class Polychaeta), oligochaetes (class Oligochaeta), and leeches (class Hirudinea). Class Polychaeta includes mainly marine worms with numerous bristles. (*Polychaites* in Greek means "many hairs.") These bristles, or **setae,** occur chiefly on appendages called **parapodia,** which function in locomotion and respiration (Figure 29.1). Setae are fewer on oligochaetes (Greek *oligos-* few). Because of the setae on polychaetes and oligochaetes, annelids are often called "bristle worms."

The leeches, class Hirudinea, lack setae. They live mostly in fresh water, but a few are terrestrial or marine. Both the oligochaetes and leeches have a swelling called a **clitellum** (Latin *clitella* pack saddle) on the back of certain segments. The clitellum is a permanent and familiar feature on some earthworms (Figure 29.14A). In most oligochaetes and leeches, however, the clitellum appears only during the period of reproduction, when it secretes a cocoon in which embryos develop. Oligochaetes and leeches are often referred to as clitellates.

Polychaeta is the largest and most diverse of the three classes and is often considered to be most like the ancestral annelids. Most polychaetes live within the upper 50 meters of oceans, but some have been found as deep as 5 km. Some have adapted to fresh water and can be found in certain lakes and streams in the

United States. Polychaetes are typically 5 to 10 cm long, but they range in length from less than 1 mm to several meters. Many are dull in appearance, but others are iridescent, luminescent, or brightly pigmented. Many are red where hemoglobin shows through the integument. Polychaetes are either freely moving (errant) predators or sedentary in tubes or burrows, where they feed on detritus or plankton. On this basis they were formerly classified into the two subclasses Errantia and Sedentaria. This taxonomy is no longer considered valid, since it does not reflect phylogeny. Still, however, it is convenient to refer to polychaetes as either errant or sedentary. In errant species all the segments tend to be similar; that is, there is little if any tagmosis. Sedentary forms have some degree of tagmosis, with the anterior metameres specialized for feeding.

Major Extant Groups of Phylum Annelida

Genera mentioned elsewhere in this chapter are noted.

Class Polychaeta POL-lee-KEY-tuh (Greek *polys* many + *chaite* long, flowing hair). Most segments with parapodia bearing numerous setae. Coelom well developed and divided into two lateral compartments per segment. Head distinct, with eyes and tentacles. No clitellum. Usually dioecious; asexual budding or fragmentation in some. Some with trochophore larvae. Mostly marine. *Amphitrite, Arenicola, Autolytus, Chaetopterus, Glycera, Nereis, Palola, Sabella, Syllis, Vanadis* (Figures 29.1A, 29.6, 29.8, 29.9, 29.10, 29.11).

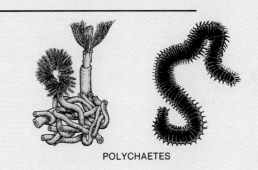

POLYCHAETES

Class Oligochaeta OH-lig-oh-KEY-tuh (Greek *oligos* few). Body divided into obvious segments, each with few setae. New segments added throughout life. Coelom well developed and compartmented. With clitellum. No parapodia or head. Hermaphroditic; asexual budding in some. No larval stage. Usually terrestrial, burrowing in moist soil. Some burrow in freshwater or marine sediments. Earthworms and other oligochaetes. *Lumbricus, Megascolides, Tubifex* (Figures 29.12, 29.13, 29.14, 29.17).

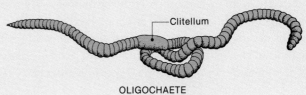

Clitellum

OLIGOCHAETE

Class Hirudinea HEAR-oo-DIN-ee-uh (Latin *hirudo* leech). A fixed number of segments (30 to 34, depending on the authority followed). Each segment divided externally into as many as 14 transverse ridges called annuli. Body usually flattened dorsoventrally. No setae (except on anterior segments of the "living fossil" *Acanthobdella*). Coelom not divided into segmental compartments; filled with spongy connective tissue and muscles. With clitellum. Usually with anterior and posterior suckers for attachment to host and for locomotion. No parapodia or tentacles. Hermaphroditic. No larval stage. Predators, ectoparasites, or scavengers. Usually in fresh water; some marine or terrestrial. Leeches. *Haemadipsa, Haementeria, Hirudo, Limnatis, Macrobdella* (Figures 29.19 and 29.20).

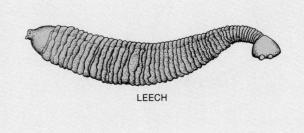

LEECH

POLYCHAETES

The clam worm *Nereis*, named for a type of sea nymph in Greek mythology, is often considered representative of polychaetes, especially the errant forms (Figure 29.1A). One member of the genus, *Nereis virens* (formerly *Neanthes virens*), is common along the Atlantic coast. It is shiny greenish with red appendages, and its more than 200 somites extend for up to 50 cm (Figure 29.1). It lives as a juvenile for two or more years along the low-tide line, then takes to the open ocean to reproduce. The juvenile spends its days in mucus-lined burrows or under rocks with the head protruding. At night it swims or crawls about, preying on other invertebrates or feeding on carrion and algae.

Feeding. To seize prey *Nereis* quickly shoots out its **pharynx**, sometimes with an audible pop (Figure 29.2). It protrudes the pharynx by contracting circular muscles in the anterior somites, which increase the pressure in the coelomic cavities of the head. (This is one advantage of having the coelom segmented into chambers.) As the pharynx retracts it turns outside in, closing the chitinous jaws and **denticles** ("teeth") on the prey. Food passes through a short esophagus into the straight **stomach-intestine**, which extends from about the 12th somite to the anus in the last somite. The gut secretes a variety of enzymes in this omnivore, including cellulase, which digests the cell walls of algae.

Nervous System. *Nereis* detects prey by receptors on the head, especially chemoreceptors (**nuchal organs**) on the prostomial palps and on a pair of prostomial tentacles. Touch receptors on the prostomial tentacles and on the four pairs of peristomial tentacles are probably also important for locating food and also shelter. *Nereis* also has two pairs of eyes. These are relatively simple pigment-cup organs, with a retina of **rhabdomeric** receptor cells (see p. 155) surrounding a spherical lens (Figure 29.3A). The lens is too close to the retina to focus an image on it, so the eyes of *Nereis* probably function primarily as light detectors. In at least two species in the family Alciopidae, however, the lens can focus images on the retina (Figure 29.3B). The general structure of these eyes shows remarkable evolutionary convergence with the eyes of vertebrates and cephalopod molluscs.

Information from the receptors on the head is integrated in the most anterior ganglion of the central nervous system, the cephalic ganglion, located dorsally in the prostomium. Two **circumpharyngeal connectives** connect the cephalic gan-

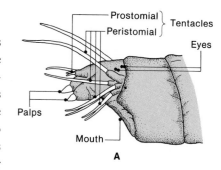

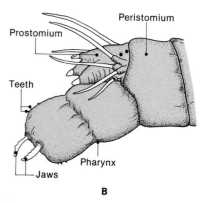

Figure 29.2
The head of Nereis virens *consists of the prostomium (Greek pro in front of + stoma mouth), and the peristomium (Greek peri- around). (A) Pharynx retracted. (B) Pharynx extended out of the mouth.*

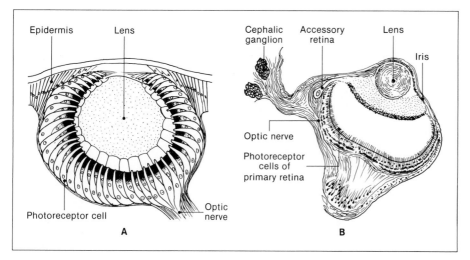

Figure 29.3
(A) One of the four eyes of Nereis. *Note that the photoreceptor cells are too close to the lens to resolve an image. (B) The eye of the deep-sea polychaete* Vanadis *is like that of vertebrates and cephalopods in having its retina in the focal plane of the lens. Also like some deep-sea fishes and cephalopods,* Vanadis *has an accessory retina that responds to wavelengths different from those of the primary retina. Since different wavelengths penetrate to different depths in the ocean, this could enable the eye to be used as a depth gauge.*

glion to the **ventral nerve cord.** As is typical in annelids, the ventral nerve cord consists of a series of ganglia joined by a pair of longitudinal **connectives** (Figure 29.1B). In *Nereis* and many other annelids the pair of connectives is fused into a single connective. There are more than 200 ganglia, each of which controls one somite. Overall behavior is coordinated by impulse traffic in nerve cells in the connective, including inhibitory impulses from the cephalic ganglion. Polychaete neurophysiology is not known in detail except for the five cells called **giant nerve fibers** that trigger escape responses. The single median giant fiber is quite sensitive to mechanical stimulation in the first 60 somites and triggers rapid contraction of the longitudinal muscles of posterior somites. The behavioral response is therefore a rapid withdrawal of the head from a potential threat in front of the animal. Two paramedial giant fibers similarly trigger a withdrawal reflex of the anterior somites when the posterior is threatened. In addition, there are two lateral giant fibers that trigger shortening of the entire body in response to strong stimulation anywhere on the body.

Movement and Support. The longitudinal muscles responsible for the withdrawal reflexes lie in four thick bands outside the **peritoneum** that surrounds the coelom (Figure 29.4). The longitudinal muscles are also responsible for rapid crawling. Those on opposite sides of the segments contract alternately, producing a wavelike wriggling. As in all annelids a layer of circular muscle surrounds the longitudinal muscles. The circular muscles are weak in polychaetes, but they are important in creating the coelomic pressure for the hydrostatic skeleton that

Figure 29.4
A cross section of Nereis, *emphasizing the muscles and parapodia. The right half of the figure represents a cut through a parapodium.*

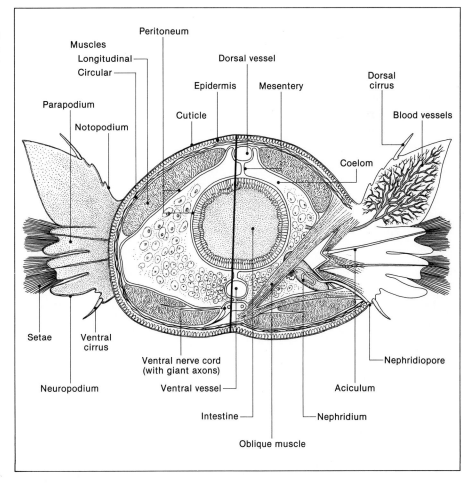

makes locomotion possible. In addition, oblique muscles to the parapodia assist in crawling.

Functions of Parapodia.   Each parapodium has a dorsal lobe, the **notopodium,** and a ventral lobe, the **neuropodium.** A stiff, chitinous **aciculum** reinforces each lobe. Each lobe also bears numerous setae and a **cirrus** that is richly endowed with sensory receptors. The notopodia of *Nereis* serve as **gills.** This function is possible because of the large number of capillaries in each notopodium. Hemoglobin, which is dissolved in the plasma, shows red through the thin integument of these dorsal lobes. (Polychaetes that have the respiratory pigment chlorocruorin have green gills.)

Circulation.   The closed circulatory system maintains an efficient supply of oxygen. The blood pressure is less than 10 mmHg in resting *Nereis* but can double in active individuals, because of increased coelomic pressure. Resting blood pressure is created primarily by peristaltic contraction of the dorsal vessel (Figures 11.3A, 29.1B, and 29.4), which pumps the blood anteriorly. Branches in each somite transport the blood to capillaries in the organs. Blood then collects in the ventral vessel, which carries it posteriorly back to the dorsal vessel.

As in most annelids, the coelomic fluid of *Nereis* carries some of the burden of transporting oxygen, nutrients, and wastes. In some errant polychaetes the coelomic fluid carries the entire burden, because there is no blood. In these worms the coelomic fluid may contain hemoglobin. One example is *Glycera*, which is commonly and paradoxically known as the bloodworm. The coelom of *Nereis* and other polychaetes is divided along the midline of each segment into a pair of coelomic cavities surrounded by the peritoneum. Where the peritonea from adjacent segments meet they form a **septum** (Figure 29.1B). Where the peritonea from the two coelomic cavities in one segment meet they form a **mesentery** along the midline (Figure 29.4). In most annelids the mesenteries and/or septa are incomplete between some coelomic cavities. In *Nereis*, for example, the septa are absent from anterior segments, forming the large chamber that everts the pharynx.

Excretion.   In addition to maintaining the hydrostatic skeleton and transporting oxygen and nutrients, the coelomic fluid is evidently important in eliminating metabolic wastes through the **metanephridia** (see Figure 14.6). The nephrostome receives the coelomic fluid of the segmental cavity anterior to the nephridium. Each nephridium expels these wastes through the nephridiopore in the body wall (Figure 29.1B). The nephridia have little to do in regulating ion and water levels in *Nereis* and most other marine polychaetes, since these animals are osmoconformers, with body fluids similar to the seawater that normally surrounds them. In estuarine and freshwater polychaetes, however, the nephridia are more highly developed and maintain a higher internal ionic concentration than occurs in the environment.

In many annelids part of the peritoneum around the gut differentiates into **chloragogue cells** (= chloragog or chloragogen cells), which assist the nephridia in eliminating nitrogenous wastes. These greenish or brownish cells remove the amine groups from amino acids and convert them into either ammonia or urea. Bits of chloragogue tissue containing these nitrogenous wastes periodically break off and either leave through the nephridia or form deposits in the body wall. Chloragogue cells also store glycogen and fats, and they gather around maturing ova and provide them with nutrients.

Sexual Reproduction. The peritoneum and nephridia are also important in reproduction. *Nereis* and other polychaetes have no distinct gonads or genital organs. Instead, the peritonea produce the gametes, and the nephridia often serve as routes for their emission. Reproduction in *Nereis* and many other polychaetes occurs during a remarkable transformation called **epitoky** (Greek *epi-* on + *tokos* birth). After living two or more years as a crawling, benthic juvenile (**atoke**), in which reproduction is inhibited by a neurosecretion from the cephalic ganglia, *Nereis* changes to a swimming, pelagic adult (**epitoke**). The body enlarges, the eyes increase in relative size, the head appendages shrink, the parapodia change shape, and muscles change from slow-twitch to fast-twitch types. The epitokes leave the shore and head for open water. Then on a night determined partly by the phase of the moon, the female epitokes release a pheromone that excites the males to swim about them in circles. As the sun rises the epitokes of both sexes release vast numbers of gametes. This synchronous mating improves the chances of fertilization. It also ensures that some of the gametes will survive predation by fishes and other animals, since there are too many gametes to be eaten at one time. (This is an example of **predator saturation;** see p. 426.) As in many flatworms, molluscs, and some other protostomes, the zygotes of *Nereis* and other polychaetes develop into trochophore larvae, characterized by a girdle of cilia (Figure 29.5).

Epitoky is different and equally notable in the palolo worm *Palola viridis* (for-

Figure 29.5
(A) A polychaete trochophore larva. (B) Development in the bloodworm Glycera. *At four weeks the parapodia are starting to develop, but the larva still uses the ring of cilia to swim. At seven weeks the larva still swims by means of its cilia. The digestive system is not developed sufficiently to permit feeding. At eight weeks* Glycera *has become a post-larva. It can now crawl with its parapodia and prey with its eversible pharynx armed with jaws. Also shown is a young worm at two months.*

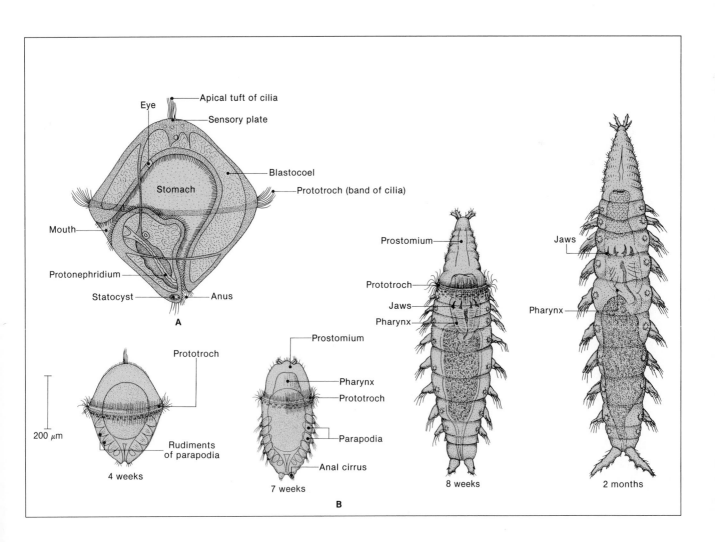

merly *Eunice viridis*). In this errant polychaete, each posterior segment becomes an individual epitoke that detaches from the anterior portion of the worm (the atoke; Figure 29.6). After detaching, the epitokes swim to the surface and release their gametes in such profusion that they cloud the water. The atokes return to burrow in coral reefs where many live to breed the following year. The swarming of the palolo worm occurs so predictably, in October or November at the start of the moon's last quarter, that the natives of Samoa schedule it as a holiday on which to gather and feast on epitokes.

Asexual Reproduction and Regeneration. Asexual reproduction occurs in some errant polychaetes by either budding or fragmentation. In budding, new individuals develop on a segment then detach. The segment that serves as the **growth zone** for buds is genetically determined, although that segment differs among individuals of the same species. In *Autolytus* buds themselves undergo budding, resulting in a chain of zooids (Figure 29.7). Some polychaetes reproduce by spontaneous fragmentation. They divide into two parts, and each fragment regenerates the parts it lacks. *Syllis*, for example, spontaneously breaks at a predetermined site between two somites. The anterior fragment grows a new tail, and the posterior fragment grows a new head. The ability to reproduce by spontaneous fragmentation is perhaps not surprising in a phylum noted for the ability to regenerate lost parts. Many annelids can regenerate a new animal from a few somites that remain after others have been lost by accident or human experimentation. In experimental fragmentation the head and tail always regenerate at the anterior and posterior ends of the fragment, respectively. That is, annelid regeneration shows **polarity,** as well as many other similarities to regeneration in *Hydra* (see pp. 515–517).

Sedentary Polychaetes. Many polychaetes are adapted to burrowing or living in tubes. The most apparent adaptations involve the parapodia and the mechanisms of feeding. Generally, the parapodia are reduced or modified for some function other than locomotion, and various food-gathering structures compensate for the inability to go out to eat. These two types of modification have led to a greater degree of tagmosis than in errant polychaetes.

Probably the most beautiful adaptations for food gathering are the plumes of tentacles in the **annelid tubeworms** in the serpulid and sabellid families. *Sabella*, for example, has a funnel-shaped plume of tentacles that gives the worm its common name "feather duster" (Figure 29.8A). Each tentacle, or **radiole,** bears numerous branches lined with cilia (Figure 29.8B). The cilia produce an upward current of water that draws particles into food grooves in the radiole. Cilia then carry the particles to the mouth. Along the way the particles are sorted according to size. Plankton are carried to the mouth and ingested, the largest particles are rejected, and medium particles of sand are stored for future tube construction.

Other sedentary polychaetes live either partially or entirely within burrows and feed on detritus rather than plankton. *Amphitrite* burrows near the low-tide line of quiet bays but feeds with its long tentacles extended over the sediment (Figure 29.9). Cilia lining the gutter-shaped tentacles carry food particles to the mouth.

The **parchment worm** *Chaetopterus* lives entirely within its U-shaped burrow (Figure 29.10). The burrow is lined with collagen (the "parchment") and lies buried in the ocean bottom down to 30 meters. The burrow has a diameter of approximately 2 cm in the middle and is often shared by crabs and other commensals. Both ends taper and extend above the sediment and thereby keep out debris. Although this living arrangement provides protection, it presents *Chaetop-*

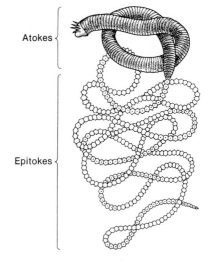

Atokes

Epitokes

Figure 29.6
The Pacific palolo worm Palola viridis (= Eunice viridis) *during epitoky. Only the posterior segments become epitokes, which break off and release the gametes within them. Each epitoke has an eyespot that guides its swimming toward the surface.*

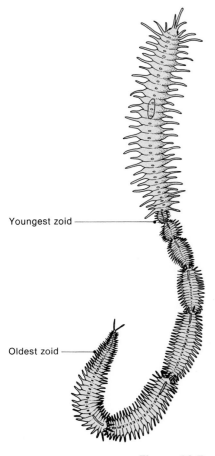

Figure 29.7
Budding in the errant polychaete Autolytus *forms a chain of zooids. Budding proceeds posteriorly, so the oldest zooids are at the rear. (Compare the formation of new segments, which proceeds anteriorly, with the oldest segments in front.)*

Youngest zoid

Oldest zoid

terus with problems of obtaining food and oxygen and of eliminating wastes. These problems are solved largely by maintaining a flow of water from anterior to posterior. Fans, consisting of the modified parapodia of segments 14 through 16, generate the flow. Particulate food drawn in by the current is trapped by a web of mucus continuously secreted by the enlarged notopodia of segment 12. The food enters a small pouch in the web, which lies within a food cup on the back of the worm, just ahead of the fans. When the mass of food within the mucous pouch reaches a diameter of about 3 mm, it is pinched off and pushed into a groove along the middle of the back. The fans then stop beating while cilia in the groove move the mucus-wrapped food to the mouth. *Chaetopterus* is clearly one of the most tagmatized of annelids.

The **lugworm** *Arenicola*, which is common in intertidal mud flats, has a less elaborate solution to the problems of living in a burrow (Figure 29.11). Peristaltic contractions of the lugworm's body draw water into the posterior end of its L-shaped burrow. The unlined burrow is closed near the worm's head; water simply percolates out through the sand or mud. The gills, which are modified notopodia on some of the segments, absorb oxygen from the water. *Arenicola* does not filter food from the water but feeds by continually ingesting sand or mud with its eversible proboscis. The caving in of sand or mud above the head forms a disc-shaped depression that, together with the mound of fecal castings around the nearby opening of the burrow, is easily recognized at low tide. By continually cycling sediments through its body, the lugworm plays a role in shallow marine environments similar to that played by earthworms in terrestrial environments. In areas where lugworms are abundant, they bring to the surface some 1900 tons of sand per acre (691,000 kg per hectare) each year and bury an equal amount of sediment. On average, all the material within the upper 50 cm of the bottom recycles through lugworms every two years.

EARTHWORMS AND OTHER OLIGOCHAETES

The origin of oligochaetes is uncertain, but it is not difficult to imagine sedentary polychaetes similar to lugworms burrowing generation after generation from the bottoms of shallow bays into the beds of freshwater streams, and eventually into moist soil far from flowing water. Such worms would have become increasingly adapted to fresh water, and their delicate tentacles, gills, and parapodia would have become reduced in the abrasive soil. In this way *Arenicola*-like polychaetes may have evolved into the roughly 3000 species of oligochaetes. A few species of oligochaetes have apparently retraced this evolutionary history and now live in burrows or tubes in freshwater or marine sediments.

One of the most common of the freshwater oligochaetes is *Tubifex tubifex*, familiar to aquarists as a fish food. These red worms are up to 10 cm long and live on the bottoms of lakes, duck ponds, and polluted streams. They keep their heads within tubes on the bottom, while waving their bright red tails in the water (Figure 29.12). In certain polluted tidal rivers, such as the Thames at London, tubifex worms are so numerous that the banks appear bright red at low tide. *Tubifex* is also common at sewage outflows and settling tanks. Often *Tubifex* is the only animal seen in its habitat, for few others are so tolerant of low oxygen levels. The ability of *Tubifex* to extract oxygen is due to the high concentration of hemoglobin dissolved in its blood and to the corkscrew motion of its tail, which draws aerated water downward. Surprisingly, *Tubifex* does not absorb oxygen through its integument, but from the lining of its intestine. To make the paradox complete, *Tubifex* obtains all its nutrients by absorbing dissolved organic material across its integument.

Figure 29.8
(A) Feather dusters (sabellids). Each worm is approximately 5 cm long and lives in a tube 10 cm long, into which it withdraws when endangered. (B) Filter feeding by Sabella. *Arrows show the direction of water currents produced by cilia on the numerous pinnules on each radiole. (C) Movement and sorting of particles in the food groove of a radiole.*

A

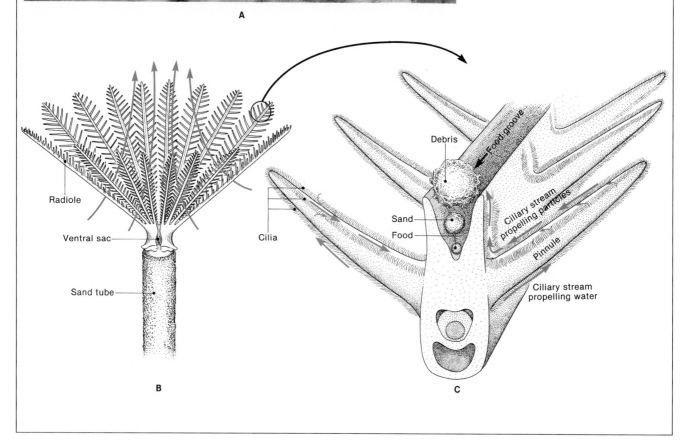

Radiole

Ventral sac

Sand tube

Cilia

B

Debris

Food groove

Sand

Food

Ciliary stream propelling particles

Pinnule

Ciliary stream propelling water

C

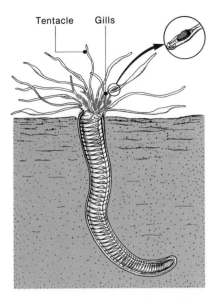

Figure 29.9
Amphitrite *feeding with its long tentacles. The inset shows food being propelled down the groove of the tentacle by means of cilia. The smaller appendages on the head are gills, which are blood-red with hemoglobin. The worm is approximately 35 cm long.*

Figure 29.10
A model of the parchment worm Chaetopterus. *Parchment worms are often three times as long as this 7-cm specimen. The smaller worm with* Chaetopterus *is the sipunculan* Phascolosoma.

The Night Crawler. The majority of oligochaetes are not aquatic like *Tubifex* but spend most of their lives burrowing through soil. These terrestrial earthworms range in length to more than 3 meters (Figure 29.13). More typical in length, however, is the night crawler *Lumbricus terrestris,* favorite of robins, fishermen, and biology teachers. *Lumbricus* (Figure 29.14) is up to 30 cm long when mature. It usually burrows within the upper 30 cm of moist soils rich in organic material, but it can descend to 2 meters to escape drying or cold. On warm, damp nights *Lumbricus* extends the anterior portion of its body out of the burrow to forage for leaves and stems or to mate. They can often be seen on the surface under these conditions, although most of them quickly retreat into the burrow if a light is shone on them. Except after a heavy rain, *Lumbricus* spends its days within the burrow. There it feeds on organic material in the soil or on the vegetable matter it has previously pulled into the burrow. The fecal castings near the entrances of the burrows are often the only clues most people have of the vast numbers of earthworms living beneath their feet. According to one estimate, 54,000 earthworms might inhabit a single acre in England.

Lumbricus and other terrestrial earthworms amply repay the soil for shelter. Charles Darwin estimated that the dry weight of soil turned over annually by earthworms was as much as 18.1 tons per acre. At the same time earthworms buried a corresponding amount of vegetable matter, forming up to 0.2 inch of humus per year on an acre of English soil. Like Darwin's previous studies of evolution and the formation of coral reefs, his studies of earthworms show how much change can be effected by small actions multiplied countless times.

Burrowing. Earthworms have two methods of burrowing. In hard soil they literally eat their way through. In soft soil and on surfaces earthworms move by means of peristaltic contractions (see Figure 10.2). This mode of locomotion uses circular and longitudinal muscles to alternately extend and thicken each segment. Four pairs of setae on each somite (except the first and last) dig into the soil and prevent backsliding. First the earthworm contracts the circular muscles of the anterior segments. (There is no head.) This narrows and lengthens the first segments, which then probe into the soil ahead. The longitudinal muscles of these segments then contract, thickening the segments and widening the channel. As the segments thicken they poke the setae into the walls of the burrow, preventing them from being pulled backward when they shorten. In the meantime the circular muscles contract in the segments farther back, narrowing and lengthening those segments so that they are pulled forward when the anterior segments shorten. The longitudinal muscles in those segments then contract and pull them forward, while circular muscles farther back contract. This sequence of alternating contractions of circular and longitudinal muscles moves rearward like a wave, pulling the worm smoothly through the soil.

Peristaltic burrowing works only because septa completely separate the coelomic cavities of most of the segments (except between segments 1 through 4, and 17 and 18). Without the septa, contraction of the circular muscles in one segment would narrow that segment but would not lengthen it, because the pressure would be dissipated over the entire length of the animal. Likewise, contraction of the longitudinal muscles of one segment would not thicken the segment to widen the burrow and anchor the setae. Segmentation is therefore essential for this type of locomotion. The setae, in spite of small sizes and numbers, contribute significantly to burrowing. The force they generate is evident from the vigorous tugging robins must use to extract an earthworm from its burrow.

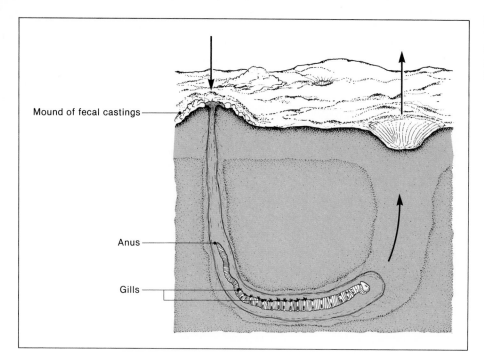

Figure 29.11
The lugworm Arenicola *in its burrow. Arrows show the direction of water movement. The worm is approximately 30 cm long.*

Mound of fecal castings

Anus

Gills

Integument. Burrowing through soil is feasible only in an animal with an integument that can resist abrasion by soil particles. The integument of earthworms has a protective outer layer of thin, collagenous cuticle secreted by the epidermis (Figures 29.14 and 29.15). Along the midline of each segment is a dorsal pore that allows coelomic fluid to leak out slowly and moisten the cuticle. The epidermis also includes numerous glands that secrete mucus that lubricates the cuticle and protects it from drying. In cuticle that has been stripped off and allowed to dry on a microscope slide, it is easy to see the pores leading to the mucous glands, and also pores leading to numerous sensory cells. These sensory cells presumably include chemoreceptors and mechanoreceptors, which would be of obvious use in a burrowing animal. Earthworms have no eyes, but there are numerous photoreceptor cells in the epidermis of each segment.

In spite of the lubricating and moisturizing secretions, the delicate integument is certainly vulnerable to injury. A more armorlike cuticle such as that in insects would, however, make the segments inflexible and would therefore interfere with burrowing. Moreover, a thicker cuticle would interfere with **integumentary respiration.** Since parapodia and gills would be even more of a liability in burrowing, earthworms must rely exclusively on integumentary respiration. The uptake of oxygen is aided by an extensive system of capillaries beneath the cuticle, by the presence of hemoglobin dissolved in the blood plasma, and by the efficient closed circulatory system with its five aortic arches ("hearts") that aid the dorsal vessel in pumping (Figure 29.16).

Digestion. Considering the amounts of food and soil that pass through them, earthworms' digestive systems are surprisingly simple. The digestive tract of *Lumbricus* starts with the mouth, which lies beneath the prostomium in the first segment. Food is pumped by the pharynx into the esophagus and is stored in the crop. (On each side of the esophagus are calciferous glands, which are thought to excrete excess calcium from the body fluids.) Following the crop in *Lumbricus* is a muscular gizzard lined on the inside with cuticle, where food is ground up. Food

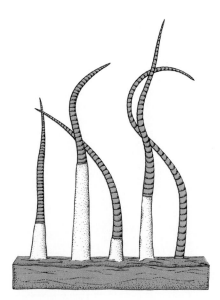

Figure 29.12
The oligochaete Tubifex tubifex, *with its head inside its tube. A spiraling motion of the tail brings oxygen-bearing water downward, and the oxygen is absorbed by the gut.*

Figure 29.13
An Australian giant earthworm Megascolides australis *being extracted from its burrow. The two men located it by hearing the gurgling sounds of burrowing.*

Figure 29.14
Structure of the earthworm Lumbricus terrestris. *(A) External structure. Some of the 115 to 200 segments are numbered here and will be mentioned later. The ventral surface is somewhat flattened and is not as dark red as the dorsal surface. (B) Cross section showing the muscles, setae, and gut, all used in burrowing.*

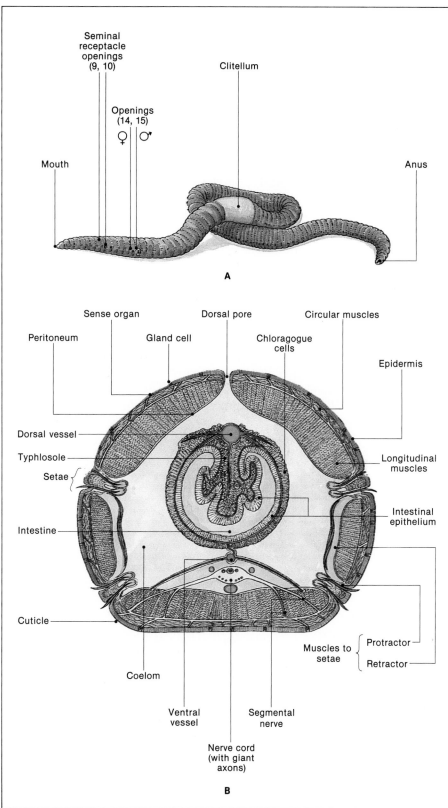

A feature characteristic of many oligochaetes is an inward fold along the length of the intestine, the **typhlosole,** which increases the area for absorption of nutrients (Figure 29.14B). Surrounding the outside of the intestine are chloragogue cells, which store nutrients and eliminate nitrogenous wastes, as described previously for polychaetes. The nephridia of freshwater and terrestrial oligochaetes are much like those of marine polychaetes except that they eliminate larger amounts of water during osmoregulation.

Reproduction. Oligochaetes clearly must have a method of reproduction different from that of their aquatic ancestors. Releasing large numbers of gametes into the environment and trusting to chance simply will not work in soil. Unlike polychaetes, which rely on the coelom and nephridia for reproduction, each oligochaete has specialized reproductive organs of both sexes. Although hermaphroditic, oligochaetes usually mate with each other rather than themselves. Pairs of *Lumbricus* mate by extending the anterior portions of their bodies from their neighboring burrows and exchanging sperm. Copulation lasts two to three hours, and an individual can copulate every three or four nights if the air is sufficiently warm and damp to allow them to stay out. During copulation the two worms are held in mutual embrace by mucus secreted from the clitellum, which lies on somites 32 through 37 (Figure 29.17). The clitellum also has specialized **genital setae** that clasp the mate by penetrating its integument. The clitellum is conspicuous and permanent in sexually mature *Lumbricus* but is temporary in earthworms that breed only during particular times of the year.

Sperm are produced by two pairs of small testes, which are in segments 10 and 11 of *Lumbricus* (Figures 29.16 and 29.18A). After maturing within the seminal vesicles (= sperm sacs), sperm pass through funnel-shaped openings into sperm ducts. During copulation cilia lining the sperm ducts propel the sperm through the male pore on the ventral surface of somite 15. The sperm from each worm are then deposited in its mate's **seminal receptacles** in somites 9 and 10. It might

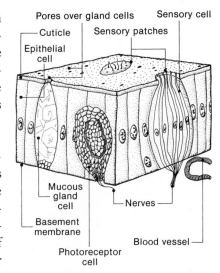

Figure 29.15
The integument of Lumbricus.

Figure 29.16
Cutaway view of the internal organs of Lumbricus. *Note the five aortic arches ("hearts") joining the dorsal and ventral vessels in segments 7 through 11.*

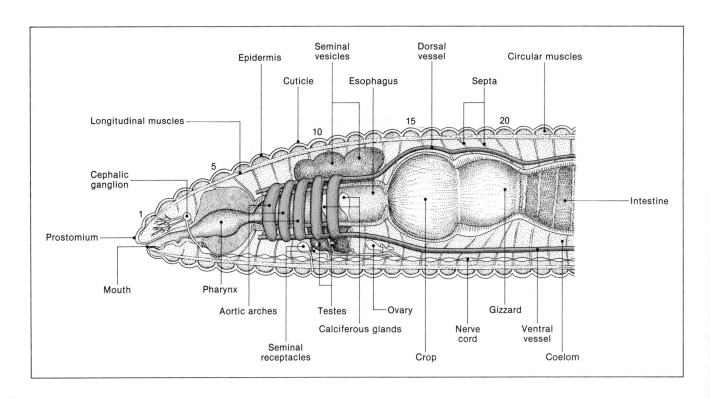

Figure 29.17
Two earthworms, Lumbricus
terrestris, *copulating.*

appear that the easiest way to exchange sperm would be for two earthworms to lie beside each other with each male pore releasing sperm directly into the mate's seminal receptacles. Many earthworms do, in fact, mate in this position. *Lumbricus,* however, copulates with the male pore opposite the clitellum of its mate, so the sperm must journey some 20 segments to reach the seminal receptacles of the mate (Figure 29.18B). The sperm pass through a **seminal groove:** a wrinkle on the ventral surface formed by contractions of certain muscles. Contractions of these muscles propel the sperm through the groove. The groove is enclosed by a tube of slime surrounding each worm, so sperm from each mate do not mix as they pass each other.

After copulating, the worms separate and return to their burrows. Fertilization and egg-laying occur a few days later, in one of the most remarkable reproductive processes among animals. First each worm secretes a sheath of mucus around its clitellum and anterior segments (Figure 29.18C). The clitellum then secretes albumen as nourishment for the eggs and envelops the mucus in a tough chitinlike material that will form a **cocoon** for the eggs. The worm then backs out of the cocoon. As the cocoon slips over segment 14 it receives an ovum from a pore leading to the oviduct. (The ovum was previously produced by an ovary in segment 13, and it matured in the coelom.) As the cocoon continues to slide, it picks up sperm from the sperm receptacles in segments 9 and 10. Once the cocoon has completely slipped off, its ends close, forming a yellowish, lemon-shaped body approximately 7 mm long. Fertilization occurs within the cocoon. Two to three weeks later one new worm emerges.

LEECHES

Like earthworms, leeches are hermaphroditic and have clitella that form cocoons. In most other ways they are quite different. Many of these differences appear to have resulted from adaptations of leeches to aquatic life, mainly in fresh water. Peristaltic burrowing is no longer appropriate, and most leeches lose their setae

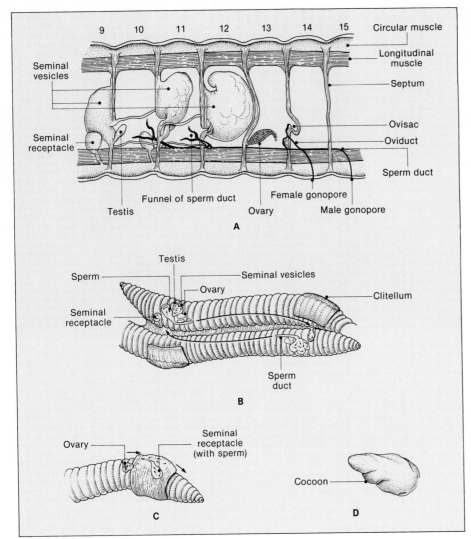

Figure 29.18
*Reproduction in Lumbricus.
(A) During copulation the fertile
male somites (10 and 11) discharge
sperm through a pore on somite 15.
Later the fertile female somite (13)
releases ova that mature in the
coelom, then exit through a pore in
somite 14. (B) Copulation. Sperm
(dashed line) must travel outside the
body to reach the seminal receptacle
of the mate. (C) Fertilization occurs
at the time the cocoon forms. As the
cocoon slides anteriorly from the
clitellum it picks up ova, then sperm
from the seminal receptacles. (D) A
cocoon.*

and intersegmental septa during embryonic development. The coelom functions
as a single large chamber, which is largely occupied by spongy tissue and dorsoventral muscles. The latter muscles, in addition to diagonal, circular, and longitudinal muscles, allow leeches to swim quite gracefully by means of dorsoventral
undulations of the body (Figure 29.19). Some leeches are adapted to terrestrial
living, at least part of the time. On solid substratum they move in inchworm
fashion, alternately attaching and detaching anterior and posterior suckers.
Leeches also differ from oligochaetes in that the segments are constant in number
and are often subdivided into numerous **annuli.** Since the fissures between annuli
look like those between segments, it is hard to determine from appearance where
the boundaries of each segment are. Internally, however, each segment can be
defined by the fact that it is innervated by one ganglion of the ventral nerve cord.

Feeding.   Because of their reputations as blood suckers, leeches do not enjoy
much esteem among humans. However, the majority of the 500 or so species of
leeches are not ectoparasites on humans. Many lack cutting or probing mouthparts
and must eat worms, invertebrate larvae, the eggs of fishes and amphibians, or
other small prey. Many also scavenge carrion. Many leeches are highly specific
for their hosts and feed on the body fluids of invertebrates (including leeches of

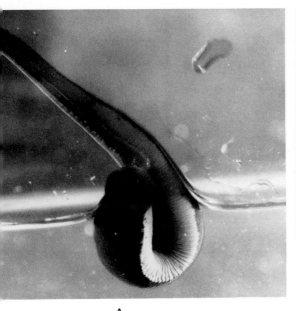

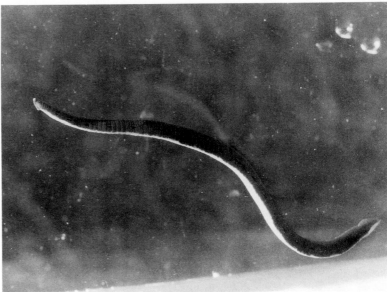

A

B

Figure 29.19
Locomotion in the American medicinal leech Macrobdella decora. *(A) Using the anterior and posterior suckers to creep up the side of an aquarium. (B) Swimming by dorsoventral undulations of the body. Extended length approximately 10 cm.*

Figure 29.20
The medicinal leech Hirudo medicinalis *removing accumulated blood (a hematoma). This European species grows up to 20 cm long.*

other species) or fishes, but not on mammals. Some leeches, however, can and do consume the blood of mammals, including humans. The best known of these is the medicinal leech, *Hirudo medicinalis* (Figure 29.20). The term "medicinal" comes from the fact that up until a century ago (and to some extent even now) leeches were used to treat a variety of human ailments. The theory was that somehow leeches could suck out disease-causing "bad blood." This ability was so prized that medicinal leeches were collected almost to extinction in Europe, where they are now protected by law. They were introduced into America, but apparently without lasting success. In spite of their relative rarity in nature, they are widely cultured in laboratories for research.

The ability of some leeches to parasitize humans, other mammals, and occasionally other vertebrates is due to a variety of adaptations that allow them to pierce or cut into the skin without notice. Leeches attach to their hosts by means of suckers. Some then pierce the skin of the host with a sharp proboscis. Others

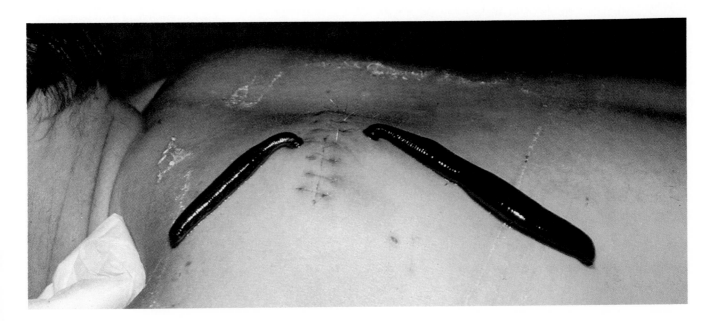

cut into the skin with sharp jaws. *Hirudo* uses three sharp-toothed jaws like circular-saw blades to make a Y-shaped incision in skin (Figure 29.21A). While cutting the skin, it secretes an unidentified local anesthetic, as well as a histamine-like substance that keeps the blood vessels open. The blood is then sucked in by the pharynx. While the blood is being swallowed, it mixes with a protein called **hirudin,** which prevents clotting by inhibiting the action of thrombin (see p. 234). Without this anticoagulant the blood clot would interfere with locomotion.

A medicinal leech takes up to 15 ml of blood during a single feeding. Much of this volume is essentially salt water, which the 17 pairs of nephridia soon eliminate. The concentrated blood is then stored within numerous diverticula of the crop. The proteins of hemoglobin and other components of the blood are then digested extremely slowly. In one study *Hirudo* took 200 days to digest a single meal, and it continued to live for another 100 days on the energy stored from that feeding. One unexpected feature of digestion in leeches is that their guts do not secrete any proteolytic enzymes. Instead, they rely on bacteria in the gut for enzymatic digestion. In *H. medicinalis* only one species of bacterium, *Pseudomonas hirudinis*, slowly digests the blood. *Pseudomonas hirudinis* also produces an unidentified antibiotic that eliminates any other bacteria from the digestive tract, allowing red blood cells to remain unspoiled within the leech intestine for more than a year. Because of the antibacterial action of *P. hirudinis*, one need not worry about getting a bacterial infection from leech bites. Some disease-causing protozoans and tapeworm cysts, however, are transmitted by leeches.

Circulation. The ability of *Hirudo* to live for many months on a single meal clearly indicates a low metabolic rate. Its circulatory system is reduced accordingly. In many leeches, including *Hirudo*, there are not even any blood vessels. The coelomic fluid, which pulsates slowly through sinuses in the tissue-filled coelom,

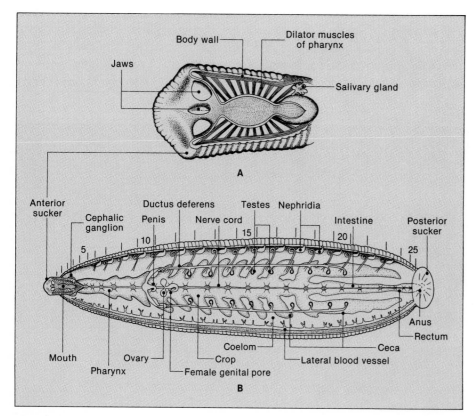

Figure 29.21
(A) The feeding apparatus of the medicinal leech Hirudo medicinalis. *The three toothed jaws slice through skin. The dilator muscles of the pharynx then contract to suck blood. (B) Ventral view of the internal structure of* H. medicinalis. *Vertical lines and numbers (top) indicate segments.*

performs the functions of blood. (Since the circulatory system is open, the coelom is also considered to be a **hemocoel.**) In *Hirudo* and many other leeches, the coelomic fluid contains hemoglobin. (It is produced by the leech rather than obtained from the host.) A few leeches possess gills, but *Hirudo* and most others exchange gases through the integument.

Nervous System.　The nervous system of leeches is typically annelidan except that there are no giant axons. Aquatic leeches generally have mechanoreceptors that respond to water currents and vibrations, and several pairs of eyes on the first segment that detect shadows of potential prey. There are also chemoreceptors and temperature receptors. Apparently hungry leeches attach to virtually any moving object detected by the mechano- and photoreceptors, but they drop off if the chemo- and thermoreceptors indicate that the object is not appropriate prey. *Hirudo medicinalis* also has well-defined receptors for touch, pressure, and injurious stimuli concentrated in the central annulus of each segment. These 14 and the other 340 or so nerve cells in each segment are connected in the central nervous system in the same way from one individual to the next. This is one reason leeches are favorite subjects of neurophysiologists. Another reason is that individual nerve cells readily regenerate.

Reproduction.　Reproduction in leeches is similar to that in oligochaetes, but some differences are worth noting. Some leeches merely deposit **spermatophores** on each other's bodies during mating, and the sperm bore through the integument, migrate through the coelom, and fertilize the eggs. In *Hirudo* and many other leeches, however, each individual has both a penis and a vagina. Fertilization can be reciprocal if two leeches happen to copulate in a head-to-tail position, but this does not always occur. Between one and nine months after copulation, *Hirudo* forms a cocoon and deposits it on land. Apparently during this time the parent also inoculates the cocoon with *P. hirudinis* bacteria that will enable the young leeches to begin their parasitism immediately. Like most oligochaetes, leeches are incapable of asexual reproduction. Unlike oligochaetes, they are incapable of regenerating lost body parts.

INTERACTIONS WITH HUMANS AND OTHER ANIMALS

Annelids attract little notice from most people, perhaps because so few of them are edible to us, and vice versa. Indirectly, however, annelids certainly have a profound effect on humans and other animals. Even if we ignore the polychaetes, about which there is little ecological knowledge, the role of annelids is impressive. The effect of earthworms in fertilizing and aerating soil has already been noted (Lee 1985; Satchell 1983). They are so beneficial that in some areas where they do not occur naturally, or where they have been eliminated by overuse of poisons, farmers pay to have them brought in. Earthworms have also been put to work reclaiming mined soils. The night crawler *Lumbricus* is also the basis for a bait-worm industry worth tens of millions of dollars. This figure does not include the thousands of entrepreneurs who dig worms in the backyard and sell them at roadside.

　　Although earthworms are not a major item in the human diet, they do feed a large number of animals that are important to us. *Turdus*, a genus that includes the American robin and the European blackbird, comes immediately to mind. Less familiar as predators on earthworms are burrowing mammals such as moles and shrews. Even some foxes eat earthworms.

Leeches have less human impact, which suits most people. In North America they are occasionally bothersome; in other areas they can be life-threatening. In areas around the Mediterranean, *Limnatis nilotica* invades the tracheae and esophagi of horses, cattle, and even humans who drink from springs and ponds that the leeches inhabit. Large numbers can cause asphyxiation. Even more distressing are land leeches in the genus *Haemadipsa*, which live in moist tropical areas. They detect groups of mammals by means of odors, vibrations, and warm air currents and actually creep toward them. They can penetrate small openings in clothing and attach to victims undetected. The worst comes after the leeches fall off, because the anticoagulant causes the wounds to bleed severely.

Even leeches have their virtues, however. Although the curative powers of the medicinal leech fell into disrepute because of excessive claims in the previous century, *Hirudo medicinalis* still has its valid medical uses. Leeches would be of even greater medical service if hirudin could be extracted in large enough quantities to prevent blood clots, but it is not practical to extract the anticoagulant from such small animals. The gene for hirudin has been cloned, however, so this problem may be solved. In the meantime, Roy T. Sawyer went to French Guiana in 1977 in pursuit of the Amazon leech *Haementeria ghilianii*, which would be large enough to supply large amounts of hirudin (Figure 29.22). Sawyer (1986) found his leeches, and eventually he and his colleagues discovered that this species produces a previously unknown anticoagulant. This enzyme, called hementin, breaks down fibrin strands. Unlike hirudin, therefore, it is effective treatment for heart attacks even after clots have already formed. Sawyer has started a leech farm in Wales that markets several useful leech products.

Figure 29.22
Dr. Roy T. Sawyer and friend, the giant Amazon leech Haementeria.

SUMMARY

Annelids include such diverse animals as polychaetes, oligochaetes, and leeches. In spite of their differences, all share the feature of segmentation, which may have evolved for burrowing or as a means of achieving larger size without all the organs growing larger. Polychaetes are either errant or sedentary. The errant polychaetes can wriggle and can swim by means of parapodia. Many prey by means of an eversible proboscis. The sedentary polychaetes, such as annelid tube worms, generally live in burrows or secreted tubes, and many feed on particulate matter by means of tentacles.

Earthworms (oligochaetes) burrow in soil either by eating their way through or by alternately contracting circular and round muscles, using their few setae to grip the sides of the burrow. Because the coelom is segmented into compartments, each segment can narrow or shorten independently of others during burrowing. Earthworms feed on vegetable material. They are hermaphroditic but generally exchange sperm. Leeches are mainly aquatic and either creep about with their suckers or swim. Many prey on smaller animals, but some suck the blood of vertebrates.

KEY TERMS

segment
tagma
seta
parapodium

clitellum
septum
mesentery
typhlosole

chloragogue cells
epitoky
radiole

SELF-TEST

1. Explain what segmentation is. What is the importance of segmentation in annelids? How is segmentation reflected in the structure of the nervous system, metanephridia, and the coelom?

2. What are the three classes of annelids? Briefly state the differences among them.

3. Describe how one member of each class obtains food. Sketch the feeding apparatus for each one.

4. Describe the means of locomotion in errant polychaetes, in earthworms, and in leeches. Explain how the coelom is adapted to locomotion in earthworms and leeches. What about locomotion in sedentary polychaetes?

5. What are some of the advantages and disadvantages of a sedentary life in polychaetes? How are the disadvantages overcome?

6. Compare sexual reproduction in a marine polychaete with that in a terrestrial earthworm. Explain how each method of reproduction is appropriate to the habitat. Why do you think leeches reproduce in a manner similar to that of earthworms?

READINGS

RECOMMENDED READINGS

Darwin, C. 1881. *The Formation of Vegetable Mould through the Action of Worms, with Observations on their Habits.*

Nicholls, J. G. and D. Van Essen. 1974. The nervous system of the leech. *Sci. Am.* 230(1):38–48 (Jan).

Stent, G. S. and D. A. Weisblat. 1982. The development of a simple nervous system. *Sci. Am.* 246(1):136–146 (Jan). (*In the leech.*)

See also relevant selections in General References at the end of Chapter 21.

ADDITIONAL REFERENCES

Lee, K. E. 1985. *Earthworms: Their Ecology and Relationships with Soils and Land Use.* Orlando: Academic Press.

Lent, C. M. and M. H. Dickinson. 1988. The neurobiology of feeding in leeches. *Sci. Am.* 258(6):98–103 (June).

Mill, P. J. (Ed.). 1978. *Physiology of Annelids.* New York: Academic Press.

Nicholls, J. G. 1984. *The Search for Connections: Studies of Regeneration in the Nervous System of the Leech.* Sunderland, MA: Sinauer Associates.

Pennak, R. W. 1989. *Freshwater Invertebrates of the United States*, Vol. 1, 3rd ed. New York: Wiley, Chap. 13.

Satchell, J. E. (Ed.). 1983. *Earthworm Ecology.* New York: Chapman and Hall.

Sawyer, R. T. 1986. *Leech Biology and Behaviour*, 3 vols. New York: Oxford University Press.

30

Introduction to Arthropods: Spiders and Other Chelicerates

Jumping spider (Marpissa mucosa).

CHAPTER OUTLINE

LEARNING OBJECTIVES

1. Why are animals as different as spiders, crayfish, and insects all considered to be arthropods?

2. How were arthropods able to adapt to life on land?

3. How is the integument constructed? How does an arthropod grow within its integument?

4. How are trilobites, spiders, harvestmen, scorpions, ticks, and mites related to each other?

5. How does a spider bite?

6. How do spiders make webs? How do they use them? What else do they do with silk?

GENERAL FEATURES OF ARTHROPODS

There are so many spiders, crustaceans, insects, and other arthropods that it will require three chapters to introduce them (Figure 30.1). This great variety seems almost like an outburst of evolutionary exuberance, as if nature were celebrating having at last perfected and completed the protostome division. It is more likely, however, that the enormous diversity and numbers of arthropods are due to their success in invading land and adapting to the great variety of terrestrial habitats.

A major key to the success of arthropods is the integument, which is made up largely of rigid **cuticle**. The integument reduces the loss of water from the body and serves as a supportive exoskeleton and protective armor. Because cuticle is rigid, however, it imposes a number of other characteristic features on arthropods. Locomotion in the manner of annelids and other soft-bodied invertebrates is out of the question, so arthropods get their mobility (and their name) from jointed appendages. Since the cuticle serves as an exoskeleton, a pressurized hydroskeleton is unnecessary, and the coelom is greatly reduced. Some other features of arthropods are found also in annelids. As in annelids, the central nervous system consists fundamentally of a ventral nerve cord with segmental ganglia. Most arthropods also have compound eyes (each with numerous lenses), which occur in a few annelids, but in no other group. Unlike annelids, however, arthropods have open circulatory systems, and they lack motile cilia (Table 30.1)

Figure 30.1
Some representative arthropods, showing the chitinous cuticle and jointed appendages. (A) The six-spotted fishing spider Dolomedes triton *on water. Leg spread approximately 3 cm. (B) A crayfish* Orconectes. *Length approximately 10 cm. (C) The centipede* Lithobius. *Approximately 3 cm long. (D) The southeastern lubber grasshopper* Romalea microptera. *Length 6 cm.*

A

B

C

D

Table 30.1 Characteristics of arthropods.

Phylum Arthropoda are-THROP-uh-duh (Greek *arthron* joint + *pod-* pertaining to the foot).

Morphology: Coelomate, although coelom reduced to portions of reproductive and excretory systems. Major body cavity is the **hemocoel,** consisting of blood-filled spaces among tissues. Integument with **chitinous cuticle** that serves as **exoskeleton.** Entire integument **molts** periodically. Body **segmented,** with many segments altered as **tagmata** (head, thorax, cephalothorax, abdomen). Each segment with a pair of primitively **uniramous or biramous appendages** that may be lost or greatly modified during development.

Physiology: Gaseous exchange through **tracheal tubes, gills, book lungs,** or **body surface.** Digestive tract complete. Central nervous system usually a chain of ganglia on a ventral nerve cord. Cerebral ganglion located dorsally, usually with inputs from various combinations of compound and simple eyes. Circulatory system open. Heart consisting of a pulsatile vessel along the dorsal midline, with blood entering through one-way **ostia.** Osmoregulation by **Malpighian tubules** and/or glands.

Locomotion: Legs jointed and moved by striated muscles attached to cuticle. Motile cilia absent except in sperm of a few groups.

Reproduction: Generally dioecious. Sexual dimorphism common. Fertilization internal. Usually oviparous.

Development: Superficial cleavage pattern (see p. 116). Protostomate. Development complete or with characteristic larval stages.

Habitat, Size, and Diversity: Crustaceans generally aquatic; others generally terrestrial. Distributed worldwide. Body length (excluding legs) from less the 0.1 mm to 60 cm. Approximately **874,000 living species** described; many more undescribed.

THE CUTICLE OF ARTHROPODS

Structure.   As previously noted, the cuticle is a major distinction of arthropods. Cuticles are also found in other phyla, such as annelids and onychophora, but they are different in composition. The arthropod cuticle consists largely of **chitin** and proteins, which are secreted by the epidermis. In crustaceans the cuticle also incorporates calcium phosphate and calcium carbonate. Chitin is a polysaccharide made of repeated subunits of *N*-acetylglucosamine that line up as crystalline, hexagonal rods. Bundles of these rods held together in a protein matrix account for the shape and strength of cuticle (Figure 30.2A,B).

Chitin occurs only in the two inner layers of cuticle, the **endocuticle** and **exocuticle** (Figure 30.2C). The difference between the two layers is that endocuticle remains pliable, while exocuticle becomes hardened, or **sclerotized,** in most areas. Sclerotization is also called **tanning.** Like the preparation of leather, the tanning of arthropod cuticle results from the cross-linking of proteins. Covering the exocuticle is a thin, sclerotized **epicuticle,** which consists of proteins and waxes (fatty acids linked to alcohols). The waxes in the epicuticle retard water loss. Generally the more arid the environment the greater the amount of wax in the epicuticle.

The cuticle of arthropods is also responsible for their often spectacular colors that serve in camouflage, recognition, and warning coloration. These colors arise in two major ways: from pigments and from the fine structure of the cuticle. Pigments in the exocuticle or in the wax are generally responsible for long-wavelength colors: brown, red, orange, and yellow. Short-wavelength colors (green, blue, violet) are generally **structural colors** (see Table 10.1, p. 199).

Running through the exocuticle and endocuticle are **pore canals** and **wax canals** that are thought to be paths by which the epidermis secretes waxes onto the epicuticle. The pore canals are also thought to be the routes by which calcium

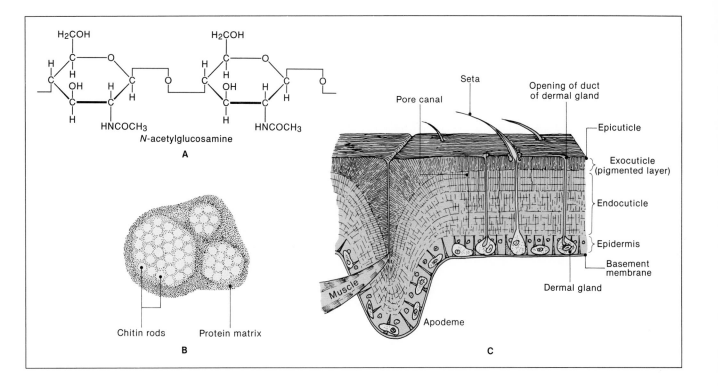

Figure 30.2
The structure of chitin and the cuticle. (A) Chitin is made of chains of N-acetylglucosamine. (B) Crystalline rods of chitin held together by protein form the main structural elements of cuticle. The cuticle is reinforced by calcium carbonate in most crustaceans. (C) Cuticle, secreted by the epidermis, has three layers: the chitinous endocuticle and exocuticle, and the lipid-rich epicuticle.

salts become incorporated into the endocuticle in the exoskeletons of crustaceans. There are also canals leading from **dermal glands,** whose function is unknown. Parts of the cuticle are modified into a variety of sensory receptors that enable the arthropod to know what is going on outside. Pores through the cuticle allow detection of chemicals by chemoreceptor cells, and various deformable plates, slits, and hairlike setae connect to mechanoreceptor cells.

The Cuticle as Exoskeleton.   Like pieces of armor, sclerotized cuticle is linked together as plates (= **sclerites**) over the body surface and as tubes around appendages (Figure 30.3A). The combination of sclerotized plates and tubes makes up the exoskeleton. Projecting internally from the exoskeleton are **apodemes,** to which muscles attach (Figure 30.2C). Movement of the body is possible because the sclerites and tubes are hinged together by a type of cuticle called **articular membrane** (= arthrodial membrane) (Figure 30.3B). Articular membrane is flexible and transparent because it lacks exocuticle.

Molting.   The arthropod exoskeleton is so useful for support and protection that one might wonder why it doesn't occur in all animals. One explanation is that it is difficult to grow while encased in a rigid exoskeleton. Molluscs solve this problem by continually enlarging their shells as they grow. Arthropods periodically replace the cuticle when they outgrow it. The process of shedding the old cuticle is called molting or **ecdysis** (EK-dis-sis; Greek *ekdysis* an escape). Insects go through a fixed number of molts until they develop into adults, which do not molt. Other arthropods, such as spiders, molt an indefinite number of times, both as juveniles and adults.

Molting is initiated in some species by environmental cues, and in others by increased pressure in the body, due to tissue growth or a meal. These triggers cause the release of the molting hormone **ecdysone** (see p. 194), which causes the epidermis to separate from the old cuticle and begin secreting a new epicuticle

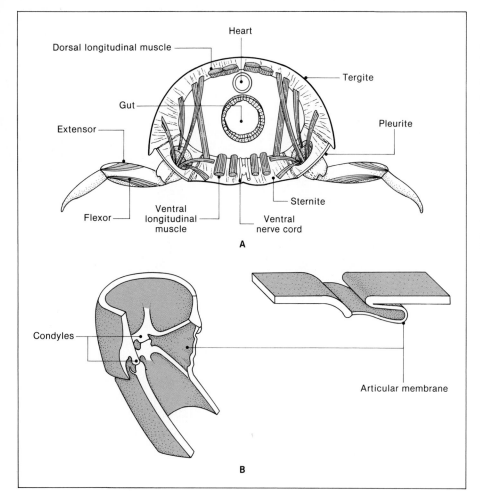

Figure 30.3
(A) Schematic representation of a primitive arthropod in cross section, showing cuticular plates (sclerites) and tubes and the attachment of some muscles. Each segment primitively has four cuticular plates: a dorsal tergite, a ventral sternite, and two lateral pleurites. Usually some of these plates combine or divide during development. Note the absence of circular muscles, which are not needed to maintain a hydroskeleton. Appendages are enclosed in cuticular tubes.
(B) Articular membranes occur between cuticular plates and at the joints of limbs. Condyles at joints compensate for the weakness of articular membranes.

(Figure 30.4). The epidermis then begins to secrete **molting fluid** containing protease and chitinase, which eat away at the old, nonsclerotized endocuticle. The products of this digestion are used in producing the new cuticle. The enzymes do not attack tanned exocuticle and new epicuticle or the nerve and muscle connections, so the animal retains its mobility and protection for as long as possible. After the old endocuticle separates from the new epicuticle the epidermis secretes a new **procuticle,** which will later differentiate into the new exocuticle and endocuticle. The old exocuticle is finally shed when the arthropod inflates itself with air or fluid, which ruptures the old exocuticle at predetermined fracture lines. Cuticles that line the gut and tracheae molt at the same time (Figure 30.4D).

It may require many minutes for the arthropod to extricate itself from its old exoskeleton. During that time it is especially vulnerable. Not only does the new unsclerotized cuticle lack protective hardness, but the arthropod must maintain an inflated posture during sclerotization in order for the cuticle to become large enough for future growth. The weakness of the exoskeleton just after molting is probably one factor that has limited the size of arthropods, especially terrestrial arthropods that do not have the benefit of buoyancy to help support their weight. The largest insect is only(!) about 30 cm long, and most are much smaller.

CLASSIFICATION

Living arthropods were traditionally divided into two classes, based largely on

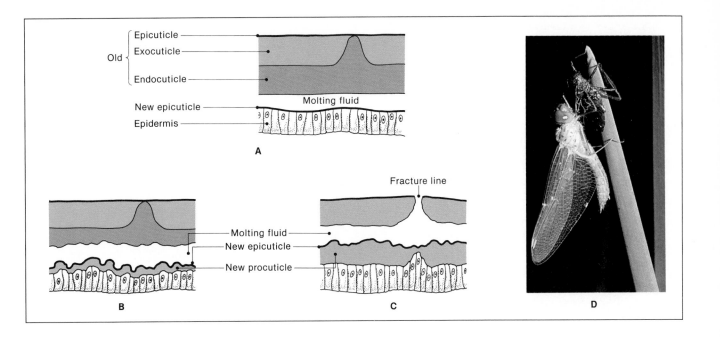

Figure 30.4
(A) Molting begins when the epidermis separates from the old cuticle, forms a new epicuticle, and secretes molting fluid. (B) Molting fluid eats away the old endocuticle. The epidermis then forms new procuticle, which will become new exocuticle and endocuticle. (C) Finally, the old cuticle ruptures. (D) A just-molted adult dragonfly clings to its old cuticle. The white strands are molted cuticle of tracheae.

differences in mouthparts. The class Chelicerata (ke-LIS-ur-AH-tuh) included the spiders and other arthropods with **chelicerae:** paired appendages near the mouth. Lacking mandibles for chewing food, most chelicerates predigest their food outside the body and suck it up as a semiliquid. The other class of arthropods was Mandibulata, which included crustaceans, insects, and other arthropods that generally have mandibles for chewing solid food.

Because of the obvious differences between crustaceans and insects, many zoologists were unhappy with lumping them together in the same class, but there did not appear to be any compelling reason to separate them until the 1960s. Then detailed study of the structure and function of arthropod appendages led S. M. Manton to divide the Mandibulata into two groups on the basis of their mouthparts. She noted that the jaws of crustaceans develop from the bases of appendages, with the distal portions of those appendages either disappearing or developing into nonchewing **palps.** In insects and some other mandibulates, on the other hand, entire appendages develop into jaws, and the tips of the appendages are used in chewing (Figure 30.5). Manton also noted that crustaceans and insects differ in the number of branches in their appendages. Crustaceans have **biramous** appendages, with essentially two branches, although one or both may be lost during development. In contrast, insects and related mandibulates have primitively **uniramous** (unbranched) appendages. Manton therefore proposed naming the latter group Uniramia, while retaining the name Crustacea for arthropods with biramous appendages.

In 1972 Manton went still further. Stressing the importance of the above differences, and noting the complete absence of any fossils linking Chelicerata, Crustacea, and Uniramia, she suggested that the three groups did not all evolve from a common ancestor that was an arthropod. In other words, she suggested that arthropods are **polyphyletic.** If that were the case the phylum Arthropoda would not be a **natural group** reflecting evolution, but simply an artificial group that zoologists find convenient. Therefore, she argued, Arthropoda should not be considered a single phylum but should be split into phylum Chelicerata, phylum Crustacea, and phylum Uniramia.

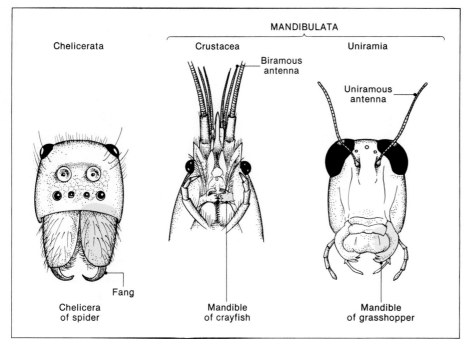

Figure 30.5
The morphological basis for arthropod classification. Spiders and other chelicerates are distinguished from both crustaceans and insects by their mouthparts, which have chelicerae instead of mandibles. The mandibles of crustaceans develop from only the proximal portions of appendages, whereas the mandibles of insects develop from entire appendages. Crustaceans are also distinguished from insects and other uniramians by the fact that their antennae and other appendages are primitively two-branched.

In the view of many zoologists, however, this is taking a good idea too far. While agreeing that the three groups differ greatly, they find the similarities among spiders, crustaceans, insects, and other arthropods sufficient evidence that they all evolved from an ancestral arthropod. If arthropods are not monophyletic, then the ancestors of spiders, crustaceans, and insects would independently have had to evolve the chitinous cuticle, compound eyes, reduced coelom, open circulation, and other arthropodan features. Molecular phylogeny and fossil evidence also show that animals as diverse as millipedes, horseshoe crabs, brine shrimp, and drosophila are indeed closely related to each other compared with annelids and many other protostomes (Briggs and Fortey 1989; Field et al. 1988). This text conforms to the view that Arthropoda is a single phylum. The force of Manton's arguments is acknowledged, however, by considering the Chelicerata, Crustacea, and Uniramia to be subphyla.

TRILOBITES

This text is concerned mainly with living animals, but one extinct group of arthropods is too important to overlook. Trilobites, usually classified in their own subphylum Trilobita, flourished during the 85 million years of the Cambrian period, then slowly declined. They finally vanished during the mass extinction at the end of the Paleozoic era, after more than 300 million years of existence. There were so many species and numbers of trilobites that they were certainly the dominant animals during much of the Paleozoic era.

Trilobites are named for the division of the chitinous carapace into three longitudinal lobes (see Figures 19.7 and 30.6). The known fossils range in length from 3 mm to almost 70 cm, but 2 to 7 cm is typical. The bodies of trilobites were clearly segmented and tagmatized into three distinct regions: the head (= **cephalon**), the thorax, and the **pygidium.** The head incorporated several fused segments and bore a pair of antennae, compound eyes, mouth, and legs. The heads of many trilobites were armed with long, sharp spines, suggesting that they were

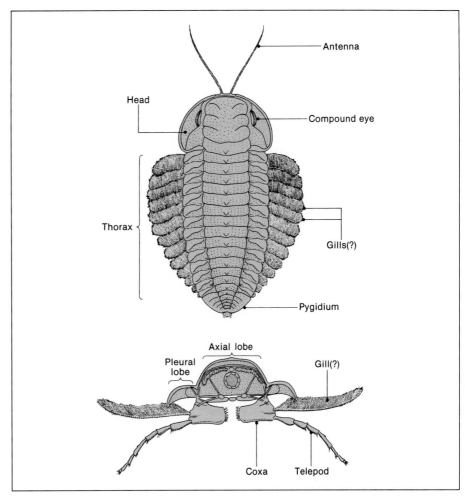

Figure 30.6
Triarthrus eatoni, *a trilobite that lived in the ocean above what is now Rome, New York, during the Ordovician period some 450 million years ago. Of the 4000 known species of trilobites, painstaking reconstructions from fossils have made this the best known. (Top) Dorsal view. (Bottom) Cross section. Body length approximately 4 cm. See also Figure 19.7.*

subject to predation. Many fossils are found with the thorax curled into a ball, as if these trilobites had been defending themselves in their final moments of life. Fossilized burrows and the shovel shape of the heads of some species suggest that some trilobites burrowed in marine sediments for food or protection. On the other hand, the dorsal position of the eyes, the adaptation of the appendages for walking, and numerous fossilized tracks indicate that most trilobites crawled. A few smaller species probably swam. The thorax consisted of 2 to 29 segments, with the number increasing as the animal grew. Each segment except the last bore a pair of biramous legs, with the ventral branch apparently adapted for locomotion. The dorsal branch was filamentous and may have served as a gill or a device for netting food.

CHELICERATES

Subphylum Chelicerata includes some 60,000 named species of horseshoe crabs, sea spiders, true spiders, scorpions, ticks, mites, and other arthropods with chelicerae. While the mandibles of other arthropods usually function as jaws for chewing solid food, chelicerae seize, pierce, or tear prey, which are then partially digested externally before being consumed as a liquid. Often feeding is aided by a pair of appendages called **pedipalps** near the mouth. Antennae are lacking in all chelicerates. Although this combination of anterior appendages defines the Chelicerata, most people do not get close enough to see them. Instead, they recognize chelicerates by the four pairs of walking legs, and the head and thorax fused into a **cephalothorax.**

Extant chelicerates are divided into three classes. Class Merostomata includes the horseshoe crabs, which are, of course, not really crabs at all. They are frequently encountered along Atlantic beaches when they come ashore to mate. Class Pycnogonida includes so-called sea spiders, which are seldom encountered except by divers. To humans and other terrestrial animals the most familiar chelicerates are members of class Arachnida: the spiders, harvestmen, scorpions, ticks, mites, and others.

MEROSTOMES

Class Merostomata is an ancient group of aquatic chelicerates, with some fossils nearly half a billion years old. One major group, subclass Eurypterida, has been extinct for about 250 million years (Figure 30.7). The only living merostomes belong to three genera of horseshoe crabs, subclass Xiphosurida.

Figure 30.7
Eurypterids from the Late Silurian period (approximately 410 million years ago) reconstructed from fossils found in New York State. Genera shown: (A) Pterygotus; (B) Stylonurus; (C) Hughmilleria; (D) Carcinosoma. Pterygotus was up to 2.3 meters long and a formidable predator.

Major Groups in Subphylum Chelicerata

Genera mentioned elsewhere in this chapter are noted.

Class Merostomata MER-oh-STOW-ma-tuh (Greek *meros* thigh + *stoma* mouth). Aquatic. Body divided into cephalothorax and abdomen, joined by thick "waist." Cephalothorax bears a pair of compound eyes and a pair of simple eyes or ocelli. Abdomen with paired appendages bearing gills, and a long tail spine (**telson**).

Subclass Eurypterida you-rip-TUR-rid-uh (Greek *eurys* wide + *pteryx* fin). Extinct (Ordovician to Permian). Cephalothorax covered dorsally by a carapace that does not extend laterally over the legs. Abdomen with 12 segments, narrowing to telson. *Carcinosoma, Hughmilleria, Pterygotus, Stylonurus* (Figure 30.7).

Subclass Xiphosurida zy-foe-SUR-id-uh (Greek *xiphos* sword + *oura* tail). Cephalothorax covered by horseshoe-shaped carapace extending laterally over legs. Abdomen unsegmented. Chelicerae with three joints each; walking legs with six joints each. Horseshoe crabs. *Limulus* (Figure 30.8).

Class Pycnogonida PICK-no-GO-nid-uh (Greek *pyknos* crowded + *gony-* knee). Cephalic somite only partly fused with thorax; abdomen vestigial. Four (sometimes five or six) pairs of walking legs with eight or nine joints each. Sucking mouth or long proboscis. Four simple eyes. Sea spiders. *Nymphon* (Figure 30.10).

Class Arachnida a-RACK-nid-uh (Greek *arachne* spider). Four pairs of legs. Abdomen usually lacking locomotory appendages. No compound eye. Mainly terrestrial. Spiders, harvestmen, scorpions, ticks, mites, and others. *Acarapis, Argiope, Boophilus, Brachypelma, Centruroides, Corythalia, Demodex, Dermacentor, Dermatophagoides, Dicrostichus, Dolomedes, Hogna, Ixodes, Latrodectus. Leiobunum, Limnochares, Loxosceles, Lycosa, Mastophora, Paruroctonus, Pandinus, Sarcoptes, Steatoda, Trombicula, Varroa* (Figures 30.1A, 30.11, 30.14, 30.15, 30.17, 30.18, 30.19, 30.20, 30.21, 30.22).

Three species of horseshoe crabs occur in the western Pacific. A fourth species, *Limulus polyphemus*, is common along the east coast of North America. This species is brown and up to 60 cm long. In early summer so many *Limulus* arrive on Atlantic beaches that they look like an invasion by midget marines, with only their helmets showing above water. The objective of this invasion is reproduction. At high tide the females burrow into the sand, and each deposits several hundred greenish eggs. One or more smaller males follow closely and deposit sperm on the eggs before the female covers them (Figure 30.8). Several weeks later larvae hatch and catch the next high tide out to sea. The larvae are called "trilobite larvae," because of their superficial resemblance to their presumed ancestors.

As any eastern beachcomber knows, many horseshoe crabs become stranded and die on beaches in their efforts to reproduce, and millions of the eggs become

Figure 30.8
A female horseshoe crab Limulus polyphemus *comes ashore to lay her eggs, closely pursued by the smaller male. The females are up to half a meter long.*

food for migrating birds. In spite of these perils, *Limulus* remains abundant, protected from predators by the hard carapace over the cephalothorax. The carapace bears a compound eye on each side and two simple eyes near the middle. In addition, there are five light-sensitive organs beneath the carapace. The abdomen can be folded beneath the animal to protect the delicate ventral appendages. The backward-pointing spines on the abdomen not only deter predators but dig into the sand and prevent waves from washing the animals out to sea as they attempt to go ashore. The tail spine, or **telson,** may also deter large predators, but it functions mainly as a lever during burrowing, and for righting the animal if it is flipped over by a wave.

Limulus can swim by flexing its abdomen, but most of the time it crawls and burrows along the bottom on five pairs of legs. Each leg of the front four pairs has a **chela** (KEE-luh; pincers) at the end. Horseshoe crabs eat molluscs and polychaete worms. This diet is more appropriate for mandibulates than for chelicerates, but horseshoe crabs make up for the lack of jaws by chewing food with their legs. Chelae on the chelicerae seize and partly macerate the food, then pass it back to the first four pairs of legs (Figure 30.9). These legs have proximal segments, called **gnathobases,** with tooth-shaped projections that enable them to function as jaws. After the gnathobases chew up the food they transfer it forward to the mouth. The fifth pair of legs do not engage in feeding and lack chelae. They provide most of the force for burrowing and also sweep the gills clean. These gills are called **book gills,** because the modified appendages that form them resemble pages. Oxygen uptake is encouraged by movement of the gills. There is also a respiratory pigment called hemocyanin, which makes *Limulus* blood blue.

SEA SPIDERS

Sea spiders are not really spiders, and some zoologists doubt that they are even chelicerates. As their names Pycnogonida (crowded knees) or Pantopoda (everywhere feet) suggest, legs are their dominant feature (Figure 30.10). Most of the approximately 600 species of pycnogonids have four pairs of long, many-jointed legs, and a few have five or six pairs. In some species the leg spread is up to 70 cm, but less than 1 cm is more typical. The males of some species also have an extra pair of legs, called **ovigerous legs** (= ovigers), with which they collect and brood the eggs they have fertilized. Pycnogonids are common in all oceans, especially cold ones. They can often be collected on sponges, hydroids, soft corals, anemones, ectoprocts, and clams. They feed on the soft parts of these animals by means of a sucking mouth at the end of a proboscis. Some species lack chelicerae. Digestion occurs in mucosal cells of the gut, which has branches radiating almost to the tips of the legs. Thanks to the enormous surface area relative to body mass, pycnogonids get by without any respiratory or excretory system.

ARACHNIDS—TERRESTRIAL PIONEERS

Most chelicerates belong to class Arachnida, which includes more than 60,000 known species of spiders, harvestmen, scorpions, ticks, mites, and a few other groups. Almost everyone is familiar with arachnids, mainly because they share our terrestrial habitat. Arachnids are, in fact, the first major group of animals discussed so far in this text that are truly adapted to a life on dry land, and they were among the first animals to abandon the habitat in which all life evolved. Scorpions, together with some crustaceans and uniramians, were apparently the first to make

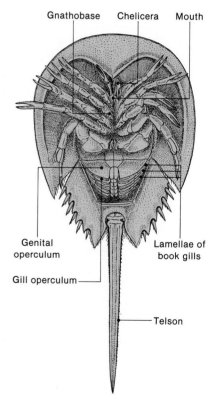

Figure 30.9
Ventral view of the horseshoe crab Limulus.

Figure 30.10
Dorsal view of a male pycnogonid,
Nymphon rubrum. *Note the unusual
location of the eyes.*

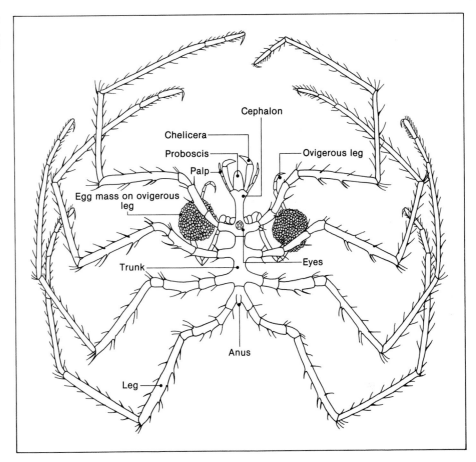

the pioneering venture out of water some 400 million years ago. We as terrestrial organisms are apt to take living on land for granted, but the transition from water to land was by no means simple. Several major adaptations were required (Table 30.2). Two of these adaptations were already present in the aquatic arthropods: an exoskeleton that can retard water loss, and the appendages that had been used for crawling or burrowing on the ocean floor.

SPIDERS

The most familiar arachnids are spiders (order Araneae), with approximately 30,000 known species. Most spiders are from 2 to 10 mm long, excluding the legs. Some female "tarantulas" have bodies up to 90 mm long (Figure 30.11). Male spiders are always smaller than females of the same species. The cuticle-covered body, divided into cephalothorax and abdomen, and the four pairs of legs are features by which most people recognize spiders quickly, and often with revulsion. The revulsion may come from an abhorrence of long, hairy legs. (If that is the case, however, why are Irish setters so popular?) The revulsion to spiders might also be due to fear of their fangs, even though only a few spiders can harm humans. The majority could not pierce human skin even if they tried.

External Structure. The external structure of spiders arises by a process that is typical for arthropods. Each metamere (= segment) of the annelid-like embryo has a pair of appendages, and there is extensive tagmosis. In spiders the anterior six metameres fuse during development to form the cephalothorax (= **prosoma**),

Table 30.2 Adaptations of arachnids and other animals for terrestrial life.
See Little (1984) for further discussion.

1. **The ability to resist drying,** primarily by means of an impermeable integument.
2. **Improved organs for maintenance of ion and water balance.**
3. **Modification of metabolic pathways with nitrogenous wastes eliminated as urea or uric acid rather than as ammonia.** (See p. 287.)
4. **The ability to obtain oxygen directly from air.**
5. **Improved ability to compensate for changes in temperature, which occur more rapidly in air than in water.** (See the discussion of poikilotherms in Chapter 15.)
6. **Modification of appendages for terrestrial locomotion, and strengthening of muscles and skeleton to compensate for the loss of buoyancy.**
7. **Modification of sensory receptors,** with less dependence on water-borne chemical information, and more dependence on airborne chemicals, vision, and vibration of substratum and air.
8. **Internal fertilization,** so that gametes remain moist. This in turn implies the development of sophisticated neural mechanisms for courtship and anatomical arrangements for copulation.
9. **Either eggs that retain water, or viviparity,** which provide a moist environment for development.
10. **Large supply of nutrients to the embryo, either by nutrient-rich eggs or viviparity,** since development must often be prolonged until the immature animal is capable of terrestrial living.

while the posterior metameres form the abdomen (= **opisthosoma**). These two tagmata are linked by a narrow **pedicel.** The six pairs of anterior appendages become the two chelicerae, the two pedipalps, and the four pairs of legs. These three kinds of structures are therefore homologous to each other (**serially homologous**).

Figure 30.11
The Mexican red-legged "tarantula" Brachypelma smithi *showing its chelicerae. Two leg-shaped pedipalps occur on each side of the chelicerae. (The name "tarantula" was erroneously applied to several large genera of the American tropics by colonists, who apparently mistook them for the true tarantulas,* Lycosa *or* Hogna, *of the Mediterranean. Both true and American tarantulas are feared, but in fact their bites, though lethal to insects and mice, are no worse for humans than wasp stings.)*

Chelicerae. In most spiders each chelicera has a fang that is usually kept within a groove, like the blade of a folding knife (Figure 30.2). During predation or defense, the fangs extend and inject venom from a poison gland into the prey or predator. Depending on species, the venom includes various neurotoxins, protein-digesting enzymes, and pain-inducing amines. After biting its prey, a spider generally backs off while the toxins kill or paralyze it. Many spiders further subdue prey by wrapping it with silk, either before or after biting. After the prey has been overcome, some spiders mash it with their chelicerae while regurgitating digestive enzymes from the gut onto it. Other spiders regurgitate enzymes into the prey through the fang holes. After allowing the enzymes to work for a few seconds, spiders suck up the semiliquid food into their mouths. Some spiders use their chelicerae not only for feeding but for carrying prey or egg cocoons, for grasping objects, or for digging burrows. In addition to the chelicerae, spiders also have a pair of pedipalps (also called simply palps) that help manipulate prey and that substitute to some extent for chewing mandibles. In adult male spiders the pedipalps are specialized as copulatory organs, as will be described.

Digestion. Spiders suck up the externally digested food by expanding the pharynx and **pumping stomach** (Figure 30.12). Most of the internal digestion and absorption of nutrients occurs in the midgut, which has numerous diverticula branching anteriorly into the bases of the legs. Digestive enzymes for both internal

Figure 30.12
Internal structure of a female spider.

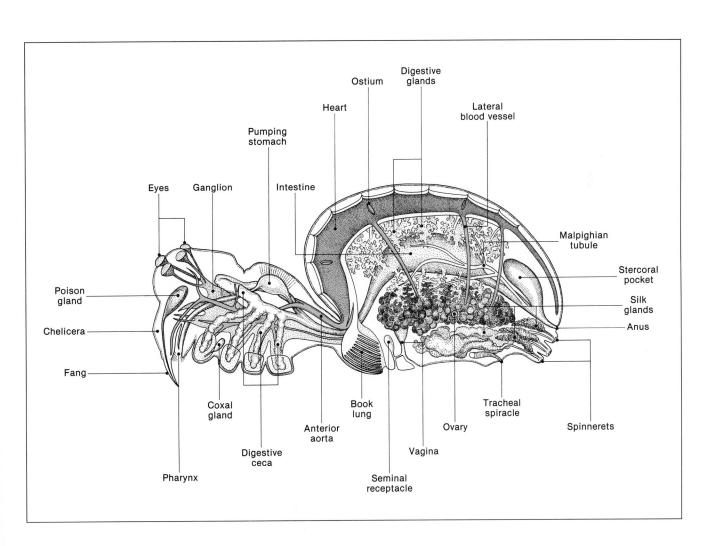

and external digestion are usually secreted by a **digestive gland,** which is unfortunately often called a "liver." Because of the highly developed digestive system and low rate of metabolism, most spiders can live for long periods without eating. "Tarantulas" can go for several months between meals, and adult black widows (*Latrodectus mactans*) can live for 200 days without eating. The abdomens of spiders can expand enormously after a meal, thanks to the arthrodial membrane linking the sclerites. Often this expansion triggers molting.

Excretion.   The main excretory organs of spiders are Malpighian tubules (see p. 286). Malpighian tubules actively transport excess ions and metabolic wastes out of the blood in the abdomen, and into fluid in the **stercoral pocket,** just dorsal to the rectum. As the fluid is excreted, rectal glands reabsorb much of the water. In addition, spiders have **coxal glands** that eliminate wastes through pores near some of the coxae (the proximal joints of the legs).

Circulation.   The coelom of spiders, and of all arthropods, is even more reduced than in molluscs. The only traces of the coelom in spiders occur in the coxal glands, gonads, and a few other places. The main body cavity is the **hemocoel:** the blood-filled space around the internal organs. The circulatory system is typically arthropodan, with a dorsal tubular heart that pumps blood (= **hemolymph**) anteriorly through an aorta. The blood percolates through the hemocoel of the thorax, returns to the abdomen through the pedicel, and is taken back into the heart through one-way valves called **ostia.** Although the circulatory system is open, spiders are capable of producing high blood pressures (up to 480 mmHg in "tarantulas"). This high blood pressure serves two functions. First, two joints in each leg lack extensor muscles, so complete extension of spider legs depends on blood pressure. (This explains why a spider's legs fold when it dies.) The second function of the high blood pressure is to transport oxygen efficiently.

Respiration.   Respiration is due to a pair of **book lungs** and to one or two pairs of tracheae. The book lungs are similar in structure to the book gills of *Limulus* and are serially homologous with the legs. Each lung consists of 15 to 20 air-filled plates within a blood-filled chamber in the abdomen. Air enters the plates through a slit in the ventral surface of the abdomen. Oxygen diffuses into the blood returning to the abdomen through the pedicel. Tracheae are cuticle-lined tubes that open externally through **spiracles** (see p. 687). Unlike the tracheae of insects, those of spiders do not branch directly to cells but bring oxygen into the blood. The pigment hemocyanin aids in oxygen transport and gives fresh spider blood a bluish hue.

Nervous System.   The central nervous systems of adult spiders are different from those of most other arthropods. Instead of numerous ganglia on a ventral nerve cord, there are two large ganglia in the cephalothorax. These ganglia are largely concerned with coordinating the legs (a major task since there are so many of them) and integrating the sensory input from eyes and numerous other receptors.

Spiders have several types of mechanoreceptor, including "hairs," receptors at the joints, and slit sensilla. The hair-shaped receptors enable spiders to hear airborne sounds, such as the buzzing of a fly. (The ability to hear is common in terrestrial arthropods and chordates but absent in all other animals.) The slit sensilla consist of slits in the cuticle arranged so that tension in a certain direction causes the slit to open or close, triggering action potentials. In spiders these slits

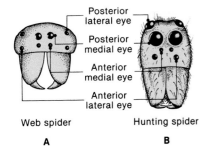

Posterior
lateral eye

Posterior
medial eye

Anterior
medial eye

Anterior
lateral eye

Web spider   Hunting spider

A   B

Figure 30.13
A comparison of the eyes of orb-weaving spiders (A) and those that stalk their prey (B).

often occur on the legs in parallel groups called **lyriform organs** (because the slits are arranged like the strings of a lyre). A spider with damaged lyriform organs has trouble finding its way back to prey once it has left it.

Typically there are six or eight simple eyes (ocelli). Arachnids, unlike trilobites, crustaceans, and insects, do not have compound eyes. Spider eyes are arranged about the head in various patterns that are useful to taxonomists for classification (Figure 30.13). The eyes cover overlapping fields of view, with virtually no blind spot. Spiders that capture prey with webs generally rely more on mechanical information from the web than on vision. Most web spiders are active at night, and blinding them by painting their eyes has no apparent effect on their ability to make webs or to capture prey trapped in them. Presumably their eyes function mainly to detect movements by daytime predators. On the other hand, hunting spiders, such as jumping spiders (family Salticidae), hunt during the day by pursuing and pouncing on prey.

Salticids commonly have a pair of large **primary eyes** with excellent resolution over a narrow angle (Forster 1982; Jackson 1985). If the smaller secondary eyes detect possible prey, the salticid orients its body so that the primary eyes are aimed at it. The eyes, of course, cannot move. The retinas, however, can move within the eyes. If potential prey moves within about 20 cm of the spider, the spider scans it by moving its retinas rather than its eyes or body. At a distance of less than 10 cm the retinas scan it rapidly to determine whether it is prey or a member of its own species. In the latter case the salticid may respond with courtship or some other appropriate social behavior.

Reproduction. Reproduction in spiders is rather conventional except that the males use their pedipalps as copulatory organs. After the final molt to adulthood, a male spider ceases to feed and devotes the rest of its abbreviated life to reproduction. It constructs a special **sperm web** on which it deposits a drop of semen, then dips the bulbs of its pedipalps into the semen to fill them with sperm (Figure 30.14A). It then goes in search of a mate. The search is often aided by pheromones that females with unfertilized ova release into the air. As in many predators, copulation is usually preceded by an elaborate courtship that helps prevent any fatal mistakes. In web spiders the male carefully approaches the female on her web, avoiding vibrations that the female could misinterpret as those of struggling prey. In some species, the male identifies himself by plucking on the web in a particular way. Hunting spiders (salticids and others) generally use visible or audible signals in courtship (Figure 30.14B). Such communication is usually successful in saving the male's life from the much larger female. Contrary to popular myth, female spiders do not routinely eat males during or after copulation. Following courtship, insemination is achieved when the male inserts the pointed tip of a pedipalp into a pore leading to the female's seminal receptacle (Figure 30.14C).

After mating, the male goes in search of another mate, while the female remains at her web. Soon she begins depositing hundreds or even thousands of fertilized eggs into silken cocoons. Female spiders often show a touching degree of care for these cocoons, and for the spiderlings that hatch from them. Many people have been startled to see a wolf spider they were about to step on divide into dozens of tiny spiderlings that the mother had been carrying on her back. Eventually all young spiders leave their mothers and disperse. A common means of dispersal is "ballooning." The young spider stands exposed to the wind, playing out a string of silk that eventually becomes long enough to carry it aloft. Ballooning spiders have been reported at altitudes up to 5 km; along with tiny insects they constitute an "aerial zooplankton."

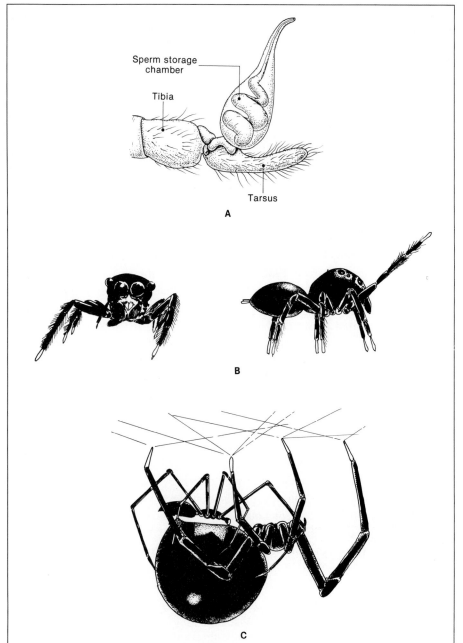

Figure 30.14
(A) A pedipalp of a male spider. These copulatory organs are extremely variable and useful in classification. (B) The jumping spider Corythalia *courts by rocking sideways and by waving its forelegs. (C) Sperm transfer by black widow spiders* Latrodectus mactans. *Contrary to myth, the male (atop the female) usually survives the encounter.*

In figure (A): Sperm storage chamber, Tibia, Tarsus, A

B

C

SPIDER SILK

Functions. Virtually every aspect of spider biology depends on the ability to produce silk. As previously noted, males produce sperm webs, females weave cocoons out of silk, juvenile spiders use silk to "balloon" to new habitats, and web spiders use silk to trap and wrap prey. In addition, most spiders lay down a **drag line** of silk as they move about. The drag line is attached to the substratum and allows the spider to recover if it falls accidentally, or if it deliberately drops to evade a predator. Silk is also used for predation by many spiders that do not weave webs. For example, trapdoor spiders (family Ctenizidae) live within burrows covered with a silk trap door. When they feel prey passing overhead, they spring out and drag the prey into the burrow. Bolas spiders (*Dicrostichus* in Australia and *Mastophora* in the Americas) produce a single, sticky strand of silk that they throw at flying insects. The usual prey of *Mastophora* are male moths, which the spiders attract with secretions that mimic the sex attractants of female moths (Stowe et al. 1987).

Composition. Spider silk consists of proteins synthesized as an aqueous solution by six types of gland in the abdomen (Figure 30.12). Each type of gland produces a different type of silk for each function: sperm web, drag line, cocoon, various parts of web, and so on. The silk is extruded from one of three pairs of **spinnerets.** The silk can be forced out by muscle contraction, or it can be pulled out by a hind leg. As the strand emerges, the proteins change from a dissolved to an insoluble, fibrous form. It had long been assumed that exposure to air triggered this change, but studies of caterpillar silk suggest that the proteins become fibrous from being pulled on. Spider silk is only 0.01 μm to a few micrometers in diameter, yet its density is so low and its strength so high that a strand would have to be 80 km long to break under its own weight. Spiders do not waste such a valuable substance. After the silk has served its function it is eaten, and the amino acids are used to synthesize new silk. In one study a web was labeled with radioactive tracer, and the spider was allowed to eat it, as it normally would. Up to 90% of the tracer appeared in a web the spider made just a half hour later.

Types of Web. Approximately half the species of spiders construct webs. There are three major types of web. **Sheet webs** are approximately horizontal and often lead to funnel-shaped retreats in which the spiders await prey. These are the webs often seen on lawns after a heavy dew. **Cobwebs** are more loosely woven than sheet webs, and they depend on sticky threads to snare prey walking beneath them (Figure 30.15). The third type of web is the **orb web,** which is constructed in a vertical plane and snares flying insects.

Orb Webs. The orb web is probably the most familiar type and is among the most fascinating constructions by animals. Most people wouldn't know how to begin constructing a web between one structure and another separated by a dis-

Figure 30.15
A cobweb spider Steatoda borealis *approaching a fly that has walked into a trap thread.*

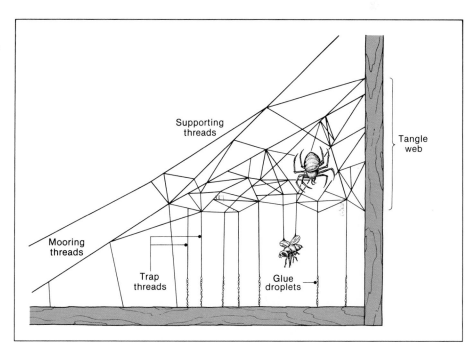

tance of tens or hundreds of body lengths. Yet many orb-weaving spiders perform this feat within a few minutes virtually every night, and they do it without vision (Figure 30.16). The portion of the web that snares prey is the **catching spiral,** which works in one of two ways, depending on the species of spider. Many spiders deposit droplets of glue on threads of the catching spiral, while others (**cribellate** spiders) construct the catching spiral of wooly threads that entangle prey. How the spider avoids getting entangled in its own catching spiral is still a largely unexplored mystery. After constructing the web the spider usually goes to the hub or to a special refuge area until prey is detected. (See Figure 30.17.) When the spider senses vibrations induced by trapped prey it rushes to the hub, if it is not already there. It then plucks the radial threads and uses their vibrations to determine the direction of the prey.

The general pattern of orb weaving varies among different species of web spiders. There are even small differences among individual spiders. These individual variations may be due in part to variations in the sizes of the front legs, which spiders use to measure lengths and angles. Webs can also be affected by many of the same drugs that affect human behavior. Peter N. Witt, a physician formerly at Duke University, first discovered this fact when asked by a zoologist to prescribe a drug that would make spiders construct webs during the daytime, when it was more convenient to observe them. Witt first tried amphetamine, which did not

Figure 30.16

Steps in the construction of an orb web by one species of spider. (A) The spider plays out a strand in the wind to form a support. (B) The spider follows the support and lays down a parallel strand then (C) pulls down at the middle, forming three radii shaped like a Y (D). (E) The spider then climbs up the stem of the Y to one of the supports, descends, and attaches a frame and another radial strand. (F–H) Additional radial and frame threads are attached. (I, J) The spider constructs an auxiliary spiral from the hub that serves as a temporary scaffolding. (K, L) It adds the catching spiral, then removes the auxiliary spiral.

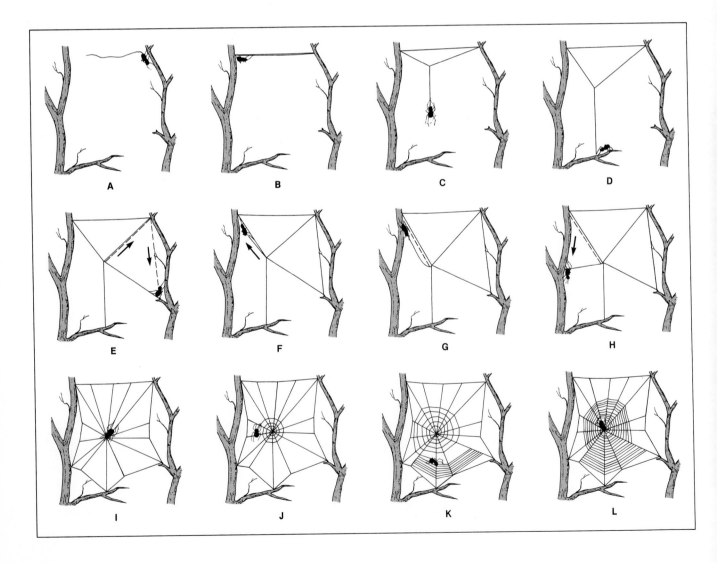

affect the time the web was built but made the spiders position the radial threads and catching spiral abnormally. Intrigued by this result, Witt spent decades studying webs and the effects on them of psychoactive drugs, such as caffeine, mescaline, and LSD. The effects of many of these drugs are so characteristic that Witt could identify a drug by examining a web constructed under its influence.

OTHER ARACHNIDS

Harvestmen. Besides spiders (order Araneae) there are several other important groups of arachnids. Harvestmen (= daddy longlegs; order Opiliones) are often conspicuous in late summer, at about harvest time (Figure 30.18). They are readily distinguished from spiders by the fused cephalothorax and abdomen, and usually by legs that are extremely long and thin relative to the body. They feed and scavenge on a variety of invertebrates and plants. A few species are among the small number of chelicerates capable of ingesting solid food. Harvestmen lack poison fangs. Perhaps for that reason most people take more kindly to them than to spiders. Harvestmen defend themselves from predators by an abdominal "stink gland," and by readily disconnecting a leg that has been grasped. Such **autotomy,** the ability to lose a leg or other appendage to save a life, is quite common among arthropods.

Scorpions. Another group of arachnids includes scorpions (order Scorpionida), which most people have heard of, but few have seen (Figure 30.19). Scorpions are found mainly in the tropics and subtropics, especially in deserts. They usually stay in underground burrows during the day and come out at night to prey on spiders and insects. Desert scorpions detect prey by sensing vibrations in sand with slit sensilla and sensory hairs on their legs. The sand scorpion *Paruroctonus mesaensis* of the Mojave Desert uses differences in the times that vibrations arrive at its eight legs to determine the direction of prey (Brownell 1984). In this way a sand scorpion can quickly and accurately home in on a cockroach up to 50 cm away. After detecting prey, a scorpion chases it down, then seizes it in the chelae of the two pedipalps.

The scorpion dismembers prey with its chelicerae and digests its externally. If the prey is large the scorpion first subdues it by injecting a paralytic venom. While holding the prey in its pedipalps it quickly arches the tail over its head, directing the stinger with great accuracy. The venoms of different species contain neurotoxins of two major types that work on the sodium channels responsible for generating action potentials (see p. 145). Although most scorpions can inject only enough venom to cause pain and swelling in humans, the stings of some species can be fatal unless treated with antibodies to the neurotoxin.

Like spiders, scorpions engage in elaborate courtship and parental behaviors. During courtship the male and female grasp each other's chelae and promenade back and forth. During the promenade the male stings the female, perhaps subduing her aggression but doing no permanent harm. The promenade may go on for days until the male locates a suitable place on which to deposit a sperm-filled capsule called a **spermatophore.** He then maneuvers the female until her external genitals flip open a lid on the spermatophore, which releases the sperm. In some species the male then flees, but in many species he makes an involuntary contribution to the nutrition of the young by being cannibalized by the female. Scorpions are viviparous, with the embryos absorbing nutrients from the mother's digestive tract. Up to 90 ova may be fertilized in one mating, and development can take more than a year in some species. Immediately after birth the young

Figure 30.17
The orange argiope Argiope aurantia *reaping the harvest of its web. The dense vertical band of silk, called the stabilimentum, is thought to deter accidental destruction by birds, or to camouflage the spider.*

Figure 30.18
The harvestman Leiobunum *on a milkweed seed pod.*

A

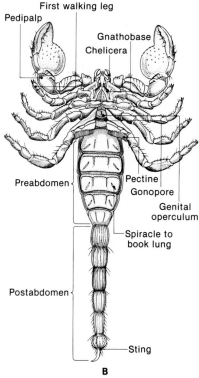

Figure 30.19
(A) Sand scorpion Paruroctonus mesaensis *feeding on a burrowing cockroach. (Photographed at night under ultraviolet.) (B) Ventral view of the scorpion* Pandinus, *the giant African scorpion. The scorpion's cephalothorax, to which the pedipalps and legs attach, is relatively short and broadly joined to the abdomen. The abdomen has two distinct regions: the preabdomen and postabdomen. On the second abdominal segment are comb-shaped pectines, which are unique to scorpions and are possibly mechanoreceptor organs.*

scorpions climb onto the mother's back and remain there until they become independent, about a week later.

Ticks and Mites. Ticks and mites (order Acari = Acarina) differ from the other arachnids in having lost all external signs of segmentation (Figure 30.20). They also differ in having a projecting mouth region, the **capitulum,** visible in Figure 30.20A. Ticks are generally much larger than mites, which are usually less than 1 mm long. Approximately 30,000 species of acarines have been described, but many authorities believe unknown species far outnumber known species. Mites are often so small that they can be identified only with the scanning electron microscope. Close examination of a small sample of leaf litter from virtually anywhere in the world will usually reveal hundreds of mites of several species.

All ticks are parasitic in their three major life stages—larva ("seed tick"), **nymph,** and adult. They feed on the blood of reptiles, birds, or mammals. Immediately after hatching in spring, the six-legged larva ascends vegetation and stays until a host passes. It then hooks its pedipalps into the skin and pierces the skin with sucking mouthparts. The larva feeds until enormously swollen, then drops off the host and molts into an immature nymph. The nymph then attaches to a new host, feeds, and molts into an adult. The adult then feeds on a new host before mating. Like ticks, many mites parasitize various tissues of terrestrial vertebrates. Other mites prey on invertebrates or feed on plants, and many are free-living.

INTERACTIONS WITH HUMANS AND OTHER ANIMALS

Benefits. Spiders, scorpions, ticks, mites, and other chelicerates evoke negative feelings in most people because we know so little about many of these animals. However, they may turn out to be beneficial in ways that we can scarcely imagine. Only in the 1960s was it discovered that the blood of horseshoe crabs clots when exposed to certain bacterial toxins. Now the blood of horseshoe crabs is widely

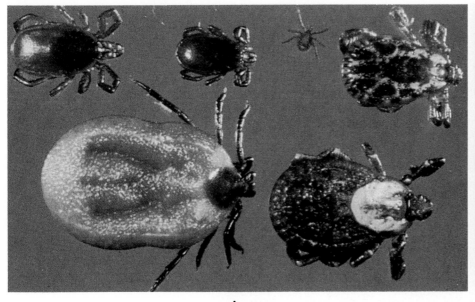

A

B

Figure 30.20

(A) Two dog ticks Dermacentor variabilis *on the right (male and female, top and bottom, respectively), and deer ticks* Ixodes dammini *at various stages of development. At top are the female (left), male (second from left), and nymph (third from left) of the deer tick. The year-old nymph, which transmits diseases to humans, is only a little larger than the period at the end of this sentence. On the lower left is a female deer tick engorged with blood. During its two-year life cycle each tick feeds just three times. As a larva and then as a nymph it usually feeds on deer mice and other small mammals; as an adult it feeds mainly on deer. If a larva feeds on a mammal carrying the bacterium responsible for Lyme disease, and then feeds as a nymph on a person, it may transmit the disease to that person. This species of tick is also responsible for a growing number of cases of human babesiosis in the United States. (B) The red freshwater mite* Limnochares americana. *This relatively large species (3 mm) can be found in ponds throughout North America. Water mites are among the few chelicerates that have returned to the aquatic habitat. Immature stages frequently live as parasites on the backs of aquatic insects. (See Figure 32.19.)*

used to screen substances for potential toxicity in humans. Even more recently horseshoe crabs were found to be essential for the survival of numerous migrating shore birds, such as the red knot (*Calidris canutus*). In early May these robin-sized sandpipers leave their wintering sites in Brazil, having fattened themselves on snails. By the time they arrive at Delaware Bay, however, they are too thin and exhausted to reach their nesting grounds in the Arctic. Fortunately, horseshoe crabs are depositing millions of eggs in the sand at this time, and red knots and other migrating birds can replenish their energy supplies by feasting on them.

Venomous Species. There is no denying that for many animals, including humans, interactions with chelicerates often have unpleasant consequences. A few species of spiders and scorpions can produce painful and even fatal bites in humans. The scorpion *Centruroides* sp. kills hundreds and perhaps thousands of people each year in Mexico alone. The most dangerous spider in North America is the female black widow *Latrodectus mactans* (Figure 30.21A). Her venom causes neuronal synapses to release the transmitter acetylcholine, resulting in muscle spasms, abdominal rigidity and cramps, sweating, salivation, high blood pressure, and sometimes convulsions. Elderly and very young people may die from the venom. Another dangerous spider is the brown recluse *Loxosceles reclusa*, which is common in the south-central United States (Figure 30.21B). It is brown with a dark violin-shaped marking on the dorsal cephalothorax. Its venom contains enzymes that destroy blood cells and induce white blood cells to attack surrounding tissues. Both spiders have an unfortunate tendency to live in homes, outhouses, and other buildings.

Vectors of Disease. Even more devastating to humans and other animals are some ticks and mites. Not only can ticks weaken terrestrial vertebrates by feeding on their blood, but they may also transmit diseases. The Rocky Mountain wood tick *Dermacentor andersoni* transmits several diseases to humans, including **Rocky Mountain spotted fever**. In spite of its name, Rocky Mountain spotted fever occurs most commonly in the eastern United States. Wild rodents serve as a reservoir for the rickettsia bacteria that cause the disease. Deer, raccoons, squirrels, mice, and other mammals serve as reservoirs for other diseases for which ticks are

Figure 30.21
(A) The female black widow Lactrodectus mactans *with a cocoon. (B) The brown recluse* Loxosceles reclusa. *Body approximately 1.5 cm long.*

A   B

vectors. A recent surge in such diseases may be due to the rise in numbers of these animals in suburbs.

The incidence of another tick-borne disease has increased even more dramatically. **Lyme disease,** named for the Connecticut town where it was first identified in 1975, is caused by a bacterium transmitted by the deer tick *Ixodes dammini* (Figure 30.20A). Thousands of people and untold numbers of dogs, cows, and horses have been infected. The first stage of the disease is often marked by a distinctive bull's-eye shaped rash and flu-like symptoms. Recovery is usually uneventful if antibiotics are administered during this stage. Thousands of people, however, develop second and third stages of the disease that may last for years. In the second stage there may be nerve damage, meningitis, partial paralysis, and heart irregularities. In the third stage there may be arthritis-like inflammation of the joints. The best treatment is prevention, by wearing long pants and sleeves outdoors in the summer in areas where the disease occurs, and by examining every square centimeter of skin and removing the tiny immature ticks (nymphs) that are the main culprits. Another approach is to kill the ticks while they are on alternative hosts, especially wild mice. Cotton laced with the poison Damminix can be left outdoors so that when mice use it in their nests the Damminix will kill the ticks. Killing mice, deer, and other alternative hosts to rid an area of ticks is probably futile.

Ticks can be just as devastating by transmitting diseases to cattle and wild animals on which people depend for food. Among the most severe tick-borne diseases of ruminants is **babesiosis,** caused by the protozoan *Babesia* (see p. 478). The cattle tick *Boophilus annulatus* is a vector for the species of *Babesia* that causes Texas cattle fever, also called red-water fever because of the bloody urine of its victims.

Many **mites** also cause severe economic losses by transmitting diseases and by parasitism. For example, several species of follicle mites (genus *Demodex;* Figure 30.22) cause mange in dogs and other skin eruptions in cattle, horses, and hogs. The tracheal bee mite *Acarapis woodi,* which inhabits the tracheae of honey bees, has recently spread over much of the United States, threatening millions of dollars

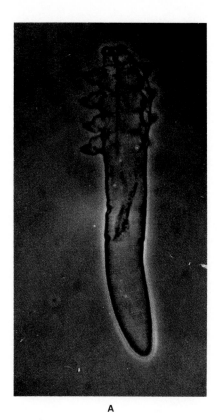

A

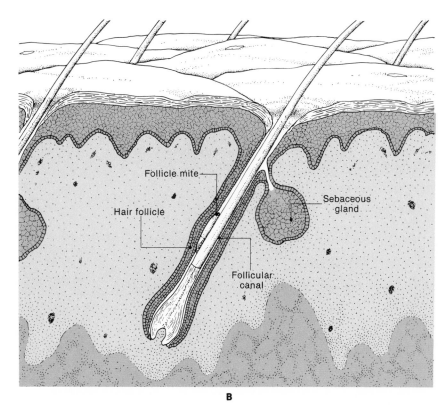

B

Figure 30.22
The follicle mite Demodex folliculorum *isolated (A) and in its normal habitat (B). This species is usually a harmless commensal in hair follicles on the human face. A related species,* D. brevis, *lives in sebaceous glands on the face. Other species of* Demodex *cause mange in dogs, and other diseases in other mammals. The genus* Demodex *includes the smallest of all known arthropods (less than 0.1 mm long).*

worth of honey and crops that depend on bees for pollination. Even more recently the ectoparasitic mite *Varroa jacobsoni* has become a serious threat to honey bees. Mites can also be a health nuisance for humans. Larvae of the chigger mite *Trombicula* cause a maddening itch when they secrete digestive enzymes into the skin and feed on the tissues. Female itch mites *Sarcoptes scabiei* tunnel through human skin, producing the disorder commonly called scabies or "the seven-year itch." Finally, feces of the house mite *Dermatophagoides farinae*, a common inhabitant of dust, is a major cause of allergies and childhood asthma.

SUMMARY

Chelicerates differ from crustaceans and uniramians in having chelicerae rather than mandibles, but in other ways they are similar to these other arthropods. Like them, chelicerates have an integument with a rigid cuticle consisting largely of chitin and protein. The cuticle is divided into endocuticle, exocuticle, and epicuticle. The exocuticle and epicuticle are sclerotized in large areas, forming an exoskeleton of plates and tubes hinged together by arthrodial membrane. The epicuticle contains waterproof waxes. In order for arthropods to grow they must molt periodically. The rigid cuticle imposes another similarity on arthropods: jointed legs. In addition, all have segmented bodies and other similarities. Chelicerates were among the first animals to live on land, thanks to the waterproof and supportive exoskeleton and numerous other adaptations.

Chelicerates include horseshoe crabs, true spiders, sea spiders, scorpions, ticks, and mites. Spiders make up the majority of named chelicerates. Their bodies are divided into two tagmata: the cephalothorax and abdomen. There are four pairs of legs. Most spiders attack prey with the fangs on their chelicerae, then feed by regurgitating digestive enzymes and ingesting the semiliquid. Spiders have open circulatory systems, and the hemocoel through which the blood circulates is the main body cavity. Oxygen enters the blood through book lungs, and to some extent through tracheae. Spiders have three or four pairs of simple eyes. Web-building spiders have small eyes and rely mainly on mechanoreceptors; hunting spiders have larger and more acute eyes.

Many spiders use silk to construct webs with which they trap prey. There are three main kinds of webs: sheet webs, cobwebs, and orb webs. Orb webs have a catching spiral that traps prey in glue or in wooly threads. Silk is also used by males for sperm webs, by females for cocoons, by juveniles for ballooning to new locations, and by all for drag lines.

KEY TERMS

cuticle
chelicera
mandible
uniramous
biramous
palp
chitin
hemocoel
endocuticle

exocuticle
epicuticle
articular membrane
molting
sclerotization
cephalothorax
telson
gnathobase
book gill

pedicel
Malpighian tubules
spinneret
cobweb
sheet web
orb web
catching spiral

SELF-TEST

1. Explain the argument in favor of dividing arthropods into three phyla. What is the argument in favor of retaining the phylum Arthropoda?

2. Describe the composition and functions of the endocuticle, exocuticle, and epicuticle.

3. List three major groups of chelicerates by their common names, and briefly describe each group. What features do they have in common, and what are the differences among them?

4. Describe four ways in which spiders are adapted to terrestrial living.

5. Describe four ways spiders use silk.

6. Describe how hunting spiders differ physiologically and behaviorally from web spiders.

7. For each of the following, state one way in which they differ from spiders: harvestmen, scorpions, ticks.

READINGS

RECOMMENDED READINGS

Barlow, R. B. Jr. 1990. What the brain tells the eye. *Sci. Am.* 262(4):90–95 (Apr). (*On vision in the horseshoe crab.*)

Brownell, P. H. 1984. Prey detection by the sand scorpion. *Sci. Am.* 251(6):86–97 (Oct).

Burgess, J. W. 1976. Social spiders. *Sci. Am.* 234(3):100–106 (Mar).

Emerton, J. H. 1961. *The Common Spiders of the United States.* New York: Dover.

Foelix, R. 1982. *Biology of Spiders.* Cambridge, MA: Harvard University Press.

Forster, L. 1982. Vision and prey-catching strategies in jumping spiders. *Am. Sci.* 70:165–175.

Gertsch, W. J. 1979. *American Spiders,* 2nd ed. New York: Van Nostrand Reinhold. (*Beautifully illustrated.*)

Habicht, G. S., G. Beck, and J. L. Benach. 1987. Lyme disease. *Sci. Am.* 257(1):78–83 (July).

Hadley, N. F. 1986. The arthropod cuticle. *Sci. Am.* 255(1):104–112 (July).

Jackson, R. R. 1985. A web-building jumping spider. *Sci. Am.* 253(3):102–115 (Sept).

Kaston, B. J. 1978. *How to Know the Spiders,* 3rd ed. Dubuque, IA: Wm. C. Brown.

Levi, H. W. 1978. Orb-weaving spiders and their webs. *Am. Sci.* 66:734–742.

Levi, H. W. and L. R. Levi, 1968. *A Guide to Spiders and Their Kin.* New York: Golden Press.

Levi-Setti, R. 1975. *Trilobites: A Photographic Atlas.* Chicago: University of Chicago Press.

McDaniel, B. 1979. *How to Know the Ticks and Mites.* Dubuque, IA: Wm. C. Brown.

Miller, J. A. 1987. Ecology of a new disease. *BioScience* 37:11–15. (*On Lyme disease.*)

Milne, L. and M. Milne, 1980. *The Audubon Society Field Guide to North American Insects and Spiders.* New York: Alfred A. Knopf.

Myers, J. P. 1986. Sex and gluttony on Delaware Bay. *Nat. Hist.* 95:68–77 (May). (*On horseshoe crab eggs as food for migrating birds.*)

Preston-Mafham, R. and K. Preston-Mafham. 1984. *Spiders of the World.* New York: Facts on File. (*Beautifully illustrated.*)

Rudloe, A. and J. Rudloe. 1981. The changeless horseshoe crab. *Nat. Geogr.* pp. 562–572 (April).

See also relevant selections in General References at the end of Chapter 21.

ADDITIONAL REFERENCES

Briggs, D. E. G. and R. A. Fortey. 1989. The early radiation and relationships of the major arthropod groups. *Science* 246:241–243.

Eberhard, W. G. 1990. Functions and phylogeny of spider webs. *Annu. Rev. Ecol. Syst.* 21:341–372.

Field, K. G. et al. 1988. Molecular phylogeny of the animal kingdom. *Science* 239:748–753.

Little, C. 1984. *The Colonisation of Land: Origins and Adaptations of Terrestrial Animals.* New York: Cambridge University Press.

Polis, G. A. (Ed.). 1990. *The Biology of Scorpions.* Stanford, CA: Stanford University Press.

Shear, W. A. (Ed.). 1986. *Spiders: Webs, Behavior, and Evolution.* Palo Alto, CA: Stanford University Press.

Stowe, M. K., J. H. Tumlinson, and R. R. Heath. 1987. Chemical mimicry: bolas spiders emit components of moth prey species sex pheromones. *Science* 236:964–967.

Witt, P. N. and J. S. Rovner. 1982. *Spider Communication: Mechanisms and Ecological Significance.* Princeton, NJ: Princeton University Press.

31

Crustaceans

Sally lightfoot crab (Grapsus grapsus).

LEARNING OBJECTIVES

1. What features of crustaceans distinguish them from chelicerates and insects?

2. Why are such diverse animals as lobsters, crabs, and barnacles all considered to be crustaceans? What other animals are considered to be crustaceans?

3. How do crustaceans, including barnacles, manage to reproduce inside the exoskeleton?

4. How do the compound eyes of crustaceans and insects work?

5. How do crustaceans tell which direction is up when they are in water?

GENERAL FEATURES OF CRUSTACEANS

Although chelicerates and crustaceans are both usually classified in phylum Arthropoda, they are so different from each other that even nonzoologists easily tell them apart. People who enjoy the sight of a lobster, crab, or shrimp on their dinner plate would faint at the sight of a spider or scorpion, even if it were boiled and served with melted butter. The reasons for this prejudice are not clear.

Zoologically there are differences between members of subphylum Chelicerata and subphylum Crustacea that are not merely matters of taste. Unlike chelicerates, which have chelicerae, crustaceans have mandibles. These mandibles develop from the bases of appendages, rather than from entire appendages, as in insects and other uniramians (see p. 637). Crustaceans have primitively biramous (two-branched) appendages and generally retain a greater number of them than do chelicerates. Crustaceans are often referred to as the aquatic mandibulates, because few have adapted to terrestrial life. In spite of these differences, most of the approximately 40,000 species of crustaceans share the important arthropod features (see Table 30.1), including segmentation, jointed appendages, and cuticular exoskeleton.

Segmentation.   Segmentation is apparent in the familiar crayfish (Figure 31.1). It is most pronounced in the abdomen, where each of six somites is covered by two plates of cuticle: a dorsal **tergite** and a ventral **sternite.** In crayfishes and many other crustaceans, segmentation is less obvious anteriorly, because somites are tagmatized as a single cephalothorax that is shielded dorsally by a **carapace.** The 13 pairs of appendages on the cephalothorax reveal, however, that this tagma originated from 13 embryonic segments.

Jointed Appendages.   The 13 pairs of jointed appendages on the cephalothorax, along with the 6 pairs of appendages on the abdomen, are all serially homologous with each other. Emerson and Schram (1990), at the San Diego Natural History Museum, have found fossil evidence that each biramous appendage originated from the fusion of two uniramous appendages at their bases. Each appendage can be thought of as developing from a generalized appendage with two branches: a

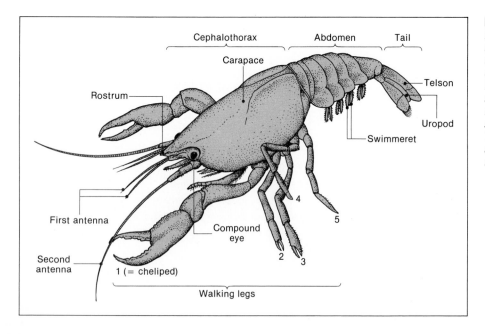

Figure 31.1
External features of a crayfish. There are several genera and approximately a hundred species of crayfishes. In some areas they are known as "crawfish" or "crawdads" and are eaten. The walking legs are numbered. Four of the five swimmerets on one side are shown. The first pair of swimmerets are adapted as copulatory organs in males.

lateral branch called the **exopod** and a medial branch called the **endopod.** These two branches are attached to a **protopod.** (See top panel in Figure 31.2.) During development the exopod may be lost from some appendages, but additional branches may grow out of protopods, endopods, or remaining exopods with each molt. In this way the appendages become modified for a variety of functions, some of which are surprising even for an arthropod.

In crayfishes and many other crustaceans, the first two pairs of segmental appendages develop into two pairs of antennae that bear receptors for touch and chemoreception. The appendages on the third through eighth segments make up a complicated assortment of mandibles, maxillae, and maxillipeds generally involved with feeding (Figure 31.3). The two mandibles crush food; the two pairs of maxillae shred food and pass it forward to the mandibles. The second pair of maxillae also have "gill bailers" that draw water anteriorly past the gills. The gills are epipods on the last pair of maxillipeds (segment 8), and on some walking legs (segments 9, 12, and 13), and are hidden beneath the carapace.

The first pair of walking legs, called **chelipeds** (segment 9), bear **chelae** (pincers) used in grasping food, in predation, and in defense. They are generally busy exploring the substratum. The second and third pairs of walking legs, but not the fourth and fifth, are also chelate. The appendages on abdominal segments 14 through 18 are called **swimmerets** (= pleopods), although they are not used for swimming in crayfish and many other crustaceans. Females use their swimmerets to carry eggs, and males have the first pair of swimmerets modified as copulatory organs. The 19th and last pair of segmental appendages are two **uropods,** which combine with the telson to make a large fin used for swimming.

Exoskeleton. Like other arthropods, crustaceans have an exoskeleton with a cuticle reinforced by chitin. Most crustaceans also have substantial deposits of calcium carbonate in the three layers of cuticle (epicuticle, exocuticle, endocuticle; Figure 31.3C). The blue crab *Callinectes sapidus*, for example, has the equivalent

Figure 31.2
Types of appendages of a crayfish and their functions. Numbers indicate the segment. All the appendages are thought to originate from a primitive biramous appendage schematically represented at top left. The protopod of this appendage has two joints, the coxa and basis. Attached to the basis is a lateral exopod and a medial endopod. During development the exopod is commonly lost, as in the mandibles and cheliped, but additional branches may sprout. A medial branch is called an endite (see the second maxilla), and a lateral branch is called an exite. An exite on a protopod is called an epipod. All the gills are epipods. Many zoologists add the sufix -ite to most of these structures: hence exopodite, endopodite, and so on.

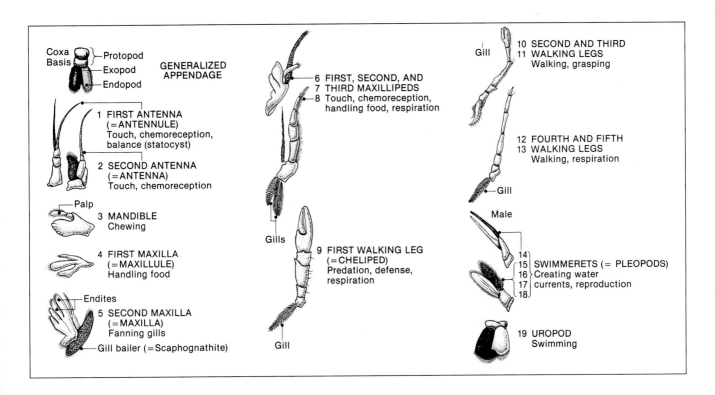

of four sticks of chalk deposited in its claws, carapace, and other hard areas of cuticle. Because of the rigid cuticle, crustaceans can grow only after molting. Molting is not only an opportunity to grow but also a chance to regenerate limbs that may have been damaged or lost before the molt. Molting is also a physiological and behavioral crisis, however. Chitin, protein, and calcium are lost and must be replaced. Even worse, the cuticle is soft just after molting, rendering the animal subject to predation. Before molting, therefore, many crustaceans must stop feeding and other activities to seek a protective shelter. While the cuticle is still soft, the lobster *Homarus americanus* and perhaps other crustaceans are also less able to compete with members of their own species and may lose territorial and other advantages. Molting is most frequently studied in the blue crab *Callinectes sapidus*, whose scientific name means "beautiful swimmer that tastes good" (Figure 31.4). Soft-shell crabs, which you may have eaten, are just-molted members of this species. One reason molting is so often studied in the blue crab is that commercial crabbers already have a great deal of knowledge of and interest in the subject (Warner 1976).

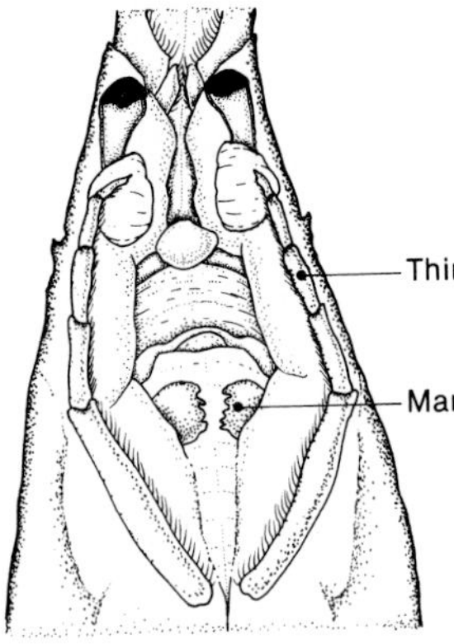

REPRODUCTION

In the blue crab and many other crustaceans, reproduction is closely tied to molting. Like most crustaceans, *Callinectes sapidus* has separate sexes and reproduces by copulation. The female can mate only after her final, or "nuptial," molt as an adult. There are several reasons for this. First, the molting hormone ecdysone is required for the production of ova. Second, copulation is physically impossible unless the female's cuticle is still soft from molting. Third, during the nuptial molt the female exchanges her narrow **apron,** which is actually the abdomen folded beneath the cephalothorax, for the broad apron required for copulation and for carrying eggs.

The nuptial molt usually occurs in autumn, in the shallow, dilute seawater of bays and estuaries. A week or so before the nuptial molt the female becomes

Figure 31.3
Mouthparts of the crayfish Cambarus.

Third maxilliped

Mandible

Figure 31.4
The edible blue crab Callinectes sapidus, *which is found all along the Atlantic coast. The blue color is restricted to the claws, which are also tinged with red, especially in females. In this and other adult crabs, the abdomen is tucked beneath the cephalothorax to form an apron that encloses the swimmerets. The rear pair of legs are modified as paddles that allow the blue crab to swim rapidly in any direction. Width 23 cm.*

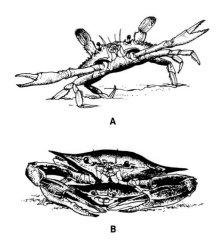

A

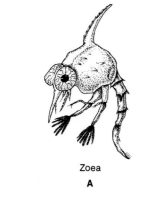

B

Figure 31.5
Stages in mating by blue crabs. (A) A male courting a female. (B) The male cradling the female until she molts.

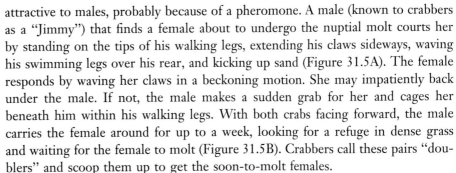

attractive to males, probably because of a pheromone. A male (known to crabbers as a "Jimmy") that finds a female about to undergo the nuptial molt courts her by standing on the tips of his walking legs, extending his claws sideways, waving his swimming legs over his rear, and kicking up sand (Figure 31.5A). The female responds by waving her claws in a beckoning motion. She may impatiently back under the male. If not, the male makes a sudden grab for her and cages her beneath him within his walking legs. With both crabs facing forward, the male carries the female around for up to a week, looking for a refuge in dense grass and waiting for the female to molt (Figure 31.5B). Crabbers call these pairs "doublers" and scoop them up to get the soon-to-molt females.

If not interrupted by crabbers, the male continues to cradle the female during the two to three hours required to molt, and during the post-molt period in which she inflates herself with water to stretch the cuticle for future growth. The male then turns the female onto her back. Then the female unfolds her broad abdomen to embrace the male and to expose two **genital pores** that lead to **sperm receptacles.** The male pulls two hollow swimmerets from beneath his narrow abdominal apron and inserts these copulatory organs into her genital pores. For the next 5 to 12 hours the male transfers two packets of sperm released from the bases of these swimmerets into the sperm receptacles of the female. This injection is essential since the sperm of crabs and most other crustaceans lack flagella and are immotile.

After copulation the male turns the female upright again and continues to cradle her for at least two more days. In this way he protects his reproductive investment until her shell hardens. The male then lets go of the female, which soon starts to swim toward the saltier waters at the mouth of the bay or estuary. Although blue crabs copulate, fertilization of the eggs occurs externally and after a long delay. In the spring following mating, the female releases a sticky mass containing some two million eggs and mixes it with one of the packets of sperm. The female broods the fertilized eggs beneath her abdominal apron. The eggs are orange at first, then turn black as the larvae within them develop.

DEVELOPMENT

In some crustaceans, including crayfish, development is direct. That is, the just-hatched young are miniatures of the adult. In blue crabs, however, the egg hatches into a larva, called a **zoea** (pronounced ZO-ee-uh; Figure 31.6A). The zoea is barely a millimeter long, has no claws, cannot actively swim, and does not feed. The zoea usually molts six times in as many weeks, developing in size and structure with each molt. The sixth-stage zoea then molts again into a **postlarva,** which, by definition, has the same complement of functional legs as the adult. In crabs the postlarva is called a **megalops** and looks more like a crayfish or shrimp than a crab (Figure 31.6B). Hordes of these megalops continue the migration to sea. (So many pass Virginia Beach on their way out of the Chesapeake Bay that the irritating nips from their tiny claws chase out the bathers, who complain of "water fleas.") In the ocean the megalops join the countless other kinds of zooplankton that form the base of most food webs. The odds of any one megalops surviving to adulthood are perhaps one in a million, but there are so many of them that blue crab populations remain stable in spite of heavy predation.

After two weeks a megalops molts into a pinhead-sized version of the adult, which migrates toward shores and into estuaries and bays. By November of the first year of its life, this adult form has molted seven or eight times and grown to a few centimeters long. It then burrows into the bottom, where it is protected

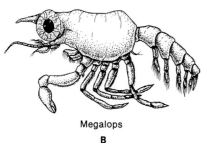

Zoea

A

Megalops

B

Figure 31.6
Stages in the life of the blue crab. (A) The egg hatches into a zoea, which usually molts seven times, increasing in size with each molt. (B) The seventh molt produces the megalops, which has claws and five pairs of legs like the adult. The megalops larva then molts into the adult form.

from winter cold. In spring it resumes its movements and molting. A blue crab may live approximately three years, grow to approximately 12 cm long, and molt more than 20 times until it is ready to complete the life cycle by mating.

In at least some members of each class of Crustacea, and especially in those considered primitive (copepods, barnacles, and some others), the first larval form is a **nauplius** (NAW-plee-us; Figure 31.7). The nauplius, with its oval, unsegmented, microscopic body, occurs only in crustaceans. Its three pairs of appendages will develop into the first and second antennae and the mandibles. Perhaps the oddest feature of the nauplius larva is the one median eye, called the **naupliar eye.** After several molts the nauplius becomes a zoea, then a postlarva. Even in crustaceans without a free-swimming nauplius, such as crayfishes and crabs, a nauplius stage is assumed to occur within the egg.

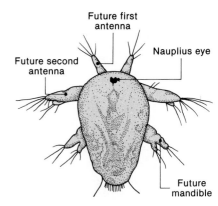

Figure 31.7
The planktonic nauplius larva of the copepod Cyclops fuscus. *The first antenna is uniramous. The biramous second antennae and mandibles are used in swimming. The nauplius does not feed.*

INTERNAL STRUCTURE AND FUNCTION

Digestion. The digestive system of crayfish and many other crustaceans includes a short esophagus connecting the mouth to the stomach, a long, straight intestine terminating at the anus, and the **digestive gland** (Figure 31.8). The final stages of digestion occur within the digestive gland, which is also called the midgut gland. The digestive gland is also called the hepatopancreas, because it also stores fats and glycogen like the vertebrate liver and secretes digestive enzymes like the ver-

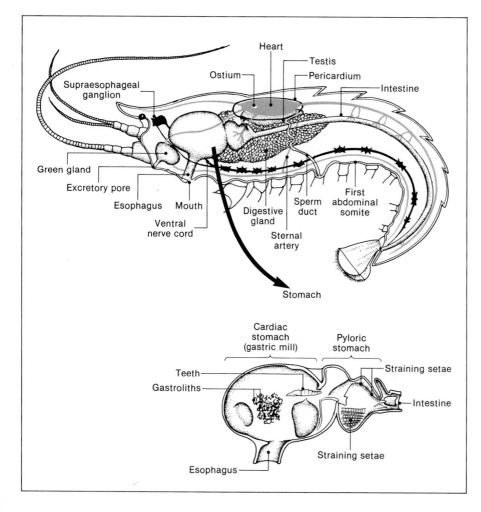

Figure 31.8
Longitudinal section of a male crayfish, showing the internal organs. The inset shows details of the stomach, which is divided into cardiac and pyloric sections.

tebrate pancreas. The stomach is divided into two chambers in crayfishes. The anterior **cardiac** chamber is used primarily for grinding and storage. Chitinous teeth attached to the muscular lining of the cardiac stomach make up a **gastric mill** that crushes and tears large pieces of food. The cardiac stomach also recovers calcium from molted cuticle and stores it as solid deposits called **gastroliths.** Food that has gone through the gastric mill passes into the smaller **pyloric stomach,** which is lined with **setae** that sort out particles small enough to pass into the intestine, and still smaller particles that go to the digestive gland.

Excretion and Osmoregulation. The crustacean organs for excretion and osmoregulation are unlike those of other arthropods. Excretion of metabolic wastes, primarily ammonia, is due to diffusion across the gills into the environment, even in terrestrial crustaceans. Little osmoregulation is required in marine crustaceans, since the body fluids have ion concentrations similar to those of seawater. In freshwater crustaceans, such as crayfishes, and in estuarine crustaceans, such as the blue crab, osmoregulation is due mainly to active transport of ions across the gill surface into the blood. Crustaceans also have **antennal glands** (= green glands) or **maxillary glands** that mainly conserve potassium and calcium and excrete excess sulfate and magnesium in most species. The two types of gland are similar to each other; the names depend on whether the organs occur at the bases of the second antennae or the second maxillae. Crustacean embryos have both antennal and maxillary glands, but one or the other is lost during development in most species.

Crayfishes have antennal glands that regulate the concentrations of certain ions and eliminate excess water (see p. 287). The antennal gland of a crayfish consists of an **end sac** (which is a rudiment of the coelom) and a **tubule** that leads into a urinary **bladder.** The antennal gland works by a combination of filtration and active transport. Pressure in the hemocoel forces hemolymph into the end sac. The filtrate then enters the green **labyrinth,** which is considered a part of either the end sac or the tubule. The function of the labyrinth is not known. As the filtrate passes through the tubule, nutrients and ions are actively reabsorbed across the lining into the hemolymph, and water is actively secreted into the urine. The urine then enters the bladder and is eventually excreted through a pore at the base of the second antenna.

Circulation and Respiration. Crustaceans, like all arthropods, are coelomates, but the coelom is rudimentary. In crayfish the coelom occurs only in the excretory organ and as a chamber surrounding the gonads. The main body cavity is the **hemocoel,** which consists of blood-filled spaces inside and around major organs. As in other arthropods, the circulatory system is open, and there is no clear distinction between the blood (= hemolymph) and the interstitial fluid. The almost colorless blood contains some dissolved respiratory pigment **hemocyanin,** as well as ameboid blood cells involved in clotting and phagocytosis. The blood is pumped under low pressure by a heart that lies within a **pericardium** along the dorsal midline (Figures 31.8 and 31.9). Blood leaves the heart through several arteries, including a **sternal artery** that connects the heart to a ventral artery. From these arteries the blood passes through the hemocoel, then collects in the **sternal sinus,** where it enters the gills. Oxygenated blood from the gills then enters the pericardium, where it returns to the heart through one-way valves called **ostia.** The gills lie on both sides of the thorax within gill chambers formed by the carapace. As noted previously, the gills are branches of the legs, and they are irrigated by the gill bailers on the second maxillae.

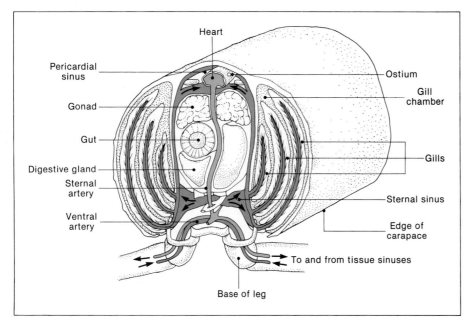

Figure 31.9
Cross section of a crayfish cephalothorax at the level of the sternal artery. (Compare with previous figure.) Blood from the heart flows through arteries (red) into sinuses within tissues, then returns to the sternal sinus (blue). From there it flows through the gills, (green) where it is oxygenated before returning to the heart through ostia.

Central Nervous System. The crustacean nervous system is typically arthropodan, with ganglia linked in series along a ventral nerve cord. In primitive crustaceans, such as fairy shrimp, each segment has a pair of ganglia side by side. The ganglia are joined to each other by connectives, forming a ladderlike array similar to that in annelids. In most crustaceans the pair of ganglia in each segment fuse into one ganglion. Ganglia from adjacent segments also tend to fuse, especially in the cephalothorax (Figure 31.10). In some crustaceans the axons in the connectives include giant axons that trigger the rapid backward movement that is familiar to anyone who has ever tried to catch a crayfish. These large-diameter giant axons are activated by mechanoreceptors that respond to movements of potential predators. The giant axons then trigger flexion of the abdomen, which causes the rapid flip of the telson that propels the crayfish backward. The speed of the reflex is due to the giant axons' ability to conduct action potentials rapidly and to electrical synapses that link the sensory nerve cells to the giant axons, and the giant axons to the motor neurons.

The largest of the fused ganglia are the **subesophageal** and **supraesophageal ganglia** in the head. The subesophageal ganglia are mainly involved in coordinating feeding movements. An important part of this coordination is the integration of information from chemoreceptors and the numerous mechanoreceptive

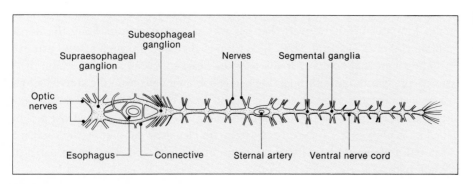

Figure 31.10
Outline of the crayfish central nervous system.

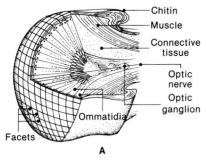

Chitin
Muscle
Connective tissue
Optic nerve
Optic ganglion
Ommatidia
Facets

A

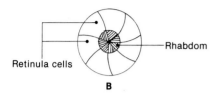

Retinula cells
Rhabdom

B

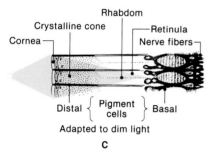

Crystalline cone
Cornea
Rhabdom
Retinula
Nerve fibers

Distal { Pigment cells } Basal

Adapted to dim light

C

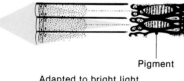

Pigment

Adapted to bright light

D

Figure 31.11
(A) The compound eye of the crayfish Cambarus, *showing some of the approximately 2500 ommatidia. (B) Cross section of an ommatidium. The seven retinula cells form the rhabdom, where light is transduced into an electrical response. (C) An ommatidium of the superposition compound eye of a crustacean adapted to dim light. (D) An ommatidium of the eye of a crayfish in bright light. Pigment from distal and basal pigment cells forms a sheath that isolates each ommatidium from light from neighboring ommatidia.*

tactile hairs on the mouthparts. The supraesophageal ganglia integrate information from mechanoreceptors and chemoreceptors on the antennae, from the statocyst organs, and from the eyes. The supraesophageal ganglia also exercise dominance over many of the other ganglia, though not to the extent implied by the term "brain" that is often used.

Vision. Most crustaceans have good vision, often with eyes on long, mobile stalks. The eyes of crustaceans (and some eyes of horseshoe crabs and insects) are called **compound eyes,** because each is composed of many subunits, called **ommatidia** (Figure 31.11A). Each ommatidium accepts light from a small area of the visual field. Photoreceptor cells in each ommatidium send action potentials to the supraesophageal ganglia, which process information about the intensity, color, and angle of polarization of the light received by the ommatidium. In this way the arthropod compound eye analyzes a visual stimulus bit by bit, without focusing an image onto a retina. The supraesophageal ganglia presumably integrate the information from all the ommatidia into some kind of picture of the visual stimulus, but it is impossible to say what such a picture looks like to an arthropod. One often sees photographs taken through compound eyes that purportedly show an arthropod's view of the world. All such photos really show, of course, is what the world would look like to us if we used arthropod eyes as contact lenses.

Each eye is covered by transparent cuticle called the **cornea,** which is divided into numerous square **facets,** with one facet covering each ommatidium (Figure 31.11A). Beneath each facet is a **crystalline cone.** The photoreceptive region of the ommatidium is the **retinula** ("little retina"), and it usually consists of seven or eight **retinula cells** (Figure 31.11B). The retinula cells have numerous parallel microvilli projecting toward the central axis of the ommatidium and constituting the **rhabdom.** The rhabdom contains the photopigments and is believed to be the site where light energy is transduced into voltage changes that evoke action potentials. (Like most other invertebrates, arthropods therefore have **rhabdomeric,** rather than ciliary eyes. See p. 155.) Many arthropods have good color vision, owing to several different kinds of photopigments that are sensitive to different wavelengths. Some mantis shrimps, in fact, have ten different photopigments, compared with the three in humans. Many arthropods can also perceive ultraviolet light, which is invisible to humans. A parallel array of microvilli in the rhabdoms also permits some arthropods to detect the angle of polarization of light. This ability allows the animal to determine the position of the sun for navigation even when the sun is hidden by clouds, the shoreline, treeline, or other objects.

The compound eyes of arthropods are generally adapted either for sharp vision in bright light or for high sensitivity in dim light. In crustaceans and insects that are active in bright light, each ommatidium is shielded from its neighbors by pigment, so it samples only a small region of the visual field. Compound eyes of this type are called **apposition eyes.** In apposition eyes the light is focused onto the retinula of each ommatidium either by the crystalline cone (in most crustaceans) or by the cornea (in insects). Apposition eyes are presumably adapted for the perception of detail. In contrast, each retinula of crustaceans and insects that are active at night, in shade, or in dark water receives light from several neighboring ommatidia, because there is less screening pigment, and neither the crystalline cone nor the cornea focuses the light. Such **superposition eyes** are more sensitive in dim light but not as sharp as apposition eyes. In bright light superposition eyes commonly reduce their sensitivity and increase their acuity with pigment that migrates around the ommatidia, isolating them as in apposition eyes (Figure 31.11C,D).

The Statocyst Organ. Within a chamber at the base of each first antenna in crayfishes and some related crustaceans is a mechanoreceptor organ, the statocyst, that allows a crayfish to tell which way is up. The statocyst organ works much like the utricle and saccule of the human inner ear (see Figure 7.16). A **statolith,** consisting of a dried secretion or sand grains stuck together, rests on hair cells that send action potentials to the central nervous system. As the orientation of the body changes with respect to gravity, the statolith rests on different hair cells, allowing the crustacean to determine its orientation. The statocyst is lined with cuticle, so the animal must replace its statolith after each molt. A crayfish cannot tell up from down until fresh sand enters the statocyst. Mischievious zoology students have been known to replace the sand in a crayfish's aquarium with iron filings, so that after molting the crayfish will flip onto its back when a magnet is held above it.

Endocrine System. Hormones play major roles in coordinating the physiology of crustaceans. The most important endocrine organ is the **X organ–sinus gland (XOSG) complex,** which is located near the optic nerve. Another important endocrine organ is the **Y organ,** which is located at the base of each maxilla. Among the hormones from the XOSG system is **molt-inhibiting hormone** (MIH). MIH blocks molting by inhibiting the secretion of **ecdysone** from the Y organs. Because of MIH, molting occurs only when certain environmental cues, such as changes in temperature or day length, inhibit the X organ, allowing the Y organs to secrete ecdysone.

Other secretions from the XOSG complex control **chromatophores,** enabling the integument to change color. One hormone causes the pigment to become more concentrated inside the red chromatophores, thereby making the integument less red. Another secretion from the XOSG system, the **crustacean hyperglycemic hormone,** is analogous to adrenaline and glucagon in vertebrates, increasing the conversion of glycogen stores into glucose. Still another secretion, the **distal retinal-pigment hormone,** aids in the adaptation of the compound eyes to dim light.

Crayfishes and related crustaceans also have **androgenic glands,** which cause masculinization. Androgenic glands are degenerate in females but normally become fully developed in adult males. If the androgenic glands are removed from males, by certain parasites, for example, the male becomes feminized in structure and behavior. If androgenic glands are implanted into females, their ovaries turn into functioning testes, and "she" resembles a male after the next molt.

MALACOSTRACANS

Crustaceans are extremely adaptable in morphology, so their taxonomy is unsettled. It does not take much searching to come up with a half dozen different schemes for classifying them. Most taxonomists, however, agree in placing the majority of crustaceans, including crayfishes, hermit crabs, "true" crabs, lobsters, shrimp, and terrestrial wood lice, in the class Malacostraca. Most of these seemingly unrelated animals share a combination of traits referred to as the "caridoid facies." The major features of the caridoid facies appear in crayfishes (Figure 31.1). These features include the particular number of somites in the cephalothorax (13) and abdomen (6), the carapace covering the cephalothorax and terminating anteriorly in a rostrum, the telson at the end of the abdomen, and the particular assortment of appendages (stalked eyes, biramous first antennae, walking legs, and swimmerets).

Figure 31.12

Some decapod crustaceans. (A) The common edible lobster Homarus americanus, *found along the Atlantic coast from Labrador to Virginia. In life the lobster is greenish black on back; it turns all red only after boiling breaks down pigment molecules. Note the difference between the crusher and the cutter claws. Length 86 cm. (B) The fiddler crab* Uca minax. *The enlarged claw of the male is used in territorial defense, sexual display, and to block its burrow after it has retreated into it. Fiddler crabs and other land crabs do not have rear legs adapted for swimming, as do blue crabs and other swimming crabs. Width approximately 4 cm. (C) Hermit crabs, such as* Pagurus, *are distinct from blue crabs, fiddler crabs, and other "true" crabs. The hermit crab does not secrete a shell but "borrows" one from a deceased mollusc. As the hermit crab grows, it must find a larger shell, which it does by inspecting any object that releases calcium but not substances that indicate the shell is already occupied (Mesce 1982). Line drawing:* Pagurus *without a shell. Approximately 4 cm long. (D) A redbacked cleaner shrimp* Hippolysmata grabhami *on a moray eel. Like certain fishes (see p. 329), these colorful shrimp remove parasites from animals that would ordinarily prey on shrimp. Length 2.5 cm.*

Decapods. Within class Malacostraca most taxonomists recognize several subclasses and superorders and more than a dozen orders. It is beyond the scope of this text to treat more than a few of the common groups. Approximately 8500 species of crayfishes, lobsters, "true" crabs, hermit crabs, "true" shrimps, and many other forms are decapods (Figure 31.12). As the name implies, the decapods

A

B

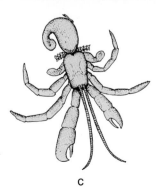

C

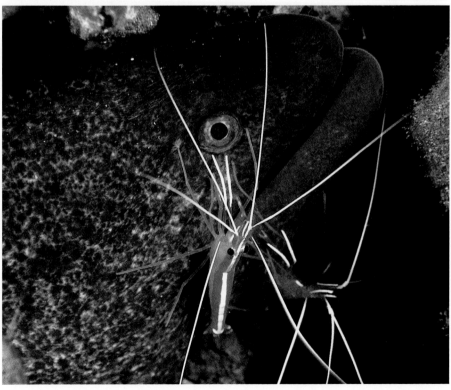

D

have five pairs of walking legs. The anterior pair is usually specialized with large chelae. Decapods are also characterized by the three pairs of maxillipeds. Crayfishes and lobsters with large claws are essentially the same, except that lobsters live in seawater and are generally larger. Spiny lobsters are differentiated from other lobsters by the absence of enlarged chelae on the first walking legs (Figure 31.21) and by their unique **phyllosoma** larvae, which are so thin that they are transparent. "True" crabs have flattened carapaces, large chelae, and reduced abdomens folded beneath the cephalothorax. Hermit crabs have coiled abdomens that allow them to back into empty mollusc shells. "True" shrimps (called prawns in Britain) are usually compressed laterally and are smaller than other decapods, but larger than the crustaceans mentioned below that are also called shrimp.

Krill. Another important malacostracan group is Euphausiacea, which includes about 90 small, shrimplike species known as "krill" (Figure 31.13). (Strictly speaking, "krill" is anything that whales eat. Often, however, the term is restricted to the Euphausiacea.) These abundant marine planktonic forms are a major item in the diets of whales, seals, penguins, cephalopods, and other consumers (see Figure 17.15). Unlike decapods, euphausids lack maxillipeds and chelae. Instead, all the thoracic appendages are biramous legs used in filter feeding. The carapace does not cover the gills. Krill are often more visible at night than in daylight because of light from a **photophore.**

Amphipods. Another group of malacostracans includes the Amphipoda, which differ from decapods and euphausids in several respects. Amphipods lack a carapace, and the abdomen is not sharply different from the thorax. The compound eyes lie flat on the sides of the head, rather than on stalks. There are two or more types of leg adapted for different functions. In most species the body is compressed laterally. Like decapods, female amphipods brood the eggs and young. Overlapping flaps from the thoracic legs form the ventral brood chamber (**marsupium**). Most amphipods live in marine or freshwater habitats, and a few are parasitic (Figure 31.14). The most familiar forms are the "beach fleas" or "sand hoppers" that are often glimpsed burrowing into sand after being uncovered at the shore line.

Isopods. Isopods resemble amphipods in the absence of a carapace and the use of a marsupium for brooding young (Figure 31.15). They differ from amphipods in being flattened dorsoventrally and in having legs of only one type (hence the name). The group Isopoda includes not only marine, freshwater, and parasitic forms but also terrestrial species. The most familiar isopods are wood lice, such

Figure 31.13
Euphausia superba, *one of the most important species among krill.*

Figure 31.14
An amphipod: the beach flea
Talorchestia *burrowing in sand.*

as the sow bug *Porcellio* and the pill bug *Armadillidium*. (The difference between sow bugs and pill bugs is that the latter roll into a ball when threatened.) These crustaceans breathe through gills and are therefore confined to moist environments, such as basements and beneath stones.

SOME OTHER CRUSTACEANS

Ostracods. In addition to the Malacostraca, several other major groups of Crustacea are generally recognized. The group (usually class) Ostracoda includes the seed shrimps, which look more like tiny clams than shrimps (Figure 31.16). Ostracods live either in seawater or fresh water on the bottom, on plants, or as plankton. Most ostracods are scavengers, but many are predators or parasites. Although small, they are quite numerous and widespread and comprise an important link in food webs between producers (algae) and consumers, such as fishes.

Figure 31.15
The common pill bug Armadillidium.

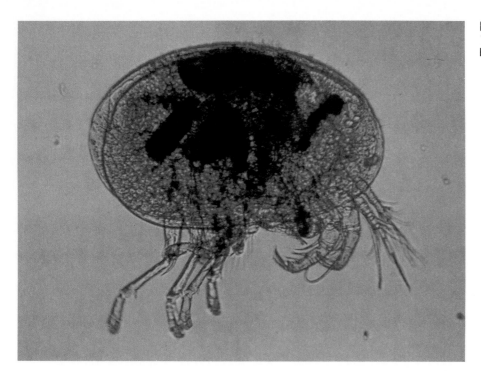

Figure 31.16
The freshwater ostracod
Entocytheria. *Diameter 1.1 mm.*

Copepods. The group Copepoda includes the copepods, which are also small but important planktonic crustaceans (Figure 31.17). Copepods vary considerably in form, but in general they lack a carapace and abdominal appendages. Usually there are four pairs of biramous appendages and a single pair of uniramous maxillipeds. Often the first antennae are the most prominent appendages and aid in locomotion. There is no compound eye, but many species retain the single median eye from the nauplius larva. The common freshwater genus *Cyclops* was named for the one-eyed monster of Greek mythology. Copepods are major food sources for whales and fishes. Like the much larger animals that feed on them, many free-living copepods are suspension feeders. *Cyclops* and some others are predatory. Many other copepods are ectoparasitic, and virtually every species of fish is host to at least one species of copepod. Each species of copepod is selective about its host species, and even about the site of attachment. Some copepods attach only to the nostrils of cod, some to the gills of salmon, and some to the spiracles of skates.

Barnacles. Until 1830 barnacles (cirripedia) were considered to be molluscs, but in fact they are crustaceans, perhaps related to copepods (Figure 31.18). Like clams, most adults in the group Cirripedia remain secluded within calcareous plates attached to solid substratum. The clue to their true identity is the nauplius larva, which allows for dispersal of the species. The nauplius develops into a **cypris larva,** named after an ostracod it resembles. The cypris larva has compound eyes and a bivalve carapace. After swimming a short time, the cypris larva glues its first antennae to substratum, using a polysaccharide cement that is the strongest adhesive known. The carapace then develops into a **mantle** that secretes the calcareous plates as the adult form develops. Inside the mantle cavity of the adult most of the typical arthropod structures are present, although inverted and greatly modified (Figure 31.18C). In nonparasitic species the legs form **cirri** (thus the name of the group, Cirripedia). The cirri draw food into the mouth except when the plates are tightly closed for protection.

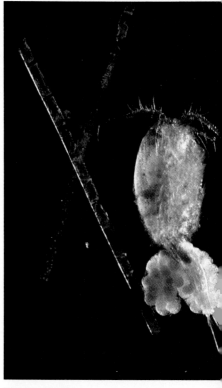

Figure 31.17
Cyclops, *a common freshwater copepod. Note the single red eye.*

A

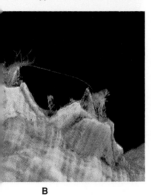

B

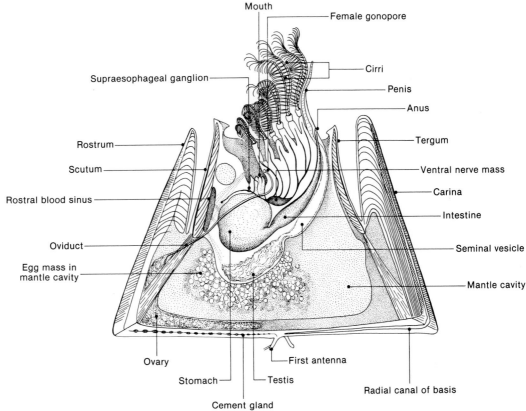

Mouth

Female gonopore

Cirri

Penis

Anus

Supraesophageal ganglion

Tergum

Rostrum

Scutum

Ventral nerve mass

Carina

Rostral blood sinus

Intestine

Seminal vesicle

Oviduct

Egg mass in mantle cavity

Mantle cavity

Ovary

First antenna

Stomach

Testis

Cement gland

Radial canal of basis

C

Figure 31.18
Barnacles. (A) The goose barnacle Lepas anatifera. The common and Latin names (anatifera = "goose-bearing") come from the resemblance of the shell and stalk to the body and neck of a goose. In the Middle Ages it was commonly believed that goose barnacles developed into geese. Note the cirri. Length 15 cm. (B) Encrusting barnacles Balanus balanoides mate by extending the penis to a neighboring barnacle. Barnacles are hermaphroditic but avoid self-fertilization. The other structures extending from the shells are cirri. Barnacles often encrust ships' hulls but also choose other animals, such as whales, molluscs, and crabs, as substratum. Height approximately 2.5 cm. (C) Internal organization of Balanus. Comparison with Figure 31.8 shows that the barnacle has typical crustacean features but is extremely modified in organization.

Rhizocephalans. The group Cirripedia also includes some even more bizarre animals, the Rhizocephala, that illustrate so well the delicate adaptations of parasites to hosts. The best-known rhizocephalan is *Sacculina*, which begins parasitism when its female cypris larva attaches to a crab and injects a mass of undifferentiated cells. These cells migrate to the intestine of the host and, like germinating seeds, develop rootlike growths that spread throughout the host's body. The cells then multiply and differentiate into the adult, reproductive form, which erupts as an **external mass** beneath the crab's apron. The term "external mass" is about as descriptive as one can get; the only resemblance of rhizocephalans to crustaceans occurs in the larvae. The position of the external mass beneath the apron is crucial. In any other position it would be wiped off, but while it is beneath the apron it is treated by a female host as if it were her own egg mass. In fact, survival of the parasite depends on the host ventilating, grooming, and protecting the external mass. If the host crab happens to be a male, the parasite feminizes it, perhaps by destroying the androgenic gland. The castrated male then develops the broad apron and brooding behavior of females. The external mass has her own brood chamber, into which a male cypris larva releases cells that become a testis. This testis then fertilizes the parasite's ova, which develop into free-swimming nauplius larvae, and then new cypris larvae.

Fish Lice and Tongue Worms. The group Branchiura includes two groups of ectoparasites that live on vertebrates. The first group, commonly called fish lice, live on marine and freshwater fishes. They have flattened bodies, compound eyes, and sucker-shaped maxillae by which they attach to hosts. Fish lice are often a problem in goldfish aquaria. The second group are the tongue worms, which are ectoparasites in the respiratory tracts of vertebrates. As might be expected from this odd habitat, the adults are unlike other crustaceans, or any other kind of animal for that matter. They are so different, in fact, that until recently they were classified in their own phylum Pentastomida (Figure 31.19). Pentastomids are up to 15 cm long, but the males are usually much smaller. The mouth lies between two pairs of retractable claws, with which the tongue worm clings to the respiratory epithelium of the host, feeding on blood, mucus, lymph, and tissue. Most pentastomids parasitize reptiles, but some live in the air sacs of sea birds, and others live in the nasal passages of canines and felines. Several species occasionally infect the nasal passages of humans, but usually without causing symptoms.

Branchiopods. Members of the group Branchiopoda include the tiny, but rather familiar, "water fleas," brine shrimp, and several other forms of so-called shrimp (Figure 31.20). The class takes its name from the leaf-shaped appendages, which exchange respiratory gases. Water fleas (group Cladocera), such as *Daphnia,* make up a large portion of freshwater zooplankton. Examination of just about any sample of pond water will reveal water fleas swimming jerkily by means of their large second antennae. One of the secrets of their success is their mode of reproduction, which is remarkably like that of rotifers (see p. 558). In the summer when conditions are good, the females parthenogenetically produce more females every two or three days. When conditions become unfavorable, the females produce some eggs that have to be fertilized, and some males to fertilize them. The fertilized eggs are extremely resistant to cold and drying and can be transported to new habitats by wind or in mud that clings to other animals.

Other branchiopods are fairy shrimp and brine shrimp (group Anostraca), which live in temporary pools. After the pools dry up the eggs have been known to survive for decades until the rains come again, or the wind carries them to a new pond. The eggs then hatch within a day or two and develop into adults. The eggs also survive shipment from San Francisco Bay or the Great Salt Lake to pet stores throughout the country, where they are sold as a convenient way to get live food for aquarium fish. They hatch into nauplius larvae. Brine shrimp eggs are also marketed to children as "sea monkeys."

Remipedians and Cephalocaridans. Two other groups deserve mention here because they are usually given the taxonomic rank of classes (Figure 31.21). Class Remipedia was discovered by Jill Yager in 1981 in submarine caves in the Baha-

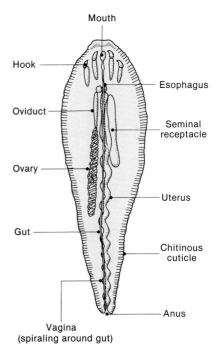

Figure 31.19
A female pentastomid, Linguatula serrata. *Pentastomids parasitize the respiratory tracts of canines, felines, and occasionally humans. Note the simple digestive system and the absence of respiratory, circulatory, and excretory organs. Approximately 10 cm long.*

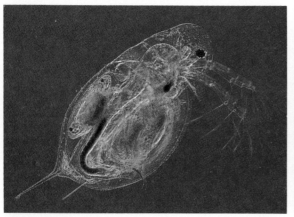

A

B

Figure 31.20
Branchiopods. (A) The female water flea Daphnia, *common in freshwater ponds. Two juveniles can be seen dorsally, developing in the brood chamber. Every few days the females molt, releasing the young and reloading the brood chamber with eggs. (B) The brine shrimp* Artemia salina *can live in ponds several times more saline than the sea. The female above is carrying eggs.*

Classes in Subphylum Crustacea According to Abele (1982)

Genera mentioned elsewhere in this chapter are noted.

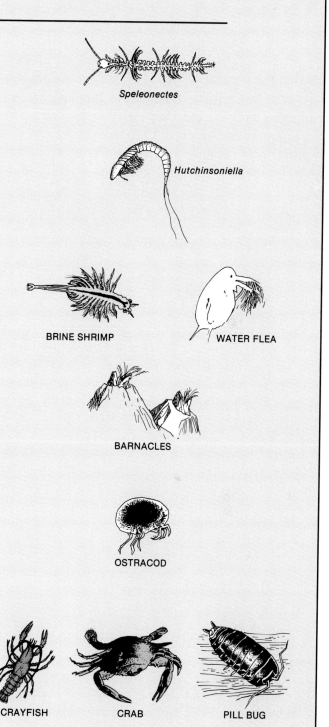

Class Remipedia rim-i-PEE-dee-uh (Latin *remipedes* oar-footed). Primitive inhabitants of marine caves. Biramous, oar-shaped trunk appendages are all similar. *Speleonectes.*

Speleonectes

Class Cephalocarida SEF-a-low-CARE-i-duh (Greek *kephale* head + *karis* shrimp). Primitive. Horseshoe-shaped head, 19-segment trunk, uniform trunk appendages. No carapace or eyes. Benthic; marine. Less than 4 mm long. *Hutchinsoniella.*

Hutchinsoniella

Class Branchiopoda BRAN-key-OP-oh-duh (Greek *branchia* gills + *podos* foot). Thoracic appendages leaf-shaped. Water fleas, fairy shrimp, and brine shrimp. *Artemia, Daphnia.*

BRINE SHRIMP   WATER FLEA

Class Maxillopoda max-ill-OP-oh-duh (Latin *maxilla* jaw). Large maxillae, used for feeding. Trunk reduced, usually to 11 segments. Fish lice, copepods, barnacles, tongue worms. *Balanus, Cyclops, Lepas, Linguatula, Sacculina* (Figures 31.11, 31.18, 31.19).

BARNACLES

Class Ostracoda os-TRACK-oh-duh (Greek *ostracodes* having a shell). Body enclosed in hinged, bivalved carapace. Swimming by second antennae and two pairs of trunk appendages. Seed shrimp. *Entocytheria.*

OSTRACOD

Class Malacostraca mal-a-KOS-tra-kuh (Greek *malakos* soft + *ostrakon* shell). Typically 19 somites (5 head, 8 thorax, 6 abdominal). Head and some thoracic somites fused; usually covered with carapace. Abdomen with appendages. Usually with stalked, compound eyes. Sand fleas, krill, "true" shrimps, crayfishes, lobsters, spiny lobsters, hermit crabs, "true" crabs, wood lice. *Armadillidium, Callinectes, Cambarus, Euphausia, Grapsus, Hippolysmata, Homarus, Limnoria, Mithrax, Pagurus, Porcellio, Talorchestia, Uca* (Figures 30.1B, 31.12, 31.13, 31.14, 31.22).

CRAYFISH   CRAB   PILL BUG

mas. Five genera of remipedes are now known, all from submerged caves. They are considered extremely primitive, because of the absence of tagmosis and the uniformity of the biramous legs, with which they swim upside down. Class Cephalocarida is also thought to be primitive on account of its uniform legs.

INTERACTIONS WITH HUMANS AND OTHER ANIMALS

Detrimental Effects on Humans. Crustaceans have a much better reputation among humans than do most other arthropods. Many crustaceans are good to eat, and few of them have poison fangs or spread diseases. One of the few crustaceans that is a vector for disease is a copepod of the genus *Cyclops*, which serves as an intermediate host for the dreaded Guinea worm (*Dracunculus*) (see pp. 552–553). Some other crustaceans present mainly economic problems. The marine isopods called gribbles, *Limnoria lignorum*, cause considerable damage to wooden pilings and boat hulls (Figure 31.22). Barnacles have long been a nuisance to shippers. A ship with a badly encrusted hull may have its top speed reduced by half, and the cost of fuel greatly increased. Until recently, the only solution was to beach the ship and scrape the hull. Hulls can now be coated with plastic to which the barnacle glue cannot adhere.

Crustaceans as Food. Crustaceans feed vast numbers of other animals, including humans. The tiny crustaceans, especially copepods, ostracods, and krill, are most important in natural food webs. These zooplankton eat phytoplankton and are eaten by fishes and other consumers. They therefore serve as links between producer and consumer organisms, including ourselves. Without them, animal populations in aquatic habitats would collapse. By reproducing efficiently these small crustaceans manage to maintain high populations in spite of heavy predation.

Krill, *Euphausia superba*, is being harvested for human consumption in the waters around Antarctica. As many as 12 tons have been netted in a single hour. It is assumed that the decline in whale populations, due largely to commercial whaling, has left a surplus of krill available for fishing. There is some doubt, however, whether the productivity of krill is really high enough to feed humans without endangering the other consumers, especially seals, penguins, and cephalopods (Ross and Quetin 1986).

Larger crustaceans, such as barnacles and decapods, tend to be protected from predators by their heavily calcified cuticles. They are, nevertheless, important food sources for a few predators. In the future, large crustaceans will be even more important components in the human diet, as techniques for ocean farming and ranching are developed. Already the harvest of spiny lobsters (= rock lobsters) on the Caribbean coast of Mexico is being increased by providing rough platforms under which they gather. There is some question, however, whether this technique increases production or merely increases the catch temporarily by attracting lobsters to the platforms (Tangley 1987).

Walter H. Adey (1987) and his staff at the Marine Systems Laboratory of the Smithsonian Institution have created artificial coral reef microcosms in the laboratory to help study the potential for increasing the productivity of edible crustaceans. They have also constructed rafts in the Caribbean as artificial reefs, providing substrata for algal growth. They suggest that edible animals such as the West Indian spider crab *Mithrax spinosissimus* would graze on the algal turfs and could be the basis for oceanic ranches that are similar in concept to cattle ranches.

Detrimental Effects of Humans. Ironically, just as we are becoming aware of the potential of crustaceans as sources of protein, we are also becoming aware of how some of our other activities are endangering that potential resource. Toxic chemicals deliberately and accidentally dumped into the oceans may be destroying the copepods, ostracods, and numerous other organisms on which fishes and other animals depend. Many of the toxic chemicals include residues of poisons that were

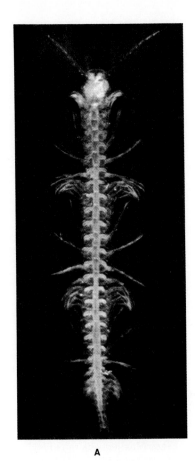

A

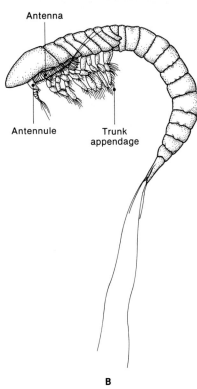

B

Figure 31.21
(A) The remipedian Speleonectes. *Approximately 2 cm long. (B) The cephalocaridan* Hutchinsoniella. *Length approximately 4 mm.*

Figure 31.22
(A) Wood-eating marine isopods,
Limnoria lignorum, *commonly called gribbles. (B) A wooden piling heavily damaged by gribbles.*

intended to kill insects, and which are most likely just as toxic to crustaceans. Deliberate and accidental oil spills may be equally devastating. A spill of fuel oil off the coast of Massachusetts in 1969 directly and indirectly killed many salt-marsh fiddler crabs *Uca pugnax* (Krebs and Burns 1977). The salt marsh had still not recovered seven years after the spill. "Burn-spot disease," which is caused by bacteria that attack chitin, has appeared in lobsters and other crustaceans in deep ocean trenches 120 miles southeast of New York City, where that city legally dumps sewage sludge.

SUMMARY

Like all arthropods, crustaceans are segmented animals with jointed appendages. Also like most other arthropods, the coelom is much reduced, and the main body cavity is a hemocoel. Crustaceans are distinguished from chelicerates in having mandibles, and from insects and other uniramians in that each mandible develops from the first segment of an appendage, rather than from an entire appendage. In addition, the appendages are primitively biramous. Most crustaceans retain many more appendages than do either chelicerates or uniramians, and these appendages are modified into a great variety of serially homologous structures, including antennae, mouthparts, walking legs, and swimmerets.

Other distinguishing features, such as gills and appendages adapted for swimming, are due largely to most crustaceans having retained an aquatic mode of life. Some crustaceans also have unique organs, such as the statocyst organ for determining the direction of gravity, the antennal gland for osmoregulation and excretion, and the X organ–sinus gland complex that coordinates molting. With some notable exceptions, such as barnacles, crustaceans are dioecious. Development is direct in many species, but many others produce a distinctive one-eyed nauplius larva.

Most familiar crustaceans, including crayfishes, hermit crabs, "true" crabs, lobsters, "true" shrimps, and wood lice, are usually placed in class Malacostraca. Other important malacostracans are krill, amphipods, and ostracods. Other classes include such important and familiar forms as copepods, barnacles, and brine shrimp.

KEY TERMS

carapace	maxilliped	statocyst
exopod	swimmeret	compound eye
endopod	gill bailer	rhabdom
protopod	cheliped	X organ–sinus gland complex
mandible	apron	antennal gland
maxilla	nauplius	

SELF-TEST

1. Describe three features that crustaceans share with other arthropods.

2. Describe three features that distinguish crustaceans from chelicerates.

3. Name five different types of crustacean appendages, and describe the function of each.

4. Diagram a compound eye, showing the location of the cornea, the retinula cells, and the rhabdoms.

5. Give the common names of five kinds of malacostracans, and briefly describe each kind.

6. Give the common names of four kinds of nonmalacostracan crustaceans, and briefly describe each kind.

READINGS

RECOMMENDED READINGS

Benson, A. A. and R. F. Lee. 1975. The role of wax in oceanic food chains. *Sci. Am.* 232(3):76–86 (Mar). (*Copepods, even more than other marine animals, store much of their energy as wax.*)

Caldwell, R. L. and H. Dingle. 1976. Stomatopods. *Sci. Am.* 234(1):81–89 (Jan).

Cameron, J. N. 1985. Molting in the blue crab. *Sci. Am.* 252(3):102–109 (May).

Fitzpatrick, J. F. Jr. 1983. *How to Know the Freshwater Crustacea.* Dubuque, IA: W. C. Brown.

Nilsson, D.-E. 1989. Vision optics and evolution. *BioScience* 39:298–307. (*Includes excellent description of various kinds of compound eyes.*)

Warner, W. W. 1976. *Beautiful Swimmers: Watermen, Crabs and the Chesapeake Bay.* Boston: Atlantic–Little, Brown. (*A popular account of the lives of crabs and crabbers.*)

Wicksten, M. K. 1980. Decorator crabs. *Sci. Am.* 242(2):146–154 (Feb).

See also relevant selections in General References at the end of Chapter 21.

ADDITIONAL REFERENCES

Abele, L. G. (Ed.). 1982. *The Biology of Crustacea*, Vol. 1. New York: Academic Press.

Abele, L. G., W. Kim, and B. E. Felgenhauer. 1989. Molecular evidence for inclusion of the phylum Pentastomida in the Crustacea. *Mol. Biol. Evol.* 6:685–691.

Adey, W. H. 1987. Food production in low-nutrient seas. *BioScience* 37:340–348.

Emerson, M. J. and F. R. Schram. 1990. The origin of crustacean biramous appendages and the evolution of Arthropoda. *Science* 250:667–669.

Krebs, C. T. and K. A. Burns. 1977. Long-term effects of an oil spill on populations of the salt-marsh crab *Uca pugnax*. *Science* 197:484–487.

Mesce, K. A. 1982. Calcium-bearing objects elicit shell selection behavior in a hermit crab. *Science* 215:993–995.

Ross, R. M. and L. B. Quetin. 1986. How productive are Antarctic krill. *BioScience* 36:264–269.

Sanders, H. L. 1963. The cephalocarida; functional morphology, larval development, comparative external anatomy. *Mem. Conn. Acad. Arts Sci.* 15:1–80.

Tangley, L. 1987. Enhancing coastal production. *BioScience* 37:309–312.

Yager, J. 1981. Remipedia, a new class of Crustacea from a marine cave in the Bahamas. *J. Crustacean Biol.* 1:328–333.

32

Insects and Other Uniramia

Blue morpho (Morpho rhetenor).

CHAPTER OUTLINE

LEARNING OBJECTIVES

1. How do insects differ from spiders and crustaceans?

2. How are insects related to centipedes, millipedes, and similar animals?

3. Why are there so many kinds of insects? Why have they not (quite) taken over the world?

4. How do terrestrial insects conserve body water?

5. How do insects communicate?

6. How do they fly?

7. How are insect mouthparts adapted for so many kinds of functions, such as biting and sucking?

8. How do insects produce so many offspring?

9. How is a colony of termites, ants, or bees organized and coordinated?

I f a probe from another planet arrived on Earth to collect random samples of four species of living organisms, chances are that three of those four species would be insects. Those insects would probably convince whoever sent the probe that our little planet is well worth a visit. With the possible exception of vertebrates, no other major group of animals has succeeded so well in exploiting the possibilities of life. Insects have mastered virtually every habitat except the ocean, and they share the skies with only a few vertebrates. Their impact on humans and other animals is impossible to calculate. If it were not for a few accidents of physiology that kept them small, insects would certainly have become even more dominant over the world, and we would probably not be here to read about them.

THE KINDS OF UNIRAMIANS

Myriapods.   Insects are the most numerous animals on Earth, but they are only one of several groups in the subphylum Uniramia. Many taxonomists divide uniramians into two superclasses, depending on whether the adults have six legs, like insects, or more than six legs. Uniramians with more than six legs constitute the superclass Myriapoda (Greek *murios* countless + *podos* foot). There are four major kinds of myriapods: centipedes, millipedes, pauropods, and symphylans. (Some zoologists also include Onychophora; see pp. 578–579.) Like all uniramians, these myriapods have uniramous appendages and chewing mandibles (see p. 637). The internal organs of myriapods resemble those of insects, as will be described later. Unlike insects, members of these four classes have legs on most of their body segments. Consequently, instead of having three tagmata as in insects (the head, leg-bearing thorax, and abdomen), the bodies of myriapods are divided into a head and a long, leg-bearing trunk.

Most of the roughly 3000 species of **centipedes** (group Chilopoda) are nocturnal predators, well adapted to chasing down insects and other small prey (see Figure 31.1C). They are seldom seen except by those who look under logs and stones. In houses where cockroaches and other insects are available as prey, however, the house centipede *Scutigera* may be encountered in bathrooms and other damp places. Centipedes have the appendages on the first trunk segment modified as poisonous fangs. Some tropical centipedes that grow up to 25 cm long can be dangerous, though seldom lethal, to humans.

On each of the other trunk segments except the last two is a pair of walking legs that extend laterally. The total number of walking legs range from 15 to 181 pairs, depending on the species. (The actual number of legs is never a hundred, as the name "centipede" implies, since the number of pairs is always odd.) The legs propel the flexible, flattened body rapidly in leaf litter. Prey are detected by the single pair of antennae on the head. In most species the eyes are clusters of simple **ocelli** that are incapable of forming an image. Centipedes reproduce in a fashion reminiscent of arachnids. After a courtship that may last several hours, the male deposits a spermatophore, which he then transfers into the female's genital opening with his mouth. In many species the female cares for the eggs and young in underground burrows.

The approximately 8000 species of **millipedes** feed mainly on decaying vegetation and are adapted for burrowing through rotting logs and humus rather than for predation. When burrowing, the head is bent beneath a dorsal shield that acts like the blade of a bulldozer. The short legs are positioned beneath the robust body and are synchronized to move in a slow, wavelike motion (Figure 32.1). In spite of the common name, no millipede has a thousand legs. The maximum

Figure 32.1
A millipede, possibly Cylindroiulus.
Approximately 3 mm long.

number is 752, and around 100 is typical. The body is usually cylindrical in cross section, and the cuticle is reinforced with calcium salts, as in crustaceans. These appear to be adaptations that prevent the body from folding as the legs push it forward. The number of places where such folding can occur has also apparently been reduced by the fusion of pairs of trunk segments into **diplosegments.** Evidence for such fusion is that each diplosegment has two pairs of openings to the heart and tracheae, two ganglia on the ventral nerve cord, and two pairs of legs. (Hence the name of their taxonomic group: Diplopoda.) The first four trunk segments, often called the thorax, remain unfused and bear either one pair of legs each or none at all.

Like centipedes, millipedes are most often restricted to moist areas under logs, stones, and leaf litter. When disturbed, some of them, called pill millipedes, curl up like pill bugs (see p. 669). Some spray or secrete toxic or irritating defensive substances. The European millipede *Glomeris marginata* secretes a substance that is chemically similar to the tranquilizer Quaalude (Carrell and Eisner 1984). Within a few hours after eating this millipede, a spider becomes totally relaxed for several days. It is not clear what good this does the millipede.

There are only about 400 species of **pauropods** (group Pauropoda), but the number of individuals is enormous. Five million may live in a hectare of forest litter, largely unnoticed because they are colorless and less than 2 mm long. They feed on dead plant and animal matter and on fungi. The trunk consists of 11 or 12 segments, with a pair of legs on all segments but the first and last (Figure 32.2). The antennae are branched. What appears to be eyes on the head are sensory organs of some other, unknown function. They have no circulatory or respiratory organs.

Symphylans (group Symphyla) number only a little over a hundred species, and they are generally confined to moist areas where they feed on living or dead vegetation. They sometimes become greenhouse pests. Although they are less than 1 cm long, they are sometimes confused with centipedes. The common "garden

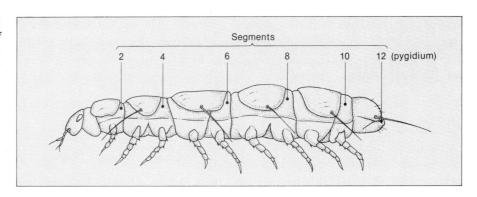

centipede" *Scutigerella* is, in fact, a symphylan (Figure 32.3). There are 12 trunk segments, each with a pair of legs. Many species reproduce from unfertilized eggs (parthenogenetically), and others reproduce sexually in a way that does not even require that the two sexes meet. Males simply leave a stalked spermatophore for a female to find by chance. The female takes the spermatophore into her mouth and stores the sperm in special pouches. To reproduce, she uses her mouth to take an egg from her genital opening, which is located on the ventral surface of an anterior segment. She deposits the egg on substratum such as moss, then smears it with stored sperm from her mouth.

Hexapods. Uniramians with six legs make up the superclass Hexapoda. Their bodies are divided into three tagmata: the head; the thorax, which bears the legs and usually two pairs of wings; and the abdomen, which retains obvious segmentation and bears the external genitalia on the terminal segments (Figures 32.4 and 32.5). The term "hexapod" is often regarded as synonymous with "insect," but taxonomists increasingly tend to include the group Entognatha along with the group Insecta.

In many ways **entognaths** are more like myriapods than insects. The main difference is that entognaths happen to have six legs, like insects. They also have mouthparts enclosed within a pouch in the head (as implied by the group name Entognatha), and they tend to have degenerate compound eyes and Malpighian tubules (excretory organs). There is little if any metamorphosis. The most common entognaths are the springtails (Figure 32.4). Some species are frequently seen hopping like fleas on the surface of snow and are therefore called "snow fleas." Others skip about on the surface of standing water. Springtails are equally common on damp vegetation but are not so easily seen there.

The rest of this chapter is devoted to the **insects,** so for now we need only compare them with the other Uniramia. Unlike spiders and many crustaceans, most members of group Insecta do not molt once they reach sexual maturity. Some insects are wingless (Apterygota), but the majority have two pairs of wings in at least some of the adults (Pterygota) (Figure 32.5).

The Pterygota are further divided into Exopterygota and Endopterygota, depending on the degree of metamorphosis. Exopterygotes change their basic form gradually or not at all as they mature; they are said to be **hemimetabolous.** The egg hatches into an immature stage called a **nymph.** (Aquatic nymphs are often called **naiads.**) The nymph molts a definite number of times, becoming a different **instar** each time. The wing buds begin to appear externally during the later instars: hence the name Exopterygota (Figures 32.6 and 32.18A). The insect becomes a winged adult during the final molt.

Endopterygotes are **holometabolous,** changing form radically usually twice in their lives (Figure 32.7). Commonly they hatch as wormlike larvae, called caterpillars, maggots, or grubs, depending on the order to which they belong. After several molts the larva becomes a **pupa** (= chrysalis). The winged adult develops during pupation and emerges by the process of **eclosion.** Metamorphosis in both Exopterygota and Endopterygota is regulated by several hormones, as described on pages 192–194.

INSECT SUCCESS

In numbers of individuals, numbers of species, and range of habitats, insects are enormously successful. More than three-quarters of a million insect species have been named so far. If this figure is too high, it won't be for long, because

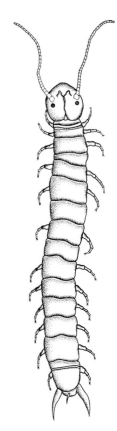

Figure 32.3
Dorsal view of the symphylan Scutigerella. *Approximately 5 mm long.*

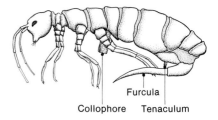

Furcula
Collophore   Tenaculum

Figure 32.4
The springtail Isotomurus, *which lives on the water surface. Unlike the rest of the body, the collophore is wettable and may serve as an anchor on water. It may also have an excretory pore. The tenaculum acts as a catch, holding the furcula while hydrostatic pressure builds up tension in it. When the tenaculum releases the furcula it launches the animal up to 30 cm. Approximately 2.5 mm long.*

Figure 32.5
The external structure of a female grasshopper, showing the body parts commonly found in insects.

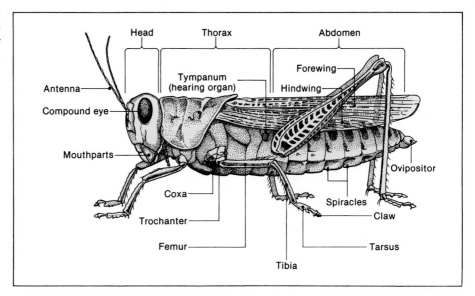

thousands of new species are described each year. The unnamed species far outnumber the named ones. Terry Erwin (1983), using a biodegradable insecticide, knocked insects out of the forest canopy in the Amazon Basin and found so many new species that he estimates the total number of insect species to be 50 million. This is approximately 35 times the number of described species of *all* living organisms. Probably only a minuscule portion of these insects will ever be described, because there are not enough taxonomists, and because species are becoming extinct so rapidly.

How can we account for such diversity? Perhaps a great deal of it is due to the insects' having become terrestrial. When the first insects evolved, during the Devonian period at least 390 million years ago, there were almost no other terrestrial animals to compete with them. They had about 40 million years to find empty habitats before reptiles came along, and at least 200 million years before the first birds and mammals. (Refer to Table 19.3.) The adaptations that allowed insects to take advantage of the many possibilities of life on land included resistance to dehydration, improved organs for the maintenance of water and ion balance, the ability to obtain oxygen from air, the ability to adjust to extreme changes in air temperature, and sensory receptors for airborne chemicals and sound. (See Table 30.2.)

Figure 32.6
Gradual metamorphosis in a typical exopterygote, the American cockroach Periplaneta americana. Shown clockwise from lower left are four nymphs representing 4 of the 13 instars of this species. Note the wing buds in the last nymph. Adult male and female, with wings. The female is carrying an egg case. Another egg case is shown in the center.

Major Groups of Uniramians

Genera mentioned elsewhere in this chapter are noted. See Figure 32.8 for common orders of insects.

Superclass Myriapoda meer-ee-AP-oh-duh (Greek *murios* countless + *podos* foot). Body divided into head and long trunk, with many trunk segments bearing legs. No compound eyes.

Class Chilopoda ky-LOP-oh-duh (Greek *chilo* lip [referring to the poison "fangs," which are actually legs]). Body usually flattened; from 18 to 184 segments in the trunk, depending on species. Most segments with one pair of walking legs. Centipedes. *Scutigera* (Figure 30.1C).

Class Diplopoda dip-LOP-oh-duh (Greek *diplos* double). Body usually cylindrical. Trunk of 9 to more than 100 diplosegments. Most diplosegments with two pairs of legs each; anterior segments have only one pair of legs or none. Millipedes. *Cylindroiulus, Glomeris* (Figure 32.1).

Class Pauropoda paw-ROP-oh-duh (Greek *pauros* few). Less than 2 mm long, with 11 or 12 segments. Nine or ten pairs of legs. No eyes; antennae three-branched. *Pauropos.*

Class Symphyla SIM-fy-luh (Greek *sim* same + *phyli* tribe). Less than 6 mm long. Twelve trunk segments, each with one pair of legs. No eyes. *Scutigerella.*

Superclass Hexapoda hex-AP-oh-duh (Greek *hex-* six). Body divided into head, thorax, and abdomen, with three pairs of legs on thorax.

Class Entognatha in-TOG-na-thuh (Greek *ento-* inside + *gnathos* jaw). Mouthparts enclosed by cranial folds. Reduced Malpighian tubules and degenerate compound eyes. Three orders: Collembola, Protura, and Diplura. Springtails and others. *Isotomurus.*

Class Insecta in-SEK-tuh (Latin *insectus* cut into [referring to the segmentation]). Mouthparts not enclosed by cranial folds. Malpighian tubules and compound eyes generally well developed. Two pairs of wings usually present.

Subclass Apterygota ap-ter-i-GO-tuh (Greek *a-* not + *pterygotos* winged). Primitively wingless. Little or no metamorphosis. Three tail appendages, long antennae, and often large eyes and scaly bodies. Up to 30 mm long. Includes order Thysanura. Bristletails and silverfish.

Subclass Pterygota ter-i-GO-tuh. Adults with wings, or wings secondarily lost. At least some metamorphosis. See Figure 32.8 for a visual synopsis of the major pterygote orders.

Superorder Exopterygota (= Hemimetabola) ex-OP-ter-i-GO-tuh (Greek *ex* external). Metamorphosis gradual, with wings appearing as buds on nymphs. Nymphs with compound eyes. True bugs,[a] mayflies, dragonflies, damselflies, stoneflies, grasshoppers, crickets, cockroaches, mantids, termites, earwigs, lice, thrips, aphids, leafhoppers, periodic cicadas, scale insects. *Abedus, Belostoma, Coenagrion, Gerris, Microtermes, Nasutitermes, Periplaneta, Phymata, Prociphilus, Schistocerca* (Figures 10.12B, 30.1D, 30.4D, 32.6, 32.14, 32.15A, 32.17, 32.18, 32.20A, 32.21C, 32.22).

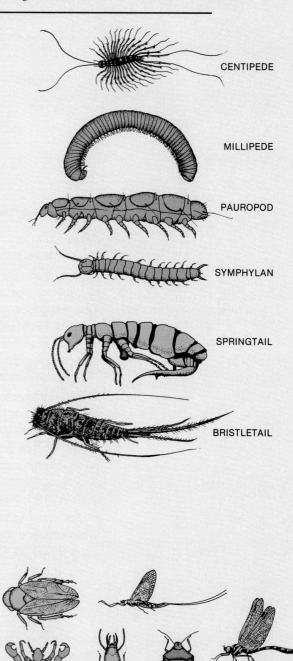

CENTIPEDE

MILLIPEDE

PAUROPOD

SYMPHYLAN

SPRINGTAIL

BRISTLETAIL

EXOPTERYGOTES

Major Groups of Uniramians *(continued)*

Superorder Endopterygota (= Holometabola) in-DOP-ter-i-GO-tuh (Greek *endo-* internal). Complete metamorphosis; larva does not resemble adult, and wings appear suddenly in adult. Larvae lack compound eyes. True flies,[a] ant lions, dobsonflies, beetles, scorpionflies, hangingflies, caddisflies, butterflies, moths, fleas, sawflies, ants, bees, and wasps. *Actias, Aedes, Anopheles, Apis, Bibio, Cactoblastis, Canthon, Chrysolina, Cryptoses, Danaus, Drosophila, Epilachna, Eurosta, Glossina, Hyalophora, Hyles, Hylobittacus, Lymantria, Megarhyssa, Metasyrphus, Morpho, Pediobius, Photuris, Pieris, Scarabaeus, Vespula* (Figures 1.7, 12.10, 16.3B, 18.7, 18.14, 18.15B, 20.12, 32.7, 32.15B, 32.16, 32.20B, 32.21A,B, 32.23, 32.24, 32.25, 32.26).

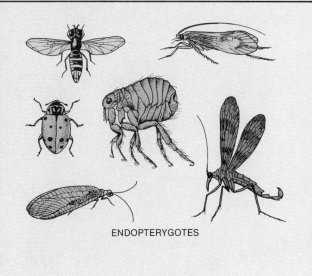

ENDOPTERYGOTES

[a] The terms "bug" and "fly" are used in the common names of many crawling or flying insects, but only insects in orders Hemiptera and Diptera, respectively, are true bugs and flies. It is good practice to indicate that an insect is not a true bug or a true fly by making "bug" or "fly" part of a longer common name. For example, the flying beetles that glow in the dark should be termed "lightningbugs" or "fireflies." However, bed bugs are true bugs, and house flies are true flies.

Figure 32.7
Complete metamorphosis in an endopterygote, the sphynx moth (= hawk moth) Hyles gallii. (A) Caterpillar. Length approximately 10 cm. (B) Pupa. Length 5 cm. (C) Adult male. Length 4 cm.

Diversity also required the ability to take advantage of different foods. The basic insect mouthparts proved to be highly modifiable for chewing, sucking, or sponging nutrients from plants, and later from animals. The ability to fly to new habitats must also have been important. From the equatorial regions of the supercontinent Pangea (see Figure 19.9), where fossil remains suggest they originated, insects spread to every terrestrial region except the poles.

A

B

C

Figure 32.8
A visual guide to the familiar orders of insects. Main features useful in identification are outlined in color. Horizontal bars show a typical body length.

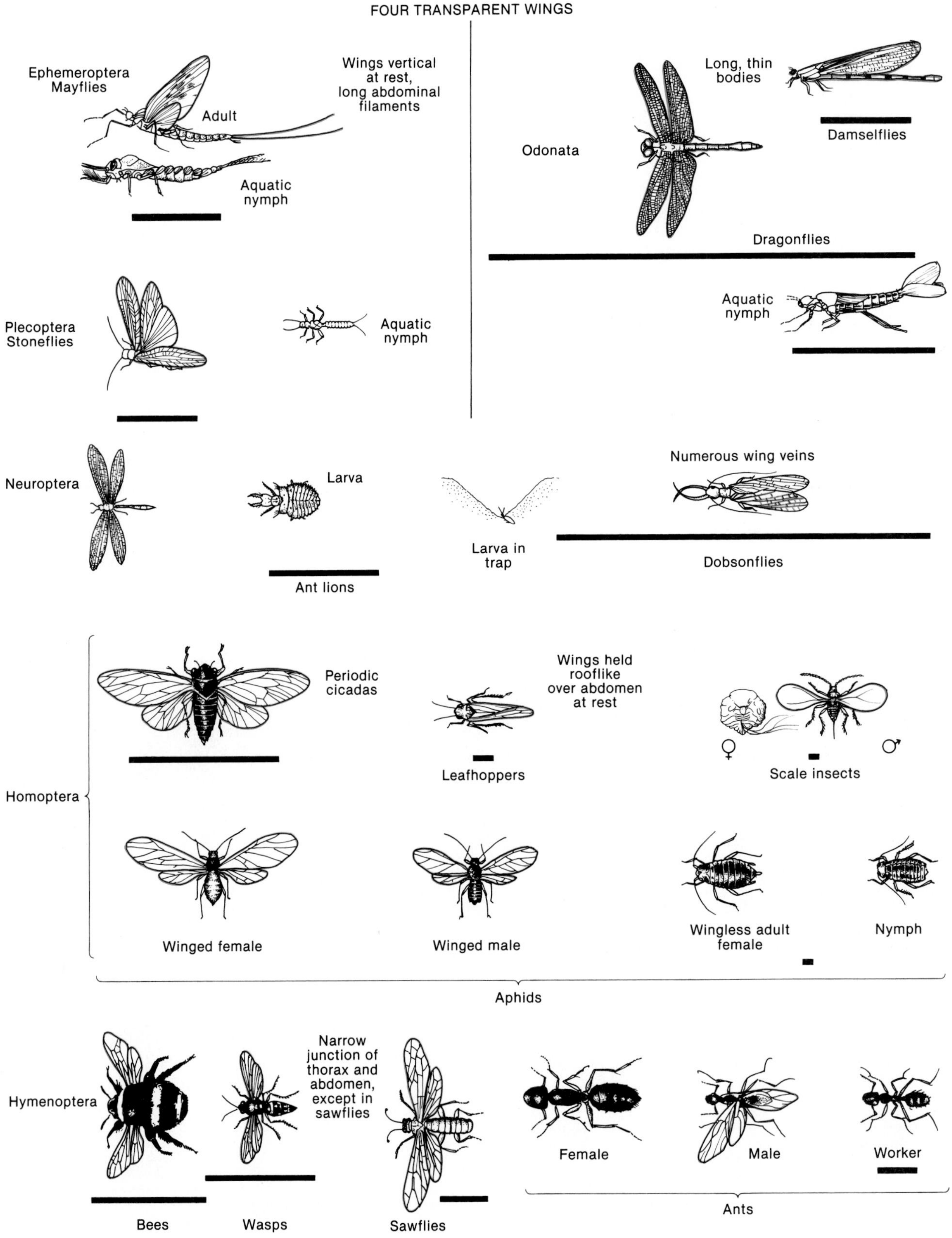

FOUR TRANSPARENT WINGS

Ephemeroptera
Mayflies

Adult

Aquatic
nymph

Wings vertical
at rest,
long abdominal
filaments

Odonata

Long, thin
bodies

Damselflies

Dragonflies

Aquatic
nymph

Plecoptera
Stoneflies

Aquatic
nymph

Neuroptera

Larva

Larva in
trap

Ant lions

Numerous wing veins

Dobsonflies

Homoptera

Periodic
cicadas

Wings held
rooflike
over abdomen
at rest

Leafhoppers

Scale insects

♀   ♂

Winged female

Winged male

Wingless adult
female

Nymph

Aphids

Hymenoptera

Narrow
junction of
thorax and
abdomen,
except in
sawflies

Female

Male

Worker

Bees   Wasps

Sawflies

Ants

684

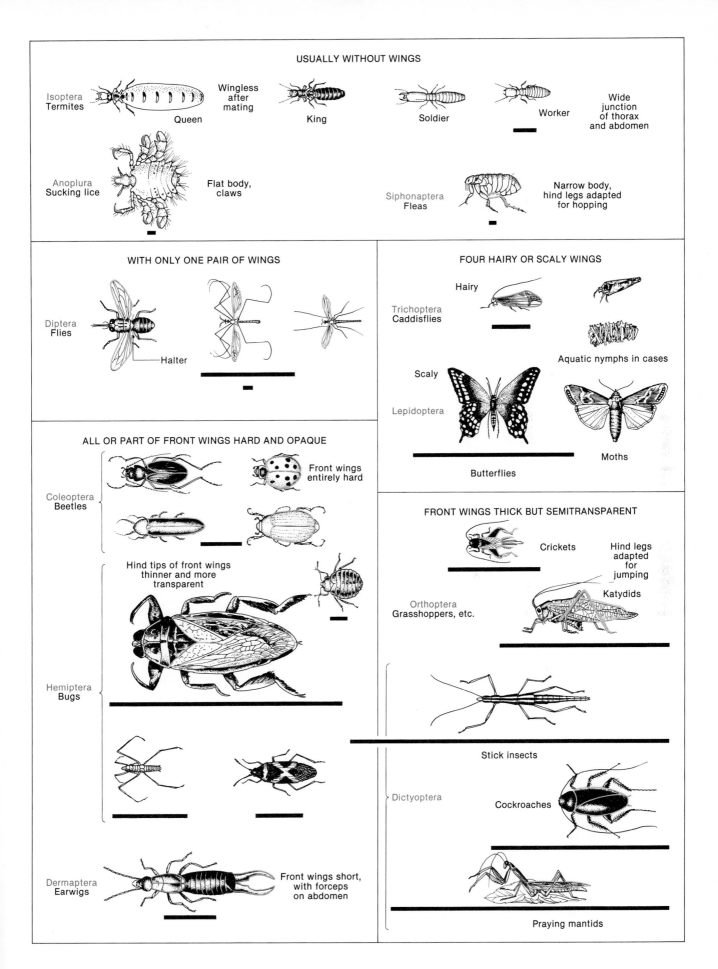

USUALLY WITHOUT WINGS

Isoptera
Termites

Queen — Wingless after mating

King

Soldier

Worker

Wide junction of thorax and abdomen

Anoplura
Sucking lice — Flat body, claws

Siphonaptera
Fleas — Narrow body, hind legs adapted for hopping

WITH ONLY ONE PAIR OF WINGS

Diptera
Flies — Halter

FOUR HAIRY OR SCALY WINGS

Trichoptera
Caddisflies — Hairy

Aquatic nymphs in cases

Lepidoptera — Scaly

Butterflies

Moths

ALL OR PART OF FRONT WINGS HARD AND OPAQUE

Coleoptera
Beetles — Front wings entirely hard

Hemiptera
Bugs — Hind tips of front wings thinner and more transparent

Dermaptera
Earwigs — Front wings short, with forceps on abdomen

FRONT WINGS THICK BUT SEMITRANSPARENT

Crickets

Hind legs adapted for jumping

Orthoptera
Grasshoppers, etc. — Katydids

Stick insects

Dictyoptera — Cockroaches

Praying mantids

685

Survival on land also required adaptations in reproduction. Fertilization had to be internal to prevent drying of gametes, and embryos had to develop within nutrient-rich eggs that let in oxygen without losing water. The rapid diversification of insects also benefited from rapid reproduction and short life cycles that enabled new habitats to be colonized even if only a small proportion of the insect pioneers were able to survive in them.

The following sections of this chapter focus on these adaptations for terrestrial success.

OSMOREGULATION AND EXCRETION

The impermeable integuments of their aquatic ancestors gave insects, along with other uniramians and arachnids, a head start toward life on land. Terrestrial life requires much more, however, than merely preventing the loss of body water. Unlike crustaceans, insects cannot simply rely on a marine environment to supply ions and water in correct proportions, and they cannot simply allow metabolic wastes to diffuse into the environment. Instead, insects depend on **Malpighian tubules** and the rectum for excretion and osmoregulation. Each insect has from two to several hundred thin Malpighian tubules, which are often yellow and contain muscles that keep them moving within the hemocoel in the abdomen. Each Malpighian tubule attaches at one end between the midgut and the hindgut (Figures 13.5B and 32.9). The other end is either unattached or attached to the rectum.

The blood pressure is so low that Malpighian tubules require an alternative means of producing filtrate. They actively absorb ions, especially potassium, from the hemolymph, and water follows the ions osmotically. Along with the water come dissolved molecules, including uric acid, the major nitrogenous waste of insects (see p. 287). The fluid then enters the hindgut, where it mixes with digestive wastes. Much of the ions may be reabsorbed into the hemolymph across the gut lining. Plants are extremely rich in potassium, and herbivorous insects excrete most of the K^+ in the feces. As the feces pass through the rectum, **rectal pads** recycle much of the water back into the hemolymph. In some insects, such as the desert locust *Schistocerca gregaria*, virtually all the water that enters the Malpighian tubules is recovered by the rectal pads.

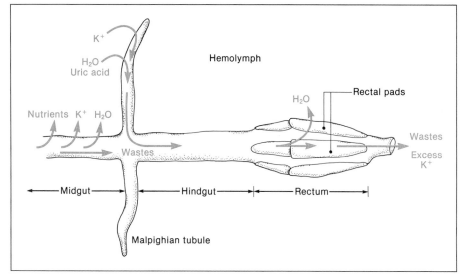

Figure 32.9
Schematic representation of the functioning of the insect excretory system. In the midgut nutrients, potassium ions (K^+), and water are absorbed from food into the hemolymph. Potassium ions are actively pumped out of the hemolymph into the Malpighian tubules. Water follows osmotically, carrying dissolved wastes such as uric acid that join the digestive wastes in the hindgut. Six rectal pads reabsorb the water and needed ions back into the hemolymph.

RESPIRATION

Tracheae. Another adaptation for terrestrial life is the system of breathing tubes, called tracheae, that allow insect tissues to obtain oxygen directly from the air. These tracheae are up to several millimeters in diameter in large insects. They are lined with glistening white cuticle that molts along with the exoskeleton (see Figure 30.4D), and they are reinforced by a spiral strand of chitin. Each trachea opens to the surface through a **spiracle** (Figure 32.10). There are two pairs of spiracles in the thorax and eight pairs in the abdomen. Each spiracle has a valve that reduces the loss of water from body fluids and keeps out parasites, particles, and liquid water. The valve opens in response to neural excitation signaling a buildup of CO_2 in the hemolymph. The large tracheae branch off into smaller and smaller tracheae, forming a pattern that varies with the species. (Compare Figures 12.10 and 32.10B.) The thinnest branches, the **tracheoles,** are generally less than 0.1 μm in diameter. Unlike the tracheae of spiders, those of insects bring air directly to tissues rather than to blood. Some tracheoles, in fact, indent the plasma membranes of cells, bringing oxygen directly to the mitochondria that require it. Because the blood does not have to supply oxygen, insects get by with open circulatory systems that operate under low pressure. A dorsal, tubular heart simply pumps blood anteriorly, then eventually draws the blood back in from the hemocoel through one-way ostia (see Figures 11.3B and 32.13).

The movement of the O_2 along the tracheal tubes is largely by diffusion. Some insects, however, aerate the tracheae by means of breathing movements that tend

Figure 32.10
The tracheal system. (A) Schematic representation of a trachea branching to tracheoles. The filter and valve in the spiracle keep out particles, parasites, and liquid water and prevent the evaporation of body water. The tracheae are reinforced by a spiral strand of chitin called the taenidium. Tracheoles bring air to within a few micrometers of most cells and actually invaginate some cells. They are usually filled with fluid at the ends, especially if the cell is not active. The largest tracheae may be several millimeters in diameter; the tracheoles are often less than 0.1 μm in diameter. (B) The tracheal system of a grasshopper. Longitudinal tracheae and air sacs allow grasshoppers, unlike many other insects, to ventilate the tracheae by breathing.

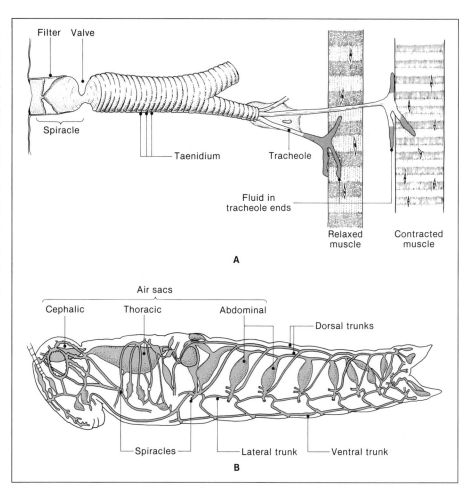

to take air into the thoracic spiracles and expel it through the abdominal spiracles. Still, the movement of gases in any long, thin tube is so slow that it may be one of the factors that has kept insects relatively small. In the largest insects, which are approximately 30 cm long, O_2 has to travel perhaps 10 cm from spiracles to the innermost tissues.

Aquatic Insects. Although insects are fundamentally terrestrial animals, some have returned to the water either as larvae or adults and obviously have some other way of obtaining oxygen. These aquatic insects have not simply "reinvented" the gills of crustaceans. Some insect larvae, such as mosquitoes, hang from the water surface, breathing air through a tube (see Figure 13.1A). Other insect larvae absorb oxygen from the water by means of **tracheal gills.** Tracheal gills are areas rich in tracheae that are especially adapted for absorbing O_2. In dragonfly larvae the tracheal gills are in the rectum, which moves water in and out much as lungs inhale and exhale air.

Other aquatic insects, such as the true bugs called water boatmen (family Corixidae) and backswimmers (family Notonectidae), have **physical gills.** These are not anatomical structures, but bubbles of air the insects carry with them beneath the water surface. These bubbles are not merely oxygen reserves. In fact, they work much better if filled with air than if filled with oxygen. As the bugs absorb oxygen from the bubbles, the partial pressure of O_2 (pO_2) in the bubble falls below that of the water. As a result, O_2 diffuses from the water into the bubble (see p. 246). If the bubble were filled with pure O_2, the pO_2 would always remain higher in the bubble than in the water, so the O_2 would not be replenished by diffusion from the water. Experiments dating back to 1915 have shown that an air-filled bubble will last a backswimmer 6 hours, while an oxygen-filled bubble will last only 35 minutes.

ADAPTATIONS FOR TEMPERATURE CHANGES

Living in air has also meant that many terrestrial insects have had to find ways to compensate for changes in temperature that are much more sudden and severe in air than in water. The majority of insects avoid the problem by living in climates where it is never cold. Others simply die after producing cold-tolerant eggs that start an entirely new generation with the return of warm weather. However, a surprisingly large number of insect larvae and adults survive winter as far north as the Arctic Circle. Some species avoid freezing by means of compounds such as glycerol and thermal hysteresis proteins, others become supercooled, and still others actually survive partial freezing. (Refer to Chapter 15 for details and references on these and other topics in this and the following paragraph.) Protected against freezing, they may remain in a state of inactivity called **torpor** for as long as the body temperature is too low to sustain normal rates of enzyme function. Other species enter a phase of inactivity called **diapause** when winter approaches. Diapause, in contrast to torpor, is not a passive response to cold but a physiological state that the insect induces.

Torpor and diapause permit survival in an inactive state, but some insects can remain active at surprisingly low environmental temperatures by means of **partial endothermy.** Partial endothermy is the ability to temporarily raise the temperature of part of the body above the environmental temperature. On cold mornings bees, moths, and many other insects bask in sunlight or shiver their flight muscles to raise the temperature of the thorax enough to permit flight. The heat is often conserved by **pile:** an insulating coat of hairlike structures.

Figure 32.11
The antennae of a male luna moth Astias luna *showing numerous branches that increase surface area for receptor cells.*

SENSORY ORGANS

Hearing. Insects also require sensory organs adapted to differences in the ways molecules move in air compared with water. The ability to detect alternating pressure changes in air—hearing—is one that insects share with only a few other animals, namely, amphibians, birds, and mammals. It has probably evolved several times among insects, judging from the variety of auditory organs. Many insects can also detect vibrations of the substratum using mechanoreceptors in or on the legs. These mechanoreceptors are associated with various cuticular structures, such as hair-shaped **setae.** Setae may also respond to wind or air movements produced by predators.

Chemoreception. Insects can also detect airborne molecules by a process comparable to olfaction in humans. Molecules pass through small pores in the cuticle, especially on antennae, and then bind to neuronal receptor cells (Figure 32.11). Among the important molecules detected are **pheromones** emitted by members of the same species in a variety of social contexts (see Figure 20.8). Many insects also have chemoreceptors for substances emitted by the plants or animals they feed on. For example, mosquitoes and some other "biting" flies have chemoreceptors in their antennae that respond to carbon dioxide, to humid convection currents, and to other volatile substances from mammals. Detection of the CO_2 increases the tendency of mosquitoes to fly. They are then apt to encounter the convection currents and volatile chemicals, which they follow to the host. Repellents work by blocking one or more of the receptors, thereby preventing the mosquitoes from locating a host.

Vision. Most insects have two compound eyes that resemble those in crustaceans, except that they tend to have hexagonal rather than rectangular facets (see p. 665). Usually there are also up to three simple ocelli. Insects are generally blind to red light but can perceive ultraviolet and also the direction of polarization.

The Central Nervous System. Other aspects of insect physiology have required less modification for terrestrial life and are generally similar to those described in the previous chapter for crustaceans. (Compare Figure 32.12 with Figure 31.8.) The central nervous system, for example, consists of a chain of ganglia on a ventral nerve cord. Each ganglion is largely autonomous and controls local activities. The thoracic ganglia generally control walking and flight. The ganglia in the head (the supra- and subesophageal ganglia) are mainly involved with the integration of sensory information from the eyes, antennae, and mouthparts and with control of feeding. The supra- and subesophageal ganglia also exercise overall control over many functions coordinated by other ganglia. For example, the subesophageal ganglion of the male mantis usually inhibits copulatory movements that are coordinated by the last abdominal ganglion. When the antennae detect pheromones from a sexually receptive female, however, the subesophageal ganglion stops inhibiting copulatory activity. Decapitating a male mantis also removes the inhibition and causes the male mantis to attempt to copulate with a finger or other object of about the same size and shape as a female mantis. Contrary to popular belief, however, female mantids do not have to bite the heads off males in order to get them to mate (Brown 1986).

FEEDING AND DIGESTION

Mouthparts. Insect mouthparts have diversified for a wide range of foods, yet they are all fundamentally the same. The mouthparts include two **mandibles,** a

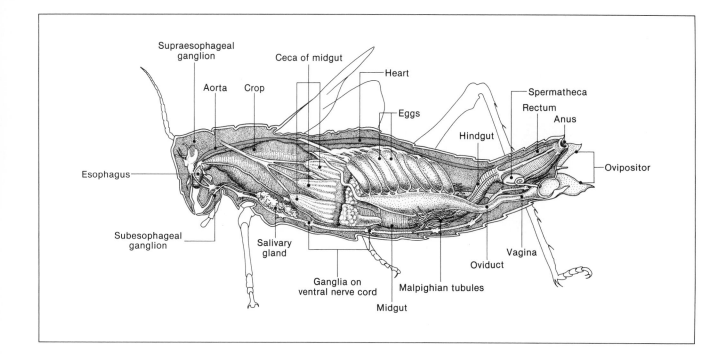

Figure 32.12
Internal structure of a female grasshopper.

Labels in figure: Supraesophageal ganglion, Ceca of midgut, Aorta, Crop, Heart, Spermatheca, Rectum, Anus, Hindgut, Eggs, Esophagus, Ovipositor, Subesophageal ganglion, Salivary gland, Ganglia on ventral nerve cord, Malpighian tubules, Midgut, Oviduct, Vagina

hypopharynx, a **labrum** and **labium,** and a pair of **maxillae** with sensory palps. In grasshoppers and other insects that chew food, the mandibles function like jaws, the hypopharynx works like a tongue, and the labrum and labium are analogous to upper and lower lips. In other insects, however, these structures vary considerably in structure and function (Figures 32.13 and 32.14).

Digestion. The digestive tract is generally a tube divided into three parts: **foregut, midgut,** and **hindgut** (Figures 13.3, 13.5B, 20.1, and 32.12). The foregut is often modified as a **crop,** in which food is stored. Some digestion may also occur in the crop from enzymes secreted by the **salivary glands** and regurgitated from the midgut. The midgut often has pouches called **ceca.** The midgut and ceca are the main sites for secretion of enzymes and absorption of nutrients into the surrounding hemocoel. The hindgut and rectum are where water and nutrients from the Malpighian tubules are reabsorbed. The guts of most insects are richly endowed with symbiotic bacteria, fungi, protozoans, and other organisms. Many are harmless commensals, and others are beneficial mutualists. Many termites and wood-eating cockroaches, for example, depend on protozoans such as *Trichomonas* and *Trichonympha* (see p. 469) to digest the cellulose in plant cell walls. However, many of the gut organisms are parasitic.

Insect Agriculture. As a final topic in the subject of insect feeding, it is worth noting that many insects have gone through the equivalent of the Neolithic Revolution, giving up the hunter-gatherer life in favor of agriculture. Many ants and termites cultivate fungi within their burrows (see Figure 1.7). Some ants guard aphids, in exchange for which the aphids let the ant "milk" them of a nectarlike secretion from the anus. Daniel Janzen has found that in tropical forests ant colonies inhabit the swollen thorns of acacia trees and feed on the nectar. The ants keep the acacias free of other herbivores and prune vegetation that would otherwise shade and kill the acacias. Acacias die without their ant colonies, and ants die without the acacia thorns.

Figure 32.13
Insect mouthparts. (A) The mouthparts of a grasshopper, which chews vegetation. (B) The mouthparts of a butterfly, adapted for sucking nectar. The proboscis (= maxilla) is unfurled by hemolymph pressure in the two galeae. The lower part of the figure shows the proboscis in cross section. (C) The enlarged labial palps of the house fly sponge up food. See also Figure 13.1A which shows the suspension-feeding mouthparts of a mosquito larva, and Figure 13.3, which shows the piercing and sucking mouthparts of an adult mosquito.

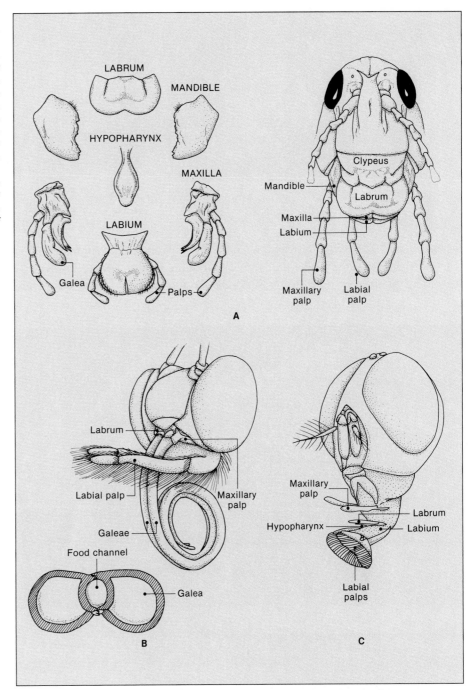

PREDATION AND DEFENSE

Insects, of course, do not merely eat but are often eaten. Many, however, have elaborate defenses against predation. The bombardier beetle puts up one of the most dramatic chemical defenses, directing a hot, explosive secretion toward potential predators (Figure 32.15). Many insects that are toxic either from their own secretions or those of host plants advertise the fact with **warning coloration.** Many other insects have appearances that **mimic** such coloration and provide protection from predation (Figures 18.14 and 32.16). Another approach to defense against predation is **camouflage** (see pp. 377–378; Figure 32.17). The sources of these colors are described on page 199.

Figure 32.14
An assassin bug (family Reduviidae) demonstrating its piercing and sucking ability on the caterpillar of a monarch butterfly. Like spiders, assassin bugs inject digestive enzymes into the prey, then suck up the nutrients. Toxins from the milkweed leaf on which the caterpillar was feeding do not deter the bug.

LOCOMOTION

Running and Jumping. Living on land enables insects to add running, jumping, and flying to the possible modes of locomotion. The majority crawl on all sixes using an **alternating tripod gait.** While the fore and hind legs on one side

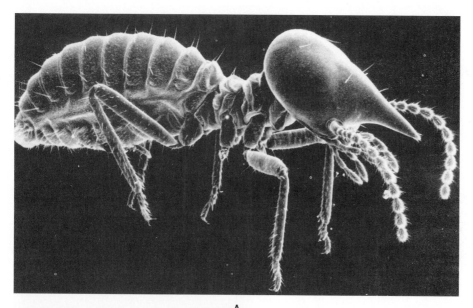

A

B

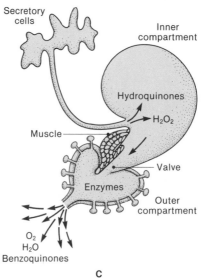

Secretory cells

Inner compartment

Hydroquinones

H_2O_2

Muscle

Valve

Enzymes

Outer compartment

O_2
H_2O
Benzoquinones

C

Figure 32.15
(A) A soldier termite Nasutitermes corniger defends its colony with a head modified as a glue gun. Soldiers of other species secrete various other adhesives and irritants. (B) Self-defense by the bombardier beetle (family Carabidae). The beetle (held in place by wax) directs a 100°C stream of benzoquinone toward a potential predator—in this case forceps pinching the left front leg. (C) This behavior has been known for a long time, but its mechanism, along with the defenses of numerous other insects, was discovered by Thomas Eisner and his colleagues at Cornell University. The beetle stores a mixture of hydroquinones and hydrogen peroxide (H_2O_2) in an inner compartment in the abdomen. When alarmed, the beetle opens a valve that allows the mixture to enter an outer compartment containing enzymes that catalyze the conversion of the H_2O_2 into H_2O and O_2, and the exothermic oxidation of hydroquinones into benzoquinones. The O_2 propels the hot benzoquinones with an audible pop.

A   B

Figure 32.16
The harmless hover fly Metasyrphus americanus *(A) resembles a yellowjacket or hornet such as* Vespula *(B). Many other dipterans, called bee flies, also derive protection from predators by mimicking bees, wasps, and yellowjackets.*

A

B

C

Figure 32.17
Defensive and aggressive camouflage. (A) The nymph of a spittlebug (Homoptera) partly exposed from its bubbly defensive secretion. (The nymph is upside down and facing left. Note the curled wing buds of this exopterygote.) (B) The wooly alder aphid Prociphilus tessellatus *covers itself in a waxy secretion that not only hides it but also sticks to the mouthparts of predators. These aphids are also guarded by ants. Thomas Eisner and colleagues (1978) have found that some predaceous lacewing larvae (order Neuroptera) disguise themselves in the "wool," which enables them to approach the aphids undetected by ants that guard them. (C) This ambush bug* Phymata *(left) was also apparently well camouflaged when the honey bee arrived to feed on goldenrod.*

and the middle leg on the other side are fixed on substratum, the other three legs advance. Insects readily vary this gait, however, if they have lost a leg or if the substratum is irregular. Some insects, notably fleas and orthopterans, have hind legs adapted for jumping. Much of our understanding of jumping and other aspects of fleas is due to Miriam Rothschild, a self-taught entomologist. Rothschild has found that before jumping the flea cocks its leg in a sort of catch, with the leg pressed against a rubbery protein called **resilin**. The flea then releases the catch, and the resilin pushes the legs back, sending the flea tumbling through the air. Grasshoppers use a different technique for cocking and releasing the hind legs. They first build up tension in both the extensor and flexor muscles of the hind legs, then suddenly relax the flexor muscles. Many insects, such as water striders, are as adept at jumping on water as they are on land (Figure 32.18). Many others, such as diving beetles and backswimmers, swim under water.

Flying. Impressive as these feats of locomotion are, it is unlikely that they could have gotten insects very far evolutionarily. The success of insects probably owes much more to the ability to fly. The origin of flight has long been a subject of controversy. It has usually been assumed that wings originated as small flaps (paranota) that first permitted insects to glide or parachute, and later became adapted for flapping flight. The idea is usually referred to as the paranotal theory, or sometimes as the flying-squirrel theory. Joel Kingsolver of Brown University and Mimi A. R. Koehl of the University of California at Berkeley, however, have shown that the paranotal theory is unlikely. Using models of insects with wings of various lengths, they found that wings provide little help in gliding until they reach a length of about a centimeter. It is a fundamental tenet of natural selection that a structure can evolve if it immediately improves the fitness of an animal, but not merely because it might improve the fitness of future generations. Therefore the structures that eventually became wings had to serve some other function initially. Kingsolver and Koehl have shown that they could first have been used as solar collectors in thermoregulation, and later for gliding as they became larger. Another theory, by Jarmila Kukalová-Peck of Carleton University in Ottawa, is that wings originated in aquatic nymphs, perhaps as gill covers and then as fins before they became adapted in the terrestrial adults for gliding.

For insects smaller than dragonflies and butterflies the viscosity of air is high relative to inertia; that is, the Reynolds number is low. (See pages 218–219.)

Figure 32.18
A common water strider Gerris *(Hemiptera) on a pond. Fine, water-repellent hairs on the legs, and the location of the sharp tarsi above the ends of the legs, keep the legs from breaking through the surface tension. The middle pair of legs do most of the rowing; the hind legs act as rudders, and the short front legs are used mainly for seizing prey. The water strider can also jump to seize prey or to escape. Prey are detected by the ripples they generate. Male water striders can also determine the sex of other water striders by the difference in ripple frequency. A population of* Gerris *generally contains some winged adults that can also fly to other habitats. (The large red spheres on the head are immature red mites that parasitize water striders.)*

Therefore most insects cannot fly the way airplanes do, by slicing through the air with smooth wings. Flying for most insects is like swimming is for us. A common pattern of wing movement, in fact, resembles the butterfly stroke of human swimmers (Figure 32.19). The wings do not flap straight up and down but are flung forward during the downstroke and backward during the upstroke. Also during downstroke the rear portions of the wings tilt up and provide forward thrust, and during the upstroke the rear portions of the wings tilt down and reduce air resistance.

The tilting of the rear portions of the wings is due to the attachment of the wings near the front edge and to the greater rigidity of the wing near the front. This rigidity is due to a greater concentration of **veins** in front. There are numerous other adaptations for flying that occur in insects but not in any other animal or machine. For example, in damselflies, butterflies, and certain moths (families Saturniidae and Noctuidae) the wings also perform a maneuver in which they clap together at the end of the upstroke, then fling apart at the start of the downstroke. This **clap and fling** pattern generates air currents that produce additional lift.

The complex movements of the wings dictate some intricate attachment of muscles. In dragonflies the flight muscles (elevators and depressors) are attached directly to the bases of the wings. (See left portion of Figure 10.3C.) A burst of action potentials from the thoracic ganglia triggers each contraction of the flight muscles. These are therefore called **synchronous muscles.** Synchronous muscles directly attached to the wings are said to be primitive, although they are good enough to permit dragonflies to hover, capture prey, and even mate on the wing. Insects considered more advanced, such as flies, beetles, bugs, and bees, have flight muscles attached to cuticular plates in the exoskeleton. (See right part of Figure 10.3C.) Deformation of these plates "clicks" the wings up and down. The flight muscles are also different, in that their contraction is not triggered by a synchronous burst of action potentials. They are therefore called **asynchronous muscles.** Asynchronous muscles are triggered to contract by being stretched, as described elsewhere (see p. 216).

The fastest flying insect is said to be a sphinx (= hawk) moth (family Sphingidae), which has been clocked at 15 m/sec—roughly 30 mph. This is about 1000

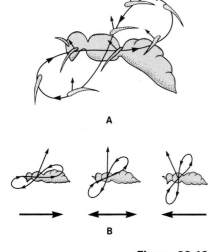

Figure 32.19
*Wing movements during flight.
(A) During the downstroke the wing
is angled downward, providing
forward thrust as well as lift.
(Arrows show direction of force on
body.) During the return stroke the
wing moves backward with respect
to the body and is angled to reduce
drag and provide additional forward
thrust. (B) By changing the
orientation of this figure-eight
pattern, many insects can hover and
fly either forward or backward.*

body lengths per second, compared with only 100 lengths per second for a jet fighter at Mach 3. The distances covered by insects are also impressive, especially in units of body length. Monarch butterflies have been known to cover 425 km in a day during migration (see pp. 421–422). From the perspective of an insect, even a few meters is a long distance to navigate. Insects use polarized light and probably other cues to help them find their way over such vast distances relative to their body sizes.

REPRODUCTION

Eggs. Another key to insect success on land is the ability to reproduce vast numbers of offspring under the hostile conditions of terrestrial life. One of the major adaptations for terrestrial reproduction is the self-sufficient or **cleidoic egg.** In contrast to the eggs of most other invertebrates, those of insects do not depend on an aqueous medium to provide water and dissolve metabolic wastes. The insect egg consists of an ovum and a nutrient yolk enclosed in a shell made of proteins and waxes. The shell offers mechanical protection and is permeable to oxygen but not to water.

After the final molt into adulthood most female insects begin egg production, **oogenesis.** Oogenesis is soon followed by **vitellogenesis,** which is the production in the **fat body** of proteins that become part of the yolk. The ovary then constructs a shell around each ovum and yolk. The release of the egg, called **oviposition,** occurs after mating. Pores (**micropyles**) provide access to sperm for fertilization of the ovum. Other pores (**aeropyles**) that are permeable to O_2 but not to H_2O allow the embryo to respire even if the egg is soaked in rain or in a puddle. Early development of the egg is described on pages 114 and 116.

Courtship. Many insects achieve a high reproductive rate through adaptations in behavior, including courtship. Courtship helps ensure that the prospective mates belong to the same species, are of opposite sex, and are physiologically ready to reproduce. Insects employ almost every imaginable sensory channel when locating mates fulfilling these requirements. The most common channel of courtship communication is by sex attractant pheromones released by one sex, usually the female, and attractive to members of the other sex in the same species. A few insects, such as cicadas, mosquitoes, bark beetles, katydids, grasshoppers, and crickets, use acoustic signals in courtship. A male cricket chirps by rubbing the medial edge of a front wing (the **scraper**) across a row of ridges (the **file**) on the inner surface of the other front wing. Chirps emitted in a particular pattern and rate attract females of the same species to the male's territory. A different pattern of chirps repels males of the same species.

Insects generally use vision only to locate and orient toward the mate at close range. (See Figure 20.3 for the use of vision by courting drosophila.) Fireflies are a striking exception. Males court in flight by flashing in specific patterns (see Figure 21.2) that sexually receptive females answer from the ground. The light is produced by a molecule called **luciferin** in the presence of ATP and the enzyme **luciferase.**

In some insects courtship also allows one member of the pair to assess the fitness of the other (sexual selection); it also prolongs copulation, thereby increasing the number of sperm transferred. In predaceous species, courtship also helps keep one mate from being eaten by the other. Extensive studies by Randy Thornhill (1980) of the University of New Mexico have demonstrated all of these functions in the black-tipped hangingfly (order Mecoptera). The male black-tipped hangingfly *Hylobittacus apicalis* begins courting by capturing another arthropod as prey, then emitting a pheromone while hanging by its forelegs. A female attracted by the pheromone and "satisfied" with the size and palatability of the prey will lower her wings as a signal that she is receptive. The female hangingfly will then hang around to feed and mate. The larger the prey, the longer copulation will last, and the more sperm will be transferred. After copulating, the male and female fight over what is left of the prey, with the male usually winning the prey and using it as bait for another female. Males may also steal prey from other males, sometimes by lowering their wings like a receptive female.

Fertilization. Fertilization occurs internally in all insects and is achieved through copulation in all but a few (Figure 32.20). The genitals of males and females generally couple by a complicated "lock and key" mechanism (see Figure 16.3B). Fertilization usually does not occur during copulation. Instead, the sperm are stored in a **spermatheca** and are released during oviposition, sometimes months or years after copulation. As the eggs move down the oviduct, the sperm enter them through the micropyles. In some species several fertilized eggs are then enclosed within a cuticular case, or **ootheca** (Figure 32.6).

Oviposition. Most insects oviposit where moisture and other physical conditions are suitable for development, or on particular plant or animal hosts that will provide food for the hatched larvae. Many flies and wasps, for example, lay their eggs in or near feces, decaying meat, or living animals (Figures 32.21A and 32.25). Many wasps inject other insects or spiders with a paralyzing but nonlethal venom, then place the helpless prey into burrows with their eggs. This behavior assures that the larvae will have fresh food when they hatch. Such larvae, and other parasites that ultimately kill the host, are called **parasitoids.**

Many insects that oviposit in plant tissues somehow induce the plants to form abnormal growths, called **galls,** around the egg. Galls protect the egg and larva and may provide food (Figure 32.21B). It is not clear whether gall-forming insects inject a growth hormone or some other substance that induces the growth, or whether the gall is an adaptation of the plant that protects its tissues from the egg. Since suitable hosts are apt to be scarce, there is a danger that so many eggs

Figure 32.20
(A) Sperm transfer in the damselfly Coenagrion puella. The male (above) is grasping the female with claspers near his anus. At the same time the female is bending her abdomen forward to receive sperm that the male has previously placed in accessory genitals on his second and third abdominal segments. (B) Almost all other insects transfer sperm directly. Here a pair of March flies (probably Bibio) copulate. Unfortunately, the male (below) has not noticed that a crab spider has killed the female.

A

B

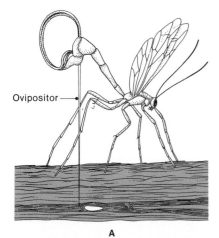

Ovipositor

A

B

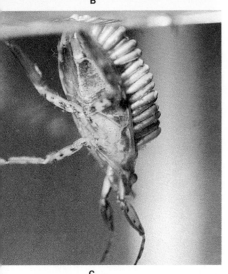

C

Figure 32.21
*Reproductive behavior. (A) A female
ichneumon wasp* Megarhyssa
macrurus *bores into wood with her
long ovipositor and deposits an egg
near the larva of a wood-boring
insect. (B) A goldenrod gall cut open
to show the pupa of the fruit fly*
Eurosta, *whose mother caused the
goldenrod to form the gall. (C) A
male giant water bug* Belostoma
with eggs on its back.

will be deposited on a single host that the nutrients will be depleted before the
larvae fully develop. Many species reduce this risk by laying conspicuous eggs or
by marking the host with a pheromone that deters other ovipositing females.

The majority of insects are egg-laying, or **oviparous.** Some females, however,
provide even further protection of the embryos from drying and predation by
retaining the eggs within themselves until they hatch into larvae. If such an insect
provides the embryo with nourishment it is said to be **viviparous,** like most mammals. In the female tsetse fly *Glossina* not only an egg but also the larva develops
within a uterus. Other insects may retain the eggs until hatching but may not
provide any nourishment. Such insects are said to be **ovoviviparous.** In aphids
there is an alternation of generations, in which a generation of females with wings
reproduce sexually and oviparously, followed by several generations of wingless,
ovoviviparous females produced from unfertilized eggs (parthenogenetically).

Parental Behavior.   Insects generally provide little parental care but produce
large numbers of offspring that ensure that at least a few survive. In contrast,
humans and some other mammals produce only a few offspring and devote a great
deal of parental care to ensure their survival. A few species of insects do, however,
provide some care for offspring, generally by brooding eggs. Almost always it is
the female that provides parental care, presumably because she has already made
the greater reproductive investment, and because it is more certain that the offspring she cares for are her own (see p. 430). Certain giant water bugs are notable
exceptions, since it is the male that carries eggs on its back (Figure 32.21C). In
Abedus herberti the male allows a female to oviposit on his back, but only if she
interrupts oviposition to copulate with him several times, thereby assuring his
paternity. (See Tallamy 1984 for discussion.)

Eusociality.   There is a variety of other behaviors that enhance reproduction,
and these are most diverse in the **eusocial insects.** Eusociality occurs when there
is cooperative rearing of the young, when infertile individuals work on behalf of
individuals that reproduce, and when there is an overlap of at least two generations
that contribute to labor in the colony. Eusociality has evolved independently in
two orders: Isoptera (termites) and Hymenoptera (ants and certain wasps and
bees). In both groups there are **castes** of sterile workers that differ morphologically from the reproductive individuals. (See Figure 32.8 for representatives of the
various castes in Isoptera and Hymenoptera.) The sterile workers generally far
outnumber the reproductives and focus much of their activity on them. This
arrangement can be compared to an individual organism, in which the functions
of most cells are directed toward maintaining optimal conditions that enable the
reproductive cells to function. In some eusocial species the reproductive individuals are so specialized that they are, in fact, little more than gonads. The female
reproductives, called **queens,** do little but lay eggs. They cannot even feed themselves (Figure 32.22). The reproductive males are also useless except to fertilize
the queens. The fact that eusociality evolved at least twice suggests that it increases
reproductive success even though the great majority of individuals lose their ability
to reproduce. A major triumph of sociobiology is that it offers a solution to this
paradox. (See pp. 423–425 and below.)

Social organization varies considerably among different species of eusocial
insects. Colonies of **termites** generally include one queen and one king that
remain together during the life of the colony. The king periodically fertilizes the
queen, who spends almost all her time laying eggs. The other members of the
colony are male and female workers. They do not reproduce as long as they

Figure 32.22
The queen termite Microtermes *has her abdomen enormously swollen with eggs. Her head and thorax are at the left in this picture. The dark dorsal stripes are terga. The helpless queen is attended by numerous workers. The king termite is at the upper right.*

remain in a colony with a functioning queen. The largest workers are soldiers that defend the colony (Figure 32.15A). One or more castes of smaller workers are specialized for foraging, brooding eggs and nymphs, building nests, and maintaining proper temperature and humidity in the nest. The nests consist of elaborate tunnels and chambers in wood or in mounds constructed of mud. Whether an individual nymph develops into a reproductive termite or one of the worker castes is determined by pheromones released by workers attending the nymphs. Periodically workers produce new queens and kings. These are the only individuals that ever have wings. After their final molt they emerge from the colony in a **nuptial flight,** pair up as queen and king, break off their wings, then start a new colony.

Social structure among many eusocial **ants** is similar to that in termites. A major difference among ants and other eusocial hymenopterans is that males never become workers. All adult males, called **drones,** are fertile, and their only function is to copulate once with a queen. The drones either die during copulation, or the workers sting them to death or exile them from the colony and let them starve. Drones cannot defend themselves, since they lack stingers (which are modified ovipositors). The queen "decides" whether to produce females or males by opening or closing a valve from the spermatheca during oviposition. If the valve is closed the eggs will not be fertilized and will develop into males. More often, however, the queen allows the eggs to be fertilized, and they develop into females. Since the eggs that develop into drones receive only one set of chromosomes from the queen, they are haploid. The females are all diploid. This system of sex determination is therefore called **haplodiploidy.**

Whether a female will develop into a new worker or a new queen depends on how well it is fed as a larva. This, in turn, depends on pheromones from the queen that inhibit workers from feeding female larvae enough to make them develop into queens. Only if the queen pheromones are in low concentrations will workers feed certain of the female larvae enough for them to develop into new queens. This can happen if the queen weakens or dies, or if the colony becomes large enough for it to establish new colonies. In many species of ants the production of various castes of workers also depends on how well the larvae are fed. Poorly fed larvae develop into members of a smaller caste of workers. The best fed larvae develop especially large heads and mandibles and become soldiers.

Social structure is as complex and diverse among other eusocial hymenopterans—certain wasps and bees—as among ants. Honey bees have been the most thoroughly studied, and several aspects of their organization have been discussed throughout this text. The accompanying box brings together and summarizes these and other aspects of honey bee life.

Honey Bees

Although the goal of zoology is to provide a complete understanding of an organism, necessity dictates that zoologists specialize in physiology, behavior, genetics, evolution, ecology, morphology, or some other discipline. It is seldom, therefore, that we can put all the pieces of information together for an integrated understanding of an animal, but we are as close to achieving that goal for the honey bee *Apis mellifera* as for any other species, even our own. Several aspects of honey bee biology are discussed elsewhere in this text, but it may be useful to summarize these and other topics in one place to gain some appreciation of this integration.

A QUEEN IS BORN

The life of a honey bee makes sense only in the context of the life of its colony. A mature colony consists of 20,000 to 80,000 workers and one queen. Generally, two pheromones from the queen inhibit workers from raising new queens. The first pheromone is **queen substance,** a fatty acid secreted from the queen's mandibular glands. The second is the **footprint pheromone,** which queens distribute as they walk about the nest ovipositing. In late spring the queen reduces her secretion of queen substance. If the colony is crowded with workers she is also unable to walk in the lower parts of the comb. The workers respond to the decline in queen substance and the absence of footprint pheromone by constructing 20 or more special queen cells near the bottom of the comb. Unlike the classic hexagonal cells in which workers store honey and pollen and raise new workers, queen cells are long and tapered (Figure 32.23). The queen then deposits eggs in these cells. After hatching, these larvae are fed a sugary substance called **royal jelly,** which is secreted from the hypopharyngeal glands in the workers' heads. Royal jelly is similar to the **worker jelly** that is fed to other larvae, but it is much richer in sugar. Sixteen days after the eggs are laid, eclosion occurs, and the new queens emerge from the royal cells.

SWARMING

In the meantime the workers have been preparing the existing mother queen to emigrate from the hive with a swarm of workers. Like trainers in a weight-loss program, the workers feed the queen less and less and jostle her more and more. This combination of enforced dieting and exercise causes the queen to lose about one-fourth of her body weight and to become restless. Finally, while the new queens are pupating, the mother queen flies out of the hive, taking about half the workers in a swarm. The swarm clusters on a limb or other substratum not far away. Worker scouts leave the swarm to find a new nest cavity, which may be a cave, hollow tree, or artificial hive. After they find a suitable site, the scouts return and communicate its location using the waggle dance (see pp. 417–419). The scouts also communicate the suitability of the site by how long they persist in dancing. "Enthusiastic" scouts will gradually convince less enthusiastic ones to inspect their sites. Over a period of several days a consensus will be reached, and the scouts will lead the entire swarm to the chosen site.

MATING

The workers in the old colony remain queenless for about eight days, until a new queen emerges. This virgin queen then takes off on a "nuptial flight" to a special "drone congregation area" where drones from nests within several kilometers gather to mate. Using her queen sub-

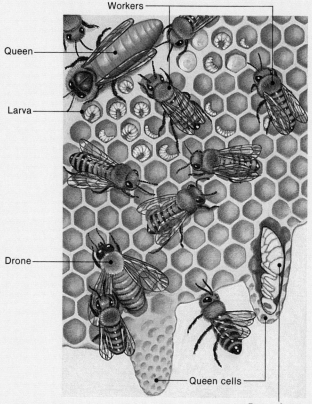

Figure 32.23
Portion of a honey bee colony. The queen (upper left) is larger than the workers. Workers continually crowd around her, feeding her and spreading chemical messages throughout the hive. Other workers tend eggs and larvae and place honey and pollen into open cells. Other open cells hold eggs and larvae. Capped cells enclose honey and pupae. A drone (lower left) is being dragged by its wing and will be killed or evicted. New queens are raised in special queen cells (lower right), one of which is shown opened to reveal the pupa inside.

(continued)

stance as a sex-attractant pheromone, the queen attracts a hundred or more of these drones, which form a comet-shaped mass as they fly after the queen. Approximately 10 or 12 drones succeed in mating with the queen, dying in the process when their genitals explode during copulation. Unsuccessful drones will either be stung to death by their sister workers or exiled from the colony and left to die. During the nuptial flight the queen accumulates more than enough sperm in her spermatheca to fertilize up to 2500 eggs per day for several years. Almost all the fertilized eggs will develop into workers.

WINTER ACTIVITIES

For the rest of the summer the workers forage for pollen and nectar, which they feed to the larvae that will develop into new workers. Workers also gather plant resins, called **propolis,** with which they seal crevices in the nest. They also bring back water with which to cool the nest by evaporation. Much of the nectar is allowed to partially evaporate to make honey. The high concentration of sugar in the honey, as well as some antibiotics, permits it to be stored in capped cells through the winter without contamination by bacteria or fungi. Stored honey and pollen allow the colony to be active through the winter. This is a major difference between honey bees and other eusocial wasps and bees. In bumble bees, for example, only the queen lives through winter, and even then she is in the inactive state of torpor. After the spring warmth arouses her, she must find a nest and raise up a brood of new workers before she can make new queens. The honey bee workers use energy from the honey and pollen to generate heat, which allows the queen to begin laying eggs at the start of winter. The pollen provides the queen with proteins for eggs. A new generation of workers will therefore be available to take advantage of the earliest flowers of spring. By late spring the colony is already producing new queens and beginning a new annual cycle.

This year-round activity of the colony is thought to have originated in the tropics, but it now allows honey bees to live in climates where the temperature drops as low as −30°C. The survival of honey bees depends on their maintaining relatively high temperatures within the nest. The workers produce heat by shivering with their flight muscles, and they conserve the heat with their hair-like pile and by huddling tightly together. The mass of workers can produce as much heat as a light bulb (40 watts or more). By clustering around the larvae the bees can warm them to around 35°C even when the outside temperature is as low as −30°C. In the northeastern United States it takes approximately 20 kg of honey to sustain a hive at this level of thermoregulation through winter. (See Heinrich 1981; Southwick and Heldmaier 1987.)

DIVISION OF LABOR

Unlike many termites and ants, in which there are several specialized worker castes, honey bees have just one worker caste. During the spring and summer each worker gradually progresses through various tasks, such as tending to larvae, cleaning cells in the comb, capping cells (putting covers on the cells containing pupae and honey), building comb, and maintaining the comb temperature and humidity. Finally, each worker becomes a forager and will remain one until she dies, usually less than two months after having become an adult. The division of labor is not rigidly scheduled but varies with environmental and genetic factors that affect the levels of juvenile hormone (Robinson et al. 1989).

FORAGING

A worker is well equipped anatomically for the task of foraging for nectar and pollen. She uses the proboscis to suck up nectar, which consists mainly of water and sucrose. The nectar is then stored in the crop, or "honey stomach," during the flight back to the nest. A valve, called the "honey stopper," prevents the nectar from entering the midgut unless the bee needs the sugars for energy. Upon returning to the nest the forager looks for an empty cell. If none is available she keeps the nectar in her crop, which causes the wax glands to redevelop, enabling her to build new cells. If she finds an empty cell she regurgitates the nectar and begins the process of making honey out of it. Along with the nectar she secretes some sucrase enzyme that digests the sucrose into the simple sugars glucose and fructose. She also secretes an enzyme that converts some of the glucose to an acid, which helps prevent bacterial contamination. The workers then fan their wings, thereby speeding up evaporation of water from the nectar, which reduces its volume and helps preserve it from microbial contamination.

Pollen accumulates on the pile, and the worker removes it with **pollen brushes** on the fore and middle legs (Figure 32.24). She then moistens the pollen with nectar and removes it from the brushes with the **pollen comb** on the opposite hind legs. Finally, she uses the **pollen packer** on the hind leg to pack the mixture of pollen and nectar into the pollen basket, which is a depression bordered by hairs. (A full pollen basket is shown in Figure 32.17C.) When the pollen baskets are full the worker flies back to the nest. Some pollen, however, gets transferred to other flowers, thereby effecting pollination. Pollination by honey bees is necessary for the survival of many plant species and ultimately, therefore, for the survival of honey bees as well.

Sources of nectar and pollen are usually located by workers sent out as **scouts.** Once a scout locates a new source of food she returns to the nest and communicates the location to other workers (see pp. 417–419). If the flowers are nearby, the scout performs a round dance while workers crowd around. The scent on the scout's body tells the **recruits** what flowers to search for. If the flowers are far away the scout performs the remarkable waggle dance mentioned previously.

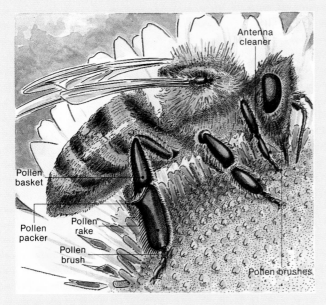

Figure 32.24
A worker honey bee, showing the structures used in foraging. Nectar is sucked into the crop by the proboscis. Pollen is scraped off the body with the pollen brushes, mixed with nectar, transferred to the pollen combs, then packed into the pollen baskets by the pollen packers.

THE ADVANTAGES OF ALTRUISM

Although many questions remain, this brief description indicates that zoologists understand quite a lot about *how* worker honey bees perform their tasks. They have even been able to suggest a satisfactory answer to the most fundamental question of all: *Why* do the workers perform those tasks? According to the theory of natural selection, a behavior should evolve only if it increases the fitness of the individual performing the behavior. That is, only those behaviors should evolve that increase the transmission of genes for that behavior into future generations. How, then, could worker altruism have evolved, since workers do not normally reproduce? Why haven't there evolved lazy workers that refuse to work, or selfish workers that eat the nectar themselves rather than donate it to the colony? Since workers retain their ovaries and can, in fact, lay eggs if neither queen substance nor larvae are present in the nest, why aren't there workers that reproduce their own young?

So far the most satisfactory answer to these questions has come from W. D. Hamilton and other sociobiologists (see pp. 423–425). Hamilton pointed out that workers carry many of the same genes as their mother and sister queens, so by helping the queens reproduce, the workers are also helping reproduce their own genes. Of course honey bees know nothing of genetics. The "reason" the workers work is that they have inherited the appropriate genes. If they inherited a gene for laziness or selfishness the queen would presumably be less successful in reproducing that gene. Such a gene would therefore disappear from the species. Likewise, if a worker inherited a gene that made her try to reproduce on her own, she would have less time to devote to maintaining the colony. Both the worker and the queen would suffer, and they would be less able to reproduce that gene.

INTERACTIONS WITH HUMANS AND OTHER ANIMALS

The Importance of Insects. The number of individual insects has been estimated at 10^{18}—a billion billion. It is not surprising, therefore, that virtually every terrestrial animal, including ourselves, interacts in some way with some insect. Numerous fishes, birds, and mammals use insects as a major source of nutrients. Humans in some cultures also eat fried grasshoppers, beetle grubs, and assorted larvae and pupae. In Western countries insects are not featured on the menus of many restaurants, although anyone who eats sausages, peanut butter, or foods made with stored grains inadvertently eats some insects. Insects also contribute to our diets as pollinators. Approximately 75% of all commercial crops, worth $20 billion per year, depend on pollination by bees.

On the other hand, insects eat a great deal of food intended for humans. In the United States alone there are approximately 800 insect species considered to be agricultural pests. Farmers—and ultimately consumers and taxpayers—spend millions of dollars each year trying to eliminate these pests. Comparable assaults are made by foresters against insects that damage trees, and by home-owners against insects that damage homes and gardens. Numerous isolated skirmishes are fought against biting flies and against cockroaches and other insects that are merely annoying. Elsewhere in the world the major enemies are flies that spread malaria, yellow fever, river blindness (see p. 553), and bacterial contamination. In parts of Africa plagues of locusts are still as common as in biblical times. If this were a battle of people against people instead of people against insects it would rank as World War III. So far the insects appear to be winning the war. Each year the mosquito *Anopheles* brings death to more people through malaria than were killed in combat in World War II. Our weapons have failed to wipe out a single species of pest insect, the cost of insect control continues to mount, and the number of victims of malaria is increasing (see pp. 478–482).

Insecticides. Few people would have bet on the insects in the 1940s and 1950s, when DDT was coming into widespread use. DDT (dichloro-diphenyl-trichlor-ethane) was the first poison that could be used practically to kill virtually any insect. In World War II it eliminated infestations of lice and fleas, two camp followers that had brought discomfort and disease to soldiers and civilians in all previous wars. Soon DDT was being used to kill mosquitoes, farm and forest pests, and even household flies and cockroaches. The very success of DDT was largely responsible for our subsequent disappointment in it and later poisons, because it raised our expectations. Before DDT we would have been content to **suppress** insects; now we expect total **eradication.** Consumers who once accepted the risk of biting into an apple and finding half a larva now reject any apple that has the slightest cosmetic damage from insects. Farmers who once had to accept that insects were going to get a portion of their crop, and in some years all of it, can no longer afford to lose part of the crop to insects. Malaria, once only one of many common diseases, seems much more serious now that antibiotics and other wonder drugs have eliminated many of those other diseases.

Another reason for disappointment in DDT and other poisons is that insects have shown that they can fight back by developing **resistance** to them. Apparently in any population of insects there are a few with genes whose products break down the chemicals or block their entry into their systems. Even if all the other insects of the species are wiped out, these few can reproduce rapidly enough to restore the population, which will inherit the resistance. Since 1970 the number of such resistant insect pests has jumped from 224 species to approximately twice that many. Because of resistance, poisons must be used in ever-increasing concentrations. On Long Island, New York, potato farmers must now spray up to 10 times per season at a cost of $300 per acre, and the heavy dose of poisons has polluted many water wells. According to one estimate, pesticides poison two million people per year worldwide and kill 14,000. When the insects get so resistant that safe and affordable levels of poisons are ineffective, new poisons must be developed. To find an effective new poison an average of 10,000 chemicals must be screened at a cost that now exceeds $45 million. This cost, of course, is passed on from the manufacturer to the farmer, and ultimately to the person who buys groceries.

In many cases chemical agents have proved not only ineffective but counterproductive. Insect pests actually increase following some applications of poisons. In one experiment cabbage plants sprayed with DDT had twice as many cabbage worms (caterpillars of the moth *Pieris rapae*) than unsprayed cabbages. The reason was that the DDT wiped out beetles that had preyed on the caterpillars.

Not only are poisons intended for insect pests harmful to beneficial insects, but many are also harmful to a wide range of other animals, as Rachel Carson pointed out in 1962. Ten years after *Silent Spring* the Environmental Protection Agency finally banned the use of DDT in the United States. Since then many people have been lulled into believing that all government-approved poisons are safe. In fact, however, only a few of the poisons in use before 1972 have even been tested to determine whether they cause cancer, birth defects, or mutations in mammals, and many developed since then are likely to be carcinogenic, teratogenic, or mutagenic to humans. A report from the National Academy of Sciences in 1987 estimates that out of every 10,000 Americans who eat a normal amount of tomatoes until age 70, as many as nine of them might get cancer from the chemicals intended to destroy insects, weeds, fungi, and other pests. Up to six of those 10,000 people might get cancer from eating beef containing such poisons, and five might get cancer from poisoned potatoes. These numbers translate into hundreds of thousands of Americans. There is little point in giving up hamburgers and french fries, however, because oranges, lettuce, apples, peaches, pork, wheat, soybeans, beans, carrots, chicken, corn, and grapes also contain high levels of carcinogenic poisons (Tangley 1987). In short, if you eat, you are at risk for cancer from poisons used to control insects and other pests. The term "poison" is therefore more accurate than "insecticide" and "pesticide," since a chemical cannot tell the difference between an insect and a human, or between a beneficial insect and one that we consider a pest.

Integrated Pest Management.   Because of the ineffectiveness, expense, and dangers of poisons for insect control, entomologists have moved toward a more sophisticated approach called Integrated Pest Management (IPM). IPM employs a growing arsenal of biological weapons against selected insects. The goal is not total eradication but suppression of insects to the point where their damage is less costly than the control. Rather than saturating an area with chemicals as soon as insects appear, IPM stresses a thorough understanding of the target insect so that a safe, effective control can be applied at the right time. Obviously such thorough understanding requires a great deal of research by many entomologists. In just a few decades since IPM has become widespread, however, such research has contributed several weapons to the arsenal against insect pests:

1. **Biological control: the use of predators, parasitoids, and pathogens.** Among the most promising weapons are parasitoid wasps that lay their eggs in the insect host (Figure 32.25). Also effective in many cases is the bacterium *Bacillus thuringiensis* (BT), which produces a protein crystal that is lethal to insects that ingest the bacteria. BT has been used successfully against tent caterpillars and aquatic larvae of mosquitoes and black flies. Through genetic engineering the gene for the protein crystal has also been introduced into soil bacteria for potential use against insects that attack plant roots.
2. **Saturating the area with male insects sterilized by radiation.** This method works for species in which the female mates only once and is therefore unable to reproduce after mating with a sterile male. The most successful use has been against several species of blow flies whose larvae, called screw worms, feed on wounds in cattle.
3. **Modifying farm methods.** Changing the spacing between rows, alternating rows of different crops (intercropping), crop rotation, and changing the time of planting are among the modifications that make it harder for insect populations to reach pest levels.
4. **Use of hormones and other substances to disrupt metabolism or behav-**

Figure 32.25
A parasitoid wasp Pediobius
foveolatus *oviposits on the larva of
the Mexican bean beetle* Epilachna
varivestis *in spite of the larva's
defensive spines.*

ior. Synthetic juvenile hormone and chemicals that disrupt JH functioning
have been applied successfully in particular cases (see pp. 192–194). Around
250 synthetic sex-attractant pheromones are sold to disrupt reproduction in
various species, but so far their main use has been as bait in insect traps.

5. **Limited use of poisons.** Poisons are still used as part of IPM, but at low
   enough levels to avoid resistance, and at times when the target insect is most
   vulnerable and beneficial insects are least endangered.

Exotic Insects. The most troublesome insects are those introduced into an area
where plants have no defenses against them, and where the insects have no natural
enemies. The classic example of such an exotic insect is the **gypsy moth** *Lymantria
dispar.* In 1869 caterpillars of this moth crawled from a house in Medford, Massachusetts, where they had been brought by a French entomologist hoping to cross
them with oriental silk worms to get a hybrid silk producer that would not have
to feed on mulberry leaves. Twenty years later the citizens of Medford were complaining of caterpillars denuding the trees, falling down their collars, and making
the sidewalks slippery with their crushed bodies. Efforts to control the outbreak
were just about to succeed when state legislators decided there was no further
need to appropriate money for the effort. As a result the gypsy moth is now
entrenched over most of the Northeast. Its caterpillars denude one to two million
acres in an average year, and ten times that area in a bad year. Massive sprayings
of DDT and other poisons have proved powerless to stop them. Nor have any of
47 imported enemies been able to keep up with the gypsy moths' prodigious
reproductive ability, and the ability of the caterpillars to spread by riding the wind
on strands of silk. Other tools of integrated pest management, such as synthetic
sex-attractant (Disparlure) and BT, are economically feasible only in limited areas.

One might think gypsy moths would have taught a valuable lesson, yet exotic
insect pests continue to enter the country. In the 1950s a geneticist in Brazil

hybridized an African race of honey bee with the common European honey bee. He had hoped to get a hybrid that would peacefully produce more honey, but instead he got a bee with the tendency of the African bee to sting with little provocation. Some of these **Africanized honey bees**, *Apis mellifera scutellata*, better known as "killer bees," escaped. In just three decades they spread throughout South and Central America, stinging hundreds of people to death and driving out European bees. Within five years after Africanized bees arrived in Venezuela in 1975, honey production plummeted from 535 metric tons per year to 78 metric tons. Many beekeepers in these areas have adapted to the Africanized bees, but in the United States increased liability insurance premiums could eliminate beekeeping as a hobby and threaten it as a business. Numerous colonies of Africanized bees have already invaded the United States but have been exterminated with pesticides at a cost of up to a million dollars each time. Eventually a colony will not be found in time to stop its spread.

In some carefully managed situations exotic insects have proved beneficial, especially in controlling other exotic species. Thousands of acres of rangeland in the American West were cleared of klamath weed by two species of *Chrysolina* beetles, and Australian rangeland was rescued from the prickly pear cactus by the moth *Cactoblastis cactorum*. Also in Australia dung beetles of several species have been introduced to deal with the cow dung that would otherwise choke the grasses and provide breeding ground for parasites. (See Figure 32.26.)

Although humans have never succeeded in deliberately extinguishing a species of insect, they have brought about the extinction of at least 33 North American species unintentionally. In the tropics the rate of extinction due to human activities is incalculably large. We will never know which of these extinct species would have controlled Africanized bees or disease-bearing mosquitoes, or which species would have brought as much pleasure as honey bees and silk worms, or which would have contributed as much to science as *Drosophila*, or which would have brightened the darkness and excited the imagination like fireflies.

Figure 32.26
The dung roller beetle Canthon imitator, *which occurs throughout North America. Females oviposit in balls of dung, then both females and males roll the balls to suitable places where they bury them. The developing larvae then feed on the dung. In the process dung beetles play an important role in eliminating feces and in recycling nutrients. The ancient Egyptians revered another dung beetle, the scarab* Scarabaeus sacer, *which they visualized as wheeling the Sun about the heavens. Dung beetles are apparently the only animals besides humans that have mastered the principle of the wheel.*

SUMMARY

Approximately three out of every four named animals is an insect. In addition to insects, several other groups make up the subphylum Uniramia. Centipedes, millipedes, pauropods, and symphylans are myriapods. Insects and entognaths (springtails and others) are hexapods. Insects are divided into the wingless apterygotes, such as silverfish, and the pterygotes. The pterygotes are further divided into the exopterygotes, in which the juveniles (nymphs) resemble the adults and have wing buds, and the endopterygotes, in which there is complete metamorphosis from larva to pupa to adult.

The reason there are so many insects may be that they were among the first to have the opportunity to diversify into the wide variety of terrestrial habitats. The impermeable integument, as well as Malpighian tubules and other organs for osmoregulation and excretion, enable them to conserve body water. The tracheal system enables their tissues to obtain oxygen directly from air. Many have special adaptations that enable them to survive cold. These include antifreeze compounds, partial endothermy, torpor, and diapause. Their sensory organs are adapted mainly for use in air, enabling them to hear and detect airborne chemicals. The mouthparts, though consisting of essentially the same structures in all species, are highly specialized for different modes of feeding, such as chewing, sucking, and sponging. Insects also have several modes of locomotion, including running, jumping, and especially flying. Finally, insects are enormously successful at reproduction, largely because of the cleidoic eggs that exchange respiratory gases but conserve water. In eusocial species of termites, ants, and bees colonies may function as superorganisms, with only a few reproducing individuals served by one or more castes of workers.

KEY TERMS

myriapod	torpor	asynchronous muscle
hexapod	diapause	paranotal theory
entognath	partial endothermy	cleidoic egg
apterygote	pheromone	fat body
pterygote	mouthparts	vitellogenesis
endopterygote	mandible	oviposition
exopterygote	hypopharynx	spermatheca
holometabolous	labrum	ootheca
hemimetabolous	labium	gall
nymph	maxilla	ovoviviparous
pupa	crop	eusocial
Malpighian tubule	midgut	haplodiploidy
rectal pad	cecum	queen substance
trachea	hindgut	royal jelly
tracheole	alternating tripod gait	altruism
spiracle	synchronous muscle	integrated pest management

SELF-TEST

1. Describe the differences between uniramians and other arthropods. Describe two kinds of myriapods and explain how they differ from insects.

2. Describe the major difference between Apterygota and Pterygota, and name one example of each.

3. Describe the differences between endopterygote and exopterygote insects. Which ones are hemimetabolous, and which are holometabolous? Which ones have nymphal stages, and which ones have pupae? Name an example of an endopterygote and an exopterygote.

4. Describe two adaptations to terrestrial life in insects, and explain how they may have contributed to the success of insects.

5. Describe how Malpighian tubules and tracheae are adapted for terrestrial living.

6. Explain how the mouthparts of insects are adapted for chewing, sucking, or sponging up food. For each of these three modes of feeding, name a kind of insect in which it occurs.

7. Briefly describe two competing theories about how flight evolved in insects.

8. Explain the functioning of an insect egg. (How is it adapted for terrestrial life? What is the role of vitellogenesis in its formation? How does the shell allow fertilization?)

9. A colony of honey bees has sometimes been compared to a "superorganism," with queens, drones, and workers performing tasks analogous to those of different organs. Elaborate on this idea. What physiological roles do the queens, drones, and various workers perform in the "superorganism"? What are some differences between a honey bee colony and a superorganism?

10. Describe an example of each of the following: parasitoid, pheromone, haplodiploidy, castes, metamorphosis, courtship.

11. Explain why simply poisoning insect pests with chemicals has proved unsatisfactory. Describe two alternatives used in integrated pest management.

READINGS

RECOMMENDED READINGS

Arnett, R. H. Jr. and R. L. Jacques Jr. 1981. *Simon and Schuster's Guide to Insects.* New York: Simon & Schuster.

Batra, S. W. T. 1984. Solitary bees. *Sci. Am.* 250(2):120–127 (Feb).

Batten, M. 1984. The ant and the acacia. *Science 84* pp. 59–67 (Apr). (*On Daniel Janzen and his studies of symbioses in tropical forests.*)

Bland, R. G. and H. E. Jaques. 1978. *How to Know the Insects,* 3rd ed. Dubuque, IA: Wm. C. Brown.

Boraiko, A. A. 1981. The indomitable cockroach. *Nat. Geogr.* 159(1):130–142 (Jan).

Borror, D. J. and R. E. White. 1970. *A Field Guide to the Insects.* Boston: Houghton Mifflin.

Camazine, S. and R. A. Morse. 1988. The Africanized honeybee. *Am. Sci.* 76:464–471.

Camhi, J. M. 1980. The escape system of the cockroach. *Sci. Am.* 243(6):158–172 (Dec).

Dethier, V. G. 1984. *The World of the Tentmakers: A National History of the Eastern Tent Caterpillar.* Amherst: University of Massachusetts Press.

Duplaix, N. 1988. Fleas: the lethal leapers. *Nat. Geogr.* 173(5):672–694 (May).

Eberhard, W. G. 1980. Horned beetles. *Sci. Am.* 242(3):166–182 (Mar).

Evans, H. E. 1985. *The Pleasures of Entomology: Portraits of Insects and the People Who Study Them.* Washington, DC: Smithsonian Institution Press.

Fabre, J. H. (Edited by E. W. Teale). 1949. *The Insect World of J. Henri Fabre.* New York: Dodd, Mead. (*One of several collections of the enjoyable writings of the 19th-century observer of insect behavior.*)

Franks, N. R. 1989. Army ants: a collective intelligence. *Am. Sci.* 77:139–145.

Funk, D. H. 1989. The mating of tree crickets. *Sci. Am.* 261(2):50–59 (Aug).

Gilbert, L. E. 1982. The coevolution of a butterfly and a vine. *Sci. Am.* 247(2):110–122 (Aug).

Gould, J. L. and C. G. Gould. 1988. *The Honey Bee.* New York: Scientific American Library.

Handel, S. N. and A. J. Beattie. 1990. Seed dispersal by ants. *Sci. Am.* 263(2):76–83 (Aug).

Heinrich, B. 1981. The regulation of temperature in the honeybee swarm. *Sci. Am.* 244(6):146–160 (June).

Hölldobler, B. and E. O. Wilson. 1983. The evolution of communal nest-weaving in ants. *Am. Sci.* 71:490–499.

Huber, F. and J. Thorson. 1985. Cricket auditory communication. *Sci. Am.* 253(6):60–68 (Dec).

Kingsolver, J. G. 1985. Butterfly engineering. *Sci. Am.* 253(2):106–113 (Aug).

Levi, H. W. and L. R. Levi. 1968. *A Guide to Spiders and Their Kin.* New York: Golden Press. (*Includes a section on myriapods.*)

Lloyd, J. E. 1981. Mimicry in the sexual signals of fireflies. *Sci. Am.* 245(1):139–145 (July).

Merritt, R. W. and J. B. Wallace. 1981. Filter-feeding insects. *Sci. Am.* 244(4):132–144 (Apr).

Milne, L. and M. Milne. 1980. *The Audubon Society Field Guide to North American Insects and Spiders.* New York: Alfred A. Knopf.

Morse, D. H. 1985. Milkweeds and their visitors. *Sci. Am.* 253(1):112–119 (July).

Nijhout, H. F. 1981. The color patterns of butterflies and moths. *Sci. Am.* 245(5):140–151 (Nov).

Prestwich, G. D. 1983. The chemical defenses of termites. *Sci. Am.* 249(2):78–87 (Aug).

Rosenthal, G. A. 1983. A seed-eating beetle's adaptations to a poisonous seed. *Sci. Am.* 249(5):164–171 (Nov).

Rosenthal, G. A. 1985. The chemical defenses of higher plants. *Sci. Am.* 254(1):94–99 (Jan).

Ryker, L. C. 1984. Acoustic and chemical signals in the life cycle of a beetle. *Sci. Am.* 250(6):112–123 (June).

Seeley, T. D. et al. 1985. Yellow rain. *Sci. Am.* 253(3):128–137 (Sept). (*How the U.S. government mistook honey bee feces for biological warfare agents.*)

Seeley, T. D. 1989. The honey bee colony as a superorganism. *Am. Sci.* 77:546–553.

Southwick, E. E. and G. Heldmaier. 1987. Temperature control in honey bee colonies. *BioScience* 37:395–399.

Stokes, D. W. 1983. *A Guide to Observing Insect Lives.* Boston: Little, Brown.

Strobel, G. A. and G. N. Lanier. 1981. Dutch elm disease. *Sci. Am.* 245(2):56–66 (Aug).

Thornhill, R. 1980. Sexual selection in the black-tipped hangingfly. *Sci. Am.* 242(6):162–172 (June).

Topoff, H. 1990. Slave-making ants. *Am. Sci.* 78:520–528.

Wootton, R. J. 1990. The mechanical design of insect wings. *Sci. Am.* 263(5):114–120 (Nov).

See also relevant selections in General References at the end of Chapter 21.

ADDITIONAL REFERENCES

Arnett, R. H. Jr. 1986. *American Insects: A Handbook of the Insects of America North of Mexico.* New York: Van Nostrand Reinhold.

Brown, S. M. 1986. Of mantises and myths. *BioScience* 36:421–423.

Carrell, J. E. and T. Eisner. 1984. Spider sedation induced by defensive chemicals of milliped prey. *Proc. Natl. Acad. Sci. USA* 81:806–810.

Dover, M. J. and B. A. Croft. 1986. Pesticide resistance and public policy. *BioScience* 36:78–85.

Eisner, T. et al. 1978. "Wolf-in-sheep's-clothing" strategy of a predaceous insect larva. *Science* 199:790–794.

Erwin, T. 1983. Beetles and other insects of tropical forest canopies at Manaus, Brazil, sampled by insecticidal fogging. In: S. L. Sutton et al. (Eds.), *Tropical Rain Forest: Ecology and Management.* Edinburgh: Blackwell Scientific Publishers, pp. 59–75.

Greany, P. D., S. B. Vinson, and W. J. Lewis. 1984. Insect parasitoids: finding new opportunities for biological control. *BioScience* 34:690–696.

Harris, P. 1988. Environmental impact of weed-control insects. *BioScience* 38:542–548.

Kukalová-Peck, J. 1978. Origin and evolution of insect wings and their relation to metamorphosis, as documented by the fossil record. *J. Morphol.* 156:53–126.

Lewin, R. 1985. On the origin of insect wings. *Science* 230:428–429.

Michelsen, A. 1979. Insect ears as mechanical systems. *Am. Sci.* 67:696–706.

Robinson, G. E. et al. 1989. Hormonal and genetic control of behavioral integration in honey bee colonies. *Science* 246:110–112.

Roitberg, B. D. and R. J. Prokopy. 1987. Insects that mark host plants. *BioScience* 37:400–406.

Seeley, T. D. 1985. *Honeybee Ecology.* Princeton, NJ: Princeton University Press.

Tallamy, D. W. 1984. Insect parental care. *BioScience* 34:20–24.

Tangley, L. 1987. Regulating pesticides in food. *BioScience* 37:452–456.

33

Echinoderms and Arrowworms

Marble sea star (Pentagonaster dubeni).

CHAPTER OUTLINE

LEARNING OBJECTIVES

1. Why are echinoderms thought by many to be related to humans and other chordates?

2. What do starfish, sea urchins, and sand dollars have in common?

3. How does the five-part radial symmetry of starfish and other echinoderms occur?

4. How do starfish and other echinoderms get about? How do they feed?

5. Do echinoderms have circulatory systems?

INTRODUCTION TO DEUTEROSTOMES

With this chapter we leave the Protostomia and begin considering four phyla that are often grouped together as Deuterostomia. In deuterostomes the mouth does not develop from the blastopore but from a second or later opening in the embryo (see p. 118). (The term "deuterostome" comes from the Greek words for "second" and "mouth.") By this criterion the deuterostomate phyla are Echinodermata (such as starfish and sea urchins), Chaetognatha (arrowworms), Hemichordata, and Chordata, such as ourselves. Some of these animals also share other developmental similarities described in Chapter 21, such as cleavage of the earliest embryonic cells in a radial rather than a spiral pattern, and the ability of those cells to form any type of tissue (**indeterminacy**). Deuterostomes are also said to be **enterocoelous,** with the coelom developing from pouches in the embryonic gut (Greek *enteron* gut).

There is absolutely no doubt that echinoderms are deuterostomate, since deuterostomy was first defined in terms of the echinoderm pattern of development. Among the other phyla often considered to be deuterostomes, however, there are many exceptions to these patterns. There are also other reasons for being skeptical of such a simple dichotomy as that of deuterostomes versus protostomes (Willmer 1990). In fact, molecular phylogeny suggests that the echinoderms and chordates are as different from each other as either of these phyla is from the protostomes (see Figures 18.16 and 34.16A). Arrowworms, which have long been difficult to classify, may also be quite different from either echinoderms or chordates. Hemichordates appear to be more closely allied to chordates and will therefore be discussed with chordates in the next chapter.

PHYLUM ECHINODERMATA

Starfish, also known as sea stars or asteroids, as well as sea urchins and sand dollars, are familiar souvenirs from trips to the beach. Other echinoderms—brittle stars, feather stars, and sea cucumbers—are seldom encountered except by divers (Figures 33.1 and 33.10). These echinoderms vary greatly in such external features as the presence or absence of arms (rays), and the position of the mouth and anus (Figure 33.2). All, however, have an organization that is fundamentally like that of starfish (Table 33.1). Starfish will therefore be used as the major example in describing the features of echinoderms.

RADIAL SYMMETRY

The most evident feature of starfish and most other echinoderms is that they tend to be radially symmetric. It is clear, however, that this is a derived rather than a primitive feature. There is a continuous fossil record of echinoderms indicating that their ancestors were bilaterally symmetric, and the larvae of echinoderms are still bilaterally symmetric. Apparently the radial symmetry of adult echinoderms is a secondary adaptation to their sessile lives on the ocean floor. In most living species of echinoderms the radial symmetry is **pentamerous,** as is evident in the five arms (or rays) of most sea stars. In other words, the bodies of most adult echinoderms can be divided into five (Greek *penta-*) similar parts (*meros*).

A

B

C

D

E

Table 33.1 Characteristics of echinoderms.

Phylum Echinodermata e-KINE-oh-DER-mah-tuh (Greek *echinos* a sea urchin or hedge-hog).

Morphology: Adults with derived, **pentamerous radial symmetry,** tending toward bilateral symmetry in some forms. Starfish, brittle stars, sea lilies, and feather stars have five (or more) arms. No head. With endoskeleton of calcareous ossicles. Integument of many forms bears spines and/or beaklike **pedicellariae.**

Physiology: Coelomic fluid resembling seawater circulated by ciliated peritoneum. Nervous system consists mainly of a plexus beneath the epithelium, together with a circumoral ring and radial nerves. No brain or other ganglia; few specialized receptor organs. No osmoregulatory organs. Exchange of respiratory gases and elimination of metabolic wastes generally through integument, especially **papulae.** Digestion by means of a simple stomach and intestine; with or without an anus.

Locomotion: Many forms move by means of **tube feet,** which may have suckers on the ends. Tube feet extend and retract by **water vascular system** and muscles. Some forms sessile; some crawl or swim by waving arms; some burrow by peristaltic contraction or movable spines.

Reproduction: Most forms dioecious, but with little external difference between sexes. Fertilization external. In at least one species of starfish, *Luidia* sp., the larvae can also reproduce asexually by fission (Bosch et al. 1989).

Development: Radial, indeterminate cleavage. Deuterostomate. Enterocoelous. Development usually indirect, with a variety of **bilaterally symmetric larvae.** Larvae are ciliated and usually free-swimming. Lost body parts readily regenerated in adults.

Habitat, Size, and Diversity: Exclusively marine. Many forms primarily intertidal or subtidal; others range in depth down to ocean trenches. None parasitic. Largest dimension of adults ranges from a few millimeters to 2 meters. Approximately **6100 living species** described.

Figure 33.1
*Representative echinoderms.
(A) Northern starfish* Asterias vulgaris *are common on rocky, northeastern shores of the United States. Diameter 20 cm. (B) A brittle star. (C) The common edible sea urchin* Echinus esculentus. *Diameter 10 cm. (D) A group of feather stars on the Great Barrier Reef off the coast of Australia. Those at the upper center (white) and right (dark) are* Oxycomanthus bennetti. *The red one at the lower center and the yellow one to its left are* Himerometra robustipinna.
(E) The sea cucumber Pseudocolochinus axiologus. *Note the feeding tentacle being withdrawn from the mouth and the warning coloration.*

Figure 33.2
Basic organization of the major kinds of echinoderms. The surface bearing the mouth is called the oral surface. The opposite side is the aboral surface. Because sea cucumbers lie on their sides with the mouth at one end, they are bilaterally symmetric.

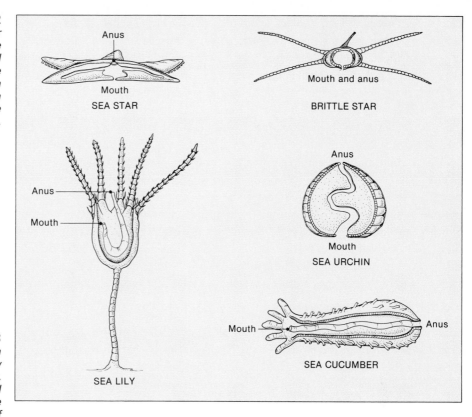

Figure 33.3
(A) Side view of starfish arm showing tube feet. (B) The hollow tube foot is filled with seawater, which enters from the water canal through the one-way valve. The tube foot is extended by contraction of muscles surrounding the ampulla. Selective contraction of retractor muscles allows the tube foot to be extended in any direction. Contraction of levator muscles causes the tube foot to apply suction. After the suckerlike endings attach to solid substratum or prey, contraction of all the retractor muscles retracts the tube foot.

TUBE FEET

Another distinctive feature of echinoderms is their tube feet, or **podia** (Figure 33.3). These structures, which are not found in any other phylum, cling to substratum and food. Starfish that live on rocky shores or coral reefs would be unable to move and would be tossed around by waves if they did not have tube feet. In these starfish the end of each tube foot is shaped like a suction cup and secretes

mucus that forms a tight contact with solid substratum. Each tube foot of the common starfish *Asterias* can lift 29 grams, so hundreds of them can exert an impressive force. Sea stars can cling to rocky shores in spite of the buffeting of heavy waves, and they can pull open the shells of clams and other prey. Some tube feet also serve as feelers and, in some other echinoderms, as tentacles.

A starfish uses its tube feet during locomotion by slightly lifting the tip of whichever arm points in the direction it is going, then extending the tube feet at the tip of that arm until they make contact. Muscles in those tube feet then contract to apply suction and to shorten the tube feet, thereby pulling the asteroid a short distance. These tube feet then release their grip while those farther from the tip of the arm repeat the sequence. The tube feet of most starfish that live on mud or sand lack suckers and simply dig into the substratum during locomotion. Starfish also use the podia to right themselves if turned upside down. A sea star that cannot right itself can neither move nor feed, since most of the tube feet and the mouth are on the oral (bottom) surface.

AMBULACRAL GROOVES

The tube feet of starfish are concentrated along ambulacral grooves, or **ambulacra,** that radiate from the mouth (Figure 33.3A). Usually there is one ambulacral groove on the oral surface of each arm, and the podia are arranged in two or four rows on both sides of each groove. In sea stars the ambulacral grooves are open, exposing the radial canals that supply water pressure to the tube feet. Within the ambulacra can also be seen radial nerves, which help coordinate movements of the arms and tube feet. Each radial nerve also conducts some visual and tactile information from the ocellus and tentacle at the end of the arm, as well as information from chemoreceptors all over the integument (Figure 33.6). This information is apparently used mainly in local reflexes responsible for defense, locomotion, righting, and feeding. There is no brain or other ganglion for overall coordination of neural activity.

THE WATER VASCULAR SYSTEM

Extension of the tube feet is due to another system unique to living echinoderms: the water vascular system (Figure 33.4). As its name suggests, the water vascular

Figure 33.4
(A) The water vascular system of a starfish. Seawater enters the madreporite on the aboral surface and passes through the stone canal into the ring canal and radial canals. The water then enters the ampullae and tube feet, as described in Figure 33.3. Polian vesicles and ampullae serve as reservoirs. Tiedemann's body is believed to produce cells (coelomocytes) in the coelom.
(B) The madreporite of the sea star Asterias.

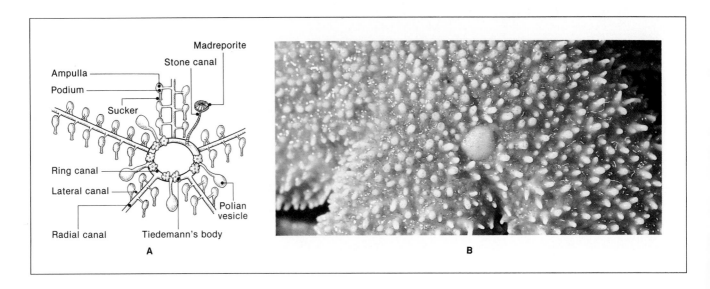

system compensates for the absence of a blood circulatory system. It comprises a network of canals filled with seawater, but also containing certain cells (**coelomocytes**), proteins, and concentrated potassium ions. The water enters through a sievelike **madreporite** on the aboral (upper) surface and passes through the **stone canal.** Cilia on the lining (peritoneum) of the canals draw the water into a **ring canal** that encircles the mouth. Five (or more) **radial canals** branch from the ring canal into the arms, where they bring water into the tube feet through valves. The water is stored in bulblike **ampullae,** which extend the tube feet when they contract.

THE COELOM

The water vascular system is one of four major parts of the coelom in adult asteroids. The largest part is the **perivisceral coelom,** which surrounds the visceral organs and branches into the arms. It is primarily responsible for transporting nutrients and perhaps respiratory gases. Another part of the coelom is the **hemal system,** which in starfish consists of poorly defined spongy tissues that lie parallel to the water vascular system. The hemal system has an aboral ring with branches to the digestive system and gonads, and an oral ring with branches radiating into the arms. Another part of the coelom, the **perihemal system**, consists of tubes and sinuses surrounding the hemal system. After more than a century of research and speculation, the functions of the hemal and perihemal systems remain a mystery.

THE INTEGUMENT

The aboral (upper) surface of a starfish is mainly adapted for defense. The **endoskeleton** plays a major role in defense against predators. The endoskeleton consists of plates called **ossicles,** each of which is a single crystal of calcium carbonate with some magnesium carbonate. The ossicles are secreted by the dermis, which lies beneath the thin epithelium that covers the animal. In sea stars, and also in brittle stars and sea urchins, the ossicles have spiny projections on the aboral surface that undoubtedly deter most predators. Starfish also have a unique **catch connective tissue** that stiffens the integument and helps defend against predation. The aboral surfaces of starfish and many other echinoderms also bear numerous beaklike structures called **pedicellariae** (Figures 33.5 and 33.11). Pedicellariae apparently keep the integument free of sponges, corals, and other encrusting organisms and are used in defense and feeding. Also important in defense may be the striking pigmentation of many echinoderms. Since echinoderms have only rudimentary eyes at best, these pigments probably serve as camouflage or warning coloration, rather than as a means of identifying members of the same species.

The aboral surface is also responsible for the elimination of wastes and the exchange of respiratory gases between the coelomic fluid and seawater. These functions do not require elaborate mechanisms, because of the low metabolic rates of these sluggish animals. Tiny projections called **papulae** provide enough surface area for the diffusion of ammonia, O_2, and CO_2. Diffusion across the papulae also maintains ionic concentrations in the coelomic fluid similar to those in seawater. Echinoderms lack organs for osmoregulation and can live only in marine habitats.

DIGESTION

Starfish have fairly simple digestive systems (Figure 33.6). The stomach is the main organ of digestion and fills most of the central disc. It has two chambers—

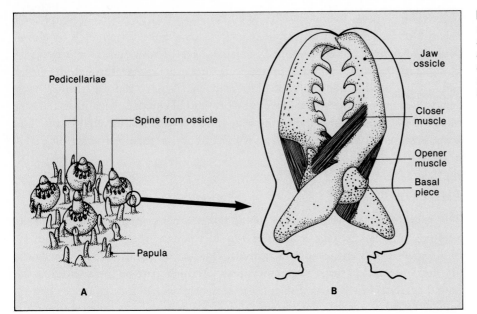

Figure 33.5
(A) Detail of the aboral surface of a
starfish, showing pedicellariae,
spines, and papulae. (B) Structure of
a pedicellaria. The two jaw ossicles
(often called valves) pivot on the
basal piece.

the **cardiac stomach** and the **pyloric stomach.** Digestive enzymes enter the stomach from a pair of **pyloric ceca** (= digestive glands) in each arm. The pyloric ceca
are also the major sites for storage of food and absorption of nutrients. The intestine and anus are either absent or too small to be of much use, so the digestive
tract is essentially incomplete. Most sea stars feed on externally digested material,
so they generate little solid waste. Those that feed on small particles use the
pedicellariae to pass the food into the mouth and directly to the stomach.

Many starfish prey on surprisingly large and well-protected animals, such as
molluscs, crustaceans, and coral polyps. While gripping the prey with their podia,
these sea stars evert their cardiac stomachs out through their mouths and partially
digest the prey externally. To feed on crustaceans and coral polyps they simply
cover the prey with their cardiac stomachs. *Asterias* and some other asteroids can
insinuate the cardiac stomach through tiny gaps between the valves of oysters and
other bivalve molluscs. A space of only 0.1 mm is all *Asterias* needs to digest an

Figure 33.6
*Internal structure of a starfish
viewed from the aboral surface.
Note the two chambers of the
stomach. The function of the rectal
ceca is unknown. The gonads
release either sperm or ova
externally through gonopores
between the arms.*

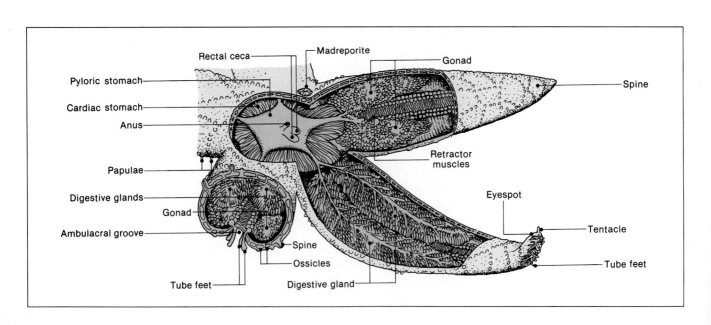

oyster inside its own shell. After partially digesting the prey the cardiac stomach retracts to bring in the nutrients.

BRITTLE STARS

With their five or more arms, brittle stars (ophiuroids) resemble sea stars in general appearance. In contrast to the lumbering movements of sea stars, however, brittle stars can scuttle rapidly across substratum and sometimes swim. Unlike the arms of starfish, those of brittle stars are slender and more flexibly attached to the central disc, enabling them to make rapid, snakelike movements. A common name for brittle stars is, in fact, serpent stars. Also contributing to the agility of brittle stars are the closely articulating ossicles that surround the arms (including the ambulacra) (Figure 33.7). These spiny ossicles enable the arms to twist and bend much like the spinal column of vertebrates; in fact, the ossicles are often called vertebrae.

Brittle stars lack suckers on their tube feet, so they are generally found in deeper water, away from wave action. Most of the time they stay hidden in dark crevices,

Figure 33.7
The structure of brittle stars. (A) Oral surface of the central disc. (See Figure 33.1B.) (B) A section through the central disc and part of one arm.

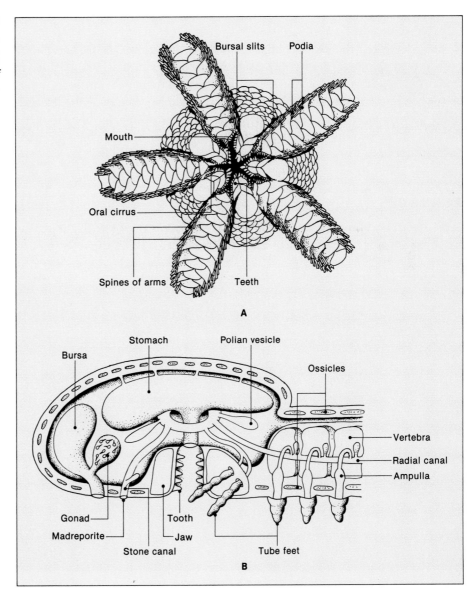

feeding on small particles that they capture with the tube feet and in mucus on the arms. Brittle stars have no pedicellariae. The mouth is armed with five toothed jaws (Figure 33.7). There is no intestine or anus. Brittle stars also lack papulae. They exchange respiratory gases by pumping water through unique slits into five **bursae** (sacs). In addition to serving as respiratory organs, the bursae also serve as attachment sites for the gonads. Either sperm or ova break through the bursae and escape through the bursal slits for external fertilization.

CRINOIDS

Sea lilies and feather stars are relics of the ancient class Crinoidea, and they are considered the most primitive echinoderms (Figure 33.8). Crinoids have five featherlike arms that are extensively branched and covered with numerous **pinnules.** The arms wave above the cup-shaped body, which is called the **calyx** (plural = calyces). In sea lilies the calyx is at the end of a long, flexible stalk whose other end is usually attached to substratum. Feather stars also have long stalks in their late larval stages, but on becoming adults the calyces break off the stalks and swim away by waving their arms. It appears, therefore, that feather stars evolved from sea lilies. Feather stars generally remain attached to substratum by means of clawlike **cirri** (visible in Figure 33.1D).

At first glance crinoids bear little resemblance to starfish. If one imagines a crinoid feeding upside down on the substratum, however, it is not so different from a starfish. Crinoids use tube feet on their arms to trap small organisms, which are passed to the mouth by the open ambulacral grooves (Figure 33.9). The pinnules greatly increase the area available for this suspension feeding. Gametes develop within the arms and pinnules and break through the epithelium for external fertilization.

ECHINOIDS

Sea urchins, heart urchins, and sand dollars, collectively known as echinoids, can be visualized as starfish with their arms raised and fused above them, so that the

Figure 33.8
(A) A "meadow" of crinoids as they appeared about 350 million years ago in the shallow sea that covered what is now the central United States. Some extinct crinoids were more than 20 meters long, but no living form is more than a meter long. (B) Ossicles from the stalks and bodies of crinoids that lived approximately 400 million years ago in what is now New York State. Such ossicles are among the most common fossils in Paleozoic limestone. Scale in centimeters.

A

B

Figure 33.9
(A) External structure of the sea lily Cenocrinus asteria, which has cirri on the stalk. (B) Prey's-eye view into the calyx of the feather star Antedon. Tube feet on the pinnules thrash about and toss small organisms into the open ambulacral grooves, where they are trapped in mucus and moved by cilia toward the mouth (broad arrows). Other cilia keep the interambulacral area clear of debris (thin arrows). Note the elevated location of the anus, which avoids fouling the mouth.

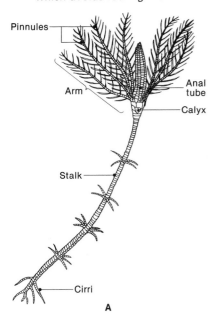

Pinnules

Arm

Anal tube

Calyx

Stalk

Cirri

A

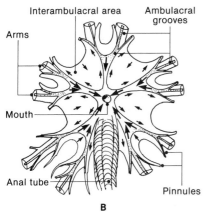

Interambulacral area

Ambulacral grooves

Arms

Mouth

Anal tube

Pinnules

B

Figure 33.10
The keyhole urchin Mellita quinquiesperforata, a common sand dollar. The "keyholes" (lunules) are believed to help prevent the sand dollar from sailing in water currents and to enable food particles to be passed from the aboral to the oral surface. The petal-shaped pattern (petaloid) shows the location of tube feet specialized for exchange of respiratory gases.

five ambulacra bend up toward the anus. Sea urchins are hemispherical in shape (if one ignores the spines) and are said to be **regular** because they are radially symmetric. Heart urchins resemble sea urchins except that they have shorter spines and have the mouth toward one edge of the oral surface. They therefore tend to be bilaterally symmetric and are said to be **irregular.** Sand dollars are disc-shaped and also irregular (Figure 33.10). In all echinoids the ossicles are close-fitting, forming tests that are often found along beaches. Living sea urchins are covered with long spines that pivot in sockets at the base. By flexing these spines, sea urchins can use them like numerous legs to creep across the ocean floor, and they can use them like pincers to aid the pedicellariae in capturing prey and deterring predators (Figure 33.11). Each spine consists of a single crystal of $CaCO_3$ that would be as brittle as chalk if it were not for certain glycoproteins in it. Heart urchins and sand dollars have shorter spines that they use instead of their tube feet for burrowing through sand as they feed on detritus.

Sea urchins generally feed on algae and encrusting animals, which they scrape up and chew with remarkable toothed jaws that make up **Aristotle's lantern** (Figure 33.12). The digestive tract is more elaborate than that of asteroids. The esophagus, stomach, and intestine make up a long tube that spirals inside the perivisceral coelom. Water that is swallowed with food bypasses the stomach and much of the intestine through a ciliated tube called the **siphon.** Heart urchins feed on organic detritus, which they pick up with tube feet as they burrow. Sand dollars generally feed by sifting sand through their short spines and passing organic material toward the mouth on cilia. In moderate currents, however, some species dig their anterior edges into sediment with the oral end facing the current to capture plankton. Like sea urchins, sand dollars have Aristotle's lanterns. Echinoids also absorb significant amounts of dissolved organic matter across the body surface.

SEA CUCUMBERS

Sea cucumbers resemble their vegetable namesake in overall shape and are so unlike other echinoderms that they were once classified in the phylum Sipuncula.

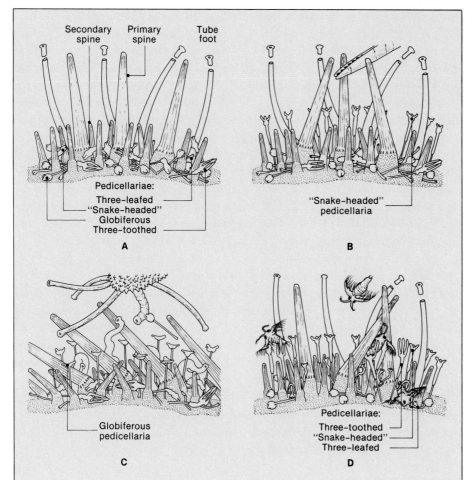

Figure 33.11
*Defense and predation by spines and pedicellariae on the aboral surface of the sea urchin Psammechinus miliaris. (A) At rest the primary and secondary spines and tube feet are extended, while the four types of pedicellaria are relaxed. The primary spines are about 5 mm long. Only parts of the much longer tube feet are shown. (B) In response to mechanical disturbance (forceps) the spines press against the object, and ophiocephalous (snake-headed) pedicellariae prepare to bite. (C) If a predaceous sea star (*Marthasterias*) threatens, the spines bend away, exposing poisonous, toothed globiferous pedicellariae that attack the starfish's tube feet. (D) Two types of pedicellaria, ophiocephalous and tridentate (three-toothed) seize plankton, such as these brine shrimp larvae (*Artemia*), while primary spines crush them. Small triphyllous (three-leafed) pedicellariae seize injured plankton that fall onto the test. (Adapted from Jensen 1966.)*

They lack arms and are elongated along the oral–aboral axis of the body. Some species lack tube feet for locomotion but burrow by contracting circular and longitudinal body muscles peristaltically. Burrowing sea cucumbers feed by trapping organic material on sticky tentacles (modified tube feet), then stuffing the tentacles one at a time into the mouth, like someone licking his fingers. Other sea cucumbers creep along substratum on tube feet equipped with suckers, grazing with the tentacles. In these species tube feet tend to be best developed on three of the five ambulacra that are usually on the bottom. These species therefore have a ventral surface (called the sole) and a dorsal surface, and they are therefore bilaterally symmetric. Podia on the dorsal side lack suckers and probably serve as feelers (Figure 33.13). The ossicles are microscopic, and there are no spines, so the integument ranges in texture from soft to leathery.

Sea cucumbers are as strange inside as they are externally. Instead of the madreporite being on the external surface, it hangs freely within the perivisceral coelom. The digestive tract is long like that of sea urchins, but it terminates in a **cloaca** before reaching the anus. The muscular cloaca serves as a respiratory organ, pumping water into and out of the **respiratory tree** that branches into the coelom. Inside the sea cucumber's anus are white, pink, or red **tubules of Cuvier.** Sea cucumbers shoot these sticky tubules out at crabs, fishes, and other potential predators, which can become fatally entangled in them (Figure 33.14). The sea cucumber readily regenerates new Cuvierian tubules for the next predator. As if this were not impressive enough, sea cucumbers also squeeze out their own digestive

Figure 33.12
(A) Internal structure of a sea urchin. (B) Internal view of Aristotle's lantern, so-called because Aristotle described it as looking like a lantern with the panes left out. (C) External view of the teeth of the purple sea urchin Strongylocentrotus purpuratus.

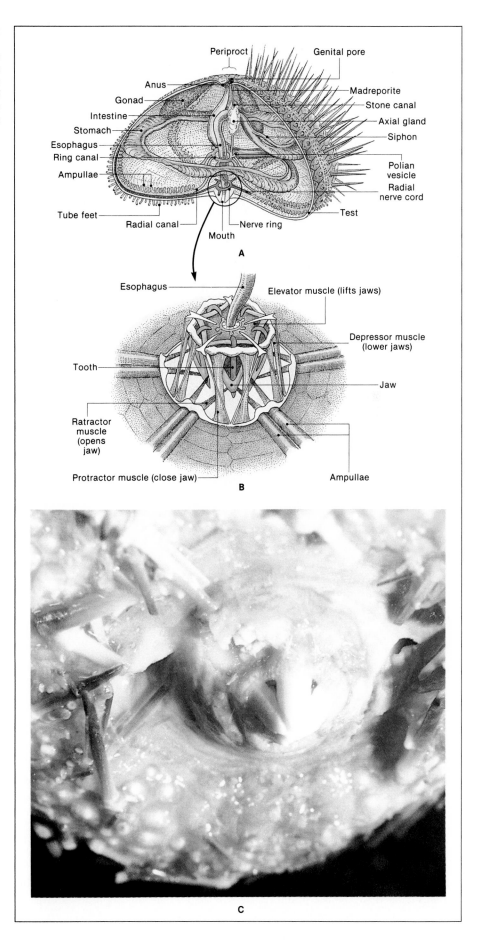

Periproct

Genital pore

Anus

Gonad

Intestine

Stomach

Esophagus

Ring canal

Ampullae

Tube feet

Radial canal

Mouth

Nerve ring

Madreporite

Stone canal

Axial gland

Siphon

Polian vesicle

Radial nerve cord

Test

A

Esophagus

Elevator muscle (lifts jaws)

Depressor muscle (lower jaws)

Tooth

Jaw

Ratractor muscle (opens jaw)

Protractor muscle (close jaw)

Ampullae

B

C

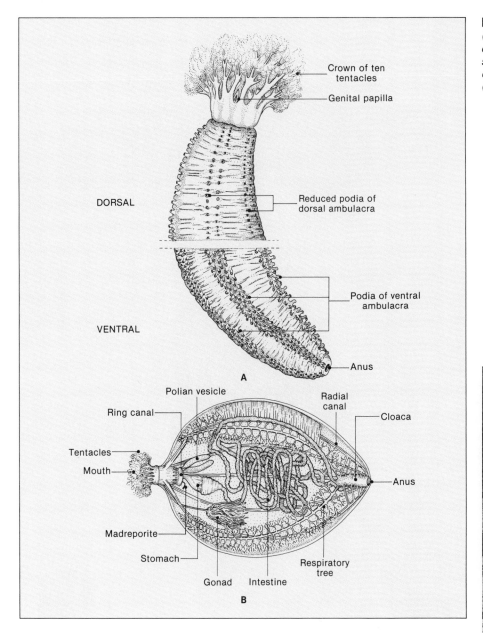

Figure 33.13
(A) External structure of the sea cucumber Cucumaria frondosa *showing the differences in tube feet on the ventral and dorsal surfaces. (B) Internal organs of* Thyone.

Figure 33.14
A sea cucumber has expelled its Cuvierian tubules from its anus toward the photographer. The Cuvierian tubules will be regenerated.

tracts, gonads, and respiratory trees under certain circumstances that apparently have nothing to do with defense. Following this self-evisceration, they regenerate the lost organs.

CONCENTRICYCLOIDS

One species of echinoderm not belonging to any of the above groups was discovered in 1986 inside wood collected more than a kilometer deep off the coast of New Zealand (Baker et al. 1986). Class Concentricycloidea now includes two species, *Xyloplax medusiformis* and *X. turnerae*. These animals have been called "sea daisies," but it remains to be seen whether they will ever become familiar enough to require a common name. Individuals are disc-shaped with a fringe of short spines around the edge (Figure 33.15). They are up to 9 mm in diameter and lack arms. They also lack a mouth, gut, or anus. The ventral surface (**velum**) of the animal may actually be the lining of the stomach, which digests food externally. The water vascular system has two concentric ring canals.

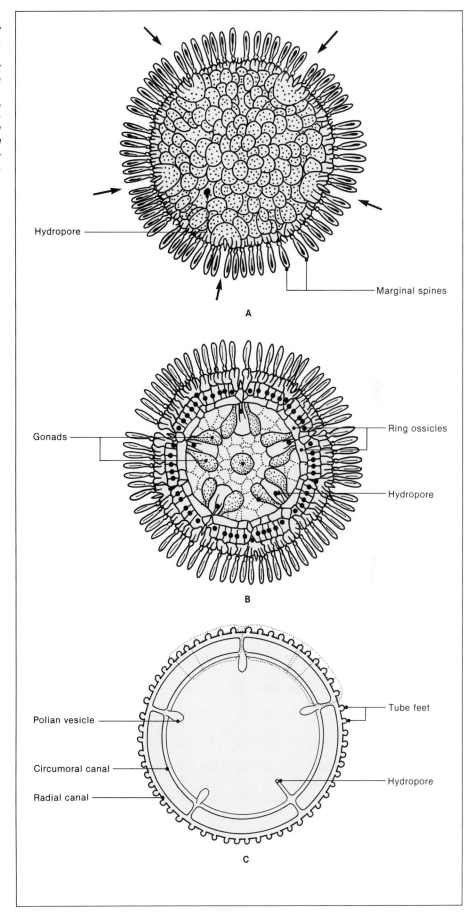

Figure 33.15
The first named species of the new class Concentricycloidea, Xyloplax medusiformis. (A) Dorsal view. Arrows help visualize the five-part symmetry. The hydropore opens into the water vascular system. (B) Ventral view. The ventral surface is a thin velum attached to the ring ossicles. Embryos and larvae in all stages of development were found within the gonads. (C) The water vascular system with its two concentric ring canals.

Hydropore

Marginal spines

A

Gonads

Ring ossicles

Hydropore

B

Polian vesicle

Tube feet

Circumoral canal

Hydropore

Radial canal

C

Extant Classes of Phylum Echinodermata

Genera mentioned elsewhere in this chapter are noted.

Class Asteroidea (AS-tur-OY-dee-uh (Greek *aster* star). Usually star-shaped, with five (or more) arms thickly joined to the central disc. Open ambulacral grooves and tube feet on oral (bottom) surface. Podia, usually with suckers, used in locomotion. Madreporite and anus (if present) on aboral surface. Starfish. *Acanthaster, Asterias, Linckia, Luidia, Pentagonaster* (Figures 33.11A, 33.18).

Class Ophiuroidea OFF-ee-your-OY-dee-uh (Greek *ophis* snake + *oura* tail). Five (or more) flexible arms clearly distinct from central disc. Closed ambulacra. Tube feet without suckers; not used in locomotion. No gut or anus. Madreporite on aboral (upper) surface. Brittle stars.

Class Crinoidea krin-OY-dee-uh (Greek *krinon* a lily). Flower-shaped, with cup-shaped body and five branched arms. Body of adult on long, attached stalk (sea lilies) or without stalk (feather stars). Feeding is by means of ciliated ambulacral grooves and podia. Anus on oral (upper) surface. No spines or madreporite. *Antedon, Cenocrinus, Himerometra, Oxycomanthus* (Figures 33.1D, 33.8, 33.9).

Class Echinoidea EK-in-OY-dee-uh. No arms; body hemispherical (sea urchins and heart urchins) or disc-shaped (sand dollars). Heart urchins and sand dollars tend toward bilateral symmetry (irregular). Ossicles form a rigid test. Mouth on bottom surface, usually with special chewing apparatus (Aristotle's lantern). Locomotion by movable spines and (in sea urchins) tube feet with suckers. Closed ambulacral grooves. *Diadema, Echinus, Mellita, Psammechinus, Strongylocentrotus* (Figure 33.10).

Class Holothuroidea HA-low-thur-OY-dee-uh (Greek *holothourion* a term Aristotle originally applied to a sea polyp). Body cucumber-shaped. Central axis elongated, with mouth and anus at opposite ends. Bilateral symmetry. No arms or spines. Podia at oral (anterior) end elongated into tentacles. Remaining tube feet have suckers used for locomotion. Sea cucumbers. *Cucumaria, Pseudocolochirus, Thyone* (Figures 33.13, 33.14).

Class Concentricycloidea con-SEN-tree-sy-KLOY-dee-uh (Latin *concentricus* concentric + *cyclus* ring). Water vascular system with two ring canals concentric on ventral surface. Disc-shaped body without arms, mouth, or anus. Ventral surface covered by velum. *Xyloplax* (Figure 33.15).

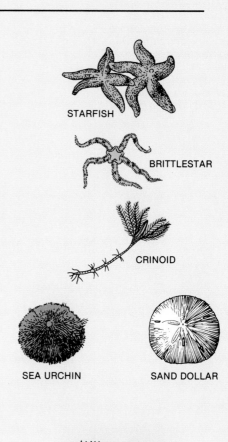

STARFISH

BRITTLESTAR

CRINOID

SEA URCHIN   SAND DOLLAR

SEA CUCUMBER

CONCENTRICYCLOID

CLASSIFICATION

As with most invertebrates, there is little fossil record of echinoderms from the Precambrian. In contrast, Paleozoic rocks are littered with the fossilized ossicles of extinct echinoderms. More than 20 classes once flourished, but most have become extinct, leaving large gaps separating the few major taxa that remain. There have been various attempts to classify the living echinoderms into subphyla, superclasses, classes, and subclasses, but it is widely agreed that none of these efforts is entirely satisfactory. For now it is just as well simply to treat each of the major groups as a class and await further research to tell us the actual taxonomic relationships among them.

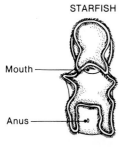

STARFISH

Mouth

Anus

Bipinnaria

SEA CUCUMBERS

Auricularia

BRITTLE STARS

Ophiopluteus

SEA URCHINS

Echinopluteus

SEA LILIES

Vitellaria

Figure 33.16
Types of echinoderm larva. Note that the forms that most resemble each other as larvae do not necessarily resemble each other as adults. Compare the ophiopluteus and the echinopluteus, for example. This is one of the sources of difficulty in echinoderm classification.

DEVELOPMENT

Echinoderms, especially sea urchins and starfish, have long been favorite subjects of developmental biologists. A major reason is that adults can be made to release millions of ova or sperm by electrically stimulating them or by injecting them with potassium chloride or acetylcholine solution. In most species the ova are relatively large (0.1 to 0.2 mm in diameter), and they have little yolk to interfere with observation of intracellular processes. Fertilization and development can therefore be watched under low magnification. (See Conway et al. 1984 for procedures.)

The early stages of development (see p. 118) are similar in almost all echinoderms. The shapes of the **larvae** vary, however, depending on the class (Figure 33.16). In some species in which there is internal fertilization and brooding, the larval stage may be abbreviated. In all other echinoderms the larvae are independent organisms competent at every function except reproduction. The larvae are better adapted to disperse the species than are adult echinoderms. After several weeks of planktonic existence, the bilaterally symmetric larva begins the remarkable metamorphosis into a radially symmetric adult. First the larva swims downward and attaches to substratum, usually with the left side down. Contrary to what one might expect, the mouth and anus of the larva usually do not migrate to their new locations on the oral and aboral sides but simply disappear. The stumps of the foregut and hindgut then reorient to the oral and aboral sides, and a new mouth and anus develop near them. Likewise, the arms of larvae do not develop into the arms of the adult but simply disappear.

REGENERATION

Many adult echinoderms have remarkable abilities to regenerate lost body parts. Asteroids often cast off an arm that has been seized by a predator, then regenerate a new one. The sea star *Asterias vulgaris* can, in fact, recover its original form from only one-fifth of the central disc attached to one arm. Some starfish can even recover their original form from only a single arm (Figure 33.17). Some species of starfish use their regenerative abilities to reproduce asexually, by splitting in half across the central disc, with each half regenerating the missing parts. Before their regenerative abilities became widely known, oyster fishermen used to try to kill starfish by cutting them in half and throwing them back into the water—a worse than useless strategy. Brittle stars and crinoids have less regenerative ability, requiring an intact central disc or calyx in order to regenerate a lost arm. Sea cucumbers lack arms but regenerate their tubules of Cuvier and viscera, as described previously.

INTERACTIONS WITH HUMANS AND OTHER ANIMALS

Crown-of-Thorns Starfish. Echinoderms almost never attack humans or transmit diseases, and only rarely does a careless diver die from handling sea urchins with especially poisonous spines or pedicellariae. On the other hand, echinoderms can indirectly have devastating effects on humans and other animals. In the late 1960s, for example, the crown-of-thorns starfish *Acanthaster planci* created near panic among people of the western Pacific and alarmed ecologists around the world (Figure 33.18). This sea star, which feeds on coral polyps, underwent a population explosion that appeared to threaten the existence of even the vast and priceless Great Barrier Reef. Numerous other coral reefs, along with the shores and islands they protect, and the fish on which the islanders depend, were also endangered. The crown-of-thorns starfish has, in fact, damaged approximately 500 km of the Great Barrier Reef and a large area of the reef off Guam. Tens of thousands of the starfish have been killed by injecting formaldehyde in an attempt to curb the damage. Although the sea stars continue to destroy large areas, it now appears that the coral can recover from the damage, and the threat no longer seems as grave as it once did.

Sea Urchins. Meanwhile, the population of another echinoderm, the black sea urchin *Diadema antillarum*, has declined throughout the Caribbean, apparently due to some infection (Lessios et al. 1984). *Diadema* is a major predator, and there was some concern that its disappearance would cause serious ecological disruption. So far, however, this appears not to have happened (Jackson and Kaufmann 1987). The major consequence has been an accumulation of algae and a resulting decline in overall productivity. The unpredictability of these population fluctuations and their effects stand as evidence of how little we understand about echinoderms and marine ecology.

We do know, however, that dumping raw sewage into the Pacific Ocean has caused some populations of sea urchins to swell. In addition to obtaining dissolved nutrients from sewage, the sea urchins also eat kelp and have destroyed several valuable kelp forests (see Figure 1.6). To stop this destruction, sea urchins have been dredged up and killed and the area "reforested" with kelp embryos raised in the laboratory. Sea urchin overpopulation has also occurred in the northwest Atlantic, due largely to the depletion of their predators, sea otters, by fur trappers. Until recently, the sea urchins were universally despised by lobster fishermen, because they prey on lobsters in the fishermen's traps. Now, however, many former lobster fishermen have become sea-urchin fishermen and sell the roe to Japan for $100 to $150 a pound.

Sea Cucumbers. In China and the Pacific islands boiled and dried sea cucumbers, called *trepang* or *bêche-de-mer*, are considered a delicacy. In the future sea

Figure 33.17
The starfish Linckia multifora *regenerating from a single arm. At this stage of regeneration the starfish is aptly called a "comet." Members of this genus are especially adept at regeneration and often reproduce asexually in this way.*

Figure 33.18
The crown-of-thorns sea star feeding on coral. Each year an individual approximately half a meter in diameter reduces over 5 square meters of coral to bare, white skeleton. In some areas of the South Pacific more than 90% of the coral has been destroyed. The starfish itself is safe from most predators because of the spines and irritating mucus that cover it. One of the few major predators is a mollusc, the giant triton (Charonia tritonis). It was originally thought that removal of hundreds of thousands of these predators by collectors was responsible for the rise in the sea star's population, but this now appears to be too simple an explanation.

cucumbers may be collected not as food but as medicine. Pacific islanders have long known that cut-up sea cucumbers can be used to poison fish in tidal pools. It now turns out that the poison, holothurin, has various effects on nerve and muscle and also suppresses the growth of certain tumors (Ruggieri 1976).

ARROWWORMS: PHYLUM CHAETOGNATHA

Arrowworms, named for their streamlined appearance, are among the most enigmatic of animals. Little is known about their daily lives except that they are important planktonic predators, feeding voraciously on copepods, small fishes, and each other. Most chaetognaths are about 4 cm long and completely transparent. If you have been in the ocean you have probably been surrounded by them without knowing it. As many as a thousand arrowworms have been found in a cubic meter of seawater.

Chaetognaths, such as those in the genus *Sagitta* (Figure 33.19), capture prey in bristles (Greek *chaite*) around the jaws (*gnathos*) (Table 33.2). They can pull a hood over the head and jaws, presumably to enhance their streamlining as they dart after prey. Bristles, jaws, and hood close around the prey instantaneously. Some chaetognaths may also paralyze prey with tetrodotoxin (p. 144; Bieri and Thuesen 1990). Swimming is due to alternating contractions of the longitudinal body muscles on each side. The fins are firmly attached and do not flap. The digestive tract is straight and suspended by a median mesentery that divides the coelom into right and left halves. The coelom is also divided by transverse septa into head, trunk, and tail regions. There is no excretory, respiratory, or circulatory organ. The nervous system includes a ventral ganglion in the trunk, a dorsal ganglion in the head, and several smaller ganglia. Sensory bristles distributed over the body detect prey, and a **ciliary loop** that passes beneath the head like a necklace is thought to detect water currents or chemicals. Two dorsal eyes (ocelli) presumably detect light and control the vertical migration at dawn and dusk.

Arrowworms are all hermaphroditic, with two ovaries in the trunk coelom and two testes in the tail coelom. The sperm circulate within the tail coelom as they mature, then enter the seminal vesicles where they are packaged into spermatophores. There is no opening from the seminal vesicles; during spawning the spermatophores simply break through the epidermis. The spermatophore then attaches to the anterior trunk of the same or a different individual, and the sperm

Figure 33.19
A chaetognath in ventral view.

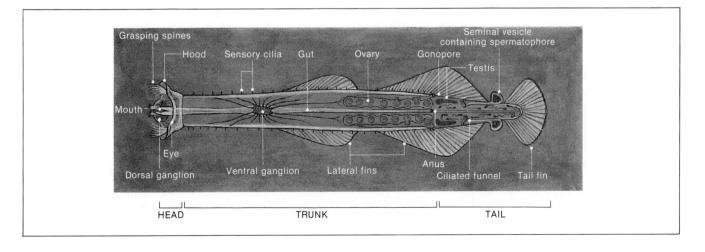

Table 33.2 Characteristics of arrowworms.

Phylum Chaetognatha key-TOG-nah-thuh (Greek *chaite* bristles + *gnathos* jaws).

Morphology: Juvenile is coelomate; peritoneum absent in adult. Body **dart-shaped**, with **bristles around jaws** and retractable hood over head. Bilateral symmetry. Not segmented. Epidermis with **cuticle**.

Physiology: Straight, complete digestive tract. Nervous system with ganglia and specialized receptors. No excretory, circulatory, or respiratory organ.

Locomotion: Most species swim by undulation of finned body. One species sessile.

Reproduction: Monoecious. Fertilization internal.

Development: Deuterostomate. Development direct.

Habitat, Size, and Diversity: All species marine, planktonic. Length of adult ranges from 0.5 to 12 cm. Body generally transparent. Approximately **100 living species** described.

migrate into the gonopore to the ovary. Fertilization occurs internally. Fertilized eggs are released in gelatinous strands that either cling to the body or float free. After approximately two days the eggs hatch into immature chaetognaths that reach maturity within another week or so.

SUMMARY

Echinoderms are deuterostomes: the mouths of larvae develop from an embryonic opening other than the blastopore, cleavage is radial and indeterminate, and the coelom is enterocoelous. All are marine animals. Most echinoderms tend to have pentamerous radial symmetry as adults. Unlike any other animal, they have tube feet arranged along ambulacral grooves, a water vascular system, and an endoskeleton of ossicles. The tube feet are used to anchor the animal to substratum, and for locomotion and feeding. They are extended by pressure from the water vascular system, which circulates seawater. Echinoderms are named for spiny protrusions on the ossicles. In addition, the integument may bear pedicellariae, which are used in predation and defense, and papulae, which exchange respiratory gases and metabolic wastes.

There are six major groups of echinoderms: starfish, brittle stars, crinoids (sea lilies and feather stars), echinoids (sea urchins, sand dollars, and others), sea cucumbers, and the little-known concentricycloids. Starfish generally have five stiff arms and feed by everting the stomach. Brittle stars resemble starfish but have longer and more flexible arms. In crinoids the arms branch from a calyx, which sits atop a long stalk in sea lilies. Echinoids have close-fitting ossicles that form a solid test and feed by means of Aristotle's lantern. Sea cucumbers are tube-shaped and have tube feet formed into tentacles.

Arrowworms are also deuterostomate, but their relationship to echinoderms, if any, is unclear. They are streamlined, transparent predators.

KEY TERMS

deuterostome
radial cleavage
indeterminacy
enterocoelous
pentamerous radial symmetry
tube foot
podium

ambulacral groove
water vascular system
madreporite
ring canal
radial canal
ossicle
pedicellaria

calyx
pinnule
Aristotle's lantern
respiratory tree
tubules of Cuvier

SELF-TEST

1. Give the common names for the five major kinds of echinoderms. Describe each of the five briefly. What shared features justify placing them all in the same phylum?

2. Explain how podia work. What is the role of the water vascular system in the functioning of podia? What is the main function of podia in sea stars?

3. Descibe each of the following: ambulacral grooves, papulae, pedicellariae, ossicles.

4. Although the two phyla discussed in this chapter bear little resemblance to each other as adults, they share several similarities in embryonic development. Describe three of these similarities. How do they differ from protostomes?

5. Describe how arrowworms capture food.

READINGS

RECOMMENDED READINGS

Bieri, R. and E. V. Thuesen. 1990. The strange worm *Bathybelos*. *Am. Sci.* 78:542–549.

Birkeland, C. 1989. The Faustian traits of the crown-of-thorns starfish. *Am. Sci.* 77:154–163.

Feder, H. M. 1972. Escape responses in marine invertebrates. *Sci. Am.* 227(1):92–100 (July). (*How sea urchins and other prey escape sea stars.*)

Inoué, S. and K. Okazaki. 1977. Biocrystals. *Sci. Am.* 236(4):82–92 (Apr). (*How a sea urchin shapes its ossicles.*)

Macurda, D. B. Jr. and D. L. Meyer. 1983. Sea lilies and feather stars. *Am. Sci.* 71:354–365.

See also relevant selections in General References at the end of Chapter 21.

ADDITIONAL REFERENCES

Baker, A. N., F. W. E. Rowe, and H. E. S. Clark. 1986. A new class of Echinodermata from New Zealand. *Nature* 321:862–864.

Bone, Q., A. Pierrot-Bults, and H. Kapps (Eds.). 1991. *The Biology of the Chaetognatha.* New York: Oxford University Press.

Bosch, I., R. B. Rivkin, and S. P. Alexander. 1989. Asexual reproduction by oceanic planktotrophic echinoderm larvae. *Nature* 337:169–170.

Conway, C. M., D. Igelsrud, and A. F. Conway. 1984. Sea urchin development. In: C. L. Harris (Ed.), *Tested Studies for Laboratory Teaching: Proceedings of the Third Workshop/Conference of the Association for Biology Laboratory Education (ABLE).* Dubuque, IA: Kendall/Hunt, Chap. 4.

Jackson, J. B. C. and K. W. Kaufmann. 1987. *Diadema antillarum* was not a keystone predator. *Science* 235:687–689.

Jensen, M. 1966. The response of two sea-urchins to the sea-star *Marthasterias glacialis* (L.) and other stimuli. *Ophelia* 3:209–219.

Lawrence, J. 1987. *A Functional Biology of Echinoderms.* Baltimore: Johns Hopkins University Press.

Lessios, H. A., D. R. Robertson, and J. D. Cubit. 1984. Spread of *Diadema* mass mortality through the Caribbean. *Science* 226:335–337.

Ruggieri, G. D. 1976. Drugs from the sea. *Science* 194:491–497.

Willmer, P. 1990. *Invertebrate Relationships.* New York: Cambridge University Press.

34

Introduction to Chordates

Sea peaches (Halocynthia pyriformis).

CHAPTER OUTLINE

LEARNING OBJECTIVES

1. What group of animals is most closely related to our own phylum?

2. What is the significance of the notochord, for which phylum Chordata is named?

3. What other features do all chordates share?

4. Aside from vertebrates, what other kinds of chordates are there?

5. What are the major kinds of vertebrates?

6. How did terrestrial vertebrates acquire their four limbs from fish ancestors?

7. How did they acquire the ability to live and reproduce on land without the embryo becoming dehydrated?

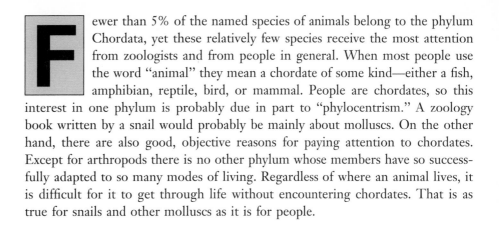

Fewer than 5% of the named species of animals belong to the phylum Chordata, yet these relatively few species receive the most attention from zoologists and from people in general. When most people use the word "animal" they mean a chordate of some kind—either a fish, amphibian, reptile, bird, or mammal. People are chordates, so this interest in one phylum is probably due in part to "phylocentrism." A zoology book written by a snail would probably be mainly about molluscs. On the other hand, there are also good, objective reasons for paying attention to chordates. Except for arthropods there is no other phylum whose members have so successfully adapted to so many modes of living. Regardless of where an animal lives, it is difficult for it to get through life without encountering chordates. That is as true for snails and other molluscs as it is for people.

PHYLUM HEMICHORDATA

One of the things most people wish to know about their own phylum is where it came from. Unfortunately, the fossil record yields few clues to the origins of chordates, but we might be able to get some idea of our humble origins by examining a living phylum—Hemichordata (Figure 34.1). Hemichordates share with chordates several similarities in development and have two features that are diagnostic of chordates. These are the pharyngeal slits and dorsal nerve cord (Table 34.1). It was once thought that, like chordates, hemichordates have dorsal skeletal rods called notochords. On that account hemichordates were considered a type of chordate. Now, however, it is recognized that what was thought to be a notochord in hemichordates is actually an extension of the buccal cavity called the **buccal diverticulum** (= stomochord), which is not homologous with the notochord. Nevertheless, hemichordates appear to have diverged close to the point at which chordates originated.

Phylum Hemichordata is divided into two classes that differ from each other greatly as adults, although they resemble each other embryologically. Class Enteropneusta includes most of the approximately 85 species of the phylum. The enter-

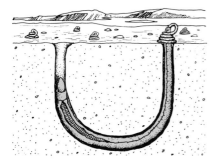

Figure 34.1
The acorn worm Balanoglossus *(phylum Hemichordata) defecating while in its U-shaped burrow in shallow seawater. The anal end of the burrows are marked by distinctive coils of castings, formed as the worm passes sediment through the gut. This species is beige or pink in color. The folds on the trunk are genital wings near the area of the gonads. Some members of this genus are up to 2.5 meters long.*

Table 34.1 Characteristics of hemichordates.

Phylum Hemichordata HEM-i-core-DAY-tuh (Greek *hemi* half + *chorda* cord).

Morphology: Coelomate (although peritoneum may be secondarily lost in some species). Body divided into proboscis, collar, and trunk. Not segmented. With **buccal diverticulum, dorsal nerve cord,** and **pharyngeal slits,** but no notochord.

Physiology: Circulatory system essentially open, with contracting dorsal and ventral vessels and heart vesicle. A network of epidermal nerve cells, concentrating in some areas to form middorsal and midventral nerve cords. Respiration through integument. Digestive system complete.

Locomotion: Peristaltic contraction of the proboscis. Larvae ciliated.

Reproduction: Sexes usually separate but similar. Some pterobranchs monoecious; some reproduction by asexual budding. Fertilization external.

Development: Deuterostomate. Coelom (when present) does not always arise in classic deuterostomate fashion, enterocoelously. Development either direct or with tornaria larvae.

Habitat, Size, and Diversity: All species marine, benthic. Length of adult enteropneusts ranges from 2.5 cm to 2.5 meters; pterobranchs generally less than 1 cm long. Approximately **85 living species** described.

Genera mentioned elsewhere in this chapter are noted.

Class Enteropneusta in-ter-op-NEW-stuh (Greek *enteron* gut + *pneustikos* for breathing). Solitary. Worm-shaped. No tentacles. Acorn worms. *Balanoglossus* (Figure 34.1).

Class Pterobranchia tear-oh-BRAN-key-uh (Greek *pteron* wing + *branchia* gills). Usually colonial, with individual zooids in secreted tubes. Collar with tentacles. *Rhabdopleura* (Figure 34.4).

---

opneusts, commonly called acorn worms, burrow in marine sediments and have wormlike bodies divided into a proboscis, collar, and trunk. Members of class Pterobranchia live in colonies inside secreted tubes.

ACORN WORMS

General Features.   Members of class Enteropneusta are commonly called acorn worms because the head and collar resemble the nut and cup of an acorn. Behind the collar is a long, slimy trunk that is more than 2 meters long in some species. Each of the three main body sections (head, collar, and trunk) has a separate coelomic compartment filled with spongy tissue that provides some mechanical support, but the body is so soft that it is easily broken by handling. Acorn worms either live in U-shaped burrows in shallow water (Figure 34.1), or they burrow through marine sediments or live under rocks or seaweed. The proboscis is the main organ of locomotion, with the trunk being pulled along passively. Many hemichordates feed by ingesting large amounts of mud or sand from which the gut extracts organic debris. Others feed by means of cilia on the proboscis, which pass food particles back toward the mouth (Figure 34.2). The food particles are bound on a mucous string and swallowed along with water.

Internal Structure.   Water swallowed during feeding exits through **pharyngeal slits** and **gill pores.** The pharyngeal slits are U-shaped openings in the pharynx that carry swallowed water into **pharyngeal pouches** and out through the gill pores (Figure 34.3). Pharyngeal slits are sometimes called gill slits, although they are much more important in feeding than in exchange of respiratory gases. There is some concentration of blood vessels surrounding the pharyngeal slits, but not enough to justify calling them gills. The body surface is the main route of respiratory exchange. The colorless blood is pumped anteriorly by a dorsal vessel into a contracting **heart vesicle** in the proboscis, and then posteriorly by a ventral vessel. The blood circulates through sinuses, making it an essentially open circulatory system. One such network of sinuses, the **glomerulus,** is assumed from its structure to be an excretory organ. The nervous system consists largely of a diffuse network in the base of the epidermis. Along the dorsal and ventral midlines this plexus is concentrated into dorsal and ventral nerve cords, which lack ganglia. In places the dorsal nerve cord is often hollow, like the dorsal nerve cords of chordates. Sensory receptors are scattered over the integument, especially on the proboscis.

Reproduction and Development.   Acorn worms have separate sexes, although there is little evident difference between males and females. The gonads are distributed in rows, often forming finlike bulges on the trunk. Each gonad has a separate gonopore, and fertilization occurs externally, apparently during mass

Figure 34.2
Method of feeding in some acorn worms. Cilia on the proboscis pass particles toward the mouth (black arrows). Sand and other dense, inorganic materials fall to the collar, while less-dense organic particles are carried into the mouth on a mucous string. Water entering the mouth during feeding exits through the pharyngeal slits and gill pores (blue arrows). The collar can be pulled forward to close the mouth.

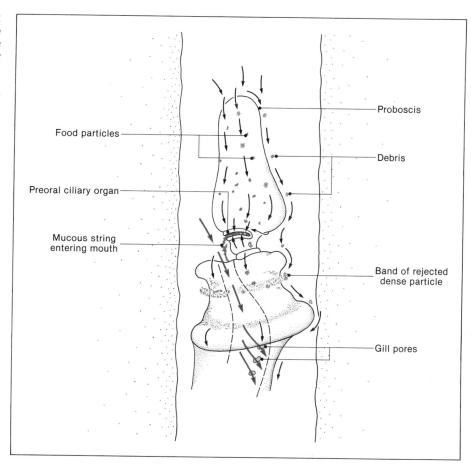

spawning initiated by the females. Early development is typically deuterostomate: cleavage is radial, the blastopore becomes the anus, and the coelom forms as an outpocketing of the archenteron. Development is direct in some species, but others produce **tornaria larvae,** so called because ciliary contraction causes them to spin (Latin *tornare* to turn on a lathe). Tornarias look much like the bipinnarias of sea stars (see Figure 33.16).

PTEROBRANCHS

The ten or so species in class Pterobranchia are organized much like the Enteropneusta, with similar development and with bodies divided into proboscis, collar,

Figure 34.3
The internal structure of the head, collar, and anterior part of the trunk of an enteropneust. Arrows show movements of water through pharyngeal slits, pharyngeal pouches, and gill pores. The buccal diverticulum was originally thought to be a notochord.

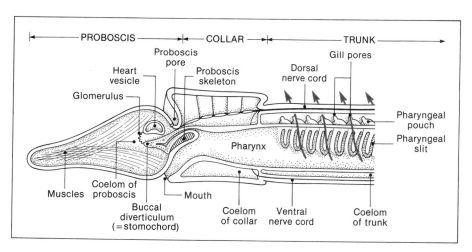

and trunk. They differ greatly in external appearance, however (Figure 34.4). They look more like ectoprocts (see pp. 565–567) than like other hemichordates, presumably because they have evolved similar adaptations for living colonially in secreted tubes. The collar expands dorsally into tentacled arms that look like the lophophores of ectoprocts. Cilia on these tentacles direct food into ciliated grooves that carry it to the mouth. Also as in ectoprocts, the alimentary canal is U-shaped, with the anus outside the fringe of tentacles. In most species there is only one pair of pharyngeal slits.

Some pterobranchs are dioecious, but as might be expected for animals confined to a tube, many are hermaphroditic. In the monoecious zooids one gonad produces sperm and the other ova. Both types of gamete are released through a gonopore near the anus. Although fertilization occurs externally, the embryos generally remain sheltered within the tubes. The sexually produced individuals then give rise to colonies by budding.

———— PHYLUM CHORDATA ————

GENERAL FEATURES OF CHORDATES

The Notochord. Phylum Chordata derives its identity from three major structural features (Table 34.2). One of these, the notochord, also gives the phylum its name. The notochord is a flexible, rodlike structure found along the dorsal midline of all chordates at some time in their lives (Figure 34.5). In most of the invertebrate chordates to be discussed in this chapter the notochord persists throughout life. In these animals the notochord keeps the body from collapsing when longitudinal muscles contract during swimming. In vertebrate chordates, except for some fishes, the notochord occurs only in the embryo, where it guides the development of vertebrae.

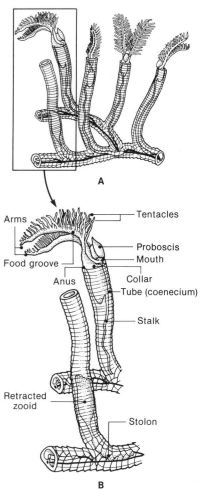

Figure 34.4
(A) A colony of Rhabdopleura. Any disturbance causes each zooid to contract the stalk and withdraw into its tube (= coenecium), as shown on the right. Afterward the ciliated proboscis climbs back up the stalk to feed. The tube is secreted by the proboscis. (B) Detail of the proboscis, tentacled collar, and trunk. Unlike most pterobranchs, members of this genus lack pharyngeal slits. Each zooid is about 1 mm long, not including the stalk.

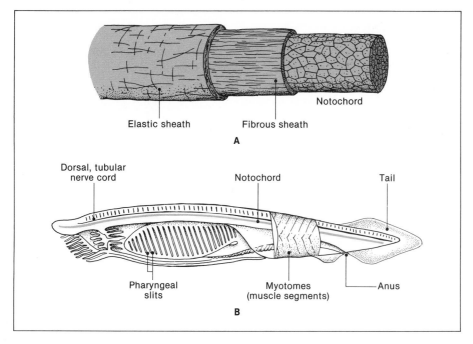

Figure 34.5
(A) Structure of the notochord. Its stiffness is due to fluid pressure in the closely packed cells and to the sheaths. (B) The amphioxus (Branchiostoma), showing the features considered diagnostic of chordates: notochord, dorsal tubular nerve cord, pharyngeal slits, and post-anal tail. Segmental muscles are not unique to chordates but are a common feature.

Table 34.2   Characteristics of chordates.

Phylum Chordata core-DAY-tuh (Latin *chorda* cord, referring to the notochord).

Morphology: With **notochord, pharyngeal slits, dorsal hollow nerve cord,** and a **post-anal tail,** at least in embryo. **Myomeres** in most species. Segmented.

Physiology: Circulatory system closed in most forms, usually with ventral heart. Central nervous system originating as dorsal, tubular nerve cord (becoming brain and spinal cord in vertebrates). Digestive system complete. Osmoregulation and excretion largely by kidneys. Respiration through gills, lungs, and/or integument. Birds and mammals homeothermic.

Locomotion: Swimming, creeping, running, or flying, using skeletal muscles attached to endoskeleton.

Reproduction: Sexes almost always separate, often with marked sexual dimorphism. Some species hermaphroditic; some parthenogenetic. Some invertebrate forms reproduce by budding. Fertilization generally external in aquatic forms; internal in terrestrial forms. Oviparous or viviparous.

Development: Cleavage pattern radial (lancelet and amphibians), bilateral (tunicates), discoidal (fishes, reptiles, birds), or rotational (placental mammals). Deuterostomate. Primitive groups enterocoelous; mammals and some other vertebrates schizocoelous. Development indirect or direct.

Habitat, Size, and Diversity: Marine, fresh water, amphibious, or terrestrial. From a few millimeters to 32 meters long. Approximately **44,000 living species** described.

Dorsal, Tubular Nerve Cord.   A second characteristic of chordates is the dorsal, hollow nerve cord. This nerve cord lies along the midline, dorsal to the notochord. In vertebrates the spinal cord forms from the dorsal nerve cord and becomes enclosed by vertebrae. The brain forms at the anterior end of the nerve cord. Many other animals, including annelids and arthropods, also have nerve cords, but in these and virtually every other bilaterally symmetric animal except hemichordates the central nervous system is ventral. The chordate nerve cord also differs from those in other phyla in being hollow, although the channel often becomes obscure during development.

Pharyngeal Slits.   A third distinguishing feature of chordates are the pharyngeal slits. These slits form in the embryo as pockets of ectoderm grow inward and fuse with pockets of endoderm lining the pharynx. In some of the chordates discussed in this chapter, as well as in fishes and some amphibians, these slits are the precursors of gills. In many terrestrial vertebrates the pharyngeal slits never completely perforate the neck, and they usually disappear before birth or hatching.

Post-anal Tail and Myomeres.   Some zoologists consider the presence of a tail extending past the anus to be a fourth hallmark of Chordata. A post-anal tail occurs in all chordates, at least in the embryo. A post-anal tail and pharyngeal slits occur in the human embryo (and occasionally in a newborn baby) and are regarded as vestiges of our evolutionary descent from aquatic vertebrates with tails. Another feature that is common in vertebrates and some invertebrate chordates is bands of segmental muscles, called **myomeres** (= myotomes).

CLASSIFICATION

Chordates are generally divided into two major groups: Craniata and Protochordata. The group Craniata (Greek *kranion* skull) comprises the subphylum Vertebrata, in which the head is well developed. The protochordates, also called

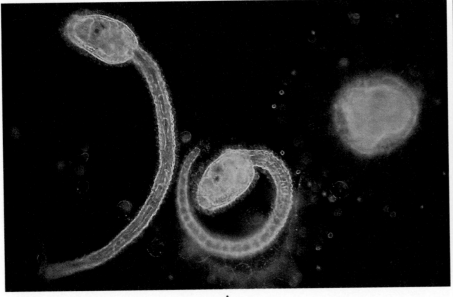

A

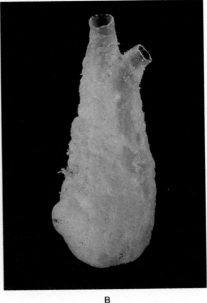

B

Acrania, are the invertebrate chordates. Protochordates are divided into two sub-phyla, Urochordata and Cephalochordata. Members of subphylum Urochordata have tadpolelike larvae with typical chordate features. As adults, however, the larg-est group of urochordates resemble sponges more than chordates at first glance. Most are sessile on the ocean floor, lack a coelom, and pump water through their hollow bodies through two siphons (Figure 34.6). A common name for them is sea squirts. Urochordates are also called tunicates, because their bodies are rein-forced with a tough **tunic.** Members of the second protochordate subphylum, the Cephalochordata, retain chordate features throughout life. The thin bodies of the adults are rather fishlike, but they have no fins, scales, or bones (Figure 34.7). Because of their resemblance to a surgeon's lancet, cephalochordates are often called lancelets.

Figure 34.6
The urochordate Ciona intestinalis. *(A) From right to left, an egg and two just-hatched tadpole larvae. Note the eyespot, and the anterior adhesive structures by which the tadpole attaches to substratum before metamorphosis into a bag-shaped, sessile adult. (Compare Figure 34.9.) (B) An adult. Water enters and leaves by the incurrent and excurrent siphons at the center and right, respectively.*

Figure 34.7
Adult cephalochordates, amphioxus (Branchiostoma). *Branchiostoma, also called the lancelet, spends much of its time partially buried in coarse gravel. Length approximately 6 cm.*

Major Groups of Phylum Chordata

Genera mentioned elsewhere in this chapter are noted.

Subphylum Urochordata (= Tunicata) YOUR-oh-core-DAY-tuh (Greek *oura* tail + *chordi* cord). Tadpole larva with notochord extending into tail. Adult usually lacks notochord, dorsal hollow nerve cord, coelom, and segmentation.

    Class Ascidiacea a-sid-ee-ACE-ee-uh (Greek *askidion* little wineskin). Sac-shaped as adults. Sessile; enclosed in well-developed fibrous tunic. With two siphons near each other. Incurrent siphon draws water into pharynx for filter feeding and gas exchange. Water exits through dorsal excurrent siphon. Sea squirts (= ascidians = tunicates). *Botryllus, Ciona, Halocynthia* (Figures 34.6, 34.8).

    Class Larvacea (= Appendicularia) lar-VASE-ee-uh (Latin *larva* ghost). Adults retain tadpole shape of larvae, including persistent notochord and nerve cord. Tunic not persistent. Pelagic (open-ocean) forms often enclosed in gelatinous "house" that traps prey. Larvaceans. *Oikopleura* (Figure 34.11).

    Class Thaliacea thal-ee-ACE-ee-uh (Greek *thalia* plenty). Adults barrel-shaped, enclosed in tunic, with siphons at opposite ends. Planktonic; either drifting in currents or swimming by jet propulsion of water through siphons. Salps. *Doliolum, Pyrosoma* (Figures 34.12, 34.13).

Subphylum Cephalochordata SEF-a-low-core-DAY-tuh (Greek *kephale* head). Notochord and nerve cord extend full length of body and persist in adult. Body of adult slender, fishlike; without scales or tunic. Lancelets. *Branchiostoma*.

Subphylum Vertebrata VER-te-BRAY-tuh (Latin *vertebratus* backboned). Vertebrae surround spinal cord. Anterior part of nerve cord becomes brain, which is enclosed in cranium.

    Superclass Agnatha (= Cyclostomata) AG-na-thuh (Greek *a-* without + *gnathos* jaws). Jawless. Fishlike, but without paired appendages. Living forms have neither scales nor bones. Notochord persists in adult. Lampreys and hagfishes. (See Chapter 35.)

    Superclass Gnathostomata NATH-oh-STOW-ma-tuh (Greek *stoma* mouth). Jaws present. Notochord usually lacking in adult. Usually with paired appendages.

        Class Chondrichthyes kon-DRIK-the-eez (Greek *chondros* cartilage + *ichthys* a fish). Fishes with cartilaginous skeletons. Scaly skin. Five to seven pairs of gills with separate, uncovered openings. Sharks, rays, and chimaeras. (See Chapter 35.)

        Class Osteichthyes OS-tee-IK-the-eez (Greek *osteon* bone). Fishes generally with bony skeletons. Scaly skin. Pairs of gills, with gill covers. (See Chapter 35.)

        Class Amphibia am-FIB-ee-uh (Greek *amphi-* double + *bios* life). Respiration by gills, integument, or lungs. Skin lacking scales. Larvae aquatic. Tetrapods (with paired, lateral appendages). Frogs, toads, newts, salamanders, and caecilians. (See Chapter 36.)

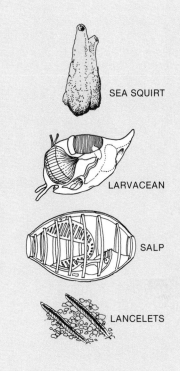

SEA SQUIRT

LARVACEAN

SALP

LANCELETS

SHARK

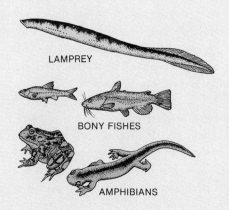

LAMPREY

BONY FISHES

AMPHIBIANS

(continued)

Class Reptilia rep-TIL-lee-uh (Latin *repere* to creep). Horny scales. Breathing through lungs. Without larval stages. Tetrapods, although legs may be absent. Snakes, lizards, sphenodonts, turtles, and crocodilians. (See Chapter 37.)

Class Aves AY-veez (Latin *aves* birds). Scales on feet only; feathers elsewhere. Tetrapod, with forelimbs modified as wings. Breathing through lungs. No larval stage. Birds. (See Chapter 38.)

Class Mammalia ma-MALE-ee-uh (Latin *mamma* breast). With mammary glands. Neither scales nor feathers; hairy skin. Tetrapod, with forelimbs modified as flippers or wings in some forms. With lungs. No larval stages. Mammals. (See Chapter 39.)

REPTILES

BIRDS

MAMMALS

TUNICATES

Ascidians. Of the approximately 2000 species in the subphylum Urochordata, most belong to the class Ascidiacea. Ascidiaceans are also the urochordates you are most likely to see, since their bag-shaped bodies are often brightly or delicately colored, or starkly black or white. The showy appearance may serve as warning coloration that deters predators, since many ascidians are highly poisonous to eat. (Julius Caesar's wife is said to have eliminated his rivals by serving them ascidians.) Individuals are up to 30 cm long, and many species form larger colonies. Along rocky shores at low tide they are hard to ignore. If you try to pick one up you may be startled by a jet of water from the excurrent siphon. This defensive reaction gives ascidians the common name sea squirts.

The current of water that flows through a sea squirt's body is not only for defense but is also necessary for most of its other vital functions. Water passes into the incurrent siphon and enters the **branchial sac,** which fills most of the upper cavity (the **atrium**) of the body (Figure 34.8). Cilia around numerous **pharyngeal slits** (often called gill slits) in the branchial sac are responsible for maintaining this flow of water. In a large ascidian the cilia can pump several liters per hour. One function of the water current is to keep the soft body from collapsing, by producing an internal pressure that maintains a **hydroskeleton.** The hydroskeleton provides body support. There is no endoskeleton, and no major coelom ever develops. Outward bulging of the body wall (**mantle**) is prevented by the tunic, which is strengthened by fibrous molecules similar to cellulose.

The branchial sac is also essential for **respiration.** Water passing through the branchial sac exchanges O_2 and CO_2 with the blood circulating in the lining. The **circulatory system** consists of a rudimentary heart and sinuses that form a hemocoel. One of the curious features of ascidians is that the direction of circulation reverses every few minutes, even if the heart is removed from the body. Studies of the possible cause and function of this reversal have continued for many years, but with no conclusion. Since there are no valves in either the vessels or the heart, reversal may occur simply because one direction is as likely as the other.

The flow of water through the branchial sac is also necessary for **filter feeding.** Small food particles in the water current become trapped in a film of mucus that is secreted by the **endostyle,** which is a groove along the ventral surface of the branchial sac. Cilia carry the mucous film along the inner lining of the branchial sac. As the mucous film collects at the dorsal surface of the branchial sac it forms

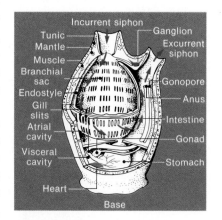

Figure 34.8
Internal structure of a solitary ascidian. Dorsal (defined by the location of the ganglion) is on the right and anterior is up. Arrows show the direction of water movement. The endostyle collects iodine and is thought by some to be homologous with the thyroid gland.

a strand. This strand, with its catch of food, then drops into the digestive tract. The digestive system, consisting of an esophagus, stomach, and intestine, lies in the **visceral cavity** (which is not a coelom). Wastes are eliminated from the anus, which is in the atrial cavity near the excurrent siphon. Periodically, the sea squirt contracts its body to clear out the feces, as well as any debris that may have accumulated in the branchial sac.

Reproduction also depends on the flow of water out the excurrent siphon. Most species are hermaphroditic, with both an ovary and a testis near the stomach in the visceral cavity. Contraction of the body forces ova and sperm out through ducts that lead to the excurrent siphon. Fertilization occurs externally in solitary species and leads to the formation of free-swimming tadpole larvae (Figure 34.6A).

A tadpole larva generally swims for only a few hours. During this time it must find a new habitat, which it does with the help of a simple eye, a statocyst (gravity receptor), and receptors for chemical and mechanical stimuli. After locating a suitable substratum, the larva attaches by means of three **adhesive papillae.** It then undergoes **metamorphosis** (Figure 34.9). During metamorphosis the larva loses some of the features that identify ascidians as chordates: the notochord, the dorsal nerve cord, and the tail. The internal organs also undergo a rotation of about 90°.

Colonial ascidians reproduce sexually like solitary ascidians, except that all the members of the colony synchronize the release of their gametes (Figure 34.10). Colonial species also form buds that reproduce asexually. Colonies often live for only a few days, but in a sense they may be immortal, since their buds are genetically identical and occupy the same site. A colony may consist of individuals much like the solitary ascidians described above, or the colony may share one circulatory system, a single tunic, and one excurrent siphon.

Figure 34.9
Metamorphosis in a solitary ascidian. (A) The free-swimming tadpole larva. (B) Immediately after settling and attaching to substratum. (C) Later in metamorphosis. (D) The adult.

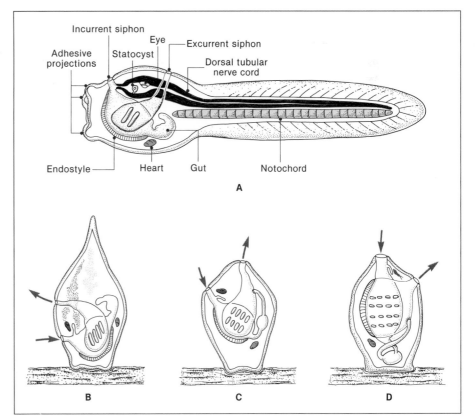

Figure 34.10
The compound colonial ascidian Botryllus schlosseri. *Each star-shaped pattern, approximately 1.8 mm long, represents a colony. Each ray of a star is the incurrent siphon of one ascidian. A single excurrent siphon in the center of the rays is shared by all the colony.*

Larvaceans. There are fewer species in the class Larvacea than in the class Ascidiacea, but in numbers of individuals larvaceans may be the most common planktonic forms in many seas. They are seldom noticed because they are only a few millimeters long and virtually transparent (Figure 34.11). Many secrete **houses** that may be up to a meter in diameter, but these are also seldom noticed because they are made of transparent mucus that disintegrates in nets. Many larvaceans and their houses glow at night, but until recently this luminescence was attributed to algae. Only since submersible research vessels have come into wide use have zoologists begun to appreciate the enormous numbers and sizes of houses and their ecological impact.

The house is used for filter feeding. When the larvacean is "at home" it continually waves its tail, drawing water through the house. Water enters at the anterior end through two **incurrent filters,** which have a mesh so fine that only plankton smaller than a few micrometers can enter. Once inside the house the plankton become trapped in a wing-shaped **feeding filter.** Houses have no opening to eliminate feces, so several times each day the larvacean abandons its home and builds a new one. Reconstruction takes only a few minutes. A larvacean will also abandon its home if threatened by a predator. Such abandoned houses, with their accumulation of plankton, may represent a significant contribution to the diets of many marine animals, which would not otherwise be able to feed on such tiny plankton.

Salps. Members of class Thaliacea are generally a few centimeters long and barrel-shaped (Figure 34.12). Rings of muscles wrapped around them like barrel hoops force water through the body. This water current is used for swimming, filter feeding, and exchange of respiratory gases. Except for this difference in the generation of the water current, salps are organized much like ascidians. The most unusual thing about salps is the alternation of sexual and asexual generations. In some species the hermaphroditic adults in the sexual generation produce larvae

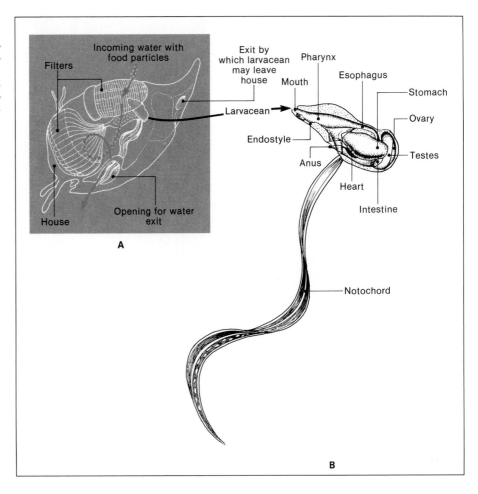

Figure 34.11
(A) The larvacean Oikopleura albicans *"at home." Arrows show the direction of the water current generated by the beating of the tail. The wing-shaped structures are feeding filters that trap tiny plankton. (B)* Oikopleura *outside its house.*

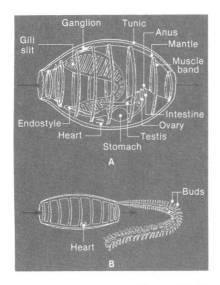

Figure 34.12
The salp Doliolum. *(A) Sexual adult. Arrows show the direction of water movement. (B) An adult in the asexual ("nurse") stage.*

externally, as in ascidians. In other species the hermaphroditic adult broods a solitary egg that develops directly. In either case the adults in the next generation are asexual. They produce a string of buds, each one of which develops into a hermaphroditic adult. Some salps form baglike colonies called **pyrosomes** that may be several meters long (Figure 34.13). Pyrosomes (Greek *pyr* fire + *soma* body) get their name from the brilliant glow they produce. This luminescence deters predators and is also a means by which individuals in the colony communicate with each other and avoid obstacles or adverse conditions.

LANCELETS

Subphylum Cephalochordata derives its name from the notochord that extends into the head. This structure leaves no room for a brain, but a dorsal, hollow nerve cord occurs in both larvae and adults. Other chordate characteristics include the numerous pharyngeal (gill) slits and the tail that projects beyond the anus (Figure 34.14). Lancelets even have several features reminiscent of vertebrates and were for many years considered representative of ancestral vertebrates. The lancelet amphioxis (genus *Branchiostoma*, formerly named *Amphioxus*) is still often used to introduce essential features of vertebrates. The vertebrate-like features of cephalochordates include the following: blocks of segmental muscles (**myomeres**); a ventral contracting blood vessel that may be homologous to the vertebrate heart; a diverticulum of the intestine that resembles a cecum in the vertebrate embryonic gut that becomes the liver; separate ventral and dorsal roots of the nerve cord;

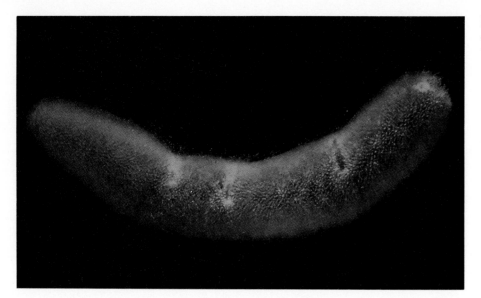

Figure 34.13
A pyrosome consisting of numerous salps of genus Pyrosoma.

and rudimentary olfactory and optic receptor organs. On the other hand, several features are quite different from those of vertebrates. One noteworthy difference is that the excretory organs drain the coelomic fluid rather than blood, and they are closed at the interior ends. They are therefore technically **protonephridia** (see pp. 285–286), which do not occur in vertebrates.

The 30 or so species of lancelets are generally found along shallow marine shores. Although they can swim by side-to-side contractions of the fin-shaped body, they are usually found with their tails buried in sand (Figure 34.7). The springy notochord prevents the body from collapsing during both swimming and burrowing. Lancelets feed and respire in much the same way as adult ascidians: cilia on the pharyngeal gill slits drive a current of water through the pharynx. Food trapped on a sheet of mucus produced by the endostyle is passed back to the cecum (the gut diverticulum mentioned before) for digestion. From the pharynx water passes through the atrium and out through the atrial opening.

Unlike tunicates, lancelets have separate sexes. As gametes ripen in the more than two dozen gonads, they burst into the atrial cavity and out through the atrial opening for external fertilization. The larvae resemble the adults but are covered by cilia that are used for propelling food toward the mouth and for swimming. As the larva grows the area of the ciliated body surface fails to keep pace with the body mass. As a result, the larva tends to sink and eventually adopts the burrowing habit of the adult.

Figure 34.14
Schematic representation of a lancelet. This is one of the few chordates in which the notochord persists in the adult.

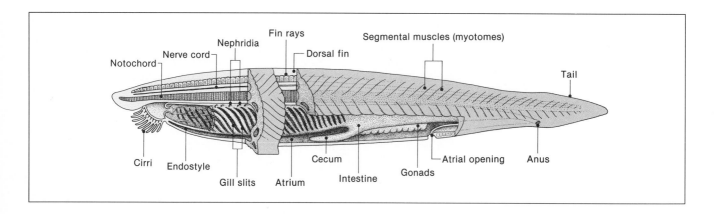

VERTEBRATES

Taxonomists now commonly divide the subphylum Vertebrata into eight classes of fishes, amphibians, reptiles, birds, and mammals. The large number of classes may indicate the success of vertebrates in being able to diversify, or it may just reflect the human tendency to see differences more easily in the familiar. Classes of vertebrates are often placed in larger informal groups according to criteria of major evolutionary significance. The jawless fishes are referred to as Agnatha; all other vertebrates have mouths with jaws and are therefore called Gnathostomata (Figure 34.15). The gnathostomes are divided into Pisces (PI-seez), which are the jawed fishes with fins, and Tetrapoda (TET-ra-PO-duh), which have two pairs of limbs for terrestrial locomotion. Fishes and amphibians, which do not develop within a fluid-filled sac (the **amnion**), are called Anamniota. As will be described shortly, reptiles, birds, and mammals develop within an amnion and are therefore referred to collectively as Amniota.

THE ORIGIN OF CHORDATES

The Arthropod–Chordate Theory.   In the 19th century, when evolution was viewed as a steady progression up to humans, zoologists assumed that an "advanced" group such as chordates must have evolved from an animal only slightly less advanced—namely, an arthropod. According to this arthropod–chordate theory, an arthropod became inverted so that its dorsal heart and ventral nerve cord became the ventral heart and dorsal nerve cord of chordates. Later,

Figure 34.15
The major groups of chordates.

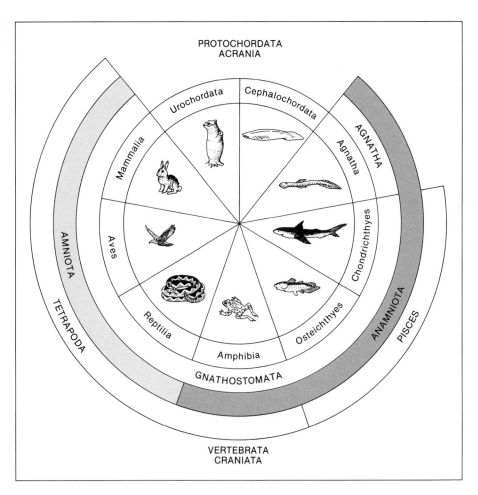

the most likely ancestor of chordates was thought to be annelids, whose hearts and nervous systems are positioned like those of arthropods. A major problem with both the arthropod–chordate and the annelid–chordate theories is that the brain would have had to undergo some tortuous rearranging in order to end up above our mouths rather than below them.

The Echinoderm Theory.   Another theory, the echinoderm theory, is based on the fact that both echinoderms and chordates are deuterostomes. This theory is weakened, however, by comparisons of ribosomal RNA, indicating that echinoderms and chordates are evolutionarily far removed from each other (Figures 18.15 and 34.16A). Moreover, the earliest chordate fossils bear no resemblance to echinoderms, or to arthropods or annelids (Figure 34.16B). In summary, we have no idea from whence our phylum came.

THE ORIGIN OF VERTEBRATES

Garstang's Theory.   The origin of vertebrates is about as uncertain as the origin of chordates. For many years amphioxus was viewed as a living relic of ver-

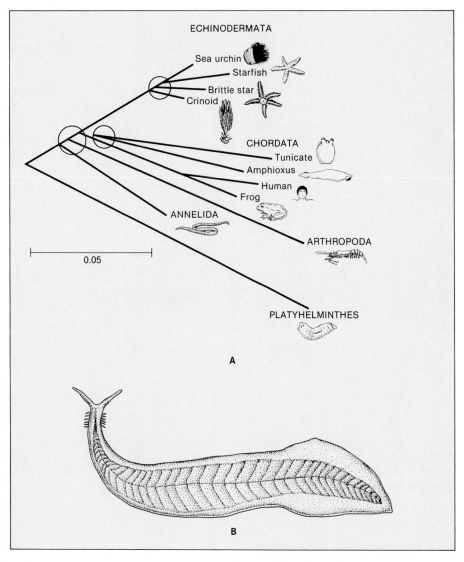

Figure 34.16
(A) Evolutionary relationship of echinoderms to chordates, according to Field et al. (1988). (See Lake 1990 for a different interpretation.) Other phyla are shown for comparison. Distance from one group to another along the line segments represents evolutionary distance, as determined by the number of differences in RNA bases in the 18S subunit of ribosomes. (Scale bar represents 0.05 base difference per position.) The method is not sufficiently precise to determine the order of branching within the circles. (B) The notochord and myomeres identify Pikaia *as one of the earliest known chordates. Its fossils are found in the Burgess shale formation, which is almost 600 million years old (see p. 395).*

tebrate ancestors. In addition to having all the earmarks of a chordate, the adult amphioxus also has several vertebrate characteristics, as noted previously. Several features of amphioxus, however, argue against its having a shared ancestry with vertebrates. Most zoologists now believe vertebrates originated from the other class of protochordates, Urochordata. It stretches the imagination to think that a sea squirt like that shown in Figure 34.6B could be ancestral to any vertebrate, especially considering that it lacks even a coelom and dorsal hollow nerve cord. The urochordate larva, however, shows more promise (Figure 34.6A). But how could a sea squirt larva evolve into anything when it can't even reproduce? William Garstang proposed an answer to this question in the 1920s, as well as in 1951 in a posthumous book of verse(!). Garstang's theory suggests that the tadpole larva of a urochordate acquired the ability to reproduce. This process, called **paedomorphosis** (see p. 134), is common among chordates, and in fact occurs in some living urochordates (larvaceans).

Theory of Northcutt and Gans. Garstang's theory remains widely accepted among zoologists, although there is little direct evidence for it. At least two other alternatives have been proposed. R.P.S. Jefferies (1986) suggests that vertebrates evolved directly from an extinct group of echinoderms called calcichordates. The evidence described in Figure 34.16A argues strongly against the calcichordate theory, however. A more likely proposal by R. G. Northcutt and Carl Gans (1983) is based on features of vertebrate development that have long puzzled embryologists. The embryonic origins of parts of the head, especially sense organs, muscles, and structures involved in gas exchange, differ from those in the rest of the body. Northcutt and Gans suggest that vertebrates evolved from an amphioxus-like protochordate in which the head was gradually elaborated during a transition from filter feeding to active predation. Crudely put, the first vertebrate consisted of a protochordate with a new head.

THE ORIGIN OF MAJOR VERTEBRATE GROUPS

Agnathans. The evolutionary origins of major groups of vertebrates are also uncertain and controversial, but we can at least sketch out some of the major transitions. It appears that the earliest vertebrates were jawless, fishlike agnathans. Fossils of agnathans are found in the United States and the Soviet Union in marine deposits more than 500 million years old (late Cambrian and early Ordovician periods). By the Silurian period millions of years later, most of these earliest agnathans had radiated into freshwater streams and lakes. Presumably the competition was less intense there, with few large animals and no other vertebrates of any kind. These agnathans were covered by bony plates and are therefore called **ostracoderms** (Greek *ostrakon* shell + *derma* skin). Judging from their tiny mouths and flat bellies, they ate small food particles on the bottoms of streams and ponds (Figure 34.17). Ostracoderms apparently had cartilaginous rather than bony skeletons, and they lacked paired fins. Ostracoderms became extinct at the end of the Devonian period, but quite different agnathans (lampreys and hagfishes) survive even now, as will be described in the next chapter. Some of the early agnathans may have evolved into various groups of fishes, both living and extinct.

Gnathostomes. During the Silurian period around 420 million years ago, one of the groups of early agnathans evolved jaws, giving rise to the first gnathostomes. The development of jaws was a pivotal event in evolution, because it enabled gnathostomes to become predatory. More important, it enabled their vertebrate

Figure 34.17
*Reconstruction of a Devonian
ostracoderm,* Hemicyclaspis. *The
two flaps behind the gills are not
paired fins. Length approximately
20 cm.*

descendants to feed on plants and on animals with skeletons, and therefore to
become terrestrial. Two of the earliest kinds of gnathostomes were the **acanthodians** and the **placoderms** (Figure 34.18). Acanthodians and placoderms were also
the first fishes with paired fins, and acanthodians also had scales rather than bony
plates. Scales and paired fins are characteristic of most cartilaginous and bony
fishes that live today.

Jaws did not evolve out of nothing but apparently originated from cartilaginous
supports called **gill arches.** (See Figure 12.7 for the location of gill arches in
modern fishes.) On each side of the head of an agnathan a pair of gill arches
apparently evolved into part of the upper and lower jaws (Figure 34.19). Unlike
some theories of vertebrate evolution, this one is supported by considerable evidence: (1) In certain primitive fishes the jaws resemble gill arches. (2) In sharks
the jaws still develop from gill arches. (3) The branching of the cranial nerve to
the jaws resembles the branching of cranial nerves to the gill arches in cartilagi-

Figure 34.18
*Two early gnathostomes. (A) An
acanthodian,* Euthacanthus.
*Acanthodians are also called spiny
sharks because of a spine on the
leading edge of each fin. Length
approximately 20 cm.
(B) Cast of the head of a placoderm,*
Dunkleosteus terrelli, *in which the
jaws were developed with a
vengeance. With a body length of
more than 10 meters, this must
have been a fearsome predator.*

A

B

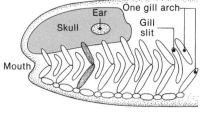

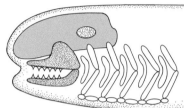

nous fishes. Initially, the jaws may have served only to close the mouth. Later they became armed with scales that evolved into teeth.

Tetrapods. An innovation that was equally important in the evolution of terrestrial vertebrates occurred during the Devonian period, when some bony fishes developed reinforcements in their fins that enabled them to support their weight on land. These were lobe-finned fishes. In addition to strong appendages, lobe-finned fishes also had lungs; both of these adaptations may have enabled them to escape drying ponds and to take advantage of new habitats. One extinct group of lobe-finned fishes, the rhipidistians, are generally credited with being the ancestors of the first tetrapods. One of the earliest tetrapods was *Ichthyostega*, whose fossil remains are found in what is now eastern Greenland, in deposits about 360 million years old (Figure 34.20). *Ichthyostega* looks very much like a lobe-finned fish, but because the skeleton shows adaptations to land it is considered to be an early amphibian.

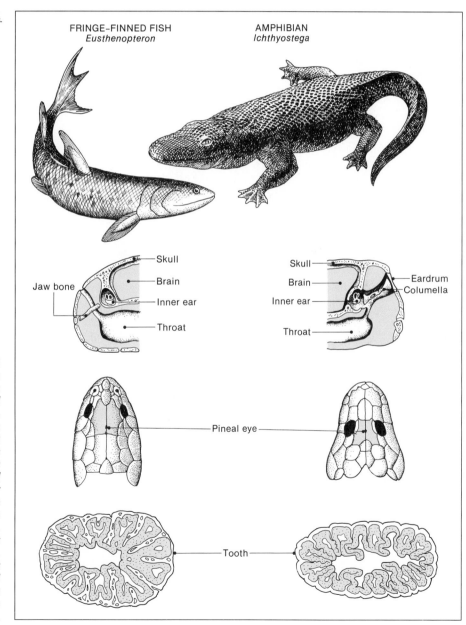

Figure 34.20
The fringe-finned fish Eusthenopteron *and the amphibian* Ichthyostega, *both of which lived during the Devonian period.* Ichthyostega *is identified as an amphibian mainly because the leg bones are adapted to support its weight on land. Each animal was about a meter long. The skulls and teeth of both animals had similar structures. The middle-ear bone (columella) of the amphibian evolved from a jaw bone of the fish. The pineal eye was a third eye that still occurs in some vertebrates.* Ichthyostega *belonged to an extinct group of amphibians called labyrinthodonts because of the labyrinthlike folding of the dentine in the teeth.*

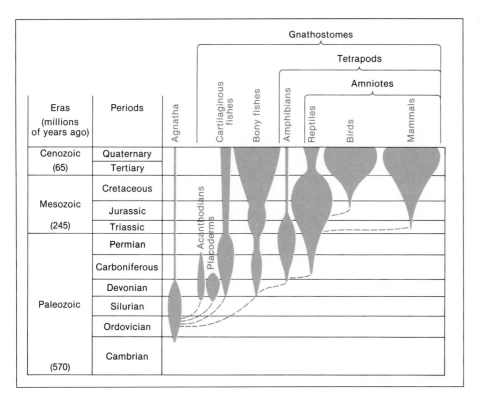

Figure 34.21
Evolution of the major groups of vertebrates. The width of each group suggests its diversity. Dashed lines are hypothetical.

Eras (millions of years ago)	Periods	Agnatha	Cartilaginous fishes	Bony fishes	Amphibians	Reptiles	Birds	Mammals
Cenozoic (65)	Quaternary							
	Tertiary							
Mesozoic (245)	Cretaceous							
	Jurassic							
	Triassic							
Paleozoic	Permian							
	Carboniferous							
	Devonian							
	Silurian							
	Ordovician							
(570)	Cambrian							

Gnathostomes / Tetrapods / Amniotes

Acanthodians / Placoderms

Amniotes. Although amphibians are adapted for living on land at least part of the time, they must return to water in order to reproduce. The water is required to keep the embryo moist and as a place into which metabolic wastes from the embryo can diffuse. The complete independence from water required a new type of egg capable of storing water and wastes, as well as nutrients for the embryo. A self-contained egg is called a **cleidoic egg.** The cleidoic egg of a vertebrate is also an **amniote egg,** because an internal water supply bathing the embryo is contained in an extraembryonic membrane called the **amnion.** Another extraembryonic membrane, the **allantois,** holds the metabolic wastes and helps a third extraembryonic membrane, the **chorion,** exchange respiratory gases. The evolution of the amniote egg was a major milestone, since it allowed the evolution of terrestrial vertebrates, beginning with reptiles in the early Carboniferous period some 340 million years ago (Figure 34.21). Mammals and then birds evolved from reptiles and retained the amnion, allantois, and chorion. The evolution of these groups will be described in more detail in later chapters.

SUMMARY

Although both echinoderms and chordates are deuterostomate, they do not appear to be closely related to each other. The phylum that does seem to be most closely related is Hemichordata. Hemichordates are also deuterostomate and have two features in common with chordates: pharyngeal slits and a dorsal nerve cord. Phylum Hemichordata comprises enteropneusts (acorn worms), which burrow in marine sediments, and pterobranchs, which are usually colonial and live in secreted tubes.

Chordates are distinguished by pharyngeal slits, a dorsal tubular nerve cord, and a notochord in at least some stage of development. In addition, a post-anal tail and myomeres are characteristic. Chordates are divided into protochordates and vertebrates. Protochordates are further divided into urochordates (sea squirts, larvaceans, and salps) and cephalochordates (lancelets). Vertebrates are divided into the agnathans, which are jawless fishlike animals, and gnathostomes. The gnathostomes are vertebrates with jaws: fishes, amphibians, reptiles, birds, and mammals. All gnathostomes except fishes are tetrapods, and all tetrapods except amphibians are amniotes. The evolution of amniotic eggs enabled reptiles, birds, and mammals to reproduce on land, since the eggs were self-contained. The four legs supported their bodies on land, and jaws enabled these animals to feed on land plants and animals.

KEY TERMS

dorsal tubular nerve cord
pharyngeal slit
notochord
post-anal tail
myomere

tunic
branchial sac
house
agnathan
gnathostome

tetrapod
amniote
cleidoic egg

SELF-TEST

1. Describe the two classes of hemichordates. Why are these dissimilar animals grouped in the same phylum?

2. Describe three features that are unique to chordates.

3. Give the common names for the three subphyla of chordates. Name a representative example in each subphylum.

4. Describe the functions of the branchial sac in adult ascidians.

5. Explain why amphioxus is so often studied in biology courses.

6. Name a common example of each of the following kinds of vertebrates: agnatha, gnathostome, pisces, tetrapod, anamniote, amniote. To which of these groups do humans belong?

7. State the essential features of Garstang's theory of the origin of vertebrates. Give the essential features of an alternative theory.

8. Explain why each of the following is considered a milestone in vertebrate evolution: jaws; tetrapody; the amniote egg.

READINGS

RECOMMENDED READINGS

Alldredge, A. 1976. Appendicularians. *Sci. Am.* 235(1):94–102 (July).

Radinsky, L. B. 1987. *The Evolution of Vertebrate Design.* Chicago: University of Chicago Press.

See also relevant selections in General References at the end of Chapter 21.

ADDITIONAL REFERENCES

American Zoologist, volume 22, number 4 (1982) includes a symposium on the developmental biology of the ascidians.

Barrington, E. J. W. and R. P. S. Jefferies (Eds.). 1975. *Protochordates.* Symposium of the Zoological Society of London, No. 36. New York: Academic Press.

Carroll, R. L. 1988. *Vertebrate Paleontology and Evolution.* New York: W. H. Freeman.

Conway Morris, S. and H. B. Whittington. 1979. The animals of the Burgess shale. *Sci. Am.* 241(1):122–133 (July).

Field, K. G. et al. 1988. Molecular phylogeny of the animal kingdom. *Science* 239:748–753.

Garstang, W. 1951. *Larval Forms and Other Zoological Verses.* Oxford: Basil Blackwell.

Gould. S. J. 1989. *Wonderful Life: The Burgess Shale and the Nature of History.* New York: W. W. Norton.

Jefferies, R. P. S. 1986. *The Ancestry of the Vertebrates.* London: British Museum (Natural History). (*Jefferies reviews his calcichordate theory and criticizes other theories of the origin of chordates.*)

Lake, J. A. 1990. Origin of the metazoa. *Proc. Natl. Acad. Sci. USA* 87:763–766.

Northcutt, R. G. and C. Gans. 1983. The genesis of neural crest and epidermal placodes: a reinterpretation of vertebrate origins. *Q. Rev. Biol.* 58:1–28.

35

Fishes

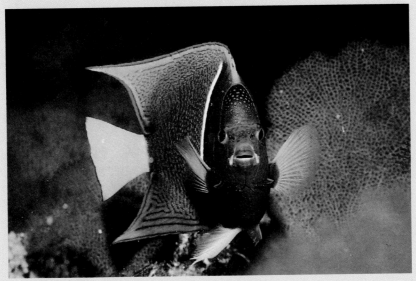

Fish at a coral reef.

LEARNING OBJECTIVES

1. How have fishes become so numerous and diverse?
2. How have sharks become formidable predators?
3. What keeps fishes from sinking?
4. How do fishes breathe?
5. Why are the fins and bodies of fishes shaped differently? Do these differences relate to differences in swimming speeds?
6. What is the function of fish scales?
7. Why do some fishes form schools?
8. Why and how do eels and salmon migrate?
9. How and why do electric fishes produce a voltage?

THE SUCCESS OF FISHES

From the sunlit surface to the deepest ocean trench, and from the clearest mountain stream to the bottom of the murkiest pond, fishes abound. The number of named species is now approximately 20,000, more than the number of all other vertebrate species. Fishermen, zoologists, and amateur divers who invade their habitat with nets, submersible research vessels, and diving gear discover approximately 100 new species of fish each year. The number of individual fishes is also greater than that of all other vertebrates. The evident success of fishes may be due to their having been among the first to take full advantage of the chordate features summarized in the previous chapter. These features are adapted in fishes in such a way that most can swim actively throughout a three-dimensional aquatic habitat, rather than settling to substratum, like many aquatic invertebrates.

One chordate distinction, the notochord, is replaced during development in most fishes by a more supportive, yet flexible **vertebral column.** This part of the skeleton provides a framework for the attachment of other skeletal structures and muscles that enable fishes to swim rapidly. A second feature of chordates, the post-anal tail, becomes the **caudal fin** that supplies most of the propulsive force in swimming. The pharyngeal slits become openings to the efficient **gills** that extract the oxygen needed to sustain a high level of activity. Finally, the dorsal hollow nerve cord is greatly enlarged at the anterior end, forming a **brain** that coordinates activity in relation to information about the environment.

KINDS OF FISHES

Agnathans.   Judging from fossils half a billion years old, the first to take advantage of the possibilities of vertebrate life were agnathans, which are considered to be vertebrates even though they lack vertebrae. These vertebrates without vertebrae or jaws are generally classified in superclass Agnatha to distinguish them from vertebrates that have jaws (superclass Gnathostomata) (see Figure 34.15; pp. 743–745). Although they look like some fishes, especially eels, agnathans are often not considered to be fishes at all. Unlike most fishes, living agnathans do not have scales, and they lack paired fins on the sides. Both modern fishes and modern agnathans evolved from quite different forms of ancient agnathans (see p. 746). The surviving forms comprise approximately 75 species of hagfishes and lampreys (class Myxini and class Cephalaspidomorphi) (Figure 35.1).

Figure 35.1
Jawless "fishes." (A) The Pacific hagfish Eptatretus stoutii. *(Length approximately 65 cm.) (B) Sea lampreys* Petromyzon marinus *parasitizing a carp,* Cyprinus carpio. *Length of each lamprey approximately 80 cm.*

A   B

Cartilaginous Fishes. Fishes with jaws are generally divided into two classes, depending in part on whether their adult skeletons consist of cartilage or mainly of bone. In modern cartilaginous fishes, class Chondrichthyes, the cartilage of the embryonic skeleton is not replaced with bone, as most of it is in other vertebrates. Fishes in class Chondrichthyes are divided into two subclasses, the Elasmobranchii and the Holocephali. The elasmobranchs include more than 300 species of sharks and more than 400 species of rays. Sharks are the most familiar cartilaginous fishes. Like most fishes, they have scales and two pairs of fins on the sides, as well as a caudal fin and unpaired fins along the dorsal and ventral midline. Rays have few scales, and their fins are highly modified in keeping with their overall flattened shape. Most elasmobranchs are predatory, and their ability to engulf prey is aided by an upper jaw that swings open like the lower jaw. The second major group of cartilaginous fishes, the subclass Holocephali, comprises approximately 30 species of chimaeras (pronounced ky-MERE-uhs; Figure 35.2). Chimaeras, commonly called ratfishes, are distinguished from elasmobranchs by an upper jaw that is fused to the skull. Their bodies are intermediate in shape between those of most sharks and most rays.

A

B

C

Figure 35.2
Cartilaginous fishes. (A) The spiny dogfish shark Squalus acanthias, *which is often used as a laboratory subject. Length approximately 1.5 meters. (B) The blue-spotted ray* Taeniura lymma. *Stingrays such as this one defend themselves with two venomous spines about midway along the tail, which can cause excruciating pain and even death to a person. Length approximately 2.4 meters. (C) Spotted ratfish (chimaera)* Hydrolagus colliei *of the Pacific. Length approximately 1 meter.*

Lobe-Finned Fishes. In bony fishes (class Osteichthyes) bones replace much of the cartilage in the skeleton during development. The oldest extant group of bony fishes are the lobe-finned fishes, named for their fan-shaped lateral fins with fleshy lobes at the base (Figure 35.3). These lobes are reinforced with bones that may be homologous to the limb bones of tetrapods. Because of this homology, and because many have lungs and can breathe air, a lobe-fin of some kind is generally thought to have been the ancestor of amphibians and other tetrapods.

Ray-Finned Fishes. Most bony fishes have fins that are not lobed but are reinforced by bony rays. Among these ray-finned fishes (subclass Actinopterygii) are almost all the familiar species that come to mind when one thinks of fish, including trout, tuna, bass, goldfish, and perch (Figure 35.3B). The ray-finned fishes will be discussed extensively later in this chapter.

AGNATHANS

Hagfishes. One group of jawless fishlike vertebrates is the class Myxini, which comprises more than 30 living species of hagfishes (Figures 35.1A and 35.4). Hagfishes inhabit temperate sea floors at depths greater than 25 meters, where they eat small invertebrates and scavenge dead and dying fishes and larger invertebrates. Using their tongues, which are toothed like snail radulas, they burrow into carcasses and feed from within. This habit and their shape may account for Linnaeus having mistaken them for intestinal worms.

Figure 35.3
Representative bony fishes. (A) Australian lungfish Neoceratodus forsteri. *Of the six living species of lungfishes, this one is most like the lungfishes of the Paleozoic era. Length approximately 1.5 meters. (B) The yellow perch* Perca flavescens, *which is often studied in zoology laboratories as a representative ray-finned fish. Length approximately 38 cm.*

A

B

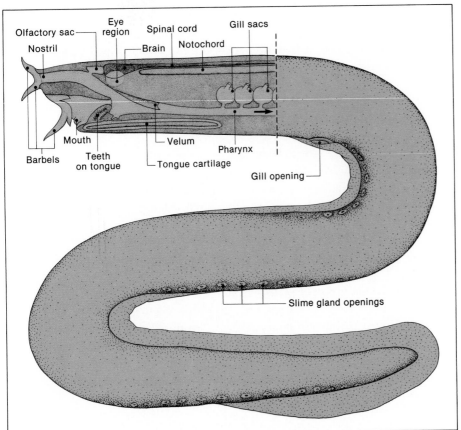

Olfactory sac
Nostril
Eye region
Brain
Spinal cord
Notochord
Gill sacs
Mouth
Teeth on tongue
Barbels
Velum
Tongue cartilage
Pharynx
Gill opening
Slime gland openings

Figure 35.4
The Atlantic hagfish Myxine glutinosa. *The eyes are degenerate and functionless, but there are light-sensitive organs in the skin. The velum pumps water from the single nostril through the gill sacs. Note the absence of vertebrae and the presence of the notochord in the adult. The "teeth" are not homologous with those of most vertebrates.*

A Classification of Fishes, According to Nelson (1984)

Genera mentioned elsewhere in this chapter are noted.

Superclass Agnatha (= Cyclostomata) AG-na-thuh (Greek *a-* without + *gnathos* jaws). Jawless vertebrates. Vertebrae rudimentary or absent. Modern forms without true appendages (paired fins), scales, or bones. Notochord persists in adult. Gill pouches empty through uncovered pores rather than slits.

   Class Myxini mix-EYE-nye (Greek *myxa* mucus). Terminal mouth with four pairs of sensory barbels. Nasal sac connecting to pharynx. Five to fifteen pairs of gill pouches. Hagfishes. *Eptatretus, Myxine* (Figure 35.1A).

   Class Cephalaspidomorphi sef-a-LASS-pid-oh-MORE-fy (Greek *kephale* head + *aspido-* shield + *morphe* shape). Sucking mouth. Nasal sac not connected to pharynx. Seven pairs of gill pouches. Lampreys. *Petromyzon* (Figure 35.1B).

Superclass Gnathostomata NATH-oh-STOW-ma-tuh (Greek *stoma* mouth). With jaws. Notochord usually absent in adult; replaced by vertebrae. Most with paired appendages. Jawed fishes, amphibians, reptiles, birds, and mammals.

   Class Chondrichthyes kon-DRIK-the-eez (Greek *chondros* cartilage + *ichthys* fish). Fishes with cartilaginous skeletons. No swim bladder or lung. Spiral valve in intestine. Male with pelvic claspers for sperm transfer. Teeth not fused to jaws; replaced continually.

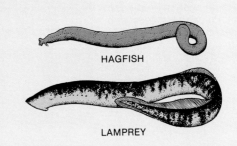

HAGFISH

LAMPREY

A Classification of Fishes, According to Nelson (1984) *(continued)*

Subclass Elasmobranchii e-LAZ-mo-BRAN-kee-eye (Greek *elasmos* plate + *branchia* gills). Five to seven pairs of gill slits without covers, and usually a pair of gill openings modified as spiracles. Upper jaw hinged to skull. Placoid scales usually present. Rigid dorsal fin. Sharks and rays (including skates). *Carcharodon, Cephaloscyllium, Cetorhinus, Isistius, Odontaspis, Pristis, Raja, Rhincodon, Squalus, Taeniura* (Figures 35.2A, 35.2B, 35.9B, 35.12).

Subclass Holocephali HO-low-SEF-al-lye (Greek *holos* whole). Upper jaw fused to skull. Four gill slits with a single cover on each side. No spiracles. Without scales. Males with club-shaped clasping organ in front of the eyes (in addition to pelvic claspers). Chimaeras. *Hydrolagus*.

Class Osteichthyes OS-tee-IK-the-eez (Greek *osteon* bone). Fishes with bony skeletons. Scaly skins. Several gills on each side with one gill cover. Usually with air sacs that function either as lungs or as swim bladders for buoyancy.

Subclass Dipneusti (= Dipnoi) dip-NEW-sty (Greek *di-* two + *pneustikos* of breathing). One or two lungs. Lobed pectoral fins. All median fins fused together. Spiral valve in intestine. Lungfishes. *Neoceratodus, Protopterus*.

Subclass Crossopterygii cross-op-ter-RIJ-ee-eye (Greek *krossoi* fringe + *pteryx* wing). With lobed pectoral fins. Two lungs in extinct species; one fat-filled lung in the living species. Spiral valve in intestine. Fringe-finned fishes. The only known living species is the coelacanth *Latimeria chalumnae*.

Subclass Brachiopterygii (= Cladistia) brake-ee-op-ter-RIJ-ee-eye (Greek *brachion* arm). Body elongated or eellike. With lungs and lobed pectoral fins. Dorsal fin divided into 5 to 18 finlets, each with a spine. Spiral valve in intestine. (Often grouped in subclass Actinopterygii, superorder Chondrostei.) Bichirs and reedfish.

Subclass Actinopterygii ACT-in-op-ter-RIJ-ee-eye (Greek *aktis* ray). Fins supported by bony rays. Swim bladders instead of lungs. No spiral valve in intestine. Ray-finned fishes. (See Figure 35.14 for examples.)

    Superorder Chondrostei kon-DROSS-te-eye. Primitive ray-finned fishes. Skeleton mostly cartilaginous. Notochords in adults. Ganoid scales. Sturgeons and paddlefishes. (Reedfish and bichirs often included.) *Acipenser* (Figure 35.15).

    Superorder Holostei ho-LOS-te-eye. "Intermediate" ray-finned fishes. Skeleton mainly bony. Swim bladder lunglike. Scales ganoid or cycloid. Bowfin and gars. *Amia, Lepisosteus* (Figure 35.16).

    Superorder Teleostei TEL-lee-OS-te-eye (Greek *teleos* complete). Skeleton almost completely bony. Scales cycloid, ctenoid, or absent. Vestigial notochord. Swim bladder maintains buoyancy. Nelson lists 35 extant orders; some of the more important ones are represented in Figure 35.14. *Alosa, Anguilla, Anomalops, Bothus, Cyprinus, Helena, Hippocampus, Monocirrhus, Oncorhynchus, Pandaka, Perca, Poecilia, Pterois* (Figures 16.1, Unit Three opener, 18.13, 20.11, 20.19, 24.15, 35.3B, 35.20, 35.23A, 35.28, 35.29).

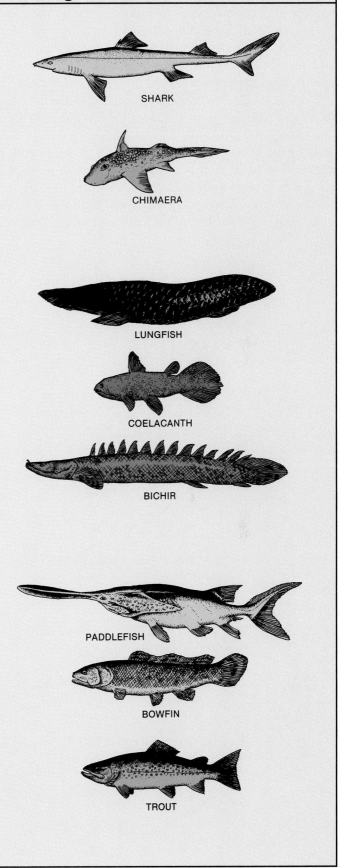

SHARK

CHIMAERA

LUNGFISH

COELACANTH

BICHIR

PADDLEFISH

BOWFIN

TROUT

Hagfishes are sometimes a nuisance to deep-sea fishermen, since they sometimes swim into a net to feed on the catch. The worst part comes when a fisherman removes the hagfish and ends up covered in **slime.** The slime comes from unique glands with openings distributed along the sides. Out of these openings come entire cells that break upon release. There are two kinds of slime gland cells. One kind releases a glycoprotein that forms clear mucus in seawater. The other kind releases protein threads up to 60 cm long, which make the mucus sticky and viscous (Downing et al. 1981). The resulting slime effectively protects hagfishes from predators but could endanger the hagfish itself by plugging the nostril through which it takes in water. The hagfish rids itself of slime by tying itself into an overhand knot at the rear, then sliding the knot forward over its head. The hagfish uses the same maneuver to slip out of the grasp of a predator or to pull off chunks of flesh when feeding (Figure 35.5).

In spite of the slime, many zoologists like hagfishes because they may provide a glimpse of the physiology of the first vertebrates. Like most marine invertebrates,

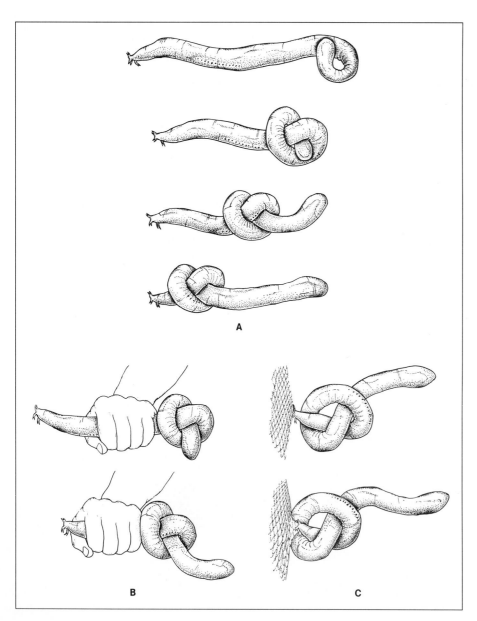

Figure 35.5
The cartilaginous body enables the hagfish to tie itself into a sliding knot to (A) clear itself of slime, (B) escape a predator, or (C) tear flesh.

but unlike most vertebrates, hagfishes have ion concentrations in their body fluids that are approximately as high as those in seawater. Thus they do not require the special osmoregulatory adaptations of most other fishes. Hagfishes are also the only vertebrates that retain in the adult both a **pronephros** and a **mesonephros:** two kinds of temporary kidney that occur only in the embryos of most vertebrates (see p. 292). Perhaps the strangest thing about hagfishes is that they have four different hearts that boost the blood pressure through gills and other areas. What little we know about the reproduction of hagfishes also appears unusual. Each individual has one long gonad, the front part of which becomes an ovary in females, and the back part of which becomes a testis in males. The female lays approximately two dozen large eggs (approximately 2 cm) with tufts at one end that anchor them to each other and to substratum. A prize for describing how the eggs get fertilized has been unclaimed ever since the Copenhagen Academy of Science offered it in 1864.

Lampreys. The remaining 41 living species of agnathans are lampreys (class Cephalaspidomorphi). Like hagfishes, lampreys lack jaws and paired fins, and have eellike bodies with cartilaginous skeletons. Lampreys differ from hagfishes in many ways, several of which reflect their different ways of feeding. Many species, such as the sea lamprey *Petromyzon marinus*, parasitize (or prey on) fishes, using the keratinized "teeth" on their tongues and **oral discs** to bore through the skin to feed on blood (Figures 35.1B and 35.6). Other lampreys have poorly developed "teeth" and are not parasitic. Parasitic lampreys locate hosts with eyes that are better-developed than those of hagfishes. Lampreys may also be able to detect electric fields generated by the muscle activity of hosts.

Figure 35.6
(A) The sea lamprey Petromyzon marinus. *(B) The oral disc of a sea lamprey. The sea lamprey uses its oral disc not only for feeding but also to move stones for nest building and to anchor itself to stones (hence the name* Petromyzon, *from the Greek words for "stone" and "suck"). Lampreys also use the oral disc to attach to other fishes for free transportation. The mouth is free to perform these functions because water flows into and out of each one of the seven pairs of gills through an individual opening.*

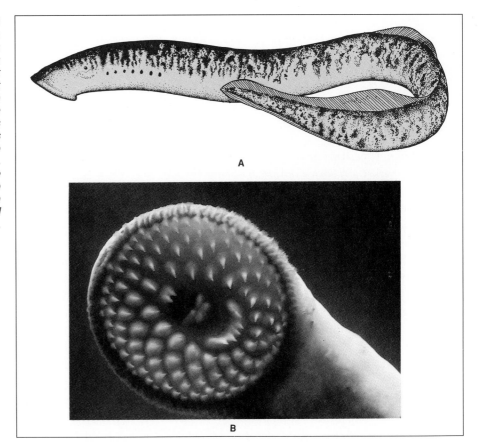

Also unlike hagfishes, lampreys spend much of their lives in fresh water, where they reproduce. Using its mouth, a male lamprey begins moving pebbles to excavate a shallow nest and is soon joined in the work by one or more females. Males then spawn repeatedly for more than a week, fertilizing numerous eggs (up to a quarter of a million for the sea lamprey). Each egg hatches into a larval form, the **ammocoete** (pronounced AM-mo-seat), that is so unlike the adult that it was long classified as a different species. The ammocoete closely resembles a lancelet internally and externally (see Figure 34.14). The ammocoete drifts out of the nest, then burrows tail first into a muddy bottom with its oral hood sticking out for suspension feeding. Generally, after four to seven years the ammocoete metamorphoses into an adult. In nonparasitic species the adults remain in the stream, living only long enough to reproduce. Some parasitic species migrate downstream to the ocean, but paradoxically the sea lamprey migrates into lakes without ever entering the sea. After a period of feeding, the parasitic species return to the streams to breed and die soon afterward.

The sea lamprey is now familiar to and despised by those who fish the Great Lakes, Lake Champlain, and many other large lakes in North America. An average sea lamprey deprives sport fishermen of approximately 20 kg of fish during the year and a half of its life as an ectoparasite. In some lakes virtually every lake trout, whitefish, salmon, carp, or other game fish has at least one lamprey on it or the scar of a lamprey. This was not always the case. In Lake Ontario, where sea lampreys were native, they had little effect on fisheries. The construction of the Welland Canal around Niagara Falls in 1829, however, provided the sea lamprey with access to Lake Erie, and from there they migrated into Lakes Huron, Michigan, and Superior. By the 1940s and 1950s they had virtually eliminated fisheries in these lakes. Those fisheries recovered after the compound TFM (3-trifluoromethyl-4-nitrophenol) was found to be toxic to ammocoetes when applied to the streams in which they live. (TFM damages gills, but it is not known how.) In the past decade, however, lampreys have begun to breed in streams where this compound cannot be applied economically or safely, and they are now threatening a $2 billion economy based on sport fishing.

SHARKS

Fins. There is perhaps no more intimidating image than the large dorsal fin of the white shark *Carcharodon carcharias* slicing through the water surface. Most sharks are too shy to live up to their reputations as vicious maneaters, but there is no denying that they are highly adapted to swiftly pursue and quickly dismember prey, including humans if the opportunity arises. The predatory abilities of sharks are due largely to their ability to swim with speed and agility, thanks to their characteristic fins. The caudal fin, which is attached to a narrow **caudal peduncle,** provides most of the forward propulsion in sharks and other fishes. In most sharks the vertebral column extends into the dorsal lobe of the caudal fin, which is larger than the ventral lobe. Such fins are termed **heterocercal** (Figure 35.7). For several decades it has been thought that a heterocercal tail tends to spin a fish head downward about its center of mass, and that this downward pitch had to be overcome by lift from the pectoral fins. Careful analysis by K. S. Thompson (1990), however, indicates that this is not necessarily the case. Sharks can apparently make small changes in the shape of the heterocercal tail to change the direction of thrust for diving, climbing, or turning. Unpaired fins along the dorsal and ventral midline keep the body swimming straight ahead as the caudal fin moves from side to side.

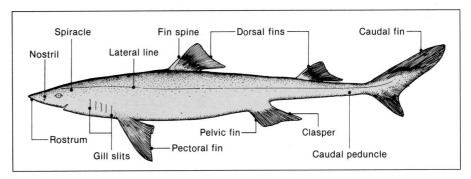

Figure 35.7
External features of a male dogfish.
(Compare Figure 35.2A.)

Spiracle   Fin spine   Dorsal fins   Caudal fin

Nostril   Lateral line

Rostrum

Gill slits   Pectoral fin   Pelvic fin   Clasper

Caudal peduncle

Neutral Buoyancy.   In sharks and other fishes swimming is aided by the maintenance of an overall density (mass divided by volume) that is nearly the same as that of water. This neutral buoyancy eliminates the need to expend energy to keep from sinking or floating to the surface. Sharks maintain neutral buoyancy mainly by means of low-density fats, especially **squalene,** in their large livers. Also contributing to neutral buoyancy in sharks is the endoskeleton of cartilage, which is less dense than bone. Unlike most other vertebrates, sharks and other elasmobranchs do not replace cartilage with bone during embryonic development, except for a few places around the teeth. This is a derived trait, since their ancestors were bony.

Skin.   The tough skin of sharks aids the cartilaginous skeleton in supporting the body and in swimming. Internal pressure against the skin serves as a **hydroskeleton** that helps keep the body from collapsing from the force of muscle contraction during swimming. The muscles used in swimming account for much of the body mass, especially in the tail. These muscles are arranged in segmental blocks (**myomeres**), each of which is shaped like a W on its side. The significance of this arrangement will be discussed in a later section on locomotion. Unlike skeletal muscles of most vertebrates, which apply their force to the skeleton, the myomeres of sharks pull on the skin, which transmits the force to the caudal fin during swimming. Another adaptation of the skin to swimming is a large number of scales with plate-shaped bases embedded in the skin. These **placoid scales** have backward-pointing spines that feel rough to the human touch but actually reduce hydrodynamic drag in water. Placoid scales are homologous with teeth, with a central pulp, dentine, and an outer covering of enamel (Figure 35.8).

Feeding and Digestion.   The structural similarity of a placoid scale to a vertebrate tooth (see Figure 13.6A) is not coincidental. The teeth of sharks are merely placoid scales that become enlarged as they migrate over the jaw (Figure 35.9). These teeth, which are generally numerous, sharp, and curved into the mouth, are all the more formidable because the loose attachment of both jaws to the cranium allows predaceous sharks to open their jaws widely to engulf larger prey. The loose attachment of the jaws also enables many sharks to extend the jaws forward so that the protruding snout does not interfere with feeding. Not all sharks use their teeth in feeding. The largest sharks feed on plankton, which they filter out of the water with their gills. These include basking sharks (*Cetorhinus*), which are up to 15 meters longs, and whale sharks (*Rhincodon*), which are up to 18 meters long. The digestive tracts of sharks are similar to those of other vertebrates (see Chapter 13), except for the **spiral valve** in the intestine, which slows the passage of food and increases the surface area for digestion and absorption (Figure 35.10).

Sensory Receptors.   Sharks locate prey using a variety of sensory receptors. The small eyes are usually not considered to be of much use in predation, although there is some debate on this point. The most important receptors appear to be chemoreceptors. There are chemoreceptors on the body surface and mouth, but the most sensitive ones are in olfactory pits in the head. Water circulates to the olfactory pits through nostrils. (The nostrils do not connect to the pharynx and therefore have no function in respiration.)

Several kinds of receptors are located in a canal called the **lateral line organ,** part of which lies just beneath the **lateral line** visible in Figure 35.7. The lateral line organ also meanders around the head. Among the kinds of receptors in the lateral line organ are electroreceptors. These electroreceptors are so sensitive they can detect the electrical activity of muscles in potential prey buried in sand. Electroreceptors may also enable sharks to detect ocean currents and to sense Earth's magnetic field for navigation. Electroreceptors will be discussed in more detail later in this chapter. Also in the lateral line organ, and elsewhere on the body surface, are mechanoreceptors called **neuromasts,** which sharks may use to detect water currents generated by prey. The neuromasts are similar to, and probably homologous with, the mechanoreceptors of the **inner ears** (see Figure 7.16C), which together with the lateral line organs constitute the **acousticolateralis system.** The inner ears of sharks consist of the **semicircular canals,** which sense acceleration of the body. Associated with the semicircular canals are **otolith organs** that respond to gravity and might also respond to low-frequency sound. Additional mechanoreceptors possibly involved in prey detection are located in **spiracles,** which are modified gill slits unique to elasmobranchs.

Respiration and Circulation.   Gills supply the oxygen needed to sustain predation and other activities of sharks. There are generally five pairs of gills, although a few species have six or seven pairs. Each gill lies in its own gill pouch and has a separate gill slit. The structure and functioning of elasmobranch gills are essentially the same as for other fishes, as described elsewhere (see pp. 249–251). Pelagic sharks, which swim in the open ocean almost constantly, ventilate the gills with water simply by holding the mouth and gill slits open. Slow-swimming and bottom-dwelling species expand and contract the gill pouches to pump water past the gills. When the mouth is busy with other matters, water enters through the spiracles. Blood circulates through the gills via branchial arteries and enters the dorsal aorta after becoming oxygenated. As in most vertebrates, red blood cells contain the respiratory pigment hemoglobin. After the blood circulates through the body tissues and becomes deoxygenated, it is pumped back to the gills by the heart, which consists of the **sinus venosus,** the **atrium,** and the **ventricle** (see Figures 11.4A and 35.10).

Excretion and Osmoregulation.   The gills of sharks and other fishes also play a role in excretion and osmoregulation. Ammonia, a nitrogenous waste from the metabolism of amino acids, diffuses across the gills into water. Sharks also produce other nitrogenous wastes, urea and methylamines, that do not diffuse across the gills. These are retained in high concentrations in the body fluids, where they serve as **osmolytes** that help maintain an osmotic concentration comparable to that of the surrounding seawater (see pp. 292–293). Although the total osmolarity of the body fluids is similar to that of seawater, the concentration of salts in the body fluids is lower than that of seawater, suggesting that sharks evolved from freshwater fishes. The ions that tend to accumulate in the body fluids from the

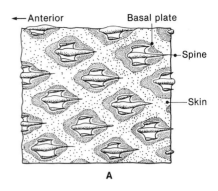

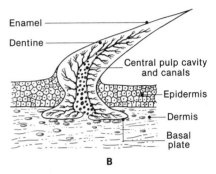

Figure 35.8
Placoid scales. (A) Surface view of shark skin. (B) Section through the spine of a single scale. The scales are homologous with the teeth of vertebrates. (Compare Figure 13.6A.)

Figure 35.9
(A) The lower jaw of a shark, showing the teeth, which are placoid scales. Like all placoid scales, the teeth of sharks are continually lost and replaced. Arrows indicate the continual outward movement of the teeth as they mature and are shed. Unlike the teeth of other vertebrates, sharks' teeth are not attached to the skeleton. (B) Head of the sand tiger shark Odontaspis taurus, *showing the teeth. This species feeds on fishes, squids, and the occasional human. Total length is over 4 meters. (C) Cookie-cutter sharks,* Isistius plutodus *and* I. brasiliensis, *have a few large teeth with which they slice out pieces of flesh from whales, porpoises, and other sharks after attaching to them with their suckerlike mouths. The sharks are approximately 40 cm long.*

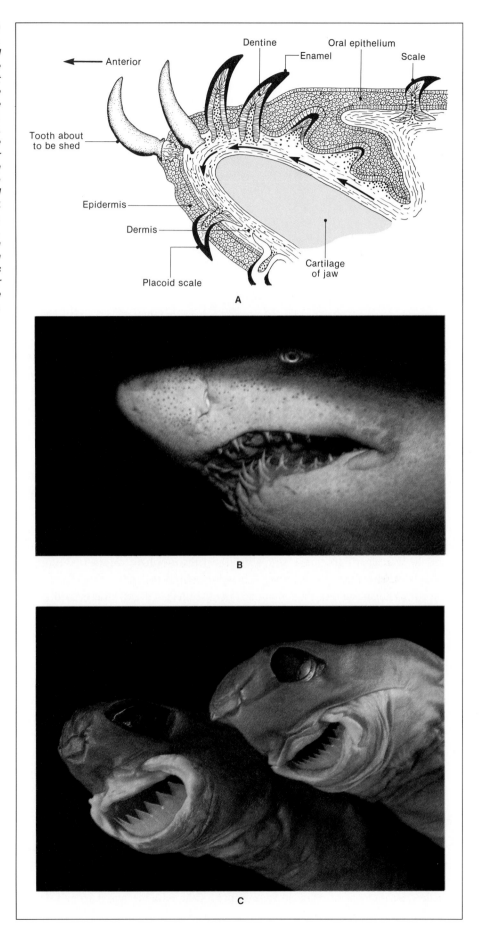

759

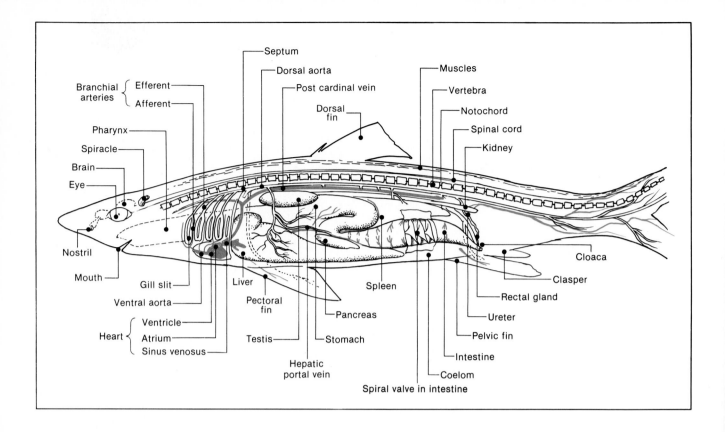

Branchial arteries {
  Efferent —
  Afferent —
}
Pharynx —
Spiracle —
Brain —
Eye —
Nostril
Mouth
Gill slit —
Ventral aorta —
Heart {
  Ventricle
  Atrium
  Sinus venosus
}
Liver
Pectoral fin
Testis —
Hepatic portal vein
Septum
Dorsal aorta
Post cardinal vein
Dorsal fin
Pancreas
Stomach
Spiral valve in intestine
Muscles
Vertebra
Notochord
Spinal cord
Kidney
Cloaca
Clasper
Rectal gland
Ureter
Pelvic fin
Intestine
Coelom
Spleen

environment are excreted mainly by the **rectal gland.** The rectal gland empties into the **cloaca,** which also receives feces and gametes. The **mesonephric kidney** is of secondary importance (see pp. 292–293).

Reproduction.   Fertilization is internal in most sharks. In males sperm are transported to the cloaca from the testes via the **mesonephric ducts** (= Wolffian ducts), which also drain the kidneys. A **clasper** behind the pelvic fins transfer the sperm from the cloaca into the female oviduct (Müllerian duct), where fertilization occurs. Some species are definitely viviparous, with a placenta-like structure connected to the embryo by an umbilical cord. In some sharks the embryo develops within the oviduct, and the mother gives birth to live young. These sharks are often said to be **ovoviviparous,** which implies that the egg is simply retained inside the oviduct during hatching, without any other support from the mother. Usually, however, there is little or no shell around the egg, and an area of the oviduct modified as a uterus provides some support to the embryo. It is therefore likely that even these live-bearing sharks are **viviparous.** A few species are indisputably **oviparous.** Their eggs are enclosed within a tough capsule in which the embryo develops externally (Figure 35.11). These embryos depend on a generous supply of yolk, which makes the egg quite large (up to 10 cm).

RAYS

About half the species of elasmobranchs are rays, including skates. Rays differ from most sharks in having few scales and in being generally adapted for feeding on bottom-dwelling animals, such as molluscs and crustaceans. Rays have a few platelike teeth adapted for crushing such prey. Their bodies are flattened dorsoventrally, enabling them to glide slowly over the bottom (Figure 35.2B). The eyes

Figure 35.10
Internal structure of a male spiny dogfish. The intestine is open to show the spiral valve.

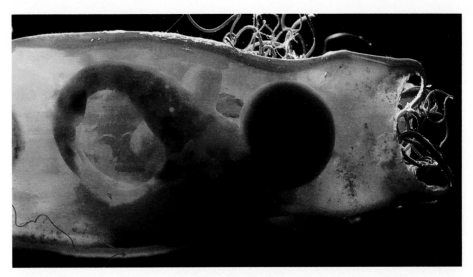

Figure 35.11
Egg capsule of the swell shark Cephaloscyllium ventriosum, containing an embryo. The tendrils at the ends may anchor the capsule to seaweed. Note the large yolk sac.

and spiracles are on top of the head. Since the mouth is often buried in bottom deposits, water for the gills enters through the spiracles on the dorsal surface and exits through gill slits on the ventral surface. The pectoral fins are greatly enlarged, and undulate gracefully during swimming. The tail fins are reduced or absent. Some rays, notably the electric rays, have electric organs used in predation, defense, and perhaps communication. These organs consist of noncontracting muscle cells whose action potentials summate to as much as 220 volts (Figure 35.28). Skates differ from other rays in being oviparous, releasing their eggs within rectangular cases—the "mermaids' purses" familiar to beachcombers. Another distinctive group of rays is the sawfishes, which have sawlike snouts with which they slash at schooling fish (Figure 35.12).

CHIMAERAS

Chimaeras are not as well known as elasmobranchs, since they are generally smaller (less than 2 meters long), are never fished, and do not eat people. Someone must have been impressed by them, however, to have named them after the chimaera of Greek myth, which was said to have a lion's head, a goat's body, and a serpent's tail. This description fits their overall shape (Figure 35.2C). Chimaeras were numerous during the Devonian period some 400 million years ago, but the surviving 30 species are now restricted to deep offshore waters, where they feed mainly on molluscs, crustaceans, and other invertebrates. Like rays, chimaeras have a few large, platelike teeth. Unlike elasmobranchs, chimaeras have the upper jaw fused to the skull (hence the name of the subclass: Holocephali).

LOBE-FINNED FISHES

Lungfishes. Three groups of bony fishes (class Osteichthyes) are often referred to as lobe-fins, because most have lateral fins with fleshy lobes at the base. These lobes are reinforced by bones that articulate with the rest of the skeleton and may be homologous with the limb bones of tetrapods. The caudal fins typically taper posteriorly to a point: a shape that is called **diphycercal** (Figure 35.3A). One group of lobe-fins, the lungfishes (subclass Dipneusti), have one or two lungs with which they can breathe air. These lungs connect to the esophagus and are inflated by swallowing air. Six species of lungfishes survive from better times hundreds of millions of years ago. Four species in the genus *Protopterus* occur in freshwater

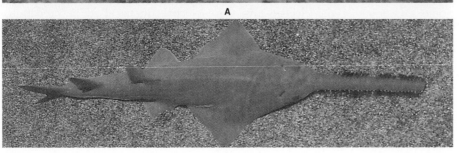

Figure 35.12
(A) The roundel skate Raja texana *from the Gulf of Mexico. The eyespots may deter predators from above. Length 60 cm. (B) The smalltooth sawfish* Pristis pectinatus, *which occurs in estuaries and shallow coastal waters from the Chesapeake Bay south. Length approximately 5 meters.*

swamps in Africa, one species lives in South America, and one inhabits shallow rivers in Queensland, Australia (Figure 35.3A). The South American and African species use their lungs to hold air reserves while burrowed into mud for up to two years to escape drought. In addition, while buried the African lungfishes enter a state called **estivation,** in which they surround themselves in a cocoon of mucus and drastically reduce their metabolic activities.

Fringe-Finned Fishes: The Coelacanth.   Some paleontologists believe that the ancestors of tetrapods were lungfishes, but many have concluded that a more likely ancestor was a fringe-finned fish (subclass Crossopterygii). (See p. 745 for the evolution of tetrapods.) Crossopterygians flourished at the same time as lungfishes and also had lungs. Rather than burrow into mud, however, they were more likely to have crawled away from drying ponds on their fins. The only crossopterygian known to have survived to the present is a marine species, *Latimeria chalumnae* (Figure 35.13). This coelacanth has only the vestige of one lung, which is filled with low-density fat that helps maintain buoyancy. Deep-sea fishermen near the Comoro Islands in the western Indian Ocean had been catching coelacanths for many years, but until 1938 scientists knew them only as fossils dating from 400 to 60 million years ago. Then a specimen reached Marjorie Courtenay-Latimer, curator of a small museum in South Africa, and J. L. B. Smith, a chemist and ichthyologist, easily identified it from a sketch (Courtenay-Latimer 1979).

Figure 35.13
A living coelacanth in its natural habitat 100 to 200 meters deep in the western Indian Ocean. The fish is approximately 1.5 meters long.

Reports of this coelacanth excited the public as much as discovery of a living Neanderthal person would. Zoologists were just as excited, for a living coelacanth would provide a rare opportunity to test theories and speculations about the evolution of vertebrates. Researchers had to wait until 1952 for a second coelacanth, but since then they have acquired around 100 specimens (enough to raise alarm about endangering the species).

Subsequent research on the coelacanth has taken some of the edge off the excitement. Molecular comparison indicates that the coelacanth is not as close to tetrapods as lungfishes are (Meyer and Wilson 1990), and in many ways the coelacanth is more like an elasmobranch that an ancestor of tetrapods (Forey 1988). Like elasmobranchs, *Latimeria* uses urea and other osmolytes in osmoregulation and has a spiral valve in the intestine. One feature not found either in elasmobranchs or modern tetrapods is a joint over the brain that allows the front part of the skull to flip upward when the mouth opens. The function of this joint is unknown. The female gives birth to live young from eggs the size of tennis balls. How these eggs get fertilized is a mystery, since the males lack copulatory organs. Unfortunately, coelacanths live for less than a day after being brought to the surface, so many details of physiology are unclear. Coelacanth behavior is a little better known since submersibles have been used to observe them several hundred meters deep off the Comoro Islands (Fricke 1988). They are nocturnal and for some reason spend minutes at a time swimming upside down, backward, or head down. Most interesting to those who regard crossopterygians as the ancestors of tetrapods is the pattern of movement by the lobed fins. Contrary to expectation, coelacanths were not seen crawling on the bottom with their fins. When they swim, however, they alternate their fins with a crawling motion, instead of moving them simultaneously like most other fishes.

Bichirs and Reedfishes. Eleven extant species of bichirs and reedfishes (subclass Brachiopterygii) constitute the third major group of lobe-fins. Bichirs and reedfishes share many features with ray-finned fishes and are sometimes classified with them. All inhabit stagnant ponds and streams in Africa, where they get most of their oxygen by gulping air into lungs, rather than by using their gills. They propel themselves along the bottom and through vegetation with their lobed fins and can also feed on land.

RAY-FINNED FISHES

Paddlefishes and Sturgeons. The largest subclass of bony fishes, the Actinopterygii, are characterized by fins reinforced by bony rays that are not attached to the rest of the skeleton. The ray-finned fishes include more than 18,000 species that have adapted in virtually every imaginable way to almost every aquatic habitat. Some of the orders of ray-finned fishes are represented in Figure 35.14. One distinctive group is the superorder Chondrostei, which comprises the paddlefishes and sturgeons (Figure 35.15). (Many taxonomists also include the bichirs and reedfishes in this group.) As a group chondrosteans are quite ancient and retain many features that are considered primitive. Most of the 25 living members have shark-like features such as ventral mouths, spiracles, spiral valves in the intestine, unfused upper jaws, and largely cartilaginous skeletons. Unlike most sharks, however, adult chondrosteans have small teeth or none at all. Sturgeons suck up invertebrates from the bottom, and paddlefishes filter feed by trapping plankton with **gill rakers** (see Figure 13.1B). Chondrosteans also differ from sharks in having a cover (**operculum**) over the gill openings.

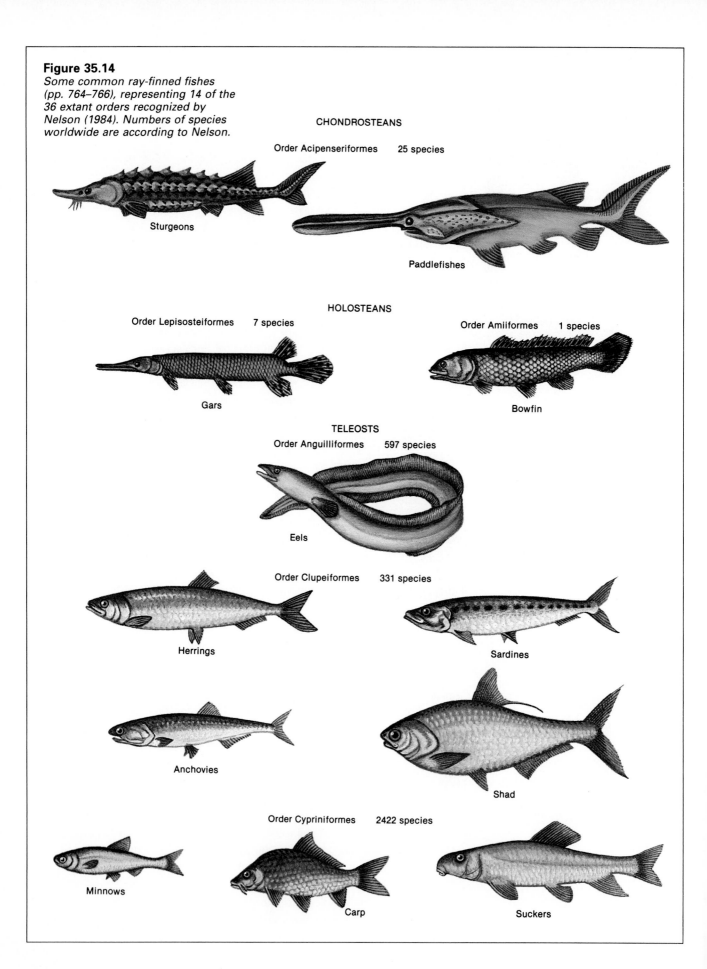

Figure 35.14
Some common ray-finned fishes (pp. 764–766), representing 14 of the 36 extant orders recognized by Nelson (1984). Numbers of species worldwide are according to Nelson.

CHONDROSTEANS

Order Acipenseriformes   25 species

Sturgeons

Paddlefishes

HOLOSTEANS

Order Lepisosteiformes   7 species

Gars

Order Amiiformes   1 species

Bowfin

TELEOSTS

Order Anguilliformes   597 species

Eels

Order Clupeiformes   331 species

Herrings

Sardines

Anchovies

Shad

Order Cypriniformes   2422 species

Minnows

Carp

Suckers

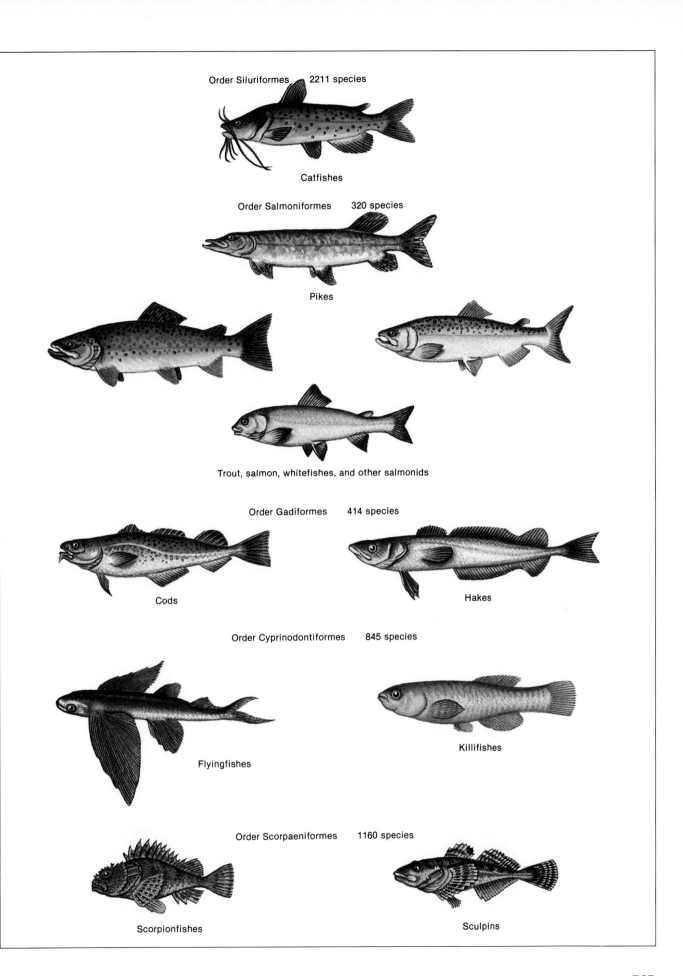

Order Siluriformes   2211 species

Catfishes

Order Salmoniformes   320 species

Pikes

Trout, salmon, whitefishes, and other salmonids

Order Gadiformes   414 species

Cods

Hakes

Order Cyprinodontiformes   845 species

Flyingfishes

Killifishes

Order Scorpaeniformes   1160 species

Scorpionfishes

Sculpins

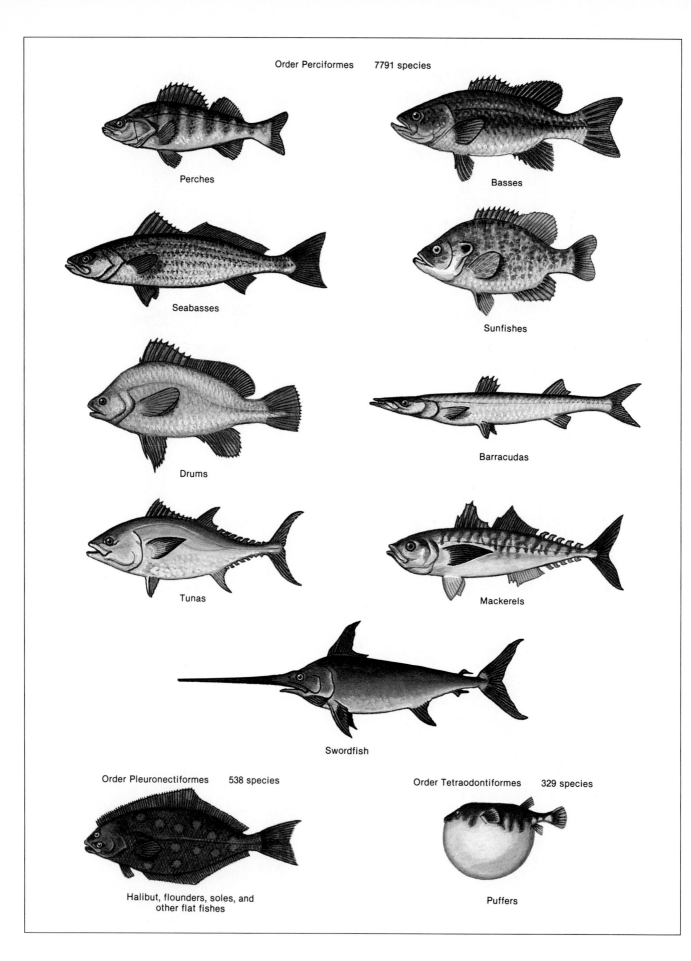

Order Perciformes   7791 species

Perches

Basses

Seabasses

Sunfishes

Drums

Barracudas

Tunas

Mackerels

Swordfish

Order Pleuronectiformes   538 species

Halibut, flounders, soles, and
other flat fishes

Order Tetraodontiformes   329 species

Puffers

Figure 35.15
A lake sturgeon Acipenser fulvescens. *These fishes attain lengths of up to 2.4 meters. Schools of them may account for persistent stories of "monsters" in Lake Champlain and other freshwater lakes.*

Bowfin and Gars. Considered somewhat less primitive are the bowfin, *Amia calva*, and gars (Figure 35.16). Both kinds of fishes have spiral valves in their intestines and can breathe air, especially when high water temperatures increase metabolic rates and decrease the amount of oxygen dissolved in water. Because they share these structures they are often referred to collectively as holosteans, but bowfins differ from gars in many ways. Bowfins can breathe by gulping air into its lung, while gars extract oxygen from a pair of swim bladders, which are not connected to the esophagus. Gars also use their swim bladders to maintain neutral buoyancy, as do most other ray-finned fishes. Bowfins and gars also differ in the kinds of scales they have. The bowfin has thin, round, overlapping **cycloid scales,** and gars have thick, nonoverlapping **ganoid scales** (Figure 35.17).

Figure 35.16
A Florida gar Lepisosteus platyrhincus *hangs motionless in a sluggish stream. Neutral buoyancy, due to the swim bladders, is essential for the "lie-in-wait" style by which gars prey on other fishes. Note the diamond-shaped ganoid scales.*

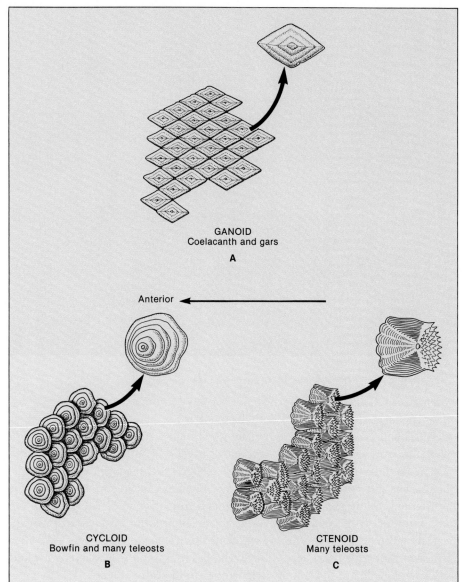

Figure 35.17
*Types of scales in ray-finned fishes.
(A) Ganoid scales are thick and
nonoverlapping and are composed
of bone overlaid with an enamellike
substance called ganoin. (B) Cycloid
scales are thin and overlapping,
permitting more flexibility. Unlike
ganoid scales, cycloid scales grow
as the fish grows and, in some
species, show annual growth rings.
(C) Ctenoid scales are essentially
cycloid scales with teeth at the
posterior edges. The teeth are
believed to reduce hydrodynamic
drag during swimming. Some
fishes, such as paddlefishes,
catfishes, and eels, sacrifice the
protection afforded by scales in
favor of the increased flexibility of
having none.*

TELEOSTS

The remaining ray-finned fishes are extremely diverse and undoubtedly polyphyletic but are commonly united under the name teleosts. There is only enough space here for a pictorial survey of some of the familiar teleosts (Figure 35.14), a more detailed discussion of the yellow perch *Perca flavescens* as a representative teleost, and comments on some of the most striking departures from the perciform model. Special adaptations of teleosts will be noted in later sections.

Fins. Like most ray-finned fishes, perches have fins supported by calcified **fin rays** (Figures 35.3B and 35.18). In the perch the fin rays are either stiff and unjointed spines, as in the anterior dorsal fin, or flexible and branching, as in the posterior dorsal fin. These dorsal fins, as well as the single anal fin, have limited mobility and mainly help keep the fish from rocking and turning when the caudal fin moves. (Some other teleosts, such as trout, also have an **adipose fin** behind the dorsal fin. The adipose fin is mainly an energy reserve.)

The pectoral and pelvic fins are more mobile and help control vertical move-

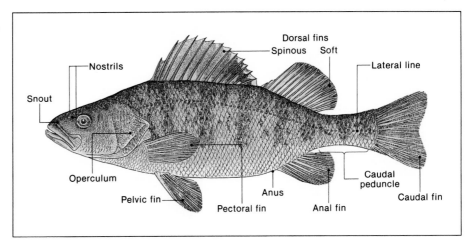

Figure 35.18
External structure of the yellow perch.

Labels in figure: Snout, Nostrils, Dorsal fins, Spinous, Soft, Lateral line, Operculum, Pelvic fin, Pectoral fin, Anus, Anal fin, Caudal peduncle, Caudal fin

ment, turning, and sudden stopping. In perches the pelvic fins are close to the pectoral fins, an arrangement that is considered more advanced than in most other teleosts, especially lobe-fins, chondrosteans, and holosteans. The pectoral fins also provide most of the forward thrust during slow swimming; the caudal fin is used for acceleration and cruising. As in most teleosts, the dorsal and ventral lobes of the caudal fin are similar in size, and the vertebral column does not extend into either lobe. Such symmetric caudal fins are termed **homocercal,** in contrast to the heterocercal caudal fins of sharks (Figure 35.7) and sturgeons, and the diphycercal caudal fins of lobe-finned fishes (Figure 35.13).

Gills. Just anterior to the pectoral fin on each side is the operculum, which covers the four gills. When swimming rapidly fishes simply open the mouth and allow "ram-jet" ventilation to drive water past the gills and out the operculi. Tunas and mackerels are, in fact, obliged to use **ram ventilation** and must cruise continually in order to obtain oxygen. When stationary or swimming slowly, perches and other teleosts ventilate their gills by **buccal pumping,** which involves compressing the **buccal** (mouth) **cavity** and the **opercular** (gill) **cavity** in sequence (see pp. 750–751).

Feeding. Perches and other teleosts also use buccal and opercular pumping to "slurp" up plankton and small invertebrates and fishes. This **suction feeding** is aided by a complicated arrangement of jaw bones that enables both the upper and lower jaws to swing open and forward with incredible speed (Figure 35.19). (Teleosts that prey on large fishes, eat coral, or scrape food off rocks generally have a firm, nonprotrusible upper jaw, as do the more primitive teleosts.) Numerous small teeth fused to the bones of the jaw enable the perch to catch and hold food. Additional teeth on the pharynx and gill rakers aid in holding and crushing.

Receptor Organs. Prey are detected by a variety of receptor organs, including the relatively large, color-sensitive eyes, and chemoreceptors in the **olfactory sacs.** In various other teleosts the olfactory sacs are also used for chemical communication by means of **pheromones.** As will be discussed later, salmon use olfaction to navigate back to their native streams for spawning. One or two nostrils on each side of the head allow water to flow into and out of olfactory sacs. In a few teleosts the nostrils connect to the pharynx, enabling the fishes to breathe while most of the body is buried in mud. Another important receptor organ is the lateral line organ (see p. 779) whose position is shown by the lateral line faintly visible on the side.

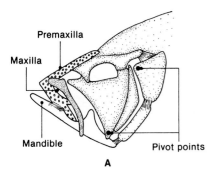

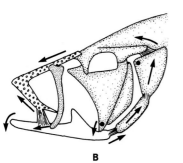

Figure 35.19
Mechanism of jaw opening and protrusion in the South American leaffish Monocirrhus polyacanthus *(order Perciformes). (A) Jaws closed. (B) Jaw opening and protrusion. Arrows show directions of movement. This is one of four mechanisms of jaw protrusion (Motta 1984).*

Labels: Premaxilla, Maxilla, Mandible, Pivot points, A, B

Integument. The entire body of the perch is covered with mucus-secreting epidermis. The mucus protects against abrasion and infection and probably reduces hydrodynamic drag. The epidermis also covers the **scales,** which grow out of the dermis, overlapping like shingles. Perches and many other teleosts have ctenoid scales, while other teleosts have cycloid scales or none at all. Unlike sharks, teleosts do not shed and replace their scales. Instead, the scales grow continually to compensate for wear. The **colors** that make fishes so attractive are not within the scales but are mainly due to special skin cells called **chromatophores** (see Table 10.1). The silvery and iridescent colors common among fishes are due to reflection from chromatophores that contain guanine. The major colors of the yellow perch are due to chromatophores that contain yellow and black pigments (carotenoid and melanin). In response to neural and hormonal signals, these pigments either condense or disperse within the chromatophores, reducing or intensifying the color. During courtship, for example, the pigments generally disperse, making the fish more intensely colored. In many teleosts the chromatophores are also used in mimicry, camouflage, or warning coloration (Figure 35.20).

The Swim Bladder. Like most teleosts, perches have a gas-filled swim bladder that maintains buoyancy (Figure 35.21). The swim bladder occupies about 7% of the body volume in the perch and other freshwater fishes, and about 5% in fishes that live in the more-buoyant salt water. The mechanism for filling the swim bladder will be described later. In many teleosts the swim bladder is close enough to the semicircular canals and otolith organs that it may help them respond to low-frequency sounds, by acting like an eardrum. In herrings a thin extension of the swim bladder contacts the semicircular canals and otolith organs, presumably transmitting sound. In some teleosts, such as carps and catfishes, certain bones

Figure 35.20
Uses of skin pigmentation in teleosts. (A) The gaudy colors of the lion fish Pterois volitans (order Scorpaeniformes) may serve as warning coloration, which predators learn to avoid after being stung by the venomous spines. This photo was taken at night. Like many coral-reef fishes, lion fishes are camouflaged during the day. Lion fishes use the feather-shaped spines to herd prey fish into crevices in coral reefs. (B) Chromatophores enable the eyed flounder Bothus mancus (order Pleuronectiformes) to camouflage itself by copying the color and pattern of the ocean floor on which it rests. Note the position of both eyes on the left side of the body. Most members of this species and most other species of flatfishes are "lefteyed" like this. Other species consist mainly of "righteyed" individuals. Young flatfishes are born bilaterally symmetric like most fishes, but gradually the mouth and eyes migrate to one side or the other, and the fish settles to the bottom.

A

B

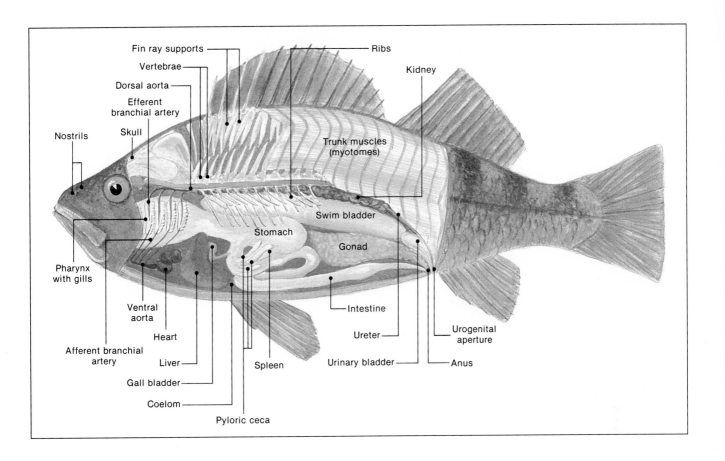

Figure 35.21
Internal structure of the yellow perch.

(**Weberian ossicles**) transmit vibrations from the swim bladder to the semicircular canals and otolith organs. Many fishes (orders Batrachoidiformes and Scorpaeniformes) also use the swim bladder to produce sounds. Drums and croakers have muscles that vibrate the swim bladder to produce sound, and grunts use the swim bladder to amplify sounds made by grinding teeth in the pharynx. These sounds are produced mainly by males during courtship and aggressive displays.

Other Internal Organs. The other internal organs of teleosts are typical for vertebrates (Figure 35.21). (Refer to the index for information from Unit Two on various aspects of vertebrate physiology.) Here we merely note several differences between perches and the sharks that were previously described. Perches and other teleosts differ from sharks in lacking spiral valves in the intestine. The yellow perch does, however, have three **pyloric ceca** that may serve the same function of increasing the duration of digestion and the area for absorption of nutrients. As a freshwater species the yellow perch also differs from the shark in its osmoregulatory problems. To compensate for ions that diffuse out of the body, the gills actively transport ions from the water into the blood. The mesonephric kidneys eliminate most of the excess water that enters through the gills and digestive tract by producing copious, dilute urine. Ammonia and other metabolic wastes are also eliminated through the urine, as well as by diffusion through the gills. (For osmoregulation among marine teleosts see p. 292.) Urine is kept in the urinary bladder until excreted through the **urogenital aperture** just anterior to the anal fin. This aperture is also used for the release of gametes. The yellow perch has separate sexes and external fertilization, but other teleosts have numerous interesting variations, some of which will be described later.

LOCOMOTION

Neutral Buoyancy. Although flounders, certain coral reef species, and a few other fishes specialize in resting or scrambling about on the substratum, most would be severely limited if they had to continually struggle to keep from sinking. The ability to exploit the entire volume of an aquatic habitat rather than only its bottom surface is due to neutral buoyancy. In teleosts neutral buoyancy is due to the swim bladder. Swim bladders probably originated in early fishes as pouches on the esophagus that served both as air reserves and as aids to buoyancy. Subsequently, these esophageal pouches specialized for one function or the other. In the evolutionary line that led to the lobe-finned fishes and to tetrapods, the pouches evolved into lungs, with inner surfaces specialized for transferring oxygen to the bloodstream. In teleosts the pouches evolved into swim bladders with linings that retain gases.

Filling the Swim Bladder. In some teleosts, such as trout, the swim bladder remains connected to the esophagus by a **pneumatic duct** and can be inflated by swallowing air. In most teleosts, however, connections to the esophagus are lost during embryonic development. How do these fishes inflate their swim bladders? How do deep-sea teleosts, especially, inflate their swim bladders under pressures of several tons per square centimeter? The answer lies in the **gas glands,** which pump gas, mainly oxygen, into the swim bladder (Figure 35.22). Each gas gland

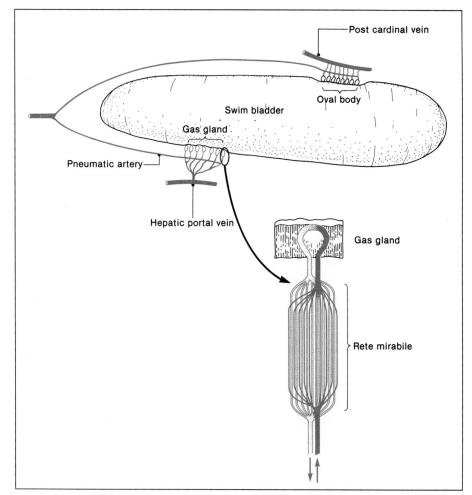

Figure 35.22
Schematic representation of the swim bladder of a perch, and mechanisms of filling and emptying it. The rete mirabile is a system of capillaries that concentrates gases in the blood, which are released into the swim bladder by the gas gland. The oval body serves as a release valve for excess pressure.

A

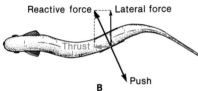

B

Figure 35.23
(A) Moray eel Helena muraena *demonstrating undulatory (anguilliform) swimming in a coral reef. Note the absence of pectoral fins, reduction of other fins, and the lateral compression of the body. Length approximately 1 meter. (B) Analysis of forces in anguilliform swimming. As a portion of the body swings left, it pushes against the water, which generates an equal and opposite reactive force on the fish. The reactive force has a lateral component that tends to push the fish sideways and a forward component—thrust—that propels it forward. A similar analysis applies to the caudal fins of other fishes.*

is supplied by a network of capillaries called a **rete mirabile** (pronounced REE-tee meer-AH-bill-lay). By means of a countercurrent mechanism (see p. 251), these retia concentrate blood gases, which the gas glands then release into the swim bladder. The gas glands increase the release of oxygen by secreting lactic acid, which causes the hemoglobin to release additional amounts of O_2. When fishes rise toward the surface the pressure on the swim bladder decreases, causing it to expand. Rupture of the swim bladder is avoided by an **oval body** that safely leaks excess gas back into the blood.

Swimming. Swimming is more than just hanging around in the water. It also requires the production of enough momentum that the viscosity of water does not stop forward movement between strokes. Some immature fishes are too short to sustain this forward momentum and are therefore either immobile or planktonic. However, even the smallest adult fish, a centimeter-long goby from the Philippines (*Pandaka pygmaea*), is long enough to swim. (See the discussion of Reynolds number, pp. 218–219.) Fishes swim in a variety of ways, and the variety of shapes of fins and bodies reflects adaptations to different styles of swimming, among other requirements for life. Eels and eel-shaped fishes use the entire body for propulsion, bending it in a series of tight undulations (Figure 35.23). Generally, the fins of such fishes are small, but the body is compressed from side-to-side and serves as one large fin.

The fins and bodies of other fishes are generally adapted for one or more types of swimming, such as maneuvering, cruising, or accelerating (Figure 35.24):

1. **Maneuvering** is important for fishes that inhabit coral reefs, weedy bottoms, and other restricted areas. Fishes adapted for maneuvering are generally short and can therefore turn easily. Forward thrust and turning are due largely to oarlike pectoral fins, rather than to the caudal fin, which is usually short. At slow speeds drag due to water pressure is negligible, so streamlining is not required. The main source of drag is the friction of water against the body surface. This friction is minimized by a spherical body shape, which

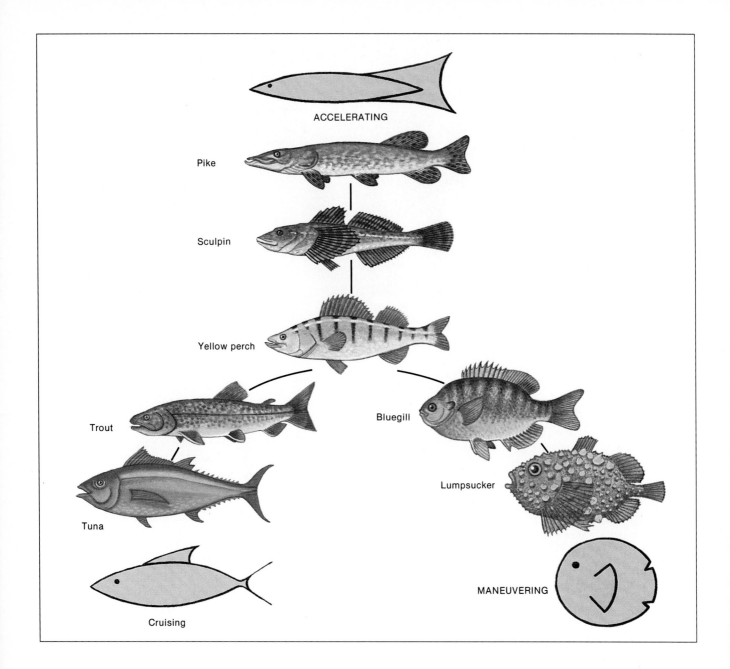

ACCELERATING

Pike

Sculpin

Yellow perch

Trout

Bluegill

Tuna

Lumpsucker

Cruising

MANEUVERING

has the least surface area for a given volume, and therefore the least friction.

2. Fishes adapted for **cruising** at steady, moderate speeds in the open ocean or lakes get most of their forward thrust from the caudal fin. However, the caudal fin and the caudal peduncle must be thin to minimize turbulence from side-to-side movements, and to reduce side-to-side movement of the body due to the lateral force. These fishes have massive, elongated bodies and large median fins that limit sideways movement during swimming. The larger surface areas of the body and fins increase drag due to friction, but friction drag during fast swimming is negligible compared to the drag due to water pressure against the front of the body. This pressure drag is reduced by streamlining.

3. Fishes adapted for **accelerating** also have streamlined bodies, but their caudal fins are larger and can therefore generate more thrust. Body mass (inertia) is reduced, permitting greater acceleration.

Figure 35.24
Specializations of fishes for maneuvering, cruising, and accelerating. Ideal shapes for each mode are shown. Most fishes, such as the yellow perch, have bodies and fins that compromise among the requirement for all three types of swimming, as well as for other functions (see Webb 1988).

Schooling.   Swimming efficiency may also be affected by schooling. The slime released from all the fish in a school appears to "lubricate" the water. It is also possible that maintaining a particular position in a school allows a fish to take advantage of water currents generated by others in the school. Fish in schools often synchronize their swimming with incredible precision, using their eyes and lateral line organs to sense the position and speeds of their neighbors. The contribution of schooling to locomotion is probably less important than its contribution to avoiding predation. A predator has a harder time selecting and catching one fish out of a school and is more likely to be spotted by a school of fish than by a single fish.

Muscles.   Most of the power for cruising comes from darker muscles that produce ATP aerobically. These muscles are made up of **slow-twitch fibers** (type I red fibers; see p. 214). In some fishes, such as mackerel, salmon, and carp, red fibers mingle with the white fibers in the myomeres. In tunas, striped bass, and some others, red fibers occur in lengthwise bundles outside the myomeres (see Figure 15.6). The power for acceleration comes mainly from the segmental blocks of muscles called **myomeres** (Figures 35.21 and 35.25). These muscles envelope the viscera, and in some fishes, such as tunas and pikes, they account for more than half the body mass. The myomeres generally consist mainly of **fast-twitch fibers** (type II white fibers; see p. 214). Although these white muscles contract rapidly, they tire quickly because they produce ATP by the anaerobic breakdown of glycogen into lactic acid.

In bony fishes the myomeres are anchored to spines and ribs on the vertebrae. When the myomeres contract the vertebral column bends, flexing the caudal peduncle and caudal fin. A horizontal septum separates the dorsal from the ventral myomeres on each side. Adjacent myomeres nest against each other and are held together by septa. (These septa dissolve when a fish is cooked, causing the myomeres to separate into flakes.) Each myomere has a complex shape, folding forward and backward in such a way that a transverse section cuts through several different myomeres. The overlapping arrangement of myomeres may allow the contraction to be distributed smoothly over a greater length of body. Within each myomere individual muscle fibers are not aligned lengthwise along the body. In fact, their orientation is so complex that it is not clear how their contraction bends the vertebral column.

MIGRATION AND HOMING

Eels.   One of the most important and fascinating uses of locomotion in fishes is in migration. Many species of fishes, such as shad, salmon, trout, and eels, are as impressive as birds in the distances over which they migrate and the precision with which they navigate. Every American eel (*Anguilla rostrata*) and European eel (*A. anguilla*) begins its life in the calm, clear Sargasso Sea south and southeast of Bermuda, migrates into freshwater streams in either America or Europe, then returns (**homes**) to the Sargasso Sea years later to reproduce (Figure 35.26). Hundreds of meters beneath the floating sargassum seaweed that accumulates there, each female deposits approximately a million and a half eggs, males fertilize them, and then the adults die. After two days each egg hatches into a hermaphroditic larva only a few millimeters long. The larva is so unlike an adult eel that it was once classified in a completely different genus, *Leptocephalus*, and it still goes by that name. Too short to swim, the leptocephalus larvae drift into currents that

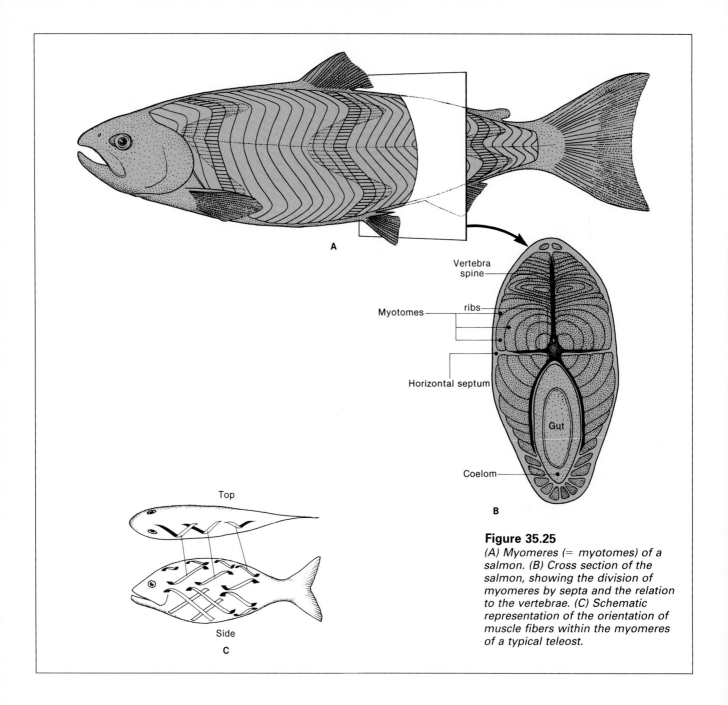

Figure 35.25
(A) Myomeres (= myotomes) of a salmon. (B) Cross section of the salmon, showing the division of myomeres by septa and the relation to the vertebrae. (C) Schematic representation of the orientation of muscle fibers within the myomeres of a typical teleost.

will eventually carry them to either the coast of America or to the coast of Europe. The American and European species mingle randomly at first, yet after a year and a half the American eels have somehow found their way to American estuaries, and approximately two years later the European eels arrive at European estuaries.

During migration the leptocephali metamorphose into more eellike juveniles called **elvers.** Upon arrival in estuaries the osmoregulatory systems of the elvers gradually adapt to less saline waters. Some elvers remain in the estuaries and after several years change from hermaphrodites to males. Other elvers, although apparently identical to those that remain in the estuaries, migrate upstream in the spring. Often they travel for several hundred kilometers, sometimes over land on moist nights, breathing through their gills and skin. Once they find a suitable location these elvers will spend the next 6 to 15 years there, eating frogs and invertebrates by day in the warm months, and remaining buried in mud during

Figure 35.26

Migrations of American and European eels. Adults of both species migrate to the Sargasso Sea where they breed and die. Leptocephalus larvae then migrate to American (blue arrows) or to European (red arrows) estuaries, where they become elvers. Female elvers migrate up streams. After several years the elvers mature into adults, which complete the cycle. The only morphological difference between the two species is in the average number of vertebrae (average of 107 in the American eel versus 114 in the European). Comparisons of mitochondrial DNA, however, show that the two species do not interbreed, even though their breeding areas in the Sargasso Sea overlap (Avise et al. 1986).

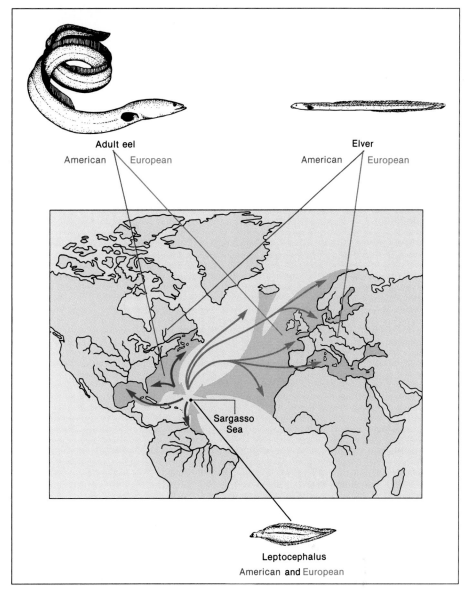

Adult eel
American   European

Elver
American   European

Sargasso
Sea

Leptocephalus
American **and** European

the cold. Finally, they all mature into females and, in the autumn, return downstream to the estuary where males join them for the return voyage to the Sargasso Sea. How eels navigate thousands of kilometers of open ocean, homing back to the Sargasso Sea, is a mystery. Some fishes can apparently determine direction from the position of the Sun, but adult eels generally migrate in the total darkness of deep water. Perhaps, like some other fishes, eels can detect Earth's magnetic field.

Eels and approximately 40 other species that hatch in seawater and migrate up freshwater streams to feed are said to be **catadromous.** For catadromous species the greatest navigational challenge is to find an area in the vast ocean where they can successfully reproduce. Eighty-seven other species, such as lampreys, shad, salmon, and some trout, are **anadromous:** They hatch in freshwater streams and migrate downstream to feed in the ocean. For anadromous species migration would appear to be easier, since all they have to do to find a breeding area is to enter the mouth of a stream and keep going against the current. This approach is risky, however, because many streams dry up, become laden with silt, or prove unsuitable for reproduction in some other way that a fish has no way of predicting.

Salmon. Perhaps because some streams prove unsuitable for reproduction, many anadromous fishes have evolved a homing ability that guides them to a stream that has proved suitable in the past—namely, the stream in which they themselves hatched. Approximately 95% of adult coho salmon *Oncorhynchus kisutch* return to spawn in the same rapid, gravel-bottomed streams in which they hatched two or three years earlier. On arriving in November, a male and female engage in a ritualized mating dance. The female then uses her caudal fin to scoop out a nest (a **redd**) into which she deposits her eggs. The male then fertilizes the eggs, and the female covers them with gravel. Exhausted, both male and female soon die. Around two months later the eggs hatch. The hatchlings, or **alevins,** remain under gravel for several months, until the nutrient supply in their yolk sac is depleted. Then the fish, now called **fry,** emerge from the nest and establish a territory in which they feed on plankton and insects that drift downstream. During a new moon in the spring of the second year of life, a surge in the levels of the thyroid hormones thyroxine and triiodothyronine triggers metamorphosis into **smolts** (Grau et al. 1981). During the smoltification process the osmoregulatory systems prepare for seawater. The smolts swim en masse downstream to the ocean, where most will feed for a year and a half. At the end of that period they stop feeding and begin to migrate back to their native streams, perhaps using geomagnetic cues to navigate. They remain at the mouth of the correct river for about a month while their osmoregulatory systems readjust to fresh water and their gonads develop. At this point they acquire a red mating coloration and distinctive hooked jaws. They then continue migrating upstream to the breeding area where they began life.

How do salmon find their way up rivers and tributaries to the correct breeding area? An answer to this question occurred to Arthur D. Hasler of the University of Wisconsin during a visit to a mountain slope he had known as a boy. "As I hiked along a mountain trail in the Wasatch Range of the Rocky Mountains where I grew up, my reflections about the migratory behavior of salmon were soon interrupted by wonderful scents that I had not smelled since I was a boy. . . . [S]o impressive was this odor that it evoked a flood of memories of boyhood chums and deeds long since vanished from conscious memory. The association was so strong that I immediately applied it to the problem of salmon homing." (Hasler and Scholz 1983, p. xii.)

Hasler began to wonder whether salmon might be **imprinted** to the odor of their native stream while they are still smolts, then home in on that odor as adults. For several decades Hasler and his colleagues and students conducted experiments to test the theory. First they showed that coho salmon could indeed be conditioned to distinguish between water from two different streams, and that they lost this ability if their olfactory sacs were destroyed, or if organic molecules were filtered from the water. Hasler and his co-workers also trapped and tagged migrating salmon in two branches of a Y-shaped stream, plugged the nostrils of some of them, then released the salmon downstream below the junction of the two branches. Nearly all the fishes with unplugged nostrils were later recovered in the same branch in which they had been trapped originally, but the salmon with plugged nostrils were much less likely to make it back to the correct branch of the stream. Conclusive evidence came from studies in which smolts from hatcheries were exposed to one of two artificial substances. When these fishes were ready to migrate, one of these substances was added to one stream, and the other substance was added to another stream. Although the salmon could have entered any one of more than a dozen streams, the great majority migrated up the stream containing the substance to which they had been imprinted as smolts.

ELECTRORECEPTION

The mechanism by which many fishes navigate over long distances may be electroreceptors that are sensitive to small voltages. Some electroreceptors can respond to a voltage gradient as small as 0.005 microvolt per centimeter, which is equivalent to connecting one terminal of an ordinary 1.5-volt battery to San Francisco Bay and the other at Victoria, British Columbia. Such sensitivity would enable the fish to navigate by detecting the voltage induced as its body moves through Earth's magnetic field. Electroreceptors could also enable a fish to detect the voltage generated by ocean currents. In addition, experiments show that electroreceptors can detect the muscle potentials generated by prey and the electrical signals certain species use to communicate or to locate nearby objects.

Electroreceptors are generally found in the lateral line organ of the fishes that have them. There are three fundamental kinds of electroreceptors (Bullock 1982). (1) The **ampulla of Lorenzini** appears to be the most primitive and is never found in teleosts. Ampullae respond to low-frequency voltage changes (less than 50 Hz), such as would be generated by geomagnetism and ocean currents. (2) Similar, but probably not homologous, ampullae occur in certain teleosts. These **ampullae of teleosts** also respond to low frequencies and might be used for navigation. (There is no evidence, however, that eels or salmon have them.) (3) Most of the teleosts with ampullae also have **tuberous organs.** Tuberous organs respond to high-frequency signals that fishes of the same species produce to locate nearby objects and to communicate.

ELECTRIC ORGANS

Fishes that can generate appreciable voltages are called electric fishes. These voltages come from electric organs that are generally derived from muscles that have lost their ability to contract. The voltage produced by each cell is an action potential no larger than that in an ordinary muscle cell (approximately 0.1 volt), but hundreds or thousands of cells in the electric organ are arranged in a series that allows the voltages to summate. In strongly electric fishes the electric organ discharge is up to several hundred volts and is used to stun or kill prey (Figure 35.27). It is not clear how these fishes avoid electrocuting themselves. Electric organs generate much lower voltages in weakly electric fishes, such as knifefishes (family Gymnotidae) and elephantfishes (family Mormyridae). Gymnotids and mormyrids are nocturnal and live in murky streams in South America and Africa, respectively. They use their electric organ discharges to locate nearby objects, by detecting the distortions in the electric fields produced by the objects. These gymnotids and mormyrids also use the electric organ discharges to communicate with other members of their own species. Although many related species may inhabit the same streams, they can distinguish one species from another, and many can discriminate age and sex on the basis of the electric organ discharge.

REPRODUCTION

One can find among fishes virtually every reproductive adaptation imaginable, and a few that are almost unimaginable. Like the yellow perch, most species have separate sexes, although it is seldom easy to distinguish them visually. Most teleosts release the eggs for external fertilization. Some, such as tuna, sardines, whitefish, and certain shads (*Alosa*), are mass spawners, releasing their gametes within large schools. Others choose particular mates, which they recognize by

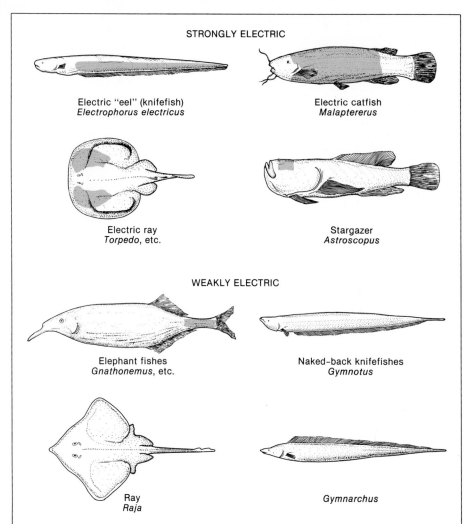

STRONGLY ELECTRIC

Electric "eel" (knifefish)
Electrophorus electricus

Electric catfish
Malaptererus

Electric ray
Torpedo, etc.

Stargazer
Astroscopus

WEAKLY ELECTRIC

Elephant fishes
Gnathonemus, etc.

Naked-back knifefishes
Gymnotus

Ray
Raja

Gymnarchus

Figure 35.27
Electric fishes. Strongly electric fishes have large electric organs that generate tens of volts used to stun or kill prey. Weakly electric fishes use their smaller electric organs for communication and location of objects. The diversity of electric organs in several distinct groups of fishes indicates that they evolved independently several times.

visual cues such as the belly of the female swollen with eggs or the bright coloration due to chromatophores. Others recognize potential mates by pheromones, and, as noted previously, males and females of some species literally electrify each other with electric organ discharges. Once potential mates recognize each other, courtship may ensue. Courtship often involves a stereotyped swimming display, such as that of the three-spined stickleback (see pp. 422–423). The display often climaxes in a nest, where the female releases her eggs and the male fertilizes them. In other species the eggs may be attached to vegetation or, as in salmon, hidden under gravel.

Following external fertilization most fishes simply swim off, but others provide some degree of support and protection for the eggs and hatchlings. In some species one or both parents remain near the nest fending off the numerous potential predators that feed on fish eggs and fry. Some aerate the eggs by fanning water currents over them with their fins. The greatest support is provided by various species that brood the eggs and sometimes the fry in or on their bodies. **Mouthbrooders** take the eggs into their mouths (see p. 430), and some allow the fry to take refuge from danger within their mouths. Others brood eggs in their gill chambers or on their skin. Perhaps the most unexpected method of brooding occurs in seahorses and some pipefishes and catfishes. In these species the sex roles seem to be reversed. The female has a penislike organ with which she depos-

Figure 35.28
Mating in the seahorse Hippocampus *(order Syngnathiformes). The female (left) is depositing eggs into the marsupium of the male.*

its eggs into a pouch (**marsupium**) on the belly of the male, which then fertilizes and broods them (Figure 35.28). In some species the male pouch even has a "placenta" with which he provides nourishment to the eggs.

Other teleosts depart radically from the dioecious, externally fertilizing "norm." Like sharks and coelacanths that were mentioned previously, some are truly viviparous, with the eggs being retained in the ovary or oviduct of the female and supplied with nutrients and oxygen. Among these live-bearing fishes are surfperches and the famous aquarium guppy *Poecilia reticulata*. In these species, fertilization is, of course, internal. Some live-bearing fishes have eliminated the males altogether and have become parthenogenetic and unisexual. Other teleosts are hermaphroditic. Some, such as most sea basses, are **simultaneous hermaphrodites.** They can release both eggs and sperm and alternate between serving as male and female several times during a single mating. Other teleosts are **sequential hermaphrodites,** which change permanently from one sex to another. The sex change is generally triggered by some environmental cue, such as a low number of males in a school (see p. 315). These fishes have generated considerable research in behavioral ecology, since they provide a means of studying the circumstances under which it is reproductively advantageous to be one sex or the other.

INTERACTIONS WITH HUMANS AND OTHER ANIMALS

Ecology. For most aquatic ecosystems it is probably safe to say that removing all the fishes would affect every other member of the community. Many fishes serve vital roles in food webs by eating plankton and channeling the energy and nutrients to higher trophic levels, often other species of fishes. Many birds, mammals, and other animals that are not members of aquatic communities also depend heavily on fishes for food. Fishes also interact symbiotically with other species. Two beautiful examples have been noted elsewhere: that of anemone fishes with sea anemones (see Figure 24.15), and cleaner fishes with various other species of fishes (see p. 329). Another fascinating symbiosis is between various species of **flashlight fishes** and the luminescent bacteria in their light organs (Figure 35.29).

Figure 35.29
The flashlight fish Anomalops katoptron *(order Beryciformes), a nocturnal species of Pacific coral reefs. Every few seconds the fish flashes the light organ by exposing it. Flashlight fishes are generally benthic or nocturnal and use the luminescent organs for communication. They may also use the light organs to attract and detect prey. The light organs are usually located near the eyes, which is the optimal location for detecting prey by means of reflected light.*

Benefits to Humans. Fishes benefit humans in a variety of ways, not least of which is the pleasure of watching and learning about them. Hobbyists spend million of dollars and countless hours annually on exotic fishes and aquarium supplies. Unfortunately, this hobby can also exact a high cost to the fishes as well, since collectors often use cyanide, which kills nine fish for every one collected. Sport fishermen derive as much pleasure and spend even more money on their hobby, again at high cost to the fish. Humans also benefit from unique fish products. One example is tetrodotoxin from pufferfish, which is widely used in neurophysiological research and as a local anesthetic (see p. 144). Researchers in the United States and the Middle East are now studying the healing properties of slime from certain catfishes. Arab fishermen have long been in the habit of rubbing this slime on cuts, and scientists have now extracted from it a variety of components that stop bleeding and promote tissue growth.

Fish as Food. Of course the most common benefit from fishes is food. Fish is becoming more popular as people become increasingly concerned about the fats and cholesterol in red meats. Americans apparently like their fish to look like fish and therefore prefer ordinary teleosts such as tuna, cod, salmon, and sardines. Sharks are also enjoyed, and Americans will even eat rays as long as the meat is cut into plugs labeled "sea scallops." Tastes are more varied elsewhere. In Europe eels are popular, and caviar (sturgeon eggs) are considered the ultimate luxury food. Fish have always been assumed to be as plentiful as . . . well, the fish in the sea, but modern fishing methods now threaten to deplete even this seemingly endless supply of food. Fishing fleets from the Far East ply the Pacific and the Atlantic with drift nets up to 30 miles long, taking various species of fishes, molluscs, and other seafoods and also killing numerous inedible turtles, birds, and other animals. Some species of sharks are now threatened by the practice of "finning," in which the fins of sharks, which contain a noodlelike cartilage much prized in Asia, are cut off, and the rest of the shark is dumped overboard to bleed to death.

Impact of People on Fishes. Considering the numerous benefits derived from fishes, one might think humans would take better care of the resource. Yet humans have always shown a disregard for fishes, mainly by showing a disregard for their aquatic habitats. When settlers arrived in America, alewives, menhaden, and other fishes were so plentiful that a thousand fish cost only a dollar, and they were used as fertilizer. Mill dams then stopped many of the spawning runs. Then came sawdust and silt from mills and mines, erratic water levels due to clearing of forests, and fertilizers and pesticides from farming. Canals and misguided stocking compounded these problems by introducing exotic species such as the sea lamprey. More recently, air pollution has made many lakes and streams too acidic to support fish, and polychlorinated biphenyls (PCBs) and other toxic wastes have either killed the fish or made them unsafe to eat. In many parts of the country people are warned not to eat more than one meal per month of locally caught fish. Even large bodies of water like the Chesapeake Bay and offshore areas, once considered bottomless dumps for garbage, toxic wastes, and sewage sludge, are now yielding reduced catches and fishes that are unhealthy and unhealthful.

Many government and other groups have attempted to eliminate some of these problems by reducing pollution and by such measures as building fish ladders to allow migration past dams. In most cases, however, it has proved cheaper and more politically acceptable to apply piecemeal remedies rather than treat underlying problems. In order to satisfy continuing demand for edible fish, **aquaculture** has become increasingly common. In the United States virtually all the rainbow trout and catfish sold now come from fish farms. Many of these fish farms employ the most advanced biological technology. Some trout farms, for example, produce only females, which look and taste better than males, since they don't waste energy in aggressive encounters. In order to get sperm for reproduction, some eggs are treated with testosterone, which causes the embryos to develop into trout that are physiologically male, though still genetically female. Since the sperm have only X chromosomes, all the fertilized eggs develop into females.

Biotechnology is also widely employed to satisfy sport fishermen and the people who depend on them economically. Often the fishes raised in hatcheries and used to stock streams bear little resemblance to native species, since they have been selected, hybridized, and genetically manipulated to produce large size, better tolerance to deteriorated habitats, and better "return" (meaning they are easier to catch). Generally, the eggs are heat-shocked or otherwise treated to interfere with meiosis, so that they are diploid and produce triploid offspring that grow up sterile. Thus if stocking turns out to damage a habitat, the damage will be reversible. Techniques of genetic engineering promise further "improvements" of this kind.

SUMMARY

Agnathans—hagfishes and lampreys—are often considered to be fishes, but they lack jaws and vertebrae. Primitive agnathans evolved into the true fishes with jaws, which are classified either in class Chondrichthyes or class Osteichthyes. Class Chondrichthyes comprises sharks, rays, and chimaeras, which have cartilaginous skeletons. Sharks are also distinguished by having placoid scales, some of which continuously form teeth, and by having open gill slits. Class Osteichthyes comprises the bony fishes, in which bone replaces much of the cartilage during development, and the gills are covered by an operculum. One bony fish, perhaps a lobe-finned fish, is believed to have evolved into the tetrapods. Most fishes—and indeed most vertebrates—are bony fishes called teleosts.

Fishes owe their success largely to being efficient swimmers. The force used in swimming comes mainly from the caudal fin, which is heterocercal in sharks, diphycercal in lobe-finned fishes, and usually homocercal in bony fishes. The pectoral fins provide some forward thrust as well, and dorsal and pelvic fins provide directional stability. The shapes of the fins and body are adapted in different fishes for cruising, accelerating, or maneuvering. Sharks and tuna tend to have narrow caudal peduncles and fins for efficient cruising. Fishes that lunge after prey, such as pike, tend to have broader tails. Fishes that maneuver among coral reefs or grasses, such as butterfly fish, tend to be spherical or disc shaped.

The shapes of scales, whether placoid, ganoid, cycloid, or ctenoid, also influence the amount of friction during swimming, as does the secretion of mucus. Neutral buoyancy is also important in swimming, since the fish does not have to expend energy to keep from sinking. Sharks maintain low body density by means of the cartilaginous skeleton and squalene, and most bony fishes use a swim bladder that is inflated by a gas gland.

Fish detect prey, communicate, and navigate by a variety of receptors, including chemoreceptors on the body surface and in olfactory pits, mechanoreceptors and electroreceptors in the lateral line organ, and the eyes. Some fishes can generate electricity for communication, navigation, and predation. Some, such as eels and salmon, can use Earth's magnetic field, electric currents, chemicals, or other cues to migrate long distances, and even to home to their native waters.

Most fishes are dioecious, fertilize the eggs externally, and provide little care for eggs and young. In some, however, fertilization is internal, and the young may be born alive. Some tend the eggs and young in the nest, or even in the mouth. Some are sequential or simultaneous hermaphrodites, and some are parthenogenetic.

KEY TERMS

vertebral column	cycloid scale	ram ventilation
caudal fin	ctenoid scale	buccal pumping
ammocoete	spiral valve	swim bladder
caudal peduncle	lateral line organ	gas gland
neutral buoyancy	operculum	catadromous
myomere	heterocercal	anadromous
placoid scale	diphycercal	ampulla
ganoid scale	homocercal	tuberous organ

SELF-TEST

1. Describe a major distinction of each of the following groups, and give the common name of one member of each group: Agnatha, Gnathostomata, Chondrichthyes, Osteichthyes, Elasmobranchii, Holocephali, Crossopterygii, Actinopterygii, chondrosteans, holosteans, teleosts.

2. For each of the following groups, describe one trait that is considered to be primitive: hagfishes, sharks, lobe-finned fishes.

3. Describe the following kind of scales and name a fish that has each kind: placoid, ganoid, cycloid, ctenoid. What advantage does each kind of scale confer to the fish?

4. Describe some major ways in which the physiology and life cycle of the sea lamprey differ from that of the American eel.

5. Why is it important for a fish to maintain neutral buoyancy? Explain how neutral buoyancy is maintained in a shark and in a bony fish.

6. For each of the fishes shown in photographs in this chapter, state on the basis of its fins and body whether it specializes in maneuvering, cruising, or accelerating.

7. Describe the differences in osmoregulation between the dogfish shark and the yellow perch.

8. Describe the structure and functioning of lateral line organs, electroreceptors, and olfactory receptors in fishes. Describe one way in which each kind of receptor organ is used.

READINGS

RECOMMENDED READINGS

Bass, A. H. 1990. Sounds from the intertidal zone: vocalizing fish. *BioScience* 40:249–258.

Boschung, H. T., Jr. et al. 1983. *The Audubon Society Field Guide to North American Fishes, Whales, and Dolphins.* New York: Alfred A. Knopf.

Brown, J. H. 1971. The desert pupfish. *Sci. Am.* 225(5):104–110 (Nov).

Coutant, C. C. 1986. Thermal niches of striped bass. *Sci. Am.* 255(2):98–104 (Aug).

Donaldson, L. R. and T. Joyner. 1983. The salmonid fishes as a natural livestock. *Sci. Am.* 249(1):51–58 (July).

Eastman, J. T. and A. L. DeVries. 1986. Antarctic fishes. *Sci. Am.* 255(5):106–114 (Nov).

Fricke, H. 1988. Coelacanths: the fish that time forgot. *Natl. Geogr.* 173(6):824–838 (June).

Hobson, E. S. 1975. Feeding patterns among tropical reef fishes. *Am. Sci.* 63:382–392. (*With beautiful color photos.*)

Leggett, W. C. 1973. The migrations of the shad. *Sci. Am.* 228(3):92–98 (Mar).

Leggett, W. C. 1990. The spawning of the capelin. *Sci. Am.* 62(5):102–107 (May).

Levine, J. S. and E. F. MacNichol Jr. 1982. Color vision in fishes. *Sci. Am.* 246(2):140–149 (Feb).

McCosker, J. E. 1977. Flashlight fishes. *Sci. Am.* 236(3):106–114 (Mar).

Partridge, B. L. 1982. The structure and function of fish schools. *Sci. Am.* 246(6):114–123 (June).

Pietsch, T. W. and D. B. Grobecker. 1990. Frogfishes. *Sci. Am.* 262(6):96–103 (June).

Policansky, D. 1982. The symmetry of flounders. *Sci. Am.* 246(5):116–122 (May).

Warner, R. R. 1984. Mating behavior and hermaphroditism in coral reef fishes. *Am. Sci.* 72:128–136. (*With beautiful color photos.*)

Webb, P. W. 1984. Form and function in fish swimming. *Sci. Am* 251(1):72–82 (July).

Wu, C. H. 1984. Electric fish and the discovery of animal electricity. *Am. Sci.* 72:598–607.

See also relevant selections in General References at the end of Chapter 21.

ADDITIONAL REFERENCES

Avise, J. C. et al. 1986. Mitochondrial DNA differentiation in North Atlantic eels: population genetics. *Proc. Natl. Acad. Sci. USA* 83:4350–4354.

Bemis, W. E., W. W. Burggren, and N. E. Kemp (Eds.). 1986. *The Biology and Evolution of Lungfishes. J. Morphol. Suppl.* 1. (Also published as a book by Liss, New York, 1987.)

Bone, Q. and N. B. Marshall. 1982. *Biology of Fishes.* London: Blackie.

Bullock, T. H. 1982. Electroreception. *Annu. Rev. Neurosci.* 5:121–171.

Courtenay-Latimer, M. 1979. My story of the first coelacanth. In: J. E. McCosker and M. D. Lagios (Eds.), *The Biology and Physiology of the Living Coelacanth.* San Francisco: California Academy of Science, pp. 6–10.

Downing, S. W. et al. 1981. Threads in the hagfish slime gland thread cells: organization, biochemical features, and length. *Science* 212:326–328.

Fischer, E. A. and C. W. Petersen. 1987. The evolution of sexual patterns in the seabasses. *BioScience* 37:482–489.

Fishman, A. P. et al. 1986. Estivation in *Protopterus. J. Morphol Suppl.* 1:237–248.

Forey, P. L. 1988. Golden jubilee for the coelacanth *Latimeria chalumnae. Nature* 336:727–732.

Grau, E. G. et al. 1981. Lunar phasing of the thyroxine surge preparatory to seaward migration of salmonid fish. *Science* 211:607–609.

Greenwood, P. H. 1986. The natural history of African lungfishes. *J. Morphol. Suppl.* 1:163–179.

Hasler, A. and A. Scholz. 1983. *Olfactory Imprinting and Homing in Salmon.* New York: Springer-Verlag.

Hasler, A. D., A. T. Scholz, and R. M. Horrall. 1978. Olfactory imprinting and homing in salmon. *Am. Sci.* 66:347–355.

Kemp. A. 1986. The biology of the Australian lungfish, *Neoceratodus forsteri* (Krefft 1870). *J. Morphol. Suppl.* 1:181–198.

Meyer, A. and A. C. Wilson. 1990. Origin of tetrapods inferred from their mitochondrial DNA affiliation to lungfish. *J. Mol. Evol.* 31:359–364.

Motta, P. J. 1984. Mechanics and functions of jaw protrusion in teleost fishes: a review. *Copeia* 1984:1–18.

Moyle, P. B. and J. J. Cech Jr. 1982. *Fishes: An Introduction to Ichthyology.* Englewood Cliffs, NJ: Prentice-Hall.

Nelson, J. S. 1984. *Fishes of the World,* 2nd ed. New York: Wiley.

Shapiro, D. Y. 1987. Differentiation and evolution of sex changes in fishes. *BioScience* 37:490–497.

Thompson, K. S. 1990. The shape of a shark's tail. *Am. Sci.* 78:499–501.

Webb, P. W. 1988. Simple physical principles and vertebrate aquatic locomotion. *Am. Zool.* 28:709–725.

36

Amphibians

Red-eyed tree frog (Agalychnis callidryas).

LEARNING OBJECTIVES

1. How do frogs and salamanders manage to live on land as well as in water?

2. Why are they still found mainly around water?

3. Why do most amphibians have thin, slimy skin? Are some of them poisonous?

4. How do frogs sing? How do they hear?

5. How do frogs catch flies?

6. How does a tadpole become a frog?

HISTORY OF AMPHIBIANS

Most people think about amphibians—if they think about them at all—as cold, slimy, and not very bright. There was a time, however, when amphibians were the most exciting animals on land. They were, in fact, the *only* animals on land except for some arthropods and an occasional lobe-finned fish. Amphibians such as *Ichthyostega* led the way for vertebrates onto land late in the Devonian period, almost 400 million years ago (see p. 394). As we saw in the previous chapter (see pp. 761–763), their ancestors were probably lobe-finned fishes that were already equipped with lungs and fleshy fins, probably as adaptations for escaping the frequent droughts of that warm, dry period. The need to escape the recently evolved jawed fishes and the lure of edible insects may also have selected for adaptations that allowed amphibians to lead double lives (Greek *amphi-* double + *bios* life).

An enormous number of other amphibians evolved between 360 and 250 million years ago—a period sometimes called the Age of Amphibians. These extinct species are often classified into two groups, the **labyrinthodonts** and the **lepospondyls.** Members of both groups had four legs and a tail. The labyrinthodonts, such as *Ichthyostega*, are distinguished by labyrinthine folds in the dentine of the teeth. The lepospondyls generally lacked labyrinthine folds, and most were more lightly built than the labyrinthodonts. There are certain other differences as well, but many paleontologists believe that the division into these two groups does not accurately reflect phylogeny.

Labyrinthodonts are thought to have evolved into reptiles, which are much better adapted to land, and probably contributed to the demise of most amphibian forms during the Mesozoic era. Class Amphibia is now reduced to some 4200 species divided among three major groups: caecilians, salamanders, and frogs (including toads) (Figure 36.1). None of these modern amphibians can be traced back any farther than about 200 million years, and the fossil record is too fragmentary to link them to the earlier amphibians or to each other.

Figure 36.1
*Representatives of the three major extant groups of amphibians.
(A) Caecilians, such as this yellow-striped caecilian* Ichthyophis katouensis, *are rarely seen because they are wormlike burrowing animals of the tropics.
(B) Salamanders have legs adapted for walking and a tail. Most remain in or close to water, but the adult Eastern newt* Notophthalmus viridescens, *known as the red eft, spends several years wandering over land. (C) Because of their enlarged hind legs, frogs walk poorly but can hop and swim well. The bullfrog* Rana catesbeiana *is one of the largest and most common frogs of North America. It spends much of its time resting in water with its green head camouflaged among algae and vegetation. The nostrils and upper halves of the eyes are usually above water, while the lower halves of the eyes watch beneath the surface of the water.*

A

B

C

CAECILIANS

Caecilians (order Gymnophiona) are little known because they are burrowing animals, and even when noticed they are easily mistaken for earthworms. Caecilians are legless, with skeletons resembling those of snakes (Figure 36.2). As a result of evolutionary convergence, they also show other similarities to snakes. Like snakes, but unlike other amphibians, some species have dermal scales as protection against abrasion. As in snakes, the lungs are long and thin, and the left one is often much smaller than the right. Although caecilians appear to be all tail, they in fact have little or no tail, since the anus is near the tip of the body. They range in length from 10 cm to more than 1.5 meters and have from 95 to 285 vertebrae. They prey on invertebrates, which they locate by protrusible sensory tentacles between the eyes and nostrils. Some 160 species occur in tropical forests of Central and South America, Africa, India, and the Seychelles islands in the Indian Ocean. Since no amphibian can survive for long in seawater, this far-flung distribution was a great mystery prior to the discovery of continental drift. Now it is understood that caecilians originated on Gondwana before it broke up (see Figure 19.9).

SALAMANDERS

Salamanders (order Caudata = Urodela) most resemble the earliest amphibians, with long post-anal tails, and legs of approximately equal size suitable for walking. The larvae of urodeles are generally aquatic, with pharyngeal gill slits, external gills, and broad tails that serve as caudal fins for swimming. Metamorphosis into a terrestrial adult involves loss of the gills and usually a switch to lungs and skin for exchange of respiratory gases (Figure 36.1B). Some terrestrial salamanders, including most in North America (family Plethodontidae), lack lungs and rely totally on the integument. Some salamanders, however, do not become terrestrial adults. In these **neotenic** species the adults remain aquatic and retain external gills and other larval traits after metamorphosis (Figures 6.33 and 36.3). There are approximately 350 species of urodeles, and they range in length from less than 15

Figure 36.2
The skeletons of caecilians, salamanders, and frogs are adapted for their different modes of locomotion. In salamanders the back-and-forth movement of the legs is assisted by flexion of the vertebral column. In frogs the urostyle transmits the force of the hind legs to the rest of the body during jumping.

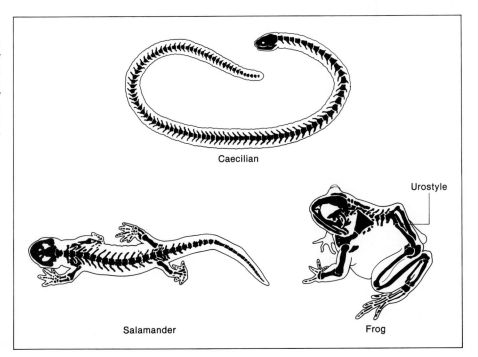

Caecilian

Urostyle

Salamander

Frog

Figure 36.3
Neotenic salamanders. (A) The adult mudpuppy Necturus maculosus, *which is common in the American Midwest. The gills, red with blood, grow larger than shown here if the mudpuppy lives in stagnant, muddy water. Length approximately 40 cm. (B) An adult female two-toed amphiuma (*Amphiuma means*) guarding her clutch of eggs. The legs, not visible here, are reduced to useless prongs. Amphiumas inhabit swamps and canals in the southeastern coastal plain of the United States. They may live for more than two decades and grow to a length of more than a meter. They are voracious predators of crayfish, frogs, and fish and can inflict a painful bite.*

cm to 1.5 meters. They occur mainly in North America, Europe, and Asia, reflecting their origins in Laurasia (see Figure 19.9). A few species have crossed land bridges into South America and Southeast Asia.

FROGS

Most amphibians are frogs, belonging to order Anura. The dominant feature of adult anurans is their large hind legs, which are adapted for hopping. The hind toes are almost as long as the upper two sections of the leg, increasing the leverage during hopping. The toes of the hind feet are also webbed for swimming. The short front legs serve mainly as props when resting, and as shock absorbers when landing after a hop. Males also use the front legs to clasp the females during mating. Frogs have fewer vertebrae than do salamanders, and several are fused into a long, thin bone (the **urostyle**) that transmits the force of the hind legs to the body (Figure 36.2). There is no tail, and the head is essentially fused to the trunk. Adult frogs rely on a combination of lungs and integumentary exchange of O_2 and CO_2. Certain frogs, known as toads, have thicker skins and are better adapted for life away from water. Adult frogs range from 1 to 30 cm long and feed mainly on insects and other invertebrates. In contrast to salamanders, most frogs undergo a dramatic metamorphosis. The tadpole larvae swim with their tails, have internal gills, and feed on algae, plants, and detritus. These adaptations for aquatic life become completely reworked during their metamorphisis into terrestrial, carnivorous adults (Figure 36.4).

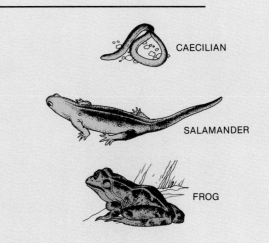
MAJOR FEATURES OF AMPHIBIANS

In many respects most amphibians are well equipped for land, at least as adults. The most obvious adaptation is lungs that enable them to breathe air, which has a much higher concentration of oxygen than does water. The lungs of amphibians are essentially sacs, although in some species folds in the walls of the lung increase the surface area for absorption of oxygen into the blood (see Figure 12.11). Some amphibians have no lungs as adults. They either retain the gills and aquatic habits of their larvae, or they rely on integumentary exchange of gases. Even amphibians with lungs exchange much of their oxygen and carbon dioxide through the skin. Adaptations to terrestrial living can also be seen in the skeleton. The bones of most amphibians are much denser and stronger than those of fishes, enabling them to support their bodies without the buoyancy of water. The paired appendages, which are only loosely attached to the rest of the skeleton in fishes, are strongly attached to pectoral and pelvic girdles in salamanders and frogs (Figure 36.2). Moreover, the pectoral and pelvic girdles are strongly attached to the vertebral column. The force generated by the legs pushing against the substratum is therefore transferred directly to the vertebral column.

In spite of these adaptations, few amphibians have become completely independent of water. Immature forms are almost always aquatic, and most adults must either live in moist habitats or frequently return to water. One reason is that, unlike arthropods, reptiles, birds, and mammals, most amphibians have an integument that is highly permeable to water. They must therefore have access to fresh water to replace the body water lost by evaporation from their integuments. Most amphibians also depend on water for reproduction, since only a few species have evolved internal fertilization, and even fewer are viviparous. Amphibian eggs also lack a shell that prevents water loss, an amnion that would enclose the embryo in fluid, or an allantois that could store metabolic wastes (see p. 120). Amphibians are also unable to adapt completely to sudden shifts in air temperature that occur in all but tropical regions. Amphibians are therefore excluded from polar regions; those in temperate climates become torpid in winter and must generally return to water or burrow in soil to avoid freezing.

Figure 36.4
Tadpole larvae of the common European frog Rana temporaria. *These two siblings are both eight weeks old, but the one on the right has already developed legs.*

SKIN

Integumentary Respiration. Most grown-ups regard frogs as ugly, cold, and slimy, largely on account of their skin. To herpetologists (specialists in amphibians and reptiles), however, the integument is one of the most beautiful features of amphibians. In contrast to the integuments of both fishes and reptiles, the skin of the adult frog or salamander lacks scales and is highly permeable to water and respiratory gases. The skin sacrifices protection from abrasion and dehydration to provide a means of exchanging O_2 and CO_2, especially in an adult frog totally immersed in water. Even on land the skin absorbs almost as much O_2 as the lungs do, and it eliminates most of the CO_2. Frogs can continue to respire through the skin even if breathing ceases to ventilate the lungs. Frogs are such convenient subjects for the study of vertebrate physiology because they go on respiring even after the central nervous system is completely destroyed (pithed) to stop movement and pain.

The respiratory function of the skin depends to a large extent on its thinness (Figure 36.5A). The **epidermis** is only a few micrometers thick, permitting easy diffusion of gases to and from the blood capillaries in the underlying dermis. Although there is an outer **cornified layer** containing the protein keratin, it is generally too thin to provide more than token protection. In toads, however, the skin tends to be thicker and more cornified, affording better protection from abrasion and dehydration. During the months when they are active, frogs shed the cornified layer periodically (every 3 to 19 days in the toad *Bufo*). The frog splits the cornified layer starting at the head, peels it off with its legs, then usually eats it.

Defense. Frog skin is unique in that it is attached to underlying tissues in only a few places. The loose attachment of the skin probably helps frogs escape from predators, especially smaller birds and other animals that cannot swallow a frog whole. A clear, slippery secretion from **mucous glands** in the dermis probably

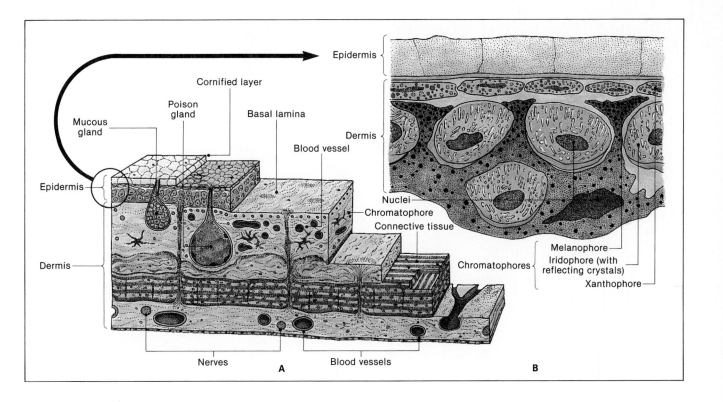

Epidermis

Cornified layer

Poison gland

Basal lamina

Mucous gland

Blood vessel

Epidermis

Dermis

Nuclei
Chromatophore
Connective tissue

Chromatophores

Melanophore
Iridophore (with reflecting crystals)
Xanthophore

Nerves

Blood vessels

A

B

Figure 36.5

(A) Schematic representation of frog skin. Total thickness is generally less than 50 μm, although the skin of many frogs, especially toads, is thicker in some places ("warts"). (B) Section from part (A), showing arrangement of chromatophores. Xanthophores contain yellow, orange, or red pigments. Blue structural color reflecting through the xanthophores from iridophores produces the green color common in frogs. Dark melanin pigments can migrate into branches of melanophores, covering the iridophores and darkening the skin.

also helps frogs slip away from predators. Another important function of the mucous glands is to keep the skin moist so that it remains permeable to respiratory gases. The skins of all amphibians also contain **poison glands** (= granular or serous glands). These glands are often concentrated in thickenings of the skin (the "warts" of toads), or in bulges behind the eyes, called **parotoid glands.** (Parotoid glands must not be confused with the parotid salivary glands. Parotoid glands can be seen in Figure 36.15C.)

When an amphibian is stressed, the poison glands secrete a frothy material containing one or more noxious substances. The secretions of some amphibians are only distasteful, but others contain toxins that are among the most potent among animals and are dangerous even to touch. Some induce paralysis by blocking ion channels or pumps in the membranes of nerve and muscle cells. Examples of these toxins are tetrodotoxin from certain newts (*Taricha*), which blocks sodium channels, and batrachotoxin from some poison-dart frogs of Central and South America, which holds the sodium channels open (see p. 144). Toxin from the cane toad *Bufo marinus* works like the neurotransmitter serotonin. (Many hallucinogenic drugs also mimic the effects of serotonin. Some people have tried licking these toads as a substitute for drugs, but they get sick rather than high.)

Colors. Many toxic amphibians are brightly colored, and potential predators soon learn to avoid them (Figure 36.6A). Red, orange, and yellow warning colors are due to pigments in certain **chromatophores,** and blue is a structural color reflected from other kinds of chromatophores (see Table 10.1; Figures 36.5B and 36.6A). Less toxic species tend to be camouflaged. Amphibians that live among green vegetation are often green (Figure 36.1C). The green color comes from structural blue reflecting through yellow pigment. Many amphibians that live on the forest floor or in trees are camouflaged with brown or gray melanin pigments in chromatophores (Figure 36.6B). These chromatophores, called **melanophores** or **melanocytes,** are also partly responsible for the darkening of the skin, which

A

B

often occurs when amphibians are on a dark substratum or in darkness. Control of the melanocytes is partly due to **melanocyte stimulating hormone** (MSH) from the pituitary. Secretion of MSH is in turn directed by the **pineal body** in the brain, which is connected by a nerve to a light-sensitive **pineal eye** (= median eye) in the top of the skull. (See Figure 34.20 for location.) The pineal body also apparently controls daily (circadian) rhythms.

INTERNAL ORGANIZATION

Lungs.   The beauty of amphibians is not merely skin deep but extends to the adaptations of their internal organs. Many of the organs are similar in structure and function to those of other vertebrates, as described in Unit Two (Figure 36.7). (See index for details on particular organs.) Here we focus on the adaptations for terrestrial life, especially in frogs. The lungs often account for more than half the oxygen taken into the blood when a frog is on land, and virtually all the oxygen when it is floating in stagnant water with only the eyes and nostrils above the surface. Small amounts of respiratory gases are also exchanged across the lining of the buccal cavity. The structure of the lungs and the pattern of breathing are described in Figure 12.11.

Circulation.   The evolution of air breathing entailed changes in the circulatory system from that of fishes. The amphibian heart has a second atrium that allows deoxygenated blood from the body tissues to be separated from oxygenated blood coming from the lungs and skin (Figures 11.4B and 36.8). Deoxygenated blood enters a chamber called the **sinus venosus** on the back of the heart, which pumps the blood into the right atrium. At the same time oxygenated blood from the skin and lungs enters the left atrium. In response to action potentials from the pacemaker region between the sinus venosus and the atria (the SA node), the two atria contract simultaneously, forcing both deoxygenated and oxygenated blood through atrioventricular valves into the one ventricle.

Figure 36.6
(A) Warning coloration in the red-backed dart-poison frog Epipedobates cainarachi *of Peru. Dart-poison frogs (family Dendrobatidiae) get their name from three species in the genus* Phyllobates, *which natives of western Colombia use to poison their blow-gun darts. (B) The wood frog* Rana sylvatica *of the northeastern United States and Canada, camouflaged against a background of leaves.*

Figure 36.7
Internal organization of a male frog.

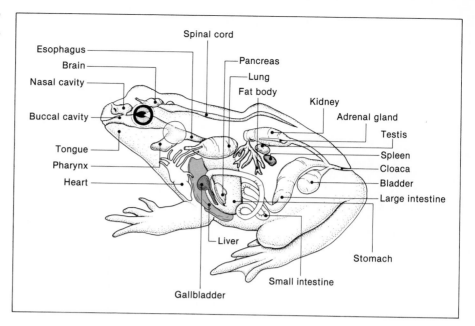

In some salamanders and frogs, such as the African clawed frog *Xenopus laevis*, the deoxygenated and oxygenated blood mix in the ventricle. In other salamanders and frogs, however, the oxygenated blood in the left side of the ventricle remains separate from deoxygenated blood in the right side of the ventricle. Amphibians that are more dependent on cutaneous respiration under water seem to have a greater degree of separation. When the ventricle contracts it forces the blood into a chamber called the **conus arteriosus**. A **spiral valve** in the conus diverts the oxygenated blood into the **carotid artery** to the head and into the **right systemic artery**. Some oxygenated blood also enters the **left systemic artery**. The two systemic arteries merge at the dorsal aorta, which supplies most of the blood to the body. Most of the deoxygenated blood is diverted by the spiral valve into the two **pulmocutaneous arteries** that go to the lungs and skin, although some also enters the left systemic artery.

Osmoregulation and Excretion.

The thin skin of an amphibian is almost useless in preventing the loss of body water when it is on the ground and is equally poor at preventing the osmotic absorption of water when it is in fresh water. The kidneys, however, compensate for the loss or gain of water through the skin. Like the kidneys of fishes, those of adult salamanders and frogs are of the mesonephric type (see p. 292). As in freshwater fishes, they produce copious amounts of dilute urine whenever the amphibian is in water. The urine passes from the kidneys through the **ureters** into the **cloaca** and may then be excreted immediately or stored in the bladder (Figures 36.7 and 36.9).

When amphibians are on land they are incapable of conserving water by producing urine more concentrated than the body fluids, but they can reabsorb any water that was previously stored in the urinary bladder. In several frogs the bladder can store urine equivalent to one-third of the body weight. Amphibians can also store large amounts of water in lymph sacs beneath the skin. Having evolved from freshwater fishes, no amphibian manages to live in seawater. Some species, however, can tolerate brackish water. The crab-eating frog *Rana cancrivora* of Southeast Asia lives in mangrove swamps that are about 80% seawater. Like sharks and some bony fishes, these frogs have high concentrations of **osmolytes** (urea and methylamines) that help prevent loss of body water by osmosis.

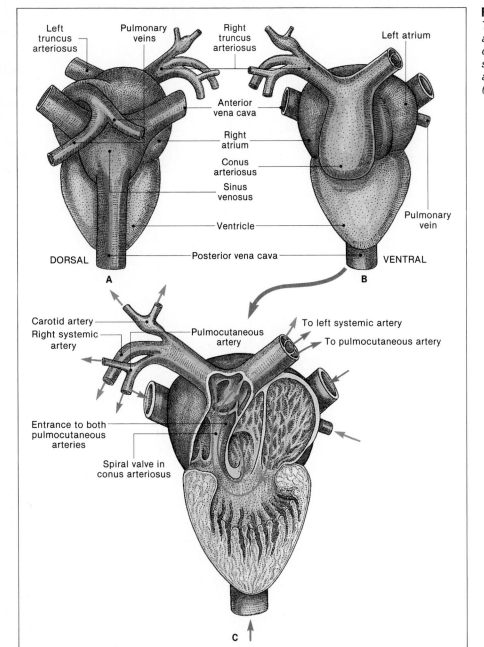

Figure 36.8
The heart of a bullfrog. (A, B) Dorsal and ventral views. (C) Ventral view of a sectioned heart showing the separation of oxygenated blood (red arrows) from deoxygenated blood (blue arrows).

When amphibians are in fresh water the skin helps compensate for the diffusion of ions out of the body by actively transporting ions in the reverse direction. (Frog skin is a favorite model system for research on ionic transport.) The kidneys help in osmoregulation by reabsorbing ions into the blood. The kidneys also eliminate nitrogenous wastes. Adult terrestrial amphibians generally excrete urea as the major nitrogenous waste. In larvae and aquatic adults ammonia is the major nitrogenous waste. Some species, such as the African clawed frog *Xenopus*, can switch from ammonia to urea as they move from water to land.

SENSORY RECEPTORS

Since pressure, chemicals, and light do not propagate in the same way through air as through water, the transition from water to land required several adaptations

Figure 36.9
Ventral view of the urogenital systems of female and male frogs. Only the right ovary is shown. Sperm enter the kidney through efferent ductules and then are stored in the ureters. Ova enter the oviducts and are stored in the "uterus."

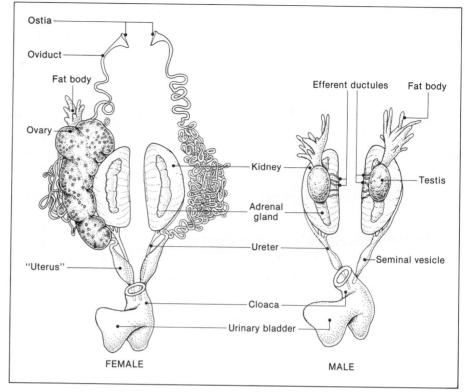

in sensory receptors. Air does not have sufficient density or electrical conductance to activate the kinds of receptors found in the lateral line organ of fishes (see p. 779), so these occur in amphibians only in aquatic larvae and adults. In most adult amphibians the function of the lateral line organ is probably largely replaced by free nerve endings in the skin that respond to touch and pressure. The absence of scales means that these receptors are much more sensitive than they would be in fishes. Free nerve endings in the skin also respond to injury and temperature. Terrestrial amphibians sense airborne molecules with olfactory epithelia in the nasal cavities, and with **Jacobson's organs** (= vomeronasal organs) in the roof of the mouth.

Like other vertebrates, amphibians have **semicircular canals** and **otolith organs** that detect position and acceleration of the head. As in fishes, the semicircular canals and otolith organs constitute the **inner ear.** There is no cochlea, which in mammals is responsible for audition. Nevertheless, in many amphibians there are mechanisms by which vibrations of water or air that we would perceive as sound do stimulate the inner ears.

1. When an amphibian is in water its body tissues are essentially "transparent" to low-frequency pressure changes, so these may pass directly to the inner ear and be "heard."
2. Airborne sound cannot penetrate body tissues, but adult frogs have external **eardrums** (= tympanic membranes) that respond just like our own eardrums to airborne vibrations (Figure 36.10). A small bone, the **columella** (= stapes), transmits the vibrations from the eardrum to the inner ear. (The columella is homologous with one of the jaw bones in fishes. See Figure 34.20.)
3. Vibrations of the substratum may be conducted through the skeleton to the inner ear.

Figure 36.10
The head of a bullfrog showing prominent structures associated with sensory reception: the tympanum (eardrum), the eyes, and the nostrils.

Vision. Eyes also do not work in quite the same way in air as in water. While an amphibian is on land the cornea aids the lens in refracting light, because of the large differences in the index of refraction of air compared with that of the cornea. Frogs can therefore accommodate (focus on) much closer objects on land than in water. Amphibians accommodate in a way that is unusual among vertebrates. Instead of changing the diameter of the lens, they move the lens forward or backward. (Compare Figures 7.13A and 36.11.) As in mammals, **lachrymal (tear) glands** and eyelids reduce abrasion and drying of the cornea in air. Frogs blink by retracting the eyes into the orbit, which causes a translucent part of the lower lid, called the **nictitating membrane,** to sweep over the eye.

The **retina** of the amphibian eye has several kinds of rods and cones, but it is not clear whether it is capable of discriminating different colors. The retina also contains other kinds of cells that apparently do most of the visual processing before they transmit information to the brain. Jerome Lettvin and others (1959) at MIT used recording electrodes on the optic nerves to study "what the frog's eye tells the frog's brain." They found that the retinas of leopard frogs (*Rana pipiens*) transmit only selected kinds of information:

1. One group of retinal cells responds only to small moving objects that are darker than the background. These cells are called **net convexity detectors** or, more descriptively, "bug detectors."
2. Some nerve cells in the retina are "wired" together in such a way that they send action potentials along axons in the optic nerve only when the edge of an object moves into range and stops. These cells continue triggering action potentials as long as the object remains in the visual field. Lettvin and his colleagues named these nerve cells **sustained contrast detectors.**

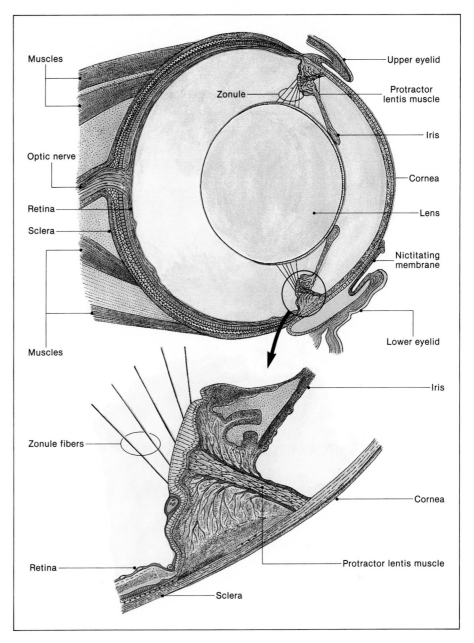

Figure 36.11
The eye of a leopard frog Rana pipiens. Inset shows detail of the mechanism for accommodation. Contraction of the protractor lentis muscle moves the lens forward, focusing nearby objects on the retina.

Muscles

Zonule

Upper eyelid

Protractor lentis muscle

Optic nerve

Iris

Cornea

Retina

Lens

Sclera

Nictitating membrane

Muscles

Lower eyelid

Iris

Zonule fibers

Cornea

Protractor lentis muscle

Retina

Sclera

3. Another system of retinal cells responds to an edge only while it is moving. These are called **moving-edge detectors.**

4. Another system responds to a sudden reduction in light intensity, such as might be due to the shadow of a potential predator. These cells make up a **net dimming detector.**

It appears, therefore, that leopard frogs see little besides the edges of objects that are or were moving, threatening shadows, and flying insects. These electrophysiological findings are consistent with behavioral observations. Leopard frogs indeed respond quickly to sudden movements, shadows, and flying insects. They apparently do not even see still insects and will starve to death surrounded by dead flies. The retinas of other frogs are apparently wired differently, however. The African clawed frog does see and will eat food that is not moving, which makes it much easier to raise in the laboratory.

FEEDING

The suction feeding employed by many fishes (see p. 769) will not work in air, but some adult frogs have a feeding method that is just as effective with flying insects: the **lingual flip.** In the two families that include bullfrogs and true toads (families Ranidae and Bufonidae), the tongue is attached to the front of the mouth and is flipped out like a whip to catch flying insects (Figure 36.12). The sticky tongue adheres to the prey, which is drawn into the mouth and swallowed whole. Many frogs have teeth on the jaw or palate (**vomerine teeth**) but use them for grasping rather than chewing prey. Adult lungless salamanders (family Plethodontidae) also capture flying insects with protrusible tongues, but the mechanism is different from the lingual flip. They extend their tongues in much the same way as humans, by contracting lateral muscles that squeeze the tongue forward. Most humans are not as good at catching flies with their tongues as salamanders are, however. Some salamanders can catch an insect as far away as 80% of their body length within 50 milliseconds.

Larger amphibians do not necessarily limit their diet to what they can catch with their tongues. Bullfrogs, for example, will eat small mammals, birds, snakes, and other frogs if the chance arises. Most other amphibian tongues are nonprotrusible but nonetheless helpful in manipulating food and in swallowing, as in humans. The African clawed frog *Xenopus* and other species in the family Pipidae have no tongues at all. These purely aquatic species usually feed on zooplankton by pumping water into and out of the mouth, and they use their fingers to stuff larger food into their mouths.

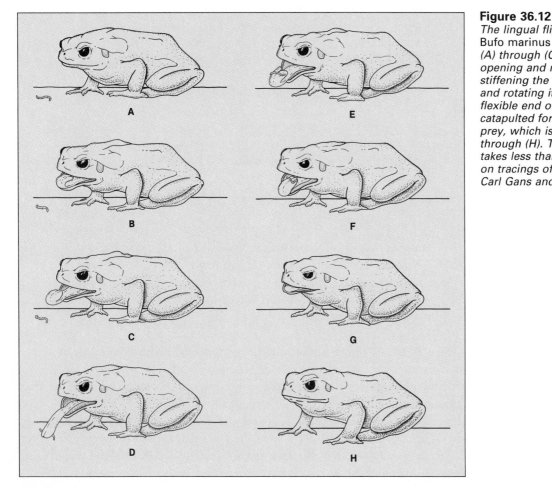

Figure 36.12
The lingual flip in a marine toad Bufo marinus *feeding on a worm. In (A) through (C) the mouth is opening and muscle contraction is stiffening the base of the tongue and rotating it forward. In (D) the flexible end of the tongue has catapulted forward and struck the prey, which is retrieved in (E) through (H). The entire sequence takes less than 0.15 second. Based on tracings of high-speed films by Carl Gans and G. C. Gorniak (1982).*

REPRODUCTION

The reproductive organs of amphibians are closely tied to the kidneys, ureters, and cloaca, making them all part of the **urogenital system** (Figure 36.9). Sperm from the testes pass through fine tubules into the kidneys and out through the ureters. They are stored in the ureters until released through the cloaca during mating. Fat bodies apparently store nutrients for the breeding season, when feeding generally ceases. During the breeding season the ovaries of females produce hundreds of eggs. These eggs spill out into the body cavity and are taken into the two oviducts. The oviducts surround the eggs with a jelly coat as they propel them toward storage chambers near the cloaca. The jelly swells in water, producing shapeless blobs (most frogs) or long strands (toads) (Figure 36.13).

Caecilians. Fertilization occurs in a variety of ways, depending largely on the degree to which the amphibian is terrestrial or aquatic. As a group caecilians are the most terrestrial. Males introduce sperm into the female by means of a protrusible copulatory organ, and fertilization occurs internally. The gametes therefore avoid desiccation and other hazards of terrestrial life. New World caecilians are mostly viviparous, so the embryos are also protected. The fetuses feed on secretions and on tissues scraped from the lining of the mother's oviduct. In oviparous species the larvae are generally aquatic, with internal gills. Little is known about courtship or many other aspects of reproduction in caecilians.

Salamanders. In about 90% of salamanders, whether aquatic or terrestrial, fertilization is internal, although copulation does not occur. Instead, the male deposits a **spermatophore,** which consists of a mass of sperm capping a conical blob of jelly secreted by a gland on the lining of the male's cloaca. Transfer of the spermatophore is achieved during courtship, in which the male captures or blocks the path of the female, then maneuvers or leads her until the spermatophore enters her cloaca (Figure 36.14). A pouch in the female's cloaca (the **spermatheca**) stores the sperm until the eggs pass through the cloaca.

In some terrestrial species the eggs hatch directly into terrestrial larvae, which do not have gills. In others the larvae develop entirely in the eggs, from which adults hatch. Still other terrestrial salamanders are viviparous. Some oviparous terrestrial salamanders breed on land but deposit the eggs in moist leaf litter, under a stone or log, or in a depression that will collect water during a rain. Still other terrestrial salamanders, such as the Eastern newt (Figure 36.1B), return to their native waters to reproduce. The aquatic salamanders spend most of their adult lives in water, where they reproduce.

Terrestrial salamanders that return to their native habitats to breed often do so by **homing** from rather long distances. Victor C. Twitty (1966) found many tagged red-bellied newts (*Taricha rivularis*) back in their native streams within a year after he had transported them across 8 km of rugged California hills. Although many amphibians are known to navigate using the position of the sun, blinding these newts did not interfere with homing. Cutting their olfactory nerves did, however. Eastern newts (*Notophthalmus*) also home, but they do so under circumstances that rule out olfaction. In the spring they orient in the direction of their home ponds even after being brought into the laboratory up to 30 km away (Phillips 1987). The directional cue appears to be geomagnetic, since their orientation can be disrupted with artificial magnetic fields. They can also be trained to orient with respect to a magnet.

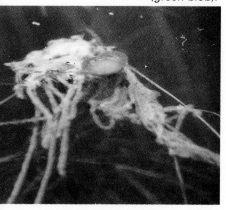

Figure 36.13
Toad eggs (long white strands) entangled on a mass of frog eggs (green blob).

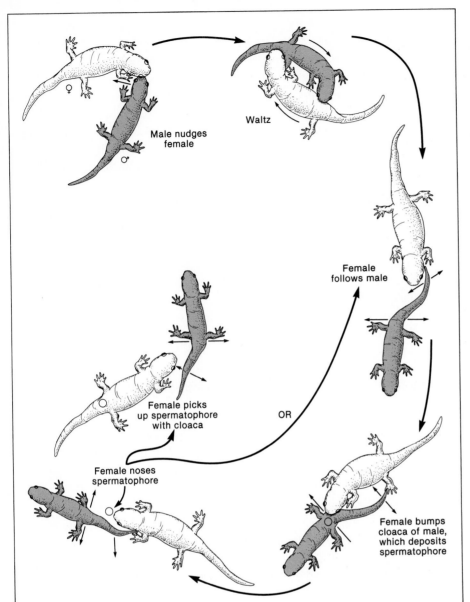

Figure 36.14
Courtship in the mole salamander Ambystoma talpoideum, *which breeds in shallow pools in the Gulf states. The male approaches and nudges a larger female, then each pushes against the other's cloaca in a circular "waltz." After one or two rounds the male pulls himself forward with his front legs while shuffling his rear from side to side and fanning his tail across the female's snout. The female follows for several minutes, then nudges the male's cloaca. He then pauses to deposit a spermatophore, then resumes his movements. After nosing the spermatophore the female walks over it. If her cloaca picks it up she leaves the male. If not, she follows the male again for another try.*

Within the figure:
Male nudges female
Waltz
Female follows male
Female picks up spermatophore with cloaca
OR
Female noses spermatophore
Female bumps cloaca of male, which deposits spermatophore
♀ ♂

Frogs. Almost all frogs are oviparous, and fertilization is usually external. This means that reproduction must occur in water or at least in a damp situation. Male frogs generally attract females during the breeding season by means of an **advertisement call.** The call is produced by exhaling air through the larynx and, in most species, is amplified by means of a **vocal sac** (Figure 36.15). In male white-lipped frogs (*Leptodactylus albilabris*) of Puerto Rico, the vocal sac may also be used in seismic communication, for it thumps the ground during calling (Narins 1990). In choice habitat there are usually many species calling simultaneously, but the calls are extremely species specific in pattern and pitch. In some species the ear of the female is narrowly tuned to the pitch of the advertising call for the male of her species, and she may not even hear the calls of other species. In other species the pattern of notes within the call is more important than the pitch. In some species the females are most attracted to the lowest-pitched calls for their species, which usually come from larger males. Males not only attract females with their advertising calls but also repel other males—usually. In some species, how-

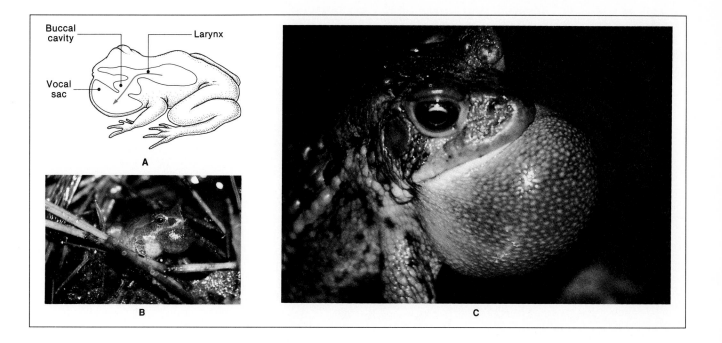

Figure 36.15

Figure 36.15
Advertising calling by male frogs. (A) Sound is produced by blowing air from the lungs into the vocal sac across the vocal cords in the larynx. The vocal sac amplifies the sound by serving as a resonating chamber and perhaps as an acoustical radiator. Females can also vocalize but do not have vocal sacs. (B) A spring peeper Hyla crucifer *calling with inflated air sac. The song is a series of notes at about 1-second intervals, with each note a high, ascending whistle. The song from a 2-centimeter male is impressive in loudness; the chorus from thousands of males in a marsh is actually painful to human ears. (C) A singing American toad* Bufo americanus. *The song is a musical trill up to half a minute long. Body length approximately 10 cm.*

ever, **satellite males** wait silently near a calling male and often intercept females he has attracted.

In most species eggs and sperm are released during a mating embrace called **amplexus,** in which the male grasps the female from behind (see Figure 16.2B). Males of many species grip the females with the help of **nuptial pads** on the fingers, arms, or chest that develop during the mating season in response to hormones. After the eggs are released and fertilized, they are usually left to hatch into free-living tadpoles on their own, with no parental care. The eggs and tadpoles are extremely vulnerable to predation and to accidents such as drying, but usually a female produces hundreds or thousands of eggs in a season.

Although this is the general pattern of mating, there are many interesting modifications that afford more protection for the offsrping and allow some independence from water. Frogs with such adaptations generally produce fewer offspring, but a greater proportion survive. One such adaptation is internal fertilization. In one species (*Ascaphus truei*) the male has a tail that serves as an intromittent organ during amplexus. In several species of African toads in the genus *Nectophrynoides* the male and female achieve internal fertilization simply by pressing their cloacae together during amplexus. Several of these species are viviparous. Many other frogs mate in the usual fashion but brood the eggs or tadpoles to protect them from drying and predation. Some construct nests, some guard the eggs, and some carry the eggs and tadpoles on their backs. In 60 species of "marsupial frogs" the females even have pouches or pits in the skin on their backs in which the eggs and tadpoles develop. Perhaps the oddest example of parental care occurs in an endangered frog from the mountains of Queensland, Australia, which broods the eggs and tadpoles in its stomach (Figure 36.16).

DEVELOPMENT

Development of the Tadpole. In ranid frogs the eggs generally begin developing immediately after fertilization. First the jelly coat surrounding them swells with water, forming masses or strands of spawn. Within a few minutes to a day,

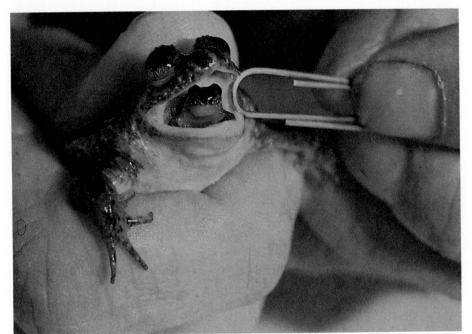

Figure 36.16
A frog is "born." The froglet
(Rheobatrachus silus, *which may be
extinct) emerges from the mouth of
its mother after having developed in
her stomach for 37 days. During this
time the mother had not eaten, and
secretion of acid and digestive
enzyme from the stomach was
suppressed. The paper clip and
thumb provide the scale.*

depending on species and temperature, the egg begins dividing repeatedly in the process called **cleavage** (Figure 36.17). Cleavage results in the formation of a hollow mass of cells called the **blastula** several hours later. The blastula then undergoes gastrulation during the next several hours. For the next several days the **gastrula** elongates and a tail bud forms. The egg then hatches into a tadpole, typically five to ten days after fertilization. At first the tadpole clings to vegetation with its sucker, obtaining nutrients from the remains of its yolk sac. Soon, however, the tadpole begins swimming with its tail fin and feeds on algae and organic particles with its toothed mouth. The gills are external at first, but soon an **operculum** grows over each gill. On the left side a **spiracle** forms as an exit for water drawn through the mouth and past the gills. The next noticeable change in the tadpole is the development of the hind legs, which are usually complete two or three weeks after fertilization. Several days later the forelimbs emerge from the gill chambers in which they have been developing.

Metamorphosis.   The emergence of the forelimbs signals one of the most remarkable processes in animals: the metamorphosis of the aquatic, herbivorous tadpole into a semiterrestrial, carnivorous frog. These changes are apparently triggered by a rise in the levels of thyroid hormones (thyroxine and triiodothyronine) and a fall in the level of prolactin. Over a period of a week or so the tadpole's mouth loses its teeth and becomes wider, jaws and a tongue develop, the gut shortens, and the tail regresses. The gills also regress, and the lungs develop. Eventually, from one to several months after fertilization, the froglet is able to emerge onto land. It may live as an amphibious juvenile for several years before it is able to reproduce.

Development in Salamanders.   Both larval and adult salamanders have legs and tails, so their metamorphosis is not as dramatic as the transition from tadpole to frog. In fact, neotenic salamanders become sexually mature with few external changes from the larval form (see pp. 133–134). They remain aquatic after met-

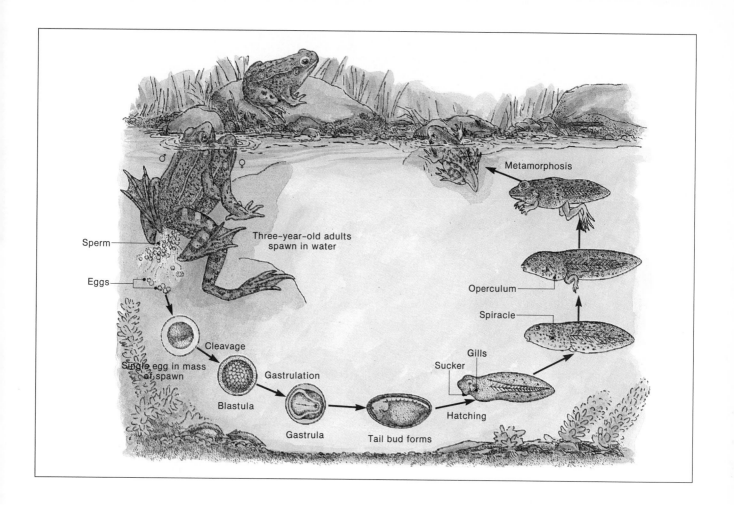

Figure 36.17
Development in anurans. For details of cleavage and gastrulation see Figures 6.13 and 6.19A. See Figure 36.4 for photographs of tadpoles.

amorphosis, and most retain external gills. Some species are neotenic regardless of environmental conditions. The mudpuppy and the eellike amphiuma are examples of such **obligate neotenes** (Figure 36.3). Other salamanders, including subspecies of tiger salamanders in the western United States and Mexico, are **facultative neotenes.** As long as there is plenty of water they retain the larval form, but if their habitats dry up they lose their gills, develop lungs, and take to land. Eastern subspecies of tiger salamanders are generally not neotenic (Figure 36.18). In other salamanders, such as the eastern newt, metamorphosis is always complete, with the adult losing its gills and becoming terrestrial (Figure 36.1B).

INTERACTIONS WITH HUMANS AND OTHER ANIMALS

Except for an occasional meal of frog legs, amphibians are not a major item in the human diet, and humans are never on an amphibian menu. Perhaps for that reason about the only people who pay much attention to amphibians are children and zoologists. It is hard to imagine a group of animals that has made a greater contribution to human welfare, however. Much of what we know about medicine originated in studies on frogs, many going back a century and more. Many of these studies have become classics in biology education. An estimated 20 million frogs per year are used in education in the United States alone. Frogs continue to contribute to new knowledge even in the age of genetic engineering. The oocytes of the African clawed frog *Xenopus laevis* are favorite cells for testing the functions of nucleic acids. RNA isolated from other species and injected in the

A
B

oocyte will often be translated into protein, enabling its function in a living cell to be studied. *Xenopus* has also joined other frogs and salamanders in a long tradition of amphibian contributions to embryology.

During the 1980s may zoologists began to report extinctions and alarming declines in various amphibians from many different habitats around the world, including the United States. Acid precipitation, deforestation, changing climate, and other environmental factors may be contributing to the decline, but so far none of these factors provides a complete explanation. Because of the thin skins of amphibians, they are extremely sensitive to environmental changes, and their decline may be regarded as a forewarning of future declines for animals that are not as sensitive. Moreover, the decline in amphibians may directly affect other species. A glimpse of what might happen can be seen in parts of India, where frogs were hunted out for sale to European restaurants. Since then the government has had to spend more on insect control than they earned from sale of the frogs.

Figure 36.18
Adults of two subspecies of the tiger salamander Ambystoma tigrinum, *one neotenic (A) and the other not (B). Like many amphibians, tiger salamanders occur throughout the United States in several different forms that may interbreed where their ranges overlap. Subspecies from separated areas will often not interbreed, however. The colors of various subspecies range from gray to yellow to olive. The neotenic forms, called axolotls, are common in the western United States. Length in (B) approximately 30 cm.*

SUMMARY

The three extant groups of amphibians are caecilians, salamanders, and frogs. Caecilians have no legs, and their skeletons are adapted for burrowing, as in snakes. In salamanders there are four approximately equal-sized legs firmly attached to the rest of the skeleton, and the vertebral column extends beyond the anus into the tail. In frogs the hind legs are greatly enlarged for hopping, and there is no tail in the adult. Most amphibians are capable of living on land as adults, although they are still dependent on fresh water for reproduction. Terrestrial forms generally have lungs, aquatic forms generally have gills, and all have thin skins capable of integumentary respiration. The skin also contains mucous glands that help protect it and keep it moist, chromatophores that provide camouflage or warning coloration, and poison glands.

Caecilians generally reproduce by copulating, salamanders generally reproduce by transferring spermatophores, and frogs reproduce by amplexus, in which the eggs are usually fertilized externally. Most male frogs attract females with advertising calls. Sounds, as well as vibrations of water or substratum, are detected by the inner ears. Vision appears to be specialized for detection of bugs, predators, and other specific images. In frogs the fertilized eggs develop into tadpoles, which are aquatic and herbivorous. Changes in hormone levels trigger their metamorphosis into terrestrial carnivores, beginning with the development of the hind legs, followed by shrinkage of the tail and the emergence of the front legs from the gill openings. In neotenic salamanders the adults remain aquatic and retain the gills or other juvenile features.

KEY TERMS

neoteny
urostyle
parotoid gland
conus arteriosus
pulmocutaneous artery
cloaca

Jacobson's organ
columella
nictitating membrane
vomerine teeth
spermatophore
spermatheca

advertisement call
vocal sac
amplexus
nuptial pad
metamorphosis

SELF-TEST

1. Discuss three of the major adaptations of amphibians that enable them to survive on land. Describe two features of amphibians that prevent their being totally independent of water.

2. Name the three modern groups of amphibians, and describe the major differences among them.

3. Describe how the amphibian skin is adapted for life in water and on land. What disadvantages are associated with this structure?

4. Describe the roles of poison glands and chromatophores in defending amphibians against predation.

5. Describe the structure of the frog's heart and the pathway of blood through it.

6. Describe two ways that sensory receptors in amphibians are adapted to air rather than water.

7. Compare the courtship and reproduction of a typical salamander with that of a typical frog.

8. Compare development in a salamander with complete metamorphosis to that of a typical frog.

READINGS

RECOMMENDED READINGS

Blaustein, A. R. and R. K. O'Hara. 1986. Kin recognition in tadpoles. *Sci. Am.* 254(1):108–116 (Jan).

Del Pino, E. M. 1989. Marsupial frogs. *Sci. Am.* 260(5):110–118 (May).

Myers, C. W. and J. W. Daly. 1983. Dart-poison frogs. *Sci. Am.* 248(2):120–133 (Feb).

Phillips, K. 1990. Where have all the frogs and toads gone? *BioScience* 40:422–424.

See also relevant selections in General References at the end of Chapter 21.

ADDITIONAL REFERENCES

Behler, J. L. and F. W. King. 1979. *The Audubon Society Field Guide to North American Reptiles and Amphibians.* New York: Alfred A. Knopf.

Dawid, I. B. and T. D. Sargent. 1988. *Xenopus laevis* in developmental and molecular biology. *Science* 240:1443–1453.

Duellman, W. E. and L. Trueb. 1986. *Biology of Amphibians.* New York: McGraw-Hill.

Gans, C. and G. C. Gorniak. 1982. How does the toad flip its tongue? Test of two hypotheses. *Science* 216:1335–1337.

Halliday, T. R. and K. Adler (Eds.). 1986. *Encyclopedia of Reptiles and Amphibians.* New York: Facts on File.

Hankin, J. 1989. Development and evolution in amphibians. *Am. Sci.* 77:336–343.

Lettvin, J. Y., H. R. Maturana, W. S. McCulloch, and W. H. Pitts. 1959. What the frog's eye tells the frog's brain. *Proc. Inst. Radio Eng.* pp. 1940–1951.

Narins, P. M. 1990. Seismic communication in anuran amphibians. *BioScience* 40:268–274.

Phillips, J. B. 1987. Laboratory studies of homing orientation in the eastern red-spotted newt, *Notophthalmus viridescens. J. Exp. Biol.* 131:215–230.

Twitty, V. C. 1966. *Of Scientists and Salamanders.* San Francisco: W. H. Freeman. (*Amusing accounts of research.*)

37

Reptiles

Double-crested basilisk (Basiliscus plumifrons).

CHAPTER OUTLINE

LEARNING OBJECTIVES

1. How are such diverse forms as turtles, lizards, snakes, and crocodiles related to each other?
2. How are they related to ichthyosaurs, plesiosaurs, pterosaurs, and dinosaurs?
3. How did the dinosaurs become extinct?
4. What adaptations enabled reptiles to become completely terrestrial?
5. How do reptiles compare with humans in intelligence and social behaviors?
6. How do snake venoms work?
7. How do sea turtles migrate for thousands of kilometers, and how do snakes slither?

THE ORIGIN OF REPTILES

No major group of animals inspires more curiosity, awe, and terror than the class Reptilia. Most of the 6200 named species of living turtles, snakes, lizards, and crocodilians are harmless, and only a few people fall victim to poisonous snakes or to crocodilians, at least in the United States. Nevertheless, the thought of encountering a snake is enough to keep many people out of the woods, and dinosaurs that have been safely extinct for more than 65 million years can still make a child sleep with the lights on. Living reptiles are also interesting enough to keep mature herpetologists up past bedtime.

The oldest known reptilian fossil dates to the Carboniferous period around 340 million years ago (see Table 19.3 and Figures 34.21 and 37.1). Reptiles therefore evolved in only some 20 million years after their amphibian ancestors originated—a remarkably short time, geologically. Reptiles quickly proved themselves so successful that we refer to the entire Mesozoic era as the Age of Reptiles. The Age of Reptiles ended abruptly and mysteriously around 66 million years ago; otherwise we would probably never have evolved to learn about it.

The amphibians and earliest reptiles that evolved from them are often referred to as **cotylosaurs,** and the stem reptiles themselves are called **captorhinomorphs.** Less than 100 million years after their first appearance, the captorhinomorphs diverged into three major lineages (Figure 37.2). One line of descent may have led to the turtles. A second led to pelycosaurs and from there to the mammallike therapsids, which then evolved into mammals, as will be described in greater detail in Chapter 39. The third line of descent was the most diverse and spectacular. It included the ancestors of living snakes, lizards, and the lizardlike *Sphenodon*, as well as extinct marine reptiles (ichthyosaurs and plesiosaurs). It also included various groups referred to as archosaurs, or ruling reptiles. Among these were the dinosaurs and flying reptiles (pterosaurs). An archosaur evolved into crocodilians, and a dinosaur subsequently evolved into birds, as will be described in the next chapter.

DIVERSITY

Paleontologists recognized these three lineages largely from the number of **temporal openings** in the skull behind the orbits of the eyes. The function of these

Figure 37.1
Reconstructed skeletons of a fossilized amphibian and a reptile, both of which lived during the Carboniferous period, and both of which were approximately 25 cm long. (A) Bruktererpeton fiebigi belonged to a group of amphibians that probably descended from the same group as reptiles. (B) Hylonomus lyelli, judged to be a reptile from details of the skeleton. This fossil was found in what is now Nova Scotia, inside a fossilized hollow tree stump where the animal is presumed to have been trapped.

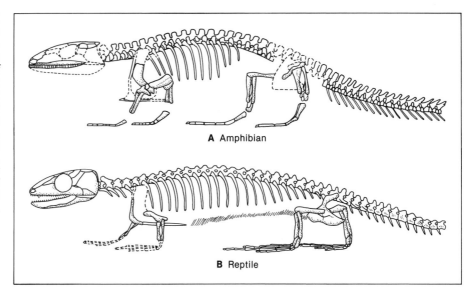

A Amphibian

B Reptile

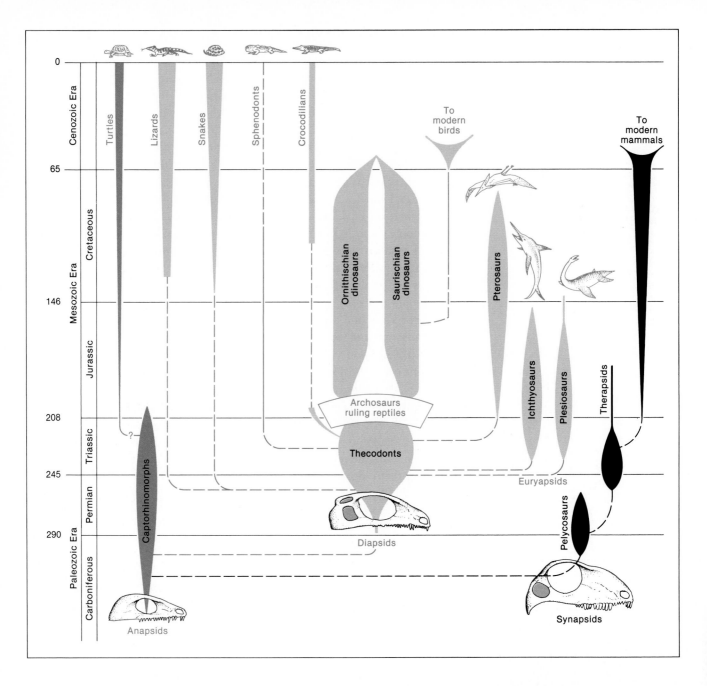

openings is not certain but may relate to the attachment of jaw muscles. Often a projecting arch develops in the skull where jaw muscles are attached. (The cheek bone of humans is an example of such an arch.) Temporal openings are often found in the skull near such arches. Captorhinomorphs and turtles lack these arches and temporal openings and are referred to as **anapsids** (from the prefix *an*, meaning without, and "apse," the arch-shaped wing that houses the altar of a cathedral). The ancestors of mammals had one pair of temporal openings and are called **synapsids** (Greek *syn* with). The early ruling reptiles, with two pairs of temporal openings, are called **diapsids** (Greek *di* two).

Turtles.   Today only a few hardy orders survive from the great Age of Reptiles. Turtles (order Chelonia) are the most distinctive from the others, being the only surviving anapsids. Most are also distinguished by a hard shell, which consists of the dorsal **carapace** and the ventral **plastron** (Figure 37.3). The shell consists of

Figure 37.2
A dendrogram representing the evolution of major groups of reptiles. Numbers on the left are millions of years before present. The three major lineages are recognized by the number of temporal openings in the skull behind the orbit of the eye. Captorhinomorphs and living turtles have no temporal openings and are called anapsids. The reptiles that were ancestral to mammals— synapsids—had one temporal opening on each side. Diapsids, with two temporal openings on each side, were the archosaurs (ruling reptiles) and their descendants. Dashed lines represent gaps in the fossil record.

pieces of dermal bone that are fused to each other and to the ribs and vertebrae. The dermal bone is covered by a layer of horny material similar to the scales of other reptiles. Only about 330 species of chelonians survive. Most occur mainly on land, but some occur only near fresh water, and some live in seawater except when breeding. The British call the terrestrial chelonians tortoises, the freshwater forms terrapins, and the marine species turtles, but Americans seldom make these distinctions.

Figure 37.3

Turtles. (A) The common eastern box turtle Terrapene carolina, *showing the carapace. The turtle can close up its shell tightly enough to retain moisture and keep out even the smallest predator. It is usually harmless, although people have died from eating box turtles that have consumed poisonous mushrooms. This and other species of turtles can live many decades and have been found with dates of more than a century ago carved on their carapaces. Length of carapace up to 22 cm. (B) The Eastern painted turtle* Chrysemys picta, *showing the plastron. This species is common in and around ponds in eastern and northern states. (See also Figure 15.1.) Faint annual growth rings on each segment of the plastron allow the age to be determined. (C) The leatherback turtle* Dermochelys coriacea *does not have a hard shell but is covered with a leathery skin with dermal plates embedded. This is the largest living turtle, reaching lengths of more than 2 meters. It wanders widely in the Atlantic and Pacific Oceans and can dive as deeply as 1.2 kilometers. Like many sea turtles, this species is endangered. This female is digging a hole in which to deposit eggs that were fertilized at sea.*

A

B

C

Lizards, Worm Lizards, and Snakes. Other living reptiles evolved from diapsids, although the number of temporal openings often changed from the original two. Most of the 3000 extant species of lizards (order Squamata, suborder Lacertilia) look like their ancestors, with four sprawling legs and long tails (Figure 37.4). There are also about 130 species of worm lizards, which are sometimes considered to be a separate suborder (Amphisbaenia). Worm lizards generally have no legs; some species have two. Contrary to appearances, snakes are closely related to lizards, and are classified in the same order but to a different suborder (Serpentes). Probably they evolved from lizardlike ancestors that took up burrowing and subsequently lost their legs. Most of the 2700 species of snakes have lost all external traces of legs, although some retain vestiges of the pelvic girdle (Figure 37.5).

Crocodilians. Order Crocodilia includes 21 species of crocodiles, alligators, caimans, and gavials (Figure 37.6). Crocodilians are more closely related to birds than to other extant reptiles, but they look like overgrown lizards, with four sprawling legs and a tail. These largest of extant reptiles use their webbed hind feet and tails with great effectiveness for swimming and live in or near water.

Figure 37.4
Lizards. (A) The common iguana, Iguana iguana, is green when young but turns brown or black with age. Some individuals grow to 2 meters long. They are nonvenomous but will defend themselves by biting, scratching, and lashing with their tails. Generally they are found in trees overhanging water, to which they escape when threatened. Common iguanas are native to Central America but have been introduced into Florida. (B) The two-legged worm lizard Bipes sp., sometimes classified in the subclass Amphisbaena. (C) A real dragon. The Komodo monitor Varanus komodoensis is up to 3 meters long and feeds on deer, pigs, carrion, and tourists. Several thousand wild Komodo dragons live on Komodo and a few neighboring islands in Indonesia.

A

B

C

A

B

Figure 37.5

Snakes. (A) The common garter snake Thamnophis sirtalis *lives in moist areas throughout the United States and well into Canada. There are numerous subspecies that vary in color and pattern. When disturbed it gives off a foul musk. It bites but is nonpoisonous. Females are up to 60 cm long and give birth to as many as 85 live young each year. Males are about one-third as long. (B) The timber rattlesnake* Crotalus horridus *of the eastern United States lives for up to 30 years and reaches lengths up to 1.9 meters. Rattlesnakes and their allies, copperheads and cottonmouths, are called pit vipers for the pit between the nostril and the eye, which leads to an infrared detector that detects warm prey. Each time the rattlesnake molts, usually two to four times per year, it adds a new segment to its rattle. (C) The highly venomous Arizona coral snake* Micruroides euryxanthus. *Another species of coral snake occurs in southeastern coastal areas of the United States, and there are more than 200 other species, mainly in the tropics. The bright colors are warning coloration that predators learn to avoid. Length up to 53 cm. (D) In areas where there are coral snakes, harmless snakes with similar coloration are presumed to be mimics. The milk snake* Lampropeltis triangulum *may be an example. Milk snakes were named for the common belief that they drank milk from cows, just one of many superstitions about snakes. Length up to 63 cm.*

C

D

A

B

C

Figure 37.6
Crocodilians. (A) The American crocodile Crocodylus acutus *lives at the extreme tip of Florida, in Central America, and on the brink of extinction. Adults may reach lengths of 7 meters, but large individuals are now rare. (B) Alligators have broader, rounder snouts than crocodiles. They commonly rest with all but their nostrils submerged. American alligators* Alligator mississippiensis *were hunted for their hides to the verge of extinction and were also sold by pet stores, but they are now common sights in the Gulf states. They reach lengths of up to 6 meters and occasionally attack swimmers. (C) After trade in alligators was prohibited, spectacled caimans* Caiman crocodilus, *named for the bony ridges around the eyes, were imported from Central and South America for sale as pets. Many were released, and some now inhabit drainage ditches in southern Florida. Length up to 2.6 meters.*

Tuataras. One final group of extant reptiles (order Sphenodonta) is also a diapsid and is represented by two surviving species. The lizardlike sphenodonts, commonly called tuataras (TOO-ah-TAH-ruz), cling to survival on more than a dozen islands of New Zealand, thanks to government protection (Figure 37.7). Until recently it was thought that there was just one species, *Sphenodon punctatus*, simply because a classification more than a century old had been ignored. Comparisons of enzyme structures and morphology have established the validity of that classification, so one small population of tuataras is now considered to be a second species, *S. guntheri* (Daugherty et al. 1990). This species is close to extinction because its taxonomic uniqueness was not recognized, and therefore the government of New Zealand did not afford it special protection. Tuataras can live more than 75 years as individuals, and they are also long-lived in evolutionary terms. They are descendants of a group that dates back more than 200 million years and was once thought to have become extinct 100 million years ago. One of the primitive features of these living fossils is the **pineal eye** (= median eye) complete with lens and retina. The pineal eye lies in the middle of the skull and regulates the activity of the pineal gland in the brain. Its ultimate function is not understood.

Extant Groups of Class Reptilia

Genera mentioned elsewhere in this chapter are noted.

Order Chelonia (= Testudines) kee-LOW-nee-uh (Greek *chelone* tortoise). No temporal openings (anapsid). Body enclosed in a two-part shell of dermal plates fused to vertebrae and ribs. Jaws without teeth. Turtles. *Caretta, Chelonia, Chrysemys, Dermochelys, Eretmochelys, Lepidochelys, Macroclemys, Pseudemys, Terrapene* (Figures 15.1, 37.3, 37.18B).

Order Squamata skwah-MAH-tuh (Latin *squama* scale). Body covered with epidermal scales that are shed with skin periodically. Paired copulatory organs. Tongue often forked and protrusible. Jaws usually with teeth.

Suborder Lacertilia (= Sauria) lay-sir-TILL-ee-uh (Latin *lacerta* lizard). Two pairs of limbs usually present. Slender body. With external ear opening and movable eyelids. Lizards and worm lizards (family Amphisbaenidae). *Amblyrhynchus, Anolis, Basiliscus, Bipes, Chamaeleo, Cnemidophorus, Draco, Heloderma, Iguana, Ptyodactylus, Tupinambis, Varanus* (Figures 37.4, 37.15, 37.18A and C, 37.21).

Suborder Serpentes sir-PEN-teez (Latin *serpens* serpent). Elongated, legless body. Tongue forked and protrusible. No external ear opening. Lidless eyes covered by transparent scale. No urinary bladder. Left lung usually vestigial. Jaws extremely flexible. Snakes. *Agkistrodon, Bungarus, Crotalus, Dispholidus, Hydrophis, Lampropeltis, Micruroides, Naja, Python, Thamnophis* (Figures 37.5, 37.17, 37.19, 37.20).

Order Crocodilia KROK-oh-DILL-ee-uh (Latin *crocodilus* crocodile). Stout body with two pairs of limbs and webbed, clawed toes. Tail laterally compressed, used as fin in swimming. Four-chambered heart. Crocodiles, alligators, caimans, and gavials. *Alligator, Caiman, Crocodylus* (Figure 37.6).

Order Sphenodonta (= Rhynchocephalia) SFEE-no-DONT-uh (Greek *spheno-* wedge + *odontos* tooth). Lizardlike. Pineal (median) eye apparent. Tuatara. Two extant species. *Sphenodon* (Figure 37.7).

TURTLE

LIZARD

SNAKE

CROCODILIAN

TUATARA

A

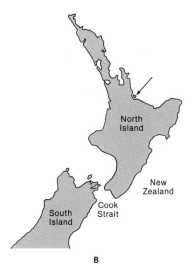

B

Figure 37.7
(A) Two species of sphenodonts, or tuataras, survive thanks to protection by the government of New Zealand. Length is approximately 70 cm. (B) Present range of tuataras.

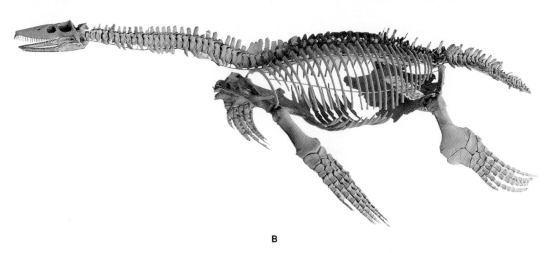

THE AGE OF REPTILES

Ichthyosaurs and Plesiosaurs. This book is mainly devoted to extant organisms, but dinosaurs and other ancient reptiles have drawn so many people to zoology that we owe them some attention. Among these were marine reptiles, ichthyosaurs and plesiosaurs. Ichthyosaurs ("fish-lizards") were shaped somewhat like sharks except that the vertebral column bent down into the lower caudal fin rather than into the upper caudal fin (Figure 37.8A). Plesiosaurs came in two forms: long-necked and short-necked (Figure 37.8B).

Figure 37.8
(A) Fossil of the ichthyosaur Stenopterygius *from the Lower Jurassic. (B) Fossilized Upper Jurassic plesiosaur* Cryptoclidus.

Figure 37.9

Part of the Age of Reptiles mural painted by Rudolph Zallinger at the Peabody Museum of Natural History at Yale University (reversed). All these species lived in North America during the Mesozoic era at different times and places. Colors are conjectural. ① Dimetrodon and ② Edaphosaurus were pelycosaurs ("finbacks"). Their dorsal fins may have absorbed or dissipated heat and may have been used in social displays. These mammal-like synapsids gave rise to the even more mammal-like therapsids, which included ③ Cynognathus. The cat-sized Saltoposuchus ④ was a crocodilian. Coelophysis ⑤, with a length of up to 2½ meters, was a rather small dinosaur. Plateosaurus ⑥ and Camptosaurus ⑦, up to 7 meters long, approached the sizes we associate with dinosaurs. Allosaurus ⑧, Stegosaurus ⑨, and Apatosaurus (= Brontosaurus) ⑩, with lengths up to 12, 9, and 21 meters, respectively, fully live up to our expectations of size. The triangular plates on Stegosaurus probably absorbed and radiated heat. Archaeopteryx ⑪ is shown soaring above. Like Edmontosaurus ⑫, a duck-billed dinosaur 13 meters long, most of the large dinosaurs up to the Cretaceous were herbivorous. Tyrannosaurus ⑬, with a length up to 12 meters, was one of the few that could take on another large reptile. Ankylosaurus ⑭, up to 11 meters long, may have defended itself against such large predators with the bony armor and club at the end of its tail. Triceratops ⑮, up to 9 meters long, protected its head and neck with horns and a bony shield. Pteranodon ⑯, a pterosaur with a wingspan of 7 meters, took to the air.

Pterosaurs. The sizes and shapes of ichthyosaurs and plesiosaurs conform to what we would expect of marine predators, but no aeronautical engineer would have designed a flying reptile like *Pteranodon* (Figure 37.9, number 16). For many years it was assumed that *Pteranodon*, with a wingspan of 7 meters and a mass of 17 kg, could only have glided from one perch to another. The discovery of fossilized fish within their fossilized ribs, however, indicated that they must have flown long distances over water. It is hard to see how they kept from crumpling their wings if they splashed into the water after prey, and even more difficult to understand how such large animals regained altitude. Perhaps they scooped up fish pelican-fashion and soared on ocean breezes. Even so, the lack of a stabilizing tail and the position of the wings behind the center of gravity made them aerodynamically unstable. The rudderlike head may have provided some lateral stability, but other pterosaurs, such as the huge *Quetzalcoatlus* (wingspan 12 meters), were even more unbalanced and lacked such rudders. Flight in *Quetzalcoatlus* has been compared to shooting an arrow backward. The aeronautical problems of pterosaurs was demonstrated in 1986 when a half-scale plastic model of *Quetzalcoatlus* with an on-board computer to correct attitude crashed on its only flight.

Dinosaurs. Of course the best known and most impressive extinct reptiles were the dinosaurs. Dinosaurs are not easily defined except as Mesozoic terrestrial reptiles with legs more upright than sprawling. They evolved from **thecodonts,** a group of relatively small, bipedal reptiles mainly from the Triassic period. Thecodonts resembled the early crocodilians, such as the one numbered 4 in Figure 37.9. Judging from differences in the structure of the pelvic bones, thecodonts gave rise to two main lines of dinosaurs: saurischians and ornithischians. The **saurischians,** or "lizard-hipped" dinosaurs, had pelvic structures similar to those of modern lizards. The pubis projected forward, and with the ischium made a fork pointing downward between the hind legs (Figure 37.10A). In **ornithischians,** or "bird-hipped dinosaurs," the pubis bent back along the ischium, as in birds (Figure 37.10B; compare Figure 38.12B). (Oddly enough, birds did not evolve from a bird-hipped dinosaur but from a lizard-hipped dinosaur similar to the species in Figure 37.10A.) All the bird-hipped dinosaurs were herbivorous and had a beaklike structure on the lower jaw. Bipeds and quadrupeds occurred among both saurischians and ornithischians. In Figure 37.9, the dinosaurs numbered 5, 6, 8, 10, and 13 were saurischians, while those numbered 7, 9, 12, 14, and 15 were ornithischians.

Dinosaur fossils were first discovered by quarrymen digging gypsum to rebuild

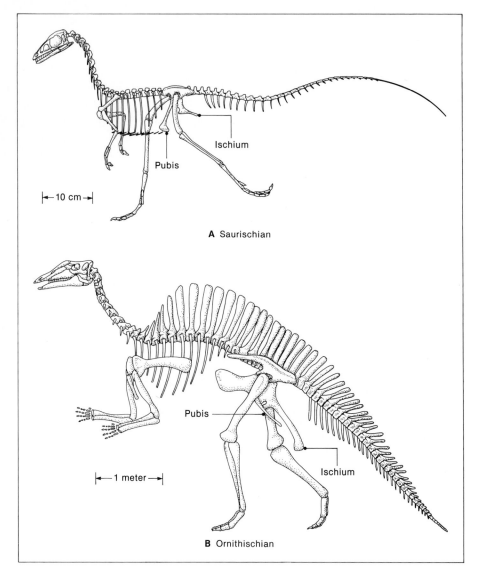

Figure 37.10
(A) A saurischian, Compsognathus, *with the pubis and ischium indicated.* Compsognathus *lived in the late Jurassic period.*
(B) Ouranosaurus, *an ornithischian from the early Cretaceous, showing the parallel pubis and ischium.*

Ischium

Pubis

|← 10 cm →|

A Saurischian

Pubis

Ischium

|← 1 meter →|

B Ornithischian

Paris after the French Revolution. Georges Cuvier (1769–1832), one of the first paleontologists, analyzed the fossils and recognized their significance. Not to be outdone, Americans later started their own explorations in the West and soon reported even more vast and exciting finds. *Tyrannosaurus rex, Apatosaurus* (formerly *Brontosaurus*), and *Diplodocus* enjoyed a few decades of fame, then were absorbed into popular culture along with Mickey Mouse. Until a few decades ago about the only people interested in dinosaurs were children and some paleontologists who were themselves sometimes suspected of being fossils. Dinosaurs were widely thought of as slow and stupid and of having become extinct from their own mass. (The term "dinosaur" is still applied disparagingly to large and cumbersome structures destined to become extinct, such as textbooks overburdened with parenthetical information.)

Today dinosaurs are usually depicted as active and intelligent, and the study of them has also become more active and intelligent. Fossilized tracks show that even large dinosaurs could run rapidly. They were also evidently capable of a high degree of social behavior, as is evident from tracks left by migrations of huge herds. Even more indicative of social behavior was the discovery by John R. Horner in 1978 of fossilized dinosaur nests, eggs, and young. The fact that young were still in the nest indicates parental care. Dinosaurs were much more like birds than lizards in their behavior, supporting a prior hypothesis, now advocated mainly by Robert T. Bakker, that dinosaurs were "warm-blooded." Bakker has argued, without much success, that dinosaurs maintained high body temperatures by producing body heat; that is, that they were endothermic.

Bakker has been more successful in erasing the image of dinosaurs as having been so slow and stupid that they simply became extinct—an idea that never made much sense in any case. But why *did* dinosaurs, after an enormous success lasting more than 100 million years, die out rather suddenly around the end of the Mesozoic era? There are as many theories about the dinosaurs' demise as there are about the decline and fall of Rome. The discovery that a large asteroid may have struck Earth around 66 million years ago is the latest contribution (see pp. 396–397). Although many paleontologists doubt that an asteroid impact was the only cause of the extinction of dinosaurs, a growing number accept it as the major cause.

AMNIOTE EGGS

Although reptiles ultimately failed to maintain dominance, they did give rise to one major adaptation that enabled their descendants, birds and mammals, to succeed as terrestrial animals. This adaptation was the amniote egg. Unlike the eggs of amphibians, the eggs of reptiles (and of other amniotes—birds and mammals) have **extraembryonic membranes** that provide water and oxygen for the embryo and remove metabolic wastes. These extraembryonic membranes include the **allantois,** the **chorion,** and the **amnion** (see p. 120). The allantois stores metabolic wastes and assists the chorion in the exchange of respiratory gases. The amnion holds amniotic fluid in which the embryo is bathed. In many species the **albumen** ("white") of the egg also serves as a water store. The eggs of amniotes are generally larger than those of fishes and amphibians, and they hatch into miniatures of the adults rather than into larvae.

The eggs of many lizards have soft shells, contain little albumen, and depend on the environment for water (Tracy and Snell 1985). The eggs of most reptiles, however, have shells more like those of birds' eggs: They contain more albumen and are totally independent of outside sources of water. Such self-contained eggs are often called **cleidoic eggs** (pronounced kly-DOH-ik). In some squamates, such as garter snakes, the eggs are even more independent of environmental supplies of water because they are incubated within the oviduct. These eggs have little or no shell, and in many and perhaps all cases the oviducts are adapted to provide some of the needs of the embryo. These reptiles are therefore viviparous.

SCALES

Another adaptation that enables most extant reptiles to survive on land are horny (keratinized) scales, which are assumed to retard the loss of body water by evaporation. Since scales occur in all living species of reptiles it is reasonable to assume that they evolved early. The earliest amphibians probably had dermal scales like their fish ancestors, but the scales of reptiles are not hand-me-downs from amphibians. They are entirely different, arising from the epidermis rather than the dermis (Figure 37.11).

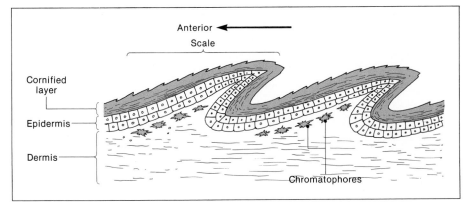

Figure 37.11
Epithelium of a reptile. Compare the epidermal scales with dermal scales of a fish (Figure 35.8). In snakes and lizards, new scales form beneath the old cornified layer of epidermis, which is shed periodically. Chromatophores often produce striking colors, to which may be added iridescent structural colors due to the fine sculpting of scales.

OSMOREGULATION AND EXCRETION

In addition to the impermeable skin of most reptiles, there are also internal adaptations for conserving water. Instead of mesonephric kidneys, as in fishes and amphibians, reptiles have metanephric kidneys like those of birds and mammals (see p. 292; Figure 37.12). Like birds, most reptiles excrete a thick, white urine with uric acid as the main nitrogenous waste. Uric acid is insoluble and therefore requires little water to eliminate. The urine empties through the cloaca, which is also the exit for feces and gametes. The external opening of the cloaca, called the **vent,** is a slit that runs lengthwise in crocodilians and turtles, and crosswise in lizards, snakes, and tuataras. Marine reptiles such as sea turtles, marine iguanas (Figure 37.16A), and sea snakes (Figure 37.17) also have **salt glands** in the head that excrete excess salt (see p. 295).

Figure 37.12
Internal structure of a male crocodile. The anatomy and physiology are similar to those of mammals, as described in Unit Two. Refer to the index for specific topics.

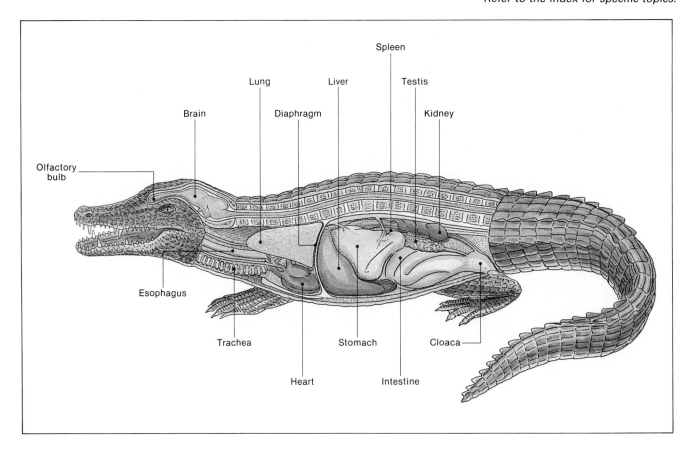

RESPIRATION

The skin of most reptiles (except sea snakes) is not only impermeable to water but also to respiratory gases. This places the entire burden for exchange of gases on the lungs. The lungs of reptiles are sacs like those of amphibians, but with larger folds in the walls that increase the surface area (Figure 37.13). Most reptiles have two lungs, but in snakes the left lung is usually vestigial. The mechanics of breathing varies widely among reptiles. Most inhale and exhale by expanding and contracting the chest cavity using the intercostal muscles between the ribs, as humans and other mammals do (see pp. 256–257). In crocodilians movement of the liver expands and compresses the lungs. During inhalation the liver is pulled backward by a muscular **diaphragm** (which is not homologous to the diaphragm of mammals). Because of their shells, turtles must use their leg muscles to expand and contract the body cavity for breathing. Aquatic turtles can also obtain oxygen by pumping water into and out of the mouth and cloaca, the linings of which function like gills. Reptiles generally have metabolic rates less than a tenth as high as those of birds or mammals of similar sizes, so their breathing rates are correspondingly lower. The respiratory tracts of most reptiles do not include a vocal apparatus, although many can make hissing noises by exhaling. In crocodilians and some lizards, however, there is a **larynx** that produces chirps, grunts, or roars.

CIRCULATION

Like amphibians, most reptiles have hearts with two atria and one ventricle (see Figure 11.4), but a partial septum in the ventricle generally separates oxygenated from deoxygenated blood. The septum is complete in crocodilians, so they have

Figure 37.13
Scanning electron micrograph of the wall of the lung of the tegu lizard Tupinambis nigropunctatus, *magnified approximately 35 times.*

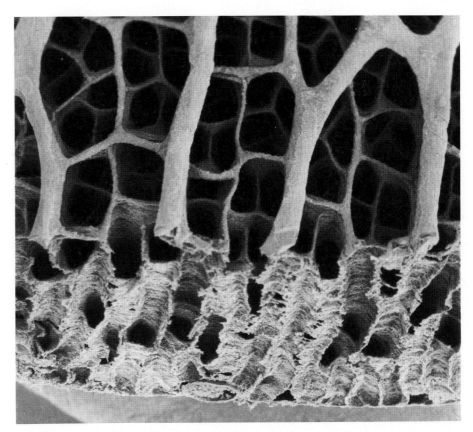

four-chambered hearts like birds and mammals. When crocodilians are submerged, however, a shunt between the ventricles allows blood to bypass the lungs. The respiratory and circulatory systems of reptiles can provide more oxygen to tissues than can those of amphibians, although less than those of birds and mammals. Nevertheless, many reptiles can move quite rapidly, using their fast-twitch muscles (type II, see p. 214), which produce ATP anaerobically. These muscles tire quickly, however, and an exhausted reptile takes hundreds or thousands of times longer than a mammal does to recover. Predaceous reptiles therefore tend to sneak up on prey rather than chase them down.

NERVOUS SYSTEM

The nervous system of reptiles approaches that of mammals in its organization, but the brain of a reptile is generally smaller than that of a mammal (or bird) of the same body size (see Figures 8.6 and 8.7). The reptilian brain also lacks a cerebral cortex, which is responsible for the most sophisticated behaviors in mammals.

Vision. Except for snakes, most reptiles that are active during the day have retinas with numerous cones, and they presumably have good color vision. Reptiles focus their eyes in a manner similar to that of mammals, by changing the shape of the lens. Most reptiles have paired eyelids, and also translucent **nictitating membranes** that sweep sideways and clear the cornea. Snakes, however, have no eyelids, which accounts for their steely gaze. Instead, their eyes are protected by transparent scales—probably an adaptation for burrowing. In *Sphenodon* and many lizards there is also a **pineal eye** on top of the head that probably detects only light intensity and may control biological rhythms. Some snakes can "see" warm prey in the dark by means of infrared receptors. Pit vipers have an infrared detector on each side of the head in a pit between the nose and eye (Figure 37.19). Pythons have as many as 13 pairs of infrared receptors around their mouths, and the related boa constrictor has infrared-sensitive scales around the mouth.

Chemoreception. Reptiles have well-developed olfactory bulbs with inputs from the olfactory epithelia in the nostrils, as well as from **Jacobson's organs** in a pair of depressions in the roof of the mouth. Jacobson's organs, also called **vomeronasal organs,** detect chemicals on the tongue. When a snake or lizard flicks out its forked tongue it is actually collecting olfactory molecules, which each tip of the tongue transfers to a Jacobson's organ.

Hearing. Most reptiles have hearing mechanisms similar to those of amphibians, with one middle-ear bone (the **columella** or **stapes**) that transfers sound from the tympanum (eardrum) to the inner ear (Figure 39.13A). In most turtles and crocodilians the tympanum is external, as in frogs, but in most lizards the tympanum is recessed in a canal. Some other lizards, as well as all snakes, lack tympani altogether. Instead, the columella touches a jawbone that picks up vibrations. This arrangement is probably responsive to vibrations of the substratum rather than to airborne sound.

REPRODUCTION

Sex Determination. Reptiles have separate sexes, although generally only they can tell the difference by casual inspection. Whether a reptile develops into a male or a female is determined genetically, but only certain squamates have sex chromosomes. In many lizards the females have two X chromosomes, and the males have one X and one Y chromosome, as in mammals. In other lizards and in many snakes, it is the female that is **heterogametic** (has differing sex chromosomes). In female heterogamety, which also occurs in birds, the sex chromosomes are designated ZW in females and ZZ in males. In some turtles, lizards, and crodocilians the genes that control sex are dependent on temperature during a critical period of incubation. Often only males develop if the incubation temperature is optimal, and only females develop if the temperature is very high or very low. The advantage of this temperature-dependent sex determination is hotly debated. It may ensure that when the population is endangered by environmental stress the more-valuable sex will be produced. It may also prevent inbreeding, since siblings tend to be of the same sex.

Copulation and Fertilization. Since most reptiles lay shelled eggs, internal fertilization is virtually dictated. Male tuataras have no copulatory organs; mating is achieved simply by pressing the cloacae together. All other male reptiles have at least one penis; snakes and lizards have two, called **hemipenes** (Figure 37.14). The male uses only one hemipenis at a time, depending on sensory information

Figure 37.14
Reproductive systems of male and female garter snakes. During erection one hemipenis of the male everts from the cloaca like the finger of a glove turned inside-out. The sex segment of the male's kidney produces seminal fluid and pheromones. In this viviparous species the oviduct also serves as a uterus.

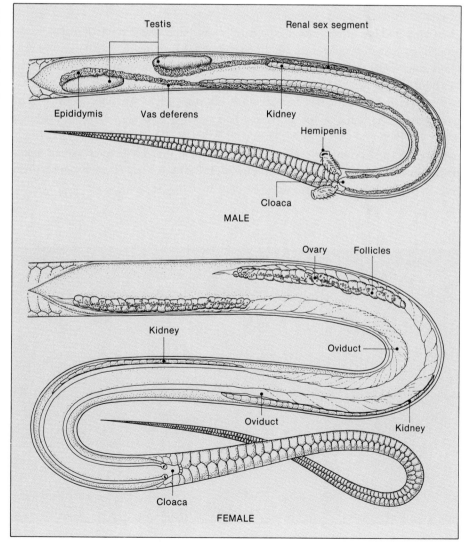

relating to which testis has the most mature sperm. The penis or hemipenis has a groove that guides the sperm into the female's cloaca during copulation. The sperm travel up the oviduct and fertilize the ova as they ripen, sometimes months or years after copulation. After fertilization the eggs pass down the oviduct, acquiring yolk, albumen (in many species), and finally the shell.

The reproductive hormones of reptiles include steroids (androgens and estrogens) from the testes and ovaries, as in mammals (see pp. 319–321). Gonadal functions are coordinated by a single hormone from the anterior pituitary instead of the two mammalian gonadotropins, FSH and LH. In most reptiles the females probably become sexually receptive while estrogen levels are high, around the time of ovulation. At this time female garter snakes, rattlesnakes, and perhaps other snakes secrete a sex-attractant pheromone from their skins, which draws males, often from great distances. In garter snakes the males do not have to travel far to find females, since thousands of both sexes den up together during the winter. The males emerge first during the spring, and dozens crowd around each female, forming a "mating ball" as she emerges. The successful male copulates by aligning his body along hers and opening her cloaca with the spines on one hemipenis. After copulation part of the ejaculate forms a plug in her cloaca that emits an **antiaphrodisiac** that discourages other males. One of the odd features of mating in garter snakes is that it occurs at a time when the gonads are regressed and sex steroid levels are low in both sexes. Apparently males fertilize eggs with sperm produced in the previous summer.

In crocodilians and lizards it is generally the male that attracts females. Male crocodiles seduce females with loud bellowing and pheromone, while lizards rely mainly on visual displays (Figure 37.15A). Perhaps the most surprising courtship behavior occurs in 15 of the 45 species of whiptail lizards (family Teiidae). What is so surprising is that courtship occurs at all, since these species consist only of parthenogenetic females. An individual behaves like a female when its estrogen levels are high and ovulation is imminent. After ovulation, when estrogen levels are low and progesterone levels are high, the same individual behaves like males do in bisexual species of whiptails (Figure 37.15B). Each female alternates between female and male behaviors several times per breeding season.

Figure 37.15
(A) Green anoles Anolis carolinensis *of the American Southeast can change color in seconds and are sold by pet stores as chameleons. Males are usually brown, especially when basking. When fighting or courting, they turn green, extend the pink dewlap on the throat, and bob the head up and down. Length up to 21 cm. (B) Pseudocopulation in desert-grassland whiptails* Cnemidophorus uniparens, *a unisexual species consisting entirely of parthenogenetic females. The upper female behaves like males of closely related bisexual species of whiptail lizards. Length approximately 20 cm.*

A

B

Parental Care. Most reptiles consider their parental duties done once they have laid fertilized eggs, or in the case of viviparous species, once the female has given birth. Crocodilians, however, are unexpectedly tender in their parental behavior. Most dig nests for the eggs. The American alligator provides the nest with wet vegetation that may generate heat by fermentation and that provides a culture for acid-secreting bacteria that help etch away the hard shell and make hatching easier. Young crocodilians often begin chirping like chicks at the time of hatching, and this encourages the mother to uncover the nest. Then, using jaws that could just as easily tear off a person's leg, she gently picks up the hatchlings and carries them to the water. Not only the mother but also the father and other adults respond to distress calls from young, but as soon as the young reach sub-adulthood adult males no longer tolerate their presence.

LOCOMOTION

Turtles. The basic pattern of reptilian locomotion is that of lizards, on four sprawling legs. Turtles retain this mode of locomotion. Unlike lizards, however, turtles are encumbered by the shell and do not have the advantage of lizards in being able to advance the legs by bending the body. A land turtle may spend its long life plodding no more than a hundred meters from where it hatched. At the other extreme, however, are **sea turtles,** especially the leatherback (Figure 37.3C), green turtles (*Chelonia*), hawksbills (*Eretmochelys*), loggerheads (*Caretta*), and ridleys (*Lepidochelys*). Many sea turtles migrate thousands of kilometers, and the females return periodically to one beach to bury their fertilized eggs in the sand. It is assumed that hatchlings become imprinted to the smell of their native beach, and that upon reaching adulthood the females home to that beach first by magnetic and celestial cues, and then by smell.

Sea turtles hatch after a month or more of incubation, then dig out of the sand and head straight toward the ocean, attracted toward the brightest part of the horizon, over the water. Once in the water they appear to be guided by mechanical stimulation from waves, and later by magnetic fields. They swim straight out for many kilometers, drawing power from the yolk located within their guts. For many years no one knew where they went during the first year at sea. Finally in 1984, after three decades of study, Archie Carr (1909–1987) and others solved the "mystery of the lost year" by tracking tagged turtles. They found that the young turtles swim out to areas where ocean currents come together, and they ride on rafts of sargassum seaweed that accumulate there. Here they find rest and shelter from predators, while they themselves prey on shrimps, crabs, jellyfish, and other animals also attracted to the rafts. Sea turtles that hatch in Florida swim out to the Gulf Stream, then float on sargassum rafts as they drift clockwise around the Atlantic. Eight months later many arrive at the Azores; then for the next year or so they continue down the west coast of Africa and back to the Caribbean. Eventually they become large enough to wander the seas actively, rather than by drifting. After 10 to 50 years they reach sexual maturity and start a new cycle.

Snakes. The locomotion of snakes is radically altered from that of their lizard ancestors and is quite adaptable to circumstances. In all forms of terrestrial locomotion snakes anchor their bodies at two or three points, then advance the portions of the body that are not anchored (Figure 37.16). At the same time the anchor points shift backward at the same speed as the snake moves forward, giving the impression that the snake is flowing effortlessly like a stream of water. Four basic styles of terrestrial locomotion can be recognized:

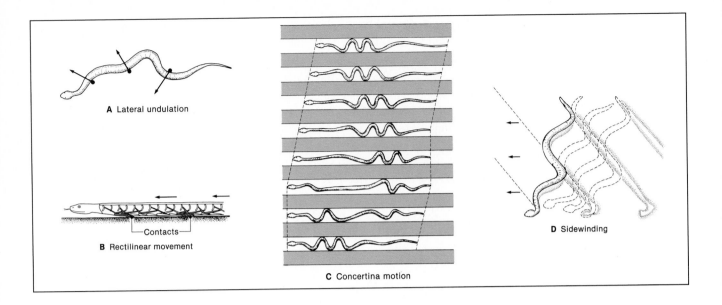

A Lateral undulation

B Rectilinear movement

—Contacts—

C Concertina motion

D Sidewinding

1. The most common is **lateral undulation,** which snakes apparently prefer when on firm and irregular substratum. In lateral undulation the snake pushes sideways against rocks, plants, or other elevated points (the optimum number is three).

2. One disadvantage of lateral undulation is that the sideways movement is easily visible to prey. While stalking, many reptiles use a less-visible **rectilinear movement** in which the body moves straight forward, using two or three points of contact between the substratum and ventral scales. The snake progresses by extending its body forward while continually reestablishing new points of contact behind the old ones.

3. Burrowing snakes use what is called **concertina motion,** in which one part of the body bends into an S shape that is braced against the walls of the burrow. The front part of the bend then straightens out to extend the body, while the bend is continually reformed at the rear. Before the bend reaches the end of the snake a new one forms behind the head.

4. Finally, many snakes on hot, loose sand use a unique **sidewinding** motion. When sidewinding, a snake twists its body into a spiral at right angles to the direction of overall motion, with two parts of the body contacting the substratum. The head is then flung forward, followed by posterior parts of the body. The advantage of sidewinding is that the body tends to dig in for better traction, and it minimizes contact with hot sand.

Miscellaneous Forms. Many squamates have interesting modifications of these locomotory patterns. Sea snakes, for example, have laterally compressed tails that adapt lateral undulation for swimming (Figure 37.17). Sea snakes can also dive effectively and stay submerged for up to eight hours. Apparently they do not inflate the lungs before diving, since that would make them too buoyant. Instead, they obtain oxygen through the integument. Some tropical snakes also leap great distances between branches and are called "flying snakes." Peculiar adaptations for locomotion can also be found among lizards. Some geckos, such as the house gecko *Ptyodactylus hasselquistii* of northern Africa, have loose folds of skin on their toes that enable them to walk on vertical walls. The flying dragons (*Draco* spp.) of the Asian tropics have flaps of skin on the sides that fill with air like parachutes, enabling these lizards to glide between trees. No less impressive are basilisks of

Figure 37.16
Modes of terrestrial locomotion in snakes. (A) In lateral undulation the sides of the body push against points of contact. Arrows show the direction of forces, which are not along the body axis but have a net direction that propels the snake forward. The forward movement does not depend on friction against the points of contact. In fact, lateral undulation is fastest if the contacts can rotate. (B) In rectilinear movement the body is aligned in the direction of movement, and ventral scales are dug in at several points (shown in color). Contractions of muscles posterior to these points pull the body forward (arrows). New points of contact are continually formed posterior to the previous ones. (C) In concertina movement successive portions of the body are bent into waves that brace against the sides of a burrow. (D) Sidewinding is used in loose sand and produces characteristic J-shaped tracks. The hook of the J is formed by the head, which is flung forward, followed by posterior portions of the body. Dashed outlines show previous positions. The body is at right angles to the direction of travel, and the tracks are diagonal to it.

Figure 37.17
The banded sea snake Hydrophis fasciatus *has a laterally compressed tail that it uses like a fin when swimming. On shore it is virtually helpless because its scales are so smooth.*

the American tropics (*Basiliscus* spp.), sometimes called "Jesus Christ lizards," because they can run on their hind feet across water at speeds up to 12 km/hr.

FEEDING

Most reptiles move slowly when foraging, feeding on plants, carrion, eggs, worms, molluscs, insect larvae, and other foods that can be caught without a chase (Figure 37.18A). The alligator snapping turtle does not have to move at all to catch fish, since it has a worm-shaped lure in its mouth (Figure 37.18B). (It is merely a myth, however, that snakes lure birds by hypnotizing them.) On the other hand, many predaceous reptiles are quite agile hunters. Crocodilians run and swim rapidly after prey, then thrash violently to tear off large chunks, which they swallow without chewing. Like all reptiles, they lack the variety of cutting, tearing, and grinding teeth found in mammals. Other carnivorous reptiles swallow their prey whole.

The inability of most reptiles to bite off pieces of flesh limits their prey to those that are smaller than the reptile's head. In snakes, however, the jaw bones are loosely hinged to each other in such a way that the mouth can accommodate prey several times wider than the head (Figure 37.19). Flexible portions of epidermis between the scales also allow the skin to stretch. Snakes do not choke on large prey because the tracheal opening is located near the front of the mouth.

VENOMS

Most predaceous reptiles simply seize prey in their mouths, which usually have sharp, recurved teeth on the jaws or palate (**vomerine teeth**) that hold the struggling victim. A few squamates, however, first subdue or kill prey. Boas and pythons (family Boidae) first suffocate their prey by wrapping themselves around the body and tightening the coils each time the prey exhales (Figure 37.20). Some squamates subdue prey with venom, which is essentially saliva containing mixtures of digestive enzymes. In addition to paralyzing and killing prey, venom also starts the digestive process from inside the prey. The enzymes in snake venoms include proteases, collagenases, and phospholipases that act on all kinds of tissues.

A

C

B

Whether blood, heart, muscles, nerves, or another tissue is attacked depends on how and where the venom is injected and the size of the prey. The common practice of classifying snake venoms as hemotoxic, neurotoxic, cardiotoxic, or myotoxic is therefore misleading (Russell 1984). It is also potentially dangerous if it encourages the assumption that an **antivenin** (blood serum containing antibodies to snake venom) of one type is useless against another.

The gila monster and a related lizard of Mexico and the southwestern United States are the only two species of venomous lizards (Figure 37.21). There are at least four families of venomous snakes. Coral snakes and cobras (family Elapidae) are the most dangerous. Indian cobras (*Naja naja*) by themselves cause an estimated 10,000 deaths each year. These snakes have two permanently erect fangs in front, each with an external groove through which venom flows into the bite. One cobra, the krait *Bungarus*, is sometimes called the "two-step," because that's about as far as a person gets after being bitten. The spitting cobra *Naja nigricollis* can also blind a victim by spitting venom into its eyes. Sea snakes (family Hydrophiidae) are similar to cobras except for their marine adaptations. Puff adders and other Old World vipers (family Viperidae) comprise a third group of venomous snakes. New World pit vipers—rattlesnakes, copperheads, and water moccasins—are also often included in this family, or in a separate family Crotalidae. Viperids and crotalids have front fangs that fold back when not in use, and each fang works like a hypodermic needle (Figure 37.19B). Family Colubridae includes most of the common nonvenomous snakes, such as garter snakes and milk snakes, but also some species, such as the African boomslang *Dispholidus typus*, with saliva so toxic that it is wise to consider it a venom. Colubrids have grooved fangs in the rear of the mouth.

Figure 37.18
(A) Marine iguanas Amblyrhynchus cristatus *bask after feeding on algae in the chilling waters off the Galápagos Islands. These harmless vegetarians reach lengths of up to 1¾ meters. (B) The alligator snapping turtle* Macroclemys temminckii *lives in the Gulf states and Mississippi area, where it lies in water with its mouth open, wiggling the wormlike fishing lure in its toothless beak. (C) The three-horned chameleon* Chamaeleo jacksoni *of eastern Africa demonstrates its hunting prowess. After anchoring itself to a branch with its tail, it suddenly inflates its tongue with blood and launches the sticky end at the target, which may be as far away as its own body length (up to 30 cm). The chameleon then hauls the prey into its mouth and coils up its tongue. This form of predation obviously requires good vision. Each of the chameleon's eyes peeps out of fused eyelids and can be moved independently of the other. Chameleons are famous for their ability to change color, which they generally do during social communication rather than for camouflage.*

Figure 37.19
Feeding mechanisms of snakes.
(A) A copperhead Agkistrodon
contortrix *swallowing a deer mouse*
(Peromyscus). Note that the opening
of the trachea (the glottis) is at the
front of the mouth, preventing
suffocation during the long process
of swallowing. (B) Jaw bones of a
rattlesnake enable the mouth to
open widely. In addition, the jaw
bones on one side can separate
from those on the other side.
(C) The mouth of a rattlesnake.

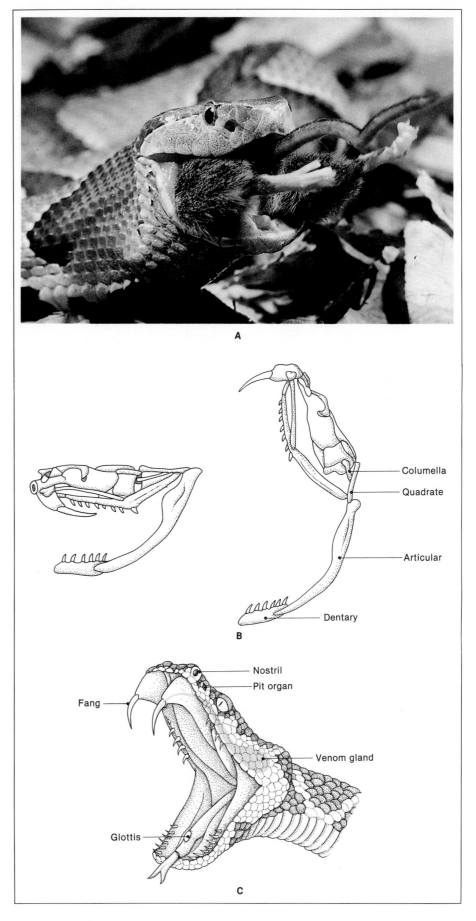

A

Columella
Quadrate
Articular
Dentary
B

Fang
Nostril
Pit organ
Venom gland
Glottis
C

INTERACTIONS WITH HUMANS

Most people's images of reptiles are colored by childhood phobias and myths. There are people in the United States who kill any snake they see because a serpent supposedly tempted Eve to eat the forbidden fruit in the Garden of Eden. (Yet they eat apples.) Many others are mistrustful of all reptiles because of the few that are dangerous. While it is true that venomous snakes are extremely dangerous in many parts of the world, the average American is more likely to be killed by another person than to be bitten by a venomous snake. Even if a person is one of the 8000 Americans bitten each year, he has a 99.8% chance of surviving.

Reptiles are much less a threat to humans than humans are to reptiles. Besides deliberately killing them, humans also destroy their habitats. The situation is especially critical for sea turtles, which have the misfortune of depending on the same sunny beaches preferred by tourists and developers. Sea turtles are also threatened by human garbage and spilled oil, which accumulate in the same ocean convergences as the sargassum rafts where sea turtles spend their first year. Sea turtles are often found with their digestive tracts blocked by plastic bags presumably mistaken for jellyfish, or their mouths glued shut by tar balls from oil spills. Thousands of sea turtles also drown in the nets of shrimp trawlers. Japan recently agreed to stop importing 18,000 hawksbill turtles per year to make turtle-shell combs, mirror handles, and trinkets.

Several species of sea turtles are so close to extinction that many people have begun to take notice. The Mexican government now has armed troops protecting the only nesting site of Kemp's ridley turtle (*Lepidochelys kempi*) from poachers, who sell the eggs for their supposed aphrodisiacal powers. The measure may be too late, however. From some 40,000 nesting females in 1947, the population of Kemp's ridley has dropped to fewer than a thousand, and the species may well be extinct by now. After years of lobbying by conservationists, U.S. lawmakers have finally taken steps to protect Kemp's ridley and other sea turtles, by requiring escape hatches (turtle excluding devices, or TEDs) on shrimp nets. Getting the shrimpers to obey the law has not been easy, however. Some 40,000 sea turtles die each year in shrimpers' nets.

Figure 37.20
The reticulated python Python reticulatus *of Southeast Asia killing a rat by constriction. The python does not exert much muscle contraction but simply locks its coils to keep the victim from inhaling. This is one of the longest snakes, reaching lengths of up to 10 meters.*

Figure 37.21
The gila monster Heloderma suspectum *of the southwestern United States and Mexico, one of only two venomous lizards. The venom comes from a gland in the lower jaw and is not injected but simply flows into the bite. The venom is seldom fatal to humans, but once a gila monster has bitten it hangs on tenaciously. Gila monsters feed mainly on eggs, young birds, small rodents, and other lizards. They store nutrients in the stout tail.*

While sea turtles have attracted the most attention, many other reptiles are also endangered by humans. Many are killed for their hides, which are often imported illegally into the United States to make shoes and handbags. Three species of sea snakes in the Philippines have become extinct from such hunting. Some reptiles, such as turtles and rattlesnakes, are hunted for food. Many other reptiles are traded legally or illegally as exotic pets. Millions of dollars in pet red-eared turtles (*Pseudemys scripta-elegans*) were sold annually in pet stores until the trade was stopped in 1975, when it was discoverd that they were transmitting salmonella infections to children. Approximately 100,000 iguanas are imported into the United States each year from Central America as pets, and most soon die from improper care. Populations are now being restored by a program of captive breeding and restoration of habitat (Cohn 1989). This is one example of what can be accomplished with timely, serious efforts. The American alligator, once nearly extinguished by poachers, is now a common tourist attraction and a source of income to hunters who can now sell the hides legally.

SUMMARY

There are three major groups of extant reptiles, distinguished mainly by the number of temporal openings in the skull. Turtles are the only extant anapsids. Synapsids now comprise tuataras, lizards, worm lizards, and snakes. Extant diapsids are the crocodilians. Reptiles were much more numerous and diverse during the Mesozoic era and included marine ichthyosaurs and plesiosaurs, flying pterosaurs, and terrestrial dinosaurs. Dinosaurs ranged in length from a few meters to 21 meters, but even the largest were probably quite agile. Their extinction was probably due largely to asteroid impact.

One of the keys to the ability of reptiles to survive on land is the amniote egg, with its extraembryonic membranes that hold water and perform other functions. Reptiles also conserve water with their scaly skins. Aside from these and some other shared features, they are extremely diverse in shape and in locomotion, feeding, reproduction, and other functions.

KEY TERMS

temporal openings
anapsid
synapsid
diapsid
carapace
plastron
amniote egg

extraembryonic membranes
amnion
allantois
chorion
albumen
cleidoic egg
salt gland

heterogametic
hemipenis
lateral undulation
rectilinear movement
concertina motion
sidewinding

SELF-TEST

1. Describe in simple terms the distinguishing features of turtles, lizards, snakes, and crocodilians.

2. For each of the major groups in the previous question, briefly describe their evolutionary origins. In what way are they related to dinosaurs?

3. Describe the amniote egg and explain how it enables most reptiles to reproduce without water.

4. Describe the integument of snakes and the functions of scales.

5. Why are the circulatory and respiratory systems of most reptiles less effective than those of humans? How can reptiles live with these less-efficient systems?

6. Describe reproduction in one species of reptile.

7. Describe the special adaptations of a rattlesnake for predation.

8. Explain how a snake is able to move on land without legs.

READINGS

RECOMMENDED READINGS

Buffetaut, E. 1979. The evolution of the crocodilians. *Sci. Am.* 241(4):130–144 (Oct).

Carr, A. 1986. Rips, FADS, and little loggerheads. *BioScience* 36:92–102.

Cole, C. J. 1984. Unisexual lizards. *Sci. Am.* 250(1):94–100 (Jan).

Crews, D. 1979. The hormonal control of behavior in a lizard. *Sci. Am.* 241(2):180–187 (Aug).

Crews, D. 1987. Courtship in unisexual lizards: a model for brain evolution. *Sci. Am.* 257(6):116–121 (Dec).

Crews, D. and W. R. Garstka. 1982. The ecological physiology of a garter snake. *Sci. Am.* 247(5):158–168 (Nov).

Gans, C. 1970. How snakes move. *Sci. Am.* 222(6):82–96 (June).

Heatwole, H. 1978. Adaptations of marine snakes. *Am. Sci.* 66:594–604.

Horner, J. R. 1984. The nesting behavior of dinosaurs. *Sci. Am.* 250(4):130–137 (Apr).

Langston, W. Jr. 1981. Pterosaurs. *Sci. Am.* 244(2):122–136 (Feb).

Lillywhite, H. B. 1988. Snakes, blood circulation and gravity. *Sci. Am.* 259(6):92–98 (Dec).

Newman, E. A. and P. H. Hartline. 1982. The infrared "vision" of snakes. *Sci. Am.* 246(3):116–127 (Mar).

Pooley, A. C. and C. Gans. 1976. The nile crocodile. *Sci. Am.* 234(4):114–124 (Apr).

See also relevant selections in General References at the end of Chapter 21.

ADDITIONAL REFERENCES

American Zoologist volume 29, number 3 (1989) has several papers on crocodilians.

Bakker, R. T. 1986. *The Dinosaur Heresies.* New York: William Morrow. (*A controversial book that brings the dinosaurs back to life.*)

Behler, J. L. and F. W. King. 1979. *The Audubon Society Field Guide to North American Reptiles and Amphibians.* New York: Alfred A. Knopf.

Carroll, R. L. 1988. *Vertebrate Paleontology and Evolution.* New York: W. H. Freeman.

Cohn, J. P. 1989. Iguana conservation and economic development. *BioScience* 39:359–363.

Daugherty, C. H. et al. 1990. Neglected taxonomy and continuing extinctions of tuatara (*Sphenodon*). *Nature* 347:177–179.

Ferguson, M. W. (Ed.). 1984. *The Structure, Development and Evolution of Reptiles.* New York: Academic Press.

Halliday, T. R. and K. Adler (Eds.). *Encyclopedia of Reptiles and Amphibians.* New York: Facts on File.

Norman, D. 1985. *The Illustrated Encyclopedia of Dinosaurs.* New York: Crescent Books. (*One of the best of a steady stream of dinosaur books.*)

Russell, F. E. 1984. Snake venoms. In: M. W. Ferguson (Ed.), *The Structure, Development and Evolution of Reptiles.* New York: Academic Press, pp. 469–480.

Thomas, R. D. K. and E. C. Olson (Eds.). 1980. *A Cold Look at Warm-Blooded Dinosaurs.* Boulder, CO: Westview Press. (*A review of Bakker's theory of homeothermic dinosaurs.*)

Tracy, C. R. and H. L. Snell. 1985. Interrelations among water and energy relations of reptilian eggs, embryos, and hatchlings. *Am. Zool.* 25:999–1008.

Wilford, J. N. 1986. *The Riddle of the Dinosaur.* New York: Alfred A. Knopf. (*Recent discoveries and the people who made them.*)

38

Birds

Crimson rosella (Platycercus elegans).

CHAPTER OUTLINE

LEARNING OBJECTIVES

1. How did feathery, warm, flying birds come to be so different from scaly, cold, creeping reptiles?
2. What are the functions of feathers? How are feathers formed, and how do they get their colors? Why are the feathers colored?
3. How do birds fly?
4. How do respiration and digestion provide enough energy to fly and maintain a warm body?
5. How do bird eggs keep the embryo alive during development? Why are the shells colored?
6. Why do birds sing?
7. How and why do birds build different kinds of nests?
8. How and why do birds care for eggs and young?

THE ORIGIN OF BIRDS

Archaeopteryx. It doesn't take a zoologist to tell the difference between a living reptile and a bird. Yet there is really only one fundamental difference between the two groups: birds have feathers. Feathers, in turn, enable birds to fly and to maintain a constant, high body temperature, unlike extant reptiles. Flight and homeothermy led in turn to other physiological and anatomical differences from reptiles. The farther one goes back in the fossil record, however, the harder it is to see a difference between birds and reptiles. The oldest known bird, *Archaeopteryx lithographica*, had a skeleton almost indistinguishable from that of certain small dinosaurs. If impressions of its feathers had not been preserved, its fossils would still be collecting dust in a museum basement, labeled as just another dinosaur (Figure 38.1). Two fossils of *Archaeopteryx* were, in fact, discovered in museum collections in just that way. Because *Archaeopteryx* had feathers, however, zoologists place it in class Aves (pronounced AY-veez). *Archaeopteryx* is generally considered to be the best fossil evidence for the evolution of one group from another—namely, birds from dinosaurs.

Other Early Birds. We know *Archaeopteryx* only because the sediment in which it was buried was fine enough to form clear impressions of the feathers, and because its bones were thicker than those of modern birds. Most sediments are too coarse to preserve the details of feathers, and most birds are so light that they float rather than get covered by sediments. Also, most birds have hollow bones that disintegrate before they can form fossils. For these reasons the fossil record is too sparse to indicate the origin of modern birds. However, fossils recently uncovered in China indicate that only 10 million years after *Archaeopteryx* there lived a sparrow-sized bird (not yet named) with a sternum for the attachment of flight muscles. It also had such modern features as a short tail and feet adapted for perching. Other early modern birds include *Hesperornis*, a large bird with a long neck and broad feet (Figure 38.2). Its sternum was too shallow to have supported large flight muscles, so its main forms of locomotion were probably swimming and diving. The second form included *Ichthyornis*, which has a deeper **sternal keel** where flight muscles were attached, and was therefore probably a good flyer. Like *Archaeopteryx*, both forms had teeth, a primitive feature not found in any modern bird.

Figure 38.1
Reconstructed skeleton of Archaeopteryx lithographica. See also Figure 18.2. Like dinosaurs, Archaeopteryx had heavy scales on its breast and gastral ribs along the belly, which may have substituted for the lack of a sternum to which flight muscles could attach. Unlike reptiles, however, Archaeopteryx had a furcula (wishbone; not visible here). The asymmetric position of the shaft of their feathers indicates that Archaeopteryx did fly (see Figure 16.5B). Archaeopteryx was first named on the basis of a single feather discovered in Germany in 1860 by quarrymen removing the fine-grained stone used in lithography. The fine sediment that formed the stone preserved the outline of the feathers. Otherwise Archaeopteryx would be indistinguishable from small dinosaurs like Compsognathus, which lived at the same time and place as Archaeopteryx. (Compare Figure 37.10A.)

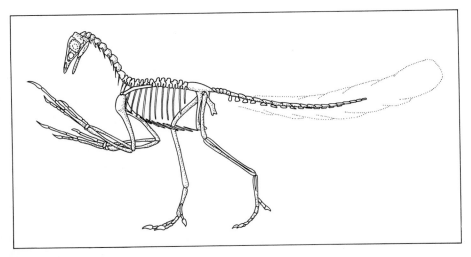

DIVERSITY

Archaeopteryx was much more dinosaurlike than *Hesperornis* and *Ichthyornis*, and is therefore placed all by itself in subclass Archaeornithes. *Hesperornis* and *Ichthyornis* are placed in subclass Neornithes with other modern birds. Beginning with the Cenozoic era around 65 million years ago, the fossil record is more nearly complete, but modern orders appear so abruptly that it is difficult to trace the relationships among them.

According to a recent count more than 8800 extant species of birds have been named, and on average two new ones are discovered each year, mainly along the remote headwaters of the Amazon. Ornithologists generally recognize 27 or 28 extant orders (Figure 38.3). The perching birds (order Passeriformes) comprise well over half of all species of birds, including the familiar song birds (suborder Passeres).

Taxonomists generally divide extant modern birds into two major groups: **ratites** and **carinates.** Ratites are large, flightless birds of the Southern Hemisphere (Figure 38.3). They include ostriches, rheas, cassowaries, emus, and kiwis, as well as the giant moa and elephant bird that were hunted to extinction by humans. Ratites generally live on islands or in other areas where they did not need to fly from predators, at least until humans arrived. Ratites have sternums shaped like flat-bottomed boats (*ratis*, in Latin), without keels for the attachment of flight muscles. Since flightlessness has evolved independently in some other large birds, such as penguins, it has often been assumed that ratites are polyphyletic. Comparisons of proteins and nucleic acids indicate, however, that they descended from a common ancestor, presumably on Gondwana (Stapel et al. 1984). The other modern birds, the carinates, have deep sternal keels (Latin *carina* ship's keel) for the attachment of large flight muscles (see Figure 38.10). All carinates are either good fliers or are descended from good fliers.

FEATHERS

Functions. T. H. Huxley called birds "glorified reptiles." If that is so, most of a bird's glory lies in its feathers. Feathers are important for three major functions: temperature regulation, flying, and camouflage or social display. The importance of feathers in temperature regulation was reviewed in Chapter 15. Birds, like mammals, are endothermic homeotherms. That is, they maintain a more or less constant body temperature by conserving the body heat produced by high metabolic rates. The majority of birds maintain core body temperatures between 40 and 42°C, which is possible only because feathers reduce thermal conductivity and radiation. Feathers also shade the thin skin from harmful solar radiation.

Figure 38.3
Representatives of 28 commonly recognized orders of birds (pp. 835–837). Many ornithologists combine orders Phoenicopteriformes and Ciconiformes.

Development. Although feathers are the essential feature by which birds are distinguished from reptiles, they do not represent a radical departure from reptilian scales. They are in fact homologous with scales. The development of a feather begins with a **papilla** that is essentially a cylindrical epidermal scale (Figure 38.4). Each papilla elongates and sinks into a **follicle** in the dermis. Blood vessels and nerves from the dermis then enter the papilla, forming a **pulp.** The pulp supplies nutrients for the surrounding epidermis to grow and produce keratin, which forms the feather as well as the sheath that protects the developing feather.

In the development of the most familiar feathers, called **vane feathers,** a **shaft** grows within the papilla, and from it numerous **barbs** grow out at an angle. These barbs are curled up within the sheath, but once the feather is fully developed the sheath bursts, and the barbs unfurl into two **vanes,** one on each side of the shaft. The pulp then dries up, leaving a hollow **quill.** The portion of the shaft extending from the quill is the **rachis** (RAY-kis). Extending from each barb are numerous tiny **barbules.** In vane feathers the barbules have hooked ends that interlock with adjacent barbs like Velcro, causing the vanes to form a single flexible surface (Figure 38.5).

Major Groups of Class Aves

Genera mentioned elsewhere in this chapter are noted.

Subclass Archaeornithes AR-kee-OR-nith-eez (Greek *archaios* ancient + *ornitho-* bird). Dinosaurlike bird with true teeth, clawed toes, and long tail with vertebrae. Finger bones unfused. No sternum. One species known: *Archaeopteryx lithographica.*

Subclass Neornithes nee-OR-nith-eez (Greek *neo-* new). No true teeth; no long bony tail; finger bones fused; sternum present.

  Superorder Odontognathae oh-don-TOG-nath-ee (Greek *odonto-* toothed + *gnathos* jaw). With toothed jaws. *Hesperornis, Ichthyornis* (Figure 38.2).

  Superorder Neognathae nee-OG-nath-ee. Modern birds. *Acrocephalus, Alcedo, Aptenodytes, Apteryx, Archilochus, Bombycilla, Bonasa, Branta, Brotogeris, Bubo, Cardinalis, Cathartes, Centrocercus, Chlamydera, Collocalia, Cuculus, Cygnus, Delichon, Ectopistes, Erithacus, Falco, Gallus, Grus, Gymnogyps, Haliaeetus, Lobipes, Melanerpes, Meleagris, Melopsittacus, Menura, Minus, Otis, Parus, Passer, Pavo, Phaethon, Phalaropus, Pharomachrus, Phasianus, Philetairus, Phoeniculus, Platycercus, Ploceus, Scolopax, Steganopus, Sterna, Struthio, Sturnus, Troglodytes, Turdus, Tyto* (Figures 20.2, 20.17, 20.18, 20.20, 38.3, 38.6B and C, 38.7, 38.11, 38.17, 38.19A, 38.20, 38.21, 38.22, 38.23, 38.24, 38.25, 38.26).

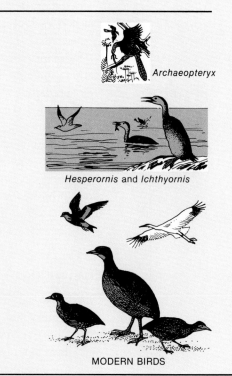

Archaeopteryx

Hesperornis and *Ichthyornis*

MODERN BIRDS

RATITES

Order Casuariiformes   **5 species**

Order Struthioniformes   **1 species (Africa)**

Order Rheiformes
2 species (S. America)

Cassowaries
(Australia and New Guinea)

Emus (Australia)

Ostriches

Rheas

CARINATES

Order Apterygiformes
3 species (New Zealand)

Order Tinamiformes
47 species (Central and S. America)

Order Sphenisciformes   **18 species**

**(Antarctica
and cold
water north
to Galapagos)**

Kiwis

Tinamous

Penguins

Order Gaviiformes
5 species (northern hemisphere)

Order Procellariiformes
104 species (worldwide)

Order Pelecaniformes
62 species (worldwide)

Loons

Shearwaters

Pelicans

Order Podicipediformes
20 species (worldwide)

(Winter)   (Summer)

Boobies

Grebes

Petrels

Cormorants

Order Ciconiiformes   **116 species (worldwide)**

Order Phoenicopteriformes
6 species (tropical and temperate)

Herons

Bitterns

Storks

Flamingos

Order Anseriformes
150 species (worldwide)

Ducks

Swans

Geese

Order Falconiformes

288 species (worldwide)

Vultures

Eagles

Hawks

Order Galliformes
236 species (worldwide)

Grouse

(Winter)

Ptarmigans

♂ ♀

Pheasants

Order Gruiformes   **223 species (worldwide)**

Rails

Coots

Cranes

Order Charadriiformes   **326 species (worldwide)**

Plovers

Sandpipers

Snipes

♀ ♂

Phalaropes

Terns

Gulls

Puffins

Order Columbiformes   **303 species (worldwide)**

Doves

Pigeons

Order Psittaciformes
340 species (tropical)

Parrots, parakeets

Order Strigiformes   146 species (worldwide)

Barn owls

Other owls

**Order Caprimulgiformes
105 species (worldwide)**

Nighthawks

**Order Apodiformes
428 species (worldwide)**

Swifts

Hummingbirds

**Order Coliiformes
6 species (southern Africa)**

Mousebirds

**Order Trogoniformes
37 species (tropical)**

Trogons

**Order Coraciiformes
195 species (worldwide)**

Kingfishers

**Order Piciformes
283 species (worldwide)**

Woodpeckers

Order Passeriformes   more than 5000 species (worldwide)

Swallows

Wrens

Thrashers

Thrushes

Nuthatches

Buntings

Vireos

Blackbirds

Finches

Starlings

Crows

Jays

Warblers

Sparrows

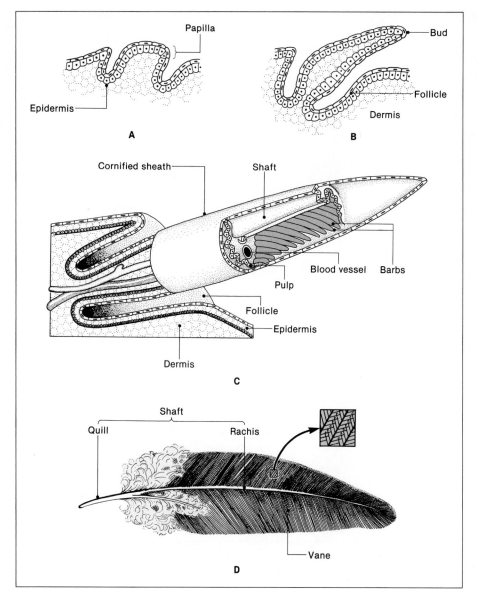

Papilla

Epidermis

A

Bud

Follicle

Dermis

B

Cornified sheath

Shaft

Blood vessel   Barbs

Pulp

Follicle

Epidermis

Dermis

C

Shaft

Quill

Rachis

Vane

D

Figure 38.4
*Development of a vane feather.
(A) The feather starts as a papilla
similar to a reptilian scale (compare
Figure 37.11). (B) The papilla forms
a bud that sinks into a follicle in the
dermis. (C) The shaft and attached
barbs of the feather develop
between a protective sheath and the
nutritive pulp. (D) When the feather
is mature the sheath bursts. The
pulp dries up, leaving a hollow quill.*

Distribution and Types. The number and types of feathers depend on species,
body size, and other factors. One ruby-throated hummingbird, *Archilochus colubris*,
was found to have 940 feathers, and a tundra swan, *Cygnus columbianus*, had 25,216.
Papillae are not distributed randomly all over the skin but lie on a few **feather
tracts** (= pterylae). In most birds vane feathers, also called **contour feathers,**
cover most of the body. (A notable exception is the lower legs, which are covered
with scales like those of reptiles.) Vane feathers also extend from the wings and
tail as **flight feathers.** Flight feathers are asymmetric, with one vane narrower
than the other. This allows flight feathers on the wings to pivot during the
upstroke, opening spaces that reduce air resistance. There are also other types of
feathers (Figure 38.6).

Color. A feather is naturally white unless it acquires color, either as pigment
from chromatophores during development or as structural color (see p. 199).
Black, brown, dull yellow, and dull red colors are due to melanin pigments. Carotenoids account for bright yellow, orange, and, in many cases, red feathers. In other

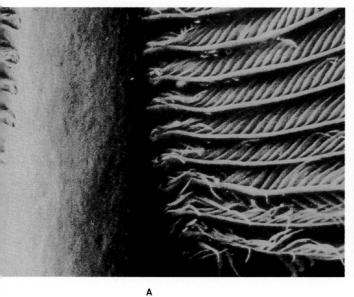

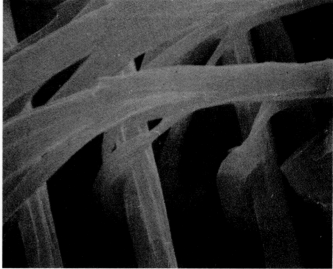

A

B

Figure 38.5
*Scanning electron micrographs of vane feathers of a robin (*Turdus migratorius, *order Passeriformes). (A) A shaft in the upper left corner has barbs that are interlocked with the barbs of another feather to the right. Such interlocking allows the feathers to form a single flexible surface. (B) Higher magnification shows the hooked barbules by which barbs of adjacent feathers interlock.*

birds bright red or green is due to porphyrin pigment. Blue color in birds is Tyndall blue, caused by the scattering of light by particles or overlapping keratin layers. The iridescent colors of peacocks and starlings, which change with the viewing angle, is a different kind of structural color due to bundles of tubules in the feathers. The activation of different combinations of chromatophores during development allows a single feather to acquire a pattern of more than one color. Color serves two important and conflicting functions in birds: camouflage and social communication. In many species juveniles and females are well camouflaged with melanin-pigmented feathers, while adult males in breeding plumage are brightly colored (Figure 38.7).

Maintenance. With so many functions depending on the feathers, it is not surprising that birds spend so much time maintaining them. One familiar maintenance behavior is **preening,** which consists of running the feathers through the beak, which zips the barbules together. While preening, most birds also oil the feathers with a water-repellent secretion from the **oil gland** (= uropygial gland) at the base of the tail. Birds also care for their feathers by bathing, both in water and in dust. Bathing apparently eliminates ectoparasites such as ticks and lice. Many birds also rid themselves of parasites by the curious practice of "anting," in which they wallow in anthills and induce the soldier ants to attack the parasites. Although the ants undoubtedly bite the bird as well, birds go into such a frenzy of excitement during anting that it is hard to believe they don't enjoy it. Orioles, jays, and starlings also pick up individual ants and tuck them into their feathers. If no ant is available they will substitute wasps, onions, vinegar, and even lit cigarettes.

Molting. In spite of maintenance, feathers do become damaged and must be replaced. In addition, feather replacement occurs during regular periods of molting. Penguins molt all their feathers over several weeks in summer. Most birds, however, undergo a gradual molt, in which they lose only a few feathers at a time. Flight feathers are molted in matched pairs from each side of the body simultaneously, so balance is not upset. Usually one pair of flight feathers is replaced before the next pair is lost, but ducks and geese lose several pairs at a time and

Figure 38.6
(A) *Varieties of feathers. Some vane feathers have aftershafts. A down feather lacks hooks on the barbules and forms an undercoat in geese and many other waterfowl. Filoplumes have barbs only at the tips. Bristles have a few barbs concentrated at the base.*
(B) *Filoplumes employed as decoration on the head of a peacock* Pavo cristatus. *Most of the other feathers are vane feathers.*
(C) *Bristles form the eyelashes of the ostrich* Struthio camelus. *Most of the other feathers have hairlike shafts without hooked barbules.*

Aftershaft

Vane with aftershaft

Down

Filoplume

Bristle

A

B

C

are therefore unable to fly during molting. Molting generally occurs once a year after the breeding season. The cue for molting may be changes in day length, which affect the levels of hormones such as thyroxine, triiodothyronine, cortisone, and the reproductive hormones. Molting may occur more than once a year in birds that depend on their feathers for long migrations, and in birds whose feathers are subject to much damage. In addition, many birds, especially males, undergo partial molts that uncover their breeding plumage.

FLIGHT

Performance. Not all birds fly, and not all flying animals are birds, but no other large group of animals has mastered flight to the extent that birds have. Birds commonly fly at speeds of 90 km/hr, and some, such as the peregrine falcon *Falco peregrinus*, reach 190 km/hr. Just as remarkably, many birds can hover at zero speed. Climbers on Mount Everest, gasping for enough oxygen to take the next

A   B

see p. 421

Figure 38.7
Functions of feather color. (A) An American woodcock, Scolopax minor *(formerly* Philohela minor*), camouflaged on its nest. Woodcocks remain motionless and usually unseen even while potential predators are only centimeters away. (B) A male resplendent quetzal* Pharomachrus mocino *sacrifices camouflage to attract females in the dark, thick foliage of Central American rainforests. The quetzal was revered as god of the air by the Maya and Aztec, who associated it with the god Quetzalcoatl. In the past many Europeans doubted its existence; because of destruction of rainforests there is now reason to doubt its future existence. The length, including the tail feathers, is more than a meter.*

step, have reported birds flying overhead, apparently unperturbed by the thin atmosphere. Many birds make aerial migrations of thousands of kilometers twice a year (see p. 421), and certain juvenile terns are thought to spend months or even years on the wing, not landing until summoned by the urge to reproduce. Birds make these feats look so easy that it is surprising how hard bird flight is to understand.

Evolution. One poorly understood aspect of bird flight is its evolution. Birds acquired the ability to fly independently rather than by inheriting wings from a pterosaur or a bat (Figure 38.8). But how? Scales or feathers do not enlarge and form wings in the "hope" that they will someday enable a descendant to fly. Changes in the forelimbs would have had to be advantageous at every stage in their evolution as wings. For more than a century there have been two major theoretical scenarios for how this evolution could have occurred: the arboreal theory and the cursorial theory. According to the **arboreal theory,** also known as the "trees-down theory," wings evolved in birds that climbed up trees hunting insects, then glided to the base of the next tree. This is a highly efficient form of hunting that is still widely used by some birds, such as woodpeckers. The **cursorial** or "ground-up theory" proposes that wings evolved in a running (cursorial) bird, perhaps as rudders that helped maintain balance, like the wings of ostriches. So far there is no compelling paleontological or aeronautical evidence for or against either theory.

Body Size. There is also much that is not understood about the mechanisms of flight, even though the basic principles have been known since the Wright broth-

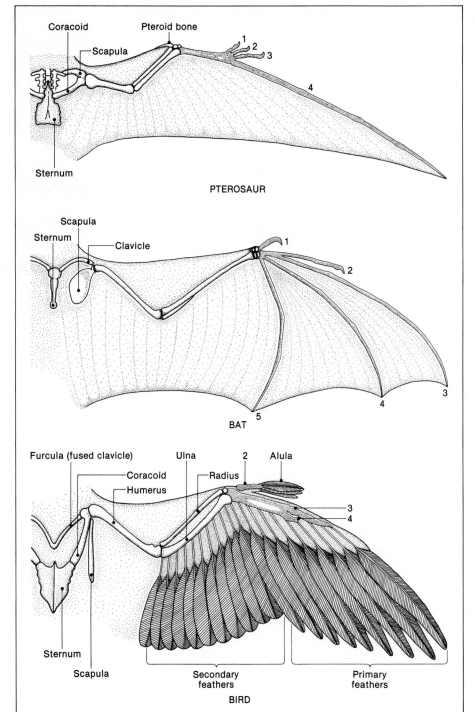

Figure 38.8
Comparison of the wings of pterosaurs, bats, and birds shows that flight evolved independently in reptiles, mammals, and birds. In pterosaurs and bats the main wing structure is skin supported by the finger bones (the fourth finger in pterosaur, and the second through fifth in bats). In birds the flight feathers form the wing surface. The primaries are attached to the fused third and fourth digits, and the secondaries are attached to the ulna of the lower arm. An auxiliary wing, the alula, consists of feathers attached to the second digit. There are also differences in the pectoral bones supporting the wings. Pterosaurs lacked clavicles (collar bones), but unlike mammals they had a coracoid that helped support the wings. In birds the clavicles are fused into a furcula (wishbone), and the coracoid provides additional support.

ers. One of the elementary principles was discovered the hard way by the early pioneers of human flight: Wings have to be large enough to generate enough **lift** to support the weight of the body. In other words, the **wing loading** (body mass divided by wing area) must be below a certain value. Small wing loading requires large wings, light bodies, or, as in most birds, both. Birds have a variety of adaptations that reduce body weight, including respiratory air sacs and hollow bones, which will be described later. In spite of these adaptations, wing loading increases as the body gets larger (see p. 219). The largest bird that can fly is the great

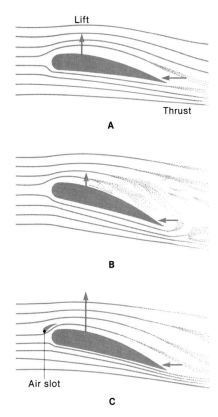

Figure 38.9
(A) The pattern of air movement over an idealized wing viewed from the end. Air forced to move more rapidly over the top has a lower pressure, which produces most of the lift. Additional lift is produced by the pressure of air against the bottom surface of the wing. This high-pressure air moving past the rear generates thrust. (B) Increasing the angle of attack decreases thrust and increases lift, up to a point. Beyond that point turbulence replaces the smooth flow of air over the wing, and there is a loss of lift (stalling). (C) Stalling is prevented by adding a wing slot to channel a fast stream of air across the upper surface. Slotting is achieved in birds by spreading the alula and primary feathers (Figure 38.11A, B).

bustard *Otis tarda*, with a mass up to 22 kg. There are compensations for being large, however, including the ability to escape predators and to conserve body heat. Ratites, penguins, and some others have given up flight in favor of these advantages. Birds must also not be too small to fly (see p. 218); all birds are, in fact, larger than the minimum size for flight.

Gliding. The generation of lift is well understood only for gliding. During gliding the wings are simply held out while the bird moves through the air, or while the air blows over the wings. The profile of a bird's wing is like that of an airplane: The front of the wing is thicker than the rear, and the rear curves downward. This forces air to flow over the wing faster than it flows beneath the wing. The faster the velocity of air, the lower its pressure, so there is a reduced pressure on the upper surface of the wing, which lifts the wing up (Figure 38.9). Some lift is also produced by the pressure of air against the bottom surface of the wing. In order to maintain forward movement contrary to the drag of friction and air pressure, the wings must also generate **thrust.** During gliding thrust is produced by air from beneath the wing escaping past the trailing edge.

Increasing the angle of attack of the wing (angling the front of the wing upward) increases the lift but decreases the thrust. Birds commonly increase the angle of attack to maintain altitude at slow speed, or to slow down their forward movement while maintaining enough lift for a gentle landing. If the angle of attack is too steep, however, the flow of air over the wing becomes turbulent, resulting in the loss of lift (**stalling**). To prevent stalling, birds do what airplane pilots do when landing: open wing slots. In birds these wing slots consist of the spread primary feathers and **alulae** (Figure 38.9C).

Flapping Flight. Flapping flight is much more complicated than gliding because the wing movements vary in so many ways, depending on the size of the bird, the direction and speed of flight, air temperature, and other factors. Presumably, however, some of the lift and thrust are produced in the same way as in gliding, except that the wing is moving actively rather than passively through air. Birds do not flap their wings straight up and down but in a figure-eight pattern, like insects (see Figure 32.20). During the downstroke the **pectoralis muscles** pull the wings downward and forward, reducing pressure on top of the wings and increasing it beneath. During the upstroke the **supracoracoideus muscles** pull the wings upward and backward (Figure 38.10). As noted previously, the asymmetry of the primary and secondary feathers causes them to pivot during the upstroke, opening spaces for air to flow between them.

Adaptations of Wing Shape. Each kind of flight makes conflicting demands on wings. The optimal wing for gliding would have the largest surface area that the bones and muscles could support. The larger the surface area, however, the greater the friction, which must be overcome by increased thrust. The wing that is most efficient for flapping flight would therefore have less surface area than a wing that is optimal for gliding. It would also be longer, since most of the lift and thrust of flapping flight come from the outer parts of the wings, which move the most. If a bird must maneuver among tree limbs, however, the wings can't be too long. Wings can produce as much lift with less area if the bird flies faster. For optimal flight at high speed the wings should be swept back to reduce pressure along the leading edge. Also, the tips of the wings should be pointed to reduce turbulence. Most birds have wings that compromise among all these demands, but some specialize in satisfying the requirements for one type of flight (Figure 38.11).

SKELETON

Flight makes enormous mechanical demands on a bird. The skeleton must be strong to withstand the stresses from the flight muscles, but it must also be light. This combination of strength and lightness is achieved by the resorption of bone marrow early in life, leaving most of the bones hollow except for internal reinforcing struts (Figure 38.12A). Contraction of the flight muscles imposes large stresses on the ribs, which avoid fracturing by being V-shaped and folding like a knife (Figure 38.12B). Excessive folding of the ribs is prevented by braces between them (**uncinate processes**).

The specialization of the forelimbs as wings means that locomotion on land is bipedal rather than quadrupedal. Like humans, therefore, birds are slower and less stable than most terrestrial vertebrates when walking and running. Birds have feet that are specialized for particular kinds of locomotion and for other functions (Figure 38.13). Bipedality also requires that the pelvic girdle support the entire body weight when the bird is standing. Consequently, the pelvic girdle of birds consists of large bones fused to the vertebrae of the back. In contrast, the vertebrae of the neck are quite flexible. This flexibility is essential for several reasons: (1) It enables the bird to use its head to maintain balance while flying and (quite noticeably in the pigeon) walking. (2) It compensates for the limited ability of birds to rotate their eyes. (3) It allows the bird's beak to reach the oil gland at the base of the tail and all the feathers for preening.

RESPIRATION

Although the physiological systems of birds are in general similar to those of reptiles and mammals (as described in Unit Two), the demands of flight and homeothermy have apparently selected for several unique adaptations. This is especially so for the respiratory system. At rest, birds generally have metabolic rates ten times higher than those of reptiles of the same size, largely because of the need to produce heat to maintain a high body temperature. While flying, the metabolic rate of a bird may increase another tenfold. Unlike the lungs of reptiles and mammals, which consist of sacs in which air can stagnate, birds' lungs consist of numerous tubules called **parabronchi** through which air flows continually in one direction (see pp. 253–255). The lungs are ventilated to a small degree by their own expansion and contraction, but mainly by the expansion and contraction of **air sacs.** During inhalation the air sacs expand, and the lungs contract somewhat. These movements draw fresh air through a central tube (the **mesobronchus**) in each lung into the posterior air sacs and force stale air from the lungs into the anterior air sacs. During exhalation the lungs expand and the air sacs contract, moving fresh air from the posterior air sacs into the lungs and emptying stale air from the anterior air sacs. Inhalation and exhalation are aided by movement of a nonmuscular **diaphragm** between the abdominal and thoracic cavities and may also be coupled to movements of the sternum and furcula during flight.

The air sacs have extensions into the skull and some other hollow bones. In fact, if one of these bones is opened to air, the trachea can be shut off completely without suffocating the bird. Normally, of course, air enters and exits through the nasal passages, trachea, and bronchi (Figure 38.14). At the upper end of the trachea is a **larynx** that, unlike the larynges (= larynxes) of mammals and some reptiles, lacks vocal cords. Instead, voice is produced by the **syrinx** at the other end of the trachea, near the bronchi. In chickens and many other birds there is one syrinx in the trachea near its junction with the bronchi. The syrinx has a pair of mem-

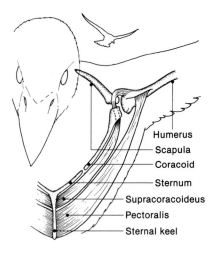

Figure 38.10
Frontal view of the pectoral girdle and attached wing muscles. The downstroke of flapping flight is due to contraction of the pectoralis muscles, which are attached to the sternal keel and to the wishbone (not shown). The upstroke is due to contraction of the supracoracoideus muscle, which is also attached to the sternal keel. The contraction of the supracoracoideus is redirected by the bending of its tendon over the scapula, like a pulley. See also Figure 38.8 for a ventral view, and Figure 38.12B for a lateral view. In chickens and most other birds that fly weakly or not at all, the flight muscles are type II white and produce ATP anaerobically. In good fliers the muscles are red and aerobic.

Humerus
Scapula
Coracoid
Sternum
Supracoracoideus
Pectoralis
Sternal keel

A

B

C

Figure 38.11
Wings adapted for different kinds of flight. (A) The turkey vulture Cathartes aura has broad wings adapted for gliding and soaring on updrafts. The slotting of the primaries allows for a steeper angle of attack, with more lift at slower speed, as described in Figure 38.9C. The wingspan is approximately 1.8 meters. (B) The white-tailed tropic bird Phaethon lepturus flies with rapid wing beats except when swooping down to catch small fish and squid. The great length compared to width (aspect ratio) minimizes friction, which reduces the amount of thrust that is required. Slotting by the alulae as well as the primaries is evident in this photo, which was taken as the bird was about to land. The long tail feathers act as stabilizers, but their main function may be in social display. Wingspan approximately 1 meter. (C) The Arctic tern Sterna paradisaea migrates 35,000 kilometers a year between the Arctic and Antarctica and thereby enjoys an endless summer. The swept-back wings and tail reduce pressure drag. The pointed wings reduce turbulence at the tips. Note slotting of the alulae and primaries. Wingspan approximately 50 cm.

branes (**tympani**) that vibrate as air blows past them. The pitch of the sound is controlled by muscles that adjust the tension on the tympani. Songbirds have essentially two syringes (= syrinxes), which enables them to sing with two voices at the same time.

CIRCULATION

The high metabolic rates of birds also place demands on the circulatory system. Like crocodilians and mammals, birds have four-chambered hearts with complete separation of oxygenated and deoxygenated blood (see Figure 11.4D). The heart rate increases with metabolic demand and is generally higher in smaller animals. The heart of a flying hummingbird beats so fast (1200 beats per minute) that it hums. The blood is similar to mammalian blood, except that the red blood cells retain their nuclei (see pp. 223–224).

OSMOREGULATION AND EXCRETION

Other physiological systems of birds are generally similar to those in reptiles and mammals. The kidneys are of the metanephric type. In marine birds **salt glands** that empty through the nostrils assist in osmoregulation (see p. 295). Birds have nephrons with short loops of Henle that help conserve water (see p. 290). Water is also conserved by the excretion of insoluble uric acid as the main nitrogenous waste (see p. 287). The white, pastelike uric acid is eliminated through the cloaca, along with feces.

THE NERVOUS SYSTEM

The Brain. The brain is highly developed, although along somewhat different lines from that of mammals. It is really not much of an insult to be called a "bird-brain," since the brains of birds are about as large as those of mammals of the same body size (see Figure 8.6). As one might expect in animals that fly so skillfully, the cerebellum is disproportionately large compared with that of a reptile or mammal (see Figure 8.7). Bulging between the cerebellum and the cerebral hemispheres are the **optic lobes.** The cerebral cortex is not as well developed as in most mammals. Instead, the core of the cerebrum, the large **corpus striatum,** appears to handle most of the complex behaviors.

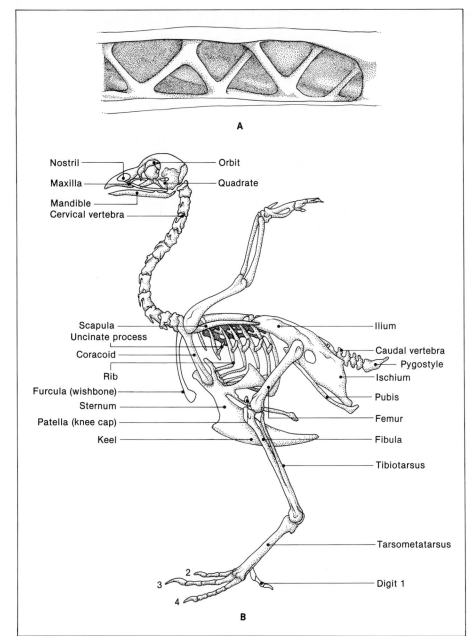

Figure 38.12
Adaptations of the skeleton for flight. (A) The bodies of birds have low density (weight divided by volume) partly because the bones are hollow. Strength is largely due to the internal struts, which resemble those engineers use in lightweight building trusses. (B) Skeleton of the chicken Gallus gallus.

Nostril
Orbit
Maxilla
Quadrate
Mandible
Cervical vertebra

A

Scapula
Uncinate process
Ilium
Coracoid
Caudal vertebra
Pygostyle
Rib
Ischium
Furcula (wishbone)
Pubis
Sternum
Patella (knee cap)
Femur
Keel
Fibula
Tibiotarsus
Tarsometatarsus
2
3
4
Digit 1

B

Learning and Instinct. Birds seem to rely on a complex combination of learning and instincts. For example, many male songbirds can make a fair approximation of their mating song without having heard it before. To sing correctly, however, they must have heard the song for their species during a certain sensitive period soon after they hatch (see p. 415). The feats of learning are often quite impressive. The marsh tit *Parus palustris*, a chickadee-like bird of Europe, can remember dozens of places where it has hidden seeds. Some birds can learn the pattern of stars in a few days, remember it for years, and can use that memory and other cues to navigate over enormous distances (see p. 421). When birds rely only on instinct, however, they can often be made to appear stupid. A male domestic turkey *Meleagris gallopavo* will try to copulate with a human hand if the hand is holding the decapitated head of a female turkey. The male Eurasian robin *Erithacus rubecula* will become enraged at a tuft of red feathers in its territory, ap-

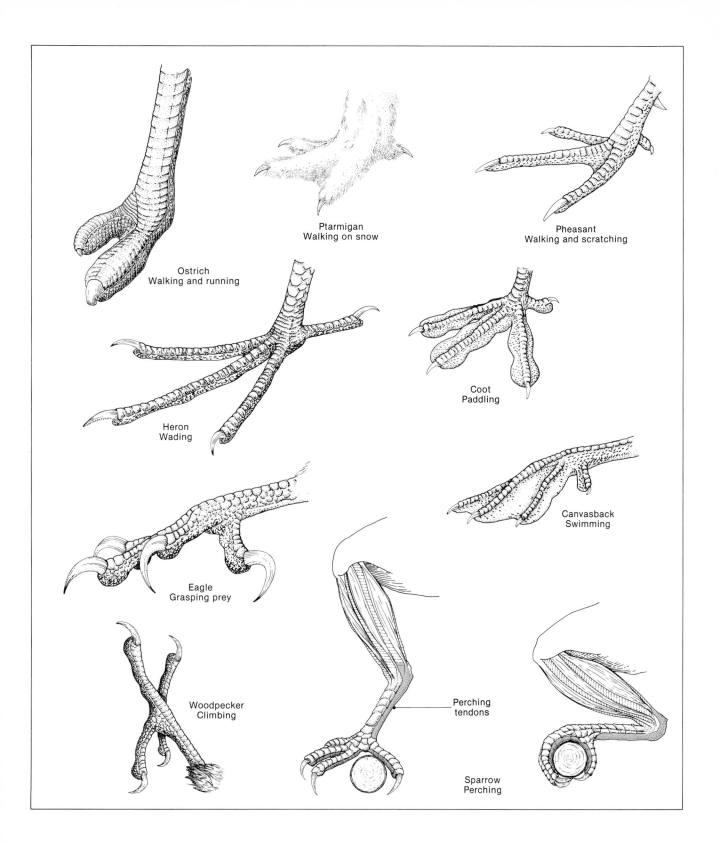

Figure 38.13
Adaptations of birds' feet. The perching tendons are shown at the lower right. When the bird's legs fold, the perching tendon automatically flexes the claws, closing them around the perch. This mechanism enables birds to sleep without falling.

parently "thinking" it is a rival male. Many other male birds will attack their own images in a mirror placed within their territories. (This is a useful trick for determining the boundaries of territories.)

Vision. Vision is the dominant sensory modality in birds. The eyes (Figure 38.15A) are about as large as the cerebrum. In hawks they are so large that they touch in the middle. In most birds the muscles that control eye movements are virtually nonexistent, presumably having been sacrificed to make room for larger eyes. Birds therefore move the head in order to change the direction of vision. Birds that are preyed upon tend to have the eyes on the sides of the head, allowing them to see in all directions. Predatory birds tend to have eyes toward the front of the head, with some overlap in the visual field permitting visual perception of depth. Each eye is protected by two ordinary lids and by a transparent **nictitating membrane.** In ducks, loons, and some other diving birds, the nictitating membrane has an area shaped like a contact lens that helps the bird focus under water. Most diurnal birds have good perception of colors, with numerous cones. In nocturnal birds, rods predominate. Since birds have smaller eyes than do humans, their vision would be expected to be poorer, because the image is compressed onto a smaller retina. In many species, however, the rods and cones in parts of the retina are packed more densely than they are in humans. Many larger birds, in fact, have a visual acuity several times better than that of humans. A vulture can resolve two closely spaced lines from three times farther away than a person can.

Hearing. Hearing is the second-most important modality in birds. As in reptiles, the inner ear of a bird has semicircular canals and otolith organs that respond to acceleration and static position, respectively (Figure 38.15B). In addition, birds have a **cochlea** specialized for hearing. Unlike the snail-shaped cochlea of a mammal, that of a bird is short and straight. Sound vibrations are picked up by the eardrum and transmitted to the cochlea by a single middle-ear bone, the stapes (columella). Birds generally can hear only sounds with frequencies below 12 kHz (12,000 cycles per second), compared with an upper limit of 20 kHz for humans. Within that range, however, some birds have an acuteness of hearing that astounds humans. Pigeons and many other birds can hear subsonic vibrations (below 20 Hz), such as those generated by wind in mountains, from distances of hundreds or thousands of kilometers, and may use such "sounds" to navigate. The barn owl *Tyto alba* can catch mice just as well when blindfolded as not. Their ear openings are located beside the eyes in the disc-shaped ruff of feathers around the face (see order Strigiformes in Figure 38.3), and the left opening is aimed downward while the right is aimed upward. By comparing the time of arrival and the intensity of a sound in the two ears, barn owls get a fix on the source of the sound in both the horizontal and vertical direction.

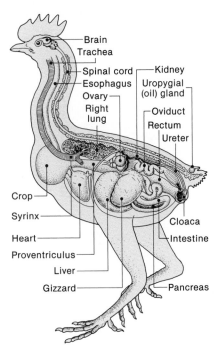

Figure 38.14
Internal organs of a domestic hen.

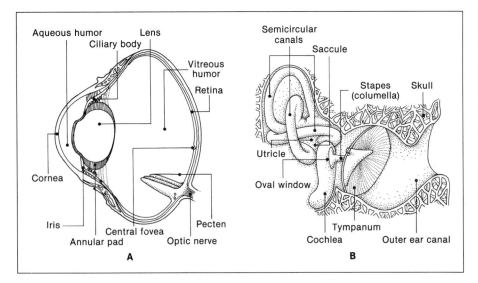

Figure 38.15
Sensory receptors of birds. Compare Figures 7.13A and 7.16. (A) The eye. The lens is shown focused close. To accommodate for distant vision the front of the lens is flattened by the annular pad. The pecten is a series of triangular ridges believed to supply oxygen and nutrients, compensating for the absence of blood vessels in the retina. The fovea is the center of focus, with the sharpest vision. Some birds have two or three foveae. (B) Frontal section of an inner ear.

Other Senses. Chemoreception in birds is much less acute than in most mammals. There is only a small olfactory epithelium and no Jacobson's organ. There are also fewer taste receptors than in mammals, and little need for any, since birds bolt down their food without chewing it. Behavioral studies show that pigeons and some other birds can detect the earth's magnetic field and changes in air pressure, although the receptors have not been identified.

DIGESTION AND FEEDING

Birds, especially small ones, need large amounts of nutrients to supply the high rate of metabolism. The 11-gram blue tit *Parus caeruleus* needs about 30% of its own weight in seeds every day. The great horned owl *Bubo virginianus* eats an average of 5% of its body weight in mice per day. (For a human that would be the equivalent of eating about 2 pounds of mice for breakfast, lunch, and dinner.) Many birds eat even larger amounts at irregular intervals. A Bohemian waxwing *Bombycilla garrulus* was estimated to eat approximately three times its own weight in berries in one day. The male emperor penguin *Aptenodytes forsteri* gorges on fish during the Antarctic summer, then marches a distance of up to 120 km to a rookery where it fasts for 115 days during the winter breeding season.

The digestive system of birds differs in several ways from that in other vertebrates. One important difference is that modern birds have no teeth, having apparently jettisoned these heavy structures to save weight. The inability to chew food is compensated by the **gizzard** (= muscular stomach), which is lined with horny plates that crush and grind hard foods. Breaking food into small pieces allows better mixing with digestive enzymes and greatly increases the rate at which the digestive system can supply nutrients. Many seed-eating birds swallow grit that aids the grinding action of the gizzard. In owls and many other birds that eat vertebrates whole, the gizzard crushes the bones, which are then regurgitated as a pellet that also contains hair and other indigestible parts. Anterior to the gizzard is the **proventriculus** (= glandular stomach), which secretes protein-digesting enzymes. In many birds the proventriculus and gizzard together are too small to store much food. Instead, an enlargement of the esophagus, called the **crop,** enables birds to store food and transport it to the young. In pigeons and doves with young the crop also secretes a "pigeon's milk" analogous to mammalian milk, which they feed to the young. Male penguins produce a similar secretion from the esophagus.

Just as the teeth of mammals are adapted for the kinds of foods they eat, the beaks of birds are similarly adapted (Figure 38.16). Even among some closely related species, such as the so-called Darwin's finches, the beaks differ in shape and size (see Figure 18.12). Those that feed on seeds tend to have heavier bills than those that feed on insects. The size of the bill is quite variable, so natural selection enables a population to adapt to changing food supplies within a few generations (see pp. 373–376). Feeding behavior is also diverse, specialized, and adaptable to circumstances (Figure 38.17).

REPRODUCTION: GENETICS AND PHYSIOLOGY

As in many snakes and lizards, the sex of a bird is determined by whether it inherits two Z chromosomes or a Z and a W. In contrast to mammals, it is the female that is **heterogametic** (has two different sex chromosomes). In many adult birds the two sexes differ greatly in coloration during the breeding season but are generally indistinguishable at other times and as juveniles. In many other species the

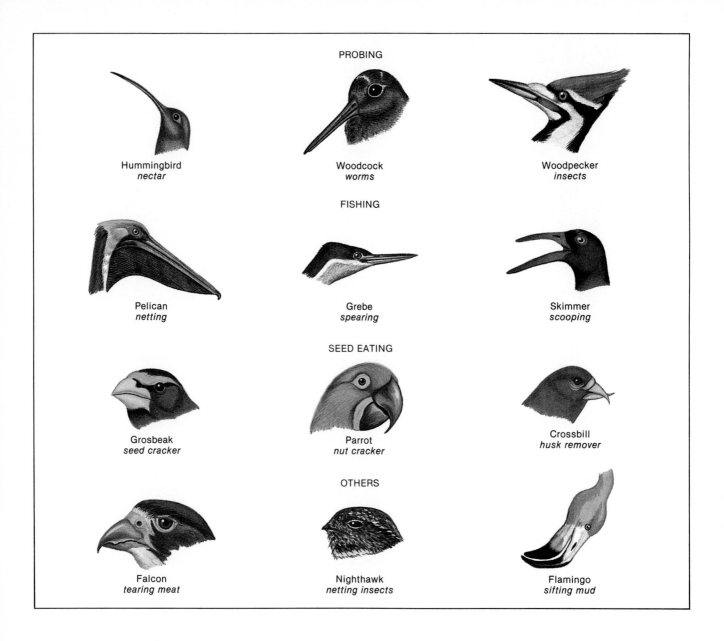

PROBING

Hummingbird
nectar

Woodcock
worms

Woodpecker
insects

FISHING

Pelican
netting

Grebe
spearing

Skimmer
scooping

SEED EATING

Grosbeak
seed cracker

Parrot
nut cracker

Crossbill
husk remover

OTHERS

Falcon
tearing meat

Nighthawk
netting insects

Flamingo
sifting mud

Figure 38.16
Adaptations of beaks for different ways of feeding.

two sexes resemble each other year round, except that they may differ in size. Fertilization is necessarily internal, since females lay eggs that are covered with shells. Males in groups considered to be primitive, such as ratites, chickens, ducks, and storks, have an erectile penis on the ventral wall of the cloaca. As in reptiles, an external groove on the penis guides sperm into the cloaca of the female. In most birds, however, internal fertilization is effected simply by the male pressing his cloaca to that of the female while treading upon her back. Generally, the testes are small except during the mating season, probably as an adaptation that reduces body weight. Females of most species save weight by having only the left ovary.

Reproductive activities are coordinated by hormones similar to those in mammals (see pp. 319–321). The anterior pituitary secretes gonadotropins (FSH and LH), which trigger gamete production and ovulation. Prolactin, which stimulates the production of milk by female mammals, also stimulates the production of "pigeon's milk" by pigeons and doves. (This "milk," however, is not homologous with mammalian milk.) Estrogens and androgens, steroid hormones from the ovaries and testes, respectively, stimulate appropriate reproductive behaviors. The

A   B   C

Figure 38.17
The Eurasian kingfisher Alcedo
atthis *catching fish. (A) Diving for
the fish, using the wings and tail for
guidance. (B) The fragile wings fold
just before impact. (C) Success. The
kingfisher beats the fish against a
branch to kill it, then swallows it
head first.*

secretion of these hormones generally follows an annual rhythm synchronized by day length. Many of the same hormones are also involved in coordinating other behaviors that must be synchronized with reproduction. For example, testosterone triggers the partial molt to breeding plumage and promotes aggression and territorial defense. Prolactin, together with cortisol from the adrenal cortex, helps coordinate migration.

EGGS

Formation.   All birds are oviparous. Why none has evolved viviparity, as many amphibians and reptiles have, is a long-standing question. The commonly offered answer is that it would be difficult for a pregnant bird to fly, but that does not seem to bother bats. Immediately after ovulation the ovum, consisting of the yolk and the **germinal spot** from which the embryo will form, enters the oviduct through the funnel-shaped **infundibulum.** As the ovum passes down the oviduct it is covered with **albumen** (egg white), a process that takes about 3 hours in the domestic hen. Farther down the oviduct the albumen becomes enclosed in two keratin membranes, which later become separated at one end of the egg to form an air pocket. The egg then enters an enlarged area of the oviduct (the "uterus"), where a protein matrix is deposited. On this matrix a shell consisting mainly of calcium carbonate crystallizes. The calcium comes from the female's bones, and much of it is absorbed by the developing chick to make its bones. Formation of the shell takes about 20 hours in a hen. The final product released from the cloaca is a self-contained (cleidoic) egg (Figure 38.18). Pores in the shell allow for the exchange of respiratory gases, assisted later by the chorion and allantois (see p. 120). Water also evaporates through the pores. The air pocket expands to fill the space left by the lost water. In the last hours of development the chick breaks into the air pocket and gets its first breath of air. Finally, the chick breaks through the egg shell using an **egg tooth** on the outside of the upper beak.

Size and Color.   Each egg is relatively large, generally from 2% to 12% of the female's body weight. In the kiwi *Apteryx* the egg is about 25% as massive as the

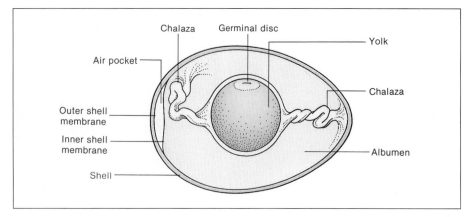

Chalaza   Germinal disc

Air pocket   Yolk

Outer shell
membrane   Chalaza

Inner shell
membrane

Shell   Albumen

Figure 38.18
A just-laid, fertilized egg. The embryo develops from the germinal spot. (Compare the later embryo in Figure 6.19B.) The chalaza, which consists of thick strands of albumen, keeps the ovum near the center of the egg and twists as the egg is turned during incubation, allowing the denser yolk (vegetal pole) to stay below the embryo (animal pole).

adult. The egg of the extinct "elephant bird" of Madagascar held 8 liters, and the aborigines used the shells as dishes.

The eggs of birds that nest in cavities are generally white. In species whose nests are exposed to potential predators and to the Sun, however, the shells are often exquisitely spotted and colored with pigments. The pigments are breakdown products of hemoglobin that are usually brown, olive, green, or blue. Unlike other kinds of pigments, these reflect infrared radiation from the Sun and thereby help prevent overheating. Experiments have established that the colors of many birds' eggs camouflage them from potential predators (Figure 38.19A). "Robin's egg blue" and other conspicuous colors are more difficult to explain. Like the colors of pearls, they might simply be incidental to some other function. On the other hand, they may enable the adult birds to detect and eject the eggs of cuckoos, cowbirds, and other brood parasites (Heinrich 1986). European cuckoos (*Cuculus* spp.) foil this adaptation by laying eggs that mimic those of the host species (Figure 38.19B).

Figure 38.19
(A) The egg of the least tern Sterna antillarum *is hardly visible even though exposed on the beach. (B) A nest of the great reed warbler* Acrocephalus arundinaceus *of Europe containing one cuckoo egg. Local populations of many cuckoos specialize in laying their eggs in the nests of one host species. Presumably through natural selection, the cuckoo eggs come to resemble those of the host species. See also Figure 20.17.*

A

B

COURTSHIP

Among the most interesting features of birds are the songs, colors, and behaviors associated with finding mates. Courtship serves several functions, such as enabling potential mates to recognize each other as being reproductively capable members of the same species and of opposite gender (see pp. 428–429). Most courtship behaviors are apparently instinctive, although they are subject to modification depending on the responses they elicit. Older birds generally have an advantage in courtship because they have learned how to modify their behavior to get the best results. Courtship in birds depends mainly on visual communication and, in songbirds, on vocalization. Some species discriminate similar species on the basis of differences in song so minute that humans can barely distinguish them (see Figure 21.1). Slight changes in appearance also often make a big difference in whether a bird is recognized as a suitable mate. Painting over the wing bar of a male warbler will often eliminate its chances of finding a mate of the same species. On the other hand, the example of the turkey mentioned above indicates that some birds can overlook quite large discrepancies in appearance.

In most species the males have the showier plumage, do all the singing, and sometimes engage in elaborate displays that attract female mates. The behavior of females is often limited to signaling receptivity by means of a particular posture (see Figure 20.2). However, in phalaropes (*Steganopus*, *Lobipes*, and *Phalaropus*; Figure 38.3, order Charadriiformes) the female is more brightly colored and attempts to attract the male. This reversal of sex roles is associated with a reversal of the usual pattern of hormone secretion. Female phalaropes secrete more androgens, and the males secrete more estrogens than in most birds. Many birds also use displays similar to those used in courtship to defend **territories** from rivals. Defense of the territory may persist only until mating is completed, or it may continue until the young are raised. In the latter case the territory must be large enough to provide all the resources needed by the family.

Displays are essential in attracting mates and defending territories, but disadvantageous in that they make birds conspicuous to predators and may interfere with flight or other functions. The tail of the lyrebird is one example of a species in which plumage that has evolved to attract females may be a handicap to the males (Figure 38.20A). According to one controversial theory, the fact that a male can survive in spite of such handicaps advertises to the female that it must have

Figure 38.20
Two male birds, both belonging to order Passeriformes and living in Australia, attract females with unusual visual and vocal techniques. Both birds are excellent singers, often mimicking other birds and even cats, dogs, sheep, and human noises. (A) The male superb lyrebird Menura novaehollandiae *has a lyre-shaped tail that probably hinders its movements through the eucalyptus forests. The ability to survive with such a tail demonstrates its fitness. Females prefer such tails to more practical ones. This male is singing to a female on a mound it has constructed. (B) The male spotted bowerbird* Chlamydera maculata *is plain looking except for a purple ruff on the back of its neck. It compensates for its drabness by constructing a bower and decorating it with bright objects, many of which it pilfers from people. Males also steal from rivals, and attempt to destroy their bowers. The ability to maintain an attractive bower is therefore evidence of the male's fitness.*

A

B

other traits that contribute to fitness. Sexual selection for showy plumage and brilliant songs may therefore augment natural selection for other adaptive traits. In some birds the males satisfy the demands of both sexual selection and natural selection by being drab in plumage but creating structures that attract females (Figure 38.20B).

MATING SYSTEMS

In some species the male and female separate immediately after copulation, and either or both then breed again with different partners. Such **promiscuity** occurs in bowerbirds and in many other species. Most species, however, are **monogamous** in the sense that they remain with each other during each breeding period, usually until the young are able to leave the nest. Even in monogamous species, however, one or both individuals usually engage in **extrapair copulations.** Using DNA fingerprinting (see p. 100), ornithologists have determined that even in monogamous species, from 10% to 70% of the chicks in a nest were not genetically related to the male that was helping to care for them. The monogamous relationship is said to be due to formation of a **pair bond** between the two mates. (An analogous relationship between humans would be attributed to love.) The advantage of the pair bond is that it commits the male to assisting in the raising of young. In **altricial** species, in which the young are born helpless and without feathers, it often requires two parents to raise a brood. Experiments have shown that in altricial species the brood fails or the chicks develop slowly if the male parent is removed. (We shall see later that males often do not have a parental role in **precocial** species, in which the young have feathers and can feed themselves soon after hatching.)

In most species the pair bond lasts for just one breeding season. In some species, however, the pair bond is thought to last for many years, and perhaps for life. Long-term data are difficult to gather, but there is good evidence that albatrosses, petrels, and some other birds mate for life. Low "divorce rates" are also reported for swans and geese in captivity. Long-term pair bonding is most common in long-lived species and has the obvious advantage of avoiding the search for a new mate each year.

Monogamy is just one of several mating systems in birds (see pp. 429–430). As in some human cultures, some species of birds practice **polygamy,** in which each bird of a gender typically has more than one mate. The mating system in which each male breeds with more than one female at the same time is called **polygyny.** Polygynous males seldom contribute to rearing the young. Polygyny is especially common in grouse, pheasants, peafowl, and chickens, which are precocial. It also occurs most often in habitats where the distribution of resources is patchy and controlled by males. In contrast to the usual situation in which the females select which males they will mate with, in most polygynous species previous competition among males determines whether a male gets to breed. The outcome of such competition is a **dominance hierarchy** or **pecking order** (see pp. 427–428).

The mating system in which each female has more than one male mate is called **polyandry.** Polyandry also occurs in patchy habitats, but the resources are dominated by the females, which are usually larger than the males. Polyandry is much less common than polygyny and occurs mainly among plovers, jacanas, phalaropes, and others in order Charadriiformes. The female lays one or more eggs fertilized by each male. Each male then incubates the eggs it has fertilized and rears the young with little help from the female.

Figure 38.21

(A) Among the most skillful nest builders are weavers of Africa, which weave oval or spherical nests out of grass and plant fibers. The black-headed weaver Ploceus cucullatus is shown here working on its nest. (B) Stages in the construction of a weaver's nest. (C) Many weavers construct nests near each other. In social weavers Philetairus socius many pairs cooperate in building one huge canopy, beneath which each pair later constructs a separate nest. (D) House martins Delichon urbica formerly built their mud-and-pebble nests beneath cliffs. Now they use the eaves of houses.

Some polygynous and polyandrous species form **leks,** areas where individuals of one sex gather and display to attract mates (see p. 429). In the polygynous sage grouse (*Centrocercus urophasianus*) the males establish leks where they put on courtship displays for females. Each successful male breeds with several females and contributes nothing to rearing the young. In the polyandrous red phalarope (*Phalaropus fulicarius*) it is the female that has the showy plumage and establishes a lek, and the males do most of the parenting. Yet another arrangement, called **cooperative breeding,** occurs among some birds, such as the acorn woodpecker *Melanerpes formicivorus* and the green woodhoopoe *Phoeniculus purpureus*. In these species only one monogamous pair in a flock mates, but other members of the flock help in rearing their young. Cooperative breeding occurs mainly in poor environments that can support only a limited number of breeding pairs (see p. 431).

NESTS

In many species reproductive behavior includes the construction of a nest in which eggs will be deposited and incubated. Nests may be built by one or both sexes and range from a mere depression in the ground to a structure that seems far more elaborate than required (Figures 38.19 and 38.21). Each species appears to follow an instinctive "blueprint," constructing nests so precisely that one can often identify the species from the nest alone. Even though each bird was raised in a nest like the one it builds, it is nevertheless amazing that it can construct a nest like those of its parents. After all, even though we have all grown up in houses, few of us are carpenters. The major function of a nest is to hide the eggs or make them inaccessible to predators. Nests also help protect the eggs and young from cold, rain, and exposure to the sun. Many birds line their nests with their own down or feathers, or with other material that insulates the eggs from variations in temperature.

INCUBATION

As with most biological processes, temperature has a profound effect on the rate of development of bird embryos. Eggs of the house wren *Troglodytes aedon*, for

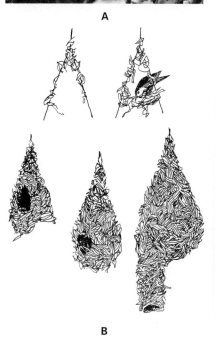

A

B

C

D

example, require 13 days to hatch at the normal temperature of 35°C, but five days longer if the temperature of the eggs is 3°C cooler. During those extra five days the eggs and parents are subject to predation and other hazards. Not surprisingly therefore, most birds incubate their eggs night and day, maintaining them at a high temperature (usually between 35 and 40°C). The source of heat is the parents' own bodies, which generally are at around 40 to 42°C. In most monogamous species both parents help incubate the eggs, or one remains on the eggs in the nest while the other brings it food. Many birds lose feathers on the belly, forming **brood patches** that transfer body heat more effectively to the eggs. The brood patch also allows the bird to sense the temperature of the egg. The urge to incubate eggs is overwhelming in many species. Penguins will even try to incubate rocks if they have lost their eggs. Even more impressive is the heroic behavior of the male emperor penguin, which stands for eight to nine weeks during the dark Antarctic winer, in temperatures down to −60°C (−77°F), incubating its one egg between its feet and abdomen (Figure 38.22).

PARENTAL CARE

Incubation is a burden on parents, but it is nothing compared with the demands placed on them after the young hatch. In altricial species the parents have to provide not only for their own high metabolic demands but also for those of their rapidly growing chicks. The body mass of a typical altricial chick increases approximately tenfold between the time it hatches and the time it leaves the nest a week or two later. This mass and more must be provided in food brought by the parents. The parents also must remove the considerable amount of feces produced by the young. Usually the feces are enclosed in a mucous coat, and the parents swallow

Figure 38.22
Male emperor penguins Aptenodytes forsteri *incubate the eggs for approximately two months during the winter, then exchange places with mates, which have been fishing.*

it or carry it away in their beaks. In addition, altricial birds are generally not capable of thermoregulation for several days, so the parents must continue to provide warmth and protection from exposure.

Even after the young leave the nest, most birds continue to feed them for some time. In addition, during the sensitive period when young birds are developing the feathers, muscles, and behavioral skills needed for flying and feeding, most adults protect them from predators (Figure 38.23). Protection takes many forms, ranging from alarm calls to the mobbing and physical attack of predators. Many birds, such as the ruffed grouse *Bonasa umbellus*, divert potential predators from the chicks to themselves by behaving as if they had broken wings. Such parental solicitude might be interpreted as love for the young, but other evidence suggests that these behaviors are rather mechanical. Many birds devote as much or more attention to **brood parasites**—offspring of different species, such as cuckoos and cowbirds—as they do to their own offspring (see Figure 20.20). Some gulls apparently do not recognize their own chicks if they stray from the nest by as little as a centimeter and will let them starve or be killed. And many of the same birds that valiantly defend a chick against a predator will do nothing to prevent the chick from being killed by a nest mate. Chicks also form a social bond to their parents, but this too appears to be an instinctive process, called **imprinting** (see p. 431). Young birds can just as easily become imprinted to humans, electric trains, or to anything else that moves and makes noise during the first days of their lives.

Figure 38.23
Wild turkey Meleagris gallopavo *defending her chicks.*

INTERACTIONS WITH HUMANS AND OTHER ANIMALS

Ecology.   Because birds are such active and mobile consumers they have enormous ecological impacts on a wide range of other animals. A small titmouse (*Parus* sp.) consumes an average of one insect every $2\frac{1}{2}$ seconds during the day in the winter, and a barn owl (*Tyto alba*) eats several mice in an average night. Most birds are herbivorous, but even they affect other animals indirectly by altering the distribution of plants. Birds are major vectors for dispersing seeds that cling to their bodies or pass intact through their digestive tracts. Many tropical plants have evolved flowers that can be pollinated only by hummingbirds, which are more reliable in rainy weather than are insect pollinators. These flowers discourage insects by producing thin nectar, having deep nectaries, having no scent, and providing no place for an insect to land. Without hummingbirds, half the plants in the Andes would become extinct within a year.

Birds and People.   The relationships between birds and humans are also wide-ranging and complex. Many early archaeological artifacts reveal wonder, envy, and fascination for the colors, songs, and flight of birds. Cave paintings, totems, and hieroglyphics frequently depict birds, often in the forms of gods. In numerous cultures even today people decorate themselves with feathers as if to borrow some of the powers of birds, or at least some of their beauty. In developed countries, our level of fascination with birds can be inferred from the billions of dollars spent by millions of people for travel, field guides, binoculars, bird feed, nest boxes, and other ornithological paraphernalia. Economically, bird watching is more important than bird hunting. Birds also account for approximately 12% of the billions of dollars in sales by pet stores, compared with 19% for dogs and 5% for cats.

Birds are valued for their practical contributions as well. Chickens and their eggs probably contribute as much protein to the human diet as any other single species except cattle. Turkey, duck, goose, pheasant, quail, and pâté de foie gras (a paste made from the livers of force-fed geese) provide rarer treats. For those who fancy the truly exotic, there is bird's-nest soup, the main ingredient of which is the dried saliva that oriental cave swiftlets (*Collocalia* spp.) use to construct their nests. Other products include goose down for clothing and comforters and feathers for hat decorations.

Exotic Species. Often our respect and appreciation for birds backfire on us, especially when we transport them out of their natural habitats. The starling *Sturnus vulgaris* and the house sparrow *Passer domesticus* were deliberately released in New York City in the 19th century, allegedly by people who thought it would be neat to have all the birds mentioned in Shakespeare. These birds now range throughout North America, competing with native songbirds, spreading diseases, and damaging crops. Other introduced species have been less troublesome. The ring-necked pheasant *Phasianus colchicus* is a valued game bird. The canary-winged parakeet *Brotogeris versicolorus* and the budgerigar *Melopsittacus undulatus* have become established around New York City and in Florida as escaped pets with apparently no adverse consequences.

The increased availability of food and nest sites in suburbs and on farms have also enabled many species to extend their ranges northward. The northern mockingbird *Mimus polyglottos* and the northern cardinal *Cardinalis cardinalis* have become northern birds only in the last few decades, apparently because of the increased presence of bird feeders and certain cultivated plants. In many areas, Canada geese *Branta canadensis* and other birds have stopped migrating south, apparently because food is plentiful throughout the winter.

Extinct and Endangered Species. On the whole the impact of people on birds has been detrimental to the birds. Birds are not nearly as numerous as they once were, and many species have become extinct. Passenger pigeons were once so plentiful in America that in the 1850s the eminent ornithologist Alexander Wilson reported one flock darkening the sky for four hours as it passed overhead. From the flight speed he calculated that the flock must have contained over two billion birds. Alfred Russel Wallace suggested in his famous evolution paper of 1858 that the passenger pigeon must be indestructible. Wallace clearly underestimated the abilities of sport and market hunters, for the last passenger pigeon died just a year after Wallace (Figure 38.24A, B).

Figure 38.24
(A) Martha, the last passenger pigeon, Ectopistes migratorius, *which died in the Cincinnati Zoo in 1914. (B) Shooting passenger pigeons for sport in Iowa in 1867. Not much skill was required. Another favorite technique was to tie one pigeon to a stool and throw a net over the hundreds that would land to investigate. This is said to be the origin of the term "stool pigeon."*

A

B

Figure 38.25
One of the last California condors Gymnogyps californianus to soar freely. Most recent deaths are believed to have been due to lead poisoning from eating wounded deer. Surviving members of the species breed in captivity and may someday be returned to the wild.

Most bird hunters in the United States now work as hard as anyone to protect the species on which they depend for sport. Nevertheless, their activities sometimes have unforeseeable consequences. Many ducks and other birds that feed on the bottoms of ponds ingest shotgun pellets and die from lead poisoning. In the area of Lake Champlain 10% of the mallards and black ducks were estimated to die from lead poisoning in 1984. Federal law now forbids lead shot in some areas, but hunters have successfully blocked efforts to extend the prohibition, because steel shot costs more, may damage gun barrels, and is thought to be less effective. Eagles and other birds that feed on wounded animals also succumb to lead poisoning. One of the victims may be the California condor, a species that now consists of fewer than 40 individuals (Figure 38.25).

The destruction of habitat is even more harmful to birds than shooting them. Millions of acres of wetlands are drained for agriculture, development, and mosquito control, depriving waterfowl and many other birds of the habitat they require for breeding. Countless species of birds are also endangered by destruction of tropical rainforests. Many of these birds are "our" familiar summer songbirds that spend their winters in the rainforests. Other ways that birds die from human-altered environments include collisions with electric lines, broadcast towers, and skyscrapers in fog and with windows that reflect the sky. Many birds also collide with automobiles and airplanes. The latter can be as fatal to humans as to birds. In September 1988 a pelican brought down one of the most sophisticated and expensive planes in the world, a B-1B bomber, killing its crew of four. To prevent such a tragedy at some commercial airports, a falcon is sent aloft to scare off other birds before each landing and takeoff.

Another hazard to birds is getting covered with some of the oil it takes to fuel all these human activities. Thousands of seabirds regularly meet this fate. In spite of the efforts of volunteers to cleanse the birds, few survive. Many birds are also still damaged by pesticides. DDT was banned in this country in 1972, after it was linked to the decline in reproduction of eagles and other birds due to egg-shell thinning. Since the ban the populations of eagles and other affected birds have started to recover, but DDT is still used in other countries. Many other pesticides undoubtedly get passed up through the food web and concentrated in top carnivores in the same way that DDT did, but most pesticides are not tested for toxicity in humans, much less in birds.

Many species of birds are threatened by trade. Hundreds of thousands of specimens of rare and endangered species of macaws, cockatoos, and other parrots are brought into the country illegally each year, many of them through Mexico. Probably an equal number die during capture and shipment. Unlike legally imported birds, which must be quarantined before sale, some of these birds have contagious diseases that can wipe out a chicken or turkey farm. Some also carry psittacosis, which can be fatal to native birds and to humans. The smuggling of some birds out of the United States is also a problem. Falcons, which are shipped to the Middle East and trained to hunt for sport, are especially at risk.

Conservation. Many people have become sufficiently alarmed to demand more conservation laws and tougher punishment for violating them. Education in the "conservation ethic" has also been effective. One now seldom hears of hunters shooting hundreds of hawks for target practice, or killing eagles in the belief that they compete for game or kill livestock. Populations of some endangered species are now recovering due partly to such efforts and partly to artificial rearing (Figure 38.26). The whooping crane, which was down to 21 individuals in 1941, now

A

C

B

numbers more than 200, thanks largely to a program of artificial insemination and the use of the related sandhill crane (*Grus canadensis*) as a foster parent. The peregrine falcon has also made a comeback thanks to artificial rearing ("hacking"). The technique, pioneered by Tom Cade at the Cornell University Laboratory of Ornithology, has led to the reintroduction of falcons into their native habitats and also into cities, where skyscrapers and bridges make fair substitutes for the cliffs preferred as nest sites. The falcons return the favor by controlling the populations of pigeons, starlings, and house sparrows. Hacking and reintroduction are now being used to restore other species of hawks, as well as owls and other birds.

Figure 38.26
Back from the brink of extinction. (A) Male and female whooping cranes Grus americana. *(B) The bald eagle* Haliaeetus leucocephalus. *(C) The peregrine falcon* Falco peregrinus *feeding on a pheasant.*

SUMMARY

Archaeopteryx was essentially a dinosaur and is considered to be a bird only because it had feathers. Feathers enabled *Archaeopteryx* and other birds to fly and to maintain high body temperatures and apparently led to the other differences now seen between extant birds and reptiles. Feathers develop from papillae that are homologous with reptilian scales. A vane feather consists of a shaft with barbs growing outward on two sides. The barbs have hooked barbules that interlock and form a continuous surface. These feathers form the area of wings used in flying and insulate the body thermally. They generally are colored in specific patterns that function in courtship and territorial defense.

During flight air flows more rapidly over the top of wings than beneath them, generating the lift that keeps the bird aloft. Thrust results from the flow of air past the rear of the wings. The angle of the wings can be changed to increase lift, but the primary feathers and alulae must then open to prevent stalling. During flapping flight the wings are depressed and raised by strong muscles attached to the sternal keel. Wings can have a variety of shapes depending on whether they are specialized for efficiency, maneuvering, speed, or hovering. The surface area of the wing must be larger than a certain minimum, which increases with body mass.

The body mass is kept low by means of air sacs and hollow bones. The air sacs pump air completely through the parabronchi of the lungs, efficiently exchanging respiratory gases, as required for the high metabolic rates of birds. Birds must also eat and digest prodigious amounts of food. The beaks of birds are generally specialized for particular foods.

All birds are oviparous, producing cleidoic eggs. Fertilization is internal and usually follows an elaborate courtship behavior involving singing and displays of plumage. In most promiscuous, polygynous, or polygamous species only one parent incubates the egg and cares for young, but in most monogamous species both parents are involved.

KEY TERMS

sternal keel	lift	polygyny
ratite	thrust	polygamy
carinate	alula	altricial
shaft	parabronchi	precocial
vane feather	air sac	lek
barb	syrinx	pecking order
barbule	crop	brood patch
rachis	promiscuity	hacking
wing loading	monogamy	

SELF-TEST

1. Summarize the evidence that birds evolved from dinosaurs.

2. How do ratites differ from carinates? Give the common names for two kinds of ratites and several carinates.

3. Sketch the structure of a vane feather, labeling the quill, rachis, and barbs. Describe the structure and function of barbules.

4. Describe three adaptations of carinates that enable them to fly.

5. Explain why wings must produce lift and thrust during flight. Explain how a wing might generate lift and thrust.

6. Describe the function of alulae and wing slotting.

7. Describe how the wings of different species of birds are specialized for soaring, speed, or maneuverability.

8. Sketch four different kinds of bird feet specialized for such functions as perching, climbing, swimming, grasping prey, and walking.

9. Explain why the metabolic rate of a bird is so much higher than that of a reptile of the same size.

10. Briefly explain how birds sing. Why do they sing?

11. Bird behaviors depend on complex interactions of instinct and learning. Describe an example of a bird behavior that appears to be mainly instinctive, one that appears to be mainly learned, and one that appears to be instinctive but modified by learning.

12. Describe the production and development of a bird egg. Describe two special adaptations of the shell.

13. Describe three mating systems in birds: monogamy, polygyny, and polyandry. Give the common name of a bird that uses each mating system. Give an ecological explanation for the occurrence of each mating system.

READINGS

RECOMMENDED READINGS

Beehler, B. M. 1989. The birds of paradise. *Sci. Am.* 261(6):116–123 (Dec).

Borgia, G. 1986. Sexual selection in bowerbirds. *Sci. Am.* 254(6):92–100 (June).

Bull, J. and J. Farrand Jr. 1977. *The Audubon Society Field Guide to North American Birds.* New York: Alfred A. Knopf. (*One of many excellent field guides. This one has photographs instead of paintings and drawings.*)

Calder, W. A. III. 1978. The kiwi. *Sci. Am.* 239(1):132–142 (July).

Davies, N. B. and M. Brooke. 1991. Co-evolution of the cuckoo and its hosts. *Sci. Am.* 264(1):92–98 (Jan).

Ehrlich, P. R., D. S. Dobkin, and D. Wheye. 1988. *The Birder's Handbook.* New York: Simon & Schuster/Fireside. (*Concise source of information on U.S. birds.*)

Feduccia, A. 1980. *The Age of Birds.* Cambridge, MA: Harvard University Press. (*Excellent account of current thinking on the evolution of birds.*)

Heinrich, B. 1986. Why is a robin's egg blue? *Audubon* 88(4):64–71 (July).

Knudsen, E. I. 1981. The hearing of the barn owl. *Sci. Am.* 245(6):112–125 (Dec).

Ligon, J. D. and S. H. Ligon. 1982. The

cooperative breeding behavior of the green woodhoopoe. *Sci. Am.* 247(1):126–134 (July).

Mock, D. W., H. Drummond, and C. H. Stinson. 1990. Avian siblicide. *Am. Sci.* 78:438–449.

Myers, J. P. et al. 1987. Conservation strategies for migratory species. *Am. Sci.* 75:18–26.

Nottebohm, F. 1989. From bird song to neurogenesis. *Sci. Am.* 260(2):74–79 (Feb). (*The brains of canaries develop new nerve cells as they learn new songs.*)

Ostrom, J. H. 1976. *Archaeopteryx* and the origin of birds. *Biol. J. Linn. Soc.* 8:91–182.

Peterson, R. T. 1980. *A Field Guide to the Birds*, 4th ed. Boston: Houghton Mifflin. (*The classic guide with paintings and text by the premier birder.*)

Rahn, H., A. Ar, and C. V. Paganelli. 1979. How bird eggs breathe. *Sci. Am.* 240(2):46–55 (Feb).

Shettleworth, S. J. 1983. Memory in food-hoarding birds. *Sci. Am.* 248(3):102–110 (Mar).

Simons, T. et al. 1988. Restoring the bald eagle. *Am. Sci.* 76:253–260.

Smith, P. and C. Daniel. 1982. *The Chicken Book*. San Francisco: North Point Press. (*Everything about the chicken.*)

Stacey, P. B. and W. D. Koenig. 1984. Cooperative breeding in the acorn woodpecker. *Sci. Am.* 251(2):114–121 (Aug).

Tickell, W. L. N. 1970. The great albatrosses. *Sci. Am.* 223(5):84–93 (Nov).

Waldvogel, J. A. 1990. The bird's eye view. *Am. Sci.* 78:342–353.

Wellnofer, P. 1990. Archaeopteryx. *Sci. Am.* 262(5):70–77 (May).

Wiley, R. H. Jr. 1978. The lek mating system of the sage grouse. *Sci. Am.* 238(5):114–125 (May).

See also relevant selections in General References at the end of Chapter 21.

ADDITIONAL REFERENCES

Stapel, S. et al. 1984. Ratites as oldest offshoot of avian system—evidence from α-crystallin A sequences. *Nature (London)* 311:257–259.

Wilford, J. N. 1986. *The Riddle of the Dinosaur*. New York: Alfred A. Knopf. (*See Chapter 12 on bird evolution.*)

39

Mammals

Black bear cub (Ursus americanus).

CHAPTER OUTLINE

LEARNING OBJECTIVES

1. How did mammals evolve from reptiles?
2. What is the significance of the fact that mammals are homeothermic and have hair and mammary glands?
3. Why are mammals so much more active and why do they consume so much more food than reptiles?
4. Why do most mammals give birth to living young, while most other vertebrates do not? Are there any egg-laying mammals? Why do kangaroos, opossums, and other marsupials rear their young in a pouch?
5. How is hair produced? Why do different species of mammals have hair of different colors and patterns?
6. Why do many mammals have claws while others have hooves? What are horns, antlers and tusks?
7. How do many bats catch insects in the night, and some whales catch prey in the dark ocean depths?
8. How are monkeys, apes, and humans related to each other? How do they differ?

ORIGINS

Mammal-like Reptiles. Like birds, mammals evolved from reptiles and gained ascendancy over them during the present Cenozoic era—the era we presume to call the Age of Mammals. Mammals evolved before birds, during the Triassic period some 300 million years ago. Our ancestors were the mammal-like reptiles, also called synapsids (see Figure 37.2). The earliest synapsids were pelycosaurs, such as *Dimetrodon* and *Edaphosaurus* (see numbers 1 and 2 in Figure 37.9). They were the most common and impressive reptiles of their day, but they were soon overshadowed by other reptiles. Before they became extinct, however, some pelycosaurs evolved into another group of synapsids called therapsids. Therapsids, such as *Cynognathus* (see number 3 in Figure 37.9), fled the evolutionary scene almost as soon as the dinosaurs made their appearance, but not without leaving descendants that would give rise to class Mammalia.

Therapsids. Although the fossil record yields no direct link between therapsids and mammals, there is little doubt that therapsids were our ancestors. One can see it in their bones (Figure 39.1). Like dinosaurs and mammals, therapsids had legs positioned beneath the body rather than sprawling, as in other reptiles. This posture provides greater support for large body weight. Many therapsids were so small, however, that this added support makes sense only if they ran for long periods at a time. Paleontologists infer, therefore, that unlike other reptiles, which can run only briefly before becoming exhausted (see p. 820), therapsids could sustain prolonged running, like mammals. Supporting this deduction is the fact that the teeth and jaws of therapsids were better adapted to process large amounts of food, which would have provided both the means and the necessity for sustained locomotion. Instead of the uniform, peg-shaped teeth of most other reptiles, which are useful only for gripping prey or tearing off leaves before swallowing them whole, therapsids had cheek teeth adapted for chewing. Chewing breaks up food, allowing digestive enzymes to process nutrients faster. The lower jaws of therap-

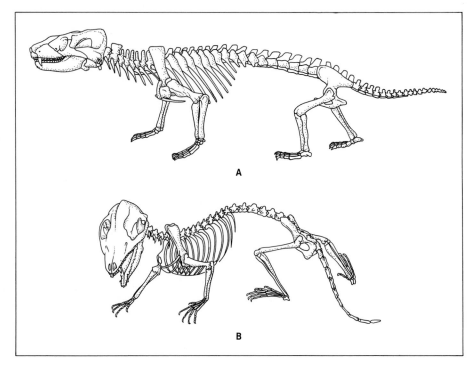

Figure 39.1
Comparison of the skeletons of a mammal-like reptile and an early mammal. (A) The skeleton of Thrinaxodon, *a therapsid reptile of the cynodont suborder, which is thought to have given rise to mammals.* Thrinaxodon *fossils are approximately* $\frac{1}{2}$ *meter long and are found in Lower Triassic deposits of South Africa and Antarctica. (B) Skeleton reconstructed from bones of two early mammals (mainly* Megazostrodon) *that lived at the same time as* Thrinaxodon, *some 200 million years ago. Note the large size of the skull relative to the rest of the body, indicating a relatively large brain that is characteristic of mammals (see Figure 5.27). This animal would have been approximately 10 cm long, not counting its tail. A primitive mammal with these teeth would have been carnivorous, probably feeding on insects. It probably resembled a mole or shrew in appearance and behavior (order* Insectivora).

sids were also stronger, consisting of one strong bone with a simple joint, rather than several bones attached by a complex joint.

Early Mammals. The oldest fossils recognized as those of early mammals are more than 200 million years old. Since hair, mammary glands, and other definitive features of extant mammals do not fossilize, paleontologists infer that these fossils are mammalian on the basis of their jaw joints and middle-ear bones. In these fossils and those of living mammals, two pairs of jaw bones have become bones of the middle ear. As will be described later, this adaptation strengthens the jaw, enabling it to chew food rather than bolt it down whole. The larger brains of these early mammals also suggest higher levels of behavioral activity. The similarity of their skeletons and teeth to those of modern moles and shrews suggests a similar life-style; early mammals probably fed voraciously on insects, most likely at night when they were safe from reptilian predators.

MAJOR FEATURES OF MAMMALS

Homeothermy. If early mammals were indeed nocturnal they were probably already homeothermic. Homeothermy accounts for most of the crucial differences between mammals and reptiles. As discussed in Chapter 15 (see pp. 307–308), the maintenance of a constant body temperature above the environmental temperature demands large expenditures of energy. The metabolic rate of a mammal, and therefore the amount of food it requires, is typically ten or more times greater than that of a reptile or other poikilotherm of the same size. Many mammals can obtain the large amounts of food required by homeothermy only because homeothermy enables them to feed at times and in places where low environmental temperatures eliminate poikilotherms from competition. In temperate regions mammals rule the night, and they share winters only with some birds. Moreover, homeothermy enables mammals to exploit wide ranges of habitat. The mice that live in the tropics are fundamentally like those of the Arctic, and temperature is no barrier to the wanderings of whales from the poles to the equator.

Hair and Milk. Homeothermy led to the evolution of two features that distinguish mammals from all other vertebrates: hair and the production of milk. Hair would have been essential for insulation in such small animals. Sweat glands would also have helped maintain a constant body temperature. It turns out that mammary glands are essentially modified sweat glands, so the production of milk also evolved indirectly as a consequence of homeothermy.

DIVERSITY

Monotremes. Class Mammalia now comprises approximately 4000 named species, with approximately one new species being discovered each year. Taxonomists generally divide them between two subclasses. The subclass Prototheria includes the egg-laying mammals, or monotremes (order Monotremata). Monotremes originated more than 130 million years ago and perhaps provide a glimpse of what early mammals were like. Only three species survive, all restricted to the Australian faunal region (see Figure 19.10), having apparently been isolated there some 50 million years ago by the breakup of Gondwana (see pp. 397–399). The most famous monotreme is the duck-bill or platypus, *Ornithorhynchus anatinus* (Figure 39.2B). The other two species, *Zaglossus bruijni* and *Tachyglossus aculeatus*, are echidnas, also called spiny anteaters (Figure 39.2A).

A

B

At first glance echidnas and the platypus seem unlikely relatives. Echidnas are terrestrial and burrowing animals, while the platypus is largely aquatic, living in the banks of streams in which it feeds. The horny beak, or **rostrum,** of echidnas is tubular and adapted for feeding on ants and termites. The rostrum of the duck-bill is adapted for feeding on crustaceans, insect larvae, and other aquatic invertebrates, which it detects by means of touch receptors and electroreceptors. The hair of echidnas is also quite different from that of the platypus. Many hairs of spiny anteaters are modified like those of porcupines as defensive weapons, while the hair of the platypus is so luxuriant that the species was nearly hunted to extinction for its fur. In spite of these differences, several similarities justify classifying the three species in the same group. In both echidnas and the platypus the males have hollow spurs on their hind legs with which they can inject poisonous venom during aggressive encounters with other males. The venom is variously described as "mildly toxic" or as being capable of killing a dog and causing excruciating pain in a human. They also are unique among mammals in having a common outlet, the **cloaca,** for feces, urine, and eggs. (Hence the name monotreme, from the Greek for "one hole.") Most important, both echidnas and the platypus share the distinction of being the only living egg-laying mammals, having apparently retained that mode of reproduction from the earliest mammals. Oviparity in monotremes will be described in greater detail later.

Marsupials. The second subclass of mammals is Theria, which includes marsupials and placental mammals. In marsupials (metatherians) most of the development of the embryo occurs outside the uterus, usually in a pouch (**marsupium**) in the mother's belly. There are now approximately 260 species of marsupials, mostly in Australia, but they were once far more numerous (Figure 39.3). The fossil record suggests that marsupials originated in what is now Canada, at a time when all the continents were united as Pangea (see Figure 19.9). The marsupials

Figure 39.2
Monotremes. (A) The short-beaked echidna Tachyglossus aculeatus. *Echidnas have long tongues and claws for feeding on ants, termites, and worms. (B) A platypus swims with its beaverlike tail and otterlike webbed feet, feeding on bottom-dwelling invertebrates with its ducklike bill. The first specimen of a platypus sent to Europe was at first dismissed as a hoax concocted by sewing together the parts of different animals. While under water the platypus keeps its eyes, nostrils, and ears closed. It rapidly scans the bottom with its bill, which contains touch receptors and electroreceptors. The latter detect the action potentials generated by the muscles of prey. The sexes are similar except that the body length is approximately 30 cm for females and 45 cm for males.*

Major Extant Groups of Class Mammalia

Genera mentioned elsewhere in this chapter are noted.

Subclass Prototheria PRO-toe-THEER-ee-uh (Greek *proto-* first + *ther* wild beast). Egg-laying mammals (monotremes). Adults lack teeth. Reproductive tract opens into cloaca. Duck-billed platypus and echidnas. *Ornithorhynchus, Tachyglossus, Zaglossus* (Figure 39.2).

Subclass Theria THEER-ee-uh. Exit of reproductive tract separate from that of alimentary tract (no cloaca).

Infraclass Metatheria MAY-tah-THEER-ee-uh (Greek *meta-* after). After an abbreviated gestation, young complete development outside the uterus, usually in a marsupium. Opossums, marsupial mice, Tasmanian devil, numbat, bandicoots, possums, wombats, kangaroos, wallabies. Marsupials. *Didelphis, Macropus, Phascolarctos, Sarcophilus* (Figures 39.3, 39.4, 39.6).

Infraclass Eutheria yew-THEER-ee-uh (Greek *eu-* true). Placental mammals. *Alces, Alopex, Balaenoptera, Bubalus, Canis, Castor, Cervus, Dasypus, Diceros, Equus, Eschrichtius, Euderma, Felis, Giraffa, Heterocephalus, Lepus, Loxodonta, Manis, Monodon, Mustela, Odobenus, Orcinus, Ovis, Phoca, Rangifer, Trachops, Tursiops, Ursus* (Figures 1.8, Unit Two opener, 20.6, 20.14, 39.5, 39.6, 39.7, 39.8, 39.9, 39.10, 39.11, 39.14, 39.15, 39.16, 39.17, 39.18, 39.19, 39.20, 39.21, 39.22).

MONOTREME

MARSUPIAL

PLACENTAL MAMMALS

then radiated into southern portions of Pangea, which later separated as Gondwana. All then became extinct in what is now North America, perhaps losing out in competition with placental mammals that invaded from the east. When the Panama land bridge reunited the Americas around three million years ago, many marsupials that had continued to evolve in South America radiated into North America. Several of these species survive, including the Virginia opossum. Placental mammals from North America likewise crossed the land bridge into South America, an event that may have led to the extinction of many marsupials there. This exchange of fauna between the Americas was first recognized by Alfred Russel Wallace and is called the Great American Interchange. Only Australia, which had become separated during the breakup of Gondwana, provided marsupials with a refuge free of competition from placental mammals except bats. Koalas, kangaroos, bandicoots, and wombats are among the unique fauna on Australia.

Many marsupials of Australia have become adapted to habitats that would otherwise be occupied by placental mammals, and some are remarkably like them. Such independent evolution of similar adaptations is referred to as **convergent evolution.** Kangaroos, for example, live on semiarid grasslands like those inhabited by antelope and buffalo elsewhere in the world, and, like these ruminants, kangaroos have multichambered stomachs containing microbes that help digest cellulose (see pp. 267–268). Many marsupials resemble their placental counterparts externally as well (Figure 39.4).

A

B

Figure 39.3
Marsupials. (A) A female opossum Didelphis virginiana *(formerly D. marsupialis). The opossum is the only extant marsupial native to North America. Following a tradition that can be traced back to an 18th-century artist, baby opossums are often portrayed riding their mothers with their tails wrapped around hers. As shown here, however, they are much less organized than that. Opossums thrive in spite of being hunted for fur, sport, and food. (B) The Tasmanian devil* Sarcophilus harrisii *has a reputation for viciousness, based largely on its behavior when confined in poor conditions. In the wild, Tasmanian devils prey on a variety of small vertebrates and invertebrates along streams, although their powerful jaws enable them to take an occasional sheep. Destruction of "the bush" eliminated them from eastern Australia, but they still live on the island of Tasmania. (C) The koala* Phascolarctos cinereus *feeding on its only food source, eucalyptus leaves. Because the eucalyptus leaves are toxic to other animals, the koala has few competitors. Its malodorous flesh also renders it safe from most predators. The species was once endangered by hunting for amusement and fur and is now threatened by destruction of eucalyptus forests by developers.*

C

Placental Mammals. Placental mammals, also called eutherians, differ from marsupials mainly in that development occurs entirely in the uterus, with support from the placenta. Placentals diverged from marsupials approximately 130 million years ago, apparently in what is now Asia, and expanded their ranges to virtually everywhere except on Antarctica and remote islands. During the Cenozoic era the diversity of placental mammals mushroomed (see Figure 34.21). They may have eliminated many of the marsupials, many other extinct groups of mammals, and numerous other kinds of animals. The diversity of placental mammals has shrunk somewhat in the last few thousand years, with the loss of some of the most spectacular species (see Figure 19.11). There are now approximately 4000 known species of eutherians, usually divided among 19 orders. Representatives of all the orders of eutherians are shown in Figure 39.5. The structure and functioning of most mammals are similar to those in humans and can be reviewed in Unit Two. Much of this chapter focuses on the adaptations of mammals that set them apart from their reptilian ancestors. Finally, we look more closely at our own species and our place among the other animals.

Figure 39.4
Evolutionary convergences of marsupial and placental mammals.

PLACENTALS

MARSUPIALS

Wolf
(*Canis*)

Tasmanian wolf
(*Thylacinus*)

Ocelot
(*Felis*)

Native cat
(*Dasyurus*)

Flying squirrel
(*Glaucomys*)

Flying phalanger
(*Petaurus*)

Ground
hog
(*Marmota*)

Wombat
(*Vombatus*)

Anteater
(*Myrmecophaga*)

Numbat
(*Myrmecobius*)

Mole
(*Talpa*)

"Mole"
(*Notoryctes*)

Mouse
(*Mus*)

Mulgara
(*Dasycercus*)

REPRODUCTION

Oviparity in Monotremes.   A major event in the history of mammals was the evolution of viviparity from oviparous ancestors. In order to appreciate what a difference this made, consider reproduction in extant oviparous mammals. Monotremes are thought to have diverged quite early, and it is possible that their reproduction is similar to that of therapsids and the earliest mammals. The female platypus produces from one to three eggs in a season, and these are fertilized by one male following an elaborate courtship and copulation. Fertilization occurs in the left oviduct. (Coincidentally, the right ovary and oviduct are nonfunctional, as in birds.) As the fertilized eggs pass down the oviduct they acquire the first of three layers of soft shell. The eggs then arrive in an enlarged area of the oviduct (the uterus), where the second shell layer is applied. While in the uterus the eggs grow, nourished by glandular secretions from the mother. Finally, the eggs acquire their third and last shell layer before entering the cloaca. Each ripe egg is oval, measuring approximately 14 mm across the middle and 17 mm from end to end. Around 20 days after mating the female retires to her burrow to lay her eggs. No one has observed oviposition, but it is supposed that the female bends so that the eggs stick to her abdomen as soon as they leave her cloaca. She then holds the eggs against her abdomen with her broad tail, incubating them for perhaps 11 days with her body heat. The young hatch out by tearing the rubbery shell with an egg tooth similar to that in birds and many reptiles. The female continues to hold the young to her abdomen as they suckle milk. Weaning occurs some four months after hatching.

Oviparity in the platypus, and presumably in early mammals, has several disadvantages. The female must invest a large supply of protein from her own body to form large, yolky eggs. To avoid predation of the eggs or herself, she has to retire to a burrow to lay and incubate the eggs. She is therefore hindered from feeding herself, even though she must soon sacrifice large amounts of fat and proteins in her milk. The incubation period is therefore necessarily brief, with the result that young are generally not well enough developed to feed or defend themselves (**altricial**). Because the young are born so early developmentally, the mother must invest even more time (four months in the platypus) before they can be weaned.

Viviparity.   In contrast to oviparous mammals, viviparous mothers can avoid predators and continue to feed themselves through almost the entire gestation period. The gestation period can therefore by prolonged, and the newborn are generally more advanced developmentally. In spite of these advantages, viviparity in placental mammals involves so many physiological adjustments that it is a wonder it ever evolved. Consider just three of the problems that had to be overcome:

1. The blastocyst had to evolve the ability to implant into the uterine lining (see pp. 112–113), and the embryo and adult jointly had to evolve mechanisms to form the placenta (see p. 324).
2. The blastocyst had to evolve some means of interrupting the normal reproductive cycles of the mother so that they could be redirected toward pregnancy. Female placental mammals are therefore sterile during the time when they are already pregnant. In placental mammals the pregnancy signal from blastocyst to mother is chorionic gonadotropin, which maintains the corpora lutea (see p. 324).
3. The immune system of the mother had to be blocked from rejecting the foreign tissue of the embryo.

Figure 39.5
Representatives of the orders of Eutheria (pp. 871–873). Numbers of species are based on Anderson and Jones (1984). The total comes to 3943 (not counting the three species of monotremes and 260 species of marsupials).

Order Insectivora 391 species

Moles

Shrews

Hedgehogs

Order Macroscelida 19 species

Elephant shrews

Order Scandentia 16 species

Tree shrews

Order Dermoptera 2 species

Flying lemurs

Order Chiroptera 917 species

Bats

Order Primates 180 species

Prosimians

Monkeys

Apes

Humans

Order Carnivora 235 species

Canids (wolves, coyotes, dogs, foxes)

Mustelids (weasels, otters, skunks, mink)

Felids (lions, tigers, cats)

Bears

Raccoons

**Order Pinnipedia
(sometimes combined with Carnivora)
34 species**

Seals

**Order Perissodactyla
(odd-toed hooved animals)
18 species**

Zebras

Horses

Walrus

Rhinoceroses

Tapirs

**Order Artiodactyla (even-toed hooved animals)
185 species**

Moose and other deer

Bison

Mountain sheep

Pronghorn

Mountain goats

Wild boars and other pigs

Giraffes

Order Proboscidea   2 species

Elephants

Order Sirenia   5 species

Manatees

**Order Hyracoidea
7 species**

Hyraxes

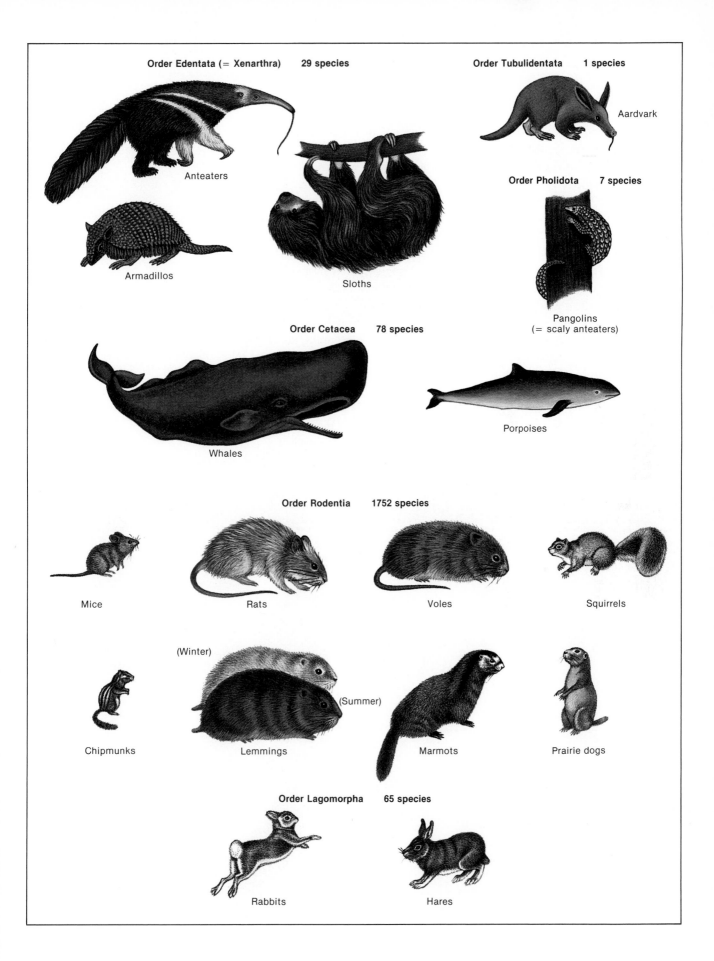

Order Edentata (= Xenarthra)   **29 species**

Anteaters

Armadillos

Sloths

Order Tubulidentata   **1 species**

Aardvark

Order Pholidota   **7 species**

Pangolins
(= scaly anteaters)

Order Cetacea   **78 species**

Whales

Porpoises

Order Rodentia   **1752 species**

Mice

Rats

Voles

Squirrels

Chipmunks

(Winter)

(Summer)

Lemmings

Marmots

Prairie dogs

Order Lagomorpha   **65 species**

Rabbits

Hares

Marsupials. How placental mammals evolved the above adaptations is unclear. It is unlikely that they went through a marsupial stage during evolution, but an examination of reproduction in marsupials may provide some insight into the selective pressures involved. Like the egg of a monotreme, that of a marsupial is large and enclosed in a shell membrane. Abundant yolk and secretions from the uterus provide nutrients for the developing embryo at first, but near the end of gestation a placental connection briefly forms.

The brevity of placental formation may be necessary to keep the embryo from triggering an immune reaction against it by the mother. Because the placenta is in place so briefly, however, there is little opportunity for the embryo to secrete chorionic gonadotropin or any other signal to the mother that she is pregnant. In the Virginia opossum *Didelphis virginiana*, and perhaps in some other marsupials, the mother's estrous cycle continues just as it would have if conception had not occurred (Harder and Fleming 1981). Its intrauterine gestation period lasts only 12.5 days and coincides with the luteal phase of the estrus cycle (see p. 322). In large kangaroos the gestation period lasts 33 days—two days less than the estrous cycle—and the mother generally conceives again shortly after birth. Thus marsupials apparently delay the problems of incubating eggs by retaining the embryo in the uterus for as long as they can. However, since they have no mechanism to redirect the reproductive physiology of the mother toward gestation, the gestation period cannot last longer than one estrous cycle.

Because the gestation period of marsupials is so brief, the neonates are still essentially embryos, comparable in developmental stage to that of a human embryo only a few weeks old (Figure 39.6A). Development must therefore con-

Figure 39.6
(A) A neonate marsupial attached to a nipple in the mother's pouch. Its developmental stage is comparable to that of a human embryo only six or seven weeks old (see Figure 6.25B). (B) Mother red kangaroo Macropus rufus *(= Megaleia rufa) with joey in her pouch (more or less).*

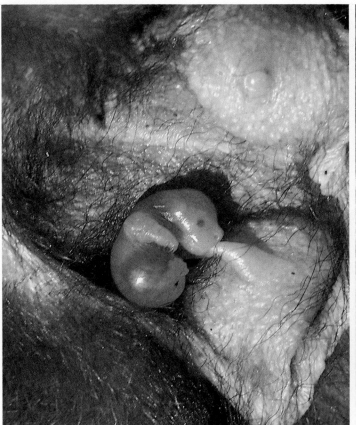

A

B

tinue outside the uterus, yet in a protected place where nourishment can be provided. In many marsupials this place is a pouch in the skin of the abdomen, which contains nipples. In a sense the marsupium serves the same function as a uterus, and the teat serves the function of a placenta. The neonate, helpless as it is, must make its own way into the marsupium without any help from the mother. In the case of the red kangaroo it is just as well that the mother does not offer to help, since she weighs 30,000 times more than the 1-gram neonate. The forelimbs of the neonate marsupial are more developed than the rest of its body, and on leaving the birth canal it struggles hand-over-hand through the hair on the mother's belly. The neonate is blind, but odor, the pull of gravity, or perhaps some unknown cue guides it to the marsupium. Once inside, the neonate finds a nipple and begins suckling. The nipple swells in the neonate's mouth, preventing it from becoming dislodged while the mother hops or runs about. Development then continues in the pouch. The young red kangaroo opens its eyes around 144 days after entering the pouch and a few days later pokes its head out for the first time (Figure 39.6B). After approximately 190 days in the marsupium the young kangaroo, now called a joey, leaves the pouch for the first time. The joey will return periodically for another 50 days to suckle from the same teat. It will remain near the mother for several months after weaning, until it is almost a year old.

While the young kangaroo has been developing in the marsupium, the mother may well have conceived again. This second embryo remains in the uterus in a dormant stage called **embryonic diapause** as long as there is already one offspring in the pouch. After about 200 days in embryonic diapause, however, the mother has evicted the joey from her pouch, and the second embryo resumes development. Thirty days later this neonate climbs into the marsupium, just as the first one did, and attaches to a second teat. The joey can still nurse from its own teat, however. Curiously the two teats produce different kinds of milk. The neonate gets a high-protein, low-fat milk, while the joey gets a high-fat, low-protein milk. Unlike most mammals, in which lactation blocks the resumption of the estrous cycle (see p. 325), the mother kangaroo might well get pregnant again while nursing both a neonate and a joey. In this way the red kangaroo can produce a steady supply of young at approximately 18-month intervals.

Reproductive Behavior. All female mammals produce milk, and this fact necessarily implies parental behavior, at least on the part of the mother. In the kangaroo, dolphin, chimpanzee, and many other mammals the males make no parental investment and generally have little to do with their offspring. These animals are often **promiscuous:** The male mates with any receptive female he encounters. A few species of mammals, such as wolves and other canoids, are **monogamous.** The male mates with only one female at a time and may make a significant parental investment. **Polygyny** is the most common mating system of mammals. It occurs among lions, gorillas, sea lions, deer, and many others. The common arrangement with polygyny is for a group occupying a territory to be dominated by one or several males that have exclusive mating privileges with all the receptive females in the group. Often, as in deer, the males compete in ritual or actual combat that establishes a **dominance hierarchy** for mating and other preferences.

Polygyny is one type of polygamy. The other type, **polyandry,** in which a female has several male mates simultaneously, is rare among mammals. One species in which polyandry does occur is the naked mole rat *Heterocephalus glaber,* a burrowing rodent of eastern Africa (Figure 39.7A). Jennifer Jarvis of the University of Cape Town in South Africa has found that the social organization of naked mole rats is remarkably similar to that of termites (see pp. 697–698). Each colony

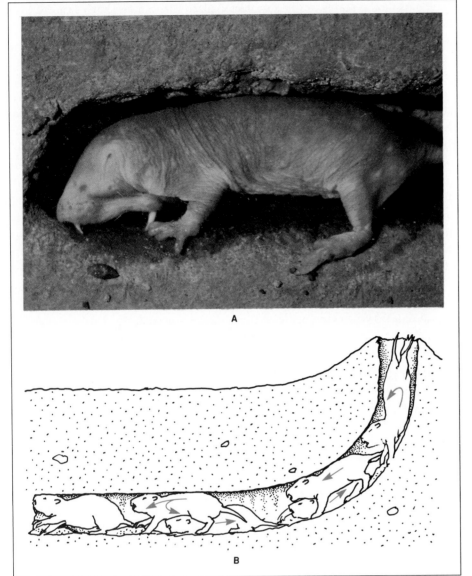

Figure 39.7
(A) A naked mole rat. Frequent workers use the incisors for burrowing and gathering roots. The lips close behind the incisors to keep dirt out of the mouth, and a flap of skin protects the nostrils. Length approximately 5 cm. (B) A digging chain of frequent workers. The digger kicks the excavated soil to the second mole rat on the bottom, which usually pushes it backward as it backs out beneath other mole rats to the mouth of the tunnel. After kicking out the dirt the mole rat then crawls over the backs of other mole rats, eventually taking its turn once more behind the digger. Periodically, the digger will be replaced.

of approximately 40 individuals has only one breeding female, and two or three castes that differ in body size and degree to which they help dig burrows.

HAIR

Functions. The next several sections describe some special adaptations of mammals, many of which are associated with the skin. One of these is hair. Many insects and some other kinds of animals produce hairlike body coverings (pile), but only mammals produce real hair. A hair is formed as epidermal cells accumulate in the root and push older cells out from the follicle (see Figure 10.1B). Like the scales of reptiles and the feathers of birds, hair therefore consists of dead, keratinized epidermal cells. Hair is not homologous with scales or feathers but appears to have evolved independently, perhaps in the synapsid reptiles. The major function of hair is to reduce the conduction of body heat to the environment. The resistance to heat conduction is increased by contraction of **arrector pili** muscles, which levers the hair outward in a process called **piloerection.** Piloerection thick-

Figure 39.8
*Color patterns in mammals. (A) The
Arctic fox* Alopex lagopus. *The thick,
white fur conserves heat and
provides camouflage, which is
essential for predation in open
terrain. In accordance with Allen's
Rule, the ears and nose are shorter
than in southern foxes, which
reduces the loss of body heat.
(B, C) The ermine, or short-tailed
weasel* Mustela erminea, *in summer
and winter. Weasels are voracious
predators and usually kill their prey
by biting through the skull.
(D) Burchell's zebra* Equus burchelli
*are clearly not camouflaged, but one
can see how confusing the stripes
would be in a stampeding herd.*

ens the fur coat (**pelage**), usually in response to cold. In many mammals, such as canids and apes, piloerection also occurs during aggressive encounters and has the effect of making the animal appear larger (see Figure 20.6). Most humans don't have enough body hair either to conserve heat or to put on an imposing display, but the arrector pili muscles still respond under similar circumstances, producing "goose bumps."

Hair also serves other functions. Often sensory neurons are attached to the hair root, allowing the hair to function as a touch receptor. The **vibrissae** ("whiskers") of mammals are especially adapted as touch receptors. Some mammals, such as spiny anteaters, hedgehogs, and porcupines, have thick, sharp hairs (**quills**) that deter predators. Contrary to popular myth, a porcupine cannot fling its quills, but it will slap at a predator or a nosy dog with its spiny tail. The quills easily detach from the skin of the porcupine, but tiny hooks at the other ends make them difficult to remove from the skin of the victim.

Color.　Many mammals that live in the Arctic are white, because the epidermal cells that form hairs do not contain pigments (Figure 39.8A). The absence of color presumably provides camouflage against the snowy landscape. The hair of most mammals contains black, brown, or grey melanins, which afford good camouflage against soil, rocks, tree bark, dry grasses, and forest shadows. Some temperate-zone animals, such as the snowshoe hare, molt twice a year, changing their color from brown to white and back to match the seasons (Figure 39.8B). In contrast, some mammals are conspicuously colored. The contrasting stripes of the zebra break up the outline of the body, perhaps making it difficult for a predator to pick out an individual from a herd (Figure 39.8C). Such patterning is referred to as **disruptive coloration.** In skunks the white stripe on the black background is undoubtedly **warning coloration.** The color patterns of many other mammals perhaps serve for social communication and recognition.

A

B

C

D

SWEAT AND MILK

Another adaptation of the skin for homeothermy in many mammals is the sweat glands. Those involved in cooling the body are the **eccrine sweat glands. Apocrine sweat glands** produce secretions containing pheromones for territorial marking and other kinds of social communication. Hair holds and concentrates pheromones used in social communication, which may explain the peculiar distribution of body hair in adult humans. Early in mammalian evolution sweat glands became adapted in a most unexpected way: as mammary glands. The derivation of mammary glands from sweat glands may seem unlikely in the adult (compare Figures 10.1B and 16.12), but their similarity can be seen in microscopical examination of embryonic tissues.

CLAWS AND HOOVES

Mammalian skin also produces hard structures that are used in defense, aggression, and for other functions. In some mammals the skin produces epidermal scales that serve as armor (Figure 39.9). In rhinoceroses and some other animals the skin is thick enough to serve as armor. In most mammals skin also produces claws, which are used in defense, capturing prey, climbing, digging, and grooming (Figure 39.10). Claws are keratinized sheaths that cover the pointed ends of the distal phalanges. In **ungulates** the claws are modified as hooves. Ungulates are classified in two orders, depending on whether each hoof has an odd number of toes (order Perissodactyla) or an even number (order Artiodactyla).

HORNS AND ANTLERS

Ungulates are now the only mammals that have horns and antlers. These two forms of head ornamentation are often confused with each other but differ in several respects:

1. An antler consists of dermal bone that is initially covered by furry skin, called velvet. A horn has a core of bone that projects partly into a hollow keratin sheath that grows out continually from patches of epidermal cells.
2. Except in caribou, antlers occur only in mature males, while horns occur in both sexes.
3. Antlers are always shed and regrown annually, while horns are generally permanent and irreplaceable. Although antlers are shed each year, they tend to get larger with age as the animal produces higher levels of sex hormones and is able to mobilize more nutrients to produce the antlers.

Cattle, sheep, goats, and antelopes (order Artiodactyla, family Bovidae) have a pair of hollow, symmetric horns that cover bone cores near the top of the head (Figure 39.11A). In most of these even-toed ungulates the horn sheath is never shed and, if lost, cannot be replaced. The American pronghorn *Antilocapra americana*, however, grows new horn sheaths inside the old ones, which are shed each year. (There is some debate over whether these should be called horns.) The horn of rhinoceroses, which are odd-toed ungulates, is very different (Figure 39.11B). A rhinoceros horn grows along the midline of the snout and consists of agglutinated fibers of keratin (sometimes erroneously said to be hair). Antlers occur in deer, caribou, and elk (order Artiodactyla, family Cervidae; Figure 39.11C). During development in the summer the antlers are sheathed in velvet. The velvet is shed prior to the breeding season, which is generally in late autumn (Figure 39.22B).

A

B

Figure 39.9
*Scaly mammals in defensive posture. (A) The cape pangolin, also called scaly anteater (*Manis temmincki), *of the African savannas. The scales are made of fused modified hairs and are shed periodically. A long tongue adds to the reptilian appearance. When not in use the tongue retracts into a sheath that extends down the abdomen to the pelvis. Like other anteaters, these are toothless and use the sticky tongue in feeding. (B) The nine-banded armadillo* Dasypus novemcinctus. *Each year millions of armadillos in the Gulf states adopt this posture as defense against automobiles, with a notable lack of success. The carapace consists of plates of dermal bone covered with keratinized epidermis.*

Figure 39.10

A bobcat Felis rufus *demonstrates the value of claws on a snowshoe hare* Lepus americanus.

ADAPTATIONS FOR FEEDING

Jaws. Because homeotherms must process much more food than reptiles, their feeding adaptations differ greatly. The chief difference is that most mammals speed up digestion by chewing food, thereby breaking it into smaller pieces and mixing it with enzymes. Birds achieve this with a gizzard, but most mammals use their teeth. Mastication places severe mechanical strain on jaws and teeth, requiring a stronger lower jaw and jaw joint. Most reptiles have a complex jaw with several small bones (quadrate, articular, and dentary; Figure 39.12A). The complex jaw joint provides little resistance to strain but is adequate in animals that swallow their food without chewing. In mammals, however, the only bone of the lower jaw is the large dentary in which the teeth are rooted (Figure 39.12B). The enlarged dentary is stronger and provides more area for the attachment of jaw muscles.

The jaw muscles attach at the other end to the **zygomatic arch** (cheek bone) and the **sagittal crest** (if present) on top of the skull. The jaw joint involves only the dentary and the squamosal bone of the skull. Although such a joint is much stronger than the complex reptilian joint, it is still mobile enough that the teeth can grind against each other in all directions, as in a cow chewing its cud. In an animal that must breathe as often as a mammal, prolonged chewing also requires that the nasal passages not open into the mouth, as they do in most reptiles. Thus mammals have a shelf of bone called the **secondary palate** that separates the nasal passages from the mouth.

Mammals did not merely discard the quadrate and articular bones during the evolution of a simpler jaw joint but incorporated them into the hearing apparatus. The quadrate and articular bones of mammal-like reptiles had already played a role in transmitting sound vibrations from the external tympanum to the inner ear. As they became smaller and lost their functions in the jaw joint, they assumed an increasing role in audition. Gradually, the quadrate and articular joined the stapes (= columella) and became the middle-ear bones now known as the **incus** and **malleus** (Figure 39.13). The presence of three middle-ear bones in mammals, rather than only the stapes, as in amphibians, reptiles, and birds, is thought to increase the efficiency of energy transfer from the air to the liquid of the inner ear (impedance matching). It may account for the fact that mammals can generally hear higher pitches than can most other terrestrial vertebrates.

Teeth. Increased efficiency of chewing in mammals also owes a great deal to adaptations in the teeth. In contrast to the uniform, peg-shaped teeth of most reptiles, which serve mainly to grip food, most mammals have a diversity of teeth that serve a variety of functions (see Figures 13.2 and 13.6). In front are the **incisors,** which are often chisel-shaped and used for biting off pieces of food. Lateral to the incisors are the **canines,** which are used by carnivores and omnivores for holding, piercing, and tearing flesh, both in feeding and in aggressive encounters. Finally, there are the **premolars** and **molars,** which tend to be flat in herbivores for crushing and grinding vegetable matter. In carnivores they tend to have sharp ridges for cutting meat. The number, size, and form of each kind of tooth vary with species and reflect diet. Since jaws and teeth are often preserved long after the rest of the skeleton has been scattered and decayed, dentition is useful in identifying carcasses. Numbers of teeth are commonly expressed by **dental formulae** (Table 39.1).

Figure 39.11
Horns and antlers. (A) Like other bovids, Dall's sheep (Ovis dalli) have a pair of symmetric horns. (B) Horns seem like overkill in a 2000-kg charging black rhino (Diceros bicornis). Ironically, the horns, which are valued in North Yemen for knife handles and in parts of Asia for "medicines," may well bring about the extinction of the species through illegal poaching. A rhino horn sells for approximately $24,000. (C) A caribou Rangifer rangifer with its antlers in velvet. (D) The bony protrusions on the head of the giraffe (Giraffa camelopardalis) are neither horns nor antlers. Their function, if any, is unknown.

Each pair of numbers gives the number of incisors, canines, premolars, or molars on each side of the upper and lower jaw (I/I, C/C, P/P, M/M). Thus the formula 2/1, 0/0, 3/2, 3/3 indicates two incisors on each side of the upper jaw and one incisor on each side of the lower jaw; no canine; three premolars on each side of the upper jaw and two on each side of the lower jaw; and three molars on each side of the upper and the lower jaw. The total number of teeth in this species would be 28.

Table 39.1   Dental formulae for some representative mammals.

Dental Formula	Feeding	Species
1/1, 0/0, 1/1, 3/3	Eating tree bark, nuts, roots	Beavers, porcupines, many squirrels, pocket mice
2/1, 0/0, 3/2, 3/3	Browsing	Hares, rabbits
0/3, 0/1, 3/3, 3/3	Grazing	Deer, bison, sheep, caribou
2/2, 1/1, 2/2, 3/3	Omnivory	Humans
3/3, 1/1, 4/4, 2/3	Predation	Dogs, foxes, wolves, bears

Reptiles' teeth do have one advantage over those of mammals: They are continually replaced. Most mammals get only two sets of natural teeth. The first set consists of **deciduous teeth** ("milk teeth"), which do not include molars. Deciduous teeth are replaced early in life by **permanent teeth.** The permanent teeth must last the mammal for the rest of its life, even though they are subject to considerable wear, especially in herbivores. Rodents compensate for such wear by having incisors that grow continually.

Some mammals have teeth adapted for functions other than biting, tearing, and chewing. In elephants the upper incisors are enlarged as **tusks** and are used to manipulate trees and other large objects. The walrus has upper canines enlarged as tusks, which it uses to haul itself onto ice and perhaps as skids or rakes as it feeds on shellfish on the ocean floor (Figure 39.14A). Most unusual is the single, spirally grooved tusk of the narwhal, which develops from an upper tooth and is up to 2.7 meters long in males (Figure 39.14B). In all three species the tusks may be as important in competition among males and in defense as they are in feeding.

Whales.   The order Cetacea is divided into two groups of whales: those with teeth and those without (Figure 39.15). The toothed whales (suborder Odontoceti) are narwhals, sperm whales, and dolphins (which are called porpoises if they have rounded rather than bottle-shaped noses). Like reptiles, most of these whales have large numbers of similar teeth, which they use for grasping prey rather than for chewing. Crushing of food occurs in a three-chambered stomach, especially in the first chamber, which contains stones, bones, and shells. Other whales (suborder Mysticeti) lack teeth. Instead, they have a curtain of **baleen** on the upper jaw that combs food out of water or sediments. For example, the gray whale (family Eschrichtidae) scoops up bottom sediments from which it filters out amphipod crustaceans (Figure 39.15B). Right whales (family Balaenidae) have baleen only on the sides of the mouth (see Figure 13.1D). They swim slowly through schools of plankton (mainly copepods), engulfing huge masses of them. They then use their large tongues to squeeze water out the sides of the mouth, through the baleen, which traps the plankton. Rorquals (family Balaenopteridae), which include the humpback, blue, and minke whales, feed on krill or fish.

Echolocation.   Sailors at least as far back as Aristotle's time have reported wails, creaks, groans, and screams from both baleen and toothed whales, and recordings have made these haunting "songs" familiar even to landlubbers. Whales apparently use them to communicate with each other, and in a few cases zoologists have been able to interpret the social context in which different songs occur. Dolphins and some other toothed whales also produce explosive sounds that may stun or even kill fish within a few meters.

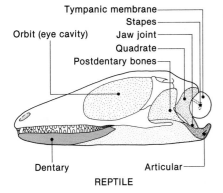

REPTILE

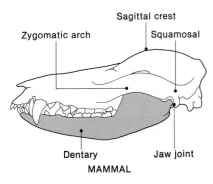

MAMMAL

Figure 39.12
The jaw of the early Triassic reptile Prolacerta *compared with that of the American opossum. See also Figure 37.19B. The dentary of the opossum and other mammals is enlarged and stronger than the combined quadrate, articular, and other bones of the reptilian jaw. Large jaw muscles of mammals attach to the sagittal crest and zygomatic arch. (The sagittal crest is clearer in Figure 39.1B.) Note also the diversity of tooth structure in the opossum.*

Besides these sounds, **toothed whales** emit pulses of ultrasound ("sound" with a frequency above the upper limit of human hearing), which are used in echolocation. Echolocation is similar to radar except that it uses ultrasound instead of radio waves. An ultrasonic beam is focused like a search light, and the reflected beam provides information about the direction, distance, speed, and shape of objects it strikes. Using echolocation, whales can find prey in the dark ocean depths and in water clouded with silt or plankton. The echolocation pulses from toothed whales do not come from the larynx. In dolphins they apparently originate in the nasal passages. A lens-shaped fatty deposit called the "melon" in the bulge of the forehead focuses the vibrations into a beam (Figure 39.16). The lower jaw picks up the returning echo and conducts it to the inner ear.

It is well known that many **bats** use echolocation to find prey, to navigate, and to avoid collisions with each other as they fly to and from their roosts. Some bats use echolocation to prey on small vertebrates (Figure 39.22), but most prey on nocturnal flying insects, availing themselves of a source of food that virtually no other kind of animal can exploit. Bats use the larynx to produce ultrasound with frequencies up to 100,000 hertz. The ultrasound emerges either from the mouth or the nose. The latter is advantageous, since a bat that has just caught prey could otherwise have difficulty navigating with its mouth full. Bats detect the echoes with their ears, as Lazaro Spallanzani demonstrated two centuries ago by plugging them with wax. Many bats have relatively huge external ears, which may help collect the echoes. Many others have bizarre ornamentation around the nose that may act like a megaphone, directing the ultrasound (Figure 39.17).

PRIMATES

Origins. Now that we are approaching the end of this book in which we have considered so many kinds of animals with scientific detachment, it might be interesting to consider our own order, Primates, with the same objectivity. There is no reason, after all, why understanding humanity should be any less a goal for zoology than it is for anthropology, sociology, psychology, history, theology, or art. In fact, none of these disciplines can provide a complete understanding of humanity if it is divorced from a zoological context. It should already be apparent that much of what we are and what we do relates to the fact that we are animals, in particular mammals, and most especially primates.

Figure 39.13
The auditory organs of the same reptile and mammal as in Figure 39.12, viewed from the rear. The quadrate and articular of the reptile have become the incus and malleus of the mammal.

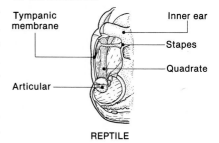
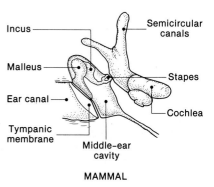

Figure 39.14
Tusks. (A) The walrus Odobenus rosmarus has enormous upper canines. The walrus is believed to use these tusks to skid along the ocean floor or to rake up clams. Body length approximately 3 meters. (B) Narwhals Monodon monoceros. A narwhal has only two teeth. In males usually the left one, rarely both, forms a tusk. Females seldom have tusks. Total body length is up to 5 meters.

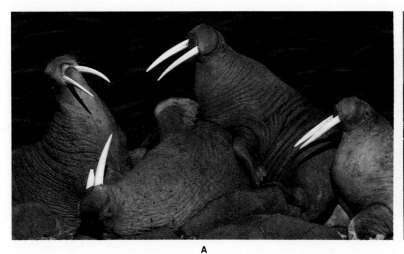
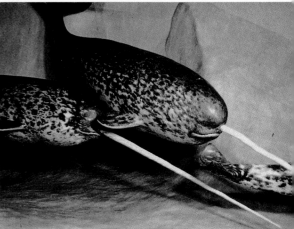

A

B

Figure 39.15
Toothed and toothless cetaceans. (A) Part of a group of approximately 30 killer whales (actually a type of dolphin) Orcinus orca *attacking a young blue whale* Balaenoptera musculus. *Of the odontocetes, killer whales are among the few that take on such large prey. They also attack seals and penguins. This blue whale was approximately 20 meters long and might have become one of the largest animals ever to live (up to 30 meters). Killer whales are up to 9 meters long. (B) The gray whale* Eschrichtius robustus *and other baleen whales have a comb of baleen on the upper jaw. Most baleen whales spend about half the year in polar waters feeding on abundant food, then migrate to warmer waters to mate and give birth. The gray whale filters crustaceans out of mouthfuls of sediment. The stirring up of sediments by the gray whale and perhaps by the walrus is believed to return minerals from the ocean floor back into nutrient cycles. Only about one-eighth of this 14-meter-long whale is shown.*

A

B

How did we get started? A single tooth found in 70-million-year-old deposits in Montana provides the earliest trace of the origins of primates. This molar has the squarish shape and blunt cusps that paleontologists generally associate with primates, though not all agree that this particular tooth belonged to a primate. Animals of this kind flourished early in the Cenozoic era. They were small, tree-climbing animals, perhaps somewhat like modern tree-shrews. The earliest undisputed fossils belong to two distinct groups dating to approximately 55 million years ago in what is now Europe and North America. In addition to molars typical of primates, they had other characteristic skull features. These include eye orbits

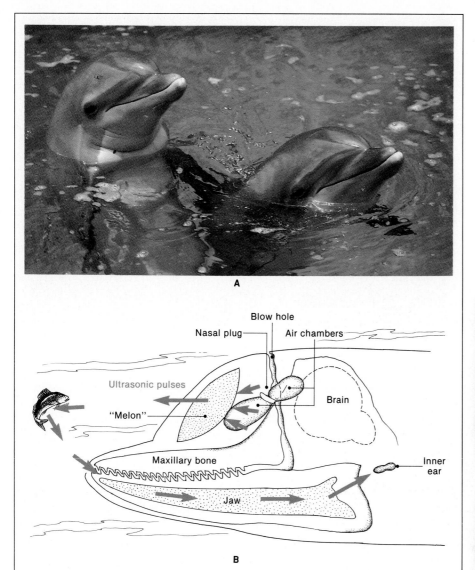

A

A

B

Figure 39.16
(A) A bottlenosed dolphin Tursiops truncatus (also commonly called a porpoise). Like other odontocetes, dolphins have only one nostril, the blow hole. (Mysticetes have two blow holes.) (B) Schematic representation of the head of a dolphin. While underwater the nasal plug shuts the blow hole. Ultrasonic vibrations appear to be produced by shuttling air between air sacs in the head, which causes them and the nasal plug to vibrate. The ''melon'' focuses the vibrations into an echolocating beam. Returning echoes are thought to be conducted to the inner ear by the lower jaw.

Figure 39.17
(A) A fringe-lipped bat Trachops cirrhosus of Panama about to capture a frog. Most bats that prey on large, immobile animals emit low-intensity echolocation pulses and are therefore said to be ''whispering bats.'' This particular species also homes in on the vocalizations of its prey. (B) The spotted bat Euderma maculatum uses high-intensity echolocation pulses to prey on moths in flight. This species is rare and, unlike many bats, is generally solitary. It roosts in rocky crevices in the southwestern United States. Body length approximately 1.1 cm.

that face forward for stereoscopic vision and a brain case that is large even for mammals. Both groups of early primates became extinct, but not before producing ancestors of the modern extant primates. One group is believed to have given rise to lemurs, lorises, and the other primates commonly called prosimians (Figure 39.18). The second group is thought to have given rise to tarsiers (Figure 39.19) and to monkeys, apes, and humans, which are commonly referred to as anthropoids.

Monkeys. The anthropoids (humanlike animals) are the monkeys, apes, and, of course, humans. Monkeys fall into two distinctive groups: the New World monkeys of South America and the Old World monkeys of Africa and Asia (Figure 39.20). Among the New World monkeys are the capuchin monkeys, howler monkeys, squirrel monkeys, spider monkeys, marmosets, and tamarins. New World monkeys lack opposable thumbs: That is, they cannot touch the tips of the fingers to the tip of the thumb. Their thumbs bend around branches on the same side as the fingers. Nevertheless, these monkeys, especially those with **prehensile tails,** are quite at home in the trees. Old World monkeys have opposable thumbs and

Figure 39.18
Prosimians (suborder Strepsirhini). (A) A ring-tailed lemur Lemur catta *with young. In its native Madagascar it is sometimes kept as a house pet. Length, including tail, 1.2 meters. (B) The slender loris* Loris tardigradus. *Lorises have no tails. They hunt insects and other small animals at night by carefully stalking them.*

A

B

Figure 39.19
*Tarsiers (*Tarsius *sp.) are extremely agile in trees, making tremendous leaps as they feed on insects, lizards, and fruit. All are nocturnal, as might be guessed from the owlish eyes. Adhesive pads on the fingertips help maintain grip. The body is about 15 cm long. The hairless tail probably assists in maintaining balance.*

big toes, and they use their tails for balance rather than for gripping tree limbs. Among the Old World monkeys are several species of macaques (including the rhesus), baboons, and mandrills.

Apes and Humans. Apes and humans appear to differ from all monkeys, since they lack tails and tend to be larger and more terrestrial than arboreal. In the kinds of characters that matter to taxonomists, however, apes and humans are more closely related to Old World monkeys than these monkeys are to New World monkeys. All three groups—Old World monkeys, apes, and humans—have opposable thumbs. They also differ from New World monkeys in having noses with high bridges and nostrils that point to the front rather than to the sides. In addition they have flat nails on all fingers and toes, whereas some of the New World primates have some nails shaped like claws. Old World monkeys, apes, and humans also share the dental formula 2/2, 1/1, 2/2, 3/3 (two incisors, one canine, two premolars, three molars on each half of both jaws), which differs from that of New World monkeys. Molecular comparisons also show that Old World monkeys, apes, and humans are more closely related to each other than any of them is to New World monkeys.

The apes include three species of African Great Apes: the chimpanzee *Pan troglodytes* (Figure 19.13A), the bonobo (formerly pygmy chimpanzee) *Pan paniscus*, and the gorilla *Gorilla gorilla* (see Figure 1.8). Other apes live in southeastern Asia: the orangutan *Pongo pygmaeus* and several species of gibbons *Hylobates* spp.

Relationship of Humans to Apes. Taxonomists generally divide humans and apes into two or three different families, reserving the family Hominidae for ourselves, and either grouping all the apes in family Pongidae or putting the gibbons into a third family Hylobatidae. Separating humans from apes has less to do with science, however, than with tradition and other considerations. When Linnaeus named *Homo sapiens* he knew that there was little scientific basis for separating

A   B

Figure 39.20
Monkeys. (A) A New World monkey, the spider monkey Ateles geoffroyi *demonstrates how handy a prehensile tail can be. (B) An Old World monkey, the rhesus monkey* Macaca mulatta, *is one of the most familiar primates. It is commonly used in medical research. Like these two, most monkeys and apes spend considerable time grooming each other. Grooming not only removes external parasites but is essential in maintaining normal social bonds.*

humans from apes. To avoid controversy he simply omitted taxonomic criteria from the description of humans, which reads in its entirety *"Homo nosce Te ipsum."* Know yourself, man. He confessed what he had done in a private letter in 1747 (quoted in Frängsmyr 1983, p. 172):

> I ask you and the whole world for a generic differential between man and ape which conforms to the principles of natural history. I certainly know of none. . . . If I were to call man ape or vice versa, I would bring down all the theologians on my head. But perhaps I should still do it according to the rules of science.

Recent studies based on molecular comparisons and on morphology support Linnaeus. For example, Miyamoto and his colleagues (1990) have reported that segments of human DNA are identical to corresponding segments of DNA from either chimpanzee or gorilla at more than 98% of their bases. These findings, and many others based on molecular phylogenetics, suggest that humans are more closely related to chimps and gorillas than either of these African apes is to the orangutan. In fact, several studies show that humans are more closely related to chimps than chimps are to gorillas. Several studies based on morphology lead to the same conclusion (Andrews 1987). Therefore if chimpanzees, gorillas, and orangutans belong in the same family, then humans belong in that family with them.

For more than 25 years a growing number of scientists, mainly of the molecular persuasion, have proposed bringing taxonomy up to date by uniting humans and the African Great Apes in the same family. The classification in the box follows the recommendation of Miyamoto and others (1990) in uniting the African Great Apes with humans in the family Hominidae and in the same subfamily Homininae. Apologies to the apes and to humans whose self-esteem depends only on the belief that they are better than apes.

A Classification of Order Primates

Genera mentioned elsewhere in this chapter are noted.

Suborder Strepsirhini STREP-sere-RINE-eye (Greek *streptos* turned + *rhin-* nose). Lemurs, lorises, galagos, indri, aye-aye, and others. *Lemur, Loris* (Figure 39.18).

Suborder Haplorhini HAP-lo-RINE-ye (Greek *haplo-* simple or single).

  Semisuborder Tarsiiformes TAR-see-eye-FORM-eez (Latin *tarsus* ankle). Tarsiers. *Tarsius* (Figure 39.19).

  Semisuborder Anthropoidea an-throw-POY-dee-uh (Greek *anthropos* man).

   Infraorder Platyrrhini PLAT-ear-RINE-eye (Greek *platys* broad, flat). Thumb not opposable. Bridge of nose flat; nostrils face to side. Dental formula 2/2, 1/1, 3/3, 3/3 or 2/2, 1/1, 3/3, 2/2. South American monkeys. *Ateles, Callithrix* (Figure 39.20A).

   Infraorder Catarrhini CAT-ar-RINE-eye (Greek *katarrhin* hook-nosed). Thumb and big toe opposable. Bridge of nose raised; nostrils toward front. Dental formula 2/2, 1/1, 2/2, 3/3.

    Superfamily Cercopithecoidea SIR-ko-pith-ee-KOY-dee-uh (Greek *kerkos* tail + *pithikos* ape). African and Asian monkeys, baboons, mandrills. *Cercopithecus, Macaca, Papio* (Figure 39.20B).

    Superfamily Hominoidea ha-min-OY-dee-uh (Latin *homines* humans).

    Family Hominidae ha-MIN-i-dee.

     Subfamily Hylobatinae HIGH-low-BATE-i-nee (Greek *hyli* forest + *bados* walk). Gibbons. *Hylobates.*

     Subfamily Homininae ha-MIN-in-ee.

      Tribe Pongini ponj-EYE-nye (West African *mpongwe* name of a people). Orangutan. *Pongo.*

      Tribe Hominini ha-min-EYE-nye.

       Subtribe Gorillina gor-ILL-in-uh (Greek *Gorillai* name of an African tribe). *Gorilla* (Figure 1.8).

       Subtribe Hominina ha-MIN-in-uh.

        Genus Pan. Chimpanzee, bonobo (Figure 19.13).
        Genus Homo. Humans.

INTERACTIONS WITH HUMANS AND OTHER ANIMALS

Ecology. Many classes of animals far outnumber mammals in both abundance and numbers of species. However, because of their high levels of activity around the clock and throughout the year and their ability to adapt to such a broad range of habitats, mammals have a disproportionate influence on other animals. Only mammalian herbivores are so large and graze and browse in such vast herds that they create and maintain entire biomes—prairies and savannas. Most of the large carnivores that dominate food webs and thereby control the populations of other kinds of animals are also mammals. Even mice, moles, shrews, and bats, although small as individuals, collectively consume huge amounts of plant material, insects, and other foods. They, in turn, are food for birds and larger carnivorous mammals. Countless examples of the ecological roles of mammals could be listed, but many are already familiar.

People and Other Mammals. The importance of other mammals in human affairs is equally familiar, although in the mechanized world we tend to overlook it. It was not so long ago, however, when much of the labor of farming and transportation was done by horses, oxen, mules, and other domesticated mammals. In many parts of the world that is still the case. Millions of people still depend on the labor of an estimated 75 million domesticated water buffalos *Bubalus bubalis.* Cattle and numerous other domesticated mammals also contribute more directly to the human food supply, as meat. In many parts of the world hunted mammals make up a major portion of the protein in the diet. In the United States and other developed countries the recreational aspects of hunting outweigh the nutritional aspects. Mammals also provide other economic benefits, such as furs and leather. Each year millions of mammals also contribute to human welfare as research animals. There is no doubt that mammals have been indispensable in finding treatments and cures for diseases in the past, and they will continue to be essential if research on cancer, AIDS, and other diseases continues.

The relationship between people and other mammals is not always one of unilateral exploitation. People often form close attachments with other mammals, and vice versa. Many mammals are bred solely for human companionship. The importance of pets to the well-being of their owners must be enormous considering the prices of dogs and cats. The quality, variety, and cost of pet food in the United States and other developed countries are further testimony of how much we value pets, especially considering how many of our own species are inadequately fed. People also value wild and captive mammals. Deer, raccoons, chipmunks, and others undoubtedly increase the allure and value of rural and suburban property. Zoos are major cultural facilities in cities that have them. Even squirrels add a touch of life to a city park. Wild animals are also cherished merely for existing, even if there is little chance of ever seeing them. People contribute millions of dollars to organizations devoted to the preservation of such mammals (Figure 39.21).

Conservation.   As with so much of nature, we most appreciate wild mammals after we have come close to destroying them. Wildlife conservation is a recent phenomenon in the United States and is not yet established in many parts of the world. In much of Africa and South America, for example, many people seem bent on repeating the depredations of the American pioneers. Ivory poachers armed with AK-47 assault rifles and chain saws have killed approximately half of the 1.35 million African elephants (*Loxodonta africana*) that were alive a decade ago. Poach-

Figure 39.21
Mother and baby harp seals, Phoca groenlandica. *In the 1960s up to 300,000 harp seals were killed each year, including pups that were clubbed to death for their white fur. Public protest forced the Canadian government to set quotas and to ban the clubbing of pups in 1985. The mother is up to 2 meters long.*

ers have killed all but a few thousand of the 60,000 black rhinoceroses (*Diceros bicornis*) alive in 1970. In Namibia wildlife authorities have had to cut off the horns of surviving black rhinos to save their lives. In many parts of Africa, and in much of South America, mammals are threatened by destruction of habitat, especially rainforests. Many undeveloped countries are so burdened by poverty and foreign debt that even if there are conservation laws, there is no money to enforce them. Often poachers and illegal developers bribe the underpaid or greedy officials who are supposed to enforce the laws.

In many places conservationists have succeeded in changing attitudes and laws regarding mammals, often by mobilizing public pressure, and increasingly by making conservation more profitable than extinction. Several African countries now get a major part of their revenue from tourists who come to see large mammals, especially following such popular films as "Out of Africa" and "Gorillas in the Mist." Some of these countries have set aside large game reserves and have cracked down on poaching and illegal trading in ivory, horns, and furs. The President of Kenya has warned that poachers will be shot on sight and has burned 12 tons of ivory in an effort to dry up the trade. In order to win support from the growing human populations in these countries, imaginative schemes have been devised so that some of the profits of tourism reach the people who would otherwise be farming the land or profiting from poaching. In South America, tourism in rainforests is just beginning, so conservationists have worked out other incentives. Many governments are allowed to sell off huge foreign debts, which they have little hope of repaying, in exchange for setting aside areas to be preserved.

Different kinds of solutions are necessary for marine mammals, which often do not come under any national jurisdiction. Faced with the extinction of many species of whales, the International Whaling Commission resolved in 1982 that all commercial whaling must end. Gradually and reluctantly all of the major whaling countries have complied with the ban except Japan, Iceland, and Norway. Taking advantage of an exception in the ban that allows whales to be killed for research, they have continued to kill several hundred whales each year. A typical "discovery" from such "research" is that pregnant female whales have higher levels of progesterone than nonpregnant females. Here also adverse publicity and changing economic incentives may end such practices. Whale-watching is already a bigger industry than whale-killing.

The populations of many species will return to healthy levels spontaneously if their habitats are preserved and they are no longer killed by humans. Some of the North American species that have recovered from near extinction are beaver (*Castor canadensis*), which were decimated to satisfy the fashion for beaver hats, and bison (*Bison bison*), which were hunted nearly to extinction for meat and to deprive Indians of food (Figure 39.22A). With the decline in farming and the restoration of habitat many other species are expanding their numbers and ranges. Among the recovering American species are the lynx (*Felis lynx*), the red wolf (*Canis rufus*), and the moose (*Alces alces;* Figure 39.22B). Other native American species require considerably more help, and their survival is in doubt (Figure 39.22C). For many species zoos and preserves must serve as a kind of "ark" until their populations increase. It often requires considerable labor and understanding of reproductive physiology and behavior to get mammals to breed in captivity. Many mammals also have to be shipped among zoos around the world to maintain genetic diversity. Even with a successful captive-breeding effort it is often questionable whether there will be a habitat in which to reintroduce a species once its numbers have recovered. This raises the further question of whether a species of wild animal that exists only in captivity can be said to exist at all.

A

B

C

Figure 39.22
Formerly and presently endangered American mammals. (A) Bison once grazed from the Rockies to as far east as Buffalo, New York, and often in herds that stretched as far as one could see. Countless numbers were killed to deprive hostile Indians of their food source, and many others were killed by market hunters who often took only the tongue. By 1900 fewer than 1000 remained. Today there are more than 30,000. Shoulder height approximately 2 meters. (B) The largest deer in the world, the moose, was nearly hunted out in the United States by 1900, but its population is now increasing throughout the northern tier of states. This bull, shedding its velvet, has a total height of approximately 2 1/2 meters. (C) A Florida panther Felis concolor coryi, subspecies of the mountain lion (= cougar or puma). Only 30 to 50 individuals survive, all in the southern tip of Florida. The survival of this subspecies is thought to be threatened by citrus farming, deer hunting (which deprives the cats of prey), disturbance due to hunting with dogs, and automobile collisions. Florida has spent $13 million for highway underpasses for the panther, and zoologists are considering test-tube fertilization as a means of restoring the population. Body approximately 2 1/2 meters long.

THE FUTURE OF HUMANS AND OTHER ANIMALS

In the preceding section, as in most of the concluding sections in this unit, I have discussed ways in which humans are affecting other animals. Often I have had to struggle to keep my personal sadness from reaching these pages as I considered the tragic results of our actions on this planet—tragic not only for other organisms, but for ourselves. Now that this book is near its close you will, I hope, forgive a brief departure from cool, scientific objectivity. Zoologists are, after all, human, and it is characteristic of humans to try to foresee the future consequences of present activities, even if it cannot be done with much confidence. This tendency to try to look into the future may well be the key to our success as a species, for it has often enabled us to avert the disastrous consequences of present actions. What follows is my own vision of the dismal future into which we seem to be rushing blindly, stampeding the other animals ahead of us.

Extinction. As noted in virtually every chapter in this unit, the success of humanity has often been at the expense of other species. Of all the consequences of our success, undoubtedly the worst is extinction. Our methods of hunting, fishing, trapping, destroying pests, domestication, studying, and otherwise dealing with other animals may raise questions about our wisdom and ethics, but their effects on a species can be reversed as long as the number of survivors is genetically viable. Extinction, though, is forever. Primarily through destruction of habitat, our species is extinguishing one other species every day on average, and threatening the continuity of three billion years of life. This should strike most people as fundamentally contrary to the ethics and principles of fairness that we like to think separates us from apes and other "lower" forms of life.

Should one species that is a relative newcomer to the planet be allowed to destroy another species? Many people say yes. Their argument takes two forms: one practical and the other abstract. In its crudest form the practical argument is that other species must justify their existence by being useful to humans. For example, during debate on a motion to weaken the Endangered Species Act one U.S. Senator said, "people should have dominion over fish, wildlife, and plants. Only where lower species are of benefit to mankind are they important." (Congressional Record, 18 July 1978, pp. S11040 and S11043. Note the word "dominion" as used in the King James translation of *Genesis* 1:28.) The practical argument has been tempered lately, and its advocates now seem more willing to let species live as long as they do not get in the way of progress. The new version of the argument is heard countless times every day, whenever a new development, highway, dam, or other construction project is proposed. It goes something like this: "We cannot let one species stand in the way of economic progress. Extinctions have always occurred, so why be so concerned about the extinction of this one species?" For some reason the same kind of reasoning is never applied in reverse. "We cannot let one project stand in the way of the continued existence of this species. Economic failures have always occurred, so why be so concerned about the failure of this one project?"

Even the moderate form of the practical argument is getting increasingly difficult to sustain in the United States and many other parts of the world as people see development making their environments increasingly unpleasant, while the promised prosperity never seems to arrive except for a few. Moreover, there appears to be an increasing appreciation of the value of **biodiversity** for its own sake. It is now widely perceived that an ecosystem is not simply a collection of isolated species, but a system that ceases to function if too many of its parts are removed. More and more people are realizing that humans are part of that system, and that they cannot go on destroying other species without endangering themselves. The extinction of one particular species may not directly endanger anyone, but little by little the losses add up to suicide.

Ironically, the abstract argument for extinction also rests on this fact. According to this argument, since humans are just one of many parts of ecosystems, we, like other species, have the right and even the necessity to exploit other organisms. There is certainly some validity to this argument. Yet the fossil record shows no other single species that has had such a devastating impact on so many others.

Overpopulation. The destruction of other organisms may not, in fact, be intrinsically wrong. Humans are perhaps no more evil or stupid than other species. The problem, however, is that the population of humans has gotten so large, and the consumption of resources by the more fortunate of us has become so great, that our one species now consumes 40% of all the organic material produced by photosynthesis. A graph showing the growth in human population since the Industrial Revolution is like what one would expect for bacteria, weeds, rats, cockroaches, and other pests that have escaped from natural controls (Figure 39.23). Indeed, an objective observer from another planet might well diagnose Earth as suffering from an infection of us.

The existence of human overpopulation is often denied, mainly by people carried away by theology or politics. An opponent of birth control has argued that the Earth cannot be overpopulated, since all five billion of us could easily fit in the state of Rhode Island. That is certainly true as long as no one needed to go to the bathroom. In the People's Republic of China the Marxist government of the 1970s discouraged birth control, arguing that the problem was not too many people but inequitable distribution of goods due to capitalism (Ding Chen 1980). In 1984 a spokesman for the U.S. government argued in Mexico City, of all places, that increasing population was good for business. In 1987 a television evangelist and candidate for President urged U.S. citizens to reproduce to keep from being outnumbered by citizens of other countries.

Others have argued that there is no cause for alarm since the human population cannot possibly continue to increase at its present rate. In fact, the rate of increase in population is already declining in most of the world. Yet at the present rate most of you will live to see the world's population twice as large as it is now. If that rate of increase were sustainable your grandchildren could live to experience a world population of more then 20 billion people. Even the most zealous theologian or politician would hesitate to argue that such a population is desirable. It is probably not even possible. Therefore, the question is not whether the population will stop increasing, but when and how. Will we rationally and gradually bring our population to a level that can be sustained? Perhaps, although there is reason to be pessimistic that it will be done in time. Fifteen years passed between the first warning about the depletion of the ozone layer by chlorofluorocarbons and the first international agreements to begin limiting their use. If we do not act in time we could overshoot the population level that can be sustained by the earth. In terms of ecological theory that means that, like the reindeer of the Pribilof Islands (see Figure 17.16B), we will overshoot the carrying capacity, destroy the

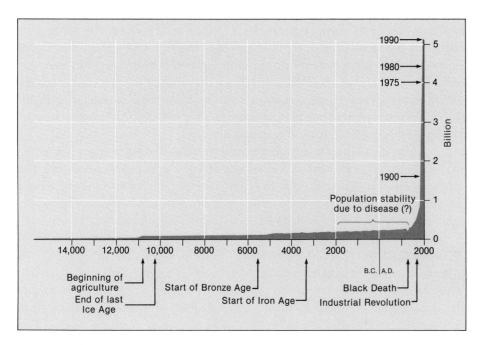

Figure 39.23
Estimated world population of humans. Population estimates are highly speculative. Population is believed to have risen at about the same time as the invention of agriculture (Neolithic Revolution) and increased gradually thereafter. Contagious diseases may have played a large role in holding populations in check. The population began its current surge around the year 1700, following the Industrial Revolution, improvements in sanitation and medicine, and more productive agricultural methods.

ability of our habitat to support us, and then experience a population collapse. In terms of reality, that means millions or even billions of people dying from famine, disease, and the consequences of political upheaval, and perhaps many more than that number will become homeless or refugees. There are those who believe we have already exceeded carrying capacity, and that we are already seeing the first signs of the coming population crash. Anyone who makes such pessimistic predictions is obviously a Cassandra. But then, Cassandra was always right.

The Importance of Zoologists. During the next few decades zoologists will play increasingly crucial roles in determining what kind of future we and our fellow animals will have, and how long that future will last. One vital role for zoologists will be to monitor the effects of human activities on other animals. Like canaries that stopped singing when the air in mines was poisoned with gas, other animals can provide early warnings that we are endangering their environments and our own. Zoologists will be the first to observe when other animals have, so to speak, stopped singing. Zoologists will also be increasingly called upon to advise on preserving species and their habitats. In many cases this preservation will have to take place in zoos, or perhaps someday even in petri dishes, until habitats recover sufficiently for reintroduction. Finally, for those species that are doomed to extinction, zoologists will be the only ones competent to memorialize them through research and to describe the many unique animals our children and grandchildren will never get to know.

SUMMARY

The most significant adaptation in the evolution of mammals from reptiles appears to have been homeothermy. The evolution of homeothermy was tied to the evolution of hair and sweat glands, which are both unique to mammals. Sweat glands evolved into mammary glands, which virtually define mammals. Homeothermy enabled mammals to spread to a wide range of climates and to remain active year round and at night. The cost for this increased activity was an extremely high metabolic rate and therefore a large demand for food. Mammals have stronger jaws, different kinds of teeth for chewing food, and an increased rate of digestion. Two of the jaw bones found in reptiles evolved into middle-ear bones in mammals. Instead of uniform, peg-shaped teeth, mammals have various combinations of incisors, canines, premolars, and molars, depending on diet.

Hair reduces the conduction of body heat, and in many mammals is colored in different patterns for camouflage, disruptive coloration, warning coloration, or social display. Skin also produces claws and hooves and horns and antlers. Horns are permanent keratinized sheaths over bony cores, while antlers are temporary formations of dermal bone.

A few mammals, the monotremes, are oviparous. Most mammals are viviparous, meaning that the young develop within the uterus and are supported by means of a placenta. The advantage of viviparity is that development can be prolonged, and the young may be able to fend for themselves soon after birth. In some mammals, the marsupials, the period in the uterus is brief, but development continues in a marsupium. Most mammals are polygynous.

Primates are generally divided into two suborders, with the prosimians (lemurs, lorises, and some others) in one, and tarsiers, monkeys, apes, and humans in the other. Old World monkeys, including baboons, are quite different from New World monkeys and are more closely related to apes and humans. The chimpanzee, the bonobo, and the gorilla are in turn more closely related to humans than to the orangutan or gibbons.

KEY TERMS

homeothermy	horn	canine
marsupium	antler	premolar
convergent evolution	zygomatic arch	molar
piloerection	sagittal crest	dental formula
pelage	secondary palate	deciduous teeth
vibrissa	incus	tusk
eccrine sweat gland	maleus	baleen
apocrine sweat gland	incisor	echolocation

SELF-TEST

1. Explain how the evolution of homeothermy is related to the evolution of hair and of mammary glands.

2. Explain how the evolution of homeothermy is related to the evolution of stronger jaws and more diverse teeth compared with those of reptiles.

3. Explain how the evolution of stronger jaws is related to the evolution of the middle-ear bones in mammals.

4. Describe the major features that distinguish monotremes, marsupials, and placental mammals from each other.

5. Explain why nonflying placental mammals are native on every large land mass except Australia and Antarctica. What was the Great American Interchange?

6. What physiological obstacles had to be overcome in the evolution of viviparity in placental mammals?

7. Explain the differences among the following: horns, antlers, tusks.

8. In Table 39.1 relate each dental formula to the diet of the mammal.

9. Describe the similarities in echolocation in toothed whales and bats. What differences are there between echolocation in these two groups?

10. Which ape is more closely related to humans than to other apes?

READINGS

RECOMMENDED READINGS

Austad, S. N. 1988. The adaptable opossum. *Sci. Am.* 258(2):98–104 (Feb).

Bekoff, M. and M. C. Wells. 1980. The social ecology of coyotes. *Sci. Am.* 242(4):130–148 (Apr).

Bronson, F. H. 1984. The adaptability of the house mouse. *Sci. Am.* 250(3):116–125 (Mar).

Burt, W. H. and R. P. Grossenheider. 1964. *A Field Guide to the Mammals.* Boston: Houghton Mifflin. (*Includes information on skulls and dentition.*)

Clark, W. C. 1989. Managing planet Earth. *Sci. Am.* 261(3):46–54 (Sept). (*Introduction to an entire issue on the global effects of humans.*)

Clutton-Brock, T. H. 1985. Reproductive success in red deer. *Sci. Am.* 252(2):86–92 (Feb).

Cronon, W. 1983. *Changes in the Land: Indians, Colonists, and the Ecology of New England.* New York: Hill and Wang. (*Many examples of the effects of humans on wild mammals.*)

Dawson, T. J. 1977. Kangaroos. *Sci. Am.* 237(2): 78–89 (Aug).

Degabriele, R. 1980. The physiology of the koala. *Sci. Am.* 243(1):110–117 (July).

Fenton, M. B. and J. H. Fullard. 1981. Moth hearing and the feeding strategies of bats. *Am. Sci.* 69:266–275.

Fossey, D. 1983. *Gorillas in the Mist.* Boston: Houghton Mifflin.

Ghiglieri, M. P. 1985. The social ecology of chimpanzees. *Sci. Am.* 252(6):102–113 (June).

Goodall [van Lawick-Goodall], J. 1971. *In the Shadow of Man.* Boston: Houghton Mifflin.

Griffiths, M. 1988. The platypus. *Sci. Am.* 258(5):84–91 (May).

Hoage, R. J. (Ed.). 1985. *Animal Extinctions: What Everyone Should Know.* Washington DC: Smithsonian Institution Press.

Kanwisher, J. W. and S. H. Ridgway. 1983. The physiological ecology of whales and porpoises. *Sci. Am.* 248(6):111–119 (June).

Macdonald, D. (Ed.). 1984. *The Encyclopedia of Mammals.* New York: Facts on File.

Marshall, L. G. 1988. Land mammals and the great American interchange. *Am. Sci.* 76:380–388.

Moehlman, P. D. 1987. Social organization in jackals. *Am. Sci.* 75:366–375.

Moore, J. A. et al. 1985. Science as a way of knowing—human ecology. *Am. Zool.* 25:375–641.

Murray, J. D. 1988. How the leopard gets its spots. *Sci. Am.* 258(3):80–87 (Mar).

Nelson, C. H. and K. R. Johnson. 1987. Whales and walruses as tillers of the sea floor. *Sci. Am.* 256(2):112–117 (Feb).

O'Brien, S. J. 1987. The ancestry of the giant panda. *Sci. Am.* 257(5):102–107 (Nov).

O'Brien, S. J., D. E. Wildt, and M. Bush. 1986. The cheetah in genetic peril. *Sci. Am.* 254(5):84–92 (May).

Reeves, B. O. K. 1983. Six millenniums of buffalo kills. *Sci. Am.* 249(4):120–131 (Oct).

Renouf, D. 1989. Sensory function in the harbor seal. *Sci. Am.* 260(4):90–95 (Apr).

Rismiller, P. D. and R. S. Seymour. 1991. The echidna. *Sci. Am.* 264(2):96–103 (Feb).

Suga, N. 1990. Biosonar and neural computation in bats. *Sci. Am.* 262(6):60–68 (June).

Terborgh, J. and M. Stern. 1987. The surreptitious life of the saddle-backed tamarin. *Am. Sci.* 75:260–269.

Tuttle, R. H. 1990. Apes of the world. *Am. Sci.* 78:115–125.

Whitaker, J. O. Jr. 1980. *The Audubon Society Field Guide to North American Mammals.* New York: Alfred A. Knopf. (*One of several excellent field guides. This one has photographs.*)

Whitehead, H. 1985. Why whales leap. *Sci. Am.* 252(3):84–93 (Mar).

Wilkinson, G. S. 1990. Food sharing in vampire bats. *Sci. Am.* 262(3):76–82 (Feb).

Wilson, E. O. (Ed.). 1988. *Biodiversity.* Washington DC: National Academy Press.

Wilson, E. O. 1989. Threats to biodiversity. *Sci. Am.* 261(3):108–116 (Sept).

Würsig, B. 1979. Dolphins. *Sci. Am.* 240(3):136–148 (Mar).

Würsig, B. 1988. The behavior of baleen whales. *Sci. Am.* 258(4):102–107 (Apr).

Zapol, W. M. 1987. Diving adaptations of the Weddell seal. *Sci. Am.* 256(6):100–105 (June).

See also relevant selections in General References at the end of Chapter 21.

ADDITIONAL REFERENCES

Anderson, S. and J. K. Jones Jr. (Ed.). 1984. *Orders and Families of Recent Mammals.* New York: Wiley.

Andrews, P. 1987. Aspects of hominoid phylogeny. In: C. Patterson (Ed.), *Molecules and Morphology in Evolution: Conflict or Compromise?* New York: Cambridge University Press, Chap. 2.

Carroll, R. L. 1988. *Vertebrate Paleontology and Evolution.* New York: W. H. Freeman.

Cohn, J. P. 1988. Halting the rhino's demise. *BioScience* 38:740–744.

Crompton, A. W. and P. Parker. 1978. Evolution of the mammalian masticatory apparatus. *Am. Sci.* 66:192–201.

Ding Chen. 1980. The economic development of China. *Sci. Am.* 243:152–165 (Sept).

Frängsmyr, T. (Ed.). 1983. *Linnaeus: The Man and His Work.* Berkeley: University of California Press.

Goodall, J. 1986. *The Chimpanzees of Gombe: Patterns of Behavior.* Cambridge, MA: Harvard University Press.

Grzimek, B. (Ed.). 1989. *Grzimek's Encyclopedia of Mammals.* New York: McGraw-Hill.

Harder, J. D. and M. W. Fleming. 1981. Estradiol and progesterone profiles indicate a lack of endocrine recognition of pregnancy in the opossum. *Science* 212:1400–1402.

King, F. A. et al. 1988. Primates. *Science* 240:1475–1482. (*Uses of primates in research.*)

Macdonald, D. W. (Ed.). 1984. *The Encyclopedia of Mammals.* New York: Facts on File.

Miyamoto, M. and M. Goodman. 1990. DNA systematics and evolution of primates. *Annu. Rev. Ecol. Syst.* 21:197–220.

Nowak, R. M. and J. L. Paradiso. 1983. *Walker's Mammals of the World,* 4th ed. Baltimore: Johns Hopkins University Press.

Vaughan, T. A. 1986. *Mammalogy,* 3rd ed. Pittsburgh: W. B. Saunders.

Würsig, B. 1989. Cetaceans. *Science* 244:1550–1557

Glossary

A band. Middle band of a sarcomere in skeletal muscle, formed by thick and thin protein filaments.

abiotic. Without life. For example, abiotic synthesis of organic molecules, or abiotic factors in ecology, such as oxygen, water, ions, and temperature.

aboral. On the side opposite the mouth, especially in an echinoderm.

acanthella. Second larval stage of acanthocephalans, which develops from an acanthor, then becomes an encysted cystacanth in the intermediate host.

acanthor. Egg containing an embryonic acanthocephalan, released into the feces of the final host.

acclimation. Gradual development of tolerance to a change in an environmental condition, such as temperature.

accommodation. Focusing of the lens of the eye.

aciculum. Stiff, chitinous reinforcement of the parapodium of a polychaete annelid.

acidity. Relative concentration of hydrogen ions in a solution.

acoelomate. Having no body cavity (neither pseudocoel nor coelom).

acontium. Tentacle-like structure attached to mesentery that extends through the mouth of a sea anemone.

acousticolateralis system. Lateral line system and inner ear of a fish or amphibian.

α (alpha) cell. Pancreatic cell that secretes glucagon.

acrosome (= acrosomal vesicle). Lysosome in sperm that contains enzymes that digest a pathway through the covering of eggs.

actin. Protein that forms intracellular microfilaments and thin filaments in muscle.

action potential. Brief, all-or-none change in the voltage across an excitable membrane of a nerve or muscle cell.

activation energy. Free energy that must be absorbed by a molecule before it can react chemically.

active site. Part of an enzyme that binds to substrate.

active transport. Energy-dependent movement of ions or molecules across a biological membrane.

adaptation. (1) Beneficial trait of an organism. (2) Reduction in sensitivity of a sensory receptor due to use.

adductor. (1) Muscle that brings a body part toward the body axis. (2) Muscle that closes the valves of a bivalve mollusc.

adenine. Nitrogenous base in nucleic acids and ATP.

adenohypophysis. Part of the pituitary gland consisting of the anterior pituitary and the pars intermedia.

adenosine. Nucleotide consisting of adenine and ribose.

adenosine triphosphate (ATP). Adenosine with three phosphate groups, which couples energy-producing processes with energy-consuming processes.

adhesion. Binding of different kinds of molecules to each other.

adhesive papilla. One of three organs by which an ascidian larva attaches to substratum prior to metamorphosis.

adipose. Pertaining to cells or tissue in which fats are deposited.

adipose fin. Finlike fatty lobe behind the dorsal fin of certain male salmonid fishes.

adrenal medulla (= chromaffin tissue). Central part of the adrenal glands, which secretes epinephrine (adrenaline) in response to stress.

advertisement call. Courtship and rivalry call of male frogs.

aeropyle. Pore in insect egg that is permeable to O_2 but not to H_2O.

aestivation (= estivation). Dormant condition during summer or dry period.

afferent. Conducting toward the center, such as a nerve cell conducting action potentials from sensory receptors into central nervous system, or a vein conducting blood to the heart. Compare efferent.

affinity. Ability of one molecule to bind another, for example, an enzyme to its substrate.

aftershaft. Secondary feather growing from the main shaft.

agglutination. Sticking together, especially of blood cells.

aggressive mimicry. Resemblance of a predator to a harmless animal or object.

air sac. Air-filled sac in the respiratory system of birds.

albinism. Inability to produce melanin, resulting in pale skin, white hair, and pink pupils.

albumen. Fluid of an amniote egg that contains large amounts of albumin. Egg white.

albumin. One of a group of soluble proteins common in blood, albumen, and other tissues.

all or none law. Principle that action potentials either occur fully or not at all.

allantois. Cavity formed by extraembryonic membrane, which stores metabolic wastes and helps exchange oxygen and carbon dioxide through the egg shell in reptiles and birds and becomes the lining of the urinary bladder in mammals.

allele. One form of a gene occurring at the same locus on a chromosome as another allele but producing a different variety of the same trait.

allergy. Immune response to an otherwise harmless antigen.

allometry (= scaling). Differential effect of changing linear, area, and volume dimensions of organisms.

allopatric. From different areas. Mainly referring to a mode of evolution in which members of a population become geographically separated and diverge into two different species.

allosteric effect. Stimulation or inhibition of an enzyme due to binding of a nonsubstrate to a site on the enzyme different from the active site.

alpha helix. Coiled secondary structure of a protein.

alpha motor neuron. Efferent neuron with large-diameter axon that directly excites skeletal muscles.

alternating tripod gait. Walking pattern in insects, in which three alternating legs support the body at any instant.

altricial. Referring to a species, especially bird or mammal, in which the very young are incapable of feeding or caring for themselves. Compare precocial.

altruism. Behavior of an animal that is disadvantageous to the animal but beneficial to another.

alula. First digit on a bird's wing.

alveolus (plural alveoli). (1) Air sac in the mammalian lung. (2) Milk reservoir in the mammary gland of a female mammal.

ambulacral groove (= ambulacrum). One of five grooves radiating from the mouth in an echinoderm, which contains canals of the water vascular system.

amebocyte. Cell capable of locomotion by formation of pseudopodia.

ameboid movement. Movement of certain cells, such as *Amoeba* and white blood cells, in which a pseudopodium slowly advances and the rest of the body flows into it.

amictic. Referring to parthenogenetic female rotifers that lay diploid, thin-shelled eggs that hatch into females. Compare mictic.

amino acid. Molecule with an amine group and an acidic carboxyl group. Subunit of proteins.

ammocoete. Larva of a lamprey.

ammonifying bacterium. Bacterium that converts nitrogen-containing molecules into ammonia.

ammonotelic. Excreting ammonia as the main nitrogenous waste.

amnion. Fluid-filled cavity formed by extraembryonic membrane, which encloses the embryo of reptiles, birds, and mammals (amniotes).

amniote. (1) Having an amnion, as in an amniote egg. (2) A vertebrate (reptile, bird, or mammal) in which the embryo has an amnion.

amphiblastula. Free-swimming larva in certain marine sponges (most calcareous sponges and some demosponges), which develops from a blastula in which flagella are pointed inward within the blastocoel.

amphid. Complex sensory organ near the heads of certain nematodes.

amplexus. Mating embrace of frogs, in which male grasps female from behind as both release gametes.

ampulla (AM-pew-luh). (1) Bulge at the end of each semicircular canal containing hair cells that sense acceleration of the head. (2) Reservoir in the water vascular system of an echinoderm. (3) Low-frequency electroreceptor of certain fishes (ampulla of Lorenzini or ampulla of teleosts).

amylase. Enzyme that digests starch into the disaccharide maltose.

anabolism. Metabolism in which organic molecules are synthesized.

anadromous. Migrating from oceans to fresh water for spawning, like lampreys, shad, trout, and salmon.

anaerobic glycolysis. Fermentation of glucose in the absence of oxygen.

analogous. Similar because of convergent evolution rather than shared ancestry.

anaphase. Stage of mitosis or meiosis when sister chromatids move to opposite poles and become chromosomes.

anaphylactic shock. Drop in blood pressure due to excessive immune response.

anapsid. A vertebrate, especially a reptile, without temporal openings in the skull.

androgen. Steroid hormone, such as testosterone, with masculinizing effects. Compare estrogen.

androgenic gland. Masculinizing gland in crayfish and related crustaceans.

aneuploidy. Difference of at least one in the normal number of chromosomes in a set.

animal pole. Part of an egg containing less yolk.

anion. Negatively charged ion such as chloride.

annulus. External subdivision of a body segment of a leech.

antennal gland (= green gland). Excretory organ in the head of certain crustaceans.

anterior pituitary. Part of the adenohypophysis that secretes several hormones in response to other hormones from the hypothalamus.

anthropomorphism. Attribution of human traits to nonhuman animals.

antiaphrodisiac. Pheromone that discourages mating, usually produced after a female has already mated.

antibody (= immunoglobulin). Protein from plasma cells that binds to a specific foreign molecule as part of the immune response.

antibody labeling. Technique in which radioactive or fluorescent antibodies bind to particular molecules and reveal their presence.

anticodon. Triplet of three bases on transfer RNA that is complementary to a codon on messenger RNA.

antifreeze. Substance that lowers the freezing point of a solution.

antigen. Molecule, usually foreign, that stimulates production of antibodies.

antiserum. Serum containing antibodies to a toxin or other antigen.

antivenin. Antiserum to venom.

antler. Bony, usually temporary, growth on head of deer, antelope, and some other even-toed ungulates.

aorta (a-OR-tuh). (1) Largest artery of vertebrates, which conducts blood from the left ventricle to body tissues. (2) Anterior part of the heart of arthropods.

aortic body. Receptor in the arch of the aorta, which senses the level of CO_2 and O_2 in the blood.

apical complex. Group of organelles characteristic of protozoans in the phylum Apicomplexa

apocrine sweat gland. Gland in the groin or armpit of a mammal that produces odiferous sweat for social communication.

apodeme. Internal projection of cuticle to which muscle attaches in arthropods.

apopyle. Opening of a radial canal into the spongocoel of a syconoid sponge, or into a flagellated chamber of a leuconoid sponge.

appendicular skeleton. Bones of the limbs, and the pectoral and pelvic girdles to which they are attached.

apron. Abdomen of certain crabs, which is folded beneath the cephalothorax.

arachnoid membrane. Membrane covering the brain, which has villi through which cerebrospinal fluid drains into blood.

arboreal. Living in trees.

archenteron. Cavity in embryo formed during gastrulation; the future lumen of the gut.

archeocyte. Type of amebocyte in sponges that wanders through the intercellular substance transporting food and differentiating into other cell types.

Aristotle's lantern. Beaklike chewing apparatus of sea urchins and related echinoderms.

arteriole. Smallest diameter artery, which conducts blood into a capillary.

artery. Blood vessel that conducts blood away from the heart.

arthrodial membrane (= articular membrane). Flexible, transparent cuticle that hinges sclerotized cuticle in arthropods.

articulamentum. Inner calcareous layer of the plates of a chiton.

articular membrane. Arthrodial membrane.

artificial selection. Breeding controlled by humans selecting for particular hereditary traits.

asconoid. Type of canal system in simple sponges in which the choanocytes line a central chamber (spongocoel). Compare syconoid and leuconoid.

asexual reproduction. Reproduction by fission, budding, or some other method not involving the fertilization of gametes.

assimilated energy. Portion of ingested energy that is available to an organism for use or storage.

association cortex. Area of the cerebral cortex that integrates information from two or more sensory modalities.

astrocyte. Type of glial cell that helps form blood–brain barrier.

asymmetric. Incapable of being bisected into similar halves.

asynchronous muscle (= fibrillar muscle). Type of flight muscle in certain insects in which contraction is not synchronized with action potentials from central nervous system.

atoke. Creeping, benthic, juvenile stage of many polychaete annelids, or the anterior, nonreproductive segments of some other polychaetes.

ATP. Adenosine triphosphate.

atrioventricular (AV) node. Tissue of the heart that is excited by action potentials in the atria, then after a delay, sends action potentials through bundles of His to ventricles.

atrioventricular (AV) valve. One-way valve that permits blood flow from an atrium into a ventricle.

atrium. (1) Chamber of the heart (auricle) that receives blood from the veins. (2) Central chamber (spongocoel) of a sponge. (3) Upper chamber in the body of an adult ascidian, which contains the branchial sac.

auricle. (1) An atrium of the heart. (2) External ear (pinna) or any similarly shaped structure.

auricularia. Larva of a sea cucumber.

autogamy. Fusion of haploid nuclei inside the cell (usually a protozoan) that produced them.

autoimmune disease. Immune reaction against an organism's own molecules.

automimicry. Protection conferred on nontoxic members of a species by their resemblance to toxic members of the species.

autonomic nervous system. Portion of peripheral nervous system of vertebrates that involuntarily controls organs.

autosome. Chromosome that occurs in the same number in both males and females. Any chromosome that is not a sex chromosome.

autotomy. Self-amputation to escape a predator.

autotroph (= primary producer). Organism that obtains its energy from inorganic sources, as in photosynthesis.

avicularium. Beak-shaped zooid of ectoprots used in defense of the colony.

avoiding reaction. Behavior in which a *Parmecium* or other ciliate backs away from a barrier, turns, then moves around it.

axial. Located along the axis.

axial skeleton. Bones of the body axis, including the skull, vertebral column, and rib cage.

axon. Long process that conducts action potentials in most nerve cells.

axoneme. Central core in a cilium or eukaryotic flagellum, usually consisting of a "9 + 2" arrangement of microtubules.

axopodium. Type of pseudopodium in protozoans in the superclass Actinopoda that is reinforced by microtubules and that shortens to draw food toward the body for phagocytosis.

β (beta) cell. Pancreatic cell that secretes insulin.

B cell. Lymphocyte that transforms into a plasma cell that produces antibodies in the immune response.

baleen. Comb-shaped structure on upper jaw of certain whales that filters food out of water or sediments.

barb. One of the thin processes from the shaft of a feather.

barbule. Hooked structure at the end of the barb of a feather that connects the barb to other barbs.

basal body (= kinetosome). Organelle to which cilia and eukaryotic flagella attach. Identical to the centriole.

basal comb. Device used by gnathostomulids to scrape up food.

basal lamina (= basement membrane). Sheet of fibrous proteins underlying epithelial tissues.

base. (1) Substance that combines with H$^+$ in solution. (2) Nitrogenous part of nucleic acids.

basement membrane. Basal lamina.

basilar membrane. Membrane in the cochlea that supports the organ of Corti and vibrates in response to sound.

basking. Exposing the body to solar radiation.

Batesian mimicry. Protection from predation conferred on a palatable species by its resemblance to a toxic species.

Bauplan (German; plural **Baupläne**). Body plan of an animal.

behavioral ecology. Approach to animal behavior that bases hypotheses on the assumption that behavior is optimized.

behaviorism. School of psychology that focuses on behavior in the laboratory and avoids inferences about internal psychological states.

benthic. At the bottom of a lake or ocean.

benthos. Benthic organisms.

bicuspid valve (= mitral valve). Left atrioventricular valve of the heart.

bilateral cleavage pattern. Pattern of cleavage in which cells move symmetrically with respect to the embryonic body axis. Characteristic of cephalopod molluscs and tunicates.

bilaterally symmetric. Capable of being bisected into two similar, mirror-image halves.

bile. Secretion from the liver that is stored in the gallbladder until released into the small intestine.

bile salt. Molecule in bile that emulsifies fats.

binary fission. Type of asexual reproduction in which the organism divides into two roughly equal parts.

bindin. Species-specific molecules on echinoderm sperm that bind to receptor molecules on the vitelline envelope of the egg of the same species.

binominal system. System of nomenclature popularized by Carolus Linnaeus in which each species is uniquely named by two words. Often called binomial.

biogenetic law. Statement by Haeckel, now discounted, which states that "the history of the fetus is a recapitulation of the history of the race [species]." Ontogeny recapitulates phylogeny.

biogeochemical cycle. Nutrient cycle, such as the carbon cycle, that involves both biological and geological processes.

biological species concept. Ernst Mayr's definition of species as "groups of actually or potentially interbreeding natural populations, which are reproductively isolated from other such groups."

bioluminescent organ. Organ that produces light.

biomass. Total mass of living tissue in an area.

biome. Terrestrial ecosystem of a particular type.

bipedal. On two feet.

bipinnaria. Ciliated, bilaterally symmetric larva of a starfish.

biradially symmetric. Radially symmetric

with some paired structures on opposite sides.

biramous. Two-branched, like the appendages of crustaceans.

bladder worm (= cysticercus). Juvenile tapeworm, consisting of an invaginated or introverted scolex in a fluid-filled cyst.

blastema. Mass of dedifferentiated cells of mesodermal origin from which a limb regenerates.

blastocoel. Cavity in the most common type of blastula (coeloblastula).

blastocyst. Preferred term for blastula in mammals.

blastomere. Embryonic cell formed during cleavage.

blastopore. Opening of the archenteron of the gastrula, which usually becomes either the anus or the mouth.

blastula. Early embryonic stage consisting of a mass of cells (blastomeres), generally enclosing a cavity (blastocoel).

blood. Fluid that moves through a circulatory system consisting of tubular vessels.

blood group. System of antigens on red blood cells that partly determines whether a donor's blood will be compatible with that of a recipient.

blood–brain barrier. Blood capillaries and certain glial cells of the brain that allow glucose and other nutrients to enter the cereobrospinal fluid but prevent the entry of many unwanted molecules.

blood–CSF barrier. Choroid plexus and arachnoid membrane, which control the concentrations of ions and some other materials in the cerebrospinal fluid.

Bohr effect. Decrease in the affinity of hemoglobin for O$_2$ in the presence of CO$_2$.

book gill. Respiratory organ of horseshoe crabs, with membranes arranged like pages of a book.

book lung. Respiratory organs of spiders, with membrane arranged like pages of a book.

bolus. Rounded mass, such as swallowed food.

Bowman's capsule. Spherical structure of each nephron of the kidney enclosing the glomerulus.

brachial. Pertaining to the arm.

brainstem. Part of brain left after removing cerebrum and cerebellum.

branchia. Thin, tentacle-like structure at the anterior end of a beard worm.

branchial. Pertaining to gills.

branchial heart. Accessory heart that pumps deoxygented blood from the body tissues to the gills in certain cephalopod molluscs.

branchial sac. Chamber in adult ascidians that produces the hydroskeleton and exchanges respiratory gases.

breathing. Ventilation of the respiratory organs.

bristle. (1) Type of feather with barbs only at the base of the shaft. (2) Short, hairlike structure.

Broca's area. Part of the human brain, usually on the left side, responsible for speech.

bronchus (BRON-kus; plural **bronchi,** pronounced BRONK-eye). Tube connecting the trachea to a lung.

bronchiole. Tubule in a lung that terminates in air sacs.

brood. (1) Eggs, larvae, or young. (2) To guard or warm eggs, larvae, or young.

brood parasite. Species such as cuckoo that depends on another species to brood its young.

brood patch. Bare spot on the belly of a bird that increases heat transfer between the bird and its egg during incubation.

brown adipose tissue (BAT = brown fat). Fatty tissue adapted for generating body heat.

brush border. Group of microvilli on a plasma membrane.

buccal cavity. Mouth cavity.

buccal ciliature. Cilia near the cytostome (mouth) that certain protozoa use in feeding.

buccal diverticulum (= stomochord). Anterior projection of the buccal cavity of a hemichordate, formerly considered a notochord.

buccal pumping. In fishes, ventilation of gills by contracting the buccal and opercular cavities.

budding. Type of asexual reproduction in which offspring form from small growths (buds) in or on the parent.

buffer. Molecule or group of molecules that resists changes in the hydrogen-ion concentration (pH) of a solution.

bulbourethral gland (= Cowper's gland). Gland that secretes clear, mucous fluid for lubrication during copulation.

bundles (of His). Fibers in the heart that conduct action potentials from the atrioventricular node to the ventricles.

bursa. Fluid-filled sac, such as one of five sacs that exchange respiratory gases in brittle stars.

byssal threads (= byssus). Filaments secreted by certain bivalve molluscs for attachment to substratum.

calcareous (kal-KARE-ee-us). Mineral structure based on calcium, especially calcium carbonate.

calmodulin. Protein that mediates the action of Ca^{2+} in cells.

calorie. Amount of heat required to warm 1 gram of water by 1°C.

calyx (plural calyces). Cup-shaped structure, such as the body of an entoproct or crinoid.

cameral fluid. Fluid in a chamber of *Nautilus.*

cancellous (= spongy = trabecular) **bone.** Bone with spaces containing marrow.

cancer (= malignancy). Tumor from which cells spread to other parts of the body.

canine. (1) Referring to a dog or other canid. (2) Sharp tooth adapted for tearing meat.

5′ cap. Structure at 5′ end of messenger RNA.

capillary. Thin-walled tubule between arteriole and venule, responsible for circulating blood in tissues.

capitulum. Projecting mouth region, as in ticks and mites.

captaculum. Ciliated tentacle of a scaphopod mollusc.

carapace. (1) Shieldlike dorsal plate over the cephalothorax of a crustacean. (2) Dorsal shell of a turtle.

carbamino compounds ($HbCO_2$). Hemoglobin with CO_2 bound to its amine groups.

carbohydrate. Organic molecule with aldehyde or ketone group and consisting of carbon, hydrogen, and oxygen in the ratio 1:2:1. Simple sugars and more complex molecules made of sugars.

carbon cycle. Biogeochemical movements of carbon among various organic and inorganic states.

carbonic anhydrase. Enzyme that catalyzes the formation of bicarbonate from carbonic acid.

carcinogen. Cancer-causing substance.

carcinoma. Cancer that begins in epithelium, especially in lungs, breasts, and intestines.

cardiac muscle. Type of muscle that forms the heart in chordates.

cardiac output. Volume of blood pumped by heart per unit of time; equals heart rate multiplied by stroke volume.

cardiac stomach. First of two chambers in the stomach of crustacean or echinoderm.

carina (= sternal keel). Long, thin process on the sternum of a bird, to which flight muscles are attached.

carnivore. Organism that feeds on animals.

carotenoid. Straight-chain hydrocarbon related to vitamin A and often found as a yellow, orange, or red pigment.

carotid body. Receptor in the sinus of the carotid artery that senses the level of CO_2 in the blood.

carrying capacity. Maximum number of individuals of a species that can be sustained in a habitat indefinitely.

cartilage. Flexible connective tissue in the skeleton of a vertebrate.

cartilaginous. Consisting of cartilage.

caste. Group of individuals within a species of eusocial insects (ants, bees, termites) that is different in appearance and behavior from other such groups.

castration. Surgical removal of ovaries or testes.

catabolism. Metabolic breakdown of organic molecules.

catadromous. Migrating from fresh water to seawater to spawn, like eels.

catastrophism. Generally discredited view that past global changes occurred suddenly, because of floods, volcanism, and so on.

catch connective tissue. Connective tissue of echinoderms that stiffens as defense against predation.

catch muscle. Type of muscle in bivalve molluscs that allows the shells to remain closed without use of energy.

catching spiral. Part of an orb web of spiders in which prey are caught.

cation. Positively charged ion, such as sodium and calcium.

caudal. Pertaining to or toward the tail.

cecum (SEE-kum; plural **ceca**). Blind pouch, for example, at the beginning of the large intestine of vertebrates, or the midgut of insects.

cell cycle. Cycle of cell activity beginning with interphase in a new daughter cell at the end of a previous mitosis and continuing through the end of the following mitosis.

cell theory. Proposal by Schwann in 1839 "that there exists one general principle for the formation of all organic productions, and that this principle is the formation of cells."

cell-adhesion molecule (CAM). Glycoprotein on cell surface responsible for adhesion of particular cells to each other.

cellular respiration (= respiration). Complete catabolism of glucose requiring oxygen and releasing carbon dioxide.

central pattern generator. Network of nerve cells that produces a rhythmic pattern of action potentials.

centriole. One of a pair of microtubular structures that give rise to mitotic spindles and to cilia or flagella. Identical with basal body.

centrolecithal. Egg with yolk concentrated in the center, as in arthropods.

centromere. Region of a chromosome where kinetochore is located, and which contains genes for kinetochore.

cephalic ganglion. Enlargement of the central nervous system in the head, often called the brain.

cephalon. Head of a trilobite.

cephalothorax. Fused head and thorax, as in spiders and many crustaceans.

cercaria. Larva of *Schistosoma* and other flukes that swims from molluscan intermediate host to final host or second intermediate host.

cercus (SIR-kus; plural **cerci** SIR-sy). Posterior antenna-like structures on certain insects.

cerebellum. Part of the vertebrate brain that coordinates learned movements.

cerebral commissure. Band of nerve fibers connecting the two hemispheres of the cerebrum.

cerebrospinal fluid (CSF). Fluid in the ventricles of the vertebrate brain and the central canal of the spinal cord.

cerebrum. Major part of the vertebrate brain responsible for perception and voluntary movement.

channel. Glycoprotein in plasma membrane that permits the passage of certain ions.

character. Any anatomical, molecular, or behavioral difference that provides a basis for classification.

character analysis. Determination of the nature and potential usefulness of a character in classification.

character displacement. Condition in which individuals with similar traits that cause them to compete with each other become less similar over succeeding generations due to evolution.

chela (KEE-luh). Pincers on an appendage of an arthropod.

chelicera (ke-LIS-ur-uh). Paired mouthpart of spiders and other chelicerates.

chelipeds (KEL-i-pedz). First pair of walking legs, with pincers, in certain crayfish, crabs, and other decapod crustaceans.

chemical synapse. Structure on a nerve cell that releases chemical transmitter onto another nerve cell or muscle cell.

chemiosmotic theory. Proposal by Peter Mitchell in 1961 that diffusion of H$^+$ across inner membrane of mitochondria produces ATP.

chemoautotrophic. Referring to bacteria that produce energy for synthesis of organic molecules by oxidizing inorganic molecules.

chitin (KY-tin). Polysaccharide that is a major component of cuticle in arthropods and occurs in integuments of many other invertebrates.

chloragogue cells (= chloragog = chloragogen). Greenish or brownish tissue in annelids that assists nephridia in eliminating nitrogenous wastes, stores glycogen and fats, and nourishes maturing ova.

chlorocruorin. Green respiratory pigment in some annelid worms.

choanocyte (ko-AN-oh-site) (= collar cell). Flagellated cell of sponges and other animals that propels water and traps food particles in its collar.

chondrocyte (KON-dro-site). Cell in cartilage that secretes collagen and proteoglycans.

chorion. Extraembryonic membrane of vertebrates that exchanges respiratory gases in the eggs of reptiles and birds and develops into part of the placenta in most mammals.

chorionic villus. Fingerlike projection of the mammalian placenta into the endometrium of the uterus.

choroid plexus. Structure that helps produce cerebrospinal fluid by actively transporting ions and certain molecules from blood.

chromatid. One of two structures formed by a dividing chromosome which becomes a new chromosome following mitosis or meiosis.

chromatin. Molecular substance (DNA and protein) composing chromosomes.

chromatophore. Cell with pigment that can be covered or uncovered, or concentrated or dispersed to change color of integument.

chromosomal aberration. Visible abnormality in structure or number of chromosomes.

chromosome. Elongated or rounded structure consisting of DNA and proteins which divides during meiosis and mitosis and carries genetic information to gametes and to dividing cells.

chrysalis. Pupa of a butterfly or moth.

chyle (pronounced kyle). Watery, digested material released by the small intestine into the large intestine.

chyme (pronounced kyme). Watery mixture of partly digested food and acid released by the stomach into the small intestine.

chymotrypsin. Pancreatic enzyme that digests proteins.

ciliary loop. Band of mechanoreceptors and/or chemoreceptors beneath the head of an arrowworm.

ciliary photoreceptor. Type of photoreceptor in vertebrates that develops around a cilium.

ciliate. Ciliated protozoan, or any protozoan in the phylum Ciliophora.

cilium. Short, hair-shaped motile structure on a cell surface that usually occurs in large numbers. Compare flagellum.

circadian rhythm. Biological rhythm with a duration of approximately one day.

circannual rhythm. Biological rhythm with a period of approximately one year.

circulatory system. Heart and vessels that transport blood to tissues and back.

cirrus. (1) Tuft of fused cilia used for locomotion by certain protozoans. (2) Copulatory organ of male flatworms and certain other invertebrates. (3) Tuft of setae on lobes of polychaete annelids and some other invertebrates. (4) Legs of barnacle adapted as brushlike structure for feeding. (5) Clawlike structures that hold feather stars to substratum.

clade. In cladistics, the taxonomic group represented by a branch on a cladogram and all the branches descended from it.

cladistics (= phylogenetic systematics). Method of classification in which outgroups are used to distinguish primitive from derived characters, and lines of descent but not times of evolution are deduced from the number of derived homologous characters (synapomorphies).

cladogram. Phylogenetic pattern derived by cladistics in which two groups branch from a single point.

clap-fling. Pattern of wing movement in some insects in which the wings come together during upstroke then peel apart to create lift.

clasper. Modified fin used as an intromittent organ in male sharks and certain other cartilaginous fishes.

class. Taxonomic level below phylum and above order.

classical conditioning (= Pavlovian = Type I conditioning). Type of learning in which a normal response to one stimulus (the unconditioned stimulus) becomes associated with a new stimulus (conditioned stimulus).

cleavage pattern. Pattern of cell movement during embryonic cleavage. See also particular types: bilateral, discoidal, radial, rotational, spiral, superficial.

cleidoic egg (kly-DOH-ik). Self-contained egg of an insect or amniote (reptile, bird, or egg-laying mammal).

climax community. Final stage of succession after which species in a community remain stable.

clitellum. External swelling on certain segments of oligochaetes and leeches that secretes mucus that forms a cocoon.

cloaca (klo-A-kuh; plural **cloacae** klo-A-see). Chamber into which gametes and excretory wastes empty.

clone. Group of genetically identical cells or organisms.

closed circulatory system. Circulatory system in which blood is confined to heart or vessels, as in vertebrates.

clot retraction. Shrinking of blood clot due to contraction of platelets.

clotting (= hemostasis = coagulation). Congealing of blood that stops bleeding.

cnida (NY-duh) (= cnydocyst). Adhesive or stinging organelle in cnidarians.

cnidocil (NY-doe-sill). Trigger that releases nematocyst of cnidarians.

cnidocyst (NY-doe-sist) (= cnida). Adhesive or stinging organelle in cnidarians.

cnidocyte (NY-doe-site). Cell that produces cnidocysts.

coated pit. Depression in a plasma membrane where receptor-mediated endocytosis occurs.

cobweb. Loosely woven spider web with sticky threads that snares prey walking beneath it.

cochlea. Tubular auditory organ of mammals, birds, and some reptiles.

cocoon. Hollow structure in which developing or resting stages occur.

codon. Triplet on messenger RNA that specifies addition of a particular amino acid, or signals the stop or start of synthesis of a protein.

coelenteron (seal-EN-ter-on) (= gastrovascular cavity). Cavity of an incomplete gut serving in both digestion and circulation in cnidarians and some other invertebrates.

coeloblastula. Blastula with a "hollow" center (blastocoel).

coelom (SEAL-um). Body cavity that develops in mesoderm and is enclosed within a peritoneal membrane.

coelomocyte. Amoebocyte in coelomic fluid, or in vascular fluid of an echinoderm.

coenzyme. Organic cofactor that transfers a functional group to or from the substrate during an enzymatic reaction.

coevolution. Interaction of two species such that each evolves adaptations to the other.

cohesion. Binding of identical molecules to each other.

collagen. Fibrous protein reinforcing skin and other flexible tissues.

collar cell. Choanocyte of sponges and some other animals.

collecting duct. Tubule in nephron of kidney in which urine is formed.

collenchyme. Mesenchyme (= mesoglea) of jellyfishes.

collencyte. Collagen-secreting amebocyte in sponges.

colligative. Property of a solution, such as freezing point and osmolarity, that depends on its concentration rather than on the nature of the solute.

colloblast (= glue cell). Cell on ctenophore tentacles that secretes sticky substance that traps prey.

colloid. Mixture of water and molecules that do not dissolve but are kept from precipitating by attraction with the water.

colloid osmotic pressure (= oncotic pressure). Hydrostatic pressure due to osmotic effect of colloid.

columella (1)Stapes. Sole middle-ear bone in an amphibian, reptile, or bird. (2) Central stalk of a gastropod shell.

columnar. Column-shaped, as in cells of columnar epithelium.

comb plate. Array of cilia used in locomotion by ctenophores.

commensalism. Association between two species in which one benefits but the other is neither helped nor harmed.

community. The species interacting within a habitat.

competitive exclusion. Exclusion of one species by another in a habitat in which both compete for the same resource.

complement. Group of proteins that attract phagocytic cells and aid antibodies in destroying bacteria.

complementary DNA (cDNA). Segment of DNA with base sequence complementary to another segment of DNA.

complete digestive tract. Digestive tract with a mouth and a separate anus.

compound eye. Arthropod eye consisting of numerous subunits (ommatidia).

concentration. Measure of the amount of a substance dissolved in a volume of solvent.

concertina motion. Type of locomotion used by burrowing snakes, in which a portion of the body is bent and braced against the sides of the burrow.

conchin (formerly conchiolin). Proteinaceous substance that forms the periostracum that covers the shells of molluscs.

conditioned stimulus. Stimulus to which an animal learns to respond through classical conditioning.

conduction. Movement, as of action potentials, heat, or electric charge.

conductivity. Ability to conduct heat, charge, and so on.

cone. Photoreceptor cell in vertebrate eye that is responsible for color vision.

congenital defect. Nonhereditary developmental defect.

conjugation. Temporary coupling and exchange of genetic material in certain protozoans (ciliates) or between bacteria.

connective. Nerve linking two ganglia in a nervous system.

connective tissue. Tissue of mesodermal origin, including cartilage, bone, blood, and connective tissue proper (ligament, tendons, fat).

consumer (= heterotroph). Organism that obtains its energy from organic materials synthesized by other organisms.

continental drift. Movement of continents on a layer of molten rock.

contour feather (= vane feather). Major type of feather, in which barbs form two vanes on opposite sides of the shaft.

contractile vacuole (= water expulsion vesicle). Osmoregulatory organelle in protozoa and freshwater sponges.

conus arteriosus. Structure in amphibian heart between ventricle and arteries.

convection. Transfer of heat through bulk movement of air or water.

convergent evolution. Evolution of similar (analogous) traits independently in two different lines of descent.

cooperative breeding. Breeding in which nonreproductive individuals assist in raising offspring of others of the same species.

copulation. Transfer of sperm by a male into a female.

copulatory plug. Secretion by snakes and certain mammals that prevents subsequent males from copulating with a female, or prevents the semen from leaking from the vagina.

copulatory sac. Storage site for sperm in certain flatworms.

cornea. (1) External transparent coat of vertebrate eye. (2) Transparent cuticle covering the eye of an arthropod.

cornified layer. Dead, keratinized cells protecting epidermis from water and abrasion.

corona. Feeding apparatus of a rotifer, consisting of a ciliated disc surrounding the mouth.

corpus luteum. Remainder of an ovarian follicle after ovulation.

cortex. Outer layer of tissue, as of cerebrum, adrenal gland, and so on.

cortical reaction. Release of the contents of cortical granules beneath the plasma membrane of an egg during fertilization, which serves as a slow block to further fertilization.

countercurrent exchange. Exchange of heat, oxygen, and so on between two fluids moving past each other in opposite directions.

courtship. Behavior by which male and female of same species prepare for sexual reproduction.

covalent bond. Chemical bond in which one or more electrons are shared by two atoms.

Cowper's gland (= bulbourethral gland). Gland that secretes a clear, mucous fluid for lubrication during copulation.

coxal gland. Excretory organ of spider, which empties through a pore near the proximal joint (coxa) of the leg.

cramp. Rigidity of muscle caused by lack of ATP.

cribellate spider. Spider with cribellum that produces wooly silk for the catching spiral of its orb web.

crista. Infolding pocket of the inner membrane of mitochondrion. Site of the respiratory chain.

critical period. Precisely timed period when an animal is susceptible to a particular effect. (Now largely replaced by "sensitive period.")

crop. (1) Anterior part (= foregut) of insect digestive tract in which food is stored and partly digested. (2) Pouch in the esophagus of a bird, in which food is stored.

cross-bridge. Connection between thin and thick filaments in muscle.

crossing over. Intertwining of the four chromatids during prophase I of meiosis, during which they exchange genetic material.

cryoprotectant. Substance that protects tissues from damage due to freezing.

cryptobiosis. Inactive state induced by dehydration.

crystalline cone. Structure in ommatidium of compound eye that, together with lens, concentrates light.

crystalline style. Rod-shaped structure in stomachs of bivalve molluscs that grinds against the gastric shield, releasing digestive enzymes.

ctenidium. Gill, especially of a mollusc.

ctenoid. With comb-shaped projections.

cuboidal. Cubical, like the cells of cuboidal epithelium.

cuticle. (1) Organic, noncellular body covering external to the epidermis, as in the integument of arthropods, annelids, and some other invertebrates. (2) Cornified layer of dead cells on vertebrate skin.

cuttlebone. Vestigial shell that aids buoyancy in cuttlefish.

cyclic nucleotide. Nucleic acid base in which phosphate is attached at two places to the ribose. Usually cyclic adenosine monophosphate (cAMP) or cyclic guanosine monophosphate (cGMP).

cycloid. Disc-shaped.

cydippid larva. Free-swimming larva of a ctenophore.

cypris larva. Larval stage following the nauplius larva of a barnacle.

cyrtocyte. Curved solenocyte (flagellated cell in the protonephridium) in a gnathostomulid.

cyst. Capsule containing a protozoan or animal in an inactive state resistant to environmental stress.

cystacanth. Embryo of an acanthocephalan that is encysted in an intermediate host and develops into adult following ingestion by final host.

cysticercus (SIS-tee-SIR-kus) (= bladder worm). Juvenile tapeworm, consisting of an invaginated or introverted scolex in a fluid-filled cyst.

cytidine. A nitrogenous base in nucleic acids.

cytochrome. Protein with iron-containing heme group; part of electron transport chain of cellular respiration.

cytokine. Interferon, interleukin, or other substance that coordinates local tissue responses such as inflammation.

cytokinesis. Division of cytoplasm during mitosis or meiosis.

cytoplasm. Cell contents, including cytosol and organelles except the nucleus and plasma membrane.

cytoplasmic membrane system. System comprising endoplasmic reticulum, Golgi complexes, and lysosomes, responsible for formation of vesicles for synthesis of new plasma membrane, formation of lysosomes, and secretion of glycoproteins.

cytoplasmic streaming. Movement of endoplasm into a pseudopodium; believed responsible for ameboid movement.

cytosine. Nucleotide consisting of cytidine linked to ribose.

cytoskeleton. Network of microfilaments, microtubules, intermediate filaments, and other protein structures that reinforce cells.

cytosol. Fluid portion of cytoplasm, which bathes cytoplasmic organelles.

cytostome. Opening through which food is ingested in many protozoa.

daily torpor. Hours-long drop in body temperature in a small bird or mammal in cold weather.

dart. Sharp structure with which certain snails stab each other during courtship.

dauer larva. Resistant larval form of some nematodes.

deciduous teeth (= milk teeth). Temporary teeth of a young mammal.

decomposer. Bacterium, fungus, or other organism that feeds on organic tissue or wastes.

defensive mimicry. Resemblance of a potential prey to a different animal or object that is not preyed upon.

definitive host (= primary or final host). Organism in which the adult form of a parasite develops.

dendrite. Branch of a nerve cell that conducts potentials toward the cell body.

dendrogram (= phylogenetic tree). Diagram representing the evolution of a group of organisms.

denitrifying bacterium. Bacterium that converts nitrite and nitrate, which can be used by plants, into gaseous nitrogen (N_2), which cannot be used.

dense bone (= compact bone). Major structural component of bone, often enclosing cancellous bone.

density-dependent factor. Ecological factor that depends on population density; for example, communicable diseases and food shortages.

density-independent factor. Ecological factor that does not depend on population density; for example, predation and natural catastrophes.

dental formula. Numerical key indicating numbers of each kind of tooth on each half of the upper and lower jaws of a mammal.

denticle. Tooth-shaped structure associated with mouth, as in pharynx of a polychaete annelid.

deoxyribonucleic acid (DNA). The material basis for genes. Nucleic acid with deoxyribose as its sugar. It generally occurs as a double helical strand.

depolarization. Reduction in the magnitude of a membrane potential.

derived. Referring to a trait that evolved as a modification of a primitive trait.

dermal bone (= membranous bone). Bone that forms from layers of embryonic connective tissue rather than by replacement of cartilage. Forms fish scales, turtle shells, antlers, and flat bones of vertebrate skulls.

dermal gland. Gland of unknown function in arthropod integument.

dermis. True skin underlying epidermis.

desmosome. Structure joining the plasma membranes of adjacent cells.

determinate cleavage (= mosaic development). Cleavage in which fates of cells are determined early. Common among invertebrates.

determination. Permanent establishment of the fate of an embryonic cell.

detorsion. Partial reversal of torsion in certain snails.

detritus feeder (= detritivore). Animal that feeds on particles of decaying organisms.

deuterostome. Animal in which the mouth originates from an opening in the embryo other than the blastopore, and usually with other distinctive patterns of embryonic development.

diadromous. Migrating either from fresh water to seawater to spawn (catadromous), or from seawater to fresh water to spawn (anadromous).

diapause. Inactive state induced as an adaptation to withstand adverse environmental condition, such as cold.

diaphragm. Thin sheet of tissue separating the thoracic and abdominal cavities of mammals, birds, and some reptiles.

diapsid. Reptile or other vertebrate with two temporal openings in the skull.

diastolic blood pressure. Minimum blood pressure just before ventricular contraction.

differentiation. Irreversible change in the structure and function of a cell during embryonic development.

diffraction colors. Structural colors formed by closely spaced grooves in integument, scales, or feathers.

diffusion. Passive movement of molecules from a region where they are highly concentrated to a region of lower concentration.

digestive gland. (1) Hepatopancreas in a crustacean, which secretes digestive enzymes and stores nutrient reserves. (2) Pyloric cecum in an echinoderm, which stores and absorbs nutrients.

dihybrid cross. Cross between two individuals heterozygous for two different genes.

dimorphism. Occurrence of two forms in the same species; for example, sexual dimorphism.

dioecious (die-EE-shus). Having separate sexes, as opposed to monoecious in which each individual has both male and female reproductive organs.

diphycercal (DIF-i-SIR-kal). Pertaining to the caudal fins of lobe-finned and some other fishes, which taper symmetrically to a point at the rear.

diploblastic. Developing from two germ layers, as may occur in cnidarians and ctenophores.

diploid. Having chromosomes occurring in homologous pairs, rather than individually (haploid).

diplosegment. An apparent segment with two pairs of legs, which results from the fusion of a pair of true segments in a millipede.

direct calorimetry. Measurement of rate of body heat production to determine metabolic rate.

direct development. Development in which juvenile resembles the adult form, without a larval stage.

disaccharide. Carbohydrate consisting of two simple sugars (monosaccharides) joined by a glycosidic link.

discoidal cleavage. Cleavage pattern in which embryo is compressed into a disc, characteristic of fishes, reptiles, and birds.

disruptive coloration. Pattern of coloration that masks outline of an animal, for example, a zebra.

distal convoluted tubule. Part of a nephron in the kidney that conducts tubular fluid to the collecting duct.

DNA hybridization. Technique for comparing DNA from two different species by combining single DNA strands from each species and determining the temperature at which they come apart.

dominance hierarchy (= pecking order). Social rank that determines which animals in a group have priority in mating, feeding, or attacking others.

dominant. Referring to an allele that is expressed in the presence of a different (recessive) allele for the same trait.

dorsal. Pertaining to the back or upper surface.

dorsal root. Dorsal branch of a nerve at the spinal cord, which contains axons from sensory receptors.

down feather. Feather in which barbs are not interconnected by barbules and are therefore fuzzy.

drag line. Silk strand played out by spiders as anchorage.

drone. Male honey bee or other hymenopteran.

ductus arteriosus. Tube connecting the pulmonary artery to the aorta in a mammalian fetus, permitting blood to bypass the lungs.

ductus deferens (plural **ductus deferentes**) (= sperm duct = vas deferens). Tube through which sperm are ejaculated.

duogland adhesive system. Combination of viscid and releasing glands used in locomotion and anchoring in flatworms and some other animals.

eccrine sweat gland. Gland that produces sweat that cools the body by evaporation.

ecdysis (EK-duh-sis) (= molting). Shedding of cuticle, skin, hair, or feathers.

ecdysone (= molting hormone). Steroid hormone of arthropods that triggers molting.

echinopluteus. Larva of a sea urchin.

echolocation. Location of prey or other objects by emission of ultrasound and reception of the echo, as in bats and whales.

eclosion. Emergence of an adult insect from the pupa.

ecological growth efficiency. Ratio of assimilated (used or stored) energy to ingested energy over a period of one year.

ecological pyramid. Diagram representing energy, biomass, or numbers of organisms at each trophic level.

ecology. The study of the interaction of organisms with their environment and other organisms.

ecosystem. Association of all of the organisms in an area, together with the physical environment.

ectoderm. (1) Outer germ layer from which external epithelium and nerve cells develop. (2) Tissues derived from ectoderm.

ectolecithal egg. Egg with yolk outside the plasma membrane of the ovum.

ectoplasm (= plasma gel). Outer, gelatinous portion of cytoplasm. Compare endoplasm.

ectotherm. Animal whose body temperature is determined primarily by the enviromental temperature.

edema (uh-DEEM-uh). Swelling.

efferent. Conducting outward, such as a nerve cell conducting action potentials out of the central nervous system, or an artery conducting blood from the heart. Compare afferent.

egg. Female gamete, or the zygote, with associated membranes, yolk, and shell (if any).

egg tooth. Temporary projection on beak of hatching birds and certain reptiles, which enables them to break through the shell.

electrical synapse. Structure by which membrane potentials conduct directly from one nerve cell to another.

electrocardiogram (ECG or EKG). Record of voltages produced by action potentials of the heart.

electron acceptor. Molecule such as oxygen that accepts electrons from an electron transport chain.

electron transport chain. System of enzymes and cytochromes involved in ATP synthesis in mitochondria.

electrophoresis. Procedure for comparing molecules by their rate of movement in an electric field, which depends on their net charges, sizes, and shapes.

electroreceptor. Receptor in many fishes and some other vertebrates that responds to electric fields.

Eltonian (ecological) pyramid. Diagram representing energy, biomass, or numbers of organisms at each trophic level.

elver. Juvenile eel.

embryo. Developing organism, especially in early stages. In humans the term is applied until the eighth week of gestation.

emulsify. To disperse large fat droplets into smaller ones.

endite. Medial minor branch of a crustacean appendage.

endocrine. Referring to a ductless gland that secretes its product (e.g., a hormone) into the blood.

endocuticle. Inner, nonsclerotized layer of arthropod cuticle.

endocytosis. Transport of material into a cell by enclosing it within a pit in the plasma membrane that becomes a vesicle.

endoderm. (1) Inner germ layer from which the lining of the gut, internal glands, and respiratory organs develop. (2) Tissues derived from endoderm.

endogenous clock. Hypothetical biological system that controls the timing of biological rhythms or compensates for movement of the sun or stars during migration.

endogenous opiate. Secretion that induces sleep and relieves pain, mimicked by morphine, heroin, and other opiates.

endolecithal egg. Egg with yolk enclosed in the plasma membrane of the ovum.

endolymph. Fluid in cochlea and vestibular apparatus.

endometrium. Inner lining of the uterus.

endoparasite. Parasite that lives within the gut or tissues of host.

endoplasm (= plasma sol). Inner, fluid portion of cytoplasm. Compare ectoplasm.

endoplasmic reticulum (ER). Cytoplasmic membranes bearing ribosomes for production of proteins (rough ER) or lacking ribosomes (smooth ER).

endopod (= endopodite). Medial branch of a biramous appendage of a crustacean.

endopterygote. Referring to an insect (superorder Endopterygota) in which the wings or wing buds are apparent only in the adult, which is therefore holometabolous.

endoskeleton. Skeleton surrounded by other body tissues, as in vertebrates.

endostyle. Groove in ventral surface of branchial sac of an ascidian, which secretes mucus that traps food particles.

endosymbiont. Organism that lives inside another organism.

endotherm. Animal whose body temperature is determined primarily by its own heat production and conservation.

endplate (= neuromuscular junction). Neuronal synapse on a skeletal muscle cell.

endplate potential (EPP). Change in voltage in plasma membrane of muscle cell caused by release of transmitter from endplate.

energy level. Energy associated with a particular position, such as distance of an electron from a nucleus.

enhancer. Portion of DNA that promotes transcription of a distant gene.

enterocoelous. Having a coelom that developed from pouches in the endoderm in the embryonic gut (archenteron).

entognath. Referring to a hexapod with mouth parts recessed, unlike those of insects.

entropy. Measure of the energy rendered unavailable for further work; equal to or greater than the amount of heat lost from a system divided by the temperature of the system. A measure of disorder.

enzyme. Organic catalyst, usually a protein, that speeds up a chemical reaction without being changed by the reaction.

ephyra. Jellyfish larva that buds asexually from a strobila.

epiblast. In placental mammals part of the inner cell mass that forms the amnion and the embryo.

epicuticle. Waxy and proteinaceous covering of arthropod cuticle.

epidermis. Outer layer of surface tissue.

epididymis. (EP-i-DID-i-mis; plural **epididymides** EP-i-DID-i-MY-deez). Structure on the testis that stores sperm.

epigenesis. Embryonic development of new structures from those that did not previously exist as such. Compare preformation.

epiglottis. Flap of cartilage at the root of the tongue that closes the opening of the trachea (the glottis) during swallowing.

epipod (= epipodite). Lateral minor branch (exite) on a protopod of a crustacean appendage.

epithelial tissue (= **epithelium**). Tissue derived from ectoderm or endoderm that forms the body covering or the lining of body cavities.

epitheliomuscle cell. Contractile cell in the epidermis of a cnidarian.

epitoky. Transformation of certain polychaete annelids from juveniles (atokes) to reproductive forms (epitokes).

equilibrium. Balanced state in which no net change occurs.

equilibrium potential. Hypothetical membrane voltage that would be required to maintain a certain ratio of concentrations of a permeant ion on each side of a membrane.

erythrocyte (= red blood cell). Cell that transports hemoglobin in blood of vertebrates.

esophagus. Tube connecting the mouth to the stomach.

essential. Required in the diet; for example, essential amino acid or essential fatty acid.

esthete. Compound receptor organ of a chiton.

estivation (= aestivation). Dormant condition during summer or dry period.

estrogen. Estradiol or related steroid hormone that promotes female behavior, morphology, and reproductive functions.

estrous cycle. Cycle of changes in the reproductive physiology of a female.

estrus. Sexual receptivity in a female vertebrate; "heat."

estuary. Area where a river empties into a sea.

ethology. Study of behavior that emphasizes mechanisms and evolution in the natural environment.

eukaryote (= eucaryote). Cell with nucleus and organelles, or an organism (protist, fungus, plant, or animal) with such cells. Compare prokaryote.

euryhaline. Able to tolerate wide shifts in environmental salinity. Compare stenohaline.

eusocial. Referring to animals such as honey bees in which a nonreproducing caste assists reproducing individuals.

eutely. The property of having a fixed number and position of somatic cells from one individual to the next within a species.

evolution. Change in the genetic makeup of an interbreeding group of organisms. Inherited change in a species.

evolutionarily stable strategy (ESS). In behavioral ecology, a behavior that persists because competing alternatives are less advantageous.

evolutionary systematics. Method of systematics in which times and patterns of evolution are deduced from comparative anatomy, paleontology, and other methods.

excitable cell. Cell capable of producing an action potential.

excurrent canal. In leuconoid sponges the canal that carries water from a flagellated chamber to the osculum.

exite. Lateral minor branch of a crustacean appendage.

exocrine. Secreted through a duct that does not empty into the bloodstream.

exocuticle. Sclerotized layer of arthropod cuticle beneath the epicuticle.

exocytosis. Transport of material out of a cell by enclosing it within a vesicle that fuses with the plasma membrane.

exon. Portion of messenger RNA or DNA that is used in translation of protein. Compare intron.

exopod (= exopodite). Lateral branch of a biramous appendage of a crustacean.

exopterygote. Referring to an insect (superorder Exopterygota) in which the wings or wing buds appear in an immature stage and which is therefore hemimetabolous.

exoskeleton. Supporting structure that encloses all or part of the body, such as in insects.

extant. Not extinct.

extensor. Muscle that extends a limb away from the body axis.

external fertilization. Fertilization of eggs outside the female's body.

extraembryonic membrane. Membrane outside the embryo of an amniote (reptile, bird, or mammal). Includes chorion, amnion, allantois, and yolk sac.

eyespot. (1) Simple photoreceptor lacking lens or other structures found in true eyes. (2) Eye-shaped pattern that may confuse potential predators.

F$_1$ generation. Hybrid offspring resulting from an experimental cross of true-breeding parents.

F$_2$ generation. Offspring from interbreeding or self-fertilizing F$_1$ hybrids.

facet. Part of the cornea over one ommatidium of a compound eye of an arthropod.

facilitated diffusion. Diffusion aided by a carrier molecule or some other mechanism.

FAD. Flavine adenine dinucleotide.

Fallopian tube (= oviduct). Tube through which eggs of mammals travel after ovulation.

family. Taxonomic level below order and above genus.

fast-twitch fiber (= type II white fiber). Muscle cell that contracts rapidly but tires easily because of reliance on glycolysis.

fat. Lipid with up to three fatty acids linked to a glycerol. Mono-, di-, or triglyceride.

fat body. Fatty, diffuse organ in insects that stores energy reserves and produces yolk.

fate map. Diagram showing structures that will be formed by parts of an embryo.

fatty acid. Chain of carbon atoms (hydrocarbon chain) with an acidic carboxyl group at one end.

faunal region (= faunal realm). Six continent-sized areas characterized by their distinctive animals.

faunal succession. Changes in kinds of animals that inhabit an area as a result of plant succession.

feather tract (= pteryla). Linear region of bird skin that produces feathers.

feature extraction. Process in which each part of a sensory system responds only to a particular feature of a stimulus.

feedback inhibition. Control mechanism in which the product of a process inhibits the process.

feral. Once-domesticated.

fermentation. Anaerobic catabolism, such as glycolysis.

fertilization. Fusion of a sperm cell with an egg, forming a zygote.

fertilization membrane. Membrane of egg after fertilization has triggered the cortical reaction that prevents further fertilization.

fetal hemoglobin. Form of hemoglobin in fetus and newborn mammal.

fetus. Unborn mammal. Developing human after the eighth week of gestation.

fever. Elevation of body temperature above normal, especially in response to infection.

fibrillar muscle (= asynchronous muscle). Muscle in certain insects in which contraction is not synchronized with action potentials.

fibrin. Blood protein that changes from globules (fibrinogen) to fibers in forming a clot.

fibroblast. Cell in connective tissue proper that secretes collagen.

fibronectin. Glycoprotein that guides cell migration and makes cells adhere.

fibrous protein. Protein in the form of a long, insoluble strand.

Fick's equation. Equation stating that the rate of diffusion is directly proportional to the surface area of a membrane and the difference in concentrations or partial pressures on each side of the membrane, and inversely proportional to the thickness of the membrane.

filopodium. Long, thin pseudopodium of certain protozoans (class Filosea).

filoplume. Type of feather with long, flexible shafts and barbs only at the tip.

filter feeding. Obtaining food by trapping small particles suspended in water or air.

fimbriae. Ciliated, finger-shaped structures that draw eggs into the end of the oviduct.

final host (= primary or definitive host). Organism in which the adult form of a parasite develops.

fission. Type of asexual reproduction in which an organism divides into two or more offspring.

fitness. (1) Measure of the average contribution to reproductive ability of one allele or genotype compared with another. (2) Loosely, how well an organism or genotype is adapted.

fixed. (1) Adjective applied to energy or matter that has been incorporated into organic molecules. (2) Preserved, as a specimen in microscopy. (3) Adjective applied to an allele that occurs in every individual in a population.

fixed action pattern (FAP). Stereotyped behavioral response to a specific signal (sign stimulus).

flagellate. (1) Bearing one or more flagella. (2) A flagellated protozoan.

flagellated chamber. In leuconoid sponges a chamber containing choanocytes.

flagellum (plural **flagella**). Long, hair-shaped motile structure on the cell surface that usually occurs individually or in small groups.

flame bulb. Hollow structure with cilia (the flame) that produces filtration pressure at closed end of certain excretory organs (protonephridia).

flame cell. Ciliated cell at closed end of an excretory organ (protonephridium) of certain animals.

flavine adenine dinucleotide (FAD). Electron carrier that is reduced to $FADH_2$ in Krebs cycle.

flexor. Muscle that brings limbs toward the body axis.

fluid mosaic model. Structure of plasma membrane in which proteins are dissolved within a bilayer of phospholipid.

fluke. (1) Parasitic flatworm in class Trematoda or Monogenea. (2) Flatfish in the order Pleuronectiformes. (3) One of two lateral extensions of a whale's tail.

follicle. (1) Structure in skin of mammal or bird that produces a hair or a feather. (2) Part of ovary in which an egg cell develops.

food chain. Linear sequence defining which organisms eat and are eaten by other organisms.

food vacuole. Vesicle formed as result of phagocytosis, which contains food for intracellular digestion.

food web. Entire trophic relationship in a community, consisting of all the interwining food chains.

foot. (1) Structure at the end of a limb adapted for terrestrial locomotion. (2) Part of mollusc adapted for locomotion or as tentacles.

footprint pheromone. Substance distributed by honey bee queen that inhibits production of other queens.

forebrain (= prosencephalon). Anterior part of the embryonic vertebrate brain that develops into the cerebrum, olfactory bulbs, thalamus, hypothalamus, pituitary gland, and other structures.

foregut (= crop). Anterior part of insect digestive tract in which food is stored and partly digested.

founder cell. (1) Amebocyte (sclerocyte) in sponges that helps form mineral spicules. (2) Stem cell. Embryonic cell from which certain other cells develop.

founder effect. Tendency of reproductively isolated populations to reflect the atypical genetic composition of the founders of the population.

fovea. Depression on the retina where vision is sharpest.

fragmentation. Type of asexual reproduction in which an individual divides into two or more parts, each of which develops into a new individual.

free energy. Potentially available energy content of a system, minus the energy that is unavailable due to increase in entropy.

freezing point depression. Reduction in freezing temperature of a solution due to concentration of solute.

fry. Recently hatched fish.

functional group. Group of atoms (hydroxyl, ketone, methyl, etc.) commonly behaving as a single unit in organic molecules.

gait. Pattern of leg movements corresponding to different speeds of locomotion.

gall. (1) Abnormal growth on plants often induced by a developing insect. (2) Bile.

gamete. Sperm or egg cell.

gametocyte. Cell specialized as a gamete.

gametogenesis. Development of gametes from primitive germ cells.

gamma motor neuron. Small-diameter nerve cell that excites muscle portion of a spindle receptor organ.

ganglion. Part of a nervous system containing cell bodies and synapses.

ganoid scale. Thick, nonoverlapping scale of gars and certain other bony fishes.

gap junction. Group of protein channels through two adjacent plasma membranes that permits movement of ions, monosaccharides, amino acids, and electrical current between cells.

gap phase. One of two periods (G_1 and G_2) in interphase of the cell cycle when DNA synthesis does not occur.

gas gland. Gland that pumps gas into the swim bladder of a fish.

gastric mill. Chitinous teeth attached to the muscular lining of the cardiac stomach of crustaceans.

gastric shield. Part of stomach lining of bivalve molluscs against which the crystalline style grinds and releases digestive enzymes.

gastrodermis. Inner cell layer of cnidarians.

gastrolith. Calcium deposit recovered from molting cuticle and stored in stomach of crustaceans.

gastrovascular cavity (= coelenteron). Cavity of an incomplete gut used for both digestion and circulation in cnidarians and some other invertebrates.

gastrula. Early embryo enclosed in two germ layers (ectoderm and endoderm) with proliferating mesoderm, and with a cavity (archenteron) opening through a blastopore.

gastrulation. Process by which a blastula develops into a gastrula.

gate. Portion of an ion channel that opens in response to a signal, increasing permeability to the ion.

gemmule. Internal bud for asexual reproduction in freshwater and some marine sponges.

gene. Genetic unit responsible for one trait or for producing one protein.

genetic code. Relationship of sequence of bases in nucleic acid to sequence of amino acids in protein.

genetic drift. Change in gene frequency in a population due to chance.

genetic mosaic. Organism in which different parts of the body are genetically different.

genital pore. Opening to a sperm receptacle in a female.

genital setae. Bristles on the clitella by which earthworms and other oligochaetes clasp each other during mating.

genome. Genetic material in a haploid set of chromosomes.

genotype. Genetic makeup of an organism. Compare phenotype.

genus (plural **genera**). Taxonomic level above species and below family.

germ cell. Cell that develops into a gamete.

germ layer. Layer of cells (usually ectoderm, endoderm, or mesoderm) formed early in embryonic development.

germ plasm. Cells responsible for reproductive transmission of heredity. Compare somatoplasm.

germinative zone. Part of a tapeworm behind the scolex where proglottids are produced asexually.

gestation. Period in which offspring of mammals are in the uterus.

giant axon. Large-diameter axon that usually triggers an escape response.

Gibbs free energy (= free energy). Potentially available energy content of a system, minus the energy that is unavailable due to increase in entropy.

gill. Organ for exchange of respiratory gases in aquatic organisms.

gill raker. Comblike portion of a fish gill used in suspension feeding.

gizzard (= muscular stomach). Posterior portion of a bird's stomach, with a horny lining that crushes food.

gladius (= pen). Vestigial shell in the back of a squid.

gland. Secretory organ.

gland cell (= renette cell). Presumed osmoregulatory cell of a nematode.

glandular stomach (= proventriculus). Anterior portion of a bird's stomach that secretes digestive enzymes.

glial cell. Nonneural cell in a nervous system that forms myelin, provides support, or forms the blood–brain barrier.

globin. Protein portion of hemoglobin, myoglobin, or cytochrome.

globular protein. Nonfibrous protein that is usually water soluble.

glochidium. Parasitic larva of certain freshwater bivalves.

glomerular filtration rate (GFR). Rate (ml/min) at which fluid is filtered through the glomeruli of the kidneys.

glomerulus. (1) Capillary in each nephron of the kidney where filtration into Bowman's capsule and the tubule occurs. (2) Sinus network in an acorn worm that is assumed to be an excretory organ.

glottis. Opening into the trachea.

glyceride (= neutral fat). Lipid consisting of glycerol with one, two, or three fatty acids attached.

glycerol. Three-carbon molecule with three —OH groups.

glycogen. Branched polysaccharide of glucose used as short-term energy store in liver, muscle, and some other cells.

glycolipid. Lipid with carbohydrate attached.

glycolysis. Partial breakdown of glucose in the absence of oxygen.

glycoprotein. Protein with carbohydrate attached.

glycosidic link. Bond joining two simple sugars, formed by an oxygen atom.

gnathobase. Toothed proximal segment of a leg of a horseshoe crab that functions as a jaw for chewing food.

goblet cell. Cell in the mucosa of the digestive or respiratory tract that secretes protective mucus.

Golgi complex (= Golgi body or Golgi apparatus). Group of hollow, disc-shaped cytoplasmic membranes responsible for preparing vesicles for synthesis of plasma membrane, or containing glycoproteins for lysosomes or for secretion.

gonad. Organ, for example, ovary or testis, that produces gametes.

gonadotropin (= gonadotropic hormone). Hormone such as FSH, LH, and hCG that stimulates gonadal activity.

Gondwana. Southern supercontinent that split from Pangea before 180 million years ago, in the Jurassic Period. It later divided into South America, Antarctica, Australia, Africa, and India.

gonocoel. Reduced coelom enclosing the gonads in a tardigrade or mollusc.

gonoduct. Duct from a gonad through which gametes are released.

G protein. Protein that binds guanosine triphosphate and helps control cellular responses to hormones or other stimuli.

gram molecular mass (= gram molecular weight). Number of grams of a substance equal to the average number of protons and neutrons in a molecule of the substance. A mole.

granular gland (= poison gland = serous gland). Poison-secreting gland in skin of an amphibian.

gravid. Bearing ripe eggs. Pregnant.

gray matter. Parts of brain and spinal cord that contain nerve cell bodies and synapses. Gray because of absence of myelin.

green gland (= antennal gland). Excretory organ in the head of certain crustaceans.

growth cone. Growing part of a nerve cell during embryonic development.

growth factor. Hormonelike substance that stimulates mitosis and maintains particular cells.

GTP. Guanosine triphosphate.

guanine. A nitrogenous base in nucleic acids and GTP.

guanosine. Nucleotide consisting of guanine and ribose.

gustation. Taste.

gynandromorph. Genetic mosaic in which part of the body is male and the rest is female.

gynogenesis. Mode of reproduction in which sperm are required for egg activation, but genes from the sperm do not contribute to the zygote.

habituation. Type of learning in which response to a stimulus declines with repetition.

hacking. Artificial rearing of a bird and reintroduction into the wild.

hair cell. Receptor cell of the cochlea and vestibular apparatus, as well as certain other mechanoreceptor organs.

halter (plural **halteres**). Club-shaped structure that occurs instead of a hind wing in a true fly.

haplodiploidy. Reproductive system in ants, bees, and other hymenopterans in which some eggs are fertilized and develop into diploid females, and others develop parthenogenetically into haploid males.

haploid. Having just one of each kind of chromosome.

Hardy–Weinberg law. Mathematical derivation showing that gene frequencies do not change in a population under equilibrium conditions.

Haversian canal. Channel in dense bone that contains vessels and nerves.

head group. Hydrophilic part of a phospholipid that is linked to the rest of the molecule through phosphate. A phosphatidyl group.

heart murmur. Blowing or roaring sound in the normal "lubb" or "dup" sound of the heart, indicating poor closure of the heart valves.

heart vesicle. Contracting vascular chamber in the proboscis of an acorn worm.

heat. (1) Energy associated with molecular vibration. (2) Estrus.

heat capacity. The amount of heat required to change the temperature of a substance by a given amount. The heat capacity of water is 1 calorie per degree Celsius.

hectocotylus arm. Arm of a male octopus specialized as a copulatory organ.

hemal system. Diffuse spongy tissue that lies parallel to the water vascular system in an echinoderm.

heme. Iron-containing, nonprotein (porphyrin) portion of hemoglobin, myoglobin, and cytochromes.

hemerythrin. Violet, iron-containing respiratory pigment of some annelids and other invertebrates.

hemimetabolous. Referring to an insect in which the immature forms (nymphs) resemble the adult. Compare holometabolous.

hemipenis. One of the two copulatory organs of a snake or lizard.

hemocoel (HE-mo-seal). Body cavity in an animal with an open circulatory system, which contains a fluid (hemolymph) that combines the functions of coelomic fluid and blood.

hemocyanin. Blue, copper-containing respiratory pigment of molluscs, crustaceans, and some spiders.

hemoglobin. Red, iron-containing respiratory pigment of vertebrates and many invertebrates.

hemolymph. Blood of animal with an open circulatory system.

hemostasis. Arrest of bleeding or blood flow.

Hensen's node. Site of cell movement during gastrulation in reptiles, birds, and mammals.

hepatic. Relating to the liver.

hepatopancreas (= digestive gland). Organ that secretes digestive enzymes and stores nutrient reserves, especially in crustaceans and molluscs.

herbivore. Animal that eats only plants.

heritability. Amount of variance that can be attributed to genetic differences.

hermaphroditic (= monoecious). Combining the reproductive functions of both genders in one individual.

heterocercal. Having a caudal fin with the dorsal lobe larger than the ventral, and with the vertebral column extending into the dorsal lobe, as in sharks.

heterochrony. Change in the timing of a developmental event; for example, neoteny.

heterogametic. Having different sex chromosomes.

heterosis (= hybrid vigor). Increased fitness of an organism due to heterozygosity (having different alleles for the same gene).

heterotherm. Homeotherm in which body temperature makes large but controlled deviations.

heterotroph (= consumer). Organism that obtains energy from organic materials synthesized by other organisms.

heterozygous. Having two different alleles for the same gene.

hexapod. Insect or other six-legged arthropod.

hexose. Simple sugar with six carbons, such as glucose and fructose.

hibernation. (1) In certain birds and mammals an adaptation to cold temperatures in which the body temperature is maintained a few degrees above freezing. (2) Loosely, torpor, diapause, or any other state of inactivity during cold.

hindbrain (= rhombencephalon). Part of the brain that forms the cerebellum, pons, and medulla.

hindgut. Part of insect digestive tract posterior to midgut.

histamine. Secretion from mast cells of injured tissues that triggers swelling and other signs of inflammation.

histone. Protein around which DNA is organized.

holdfast. (1) Structure for attachment of an animal to substratum. (2) The scolex of a tapeworm.

holoblastic. Complete cleavage of an egg, as opposed to meroblastic cleavage in which yolk prevents or delays cleavage of part of the egg.

holometabolous. Referring to an insect in which the immature forms do not resemble the adult.

home range. Area shared by a group of social animals of the same species.

homeo box. Sequence of 180 base pairs found in many homeotic genes of insects and certain other animals.

homeostasis. Maintenance of certain conditions in a steady state.

homeothermy. Maintenance of stable body temperature.

homeotic mutation. Mutation in a homeotic gene, which changes the identity of a segment, causing its appendages to develop into the homologous appendages of a different segment.

homocercal. Having a caudal fin with dorsal and ventral lobes of approximately equal size, and with the vertebral column not extending into either lobe, as in most bony fishes.

homologous. (1) Similar because of shared ancestry. (2) Referring to chromosomes that are similar in shape and carry identical genes. (3) Referring to different parts (e.g., legs and antennae) that develop from similar structures in different segments of segmented animals. Serially homologous.

homozygous. Having two identical alleles for a gene.

hormone. Secretion from a specific tissue that travels through the bloodstream and evokes a response in distant cells.

horn. Keratinous growth, usually permanent, on the heads of cattle and certain other hoofed mammals.

hybrid. Offspring of genetically different parents. A heterozygote.

hybrid vigor (= heterosis). Increased fitness of an organism due to heterozygosity (having different alleles for the same gene).

hybridogenesis. Fertilization of a primarily parthenogenetic species by males of a related species.

hydride (H:⁻). Hydrogen atom with an additional electron.

hydrogen bond. Attraction of a strongly electronegative atom, especially O or N, for a hydrogen atom that is covalently bonded to a strongly electronegative atom.

hydroid. A cniarian polyp, such as a hydra.

hydrologic cycle (= water cycle). Biogeochemical movement of water among various organic and inorganic states.

hydrolysis. Chemical reaction in which a molecule is decomposed by the addition of water.

hydrophilic. Having a high solubility in water, like most polar molecules.

hydrophobic. Having a low solubility in water, like fats.

hydrostatic skeleton (= hydroskeleton). Internal hydrostatic pressure that supports the body of certain animals.

hyperosmotic. Referring to a solution with a higher osmotic concentration than another solution.

hyperpolarization. Increase in the magnitude of a membrane voltage.

hypertension. High blood pressure.

hypertonic. Referring to a solution that causes a cell or organism to lose water and shrink.

hypnozoite. Dormant stage of malaria-causing *Plasmodium* in liver.

hypoblast. Part of the inner cell mass of a mammalian embryo that forms the yolk sac.

hypodermic impregnation. Insemination by inserting the penis through the integument, as occurs in certain flatworms.

hypodermis. Cuticle-secreting epidermis in animals such as annelids and arthropods.

hypopharynx. Tonguelike structure in mouthparts of an insect.

hypophysis (= pituitary gland). Compound gland that secretes several hormones in response to signals from the nearby hypothalamus in the brain.

hyposmotic (= hypo-osmotic). Referring to a solution with a lower osmotic concentration than another solution.

hypothalamus. Part of the brain that helps regulate feeding, water and ion balance, body temperature, and the secretion of hormones from the pituitary gland.

hypotonic. Referring to a solution that causes a cell or organism to gain water and swell.

Ia afferent. Large-diameter neuron that conducts action potentials from spindle receptors in muscles to the spinal cord.

I band. Band at the end of a sarcomere in skeletal muscle, formed by thin protein filaments.

ice age. Period of reduced mean temperature when annual snow accumulation exceeds melting, causing the spread of glaciers.

ice era. Period lasting about 50 million years during which ice ages occur.

imaginal disc. Structure in insect larva that becomes a particular adult structure after metamorphosis.

imago (im-A-go). Adult form, especially of an insect.

immunity. Specific defense reaction against a parasite or foreign molecule.

immunoglobulin (= antibody). Blood protein from plasma cells that binds to specific foreign molecules.

immunological memory. Ability of the immune system to recognize and quickly respond to a second exposure to a particular foreign molecule.

implantation. Process by which embryo enters the lining of the uterus.

imprinting. (1) Social bonding between a young animal (usually a bird) and a parent or other object. (2) Long-term memory of the odor of a native stream used by certain fishes to migrate to the same stream for spawning.

inbreeding. Breeding between closely related organisms.

incisor. Chisel-shaped tooth adapted for cutting.

inclusive fitness. Individual fitness (ability to reproduce) plus the sum of the individual fitnesses of relatives multiplied by their relatedness.

incomplete digestive tract. Digest tract lacking an anus.

incurrent canal. In leuconoid sponges, a channel leading water into a flagellated chamber.

independent assortment. Inheritance of one allele regardless of whether another allele at a different locus is also inherited.

indeterminate cleavage (= regulative development). Cleavage in which fates of cells (blastomeres) are determined later in cleavage.

indirect calorimetry. Determination of metabolic rate by measuring rate of O_2 consumption or CO_2 production.

indirect development. Development in which there is a larval form.

induction. During embryonic development, the influence of some cells (organizer or morphogen) on the development of others.

industrial melanism. Darkening of the integument of insects in areas where soot has accumulated. Believed to be an evolved camouflage.

initiator. Mutagen or oncogenic virus that initiates cancer.

inflammation. Reaction to tissue damage involving swelling, pain, heat, redness, and itching.

infundibulum. Funnel-shaped opening through which eggs enter the oviduct of a bird.

ink. Dark or luminescent secretion released by cephalopods and some other molluscs in response to a threat.

inner cell mass. Cells in a developing reptile, bird, or mammal that form the embryo and extraembryonic membranes.

instar. Stage between molts in an immature arthropod.

integument. The body surface of an animal.

intercalated disc. Electrically conducting junctions between muscle cells in the heart.

intercostal muscle. Skeletal muscle between the ribs that is partly responsible for breathing.

interference color. Structural colors due to overlapping structures that cause incident and reflected rays of light to interfere with each other.

interferon. Secretion (a cytokine) from virus-infected cells that interferes with the ability of virus to reproduce in other cells.

interleukin. Chemical (a cytokine) that stimulates immune responses.

intermediate filament. Part of the cytoskeleton consisting of protein strands between microtubules and microfilaments in diameter (7 to 11 nm).

internal cellularization. Hypothetical process by which metazoans originated from a multinucleate protozoan in which the nuclei became separated by intracellular membranes.

internal fertilization. Fertilization of ova inside the mother's body.

interneuron. Neuron with both synaptic inputs and outputs to other neurons.

interphase. Period in the life cycle of a cell when mitosis and meiosis are not occurring, or between telophase I and prophase II of meiosis.

interstitial cell (= Leydig cell). Cell of the testis that produces androgens.

interstitial fluid. Fluid surrounding cells.

intertidal zone (= seashore). Part of the continental-shelf ecosystem between high and low tide levels.

intracellular digestion. Digestion of food inside cells, as in many protozoans, sponges, cnidarians, flatworms, rotifers, bivalve molluscs, and primitive chordates.

intron (= intervening sequence). Portion of messenger RNA that is removed during processing and does not code for a protein. Also the DNA coding for that portion of the mRNA.

introvert. Retractable anterior portion of a sipunculid.

inversion. Reversal of a portion of a chromosome.

ion. Atom or molecule with a net charge due to gain or loss of one or more electrons.

ionic bond. Chemical bond in which an electron from one atom is transferred to another. Characteristic of atoms that form ions.

iridophore. Type of chromatophore that contains guanine or other purine and produces silvery and iridescent colors by reflecting light.

isoenzyme (= isozyme). Form of an enzyme that differs from another form of the same enzyme.

isolecithal. Egg with sparse yolk that is evenly distributed.

isosmotic (= iso-osmotic). Having the same osmotic concentration as another solution.

isotonic. Referring to a solution that does not cause a cell or organism to shrink or swell.

isotope. Atom that differs in number of neutrons in the nucleus compared with another in the same element.

Jacobson's organ (= vomeronasal organ). Olfactory organ in the roof of the mouth of an amphibian, reptile, or mammal.

jelly layer. Covering of echinoderm egg.

jumping gene (= transposing element). Segment of a chromosome that moves to another chromosome or to a different position on the same chromosome.

juxtaglomerular apparatus (JGA). Structure near the glomerulus of the kidney that secretes renin.

karyotype. Chromosomes of an organism as they appear in the microscope.

keratin. Complex of globular and fibrous proteins held together by disulfide bonds. Replaces cytoplasm in epidermal cells as they become cornified and forms nails, claws, and horns.

kidney. Osmoregulatory organ of vertebrates, or analogous organ in molluscs.

killer cell. Cell of the immune system that attacks foreign cells.

kinesis (plural **kineses**). Nondirectional change in rate of movement in response to a stimulus. Compare taxis.

kinetochore. Protein on centromere of chromosome by which spindle fibers attach during mitosis and meiosis.

kingdom. Highest taxonomic level, such as kingdom Animalia.

kin recognition. Recognition by one animal that another is related.

kin selection. Preferential behavior by one animal toward related animals.

kinetosome (= basal body). Organelle to which a cilium or eukaryotic flagellum attaches.

kinin. Peptide that relaxes arterial smooth muscle and increases the flow of blood out of capillaries.

kleptocnida (KLEP-to-NY-duh). Nematocyst of a cnidarian ingested by a predator and used for the predator's own defense.

knuckle-walking. Locomotion of gorillas and chimpanzees in which the weight of the upper body rests on the backs of the fingers.

Krebs citric acid cycle (= tricarboxylic acid or TCA cycle). Cycle of reactions in mitochondria by which pyruvate from glycolysis is oxidized to carbon dioxide during cellular respiration.

krill. Filtered food of toothless whales. Especially small crustaceans in order Euphausiacea.

labium. (1) Liplike fold of tissue, such as an inner or outer lip of the female genitals. (2) Lower lip of insect or crustacean mouthparts.

labrum. Upper lip of insect or crustacean mouthparts.

lachrymal gland. Tear gland.

lactate. Waste product of anaerobic glycolysis.

lactation. Production of milk.

lacteal. Lymph vessel in the villus of the small intestine that absorbs digested fats.

Lamarckism. Inheritance of acquired characteristics, one component of Lamarck's theory of evolution.

lamella. Thin, layered structure, such as in dense bone.

larva. Juvenile that is sharply different in form from the adult.

larynx. Enlargement of the entrance of the vertebrate trachea; voice-producing organ of amphibians, reptiles, and mammals.

lateralization. Unequal distribution of functions on each side of the brain, with each side generally controlling functions of the other side of the body.

lateral line organ. Canal or line bearing mechanoreceptors on each side of a fish.

lateral undulation. Most common mode of locomotion in snakes, in which the body pushes sideways against several supports.

Laurasia. Northern supercontinent that split off from Pangea before 180 million years ago, in the Jurassic Period. Later divided into North America, Europe, and most of Asia.

lek. Area where animals of one sex display to attract mates.

lentic. Referring to water with little or no current, as in a pond or swamp.

leptocephalus. Leaf-shaped larva of an eel.

lethal limit. Upper or lower extreme for a limit of tolerance. Maximum or minimum temperature, oxygen level, or other abiotic factor for an organism.

leuconoid. Type of canal system in most sponges, in which choanocytes are located in numerous flagellated chambers and there is no spongocoel.

leukemia. Cancer of white blood cells.

leukocyte (= white blood cell). One of several types of blood cell important to immunity.

lift. Force generated during flight that overcomes the force of gravity.

limax form. Type of locomotion in some amebas in which the whole cell creeps like a single pseudopodium.

limbic system. Part of the cerebral cortex that controls emotional behavior.

limit of tolerance. Range of temperature, oxygen level, ionic concentration, or other abiotic factor in which an organism can live.

limnetic zone. Surface waters away from shore.

lingual flip. Flip of tongue by which certain frogs capture prey.

linkage group. Group of genes that tend to be inherited together because all are found on the same chromosome.

linkage map. Diagram showing the locations of genes on a chromosome.

lipase. Enzyme that digests lipids.

lipid. Hydrophobic molecule; a neutral fat, oil, steroid, or wax.

lipid bilayer. Double layer formed spontaneously and in cell membranes, with hydrophobic tails pointing toward each other and hydrophilic head groups pointing outward.

littoral zone. Shallow water along a shore.

lobopodium. (1) Lobe-shaped pseudopodium of amebas and some other protozoans (classes Lobosea and Eumycetozoea). (2) Leglike appendage of an onychophoran.

locus (LOW-kus; plural **loci**, LOW-sy). Position of a gene on a chromosome.

logistic equation. An equation describing idealized exponential population growth, leveling off at the carrying capacity.

loop of Henle. Bend in the tubule of a nephron of the kidney that is important in water reabsorption.

lophophore. Arch of hollow (coelomate) tentacles enclosing the mouth but not the anus.

lorica. Girdlelike enclosure of some protozoans, rotifers, and other invertebrates.

lotic. Referring to rivers, streams, and other moving fresh water.

luciferase. ATPase that catalyzes emission of light in fireflies.

luciferin. Molecule in fireflies that emits light when activated by ATP and luciferase.

lumen. Cavity of an organ such as the gut or a blood vessel.

lunar cycle. Period of a biological rhythm that is determined by movement of the Moon through its effect on tides.

lung. Organ for exchange of respiratory gases in terrestrial animals.

lymph. Fluid that accumulates in the lymphatic system from interstitial fluid.

lymphatic capillary. End of a lymph vessel, where lymph collects from interstitial fluid.

lymphatic vessel. Tubular channel that transports lymph into the blood.

lymph node. Lymphoid tissue, especially in the groin, armpits, and neck, through which all lymph is filtered.

lymphocyte. Cell of the *B* or *T* type responsible for immunity.

lymphoid organ. Tissue that produces lymphocytes, including lymph nodes, tonsils, thymus, and spleen.

lymphoma. Cancer of solid tissues associated with the lymphoid tissue.

lyriform organ. Parallel array of mechanoreceptive slits in cuticle of spiders.

lysosome. Spherical, membrane-enclosed organelle containing enzymes that digest materials absorbed by endocytosis.

lysozyme. Enzyme in saliva, tears, and other secretions that breaks down bacterial cell walls.

macroevolution. Evolutionary change large enough to be recognizable at the level of species, genus, or higher taxon.

macromere. Blastomere that is larger than other blastomeres (micromeres).

macronucleus. Larger of two types of nucleus in certain protozoans (ciliates), which directs metabolism, development, and the physical traits of the cell but not reproduction.

macrophage. Phagocytic white blood cell.

madreporite. Sievelike input of water vascular system of an echinoderm.

magnetoreceptor. Receptor in insects, salamanders, birds, and perhaps mammals that responds to magnetic fields.

Malpighian tubule. Excretory organ of insect or spider that is attached at one end between the midgut and hindgut.

mammary gland. Milk-producing gland of a mammal.

mandible. Jaw of a vertebrate, or jawlike part of arthropod mouthparts.

manometer. Fluid-filled tube for measuring pressure.

mantle (= pallium). (1) Part of the visceral mass of molluscs and brachiopods consisting of two flaps of skin on the dorsal surface of the body that secretes the shell and forms the mantle cavity. (2) Structure in barnacles analogous to the mantle of molluscs. (3) Thin body wall of tunicates.

mantle cavity. Major body cavity of molluscs and brachiopods.

marsupium. In amphipods, marsupial mammals, and some fishes and other animals, a receptacle in which eggs and young are brooded.

mass extinction. Episode in which many species become extinct at about the same time.

mass spawning. Simultaneous release of gametes by numerous animals of same or different species.

mastax. Pharynx of a rotifer modified with crushing jaws (trophi).

maternal factor. Substance in egg that affects embryonic development.

mating system. Proportion of males to females involved simultaneously in reproduction. Monogamy, polyandry, or polygyny.

mating type. Group of ciliate protozoans incapable of conjugating with each other but able to with members of other mating types.

matrix. Central space within mitochondrion, enclosed by the inner membrane.

maxilla. (1) Upper jaw bone of a vertebrate. (2) Part of arthropod mouthparts.

maxillary gland. Gland in crustacean that regulates ion concentrations.

maxilliped. Appendage associated with the mouth of a crustacean.

median eye (= pineal eye). Third eye of sharks and certain other vertebrates.

medulla oblongata. Part of the brainstem that regulates circulation, breathing, coughing, sneezing, and swallowing.

medusa. Jellyfish or any similar free-swimming cnidarian.

megalops. Larval stage of a true crab.

meiofauna. Small animals that live among gravel and sand grains.

meiosis. Process by which chromosomes in a diploid cell duplicate, and the cell divides twice into four haploid cells that form gametes.

melanin. Black, gray, or brown pigment in skin, hair, feathers, retina, and other structures.

melanophore (= melanocyte). Type of chromatophore containing melanin pigment.

membranelle. Plate-shaped buccal ciliature of certain protozoans (ciliates).

membrane potential. Voltage across a membrane.

membranous bone (= dermal bone). Bone that forms from layers of embryonic connective tissue rather than by replacement of cartilage. Forms fish scales, turtle shells, antlers, and flat bones of vertebrate skulls.

memory lymphocyte. Lymphocyte responsible for triggering immune response to second invasion by a particular foreign cell or molecule.

menstrual cycle. Cycle of changes in reproductive function (estrus cycle) of women and other primates that menstruate.

meridional. Cleavage furrow that is linear from pole to pole, like a meridian on a globe.

meroblastic. Term applied to an egg in which cleavage in one area is delayed or prevented by yolk.

merozoite. Stage of certain protozoans, such as *Plasmodium*, resulting from multiple fission (schizogony) by the schizont.

mesencephalon (= midbrain). Middle part of the brain between the cerebrum and brainstem (forebrain and hindbrain).

mesenchyme. Middle layer of connective tissue consisting of cells and cell products in a jellylike matrix (mesoglea).

mesentery. (1) Thin fold of peritoneum that holds viscera in place. (2) Septum of a sea anemone.

mesobronchus (plural **mesobronchi**) Central tube leading to parabronchi in the lung of a bird.

mesoderm. (1) Germ layer that gives rise to blood, bone, and other connective tissues, as well as muscle. (2) Tissue derived from mesoderm.

mesoglea. Jellylike substance in mesenchyme. Synonymous with mesenchyme in cnidarians and ctenophores.

mesohyl. Cellular layer between inner and outer layers of a sponge. Often called mesenchyme or mesoglea.

mesolecithal. Egg with moderate amount of yolk that is unevenly distributed. Moderately telolecithal.

mesonephros. Kidney in adult fishes and amphibians and in embryos of other vertebrates. Develops into epididymis, vas deferens, and seminal vesicle in male mammals.

messenger RNA (mRNA). Ribonucleic acid transcribed from DNA that determines the sequence of amino acids during synthesis of proteins.

metabolic rate (MR). Rate of overall metabolism; rate of production of body heat.

metabolism. Synthesis of new organic molecules from nutrients (anabolism) and breakdown of nutrient molecules to extract energy (catabolism)

metacercaria. Juvenile fluke that develops when the cercaria loses its tail and which encysts in second intermediate host.

metachronal wave. Coordinated wave of contraction of cilia.

metamere (= somite = segment). Linearly repeated subdivision of a segmented animal (Pogonophora, Annelida, Arthropoda, and Chordata).

metamerism (= segmentation). Division of body into segments (metameres).

metamorphosis. Development of a larva into the adult stage.

metanephridium. (Often called simply nephridium.) Excretory organ of many invertebrate coelomates, consisting of a tube open at both ends that drains coelomic fluid.

metanephros. Kidney of adult reptiles, birds, and mammals.

metaphase. Stage of mitosis or meiosis in which chromosomes are visible and line up between the cellular poles.

metazoon (plural **metazoa**). An animal (multicellular).

microclimate. Atmospheric condition in a restricted space.

microevolution. Small change in gene frequency, or minor heritable alteration of a species.

microfilament. Thinnest (6 nm) of three fibers in the cytoskeleton, consisting of the protein actin.

microfilaria. First larval stage of a filarial worm (phylum Nematoda).

micromere. Blastomere that is smaller than others (macromeres).

micronucleus. Smaller of two types of nucleus in certain protozoans (ciliates), which transmits genetic information during reproduction.

micropyle. Pore in insect egg that admits sperm for fertilization.

microthrix (plural **microtriches**). Microvillus-like projection on tegument of tapeworm that increases surface area for absorption of nutrients from gut of host.

microtubule. Thickest (22 nm) of the three fibers in the cytoskeleton, constructed of subunits of tubulin and other proteins.

microvillus. Submicroscopic projection of the plasma membrane.

mictic. Referring to female rotifers that lay haploid eggs that hatch into male rotifers. Compare amictic.

midbrain (= mesencephalon). Middle part of the brain between the cerebrum and brainstem (forebrain and hindbrain).

middle-ear bones. Small bones (ossicles) that transmit vibrations from the eardrum to the cochlea.

middle piece. Part of sperm cell between head and flagellum that contains mitochondria.

midgut. Part of insect digestive tract between crop and hindgut, where most digestion and absorption occur.

milk let-down. Release of milk during suckling.

miracidium. Ciliated larva of a fluke, released in egg from final host and infectious to molluscan intermediate host.

mitochondrion (plural **mitochondria**). Organelle in which Krebs cycle and the respiratory chain occur. Major source of ATP.

mitosis. Process by which chromosomes duplicate and separate prior to cell division.

mitral valve (= bisupid valve). Left atrioventricular valve.

molar. (1) Tooth with several cusps adapted for grinding. (2) Unit of measure of concentration. One mole per liter.

molarity. Measure of concentration. Number of moles of a substance dissolved in a liter of solution.

mole. 6.02×10^{23} molecules (Avogadro's number) of a substance; one gram molecular mass.

molecular phylogenetics. Determination of evolutionary relationships of organisms by comparing structures of proteins, nucleic acids, or other molecules.

molting (= ecdysis). Shedding of cuticle, skin, hair, or feathers.

molting fluid. Secretion of arthropod epidermis that digests endocuticle of old exoskeleton that is to be molted.

monoecious (mon-EE-shus) (= hermaphroditic). Combining both male and female reproductive organs in one individual.

monogamy. Mating system in which one male pairs with one female at a time in reproduction.

monohybrid cross. Cross between two individuals with different alleles for a given gene.

monophyletic. Referring to a natural taxonomic group: one that has evolved from only one ancestral species. Compare polyphyletic.

monosaccharide. Simple sugar molecule, such as glucose.

monosynaptic reflex. Simple reflex in which there is only one synapse connecting sensory input to motor output.

morphogen. Substance from one part of an embryo that induces differentiation in another part.

morphogenesis. Development process by which germ layers form and body pattern is established.

morphology. (1) Structure. (2) The study of structure.

mosaic development (= determinate cleavage). Development in which fates of cells are determined in early cleavage. Common among invertebrates.

motor unit. One motor axon and all the muscle fibers it controls.

mouthbrooder. Fish that broods eggs or young in its mouth.

mucosa. Mucus-secreting tissue layer, such as the inner lining of the gut.

mucus. Glycoprotein solution that protects and lubricates tissues.

Müllerian duct. Embryonic duct that develops into the oviduct of a female.

Müllerian mimicry. Resemblance of two or more toxic species to each other, which presumably increases the deterrence to predation.

multiple fission. Type of asexual reproduction in which an organism divides into many offspring simultaneously.

muscular stomach (= gizzard). Posterior portion of a bird's stomach, with a horny lining that crushes food.

mutagen. Agent such as a chemical, ionizing radiation, or ultraviolet radiation that induces genetic mutations.

mutation. Hereditary change, especially in the sequence of bases in DNA.

mutualism. Type of symbiosis in which both symbionts benefit.

myelin. Layers of membrane surrounding an axon, periodically interrupted by nodes of Ranvier.

myenteric plexus. Network of nerve cells in the gut that coordinates motility.

myocyte. Contractile cell encircling an osculum or channel in a sponge.

myofibril. Longitudinal contractile subunit within a muscle fiber.

myogenic. Referring to hearts that contract spontaneously, without neural excitation.

myoglobin. Respiratory pigment that transports oxygen in muscle.

myomere (= myotome). Block of segmental muscles in a chordate.

myometrium. Smooth-muscle layer of the uterus.

myosin. Protein that forms thick filaments in muscle.

myotome. (1) Mesodermal segment of vertebrate embryo from which a block of muscles (myomere) develops. (2) A myomere.

nacre (= nacreous layer). Inner layer of mollusc shell.

NAD. Nicotinamide adenine dinucleotide.

naiad. Immature aquatic insect (nymph).

Na$^+$–K$^+$ ATPase (= sodium–potassium

pump). Enzyme in plasma membranes responsible for coupled transport of sodium and potassium.

naris (plural **nares;** NARE-eez). Opening of a nasal cavity.

natural group. Taxonomic group that corresponds to an actual evolutionary group.

natural selection. Differential survival and reproduction of one phenotype compared with another.

nauplius. Free-swimming microscopic larva of a copepod, barnacle, or other crustacean with one median eye (naupliar eye).

nekton. Aquatic organisms that move actively through water, rather than drifting or floating passively like plankton.

nematocyst. Organelle in cnidarians that shoots out stinging thread for predation or defense. Most common type of cnida.

nematocyte. Cell that produces nematocysts.

neocortex. Outer few millimeters of the cerebral cortex in mammals, which contains most synapses.

neo-Darwinian theory (= synthetic theory). Current theory that combines Darwin's theory of natural selection with modern knowledge of genetics to explain most evolution.

neoteny. Occurrence of larval form in an adult, resulting from either paedomorphosis or progenesis.

nephridiopore. Opening through which a nephridium (protonephridium or metanephridium) excretes wastes and excess water.

nephridium. Tubular osmoregulatory and excretory organ of many invertebrates; usually a metanephridium.

nephron. Functional subunit responsible for osmoregulation and excretion in the vertebrate kidney, consisting of a glomerulus, Bowman's capsule, and a tubule.

nephrostome. Funnel-shaped, internal opening of a metanephridium through which cilia filter coelomic fluid for osmoregulation and excretion.

nerve net. Diffuse network of nerve cells forming the nervous system of cnidarians, ctenophores, and echinoderms.

nerve ring. Circular nerve around the pharynx that forms a major part of the nervous system of many small invertebrates.

net primary production. Increase in biomass or energy in an area at the end of a given period.

neural crest. Ectoderm that develops into sensory nerve cells, sympathetic ganglia, adrenal medulla, dentine of teeth, and connective tissues of head.

neural fold. One of two folds of ectoderm that form the embryonic neural tube from which the brain and spinal cord develop.

neural tube. Hollow cylinder of ectoderm from which the brain and spinal cord develop.

neurogenic. Referring to hearts that are stimulated to contract by the nervous system.

neurohypophysis (= posterior pituitary). Part of the pituitary gland that releases hormones from neurosecretory nerve endings.

neuroid cells. Syncytial amebocytes forming a nervelike network in certain sponges.

neuromast. Mechanoreceptor cell in lateral line organ of fishes.

neuromuscular junction (= endplate). Junction between a nerve cell and a muscle cell.

neuron. Nerve cell.

neuropodium. Ventral lobe of a parapodium of a polychaete annelid.

neurosecretory cell. Nerve cell that secretes hormone or other substance rather than synaptic transmitter.

neurula. Stage of vertebrate development in which the neural tube and neural crest form.

neurulation. Formation of neural tube and neural crest in the embryo.

neutral. Having a pH of approximately 7, equal to that of pure water.

neutral buoyancy. Having a density (mass divided by volume) equal to that of the external medium and therefore being able to float.

neutral fat (= glyceride). Lipid consisting of glycerol with one, two, or three fatty acids attached.

neutral theory of evolution. Theory that most evolution results from mutations that have no immediate effect on the ability of organisms to survive and reproduce, and are therefore not subject to natural selection.

niche. Abiotic and biotic factors that define the role of a species in a community.

nicotinamide adenine dinucleotide (NAD^+). Coenzyme that is reduced to NADH during glycolysis and cellular respiration.

nictitating membrane. Translucent eyelid of many vertebrates.

El Niño. Invasion of warm Pacific water that prevents normal upwelling of cold water off west coast of South America.

nitrate bacterium. Bacterium that converts nitrite NO_2 into nitrate NO_3^-.

nitrite bacterium. Bacterium that converts ammonia into nitrite NO_2^-.

nitrogen cycle. Biogeochemical movements of nitrogen among various inorganic and organic states.

nitrogenous base. Adenine, guanine, cytosine, thymine, or uracil; the chemical structures that determine the genetic information in nucleic acids.

nitrogenous waste. Substance, usually ammonia, urea, or uric acid, by which an animal eliminates nitrogen.

node of Ranvier (RAHN-vee-ay). Gap between adjacent sections of myelin.

nomenclature. Branch of systematics dealing with the naming of groups of organisms.

nondisjunction. Failure of chromatids to separate during meiosis, leading to gain or loss of chromosomes in gametes.

notochord. Stiff rod of tissue along the dorsal midline of lancelets and vertebrate embryos.

notopodium. Dorsal lobe of a parapodium of a polychaete annelid.

nuchal organ. Anterior chemoreceptor of a polychaete annelid.

nuclear envelope. Porous, double membrane system that encloses the cell nucleus.

nucleator. Substance that favors the formation of ice crystals.

nucleic acid. DNA or RNA, both of which consist of chains of nucleotides.

nucleolus. Dark-staining part of cell nucleus that synthesizes ribosomes.

nucleosome. Fundamental subunit of chromatin consisting of a length of DNA partly wrapped around histones.

nucleotide. Nucleic acid base attached to ribose or deoxyribose, to which is attached a phosphate.

nucleus (plural **nuclei**). (1) Cluster of cell bodies and synapses in the brain. A ganglion. (2) Organelle in eukaryotic cells that contains the hereditary material.

numerical taxonomy (= numerical phenetics = phenetics). Method of classification in which numerical values are assigned to quantifiable differences as a basis for grouping organisms.

nuptial pad. Thickened skin on hands or other parts of male frog by which it grasps female during mating.

nuptial swarm. Aggregation of animals, especially insects, during mass mating.

nurse cell. Cell that attends developing larvae in a sponge or other animal.

nutritive muscle cell. Contractile cell in the gastrodermis of a cnidarian.

nymph. (1) Immature insect that is hemimetabolous (gradually metamorphic, without a pupal stage); (2) Legless, immature stage of a pentastomid.

ocellus. Simple eye of an arthropod or other invertebrate.

odontophore. Cartilaginous support for the radula of a snail.

oil gland (= uropygial gland). Gland at the base of the tail of a bird that produces oil for preening.

olfaction (= smell). Chemoreception of airborne chemicals.

olfactory bulbs. Pair of brain structures just above the olfactory epithelium that are responsible for perception of odors.

olfactory epithelium. Epithelium lining the roof of the nasal cavity and bearing olfactory receptors.

ommatidium. Subunit of arthropod compound eye.

omnivore. Animal that eats both plants and other animals.

oncogene. Mutated proto-oncogene or gene acquired by viral infection that initiates cancer.

oncomiracidium (ON-ko-MERE-a-SID-ee-um). Ciliated larva of flatworm in class Monogenea.

oncosphere. Egg-enclosed larva produced by proglottid of a tapeworm.

oncotic pressure (= colloid osmotic pressure). Hydrostatic pressure due to osmotic effect of a colloid.

ontogeny. Development of an individual.

oocyst (OH-uh-sist). Cyst enclosing zygote of malaria-causing *Plasmodium* and certain other protozoans.

oocyte (OH-uh-site). Cell in meiosis during development into an ovum.

oogenesis (oh-uh-GEN-uh-sis) Development of egg cells.

oogonium (oh-uh-GO-ne-um) (= primordial germ cell). Cell that will develop into an oocyte, then an ovum.

ootheca (oh-uh-THEEK-uh). Protective case containing fertilized eggs.

open circulatory system. Circulatory system in insects and some other invertebrates, in which blood is not always confined within the heart or vessels.

operant conditioning (= Skinnerian, instrumental, or Type II conditioning). Learning of a behavior due to the rewarding (reinforcement) of closer and closer approximations to that behavior.

opercular chamber. Cavity enclosing gills of a fish.

operculum. Covering of a chamber, such as the gill chamber of a fish, the nematocyst of a cnidarian, or the shell of a snail.

ophiopluteus. Larva of a brittle star.

opisthaptor. Posterior adhesive organ of a flatworm in the class Monogenea.

opisthosoma. Segmented posterior portion of a pogonophoran or an arachnid.

optimum. Level of temperature, ion concentration, or other abiotic factors at which an organism lives best.

oral disc. Double ring of tentacles around the mouth of a sea anemone.

orb web. Spider web constructed in a vertical plane to snare flying insects.

order. Taxonomic level below class and above family.

organ of Corti (KOR-tee). Mechanoreceptive tissue in the cochlea responsible for hearing.

organelle. Discrete structure in a cell that performs a specific function.

organic molecule. Any carbon-based molecule.

organogenesis. Embryonic development of organs.

osculum. Opening by which water exits a sponge.

osmoconformer. Animal, generally a marine invertebrate, that does not regulate the solute concentration of its body fluids when the concentration in the environment changes.

osmolarity. Osmotic effect of a given solution. The concentration that would produce the same osmotic effect as that of the solution. Osmoles per liter.

osmolyte. Organic molecule, such as urea, that increases the osmolarity of body fluids and helps prevent loss of water due to osmosis.

osmometer. Device for measuring osmolarity.

osmoregulator. Animal that regulates the solute concentration of its body fluids as the concentration in the environment changes.

osmosis. Movement of water from a low solute concentration into a higher solute concentration.

osmotic pressure. Hydrostatic pressure due to osmosis.

osphradium. Sense organ that samples incoming water of bivalve molluscs and aquatic snails.

ossicle. (1) Small calcareous plate forming part of the endoskeleton of an echinoderm. (2) Any middle-ear bone.

osteoblast. Cell that forms bone by depositing minerals.

osteoclast. Cell that dissolves minerals in bone.

ostium. An opening, especially (1) a microscopic pore through which water enters a sponge, or (2) an opening by which blood enters the arthropod heart.

otolith (= **otoconium**). Mineral deposit that deflects hair cell receptors in the utricle and saccule (the otolith organs) of vertebrates or in the statocyst of certain invertebrates for perception of orientation.

outcrossing. Mixing of alleles due to interbreeding between two genetically different populations.

outgroup. Group of organisms used in cladistics as a reference in determining whether characters in other groups are primitive or derived.

oval body. Organ that leaks excess gas out of the swim bladder of a fish.

ovary. Organ of females that produces eggs.

ovicell. Zooid of an ectoproct colony specialized as an egg-brooding chamber.

oviduct. Tube through which eggs pass after release from the ovary.

oviger (= ovigerous leg). Leg of a male sea spider (Pycnogonida) that is specialized to gather and brood fertilized eggs.

oviparity. Egg laying.

oviposition. The act of egg laying.

ovoviviparity. Retention of fertilized egg in oviduct until hatching, without nutritional support from the mother.

ovulation. Release of a female gamete from the ovary.

ovum. Mature female gamete.

oxidation. Loss of electrons by a molecule.

oxidation–reduction reaction (= redox reaction). Reaction involving loss of electrons from one molecule, which is oxidized, to another, which is reduced.

oxygen cycle. Biogeochemical movement of oxygen among various inorganic and organic states.

oxygen dissociation curve (= oxygen equilibrium curve). Graph showing the degree of oxygenation of a respiratory pigment at different levels of O_2 in blood.

oxyntic cell (= parietal cell). Cell of the stomach lining that secretes H^+.

pacemaker. Nerve or muscle cells that trigger activity in an organ, especially the heart.

paedomorphosis. Occurrence of features usually characteristic of adults in an animal with a juvenile form. Results from neoteny or progenesis.

pallium. Mantle of a mollusc or brachiopod.

palp. Small appendage usually involved in feeding.

pancreas. Organ that secretes insulin and glucagon into the blood, and digestive enzymes into the small intestine in vertebrates. In some invertebrates, a digestive gland.

Pangea. Supercontinent comprising entire land mass of Earth approximately 250 million years ago.

pangenesis. Obsolete theory that genetic traits come from all parts of the bodies of parents.

papilla. Small nipple-shaped projection, such as a taste papilla on the tongue or a feather bud on a bird's skin.

papula. Projection on integument of echinoderms that exchanges respiratory gases and nitrogenous wastes.

parabiosis. Experimental procedure in which bloodstreams of two living animals are interconnected.

parabronchus (plural **parabronchi**). Tube that exchanges O_2 and CO_2 in a bird's lung.

parallel processing. Simultaneous analysis of different features of a stimulus, such as color and movement, by different parts of the nervous system.

parapodium. Appendage used by polychaete annelids for locomotion and exchange of respiratory gases.

parasitism. Symbiosis in which one organism (the parasite) lives at the expense of the other (host).

parasitoid. Parasite, especially an insect larva, that consumes and ultimately kills the host.

parasympathetic division. Part of autonomic nervous system active during rest.

parathyroid glands. Glands on the thyroid that secrete parathyroid hormone.

parenchyme (= parenchyma). Mesenchyme containing densely packed cells, especially in acoelomates.

parenchymula. Free-swimming larva of some types of sponges, which is solid and covered with flagella.

parental investment theory. Body of theory in sociobiology that attempts to explain the evolutionary advantages of parental behavior.

parietal cell (= oxyntic cell). Cell of the stomach lining that secretes H^+.

parotoid gland. Poisonous swelling behind eye of some frogs.

parthenogenesis. Development of an embryo from an unfertilized ovum.

partial endothermy. Temporary elevation of body temperature of a poikilotherm by production, absorption, and conservation of heat.

partial pressure. Portion of the pressure in a fluid due to one substance in the fluid.

parturition. Process of giving birth to young.

pecking order (= dominance hierarchy). Social rank that determines which animals in a group have priority in mating, feeding, or attacking others.

pectoral girdle. Bones that attach the bones of the forelimbs to the axial skeleton.

pedalium. Rudderlike appendage at each corner of a cubozoan.

pedicel. A small stalk, especially the narrow "waist" between the thorax and abdomen of a spider, ant, or other arthropod (also called a petiole); or the second segment of an insect antenna.

pedicellaria. Sharp or beaklike defensive structure on integument of a sea urchin or other echinoderm.

pedicle (= peduncle). Stalk, such as that by which a brachiopod attaches to substratum or the caudal fin attaches to a fish.

pedipalp. One of the second pair of appendages in arachnids, near the mouth.

peduncle. A pedicle.

pelage. Coat of hair on a mammal.

pelagic. Referring to open ocean.

pellicle. Plasma membrane and associated fibrous cytoplasm in ciliate and flagellate protozoans.

pelvic girdle. Bones by which bones of the hindlimbs attach to the axial skeleton.

pen (= gladius). Vestigial shell in the back of a squid.

penis (plural **penes**). Male organ used for intromission and sperm transfer.

penis bulb. Organ in certain flatworms that stores sperm until copulation.

pentamerous radial symmetry. Five-part radial symmetry, characteristic of starfish and other echinoderms.

pentose. Simple sugar with five carbons, such as ribose or deoxyribose.

pentose shunt. Series of reactions responsible for metabolism of five-carbon sugars using three- and six-carbon sugars in glycolysis.

pepsin. Protein-digesting enzyme secreted by the stomach.

peptide. Molecule consisting of a few amino acids.

peptide bond. Linkage between two amino acids, formed by removal of one hydrogen from the amino group of one amino acid and the OH from the carboxyl group of the other amino acid

pericardial sinus. Sinus in body cavity of arthropod that contains the heart.

pericardium. Sac enclosing a heart.

periostracum. Outer protective layer of a mollusc shell.

peristalsis. Wavelike contraction, such as occurs in the esophagus during swallowing.

peristaltic progression. Type of locomotion in earthworms and some other animals involving wavelike contraction of body segments.

peritoneum (= peritoneal membrane). Membrane enclosing the coelom.

peritubular capillary. Capillary surrounding the tubule of a nephron in the kidney.

permeable. Permitting diffusion, as in a membrane that is permeable to ions.

permeant. Able to diffuse through, as an ion through a membrane.

pH. Measure of acidity, equal to the negative of the exponent of the H^+ concentration expressed as a power of 10.

phagocytosis. Transport (endocytosis) of solid material into a cell by formation of an enclosing vesicle.

pharyngeal pouch. Sac for filter feeding by acorn worms.

pharyngeal slits. Openings in the neck characteristic of chordates.

pharynx (plural **pharynges**). Part of the digestive tract between the mouth and the esophagus. In birds and mammals, the place where the respiratory and digestive tracts cross.

phasmid. Posterior sensory organ in certain nematodes.

phenetics (= numerical phylogeny = numerical phenetics). Method of classification in which numerical values are assigned to quantifiable differences as a basis for grouping organisms.

phenotype. Expressed characteristics of an organism, especially those genetically influenced. Compare genotype.

pheromone. Airborne chemical secreted by one animal that influences the behavior of another in the same species.

phosphagen. Molecule, usually creatine phosphate or arginine phosphate, that stores phosphate bonds for ATP or that shuttles phosphate from mitochondria to make ATP.

phospholipid. Molecule consisting of glycerol to which are linked two fatty acids and one phosphate with a polar head attached.

phosphorylation. Chemical addition of a phosphate group.

photophore. Light-emitting organ.

phylogenetic systematics (= cladistics). Method of classification in which outgroups are used to distinguish primitive from derived characters, and lines of descent but not times of divergence are deduced from the number of derived homologous characters (synapomorphies).

phylogenetic tree (= dendrogram). Diagram representing the evolution of a group of organisms.

phylogeny. Evolutionary history of an organism or taxon.

phylum (plural **phyla**). Taxonomic level below kingdom and above class.

physical gill. Bubbles of air that function as a gill in an aquatic insect.

physiological saline. NaCl solution osmotically balanced to substitute for a body fluid.

phytoplankton. Small algae and plants suspended in water.

pile. Hairlike covering on arthropod integument.

pilidium. Free-swimming larva of nemertine worms, enclosed in plates arranged like a cap with ear flaps.

piloerection. Erection of hair on a mammal due to contraction of the arrector pili muscles.

pinacocyte. Cell lining the external surface (pinacoderm) or a nonflagellated channel in a sponge.

pinacoderm. Epidermis of a sponge.

pineal eye (= median eye). Third eye of sharks, amphibians, and certain other vertebrates.

pineal gland (= pineal body). Gland in vertebrate brain derived from the pineal eye, which appears to be important in biological rhythms.

pinnule. Small projection, as on the arm of a crinoid.

pinocytosis. Transport (endocytosis) of fluid and dissolved material into a cell by formation of an enclosing vesicle.

pituitary (= hypophysis). Compound gland that secretes several hormones in response to signals from the nearby hypothalamus in the brain.

placenta. Organ for exchange of oxygen, nutrients, wastes, or other substance between a viviparous mother and the embryo.

placoid. Plate-shaped.

plankton. Small aquatic organisms suspended in water. Includes phytoplankton (algae and small plants) and zooplankton (larvae and small animals). Compare nekton.

planula. Free-swimming ciliated larva of a cnidarian.

plasma. Fluid portion of blood that remains after cells are removed.

plasma cell. Cell formed from a *B* lymphocyte that makes antibodies.

plasma gel (= ectoplasm). Outer, gelatinous portion of cytoplasm. Compare plasma sol.

plasma membrane. Membrane enclosing a cell.

plasma sol (= endoplasm). Inner, fluid portion of cytoplasm. Compare plasma gel.

plasmid. Loop of DNA that bacteria exchange among themselves naturally.

plastron. Ventral shell of a turtle.

plate tectonics. Branch of geology relating to the movement of plates of Earth's crust due to sea floor spreading and subduction at continental margins.

platelet. Fragment of a white blood cell (megakaryocyte) in blood that initiates clotting.

pleopods (= swimmerets). Appendages on abdominal segments 14 through 18 of decapod crustaceans.

pleura. Membrane enclosing the lung.

pleurite. Cuticular plate (sclerite) on the side (pleuron) of an arthropod.

pneumatic duct. Connection from the esophagus to the swim bladder or lung of a bony fish.

podium (= tube foot). Hollow grasping organ of an echinoderm, controlled by the water vascular system and muscles.

poikilotherm. Animal that does not regulate its body temperature.

point mutation. Change in one base in DNA.

polar. (1) Property of certain molecules, such as water, in which positive and negative charges are not uniformly distributed. (2) Associated with a pole, such as the north or south pole, or the animal or vegetal pole of an egg.

polar body. One of three functionless haploid cells formed as a by-product of meiosis in the development of an ovum.

polarity. (1) Property of certain regenerating tissues in which the orientation (e.g., head or tail) is preserved. (2) Direction with reference to poles.

polarized. Oriented with respect to opposite extremes or poles.

pole. One of a pair of opposites, such as the animal or vegetal pole of an egg.

pollen brush, comb, packer. Structures on legs of bees used to transport pollen.

poly-A tail. Sequence of 150 to 200 adenosine monophosphates attached to messenger RNA during processing.

polyandry. Mating system in which one female has more than one male mate at a time.

polygamy. Having more than one mate at a time. Either polygyny or polyandry.

polygyny (puh-LIJ-uh-nee). Mating system in which one male has more than one female mate at a time.

polymorphism. Variety in a genetically determined trait within a population.

polyp. Sessile stage of a cnidarian.

polypeptide. Molecule consisting of many amino acids.

polyphosphoinositide lipid. Substance such as inositol triphosphate (IP$_3$) or diacylglycerol (DAG) that functions as a second messenger in a cellular response to a hormone or other signal.

polyphyletic. Referring to a taxonomic group that combines two or more distinct lines of evolution. Compare monophyletic.

polypide. Fleshy part of an ectoproct or other zooid.

polyploidy. Occurrence of three or more sets of chromosomes rather than the normal diploid set.

polysome. Messenger RNA with ribosomes attached during translation.

polysaccharide. Carbohydrate consisting of many simple sugars joined by glycosidic links.

polyspermy. Fertilization of an egg by more than one sperm cell.

pons. Part of the brainstem with horizontal axons that bridge the two halves of the brain.

population. (1) Number of individuals in an area. (2) Group of interbreeding organisms.

population cycle. Periodic increase or decrease in population around the carrying capacity.

pore canal. Channel through arthropod cuticle thought to conduct waxes to the surface.

porocyte. Cell (pinacocyte) surrounding a pore (ostium) in a sponge.

portal vessel. Any blood vessel joining two capillary systems, such as the vessel connecting the hypothalamus with the pituitary or the small intestine with the liver.

post-anal tail. Tail that extends past the anus, as in chordates.

posterior pituitary (= neuro-hypophysis). Part of the pituitary that releases hormones as neurosecretions.

postsynaptic potential. Change in voltage of a nerve cell membrane in response to synaptic transmitter.

powder down. Feathers that continually grow and distintegrate into talcumlike powder in certain birds.

precapillary sphincter. Band of smooth muscle that closes and diverts blood into a capillary system.

precocial. Referring to a species, especially bird or mammal, in which the young are able to feed and care for themselves soon after hatching or birth. Compare altricial.

preening. Running the feathers through the beak of a bird to restore the interconnections among barbs.

preferred range. Range of temperature, oxygen concentration, and other levels of abiotic factors in which the highest population density of a species is found.

preformationism. Obsolete view that each organism was already preformed within either the ovum or the sperm. Compare epigenesis.

prehensile. Adapted for grasping, as in the prehensile tails of New World monkeys.

premolar. Two-pointed mammalian tooth adapted for grinding.

pressure. Force divided by the area over which the force is distributed.

primary consumer. Animal that eats autotrophs.

primary host (= final or definitive host). Organism in which the adult form of a parasite develops.

primary producer (= autotroph). Organism that obtains its energy from inorganic sources, as in photosynthesis, and thereby makes it available to other organisms.

primary production. Fixation of matter and energy in organic molecules by primary producers.

primary succession. Establishment and subsequent changes in species composition in an area previously devoid of life.

primary structure. Sequence of amino acids in a protein.

primary transcript (= mRNA precursor). Messenger RNA before it has been processed by removing introns and attaching the poly-A tail and 5′ cap.

primitive. (1) Trait that is ancestral to some other (derived) trait. (2) Organism with many primitive traits, therefore resembling an ancestral organism.

primitive streak. Area along the dorsal midline of the vertebrate embryo where the development of the central nervous system occurs.

primordial germ cell. Cell that develops into a gamete.

prismatic layer. Chalky middle layer of mollusc shell.

proboscis (plural **proboscides**). Snout or other projection on the head, especially associated with the nasal portion of vertebrates or with the mouth in planarians, leeches, and insects.

procuticle. Newly formed cuticle immediately after molting, before sclerotization.

profundal zone. Deepest water of a lake or pond, beneath the limnetic zone.

progenesis. Accelerated development of some adult features in an organism that otherwise retains its juvenile form.

proglottid. Reproductive subunit of a tapeworm.

prokaryote (= procaryote). Organism, mainly bacterial, without cell nuclei and other organelles.

promiscuity. Breeding with several individuals without forming bonds with any. Compare polygamy.

promoter. (1) Sequence of DNA that initiates transcription of a nearby gene. (2) Anything that promotes proliferation of tumor cells after they have been initiated by an oncogene.

pronephros. Primitive kidney in the embryo of a vertebrate.

pronucleus. (1) Haploid nucleus of a gamete after fertilization but before fusion into the zygote nucleus. (2) Haploid product of meiosis of a micronucleus exchanged during conjugation by certain ciliate protozoans.

prophase. First stage of mitosis or meiosis, in which chromosomes become visible.

propolis. Plant resins with which honey bees seal the hive.

prosencephalon (= forebrain). Anterior part of the vertebrate brain including the cerebrum, olfactory bulbs, thalamus, hypothalamus, pituitary gland, and other structures.

prosoma (= cephalothorax). Fused head and thorax of a spider or other arthropod.

prosopyle. Opening of an incurrent canal into a radial canal of a syconoid sponge, or of a flagellated chamber into an excurrent canal in a leuconoid sponge.

prostaglandin. Hormonelike secretion that triggers contraction in uterine smooth muscle and affects blood pressure, blood clotting, inflammation, and numerous other processes.

prostate gland. Organ that produces milky fluid of semen in mammals.

prostomium. Part of the head projecting in front of the mouth, particularly in annelids and certain molluscs.

protandry. Sequential hermaphroditism in which gender changes from male to female.

protein. Molecule consisting of amino acids linked to each other by peptide bonds.

protein family. Group of similar proteins believed to have evolved by duplication and mutation of a single gene.

protein sequencing. Determining the sequence of amino acids in a protein.

proteoglycan. Complex of protein and carbohydrate that together with collagen makes up a fibrous network in cartilage.

prothrombin. Protein in blood that changes to thrombin as part of the clotting process.

protogyny. Sequential hermaphroditism in which gender changes from female to male.

protonephridium. Organ for osmoregulation and execretion, consisting of branched tubule with closed ends internally. Occurs in some acoelomates and pseudocoelomate invertebrates, and in the chordate amphioxus.

proton pump. Membrane device responsible for active transport of hydrogen ions.

proto-oncogene. Normal gene that can mutate and become a cancer-initiating oncogene.

protopod (= protopodite). Base of a biramous appendage of a crustacean.

protostome. Animal in which the mouth originates from the blastopore, and usually with other characteristic patterns of embryonic development.

proventriculus. (1) Anterior portion of a bird's stomach (glandular stomach) that secretes digestive enzymes. (2) Gizzardlike structure between the crop and the midgut of an insect.

proximal convoluted tubule. Part of the tubule of a nephron between Bowman's capsule and the loop of Henle.

pseudocoel (= pseudocoelom). Major body cavity that is similar to a coelom but not derived from mesoderm or not enclosed by a peritoneum.

pseudocoelomate. Animal with a pseudocoel.

pseudogene. Segment of DNA that resembles a gene but is not transcribed.

pseudopodium (= pseudopod). Extension from a protozoan or ameboid cell used in locomotion and endocytosis.

pulmonary. Referring to a lung, as in pulmonary circulation.

pulmonate. Having a lung, as in snails.

pulp. Nerves, blood vessels, and other soft tissue in a tooth or feather.

pump. Structure in membranes responsible for active transport of ions and molecules.

punctuated equilibrium. Pattern of evolution in which long periods of little change are punctuated by brief periods (lasting tens of thousands of years) in which there is rapid evolution.

Punnett square. Diagrammatic device for determining the expected proportions of each genotype resulting from a cross.

pupa. Stage of development in certain insects (holometabolous) between the larva and the adult.

pupil. Opening in the iris of the eye that admits light.

Purkinje fiber. Fiber in the heart that conducts action potentials to the ventricle.

pygidium (py-JI-dee-um). Nonsegmented posterior of a trilobite or other segmented animal.

pyramid (of energy, mass, numbers) (= Eltonian pyramid). Diagram representing energy, biomass, or numbers of organisms at each trophic level.

pyrogen. Substance that induces fever.

pyrosome. Large, luminescent baglike colony of salps.

pyruvate. Three-carbon molecule produced in glycolysis and oxidized in the Krebs cycle.

Q$_{10}$. Measure of the effect of temperature on the rate of a process. Calculated by dividing the rate at one temperature by the rate at a temperature ten degrees lower.

quadrupedal. On four feet.

quaternary structure. Combination of two or more protein subunits into a functional unit.

queen. Reproductive female termite, ant, or bee.

queen substance. Pheromone secreted from honey bee queen's manidibular glands that inhibits production of new queens.

quill. (1) Sharp, stiff defensive hair of a mammal such as a porcupine. (2) Hollow portion of the shaft of a feather.

rachis (RAY-kis). Portion of the shaft of a feather that is not hollow.

radial canal. Part of echinoderm water vascular system that branches into an arm.

radial cleavage pattern. Cleavage pattern in which newly divided cells are aligned with the central axis. Characteristic of echinoderms, amphibians, and some protostomes.

radially symmetric. Forming two similar halves when bisected at any angle along the axis.

radiation. (1) Radial movement from a source. (2) Emission of electromagnetic or ionizing radiation.

radioactive dating. Technique for determining the age of materials by measuring quantities of decay products of radioactive isotopes.

radioimmunoassay (RIA). Technique for measuring small concentrations of hormones or other molecules.

radiole. Branched, ciliated tentacle of annelid tube worms.

radula (RAD-ju-luh). Rasping tonguelike organ of a snail or other mollusc.

ram ventilation. Forcing water over gills by swimming with the mouth open.

ratite. Having a breastbone without a sternal keel, as in flightless birds.

reabsorption. Active transport from filtrate into body fluid by an organ of osmoregulation and excretion.

reaggregation. Re-adhesion of cells after they have been separated, as in sponges.

recapitulation. Summary repetition, especially the repetition of the pattern of evolution of an organism in its pattern of development.

receptive field. Group of receptors that directly affect a nerve cell's electrical activity.

receptor-mediated endocytosis. Type of endocytosis in which receptors binding a particular substance migrate to a coated pit on the plasma membrane, where a vesicle is formed that transports the material into the cell.

receptor molecule. Molecule in or on a cell that binds a particular hormone, synaptic transmitter, or other substance.

receptor potential. Change in voltage across membrane of a sensory receptor cell during stimulation.

recessive. Referring to an allele that is not expressed in the presence of a different homologous (dominant) allele.

reciprocity. Behavioral assistance by one organism that is returned by another.

recombinant DNA. DNA combined from two different chromosomes during meiosis, combined from two different individuals during fertilization, or combined by other natural or artifical means.

recruitment. Increasing the number of nerve cells activated, especially in increasing the force of contraction of a muscle.

rectal gland. Osmoregulatory organ of sharks and rays that secretes concentrated NaCl solution into the posterior intestine.

rectal pad. Structure on insect rectum that recycles water back into the hemolymph.

rectilinear movement. Locomotion used by snakes when stalking, in which the body lies in the direction of movement and creeps forward without twisting.

rectum. Terminal portion of a digestive tract.

red blood cell (= erythrocyte). Cell that transports hemoglobin through blood of vertebrates.

redd. Circular depression in gravel used as a nest by salmonid fishes.

redia. Larval form of a fluke that is produced by sporocysts and feeds in an intermediate host.

redox reaction (oxidation–reduction reaction). Reaction involving loss of electrons from one molecule, which is oxidized, to another, which is reduced.

reduction. Gain of electrons by a molecule.

reductionism. Term applied, usually disparagingly, to attempts to explain complex biological phenomena in terms of simpler phenomena, especially physical and chemical.

reflex. Simple, stereotyped response to a stimulus.

reflex ovulation. Ovulation triggered by copulation.

regeneration. Development of a new limb or organ that replaces a lost one.

regulative development (= indeterminate cleavage). Development in which fates of cells (blastomeres) are determined later in cleavage.

relatedness. Degree of kinship, expressed as a number between 0 (no kinship) and 1 (genetic identity). The probability that two individuals have inherited the same gene from the same ancestor.

releaser. Sign stimulus that triggers a stereotyped social response (fixed action pattern) in a member of the same species.

releasing gland. Part of the duogland system of a flatworm that dissolves the adhesive released by the viscid gland for attachment and locomotion.

releasing hormone. Peptide from hypothalamus that, together with release-inhibiting hormones, regulates secretion from the anterior pituitary.

renal. Pertaining to the kidney.

renette cell (= gland cell). Presumed osmoregulatory cell of a nematode.

rennin (REN-in). Stomach enzyme of infant mammals that coagulates milk.

replication. Duplication of a cell's DNA prior to mitosis.

reproductive isolation. Separation of two populations by geographic barriers or by differences in behavior or appearance so that they do not interbreed.

resilin. Rubbery protein that contributes to jumping in fleas.

respiration (= cellular respiration). Complete catabolism of glucose requiring oxygen and producing carbon dioxide.

respiratory chain. Oxygen-requiring sequence of reactions in mitochondria involving transfers of protons and electrons, resulting in production of ATP.

respiratory circulation. Circulation of blood through lungs, gills, or other organs for exchange of O_2 and CO_2.

respiratory pigment. Colored molecule such as hemoglobin that increases the capacity of blood to transport oxygen.

respiratory tree. (1) System of branching bronchioles in mammalian lungs. (2) Branched respiratory organ in the body cavity of a sea cucumber.

resting membrane potential. Voltage across the plasma membrane of a cell except when the potential is specifically changed, as by an action potential.

restriction enzyme. Enzyme produced by bacteria that cleaves DNA at particular base sequences.

restriction fragment length polymorphism (RFLP). Segment of DNA of varying length produced by restriction enzyme. Used as a marker in mapping chromosomes.

rete mirabile (REE-tee mir-AH-bi-lay; plural **retia mirabiles**). Intertwined veins and arteries conducting blood in opposite directions for the concentration of oxygen, heat, or ions.

reticular formation. Part of the brainstem that maintains wakefulness.

reticulopodium. Type of pseudopodium in certain protozoans (foraminiferans, class Granuloreticulosea) that forms a food-trapping net.

retina. Structure of an eye that bears photoreceptors and other nerve cells for vision.

retinula. Photoreceptive region of an arthropod compound eye usually consisting of seven or eight retinula cells.

retrocerebral sac. Mucus-secreting structure of rotifers.

reuptake. Active transport of synaptic transmitter back into a synaptic knob.

Reynolds number (Re). Dimensionless number representing the ratio of inertial to viscous forces in swimming or flying. It increases with speed and body length.

R group. Part of an amino acid that differs from one amino acid to another.

rhabdite. Rod-shaped structure in epidermis of a turbellarian flatworm that helps produce mucus.

rhabdom. Microvilli in retinula of arthropod compound eye that contain the photopigment.

rhabdomeric photoreceptor. Type of photoreceptor cell in most invertebrates organized around microvilli or membrane lamellae, rather than cilia as in vertebrates.

Rh factor. Protein antigen on red blood cells that partly determines whether donated blood is compatible. Presence or absence designated by + or − following the ABO type.

rhodopsin. Visual pigment in vertebrate rods.

rhombencephalon (= hindbrain). Part of the brain that forms the cerebellum, pons, and medulla.

rhopalium (= tentaculocyst). Rod-shaped structure enclosing a statocyst organ and sometimes ocelli in a jellyfish.

rhynchocoel (RING-ko-SEAL). Cavity that encloses the proboscis of a nemertean.

ribonucleic acid (RNA). Nucleic acid with ribose as its sugar. It is generally single stranded and occurs as messenger RNA, ribosomal RNA, and transfer RNA.

ribose. Five-carbon sugar in ribonucleic acid, ATP, and cyclic AMP.

ribosome. Organelle consisting of protein and ribosomal RNA (rRNA) in the cytoplasm and on rough endoplasmic reticulum, which organizes the synthesis of proteins.

rigor. Cramping of muscle due to lack of ATP; especially rigor mortis, which occurs after death.

ring canal. Part of echinoderm water vascular system that encircles the mouth.

Ringer's solution. Solution containing balanced concentrations of Na^+, K^+, Ca^{2+}, and other ions for replacement of body fluids. Especially such a solution for use with amphibians.

rod. Vertebrate photoreceptor cell responsible for vision in dim light.

rotational cleavage pattern. Cleavage pattern in which one pair of cells formed by the second cleavage lies at right angles to the other pair of cells. Characteristic of placental mammals.

rough endoplasmic reticulum (rough ER). Cytoplasmic membranes bearing ribosomes.

royal jelly. Sugary substance that worker honey bees feed to female larvae to induce them to develop into queens.

rumen. Large fore-stomach of a ruminant in which plant fibers are initially digested.

ruminant. Animal such as cow, sheep, and deer that has a rumen.

rumination. Regurgitation of food from the stomach for further chewing.

saccule. Mechanoreceptor in inner ear that detects position of head.

sagittal. Along the vertical midline of the head; e.g., sagittal crest.

salinity. Measure of total solute concentration, especially in seawater.

saliva. Secretion from salivary glands into the mouth.

saltatory conduction. Conduction in myelinated axons, in which action potentials "jump" from one node of Ranvier to the next.

salt gland. Gland in marine vertebrates that secretes excess salt, usually from the eye sockets or nostrils.

sarcodine. Protozoan that characteristically forms pseudopodia.

sarcoma. Cancer that arises from bone, cartilage, fat, muscle, or other tissues of mesodermal origin.

sarcomere. Subunit of a skeletal muscle fiber representing the structure between two Z discs.

sarcoplasmic reticulum. Membranous organelle of muscle that controls contraction by releasing and storing Ca^{2+}.

satellite male. Male, especially frog, that stays near a courting male and intercepts females he has attracted.

saturated fat. Neutral fat lacking double bonds and thus having two hydrogens bonded to each carbon in the fatty acid chain.

saturation effect. Advantage of group living or mass hatching resulting from the inability of predators to eat more than a small proportion of the group at one time.

scalid. Circle of spines on the proboscis in certain invertebrates.

scaling (= allometry). Differential effects of changing linear, area, and volume dimensions in organisms.

schizocoelous. Having a coelom formed by division of mesoderm.

schizogony. Multiple fission, especially in certain protozoans.

schizont. Multinuclear form of *Plasmodium* or other protozoan that forms from a sporozoite and undergoes multiple fission (schizogony).

sclerite. Platelike piece of sclerotized cuticle.

sclerocyte (= scleroblast). Amebocyte that forms spicules in a sponge.

sclerotization (= tanning). Hardening and darkening of arthropod cuticle.

scolex (= holdfast). Adhesive organ of a tapeworm.

scute. Horny or bony scale, such as that on a turtle or other reptile.

scyphistoma. Polyp of a jellyfish just after settling.

sebaceous gland. Skin gland that produces oily sebum.

second messenger. Intracellular chemical that mediates action of a hormone or other stimulus in a cell.

secondary consumer. Animal that consumes a primary consumer.

secondary palate. Bony structure separating nasal and oral passages in mammals.

secondary structure. Conformation of part of a protein due to hydrogen bonding between adjacent parts of a protein strand. Secondary structure takes two common forms: α-helix and β-sheet.

secondary succession. Changes in species composition in an area already inhabited.

secretion. (1) Active transport of a substance from body fluid into an organ for excretion. (2) Movement of a substance out of a gland.

sedentary. Not moving.

segment (= somite = metamere). Linearly repeated subdivision of a body in Pogonophora, Annelida, Arthropoda, and Chordata.

segmentation (= metamerism). Division of body into segments.

segmentation mutation. Mutation in insects that affects the number or pattern of segments.

segregation. Independence of expression of one allele regardless of presence of other alleles.

semen. Fluid containing sperm.

semicircular canals. Three canals in inner ear that respond to rotational acceleration of the head.

semiconservative replication. Duplication of DNA in such a way that each new molecule conserves one strand of the old DNA.

semilunar valve. Valve in the aorta or pulmonary artery that prevents regurgitation of blood into the ventricle.

seminal vesicle. Organ of male vertebrates that secretes fructose-rich fluid into semen.

seminiferous tubule. Tubule in testis that produces sperm.

semipermeable. Referring to a membrane that is permeable to some substances but not to others.

semiplume. Feather intermediate between a vane feather and down, lacking hooks on the barbules but with a rachis.

sensitive period. Period during which an organism is subject to a certain effect.

sensory neuron (= afferent neuron). Nerve cell that conducts action potentials from sensory receptors into central nervous system.

septum. Sheet of tissue dividing two chambers, such as two ventricles of the heart.

sequencing. Determining the sequence of amino acids in a protein or bases in a nucleic acid.

serial homology. Derivation of structures from corresponding parts of different segments in an animal. For example, the antenna of an insect is serially homologous to a leg.

serous gland (= poison gland = granular gland). Poison-secreting gland in skin of amphibians.

serum. Fluid part of clotted blood that remains after cells and the clot are removed.

serum albumin. Proteins in blood that are largely responsible for its osmotic pressure.

servomechanism. System that uses its own output for self-correction.

sessile. Attached to substratum.

seta (plural **setae**). Bristle, usually chitinous, on an annelid, arthropod, or other invertebrate.

set point. Hypothetical temperature, pH, or other level that is maintained by homeostasis.

settling. Process by which a free-swimming larva attaches to substratum prior to developing into a sedentary adult.

sex. (1) Reproductive process by which a new individual develops from gametes produced by two parent individuals. (2) Gender (male or female).

sex chromosome. Chromosome that occurs in different numbers in different sexes. Chromosome that is not an autosome.

sex linkage. Location of a gene on a sex chromosome, making it likely to be transmitted differently to different sexes.

sexual selection. Differential reproduction due to selection of one mate by the other.

sheet web. Spider web built horizontally and often leading to a funnel-shaped retreat in which the spider awaits prey.

sickle cell. Deformed red blood cell resulting from a point mutation in a gene for hemoglobin. Causes sickle cell trait in heterozygotes, and sickle cell anemia in homozygotes.

sidewinding. Mode of locomotion used by snakes on sand, in which the body is formed into a spiral at right angles to the direction of locomotion.

sign stimulus. Signal from one animal to another that evokes a stereotyped behavioral response (fixed action pattern).

siliceous. Containing silica (silicon dioxide).

simple epithelium. Epithelium with one cell layer.

simple eye. Light-detecting organ with a lens that concentrates light without the ability to resolve an image.

sinoatrial (SA) node. Cluster of cells in the right atrium that normally functions as the pacemaker of the vertebrate heart.

sinus venosus. Chamber between the vena cava and the right atrium of most vertebrates and in fetus of birds and mammals.

siphonoglyph. Ciliated groove in the mouth of sea anemones that generates hydrostatic pressure.

siphuncle. Strand of tissue that empties fluid from the chambers of *Nautilus*.

skin. (1) Integument of vertebrates, consisting of dermis and epidermis. (2) Soft integument of any animal.

sliding filament theory. Theory that thin and thick muscle filaments slide among each other and that the force of muscle contraction comes from the interaction of the two types of filament.

slit sensillum. Mechanoreceptor in the cuticle of arachnids arranged so that tension in a certain direction generates a response by opening or closing a slit.

slow-twitch fiber (= type I red fiber). Muscle fiber that contracts slowly and metabolizes aerobically.

smolt. Young salmonid fish at time of migration from native stream.

smooth endoplasmic reticulum (= smooth ER). ER without ribosomes, generally responsible for packaging glycoproteins from rough ER into vesicles to be processed by Golgi complex.

smooth muscle. Involuntary muscle that lacks striations.

sociobiology. Biological study of animal

behavior that emphasizes the evolution of social behavior.

solenocyte. Flagellated type of flame bulb that drives filtrate in certain excretory organs (protonephridia).

solubility. Amount of a substance that can be dissolved in a given volume of solvent.

solute. Substance dissolved in another substance.

solution. Liquid in which something is dissolved.

solvent. Liquid such as water that dissolves another substance.

somatoplasm. Body of an organism exclusive of the germ plasm that is responsible for transmission of hereditary information.

somatosensory (= somesthetic). Referring to touch, pain, and other body sensations.

somesthetic. Somatosensory.

somite (= segment = metamere). (1) Linearly repeated subdivision of a body in a segmented animal (Pogonophora, Annelida, Arthropoda, or Chordata). (2) In vertebrates, one of 40 masses of embryonic mesoderm.

spat. Juvenile oyster with protective shells before attachment to substratum.

speciation. Evolution of a new species.

species (singular and plural). (1) For sexual organisms, "groups of actually or potentially interbreeding natural populations, which are reproductively isolated from other such groups" (Ernst Mayr). For asexual organisms, groups of organisms that are similar to each other but recognizably different from other such groups. (2) Taxonomic category below the genus or subgenus level, which may or may not correspond to the biological definition above.

sperm (singular and plural; = **spermatozoon**). One or more male gametes.

spermatheca. Sperm reservoir in a female.

spermatid. Haploid, immature sperm cell that has not acquired its flagellum.

spermatocyte. Sperm cell in the process of meiosis.

spermatogenesis. Development of sperm.

spermatogonium. Primordial germ cell that eventually develops into sperm.

spermatophore (sperm packet). Packet or globule of sperm produced by certain male invertebrates and salamanders.

spermiogenesis. Maturation of a spermatid into a sperm cell.

sperm receptacle. Reservoir for sperm in a female crab.

sperm web. Special web on which a male spider deposits a drop of semen to fill the bulbs of its pedipalps.

S phase. Part of interphase in the cell cycle during which DNA is replicated.

sphincter. Ring of muscle that controls flow through an aperture or tube.

spicule. Needle-shaped mineral particle that helps form the endoskeleton of a sponge or sea cucumber.

spindle apparatus. Arrangement of microtubules radiating from centrioles to chromatids during mitosis and meiosis.

spinneret. Tubule on abdomen of a spider from which silk is extruded.

spiracle. (1) External opening of an insect trachea. (2) Modified first gill opening of a shark. (3) Excurrent channel for tadpole gills.

spiral cleavage. Cleavage pattern in which daughter cells of cleavage move spirally into the furrows between other cells. Characteristic of many invertebrates other than arthropods and echinoderms.

spiral valve. (1) Helical membrane in intestine of sharks and primitive fishes. (2) Membrane that separates oxygenated from deoxygenated blood in conus arteriosus of amphibian heart.

spongin (SPUN-jin). Fibrous protein that helps form the endoskeleton of a sponge.

spongocoel (SPUN-jo-seal) (= atrium). Central chamber in an asconoid or syconoid sponge.

spongy (= cancellous = trabecular) **bone.** Bone with spaces containing marrow.

spontaneous generation. Supposed production of an organism without parents.

spore. Inactive, resistant form of a zygote. In certain protozoans, the product of an oocyst.

sporocyst. Larval form of *Schistosoma* and other flukes that lacks a mouth and digestive tract and absorbs nutrients from the intermediate host.

sporogony. Development of numerous spores or sporozoites within oocyst of malaria-causing *Plasmodium* and certain other protozoans.

sporozoan. Protozoan that reproduces from spores.

sporozoite. Form of malaria-causing *Plasmodium* and some other protozoans that may form from a spore and develop into a schizont.

squalene. Low-density fat that maintains buoyancy in sharks.

squamous epithelium. Epithelium with flattened cells.

stapes (stay-peez). A middle-ear bone. The columella in amphibians and reptiles.

Starling effect. Balance of blood pressure and colloid osmotic pressure in capillaries that prevents fluid from accumulating in tissues.

statoblast. Resistant form from which a freshwater ectoproct reproduces a new zooid asexually.

statocyst. In crustaceans and certain other aquatic animals a hollow mechanoreceptor organ containing granules that determines direction of gravity.

statolith. Dried secretion or sand grains in statocyst organs, which deflect hair cells in response to gravity.

stem cell (= founder cell). Embryonic cell from which certain other cells develop.

stenohaline. Able to tolerate slight changes in environmental osmolarity. Compare euryhaline.

stercoral pocket. Receptacle for metabolic wastes just dorsal to the rectum in a spider.

stereoblastula. Solid blastula that lacks a blastocoel.

sternal keel (= carina). Long, thin process on the sternum of a bird, to which flight muscles are attached.

sternum. (1) Vertebrate breastbone. (2) Ventral surface of an arthropod.

sternite. Cuticular plate (sclerite) on ventral surface (sternum) of an arthropod.

steroid. Four-ringed lipid such as cholesterol, vitamin D, or a steroid hormone.

stigma (plural **stigmata**). (1) "Eyespot" of *Euglena* or other protozoan. (2) Type of spiracle on certain insects.

stoma (plural **stomata**). A mouth.

stomach. Enlargement of the digestive tract for storage and digestion of food.

stomochord (= buccal diverticulum). Anterior projection of the buccal cavity of a hemichordate, formerly considered a notochord.

stone canal. Tube that brings water from the madreporite into the rest of the water vascular system in an echinoderm.

stratified epithelium. Epithelium with more than one cell layer. Compare simple epithelium.

stress. Activation of the adrenal cortex and sympathetic nervous system.

stretch receptor. Mechanoreceptor that responds to stretch, especially in skeletal muscle.

stretch reflex. Self-correcting contraction of a skeletal muscle in response to sudden stretching.

striated muscle. Skeletal muscle, which has repeating subunits producing a striped pattern.

strobila. (1) Stage in development of jellyfish that forms a series of free-swimming larvae (ephyrae). (2) Chain of tapeworm proglottids.

stroke volume. Volume of blood pumped by one contraction of the left ventricle heart.

structural color. Color due to structure, rather than to pigment, that alters the transmission or reflection of light.

structural white. White due to dust-sized, transparent particles that scatter all wavelengths equally.

stylet. Sharp projection, such as on the proboscis of a nemertean or tardigrade.

subepidermal nerve plexus. Portion of the nervous system of a flatworm that is independent of the cerebral ganglia.

subjunctional fold. Fold in the plasma membrane of a muscle cell opposite the neuromuscular junction.

submucosal plexus. Network of nerve cells in the gut that coordinates motility.

substrate. (1) Molecule that an enzyme binds to and catalyzes a reaction in. (2) Substratum.

substrate-adhesion molecule (SAM).

Glycoprotein such as fibronectin that guides movement of cells during development.

substratum (= substrate). Surface on which something rests.

succession. Change in species inhabiting an area.

summation. Additive effect of multiple postsynaptic potentials or of muscle twitches.

supercooling. Reduction in temperature below the nominal freezing point without formation of ice crystals.

superficial cleavage pattern. Cleavage pattern in which blastomeres are confined to the periphery of the egg. Characteristic of arthropods.

suppressor gene. Gene whose product normally prevents abnormal growth of tissue (a tumor).

suprachiasmatic nucleus (SCN). Part of the hypothalamus above the optic chiasm that apparently controls the pineal gland's influence on biological rhythms in a mammal.

suspension feeding. Feeding on particles or small animals suspended in water.

swelling (= edema). Accumulation of fluid in tissues due to inflammation, tissue damage, or other cause.

swim bladder. Gas-filled chamber that maintains buoyancy in many fishes.

swimmerets (= pleopods). Appendages on abdominal segments 14 through 18 of decapod crustaceans.

syconoid. Type of canal system in a sponge in which choanocytes do not line the spongocoel but are found within radial canals.

symbiosis. Close association between members of different species. Includes mutualism, parasitism, and commensalism.

symmetry. Complementarity of shape such that an animal can be bisected into two similar halves.

sympathetic divison. Part of the autonomic nervous system that is active during stress ("fight or flight") situations.

sympatric. In the same area. Referring especially to evolution in which a population becomes reproductively divided into two species even though they are not geographically separated.

symplesiomorphy. In cladistics a primitive character shared by two groups that is useless in classifying the two groups because it is also shared by the outgroup.

synapomorphy. In cladistics a trait not found in the outgroup but homologous in two groups being classified. The only traits useful for classification in cladistics.

synapse. Structure by which a nerve cell excites or inhibits another nerve cell or a muscle cell.

synapsid. Reptile or other vertebrate with one temporal opening.

synapsis. Alignment of homologous chromosomes side by side during meiosis.

synaptic knob. Enlarged terminal of a nerve cell associated with a chemical synapse.

synaptic transmitter (= neurotransmitter). Chemical released from a synapse that tends to either excite or inhibit a nerve or muscle cell.

synchronous muscle. Muscle in insects in which contraction is synchronous with action potentials in the muscle. Compare asynchronous muscle.

syncytium (sin-SISH-ee-um). Multinucleate cell, usually resulting from the fusion of two or more cells.

syngamy. Fertilization of one gamete by another in sexual reproduction, especially in protozoans.

synthetic theory (= neo-Darwinian theory). Current theory that combines Darwin's theory of natural selection with modern knowledge of genetics to explain most evolution.

syrinx (plural **syringes** or **syrinxes**). Voice-producing structure in the trachea of a bird.

systemic circulation. Circulation of blood to metabolizing tissues, as opposed to respiratory circulation.

systolic blood pressure. Peak blood pressure, which follows ventricular contraction.

taenidium. Spiral strand of chitin that reinforces insect trachea.

tagma (plural **tagmata**). Group of body segments forming a distinct region, such as head, thorax, or abdomen.

tagmosis (= tagmatization). Development of distinct body regions (tagmata) in segmented animals.

tanning (= sclerotization). Hardening and darkening of arthropod cuticle.

tapetum. Reflecting layer in the eyes, especially in nocturnal animals.

taste. Chemoreception of substances not airborne.

taste bud. Structure on tongue bearing chemoreceptors for taste.

taxis (plural **taxes**). Orientation of an animal with respect to direction of a stimulus.

taxon (plural **taxa**). Category such as phylum, order, or species in which organisms are classified.

T **cell.** Type of lymphocyte responsible for destroying damaged cells.

tectorial membrane. Tissue in organ of Corti that deflects hair cells in response to sound.

tegument. Outer covering of parasitic flatworm (tapeworm or fluke) or acanthocephalan, consisting of hardened portions of living cells partially submerged within parenchyme.

telolecithal. Egg with abundant, unevenly distributed yolk.

telophase. Final stage of mitosis or meiosis in which chromosomes are separated to opposite poles and begin to disperse prior to cell division.

telson. (1) Posterior projections of a decapod crustacean that combines with the uropod to make a fin. (2) Spinelike posterior projection of a horseshoe crab.

temperature. Measure of the average energy associated with the thermal agitation of a molecule.

tendon. Connective tissue that attaches muscle to bone.

tentacle. Flexible, elongated, unsegmented structure usually used in feeding.

tentaculocyst (= rhopalium). Rod-shaped structure enclosing a statocyst organ and sometimes ocelli in a jellyfish.

teratogen. Substance that causes congenital defects.

tergite. Cuticular plate (sclerite) on dorsal surface (tergum) of an arthropod.

tergum. Dorsal surface of an arthropod.

termination sequence. DNA base sequence that stops transcription.

territory. Area from which one animal excludes others of the same species.

tertiary structure. Shape of a protein determined by interactions among R groups.

test. A hard case, such as the shell of a sea urchin.

testis (plural **testes**). Organ that produces sperm.

testis determining factor (TDF). Substance produced by a gene on the Y chromosome that induces development of testes rather than ovaries.

tetrad Four chromatids joined in synapsis during prophase I of meiosis.

tetrapod. Vertebrate belonging to a group that generally has four limbs (amphibian, reptile, bird, or mammal).

thalamus. Part of midbrain consisting of nuclei and neural tracts linking parts of the cerebrum with each other and with other parts of nervous system.

theca (THE-kuh; plural **thecae** THE-see). (1) A case, covering, receptacle, or sheath. (2) Overwintering cyst of a hydra.

thermal hysteresis protein (THP). Protein in insects that lowers the freezing point of tissues without affecting the melting point.

thermal stratification. Layering of temperate lakes resulting from differences in water density due to temperature.

thermogenesis. Production of body heat, especially to raise body temperature.

thermoneutral zone. Range of environmental temperatures in which the metabolic rate of a homeotherm is at minimum.

thickener cell. Sclerocyte that helps form spicules in a sponge.

thick filament. Filament of the protein myosin in muscle.

thin filament. Filament of the protein actin in muscle.

threshold. Minimum stimulus required for a response such as an action potential.

thrombin. Protein made from prothrombin during the process of blood clotting.

thrombosis. Blood clotting that interferes with circulation.

thrust. Force generated during flight that pushes an animal forward against friction and pressure drag.

thymine. Nitrogenous base in DNA.

tight junction. Band joining one cell to surrounding cells that blocks the passage of substances between them.

tissue. Group of interconnected cells structurally and functionally related to each other.

tissue culture. Technique of maintaining cells outside the body (*in vitro*).

tonofilament. Intermediate filament of keratin that anchors a desmosome to the cell contents.

top carnivore. Carnivorous animal that is not, itself, prey to another carnivore.

tornaria larva. Larva characteristic of acorn worms, which spins as it swims.

torpor. State of inactivity due to low body temperature.

torsion. Twisting of veliger larva of a snail or other gastropod mollusc into a U shape.

toxin. Substance released by one organism that is harmful to another.

trabecular (= cancellous = spongy) **bone.** Bone with spaces containing marrow.

trace element. Inorganic nutrient required in minute amount.

trachea (plural **tracheae**). (1) Tube connecting mouth and nose with the bronchi in a vertebrate. The windpipe. (2) In insects and spiders, a tube through which air diffuses to tissues.

tracheal gill. Area of aquatic insect that is rich in tracheae adapted for absorbing O_2 underwater.

tracheole. Smallest branch of a trachea in tissue of an insect or spider.

transcription. Production of messenger RNA with a base sequence complementary to one strand of DNA in a gene.

transducin. Protein that removes cGMP during the response of rods and cones to light.

transfer RNA (tRNA). RNA that binds a particular amino acid and inserts it where called for by messenger RNA during protein synthesis.

translation. Synthesis of protein as directed by messenger RNA.

translocation. Change in location, such as movement of part of a chromosome to another location on the same or another chromosome.

transposition. Movement of a submicroscopic segment of a chromosome (transposing element = jumping gene) to another chromosome or to a different position on the same chromosome.

triacylglycerol (= triglyceride = neutral fat). Lipid consisting of three fatty acids linked to a glycerol.

tricarboxylic acid cycle (= TCA cycle = Krebs citric acid cycle). Cycle of reactions in mitochondria by which

pyruvate from glycolysis is oxidized to carbon dioxide during cellular respiration.

trichocyst. Hair-shaped defensive structure discharged from the pellicle of certain protozoans (ciliates).

tricuspid valve. Right atrioventricular valve of the heart.

triglyceride (= triacylglycerol = neutral fat). Lipid consisting of three fatty acids linked to a glycerol.

triose. Simple sugar with three carbons.

triploblastic. Developing from three germ layers, usually ectoderm, mesoderm, and endoderm.

trochophore larva. Free-swimming larva with cilia encircling the mouth. Characteristic of molluscs and some marine flatworms, annelids, brachiopods, and others.

trophic level. Level of an animal in a food web or food chain that depends on the number of energy transfers between the animal and the primary producers. Examples: primary consumer, secondary consumer.

trophoblast. Cells surrounding the blastocyst that are responsible for implantation into the uterine lining.

trophosome. Chamber within certain pogonophorans containing symbiotic chemoautotrophic bacteria.

trophozoite. Mature, feeding stage of a parasitic protozoan.

tropomyosin. Protein on actin that blocks cross-bridge formation in vertebrate skeletal muscle.

troponin. Protein on actin that binds Ca^{2+} and triggers contraction in vertebrate skeletal muscle.

trypsin. A protein-digesting enzyme.

t tubule. Transverse tubule in muscle that conducts action potentials into sarcomeres and triggers contraction.

tube foot (= podium). Hollow grasping organ of an echinoderm controlled by the water vascular system and muscles.

tuberous organ. High-frequency electroreceptor of certain fishes.

tubules of Cuvier. Sticky strands ejected by sea cucumbers when disturbed.

tubulin. A protein in microtubules.

tumor. Mass of cells resulting from inappropriate cell division.

tun. Barrel-shaped body of a tardigrade in cryptobiosis.

tunic. (1) Collagenous sheath reinforcing radial and circular muscles of a cephalopod mollusc. (2) Sheath enclosing the body of a tunicate.

tusk. Greatly enlarged tooth of a mammal such as an elephant.

twitch. Response of a muscle to a single action potential.

tympanum. An eardrum or other vibrating membrane.

Tyndall blue. Structural color caused by scattering of light by submicroscopic particles.

typhlosole. Longitudinal infolding of the intestine of an earthworm.

ultrafiltrate. Fluid filtered across a capillary wall.

umbo (plural **umbones** um-BO-neez). Bulge in each valve at the hinge of a bivalve mollusc, or the beak of a brachiopod.

unconditioned stimulus. In classical conditioning, a stimulus to which an animal normally responds.

undulating membrane. (1) Long, finlike group of cilia (buccal ciliature) near the cytostome (mouth) of a protozoan. (2) Finlike fold in pellicle and the flagellum of trypanosomes and related protozoa.

unequal crossing over. Failure of homologous chromosomes to break in exactly the same place during crossing over in meiosis, resulting in gain or loss of genes in gametes.

uniformitarianism. Concept that geological changes in the past occurred in the same way as those occurring now.

uniramous. Unbranched, like legs of insects and other members of subphylum Uniramia.

unisexual species. Species in which only parthenogenetic females occur.

upwelling. Rise of cold, nutrient-rich water to the surface due to prevailing local winds or water currents.

urea. Nitrogenous waste excreted mainly by mammals and other viviparous animals.

ureotelic. Excreting urea as the major nitrogenous waste.

ureter. Tube conducting urine from the kidney to the urinary bladder.

urethra. Tube through which urine is voided from the bladder.

uric acid. Main nitrogenous waste in egg-laying terrestrial animals.

uricotelic. Excreting uric acid as the main nitrogenous waste.

urogenital aperture. Aperture through which urine and gametes are released by certain fishes.

urogenital system. Excretory and reproductive systems with shared parts.

uropods. Last pair of segmental appendages of a decapod crustacean, which combine with the telson into a fin.

uropygial gland (= oil gland). Gland at the base of the tail of a bird that produces oil for preening.

urostyle. Fused vertebrae of a frog.

uterus (= womb). (1) Cavity in which embryos develop in a viviparous animal. (2) In some nonviviparous animals, a cavity in which eggs are retained temporarily.

utricle. Mechanoreceptor organ in inner ear that detects head position.

vaccination (= inoculation). Stimulating immunity by exposing an animal to the antigen of a virus or bacterium.

vagal inhibition. Reduction in heart rate by the vagus nerve.

vagina. Receptacle for the penis during copulation.

vagus nerve. Major branch of the

parasympathetic division of the autonomic nervous system.

valve. (1) One of two shells in a bivalve mollusc or brachipod. (2) Device that controls flow of a fluid, such as a valve in the heart.

vane feather (= contour feather). Major type of feather, in which barbs form two vanes on opposite sides of the shaft.

vaporization. Conversion of a substance from a liquid to a gas phase.

variety. Race or other genetically different group within a species.

vas deferens (plural **vasa deferentia**) (= sperm duct = ductus deferens). Tube through which sperm are ejaculated.

vasectomy. Sterilization by cutting and tying the vasa deferentia.

vas efferens (plural **vasa efferentia**). In flatworms and some other invertebrates, a tubule conducting sperm into the vas deferens.

vasoconstriction. Narrowing of arteries due to contraction of smooth muscles.

vasodilation. Widening of arteries due to relaxation of smooth muscles.

vegetal pole. Part of ovum containing more yolk than the animal pole.

vein. (1) Blood vessel that circulates blood from capillaries to heart. (2) Tubule in insect wing that circulates blood and contains nerves.

veliger. Larval form of certain molluscs that occurs between the trochophore larva and the adult.

velum. (1) Infolding flap on the edge of the bell of a hydrozoan medusa. (2) Ciliated swimming organ of a molluscan veliger larva. (3) Ventral surface of *Xyloplax* (Echinodermata: Concentricycloides). (4) Muscular membranes that pump water into gills of a jawless fish.

vena cava. Largest vein in the vertebrate, which returns blood to the heart.

venous pump. Forcing of blood toward heart due to one-way valves in veins and contraction of skeletal muscle.

vent. External opening of a cloaca.

ventral. Pertaining to the lower or front surface. Compare dorsal.

ventral root. Ventral branch of nerve from spinal cord, which contains motor neurons.

ventricle. (1) Chamber of the heart that pumps blood into an artery. (2) Chamber in the brain filled with cerebrospinal fluid.

venule. Small vein connected to a capillary.

vermiform. Worm-shaped.

vertical migration. Daily rhythm of vertical movements of zooplankton, usually down at dawn and to the surface at night.

vertical stratification. Systematic vertical differences in distribution of abiotic factors, such as light, humidity, temperature, and wind speed.

vesicle. Spherical, membrane-bound structure enclosing material being transported into a cell by endocytosis, or material (digestive enzyme, synaptic

transmitter, or hormone) being transported out of the cell by exocytosis.

vestibular apparatus. Mechanoreceptive organs that detect head motion and orientation, consisting of semicircular canals and otolith organs (utricle and saccule).

vestigial. Referring to a functionless structure that may have evolved from a useful structure in an ancestral species.

vibrissa (= whisker). Stiff facial hair adapted as a touch receptor.

vicariance. Reproductive isolation of two populations in a species by geological change or by extinction of members of species between the two populations.

villus. Finger-shaped microscopic projection of a tissue layer, especially of the small intestine or the placenta.

visceral mass. Fleshy part of a mollusc consisting of the foot and mantle.

viscid gland. Part of the duogland system of flatworms that secretes an adhesive for attachment and locomotion.

vitalism. Belief that organisms have powers unique to life.

vitamin. Organic molecule required in the diet for metabolism, but which is not itself metabolized.

vitellaria. (1) Larvae of sea lilies. (2) Plural of vitellarium.

vitellarium (= yolk gland). Gland that produces yolk and, in many flatworms, eggshell.

vitelline envelope. Covering of echinoderm egg between the plasma membrane and the jelly coat.

vitellogenesis. Production of proteins (vitellogenins) and other constitutents of yolk.

viviparity. Birth of live young that have been nurtured during development within the reproductive tract of the mother. Compare ovoviviparity and oviparity.

vocal sac. Skin pouch on a male frog that amplifies the advertising call.

vomerine. Relating to the flat bone that forms part of the septum between the nostrils; for example, the vomerine teeth on the palate of some fishes, amphibians, and reptiles.

vomeronasal organ (= Jacobson's organ). Olfactory organ in the roof of the mouth of an amphibian, reptile, or mammal.

von Baer's law. Proposal that embryonic development in vertebrates goes from general forms common to all vertebrates to increasingly specialized forms characteristic of classes, orders, and lower taxonomic levels.

vulva. Outer chamber leading to female reproductive tract.

waggle dance. Form of tactile communication by which a scout honey bee communicates the distance and direction of food to other worker bees.

warning coloration. Highly visible coloration of toxic species that deters predation.

water cycle (= hydrologic cycle). Biogeochemical cycling of water among various organic and inorganic states.

water expulsion vesicle (WEV; =contractile vacuole). Osmoregulatory organelle in protozoa and freshwater sponges.

water vascular system. Network of water-filled canals in an echinoderm.

wax. Long-chain fatty acid linked to a long-chain alcohol or carbon ring.

wax canal. Channel through arthropod cuticle believed to transport wax to the surface.

Weberian ossicle. Bone by which swim bladder is believed to transmit sound to semicircular canals and otolith organs of some fishes.

white blood cell (= leukocyte). Blood cell important in immunity.

white matter. Portions of the brain or spinal cord consisting largely of myelinated axons.

wing loading. Body mass divided by wing area.

withdrawal reflex. Simple reflex that pulls a limb away from injury.

Wolffian duct. Duct of the embryonic kidney (mesonephros) that develops into part of the reproductive system in males.

worker jelly. Sugary substance that worker honey bees feed to larvae.

xanthophore. Type of chromatophore that contains red, orange, or yellow pigment.

X organ–sinus gland (XOSG) complex. Glands in crustacea that secrete molt-inhibiting hormone and several other hormones.

yolk. Nutrient deposit in an egg.

yolk sac. Membrane enclosing the yolk in developing vertebrates.

Y organ. Gland in crustaceans that secretes molting hormone.

Z disc. Structure at each end of a sarcomere in skeletal muscle.

zoea. (ZO-ee-uh). Larval stage of certain crabs.

zoecium (zo-EE-she-um). Cuticular chamber enclosing an individual zooid (polypide) of an ectoproct.

zona pellucida. Glycoprotein covering a mammalian egg.

zona reaction. Response to fertilization triggered by the cortical reaction, which blocks polyspermy by hardening the zona pellucida and preventing the binding of sperm.

zooid (ZO-oyd). Individual in a colony of animals or protozoans.

zooplankton (singular **zooplankter**). Aquatic larvae and animals suspended in water.

zooxanthella (ZO-oh-zan-THELL-uh). Photosynthesizing protist (dinoflagellate) that lives as a mutualist within a sponge, coral, clam, or other animal.

zygote. Fertilized ovum.

Illustration Credits

Frontmatter

Page ii: Doug Allari/Oxford Scientific Files/Animals Animals; *Page vii (top):* Richard G. Kessel and R. H. Kardon. *Tissues and Organs: A Text-Atlas of Scanning Electron Microscopy.* W. H. Freeman, p. 37; *Page vii (center):* Tom J. Ulrich/VU[*]; *Page vii (bottom):* Denise Tackett/TS[†]; *Page viii:* Dave B. Fleetham/TS; *Page ix (top):* C. Leon Harris; *Page ix (bottom):* Reproduced from K. Tanaka, A. Iino, and T. Naguro (1976). *Arch. Histol. Jap.* 39:165–175 by permission of the Japan Society of Histological Documentation; *Page x (center):* Robert Winslow/TS; *Page xi (top):* Carlo Bevilacqua/CEDRI; *Page xiii (bottom):* Dr. Warren W. Burggren, Department of Zoology, University of Massachusetts, Amherst; *Page xiv (top):* Dwight R. Kuhn; *Page xv:* C. Leon Harris; *Page xvi (top):* Allan Roberts; *Page xvii:* Painting by Rudolf Freund/Courtesy Carnegie Museum of Natural History, Pittsburgh; *Page xviii (bottom):* Jeff Foott/Bruce Coleman Inc.; *Page xix (top):* Alan Blank/Bruce Coleman Inc.; *Page xix (bottom):* Reproduced from L.E. Roth, D.J. Pihlaja, and Y. Shigemaka (1970). *J. Ultrastruct. Res.* 30:7–37 by copyright permission of Academic Press, Inc., Orlando, Fl.; *Page xx (top):* Larry Roberts/VU; *Page xx (bottom):* The University of Alberta Archives, D. M. Ross Collection; *Page xxi (top):* Ed Robinson/TS; *Page xxi (bottom):* R. Calentine/VU; *Page xxii (top):* J.D. Wrobel/Monterey Bay Aquarium/Biological Photo Service; *Page xxiii (top):* William H. Amos; *Page xxiii (bottom):* Richard Walters/VU; *Page xxiv (top):* Joe McDonald/TS; *Page xxv (bottom):* Dwight R. Kuhn; *Page xxvi:* Ed Robinson/TS; *Page xxviii (top):* D.G. Barker/TS.

Line Drawings

p. 39, Figure 2.21: From *Life: An Introduction to Biology,* Third Edition, by William S. Beck, Karel F. Liem and Anne Roe Simpson. Copyright © 1991 by William S. Beck, Karel F. Liem and Anne Roe Simpson. Reprinted by permission of HarperCollins Publishers Inc.; *p. 67, 4.2:* From *Life: An Introduction to Biology,* Third Edition, by William S. Beck, Karel F. Liem and Anne Roe Simpson. Copyright © 1991 by William S. Beck, Karel F. Liem and Anne Roe Simpson. Reprinted by permission of HarperCollins Publishers Inc.; *p. 74, 4.7:* From *Life: An Introduction to Biology,* Third Edition, by William S. Beck, Karel F. Liem and Anne Roe Simpson. Copyright © 1991 by William S. Beck, Karel F. Liem and Anne Roe Simpson. Reprinted by permission of HarperCollins Publishers Inc.; *p. 76, 4.9B:* After *Genetics,* Second Edition, by Farnsworth. Reprinted by permission of HarperCollins Publishers; *p. 78, 4.12:* From *Life: An Introduction to Biology,* Third Edition, by William S. Beck, Karel F. Liem and Anne Roe Simpson. Reprinted by permission of HarperCollins Publishers Inc.; *p. 91, 5.8A–B:* After *Principles of Cell Biology* by Kleinsmith and Kish. Reprinted by permission of HarperCollins Publishers; *p. 92, 5.9:* From *Principles of Cell Biology* by Kleinsmith and Kish. Reprinted by permission of HarperCollins Publishers; *p. 99, 5.14:* After *Genetics,* Second Edition, by Farnsworth. Reprinted by permission of HarperCollins Publishers; *p. 108, 6.3:* After *Principles of Cell Biology* by Kleinsmith and Kish. Reprinted by permission of HarperCollins Publishers; *p. 113, 6.9B:* From *Life: An Introduction to Biology,* Third Edition, by William S. Beck, Karel F. Liem and Anne Roe Simpson. Copyright © 1991 by William S.

Beck, Karel F. Liem and Anne Roe Simpson. Reprinted by permission of HarperCollins Publishers Inc.; *p. 123, 6.21B:* After Figure 1, Chapter 6, *Developmental Biology,* Second Edition by S. F. Gilbert. Sunderland, MA: Sinauer Associates, Inc., 1988. Reproduced by permission; *p. 124, 6.22A:* From *Life: An Introduction to Biology,* Third Edition, by William S. Beck, Karel F. Liem and Anne Roe Simpson. Copyright © 1991 by William S. Beck, Karel F. Liem and Anne Roe Simpson. Reprinted by permission of HarperCollins Publishers Inc.; *p. 124, 6.22C:* Human embryo after *Developmental Stages in Human Embryos* by R. O'Rahilly and F. Muller. Publication 637, p. 97, Fig. 10.4, Specimen 7S-6330. Baltimore: Carnegie Institution of Washington, 1987. Reproduced by permission of the Carnegie Institution of Washington and R. O'Rahilly, M.D.; *p. 131, 6.29:* After *Genetics,* Second Edition by Farnsworth. Reprinted by permission of HarperCollins Publishers; *p. 141, 7.1B:* After *Introduction to the Human Body: The Essentials of Anatomy and Physiology,* Second Edition, by Gerard J. Tortora. Copyright © 1991 by Biological Sciences Textbooks, Inc. and A and P. Textbooks, Inc. Reprinted by permission of HarperCollins Publishers Inc.; *p. 156, 7.13A:* After *Introduction to the Human Body: The Essentials of Anatomy and Physiology,* Second Edition, by Gerard J. Tortora. Copyright © 1991 by Biological Sciences Textbooks, Inc. and A and P. Textbooks, Inc. Reprinted by permission of HarperCollins Publishers Inc.; *p. 156, 7.13B:* From *Life: An Introduction to Biology,* Third Edition, by William S. Beck, Karel F. Liem and Anne Roe Simpson. Copyright © 1991 by William S. Beck, Karel F. Liem and Anne Roe Simpson. Reprinted by permission of HarperCollins Publishers Inc.; *p. 157, 7.14:* After *Introduction to the Human Body: The Essentials of Anatomy and Physiology,* Second Edition, by Gerard J. Tortora. Copyright © 1991 by Biological Sciences Textbooks, Inc. and A and P. Textbooks, Inc. Reprinted by permission of HarperCollins Publishers Inc.; *p. 158, 7.15:* From *Life: An Introduction to Biology,* Third Edition, by William S. Beck, Karel F. Liem and Anne Roe Simpson. Copyright © 1991 by William S. Beck, Karel F. Liem and Anne Roe Simpson. Reprinted by permission of HarperCollins Publishers Inc.; *p. 159, 7.16A,C,D:* After *Introduction to the Human Body: The Essentials of Anatomy and Physiology,* Second Edition, by Gerard J. Tortora. Copyright © 1991 by Biological Sciences Textbooks, Inc. and A and P. Textbooks, Inc. Reprinted by permission of HarperCollins Publishers Inc.; *p. 161, 7.17:* After *Introduction to the Human Body: The Essentials of Anatomy and Physiology,* Second Edition, by Gerard J. Tortora. Copyright © 1991 by Biological Sciences Textbooks, Inc. and A and P. Textbooks, Inc. Reprinted by permission of HarperCollins Publishers Inc.; *p. 162, 7.18:* From *Life: An Introduction to Biology,* Third Edition, by William S. Beck, Karel F. Liem and Anne Roe Simpson. Copyright © 1991 by William S. Beck, Karel F. Liem and Anne Roe Simpson. Reprinted by permission of HarperCollins Publishers Inc.; *p. 163, 7.19A–B:* From *Life: An Introduction to Biology,* Third Edition, by William S. Beck, Karel F. Liem and Anne Roe Simpson. Copyright © 1991 by William S. Beck, Karel F. Liem and Anne Roe Simpson. Reprinted by permission of HarperCollins Publishers Inc.; *p. 168, 8.1A:* After *Introduction to the Human Body: The Essentials of Anatomy and Physiology,* Second Edition, by Gerard J. Tortora. Copyright © 1991 by Biological Sciences Textbooks, Inc. and A and P. Textbooks, Inc.

Reprinted by permission of HarperCollins Publishers Inc.; *p. 170, 8.2:* After *Introduction to the Human Body: The Essentials of Anatomy and Physiology,* Second Edition, by Gerard J. Tortora. Copyright © 1991 by Biological Sciences Textbooks, Inc. and A and P. Textbooks, Inc. Reprinted by permission of HarperCollins Publishers Inc.; *p. 172, 8.4:* From *Life: An Introduction to Biology,* Third Edition, by William S. Beck, Karel F. Liem and Anne Roe Simpson. Copyright © 1991 by William S. Beck, Karel F. Liem and Anne Roe Simpson. Reprinted by permission of HarperCollins Publishers Inc.; *p. 175, 8.8:* After *Introduction to the Human Body: The Essentials of Anatomy and Physiology,* Second Edition, by Gerard J. Tortora. Copyright © 1991 by Biological Sciences Textbooks, Inc. and A and P. Textbooks, Inc. Reprinted by permission of HarperCollins Publishers Inc.; *p. 178, 8.10B:* From *Life: An Introduction to Biology,* Third Edition, by William S. Beck, Karel F. Liem and Anne Roe Simpson. Copyright © 1991 by William S. Beck, Karel F. Liem and Anne Roe Simpson. Reprinted by permission of HarperCollins Publishers Inc.; *p. 198, 10.1:* After *Introduction to the Human Body: The Essentials of Anatomy and Physiology,* Second Edition, by Gerard J. Tortora. Copyright © 1991 by Biological Sciences Textbooks, Inc. and A and P. Textbooks, Inc. Reprinted by permission of HarperCollins Publishers Inc.; *p. 201, 10.3C:* From *Life: An Introduction to Biology,* Third Edition, by William S. Beck, Karel F. Liem and Anne Roe Simpson. Copyright © 1991 by William S. Beck, Karel F. Liem and Anne Roe Simpson. Reprinted by permission of HarperCollins Publishers Inc.; *p. 203, 10.5:* From *Life: An Introduction to Biology,* Third Edition, by William S. Beck, Karel F. Liem and Anne Roe Simpson. Copyright © 1991 by William S. Beck, Karel F. Liem and Anne Roe Simpson. Reprinted by permission of HarperCollins Publishers Inc.; *p. 206, 10.8:* After figure by Bunji Tagawa from "The Response to Acetylcholine" by Henry A. Lester, *Scientific American,* February 1977, p. 108, a and b. Copyright © 1977 by Scientific American, Inc. All rights reserved; *p. 218, 10.18:* After figure by V. A. Tucker, *American Scientist,* Vol. 63, pp. 413–419. Reprinted by permission of *American Scientist,* Journal of Sigma Xi, The Scientific Research Society; *p. 219, 10.19:* After figure by J.H. Marden, *Journal of Experimental Biology,* Vol. 130, pp. 235–258, 1987. Reproduced by permission of Company of Biologists Ltd.; *p. 228, 11.5:* From *Life: An Introduction to Biology,* Third Edition, by William S. Beck, Karel F. Liem and Anne Roe Simpson. Copyright © 1991 by William S. Beck, Karel F. Liem and Anne Roe Simpson. Reprinted by permission of HarperCollins Publishers Inc.; *p. 228, 11.6:* After *Introduction to the Human Body: The Essentials of Anatomy and Physiology,* Second Edition, by Gerard J. Tortora. Copyright © 1991 by Biological Sciences Textbooks, Inc. and A. and P. Textbooks, Inc. Reprinted by permission of HarperCollins Publishers Inc.; *p. 232, 11.9A:* After *Introduction to the Human Body: The Essentials of Anatomy and Physiology,* Second Edition, by Gerard J. Tortora. Copyright © 1991 by Biological Sciences Textbooks, Inc. and A. and P. Textbooks, Inc. Reprinted by permission of HarperCollins Publishers Inc.; From *Life: An Introduction to Biology,* Third Edition, by William S. Beck, Karel F. Liem and Anne Roe Simpson. Copyright © 1991 by William S. Beck, Karel F. Liem and Anne Roe Simpson. Reprinted by permission of HarperCollins Publishers Inc.; *p. 235, 11.12:* After *Introduction to the Human Body: The Essentials of Anatomy and Physiology,* Second Edi-

[*]The abbreviation VU is used for Visuals Unlimited throughout the credits.
[†]The abbreviation TS is used for Tom Stack & Associates throughout the credits.

Photographs

p. 1, Chapter 1 opener: Wendy Shattil & Bob Rozinski/TS; *p. 2, Figure 1.1:* Trans. no. 2373(2). Courtesy Department of Library Services, American Museum of Natural History; *p. 3, 1.2:* C. Leon Harris; *p. 3, 1.3:* C. Leon Harris; *p. 8, 1.5:* David M. Phillips/VU; *p. 9, 1.6:* John D. Cunningham/VU; *p. 9, 1.7:* Neal A. Weber; *p. 10, 1.8:* George Holton/Photo Researchers; *p. 16, Unit One opener:* Kessel, R.G. and R.H. Kardon. *Tissues and Organs: A Text-Atlas of Scanning Electron Microscopy*, W.H. Freeman, P. 37; *p. 22, 2.3A:* C. Leon Harris; *p. 36, 2.18A–D:* C. Leon Harris; *p. 36, 2.18E:* T. Fujita and D.W.Fawcett/VU; *p. 36, 2.18F:* Reproduced from *A Textbook of Histology*, W. Bloom and D. W. Fawcett, 10th ed., © 1975 by permission of W.B. Saunders Company, Philadelphia; *p. 36, 2.18G–H:* Reproduced from A. L. Beyer, O. L. Miller, Jr., and S. L. Knight (1980). *Cell* 20:75–84 by copyright permission of The Massachusetts Institute of Technology, Boston; *p. 38, 2.20A–D:* Reproduced from *Molecular Biology of the Cell*, B. Alberts, D. Bray, J. Lewis, M. Raff, K. Roberts and J. D. Watson, © 1983, Garland Publishing, New York; *p. 38, 2.20E–F:* Reproduced from *A Textbook of Histology*, W. Bloom and D. W. Fawcett, 10th ed., © 1975 by permission of W. B. Saunders Company, Philadelphia; *p. 44, 2.27:* Reproduced from E. Lazarides and K. Weber (1974). *Proc. Natl. Acad. Sci. USA* 71:2268–2272, by courtesy of the National Academy of Sciences; *p. 45, 2.28B:* Courtesy, M. McGill; *p. 46, 2.29A:* Courtesy, R. S. Decker; *p. 46, 2.29B:* Reproduced from D. S. Friend and N. B. Gilula (1972). *J. Cell Biology* 53:758–776 by copyright permission of The Rockefeller University Press, New York; *p. 46, 2.29C:* Courtesy, R. S. Decker; *p. 47, 2.30A and C:* Reproduced from K. Tanaka, A. Iino, and T. Naguro (1976). *Arch. Histol. Jap.* 39:165–175 by permission of the Japan Society of Histological Documentation; *p. 47, 2.30B:* Reproduced from *The Cell*, D. W. Fawcett, 2nd ed. © 1981, by permission of W. B. Saunders Company, Philadelphia; *p. 47, 2.30D:* Reproduced from M. Bielinska, G. Rogers, T. Rucinsky, and I. Boine (1979). *Proc. Natl. Acad. Sci. USA* 76:6152–6156, by courtesy of the National Academy of Sciences; *p. 48, 2.31:* Reproduced from D. S. Friend (1965), *J. Cell Biol.* 25:563–576 by copyright permission of The Rockefeller University Press, New York; *p. 49, 2.33:* Dr. Daniel Branton; *p. 51, Chapter 3 opener:* Steve McCutcheon/VU; *p. 58, 3.5A:* Courtesy, Pierre and Nina Favard, Centre National de la Recherche Scientifique, France; *p. 65, Chapter 4 opener:* Robert Winslow/TS; *p. 66, 4.1:* Culver Pictures; *p. 72, 4.5A–E:* Supplied by Carolina Biological Supply Company; *p. 73, 4.6:* Courtesy, S. M. Gollin and W. Wray; *p. 75, 4.8:* Reproduced with permission from *The Meiotic Mechanism*, B. John and K. R. Lewis, 2nd ed. 1984. © Carolina Biology Reader Series. Supplied by Carolina Biological Supply Company; *p. 76, 4.9A:* Photo by Calvin Bridges. Reproduced from *American Zoologist* 26:386(1986); *p. 81, Chapter 5 opener:* Alvin E. Staffan/Photo Researchers; *p. 83, 5.2:* Photographer: A. C. Barrington Brown From J. D. Watson, 1968, *The Double Helix*, New York: Atheneum, p. 215. © by J. D. Watson; *p. 85, 5.3B:* Courtesy, U. K. Laemmli; *p. 94, 5.11:* Children's Hospital, Denver; *p. 95, 5.12:* Reprinted by permission of John Wiley & Sons Inc. from J. de Grouchy and C. Turleau, *Clinical Atlas of Human Chromosomes*, © 1977 by John Wiley & Sons, Inc., New York; *p. 103, 5.15:* R. L. Brinster and R. E. Hammer, School of Veterinary Medicine, University of Pennsylvania; *p. 105, Chapter 6 opener:* Dwight R. Kuhn; *p. 106, 6.1:* From N. Hartsoecker, *Essai de Dioptrique*, Paris, 1694; *p. 110, 6.5A:* From Lennart Nilsson, *Behold Man*, Lennart Nilsson/Bonnier Fakta © 1974 by Albert Bonniers Forlag and Little, Brown and Co. Canada, Ltd.; *p. 110, 6.6:* Reproduced from H. Shatten and G. Shatten (1980), *Developmental Biology* 78:435–449, by copyright permission of Academic Press, Inc., Orlando, Fl; *p. 111, 6.7:* Reproduced from M. J. Tegner and D. Epel (1973). *Science* 129:685–688, by copyright permission of the AAAS; *p. 113, 6.9A–E:* From J. G. Mulnard, 1967, Analysé microcinématographique du développement de l'oeuf de souris du stade II au blastocyte. *Arch. Biol.* (Liege) 78:107–138; *p. 115, 6.12A–G:* Supplied by Carolina Biological Supply Company; *p. 117, 6.14A–B:* Einhard Schierenberg, Zoologisches Institut der Universität Köln; *p. 120, 6.17A:* Supplied by Carolina Biological Supply Company; *p. 124, 6.22B:* From R. O'Rahilly and F. Muller, 1987, *Developmental Stages in Human Embryos.* Washington, D.C.: Carnegie Institution of Washington, Publication 637. Page 94, Figure 10-1A; *p. 125, 6.23A:* From R. O'Rahilly and F. Muller, 1987. *Developmental Stages in Human Embryos.* Washington, D.C.: Carnegie Institution of Washington, Publication 637. Page 158, Figure 14-1D; *p. 125, 6.23B:* Carlo Bevilacqua/CEDRI; *p. 127, 6.25:* Carlo Bevilacqua/CEDRI; *p. 129, 6.26:* From E. B. Lewis, *American Zoologist* 3,33–56, 1963; *p. 130, 6.28:* Courtesy of S. M. Rothman. From W. M. Cowan, 1979, *Scientific American* 241(Sept.):113–130. Copyright © Scientific American, Inc. All rights reserved. Reproduced with permission; *p. 136, Unit Two opener:* Tom J. Ulrich/VU; *p. 139, Chapter 7 opener:* John D. Cunningham/VU; *p. 141, 7.1A:* Courtesy Dr. Christine Gall, University of California at Irvine; *p. 146, 7.6A:* Cedric S. Raine, Albert Einstein College of Medicine; *p. 146, 7.6B:* S. Tsukita and H. Ishikawa, *J. Cell Biology.* 84:513 (1970), by copyright permission of The Rockefeller University Press; *p. 148, 7.7A:* E. R. Lewis, Y. Y. Zeevi, T. E. Everhart, University of California/Biological Photo Service; *p. 148, 7.7B:* Courtesy of J. E. Heuser. From H. A. Lester (1977). *Scientific American* 236 (Feb.):106–118. Copyright © Scientific American, Inc., All rights reserved. Reproduced with permission; *p. 166, Chapter 8 opener:* Thomas Kitchin/TS; *p. 179, 8.10C:* Reproduced by permission of Dr. William Feindel, Montreal Neurological Institute; *p. 182, Chapter 9 opener:* S. Maslowski/VU; *p. 193, 9.11B:* C. Leon Harris; *p. 194, 9.12B:* William S. Bowers, Department of Entomology, University of Arizona; *p. 196, Chapter 10 opener:* Russell C. Hansen/Peter Arnold, Inc.; *p. 206, 10.8A:* T. Reese and D. W. Fawcett/VU; *p. 208, 10.10A:* C. Leon Harris; *p. 211, 10.12A:* C. Leon Harris; *p. 211, 10.12B:* H. E. Huxley, In Murray, J. M. A. Weber, 1974. The cooperative action of muscle proteins. *Sci. Am.* 230(2):58(top)(Feb.) and copyright © Scientific American, Inc. All rights reserved; *p. 216, 10.16:* R. D. Allen; *p. 217, 10.17A:* Dr. William E. Barstow; *p. 217, 10.17B:* Sidney L. Tamm, In Tamm, S. L. and G. A. Horridge. 1970. *Proc. Roy. Soc. London B* 175:219–233; *p. 217, 10.17C:* David M. Phillips/VU; *p. 222, Chapter 11 opener:* Michael Fogden/Bruce Coleman Inc.; *p. 237, 11.13:* SEM by Dr. Gilla Kaplan, The Rockefeller University, New York; *p. 244, Chapter 12 opener:* Nuridsany et Pérennou/Photo Researchers; *p. 250, 12.7C:* Dr. Warren W. Burggren, Department of Zoology, University of Massachusetts, Amherst; *p. 259, Chapter 13 opener:* Dwight R. Kuhn; *p. 266, 13.3A:* Rod Planck/Tom Stack & Associates; *p. 271:* The Bettmann Archive; *p. 280, Chapter 19 opener:* Thomas Kitchin/TS; *p. 297, Chapter 15 opener:* Wolfgang Bayer/Bruce Coleman Inc.; *p. 298, 15.1:* C. Leon Harris; *p. 302:* C. Leon Harris; *p. 313, Chapter 16 opener:* Hans Pfletschinger/Peter Arnold, Inc.; *p. 315, 16.1:* Warren Williams/Planet Earth Pictures; *p. 316, 16.2A:* Marty Snyderman/VU; *p. 316, 16.2B:* Allan Roberts; *p. 328, Unit Three opener:* Denise Tackett/TS; *p. 331, Chapter 17 opener:* Joe McDonald/TS; *p. 361, Chapter 18 opener:* Leonard Lee Rue III/TS; *p. 365, 18.2A:* John H. Ostrom, Yale University; *p. 365, 18.2B:* Painting by Rudolf Freund. Courtesy Carnegie Museum of Natural History, Pittsburgh; *p. 368, 18.4:* Neg. no. 326795, Courtesy Department of Library Services, American Museum of Natural History; *p. 368, 18.5: Alfred Russel Wallace* by Thomas Sims (original photograph), © 1863–66. National Portrait Gallery, London; *p. 368, 18.6: Sir Thomas Lyell* by L. Dickinson (replica), 1883. National Portrait Gallery, London; *p. 370, 18.8:* By permission of the Darwin Museum, Down House; *p. 376, 18.12A:* T. W. Pietsch and D. B. Grobecker; *p. 376, 18.12B:* Steinhart Aquarium; *p. 378, 18.14A–B:* Photograph by H. B. D. Kettlewell. Courtesy of David Kettlewell; *p. 386, Chapter 19 opener:* Wolfgang Bayer/Bruce Coleman Inc.; *p. 389, 19.1A:* Manfred Gottschalk/TS; *p. 391, 19.3:* Andrew H. Knoll, Harvard University; *p. 393, 19.5:* From P. Cloud and M. F. Glaessner, 1982. The Ediacaran Period and System: Metazoa Inherit the Earth, *Science* 218:783–792; *p. 395, 19.7:* E. N. K. Clarkson, Grant Institute of Geology, Edinburgh; *p. 400, 19.11A:* Trans. No. 3148(3). Courtesy Department of Library Services, American Museum of Natural History; *p. 400, 19.11B:* Trans. no. 2435(4). Courtesy Department of Library Services, American Museum of Natural History; *p. 401, 19.13A:* H. Albrecht/Bruce Coleman Inc.; *p. 403, 19.15A:* Science VU/VU; *p. 403, 19.15B:* Jay Matternes; *p. 407, Chapter 20 opener:* C. J. Smale/Frank Lane/Bruce Coleman Inc.; *p. 409, 20.2:* Dr. Andrew P. King, Indiana Univ.; *p. 414, 205A:* C. Leon Harris; *p. 414, 20.5B: John H. Ostrom, Yale University; p. 414, 20.5C:* © copyright 1991 John Reader; *p. 415, 20.6:* Ronald M. Schassburger, Ph.D., Natural History Society of the Finger Lakes Region, Ithaca, NY; *p. 416, 20.7:* James E. Lloyd; *p. 420, 20.11:* Jeff Foott/Bruce Coleman Inc.; *p. 421, 20.12A:* C. Leon Harris; *p. 421, 20.12B:* Photograph by Biana Lavies © 1978 Copyright, National Geographic Society; *p. 425, 20.14:* John Gerlach/TS; *p. 427, 20.15:* George W. Barlow, Courtesy University of California, Berkeley; *p. 429, 20.17:* Diana L. Stratton/TS; *p. 430, 20.18:* Tim Davis/Photo Researchers; *p. 430, 20.19:* Jane Burton/Bruce Coleman Inc.; *p. 431, 20.20:* Roger Wilmshurst/Bruce Coleman Inc.; *p. 431, 20.21:* Photo by Nina Leen © 1965/Life Magazine Time Warner Inc.; *p. 432, 20.22:* Martin Rogers/Tony Stone Worldwide; *p. 436, Unit Four opener:* Dave B. Fleetham/TS; *p. 439, Chapter 21 opener:* Alan Blank/Bruce Coleman Inc.; *p. 440, 21.1A:* Photographed by Mike Hopiak for the Cornell Laboratory of Ornithology; *p. 440, 21.1B:* Robert C. Simpson/TS; *p. 441, 21.2A:* Robert C. Simpson/TS; *p. 441, 21.2B:* S. Maslowski/VU; *p. 460, Chapter 22 opener:* A. M. Siegelman/VU; *p. 463, 22.2A:* D. M. Phillips/VU; *p. 463, 22.2B:* Walne, P. L. and Arnott, H. J. 1967. *Planta* 77:325–354; *p. 465, 22.3A:* Karl Aufderheide/VU; *p. 469, 22.4A:* David M. Phillips/VU; *p. 469, 22.4B:* Manfred Kage/Peter Arnold, Inc.; *p. 470, 22.6:* K. W. Jeon/VU; *p. 471, 22.7:* Reproduced from L. E. Roth, D. J. Pihlaja, and Y. Shigemaka (1970). *J Ultrastruct. Res.* 30:7–37, by copyright permission of Academic Press, Inc., Orlando, Fl; *p. 475, 22.12:* Manfred Kage/Peter Arnold, Inc.; *p. 477, 22.13:* Armed Forces Institute of Pathology; *p. 477, 22.14:* Jerome Paulin/VU; *p. 485, Chapter 23 opener:* Nancy Sefton/Photo Researchers; *p. 488, 23.1A:* Larry Roberts/VU; *p. 490, 23.4A:* Biological Photo Service; *p. 491, 23.5A:* Field Museum of Natural History, Chicago; *p. 492, 23.6A:* Glenn Oliver/VU; *p. 492, 23.6B:* Patricia R. Bergquist, Department of Zoology, The University of Auckland; *p. 493, 23.7A:* Courtesy of the Peabody Museum of Natural History, Yale University; *p. 494, 23.8A:* C. Leon Harris; *p. 501, Chapter 24 opener:* Brian Parker/TS; *p. 502, 24.1A:* John D. Cunningham/VU; *p. 502, 24.1B:* Fred Bavendam/Peter Arnold, Inc.; *p. 505, 24.4:* Supplied by Carolina Biological Supply Company; *p. 506, 24.5A:* Ed Reschke/Peter Arnold, Inc.; *p. 507, 24.6:* John D. Cunningham/VU; *p. 508, 24.7A:* John D. Cunningham/VU; *p. 510, 24.8A:* Neville Coleman/VU; *p. 510, 24.9A:* Daniel W. Gotshall/VU; *p. 510, 24.9B:* James R. McCullagh/VU; *p. 511, 24.10A:* Peter Parks/Oxford Scientific Films/ANIMALS ANIMALS; *p. 511, 24.10C:* Larry Tackett/TS; *p. 511, 24.11A:* Fred Bavendam/Peter Arnold, Inc.; *p. 515, 24.13A–F:* The University of Alberta Archives, D. M. Ross Collection; *p. 518, 24.15:* Photo by Phillip Playford from P. E. Playford, 1980, Ordovician Great Barrier Reef of the Canning Basin, Western Australia. *Petroleum Geologists* 64: 814–840; *p. 518, 24.16:* Don and Pat Valenti; *p. 519, 24.17:* Laurence P. Madin, Woods Hole Oceanographic Institution; *p. 524, Chapter 25 opener:* Scott Johnson/ANIMALS ANIMALS; *p. 527, 25.1A:* Ed Robinson/TS; *p. 527, 25.1B:* Peter J. Bryant/Biological Photo Service; *p. 529, 25.3:* Dwight R. Kuhn; *p. 534, 25.8B:* Harvey D. Blankespoor, Hope College, Holland, Michigan; *p. 536, 25.9A:* Allan Roberts; *p. 536, 25.9B:* Stanley Flegler/VU; *p. 536, 25.9C:* John D. Cunningham/VU; *p. 537, 25.10:* From Ralph Buchsbaum et al., ANIMALS WITHOUT BACKBONES, third edition, 1987, University of Chicago Press, page 201. Photo courtesy of the Department of Pathology, The University of Chicago; *p. 543, Chapter 26 opener:* R. B. Taylor/Science Photo Library/Photo Researchers; *p. 550, 26.3:* N. Allin and G. L. Barron, University of Guelph; *p. 552, 26.5:* R. Calentine/VU; *p. 553. Figure 26.6B:* Courtesy, Institute of Public Health Research, Teheran University School of Public Health; *p. 554, 26.7:* Armed Forces Institute of Pathology; *p. 554, 26.8:* Dwight R. Kuhn; *p. 555, 26.9B:* Courtesy of the Trustees of the British Museum (Natural History); *p. 558, 26.13:* John Gilbert, Dartmouth College; *p. 564, Chapter 27 opener:* Michael Fogden/ANIMALS ANIMALS; *p. 565, 27.1:* Kjell B. Sandved/VU; *p. 566, 27.3:* C. Leon Harris; *p. 568, 27.4A:* J. D. Wrobel, Monterey Bay Aquarium/Biological Photo Service; *p. 570, 27.5A:* Allan Roberts; *p. 571, 27.6:* Kjell B. Sandved/VU; *p. 573, 27.8A:* John D. Cunningham/VU; *p. 576, 27.10:* Courtesy of Diane R. Nelson, East Tennessee State University; *p. 578, 27.12:* C. B. and D. W. Firth/Bruce Coleman Inc.; *p. 581, 27.15:* D. Foster/Science VU/VU; *p. 583, Chapter 28 opener:* Alex

Kerstitch/VU; *p. 584, 28.1A:* Kjell B. Sandved/VU; *p. 584, 28.1B:* Jeff Foott/TS; *p. 584, 28.1C:* C. Leon Harris; *p. 584, 28.1D:* C. R. Wyttenbach, University of Kansas/Biological Photo Service; *p. 584, 28.1E:* Gary Milburn/TS; *p. 585, 28.2:* James L. Amos/Photo Researchers; *p. 587, 28.4A:* C. Leon Harris, *p. 588, 28.5B:* neg. 3830, Field Museum of Natural History, Chicago; *p. 590, 28.7A–B:* C. B. Calloway; *p. 595, 28.12A:* William H. Amos; *p. 595, 28.12B:* James H. Carmichael/Bruce Coleman Inc.; *p. 595, 28.12C:* Scott Johnson/Animals Animals; *p. 595, 28.12D:* Matt Bradley/TS; *p. 595, 28.12E:* Brian Parker/TS; *p. 595, 28.13:* Kjell B. Sandved/VU; *p. 596, 28.14:* Milton Rand/TS; *p. 598, 28.18B:* Alex Kerstitch/VU; *p. 599, 28.19A:* Geoff du Feu/Planet Earth Pictures; *p. 600, 28.20A:* William C. Jorgensen/VU; *p. 600, 28.20B:* Dave B. Fleetham/TS; *p. 602, 28.22:* Dave B. Fleetham/TS; *p. 603, 28.24A:* Douglas Faulkner/Photo Researchers; *p. 605, 28.26:* Douglas Faulkner; *p. 609, Chapter 29 opener:* William H. Amos; *p. 611, 29.1A:* R. DeGoursey/VU; *p. 620, 29.8A:* Dave Woodward/TS; *p. 621, 29.10:* Neg. No. 33717. Courtesy Department of Library Services, American Museum of Natural History; *p. 623, 29.13:* Globe Photo; *p. 625, 29.17:* C. Leon Harris; *p. 627, 29.19A–B:* C. Leon Harris; *p. 627, 29.20:* St. Bartholomew's Hospital/Science Photo Library/Photo Researchers; *p. 630, 29.22:* T. Branning; *p. 632, Chapter 30 opener:* Kim Taylor/Bruce Coleman Inc.; *p. 633, 30.1A–D:* C. Leon Harris; *p. 637, 30.4D:* C. Leon Harris; *p. 640, 30.7:* Slide # GE080819.1, Field Museum of Natural History, Chicago; *p. 641, 30.8:* Lynn M. Stone; *p. 644, 30.11:* Dick George/TS; *p. 651, 30.17:* C. Leon Harris; *p. 651, 30.18:* C. Leon Harris; *p. 652, 30.19A:* Philip Brownell, Department of Zoology, Oregon State University; *p. 653, 30.20A:* Ed Bosler, Ph.D., State of New York Department of Health; *p. 653, 30.20B:* William H. Amos; *p. 654, 30.21A:* W. J. Weber/VU; *p. 654, 30.21B:* Richard Walters/VU; *p. 655, 30.22A:* Mite from and photo by Clifford E. Desch, Jr., Department of Ecology and Evolutionary Biology, University of Connecticut at Hartford; *p. 657, Chapter 31 opener:* Joe McDonald/VU; *p. 660, 31.4:* Lynn M. Stone; *p. 667, 31.12A:* Fred Bavendam/Peter Arnold, Inc.; *p. 667, 31.12B:* Joe McDonald/TS; *p. 667, 31.12C:* Paulette Brunner/TS; *p. 667, 31.12D:* Dave B. Fleetham/TS; *p. 668, 31.13:* Jack Stein Grove/TS; *p. 669, 31.14:* William H. Amos; *p. 669, 31.15:* C. Leon Harris; *p. 670, 31.16:* John D. Cunningham/VU; *p. 670, 31.17:* Dwight R. Kuhn; *p. 671, 31.18A:* Tom Stack/TS; *p. 671, 31.18B:* Heather Angel/Biofotos; *p. 672, 31.20A:* Paulette Brunner/TS; *p. 672, 31.20B:* Ed Degginger/Bruce Coleman Inc.; *p. 674, 31.21A:* Jill Yager, Antioch University; *p. 675, 31.22B:* D. P. Wilson; *p. 677, Chapter 32 opener:* Alex Kerstitch/VU; *p. 679, 32.1:* C. Leon Harris; *p. 681, 32.6:* C. Leon Harris; *p. 683, 32.7A–C:* C. Leon Harris; *p. 689, 32.11A:* Dwight R. Kuhn; *p. 692, 32.14:* C. Leon Harris; *p. 692, 32.15A:* Photo by Barbara L. Thorne, Museum of Comparative Zoology, Cambridge, Mass.; *p. 692, 32.15B:* Thomas Eisner and Daniel Aueshansley; *p. 693, 32.16A–B:* C. Leon Harris; *p. 693, 32.17A–C:* C. Leon Harris; *p. 694, 32.18:* C. Leon Harris; *p. 696, 32.20A:* Hans Pfletschinger/Peter Arnold, Inc.; *p. 696, 32.20B:* C. Leon

Harris; *p. 697, 32.21B–C:* C. Leon Harris; *p. 698, 32.22:* Dr. Edward S. Ross, California Academy of Sciences; *p. 704, 32.25:* Photo by B. M. Shepard, G. R. Carner and J. Hudson; *p. 705, 32.26:* Allan Roberts; *p. 708, Chapter 33 opener:* Alex Kerstitch/VU; *p. 710, 33.1A:* Dwight R. Kuhn; *p. 710, 33.1B:* Brian Parker/TS; *p. 710, 33.1C:* Tom Stack/TS; *p. 710, 33.1D:* David L. Meyer; *p. 710, 33.1E:* Brian Parker/TS; *p. 711, 33.3A:* Allan Roberts; *p. 712, 33.4B:* C. Leon Harris; *p. 716, 33.8A:* Slide # 80871.1, Field Museum of Natural History, Chicago; *p. 716, 33.8B:* C. Leon Harris; *p. 717, 33.10:* David S. Addison/VU; *p. 719, 33.12C:* C. Leon Harris; *p. 720, 33.14:* Ed Robinson/TS; *p. 724, 33.17:* Brian Parker/TS; *p. 724, 33.18:* Jack Stein Grove/TS; *p. 728, Chapter 34 opener:* Fred Bavendam/Peter Arnold, Inc.; *p. 734, 34.6A:* Robert Brons/Biological Photo Service; *p. 734, 34.6B:* Biological Photo Service; *p. 734, 34.7:* Heather Angel/Biofotos; *p. 738, 34.10:* Ed Robinson/TS; *p. 740, 34.13:* Gary R. Robinson/VU; *p. 744, 34.17:* Slide # 82670, Field Museum of Natural History, Chicago; *p. 744, 34.18A:* Slide # 82664, Field Museum of Natural History, Chicago; *p. 744, 34.18B:* Neg. # GE082014, Field Museum of Natural History, Chicago; *p. 748, Chapter 35 opener:* Marty Snyderman/VU; *p. 749, 35.1A:* Biological Photo Service; *p. 749, 35.1B:* Tom Stack/TS; *p. 750, 35.2A:* Tom McHugh/Photo Researchers; *p. 750, 35.2B:* Dave B. Fleetham/TS; *p. 750, 35.2C:* Dave B. Fleetham/TS; *p. 753, 35.3A:* Tom Stack/TS; *p. 753, 35.3B:* Patrice Ceisel/VU; *p. 755, 35.6B:* Patrice Ceisel/TS; *p. 759, 35.9B:* William H. Amos; *p. 759, 35.9C:* Leonard J. V. Compagno, Shark Research Center, South African Museum; *p. 761, 35.11:* David J. Wrobel/Biological Photo Service; *p. 762, 35.12A:* Miami Seaquarium; *p. 762, 35.12B:* Ken Lucas/Biological Photo Service; *p. 763, 35.13:* J. Schauer and H. Fricke, reprinted by permission from *Nature,* 329, cover, copyright © 1987, Macmillan Magazines Ltd.; *p. 767, 35.15:* Tom Stack/TS; *p. 767, 35.16:* Patrice Ceisel/VU; *p. 770, 35.20A:* Denise Tackett/TS; *p. 770, 35.20B:* Denise Tackett/TS; *p. 773, 35.23A:* Tom McHugh/Photo Researchers; *p. 781, 35.29:* Alex Kerstitch/VU; *p. 785, Chapter 36 opener:* Stephen Dalton/Animals Animals; *p. 786, 36.1A:* D. G. Barker/TS; *p. 786, 36.1B–C:* C. Leon Harris; *p. 788, 36.3A:* Allan Roberts; *p. 788, 36.3B:* Allan Blank/Bruce Coleman Inc.; *p. 790, 36.4:* Jane Burton/Bruce Coleman Inc.; *p. 792, 36.6A:* D. G. Barker/TS; *p. 792, 36.6B:* C. Leon Harris; *p. 796, 36.10:* C. Leon Harris; *p. 799, 36.13:* C. Leon Harris; *p. 801, 36.15B–C:* C. Leon Harris; *p. 802, 36.16:* Copyright M. J. Tyler; *p. 804, 36.18A:* Victor Hutchinson/VU; *p. 804, 36.18B:* Allan Roberts; *p. 806, Chapter 37 opener:* Brian Parker/TS; *p. 809, 37.3A:* Dwight R. Kuhn; *p. 809, 37.3B:* Don and Pat Valenti; *p. 809, 37.3C:* John C. Murphy/TS; *p. 810, 37.4A:* Dwight R. Kuhn; *p. 810, 37.4B:* David M. Dennis/TS; *p. 810, 37.4C:* John Nees/Animals Animals; *p. 811, 37.5A:* Lynn M. Stone; *p. 811, 37.5B:* Zig Leszczynski/Animals Animals; *p. 811, 37.5C:* Alan and Sandy Carey; *p. 811, 37.5D:* Lynn M. Stone; *p. 812, 37.6A:* Lynn M. Stone; *p. 812, 37.6B:* David R. Frazier Photolibrary; *p. 812, 37.6C:* Gary Milburn/TS; *p. 813, 37.7A:* John Cancalosi/TS; *p. 814, 37.8A:* Slide # GE084968, Field Museum of

Natural History, Chicago; *p. 814, 37.8B:* Institut und Museum fur Geologie und Palaontologie, Universitat Tubingen; *pp. 814–815, 37.9:* Painted by Rudolph F. Zallinger. Courtesy of the Peabody Museum of Natural History, Yale University; *p. 819, 37.13:* Daniel Luchtel and Michael Hlastala; *p. 822, 37.15A:* J. H. Robinson/Animals Animals; *p. 825, 37.17:* Jack S. Grove/TS; *p. 826, 37.18A:* William H. Amos; *p. 826, 37.18B:* Mary Clay/TS; *p. 826, 37.18C:* Dwight R. Kuhn; *p. 827, 37.19A:* Joe McDonald/VU; *p. 828, 37.20:* C. B. Frith/Bruce Coleman Inc.; *p. 828, 37.21:* Tom Brakefield/Bruce Coleman Inc.; *p. 831, Chapter 38 opener:* R. H. Armstrong/Animals Animals; *p. 833, 38.2:* From Richard Moody, *Nature Library: Prehistoric Life,* 1983, Exeter Books, New York; *p. 839, 38.5A:* David M. Phillips/VU; *p. 839, 38.5B:* S. D. Carlson; *p. 840, 38.6B–C:* C. Leon Harris; *p. 841, 38.7A:* C. Leon Harris; *p. 841, 38.7B:* Michael Fogden/VU; *p. 845, 38.11A:* Lynn M. Stone; *p. 845, 38.11B:* James H. Carmichael/Bruce Coleman Inc.; *p. 845, 38.11C:* Dwight R. Kuhn; *p. 851, 38.17A,C:* Silvestris & Gross/Peter Arnold, Inc.; *p. 851, 38.17B:* Fritz Polking/Peter Arnold, Inc.; *p. 852, 38.19A:* Lynn M. Stone; *p. 852, 38.19B:* Jim Markham/Bruce Coleman Inc.; *p. 853, 38.20A:* C. B. & D. W. Frith/Bruce Coleman Inc.; *p. 853, 38.20B:* C. B. Frith/Bruce Coleman Inc.; *p. 855, 38.21A:* John K. Nakata/Terraphotographics/Biological Photo Service; *p. 855, 38.21C:* David L. Pearson/VU; *p. 855, 38.21D:* Stephen Dalton/Photo Researchers; *p. 856, 38.22:* Kjell B. Sandved/VU; *p. 857, 38.23:* John L. Ebeling; *p. 858, 38.24A:* National Audubon Society photo/Photo Researchers; *p. 858, 38.24B:* The Bettmann Archive; *p. 859, 38.25:* John Borneman/Photo Researchers; *p. 860, 38.26A:* Wendy Shattil and Bob Rozinski/TS; *p. 860, 38.26B:* Ben Goldstein/Don and Pat Valenti; *p. 860, 38.26C:* Alan & Sandy Carey; *p. 863, Chapter 39 opener:* Thomas Kitchin/TS; *p. 866, 39.2A:* J. Alcock/VU; *p. 866, 39.2B:* Dave Watts/TS; *p. 868, 39.3A:* Allan Roberts; *p. 868, 39.3B:* Dave Watts/TS; *p. 868, 39.3C:* Lynn M. Stone; *p. 874, 39.6A:* G. R. Roberts; *p. 874, 39.6B:* Lynn M. Stone; *p. 876, 39.7A:* John Visser/Bruce Coleman Inc.; *p. 877, 39.8A:* Alan and Sandy Carey; *p. 877, 39.8B:* Dwight R. Kuhn; *p. 877, 39.8C:* Dwight R. Kuhn; *p. 877, 39.8D:* Jonathan Scott/Planet Earth Pictures; *p. 878, 39.9A:* Jane Thomas/VU; *p. 878, 39.9B:* Leonard Rue Enterprises; *p. 879, 39.10:* Marty Stouffer/Animals Animals; *p. 880, 39.11A:* Alan and Sandy Carey; *p. 880, 39.11B:* Gerald Corsi/TS; *p. 880, 39.11C:* John Shaw/TS; *p. 880, 39.11D:* Joe McDonald/TS; *p. 882, 39.14A:* Jeff Foott/TS; *p. 882, 39.14B:* Science VU/VU; *p. 883, 39.15A:* Copyright © 1991 Sea World, Inc. All rights reserved. Reproduced by permission; *p. 883, 39.15B:* F. Gohier/Photo Researchers; *p. 884, 39.16A:* Brian Parker/TS; *p. 884, 39.17A–B:* Dr. Merlin D. Tuttle/Bat Conservation International; *p. 885, 39.18A:* Lynn M. Stone; *p. 885, 39.18B:* Ron Garrison/Zoological Society of San Diego; *p. 885, 39.19:* Gary Milburn/TS; *p. 886, 39.20A:* Jack Swenson/TS; *p. 886, 39.20B:* Bruce Coleman Inc.; *p. 888, 39.21:* William R. Curtsinger/Photo Researchers; *p. 890, 39.22A:* Don and Pat Valenti; *p. 890, 39.22B:* Thomas Kitchin/TS; *p. 890, 39.22C:* Lynn M. Stone.

Index

Note: Page numbers in **boldface** indicate major discussion.

representative, 633
respiration, 252
visual communication, 415
Arthropod–chordate theory, 741–742
Arthropoda, 638
Articulamentum, 591
Articular, 879, 881–882
Articular membrane, 635
Articulata, 569
Artiodactyla, 267, 872, 878
Ascaphus, 789, 801
Ascaris, 71, 545–547, **550–551**
Ascetospora, 467
Aschelminthes, 544
Ascidiacea, 735–736
Ascidian. *See* Sea squirt
Asconoid canal system, 487, 489
Ascorbic acid. *See* Vitamin C
Asio, 166
Aspirin, 130, 184
Asplanchna, 558
Assimilation, 348
Association cortex, 176
Astacus, 287
Asterias, 710
    classification, 722
    feeding, 714
    ion concentrations in, 282
    madreporite, 712
    osmoconformity, 284
    regeneration, 723
    tube foot, 712
Asteroid. *See* Starfish
Asteroidea, 722
Asteroid impact, 387, **396,** 817
Asthma, 257
Astrocyte, 172
Astrorhiza, 490, 493
Astroscopus, 780
Asymmetry, 450
Asynchronous muscle, 216
Ateles, 886, 887
Atmosphere, 387
Atoke, 617
Atoll, 357
Atom, 20
ATP
    function, 52–55
    production, **55–62,** 263
    regulation of metabolism, 62
    structure, 52–53
    use by muscle, 210, 213
    from various nutrients, 261
ATPase, 41, 61, 210–211
Atrial natriuretic hormone, 187, 291
Atriopeptin. *See* Atrial natriuretic hormone
Atrioventricular node, 228
Atrioventricular valve, 227–229
Atrium
    of heart, 225–228
    of sea squirt, 736
Atta, 9
Attachment, of sperm, 110
Audition. *See* Cochlea; Hearing
Auditory cortex, 177
Auditory nerve, 160
Aurelia, 506–509
Auricle
    of heart. *See* Atrium, of heart
    of planaria, 530
Auricularia larva, 723
Auriculin. *See* Atrial natriuretic hormone
Australopithecus, 402, 404
Autogamy, 463, 473
Autoimmune disease, 240
Autolytus, 613, 618–619

Autonomic nervous system, 171–172
    control of copulation, 318
    control of digestion, 273
    effect on circulation, 231
    effect on smooth muscle, 215
    effect on thermoregulation, 308
Autosome, 77
Autotomy, 651
Autotroph, 333
Aves, 736, 834
Avicularium, 567
Avogadro's number, 23
Avoiding reaction, 470–472
Axial cell, 498
Axis, developmental, 128
Axolotl, 804
Axon, 141–145
Axoneme, 217
Axopodium, 468, 471
Aye-aye, 887
Aysheaia, 395

β. *See* Beta *entries*
Babesia, 467, 478, 654
Baboon, 428, 887
Bacillus, 703
Backswimmer, 688
Bacteria, 8
    ammonifying, 346
    chemoautotrophic, 581
    as decomposers, 345, 354
    defenses against, 236, 256, 309, 325
    denitrifying, 348
    DNA of, 84
    in flashlight fishes, 357, 781
    as insecticide, 703
    in lakes, 337
    in large intestine, 273
    in leeches, 628–629
    molecular phylogeny, 8, 379
    nitrate, 346
    nitrite, 346
    nitrogen-fixing, 346
    and origin of eukaryotes, 392
    sulfur-metabolizing, 581
Badger, 871
Baer, Karl Ernst von, 132–133, 363
Bakker, Robert, 307, 817
Balaenidae, 881
Balaenoptera, 218, 867, 883
Balaenopteridae, 881
Balance, 160
Balanoglossus, 729–730
Balanophylia, 340
Balanus, 340, 671, 673
Baleen, 264–265, 881, 883
Bandicoot, 449, 867
Bankia, 593, 599
Baobab tree, 336
Barbus, 298
Barnacle, 670–671, 673–674
    freeze tolerance, 300
    in tidal pool, 340
Barracuda, 766
Barrier reef, 357
Basal body, 217, 465
Basal lamina. *See* Basement membrane
Basal plate, 538
Base, 24
    nitrogenous, 82–83, 91
Basement membrane, 122, 124, 198
Basilar membrane, 160–161
Basiliscus, 806, 813, 825

Basilisk, 806, 824–825
Basis, 659
Basking, 298, 301, 306
Basophil, 224
Bass, 766
Basslet, 315
Bat, 871, 882
    echolocation, 884
    flight, 219–220
    fringe-lipped, 884
    hibernation, 309
    homologous limbs, 364
    reciprocity, 425
    spotted, 884
    vampire, 425
    whispering, 884
    wing, 842
Bates, Henry Walter, 369, 377
Batesian mimicry, 377
Batrachoidiformes, 771
Batrachotoxin, 145, 791
Bauplan. See Body plan
Bayliss, W. M., 185
B cell, 237–239
B–c1 complex, 61
Beach flea, 668–669
Beagle, 369–371
Beak, 198, 376, **849–850**
Bear, 334–336, 863, 871
Beard worm, 264. *See also* Pogonophora
Beaumont, William, 271
Beaver
    habitat, 335
    pond production, 337
    recovery, 889
    and succession, 344
    vasectomy, 323
Bêche-de-mer, 724
Bee, 684
    bumble, 301, 700
    classification, 683
    eusociality, 697
    flight, 201
    honey. *See* Honey bee
    "killer," 705
    muscle, 216
    partial endothermy, 301, 304
Beehler, Bruce, 429
Beetle, 685
    Asiatic lady, 369
    bombardier, 691–692
    brood parasite, 417
    carrion, 345–346
    classification, 683
    dung, 345–346, 705
    flight, 201
    flour, 342
    larva, 346
    Mexican bean, 704
    muscle, 216
    scarab, 705
    variations within species, 369
Behavior, **407–435**
    evolution, 12, 412–414
    fossilized, 412, 414
    and genetics, **410–413,** 471
    human, 413
    innate, 409–410
    modification by parasites, 559–560
    neural control, 408–409, 411
    of *Paramecium*, 470–472
    parental, 430–431
    of protozoa, 470–472
    sexual differences, 319, 321
    social, 423
Behavioral ecology, 425–426
Behaviorism, 408
Békésy, Georg von, 160

Belostoma, 682
Beneden, Edouard von, 71, 549
Benthos, 357
Beriberi, 263
Bering Strait, 400
Bernard, Claude, 11
Beroe, 519, 521–522
Beryciformes, 781
Beta cell, 275
Beta sheet, 33, 34
Beta thalassemia, 99, 483
Bibio, 683, 696
Bicarbonate, 24
    in blood, 249
    pancreatic, 270, 273–274
Bichir, 753, 763
Bicuspid valve, 227
Bilateral cleavage, 114
Bilateral symmetry, 450, 525, 549
Bilateria, 525
Bile, 271–274
Bile salt, 272–273
Bilharziasis. *See* Schistosomiasis
Bilirubin, 272
Binary fission
    in fossils and bacteria, 391
    in Placozoa, 496
    in protozoa, 463, 472–473
Bindin, 110
Binding, of sperm, 110
Binkley, Sue, 420
Binomen, 443
Binominal system, 442–443
Biodiversity, 891
Biogenetic law, 133
Biogeochemical cycle. *See* Nutrient cycle
Biolistics, 101
Biological clock, 419–420. *See also* Circadian rhythm
Biological compass, 421
Biological species concept, 373, 382, 440–441
Bioluminescence
    bacteria, 357, 781
    firefly, 415
    jellyfish, 211
Biomass, 348–349
Biome, 333–336. *See also particular biomes*
Biosystematics. *See* Systematics
Biotic, 333
Biotin, 263
Bipalium, 526–527
Bipedal locomotion, 401–403, 412, 844
Bipes, 810, 813
Bipinnaria larva, 723
Bipolar affective disorder, 151
Bipolar cell, 156–157
Biradial symmetry, 450, 502
Biramous appendage, 637, 658–659
Bird, **831–862**
    beak, 376, **849–850**
    biological clock, 420
    brain, 174, 845
    chromosome number, 96
    circulation, 226–227, 845
    classification, 833–834
    cleavage pattern, 114
    cochlea, 848
    color, 838–839, 841
    conservation, 859–860
    cooperative breeding, 431
    courtship, 853–854
    and culture, 857
    dependence on tropical rainforests, 353

Pressure
  atmospheric, 246
  blood. *See* Blood pressure
  colloid osmotic, 25, 233
  intracellular, 44
  osmotic, 25–26
  partial, 246
Pre-urine, 290
Prey, 341
Priapulida, 395, 555, **557–558**
Primary structure, 33–34
Primary transcript, 89
Primate, **882–887**
  brain, 174
  love, 432
  menstruation, 321
Primates, order, 871, 882, 887
Primitive, 444, 450
Primitive streak, 121
Primordial germ cell, 107
Primordial soup, 387
Priscoan eon, 388
Prismatic layer, 587
Pristis, 753, 762
Problognathia, 539
Proboscidea, 872
Proboscis, 572, 691, 730–732
Procellariiformes, 835
Processing of messenger RNA, 89
Prociphilus, 682, 693
Proconsul, 402
Procuticle, 636–637
Production
  of coral reefs, 358
  in marine ecosystems, 339
  net primary, 333
  primary, 333–334
  in tropical rainforest, 353–354
Profundal zone, 337
Progenesis, 134
Progesterone, 187, 189, **321–325**
Proglottid, 536
Progress, evolutionary, 380
Prokaryote, 7, 390
Prolacerta, 881
Prolactin
  effect on birds, 851
  function, 325
  and metamorphosis, 802
  source, 185, 191
Prolactin release-inhibiting hormone, 191
Proline, 300
Promiscuity, **429**, 854, 875
Promoter, 89, 103
Pronephros, 292, 755
Pronghorn, 872, 878
Pronucleus, 112, 472
Proofreading in DNA replication, 88
Prophase, 73, 75
Prophase I, **75**, 79, 95
Propolis, 700
Prosencephalon, 173–174
Prosimian, 871, 884–885
Prosobranchia, 593–595
Prosoma, 643
Prostaglandin, **183–184**, 318, 323
Prostate gland, 318
Prostoma, 573
Protandry, 315
Protease, 270
Protein
  as buffer, 24
  comparison, 378
  function, 33
  metabolism, 62

as nutrient, 261–262
in plasma membrane, 35, 39
polymorphism, 382
sequencing, 378
structure, 31, 33–34
synthesis, 93, 262
Protein family, 97
Proteoglycan, 202
Proterozoic eon, 388, **390–391**, 394
Prothoracic gland, 193–194
Prothoracicotropin, 193–194
Prothrombin, 234, 263
Protist, 8, 379, 463
Protochordate, 733–734
Protoeukaryote, 391–392
Protogyny, 315
Proton, 20, 59
Proton pump, 59, 61, 270
Protonephridium, **285–286**, 529, 740
Proto-oncogene, 131
Protopod, 659
Protopterus, 753, 762
Protostome, 118, 451–452, 565
Prototheria, 865, 867
Protozoa, **460–484**
  behavior, 470–471
  characteristics, 462
  cilia, 464–465
  classification, 8, 463–464, 466–467
  defenses against, 236
  diseases caused by, 477–479
  ecology, 476
  feeding, 462
  interactions with humans and other animals, 476–479
  light detection, 155
  locomotion, 462, 464–465
  osmoregulation, 285
  reproduction, 463, 472–473
  symbioses, 477–478
Protractor lentis, 797
Protura, 682
Proturan, 346
Proventriculus, 267, 849
Proximal convoluted tubule, 288
Psammechinus, 718, 722
Pseudacris, 316
Pseudemys, 813, 829
Pseudoceros, 524, 526
Pseudocoel, 452, 544–546
Pseudocoelomate, 451, 544. *See also particular groups*
Pseudocolochinus, 710, 722
Pseudocopulation, 822
Pseudogene, 97, 380
Pseudomonas, 628
Pseudopod, 129
Pseudopodium, 216, 468, 471
Psittaciformes, 836
Psittacosis, 859
Psychological disorders, 151
Psychology, 408
Psychosis, 151
Ptarmigan, 334, 836
Pteranodon, 815
Pteridine, 199
Pterobranchia, 730–732
Pterois, 753, 770
Pterosaur, 815, 842
Pterygota, 680, 682
Pterygotus, 640, 641
Pteryla, 838
Ptyodactylus, 813, 824
Puffer fish, 144, 766
Puffin, 836
Pulmocutaneous artery, 226, 793
Pulmonary artery, 227–228, 256

Pulmonary circulation, 226
Pulmonary vein, 226–228
Pulmonata, 593–594
Pulmonate. *See also* Slug; Snail
  eye, 606
  reproduction, 599
  respiration, 252, 588
Puma, 890
Pump, ionic, 41
Pumping stomach, 645
Punctuated equilibrium, 396
Punnett square, 70, 77
Pupa, 680, 683
Pupil, 156
Purine, 83, 199
Purkinje fiber, 228
Pycnogonida, 641–642
Pygidium, 638
Pyloric cecum, 714, 771
Pyloric sphincter, 269
Pyloric stomach, 662–663, 714
Pyramid, ecological, 349–350
Pyridoxine, 263
Pyrimidine, 83
Pyrogen, 309
Pyrosoma, 735, 740
Pyrosome, 739–740
Pyruvate, 56–60
Python
  classification, 813
  constriction, 825, 828
  infrared receptor, 155, 820

Qiviut, 137
Q_{10}, 299
Quadrate, 879, 881–882
Quality, sensory, 152
Quaternary period, 399
Quaternary structure, 33–34
Queen, 697
  ant, 9
  bumble bee, 301
  honey bee, 699
  termite, 698
Queen substance, 699
Quetzal, 841
Quetzalcoatlus, 815
Quill, 837, 877
Quinine, 482

Rabbit, 322, 336, 873
Raccoon, 337, 871
Radial canal, 487, 507, 712–713
Radial cleavage, 114–115
Radial symmetry, 450, 709
Radiation, thermal, 305–306
Radioactive dating, 389
Radioimmunoassay, 185
Radiolaria, 476
Radiole, 618, 620
Radula, 588, 598
Rail, 836
Rainfall, 335
Rainforest, 337
Raja, 753, 762, 780
Ramapithecus, 402
Rana
  camouflage, 792
  cancrivora, 793
  catesbeiana, 786
  classification, 789
  eye, 797
  heart rate, 299
  neuromuscular synapse, 148, 206
  pipiens, 796
  respiration, 253
  sylvatica, 792
  tadpole, 790

temporaria, 790
  vision, 796
Rangifer, 352, 867, 880
Ranidae, 798
Rat
  behavioral studies, 408
  brain, 174
  classification, 873
  effect of reproductive hormones, 321
  learning, 410
  mole, 875–876
Ratfish. *See* Chimaera
Ratite, 833, 835
Rattlesnake
  aggressive behavior, 428
  classification, 826
  infrared detection, 155
  jaw bones, 827
  mouth, 827
  predation, 428
  rattlers, 428
  social hierarchy, 428
  timber, 811
Ray, **760–762**
  blue-spotted, 750
  classification, 753
  electric, 780
  osmoregulation, 292–293
  skeleton, 200
Reabsorption, 285, 290
Reaction
  cortical, 111–112
  coupled, 53
  dehydration, 30
  hydrolysis, 30
  oxidation-reduction, 57
  zona, 112
Reading frame, 98
Reaggregation, 492–494
Receptive field, 177
Receptor, sensory, 152–164. *See also particular sensory receptors; "nervous system" or "sensory receptors" under particular groups*
  chemical, 161–164
  crayfish stretch, 153
  mechanical, 157–161
  photoreceptor organs, 155–157
  retinal, 154–155
  somesthetic, 157–160
Receptor molecule
  hormonal, 188–189
  for insulin, 189
  on leukocytes, 237
  for synaptic transmitter, 147, 183
Receptor-mediated endocytosis, 42–43
Recessivity, 67–69, 98
Reciprocity, 425
Recombinant DNA, 101–103
Recombination
  artificial, 101–103
  by crossing over, 79, 383
  in evolution, 362
  by sexual reproduction, 314, 383
Recruit, honey bee, **417–419**, 701
Rectal cecum, 714
Rectal gland, 294, 760
Rectal pad, 267, 287, 686
Rectilinear movement, 824
Rectum
  human, 272, 318, 320
  insect, 267
Red blood cell. *See* Erythrocyte

Snake (*Continued*)
  pelvic bones, 363
  sea, 295, 824–826, 829
  venom, 825–826
Sneezing, 175
Snipe, 836
Social behavior, 423–426, 697–698
Sociobiology, 423–425
Sodium, 23
  in action potential, 143
  cellular concentrations, 41, 140, 142
  concentrations (table), 282
  in diet, 295
  effect on blood pressure, 292
  and membrane potential, 140, 143
  as nutrient, 260
  reabsorption in kidney, 291
  and synaptic function, 147
  and taste, 164
  transport by gills, 292
  transport in kidney, 290
Sodium bicarbonate, 24
Sodium chloride, **23**, 260, 295. *See also* Salt
Sodium–potassium pump, 41, **142–144**, 273
Soil, 354
Sol, 217
Soldier ant, 9
Sole, 766
Solea, 356
Solenocyte, 286, 538–539
Solenopsis, 417
Solute, 23
Solution, 23
Solvent, 22–23
Somatoplasm, 71
Somatosensory cortex, 176–178
Somatostatin, 191
Somatotopic organization, 178
Somesthetic receptor, 157
Somite, 125. *See also* Segmentation
Song
  communication, 414–415
  frog, 801
  function, 427
  innate response to, 409
  innate versus learned, 415, 846
  whale, 881
Sour, 164
Sow bug, 669
Spallanzani, Lazaro, 882
Sparrow, 415, 837, 858
Spat, 600
Speciation, 382
Species
  defined, 373, 440–441
  evolution, 362
  nomenclature, 443
  unisexual, 316
Speech, 179
Speleonectes, 673, 674
Spemann, Hans, 129
Spencer, Herbert, 380
Sperm
  ameboid, 547
  development of, 107
  discovery of, 106
  effect on pattern formation, 128
  in fertilization, 109–112
  number of, 318
  production, 108
  structure, 109
Spermatheca, 317, 696, 799

Spermatid, 108
Spermatocyte, 108
Spermatogenesis, 108, 130
Spermatogonia, 107–108
Spermatophore, 316
  arrowworm, 725
  leech, 629
  octopus, 606
  salamander, 799–800
  scorpion, 651
Spermicide, 323
Spermiogenesis, 108
Spermophilus, 425
Sperm receptacle, 661
Sperm web, 647
S phase, 84, 86, 112
Spheciospongia, 487, 491, 495
Spheniscimorformes, 835
Sphenodon, 812–813
Sphenodonta, 812–813
Sphere, in mitochondrion, 59, 61
Spheroides, 144
Sphygmomanometer, 229
Spicule, 118, 200, 487–490
Spider, **643–655**
  antifreeze, 300
  black widow, 646, 648, 653–654
  bolas, 648
  brown recluse, 653–654
  characteristics, 643
  circulation, 646
  cobweb, 649
  copulation, 316, 647
  crab, 696
  cribellate, 650
  digestion, 645–646
  excretion, 646
  external structure, 643–644
  eyes, 647
  feeding, 645
  fishing, 633
  garden, 651
  internal structure, 645
  jumping, 632, 647–648
  nervous system, 646–647
  osmoregulation, 286
  reproduction, 647–648
  respiration, 252, 646
  silk, 648–649
  "tarantula," 644, 646
  venom, 653
  web, 649–651
Spinal cord, 176
  anatomy, 167–168
  and autonomic nervous system, 171
  development, 124–125
  and reflexes, 167, 169
Spindle apparatus, 73, 75
Spindle receptor, 169–170
Spinneret, 649
Spiny anteater. *See* Echidna
Spiny-headed worm, 559–561
Spiracle
  of insect, 252, 687
  of ray, 761
  of shark, 758
  of spider, 646
  of tadpole, 802
Spiral cleavage, 114, 565
Spiral valve
  of amphibian heart, 793
  in fish intestine, 757, 760, 763
Spirontocaris, 340
Spittlebug, 693
Spleen, 236, 260
Split-brain, 179–180

Sponge, **485–496**
  bath, 490, 495–496
  behavior, 170
  boring, 358, 495
  calcareous, 490
  canal system, 487–489
  cells, 490–491, 493
  characteristics, 487
  chromosome number, 96
  classification, 490–491
  cleavage pattern, 114
  commensals, 495
  competition with corals, 495
  coralline, 490
  in coral reef, 357
  crumb-o'-bread, 492
  cyanobacteria mutualists, 495
  development, 494–495
  elephant ear, 496
  endoskeleton, 200, 488–490
  external fertilization, 316
  feeding, 488
  freshwater, 494
  genome size, 97
  glass, 490
  interactions with humans and other animals, 495–496
  larvae, 494–495
  loggerhead, 487, 495
  osmoregulation, 285
  predators, 495
  reaggregation, 492–494
  red, 488
  redbeard, 493
  red encrusting, 340
  reproduction, 494
  "smoking," 316, 494
  structure, 488
  vase, 485
  yellow, 488
Spongia, 490–491, 495–496
Spongicola, 495
Spongilla, 491, 494
Spongin, 200, 488, 493
Spongocoel, 486, 488
Spongocyte, 493
Spongy bone, 202
Spoon, D. M., 468
Spoon worm, 574–576
Spore, 467, 474. *See also* Cyst
Sporocyst, 532–534
Sporogony, 481
Sporozoa, 463–464
Sporozoea, 467
Sporozoite, 480, 482
Spriggina, 393
Spring overturn, 338
Spring peeper, 801
Springtail, 346, 680, 682
Squalene, 757
Squalus, 750, 753
Squamata, 810, 813
Squamosa, 879
Squid, **602–604**
  in Antarctic food web, 350
  giant axon, 143
  hormones, 193
Squirrel, 873
  body temperature, 309
  ground, 309
  habitat, 335–336
  teeth, 265
Stabilimentum, 651
Stability, population, 351, 353
Stalling, 843
Stapes, 795, 820, 879
Starch, 29, 261
  digestion, 268–269, 273
Starfish, **710–715**
  classification, 722

  crown-of-thorns, 517, 723–724
  ion concentrations in, 282
  marble, 708
  osmoconformity, 284
  predation by, 515
  regeneration, 723–724
Stargazer, 780
Starling, 835, 858
Starling, E. H., 185, 233
Starling effect, 233
Stars, 421
Statoblast, 567
Statocyst organ
  in Cnidaria, 507, 514
  in crustacea, 665–666
  in Ctenophora, 519
  in Turbellaria, 530
Statolith, 521, 666
Steatoda, 641, 649
Steganopus, 834, 853
Stegosaurus, 815
Stem cell, 117, 224
Stenohaline, 284
Stenolaemata, 567
Stenopterygius, 814
Stentor, 461, 469
  classification, 467
  feeding, 465
Stercoral pocket, 646
Stereoblastula, 494
Stereocilium, 158
Sterna, 834, 845, 852
Sternal artery, 663–664
Sternal keel, 832–833, 844
Sternal sinus, 663–664
Sternite, 658
Steroid
  absorption, 268, 274
  anabolic, 319
  diffusion across membrane, 39–40
  hormones, 188–189
  structure, 31
Steroid hormone, 188–189
Stick insect, 685
Stickleback, 407, 422–423, 430
Stigma, 463, 470
Stinger, 698
Stingray. *See* Ray
Stomach
  history of research on, 271
  hormone from, 187
  human, 269, 272
Stomach-intestine, 614
Stomochord, 729
Stomphia, 509, 514–515
Stone canal, 712–713
Stonefly, 682, 684
Stork, 835
Stream, 338
Strepsirhini, 885, 887
Stress, physiological
  autonomic nervous system in, 171
  effect on digestion, 273
  hormones, 191
  in subordinate baboon, 428
Stress fiber, 44
Stretch receptor, 153, 169–170
Striated muscle. *See* Skeletal muscle
Strigiformes, 837, 848
Strobila, 508, 536
Stroke, 229
Stroke volume, 231
Stromatolite, 391
Strongylocentrotus, 340, 719, 722
Structural color, 199
Structural white, 199

Yolk sac, 120–121
Y organ, 193, 666

Zaglossus, 865, 867
Zallinger, Rudolph, 815
Zebra
    classification, 872
    community, 342–343
    disruptive coloration, 877
    habitat, 336

Zeiss, Carl, 37
Zinc, 260
Z line (or disc), 210, 213
Zoea larva, 661
Zoecium, 566
Zona pellucida, 110
Zona reaction, 112
Zonotrichia, 415
Zoo, 889

Zooid
    of *Autolytus*, 618–619
    of ectoproct, 566
    of entoproct, 561
    of flatworm, 530
    of Portuguese man-of-war,
        506–507
Zoology, 2–5, 7, 10
Zoomastigophorea, 466
Zooplankton, 337, 355

Zooxanthellae
    and coral reefs, 358
    of Cnidaria, 503, 511, 517
    symbiosis, 262, 477
ZP3, 110–111
Zygomatic arch, 879, 881
Zygote, 107, 112
    diploidy of, 71
    of frog, 115

Common roots of zoological terms

Each root is followed by its meaning and example(s). The root is from Greek unless noted by (L) as being from Latin.

A-, an- not, without (acoelomate, anaerobic)
ab- (L) off, from, away (aboral)
acanth thorn (acanthocephalan)
acti ray (Actinopterygii)
ad- (L) to, toward (adoral)
alb (L) white (albino)
allo- different, other (allele)
amphi around, on both sides, double (amphibian)
andr male (androgen)
aqua (L) water (aquatic)
arche ancient, first (archenteron)
astr, aster star (astrorhizae)
auto self (autotrophic)
axo, axi axis (axial)

Bi- (L) two, double (bilateral)
blast sprout (blastula, fibroblast)
brach (L) arm (brachiopod)
branch gill (nudibranch)

Card heart (cardiac)
caud (L) tail (caudal)
centr center (centromere)
cephal head (cephalopod)
chaet bristle, hair (polychaete)
chondr cartilage (Chondrichthyes)
chord string (notochord)
chori membrane (chorion)
chrom color (cytochrome)
clad branch (cladoceran)
coel hollow (coelom, schizocoel)
corn (L) horn (unicorn)
cyst bladder (statocyst)
cyt cell (cytoplasm, erythrocyte)

Dactyl finger or toe (pentadactyl)
de- (L) from, down, out (defecate)
dendr tree (dendrite)
derm skin (epidermis, placoderm)
di- two, double, separate (diploid)
dia- across, through (diaphragm)
diplo two, double (diploid)
dist (L) distant (distal)

Echin hedgehog, i.e. spiny (echinoderm)
ect outside (ectoparasite)
en-, end-, ent- inside (endoparasite)
enter intestine, gut (coelenterate)
entomo insect (entomology)
epi upon, over, beside (epidermis)
erythr red (erythrocyte)
eu- true, good (eukaryotic)
eury wide (euryhaline)
ex- (L) out, from (excrete)

Fer (L) bear (porifera)

Gam marriage (gamete)
gast stomach, belly (gastrula)
gen produce (carcinogen)
gene origin, birth (genetics)
gnath jaw (Agnatha)
gon seed, generation, offspring (gonad)
gyn female (gynandromorph)

Helmin worm (helminthes)
hem blood (hemoglobin, uremia)
hemi- half (hemichordate)

hetero- other, different (heterozygote)
hex six (hexactinellid)
hol whole (holoblastic)
hom (L) human (hominid)
homeo-, homo- same, alike (homeostasis)
hyal glass (hyaline)
hydr water (hydrated)
hyper- over, excessive (hyperthermia)
hypo- under (hypotonic)

Ichthy fish (Osteichthyes)
inter- (L) between (intercellular)
intr- (L) inside (intracellular)
iso equal (isotonic)

Leuc, leuk white (leukocyte)
lob lobe (lobose)
loph crest, tuft (lophophorate)
lys, lyt loose (proteolytic)

Macro large (macromolecule)
melan black (melanin)
mer part (sarcomere, meroblastic)
meso middle (mesoderm)
meta after (metazoan)
micro small (microtubule)
mono one, single (monoecious)
morph form (metamorphosis)
myo muscle (myofibril)

Naut ship, sail (nautiloid)
nem thread (nematode)
neo new (neo-Darwinism)

Oo pronounced OH-oh; egg (oocyte)
ophi snake (Ophiuroidea)